W9-BUO-622

GEOCHEMISTRY, GROUNDWATER AND POLLUTION

SECOND EDITION

GEOCHEMISTRY, GROUNDWATER AND POLLUTION, 2ND EDITION

C.A.J. APPELO
Hydrochemical Consultant, Amsterdam, the Netherlands

D. POSTMA
Environment & Resources DTU, Technical University of Denmark, Kgs. Lyngby, Denmark

A.A. BALKEMA PUBLISHERS Leiden/London/New York/Philadelphia/Singapore

Library of Congress Cataloging-in-Publication Data
Applied for

Cover Illustration: Ben Akkerman – Landscape

Published by: A.A. Balkema Publishers, Leiden, The Netherlands a member of
Taylor & Francis Group plc
www.balkema.nl and www.tandf.co.uk

ISBN 04 1536 421 3 Hardback
ISBN 04 1536 428 0 Student paper edition

Printed in Great Britain

Preface

The first edition of *Geochemistry, groundwater and pollution* has become a popular textbook that is used throughout the world in university courses and as a reference text. The new developments in our science during the last decade have inspired us to update the book. There are now more data available, more experiments have been done, new tools were developed, and of course, new problems emerged. An important catalyst for rewriting the book was the advancement of the computer code PHREEQC. From the start in 1980, this model was developed to simulate field data and laboratory experiments and at present it encompasses over 20 years of experience in modeling water quality. The code is used all over the world in research and to predict the long-term effects of pollution and radioactive waste storage. Modeling with PHREEQC is an integrated part of the presentation in the second edition. It is used to elucidate chemical concepts when they become complicated and in particular to analyze the interactions between chemical processes and transport. Numerous examples in this book show PHREEQC applications with input files listed in the text. The more comprehensive ones can be downloaded from the net and are useful templates for similar problems.

The basic structure of the first edition is maintained. However, flow and transport in aquifers is introduced already in Chapter 3 since it is important to become familiar with the principles that govern water flow and residence times in aquifers at an early stage. The transport equations developed in this chapter are applied throughout the text when transport and chemical reactions interplay, for example to calculate retardation and chromatographic patterns. The behavior of heavy metals is treated more extensively in a separate chapter with emphasis on surface complexation theory. A chapter on organic pollutants has also been added, discussing the evaporation and leaching of these chemicals as well as inorganic and microbial degradation. The last chapter introduces numerical transport models and provides application examples of organic pollution, acid mine water, in-situ aquifer remediation, arsenic in groundwater, and more. Similar to the previous edition, examples serve to illustrate the application of theory. In addition to the problems at the end of each chapter, questions have been interspersed in the text to focus the reader and to provide diversion with simple calculations that elucidate the concepts.

We like to thank colleagues and friends for their contributions to the second edition. First and foremost, we owe much to David Parkhurst who initiated PHREEQC, solved many of our problems, and reviewed Chapter 11. Vincent Post developed the graphical interface for PHREEQC that is used throughout this book. Individual chapters were read and commented by Martin S. Andersen, Philip Binning, Peter Engesgaard, Rien van Genuchten, Pierre Glynn, Jasper Griffioen, Rasmus Jakobsen, David Kinniburgh, Claus Kjøller, Niel Plummer, Ken Stollenwerk and Art White. We also appreciate the contributions and comments of Boris van Breukelen, Liselotte Clausen, Flemming Larsen, Uli Mayer, Hanne Dahl Pedersen and Judith Wood. We are indebted and immensely thankful for the time spent by all. Projects and studies with Paul Eckert, Laurent Charlet,

Christoph König and Alain Dimier have helped to practize ideas and provided financial support that is gratefully acknowledged (TA). Dieke thanks Philip Binning and the University of Newcastle N.S.W., Australia, for hospitality and financial support during a sojourn in the winter of 2002. Finally, Tony could not have imagined to have started the work without having Dorothée, Els and Maria close-by for inspiration, wisdom and love, and Tobias for surviving.

January 2005.
Amsterdam, Tony Appelo
Hanoi, Dieke Postma

Contents

List of Examples

Notation

Greek characters

α	Speciation factor, fraction of total concentration (−), *or* isotope fractionation factor, see below (−), *or* solid solution fractionation factor, the activity ratio of two elements in a solid, divided by the activity ratio in solution (−), or exchange factor between mobile and immobile regions (1/s), *or* spatial weighting factor in numerical scheme (−)
$\alpha^{18}O_{l/g}$	isotope fractionation factor (−), here for ^{18}O in liquid/gas equilibrium written in the form $H_2{}^{18}O_{(g)} + H_2{}^{16}O_{(l)} \leftrightarrow H_2{}^{18}O_{(l)} + H_2{}^{16}O_{(g)}$
α_L	longitudinal dispersivity (m); subscript $_T$ for transversal dispersivity
α_i	equivalent fraction of ion *i* in water (−)
β_i	equivalent fraction of *i* on exchanger (−)
β_i^M	molar fraction of *i* on exchanger (−)
δ	difference of isotopic ratio with respect to a standard (‰)
γ_i	activity-coefficient of species *i* in water (−)
ε	porosity, a fraction of total volume (−), *or* isotopic enrichment factor (−), notation follows example $\alpha^{18}O_{l/g}$, *or* dielectric constant (F/m)
ε_w	water filled porosity, a fraction of total volume (−); subscript $_g$ for gas filled porosity
ε_m	mobile water filled porosity, a fraction of total volume (−); subscript $_{im}$ for immobile water
η	viscosity (g/m/s)
θ	tortuosity, the ratio of actual path over straight line path (−), *or* water content (g/g)
$1/\kappa$	Debye length (m)
Λ_m	molar electrical conductivity ($S \cdot m^2/mol$)
λ	radioactive decay constant (1/s), *or* activity coefficient in solid solution (−), *or* correlation length (m)
μ	reduced mass (g/mol)
μ_{max}	specific degradation rate (1/s)
ν	stoichiometric coefficient in reaction (−)
ρ	charge density (C/m^3), *or* solid or liquid density (g/cm^3)
ρ_b	bulk density of solid (with air filled pores) (g/cm^3)
σ	charge density at solid surface (eq/kg solid); subscripts $_0$ for structural charge, $_H$ for protons, $_M$ for metals, $_A$ for anions, $_D$ for double layer charge
σ_{DL}	double layer charge (C/m^2 solid)
σ^2	variance in a set of parameters; of the diffusion curve (units depend on parameter)
τ	residence time (s)
Φ_m	mobile fraction of pore water (−), $\varepsilon_m/\varepsilon_w$

χ	mole fraction in solid or liquid solution (−)
ψ	potential in a double layer (V)
ψ_0	potential at a solid surface (V)
Ω	saturation state or ratio, *IAP/K* (−)
ω	weighting factor for time in numerical scheme (−)

Capital letters

A	surface area (m^2), *or*
	temperature dependent coefficient in Debye-Hückel theory (−)
A_0	initial surface area (m^2), *or*
	total cation concentration (meq/L)
A_{11}	longitudinal macrodispersivity (m)
Alk	Alkalinity (meq/L)
CEC	cation exchange capacity (meq/kg)
B	number of bacterial cells as dry biomass (mg/L, mol C/L), *or*
	temperature dependent coefficient in Debye-Hückel theory (1/m)
D	diffusion coefficient (m^2/s), *or*
	dry (atmospheric) deposition ($\mu g/m^2/yr$), *or*
	aquifer thickness (m)
D_e	effective diffusion coefficient in a porous medium (m^2/s)
D_f	diffusion coefficient in free water (m^2/s); subscript $_a$ in free air
D_L	longitudinal dispersion coefficient (m^2/s); subscript $_T$ for transversal dispersion coefficient
DDL	diffuse double layer
E	evapotranspiration (m), *or*
	electrical field strength (V/m), *or*
	kinetic energy ($kg \cdot m^2/s^2$), *or*
	efficiency of in situ iron removal (−)
E.B.	electrical balance of water analysis (%)
EC	electrical conductivity of water sample ($\mu S/cm$)
Eh	redox potential relative to the H_2/H^+ half reaction (V)
E^0	standard (redox) potential (V)
ESR	exchangeable sodium ratio (−)
F	flux (mass/m^2/s)
F_{Na}	fractionation factor with respect to Na^+, used when comparing concentrations in rainwater and seawater (−)
ΔG_r^0	Gibbs free energy of a reaction (J/mol)
$\Delta G_{f(i)}^0$	Gibbs free energy of formation of *i* from elementary *i* (J/mol)
ΔH_r^0	reaction enthalpy (heat) (J/mol)
I	ionic strength (−)
IAP	ion activity product (−)
K	mass action constant (−) or solubility product (−)
K_a	acid dissociation constant (−)
$K_{i\backslash j}$	exchange constant, the subscript ions indicate solute ions in the order of appearance in the mass-action equation: $i + j\text{-X} \leftrightarrow i\text{-X} + j$, Gaines and Thomas convention (−)
$K_{i\backslash j}^G$	idem, Gapon convention (−)
K_d'	distribution coefficient (mg sorbed/kg solid)/(mg solute/L pore water) (L/kg)
K_F	constant in Freundlich sorption isotherm (units depend on concentration units and on exponent)
K_H	Henry's constant for liquid ↔ gas equilibrium (atm · L/mol)
K_H'	Henry's constant with the gaseous concentration in mol/L gas (L water/L gas); $K_H' = K_H/RT$
K_d	distribution coefficient (mg sorbed/L porewater)/(mg solute/L pore water) (−)
K_L	constant in Langmuir sorption isotherm (mol/L)
K_{oc}	distribution coefficient among 100% organic carbon and water (L/kg)
K_{ow}	idem, among octanol and water (L/kg)
L	length of column or flowtube (m)
M	molality (mol/kg H_2O)
M	mass (mol, kg)

N normality (eq/L), *or*
N a number of molecules (−)
P gas pressure (atm), *or*
net rate of precipitation (groundwater infiltration) m/yr
Pe Peclet number (−)
PV pore volumes injected/displaced, V/V_0 (−)
Q discharge (m^3/s)
R gas constant (8.314 J/K/mol), *or*
retardation (−), *or*
rate of dissolution/precipitation (mol/L/s), *or*
resistance (Ω, V/A), *or*
isotopic ratio (−)
S substrate concentration (mg/L, mol/L)
SR saturation state or ratio, IAP/K (−)
SAR sodium adsorption ratio $(mmol/L)^{0.5}$
SI Saturation Index, $\log(IAP/K)$ (−)
U potential energy in Boltzmann equation (J/atom)
V volume (m^3)
V_0 pore volume of a column (m^3)
X soil exchanger with charge number −1
Y yield factor, transforming substrate (S) into biomass (B)

Lower case letters
a ion radius (m), *or*
ion size in Truesdell-Jones Debye-Hückel equation (m), *or*
stagnant body size (m)
å ion size in Debye-Hückel equation (m)
b coefficient in Truesdell-Jones Debye-Hückel equation (−)
c concentration (mol/L, mol/kg, mg/L, etc)
c_0 concentration at column inlet ($x = 0$) (mol/L)
c_i initial concentration (mol/L)
d differential symbol
d depth (m), *or*
deuterium excess in rain (‰)
f a fraction (−), *or*
dilution factor (−)
f_{oc} fraction of organic carbon in soil (−)
h hydraulic head or potential for water flow (m)
k hydraulic conductivity (m/day), *or*
rate constant in kinetic rate law (units depend on mechanism)
$k_{1/2}$ substrate concentration giving half of the maximal Monod rate (mol/L)
k_{max} maximal Monod rate (mol/L/s)
m molality (mol/kg H_2O), *or*
temporal or spatial moment (units depend on number)
m/m_0 mass fraction (−)
$meq_{I\text{-}X}$ exchangeable cation I (eq/kg)
p in pH, pK, etc. indicates −log, the negative, decimal logarithm
q sorbed concentration (on solid), expressed *per* L porewater (mol/L, mg/L, etc.)
r specific kinetic rate ($mol/m^2/s$ for solids, mol/s otherwise), *or*
radial distance (m)
s sorbed concentration (on solid) (mol/kg, mg/g, etc.); $q = (\rho_b/\varepsilon)s$
u ionic mobility ($m^2/s/V$)
v velocity (m/s)
v_D specific discharge, or Darcy velocity, $v_D = v_{H_2O}/\varepsilon_w$ (m/s)
v_{H_2O} pore water flow velocity (m/s)
v_c velocity of a given concentration (m/s)

z_a	boundary film thickness of air (m); subscript $_w$ for water
z_i	charge number of ion i (−)

General data

N_a	Avogadro's number (6.022×10^{23}/mol)
F	Faraday constant (96485 C/mol or J/Volt/g eq)
R	gas constant (8.314 J/K/mol)
T	temperature in K (= °C + 273.15)
q_e	charge of the electron (1.602×10^{-19} C)
k	Boltzmann constant (1.38×10^{-23} J/K)

1 cal = 4.184 J
1 atm = 1.013×10^5 Pa
at T = 298.15 K:

RT = 2.479 kJ/mol
RT/F = 25.69 mV; 2.303 RT/F = 59.16 mV
$V_{gas} = RT/P$ = 24.79 L/mol at 1 atm
η_{H_2O} 0.891 g/m/s

Global data

Earth surface: 5.1×10^{14} m^2
Ocean surface: 3.6×10^{14} m^2
Land surface: 1.5×10^{14} m^2

Water in

Oceans: 13700×10^{20} g
Ice caps and glaciers: 290×10^{20} g
All groundwater: 95×10^{20} g
Shallow groundwater < 100 m: 5.6×10^{20} g
Atmosphere: 0.13×10^{20} g
Global precipitation 4.7×10^{20} g H_2O/yr

Prefixes

p	n	μ	m	c	d	k	M	G
pico	nano	micro	milli	centi	deci	kilo	mega	giga
10^{-12}	10^{-9}	10^{-6}	10^{-3}	10^{-2}	10^{-1}	10^3	10^6	10^9

1

Introduction to Groundwater Geochemistry

Groundwater geochemistry is the science that explores the processes controlling the chemical composition of groundwater, the *groundwater quality*. The groundwater quality influences the use of this resource. Groundwater may contain hazardous substances that affect health when consumed or which deteriorate the environment when the water is thoughtlessly spilled at the surface. The groundwater quality may change during the exploitation or it may be affected by human activities of which the impact is not always immediately evident.

The interest of society in groundwater geochemistry is mainly to ensure good quality drinking water. Although drinking water can be manufactured, for example by desalinization, this still is a costly affair, and to surrender to this option is in conflict with our desire to utilize groundwater as a sustainable resource, refreshingly and cleanly flowing from a well. Preservation of good groundwater therefore has a high priority for environmental authorities.

This chapter introduces basic subjects such as the concentration units, water sampling techniques, and how to examine the accuracy of a chemical analysis. We begin with the concentration limits for drinking water.

1.1 GROUNDWATER AS DRINKING WATER

Access to clean freshwater will be one of the biggest global resource problems of the coming decades. One billion people had no access to clean drinking water from public supply in the year 2003. Probably, between 2 and 7 billion people will live in water scarce countries in the middle of this century. Recent estimates suggest that climate change will account for about 20 percent of the increase in global water scarcity (www.unesco.org). Lack of fresh water presents a global problem of huge dimensions and a major effort is required to ensure good quality drinking water for the world population. Much of the drinking water is derived from surface waters but, particularly in developing countries, groundwater is often preferred because it needs less treatment and has a better bacteriological quality which helps to minimize the spread of water-born diseases like cholera.

The general public is well aware of the value of good quality drinking water and consequently, the sale of bottled waters is world-wide a big business. Figure 1.1 shows the labels of various mineral waters, and they seem to suggest that the composition of the bottled water has a favorable health aspect.

1.1.1 *Standards for drinking water*

Table 1.1 shows the maximal admissible concentration in drinking water for a number of common constituents imposed by most national authorities. Generally, the concentration limits in Table 1.1 comply with the guidelines of the World Health Organization (WHO).

Figure 1.1. Labels of mineral water bottles, advertizing beneficial effects of drinking mineral water because of their composition.

Interestingly, bottled mineral waters are not subject to these drinking water limits since, from a legislation point of view, they are considered a medicine rather than drinking water and the comparison of the analyses printed on labels of mineral water bottles with the values in Table 1.1 may be an interesting exercise. Table 1.1 contains both elements with a natural origin and constituents mainly derived from pollution (like nitrate and nitrite) or resulting from water treatment and distribution systems (Cu, Zn). For pesticides the concentration of each pesticide should be less than 0.0001 mg/L while the sum of the pesticides should not exceed 0.0005 mg/L.

Table 1.1. Standards for the composition of drinking water and contribution of drinking water to the intake of elements in nutrition. The European Union drinking water limits can be found on www.europa.eu.int in the Official Journal L330 of December 5, 1998. For the USA look at www.epa.gov/safewater/. For WHO guidelines see www.who.int/entity/dwq/en.

Constituent	Contribution to mineral nutrition (%)	Maximal admissible concentration (mg/L)	Comment
Mg^{2+}	3–10	50	Mg/SO_4 diarrhea
Na^+	1–4	200	
Cl^-	2–15	250	taste; safe < 600 mg/L
SO_4^{2-}		250	diarrhea
NO_3^-		50	blue baby disease
NO_2^-		0.5	
F^-	10–50	1.5	lower at high water consumption
As	≈30	0.01	black-foot disease, skin cancer
Se		0.01	
Al	..	0.2	acidification/Al-flocculation
Mn		0.05	
Fe		0.2	
Ni		0.02	allergy
Cu	6–10	2	3 mg/L in new piping systems
Zn	negligible	0.1	5 mg/L in new piping systems
Cd	..	0.005	
Pb	..	0.01	
Cr	20–30	0.05	
Hg		0.001	

The average contribution of drinking water to the daily elemental intake with nutrition is also listed (Safe Drinking Water Comm., 1980). For example for Na^+, the intake through drinking water is insignificant as the Na^+ intake completely is dominated by the addition of salt (NaCl) to our food. However, for elements like F^- and As, the intake through drinking water is highly significant, and a too high concentration of these two elements is a severe health threat affecting millions of people in many places around the world. The intake of excess fluoride leads to painful skeleton-deformations termed fluorosis; it is a common disease in African Rift Valley countries such as Kenya and Ethiopia, where volcanic sources of F^- are important, and in India and West Africa where salts and sedimentary F-bearing minerals are the primary sources. In India alone about 67 million people are at risk of developing fluorosis (Jacks et al., 2000).

Groundwater high in arsenic may be found in inland basins in arid and semi-arid climate (Welch et al., 2000). Another typical setting for high arsenic groundwater is in alluvial plain sediments where strongly reduced groundwater, often with a high Fe^{2+} concentration, is present. In both cases the aquifers are situated in geologically young sediments and in rather flat, low-lying areas were the extent of flushing with groundwater is low (Smedley and Kinniburgh, 2002). Arsenic is also associated with sulfide minerals, and may therefore be high in mine waste waters. Chronic arsenic poisoning can result in skin lesions, hyperkeratosis (thickening, hardening and cracking of palms and soles), skin cancer and liver disease (Karim, 2000; Figure 1.2). In Bangladesh an estimated 20–70 million people are at risk from drinking high arsenic groundwater and it probably presents the biggest case of mass poisoning in history (Halim, 2000).

1.2 UNITS OF ANALYSIS

The concentration of dissolved substances in water can be reported in different units, depending on the purpose of the presentation and also on tradition. Some common units are listed in Table 1.2.

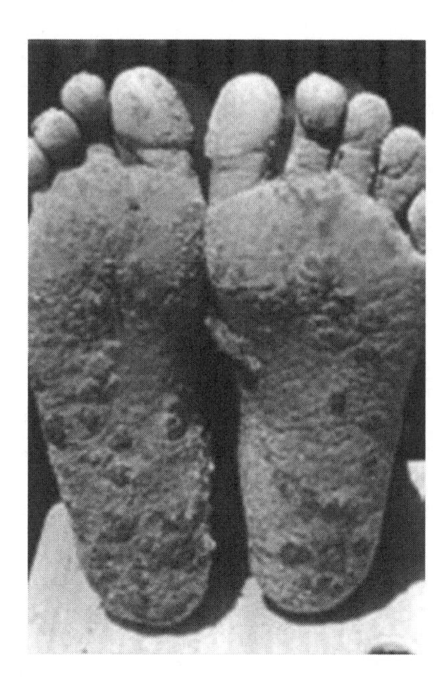

Figure 1.2. Skin lesions (black foot disease) as result of drinking water with high concentrations of arsenic.

What is the best unit to use? The labels of bottled mineral waters in Figure 1.1 suggest that milligrams per liter is the obvious unit to express the analysis and mg/L is used by most laboratories to report their analytical results. Other common units are parts per million (ppm = 1 mg/kg) and parts per billion (ppb = 1 μg/kg). These are numerically equal to mg/L and μg/L, respectively, when the density of the water sample is 1 kg/L, as is the case for dilute fresh waters. However, the density of evaporation brines is much higher and a correction is required when converting the results reported per volume to concentrations by weight.

Table 1.2. Concentration units for dissolved substances.

mg/L	milligrams per liter sample
μg/L	micrograms per liter sample
ppm	parts per million by weight of sample
ppb	parts per billion by weight of sample
mmol/L	millimoles per liter sample
μmol/L	micromoles per liter sample
meq/L	milliequivalents per liter of sample
$mmol_c$/L	milliequivalents per liter of sample
epm	equivalents per million, by weight of sample
M	molality, moles per kg of H_2O
mM	millimoles per kg of H_2O
N	normality, equivalents per liter

The unit molarity (mol/L) reflects the number of molecules of a substance (Avogadro's number: 1 mol = 6.022×10^{23} molecules), rather than their weight, and has the advantage that changes in concentration can be related directly to the coefficients in reaction equations. For example, if 1 mmol gypsum/L dissolves (the chemical formula of gypsum is $CaSO_4 \cdot 2H_2O$), the Ca^{2+} concentration increases by 40 mg/L and the SO_4^{2-} concentration by 96 mg/L, while the increase for both is simply 1 when expressed in mmol/L.

The official SI-unit is moles/m^3, but this unit is not commonly applied. Numerically, moles/m^3 are equal to mmol/L, the unit we often use in this book. To recalculate an analysis from mg/L to mmol/L, the numbers are divided by the gram formula weight for each species. The gram formula weight is the weight in grams of 1 mol of atoms or molecules (Example 1.1). The unit milliequivalent per liter (meq/L or $mmol_{charge}$/L, abbreviated as $mmol_c$/L) is the molar concentration multiplied by the charge of the ions (1 mmol Na^+/L = 1 meq Na^+/L; 1 mmol Ca^{2+}/L = 2 meq Ca^{2+}/L). Milliequivalents per liter are useful for checking the charge balance of chemical analyses.

Table 1.3. Recalculation of analysis units.

$$\text{mmol/L} = \text{mg/L} / (\text{gram formula weight})$$
$$\text{mmol/L} = \text{ppm} \cdot (\text{density of sample}) / (\text{gram formula weight})$$
$$\text{mmol/L} = \text{meq/L} / (\text{charge of ion})$$
$$\text{mmol/L} = \text{molality} \times \text{density} \times \frac{(\text{weight solution} - \text{weight solutes})}{(\text{weight solution})} \times 1000$$

The unit molality gives the number of moles per kg of H_2O and is generally used in physical chemistry. The advantage is that this concentration unit is independent of a change in density with temperature, or a change in concentration of other constituents. In normal fresh to slightly brackish water, molality is identical with molarity (mol/L). The density of sea water at 25°C is 1.023 kg/L as compared to 0.997 kg/L for air-free pure water so that the difference between molality and molarity amounts to approximately 2.5%. Table 1.3 presents equations for the conversion between different units and some examples are given in Example 1.1.

EXAMPLE 1.1. *Recalculation of concentration units*

1. Gram formula weights are calculated from the periodic system as reproduced in Table 1.4 from the *Handbook of chemistry and physics*.
   The mass of 1 mol Ca is 40.08 grams.
   1 mol SO_4^{2-} weighs: 32.06 grams from sulfur + 4 × 15.9994 grams from oxygen, in total: 96.06 grams.
2. Conversion of mg/L to mmol/L is obtained by dividing by the weight of the element or molecule.
   Thus, a river water contains 1.2 mg Na^+/L;
   This corresponds to 1.2 / 22.99 = 0.052 mmol Na^+/L;
   The sample also contains 0.6 mg SO_4^{2-}/L;
   This equals 0.6 / 96.06 = 0.006 mmol SO_4^{2-}/L.
3. The term mmol/L indicates the number of ions or molecules in the water when multiplied by Avogadro's number. For Na^+ in the river water sample it amounts to $0.052 \times 10^{-3} \times 6.022 \times 10^{23} = 3.1 \times 10^{19}$ ions of Na^+ in 1 liter of water. (Quite a lot really!)
4. Ions are electrically charged, and the sums of positive and negative charges in a given water sample must balance. This condition is termed the electroneutrality or electrical balance of the solution. Since mmol/L represents the number of molecules, it should be multiplied by the charge of the ions to yield their total charge in meq/L. Thus:
   0.052 mmol Na^+/L × 1 = 0.052 meq/L;
   1.8 mmol Ca^{2+}/L × 2 = 3.6 meq/L;
   0.41 mmol SO_4^{2-}/L × −2 = −0.82 meq/L.

Table 1.4. Periodic table of the elements. Reprinted with permission from *Chem. Eng. News*, February 4, 1985, 63(5), p. 26–27. Published in 1985 by the American Chemical Society.

New notation / Previous IUPAC form / CAS version

(Numbers in parentheses are mass numbers of most stable isotope of that element)

KEY TO CHART

Atomic Number →	50 +2	← Oxidation States
Symbol →	Sn +4	
1987 Atomic Weight →	118.71	
	18 18 4	← Electron Configuration

1 Group IA	2 IIA	3 IIIA IIIB	4 IVA IVB	5 VA VB	6 VIA VIB	7 VIIA VIIB	8	9 VIIIA VIII	10	11 IB	12 IIB	13 IIIB IIIA	14 IVB IVA	15 VB VA	16 VIB VIA	17 VIIB VIIA	18 VIIIA	Orbit
1 +1 −1 H 1.00794 1																	2 0 He 4.0020602 2	K
3 +1 Li 6.941 2-1	4 +2 Be 9.012182 2-2											5 +3 B 10.811 2-3	6 +2 +4 −4 C 12.011 2-4	7 +1 +2 +3 +4 +5 −1 −2 −3 N 14.00674 2-5	8 −2 O 15.9994 2-6	9 −1 F 18.9984032 2-7	10 0 Ne 20.1797 2-8	K-L
11 +1 Na 22.989768 2-8-1	12 +2 Mg 24.3050 2-8-2											13 +3 Al 26.981539 2-8-3	14 +2 +4 −4 Si 28.0855 2-8-4	15 +3 +5 −3 P 30.97362 2-8-5	16 +4 +6 −2 S 32.066 2-8-6	17 +1 +5 +7 −1 Cl 35.4527 2-8-7	18 0 Ar 39.948 2-8-8	K-L-M
19 +1 K 39.0983 -8-8-1	20 +2 Ca 40.078 -8-8-2	21 +3 Sc 44.955910 -8-9-2	22 +2 +3 +4 Ti 47.88 -8-10-2	23 +2 +3 +4 +5 V 50.9415 -8-11-2	24 +2 +3 +6 Cr 51.9961 -8-13-1	25 +2 +3 +4 +7 Mn 54.93085 -8-13-2	26 +2 +3 Fe 55.847 -8-14-2	27 +2 +3 Co 58.93320 -8-15-2	28 +2 +3 Ni 58.69 -8-16-2	29 +1 +2 Cu 63.546 -8-18-1	30 +2 Zn 65.39 -8-18-2	31 +3 Ga 69.723 -8-18-3	32 +2 +4 Ge 72.61 -8-18-4	33 +3 +5 −3 As 74.92159 -8-18-5	34 +4 +6 −2 Se 78.96 -8-18-6	35 +1 +5 −1 Br 79.904 -8-18-7	36 0 Kr 83.80 -8-18-8	-L-M-N
37 +1 Rb 85.4678 -18-8-1	38 +2 Sr 87.62 -18-8-2	39 +3 Y 88.90585 -18-9-2	40 +4 Zr 91.224 -18-10-2	41 +3 +5 Nb 92.90638 -18-12-1	42 +6 Mo 95.94 -18-13-1	43 +4 +6 +7 Tc (98) -18-13-2	44 +3 Ru 101.07 -18-15-1	45 +3 Rh 102.90550 -18-16-1	46 +2 +4 Pd 106.42 -18-18-0	47 +1 Ag 107.8682 -18-18-1	48 +2 Cd 112.411 -18-18-2	49 +3 In 114.82 -18-18-3	50 +2 +4 Sn 118.710 -18-18-4	51 +3 +5 −3 Sb 121.75 -18-18-5	52 +4 +6 −2 Te 127.60 -18-18-6	53 +1 +5 +7 −1 I 126.90447 -18-18-7	54 0 Xe 131.29 -18-18-8	-M-N-O
55 +1 Cs 132.90543 -18-8-1	56 +2 Ba 137.327 -18-8-2	57* +3 La 138.9055 -18-9-2	72 +4 Hf 178.49 -32-10-2	73 +5 Ta 180.9479 -32-11-2	74 +6 W 183.85 -32-12-2	75 +4 +6 +7 Re 186.207 -32-13-2	76 +3 +4 Os 190.2 -32-14-2	77 +3 +4 Ir 192.22 -32-15-2	78 +2 +4 Pt 195.08 -32-16-2	79 +1 +3 Au 196.96654 -32-18-1	80 +1 +2 Hg 200.59 -32-18-2	81 +1 +3 Ti 204.3833 -32-18-3	82 +2 +4 Pb 207.2 -32-18-4	83 +3 +5 Bi 208.98037 -32-18-5	84 +2 +4 Po (209) -32-18-6	85 At (210) -32-18-7	86 0 Rn (222) -32-18-8	-N-O-P
87 +1 Fr (223) -18-8-1	88 +2 Ra 226.025 -18-8-2	89** Ac +3 227.028 -18-9-2	104 Unq +4 (261) -32-10-2	105 Unp (262) -32-11-2	106 Unh (263) -32-12-2	107 Uns (262) -32-13-2												OPQ

															Orbit
*Lanthanides	58 +3 +4 Ce 140.115 -20-8-2	59 +3 Pr 140.90765 -21-8-2	60 +3 Nd 144.24 -22-8-2	61 +3 Pm (145) -23-8-2	62 +2 +3 Sm 150.36 -24-8-2	63 +2 +3 Eu 151.965 -25-8-2	64 +3 Gd 157.25 -25-9-2	65 +3 Tb 158.92534 -27-8-2	66 +3 Dy 162.50 -28-8-2	67 +3 Ho 164.93032 -29-8-2	68 +3 Er 167.26 -30-8-2	69 +3 Tm 168.93421 -31-8-2	70 +2 +3 Yb 173.04 -32-8-2	71 +3 Lu 174.967 -32-9-2	NOP
**Actinides	90 +4 Th 232.0381 -18-10-2	91 +5 +4 Pa 231.03588 -20-9-2	92 +3 +4 +5 +6 U 238.0289 -21-9-2	93 +3 +4 +5 +6 Np 237.048 -22-9-2	94 +3 +4 +5 +6 Pu (244) -24-8-2	95 +3 +4 +5 +6 Am (243) -25-8-2	96 +3 Cm (247) -25-9-2	97 +3 +4 Bk (247) -27-8-2	98 +3 Cf (251) -28-8-2	99 +3 Es (252) -29-8-2	100 +3 Fm (257) -30-8-2	101 +2 +3 Md (258) -31-8-2	102 +2 +3 No (259) -32-8-2	103 +3 Lr (260) -32-9-2	OPQ

Some terms often used in relation to groundwater chemistry are explained in Table 1.5. Apart from standard chemical terminology, like pH and *Eh*, it also contains specific items from the water world. For example, hardness of water is among other things used to dose the amount of soap needed in a washing machine.

Table 1.5. Common parameters used in the water world (cf. Standard Methods, 1985; Hem, 1985).

Hardness	Sum of the ions which can precipitate as "hard particles" from water. Sum of Ca^{2+} and Mg^{2+}, and sometimes Fe^{2+}. Expressed in meq/L or mg $CaCO_3$/L or in hardness degrees. 100 mg $CaCO_3$/L $\cong$ 1 mmol $CaCO_3$/L $\cong$ 2 meq Ca^{2+}/L
Hardness degrees	1 german degree = 17.8 mg $CaCO_3$/L 1 french degree = 10 mg $CaCO_3$/L
Temporary hardness	Part of Ca^{2+} and Mg^{2+} concentrations which are balanced by HCO_3^- (all expressed in meq/L) and can thus precipitate as carbonate
Permanent hardness	Part of Ca^{2+} and Mg^{2+} in excess of HCO_3^- (all expressed in meq/L)
Color	Measured by comparison with a solution of cobalt and platinum
EC	Electrical Conductivity, in μS/cm (= μmho/cm), $EC \approx 100 \times$ meq (anions or cations)/L
pH	$-\log[H^+]$, the log of H^+ activity (dimensionless).
Eh	Redox potential, expressed in Volt. measured with platinum/reference electrode
pe	Redox potential expressed as $-\log[e^-]$. $[e^-]$ is "activity" of electrons. pe = $Eh/0.059$ at 25°C.
Alkalinity (*Alk*)	Acid neutralizing capacity. Determined by titrating with acid down to a pH of about 4.5. Equal to the concentrations of $m_{HCO_3^-} + 2m_{CO_3^{2-}}$(mmol/L) in most samples.
Acidity	Base neutralising capacity. Determined by titrating up to a pH of about 8.3. Equal to H_2CO_3 concentration in most samples except when Al^{3+} or Fe^{3+} are present
TIC	Total inorganic carbon
TOC	Total organic carbon
COD	Chemical oxygen demand. Measured as chemical reduction of permanganate or dichromate solution, and expressed in oxygen equivalents
BOD	Biological oxygen demand.

1.3 GROUNDWATER QUALITY

The classical use of water analyses in hydrology is to show the regional distribution of the water compositions in a map. Such maps serve environmental authorities, water resource managers, drilling operators and other practitioners to identify aquifers with good quality groundwater, but they are also useful for a first assessment of the relation between the aquifer mineralogy and groundwater composition. For example, groundwater from a limestone aquifer is likely to contain enhanced calcium and bicarbonate concentrations.

A standard groundwater chemical analysis will as a minimum comprise values for temperature, *EC*, pH, the four major cations (Na^+, K^+, Mg^{2+}, Ca^{2+}) and four major anions (Cl^-, HCO_3^-, SO_4^{2-}, NO_3^-), in other words eleven variables. When the number of available water analyses in an area accumulates, it becomes increasingly more difficult to overview all the numbers. Therefore graphical methods have been developed to display the chemical composition of the main components in groundwater at a glance. Figure 1.3 shows examples of two common diagram types, the bar diagram and the circle diagram. The bar diagram consists of two adjacent columns; the left column for cations and the right column for anions. Each column shows the stacked contribution of the various ions with concentrations expressed in milliequivalents per liter. Since the sum of positive charges should be balanced by the sum of negative charges, the height of the two columns should be the same for a good water analysis.

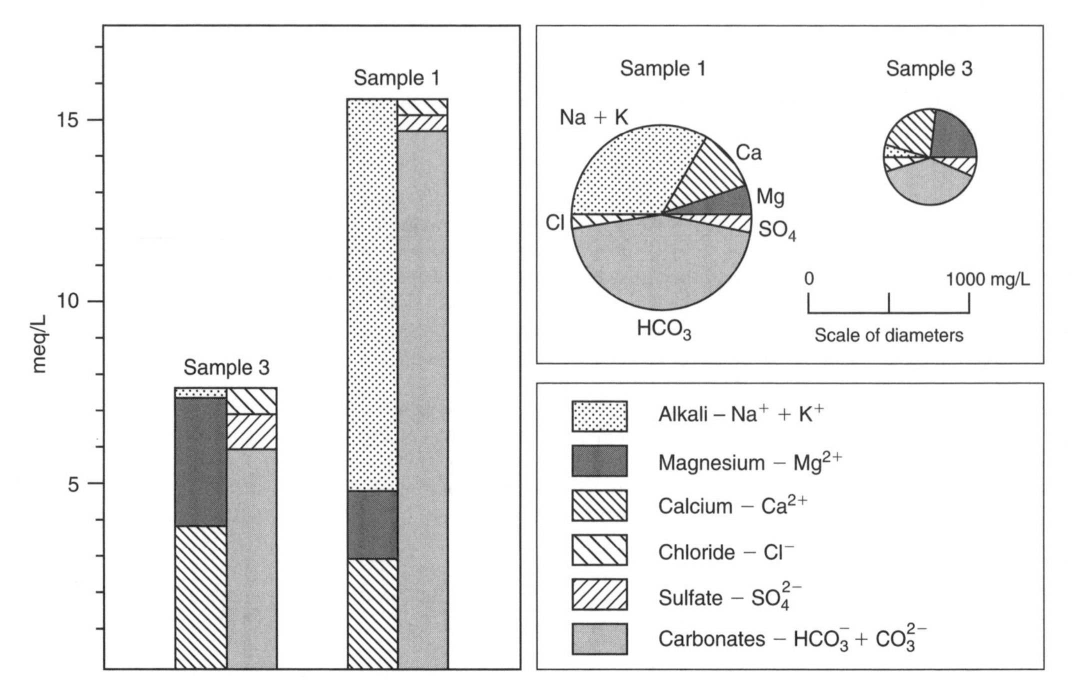

Figure 1.3. Bar and circle diagrams, two graphical methods to display the bulk chemical composition of groundwater (Domenico and Schwartz, 1997).

In the circle diagram the size of the circle is proportional to total dissolved solids. The circle is then subdivided in an upper half showing the relative composition of the cations, and a lower half with the anions. Again, concentrations are expressed in milliequivalents per liter. In both bar and circle diagrams, waters with a similar chemical composition are quickly recognized.

A further step in visualizing is the Stiff diagram. It consists of three to four horizontal axes displaying selected components. On each axis a cation is plotted to the left and an anion to the right, again in milliequivalents per liter. The uppermost axis has Na^+ to the left and Cl^- to the right and reflects a possible marine influence since in seawater NaCl is the dominant salt. The second axis has Ca^{2+} to the left and HCO_3^- to the right and this axis is meant to display the dissolution of $CaCO_3$. The third axis has Mg^{2+} to the left and SO_4^{2-} to the right, presenting the remaining two major components in most waters. The fourth axis is optional and its variables may change from study to study. The values on each axis are connected by lines and a typical shape emerges for a given water composition. In Figure 1.4, sample 1 is rainwater with a low concentration of all the components. Sample 4 is from a limestone aquifer where $CaCO_3$ dissolution is the predominant process yielding a typical bird-like shape. Sample 3 is from an aquifer where dolomite, $CaMg(CO_3)_2$, is the dissolving mineral and therefore the concentrations of Ca^{2+} and Mg^{2+} are similar, the bird seems to have lost half of its left wing.

A large number of chemical analyses can be compiled in the so-called Piper diagram (Figure 1.5). The Piper diagram contains two triangular charts for depicting the proportions of cations and anions, expressed in meq/L. The triangle for cations has 100% Ca^{2+} in the left corner, 100% $Na^+ + K^+$ towards the right and 100% Mg^{2+} upwards (traditionally, Na^+ and K^+ are combined because these ions were analyzed jointly in a precipitate with uranium). The sum of the concentrations of the three ions, in milliequivalents per liter, is recalculated to 100% and the relative composition is plotted in the triangle. For anions the triangle has 100% carbonate to the left, Cl^- to the right and SO_4^{2-} on top.

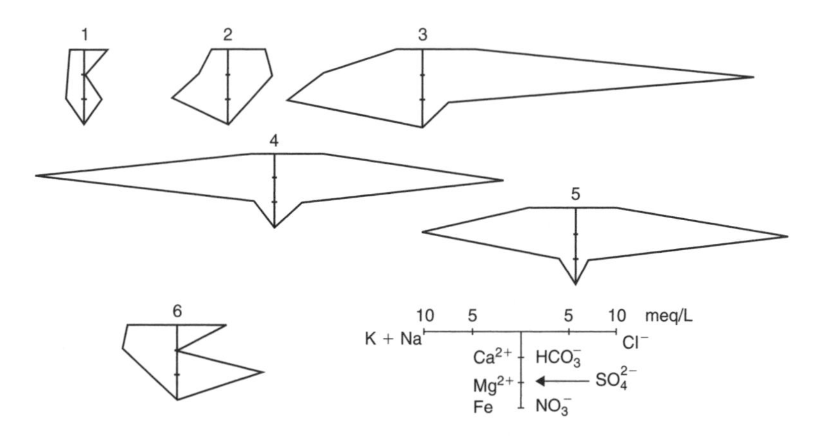

Figure 1.4. Stiff-diagrams reflecting the chemical composition of groundwaters. For explanation see text. Free software to construct Stiff diagrams can be downloaded from http://www. twdb.state.tx.us/publications/reports/GroundWaterReports/Open-File/Open-File_01-001.htm

The two data points in the triangles are joined by drawing lines parallel to the outer boundary until they unite in the central diamond shape (Figure 1.5). The relative chemical composition of the water sample is now indicated by a single point. The diamond diagram often forms the background for a descriptive terminology of the chemical composition of groundwater. For example a calcium-bicarbonate water type must plot near the lefthand corner of the diamond while a sodium-bicarbonate type will be located closer to the bottom corner.

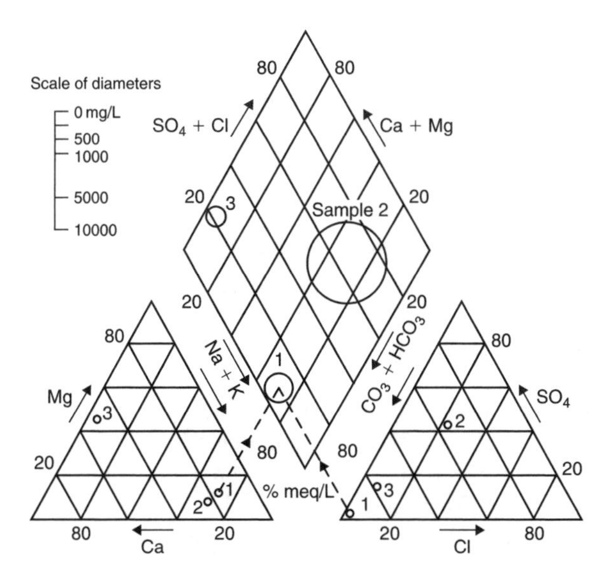

Figure 1.5. The Piper diagram, used for graphically displaying the bulk chemical composition of groundwaters. The two triangular plots give the relative composition of cations and anions, expressed in percentage of total cation or anion meq/L. From the triangular diagrams the points are conducted to the diamond shaped diagram parallel to the outer side of the diagram until they intersect (dotted lines for sample 1). The size of the circle in the diamond indicates the total dissolved solid concentration (Domenico and Schwartz, 1997). Free software for constructing Piper diagrams can be downloaded from http://water.usgs.gov/nrp/gwsoftware/GW_Chart/GW_Chart.html

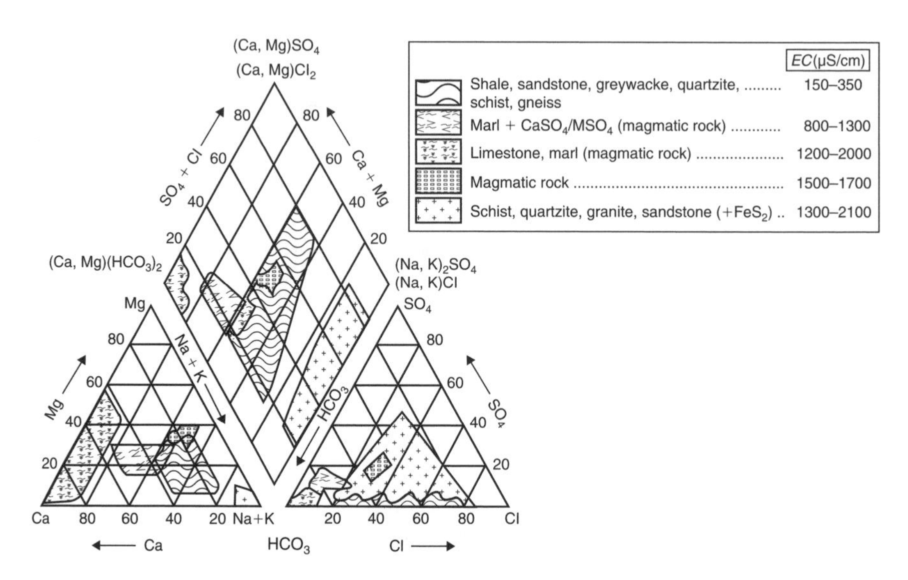

Figure 1.6. A Piper plot of European bottled mineral waters and their relation to the rock type from which the water was extracted (modified from Zuurdeeg and Van der Weiden, 1985).

Figure 1.6 shows a Piper diagram with the composition of mineral waters from Europe and their relation to the rock type from which they have been extracted (Zuurdeeg and Van der Weiden, 1985). Note that mineral waters are retrieved from almost any rock type and accordingly there is also a large variation in their chemical composition. The diagram shows that limestones and marls produce a Ca, Mg-HCO_3 type of water, whereas groundwater in metamorphic rocks (schists, sandstones, etc.) or igneous rocks (granites and magmatic rocks) also contain high concentrations of elements such as Na^+, K^+, and Cl^-. The electrical conductivity (*EC*) of the mineral waters indicate that appreciable amounts of salts are present that may exceed the limit for a regular drinking water supply (Table 1.1).

QUESTION:
The composition of sample 1 in Figure 1.5 is (mg/L): Na^+ 245; Mg^{2+} 21.9; Ca^{2+} 61.7; Cl^- 11.3; HCO_3^- 906; SO_4^{2-} 21.2. Calculate the cation and anion percentages, and compare with Figure 1.5.

1.4 SAMPLING OF GROUNDWATER

Both the sampling and the chemical analysis of groundwater are expensive and particularly the cost of drilling new boreholes is high. It is therefore important to evaluate beforehand which data are required to solve a specific problem. Furthermore, in the interpretation of the chemical analyses, both the well construction and the sampling procedures should be considered when judging the representativeness of the samples for the actual groundwater composition.

1.4.1 *Depth integrated or depth specific sampling*

Figure 1.7 shows two different approaches towards sampling of groundwater. The borehole on the left has a long screen and the sample obtained is integrated over the length of the screen as well as over the permeability of the formation. The screen interval with fine grained sand will contribute relatively less to the total yield than the coarse grained part with a much higher permeability.

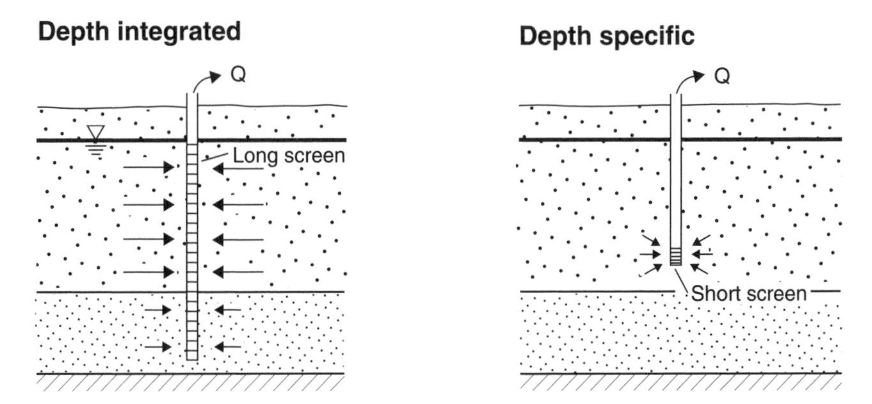

Figure 1.7. Depth integrated versus depth specific groundwater sampling. Coarse dotted areas represent coarse grained sand and densely dotted areas fine grained sand. The size of the arrows reflects the flow rate (modified from Cherry, 1983).

The borehole to the right has a short screen and will provide depth specific information on the chemical composition of the groundwater in a distinct sediment layer. Depth specific sampling is usually required for detailed studies of the chemical processes in the aquifer.

In many cases groundwater compositions show major variations with depth even on a small scale. Integrated samples from a screen interval of several meters may accordingly represent mixtures of waters with different concentrations and the mixing process may even induce chemical reactions during sampling. Figure 1.8 displays a groundwater chemistry profile, obtained by depth specific sampling in a sandy aquifer with a well defined oxic zone on top of an anoxic zone containing Fe^{2+}.

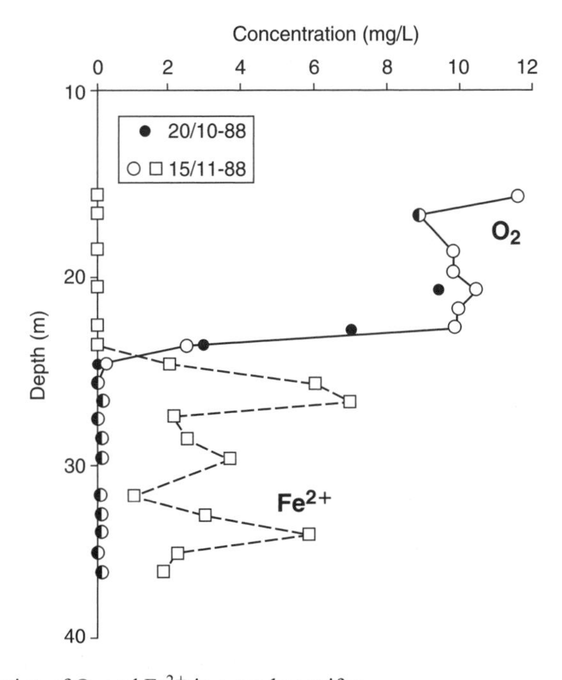

Figure 1.8. The distribution of O_2 and Fe^{2+} in a sandy aquifer.

Suppose the borehole has a screen installed that draws water from both the oxic and anoxic zones. The water with O_2 will mix and react with Fe^{2+} containing water and the concentrations of O_2, Fe^{2+} and pH or alkalinity will alter by the reaction:

$$2Fe^{2+} + \frac{1}{2}O_2 + 3H_2O \rightarrow 2FeOOH + 4H^+ \quad (1.1)$$

The composition of the resulting sample depends on the extent of mixing and reaction. In principle even a sample containing both Fe^{2+} and O_2 might be captured although such a water quality is not present in the subsoil. In any case, the resulting sample will not reflect the *in situ* conditions in the aquifer very well. Similar reactions occur in shallow, open wells with a large body of stagnant water in contact with air. Production wells, usually equipped with screens of tens of meters, may be pumped sectionwise in an attempt to obtain more detailed, depth specific information (Lerner and Teutsch, 1995).

1.4.2 *Procedures for sampling of groundwater*

The first concern in sampling is the contamination and disturbance of natural conditions caused by drilling operations. It may take a long time before the influence of new materials brought into the aquifer, including drilling fluids, gravel pack or casing materials, is sufficiently diminished to allow for representative sampling. Water from boreholes where drilling mud has been used or where clay layers are sealed with bentonite (a swelling clay mineral) often displays cation exchange with the clay. If trace metals or trace organics are going to be analyzed, special care is required to avoid contamination by or sorption onto the sampling materials. The time period that should pass before representative samples can be obtained depends on the groundwater flow rate, the ion exchange capacity etc., and may be in the order of two to three months when pumping on the screen is limited (as is often the case with depth specific samplers).

Wells which have been out of production for some time may yield a water chemistry that is different from the composition during pumping. The main reason is the presence of stagnant water above the screen in the well and it is therefore necessary to empty the well for a number of volumes. On the other hand, excessive pumping may draw waters with a different composition towards the screen and cause mixing of waters. Thus, a balance has to be found between these two aspects.

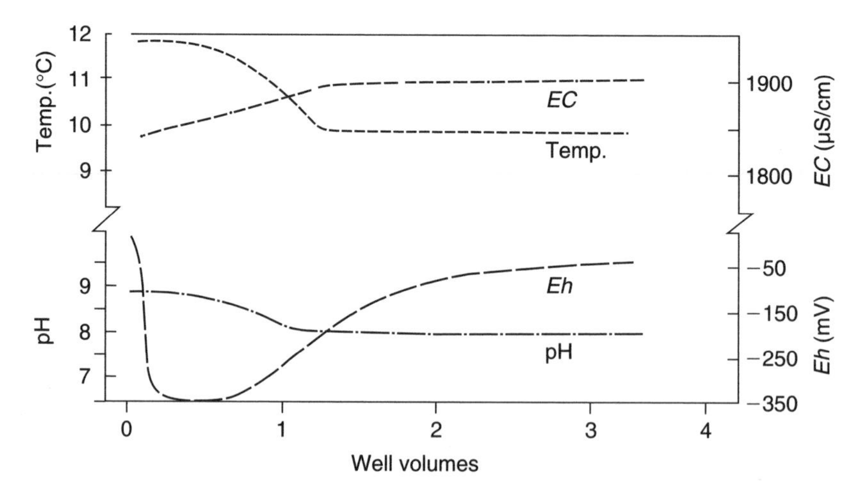

Figure 1.9. The change in chemical composition of discharge during well flushing (modified from Lloyd and Heathcote, 1985).

Estimates for the number of well volumes to be emptied vary between 2 and 10 times and depend on the local hydrological conditions (Barber and Davis, 1987; Robin and Gillham, 1987; Stuyfzand, 1983). In most cases 2 to 4 times seems sufficient.

To accomplish effective flushing, the well should be pumped from just below the air-water interface (Robin and Gillham, 1987). The best way to evaluate the degree of flushing needed is to monitor easily measured field parameters such as the electrical conductivity (*EC*) or pH over time. The example shown in Figure 1.9 demonstrates that in this case stationary conditions are obtained after emptying about two well volumes.

Problems with sampling from existing wells are often related to faulty completion of wells and are not always easy to detect. Some examples of commonly encountered problems are illustrated in Figure 1.10. In Figure 1.10A, the screen length is too large causing drainage of water from an upper sandy layer, with a higher pressure, into a lower sandy layer with a lower hydraulic potential. Case B displays an improperly placed bentonite seal which results in leakage from the upper part of the aquifer into the lower part. In C, the gravel pack functions as a drain which also may cause short circuiting. There are examples of the application of herbicide to well fields, intended to make things look pleasant at the surface, but resulting in downward seepage of the chemical along the outer side of the well casing and polluting the well. Finally, case D displays the effect of leaky coupling of the casing.

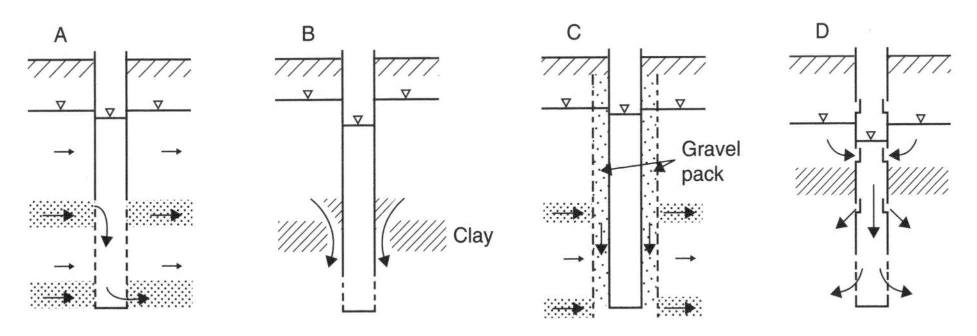

Figure 1.10. Different forms of leakage and short circuiting during groundwater sampling from screened wells (modified from Stuyfzand, 1989). For explanation see text.

For proper depth specific sampling, dedicated systems are required (Figure 1.11). In shallow aquifers a drive-point piezometer, hammered or pushed down, may provide high-resolution profiles of the groundwater chemical composition (Jakobsen and Postma, 1999). Towards greater depth, a hollow stem auger equipped with a water sampling system may be useful (Figure 1.11), especially when combined with geophysical sensors to delineate the geological variations (Sørensen and Larsen, 1999). It is also possible to take cores, which can be centrifuged or pressurized to procure water samples from clay layers. The disadvantage of the drive-point piezometer is that resampling the groundwater needs renewed augering, and therefore permanent installations might be preferred for monitoring purposes. Sampling points can be installed individually, or in a multiple-level, single borehole piezometer (Obermann, 1982; Lerner and Teutsch, 1995).

Single hole multiple piezometers can be introduced as bundles (Cherry et al., 1983; Appelo et al., 1982; Stuyfzand, 1983; Leuchs, 1988) or enclosed in a casing (Pickens et al., 1978; Postma et al., 1991). While the first approach is by far the cheapest, it requires cohesionless aquifer material to ensure sealing between sampling points, and the risk of short circuiting between different sample levels through the bundles of tubing always remains even when bentonite is applied in-between.

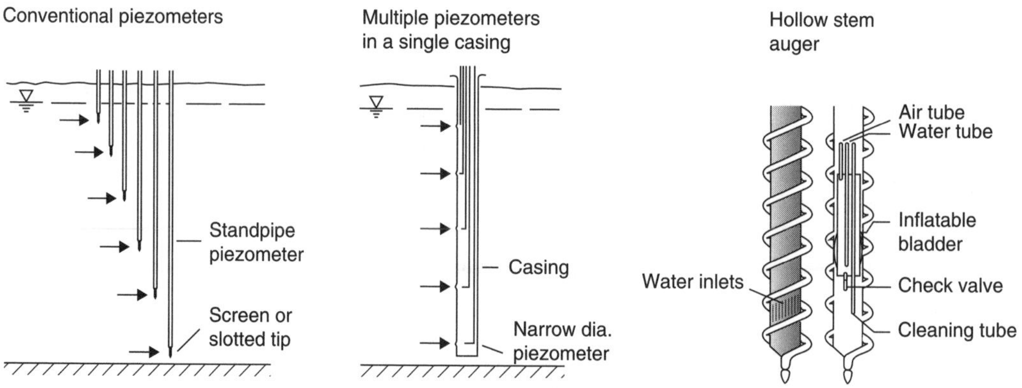

Figure 1.11. Three approaches for depth specific groundwater sampling; left, multiple-borehole piezometer nest; middle, multiple piezometers enclosed in a casing (Cherry, 1983); right, hollow stem auger with a water sampling device (Sørensen and Larsen, 1999).

This risk increases with the number of sample levels used and the difference in hydraulic head in the layers which have been penetrated.

Sample retrieval is done most easily by suction if the water table is less than 9 m below the surface. However, the application of vacuum may result in degassing of the sample (Suarez, 1987; Stuyfzand, 1983) and produce erratic results for dissolved gases, pH and volatile organics. Alternative approaches are downhole pumps (Gillham and Johnson, 1981), or double line gas driven sampling devices (Appelo et al., 1982). In the latter approach (Figure 1.12), a downhole reservoir being filled with groundwater, is emptied by pressurizing the sampler with an inert gas like N_2 or Ar.

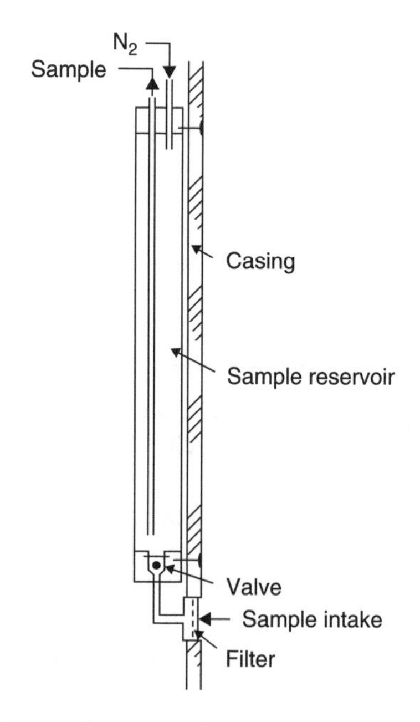

Figure 1.12. Sampling of groundwater using a downhole reservoir with a check valve. Pressurizing the reservoir with N_2 or Ar, closes the check valve and the sample is transported towards the surface.

A check valve at the bottom of the reservoir is then closed and the sample transported to the surface. Upon (gentle) release of the pressure, the reservoir is filled again. There is, however, still a risk for gas exchange at the interfaces between the sample and the pressurizing gas. Samples should therefore preferably be taken in the middle part of the sampling volume.

1.5 CHEMICAL ANALYSIS OF GROUNDWATER

There is probably no such thing as a perfect chemical water analysis carried out by routine procedures. For specific purposes both the sampling and analytical procedures have to be adjusted to obtain the best possible results for the critical parameters in the given study.

The analytical chemistry of groundwater in general is beyond the scope of this book. In most cases water analyses are carried out by standard procedures such as described in handbooks like "Standard Methods for the examination of water and wastewater" (1985). However, some aspects concerning collection and conservation of water samples, and the evaluation of the quality of chemical analysis deserve special attention.

1.5.1 *Field analyses and sample conservation*

Water samples may contain variable concentrations of suspended solids or colloids that need to be removed before the analysis of the "real" solute ions can begin. The size of the colloids ranges from 10^{-9} to about 10^{-4} m, and actually, the scale from solutes to colloids is a sliding one (Stumm and Morgan, 1996). The standard procedure is to filtrate through 0.45 μm membrane filters. However, finely dispersed Fe- and Al-oxyhydroxides may pass through 0.45 μm filters, and 0.1 μm is a better choice and able to remove viruses as well (Kennedy and Zellweger, 1974; Laxen and Chandler, 1982). Pressure filtration, using an inert gas is preferred since vacuum application may degas the sample. Disposable syringes and filter cartridges are often handy for sample manipulation and to keep the sample out of contact with the atmosphere. Access of atmospheric oxygen may result in the oxidation of components like Fe^{2+}, H_2S, etc. which are often present in anoxic groundwater. Furthermore, degassing of CO_2 may occur, causing changes in pH, alkalinity and total inorganic carbon, which may also induce carbonate precipitation (Suarez, 1987). Also degassing of methane may occur.

Precautions to avoid changes in the chemical composition of the sample before analysis consist of either measuring sensitive components in the field or the conservation of samples for later analysis in the laboratory. An overview of the required treatment is presented in Table 1.6. Conservation is in most cases done by adding acid to the sample until the pH is <2 (0.7 mL of 65% HNO_3 is usually sufficient to neutralize alkalinity and to acidify 100 mL sample). Acidification stops most bacterial growth, blocks oxidation reactions, and prevents adsorption or precipitation of cations. The effect of improper handling on the analytical results is illustrated in Example 1.2.

EXAMPLE 1.2. *Effect of iron oxidation on analytical results*
A groundwater sample contains 20 mg Fe^{2+}/L. How does the alkalinity change when this iron oxidizes and precipitates as $Fe(OH)_3$ in the sample bottle?
The reaction in the bottle is:

$$2Fe^{2+} + 4HCO_3^- + \frac{1}{2}O_2 + 5H_2O \rightarrow 2Fe(OH)_3 + 4H_2CO_3$$

When 20 mg Fe/L precipitates, this corresponds to 20 / 55.8 = 0.36 mmol/L. The reaction consumes twice this amount of HCO_3^-, and the alkalinity is expected to decrease with $2 \times 0.36 = 0.72$ meq/L.

Table 1.6. Conservation of chemical parameters of water samples.

Parameter	Conservation/field analysis
Na^+, K^+, Mg^{2+}, Ca^{2+} NH_4^+, Si, PO_4^{3-}	Acidify to pH <2 in polyethylene container (preferably HNO_3 for AAS or ICP-analysis).
Heavy metals	Acidify to pH <2 in glass or acid rinsed polypropylene container.
SO_4^{2-}, Cl^-	Cool to 4°C.
NO_3^-, NO_2^-	Store cool at 4°C and analyze within 24 hours or add bactericide like thymol. (Note that NO_3^- may form from NH_4^+ in reduced samples. NO_2^- may self-decompose even when a bactericide is added.)
H_2S	To avoid degassing, collect sample in a Zn-acetate solution, precipitating ZnS. Spectrophotometry in the field or later in the laboratory.
TIC	Dilute sample to $TIC < 0.4$ mmol/L. (This effectively reduces CO_2 pressure, and prevents the escape of CO_2).
Alkalinity	Field titration with the GRAN method (Stumm and Morgan, 1996)
Fe^{2+}	Spectrophotometry in the field. Alternatively determined as Fe-total in an acidified sample.
pH, Temp., *EC*, O_2	Field measurement in a flow cell.
CH_4	Unfiltered sample collected avoiding degassing, then acidified.

Field analyses are usually carried out for parameters like pH, *EC*, *Eh*, O_2, which are measured by electrode, and sometimes also for alkalinity, Fe^{2+} and H_2S. Electrode measurements should preferably be carried out in a flow cell mounted directly on the well head in order to prevent air admission. *EC* measurements are particularly useful as a control on analysis and conservation of samples. *Eh* measurements only give a qualitative indication of the redox conditions and should be made as sloppy as possible, so you will not be tempted to relate them to anything quantitative afterwards (Chapter 9).

Figure 1.13 compares pH measurements carried out immediately in the field with later laboratory measurements and shows substantial differences. These differences are even more significant when it is remembered that pH is a logarithmic unit ($pH = -\log[H^+]$). Since the pH is of major importance in geochemical calculations, care should be taken to obtain reliable measurements.

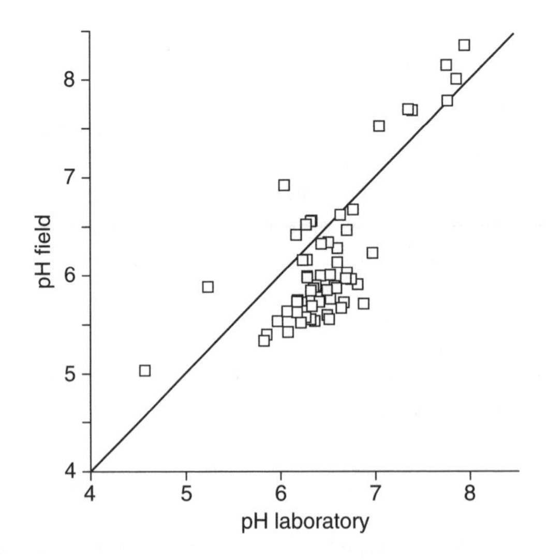

Figure 1.13. Comparison of field measurements of pH, in a carbonate-free sandy aquifer, with those performed in the laboratory (Postma, unpublished results).

While a pH measurement in principle is a simple operation, there are a number of potential sources of error. Some problems are related to the removal of groundwater from its position in the subsoil, usually anoxic and possibly at a high hydrostatic pressure. Most problems can be overcome by careful sampling and field measurement procedures. In particular, the use of in-line flow cells in a pressurized sampling system helps to minimize the problems of pH measurement related to the change of environment. Another source of error in low ionic strength groundwaters is the liquid junction between the reference electrode and the solution to be measured. The fundamental problem is that the liquid junction potential across the porous ceramic plug of the reference electrode will vary with the solution composition (Bates, 1973). Liquid junction problems in low ionic strength waters have been recognized (Illingworth, 1981; Brezinski, 1983; Davison and Woof, 1985) and may amount to several tens of a pH unit. A tricky aspect of this error is that it is not revealed by a standard two buffer calibration (Illingworth, 1981; Davison and Woof, 1985), since commercial buffers are all high ionic strength solutions. Electrode performance should therefore be checked in dilute solutions, for example a solution of 10^{-4}M HCl or 5×10^{-5}M H_2SO_4, which should yield a pH of 4.00 ± 0.02, cf. Davison (1987) and Busenberg and Plummer (1987).

1.5.2 *Accuracy of chemical analysis*

In general, two types of errors are discerned in chemical analyses:

Precision or statistical errors which reflect random fluctuations in the analytical procedure.
Accuracy or systematic errors displaying systematic deviations due to faulty procedures or interferences during analysis.

The precision can be calculated by repeated analysis of the same sample. It is always a good idea to collect a number of duplicate samples in the field as a check on the overall procedure. Systematic errors can be tested only by analyzing reference samples and by interlaboratory comparison of the results. At low concentrations, duplicate analyses may show large variations when the sensitivity of the method is insufficient. The accuracy of the analysis can for major ions be estimated from the electrical balance (*E.B.*) since the sum of positive and negative charges in the water should be equal:

$$\text{Electrical Balance } (E.B., \%) = \frac{(\text{Sum cations} + \text{Sum anions})}{(\text{Sum cations} - \text{Sum anions})} \times 100 \qquad (1.2)$$

where cations and anions are expressed as meq/L and inserted with their charge sign. The sums are taken over the cations Na^+, K^+, Mg^{2+} and Ca^{2+}, and anions Cl^-, HCO_3^-, SO_4^{2-} and NO_3^-. Sometimes other elements contribute significantly, for example ferrous iron (Fe^{2+}) or NH_4^+ in reduced groundwater, or H^+ and Al^{3+} in acid water. The presence of the last two substances in significant amounts requires more accurate calculations using a computer speciation program like PHREEQC which will be presented in Chapter 4. Differences in *E.B.* of up to 2% are inevitable in almost all laboratories. Sometimes an even larger error must be accepted, but with deviations in excess of 5% the sampling and analytical procedures should be examined.

EXAMPLE 1.3. *Estimating the reliability of water analyses*
Your laboratory returns the following water analysis:

pH = 8.22 *EC* = 290 μS/cm

Na^+	K^+	Mg^{2+}	Ca^{2+}	Cl^-	HCO_3^-	SO_4^{2-}	NO_3^-
13.7	1.18	3.2	42.5	31.2	79.9	39	1.3 mg/L

Is this a reliable analysis?

A first requirement for a water analysis is a reasonable charge balance of cations and anions: the solution should be electrically neutral. The analysis is recalculated from mg/L to mmol/L and then to meq/L.

	mg/L		Formula wt. (gram/mol)		mmol/L		charge		meq/L
Na^+	13.7	/	22.99	=	0.60	×	1	=	0.60
K^+	1.18	/	39.1	=	0.03	×	1	=	0.03
Mg^{2+}	3.21	/	24.31	=	0.13	×	2	=	0.26
Ca^{2+}	42.5	/	40.08	=	1.06	×	2	=	2.12
								Σ =	3.01
Cl^-	31.2	/	35.45	=	0.88	×	−1	=	−0.88
HCO_3^-	79.9	/	61.02	=	1.31	×	−1	=	−1.31
SO_4^{2-}	39	/	96.06	=	0.41	×	−2	=	−0.82
NO_3^-	1.3	/	62.0	=	0.02	×	−1	=	−0.02
								Σ =	−3.03

The difference between cations and anions is only 0.02 meq/L. The electrical balance is:

$$E.B. = (3.01 + (-3.03))/(3.01 - (-3.03)) \times 100 = -0.02/6.04 = -0.3\%.$$

which is very good indeed.

Another useful technique is to compare the calculated and measured electrical conductivity. Different ions have a different *molar conductivity* Λ_m which is defined as:

$$\Lambda_m = EC_i/m_i \tag{1.3}$$

where EC_i is the electrical conductivity (Siemens/m, S/m), contributed by ion i with a concentration of m_i mol/L. The Λ_m has accordingly the units of $S \cdot m^2/mol$. The conductivity is related to the mobility of the ion and its charge, and varies for different electrolyte ions as noted in Table 1.7.

Table 1.7. Mobility and conductivity for ions at trace concentration in water at 25°C. Conductivities from Landolt and Bornstein (1960).

	Na^+	K^+	Mg^{2+}	Ca^{2+}	Cl^-	HCO_3^-	SO_4^{2-}	
Conductivity	50.1	73.5	106	119	76.35	44.5	160	$S \cdot cm^2/mol$
Mobility	5.19	7.62	5.46	6.17	7.91	4.61	8.29	$\times 10^{-8} m^2/s/V$

The unit for the molar conductivity in the table is $S \cdot cm^2/mol$ which is equal to $\mu S/cm \cdot L/mmol$. The conductivity of a solution is, at least for trace concentrations, obtained by multiplying the conductivity listed in Table 1.7 with the concentrations of the ions in mmol/L and summing up. For example, the *EC* of a solution of 0.1 mM NaCl is $50.1 \times 0.1 + 76.35 \times 0.1 = 12.6\ \mu S/cm$, and of 0.3 mM $Ca(HCO_3)_2$ is $119 \times 0.3 + 44.5 \times 2 \times 0.3 = 62.4\ \mu S/cm$, both at 25°C. The *ion mobility* is obtained by dividing the conductivity by the charge number $|z|$ of the ion and the Faraday constant, $F = 96485$ C/mol:

$$u = \Lambda_m/(|z| \cdot F) \tag{1.4}$$

where u is the mobility in $10^{-8} m^2/s/V$. The ion mobilities are related to the diffusion coefficients, as we will see in Chapter 3.

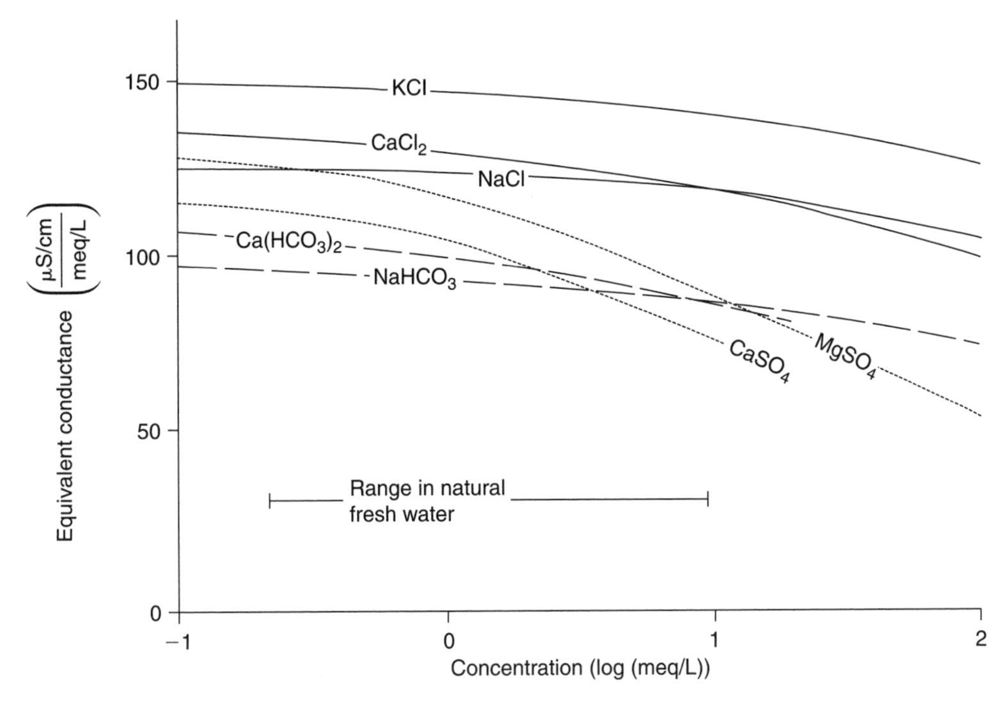

Figure 1.14. The relation between equivalent electrical conductivity and concentration for different salts.

The molar conductivity generally decreases with increasing concentration as is illustrated in Figure 1.14 for various salts. Here, the equivalent conductivity (or conductance) is plotted, i.e. the molar conductivity is divided by the sum of the charge numbers of the ions in the salt. The decrease of the conductivity is due to ion-association and electrostatic shielding effects which increase with concentration. Figure 1.14 shows that a $Ca(HCO_3)_2$ fresh water will have an equivalent conductivity of close to 100 μS/cm per meq/L. Therefore, at 25°C, the *EC* divided by 100, will in most cases give a good estimate of the sum of anions or cations (both in meq/L):

$$\Sigma \text{ anions} = \Sigma \text{ cations (meq/L)} \approx EC/100 \text{ } (\mu\text{S/cm}) \tag{1.5}$$

This relation is valid for EC-values up to around 1500 μS/cm. More complicated expressions have been derived by Rossum (1975) and Stuyfzand (1987). Also relationships between the *EC* and total dissolved solids (*TDS* in mg/L) have been suggested (Hem, 1985), but these are less reliable since they contain an additional assumption about average molecular weight of the conducting ions.

The calculation of the *E.B.*, and comparison with the *EC*, as a check on the chemical analyses, is only applicable for major elements. The accuracy for minor elements is much more difficult to estimate. Sometimes the incompatibility of different elements found together in one sample can be a warning that something went wrong. Thus, it is unlikely that O_2 or NO_3^-, which indicate oxidizing conditions, are found together with appreciable concentrations of Fe^{2+} in natural water (see Figure 1.8). Ferrous iron (Fe^{2+}) only occurs in appreciable concentrations (more than 1 μmol/L) in reduced environments. The solubility of ferric iron (Fe^{3+}) is low in water with a pH between 3 and 11, and it precipitates rapidly as $Fe(OH)_3$. The total iron concentration should therefore also be low in water where O_2 or NO_3^- is present. Also, the solubility of aluminum is low in the pH range 5 to 10, and it is unlikely to find an Al concentration of more than 1 μmol/L in water having a near neutral pH.

EXAMPLE 1.4. *Analytical errors due to precipitation in the sampling bottle*
In samples brought from Portugal to the laboratory in Amsterdam a large charge imbalance was found. A sample consisted of 100 mL water acidified to pH <2 for the analysis of cations, and 200 mL water for the analysis of HCO_3^-, Cl^-, SO_4^{2-}, NO_3^- as well as pH and *EC*. The samples were transported by car, and were subjected to significant temperature variations. A typical example is given below:

temp. = 18°C
pH_{field} = 7.00 $pH_{lab.}$ = 7.62
EC_{field} = 720 $EC_{lab.}$ = 558 μS/cm

Na^+	K^+	Mg^{2+}	Ca^{2+}	Cl^-	HCO_3^-	SO_4^{2-}	NO_3^-
0.55	0.01	0.35	3.00	0.64	4.31	0.16	0.12 mmol/L

Sum cations = 7.36; Sum anions = −5.38 meq/L

The sum of anions and cations in this sample differs by +16%, and could not be improved by duplicate analyses. It was realised, however, that cations and anions were analyzed in different bottles. When the cations were analyzed in the unacidified sample bottle, lower concentrations of Ca^{2+} were found. The reason for the imbalance was clearly that calcite had precipitated in the bottle without acid, thus decreasing the HCO_3^- concentration.

The reactions in the bottle can also be discovered when field measurements are compared with laboratory measurements. *EC* of the samples measured in the field was consistently higher by about 100 μS/cm than *EC* measured in the laboratory. Note the good agreement between $EC_{field}/100$ and the sum of cations.

PROBLEMS

1.1. An analysis of groundwater gave as results:

pH	Na^+	K^+	Mg^{2+}	Ca^{2+}	Cl^-	*Alk*	SO_4^{2-}	mg/L
8.5	62	0.1	9	81	168	183	–	

Alkalinity is expressed as HCO_3^-

a. Is the analysis correct?
b. Calculate total hardness, in degrees german hardness, and non-carbonate hardness.

1.2. Analyses of three water samples are listed below. Check the analyses and indicate where they might be wrong (concentrations in mmol/L).

	1	2	3
pH (−)	8.01	7.2	7.5
EC (μS/cm)	750	150	450
Na^+	2.3	0.3	0.5
K^+	0.1	0.08	<0.001
Mg^{2+}	1.1	0.2	0.2
Ca^{2+}	1.4	1.4	1.5
Cl^-	2.0	0.3	0.5
HCO_3^-	0.3	0.7	1.5
SO_4^{2-}	0.9	0.2	0.5
NO_3^-	1.8	0.1	1.5
Al	<0.001	0.2	<0.001
Fe	<0.001	<0.001	0.3

1.3. The analyses listed below reflect a range of water compositions in Denmark (concentrations in mg/L).

A = St. Heddinge waterworks, C = Harald's Mineralwater, B = Maarum waterworks, D = Rainwater Borris

Sample	A	B	C	D
pH	7.37	7.8	5.5	4.4
EC (μS/cm)	750.0	1900.0	88.0	
Na^+	18.0	440.0	9.0	7.2
K^+	6.3	14.0	1.0	0.63
Mg^{2+}	14.0	29.0		0.78
Ca^{2+}	121.0	42.0		1.4
NH_4^+	0.012	3.5		1.23
Cl^-	29.0	201.0	17.0	12.0
Alk (mg HCO_3^-/L)	297.0	1083.0		0.0
SO_4^{2-}	61.0	0		4.48
NO_3^-	35.0	2.0	5.0	4.52
PO_4	0	0.02		
F^-	0.32	1.7		0

a. Compare the water analyses with the drinking water limits in Table 1.1. Are these waters suitable as drinking water?
b. Check the quality of each analysis by setting up a charge balance. Compare the total amount of equivalents of ions with the electrical conductivity (*EC*) and identify possible errors in the analysis. Analysis C (Harald's Mineralwater) is incomplete. However, the hardness in german degrees is known to be 1°dH. Use this information to calculate the electrical conductivity. Note that this also sets limits to the missing anions.
c. Analysis D is rainwater from Borris (C. Jutland). Calculate the contribution of rainwater ions to the groundwater composition, assuming that chloride behaves like a conservative element. When would this assumption become dubious?
d. Draw a Stiff diagram for each analysis and discuss which processes have influenced the water composition. Consider what kind of geological deposit the analysis has been extracted from.

REFERENCES

Appelo, C.A.J., Krajenbrink, G.J.W., Van Ree, C.C.D.F. and Vasak, L., 1982. *Controls on groundwater quality in the NW Veluwe cacthment* (in Dutch). Soil Protection Series 11, Staatsuitgeverij, Den Haag, 140 pp.

Barber, C. and Davis, G.B., 1987. Representative sampling of ground water from short-screened boreholes. *Ground Water* 25, 581–587.

Bates, R.G., 1973. *Determination of pH; theory and practice*, 2nd ed. Wiley and Sons, New York, 479 pp.

Brezinski, D.P., 1983. Kinetic, static and stirring errors of liquid junction reference electrodes. *Analyst* 108, 425–442.

Busenberg, E. and Plummer, L.N., 1987. *pH measurement of low-conductivity waters.* U.S. Geol. Surv. Water Resour. Inv. Rep. 87–4060.

Cherry, J.A., 1983. Piezometers and other permanently-installed devices for groundwater quality monitoring. Proc. conf. on groundwater and petroleum hydrocarbons: protection, detection and restoration, IV-1 IV-39.

Cherry, J.A., Gillham, R.W., Anderson, E.G. and Johnson, P.E., 1983. Migration of contaminants in groundwater at a landfill: a case study. 2. Groundwater monitoring devices. *J. Hydrol.* 63, 31–49.

Davison, W., 1987. Measuring pH of fresh waters. In A.P. Rowland (ed.), *Chemical analysis in environmental research.* ITE symp. no 18, 32–37, Abbots Ripton.

Davison, W. and Woof, C., 1985. Performance tests for the measurement of pH with glass electrodes in low ionic strength solutions including natural waters. *Anal. Chem.* 57, 2567–2570.

Domenico, P.A. and Schwartz, F.W., 1997. *Physical and chemical hydrogeology*, 2nd ed. Wiley and Sons, New York, 506 pp.

Gillham, R.W. and Johnson, P.E., 1981. A positive displacement ground-water sampling device. *Ground Water Monitor. Rev.* 1, 33–35.

Halim, N.S., 2000. Arsenic mitigation in Bangladesh, *The Scientist*, March 6, 14–15.

Hem, J.D., 1985. *Study and interpretation of the chemical characteristics of natural water*, 3rd ed. U.S. Geol. Survey Water Supply Paper 2254, 264 pp.

Handbook of Chemistry and Physics. Am. Rubber Cy. (CRC Press)

Illingworth, J.A., 1981. A common source of error in pH measurement. *Biochem. Jour.* 195, 259–262.

Jacks, G., Bhattacharya, P. and Singh, K.P., 2000. High-fluoride groundwaters in India. In P.L. Bjerg, P. Engesgaard and Th.D. Krom (eds), *Groundwater 2000*, 193–194, Balkema, Rotterdam.

Jakobsen, R. and Postma, D., 1999. Redox zoning, rates of sulfate reduction and interactions with Fe-reduction and methanogenesis in a shallow sandy aquifer, Rømø, Denmark. *Geochim. Cosmochim. Acta* 63, 137–151.

Karim, M. M., 2000. Arsenic in groundwater and health problems in Bangladesh. *Water Res.* 34, 304–310.

Kennedy, V.C. and Zellweger, G.W., 1974. Filter pore-size effects on the analysis of Al, Fe, Mn, and Ti in water. *Water Resour. Res.* 10, 785–790.

Landolt and Bornstein, 1960. *Zahlenwerte und Funktione*, 7th Teil. Springer, Berlin.

Laxen, D.P.H. and Chandler, I.M., 1982. Comparison of filtration techniques for size distribution in freshwaters. *Anal. Chem.* 54, 1350–1355.

Leuchs, W., 1988. *Vorkommen, Abfolge und Auswirkungen anoxischer Redoxreaktionen in einem pleistozänen Porengrundwasserleiter*. Bes. Mitt. dt. Gewässerk. Jb. 52, 106 pp.

Lerner, D. and Teutsch, G. 1995. Recommendations for level-determined sampling in wells. *J. Hydrol.* 171, 355–377.

Lloyd, J.W. and Heathcote, J.A., 1985. *Natural inorganic hydrochemistry in relation to groundwater*. Clarendon Press, Oxford, 296 pp.

Obermann, P., 1982. *Hydrochemische/hydromechanische Untersuchungen zum Stoffgehalt von Grundwasser bei landwirtschaftlicher Nutzung*. Bes. Mitt. dt. Gewässerk. Jb. 42.

Pickens, J.F., Cherry, J.A., Grisak, G.E., Merrit, W.F. and Risto, B.A., 1978. A multilevel device for groundwater sampling and piezometric monitoring. *Ground Water* 16, 322–323.

Postma, D., Boesen, C., Kristiansen, H. and Larsen, F., 1991. Nitrate reduction in an unconfined sandy aquifer: Water chemistry, reduction processes, and geochemical modeling. *Water Resour. Res.* 27, 2027–2045.

Robin, M.J.L. and Gillham, R.W., 1987. Field evaluation of well purging procedures. *Ground Water Monitor. Rev.* 7, 85–93.

Rossum, J.R., 1975. Checking the accuracy of water analyses through the use of conductivity. *J. Am. Water Works Ass.* 67, 204–205.

Safe Drinking Water Comm., 1980. *Drinking water and health*, Vol. 3. Nat. Acad. Press, Washington, 415 pp.

Smedley, P.L. and Kinniburgh, D.G., 2002. A review of the source, behavior and distribution of arsenic in natural waters. *Appl. Geochem.* 17, 517–568.

Sørensen, K. and Larsen, F., 1999. Ellog auger drilling: Three-in-one method for hydrogeological data collection. *Ground Water Monit. Rev.* 17, 97–101.

Standard Methods for the examination of water and wastewater, 1985. Joint publication of APHA, AWWA, WPCF, 16th ed. Am. Publ. Health Ass., Washington, 1268 pp.

Stumm, W. and Morgan, J.J., 1996. *Aquatic chemistry*, 3rd ed. Wiley and Sons, New York, 1022 pp.

Stuyfzand, P.J., 1983. Important sources of errors in sampling groundwater from multilevel samplers (in Dutch). *H_2O* 16, 87–95.

Stuyfzand, P.J., 1987. An accurate calculation of the electrical conductivity from water analyses (in Dutch). KIWA SWE-87.006, Rijswijk, 31 pp.

Stuyfzand, P.J., 1989. Hydrochemical methods for the analysis of groundwater flow (in Dutch). *H_2O* 22, 141–146.

Suarez, D.L., 1987. Prediction of pH errors in soil-water extractors due to degassing. *Soil Sci. Soc. Am. J.* 51, 64–67.

Welch, A.H., Westjohn, D.B., Helsel, D.R. and Wanty, R.B., 2000. Arsenic in ground water of the United States: occurrence and geochemistry. *Ground Water* 38, 589–604.

Zuurdeeg, B.W. and Van der Weiden, M.J.J., 1985. Geochemical aspects of European bottled waters. In *Geothermics, Thermal-mineral waters and hydrogeology*, 235–264. Theophrastus Publ., Athens.

2

From Rainwater to Groundwater

Rainwater is the source of most groundwater and a logical starting point for the study of groundwater geochemistry. Continental rainwater is dominated by oceanic vapor and it does indeed resemble strongly diluted seawater. However, natural and anthropogenic dusts and gases modify the composition. Before the rain turns into groundwater, various processes in the soil may affect the concentrations. Dry-deposited dust particles and gases will dissolve. Evapotranspiration concentrates the solutes, and vegetation selects essential elements to store them temporarily in the biomass. Particularly the weathering of minerals in the soil is of importance in changing concentrations, and all the soil processes together often generate already much of the groundwater chemistry. In this chapter, we will follow the evolution in water chemistry from rain, via soil, to the aquifer.

Mass balance calculations are useful to obtain first insight in the water chemistry. The mass balance approach is fundamental and universally applicable, and we will use it to decipher the hydrological cycle, the sources of the constituents in rainwater, the isotopic evolution of rainwater, cycling of elements in the biosphere, and to derive the groundwater quality below farmland.

2.1 THE HYDROLOGICAL CYCLE

The inventory of the global water cycle in Figure 2.1 shows the amounts of water in the various reservoirs. The oceans contain $13{,}700 \times 10^{20}$ g H_2O, or 97% of the total amount of water on earth. The water in the ice caps and glaciers amounts to 2% of the total. If all the ice would melt, the depth of the oceans would increase by 2% and the sea level would rise by 2% of the average depth, or about 80 m. Groundwater is placed third with about 1% of the total amount of water on earth, and it is many times larger than water in rivers and lakes (together about 0.009%). Finally the amount of water present in the atmosphere constitutes only 0.001% of the global total.

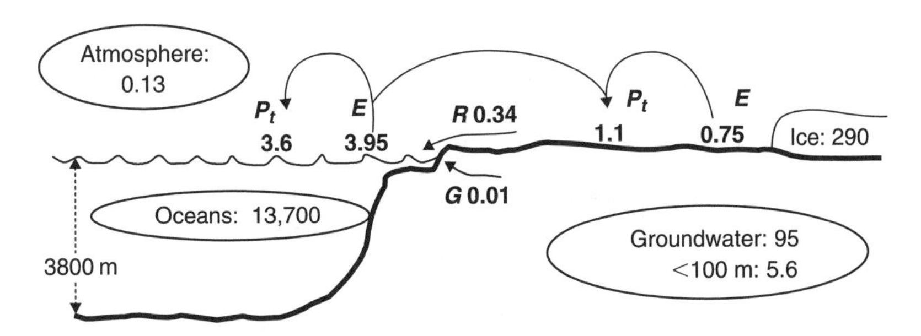

Figure 2.1. The global hydrologic cycle showing amounts in 10^{20} g H_2O (encircled entities), and mass transfers among the reservoirs in 10^{20} g H_2O/yr (numbers associated with arrows). Based on data in Berner and Berner, 1996; Holland and Petersen, 1996; Budyko, 1982.

Figure 2.1 also indicates the mass transfers among the reservoirs. The numbers are based on measurements and calculations from mass and energy balances and contain a variable uncertainty. Well measured are the continental precipitation (P_t) and the continental surface runoff (R) into the oceans. The amount of groundwater draining directly into the oceans (G) is by comparison rather small. From these numbers (cf. Figure 2.1), we can estimate that over land the evapotranspiration ($E = P_t - R - G = 1.1 - 0.34 - 0.01 = 0.75 \times 10^{20}$ g H_2O/yr) amounts to 68% of the precipitation.

The oceanic evaporation is estimated with the energy budget method, and the precipitation over the oceans is subsequently obtained from mass balance (Budyko, 1982). Over the oceans, evaporation exceeds precipitation and a vapor transport from the ocean to the continent balances the continental surface runoff and the groundwater outflow into the oceans. Water evaporated over the oceans contributes with 32% to the continental precipitation on average, but the contribution varies depending on the distance from the coast and the meteorological conditions. The proportion of land derived vapor or *recycled precipitation* (Eltahir and Bras, 1996) can be estimated from a mass balance, which along a transect accounts for precipitation and evapotranspiration (Budyko, 1982). The atmospheric vapor transfer at any point x (Figure 2.2) is the sum of the inflow at the border ($x = 0$), the loss by precipitation, and the gain from evapotranspiration:

$$I_x = I_0 - P_t \cdot x + E \cdot x \tag{2.1}$$

Here I_x is the vapor mass transfer (m $H_2O \times$ m/yr) at distance x from the coast, I_0 is the atmospheric input at the coast, P_t is precipitation and E is evapotranspiration along the transect (both in m H_2O/yr). For the global data in Figure 2.1, $P_t / E = 1.1 / 0.75 = 1.47$, or simplified $P_t = 1.5E$. Substituted in Equation (2.1) this gives $I_x = I_0 - 0.5E \cdot x$. The ratio of the land derived vapor ($E \cdot x$) to the total vapor transfer (I_x) then is:

$$\frac{E \cdot x}{I_x} = \frac{E \cdot x}{I_0 - 0.5\, E \cdot x} \tag{2.2}$$

I_0 is the product of atmospheric vapor content (u) and the air flow velocity (v) across the sea-land boundary. To calculate Figure 2.2 we used u = (atmospheric vapor content) / (density of water $\times$ earth's surface) $= 0.13 \times 10^{20}$ g H_2O / (10^6 g/m^3 $\times$ 5.1×10^{14} m^2) $= 0.025$ m H_2O. Furthermore, a wind velocity $v = 4$ m/s directed eastward, the average for Europe at 50°N (Peixóto and Oort, 1983), gives $I_0 = 0.025$ m $\times$ 4 m/s $= 3.15 \times 10^6$ m^2/yr. And lastly, E = (g H_2O/yr) / (density of water $\times$ earth's land surface) $= 0.75 \times 10^{20} / (10^6 \times 1.5 \times 10^{14}) = 0.5$ m/yr.

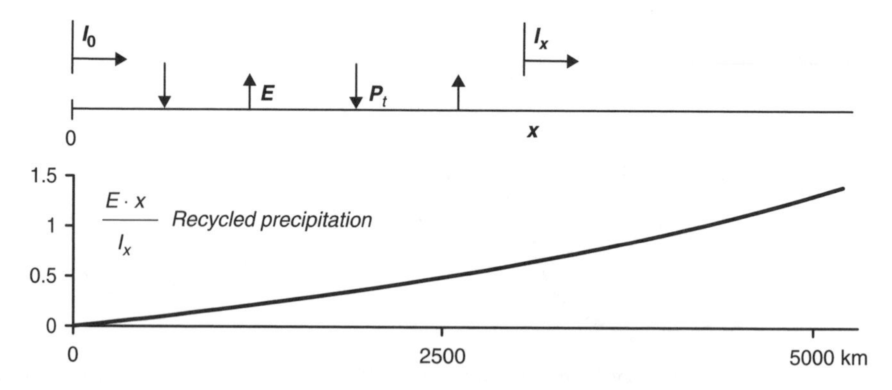

Figure 2.2. Recycled precipitation, the ratio of land derived vapor ($E \cdot x$) over the total vapor transfer (I_x) in the atmosphere as a function of the distance to the coast.

Figure 2.2 shows that the importance of land-derived vapor increases as a function of distance to the coast. However, first at 2500 km from the coast does the land-derived vapor exceed 50% of the atmospheric water vapor. The integrated amount over this distance gives an even smaller percentage (Problem 2.2), illustrating that the chemistry of rainwater on the continents, including its isotopes, is very strongly influenced by oceanic water vapor.

Assuming constant volumes of water in each reservoir and constant mass transfers among them (steady state conditions) we can calculate the *residence time*, which is the volume divided by the mass transfer:

$$\tau = \frac{V}{Q} \tag{2.3}$$

where τ is the residence time (s), V is volume of the reservoir (g H_2O) and Q is discharge (g H_2O/yr) into or out from the reservoir.

For the oceans, the residence time based on surface runoff and groundwater outflow is 13,700 / (0.34 + 0.01) = 39,000 years. For the atmosphere, based on precipitation, the residence time is 0.13 / (3.6 + 1.1) = 10 days. For groundwater, the residence time based on direct groundwater discharge into the ocean would be 95 / 0.01 = 9,500 years. However, most groundwater discharges via rivers and a minimum estimate of the residence time would rather be 95 / (0.34 + 0.01) = 271 years (all the basic data are in Figure 2.1). The residence time indicates the timescale that the subsystem operates on and marks its susceptibility towards changes. Clearly, the oceans are large, slowly responding systems, while the atmosphere is highly dynamic. Groundwater reservoirs are intermediate and operate on timescales of at least hundreds of years.

The global water cycle averages the existing variations with lattitude and ignores the large differences among the continents. Figure 2.3 shows global net recharge variations as a function of lattitude. Precipitation is highest in the tropics and in the temperate climate. Evaporation is highest in arid regions with a strong heat source (sun radiation) and low air moisture. In the subtropics the potential evaporation exceeds precipitation. Over the continents this results in the formation of deserts like the Sahara and the Great Desert of Australia.

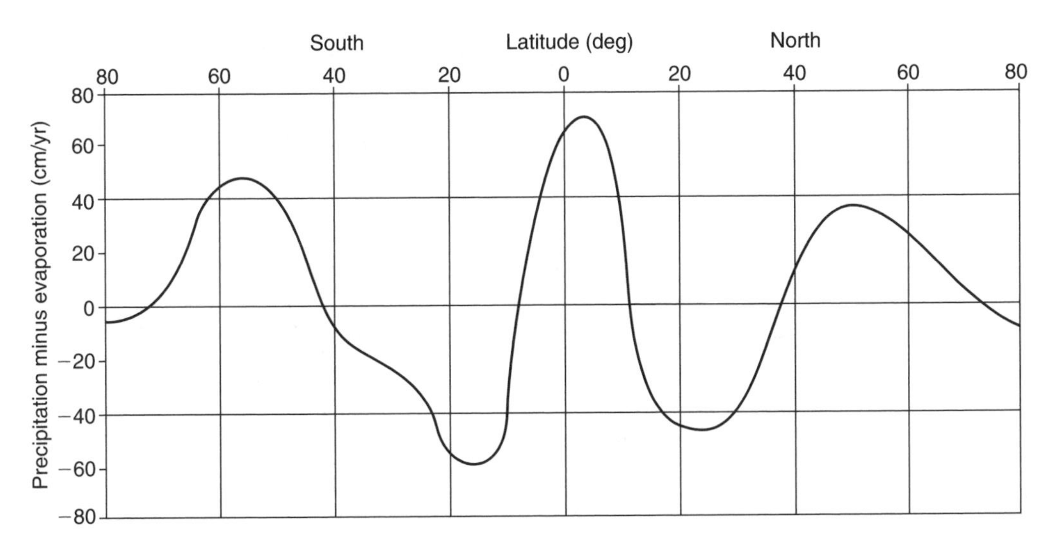

Figure 2.3. Net precipitation (precipitation minus evaporation) as a function of lattitude. Positive values indicate net infiltration and negative values potential evaporation (Peixóto and Keitani, 1973).

QUESTIONS:
Estimate the surface area of the oceans, using Figure 2.1.

ANSWER: $1.37\times10^{18}\,m^3/3800\,m = 3.6\times10^{14}\,m^2$

About 6% of groundwater resides in shallow aquifers <100 m with 20% porosity. Estimate the percentage land on earth with shallow aquifers.

ANSWER: $5.6\times10^{14}\,m^3/0.2/100\,m = 2.8\times10^{13}\,m^2$ / (earth's land surface = $1.5\times10^{14}\,m^2$) $\times\ 100 = 18.7\%$

Estimate the residence time of groundwater in shallow aquifers.

ANSWER: $Q_{in} = 0.35\times10^{20}\,g\ H_2O/yr \times 0.187 = 0.065\times10^{20}\,g\ H_2O/yr.$

$\tau = 5.6\times10^{20}\,g\ H_2O/0.065\times10^{20}\,g\ H_2O/yr = 86\,yr.$

2.2 THE COMPOSITION OF RAINWATER

The composition of rainwater is determined by the source of the water vapor and by the ions which are acquired (or lost) by water on its journey through the atmosphere. Over the oceans, rainwater resembles strongly diluted seawater with a chloride concentration of 10–15 mg/L. Given the dominance of oceanic vapor in the atmosphere (Figure 2.2), this composition will form the basis of inland rainwater quality thousands of kilometers away from the sea. The sea-salt content originates from marine aerosols that form from the liquid droplets, produced when gas bubbles burst at the sea surface (Monahan, 1986; Blanchard and Syzdek, 1988). Evaporation of the water from the droplets produces aerosols with different size-ranges.

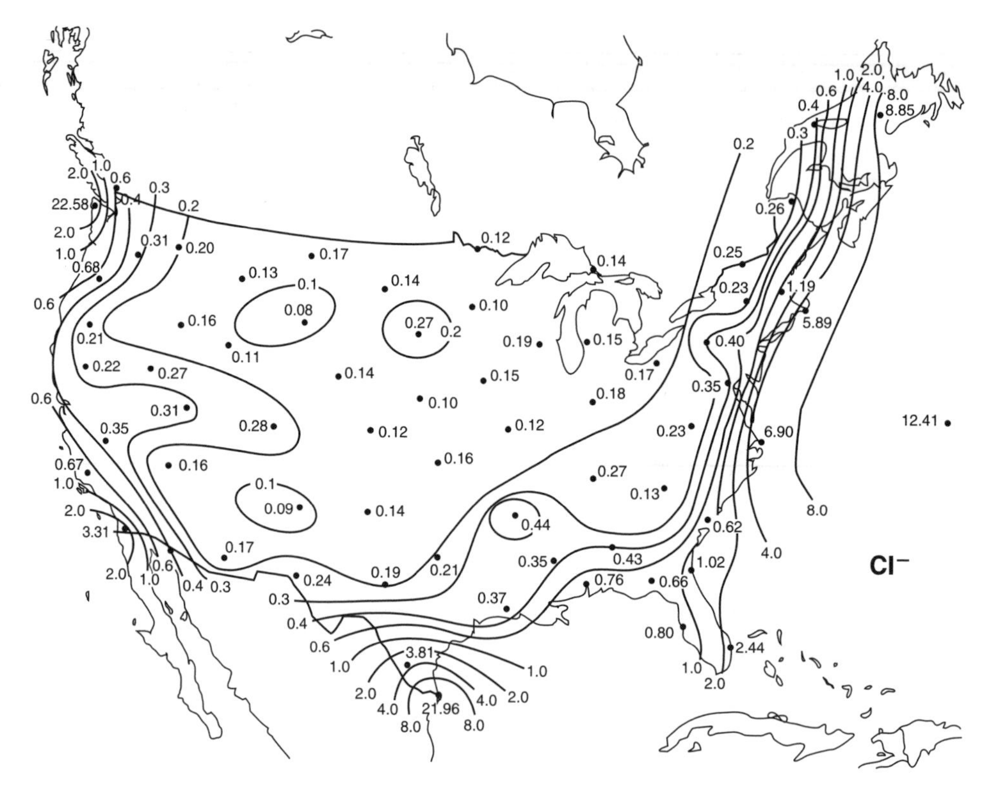

Figure 2.4. Average chloride concentrations (mg/L) in precipitation over the United States (Junge and Werby, 1958).

The coarsest particles have diameters exceeding 10 μm and retain a composition that is closely akin to seawater. From smaller particles, chloride evaporates presumably in the form of HCl, which results in Cl^- depletions relative to Na^+ of up to 40% (Raemdonck et al., 1986; Möller, 1990). Rainout of Cl^- depleted particles produces a Cl^- depleted rain, but since the evaporated HCl also contributes to the total rainwater chemistry, the net effect is that the Na/Cl ratios remain similar to the ratio in seawater, as found in coastal areas of Europe (Möller, 1990), the United States (Wagner and Steele, 1989), and in the Amazonas (Andreae et al., 1990). The importance of the sea-salt contribution is easily traced in the chloride concentration of rainwater. Figure 2.4 shows the distribution of chloride in the rain over the USA. The highest concentrations are found near the coast as a result of rapid washout of the largest sea salt particles, and the concentrations decrease exponentially inland to a background value of 0.1–0.2 mg/L at about 600 km from the coast. Over the continent, there are additional Cl^- sources in industrialized areas, like burning of waste plastics (Lightowlers and Cape, 1988), coal fired power plants (Wagner and Steele, 1989), or even the evaporation of salt water used as a coolant in the steel industry (Vermeulen, 1977).

The relative composition of rainwater and seawater can be compared by using a *fractionation factor*. For example, for Cl^- relative to Na^+ the fractionation factor F_{Na} is:

$$F_{Na} = \frac{(Cl/Na)_{rain}}{(Cl/Na)_{seaw.}} \qquad (2.4)$$

Since we are comparing ratios, the Cl^- and Na^+ concentrations in rain or seawater can be expressed in any concentration-units. Table 2.1 compiles fractionation factors for rainwater on the oceanic island of Hawaii, which for major cations will be similar to other near-coastal stations.

Table 2.1 indicates that the major elements are not fractionated very much. For Cl^-, the fractionation factor ranges from 0.93 to 1.0, and perhaps the lower value is the result of some volatilization. For Ca^{2+}, the fractionation factor ranges from 0.97 to 1.22, indicating some enrichment of Ca^{2+} with respect to Na^+, which suggests the presence of a land-based Ca^{2+} source. Bromide and organic-N have fractionation factors much higher than 1. These elements are part of lipid containing organic matter that is present in the surface water layer. When gas bubbles from below the water surface burst into the air, the interfacial foam is dragged along and the associated elements are enriched in the atmosphere and in rainwater. For major components the absence of extensive fractionation leads to the conclusion that rainwater of marine origin is basically strongly diluted seawater.

During transport over land, the air masses and clouds pick up continental dust and gases from natural or industrial origin, which modifies the composition of rainwater. Figure 2.5 shows the distribution of Ca^{2+} in rain over the USA. The concentration pattern is the reverse of that for Cl^-. The lowest Ca^{2+} concentrations are found near the coast and the concentrations are particularly high in the arid southwestern part of the US. Soil dust, containing $CaCO_3$ particles, is the source of most Ca^{2+} in rain, together with dust from cement industry, and fuel and waste burning. Na^+ in rainwater may also originate from soil dust particles in arid climates (Nativ and Issar, 1983), and roadspray used for de-icing may increase the Na^+/Cl^- ratio with respect to seawater (Baisden et al., 1995).

Table 2.1. Fractionation factors, given as a range, of seawater elements in marine aerosols. Fractionation factors expressed with respect to Na (F_{Na}). (Data from Duce and Hoffman, 1976.)

		Mg^{2+}	Ca^{2+}	K^+	Sr^{2+}	Cl^-	SO_4^{2-}	Br^-	Org.N
F_{Na}	upper limit	1.07	1.22	1.05	0.89	1.0	?	12	1×10^6
	lower limit	0.98	0.97	0.97	0.84	0.93	1	1	2×10^4

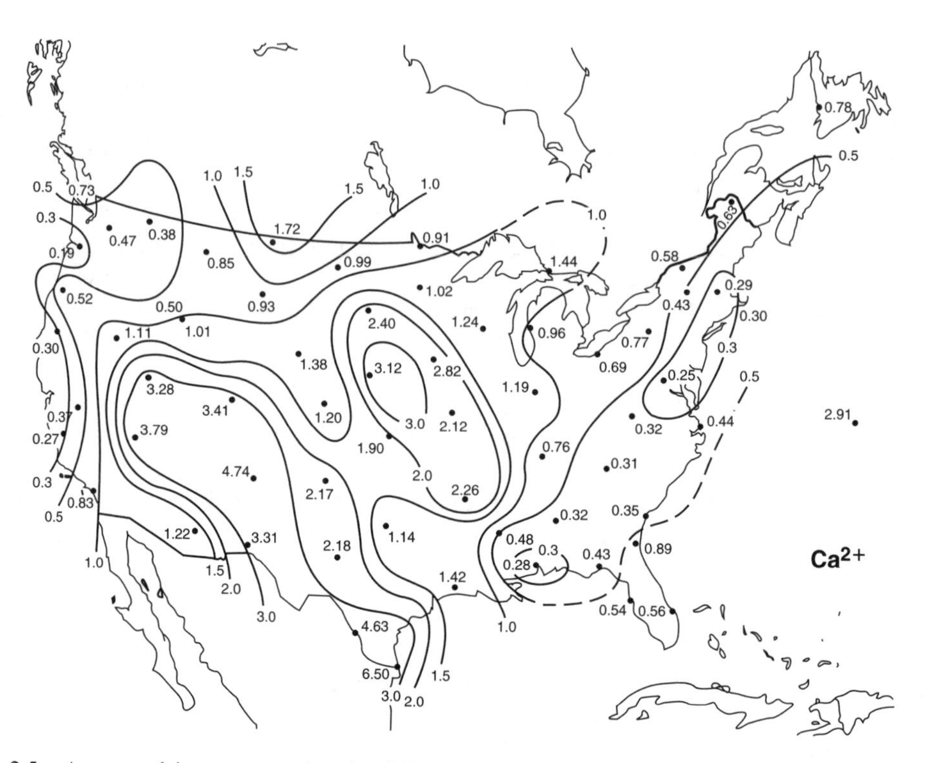

Figure 2.5. Average calcium concentrations (mg/L) in precipitation over the United States (Junge and Werby, 1958).

Table 2.2 contains a number of rainwater analyses from different parts of the world that illustrate the compositional variations. The older analyses are all from bulk rain samplers that are permanently open. However, for the stations Beek (Netherlands) and Delhi (India), more recent analyses have been included of 'wet-only' samples collected with samplers that open only when it rains, which excludes dry deposition (Section 2.4) and generally results in lower concentrations.

Table 2.2. Rainwater analyses (values in μmol/L).

Period	a Kiruna Sweden '55–'57	b Hubbard Brook,US '63–'74	c De Kooy Neth. '78–'83	c Beek Neth. '78–'83	d Beek wet-only '97–'98	e Thumba India 1975	e Delhi India 1975	f Delhi wet-only 1997
pH	5.6	4.1	4.40	4.75	5.5			6.3
Na^+	13	5	302	43	26	200	30	9
K^+	5	2	9	7	3	6	7	9
Mg^{2+}	5	2	35	9	4	19	4	2
Ca^{2+}	16	4	19	47	10	23	29	13
NH_4^+	6	12	78	128	74			21
Cl^-	11	7	349	54	29	229	28	13
SO_4^{2-}	21	30	66	87	29	7	4	10
NO_3^-	5	12	63	63	35			9
HCO_3^-	21	–	–	9				25

(a) Granat, 1972; (b) Likens et al., 1995; (c) KNMI / RIV, 1985; (d) Boschloo and Stolk, 1999; (e) Sequeira and Kelkar, 1978; (f) Jain et al., 2000.

To facilitate the interpretation of these analyses and to illuminate supplementary sources of ions, the sea salt contribution can be subtracted from the rainwater analyses. Chloride is chosen as an inert reference component in the calculation because, in an overall sense, it shows little fractionation. For Na^+ the calculation proceeds as follows (Table 2.3). In seawater the Na^+ concentration is 485 mmol/L and the Cl^- concentration is 566 mmol/L. The ratio of Na / Cl in seawater is therefore 0.86. In Kiruna rain, the Cl^- concentration is 11 μmol/L and the part of Na^+ that is derived from seawater amounts therefore to $0.86 \times 11 = 9$ μmol/L. The observed Na^+ concentration in Kiruna rainwater is 13 μmol/L, and $13 - 9 = 4$ μmol Na^+/L must have a non-marine origin. This difference (which indicates either a continental source of Na^+, or a depletion of Cl^-) is, however, barely significant compared to the analytical accuracy. Similar calculations were performed for all the rainwater analyses in Table 2.2, and the results are given in Table 2.3. After subtracting the seawater contribution, the precipitation of the near coastal stations, De Kooy or Thumba, shows low residuals of Na^+, K^+ and Mg^{2+} ions, which indicates that the content of these ions in rain primarily is derived from seawater. On the other hand, precipitation in Delhi is enriched in Ca^{2+} derived from soil dust and if 1975 and 1997 numbers are comparable, more than half comes down as dry deposition.

A conspicuous enrichment of the anions NO_3^- and SO_4^{2-} is found in the rain of the United States and Europe (Table 2.3, Hubbard Brook and Beek). These ions originate mainly from industrial and traffic fumes containing gaseous NO_x and SO_2. The gases become oxidized in the atmosphere, producing the strong acids HNO_3 and H_2SO_4. Dissociation of these acids lowers the pH of rainwater down to values of 4 and are the cause of *acid rain* (Berner and Berner, 1996). In Europe the emission of SO_2 has decreased substantially during the last decade and the increase in the pH of rainwater in Beek from 4.75 during the period 1978–1983 to 5.5 in 1998 looks promising.

The pH of rainwater is also affected by base producing processes which vary over time. The intensification of animal husbandry increasingly produces the neutralizing base NH_4OH, which evaporates from manure that is spread over agricultural land. Its influence can be traced again in the analysis of Beek, located in an area of the Netherlands where factory farming has grown rapidly since 1970. Not much NH_4^+ is present in the rain at Hubbard Brook and the pH is therefore lower at this site. Cement industry, and waste burning, also emit particles that contribute with base. The analysis of Beek rainwater shows a large excess of Ca^{2+} when the sea salt contribution has been subtracted (Table 2.3) due to the presence of a local cement industry. Ironically, the increased implementation of filters on chimneys has reduced the emission of base producing dust, which has led to a sharp decline in base cations in rain and counteracted the effect of pH increase that would result from the reduction in SO_2 emission (Hedin et al., 1994).

Table 2.3. Concentrations in rainwater from sources other than seawater. For the Kiruna station both the contribution from seawater and from additional sources is shown while for the other stations only the additional sources are tabulated. (Units are μmol/L, except for seawater).

	seaw. (mmol/L)	Kiruna			Hub. Br.	De K.	Beek	Beek wet-only	Thum	Delhi	Delhi wet-only
		rain	seaw. contr.	other src.							
Na^+	485	13	9.4	4	−1	3	−3	1	4	6	−2
K^+	10.6	5	0.2	5	2	3	6	2	2	6	9
Mg^{2+}	55.1	5	1	4	1	1	4	1	−3	1	0
Ca^{2+}	10.7	16	0.2	16	4	12	46	9	19	28	13
NH_4^+	2×10^{-6}	6	0	6	12	78	128	74			21
Cl^-	566	11	11	–	–	–	–	–	–	–	–
SO_4^{2-}	29.3	21	0.6	20	30	48	84	27	−5	3	9
NO_3^-	5×10^{-6}	5	0	5	12	63	63	35			9
HCO_3^-	2.4	21		21	0	−1	9				24

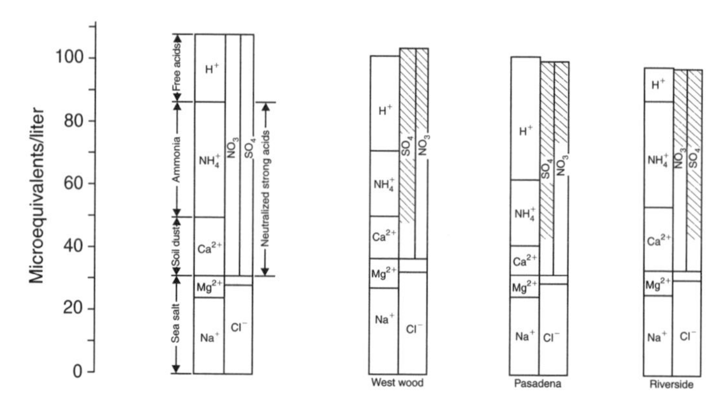

Figure 2.6. Composition of rain in southern California, 1978–79, interpreted in terms of input components and source type (Morgan, 1982). Hatched areas in the SO_4^{2+} and NO_3^- fields represent immobile sources, unhatched is the contribution from traffic.

The net effect of different acid and base producing processes on the rainwater pH is illustrated in a bar diagram by Morgan (1982) showing rainwater analyses from California (Figure 2.6). The bottom part shows the sea salt component with the anions Cl^- and part of the SO_4^{2-}, and the cations Na^+ and Mg^{2+}. The remainder of SO_4^{2-} and NO_3^- come from industrial sources and were originally present as sulfuric and nitric acid but have become partly neutralized by $CaCO_3$ from continental dust and NH_4OH from manure evaporation. The unneutralized part of the acids is present as free H^+, and determines the pH of rainwater.

QUESTION:
Calculate the concentration of free acids in Hubbard Brook and Delhi rain (μeq/L) from Table 2.2.

ANSWER: Hubbard Brook: $10^{-4.1} \times 10^6 = 79$ μeq/L. Delhi: $10^{-6.3} \times 10^6 = 0.5$ μeq/L.

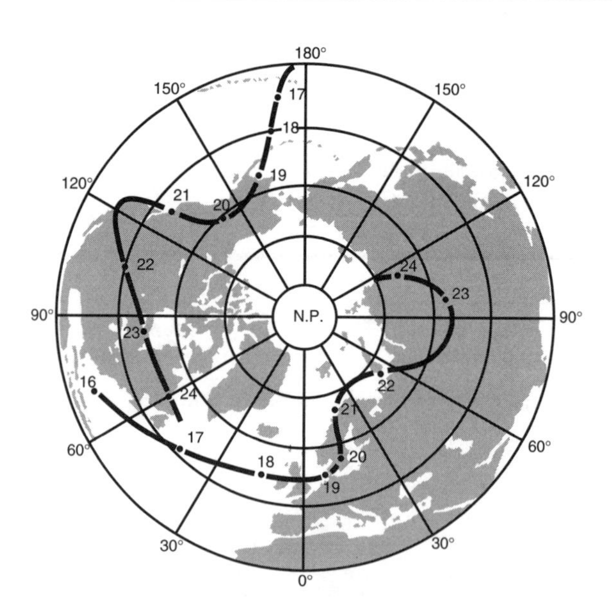

Figure 2.7. Tracks of air parcels around the world. The numbers represent successive days in April, 1964. (Modified from Newell, 1971.)

2.2.1 *Sources and transport of atmospheric pollutants*

Atmospheric pollution operates on an international scale due to the rapid dispersion and world-wide transport of local emissions through the atmosphere. The transport velocity in the atmosphere has been demonstrated in trajectory studies in which atmospheric particles (or air parcels) are traced back in time, based on meteorological observations. Figure 2.7 presents the results of an early study (Newell, 1971) and shows that particles or gases travel the distance from the United States to Europe in only about 4 days. The industrial pollution of the atmosphere is clearly not limited by national frontiers.

Mass balances for the acidifying gases NO_x and SO_2 have been constructed for individual countries, and an example for the Netherlands is shown in Figure 2.8 for the year 1980. The figure shows that of 550,000 tons of NO_x emitted (expressed as NO_2), only 90,000 tons were deposited within the Netherlands. The remainder was dispersed, in part to the sea (60,000 tons), and in part to other countries (400,000 tons). The exported amounts exceeded the 150,000 tons deposited as import from foreign sources. Similar numbers applied for SO_2 before 1980 but the emission of this acidifying oxide has been forcefully reduced in later years and amounted to about 70,000 tons in 1999.

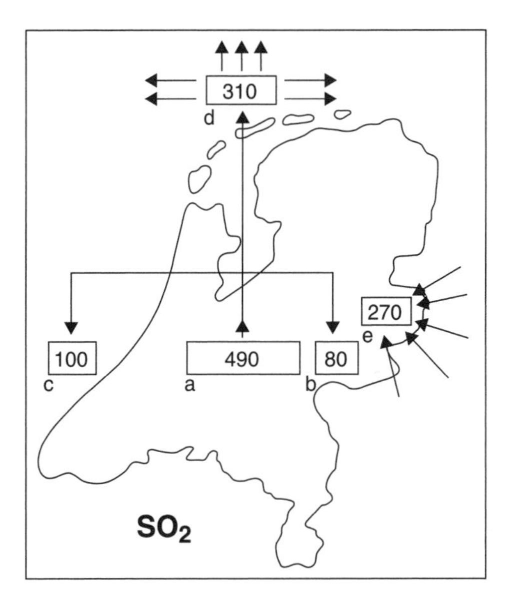

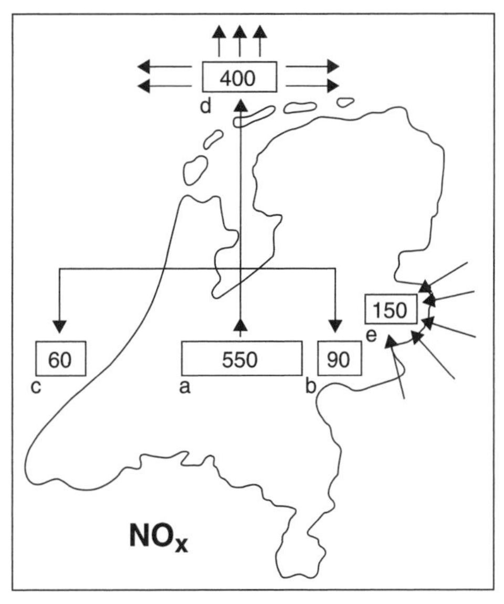

Figure 2.8. Mass balances for SO_2 and NO_x in the Netherlands in 1980. The values are given in 1000 tons/year. (a) is total industrial emission from the Netherlands, of which (b) is deposited in the Netherlands, (c) in sea, (d) in other countries and (e) is deposited in the Netherlands from foreign sources.

2.3 STABLE ISOTOPES IN RAIN

Many elements, including hydrogen and oxygen, exist in nature with different mass numbers, also called isotopes. For example, hydrogen occurs in nature as a mixture of the isotopes ^1H and ^2H (deuterium) while oxygen is found as isotopes with the atomic masses ^{18}O, ^{17}O and ^{16}O. Ocean water contains two ^{18}O atoms for every thousand ^{16}O atoms (Table 2.4), but the ratio is different in fresh water. ^1H, ^2H, ^{18}O and ^{16}O do not engage in nuclear transformations and are therefore termed *stable isotopes*. In contrast, *radioactive isotopes* like tritium (^3H) or radiocarbon (^{14}C) will decay over time and can be used for dating (Section 3.3).

Due to the difference in mass, stable isotopes behave slightly differently in physical, chemical and biological processes. For example, during evaporation of water and again during rainout from the atmosphere, the stable isotopes ^2H / ^1H and ^{18}O / ^{16}O become fractionated. The resulting small variations in isotopic concentrations may yield information on the climate at the point of infiltration or the provenance of the water. Also for other elements, isotope fractionation may help to elucidate the geochemical processes and reactions which have been creating a given water composition (Clark and Fritz, 1997).

Table 2.4. The stable environmental isotopes.

Isotope	Standard	Ratio *R*	Measured phases/process	δ-range (‰)
^2H	VSMOW[1]	$^2H / ^1H = 1.5576\times10^{-4}$	H_2O, CH_2O, CH_4, H_2, OH minerals	−450 (CH_4), −200 (arctic ice), +50 (evap. lakes)
^3He	Air	$^3He / ^4He = 1.3\times10^{-6}$	Gas-sources, dating	
^6Li	NBS[2] L-SVEC	$^6Li / ^7Li = 0.0832$	Water, rocks	
^{11}B	NBS 951	$^{11}B / ^{10}B = 4.0436$	Water, pollution from waste sites	
^{13}C	VPDB[3]	$^{13}C / ^{12}C = 0.011237$	*TIC*, CH_4, carbonates, ^{14}C dating	−120 (bact. CH_4), +20 (non-marine carb.)
^{15}N	Air	$^{15}N / ^{14}N = 3.677\times10^{-3}$	NO_3^-, N_2, NH_4^+, N-organics	
^{18}O	VSMOW	$^{18}O / ^{16}O = 2.005\times10^{-3}$	H_2O, CH_2O, HCO_3^-, SO_4^{2-}, NO_3^-, O minerals	−60 (arctic ice) +40 ($CO_{2(g)}$)
^{34}S	CDT[4]	$^{34}S / ^{32}S = 0.045$	SO_4^{2-}, S^{2-}	
^{37}Cl	SMOC[5]	$^{37}Cl / ^{35}Cl = 0.32398$	Water, salts, diffusion	−1.5 to +1.0
^{87}Sr	–	$^{87}Sr / ^{86}Sr \approx 0.710$	Water, minerals, weathering	

[1] Vienna Standard Mean Ocean water; [2] National Bureau of Standards; [3] Vienna Pee Dee Belemnite; [4] Cañon Diablo Troilite; [5] Standard Mean Ocean Chloride, ratio from natural abundances (Wedepohl, 1969).

2.3.1 *Isotopic ratios and the δ notation*

The concentration of stable isotopes is normally given as the ratio of the least abundant isotope over the most abundant isotope and expressed relative to a standard. For water the internationally agreed standard is called "Vienna Standard Mean Ocean Water" (VSMOW). According to Table 2.4, in VSMOW the isotopes of oxygen and hydrogen are present in the ratios $^{18}O / ^{16}O = 2.005\times10^{-3}$ and $^2H / ^1H = 1.56\times10^{-4}$. Standards for stable isotopes of other elements that are commonly used in geochemical studies are also listed in Table 2.4.

The variations of isotopic ratios in nature are small and are studied more easily by using the δ notation. The δ notation expresses the deviation of the isotopic ratio *R* in the sample with respect to the ratio in the standard:

$$\delta_{sample} = \frac{R_{sample} - R_{standard}}{R_{standard}} \cdot 1000 \tag{2.5a}$$

or:

$$\frac{R_{sample}}{R_{standard}} = \frac{\delta_{sample}}{1000} + 1 \tag{2.5b}$$

The least abundant isotope is specified with the δ symbol, for example the ratio $^{18}O / ^{16}O$ in rain is described as $\delta^{18}O_{rain}$.

It follows from Equation (2.5) that the δ value of a mixture of n waters can be calculated by adding the δ values of the contributing waters multiplied by the mass fraction of the reference isotope, and dividing by the concentration of the reference isotope in the mixture:

$$\delta_{mixture} = \frac{\delta_1(x_1 \cdot {}^{ref}m_1) + \cdots + \delta_n(x_n \cdot {}^{ref}m_n)}{(x_1 \cdot {}^{ref}m_1) + \cdots + (x_n \cdot {}^{ref}m_n)} = \frac{\sum_{i=1}^{n} \delta_i(x_i \cdot {}^{ref}m_i)}{{}^{ref}m_{mixture}} \quad (2.6)$$

where x_i is the fraction of water i, which has concentration ${}^{ref}m_i$ of the reference isotope (e.g. ^{16}O, or ^{12}C).

Usually to a good approximation, the total concentration of the element m_i ($= {}^{ref}m_i + {}^{j}m_i$, where j is the isotope of interest) can be used instead of the reference isotope:

$$\delta_{mixture} = \frac{\delta_1(x_1 \cdot m_1) + \cdots + \delta_n(x_n \cdot m_n)}{(x_1 \cdot m_1) + \cdots + (x_n \cdot m_n)} = \frac{\sum_{i=1}^{n} \delta_i(x_i \cdot m_i)}{m_{mixture}} \quad (2.6a)$$

For example, when a water (1) with $TIC_1 = 10\,\text{mM}$ and $\delta^{13}C_1 = 0‰$ is mixed 1:1 with a water (2) with $TIC_2 = 2\,\text{mM}$ and $\delta^{13}C_2 = -18‰$, Equation (2.6a) gives $\delta^{13}C_{\text{mixture}} = -3‰$, while the exact answer is $-3.0005‰$ according to Equation (2.6).

QUESTIONS:

Calculate the ratio of ^{13}C and ^{12}C in water (1) with $\delta^{13}C_1 = 0‰$.

ANSWER: $^{13}C\,/\,^{12}C = 0.011237$

Calculate the concentrations of ^{13}C and ^{12}C in water (1) with $TIC = 10\,\text{mM}$ and $\delta^{13}C = 0‰$.

ANSWER: $^{12}m_1 = 10\,/\,(1 + 0.011237) = 9.888879\,\text{mM}$, $^{13}m_1 = 0.111121\,\text{mM}$.

Calculate the concentrations of ^{13}C and ^{12}C in water (2) with $TIC = 2\,\text{mM}$ and $\delta^{13}C = -18‰$.

ANSWER: $^{12}m_2 = 2\,/\,(1 + 0.011035) = 1.978171\,\text{mM}$, $^{13}m_2 = 0.021829\,\text{mM}$.

Calculate $\delta^{13}C$ when waters (1) and (2) are mixed 1:9.

ANSWER: $\delta^{13}C = -11.57‰$

Calculate $\delta^{18}O$ of a 1:1 mixture of river Rhine water with $\delta^{18}O = -9‰$ and seawater.

ANSWER: $-4.5‰$

2.3.2 *The Rayleigh process*

Rayleigh originally derived a formula to calculate the products of distillation for a mixture of liquids with different boiling points. The principles are, however, applicable to any separation process with a constant fractionation factor, as is the case for stable isotopes. For example during condensation from water vapor, the heavier molecule $H_2{}^{18}O$ condenses more readily than $H_2{}^{16}O$, producing liquid water that is isotopically heavier than the vapor. The remaining water vapor must become increasingly enriched in the light isotope ^{16}O and the change in composition can be calculated with the Rayleigh formula. In the Rayleigh process, the product is removed continuously from the reactant, and further interactions are eliminated. This condition applies to the situation where a raindrop falls out of a cloud, or where isotopes fractionate between water and a solid phase along a flowline, for example in partitioning of ^{13}C and ^{12}C during calcite precipitation.

Consider a small number, dN, molecules of light $H_2{}^{16}O$ that condense and are accompanied by dN_i molecules of the heavy $H_2{}^{18}O$. The ratio $dN_i\,/\,dN$ in the condensate will be higher than the ratio $N_i\,/\,N$

in the vapor:

$$\frac{dN_i}{dN} = \alpha \frac{N_i}{N} \tag{2.7}$$

where α is the (*isotope*) *fractionation factor* (no dimension, the fractionation process is similar to element fractionation among rain and seawater, Equation 2.4). Going from vapor (g) to liquid (l) the fractionaton factor is $\alpha^{18}O_{l/g} = 1.0098$ at 20°C.

The derivative of the equation $y = (\ln x)$ is $d(y) / dx = 1 / x$, or $d(\ln x) / d x = 1 / x$. Hence $d(\ln x) = d(x) / x$. Thus, in Equation (2.7) we have $d(\ln N_i) = dN_i / N_i$ and $d(\ln N) = dN / N$, which we substitute:

$$\frac{d \ln N_i}{d \ln N} = \alpha \tag{2.8}$$

Substracting 1 from the left and right hand sides and recalling that $\ln a - \ln b = \ln(a / b)$, leads to:

$$\frac{d \ln N_i}{d \ln N} - 1 = \frac{d \ln N_i}{d \ln N} - \frac{d \ln N}{d \ln N} = \frac{d \ln (N_i / N)}{d \ln N} = \alpha - 1$$

and inserting the isotopic ratio $R = N_i / N$ yields:

$$\frac{d \ln R}{d \ln N} = \alpha - 1 \tag{2.9}$$

Integration of Equation (2.9) from the initial values R_0 and N_0:

$$\int_{R_0}^{R} d \ln R = (\alpha - 1) \int_{N_0}^{N} d \ln N$$

solves to:

$$\ln R - \ln R_0 = (\alpha - 1) \cdot (\ln N - \ln N_0) \quad \text{or} \quad \frac{R}{R_0} = \left(\frac{N}{N_0}\right)^{(\alpha-1)} \tag{2.10}$$

Since N_i is small compared to N, $N \approx N + N_i$, and the ratio N / N_0 is almost equal to the remaining fraction f of the vapor. Therefore:

$$\frac{R_{\text{vapor}}}{R_{\text{vapor, 0}}} = f^{(\alpha-1)} \tag{2.11}$$

With Equation (2.11) we can then calculate how R in atmospheric vapor changes during a rain-out. The isotopic ratio in the rain follows straightforwardly from Equation (2.7):

$$R_{\text{rain}} = dN_i / dN = \alpha R_{\text{vapor}} \tag{2.12}$$

and substitution in Equation (2.11) gives:

$$R_{\text{rain}} = \alpha R_{\text{vapor, 0}} f^{(\alpha-1)} = R_{\text{rain, 0}} f^{(\alpha-1)} \tag{2.13}$$

We can rewrite Equation (2.13) for ^{18}O, inserting δ instead of R according to Equation (2.5b):

$$\left(\frac{\delta^{18}O_{rain}}{1000}+1\right)=\left(\frac{\delta^{18}O_{rain,0}}{1000}+1\right)\cdot f^{(\alpha-1)} \tag{2.14}$$

The term $\delta / 1000$ will normally be small (<0.01), and therefore $\ln(\delta / 1000 + 1) \approx \delta / 1000$ (check this for $\delta = 36$ and $\delta = -24$). Taking logarithms of Equation (2.14) and substituting $\ln(\delta / 1000 + 1) = \delta / 1000$ yields:

$$\delta^{18}O_{rain} = \delta^{18}O_{rain,0} + 1000 \cdot (\alpha^{18}O_{l/g} - 1) \cdot \ln f \tag{2.15}$$

For shorthand notation the *enrichment factor*, $\varepsilon = 1000 \cdot (\alpha - 1)$, is used:

$$\delta^{18}O_{rain} = \delta^{18}O_{rain,0} + \varepsilon^{18}O_{l/g} \cdot \ln f \tag{2.16}$$

We can also rewrite Equation (2.12) in δ notation:

$$\delta_{rain} = \delta_{vapor} + 1000 \cdot \ln \alpha \tag{2.17}$$

Since α is very close to 1, $\ln \alpha \approx (\alpha - 1)$, and therefore:

$$\delta_{rain} \approx \delta_{vapor} + 1000 \cdot (\alpha - 1) = \delta_{vapor} + \varepsilon_{l/g} \tag{2.18}$$

The enrichment factor is very helpful since it immediately gives the isotopic composition of two different phases at isotopic equilibrium. For example, at 20°C the equilibrium enrichment $\varepsilon^{18}O_{l/g} = 9.8 \approx (\delta^{18}O_l - \delta^{18}O_g)$. For $\delta^{18}O_{vapor} = -15.8‰$ the isotopic composition of the rain is $\delta^{18}O = -6‰$. Note that fractionation can be large for the hydrogen isotopes, in which case the fractionation should be determined for the ratio $^2H / {}^1H$, followed by calculation of the δ^2H value.

QUESTION:
Derive the equation to describe δ_{vapor} as function of f?
ANSWER: Combine Equations (2.16) and (2.18): $\delta^{18}O_{vapor} = \delta^{18}O_{vapor,0} + \varepsilon^{18}O_{l/g} \cdot \ln f$

EXAMPLE 2.1. *Calculate $\delta^{18}O$ of rain condensing from vapor* with, initially, $\delta^{18}O = -12.2‰$, as a function of the remaining vapor, temperature is fixed at 20°C. $\varepsilon^{18}O_{l/g} = 9.8‰$ at 20°C.
We consider three possibilities,

1. the rain is separated continuously from the vapor (Rayleigh condensation),
2. the rain is separated from the vapor but collected in a vessel,
3. the rain is collected and remains in contact with the vapor for all fractions.

Ad 1. For the rain in a Rayleigh process, we use Equation (2.16) and combine with (2.18):

$$\begin{aligned}\delta^{18}O_{rain} &= \delta^{18}O_{rain,0} + \varepsilon^{18}O_{l/g} \cdot \ln f = (\delta^{18}O_{vapor,0} + \varepsilon^{18}O_{l/g}) + \varepsilon^{18}O_{l/g} \cdot \ln f \\ &= -2.4 + 9.8 \times \ln f\end{aligned}$$

Ad 2. We calculate the vapor composition according to a Rayleigh process,

$$\delta_{\text{vapor}} = \delta_{\text{vapor, 0}} + \varepsilon_{\text{l/g}} \cdot \ln f$$

and combine with the mass balance (Equation 2.6a),

$$f\delta_{\text{vapor}} + (1 - f)\delta_{\text{rain}} = 1 \cdot \delta_{\text{vapor, 0}}$$

to obtain:

$$\begin{aligned} \delta^{18}\text{O}_{\text{rain}} &= \{\delta^{18}\text{O}_{\text{vapor, 0}} - f\delta^{18}\text{O}_{\text{vapor, 0}} - f \cdot \varepsilon_{\text{l/g}} \cdot \ln f\} \,/\, (1 - f) \\ &= -12.2 - 9.8 \times f \times \ln f/(1 - f) \end{aligned}$$

Ad 3. We combine the mass balance (Equation 2.6a),

$$f\delta_{\text{vapor}} + (1 - f)\delta_{\text{rain}} = 1 \cdot \delta_{\text{vapor, 0}}$$

and the equilibrium relation (Equation 2.18),

$$\delta_{\text{rain}} = \delta_{\text{vapor}} + \varepsilon_{\text{l/g}}$$

which gives:

$$\delta_{\text{rain}} = \delta_{\text{vapor, 0}} + \varepsilon_{\text{l/g}} \cdot f = -12.2 + 9.8 \times f$$

The resulting δ^{18}O's in rain for the three options are shown in Figure 2.9, as a function of the fraction of remaining vapor, f. Also shown is the composition of the vapor which loses water in the Rayleigh process (options 1 and 2).

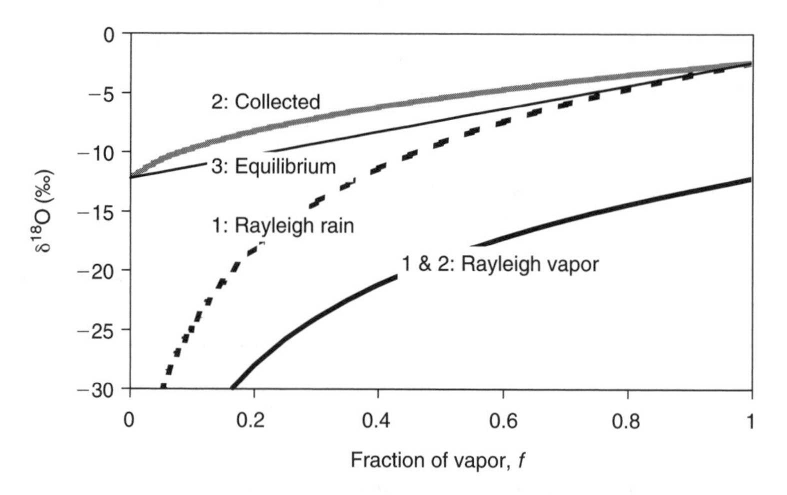

Figure 2.9. Isotopic composition of rain for three options of separating and equilibrating rain from vapor.

QUESTION:
Draw δ^{18}O of vapor for option 3?

ANSWER: a line parallel to case 3 but shifted −9.8‰

It is difficult to predict the value of a fractionation factor (cf. Chacko et al., 2001; Thorstenson and Parkhurst, 2004), but in general, fractionation is related to the movement of individual molecules and it decreases with increasing temperature. At high temperature, $\varepsilon = 0$ for any isotope pair. Also, qualitative reasoning suggests that the denser phase (liquid compared to gas) or the denser compound ($CaCO_3$ *vs* CO_2) will contain more of the heavy isotope. Furthermore, heavier isotopes are discriminated in kinetic and diffusion processes, and therefore in biological activities as well.

2.3.3 *The isotopic composition of rain*

The rainout of atmospheric water vapor is driven by the cooling of air during transport to higher latitude or altitude. The change in isotopic composition of rain and snow, as a function of the remaining fraction *f*, and indirectly of temperature, was calculated according to option 1 of Example 2.1 and is shown in Figure 2.10 (Clark and Fritz, 1997). Note that the enrichment factor increases when temperature decreases, and that Clark and Fritz used a slightly larger value at 20°C than in Example 2.1.

Figure 2.11 shows the isotopic composition of rainwater from various locations on earth in a plot of δ^2H *vs* $\delta^{18}O$. It is found that the stable isotopes for hydrogen and oxygen in rainwater obey the empirical relation:

$$\delta^2H = 8\ \delta^{18}O + 10 \tag{2.19}$$

This equation is known as the *meteoric water line*. The slope of the line $d(\delta^2H) / d(\delta^{18}O) = 8$, and has its origin in the way most precipitation is generated, *viz.* derived from evaporated oceanic water in the tropics, then transported to higher latitudes where it cools and rains out.

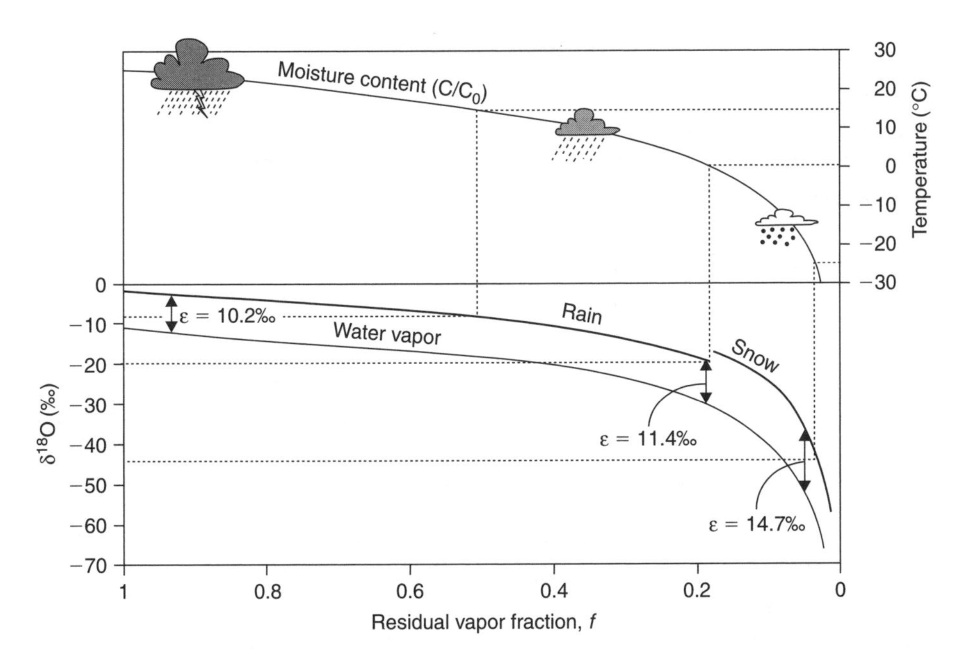

Figure 2.10. Isotopic composition of rain and snow during progressive cooling and rainout. Initial temperature of 25°C and $\delta^{18}O_{vapor} = -11$‰. Dashed lines link $\delta^{18}O$ of precipitation with temperature of condensation. (From Clark and Fritz, 1997, or calculated with PHREEQC, Chapter 11).

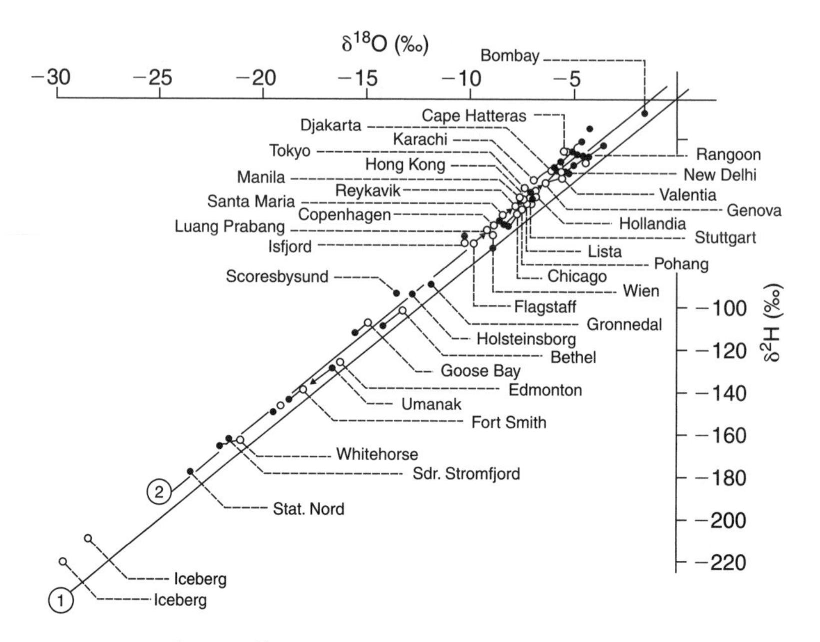

Figure 2.11. The relation of δ^2H and $\delta^{18}O$ in rain for various locations (Dansgaard, 1964). Line 2 is the meteoric water line, line 1 has the same slope but an intercept of zero.

Comparison with the pattern expected for Rayleigh distillation proceeds as follows. The Rayleigh equation for $\delta^{18}O_{rain}$ is:

$$\delta^{18}O_{rain} = \delta^{18}O_{rain,\,0} + \varepsilon^{18}O_{l/g} \cdot \ln f \tag{2.20}$$

Differentiation towards f, we find $d(\delta^{18}O_{rain}) / df = (\varepsilon^{18}O_{l/g}) / f$ in the rain. (The slope of the equation $y = a + b \ln x$ is equal to $dy / dx = b/x$). A similar relation applies for δ^2H. The slope among δ^2H and $\delta^{18}O$ now becomes:

$$\frac{d\,\delta^2H / df}{d\,\delta^{18}O / df} = \frac{\varepsilon^2H_{l/g}}{\varepsilon^{18}O_{l/g}} = 8.6\ (18°C) \tag{2.21}$$

The slope calculated for the Rayleigh process (Equation 2.21) is close to the meteoric water line (Equation 2.19) and indicates that rainout is an equilibrium process for isotope fractionation. The slope for the Rayleigh proces is slightly higher because we have neglected that the initial values for $\delta^2H_{rain,\,0}$ and $\delta^{18}O_{rain,\,0}$ are non-zero when substituting the δ notation for isotopic ratios in the Rayleigh equation (Problem 2.4).

In contrast, evaporation is a *kinetic* fractionation process. Kinetic fractionation produces a vapor that is lighter than predicted by equilibrium fractionation. Moreover, kinetic fractionation is stronger for ^{18}O than for ^2H, and the vapor becomes, relative to the equilibrium process, more depleted in ^{18}O than in ^2H. For example, oceanic vapor at 25°C would have $\delta^{18}O = -9.4$‰ and $\delta^2H = -85$‰ at equilibrium, but the observed values are close to −13‰ and −95‰, respectively. As the result, rain obtains its isotopes by equilibrium fractionation during condensation from a vapor that is depleted in ^{18}O compared to the original liquid from which the vapor was generated. Since the depletion is stronger for ^{18}O than for ^{2}H, the rain has a *deuterium excess*, which is calculated from analyzed values by:

$$d = \delta^2H_{rain} - 8\ \delta^{18}O_{rain} \tag{2.22}$$

The meteoric water line (Equation 2.19) gives $d = 10‰$ for average precipitation. A higher deuterium excess may indicate a contribution from evaporated water of an arid climate with low relative humidity in the atmosphere (Gonfiantini, 1986; Clark and Fritz, 1997), or a contribution from evaporated inland water to the rain.

The temperature is the driving force in cooling and condensing atmospheric vapor and results in the global variations of the isotopic composition of rain (Figures 2.12 and 2.13). In coastal areas the relation between $\delta^{18}O$ and temperature is (Dansgaard, 1964):

$$\delta^{18}O = 0.695\ t_c - 13.6‰ \tag{2.23a}$$

and for deuterium:

$$\delta^2H = 5.6\ t_c - 100‰ \tag{2.23b}$$

where t_c is the average yearly temperature in °C. For continental stations the slope in Equation (2.23a) is less, for example 0.58‰/°C instead of 0.695‰/°C (Rozanski et al., 1993). There is also a *latitudinal* effect, indicating that rain is generally more depleted at higher and colder latitudes (Figure 2.13). *Altitude* is another temperature-related effect that depletes $\delta^{18}O$ by -0.1 to $-0.5‰/100\,m$.

The temperature dependence also leads to seasonal fluctuations of the isotopic composition of rain, which are shown for various stations in Figure 2.12. As expected, the rain turns isotopically lighter in the colder season. The *seasonal* effect introduces a greater range of isotope compositions for continental stations with winter-snow (The Pas) than for near-coastal stations (Stanley).

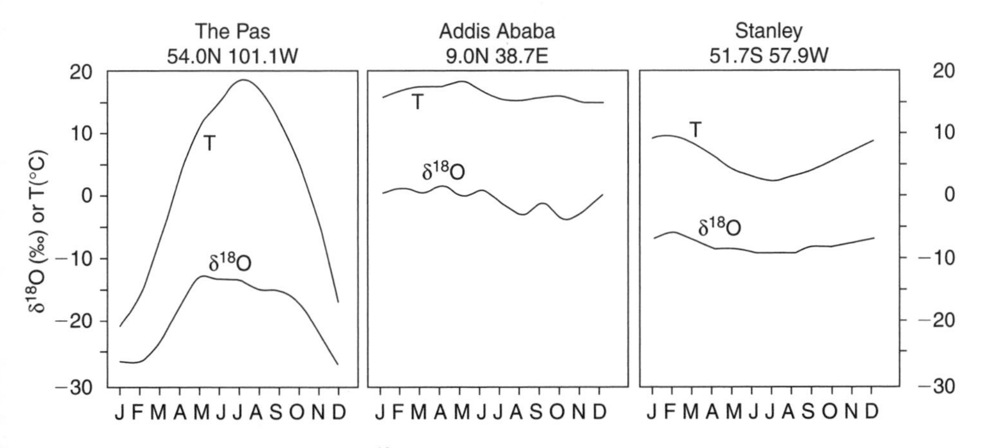

Figure 2.12. Seasonal variations in $\delta^{18}O$ and temperature in The Pas (Canada), Addis Ababa (Ethiopia) and Stanley (Falkland Islands) (Rozanski et al., 1993).

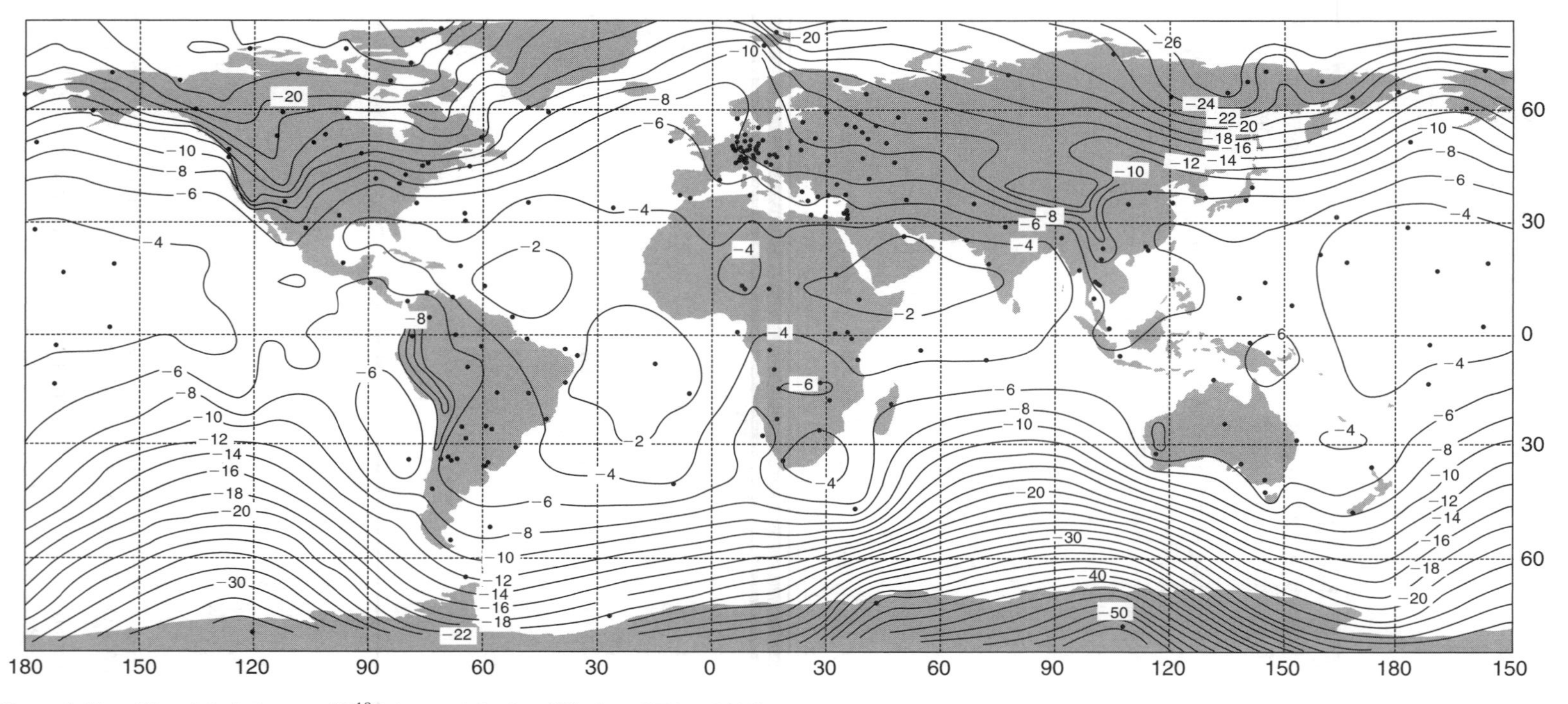

Figure 2.13. The global picture of $\delta^{18}O$ in precipitation (Clark and Fritz, 1997).

Lastly, a *continental* effect makes the isotopic composition of rain lighter, first because the atmospheric vapor needs progressively lower temperatures for rainout, second because evaporation of inland waters is a kinetic process which lightens the atmospheric vapor as was discussed before. During transpiration, on the other hand, plants take up water indiscriminately, and although some fractionation may occur in the leaves, steady state requires that the water that evaporates from the leaves has the same isotopic composition as soil water extracted by the roots.

QUESTIONS:
Estimate $\delta^{18}O$ and $\delta^{2}H$ of rain in your home town (e.g. Amsterdam, avg. temp. 10°C).
ANSWER: Amsterdam: $\delta^{18}O = -6.65‰$, $\delta^{2}H = -44‰$.
Find the deuterium excess in your home town's rain.
ANSWER: (Amsterdam) $d = 9.2‰$.
What is the meteoric water line according to Equations (2.23a + b)?
ANSWER: $\delta^{2}H = 8.06\ \delta^{18}O - 9.58‰$.
What will be the slope of $\delta^{2}H$ with temperature in continental rain according to Rozanski?
ANSWER: $8 \times 0.58 = 4.6‰/°C$.

2.4 DRY DEPOSITION AND EVAPOTRANSPIRATION

Dry deposition comprises both the deposition of particulate aerosols and of atmospheric gases by adsorption. Since the Cl^- concentration in groundwater is being used as a conservative parameter for estimating the recharge (Schoeller, 1960; Eriksson, 1960; Lerner et al., 1990; Edmunds and Gaye, 1994; Wood and Sanford, 1995; Wood, 1999), it is important to estimate the contribution of dry deposition to the total atmospheric input of Cl. The importance of dry deposition shows up as the difference between the composition of rainwater collected by *wet-only* rain gauges, open only during rain, and of rainwater collected by *bulk* samplers which remain open all the time (Galloway and Likens, 1978; Table 2.2). During dry periods, bulk samplers collect aerosols as well as anomalies such as bird droppings on the funnel that wash into the sampling bottle with the next rain. In the temperate climate of the Netherlands where 800 mm rain is distributed regularly over the year, bulk samplers overestimate the wet input of the major ions by a factor of about 1.3 (Ridder et al., 1984). The difference between wet-only and bulk sampling is also fundamental when estimating the atmospheric input of different elements to the aquifer, since dry deposition depends on the surface characteristics, vegetation, wind direction etc., all factors that cannot be easily extrapolated from the rain gauge to the natural surface.

The large variability in dry deposition is illustrated in Figure 2.14. It shows the sum of anions, basically chloride, in soil water below a forest in Denmark. At the windward margin of the forest, a high rate of dry deposition produces high ion concentrations in the unsaturated zone. Further towards the center of the forest, the air is filtered and the concentration decreases by up to a factor of 10.

Dry deposition is an important source of elements, as shown already by Garrels and Mackenzie (1972). Aluminum in dry deposition is mostly derived from soil dust and its dry deposition flux ranges from 13–380 mg/m^2/yr at various locations (Shahin et al., 2000). Associated with aluminum are trace elements in the same concentration ratio as in the soil and in many cases there are further increases as the result of anthropogenic emissions.

Over the North Sea, the dry deposition of trace elements such as Cd, Cu, Pb and Zn is about equal to the wet deposition. The total atmospheric input of Pb into the North Sea is about equal to the total riverine input, while the atmospheric input of Cd and Zn amounts to about ⅓ of the riverine input (Van Aalst et al., 1983). Dry deposition also contributes with significant amounts of NO_x and SO_2, and since gaseous adsorption is relatively important for these components, dry deposition is normally greater for conifer than for deciduous canopies (Fowler, 1980).

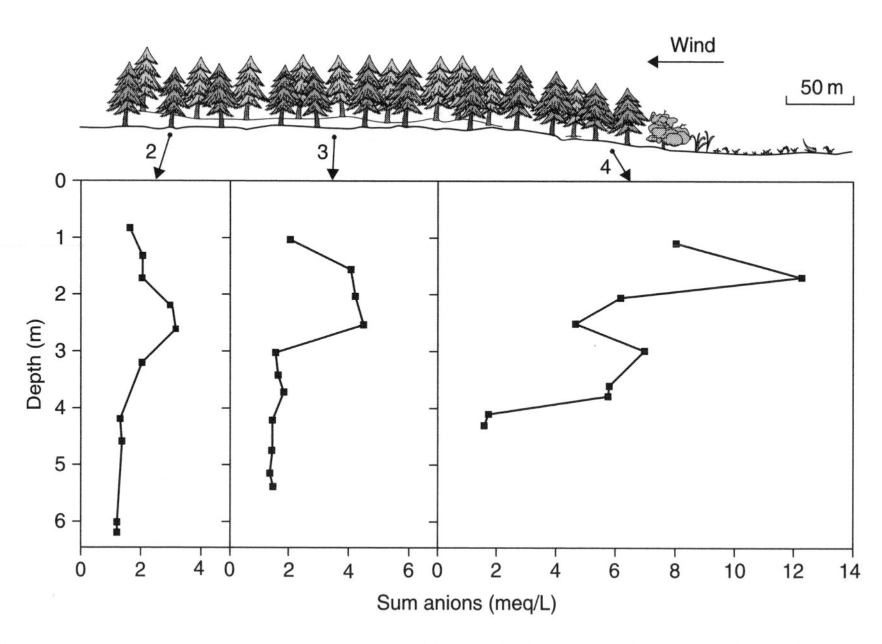

Figure 2.14. Dry deposition and total anion concentrations in the unsaturated zone of a sandy aquifer. At the windward margin, boring 4, the concentrations are high and decrease then further inward into the forest (Hansen and Postma, 1995).

Unfortunately, dry deposition on vegetation is notoriously difficult to measure. Samples of throughfall under the canopy consist of wet input affected by *canopy-exchange.* Ions may have been lost by adsorption to the canopy, or concentrations may have increased by admixture of flushing of salts deposited during the previous dry period and leachates from the vegetation (e.g. Ulrich et al., 1979; Miller and Miller, 1980). Some idea of the importance of dry Cl^- deposition can be obtained from chemical balances of watersheds or lysimeters in which both input and output have been measured over a period of several years. Figure 2.15 shows the ratio of input and output for a number of ecosystems with different vegetations. The input is the product of the Cl^- concentration in precipitation multiplied by the amount of precipitation over the study area (g/year); the output is estimated from either the concentrations in streamflow water and discharge measurements, from groundwater analyses and precipitation-surplus, or from lysimeter studies (also in units of g/year). The ratio of output over input is plotted against Cl^- concentration in precipitation since a higher Cl^- concentration in rain should be associated with a higher abundancy of Cl in aerosols, leading to increased dry deposition.

Figure 2.15 shows a ratio close to 1 in areas covered with heather or low shrubs, so that the output is balanced by the input measured with bulk rain samplers, without much contribution from additional dry deposition. In forests with very low Cl^- concentrations in precipitation, ratios less than 1 are found, indicating an actual loss of Cl^- in the watershed. This could possibly be due to biological uptake (Oberg, 1998) or alternatively to anion adsorption on clay minerals (Feth et al., 1964), although this has not been documented in the field studies compiled in Figure 2.15. When the average Cl^- concentration in wet deposition of forested areas exceeds 50 μmol/L, a ratio of around 2 is found. This suggests a total Cl^- deposition that is about twice the amount supplied with rain and underlines the importance of land use for dry deposition. In Figure 2.15 there is no clear difference between deciduous and coniferous forests, except for the 3 lysimeters at a coastal site in the Netherlands with a 40 year long record (Stuyfzand, 1984) that plot at the extreme right of Figure 2.15. Here, the pine forest has a ratio exceeding 5, which shows that its perennial canopy effectively collects aerosols during winter storms.

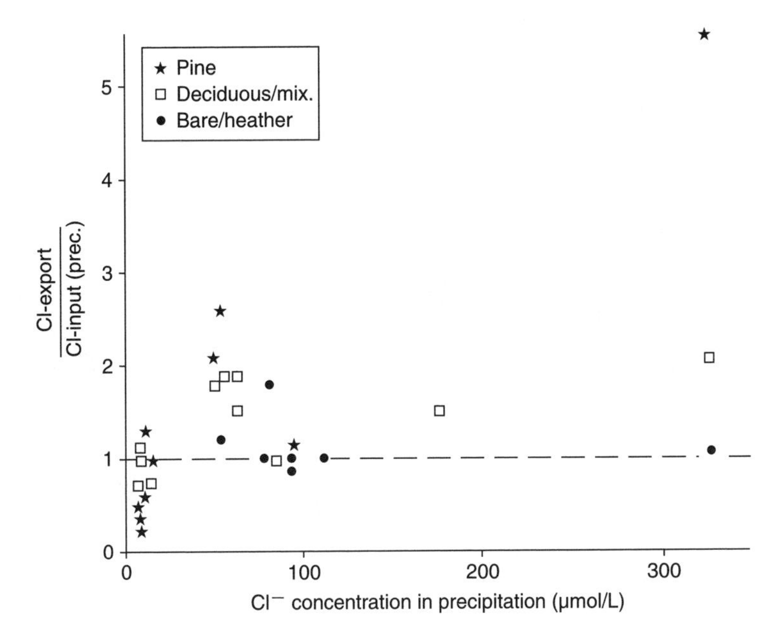

Figure 2.15. Ratio of Cl output/input in watersheds and lysimeters against Cl^- concentration in rain. The Cl^- input was measured with bulk rain samplers (Appelo, 1988).

The dry deposition derived from a chemical balance can be compared with a physical model (Eriksson, 1959). Dry deposition can be calculated from:

$$D = 3.15 \times 10^5 \, v_d c \tag{2.24}$$

where D is dry deposition (μg/m²/yr), v_d is the dry deposition velocity (cm/s), c is concentration in (μg/m³), and 3.15×10^5 recalculates cm/s into m/yr. The value of v_d depends, according to Stokes' law, on the aerosol particle diameter, and the various size fractions can be summed to obtain the operational dry deposition (Holsen and Noll, 1992). The size of aerosol particles often shows a bimodal character, with one fraction between 0.1 and 0.8 μm, and a second fraction between 10 and 100 μm. The larger particles result from an equilibrium of production from seasalt, soil dust, soot, etc. and sedimentation, while the fine particles are produced by coagulation and condensation of particles, which become strongly hygroscopic when finer than 0.1 μm (Whitley et al., 1972). The small-size aerosols, which are the most important for transport over more than 10–100 kms, have a considerably higher deposition velocity than calculated from Stokes' law, probably because the particles are filtered out by adsorption on fixed surfaces. Their velocity of deposition is estimated to range from 0.2 to 2 cm/s (Mészáros, 1981; Möller, 1990). If we take $v_d = 1$ cm/s, and $c_{Cl} = 5$ μg/m³, which is the value observed at 600 m high above the sea (Blanchard and Syzdek, 1984), then $D = 1.58 \times 10^3$ mg/m²/yr is the continental dry deposition rate of Cl in areas where vegetation scavenges the small aerosol particles. This value translates to 2.0 mg/L in 800 mm rain per year and dry deposition then approximately doubles the wet-input in forested areas situated at more than about 10 km from the coast. This is in reasonable agreement with estimates derived from chemical balances for forests in temperate climate (Figure 2.15; Matzner and Hesch, 1981; Appelo, 1988; Mulder, 1988; Rustad et al., 1994). The model assumes a linear relation among dry deposition and atmospheric concentration that may be correct for stagnant air in forests, but becomes somewhat questionable for fields with low shrubs or grass vegetations where air turbulence apparently reduces dry deposition. For such areas, the total atmospheric input equals the input measured with bulk rain samplers.

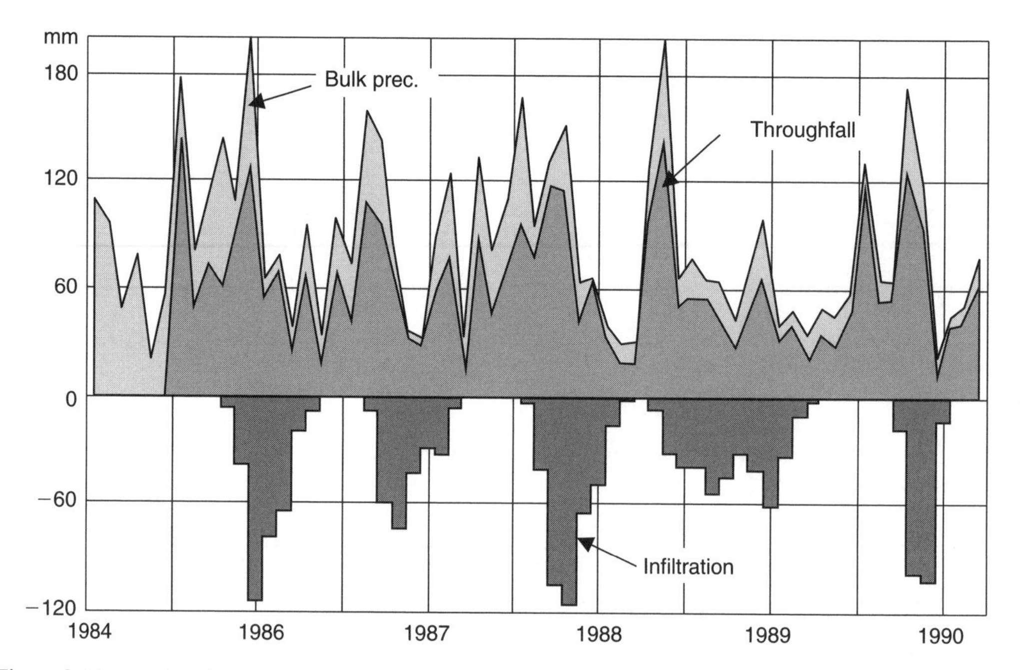

Figure 2.16. Hydraulic budgets in the temperate climate, conifer forest at Klosterhede, Denmark. (Modified from Rasmussen, 1988).

Only part of the precipitation will ultimately infiltrate into the aquifer. Figure 2.16 shows the annual water balance in a conifer forest in a temperate climate with an annual precipitation of 800 mm/yr and an infiltration of 400 mm/yr. The difference is lost by evapotransporation from vegetation and soil. Throughfall, water dripping from the canopy, is already seen to be significantly less than the bulk precipitation, while infiltration in most years starts during the fall and stops again in the spring so that there is no net infiltration during the summer. Any salt deposition by rain or dry deposition will accumulate during summer and be flushed down with the autumn rain.

The evapotranspiration in an area can be calculated by comparing the Cl^- concentrations in soil- or groundwater and in rainwater:

$$E = P_t - \frac{c_{\mathrm{Cl,\,rain}} \times P_t + D}{c_{\mathrm{Cl,\,soil}}} \tag{2.25}$$

where E is evapotranspiration (mm/yr), P_t is total precipitation (mm/yr), c_{Cl} is the Cl^- concentration in precipitation (mg/L) and D is 1575 mg/m²/yr for deciduous or mixed forests and 0 for bare or grass-covered land. When applying Equation (2.25), the precipitation chemistry from bulk samplers should be used, and the soil- or groundwater concentrations must be averaged. Also, the relation becomes invalid when other sources of Cl^- are present, such as manure or waste disposal, or when the chloride mass transfer has changed over time, for example by changes in land use (Wood, 1999).

EXAMPLE 2.2. *Calculate recharge using the Cl^- mass balance* and estimate the flow velocity in the unsaturated zone. Gehrels (1999) sampled and analyzed soil water Cl^- in the Netherlands' Veluwe area below grassland (Molinia sp.) and Picea/Prunis forest. He obtained bulk rain and dry deposition rates of Cl^- for the two vegetation types in the period 1993–1994, when total precipitation was $P_t = 975$ mm/yr.

Precipitation	*grass*	*forest*	
bulk	0.099	0.099	mmol Cl^-/L
dry	0	99	mmol $Cl^-/m^2/yr$
total	0.099	0.201	mmol Cl^-/L
soil, below root	0.19	0.45	mmol Cl^-/L
ε_w	0.089	0.072	m^3/m^3
Result			
P	508	435	mm *net recharge*/yr
v	5.7	6.0	m/yr

The dry deposition of 99 mmol $Cl^-/m^2/yr$ is equivalent to 99/975 = 0.102 mmol Cl^-/L, which is added to the concentration in bulk rain. The net recharge $P = P_t \times m_{Cl,\ total\ rain} / m_{Cl,\ soil}$ and the flow velocity $v = P/\varepsilon_w$ are given in the table.

Finally, the importance of evaporation for the chemical composition of surface waters is illustrated in Figure 2.17. It shows the distribution of total dissolved solids (*TDS*) in surface waters of the USA with the highest concentrations occurring in the central parts with a high evaporation, and the lowest concentrations in the marginal parts of the continent. This pattern is the inverse of the distribution of Cl^- in rainwater (Figure 2.4) and underpins the importance of evaporation.

QUESTION:
The dry deposition of Cd is 1 mg/m^2/yr. Calculate the concentration in 1000 mm/yr of rain.
ANSWER: 1 μg/L

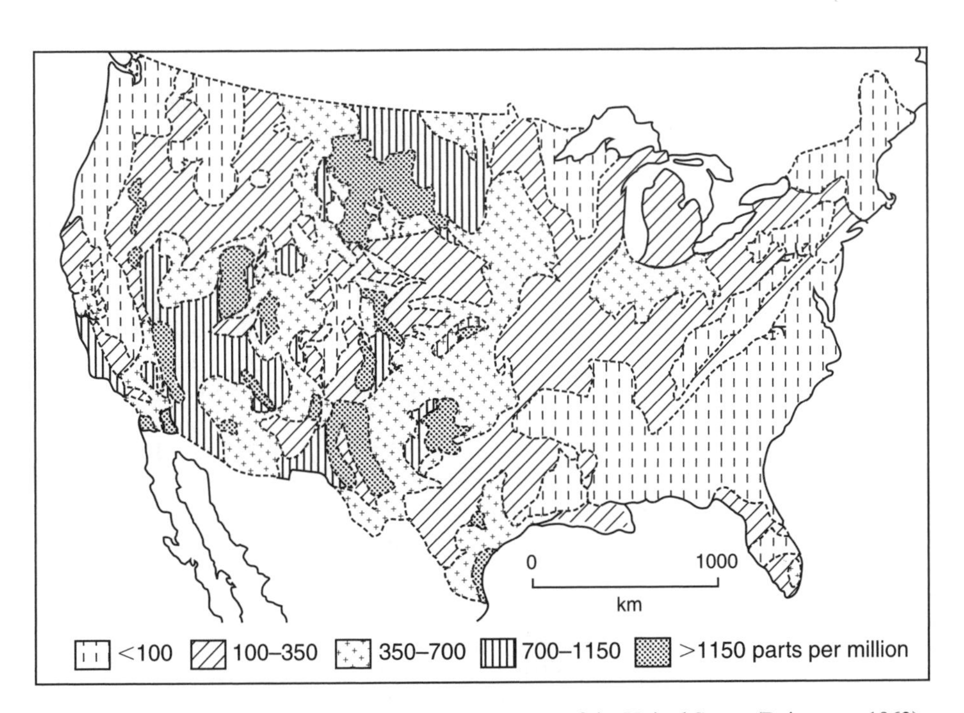

Figure 2.17. Dissolved solids concentrations in stream waters of the United States (Rainwater, 1962).

2.5 MASS BALANCES AND ECOSYSTEM DYNAMICS

Mass balances can be used to estimate the water quality in an area. They consist of an inventory of all the sources and sinks for the elements of interest, including water, and the calculation of the water composition from the resulting net output. For agricultural areas the most important terms in the balance are related to harvesting and the application of manure and fertilizers. The composition of crops and the manure from various animals is continuously analyzed by various institutions and some uptake and release rates are listed in Table 2.5. The numbers allow to assess the groundwater quality below agricultural land as demonstrated in Example 2.3.

Table 2.5. Elemental uptake rates by crops and release rates from animal manure. (1 ha = $10^4 m^2$)

	Water	N	P	K	Na	Mg	Ca	Cl	S	
Crop										
Grass, 4 cuttings	4000 (m^3/ha/yr)	15.4	0.9	5.4	0.1	0.5	1.0	2	0.7	(kmol/ha/yr)
Green maize	3000	13.4	1.1	4.9	0.1	0.8	0.9	1.7	0.43	
Barley	3000	6.1	0.6	1.8	0.05	0.3	0.3	0.5	0.2	
Animal manure										
Dairy cow	18.1 (m^3/yr)	6.2	0.56	2.1	0.64	0.50	0.72	1.7	0.45	(kmol/yr)
Fatting calf	2.2	0.46	0.04	0.11	0.04	0.07	0.06	0.16	0.01	
Porker	1.7	0.90	0.12	0.15	0.06	0.05	0.11	0.08	0.06	
Breeding pig	3.7	2.0	0.26	0.34	0.13	0.10	0.25	0.17	0.12	
1000 chickens	3.7	16	3.0	4.0	1.1	1.3	3.2	1.6	0.99	
1000 laying-hens	27	36	11	7.6	2.6	2.5	17	4.0	2.0	

EXAMPLE 2.3 *Estimate the Cl^- and NO_3^- concentration in groundwater below agricultural land*

We consider a farm run by a family of four, with 10 ha grassland and 40 cows. The input consists of precipitation, waste water, manure and fertilizer. The export is by vegetation uptake, harvesting and groundwater outflow. Concentrations in the groundwater are obtained dividing the net output of an element by the net output of water.

	H_2O		N	Cl	
Yearly input					
Rain	840	mm	0.18	0.08	mmol/L
1 person	36.5	m^3	6	3	mmol/L
1 cow	18.1	m^3	340	95	mmol/L
Yearly output					
Grass (4 cuttings)	400	mm	15.4	2	kmol/ha
Mass transfer for 10 ha					
Rain	8.40×10^4 m^3/yr		15.1	6.7	kmol/yr
4 persons	1.46×10^2		0.88	0.44	
40 cows	7.24×10^2		246	68	
Total	**8.49×10^4**		**262**	**75.9**	
Grass	-4.00×10^4		−154	−20	
Net	**4.49×10^4 m^3/yr**		**108**	**55.9**	**kmol/yr**
Concentration			2.4	1.2	mmol/L

The concentrations calculated in Example 2.3 can be compared with drinking water limits, which are 0.4 and 5.5 mmol/L for NO_3^- and Cl^- (Table 1.1), respectively. The estimated N concentration by far exceeds the drinking water limit, but fortunately not all the nitrogen will be in the form of NO_3^-. Most of the N is introduced as ammonium and amines associated with the organic waste and will only partly be transformed into nitrate by microbes under aerobic conditions. A part will evaporate already as ammonia during field application, and some of the nitrate in the soil will be denitrified to N_2 gas and escape to the atmosphere. The various processes which affect nitrogen conversions can be calculated for agricultural soils with computer programs (Rijtema and Kroes, 1991; Wu and McGechan, 1998). A rough estimate for an aerobic soil with a neutral pH is, that about half of N will persist in the form of nitrate and thus will be flushed. However, note that the resulting 1.2 mmol NO_3^-/L in Example 2.3 is still three times higher than the drinking water limit.

The comparison of observed and calculated concentrations will rapidly reveal whether processes have been neglected in the mass balance. For example, the influence of agricultural practices on water quality is illustrated in Figure 2.18. It shows the differences in groundwater composition with depth in a borehole equipped with multilevel samplers. The high NO_3^- concentrations in the upper part of the boring are due to heavy application of manures and fertilizers to fields upstream, and are close to what was calculated in Example 2.3.

The mass balance calculation is an overall summary in which concentration fluctuations in time and space are averaged and internal biogeochemical cycles are neglected. However, uptake and cycling of elements by vegetation and storage in accreting biomass, or the release from decaying organic matter can profoundly influence the concentrations of elements in water leached from the soil.

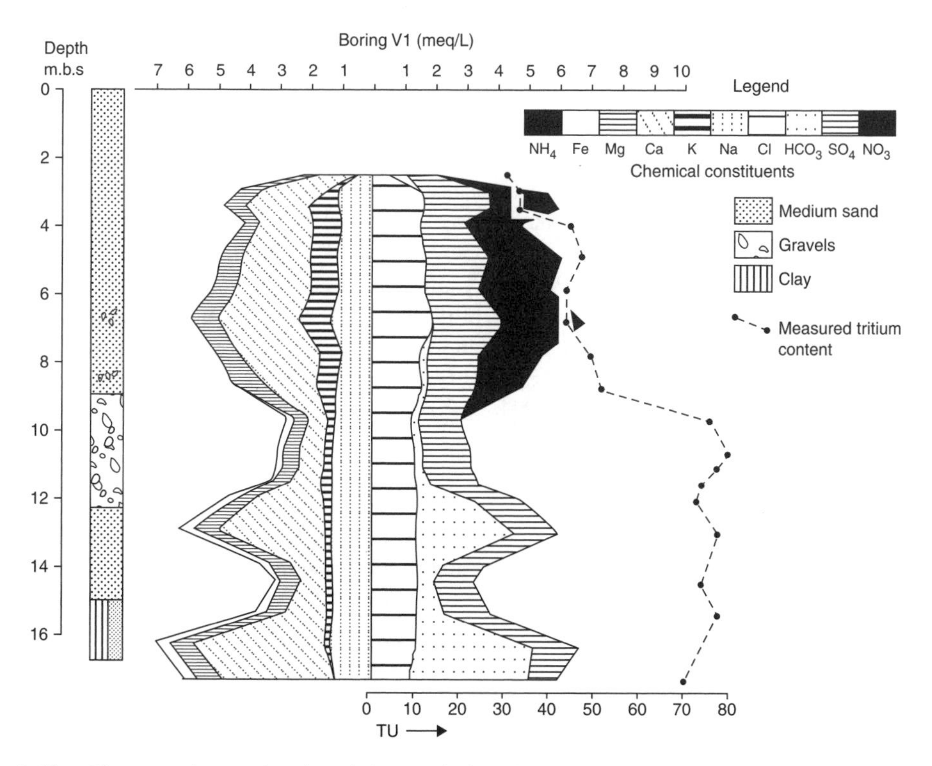

Figure 2.18. The groundwater chemistry below agricultural land. Concentrations of cations are plotted cumulatively to the left, and anions cumulatively to the right (Appelo, 1988).

Figure 2.19 illustrates the transfers of Ca^{2+} in the forested ecosystem developed on gneiss and till deposits at Hubbard Brook (Likens and Bormann, 1995). The input of Ca^{2+} into the ecosystem comes mainly from the weathering of minerals and for only 10% from precipitation. The outflow of Ca^{2+} with the runoff is about half of the total input. The difference is due to the uptake of Ca^{2+} in the biomass of growing trees. Furthermore, the Ca^{2+} uptake by the biomass is 3 times the Ca^{2+} weathering flux, implying that Ca^{2+} is extensively recycling in the soil-biomass system.

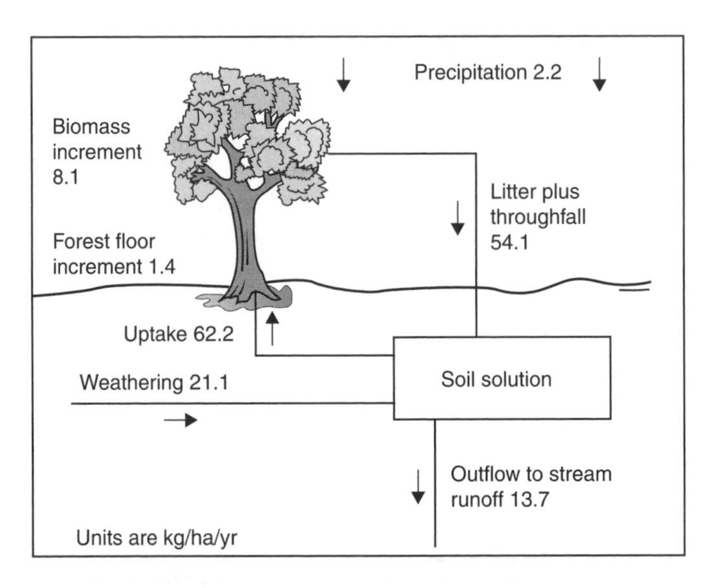

Figure 2.19. Cycling of Ca^{2+} in a forested ecosystem at Hubbard Brook (Drever, 1997).

Table 2.6 displays the transfers for other elements in the Hubbard Brook ecosystem and shows significant difference among the elements. For example, Mg^{2+} is recycled in the biomass to a much smaller extent than Ca^{2+}. On the other hand K^+ is very strongly conserved within the soil and biomass. The vegetation uptake of K^+ is 64 kg/ha/yr, whereas the streamwater output is only 1.9 kg/ha/yr. A nutrient like P is even more strongly recycled and withhold in the biomass. Natural vegetations are, as may be expected from an evolutionary standpoint, masterful in storing essential elements with a limited availability. However, this also implies that a small disturbance in the flux through living matter, for example by logging trees or bush fires, may greatly disturb the much smaller flux of dissolved constituents leaving the biosphere.

Table 2.6. Elemental fluxes in the biomass compared to fluxes with rain, weathering and streamwater output in the Hubbard Brook ecosystem (fluxes in kg/ha/yr from Likens and Bormann, 1995).

	Ca	Mg	Na	K	N	S	P	Cl
Bulk precipitation input	2.2	0.6	1.6	0.9	6.5	12.7	0.04	6.2
Streamwater output	13.7	3.1	7.2	1.9	3.9	17.6	0.01	4.6
Vegetation uptake	62	9	35	64	80	25	9	small
Root exudates	4	0.2	34	8	1	2	0.2	1.8
Weathering release	21	4	6	7	0	1	?	small

Table 2.6 shows for N and S that precipitation constitutes the main input into the ecosystem. This is partly due to adsorption of gases like SO_2, NH_3 and NO_2 on the vegetation. Some of these gases are taken up directly in the plant, while the remainder is flushed away with the rain.

Vegetation uptake is usually determined by considering above-ground biomass only, neglecting the growth and production of roots. However, the proportion of below-ground production of the total net primary production varies from about 0.2–0.3 for forests to 0.6 for grasses (Gower et al., 1999) and is therefore highly significant in the chemical balance, but also quite difficult to analyze.

QUESTIONS:

Estimate the Cl^- concentration in the urine of a dairy cow and its calf.

ANSWER: From Table 2.5, cow: 1700 mol/yr / 18.1 m^3/yr = 94 mM; calf: 73 mM.

A dairy cow produces about 50 L milk/day. Estimate the daily water consumption of the animal.

ANSWER: 50 L/day (milk) + 50 L/day (urine, Table 2.5) = 100 L/day.

Calculate the *E.B.* of a dairy cow's faeces, using Table 2.5

ANSWER: $\Sigma+ = 5.18$ keq/yr. $\Sigma- = -2.6$ keq/yr (Cl^- and SO_4^{2-} only); estimated organic acids -2.58 keq/yr; N is mainly present in amines.

2.5.1 *Water quality profiles in the unsaturated soil*

The effects of variations in rainwater composition and dry deposition, and of seasonal fluctuations in infiltration and evapotranspiration on the groundwater chemistry are seen most clearly in the unsaturated zone, and particularly when the rocks contain no highly reactive minerals such as carbonates. This is illustrated by a water chemistry profile through the unsaturated zone of a sandy deposit near the coast in Denmark in Figure 2.20.

Very large variations in the total dissolved ion concentrations are mostly due to variations in the concentrations of Na^+ and Cl^- which both have a marine origin. At this site the average water transport rate through the unsaturated zone is about 5 m/yr, so that the whole profile represents roughly one year of infiltration, although net infiltration occurs only from October to May.

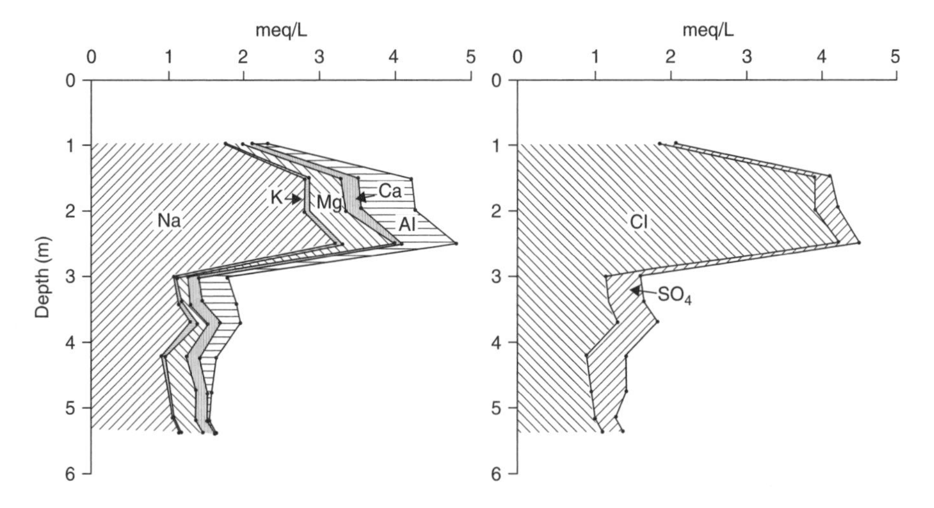

Figure 2.20. The groundwater chemistry in the unsaturated zone of a sandy, carbonate free sediment under a conifer forest in Denmark. Concentrations are plotted cumulatively. The annual precipitation is 860 mm/yr, infiltration 400 mm/yr, and unsaturated zone pore water velocity 4–5 m/yr (Hansen and Postma, 1995).

During the fall, the salts accumulated in the soil over the previous summer period start to be washed down into the unsaturated zone. During the same season westerly storms are frequent and may deposit large amounts of sea salt. The effects of these two processes are not easily separated and in addition, transpiration may also contribute to the maximal concentrations observed in Figure 2.20 in the upper meters.

The influence of dry deposition, and differences in evapotranspiration on the Cl^- content of water in the unsaturated zone of a sandstone in England are illustrated in Figure 2.21. Below heathland, Cl^- concentrations are low and show little variation. Below the forest, chloride concentrations are much higher and vary widely, although generally lower concentrations are found below the rooting depth. The high Cl^- concentrations below the forest can partly be explained by the effect of dry deposition (Figure 2.15). Moreover, evapotranspiration varies distinctly between heath/grassland and forested areas. In particular interception (evaporation from wetted surfaces) is much higher in forested areas and may amount to about 40% of the annual precipitation (Calder, 1990, 2003). The concentration of dissolved ions becomes correspondingly higher. Moss and Edmunds (1989) found that the recharge rate below heathland was roughly 3 times higher than below the forest.

Unsaturated profiles often display the characteristic shape of Figures 2.20 and 2.21 with higher concentrations present in the rooted, upper part than at greater depth. In most cases these profiles are stable throughout the year in both arid and temperate climate (Sharma and Hughes, 1985; Gehrels, 1999). This indicates that the rain is short-circuiting to greater depths via root channels and along water-repellent, humic surfaces of soil cutans, and only partly exchanges with the more concentrated soil solution in the stagnant structure. Short-circuiting implies that the estimates of net recharge with the Cl mass balance method must be based on water quality from below the root zone. The bypass flow also explains why seasonal signals in isotope and Cl^- concentrations in rain are quickly lost in the soil solution.

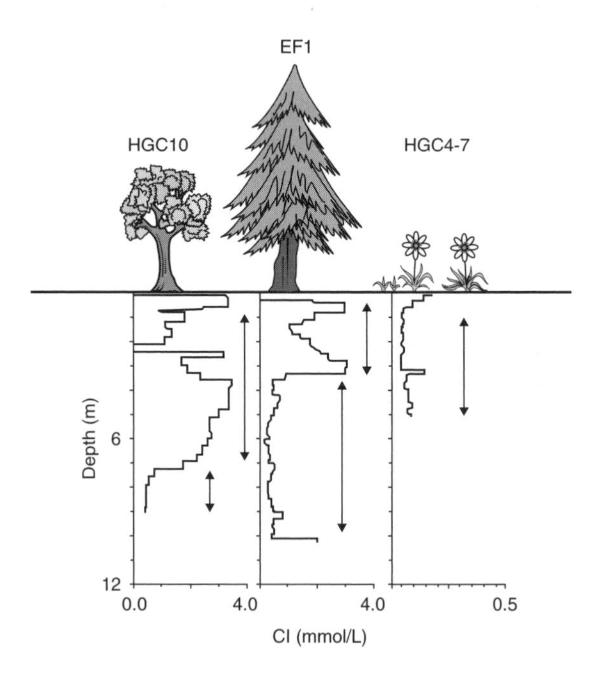

Figure 2.21. Chloride profiles in the unsaturated zone of the Sherwood sandstone below birch woodland, Cyprus Pine and heathland (Moss and Edmunds, 1989).

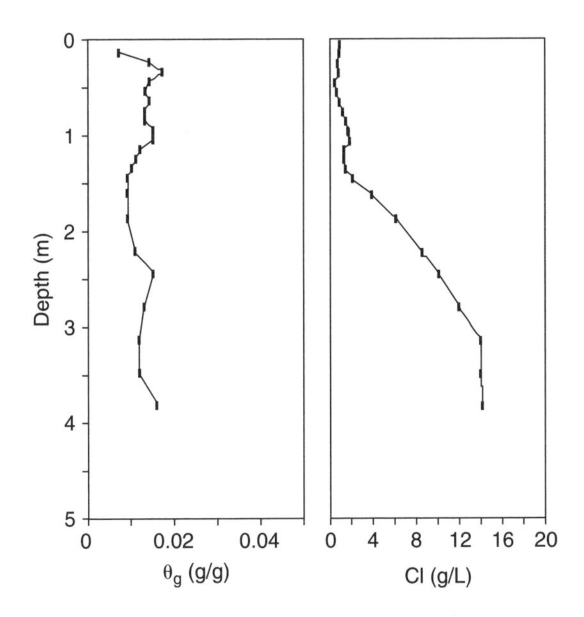

Figure 2.22. Profiles of gravimetric water content (left, grams per gram) and Cl^- concentration in the soil solution below an Eucalyptus stand in a semi-arid climate, Murbko, S. Australia. Annual precipitation is 260 mm/yr, potential evaporation 1800 mm/yr and the recharge rate 0.1 mm/yr (Cook et al., 1994).

In arid or semi-arid areas where evapotranspiration is high, very high concentrations of salts may accumulate in the soil solution. Figure 2.22 shows the Cl^- distribution in a soil solution profile below a stand of Eucalyptus in S.Australia. As the water moves downward through the soil, roots selectively remove water molecules and the Cl^- concentration increases. First at a depth of about 3 m the effect of evaporation ceases and the Cl^- concentration, at 14 g/L, becomes constant over depth. If the chloride mass balance method is used to calculate the recharge rate, clearly the Cl^- concentration at depths >3 m should be used in this case. The calculated recharge rate was 0.1 mm/yr which can be compared with the mean annual precipitation of 260 mm/yr. The low recharge rates will also give long reponse time of the Cl^- profiles upon changes in landuse or climate before a new steady state situation is attained (Allison et al., 1994; Phillips, 1994).

2.6 OVERALL CONTROLS ON WATER QUALITY

The chemical composition of fresh groundwaters of the USA is shown in a frequency plot in Figure 2.23. The most abundant cations are Ca^{2+}, Na^+ and Mg^{2+}. The main anions are HCO_3^-, SO_4^{2-} and Cl^-. The data in Figure 2.23 are older than 1960, before groundwater became polluted with NO_3^-. More recent data would place NO_3^- together with the other major anions. Minor components are Fe^{2+}, Mn^{2+}, and many trace metals, as well as anions like F^- and boron, are also plotted. For some elements, such as Na^+, SO_4^{2-}, and Cl^- the distributions approach the sigmoidal shape of a normal distribution. This suggests that the concentration of these species depends on their availability in rocks, very slowly dissolving minerals, or control by biological processes, which have a random character on the continental scale. For other components, like Ca^{2+}, HCO_3^-, SiO_2, K^+ and F^-, the gradient steepens at higher concentrations which could indicate that the solubility of a mineral places an upper limit on the maximal concentration of the species in natural waters.

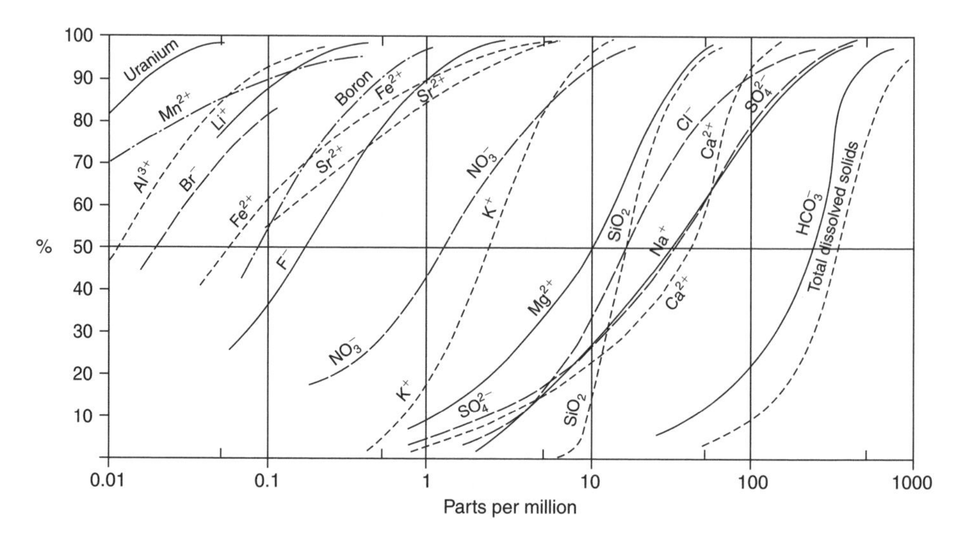

Figure 2.23. Concentrations of different species in groundwaters in the USA displayed in a frequency plot (Davis and De Wiest, 1966). Reprinted by permission of John Wiley & Sons, Inc.

Zooming in from the continental scale to a single watershed, the complexity increases and a myriad of processes may affect the groundwater chemistry, as is outlined schematically in Figure 2.24. We have already discussed the importance of the external factors like rainwater chemistry, evapotranspiration and uptake by vegetation in some detail. What follows is a concise introduction to the processes that affect water quality in the soil and groundwater zone.

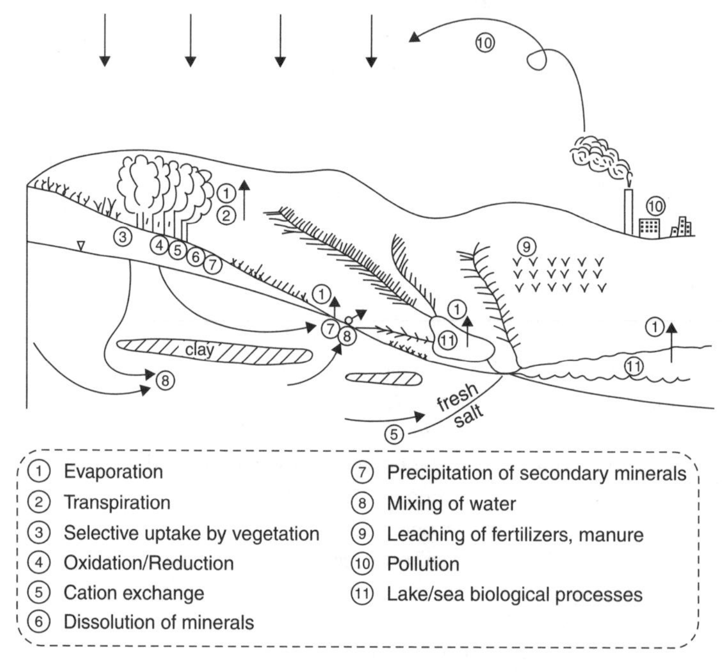

Figure 2.24. An overview of processes that affect the water quality in a watershed.

Oxidation/reduction processes are of importance and often related to the decay of organic matter. It is the reverse process of uptake and storage of elements in growing plants. Decay of organic matter is an oxidation reaction. The process uses oxygen and produces carbonic acid:

$$CH_2O + O_2 \rightarrow H_2O + CO_2 \tag{2.26}$$

In this reaction a carbohydrate, CH_2O, is used as a simple representative of the, in reality, highly complex organic compounds in organic material.

The oxidation of organic matter may occur in the soil but also within aquifers where fossil organic matter is present as peat, lignite, etc. Once oxygen becomes exhausted other electron acceptors, like nitrate, iron oxides or sulfate may mediate the oxidation of organic matter. The reduction of iron oxides results in high iron concentrations in groundwater, the reduction of sulfate in badly smelling hydrogen sulfide and finally, fermentation of organic matter may result in groundwaters containing methane. The production of CO_2 as result of organic matter oxidation (Equation 2.26) may also enhance the weathering of carbonate and silicate minerals.

Weathering and dissolution of minerals releases elements to the water. The relation between rock type and water composition is illustrated in the composition of the spring waters of the Ahrntal (Zillertal Alps) in N. Italy. The geology displays a typical central alpine sequence of micaschists, limestones, dolomites, gypsiferous layers, and serpentinite rock. The rocks dip vertically and are incised by alpine trough valleys (Figure 2.25). Table 2.7 lists the characteristic chemical composition for each of the four rock types; Group 1 consists of calcareous schist, a metamorphic rock containing $CaCO_3$ as its most reactive component and Ca^{2+} and HCO_3^- are the predominant ions in the spring water.

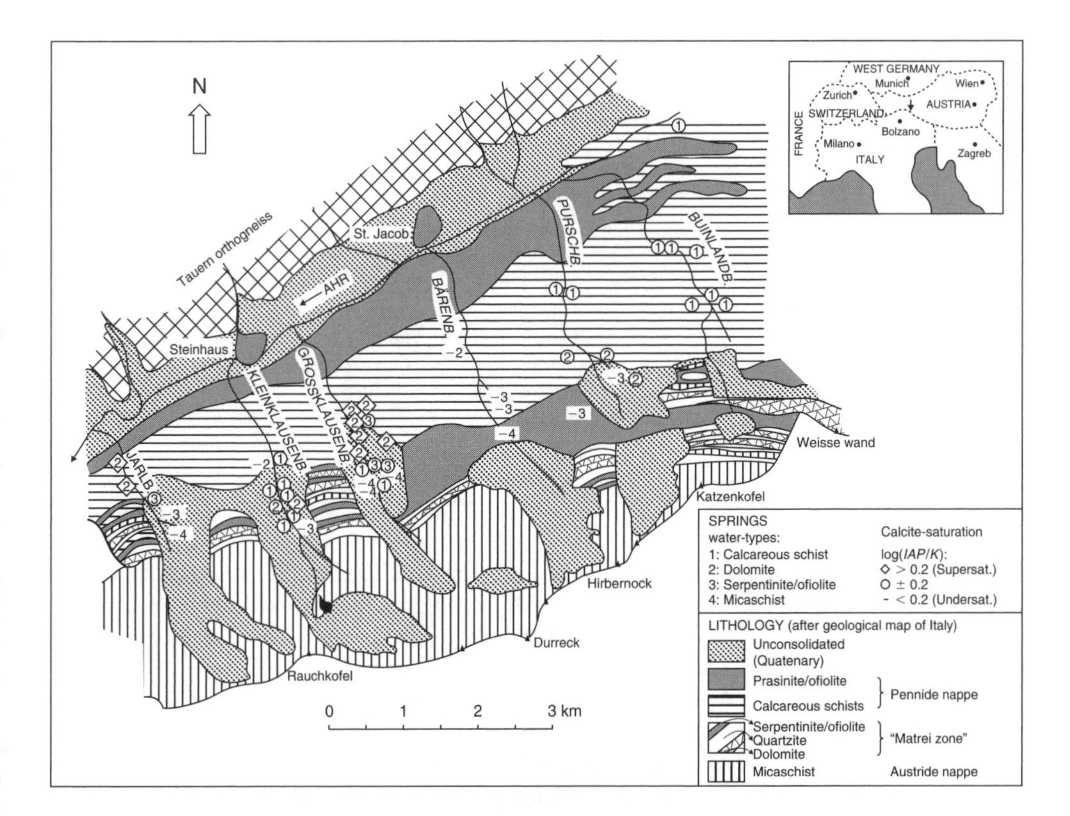

Figure 2.25. The geology of the Ahrntal (N. Italy) with location and classification of springs.

Table 2.7. Typical composition of spring-waters. "mcs" micaschist; "serp" = serpentinite; "pras" = prasinite/ofiolite; "dol" = dolomite; "cc" = calcareous schist (Appelo et al., 1983).

Group code Geol. unit	4 "mcs"	4 "mcs"	3 "serp"	3 "pras"	2 "dol"	2 "dol"	1 "cc"
pH	7.63	7.70	8.47	7.41	8.00	7.93	8.12
EC (μS/cm)	52	100	152	610	325	505	250
Temp (°C)	3.7	4.0	4.0	4.8	5.5	7.5	2.0
Na^+ (mmol/L)	0.04	0.03	0.02	0.06	0.04	0.04	0.04
K^+	0.016	0.026	0.011	0.015	0.011	0.008	0.011
Mg^{2+}	0.11	0.12	0.44	2.30	0.67	0.93	0.31
Ca^{2+}	0.17	0.35	0.33	1.20	0.77	1.95	0.94
Cl^-	0.05	0.05	0.06	0.05	0.04	0.06	0.05
HCO_3^-	0.37	0.86	1.35	5.54	2.10	3.34	2.23
SO_4^{2-}	0.08	0.04	0.05	0.72	0.14	1.19	0.15
Si	0.05	0.05	0.06	0.13	0.06	0.09	0.05

Group 2 comprises rock with the mineral dolomite ($CaMg(CO_3)_2$) and the water contains, apart from Ca^{2+} and HCO_3^-, significant amounts of Mg^{2+}. At some places the mineral gypsum ($CaSO_4 \cdot 2H_2O$) is present and the water becomes enriched in SO_4^{2-}. Group 3 contains magmatic rocks with Mg-minerals like serpentine and talc. These weather easily and produce waters enriched in Mg^{2+}, HCO_3^- and silica. Finally group 4 consists of metamorphic micaschists. The minerals in this rock have a low reactivity and the solute concentrations are low. Table 2.8 lists the major elements and their main mineral source (as well as some additional sources), and the range of concentrations which may be expected in fresh waters.

Table 2.8. Normal ranges of concentrations in unpolluted fresh water and the sources of elements.

Element	Concentrations (mmol/L)	Source
Na^+	0.1–2	Feldspar, Rock-salt, Zeolite, Atmosphere, Cation exchange
K^+	0.01–0.2	Feldspar, Mica
Mg^{2+}	0.05–2	Dolomite, Serpentine, Pyroxene, Amfibole, Olivine, Mica
Ca^{2+}	0.05–5	Carbonate, Gypsum, Feldspar, Pyroxene, Amfibole
Cl^-	0.05–2	Rock-salt, Atmosphere
HCO_3^-	0–5	Carbonates, Organic matter
SO_4^{2-}	0.01–5	Atmosphere, Gypsum, Sulfides
NO_3^-	0.001–0.2	Atmosphere, Organic matter
Si	0.02–1	Silicates
Fe^{2+}	0–0.5	Silicates, Siderite, Hydroxides, Sulfides
PO_4	0–0.02	Organic matter, Phosphates

Different mechanisms of rock weathering can be deducted from Figure 2.26. It shows the maximal dissolved solids concentration in streams from small catchments, located in different rock types, as a function of the mean annual runoff (Walling, 1980). It is instructive to interpret the mean annual runoff in terms of the residence time of water. The average catchment will have a soil that is 1 m thick with a water filled porosity of 10%. The soil profile then contains 0.1 m water. With a precipitation surplus (rain – evapotranspiration) of 0.1 m/yr, it will take 1 year for the water to travel through the soil. If the precipitation surplus were 0.2 m/yr, the water travel time would be half a year, etc. The annual runoff is therefore inversely related to the residence time of water in the soil; the higher the annual runoff, the shorter the contact time between water and rock.

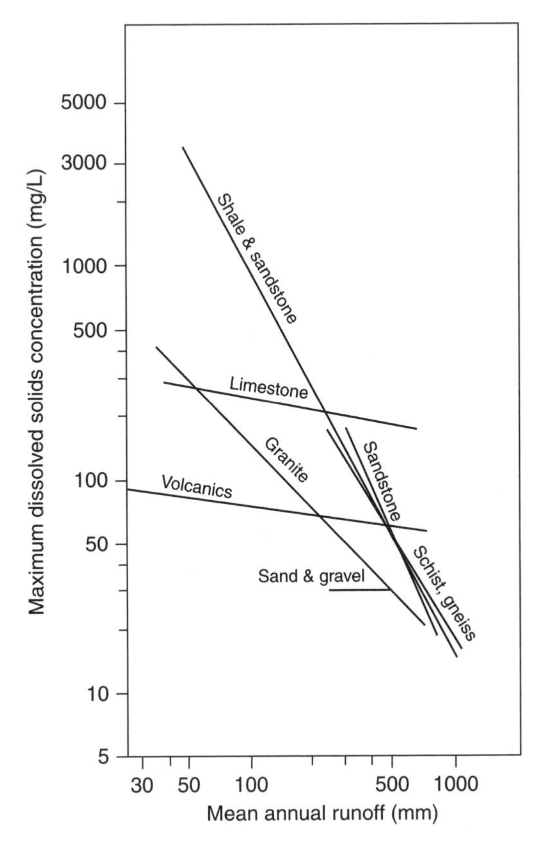

Figure 2.26. Differences in total dissolved solid concentration in surface runoff as a function of rock type and annual runoff (Walling, 1980). Reprinted by permission of John Wiley & Sons, Ltd.

Two trends can be identified in Figure 2.26. First, the content of total dissolved solids in the water derived from limestone, volcanics, sand and gravel deposits is almost independent of the amount of runoff, which is not the case for the other rock types. Second, for a specific value of the annual runoff, the total dissolved concentration varies for different rocks. The differences are related to the *solubilities* of the minerals present in the parent rock, and the *rate of dissolution* of these minerals. Gravel consists mainly of insoluble quartz pebbles which are highly resistant to dissolution. Accordingly, the concentration level remains low, and the composition of stream water is mainly controlled by rainwater chemistry and evapotranspiration. On the other hand, volcanics and limestones contain more soluble minerals so that the solute concentration in runoff increases. The dissolved solute concentration here is nearly constant and independent of runoff. This indicates that the dissolution rate of the solids is fast compared to the residence time of the water in the drainage area, and the concentration level may well represent the solubility of the minerals.

On the other hand, streamflow from granites and sandstones show concentration levels that depend strongly on the annual runoff. This suggests that the dissolution rate of the minerals is slow compared to the residence time and that the concentration level is determined by the latter. The dominant minerals in granite and sandstone are feldspars, micas and quartz and, as is shown in Chapter 8, these minerals exhibit indeed very slow dissolution kinetics. In other words, Figure 2.26 demonstrates that in some cases, as for limestones, an equilibrium approach is appropriate, while in other situations, as with most silicate rocks, dissolution kinetics must also be considered.

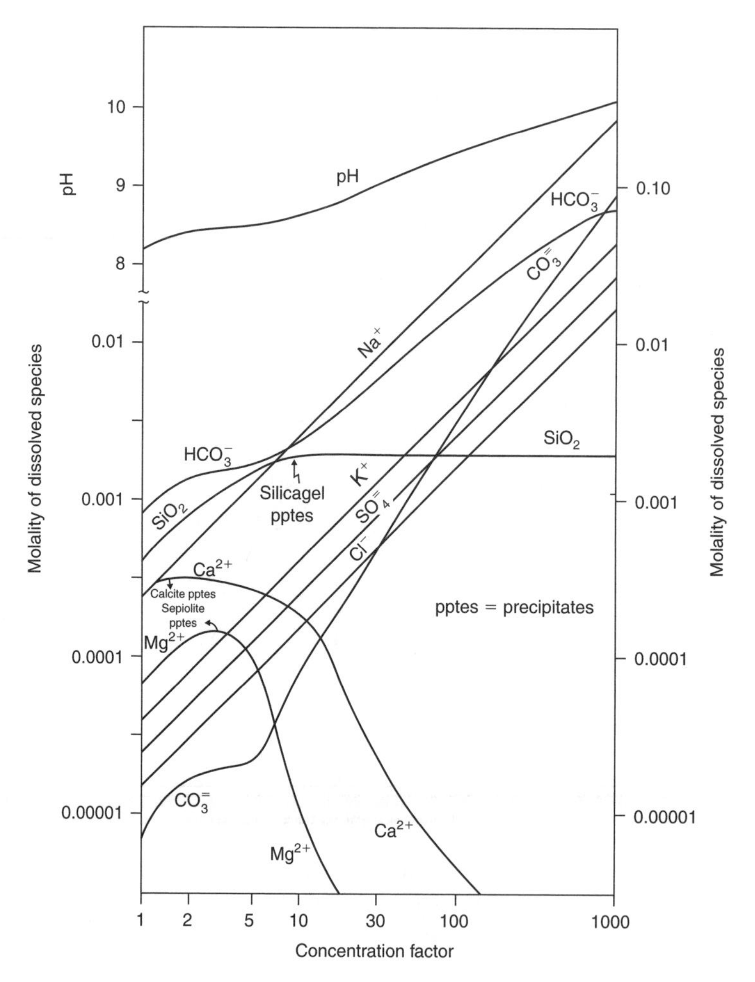

Figure 2.27. Calculated results of evaporation of Sierra Nevada (USA) spring water (modified from Garrels and Mackenzie, 1967).

Minerals may also precipitate in soils or aquifers. Weathering of silicate minerals leads generally to the formation of secondary clay minerals. The type of clay mineral that forms depends on the composition of the parent rock and the weathering stage. Basically, clay minerals consist of Al-silicates with or without other cations. As weathering proceeds, the clay minerals are stripped first of cations and then of silicon, until the sparingly soluble Al-hydroxide remains. If iron is present in the parent rock, the formation of Fe-oxides can also be expected.

In arid climates, evaporation may lead to the precipitation of a suite of minerals in the soil. These include calcite, gypsum, and chloride-salts. The changes in solute concentration and the sequence of minerals that precipitate during evaporation can be calculated for a given initial water composition, as shown for the Sierra Nevada spring waters in Figure 2.27. In this case, the initial water stems from springs draining a granitic terrain, and the minerals which precipitate are sepiolite (a Mg-silicate), calcite, and silica-gel. The final water is a Na-Cl + HCO_3 brine which is commonly found in soda lakes.

Ion exchange reactions, between dissolved cations and those sorbed to mineral surfaces may profoundly change the water chemistry when the infiltrating water differs from what is already present. Examples are found in fresh and salt water displacements in coastal aquifers, pollution plumes spreading out from waste sites, and acid rain moving downward through soil. Ion exchange tends to smoothen concentration gradients which are the result of changing conditions in the groundwater reservoir, but it can also produce concentration peaks in a chromatographic sequence. The sediment exchange complex in a soil consists of clay minerals, iron oxides, weathered primary minerals and organic material. Per volume of wet sediment, the exchange complex can easily contain an amount of in essence mobile cations that is 300 times larger than is present in soil moisture!

Mixing of different water qualities occurs in coastal zones between fresh water and seawater, but also in groundwater seepage zones, and near springs. An upland area with locally different characteristics (geology, vegetation) can discharge different water qualities to the same spring. Mixing of different waters by dispersion is also important for transport of pollutants in aquifers. Mixing of waters can lead to subsaturation with respect to calcite, even if the original waters were at saturation before mixing.

Anthropogenic activities can thoroughly change water quality. These contributions may be of any kind, from NaCl used in households to heavy metals or poisonous organic constituents leached from tip heaps. Air pollution and acid rain already have been mentioned. The large amounts of NO_3^- and pesticides leaching from agricultural fields constitute a serious problem in many countries.

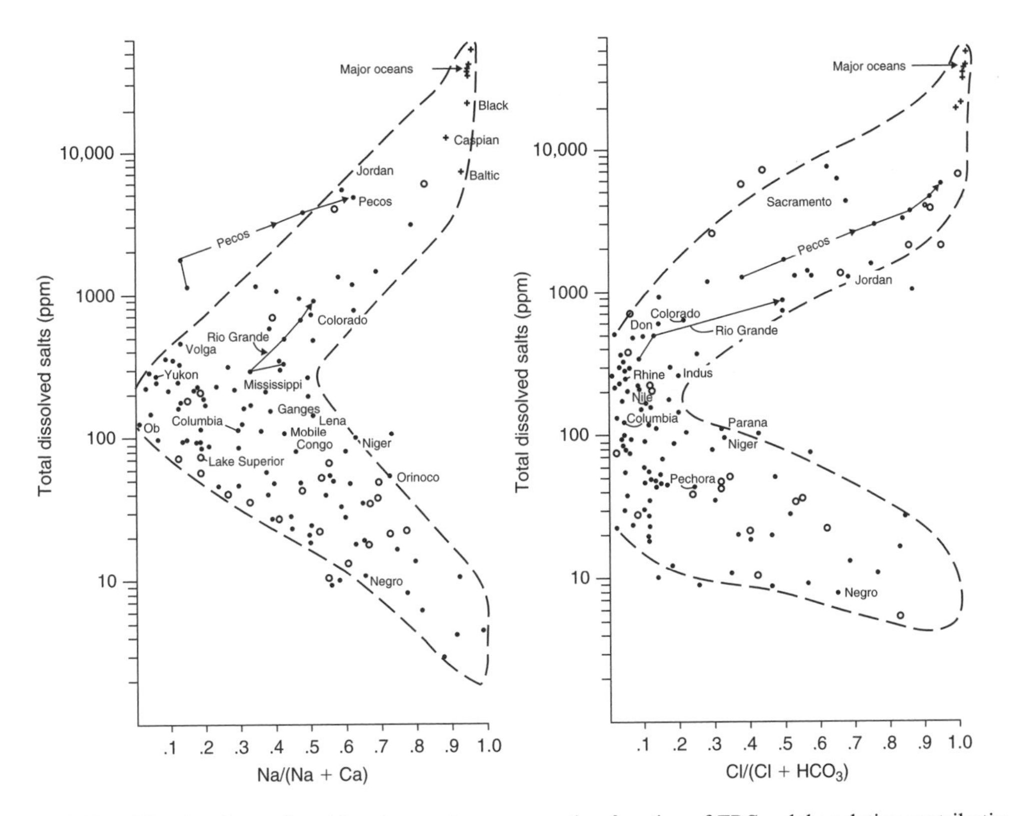

Figure 2.28. The chemistry of world surface waters expressed as function of *TDS* and the relative contribution of Na^+ and Ca^{2+}, and of Cl^- and HCO_3^- (Gibbs, 1970).

This chapter started with the hydrological cycle, noting that groundwaters mostly discharge to rivers and surface waters. Thus, surface runoff will present an overall average of groundwater compositions in the drainage basin (although quality changes are retarded by the long groundwater residence times). Figure 2.28 assembles surface water compositions from all over the world, graphing total dissolved solids as a function of the dominance of Na^+ or Ca^{2+} and of Cl^- or HCO_3^-. The figure illustrates the processes in groundwater quality in a nutshell. At low dissolved concentrations, the dominant ions are Na^+ and Cl^-, contributed by rainwater without much geochemical reaction. Contact with rapidly dissolving calcite and Ca-silicates increases the relative content of Ca^{2+} and HCO_3^-. If subsequent evaporation concentrates the solution, Ca^{2+} and HCO_3^- are lost by precipitation of $CaCO_3$ and the water composition moves upwards in the figures and returns to the seawater composition dominated by Na^+ and Cl^-.

Thus, the overall picture of groundwater quality can be understood with simple, qualitative reasoning. However, for the details a variety of chemical, physical and biological processes must be considered, which are discussed in the remainder of this book.

PROBLEMS

2.1. Estimate the distance from the coast where the global averaged proportion of recycled precipitation equals 1.

2.2. Estimate the integrated proportion of recycled precipitation from 0 to 2500 km from the coast, using Equation 2.2.

2.3. The composition of rain at 3 locations in the Netherlands is listed in the table (Summer-averages, 1979, concentrations in μmol/L, except pH).

	pH	NH_4^+	Na^+	K^+	Ca^{2+}	Mg^{2+}	Cl^-	NO_3^-	SO_4^{2-}
Vlissingen	4.00	49	265	8	39	34	297	92	85
Deelen (Veluwe)	3.95	119	30	3	16	5	37	90	..
Beek (Limburg)	....	143	23	11	50	5	34	87	96
Sea-water (mmol/l)	–	–	485	10.6	10	55	566	–	29.5

a. Check the quality of the Vlissingen-analysis. Estimate missing parameters in the two other analyses.
b. Calculate seawater-influence in the three samples.
c. Which other influences can be discerned?

2.4. Calculate the MWL slope by differentiating Equation (2.11) with respect to *f* after substituting δ for *R*. Use $\delta_0{}^2H = -95‰$ and $\delta_0{}^{18}O = -12.2‰$ (values for oceanic vapor), $f = 0.9$, $\alpha_{l/g}{}^2H = 1.0876$ and $\alpha_{l/g}{}^{18}O = 1.0099$ (values for 18°C). Also calculate the deuterium excess.

2.5. A sample contains $^{16}O = 55.45$ mol/L. Calculate $\delta^{18}O$ when $^{18}O = 0.11117, 0.11118$ and 0.11 mol/l. Also find the number of digits accuracy for ^{18}O, to obtain 0.05‰ accuracy in the δ value.

2.6. In unsatured profiles, Gehrels (1999) measured Cl^- concentrations of 0.7 and 0.19 mmol/L in the root-zone and below, respectively. The water filled porosity was $\varepsilon_{rz} = 0.25$ in the root zone and $\varepsilon_w = 0.089$ below. Estimate the concentration in the stagnant part of the root zone.

2.7. How does the ratio of Na- over Ca-concentration change in rain from the coast to further inland locations?

2.8. How does the Na / Ca-ratio change when surface water evaporates?

2.9. Which elements are enriched in vegetation when compared with soil moisture?

2.10. What is the effect of a larger discharge on runoff composition from:

a. gravel; b. calcareous rocks; c. granite.

2.11. Find the change of δ^2H in water condensing in a closed bucket cooling from 20°C to 0°C. Initially there is only vapor in the bucket, $\delta^2H_{vapor,0} = -80‰$. $\varepsilon^2H_{l/g} = 91.3‰$ (at 15°C).

REFERENCES

Allison, G.B., Gee, G.W. and Tyler, S.W., 1994. Vadose-zone techniques for estimating groundwater recharge in arid and semiarid regions. *Soil Sci. Soc. Am. J.* 58, 6–14.

Andreae, M.O., Talbot, R.W., Berresheim, H. and Beecher, K.M., 1990. Precipitation chemistry in Central Amazonia. *J. Geophys. Res.* 95, 16987–16999.

Appelo, C.A.J., 1988. *Water quality in the Hierdensche Beek watershed* (in Dutch). Free University, Amsterdam, 100 pp.

Appelo, C.A.J., Groen, M.M.A., Heidweiller, V.M.L. and Smit, P.M.H., 1983. Hydrochemistry of springs in an alpine carbonate/serpentinite terrain. Int. Conf. Water-Rock Interaction, Misasa, Japan, Extend. Abstr. 26–31.

Baisden, W.T., Blum, J.D., Miller, E.K. and Friedland, A.J., 1995. Elemental concentrations in fresh snowfall across a regional transect in the northeastern US – apparent sources and contribution to acidity. *Water Air Soil Poll.* 84, 269–286.

Berner, E.K. and Berner, R.A., 1996. *Global environment: water, air, and geochemical cycles*. Prentice-Hall, Englewood Cliffs, 376 pp.

Blanchard, D.C. and Syzdek, L.D., 1988. Film drop production as a function of bubble size. *J. Geophys. Res.* 93, 3649–3654.

Boschloo, D.J. and Stolk, A.P., 1999. *National network for analysis of rainwater chemistry* (in Dutch). RIVM Rep. 723101049/054.

Budyko, M.I., 1982. *The earth's climate: past and future*. Academic Press, New York, 307 pp.

Calder, I.R., 1990. *Evapotranspiration in the uplands*. Wiley and Sons, New York, 148 pp.

Calder, I.R., Reid, I., Nisbet, T.R. and Green, J.C., 2003. Impact of lowland forests in England on water resources: Application of the hydrological land use change (HYLUC) model. *Water Resour. Res.* 39 (11), art. 1319.

Chacko, T., Cole, D.R. and Horita, J., 2001. Equilibrium oxygen, hydrogen and carbon isotope fractionation factors applicable to geologic systems. *Rev. Mineral.* 43, 1–81.

Clark, I.D. and Fritz, P., 1997. *Environmental isotopes in hydrogeology*. CRC Press, Boca Raton, 328 pp.

Cook, P.G., Jolly, I.D., Leaney, F.W., Walker, G.R., Allan, G.L., Fifield, L.K. and Allison, G.B., 1994. Unsaturated zone tritium and chlorine 36 profiles from southern Australia: Their use as tracers of soil water movement. *Water Resour. Res.* 30, 1709–1719.

Dansgaard, W., 1964. Stable isotopes in precipitation. *Tellus* 16, 436–468.

Davis, S.N. and DeWiest, R.C.M., 1966. *Hydrogeology*. Wiley and Sons, New York, 463 pp.

Drever, J.I., 1997. *The geochemistry of natural waters*, 3rd ed. Prentice-Hall, Englewood Cliffs, 436 pp.

Duce, R.A. and Hoffman, E.J., 1976. Chemical fractionation at the air/sea interface. *Ann. Rev. Earth Planet. Sci.* 4, 187–228.

Edmunds, W.M. and Gaye, C.B. 1994. Estimating the spatial variability of groundwater recharge in the Sahel using chloride. *J. Hydrol.* 156, 47–59.

Eltahir, E.A.B. and Bras, R.L., 1996. Precipitation recycling. *Rev. Geophys.* 34, 367–378.

Eriksson, E., 1959. The yearly circulation of chloride and sulfur in nature; meteorological, geochemical and pedological implcations. Part I. *Tellus* 11, 375–403.

Eriksson, E., 1960, The yearly circulation of chloride and sulfur in nature; meteorological, geochemical and pedological implications, Part II. *Tellus* 12, 63–109.

Feth, J.H., Roberson, C.E. and Polzer, W.L., 1964. *Sources of mineral constituents in water from granitic rocks, Sierra Nevada, California and Nevada*. US Geol. Surv. Water Supply Pap. 1535 I, 70 pp.

Fowler, D., 1980. Removal of sulphur and nitrogen compounds from the atmosphere in rain and by dry deposition. In D. Drabløs and A. Tollan (eds), *Ecological impact of acid precipitation*, 22–32, SNSF Project, Oslo.

Galloway, J.N. and Likens, G.E., 1978. The collection of precipitation for chemical analysis. *Tellus* 30, 71–82.

Garrels, R.M. and Mackenzie, F.T., 1967. Origin of the chemical compositions of some springs and lakes. In W. Stumm (ed.), *Equilibrium concepts in natural water systems*. Adv. Chem. Ser. 67, 222–242.

Garrels, R.M. and Mackenzie, F.T., 1972. *Evolution of sedimentary rocks*. Norton, New York, 397 pp.

Gehrels, J.C., 1999. *Groundwater level fluctuations*. Ph.D. thesis, Free University, Amsterdam, 269 pp.

Gibbs, R., 1970. Mechanisms controlling world water chemistry. *Science* 170, 1088–1090.

Gonfiantini, R., 1986. Environmental isotopes in lake studies. In P. Fritz and J.Ch. Fontes (eds), *Handbook of environmental isotope geochemistry*, 113–168, Elsevier, Amsterdam.

Gower, S.T., Kucharik, C.J. and Norman, J.M., 1999. Direct and indirect estimation of leaf area index, f_apar, and net primary production of terrestrial ecosystems. *Remote Sens. Env.* 70, 29–51.

Granat, L., 1972. On the relation between pH and the chemical composition in atmospheric precipitation. *Tellus* 24, 550–560.

Hansen, B.K. and Postma, D., 1995. Acidifcation, buffering, and salt effects in the unsaturated zone of a sandy aquifer, Klosterhede, Denmark. *Water Resour. Res.* 31, 2795–2809.

Hedin, L.O., Granat, L., Likens, G.E., Buishand, T.A., Galloway, J.N., Butler, T.J. and Rodhe, H., 1994. Steep declines in atmospheric base cations in regions of Europe and North America. *Nature* 367, 351–354.

Holland, H.D. and Petersen, U., 1996. *Living dangerously*. Princeton Univ. Press, 490 pp.

Holsen, T.M. and Noll, K.E., 1992. Dry deposition of atmospheric particles – application of current models to ambient data. *Env. Sci. Technol.* 26, 1807–1815.

Jain, M., Kulshrestha, U.C., Sarkar, A.K. and Parashar, D.C., 2000. Influence of crustal aerosols on wet deposition at urban and rural sites in India. *Atmos. Env.* 34, 5129–5137.

Junge, C.E. and Werby, R.T., 1958. The concentration of chloride, sodium, potassium, calcium and sulfate in rain water over the United States. *J. Meteorol.* 15, 417–425.

KNMI/RIVM, 1985. *Chemical composition of precipitation over the Netherlands*. Ann. Rep. 1983. KNMI, De Bilt, the Netherlands.

Lerner, D.N., Issar, A.S. and Simmers, I., 1990. *Groundwater recharge*. Int. Contrib. Hydrogeol., vol 8, IAH, 345 pp.

Lightowlers, P.J. and Cape, J.N., 1988. Sources and fate of atmospheric HCl in the UK and Western Europe. *Atmos. Env.* 22, 7–15.

Likens, G.E. and Bormann, F.H., 1995. *Biogeochemistry of a forested ecosystem*, 2nd ed. Springer, New York, 159 pp.

Matzner, E. and Hesch, W., 1981. Beitrag zum Elementaustrag mit dem Sickerwasser unter verschieden Ökosystemen im Nordwestdeutschen Flachland. *Z. Pflanzenern. Bodenkd.* 144, 64–73.

Mészáros, E., 1981. *Atmospheric chemistry, fundamental aspects*. Elsevier, Amsterdam, 201 pp.

Miller, H.G. and Miller, J.D., 1980. Collection and retention of atmospheric pollutants by vegetation. In D. Drabløs and A. Tollan (eds), *Ecological impact of acid precipitation*, 33–40, SNSF Project, Oslo.

Möller, D., 1990. The Na/Cl ratio in rainwater and the seasalt chloride cycle. *Tellus* 42B, 254–262.

Monahan, E.C., 1986. The ocean as a source for atmospheric particles. In P. Buat-Menard (ed.), *The role of air-sea exchange in geochemical cycling*, 129–163, D. Reidel, Dordrecht.

Morgan, J.J., 1982. Factors governing the pH, availability of H^+, and oxidation capacity of rain. In E.D. Goldberg (ed.), *Atmospheric chemistry*, 17–40. Springer, Berlin.

Moss, P.D. and Edmunds, W.M., 1989. Interstitial water-rock interaction in the unsaturated zone of a Permo-triassic sandstone aquifer. In D.L. Miles (ed.), *Proc. 6th water-rock bnteraction symp.*, 495–499, Balkema, Rotterdam.

Mulder, J., 1988. Impact of acid atmospheric deposition on soils. Ph.D. thesis, Wageningen.

Nativ, R. and Issar, A., 1983. Chemical composition of rainwater and floodwaters in the Negev desert, Israel. *J. Hydrol.* 62, 201–223.

Newell, R.E., 1971. The global circulation of atmospheric pollutants. *Scientific American* 224, 32–43.

Oberg, G., 1998. Chloride and organic chlorine in soil. *Acta Hydrochim. Hydrobiol.* 26, 37–44.

Peixóto, J.P. and Kettani, M.A., 1973. The control of the water cycle. *Scientific American* 228, 46–61.

Peixóto, J.P. and Oort, A.H., 1983. The atmospheric branch of the hydrological cycle and climate. In A. Street-Perrott et al. (eds), *Variations in the global water budget*, 5–65. D. Reidel, Dordrecht.

Phillips, F.M., 1994. Environmental tracers for water movement in desert soils of the American Southwest. *Soil Sci. Soc. Am. J.* 58, 15–24.

Raemdonck, H., Maenhout, W. and Andreae, M.O., 1986. Chemistry of marine aerosol over the tropical and equatorial Pacific. *J. Geophys. Res.* 91, 8623–8636.

Rasmussen, L., 1988. Sur nedbørs effekt på ionbalancen og udvaskningen af metaller og anioner i danske nåleskovøkosystemer i perioden 1983–1987. Report DTH, Lyngby.

Rainwater, F.H., 1962. *Stream composition of the conterminous United States*. US Geol. Surv. Hydrol. Inv. Atlas HA 61.

Ridder, T.B., Baard, J.H. and Buishand, T.A., 1984. *The influence of sampling methods on the quality of precipitation* (in Dutch). KNMI Techn. Rep. 55, 40 pp.

Rijtema, P.E. and Kroes, J.G., 1991. Some results of nitrogen simulations with the model ANIMO. *Fert. Res.* 27, 189–198.

Rozanski, K., Araguás-Araguás, L. and Gonfiantini, R., 1993. Isotopic patterns in modern global precipitation. In P.K. Swart et al. (eds), *Continental isotope indicators of climate*, AGU Monogr. 78, 1–36.

Rustad, L.E., Kahl, J.S., Norton, S.D. and Fernandez, I.J., 1994. Underestimation of dry deposition by throughfall in mixed northern hardwood forests. *J. Hydrol.* 162, 319–336.

Schoeller, H., 1960. Salinity of groundwater, evapotranspiration and recharge of aquifers (in French). IASH Pub. 52, 488–494.

Sequeira, R. and Kelkar, D., 1978. Geochemical implications of summer monsoonal rainwater composition over India. *J. Appl. Meteor.* 17, 1390–1396.

Shahin, U, Yi, S.M., Paode, R.D. and Holsen, T.M., 2000. Long-term elemental dry deposition fluxes measured around Lake Michigan with an automated dry deposition sampler. *Env. Sci. Technol.* 34, 1887–1892.

Sharma, M.L. and Hughes, M.W., 1985. Groundwater recharge estimation using chloride, deuterium and oxygen-18 profiles in the deep coastal sands of Western Australia. *J. Hydrol.* 81, 93–109.

Stuyfzand, P.J., 1984. Effects of vegetation and air pollution on groundwater quality in calcareous dunes near Castricum: measurements from lysimeters (in Dutch). *H_2O* 17, 152–159.

Thorstenson, D.C. and Parkhurst, D.L., 2004. Calculation of individual isotope equilibrium constants for geochemical reactions. *Geochim. Cosmochim. Acta* 68, 2449–2465.

Ulrich, B., Mayer, R. and Khanna, P.K., 1979. *Deposition of air pollutants and their effects on forest ecosystems in Sölling* (in German). Sauerländer, Frankfurt a.M., 291 pp.

Van Aalst, R.M., Van Aardenne, R.A.M., De Kreuk, J.F. and Lems, Th., 1983. *Pollution of the North Sea from the atmosphere.* TNO-Report CL 82/152., TNO, Delft, 124 pp.

Vermeulen, A.J., 1977. *Immision measurements with rain gauges* (in Dutch). Prov. North Holland, Haarlem.

Wagner, G.H. and Steele, K.F., 1989. Na^+/Cl—ratios in rain across the USA, 1982–1986. *Tellus* 41B, 444–451.

Walling, D.E., 1980. Water in the catchment ecosystem. In A.M. Gower (ed.), *Water quality in catchment ecosystems*. 1–48. Wiley and Sons, New York.

Wedepohl, K.H. (ed.), 1969–1976. *Handbook of geochemistry*. Springer, Berlin.

Whitley, K.T., Husar, R.B. and Liu, B.Y.H., 1972. Aerosol size distribution of Los Angeles smog. *J. Coll. Interf. Sci.* 39, 177–204.

Wood, W.W., 1999. Use and misuse of the chloride-mass balance method in estimating groundwater recharge. *Ground Water* 37, 2–3.

Wood, W.W. and Sanford, W.E., 1995. Chemical and Isotopic methods for quantifying ground-water recharge in a regional, semiarid environment. *Ground Water* 33, 458–468.

Wu, L. and McGechan, M.B., 1998. A review of carbon and nitrogen processes in four soil nitrogen dynamics models. *J. Agr. Eng. Res.* 69, 279–305.

3

Flow and Transport

In order to understand the spatial and temporal variability of groundwater quality, it is imperative to have a perception of the groundwater flowpath and the travel time from the site of infiltration to the sampling point. Along the flowline, chemicals may become sorbed which retards their transport compared to the velocity of water. Furthermore, physical processes like diffusion and dispersion cause mixing and smoothen concentration changes. To achieve a conceptual understanding of the processes, we will in this chapter focus on hand calculations that provide insight in flowlines and residence times, in retardation, and in diffusion and dispersion.

3.1 FLOW IN THE UNSATURATED ZONE

Water in the unsaturated zone percolates vertically downward along the maximal gradient of the soil moisture potential when the relief is moderate. The rate of percolation in an unsaturated profile can be derived from a mass balance, dividing the precipitation surplus by the water filled porosity of the soil (Figure 3.1; Allison et al., 1994; Phillips, 1994):

$$v_{H_2O} = P / \varepsilon_w \tag{3.1}$$

where v_{H_2O} is the velocity of water (m/yr), P is the precipitation surplus (m/yr) and ε_w the water-filled porosity (m^3/m^3). This relationship has been verified in the field by Andersen and Sevel (1974) and Engesgaard et al. (1996) by analyzing tritium transport in a 20 m thick unsaturated zone in sandy sediments.

A water velocity that is simply determined by mass balance, means that infiltrating water pushes the old water ahead, a type of flow adequately termed *piston flow*. Figure 3.2 shows how the tritium peak in the rain from 1963 is translated by piston flow in the unsaturated zone of chalk at 3 locations in England. The velocity of water in the profile measure in 1968 is v_{H_2O} = 4.4 m / 5 yr = 0.9 m/yr, and similar at the other sites. The tritium peak decreases with time due to radioactive decay (Section 3.3).

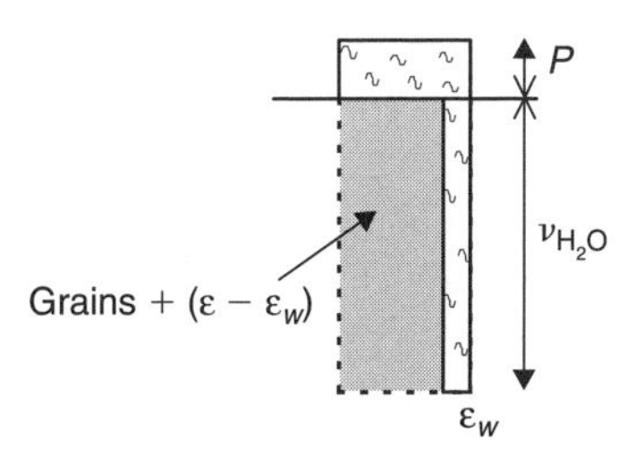

Figure 3.1. A precipitation surplus of 100 mm/yr translates to v_{H_2O} = 500 mm/yr in a soil with ε_w = 0.2.

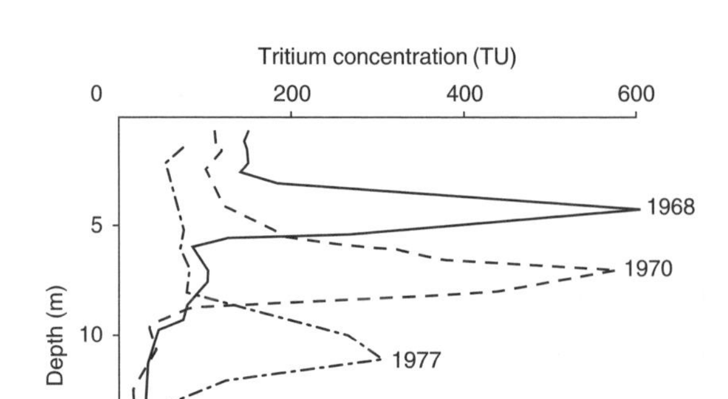

Figure 3.2. Tritium variation in rain and piston flow displacement at three sites in the English chalk, measured in 3 different years (Foster and Smith-Carrington, 1980).

However, the observed displacement of about 1 m/yr was too low compared with the precipitation surplus and the water filled porosity of the chalk. Part of the infiltration was therefore believed to proceed along preferential flowpaths through macro-fissures (Foster and Smith-Carrington, 1980). This rapid bypass flow is also observed in soil moisture in the root zone (Section 2.5.1).

QUESTION:
Estimate the age of water at 3 m depth in Figure 2.22: average water content $\theta_g = 0.012$ g/g, $P = 0.1$ mm/yr. (Recalculate θ_g from g/g into water filled porosity ε_w using a bulk density of 1.8 g/cm^3).

ANSWER: Average $\theta_g = 0.012$ g/g, or 0.022 g H_2O/cm^3. Hence $v_{H_2O} = 0.1 / 0.022 =$ 46 mm/yr.
Age: 3000 mm / 46 mm/yr = 65 yr.

3.2 FLOW IN THE SATURATED ZONE

Consider the landfill site shown in cross section in Figure 3.3. Leachate from the waste is percolating into the aquifer and the question is, how fast the various chemicals travel through the subsoil, and where and when they pollute drinking water wells further downstream. For an answer, we must define the flowlines of groundwater in the aquifer and derive the travel time. The detailed calculation of flow patterns is usually done with numerical models (Zheng and Bennett, 2002; Chiang and Kinzelbach, 2001). However, a first guess can be readily obtained by hand calculations.

3.2.1 *Darcy's law*

The groundwater levels measured in wells (the *piezometric level*) can be compiled in a map of *equipotentials* (*isohypses*), the lines that connect points with the same groundwater elevation or potential (Figure 3.4). Water flows from a high to a low potential and the groundwater flow direction is at right angles to the isohypses if the aquifer is *isotropic*, which means that the hydraulic properties are equal in all directions. Hydraulic gradients follow primarily the local relief, but are also influenced by recharge and discharge rates, and groundwater withdrawal (Freeze and Cherry, 1979; Domenico and Schwartz, 1997; Fetter, 1994).

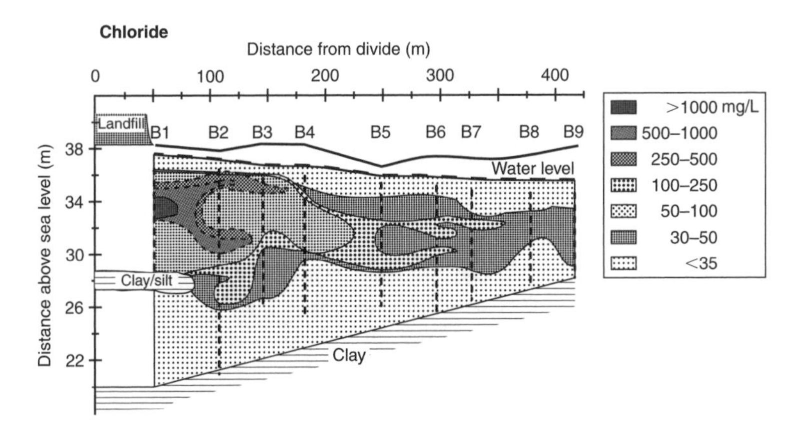

Figure 3.3. Cl^- concentration contours in the aquifer below the Vejen waste site. The small dots on the vertical borings (B1–B9) represent sample points (Lyngkilde and Christensen, 1992).

Groundwater flow thus depends on the hydraulic gradient, but it depends also on the hydraulic conductivity of the subsoil. These two parameters are combined in Darcy's law:

$$v_D = -k \, \mathrm{d}h / \mathrm{d}x \tag{3.2}$$

where v_D is the specific discharge or *Darcy flux* (m/day), k is the hydraulic conductivity (m/day), and $\mathrm{d}h / \mathrm{d}x$ is the hydraulic gradient.

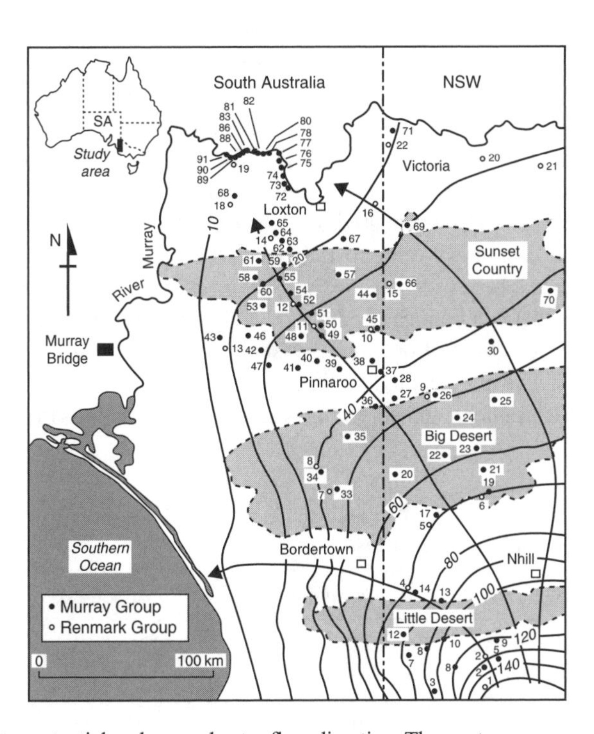

Figure 3.4. Groundwater potential and groundwater flow direction. The contours represent the hydraulic head (isohypses), and the arrows indicate the inferred direction of groundwater flow (Dogramaci and Herczeg, 2002).

The piezometric level of groundwater gives, after corrections for density differences, the potential (h) for groundwater flow. The discharge is obtained by multiplying the specific discharge with the surface area perpendicular to the flow:

$$Q = v_D \cdot A = -kA \cdot \mathrm{d}h / \mathrm{d}x \tag{3.3}$$

where Q is discharge (m^3/day), and A is the *total* surface area which includes pore space and grain-skeleton together.

The hydraulic conductivity of a sediment can be determined in a permeameter and is calculated with Equation (3.3) from Q, A and $\Delta h / \Delta x$. The Darcy flux is defined as specific discharge for a medium containing both grains and water. However, only the pore space contributes to the water flow, so that the actual velocity of water through the pores must be larger. Mass balance applies as for the unsaturated zone, and the velocity is:

$$v_{H_2O} = v_D / \varepsilon_w = -k / \varepsilon_w \cdot \mathrm{d}h / \mathrm{d}x \tag{3.4}$$

The water velocity can also be calculated from the discharge and the cross section $A \cdot \varepsilon_w$:

$$v_{H_2O} = Q / (A \cdot \varepsilon_w) \tag{3.5}$$

Some information on groundwater flow velocities can be obtained using hydraulic conductivities and porosities of sediments presented in Table 3.1. For example, the actual distance traveled by groundwater in a sandy aquifer with porosity $\varepsilon_w = 0.3$, hydraulic conductivity $k = 50$ m/day, and potential gradient $\mathrm{d}h / \mathrm{d}x = 0.001$ (1 m per 1 km), amounts to 60 m/yr. The hydraulic conductivity varies by many orders of magnitude for different sediments, while the porosity only varies from 0.2 for coarse, unsorted sands to 0.65 for clay.

Layers of clay, peat and loam, which are considered impermeable, are termed *aquicludes* or, when almost impermeable, *aquitards*. Such layers can separate more permeable (sandy) *aquifers*. If the watertable in the aquifer is higher than the boundary with the aquiclude, the aquifer is *confined*, otherwise the aquifer is *unconfined* and possibly unsaturated in the upper part. If the uppermost aquifer extends right to the earth's surface it is called a *phreatic* aquifer.

The groundwater velocity can, in principle, be calculated with Darcy's law (Equation 3.2), when the hydraulic gradient and the hydraulic conductivity are known. However, the large variation in hydraulic conductivities makes the calculated velocity rather uncertain. Often, the water balance based on Equation (3.5) may provide a better idea of the average, or *effective* parameters. Effective parameters may differ from the ones that can be measured directly. The effective porosity comprises the porosity containing water that participates in flow, in contrast to the total porosity which includes the pores filled with stagnant water. The effective hydraulic conductivity is similarly obtained from the quotient of discharge and hydraulic gradient, and so averages small scale variations in hydraulic conductivity.

Table 3.1. Hydraulic conductivity and porosity of different sediments.

	k (m/day)	ε (fraction)
Gravel	200–2000	0.15–0.25
Sand	10–300	0.20–0.35
Loam	0.01–10	0.30–0.45
Clay	10^{-5}–1	0.30–0.65
Peat	10^{-5}–1	0.60–0.90

In numerical models, the water balance is, together with the measured groundwater potential, used to optimize the *transmissivity* (the product of hydraulic conductivity and aquifer thickness). If only the quantity of (horizontal) water flow is important, the transmissivity is an adequate parameter. However, increased pollution has aroused interest for unraveling the actual flowlines in the aquifer. It has proven difficult to assess flowlines exactly using numerical hydrological models because permeability variations have a large influence on flowpaths but only small effect on hydraulic heads, and thus do not show up in the traditional hydrological measurements.

3.2.2 *Flowlines in the subsoil*

The cross section through a Canadian prairie landscape in Figure 3.5 illustrates how water may flow along curved flowlines from recharge areas to discharge areas, often near creeks. The vertical scale is grossly exaggerated and in reality the flowpaths are more horizontal. Nevertheless, the figure illustrates the important points that the flowpaths are at right angles to the isolines for the hydraulic potential (dashed lines in Figure 3.5, note that scaling *does* affect the intersection angle in a graph), and that the intersection of the isoline with the water table defines its potential in the terms of hydraulic head. By specifying the shape of the water table with a mathematical formula, it is possible to solve the potential field in the cross section and to obtain the flowlines (if the flow domain is homogeneous, otherwise numerical models need to be used) (Tóth, 1962; Freeze and Witherspoon, 1966; Domenico and Schwartz, 1997).

It is easier to use the water balance to visualize how groundwater flows in aquifers (Vogel, 1967; Ernst, 1973; Gelhar and Wilson, 1974; Dillon, 1989). Let us take a segment of the cross section of Figure 3.5, from Kneehills Creek in the West to the divide at the topographic high, and redraw it as the phreatic aquifer shown in Figure 3.6. The sediment is homogeneous so that porosity and hydraulic conductivity are the same everywhere. In the rectangular aquifer of Figure 3.6 the velocity is (very nearly) equal at all depths along a vertical line. (Why does an aquifer not show the velocity profile of a river?). The point along the upper reach where water infiltrates, and depth in the aquifer are then related proportionally:

$$\frac{x}{x_0} = \frac{D}{D - d} \tag{3.6}$$

where D is the thickness of the aquifer. Water infiltrated at a point x_0, upstream of x, is at a given time, found at depth d in the aquifer. Above d flows water that infiltrated between point x_0 and x, while water at greater depths infiltrated further upstream of x_0.

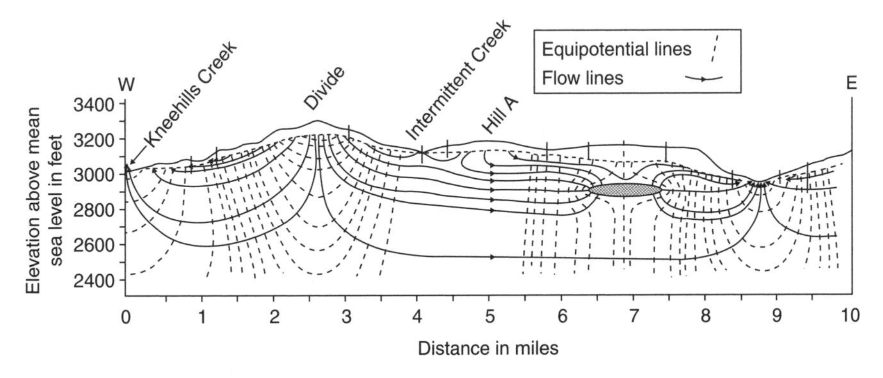

Figure 3.5. Groundwater potential distribution and flow pattern, and the effect of a highly permeable body, near 7 miles, on the flow system (Tóth, 1962).

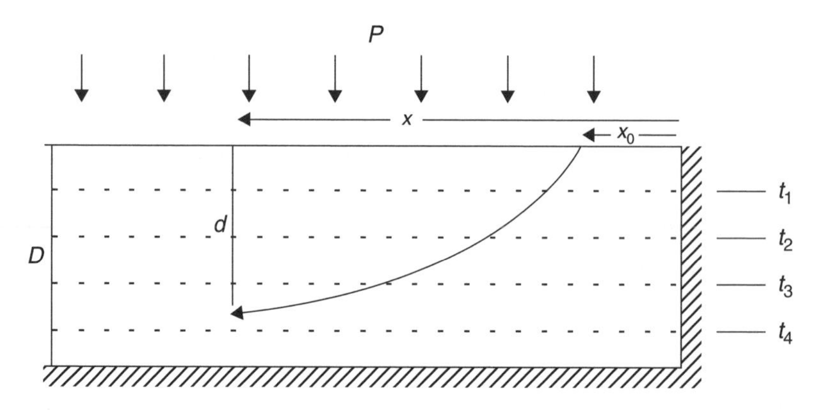

Figure 3.6. A profile through a homogeneous, phreatic aquifer.

The flow velocity in the aquifer can also be calculated. If a precipitation surplus of P m/yr enters the aquifer, $Q = P \cdot x$ m^2/yr must be discharged through the aquifer at point x. Using Equation (3.5), the flow velocity at point x is:

$$v = \frac{\mathrm{d}x}{\mathrm{d}t} = \frac{Px}{D\varepsilon_w} \tag{3.7}$$

The thickness D has replaced the surface area A, since we consider flow in a profile (for example per m width). The groundwater velocity increases with distance from the divide, since more precipitation must be discharged through the same layer. The increase of the flow velocity requires that the hydraulic gradient increases with the square of distance.

Integration of (3.7) gives, when water infiltrates at $t = 0$ at $x = x_0$:

$$\ln \frac{x}{x_0} = \frac{P}{D\varepsilon_w} t \tag{3.8}$$

or the distance x reached in time t:

$$x = x_0 \operatorname{cxp}\left(\frac{Pt}{D\varepsilon_w}\right) \tag{3.9}$$

We may also substitute the proportionality relationship (3.6) into (3.8), which yields:

$$\ln \frac{D}{D - d} = \frac{Pt}{D\varepsilon_w} \tag{3.10}$$

or:

$$d = D\left(1 - \exp\left[\frac{-Pt}{D\varepsilon_w}\right]\right) \tag{3.11}$$

This is a remarkable result. It demonstrates that in a homogeneous aquifer with uniform infiltration, water at a specific depth is all of the same age, independent of the location. In other words, the *isochrones* are horizontal, water resides in the aquifer in planes of equal age.

EXAMPLE 3.1. *Calculate the time for water to flow from midway in the Vejen waste site to 125 m downstream* (Figure 3.3), and the thickness of the plume at that point. The waste site is located on the groundwater divide for the clay/silt layer and is 50 m long, the aquifer is 9 m thick. The precipitation surplus is 0.4 m/yr, the porosity 0.35 (Brun et al., 2002).

ANSWER:
The midpoint of the site is at $x_0 = 25$ m, $x = 125$ m from the divide. With Equation (3.8), we obtain $t = 9 \times 0.35 / 0.4 \times \ln(125 / 25) = 12.7$ yr. The percolate from 50 m waste is distributed proportionally over depth, and the thickness of the plume is $50 / 125 \times 9 = 3.6$ m. Note that the actual plume is much more spread out in the vertical as a result of the hydraulic conductivity variations in the subsoil, the irregular filling of the site, and probably because density flow occurs (Christensen et al., 2001).

The depth/time relation is valid for an aquifer of infinite length, or in practice, not too close to the point of discharge of the aquifer (in formula: $0 < x < (L - D)$, where L is the distance from divide to drain). Ernst (1973) and Gelhar and Wilson (1974) have derived formulas which can be used near the discharge point, assuming radial upward flow into the spring or seepage zone.

In principle, seepage occurs where the water table intersects the topography and from where a drainage network develops (Ernst, 1978; De Vries, 1995). The drainage distance can be estimated from Darcy's law for the homogeneous aquifer:

$$Q = P \cdot x = -kD \, \mathrm{d}h / \mathrm{d}x \tag{3.12}$$

Integration from h_m at $x = 0$ at the divide to h at the discharge point at $x = L$ gives:

$$L = \sqrt{\frac{2kD}{P}(h_m - h)} \tag{3.13}$$

Applied to the aquifer in Figure 3.5 between the divide and Kneehills Creek in the West, the hydraulic conductivity k can be calculated using Equation (3.13). With $L = 4216$ m (2.62 miles), $(h_m - h) = 62.8$ m (206 ft), $D = 185$ m (600 ft) and $P = 0.03$ m/yr, we obtain $k = 0.06$ m/day. (Tóth, pers. comm., estimated $k = 0.08$ m/day for the area).

EXAMPLE 3.2. *Calculate the water level in the Vejen river, 1 km downstream from the Vejen waste site* and estimate the travel time of waste leachate to the river. The hydraulic conductivity of the aquifer is 35 m/day (Brun et al., 2002), the other data are given in Example 3.1.

ANSWER:
The water level at the confinement of the waste (50 m from the divide) is 38 m above sea level. We integrate Equation (3.12) from $h_{x=50\,\mathrm{m}} = 38$ m to $h_{x=1000\,\mathrm{m}}$:

$$\tfrac{1}{2}(1000^2 - 50^2) = -(35 \times 9 / (0.4 / 365)) \, (h_{x=1000\,\mathrm{m}} - 38)$$

which solves to $h_{x=1000\,\mathrm{m}} = 36.3$ m. The travel time from the border of the waste is 24 yrs.

Actually, the water level in the river is about 32 m. There is a greater hydraulic gradient along the flowline towards the river, indicating that the hydraulic conductivity decreases, or that the aquifer thins out, or that the clay layer disappears allowing water from below to enter the aquifer. Figure 3.3, but also the regional hydrogeology suggest the importance of the two latter factors, implying that the travel time will be less than estimated by our calculation.

QUESTIONS:

Why does an aquifer not show the velocity profile of a river?

ANSWER: The flow profile in water is determined by the resistance at the boundaries: in the river at the river bed and the air, in the aquifer by the resistance at the pore boundaries which remain the same over depth.

Estimate the flowtime from midway in the Vejen waste site ($x_0 = 25$ m) to the river?

ANSWER: less than 29 yr (note that the last 850 m are traveled in the same time as the first 150 m).

In Example 3.2, consider the effect of the decrease of water filled thickness by the gradient of h with x?

ANSWER: Take $D = 7.3$ m, then $h_{x=1000\,\mathrm{m}} = 35.9$, still higher than observed; the travel time decreases slightly: 19.4 yrs.

3.2.3 *Effects of non-homogeneity*

Most aquifers are not homogeneous and large differences in hydraulic conductivity are usual. The distribution of flowlines can be calculated by proportioning the volume of flow according to the transmissivity difference. The correction is performed in the vertical plane as shown in Figure 3.7. For two layers of thickness D_1 and D_2, porosities ε_1 and ε_2, and with hydraulic conductivities k_1 and k_2 (m/day), the discharge through the vertical section can be divided in two portions Q_1 and Q_2. The ratio of the two is (assuming equal hydraulic gradient in both layers):

$$\frac{Q_1}{Q_2} = \frac{k_1 D_1}{k_2 D_2} \tag{3.14}$$

and the velocities in the layers are:

$$\frac{v_1}{v_2} = \frac{k_1}{k_2}\,\frac{\varepsilon_2}{\varepsilon_1} \tag{3.15}$$

Various examples of flowpatterns affected by inhomogeneities are illustrated in Figure 3.37 and Problem 3.7, and can further be found in Freeze and Cherry (1979). Analytical solutions for various cases may be found in Bruggeman (1999) and more flexible computer programs for calculating and visualizing flowlines in 2D profiles are TopoDrive&ParticleFlow by Hsieh (2001) which can be downloaded from water.usgs.gov/nrp/gwsoftware/tdpf/tdpf.html and Flownet by Van Elburg, Hemker and Engelen which can be downloaded from www.microfem.nl.

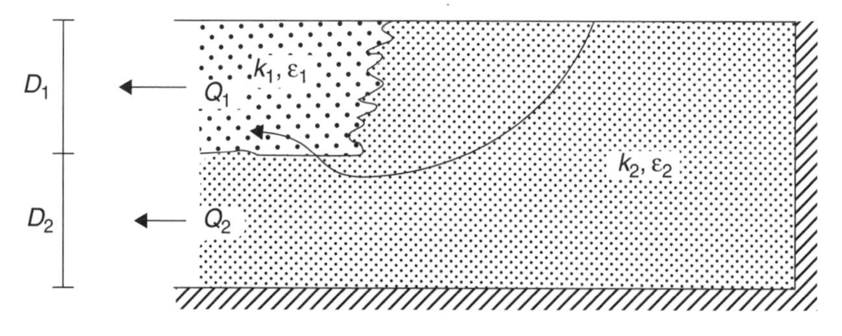

Figure 3.7. Effect of hydraulic conductivity-changes on flowlines.

3.2.4 *The aquifer as a chemical reactor*

Concentrations of dissolved substances were so far considered as depth related entities which can be measured using depth specific samplers (Chapter 1). On the other hand, a large pumping well with a long screen yields an average of the concentrations present along the length of the screen. Assume, for example in Figure 3.6, that the concentration of NO_3^- in infiltrating water increases from an initial c_i at time $t < 0$, to c_0 at time $t = 0$. When the concentration front has arrived at depth d, the average concentration is:

$$c = \frac{c_0 d + c_i(D - d)}{D} \tag{3.16}$$

Now, assume that the initial concentration in the aquifer $c_i = 0$, and use (3.11) to obtain:

$$c = c_0 \frac{d}{D} = c_0(1 - e^{-Pt/D\varepsilon_w}) \tag{3.17}$$

This shows that the concentration in a fully penetrating well (screen length equals D) increases with time to the final value by an exponential function. This function is similar to the one used for an "ideal mixed reactor" in chemical engineering (Levenspiel, 1972). For such a reactor the *residence time* is defined as:

$$\tau = \frac{Total\ mass}{Mass\ exchange} = \frac{M_{tot}}{\mathrm{d}M_{in} / \mathrm{d}t} \tag{3.18}$$

where τ is the residence time (s). $\mathrm{d}M_{in} / \mathrm{d}t$ can be replaced with $\mathrm{d}M_{out} / \mathrm{d}t$, since the M_{tot} does not change. Replacing mass with volume of water, the residence time becomes:

$$\tau = V / Q \tag{2.3}$$

which we used in Chapter 2 for calculating the residence time of water.

When a step input c_0 is introduced at time $t = 0$ in the input stream, the concentration in the exit stream of the reactor increases as:

$$c = c_0(1 - e^{-t/\tau}) \tag{3.19}$$

Equation (3.17) is identical with (3.19) when $\tau = D\varepsilon_w / P$, and this is correct since in Equation (2.3) $V = D\varepsilon_w x$, and $Q = Px$. The comparison is logical since a fully penetrating well samples a mixture of waters which infiltrated from time zero in the uppermost part to successively earlier times deeper down. With uniform flow (no velocity variations with depth) the averaging in the groundwater well is similar to the mixing in the reactor.

A difference between a stirred chemical reactor and the phreatic aquifer is that the reactor is well mixed to let all the reactions elapse in the same way everywhere. In the phreatic aquifer, mixing only takes place in the well, while the chemical reactions among water and solids are variable in time and distance along the flowlines through the aquifer. Nevertheless, the chemical reactor equations are useful when reactions between water and sediment are unimportant. Equation (3.19) has been used to derive characteristics of the groundwater reservoir from tritium measurements in spring water (Yurtsever, 1983; Maloszewski and Zuber, 1985). When the residence time is known, either from tracer measurements or from hydrogeological conditions, a first estimate of the change in concentration over time is easily obtained with Equation (3.19), for example of the NO_3^- concentration in production wells after the application of excess fertiliser on the surface has been banned.

EXAMPLE 3.3. *Flushing of NO_3^- from an aquifer*
Estimate the time required to reduce the NO_3^- concentration from 150 mg/L to 25 mg/L (the drinking water limit) in an aquifer after fertilizer input has been stopped. $D = 10$ m, $\varepsilon_w = 0.3$, $P = 0.6$ m/yr.

ANSWER:
The residence time in the aquifer is $\tau = D \cdot \varepsilon / P = 5$ yr. From (3.16) and (3.17) we obtain:

$$c = c_0(1 - e^{-t/\tau}) + c_i e^{-t/\tau}$$

Here $c_0 = 0$ mg/L, $c_i = 150$ mg/L. From $c = c_i\, e^{-t/\tau}$ find $25 / 150 = e^{-t/\tau}$, or $t = -\tau \cdot \ln(25 / 150) = 9$ yrs.

3.3 DATING OF GROUNDWATER

Calculated groundwater flow patterns and residence times can be verified by groundwater dating. There is a range of dating methods available based on radioactive decay (Cook and Herczeg, 2000). These comprise tritium (^3H), $^3H / ^3He$ and ^{85}Kr for young groundwaters of up to about 50 years, radiocarbon (^{14}C) for older groundwaters up to 30,000 years and finally isotopes like cosmogenic ^{81}Kr and ^{36}Cl for even older groundwaters. Using the latter method, groundwater with an age of 400,000 years has been dated (Collon et al., 2000).

For all radioactive decay processes, the rate equals a decay constant, λ, times the concentration of the reactant. For example, tritium (^3H) disintegrates to ^3He and β^- at a rate:

$$\frac{d^3H}{dt} = -\lambda \cdot {}^3H \tag{3.20}$$

The tritium concentration is expressed in tritium units (TU). One TU equals 10^{-18} mol ^3H/mol H, and is equivalent to 3.2 pCi/L from the emitted β particles. The decay constant $\lambda = 0.0558$/yr. The equation can be integrated from the initial value $^3H_0$ at time $t = 0$ and gives:

$$\ln({}^3H / {}^3H_0) = -\lambda \cdot t \tag{3.20a}$$

The time to let tritium decay to half of its initial concentration is called the half-life, $t_{½} = \ln(0.5) / -\lambda = 12.43$ years.

Figure 3.8 shows the distribution of ^3H and its daughter ^3He in an aquifer. The peak in ^3H and ^3He at 10 m depth corresponds to 1963, the year of maximal hydrogen bomb testing in the atmosphere (cf. Question). Tritium/Helium offers an absolute dating method because the initial $^3H_0$ is given by the sum of ^3H and ^3He in the sample. However, the resulting age is affected by dispersion (Schlosser et al., 1989) and by different diffusion rates of ^3He and ^3H (Chapter 11). Moreover, radiogenic and atmospheric ^3He contributions must be subtracted, and some tritiogenic ^3He may escape into the unsaturated zone, which can give substantial corrections in young or old (infiltrated before 1950) groundwater (Schlosser et al., 1989; Cook and Solomon, 1997).

Krypton 85 (^{85}Kr) is a radioactive, and chemically inert, noble gas that decays to stable ^{85}Rb with a half-life of 10.7 years. Almost all ^{85}Kr in the environment is of anthropogenic origin. It is a fission product of uranium and plutonium and released by nuclear weapon testing, nuclear reactors and nuclear reprocessing plants. The activity of ^{85}Kr in the atmosphere has steadily increased except for a few years around 1970 (Figure 3.9) and may be used for groundwater dating. ^{85}Kr is measured as specific activity, i.e. as the ratio of ^{85}Kr to stable Kr (in disintegrations per minute of ^{85}Kr per volume Kr, dpm/cm^3).

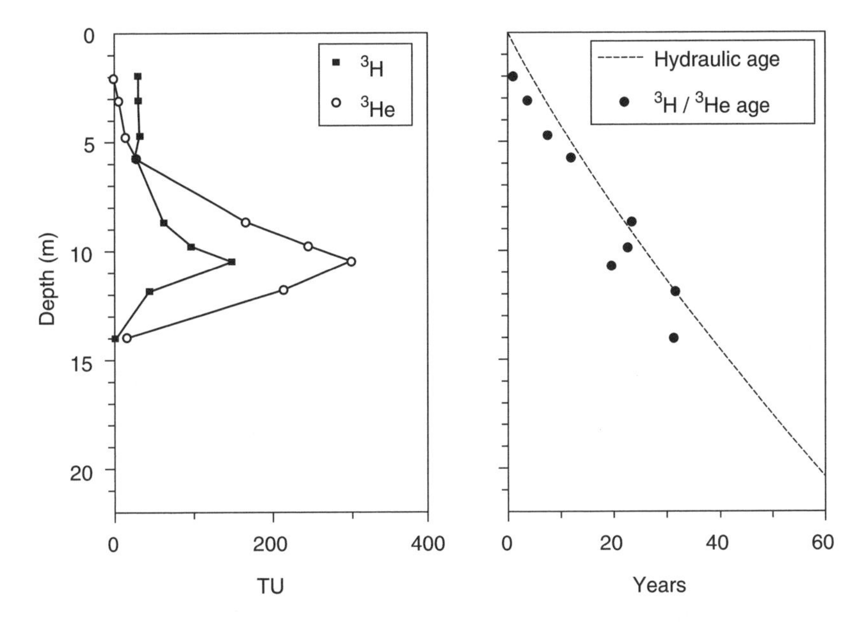

Figure 3.8. Groundwater ^{3}H and ^{3}He concentrations and ^{3}H / ^{3}He age profiles at Sturgeon Falls, Ontario (Solomon et al., 1993). The ^{3}He concentration is expressed in TU (10^{-18} mol/mol H).

Because the ratio of ^{85}Kr over stable Kr is measured, the Kr concentration in the recharge water, which depends on krypton solubility and temperature, does not have to be known (Ekwurzel et al., 1994). Groundwater dating proceeds by fitting downward decreases in groundwater ^{85}Kr to the historical atmospheric record.

Carbon 14 is a natural radioactive compound that forms in the atmosphere. The decay constant for ^{14}C is $\lambda = 1.21 \times 10^{-4}$/yr, and the half-life is 5730 yr. It is therefore suitable for older groundwaters. The ^{14}C activity of a sample is expressed as percent modern carbon (pmc or ^{14}A) given by the ratio:

$$^{14}A = \frac{(^{14}\mathrm{C} / \mathrm{gC})_{sample}}{(^{14}\mathrm{C} / \mathrm{gC})_{standard}} \frac{(^{13}\mathrm{C} / {}^{12}\mathrm{C})^2_{standard}}{(^{13}\mathrm{C} / {}^{12}\mathrm{C})^2_{sample}} \times 100 \tag{3.21}$$

where the ^{14}C / gramC in the standard equals the ratio in equilibrium with a "modern", pre-nuclear bomb testing atmosphere. The squared ratio $(^{13}\mathrm{C} / {}^{12}\mathrm{C})^2$ accounts for fractionation of ^{14}C. The radiocarbon age of groundwater from simple radioactive decay is $t = \ln(^{14}A / {}^{14}A_0) / -1.21 \times 10^{-4}$ yr (the "conventional" age). However, the measured ^{14}C concentration may have been diluted by 'dead' carbon derived from carbonate dissolution or degradation of organic carbon. In that case a number of corrections must be applied before ^{14}C can be used for dating (Chapter 5).

Chlorofluorocarbons (CFCs) and sulfurhexafluoride (SF_6) have been introduced in the atmosphere and hydrosphere during the last couple of decades. CFCs, also called freons, are stable synthetic organic compounds produced for use as refrigerants, aerosol propellants, solvents etc. (Busenberg and Plummer, 1992). Gaseous SF_6 is a product of industrial processes (Bauer et al., 2001). CFCs have received much attention because of the detrimental effects on the atmospheric ozone layer. The increase in the atmospheric concentration of various CFCs is also shown in Figure 3.9.

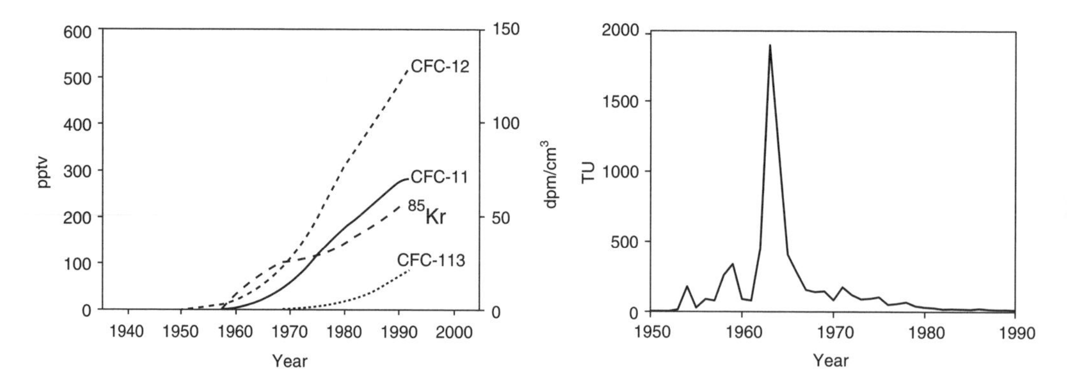

Figure 3.9. Left: Atmospheric concentrations of CFC-11, CFC-12 and CFC-113 (part per trillion by volume (pptv)) and ^{85}Kr (disintegration per minute per cubic centimeter) (Cook and Solomon, 1997); Right: Tritium concentrations in rain in the Netherlands.

The steady increases in both CFCs and ^{85}Kr concentration have been used for dating of young groundwaters. While for the solutes ^{36}Cl and ^{3}H the clock is set at the time of infiltration, for the atmospheric gases like CFCs the relation is more complex. For gases, the temperature-dependent solubility and transport through the unsaturated zone plays a role (Cook and Solomon, 1997). Dating of young groundwater using CFCs proceeds by fitting downward increases in measured CFC concentration to the historical atmospheric record and thereby interpreting the age of the groundwater. Such an age interpretation over depth is for three different CFCs given in Figure 3.10. For comparison the position of the 1963 tritium peak in 1991 is also shown, presenting in general a good agreement with the CFC ages.

Comparison (Figure 3.10) of CFC-11 with CFC-12 and CFC-113, suggests consistently higher ages derived from CFC-11. Probably this is an artefact due to degradation of CFC-11.

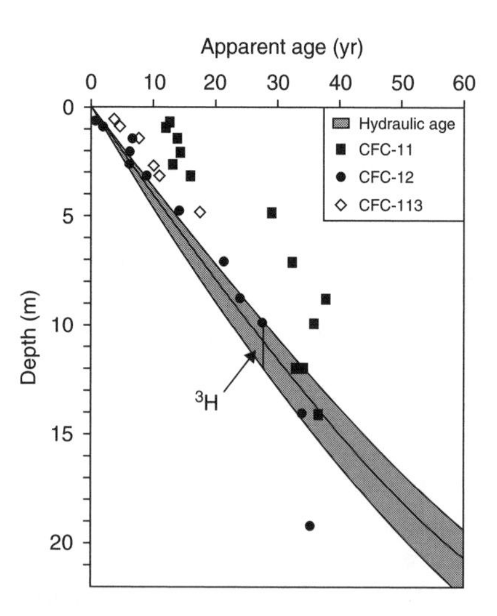

Figure 3.10. Apparent groundwater age profiles estimated from CFC's and ^{3}H. The error bar depicts the position of the ^{3}H peak in 1991. The shaded region depicts the hydraulic age profile calculated with Equation (3.11) and a recharge rate derived from the position of the tritium peak (Cook et al., 1995).

It appears that CFC-11 is readily degraded under anoxic conditions, while CFC-12 degrades at a ten times slower rate (Oster et al., 1996). CFC-113 on the other hand has been reported to be slightly adsorbed to sediment particles (Bauer et al., 2001). Thus, care should be taken in the interpretation of CFC concentrations for groundwater dating and SF_6 may in some cases be a better option (Bauer et al., 2001).

QUESTIONS:
Calculate the maximal tritium concentration in rain for Figure 3.8?
ANSWER: The maximum is 306 (^3He) + 165 (^3H) = 471 TU.
Estimate the decay constant for ^3H for this maximum?
ANSWER: $\lambda_{est} = 0.055$/yr (for the plotted age of 19 yr).
In Figure 3.8, find the depth where ^3H and ^3He have equal concentration, $D = 35$ m, $\varepsilon_w = 0.35$, $P = 0.16$ m/yr?
ANSWER: $t = 12.43$ yr, $d = 5.24$ m
Two groundwater samples have ^{14}A of 50 and 25%, respectively. Calculate the uncorrected (conventional) ages?
ANSWER: 5730 and 11460 yrs.

3.4 RETARDATION

Again, the aquifer downstream from the Vejen waste site is depicted, showing the Na^+ concentrations in Figure 3.11. Probably, most of the Na^+ is derived from dissolution of salt that is present in the waste, which implies that Na^+ and Cl^- are released in equal (molar) quantities to the percolate (but the rain may have the ratio of Na/Cl from seawater). When the distribution of the two ions is compared, it is evident that the Na^+ plume has traveled less far than Cl^- (Figure 3.3). Part of Na^+ is sorbed to the sediment, while sorption of Cl^- is negligible in most situations.

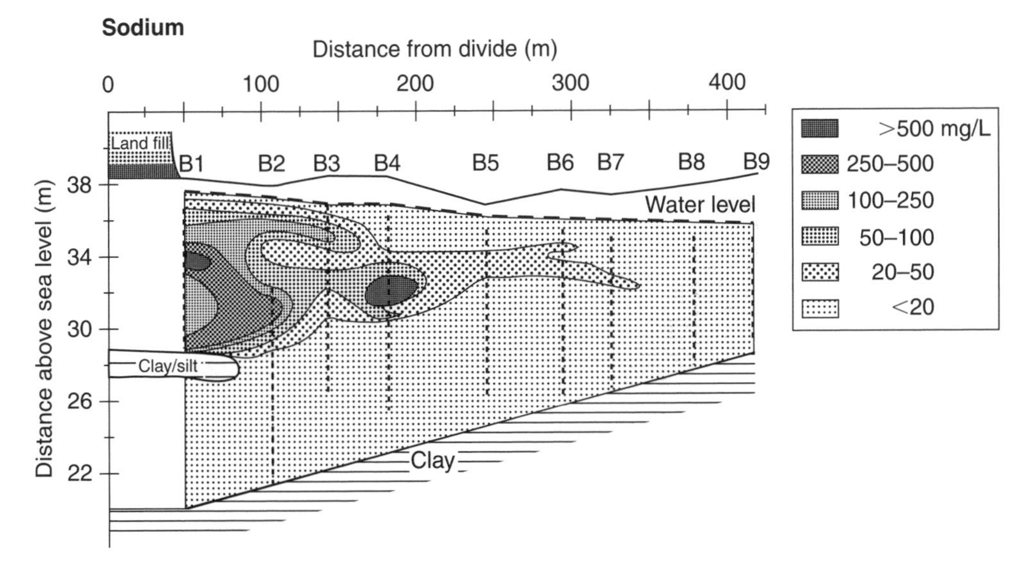

Figure 3.11. Profile through the Vejen aquifer showing Na^+ concentrations (Christensen et al., 2001).

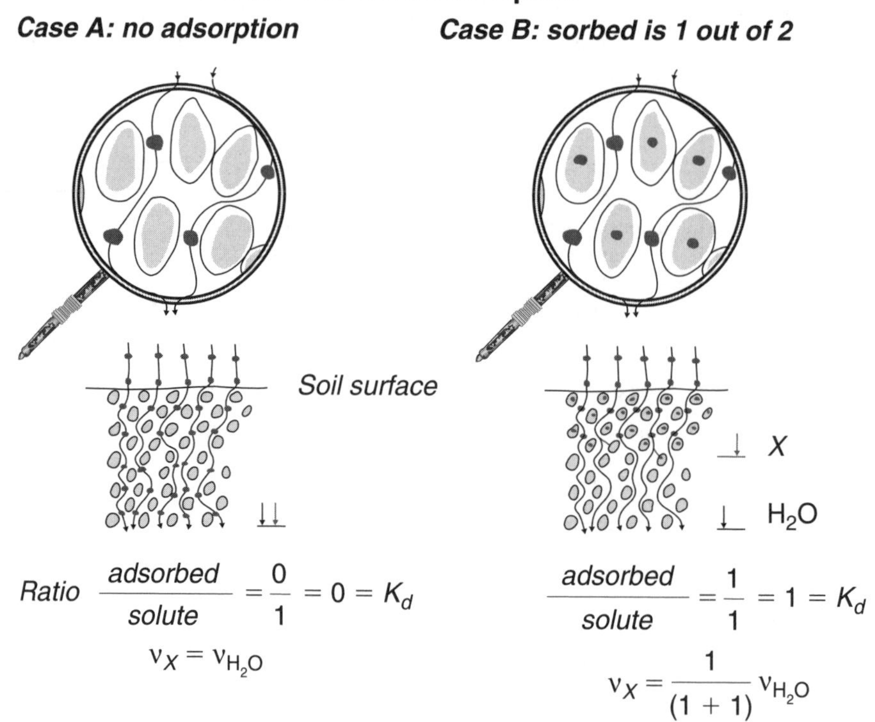

Figure 3.12. Transport of chemicals is retarded by sorption. Case *A*: no sorption; retardation is $1 + 0 = 1$, relative to water flow. Case *B*: sorption of half of total mass, and the chemical migrates at half the velocity of water; retardation is $1 + 1 / 1 = 2$.

The effect of sorption on the transport velocity of solutes is illustrated in Figure 3.12. In case (A) the solute does not sorb to the sediment grains. Consequently this solute is transported with the velocity of water. In case (B), one out of two solute molecules is sorbed. Thus, half the mass of chemical is lost from solution and therefore it will travel with half the velocity of water. The solute is *retarded* compared to water migration.

3.4.1 *The retardation equation*

For quantifying retardation, we must consider the effects of sorption processes on transport. First, examine the left part of Figure 3.13 which shows the concentration profile of a solute, sampled at time t_1 along a flowline. If we come back at a later time t_2, the solute has migrated downstream and the concentration profile has shifted along x, but also the shape has changed. The other way to observe the migration of the solute is to sit down at a single well, for example located at x_1 and to measure the concentration changes over time. The corresponding result is shown in Figure 3.13B. The slopes of the lines in Figure 3.13A, or the gradients $\mathrm{d}c / \mathrm{d}x$, are denoted by partial signs ∂, which indicates that c not only depends on x, but also on t. The time is fixed in $(\partial c / \partial x)_t$, and likewise, x is fixed in Figure 3.13B where the gradient $(\partial c / \partial t)_x$ is given by the slope of the c / t plot. Often, the subscripts are omitted when the fixed dimension is obvious.

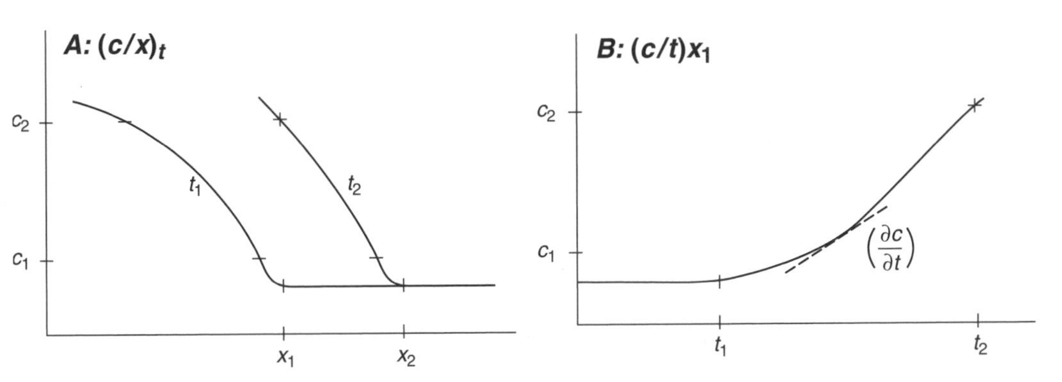

Figure 3.13. *A*: A concentration profile over distance, profiles are given for times t_1 and t_2, *B*: The change in concentration over time at point x_1. Dotted line in *B* shows the gradient $(\partial c / \partial t)_x$.

QUESTION:
Can you indicate the slope $(\partial c_2 / \partial x)_{t_1}$ in Figure 3.13?
Draw the c / t profile at x_2.

ANSWER: Note that the $(c / t)_{x_2}$ profile steepens at x_2 for c_2 because $(\partial c / \partial x)$ is steeper at t_2.

Next, consider a small volume at position x_1 with dimensions $dx \cdot dy \cdot dz$ as in Figure 3.14, and calculate the mass balance for solute c when water passes the volume with a fixed velocity v_{H_2O} in the x direction.

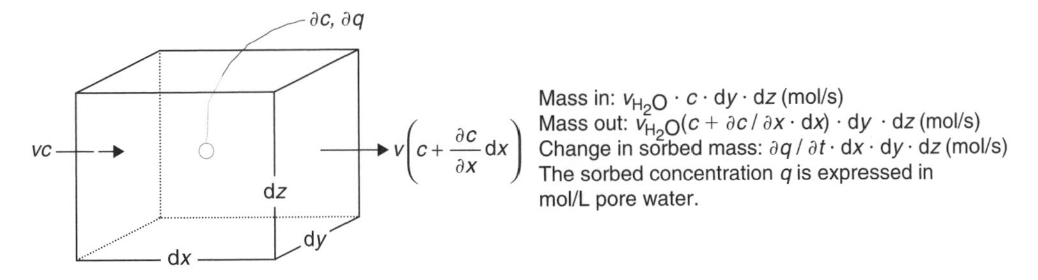

Figure 3.14. Mass balance of solute transport through a cube.

The mass balance for the solute in the volume is:

$$\frac{\partial M}{\partial t} = \text{In} - \text{Out} - \text{Sorbed} = -v_{H_2O} \frac{\partial c}{\partial x} dx \cdot dy \cdot dz - \frac{\partial q}{\partial t} dx \cdot dy \cdot dz \qquad (3.22)$$

Dividing the mass M by the volume $dx \cdot dy \cdot dz$ gives the concentration c, and thus:

$$\frac{\partial c}{\partial t} = -v_{H_2O} \frac{\partial c}{\partial x} - \frac{\partial q}{\partial t} \qquad (3.23)$$

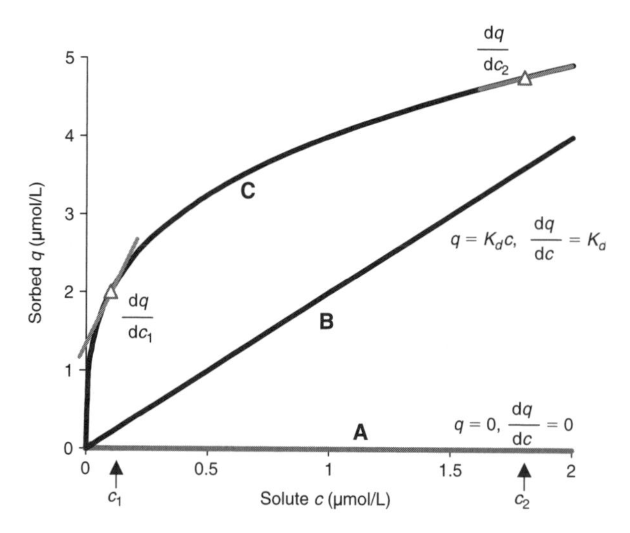

Figure 3.15. Examples of sorption isotherms: A: no sorption, B: linear sorption, C: non-linear sorption.

The third step is to combine the time gradients of q and c. We expect that, when the concentration of solute c increases, also the concentration of the sorbed species q will increase, but the exact relationship will depend on the chemistry of the system. Figure 3.15 shows three examples of *sorption isotherms* which couple q and c.

In case A (Figure 3.15), there is no sorption ($q = 0$) and $\mathrm{d}q / \mathrm{d}c = 0$ for all c (the total derivative is used here because $\mathrm{d}q / \mathrm{d}c$ is independent of x and t). This case applies to Cl^- which conserves all the mass in solution (and is hence called a *conservative* substance). In case B, q increases linearly with c, with $q = 2c$ and slope $\mathrm{d}q / \mathrm{d}c = 2$. When the slope is identical for all c, the *distribution coefficient*, $K_d = \mathrm{d}q / \mathrm{d}c = q / c$ is constant. Usually, sorption diminishes as solute concentration increases, and the slope $\mathrm{d}q / \mathrm{d}c$ decreases as c increases (the slope is smaller for $c_2 > c_1$, case C in Figure 3.15). The isotherm of case C is defined by $q = 3.5c^{0.5}$ and $\mathrm{d}q / \mathrm{d}c = 1.75c^{-0.5}$. The isotherm is conform with the general equation $q = K_F c^n$, also called a *Freundlich isotherm*.

The slope of the sorption isotherm can be introduced into Equation (3.23) by replacing $\partial q / \partial t$ with $(\mathrm{d}q / \mathrm{d}c) \cdot (\partial c / \partial t)$ which gives:

$$\left(1 + \frac{\mathrm{d}q}{\mathrm{d}c}\right)\left(\frac{\partial c}{\partial t}\right) = -v_{H_2O} \frac{\partial c}{\partial x} \tag{3.24}$$

Multiplying both sides with $\partial x / \partial c$ yields:

$$\left(1 + \frac{\mathrm{d}q}{\mathrm{d}c}\right)\left(\frac{\partial c}{\partial t}\right) \cdot \left(\frac{\partial x}{\partial c}\right) = -v_{H_2O}\left(\frac{\partial c}{\partial x}\right) \cdot \left(\frac{\partial x}{\partial c}\right) \tag{3.25}$$

The product $(\partial c / \partial x) \cdot (\partial x / \partial c) = 1$ (the value of a derivative multiplied by its inverse is 1). But the product $(\partial c / \partial t) \cdot (\partial x / \partial c)$ is not simply $(\partial x / \partial t)$! As you can see in Figure 3.13, a negative slope $(\partial c / \partial x)$ translates into a positive slope $(\partial c / \partial t)$, and the product of the two terms must be negative. More formally it is derived by implicit differentiation of a constant concentration:

$$\mathrm{d}c = 0 = \frac{\partial c}{\partial t} \cdot \mathrm{d}t + \frac{\partial c}{\partial x} \cdot \mathrm{d}x \tag{3.26}$$

which gives:

$$\frac{\partial c}{\partial t} \cdot \frac{\partial x}{\partial c} = -\left(\frac{\mathrm{d}x}{\mathrm{d}t}\right)_{\text{constant concentration}} = -\left(\frac{\partial x}{\partial t}\right)_c \tag{3.27}$$

We have implied that the concentration is constant, and therefore $(\mathrm{d}x / \mathrm{d}t)$ is a partial differential for a given, constant concentration. The partial differential $(\partial x / \partial t)_c$ (m/s) is equal to the velocity v_c of that specific concentration. If we combine Equations (3.27) and (3.25):

$$v_c = \frac{v_{H_2O}}{1 + \dfrac{\mathrm{d}q}{\mathrm{d}c}} = \frac{v_{H_2O}}{R_c} \tag{3.28}$$

Here, the retardation is defined as:

$$R_c = 1 + \frac{\mathrm{d}q}{\mathrm{d}c} \tag{3.29}$$

which varies according to isotherm slope. Equation (3.28) is called the retardation equation. The retardation equation shows that the transport velocity is different for different concentrations when the slope of the sorption isotherm is variable. It was first derived by DeVault (1943) and presented by Sillén (1951) in a particularly insightful manner.

QUESTIONS:

In Figure 3.13A,

Compare the distances Δx traveled by c_1 and c_2 in time $(t_2 - t_1)$.

ANSWER: Δx is larger for c_2 than for c_1.

Compare the flow velocities $v = \Delta x / (t_2 - t_1)$ for c_1 and c_2.

ANSWER: v is greater for c_2 than for c_1.

Compare the retardation for c_1 and c_2.

ANSWER: the retardation is smaller for c_2 than for c_1.

Sketch the isotherm for the chemical transported in Figure 3.13.

ANSWER: the isotherm is convex, i.e. everywhere the slope lies above the line, and the slope is smaller for c_2 than for c_1 (case c in Figure 3.15).

3.4.2 *Indifferent and broadening fronts*

The retardation equation yields v_c for a specific concentration c and we can evaluate how retarded solutes move compared to water. We can make time-distance graphs for a concentration using

$$x_c = v_c \cdot t = \frac{v_{H_2O}}{R_c} t \tag{3.30}$$

We can also plot concentrations as a function of distance for a fixed time:

$$x_c = \frac{x_{H_2O}}{R_c} \tag{3.31}$$

or we can chart them as a function of time for a fixed position:

$$t_c = t_{H_2O} \cdot R_c \tag{3.32}$$

Let us explore the effect of the isotherm slope dq / dc on the retardation of different concentrations in Example 3.4.

EXAMPLE 3.4. *Retardation and isotherm slope*
The first chemical to be considered shows linear adsorption with $q = c$ and therefore dq / dc = 1. The initial concentration along the flowline c is 0.8 mol/L which is flushed by a 0.1 mol/L solution. The sorbed concentration q is expressed in mol/L pore water. Calculate x_c and t_c for $x_{H_2O} = 20$ m and $t_{H_2O} = 1$ yr. Plot the concentrations in c / x and c / t graphs.

Table 3.4.1. Linear retardation, linear isotherm $q = c \cdot \mathrm{d}q / \mathrm{d}c = 1$.

c	q	dq / dc	R	x_{H_2O}	x_c	t_{H_2O}	t_c
0.8	0.8	1	2	20	10	1	2
0.45	0.45	1	2	20	10	1	2
0.1	0.1	1	2	20	10	1	2

The second chemical shows a convex adsorption isotherm with $q = c^{0.5}$ (a Freundlich type isotherm). Do the same calculation as above and add the calculated curves to the previous plots.

Table 3.4.2. Broadening front, convex isotherm, $q = c^{0.5}$. dq / dc = $0.5c^{-0.5}$.

c	q	dq / dc	R	x_{H_2O}	x_c	t_{H_2O}	t_c
0.8	0.894	0.559	1.559	20	12.829	1	1.559
0.45	0.671	0.745	1.745	20	11.459	1	1.745
0.1	0.316	1.581	2.581	20	7.749	1	2.581

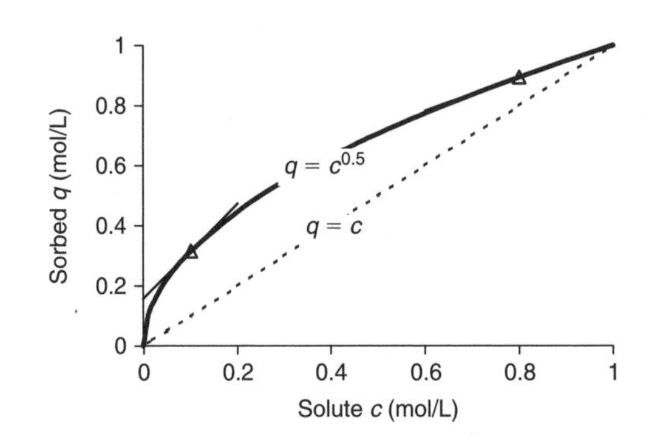

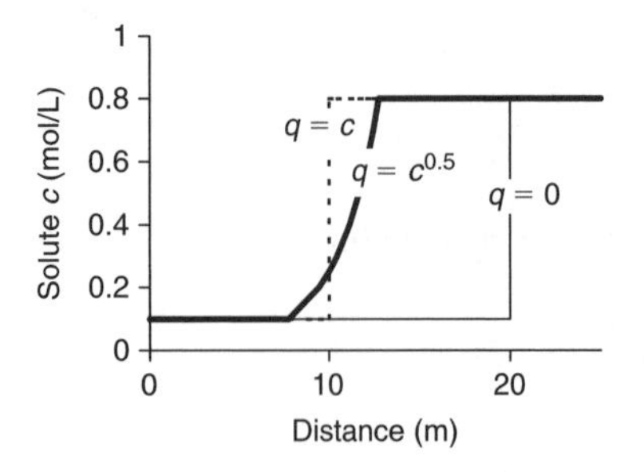

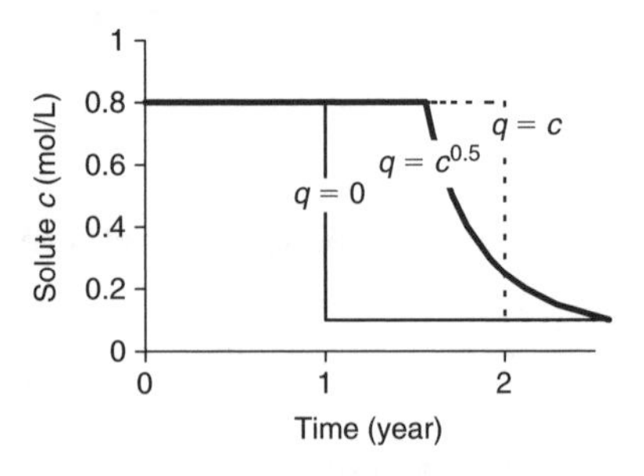

Figure 3.16. Isotherms and concentration/distance and concentration/time plots for Example 3.4.

For linear adsorption ($q = c$ in Figure 3.16) the sorbing chemical is delayed compared to a non sorbing chemical ($q = 0$) while the shape of the concentration front remains unchanged (indifferent front). When the adsorption isotherm is convex, a high concentration gives a relatively low dq / dc, thereby a low retardation and consequently, the concentration moves fast. When the concentration is low, the slope dq / dc is relatively high, retardation is high, and the concentration moves slower. This results in a broadening front when the chemical is eluted.

EXAMPLE 3.5. *Analytical modeling of column elution*
A 4 cm long sediment column has been spiked with 37 µg β-HCH/L (β-hexachlorocyclohexane). The sorption isotherm is $q = 7c^{0.7}$ µg/L. Estimate the elution of β-HCH with respect to conservative breakthrough.

ANSWER:
We relate the breakthrough time of β-HCH to Cl^- by:

$$\frac{t_{\beta-\mathrm{HCH}}}{t_{\mathrm{Cl}}} = R_c = \left(1 + \frac{dq}{dc}\right) = 1 + 4.9c^{-0.3}$$

The breakthrough of Cl^- shows dispersion (see Section 3.6) and $t_{\beta\text{-HCH}}$ is calculated relative to the concentrations of the dispersed Cl^- curve. The breakthrough time is expressed in 'pore volumes' injected in the column: after 1 pore volume the resident water (with Cl^-) is displaced and Cl^- drops to zero (when dispersion is absent). Two pore volumes need twice more time than one, etc. Figure 3.17 compares measured β-HCH concentrations in a column experiment with the calculated breakthrough curve ($R_c = (V^* + V_c) / V_c$, where $V^* = dq / dc$ indicates the variable retardation) and also shows results from a numerical model which accounts for dispersion (see Chapter 11) (Appelo et al., 1991). Clearly, the simple formula provides very good first estimates, although numerical modeling is essential for fine tuning.

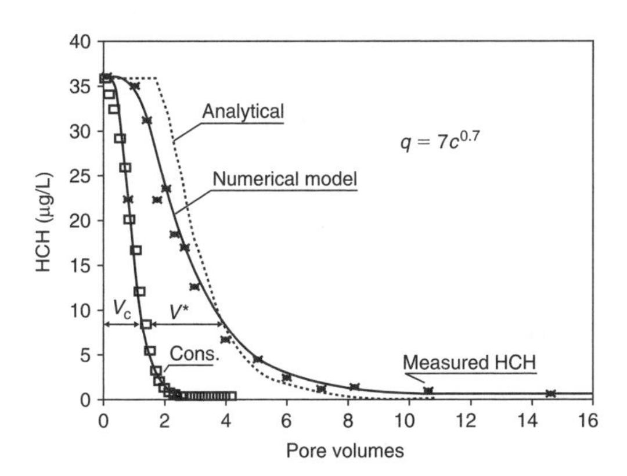

Figure 3.17. Elution of a sediment column spiked with 37 µg β-HCH/L. Thin lines are results from numerical model, dotted line is calculated with the retardation equation. The Cl^- breakthrough curve is scaled to the maximal concentration of β-HCH.

QUESTION:
For one year, a sorbing chemical is spilled in soil at a concentration of 1 mM. The isotherm is linear, $dq / dc = 4$; $v_{H_2O} = 10$ m/yr.

1) After 10 yrs, where is the pollutant forefront?
   ANSWER: at 100 / 5 = 20 m

2) After 10 yrs, where is the pollutant hindfront?

ANSWER: at 90 / 5 = 18 m

3) What is the solute concentration in the soil water?

ANSWER: 1 mM

4) What is the sorbed concentration in the soil (expressed per L pore water)?

ANSWER: 4 mM (Note the mass balance 1 mM × 10 m = (1 + 4) mM × 2 m)

5) When arrives the chemical at 30 m?

ANSWER: after 15 years

6) When is the chemical flushed at 30 m?

ANSWER: after 16 years

Estimate the pore volumes needed to lower the concentration of β-HCH to below 1 μg/L in Example 3.5.

ANSWER: 10.8 in case dispersion is absent.

3.4.3 *Sharpening fronts*

In Example 3.4, the solution with a low concentration flushed the solution with a high concentration. Now let us look at the reverse situation where a high concentration (0.8 mol/L) displaces a low concentration (0.1 mol/L) of the chemical with the convex isotherm. The slope dq / dc at 0.1 mol/L is steeper than at 0.8 mol/L and the retardation for 0.1 mol/L is larger than for 0.8 mol/L. Thus, the concentration 0.8 mol/L travels faster than 0.1 mol/L. We may draw a c / x diagram in Figure 3.18, which shows that the high concentration will tend to overtake the low concentration.

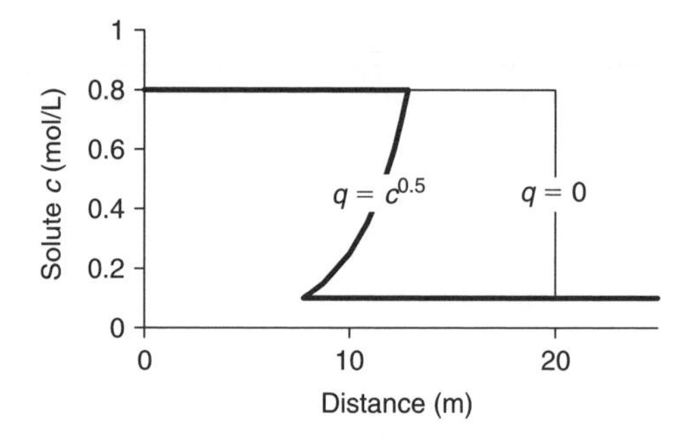

Figure 3.18. A chemical with convex isotherm $q = c^{0.5}$ enters the flowline.

This situation is, of course, impossible: at 10 m a choice of 3 concentrations can be made of, 0.8, or 0.1 or about 0.2 mol/L. What happens is, that any contact of $c = 0.8$ mol/L with soil that contains sorbed chemical in equilibrium with a lower solute concentration, will result in sorption and thus lower the solute concentration. Accordingly, the higher migration rate of the high concentration is arrested to the lower velocity of the smaller concentration. The result is a sharp front where the solute concentration jumps directly from 0.1 to 0.8 mol/L, accompanied by a corresponding jump in sorbed concentrations. Over the front, mass balance applies in which the change Δq of sorbed mass is connected with the change Δc in solution. Obviously, retardation is stronger when the ratio $\Delta q / \Delta c$ is greater. In addition to the retardation equation that is based on the slope of the isotherm (Equation 3.29), we can formulate the *retardation equation for a sharp front*:

$$R_{jump} = 1 + \frac{\Delta q}{\Delta c} \tag{3.33}$$

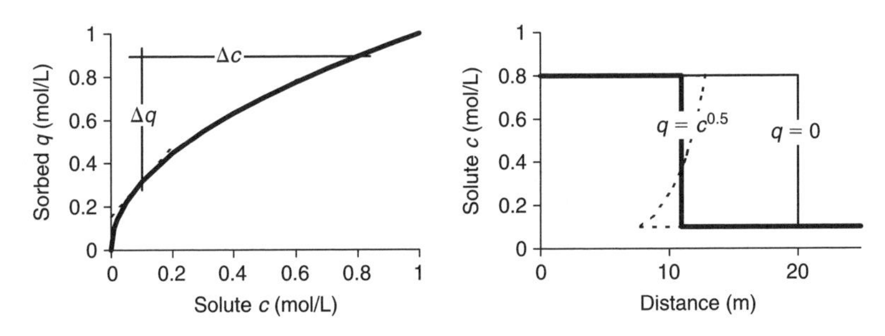

Figure 3.19. Convex isotherm gives a sharp front when concentrations increase, here from 0.1 to 0.8 mol/L.

The sharp front equation is illustrated in Figure 3.19. The differences between the initial and the final sorbed and solute concentrations are sufficient for calculating the retardation of the front.

To discern whether a front is sharpening, we look at the sorption isotherm. A slope dq / dc that is smaller for the final concentration than for the initial solution, will give a jump-like concentration change, a *sharpening front*, or a shock. On the other hand, a slope that is greater for the the final concentration will yield a *broadening front*, or a *wave*. When the slope is constant (linear isotherm) the front is not affected by concentration dependent retardation and we have an *indifferent front*. The retardation of an indifferent front is also calculated with Equation (3.33). This reasoning can be applied piecewise along isotherms with variable shape and any slope dq / dc (Problem 3.11).

EXAMPLE 3.6. *Retardation of a sharp front*

In continuation of Example 3.4 we calculate the fronts for the chemical with $q = c^{0.5}$ when the solute concentration increases from 0.1 to 0.8 mol/L. Moving along the sorption isotherm from 0.1 to 0.8 mol/L, the slope dq / dc decreases and we are dealing with a sharp front. The retardation is then calculated from $R = 1 + (q_i - q_0) / (c_i - c_0)$, using the initial and the final concentration, and is identical for all concentrations in between. The sorbed concentration q is expressed in mol/L pore water.

Table 3.6.1. Sharpening front, convex isotherm $q = c^{0.5}$. $dq / dc = 0.5c^{-0.5}$.

c	q	dq / dc	R	x_{H_2O}	x_c	t_{H_2O}	t_c
0.1	0.32	1.58	1.82	20	10.95	1	1.82
0.45	0.67	0.75	1.82	20	10.95	1	1.82
0.8	0.89	0.56	1.82	20	10.95	1	1.82

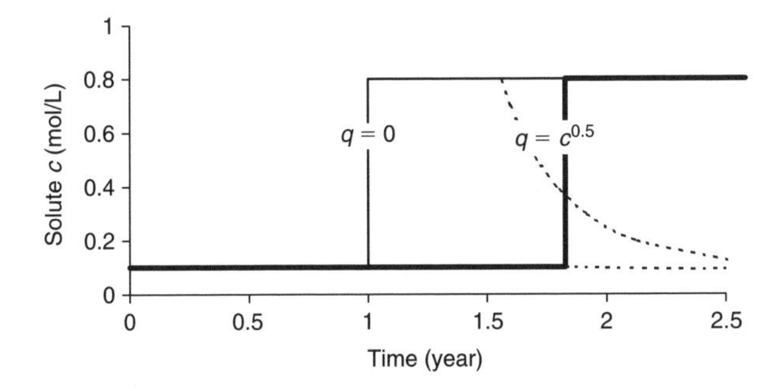

Figure 3.20. A c / t plot for a sharp front, Example 3.6 (the c / x plot is shown in Figure 3.19).

Sharp fronts are found when chemicals with convex isotherms enter an aquifer, or when several ions or species compete for the same sorption site, and the displacing species has a stronger affinity for the site than the displaced one. Sharpening fronts counteract the tendency of concentration changes to become more diffuse with time or distance. Retardation is a matter of mass transfer between the solid and the solution only and is therefore not confined to sorption. Also mineral dissolution and precipitation yield retardation and when the dissolution/precipitation reactions are fast enough to reach equilibrium, the associated fronts will be sharp as well.

QUESTION:
Calculate the retardation when 36 μg β-HCH/L enters a pristine aquifer (cf. Example 3.5). First consider, is the front sharp?

ANSWER: The front is sharp. $R = 1 + (7 \times 36^{0.7} - 0) / (36 - 0) = 3.4$.

3.4.4 *Solid and solute concentrations*

Chemical analysts usually express the concentration of substances in solids as concentration per mass of solid, mol/kg, g/kg, ppm by weight, etc. These numbers can be converted to equivalent solute concentrations, by multiplying with the bulk density ρ_b and dividing by the water filled porosity ε_w, cf. Example 3.7. The factor $\rho_b / \varepsilon_w \approx 6$ kg/L for sandy aquifers.

EXAMPLE 3.7. *Recalculate 10 ppm As in (unsaturated) quartz sand to an equivalent solute concentration in mg/L pore water.* Total porosity $\varepsilon = 0.3$, water filled porosity $\varepsilon_w = 0.2$. Density of quartz is $\rho_{qu} = 2.65$ g/cm^3.

ANSWER:
The bulk density of the sediment is the density with air filled pores:

$$\rho_b = \rho_{qu}(1 - \varepsilon) + 0(\varepsilon) = 2.65 \times 0.7 = 1.855 \text{ g/cm}^3$$

Take 1 kg aquifer, the volume is $1 / \rho_b = 0.54$ L. Hereof, $\varepsilon_w = 0.2$ is filled with water, or 0.108 L. The 1 kg aquifer contains 10 mg As, which is $10 / 0.108 = 93$ mg As/L pore water.

The concentrations expressed per liter pore water always appear surprisingly high compared to those expressed relative to the solid phase. This is a simple consequence of the fact that most of the mass and volume in the aquifer is located in the solid phase. The dimensionless ratio of the concentrations, or the slope of the isotherm when the ratio varies, allows for a readily estimate of the mobility of chemicals. In Example 3.8 a comparison is given of the calculation of the herbicide diuron's travel time in an aquifer using both the dimensional and the non-dimensional approach.

EXAMPLE 3.8. *Travel time of diuron in a soil*
In a specific soil, diuron (a herbicide) was found to be sorbed 6 μg/g soil at a solute concentration of 1 μg/mL (Nkedi-Kizza et al., 1983). Estimate the arrival time of 1 mg diuron/L at 1 m depth when irrigation water percolates with 0.3 m/yr, $\rho_b = 1.8$ g/cm^3, $\varepsilon_w = 0.1$.

ANSWER:
By full mass balance: We take a soil column $1\,m^2 \times 1\,m$ depth. When diuron has arrived at 1 m depth, the column contains $0.1\,m^3$ water with 1 g diuron/m^3 and 1800 kg soil with 6 g diuron/kg. Total is $0.1 \times 1 + 1800 \times 6 = 10.9$ g. Irrigation water is applied at 0.3 m^3/yr and contains 1 g/m^3. Thus, it takes $10.9 / 0.3 = 36.3$ yr.

By pore volume and retardation: Express sorbed diuron in μg/mL pore water, $q = 6\,\mu g/g \times \rho_b / \varepsilon_w = 108\,\mu g/mL$. The retardation is $R = 1 + \Delta q / \Delta c = 1 + 108 / 1 = 109$. It takes water $0.1 / 0.3 = 0.33$ yr to arrive at 1 m depth and to flush one pore volume. Hence, the diuron arrives after $0.33 \times 109 = 36.3$ yr.

The effect of the curved sorption isotherm on transport time becomes clear when transport is calculated for a concentration pulse (Scheidegger et al., 1994). For diuron the following sorption isotherm was determined: s (μg/g soil) $= 6c^{0.67}$ (Nkedi-Kizza et al., 1983). If in Example 3.8, diuron was applied for only one year, while everything else remains the same, then the front of 1 mg diuron/L still has a retardation of 109 because it is sharp. The zone with 1 mg diuron/L compacts to the distance traveled by water in one year, divided by the retardation (Equation 3.31) or $3\,m / (R_{jump} = 109) = 0.0275$ m and arrives at 1 m depth after $0.33 \times 109 = 36.3$ yr. However, a concentration of 0.1 μg/L has a retardation of 2257, and it requires $0.33 \times 2257 = 752$ yr before the concentration at 1 m depth is reduced to 0.1 μg/L (the drinking water limit). We have neglected dispersion and decay of the chemical here, but the essential feature still emerges that transport in the soil and subsoil is slow indeed, and is slowed down even further by sorption and chemical reactions.

Let us return once more to the Vejen aquifer, Figures 3.3 and 3.11. Can we estimate now how transport of Cl^- and Na^+ are interrelated? Figure 3.21 shows averaged concentrations of Cl^- and Na^+ in the plume downstream of the waste site and a model line for Na^+ that is calculated from the trendline for Cl^- as follows. The background concentrations for Na^+ and Cl^- are 11 and 20 mg/L respectively. (The molar Na/Cl ratio in background water is 0.85, closely equal to the ratio in seawater as expected). From the waste site, NaCl is leached with a molar ratio of 1. Thus, the concentration of Na^+ can be calculated from the concentration of Cl^-:

$$c_{\mathrm{Na}} = 11 + \frac{23\,\mathrm{g\ Na/mol}}{35.45\,\mathrm{g\ Cl/mol}}(c_{\mathrm{Cl}} - 20)$$

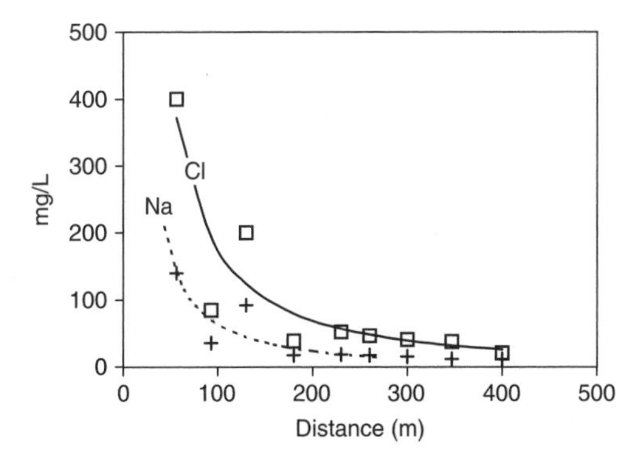

Figure 3.21. Concentrations of Na^+ (crosses) and Cl^- (squares) in the plume of the Vejen waste site, averaged over depth from the multilevel samplers. Dotted line for Na^+ shows estimated concentrations based on the Cl^- trendline with a retardation of 1.5.

However, Na^+ is retarded by ion exchange. The retardation factor is almost constant 1.5 for the concentrations in the aquifer (ion exchange is discussed in Chapter 6, cf. Problem 6.17). Thus, $x_{Na} = x_{Cl} / 1.5$ was used to obtain the travel distance of Na^+, shown in Figure 3.21, dotted line. We observe a concentration pattern that is irregular, even when averaged over depth. This will be related to irregularities in the filling of the site, the subsequent leaching of the waste and, most importantly, be influenced by the hydraulic conductivity variations in the aquifer as was noted before. Finally, the low concentrations are running ahead of the main front, and therefore we must also discuss front spreading and dispersion.

3.5 DIFFUSION

A concentration difference between two points in a stagnant solution will be leveled out in time by the random Brownian movement of molecules. The process is called *molecular diffusion* and the movement of molecules by diffusion is described by Fick's laws. Fick's first law relates the flux of a chemical to the concentration gradient:

$$F = -D\frac{\partial c}{\partial x} \tag{3.34}$$

where F is the flux (mol/s/m^2), D the diffusion coefficient (m^2/s), and c the concentration (mol/m^3).

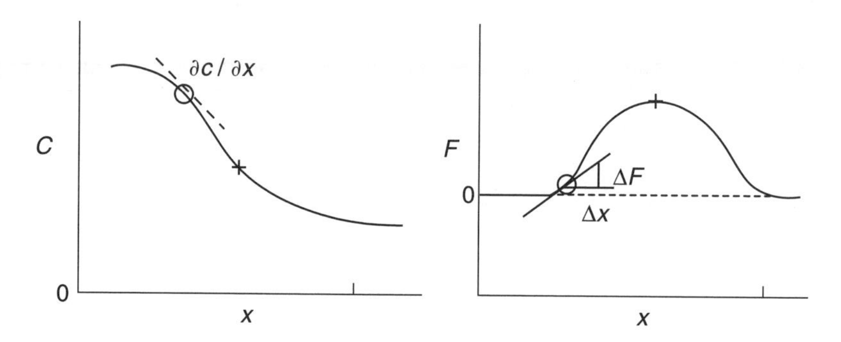

Figure 3.22. Diffusion develops as a result of a concentration gradient.

Figure 3.22 illustrates how the flux depends on the concentration gradient. Diffusion changes the concentration at a given x and the flux is therefore time dependent.The concentration change can be found by constructing a mass balance for a small volume, as shown in Figure 3.23, similar to Figure 3.14 and Equation (3.22). The mass that enters through the left side is:

$$F_1 \times \mathrm{d}z\mathrm{d}y = -D\frac{\partial c}{\partial x}\mathrm{d}z\mathrm{d}y \tag{3.35}$$

and the mass leaving through the right side is:

$$F_2 \times \mathrm{d}z\mathrm{d}y = \left(F_1 + \frac{\partial F}{\partial x}\mathrm{d}x\right)\mathrm{d}z\mathrm{d}y = -D\left(\frac{\partial c}{\partial x} + \frac{\partial}{\partial x}\frac{\partial c}{\partial x}\mathrm{d}x\right)\mathrm{d}z\mathrm{d}y \tag{3.36}$$

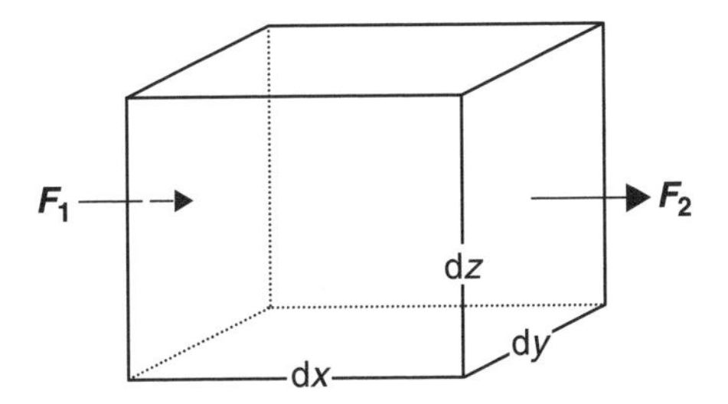

Figure 3.23. Concentration changes in a cube as result of diffusion.

The change in number of moles M in the cube is the difference:

$$\frac{\partial M}{\partial t} = (F_1 - F_2) \times \mathrm{d}z\mathrm{d}y = \left(-D\frac{\partial c}{\partial x} + D\frac{\partial c}{\partial x} + D\frac{\partial^2 c}{\partial x^2}\mathrm{d}x\right)\mathrm{d}z\mathrm{d}y$$
$$= D\frac{\partial^2 c}{\partial x^2}\mathrm{d}x\mathrm{d}y\mathrm{d}z \tag{3.37}$$

Dividing by the volume gives the change in concentration: $\delta c = \delta M / (\mathrm{d}x \cdot \mathrm{d}y \cdot \mathrm{d}z)$. We therefore obtain:

$$\frac{\partial c}{\partial t} = D\frac{\partial^2 c}{\partial x^2} \tag{3.38}$$

Equation (3.38) is known as Fick's second law of diffusion.

3.5.1 *Diffusion coefficients*

For solute ions the diffusion coefficient, D, will be related to their mobility, u, as derived from the molar electrical conductivity in m/s per V/m (m^2/s/V) given in Table 1.7 (Atkins and de Paula, 2002; Lyman et al., 1990). Differences in mobility are a function of the friction ions experience when traveling through water, and are related to the ion-size and the viscosity:

$$u = |z|q_e / (6\pi\eta a) \tag{3.39}$$

where $|z|$ is the absolute charge of the ion, q_e is the charge of the electron (1.602×10^{-19} C), η the viscosity of water (0.891×10^{-3} kg/m/s at 25°C), and a the radius of the ion. The radius of the solvated ion includes water of hydration and is found by combining the molar conductivity (Equation 1.4) with Equation (3.39). For example, the mobility of Na^+ is $5.19\times10^{-8}\,m^2 \cdot$ s/V (Table 1.7), which gives $a = 0.164$ nm while the radius of unhydrated Na^+ is 0.10 nm.

Similarly, the diffusion coefficient is given by the Stokes-Einstein equation:

$$D = kT / (6\pi\eta a) \tag{3.40}$$

where k is the Boltzmann constant (1.3805×10^{-23} J/K) and T the absolute temperature (°C + 273.15). The combination of Equations (3.39) and (3.40) relates the diffusion coefficient to the ionic mobility:

$$D = u \cdot kT / (|z|q_e) \tag{3.41}$$

Diffusion coefficients, calculated using (3.41), are listed in Table 3.2 and vary by a factor of 3. There will be some interaction between different diffusing species because charge must be conserved.

When NaCl is diffusing in distilled water, the more mobile Cl^- ions will drag along the Na^+ ions and accelerate Na^+ slightly, while the movement of Cl^- will be impeded somewhat. Similarly for a $Ca(HCO_3)_2$ solution, some acceleration results for HCO_3^-, and some retardation for Ca^{2+}. For a symmetric salt (in which cation and anion have the same charge), the resulting diffusion coefficient can be calculated with the Nernst formula, which emphasizes the smaller value of the pair of ions. For multicomponent solutions, an average diffusion coefficient may be used which, for the ions in Table 3.2, is $D_f = 1.3\times10^{-9}$ m²/s at 25°C. Values at other temperatures can be obtained using Equation (3.40):

$$D_{f,T} = D_{f,298} \cdot T \cdot \eta_{298} / (298 \cdot \eta_T) \tag{3.42}$$

which yields $D_{f,10} = 0.84\times10^{-9}$ m²/s at 10°C (cf. also Li and Gregory, 1974). Finally, the ionic mobility listed in Table 3.2 is for ions at trace concentration, and decreases towards higher concentrations (cf. Figure 1.13). Accordingly, the diffusion coefficients decrease slightly with increasing concentration for simple salt solutions (Lasaga, 1998, p. 317), while the relations become quite complicated for multicomponent solutions (Felmy and Weare, 1991).

Table 3.2. Diffusion coefficients (D_f 10^{-9} m²/s) in "free" water as calculated from ionic mobilities at 25°C (obtained from molar conductivities, Table 1.7).

	Na^+	K^+	Mg^{2+}	Ca^{2+}	Cl^-	HCO_3^-	SO_4^{2-}	Average
Mobility	5.19	7.62	5.46	6.17	7.91	4.61	8.29	6.46×10^{-8} m²/s/V
Diffusion coefficient	1.33	1.96	0.70	0.79	2.03	1.18	1.06	1.30×10^{-9} m²/s

Overall, the differences in D_f remain small and in most cases it is sufficient to use a constant diffusion coefficient in "free" water of $D_f \approx 10^{-9}$ m²/s for simple electrolytes. For neutral organic molecules, the diffusion coefficient can be estimated with (Schwartzenbach, et al., 1993; Lyman et al., 1990):

$$D_{f,298} = 2.8\times10^{-9}\, V_l^{-0.71} \tag{3.43}$$

where V_l is the liquid molar volume of the substance (cm³/mol; V_l = (Mol. wt, g/mol) / (ρ_l, g/cm³).

Compared to diffusion in "free" water, solutes diffusing through a sediment-water system must travel an extra distance because they have to circumnavigate the sediment grains. The *effective diffusion coefficient* D_e corrects for the additional pathway. There is considerable controversy on how the effective diffusion coefficient can be calculated from the free water value and related to sediment properties (Lerman, 1979; Boudreau, 1997). In the Kozeny theory (Childs, 1969), it is related to the *tortuosity* which is defined as the length of the actual travel path taken by a solute in a porous medium, divided by the straight line distance:

$$D_e = D_f / \theta^2 \tag{3.44}$$

where θ is the tortuosity of the porous medium. The quadratic dependence of diffusion on tortuosity results from the following reasoning. Imagine a column which has its ends in two different solutions, so that a steady state flux of some ion passes through the column. The sediment in the column has pores which are connected by tortuous capillaries of length L_a. Now mould the column in such a way that the capillaries are straightened out to length L, without affecting the volume of the pore space, the pore geometry, or the flux of the ion through the column.

The concentration gradient in the moulded column will be smaller by (L_a / L). The volume of the column has not changed, and therefore the cross-sectional area will also be smaller by (L_a / L). Since the total flux through the column did not change either, the apparent diffusion coefficient in the straightened capillaries must be larger by $(L_a / L)^2 = \theta^2$. A tortuous path, at an average angle of 45° with the straight line, obviously gives a tortuosity $\theta = L_a / L = 2 / \sqrt{2} = 1.4$, and hence $D_e = D_f / 2$.

Alternatively, the effective diffusion coefficient can be related to the *formation factor F*, which is the ratio of the specific electrical resistance of a sediment with a solution, and the specific resistance of the free solution. The formation factor is thus defined for a unit area of porous medium, i.e. pore volume and grain skeleton together. The formation factor has the advantage that it can be measured directly. The formation factor itself is also a function of porosity. Empirical relationships between formation factor and porosity take the form of Archie's law:

$$F = 1 / \varepsilon_w^n \tag{3.45}$$

where the exponent n varies from 1.4 to 2.0 (McNeil, 1980). When the sediment grains are perfect insulators and only the pore space conducts electricity, the tortuosity is related to the formation factor as (Childs, 1969):

$$\theta^2 = F \cdot \varepsilon_w \tag{3.46}$$

Substitution of $F = \varepsilon_w^{-2}$ yields $\theta^2 = \varepsilon_w^{-1}$, and the effective diffusion coefficient follows from:

$$D_e = D_f / \theta^2 = D_f \cdot \varepsilon_w \tag{3.47}$$

With $\varepsilon_w = 0.5$, the relation gives $D_e = 0.5D_f$, the same value as derived from tortuosity *per se*. Other values of ε_w and n give of course a slightly different result; Boving and Grathwohl (2001) found $\theta^2 = \varepsilon_w^{-1.2}$ for sandstones and limestones, which yields $D_e = 0.44D_f$ for $\varepsilon_w = 0.5$. In dry soils, the diffusion coefficient may be further reduced when water resides only in very thin films or unconnected pores (Hu and Wang, 2003).

When applying diffusion coefficients for "free water" to sediment-water systems, one should remember that only the water filled porosity, ε_w, participates in the diffusive flux. Thus, the flux must be multiplied with ε_w to obtain the mass transfer for a surface composed of grains and water (compare with the difference among water velocity and Darcy flux). Some authors incorporate the correction in the value of D_e, which has led partly to the confusion noted by Lerman (1979) and Bear (1972).

3.5.2 *Diffusion as a random process*

Fick's second law (3.38) gives the change of c in space (x) and time (t). For comparison with experimental data we are normally more interested in the actual concentration at a given time and location, which requires integration of Equation (3.38). By differentiation with respect to t and x, we can verify that the integration of (3.38) yields a solution such as:

$$c = A \cdot t^{-1/2} \exp\left(\frac{-x^2}{4Dt}\right) \tag{3.48}$$

Here A is a constant that must be found from the initial and boundary conditions. Consider the initial condition where no chemical is present at time $t < 0$, while at $t = 0$, N moles are injected at the origin, $x = 0$.

This is known as a *single shot input* or *Dirac delta function*; as $t \rightarrow 0$, $c = 0$ everywhere except at the origin where $c \rightarrow \infty$. Mass conservation at any time t requires:

$$N = A \cdot t^{-1/2} \int_{-\infty}^{\infty} \exp\left(\frac{-x^2}{4Dt}\right) dx \tag{3.49}$$

Substitution of $x^2 / (4Dt) = s^2$, and the associated derivative $dx = \sqrt{(4Dt)} \cdot ds$ gives:

$$N = A \cdot \sqrt{4D} \int_{-\infty}^{\infty} e^{-s^2} ds \tag{3.50}$$

The integral now contains a form closely akin to the error function erf(z):

$$\text{erf}(z) = \frac{2}{\sqrt{\pi}} \int_{0}^{z} e^{-s^2} ds \tag{3.51}$$

The error function is tabulated, and numerical approximations are known, as given in Table 3.3. The error function is 0 for $z = 0$, and 1 for $z = \infty$. Furthermore, the error function is symmetric around 0, so that erf($-z$) $= -$erf(z). We can thus evaluate the integral in (3.50) as:

$$\int_{-\infty}^{\infty} e^{-s^2} ds = 2 \frac{\sqrt{\pi}}{2} \, erf(\infty) = \sqrt{\pi} \tag{3.52}$$

The constant A in (3.50) then becomes:

$$A = N/(4\pi D)^{1/2}$$

and the solution of Equation (3.38) for the initial condition stated is:

$$c(x, t) = \frac{N}{\sqrt{4\pi Dt}} \exp\left(\frac{-x^2}{4Dt}\right) \tag{3.53}$$

where N is the input mass (moles) at time $t = 0$ at $x = 0$. $c(x, t)$ is expressed in mol/m since we consider only one dimension. Fundamentally, (3.53) is equal to the normal density function (the Gauss curve):

$$n(x) = \frac{N}{\sqrt{2\pi\sigma^2}} \exp\left(\frac{-(x - x_0)^2}{2\sigma^2}\right) \tag{3.54}$$

where x_0 is the average location (in Equation (3.53), $x_0 = 0$), and σ^2 is the variance of the distribution. Diffusion can therefore be considered as a statistical process. Figure 3.24 illustrates the concept as spreading of particles moving in random steps. One thousand particles start at $x = 0$ and make steps of random size within the range -2.5 to 2.5. The curve above the swarm sums the number of particles for each x. In theory, a Gauss curve results with a variance that increases with each round (i) of random steps as:

$$\sigma^2 = i \cdot 25/12 \tag{3.55}$$

Table 3.3. Values and approximations for the error function (erf) and the error function complement (erfc = 1 − erf).

	z	erf(z)	erfc(z)
	−3.00	−1.0000	2.0000
	−2.50	−0.9996	1.9996
	−2.00	−0.9953	1.9953
	−1.00	−0.8427	1.8427
	−0.50	−0.5205	1.5205
	0.00	0.0000	1.0000
	0.10	0.1125	0.8875
	0.20	0.2227	0.7773
	0.30	0.3286	0.6714
	0.40	0.4284	0.5716
	0.50	0.5205	0.4795
	0.60	0.6039	0.3961
	0.70	0.6778	0.3222
$z = \sqrt{½} = \sigma$	0.7071	0.6827	0.3173
	0.80	0.7421	0.2579
	0.90	0.7969	0.2031
	1.00	0.8427	0.1573
	1.10	0.8802	0.1198
	1.20	0.9103	0.0897
	1.30	0.9340	0.0660
	1.40	0.9523	0.0477
	1.60	0.9763	0.0237
	1.80	0.9891	0.0109
	2.00	0.9953	0.0047
	2.25	0.9985	0.0015
	2.50	0.9996	0.0004
	2.75	0.9999	0.0001
	3.00	1.0000	0.0000

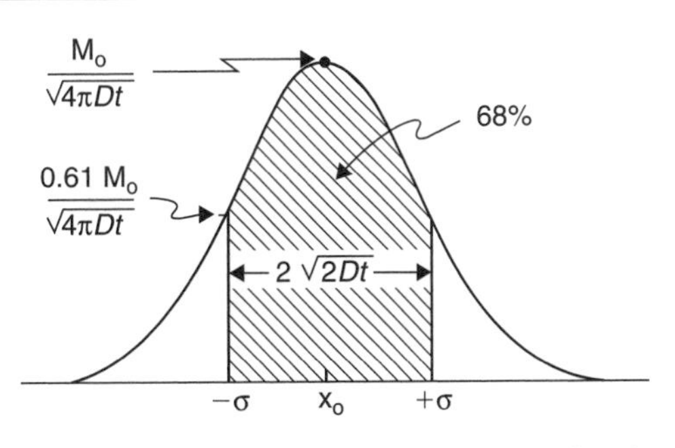

Figure 3.24. A Gaussian (normal) density function.

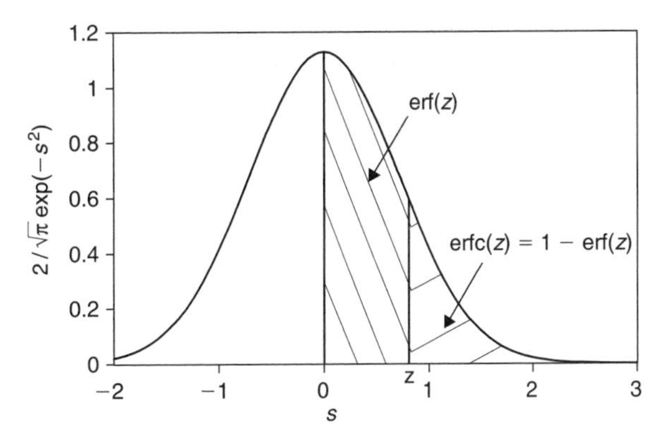

Figure 3.25. The error function, erf(z), is the integral of the Gaussian density function from 0 to z. The error function complement, erfc(z) = 1 − erf(z), is the integral from z to ∞.

Approximations for erf(z) (Abramowitz and Stegun, 1964):

$$\mathrm{erf}(z) = (1 - e^{-(4z^2/\pi)})^{1/2}\ (\pm 0.007)$$

$$\mathrm{erf}(z) = 1 - (1 + a_1 z + a_2 z^2 + a_3 z^3 + a_4 z^4)^{-4}\ (\pm 0.0005)$$

a_1	a_2	a_3	a_4
0.278393	0.230389	0.000972	0.078108

$$\mathrm{erf}(z) = 1 - (a_1 b + a_2 b^2 + a_3 b^3 + a_4 b^4 + a_5 b^5)\exp(-z^2)\ (\pm 1.5\times 10^{-7}) \qquad b = 1/(1 + a_6 z)$$

a_1	a_2	a_3	a_4	a_5	a_6
0.254829592	−0.284496736	1.421413741	−1.453152027	1.061405429	0.3275911

Note that the error function is symmetric around 0, and that erf(−z) = −erf(z).

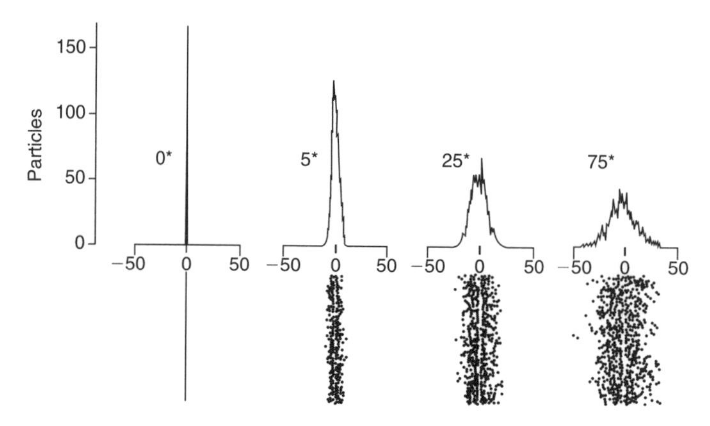

Figure 3.26. Computer simulation of one thousand particles that do 5, 25 or 75 steps with a random size in the range −2.5 to 2.5.

As the enlargement after 75 steps in Figure 3.27 indicates, the thousand particles portrayed in Figure 3.27 still only roughly approximate a smooth Gauss curve. Computed random motions require many thousands of particles before a reasonable fit is obtained within a few percent deviation. Computer programs which use the "random walk" based on particle tracking in combination with a random step (e.g. Uffink, 1988; Kinzelbach, 1992, Sun, 1999), are known as devourers of computer time.

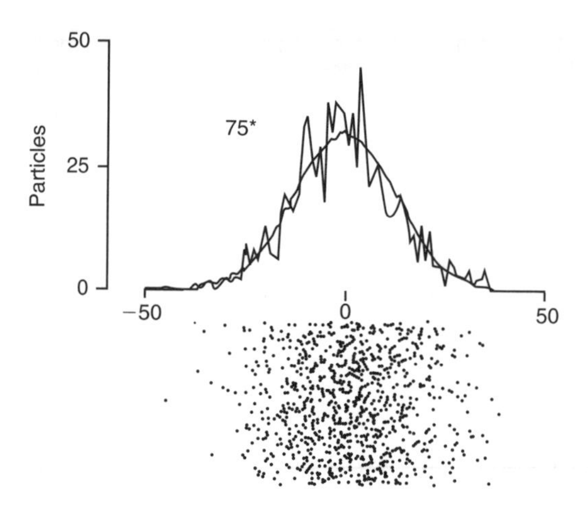

Figure 3.27. Enlargement of Figure 3.26 for 75 random steps and comparison with the ideal Gauss curve.

Since Equations (3.53) and (3.54) fundamentally are the same, the variance σ^2 is related to the diffusion coefficient by:

$$\sigma^2 = 2Dt \tag{3.56}$$

where σ has the dimension of length. This simple formula, first derived by Einstein, provides a rapid estimate of the mean diffusion length, not only of chemicals in aquifers but in gases and in minerals as well (Moore, 1972). As shown in Figure 3.24, 68% of the mass that is spreading is contained in the area from $-\sigma$ to $+\sigma$, after a given period of time.

For diffusion in three dimensions the variances of the extra dimensions must be added. Thus, for diffusion in water where $\sigma_x^2 = \sigma_y^2 = \sigma_z^2$, the sphere where 68% of a point source is located has a radius $r = \sigma_{xyz}$:

$$\sigma_{xyz} = \sqrt{(\sigma_x^2 + \sigma_y^2 + \sigma_z^2)} = \sqrt{6D_f t} \tag{3.57}$$

3.5.3 *Diffusive transport*

When is diffusive transport important as compared to advective transport? Using Equation (3.56) diffusive spreading can be calculated straightforwardly. Table 3.4 shows results for different time periods, and for comparison also the traveled distance L, during the same time, by an advective flow of 10 m/year. Over short time periods diffusive transport equals advective transport, but over longer time periods (and hence also over larger distances) advective flow becomes more and more important.

Table 3.4. Spreading of a point source through diffusion ($D_f = 10^{-9}\,m^2/s$), compared with advective transport ($v = 10\,m/year$).

t (s)	Diffusion, σ (cm)	Advection, L (cm)
60 (=1 min)	0.03	0.002
3600 (=1 hr)	0.27	0.11
21600 (=6 hr)	0.66	0.68
68400 (=1 day)	1.3	2.7
32×10^6 (=1 year)	25	1000

Fick's second law can be integrated for the boundary conditions:

$$c(x, t) = c_i, \quad \text{for } x > 0,\ t = 0 \quad \text{and} \quad \text{for } x = \infty,\ t > 0;$$
$$c(x, t) = c_0, \quad \text{for } x = 0,\ t > 0$$

The solution for these boundary conditions can be obtained with similar methods as applied for Equation (3.53) (Lasaga, 1998, p. 368) or via Laplace transforms (Boas, 1983, p. 676; Van Genuchten, 1981) and is:

$$c(x,t) = c_i + (c_0 - c_i)\,\mathrm{erfc}\left(\frac{x}{\sqrt{4D_e t}}\right) \tag{3.58}$$

where erfc[z] is the error function complement (i.e. 1 − erf[z], Table 3.3). This equation has been used to model field data of diffusion processes. In the Champlain Sea clay deposits, the Cl^- profiles reflect diffusion out of the clay over a period of 10,000 years since the regression of the sea (Desaulniers and Cherry, 1988). The model considered an initial homogeneous Cl^- distribution (⅓ of seawater concentration) and an upper weathered zone, where the salt is flushed. The Cl^- distribution was succesfully modeled using a diffusion coefficient D_e of $2 \times 10^{-6}\,cm^2/s$. This effective diffusion coefficient can be compared with a "free" water diffusion coefficient of about $8 \times 10^{-6}\,cm^2/s$ at 10°C that we found in Section 3.5.1.

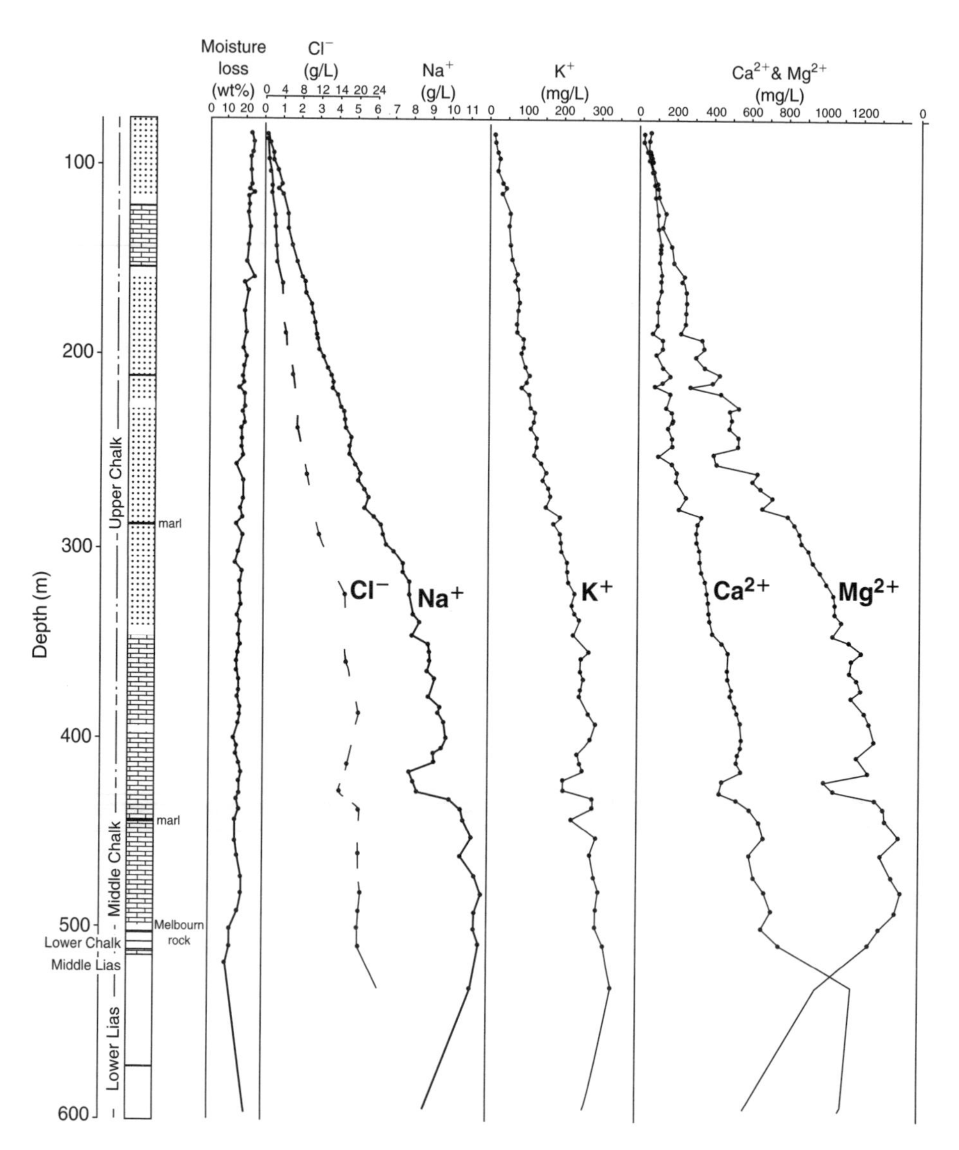

Figure 3.28. The chemical composition of pore waters in Upper Cretaceous Chalk deposits, UK. The upward decrease indicates an upward loss of ions by mainly diffusion (Bath and Edmunds, 1981).

Similar effective diffusion coefficients have been used by Volker (1961), Sjöberg et al. (1983), Groen et al. (2000), Hendry and Wassenaar (2000) and Beekman (1991) in clays where a geologic change induced a salinity jump and Cl^- diffusion occurred over periods of about 300 to 500 years.

Figure 3.28 shows the water composition in connate waters in chalk deposits (Bath and Edmunds, 1981). Upward diffusion is thought to be the main driving force behind the concentration profiles. However, the upper part of the chalk is fissured and here advective flow removes solutes, and the transport mechanism may be somewhat hybrid. Equation (3.58), assuming a diffusion coefficient D_e of $5 \times 10^{-10} m^2/s$, gives a transport time of around 3 million years in this case, since late Tertiary. Even longer diffusion times of over 50 million years were used to model Cl^- and δ^2H profiles in clays which are considered as repositories for the storage of nuclear waste (Patriarche et al., 2004).

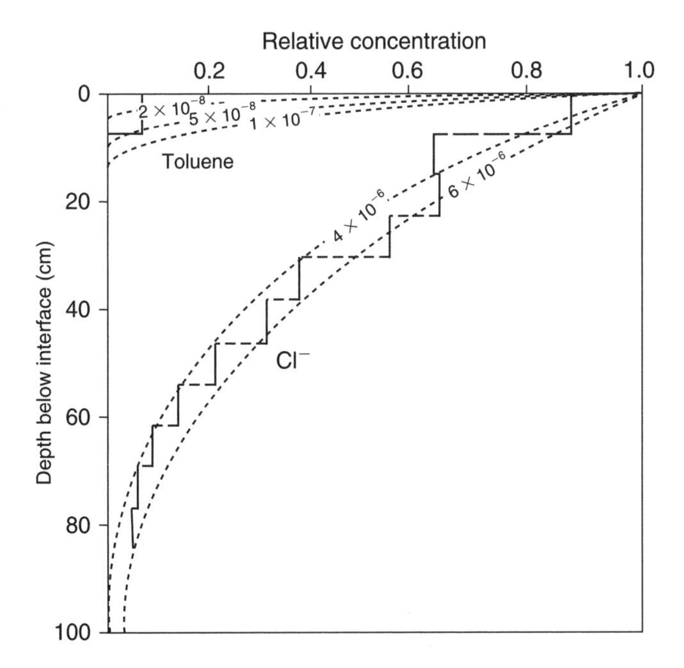

Figure 3.29. Chloride and toluene concentrations in interstitial solutions obtained from a clay core below a waste site. Simulation data for 1800 days and for indicated diffusion coefficients. Reprinted with permission from Johnson, et al., 1989. Copyright American Chemical Society.

Advective flow in clays or peats (aquicludes and aquitards) is often negligible and diffusion is then the only transport mechanism for chemicals. An example, showing diffusion of Cl^- and toluene in thick clays underlying a waste deposit, is given in Figure 3.29. The diffusion coefficients used in the simulations are indicated in the figure; the best fit for Cl^- and toluene is obtained with 5×10^{-6} and 5×10^{-8} cm²/s, respectively. For linearly retarded chemicals D_e is replaced by D_e / R, which indicates that $R \approx 100$ for toluene.

EXAMPLE 3.9. *Calculation of a diffusion profile*
Calculate the relative concentration profile of Cl^- diffusing in the clay after 1800 days (Figure 3.29). $D_f = 1.1\times10^{-9}$ m²/s for Cl^-, $\varepsilon_w = 0.5$.

ANSWER:
The effective diffusion coefficient is $D_e = D_f \cdot \varepsilon_w = 1.1\times10^{-9} \times 0.5 = 5.5\times10^{-10}$ m²/s. Time is 1.56×10^8 s. We use Equation (3.58) with $\sqrt{(4D_e t)} = 0.585$ m, $c_i = 0$ and $c_0 = 1$ mol/L. Hence:

$$c(x, 1800 \text{ days}) = 0 + (1 - 0) \times \operatorname{erfc}(x / \sqrt{(4D_e t)}) = 1 - \operatorname{erf}(x / 0.585)$$

The simple approximation for $\operatorname{erf}(z) = (1 - \exp(-4z^2 / \pi))^{0.5}$ provides sufficient accuracy here, and gives $c(0.4\,\text{m}, 1800 \text{ days}) = 0.33$ mol/L, etc.

QUESTION:
Calculate c at 0.6 m
ANSWER: 0.14 mol/L

Figure 3.29 shows that adsorption initially can slow down the diffusive flux of a pollutant. However, once all the sorption sites are filled, retardation stops, and the diffusive fluxes of conservative elements and adsorbed elements become the same. When, for example, a 1 m thick clay liner is used to act as a barrier against downward percolation of leachates from a waste site, the steady state flux may become considerable, and cause extensive pollution of groundwater in the aquifer below the site (Example 3.10).

EXAMPLE 3.10. *Diffusive flux through a clay barrier (after Johnson et al., 1989)*
Estimate the steady state flux of benzene through a 1 m thick clay liner below a $10^4\,m^2$ waste site. The concentration of benzene in the waste site is 1 g/L, the groundwater below the liner has a concentration at 0.01 g/L. The clay has a porosity $\varepsilon_w = 0.5$, the free water diffusion coefficient of benzene $D_f = 7\times10^{-6}\,cm^2/s$. Also estimate the time needed to arrive at steady state when benzene is sorbed linearly by the clay, $K_d = 2.08$.

ANSWER:
An effective diffusive coefficient is calculated from Equation (3.47): $D_e = D_f\varepsilon_w = 7\times10^{-6} \times 0.5 = 3.6\times10^{-6}\,cm^2/s$. From Fick's first law obtain a flux

$$F = -D_e \cdot \varepsilon_w \cdot (\partial c/\partial x) = -3.6\times10^{-6} \times 0.5 \times (1 - 0.01) \,/\, -100 = 1.8\times10^{-8}\ mg/s/cm^2.$$

This amounts to 56 kg benzene/yr, a quantity which has the potential to contaminate 11.2×10^9 liters of water at the EPA drinking water limit of 0.005 mg/L. Note that we multiplied the flux with porosity to obtain the flux through the open, pore space.

The maximum time necessary to reach steady state can be calculated from amount of adsorption that must occur. Adsorbed with 1 g/L and 0.01 g/L in solution are 2.1 and 0.02 g/L pore water, or the average for the whole clay liner: $1.05\times10^4 \times 0.5 = 5252$ kg benzene. Add the amount in solution, which is 2525kg, to find a total quantity of 7777 kg benzene in the clay liner (one may wonder whether this is available in the waste). This amount is carried into and through the clay in about 140 years by the steady state flux. The initial diffusion profile is concave, and consequently exhibits a larger gradient, and a larger flux; the calculated time space to reach steady state is therefore a maximal one. The time for steady state can also be calculated with an analytical solution (Johnson et al., 1989).

Diffusion also has important consequences for transport in fractured aquifers, where the main flow occurs along faults or fissures, but solutes exchange through diffusion with the stagnant water in the bulk rock. The phenomenon was described in Section 3.1 in the discussion of tritium transport in the unsaturated zone of chalk (Figure 3.1). A porous formation with these characteristics is termed a *dual porosity* medium, or a rock with *mobile–immobile* zones. Aggregated soils in which transport occurs both by advective flow along shrinkage fissures, and by diffusion into the peds, have similar transport characteristics. It is typical to observe tailing of the concentrations when a pulse injection has passed the observation point, due to back-diffusion, again out of the matrix into the channels which contain the mobile fluid. These processes generally require numerical models as applied in Chapters 6 and 10, and discussed in Chapter 11.

3.5.4 *Isotope diffusion*

For perfect gases, the diffusion coefficient of gas 1 into gas 2 is given by kinetic theory (Bird et al., 1960; Tabor, 1991):

$$D_{1,2} = \frac{\sqrt{v_1^2 + v_2^2}}{12\pi(c_1 + c_2)(a_1 + a_2)^2} \tag{3.59}$$

where v is the molecular velocity (m/s), c the concentration (mol/m^3) and a is the radius of the molecule. With equal kinetic energy $E = \frac{1}{2}mv^2$ for all the molecules, and equal radii for isotopes a and b of gas 1, we find the ratio of the diffusion coefficients for isotope tracer-diffusion in gas 2:

$$\frac{D_{1a,2}}{D_{1b,2}} = \sqrt{\frac{v_{1a}^2 + v_2^2}{v_{1b}^2 + v_2^2}} = \sqrt{\frac{\left(\frac{m_{1a} + m_2}{m_{1a}m_2}\right)}{\left(\frac{m_{1b} + m_2}{m_{1b}m_2}\right)}} = \sqrt{\frac{m_{1b}(m_{1a} + m_2)}{m_{1a}(m_{1b} + m_2)}} \tag{3.60}$$

For example, for diffusion of $^{13}CO_2$ and $^{12}CO_2$ in air, the diffusion coefficients show the proportion:

$$\frac{D_{^{12}CO_2,\,air}}{D_{^{13}CO_2,\,air}} = \sqrt{\frac{45 \times (44 + 28.8)}{44 \times (45 + 28.8)}} = 1.0044 \tag{3.61}$$

where 28.8 g/mol is the molecular weight of air (with 79% N_2 and 21% O_2). The ratio of 1.0044 indicates that the light molecule $^{12}CO_2$ will diffuse a certain distance quicker than $^{13}CO_2$. The ratio $m_1m_2 / (m_1 + m_2)$ is called the *reduced mass* $\mu_{1,2}$ (g/mol).

The diffusion coefficients for water molecules show a similar mass effect, as illustrated in Table 3.5 for water labeled with tritium and deuterium. Also, the self-diffusion coefficients of $H_2{}^{16}O$ and $H_2{}^{18}O$ show a ratio of $1.054 (= \sqrt{20/18})$ as expected for perfect gases (Tyrrell and Harris, 1984). Apparently, water behaves here according to simple kinetic theory and we can estimate the tracer-diffusion coefficient for isotopes of solute ions with Equation (3.60). For the solute ions it is necessary to include the water molecules that adhere to the solute ion (the hydration number is approximately given by the Debye-Hückel å values in 10^{-10} m, cf. Chapter 4). For example for Cl^-, $å / 10^{-10} = 3$, which gives a mass $m_{^{37}Cl} = 91$ g/mol and $\mu_{^{37}Cl,H_2O} = 15.0275$ g/mol. Similarly for $^{35}Cl^-$, we have $m_{^{35}Cl} = 89$ g/mol and $\mu_{^{35}Cl,H_2O} = 14.972$ g/mol. The resulting ratio of the diffusion coefficients is $D_{^{37}Cl,H_2O} / D_{^{35}Cl,H_2O} = (\mu_{35} / \mu_{37})^{1/2} = 0.9981$. Desaulniers et al. (1986) optimized a ratio of 0.9988 for a diffusion profile in clay, and Eggenkamp et al. (1994) obtained 0.9977 for a diffusion profile in a marine sediment.

Table 3.5. Tracer diffusion coefficients ($D_f / 10^{-9}$ m^2/s) of tritium (T) and deuterium (D) labeled water in ordinary water ($H_2{}^{16}O$) (Mills and Harris, 1976).

°C	HTO	HDO	Ratio	$(\mu_{HDO} / \mu_{HTO})^{1/2}$
5	1.272	1.295	0.982	
25	2.236	2.272	0.984	$(9.2432 / 9.4737)^{1/2} = 0.988$
45	3.474	3.532	0.984	

Beekman (1991) investigated a site flooded by seawater where Cl^- diffused over a period of 400 years downward into a freshwater clay. In 1932 a dike was built that isolated the area from the sea and a freshwater lake developed. From 1932 onwards, Cl^- therefore diffused backward into the freshwater. Figure 3.30 shows the profiles for Cl^- and $\delta^{37}Cl$ as calculated by a numerical model. The solid lines depict the situation after 400 years of downward Cl^- diffusion and give a downward concave Cl^- profile. When backward Cl^- diffusion into freshwater is included, the result is the dotted line marked 1987 (Figure 3.30). Chloride is diffusing out of the upper part and a Cl^- maximum forms.

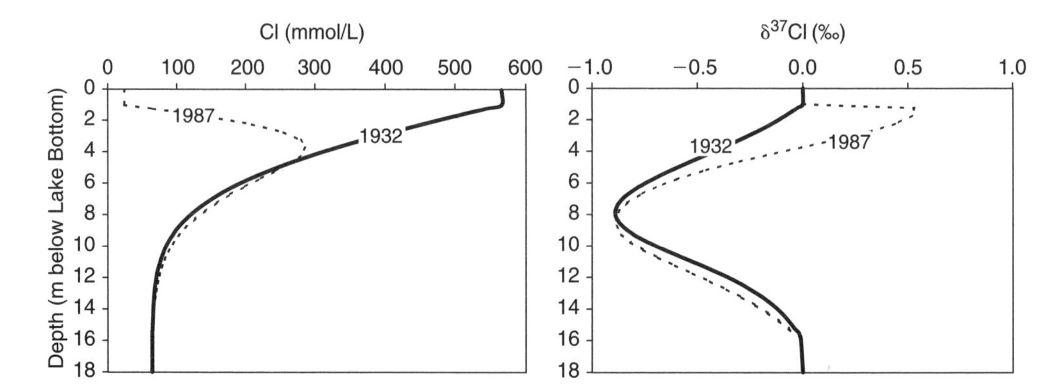

Figure 3.30. Calculated concentrations of Cl^- and $\delta^{37}Cl$ in the Markermeer sediment in 1932, at the termination of salt water diffusion. Solid lines are for 400 years of downward Cl diffusion until 1932. Dotted lines include backward Cl diffusion from 1932 to 1987.

When seawater diffuses down in the sediment, ^{37}Cl migrates slightly slower than ^{35}Cl, therefore negative $\delta^{37}Cl$ values are expected. The model isotope ratio (Figure 3.30, solid line) has zero enrichment at 0 and 18 m depth, the boundary locations where seawater and initial water are found. In between, a pronounced minimum is calculated for $\delta^{37}Cl$ at 8 m. When the Cl^- diffuses out of the sediment, ^{37}Cl is again slower than ^{35}Cl, and it becomes enriched in the remaining salt. This enrichment results in a distinct $\delta^{37}Cl$ maximum at 2 m depth (Figure 3.30, dotted line).

Figure 3.31 shows the measured profiles for Cl^- and $\delta^{37}Cl$ together with model calculated lines. Scenario A corresponds to the 1987 dotted lines in Figure 3.30. The Cl^- concentration approximates the observed concentrations for depths greater than 4 m reasonably well.

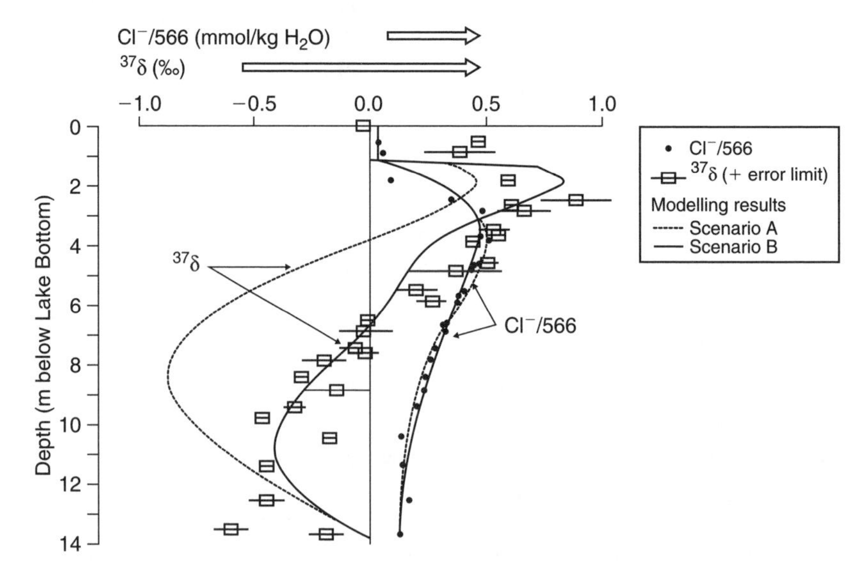

Figure 3.31. Profiles for Cl and $\delta^{37}Cl$ in the Markermeer sediment measured in 1987. Scenario A corresponds to the dotted model line in Figure 3.30. Scenario B included a sedimentation/erosion component. (Beekman, 1991; Eggenkamp, 1994).

The Cl^- concentration in the upper 4 m has decreased due to backward Cl diffusion, and the observed concentrations are also well matched by the computed line. For $\delta^{37}Cl$, the field data do show a maximum and a minimum as predicted by the model. However, the modeled minimum is much greater than measured, while the modeled maximum is smaller. Beekman (1991) concluded that the bottom sediments were homogenized during storms and gradually accumulated in the period of downward salt diffusion. Thus, more ^{37}Cl was brought into the profile than with diffusion only. The full lines, scenario B in Figure 3.31, are for a sedimentation/erosion model which combines spasmodic events and calm periods in-between when the concentration variations are smoothed out again.

EXAMPLE 3.11. *Chloride isotope fractionation during diffusion*
Calculate $\delta^{37}Cl$ at 8 m depth in the salt water diffusion profile shown in Figure 3.30. Diffusion time is 362 yr, $D_{^{35}Cl} = 8.4\times10^{-10}$ m²/s, $D_{^{37}Cl} / D_{^{35}Cl} = 0.9988$, $Cl_i = 65$ mmol/L and $Cl_0 = 566$ mmol/L. The upper meter sediment is homogenized with seawater due to storm events and bioturbation (Cl_0 extends to 1 m bsl). Both fresh water and seawater have $\delta^{37}Cl = 0$‰.

ANSWER:
The Cl^- isotope composition of seawater is 24.47% ^{37}Cl and 75.53% ^{35}Cl (Table 2.4). Thus, $^{35}Cl_i = 0.7553 \times 65 = 49.0945$ mmol/L and $^{35}Cl_0 = 0.7553 \times 566 = 427.4978$ mmol/L. We calculate the concentration after 1.14×10^{10} seconds at $x = 7$ m (corresponding to 8 m depth) with Equation (3.58), hence for $z_{35} = x / \sqrt{(4Dt)} = 7 / (4 \times 8.4\times10^{-10} \times 1.14\times10^{10})^{½} = 1.13024$. It gives, with the accurate approximation of the error function:

$$^{35}Cl = 49.0945 + (427.4978 - 49.0945) \times \mathrm{erfc}(1.13024) = 90.702 \text{ mmol/L}$$

Similarly for ^{37}Cl, with $z_{37} = z_{35} / \sqrt{0.9988} = 1.13092$:

$$^{35}Cl = 15.9055 + (138.5022 - 15.0945) \times \mathrm{erfc}(1.13092) = 29.359 \text{ mmol/L}$$

The total Cl concentration is $^{35}Cl + {}^{37}Cl = 90.7 + 29.4 = 120$ mmol/L. $\delta^{37}Cl$ is the ratio in the sample normalized to the ratio in the standard:

$$\delta^{37}Cl = \left(\frac{(^{37}Cl / {}^{35}Cl)_{sample}}{(^{37}Cl / {}^{35}Cl)_{standard}} - 1 \right) \times 1000 = \left(\frac{29.359 \ / \ 90.702}{24.47 \ / \ 75.53} - 1 \right) \times 1000 = -0.883$$

QUESTION:
Calculate $\delta^{37}Cl$ at $x = 2$ m (3 m depth)
ANSWER: $^{35}Cl = 235.77$, $^{37}Cl = 76.35$, $\delta^{37} = -0.404$‰

3.6 DISPERSION

Groundwater flowing through a sand layer is forced to move around the sediment grains as illustrated in Figure 3.32. The resulting spreading of a concentration front is called *dispersion*. There are two types of dispersion (Figure 3.32); differences in travel time along flowlines around grains cause *longitudinal* dispersion (D_L), whereas *transverse* dispersion (D_T) is due to stepover onto adjacent flowlines by diffusion. Also indicated in Figure 3.32 is the statistical nature of dispersion; the choice just before a grain to go left or right leads automatically to a probabilistic interpretation of the process exactly as for diffusion.

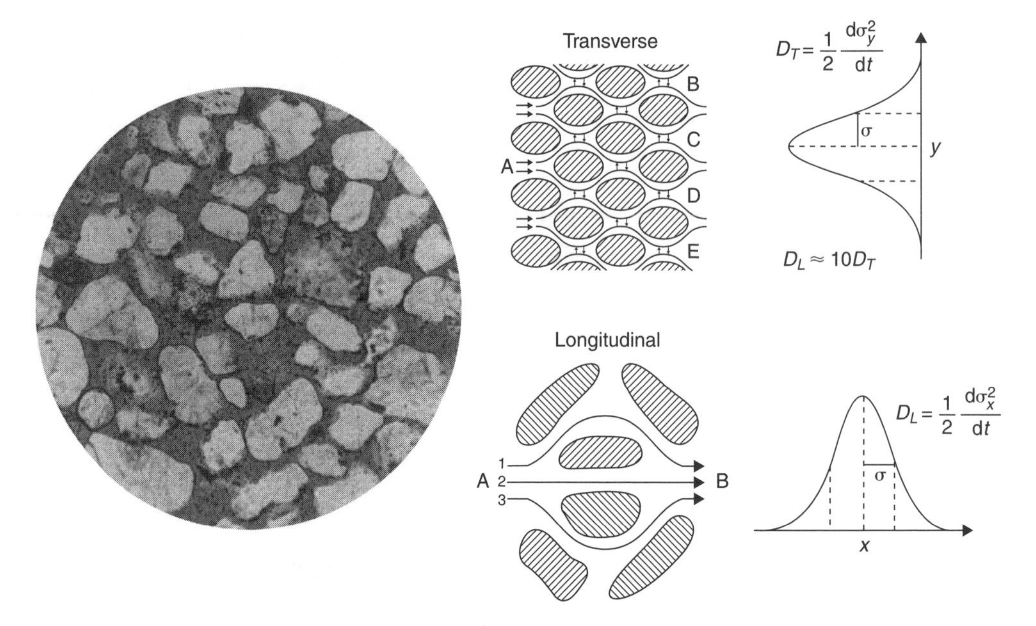

Figure 3.32. Longitudinal and transverse dispersion viewed at the microscopic scale.

While the physical process behind dispersion is different from diffusion, mathematically we can still use Fick's laws to quantify the spreading of concentration fronts. Combining the reactive transport and the diffusion equation (Figures 3.14 and 3.23) we obtain:

$$\frac{\partial c}{\partial t} = -v\frac{\partial c}{\partial x} - \frac{\partial q}{\partial t} + D_L \frac{\partial^2 c}{\partial x^2} \tag{3.62}$$

The longitudinal dispersion coefficient, D_L, is also termed the *hydrodynamic dispersion coefficient*. Equation (3.62) contains three terms on the right hand side; the first describes *A*dvective flow, the second chemical *R*eactions and the third *D*ispersion, therefore it is also called the *ARD* equation. This sequence of the three terms may be the order of their relative importance in controlling groundwater quality in most situations. The equation requires considerable mathematical background to solve under various conditions (Bear, 1972). However, as in the case of diffusion, front spreading by dispersion can be calculated with the statistical approach. Figure 3.33 shows how a Gaussian density function migrates down a flowline and is broadening due to dispersion. The dispersion can be calculated using Equation (3.53) with the difference that x_0, the initial location of the point source, is not zero, as for diffusion, but increases with the distance covered by the moving fluid, i.e. $x_0 = vt$.

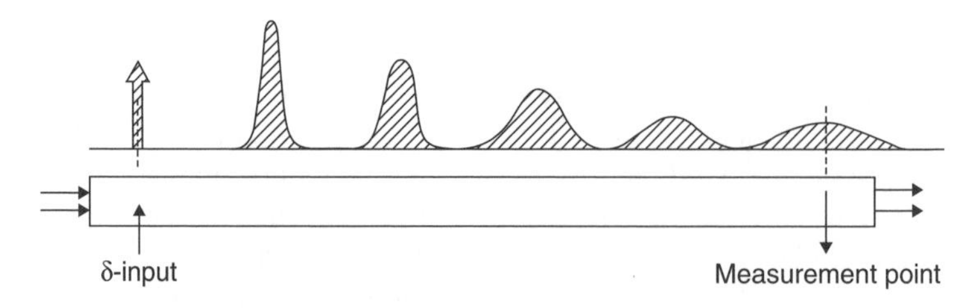

Figure 3.33. Propagation of a Gaussian density function along a flowline (Levenspiel, 1972).

Methods for determining the value of σ, and hence of D_f or D_L, from measured data, are illustrated with the Gaussian density function in Figure 3.24. A value of 2σ gives the distance comprising 68% of the original mass. When a variance σ^2 can be calculated from a series of observation points, either on time- or on distance-basis, the value of the dispersion coefficient can be calculated from Equation (3.56). A time-based variance can be recalculated to become dimensionless, and to a distance basis, or vice versa as illustrated in Example 3.12.

EXAMPLE 3.12. *Dispersion coefficient from a single shot input*

Calculate the dispersion coefficient from tracer output concentration at the end of a column measured as a function of t. This example is from Levenspiel (1972).

Time t (min)	Tracer output Concentration (g/L)
0	0
5	3
10	5
15	5
20	4
25	2
30	1
35	0

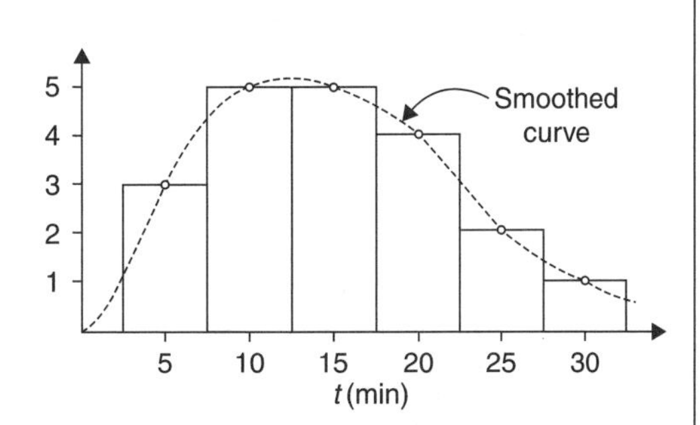

ANSWER:

The variance of the measured output distribution can be calculated as the variance of a number of measurements,

$$\frac{(\text{individual measurements})^2}{(\text{no. of observations})} - (\text{average})^2 = \sigma^2$$

Hence:

$$\sigma_t^2 = \frac{\sum t_i^2 c_i}{\sum c_i} - \bar{t}^2 = \frac{\sum t_i^2 c_i}{\sum c_i} - \left[\frac{\sum t_i c_i}{\sum c_i}\right]^2$$

$$\sum c_i = 3 + 5 + 5 + 4 + 2 + 1 = 20$$

$$\sum t_i c_i = (5 \times 3) + (10 \times 5) + \cdots + (30 \times 1) = 300 \text{ min}$$

$$\sum t_i^2 c_i = (25 \times 3) + (100 \times 5) + \cdots + (900 \times 1) = 5450 \text{ min}^2$$

Therefore:

$$\sigma_t^2 = \frac{5450}{20} - \left(\frac{300}{20}\right)^2 = 47.5 \text{ min}^2$$

The time-based variance is normalized by dividing by the square of the average arrival time:

$$\sigma^2 = \frac{\sigma_t^2}{\bar{t}^2} = \frac{47.5}{(15)^2} = 0.211$$

The variance can be expressed on distance basis: $x = v \cdot t$, for example when $v = 1$ cm/min, $x = 1 \times 15 = 15$ cm, and

$$\sigma_x^2 = \sigma_t^2 \,/\, \bar{t}^2 \cdot x^2 = 0.211 \times 225 = 47.5 \text{ cm}^2$$

Since $\sigma_x^2 = 2\, D_L\, t$, $D_L = \sigma_x^2 / (2t) = 47.5 / (2 \times 15) = 1.58$ cm²/min.

The transformation can only be made when dispersion coefficients are small (and observations obey, anyhow, the normal distribution, i.e. are symmetric around midpoint breakthrough).

In Example 3.12 the distribution of breakthrough concentrations was characterized by its variance. Temporal or spatial moments of the breakthrough distribution also serve that purpose. The temporal moments $m_0, m_1, \ldots, m_i$ are defined as:

$$m_0(L) = \int_0^\infty t^0 c_{L,t}\, \mathrm{d}t = \int_0^\infty c_{L,t}\, \mathrm{d}t \tag{3.63a}$$

$$m_1(L) = \int_0^\infty t^1 c_{L,t}\, \mathrm{d}t \tag{3.63b}$$

or in general:

$$m_i(L) = \int_0^\infty t^i c_{L,t}\, \mathrm{d}t \tag{3.63c}$$

The zeroth moment $m_0(L)$ is the total mass of chemical, integrated over time at point L (mass · s/L). The first moment divided by the zeroth moment gives the mean arrival time $\bar{t} = m_1 / m_0$ (s). In Example 3.12, we calculated the variance $\sigma_t^2 = m_2 / m_0 - (m_1 / m_0)^2$ (s²). The third moment gives the skewness of the distribution. Besides characterizing empirical distributions, moment analysis is used as a mathematical tool to derive system properties in the Laplace space, and to compare physical behavior such as dispersion, interaction among mobile and stagnant zones, and kinetic chemical reactions (Valocchi, 1985; Goltz and Roberts, 1987; Harvey and Gorelick, 1995; Garabedian et al., 1991).

3.6.1 *Column breakthrough curves*

When a conservative tracer is injected into a column at concentration $c = c_0$, the front will move with the average water flow velocity through the column. At the same time the front disperses, as illustrated in Figure 3.34. In the figure, the concentrations are scaled from 0 to 1, by dividing all values with c_0. The concentrations are found by integrating Equation (3.62) for $\mathrm{d}q = 0$ which becomes:

$$\left(\frac{\partial c}{\partial t}\right)_x = -v\left(\frac{\partial c}{\partial x}\right)_t + D_L\left(\frac{\partial^2 c}{\partial x^2}\right)_t \tag{3.64}$$

To facilitate the solution, we make the column infinite and admit a non-zero initial concentration c_i:

$$c = c_i \quad \text{for } x > 0;\ t = 0, \quad \text{and} \quad \text{for } x = \infty;\ t > 0$$

$$c = c_0 \quad \text{for } x \le 0;\ t = 0, \quad \text{and} \quad \text{for } x = -\infty;\ t > 0$$

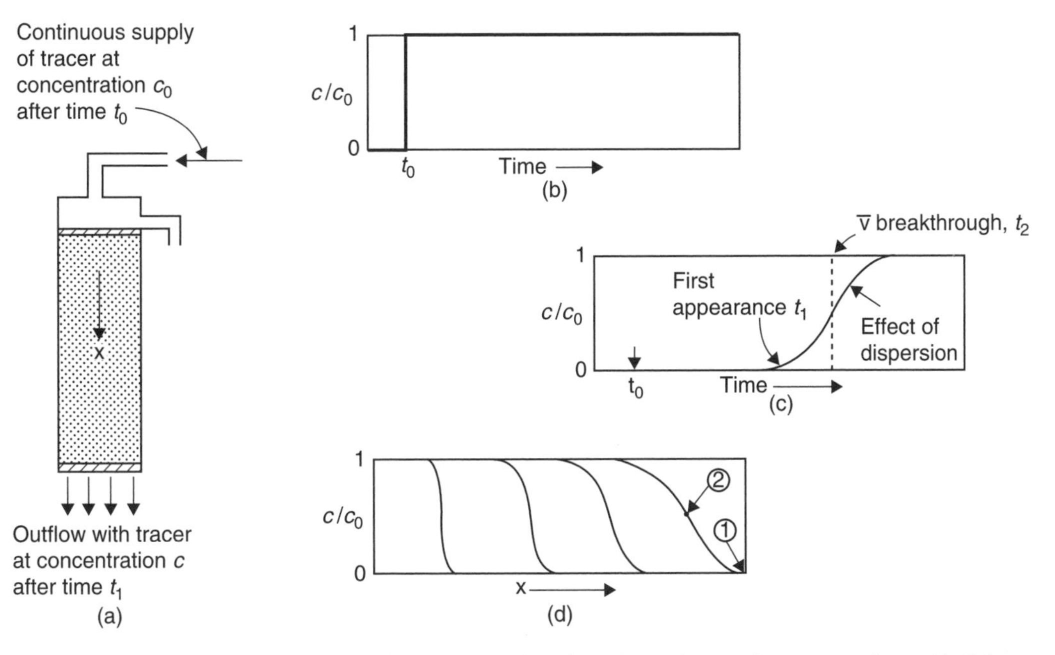

Figure 3.34. Longitudinal dispersion of a tracer passing through a column of porous medium. (a) Column with steady flow and continuous supply of tracer after time t_0; (b) step-function-type tracer input; (c) relative tracer concentration in outflow from column (dashed line indicates plug flow condition and solid line illustrates effect of mechanical dispersion and molecular diffusion); (d) concentration profile in the column at various times. Reprinted with permission from Freeze and Cherry, 1979. Copyright Prentice Hall, Englewood Cliffs, New Jersey.

The solution is similar to Equation (3.58), but x is taken relative to the advective front $v \cdot t$:

$$c(x,t) = c_i + \frac{(c_0 - c_i)}{2} \mathrm{erfc}\left(\frac{x - vt}{\sqrt{4D_L t}} \right) \tag{3.65}$$

Note that erfc$[-z] = 1 + \mathrm{erf}[z]$, so that Equation (3.65) is symmetric around $x_{0.5} = vt$. The concentrations from c_i to $(c_0 - c_i) / 2$ run ahead of the average front position at $x_{0.5} = vt$, and the argument of the error function complement is positive. The concentrations from $(c_0 - c_i) / 2$ to c_0, on the other hand, lag behind $x_{0.5}$, and the argument is negative.

EXAMPLE 3.13. *Front dispersion in a column*
Calculate the Cl^- concentration front in a column injected with a 1 mmol/L NaCl solution. The water velocity is 2.8×10^{-4} cm/s, $D_L = 10^{-4}$ cm^2/s; the initial Cl^- concentration in the column $c_i = 0$ mmol/L. Calculate the concentrations after 5 hr injection time at the advective front, and at $x = vt \pm \sqrt{2D_L t}$.

ANSWER:
The advective front has arrived at $v \cdot t = 2.8 \times 10^{-4} \times (5 \times 60 \times 60) = 5.04$ cm. The concentration at this point, where $x = vt$, is ½ erfc[0] = 0.5 mmol Cl^-/L (clearly just half of initial and input concentrations).

The concentration at $x - vt = \sqrt{2D_L t}$ is:

$$0.5\ \mathrm{erfc}\left[\sqrt{2D_L t}\ /\ \sqrt{4D_L t}\right] = 0.5 \times \left(1 - \mathrm{erf}\left[\sqrt{½}\right]\right) = 0.5 - 0.5 \times 0.68 = 0.16$$

where the value of $\mathrm{erf}\left[\sqrt{½}\right]$, as found from Table 3.3, is 0.68.
The concentration at $x - vt = -\sqrt{(2D_L t)}$ is similarly:

$$0.5\ \mathrm{erfc}\left[-\sqrt{2D_L t}\ /\ \sqrt{4D_L t}\right] = 0.5 \times \left(1 + \mathrm{erf}\left[\sqrt{½}\right]\right) = 0.5 + 0.5 \times 0.68 = 0.84$$

The spread between $c = 0.16$ and $c = 0.84$ is obtained over a length exactly equal to 2σ, and encompasses 68% of the concentrations ($0.84 - 0.16 = 0.68$) around the midpoint breakthrough, as illustrated in Figure 3.35. The distances from the midpoint breakthrough are $\pm\sqrt{2D_L t} = \pm 1.9$ cm.

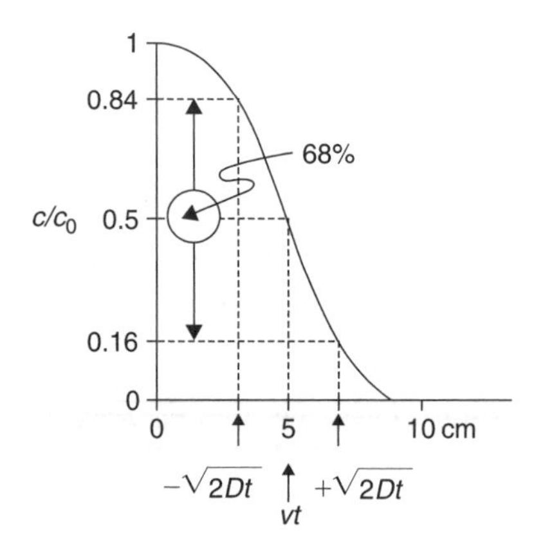

Figure 3.35. Dispersion at a concentration front and its relation to the variance.

QUESTION:
Calculate the concentration at $x = \infty$ in the column, using Equation (3.65).
ANSWER: $\mathrm{erfc}(\infty) = 0$, hence $c(\infty, t) = c_i = 0$.

The solution for a dispersion front, given in Equation (3.65), is for an infinitely long column, $-\infty < x < \infty$, and for *resident* concentrations, c_r, within the column. If we instead consider a semi-infinite column from $x = 0$ to $x = \infty$, and define a constant concentration boundary at $x = 0$ ($c_r(0, t \geq 0) = c_0$), the concentrations in the column are (Lapidus and Amundson, 1952):

$$c_r(x,t) = c_i + \frac{(c_0 - c_i)}{2}\left[\mathrm{erfc}\left(\frac{x - vt}{\sqrt{4D_L t}}\right) + \exp\left(\frac{vx}{D_L}\right)\mathrm{erfc}\left(\frac{x + vt}{\sqrt{4D_L t}}\right)\right] \qquad (3.66)$$

This equation adds an additional term to (3.65) that describes the diffusive flux at the entrance of the column. Resident concentrations in the column can be obtained using probes that monitor radioactive or color tracers, or by dissecting the column when the front has penetrated a certain length (Bond and Phillips, 1990). However, it is more common to collect the effluent at the column outlet.

The concentrations obtained in this way are *flux averaged* and differ from the resident concentration in that dispersion into or out of the column is discontinued. For a flux vc_0 injected in a column initially free of the chemical, the flux into the column, at the inlet, consists of an advective part $vc_{r,x=0}$ and a dispersive part $-D_L(\partial c_r / \partial x)_{x=0}$:

$$vc_0 = vc_{r,x=0} - D_L \left(\frac{\partial c_r}{\partial x} \right)_{x=0} \tag{3.67}$$

Consequently, the resident c_r is smaller than the flux-averaged concentration c_0:

$$c_r = c_0 + \frac{D_L}{v} \left(\frac{\partial c_r}{\partial x} \right) \tag{3.68}$$

(Note that $(\partial c_r / \partial x)$ is negative because initially the concentration in the column was zero.)

Now, we can define a flux averaged concentration at the inlet of the column ($c_f(0,t) = c_0$) and then Equation (3.66) will give the flux averaged concentration $c_f(x,t)$ inside the column (Kreft and Zuber, 1978; Van Genuchten and Parker, 1984; it is assumed that $\partial c_r / \partial x = \partial c_f / \partial x$). But, these flux averaged concentrations are for effluent that is sampled at the end of the column. Therefore, although Equation (3.66) was derived for a semi-infinite column, it is valid as well for a short laboratory column with flux boundary conditions at the inlet and outlet points.

Equation (3.66) can be used to obtain values for the dispersion coefficient from breakthrough experiments. An excellent program, CXTFIT2 by Toride et al. (1995), can do the job with a least squares fit and can be obtained from the US Soil Salinity Lab, Riverside, www.ussl.ars.usda.gov. Sometimes it is advantageous to have a visual fit as well, when some data points are considered less reliable. A visual fit can be obtained by programming (3.66) in a spreadsheet program, and comparing results for various D_L with observed concentrations. For retarded chemicals that show linear retardation, Equations (3.65) and (3.66) can still be used, replacing t with t / R. For non-linear retardation a numerical solution of Equation (3.62) is required as presented in Chapter 11.

3.6.2 *Dispersion coefficients and dispersivity*

In Figure 3.36 values of dispersion coefficients are plotted as function of the non-dimensional Peclet number, *Pe* (Bear, 1972). The Peclet number was defined for this figure as:

$$Pe = vd / D_f \tag{3.69}$$

where d is diameter of the particles packed in the column (m). By varying the water velocity v, the Peclet number was varied for a given packing and by injecting a tracer the dispersion coefficient was determined for the Peclet number. Figure 3.36 plots the ratio of the dispersion-coefficient D_L over the diffusion coefficient D_f in pure water on the vertical axis. Two domains can be discerned in the figure (Bear, 1972; Plumb and Whitaker, 1988). At low Peclet numbers (=low water velocity), the ratio of D_L / D_f is constant. At intermediate Peclet numbers from 1 to 10^5, the value of D_L / D_f increases linearly with the Peclet number.

The interpretation of the dispersion domains is as follows. At a $Pe < 0.5$ (Figure 3.36) the ratio $D_L / D_f = 10^{-0.3}$, or $D_L = D_f / 2$. This value corresponds with the effective diffusion coefficient that we found when correcting the free water diffusion coefficient for the tortuosity of a sediment.

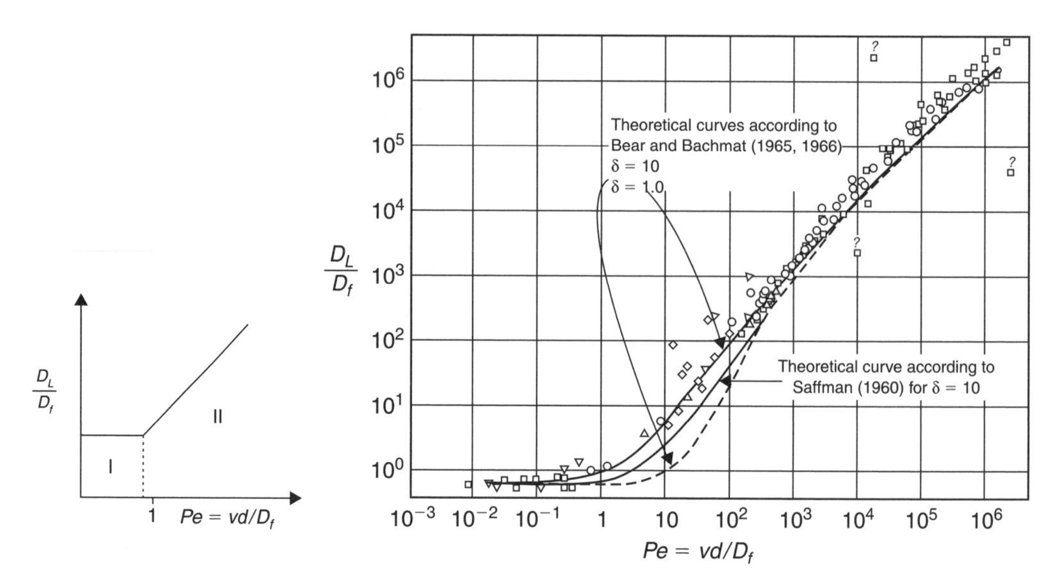

Figure 3.36. A plot of the ratio of dispersion coefficient/diffusion coefficient versus Peclet number in packed laboratory columns (Bear, 1972).

At a $Pe > 1$, the ratio D_L / D_f increases in the proportion 1 : 1 with the Peclet number and therefore:

$$D_L = D_f \cdot Pe = dv \tag{3.70}$$

For low Peclet numbers where D_L is constant, it follows that $\sigma^2 = 2D_L t = \text{constant} \times t$, or $\sigma^2 \equiv$ constant/v (over a given travel length). The variance increases with time, or equivalently, the front spreading at the column end increases with decreasing velocity. In other words, diffusion is the dominant process. The upper boundary for diffusion controlled spreading is found at a Peclet number of 0.5. For d values in the range 0.4 to 2 mm (medium to coarse sands), a flow velocity lower than 37 to 8 m/yr, respectively, will result in diffusion-controlled spreading.

For higher Peclet numbers, or for flow velocities higher than about 50 m/yr, the variance $\sigma^2 = 2D_L t = 2d \cdot vt = 2dx$, or $\sigma^2 \equiv$ constant (over a given travel length). Front spreading is now independent of the water velocity and depends only on the travelled length. In other words, dispersion becomes a characteristic property of the porous medium denoted by the *dispersivity*:

$$\alpha_L = D_L / v \tag{3.71}$$

where α_L is the dispersivity (m). It can be subdivided into a longitudinal (α_L) and transverse (α_T) dispersivity. The variance is related to dispersivity by:

$$\sigma^2 = 2D_L t = 2\alpha_L \cdot vt = 2\alpha_L \cdot x \tag{3.72}$$

This important formula is very efficient for calculating the spread of pollutant fronts.

The contributions of diffusion and dispersion are combined in the hydrodynamic, longitudinal dispersion coefficient:

$$D_L = D_e + \alpha_L v \tag{3.73}$$

The dispersivity in columns, as shown in Figure 3.36, is equal to the representative grain-diameter used in packing the column. When aquifer sediments are used in the column, the grain-diameter is taken as d_{10}, the diameter below which 10% of particles fall, and Perkins and Johnston (1963) suggest the relationship:

$$\alpha_L = 3.5\, d_{10} \tag{3.74}$$

Here 3.5 is a shape factor, which increases somewhat with smaller grain size. In aquifers the dispersivity is generally much higher than in packed laboratory columns, and the diffusion contribution in Equation (3.73) becomes already negligible at a water velocity higher than 1 m/yr.

QUESTIONS:
Compare the effect of dispersion in c/x plots of a pulse injection of two chemicals, one conservative, and the other with linear retardation R, at times t_1 and $t_1 \times R$?

ANSWER: The front of the retarded chemical arrives R times later, both the width and the variance of the pulse of the retarded chemical are R times smaller. For small pulses this results in some broadening of the retarded pulse and also in some peak reduction.

And the same question, comparing a c/t plot?

ANSWER: The width and the variance of the pulses are exactly the same, only shifted in time by R. However, small retarded pulses will show broadening and peak reduction.

Are the answers different for diffusion, cq. dispersion dominated spreading?

ANSWER: No

Calculate the number of grains in a column packed with 0.5 kg, 2 mm diameter quartz sand, density 2.65 g/cm^3?

ANSWER: 45044

3.6.3 *Macrodispersivity*

Looking at the Cl^- concentrations downstream the Vejen waste site in Figures 3.2 and 3.21, we have noted the irregular distribution pattern, even when the concentrations are averaged over depth. The larger spreading in aquifers, compared to laboratory columns, is due to the heterogeneous structure of natural sediments with alternating sands of different hydraulic conductivity, and with intercalated layers of clay, silt or gravel. It is no longer the individual grain size, but rather the variation in flow length around low-conductivity bodies that determines the dispersivity. This effect is known as *macrodispersivity*.

We have considered dispersion as a statistical process, where flow is redirected around obstacles so many times, that a normal (or Gaussian) distribution of flowpaths is obtained. Several thousands obstacles are required to approach a normal distribution and this is easily reached in a column with homogeneously packed grains, but in an aquifer this is not the case.

Figure 3.37 illustrates the effect of geological structures on the Dirac delta function ("the single shot input"), and also shows how the sampling method influences the measured field dispersivity. When filter screens are longer than 1 m, mixing of water from different layers is inevitable and dispersivity is increased. Field dispersivities are therefore lowest when small filters or depth specific sampling techniques are used (Schröter, 1984). It is possible to obtain dispersivity values in an aquifer that are as small as in column experiments, when a point sampling method is employed and a single flowline is monitored. Pickens and Grisak (1981) found a dispersivity of 7 mm which remained constant over distances from 0.36 to 4 m. Taylor and Howard (1987), who used gamma-ray counting of a radiotracer in dry access tubes, even found a constant dispersivity of 1.6 cm over a distance of up to 40 m.

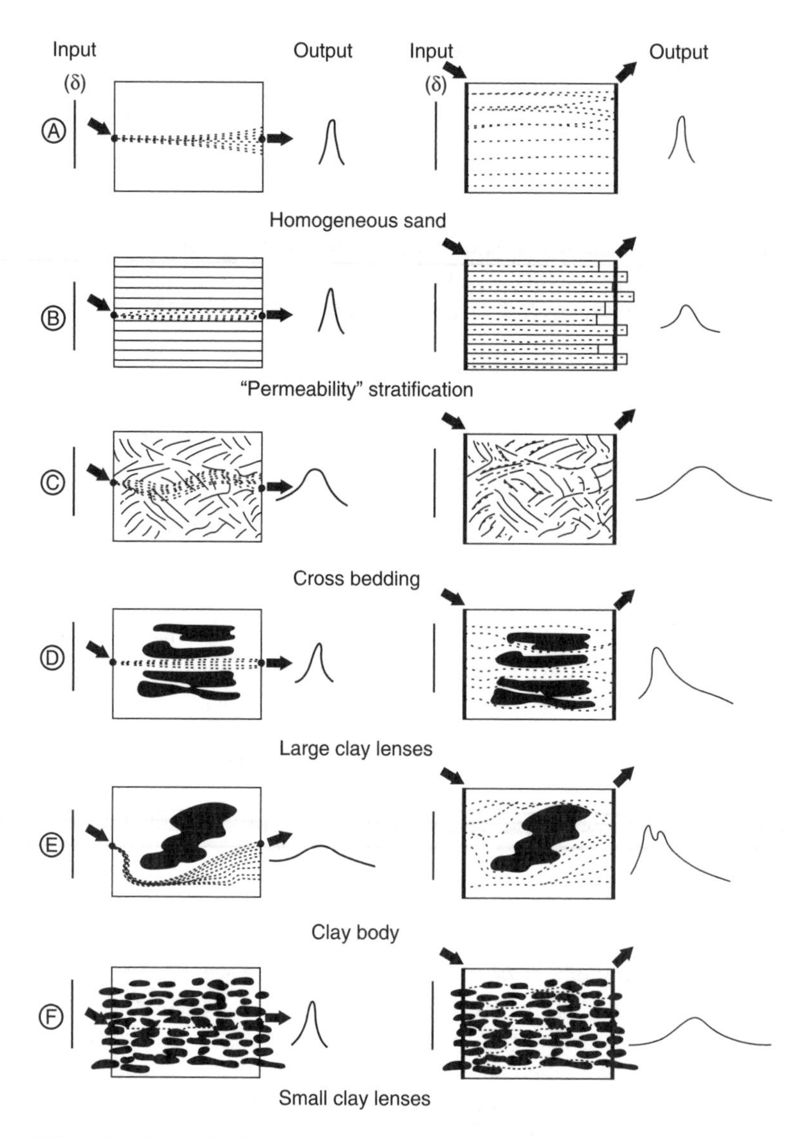

Figure 3.37. Effect of sedimentological structures on dispersion of a shot inpulse. At left: point specific sampling, at right: depth integrated sampling in the same aquifer.

Depth integrated sampling includes the effects of the geological heterogeneity, and is thus more apt to give an overall picture of transport properties of an aquifer. Figure 3.38 shows that depth integrated dispersivity in aquifers may be orders of magnitude larger than in columns. Even more important is that dispersivity appears to increase with distance. In other words it becomes a *scale dependent* property, while before dispersivity was considered as a constant. The general trend in Figure 3.38 is that the macrodispersivity is about 10% of the traveled distance:

$$A_{11} = 0.1\,x \tag{3.75}$$

where A_{11} is the longitudinal macrodispersivity. However, the increase lessens for travel distances above 1000 m.

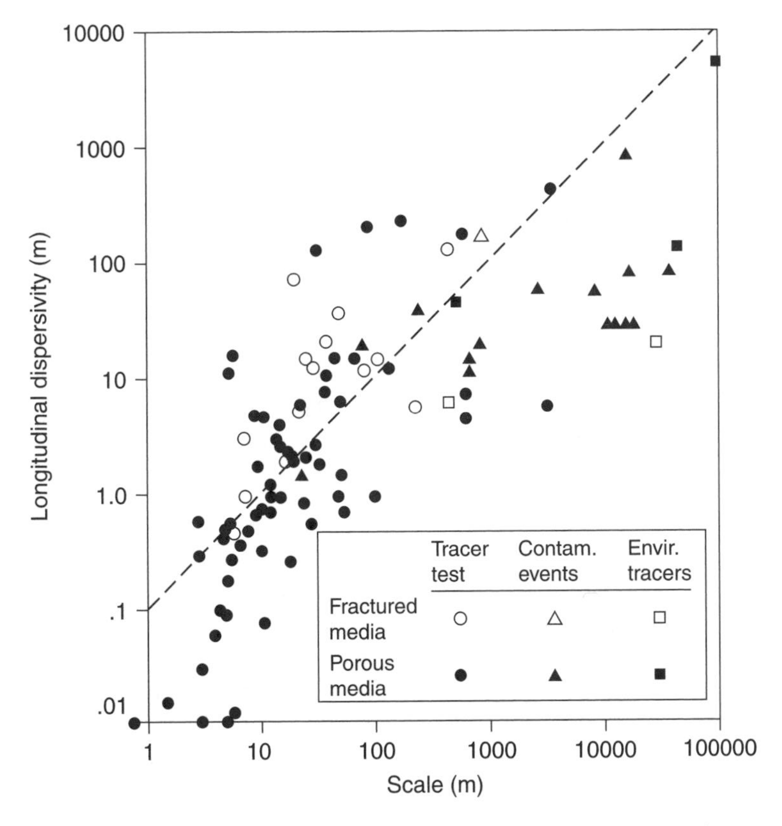

Figure 3.38. Values for dispersivities obtained in field experiments (Gelhar et al., 1985).

If we combine Equations (3.75) and (3.72) we obtain:

$$\sigma^2 = 2A_{11} \cdot x = 2 \cdot 0.1x \cdot x = 0.2x^2 \tag{3.76}$$

for the variance of a pollutant front in aquifers (cf. Example 3.14).

EXAMPLE 3.14. *Pollutant spreading during transport in an aquifer*
An aquifer has thickness $D = 50$ m; $P = 0.3$ m/yr; porosity $\varepsilon_w = 0.3$. Calculate the area which contains 68% of a point source pollutant, infiltrated 1 km from the divide, after 1 km flow. Also estimate for the point at 2 km from the divide, the time space during which 68% of the pollutant mass with highest concentrations passes.

ANSWER:
Assume that dispersivity is 10% of travel length, and obtain $\sigma^2 = 0.2x^2$ (Equation 3.76); with $x = 1000$ m, $\sigma = 447$ m. Hence we estimate that 68% of our pollutant is within 2447 and 1553 m (from the divide). We next want the time period during which the highest concentrations pass, covering 68% of the pollutant mass. This is the period between the arrival time of $(\sigma + x)$ at 2000 m, and the time when $(x - \sigma)$ passes the point. Now, our estimated $\sigma = x\sqrt{0.2}$, and we find that $x = 2000 / (1 + \sqrt{0.2}) = 1382$ m. We use Equation (3.8) and find that this distance is attained in $50 \times \ln[1382 / 1000] = 16.2$ years. Similarly, we find from $x - \sigma = 2000$, that $x = 3618$ m, and the time necessary is 64.3 years. Hence the time period for 68% of pollutant mass, with highest concentrations, to pass the 2 km point is $64.3 - 16.2 = 48.1$ years.

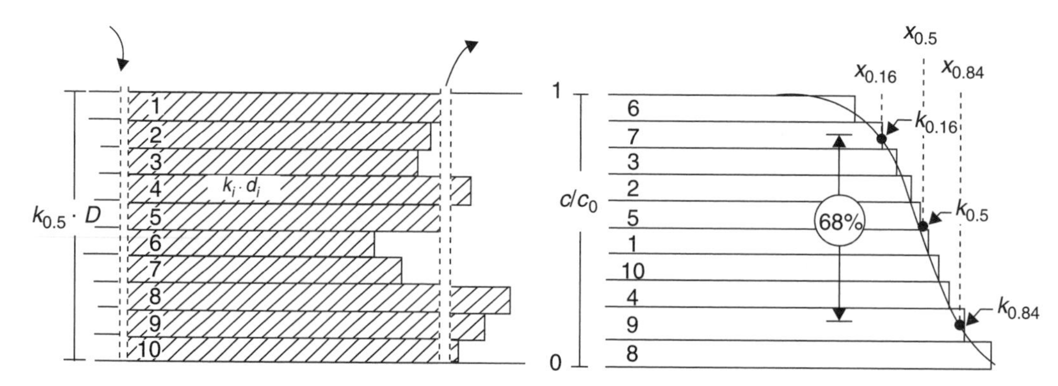

Figure 3.39. A hydraulic-conductivity stratified aquifer in which the layers contribute successively to tracer breakthrough (left); ordering the layers according to hydraulic conductivity shows more clearly that travel distance follows the conductivity distribution (right).

The larger field dispersivity is a result of the variation in arrival time at the observation well, which means that the hydraulic conductivity distribution in the aquifer lies at the heart of the process. Mercado (1967; 1984) developed the concept of a conductivity stratified aquifer. An aquifer, consisting of stacked layers with different hydraulic conductivities, is depicted in Figure 3.39. Assume that flow follows the layers only (transverse dispersion is absent). Tracer from the (fully penetrating) injection well arrives at the observation well in a stepwise fashion depending on the hydraulic conductivity of the layer. In Figure 3.39 layers 1, 4, 8, 9 and 10 are contributing tracer to the observation well, and layer 5 has just started to contribute. We may reshuffle the layers, and order them from top to bottom by increasing hydraulic conductivity (Figure 3.39, right). This orders at the same time the distance traveled by the chemical according to the conductivity distribution. An average transmissivity is obtained from:

$$k_{0.5}D = \sum k_i \, d_i \tag{3.77}$$

where the individual layers have thickness d_i and conductivity k_i, and D is the total thickness of the aquifer, with average hydraulic conductivity $k_{0.5}$. It is assumed that the porosity ε_w is equal for all the layers (cf. Equation 3.14). Now suppose that hydraulic conductivity variations among the layers follow a normal distribution, with standard deviation:

$$\sigma_k = k_{0.84} - k_{0.5} \tag{3.78}$$

where $k_{0.84}$ is the hydraulic conductivity which is larger than 84% of all values. According to Darcy's law, the average distance covered by tracer is:

$$x_{0.5} = k_{0.5} \, (-\mathrm{d}h \,/\, \mathrm{d}x \cdot \Delta t \,/\, \varepsilon) \tag{3.79}$$

while the distance for 84% of the layers is less than

$$x_{0.84} = k_{0.84} \, (-\mathrm{d}h \,/\, \mathrm{d}x \cdot \Delta t \,/\, \varepsilon) \tag{3.80}$$

If the potential gradient $(-\mathrm{d}h \,/\, \mathrm{d}x)$ and the porosity ε_w are equal for all the layers, then

$$x_{0.84} - x_{0.5} = (k_{0.84} - k_{0.5}) \cdot x_{0.5} \,/\, k_{0.5} = \sigma_k \cdot x_{0.5} \,/\, k_{0.5} \tag{3.81}$$

The tracer front is spread according to the distances covered in the individual layers, and the spread is related to the dispersivity:

$$\sigma_x = x_{0.84} - x_{0.5} = \sqrt{2A_{11}\, x_{0.5}} \tag{3.82}$$

Combining (3.81) and (3.82) yields the longitudinal dispersivity as:

$$A_{11} = \frac{1}{2}\left(\frac{\sigma_k}{k_{0.5}}\right)^2 x_{0.5} \tag{3.83}$$

Recall that $x_{0.5}$ is the distance covered by the tracer in half of the layers, which is equal to vt. Equation (3.83) therefore suggests that dispersivity increases linearly with distance, as also was indicated by Figure 3.38. If we assume that the hydraulic conductivity is log-normally distributed we obtain in a similar way:

$$A_{11} = \tfrac{1}{2}(\exp[\sigma_{\ln(k)}] - 1)^2\, x_{0.5} \tag{3.84}$$

where $\sigma_{\ln(k)}$ is the standard deviation of the log-normally distributed conductivity.

It is of interest that the value for $\frac{1}{2}(\sigma_k / k_{0.5})^2$ can be around 0.1 (Pickens and Grisak, 1981; Molz et al., 1983; random sample of Smith, 1981), conform with Equation (3.75). For log-normally distributed conductivities, the value of $\sigma_{\ln(k)}$ is 0.4 for the Borden aquifer (Sudicky, 1986) and 0.49 for the Cape Cod aquifer (Garabedian et al., 1991), which give $A_{11} = 0.1x_{0.5}$ and $0.2x_{0.5}$, respectively.

A remarkable point of the conductivity stratified aquifer is that dispersion diminishes again when flow direction is inverted (a tracer would return in the same step form during pumping as it was injected in the well). This is different from the diffusion concept, where spreading increases with the square root of time irrespective of flow inversion. However, the Mercado concept neglects that the layers have limited extent in the real aquifer and that mixing occurs where a high conductivity layer ends. Consequently, the increase of the dispersivity with distance becomes smaller and will level off to a constant value. More intricate models have been developed in which horizontal space variability is allowed, as well as transverse dispersion (Gelhar and Axness, 1983; Dagan, 1989). The final, asymptotic value of the macrodispersivity can be well approximated by (Gelhar, 1997):

$$A_{11} = (\sigma_{\ln(k)})^2 \cdot \lambda \tag{3.85}$$

where $(\sigma_{\ln(k)})^2$ is the variance of the log-normally distributed conductivities, and λ is the correlation length of the conductivity. The correlation length indicates the continuity of the layers and can be obtained from spatially distributed data (De Marsily, 1986; Domenico and Schwartz, 1997). Correlation lengths of 2 m, observed in the Borden and Cape Cod aquifers, lead to values of A_{11} of 0.3 and 0.6 m, respectively. In these cases, the asymptotic dispersivity is attained already within a flowlength of 10 m. The Borden and Cape Cod studies are for single aquifers and flow-paths of a few 100 m. Larger units will have a larger macrodispersivity and a larger correlation length, and so on and on, which can be introduced in flow models (Elfeki et al., 1999; Rubin and Bellin, 1998).

Transverse dispersivity has been less studied, but it is smaller than the longitudinal dispersivity. Measurements from tracer studies indicate that transverse dispersivity is about 10% of the longitudinal dispersivity in the bedding plane, and about 1% at straight angles, corresponding to transverse horizontal and vertical dispersivity, respectively (Gelhar, 1997). Klenk and Grathwohl (2002) found transverse vertical dispersivities that were determined mostly by diffusion.

EXAMPLE 3.15. *Longitudinal dispersivity in the Borden aquifer*
A point injection of a number of chemicals in an 8 m thick aquifer was followed over two years with a dense network of depth specific samplers (Mackay et al., 1986). Figure 3.40 shows the *depth averaged* Cl-cloud after 462 days, with a midpoint at 41 m from the injection point. Estimate the longitudinal dispersivity.

ANSWER:
The cross section suggests approximate Gaussian behaviour, and we may apply estimation techniques from Figure 3.24 in addition to a specific calculation of variance as in Example 3.12. If we take the value of (0.61 · top) as border of the 68% area which encompasses 2σ, we obtain $2\sigma = 9$ m, and $A_{11} = 0.25$ m.

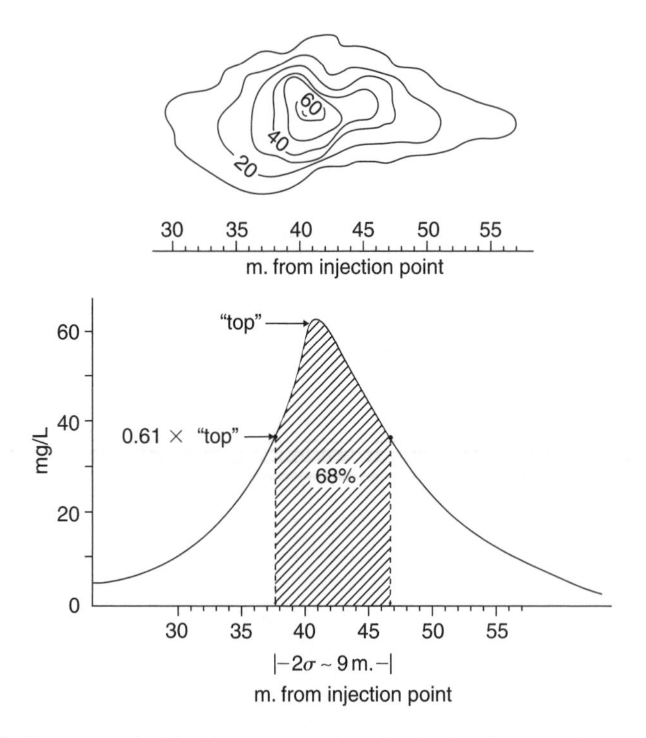

Figure 3.40. Vertically averaged chloride concentrations in the Borden experiment, 462 days after injection, and cross section over the plume. (Plume delineation after Sudicky, 1986).

Calculating the variance as in Example 3.12 gives:

$$\sigma^2 = \frac{\sum x_i^2 c_i}{\sum c_i} - \left(\frac{\sum x_i c_i}{\sum c_i}\right)^2 = 25 \text{ m}^2 = 2D_L t$$

where x_i is the distance (m) traveled by concentration c_i.

With $\sigma^2 = 2A_{11} \cdot vt = 2A_{11} \cdot 41$, we obtain $A_{11} = 0.3$ m. This value should be compared with 0.36 m, given by Freyberg (1986). It is of interest to show here the development of the longitudinal spread σ_{11}^2 and the transverse spread σ_{22}^2 as observed when the plume was transported with groundwater flow. The dashed curves are calibration fits by Freyberg (1986) (note that the longitudinal variance increases linearly with time, and hence with travel distance; the asymptotic value of the macrodispersivity is rapidly reached), and the solid lines indicate model calculations based on hydraulic conductivity variations in the aquifer by Sudicky (1986).

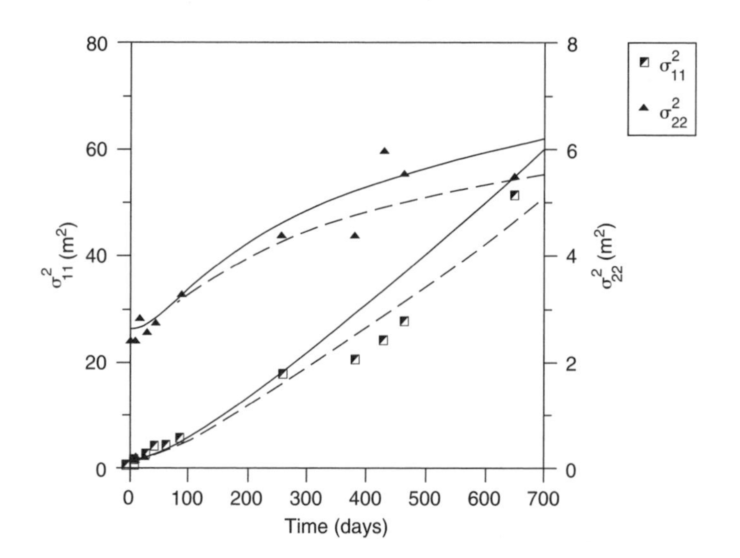

Figure 3.41. Development of longitudinal and transverse variance of the spread of Cl^-, injected in the Borden aquifer (Sudicky, 1986).

QUESTION:
The Vejen waste site is 50 m wide and is located 1 km from the Vejen river. Estimate the dilution of waste leachate.

ANSWER: 20 times (e.g. 1000 mg Cl^-/L will dilute to 50 mg/L in the seepage zone).

PROBLEMS

3.1. Calculate the downward velocity in chalk, when $P = 0.3$ m/yr, and $\varepsilon_w = 0.1$. Compare results with Figure 3.2.

3.2. An aquifer has thickness $D = 50$ m; $P = 0.3$ m/yr; $\varepsilon_w = 0.3$. At 900–1100 m from the divide infiltrates polluted water in the aquifer, and a private well at 2000 m from the divide may be affected. Calculate the thickness of the tongue of polluted water, its mean depth, and arrival time at the well point.

3.3. Draw a sketch of the hydrological situation described in Problem 3.2.
   a. Calculate pore water velocity and Darcy velocity in the aquifer at 1000 and 2000 m.
   b. Calculate the travel time of water between these two points.
   c. What is the age of water at $d = 10, 20, 30, 40$ m below the phreatic surface?
   d. Compare the result of c. with the formula $t = d \cdot P / \varepsilon_w$ which is valid for an unsaturated zone (but gives approximate results for the saturated zone as well).

3.4. Derive the time/distance and time/depth relationships for groundwater in a uniform phreatic aquifer with radial outward flow (e.g. below a hill).

3.5. Derive the time/distance relationship for groundwater in a uniform phreatic aquifer towards a pumping well with a radius of influence of r m.

3.6. A municipal well in a 100 m thick phreatic aquifer, $\varepsilon = 0.3$, $P = 0.3$ m/yr, has an influence up to $r = 5$ km. Calculate the traveltime for water from a factory situated at 1 km from the well.

3.7. Consider an impermeable clay layer that divides the aquifer in a phreatic part and a confined part. The flow pattern in the confined part remains intact when the confining layer is strictly impermeable (Figure 3.42). For the upper, phreatic aquifer the Equations (3.6) to (3.10) apply again. For the confined part of the aquifer the age distribution with depth is conserved. The flowtime at point x^* in the confined part is (see Figure 3.42 for explanation of symbols):

$$t = \frac{D\varepsilon_w}{P} \ln\left(\frac{D}{D - d - D^*}\right) + \frac{(x^* + x_0 - x)(D - D^*)\varepsilon_w}{Px_0}$$

Redraw the Figure on transparant paper.
a. Draw the hydraulic gradients in the upper and lower aquifer.

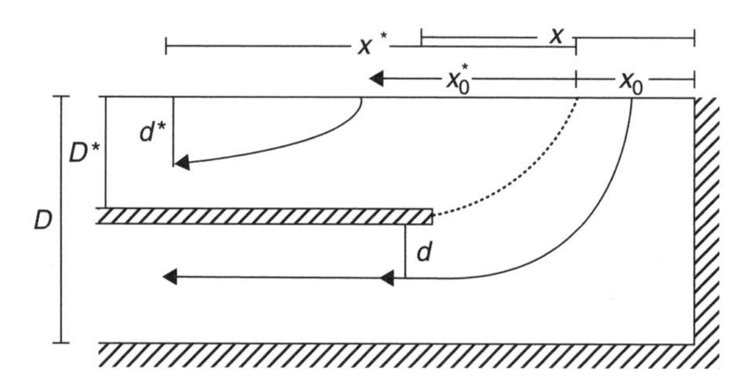

Figure 3.42. A clay layer separates an aquifer in two parts.

b. Indicate the direction of water flow through the confining layer, when it is leaky.
c. Draw isochrones (lines connecting water with equal age) in the confined aquifer.

3.8. The clay layer in the aquifer of Figure 3.42 starts at 1 km from the divide, and is 1 km long. $D^* = 2/3 \cdot D$; what is x_0? The two aquifers merge again downstream from the clay layer. Compare travel times for water just above and below the clay layer.

3.9. An aquifer sediment is made of quartz grains, with density 2.65 g/cm^3; porosity is 0.2. Calculate the bulk density of the sediment, and express sorbed concentration of Cd, $s_{Cd} = 2$ ppm, as mg/L pore water.

3.10. Bulk density of a limestone is 2.0 g/cm^3; calcite, the only mineral in the rock, has density 2.7 g/cm^3. Calculate porosity of the rock.

3.11. The sorption isotherm has the shape shown in the figure.
a. Draw c/x and c/t diagrams when a solution with concentration c_2 displaces c_1.
b. And also when a solution with c_1 displaces c_2.

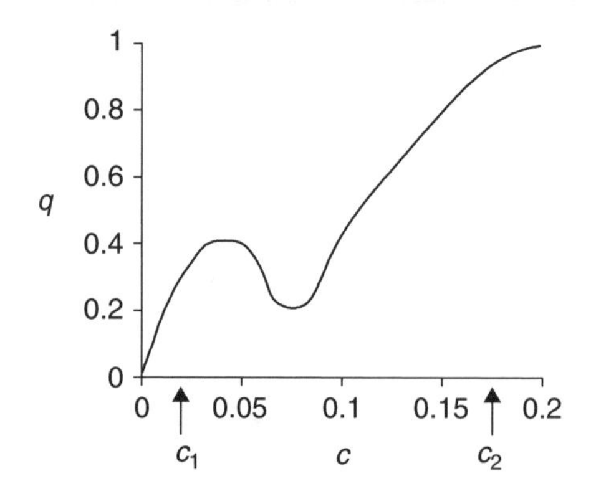

3.12. Calculate the distance covered by diffusion in 10 year in clay, for $D_e = 10^{-6}$ and 10^{-5} cm^2/s.

3.13. A single shot of 10 g. NaCl is injected in a clay, there is no water flow. Calculate the mass of Cl^- between $x = x_0$, and $x = x_0 + 1$ m after 10 years; $D_e = 10^{-9}$ m^2/s. *Hint*: Integrate Equation (3.53), and substitute the error function.

3.14. Derive the formula for the dispersivity in a permeability stratified aquifer with log-normally distributed permeabilities (Equation 3.84). *Hint*: $\sigma_{\ln(k)} = \ln(k_{0.84}) - \ln(k_{0.5})$, use $x_{0.5} = \exp[\ln(k_{0.5})] \, (-dh / dx \cdot \Delta t)$ and similar for $x_{0.84}$.

3.15. Calculate the value of the dispersion coefficient for a column with dispersivity $\alpha_L = 5$ mm, for $v_{H_2O} = 10^{-4}, 10^{-5}, 10^{-6}$ m/s while $D_f = 10^{-9}$ m^2/s, porosity is 0.3.

3.16. Give an estimated macro-dispersivity for an aquifer, 50 m thick, $k_{0.5} = 50$ m/day, $\sigma_k = 10$ m/day. Also for an aquifer with log-normally distributed k, $\sigma^2_{(\ln/k)} = 0.3$.

3.17. Estimate the diffusion coefficient of NaCl and $Ca(HCO_3)_2$ in water at 10°C, $\eta_{298} = 0.891$ cP, $\eta_{283} = 1.301$ cP.

REFERENCES

Abramowitz, M. and Stegun, I.A. 1964. *Handbook of mathematical functions*, 7th print. Dover, New York, 1064 pp.

Allison, G.B., Gee, G.W. and Tyler, S.W., 1994. Vadose-zone techniques for estimating groundwater recharge in arid and semiarid regions. *Soil Sci. Soc. Am. J.* 58, 6–14.

Andersen, L.J. and Sevel, T., 1974. Six years' environmental tritium profiles in the unsaturated and saturated zones, Grønhøj, Denmark. In *Isotope techniques in groundwater hydrology*, vol. 1, 3–18. IAEA, Vienna.

Appelo, C.A.J., Bäer, T., Smit, P.H.H. and Roelofzen, W., 1991. *Flushing a column spiked with HCH* (in Dutch). Report, Free University, Amsterdam.

Atkins, P.W. and de Paula, J., 2002. *Atkins' Physical chemistry*, 7th ed. Oxford Univ. Press, 1149 pp.

Bath, A.H. and Edmunds, W.M., 1981. Identification of connate water in interstitial solution of chalk sediment. *Geochim. Cosmochim. Acta* 45, 1449–1461.

Bauer, S., Fulda, C. and Schäfer, W., 2001. A multi-tracer study in a shallow aquifer using age dating tracers ^{3}H, ^{85}Kr, CFC-113 and SF_6-indication for retarded transport of CFC-113. *J. Hydrol.* 248, 14–34.

Bear, J., 1972. *Dynamics of fluids in porous media.* Elsevier, Amsterdam, 764 pp.

Beekman, H.E., 1991. *Ion chromatography of fresh- and seawater intrusion.* Ph.D. thesis, Free University, Amsterdam, 198 pp.

Bird, R.B., Stewart, W.E. and Lightfoot, E.N., 1960. *Transport phenomena.* Wiley and Sons, New York, 780 pp.

Boas, M.L., 1983. *Mathematical methods in the physical sciences*, 2nd ed. Wiley and Sons, New York, 793 pp.

Bond, W.J. and Phillips, I.R., 1990. Cation exchange isotherms obtained with batch and miscible displacement techniques. *Soil Sci. Soc. Am. J.* 54, 722–728.

Boudreau, B.P., 1997. *Diagenetic models and their implementation.* Springer, Berlin, 414 pp.

Boving, T.B. and Grathwohl, P., 2001. Tracer diffusion coefficients in sedimentary rocks: correlation to porosity and hydraulic conductivity. *J. Contam. Hydrol.* 53, 85–100.

Bruggeman, G.A., 1999. *Analytical solutions of geohydrological problems.* Elsevier, Amsterdam, 959 pp.

Brun, A., Engesgaard, P., Christensen, T.H. and Rosbjerg, D., 2002. Modelling of transport and biogeochemical processes in pollution plumes: Vejen landfill, Denmark. *J. Hydrol.* 256, 228–247.

Busenberg, E. and Plummer, L.N., 1992. Use of chlorofluorocarbons (CCl_3F and CCl_2F_2) as hydrologic tracers and age-dating tools. *Water Resour. Res.* 28, 2257–2283.

Chiang, W.-H. and Kinzelbach, W., 2001. *3D-Groundwater modeling with PMWIN.* Springer, Berlin, 346 pp.

Childs, E.C., 1969. *The physical basis of soil water phenomena.* Wiley and Sons, New York, 493 pp.

Collon, P., Kutschera, W., Loosli, H.H., Lehmann, B.E., Purtschert, R., Love, A., Sampson, L., Anthony, D., Cole, D., Davids, B., Morrissey, D.J., Sherrill, B.M., Steiner, M., Parso, R.C. and Paul, M., 2000. ^{81}Kr in the Great Artesian Basin, Australia: a new method for dating very old groundwater. *Earth Planet. Sci. Lett.* 182, 103–113.

Cook, P.G. and Herczeg, A.L. (eds), 2000. *Environmental tracers in subsurface hydrology*. Kluwer, Boston, 529 pp.

Cook, P.G. and Solomon, D.K., 1997. Recent advances in dating young groundwater: chlorofluorcarbons, $^{3}H/^{3}He$ and ^{85}Kr. *J. Hydrol.* 191, 245–265.

Cook, P.G., Solomon, D.K. and Plummer, L.N., 1995. CFC's as tracers of groundwater transport processes. *Water Resour. Res.* 31, 425–431.

Christensen, T.H., Kjeldsen, P., Bjerg, P.L., Jensen, D.L., Christensen, J.B., Baun, A., Albrechtsen, H.-J. and Heron, G., 2001. Biogeochemistry of landfill leachate plumes. *Appl. Geochem.* 16, 659–718.

Dagan, G., 1989. *Flow and transport in porous formations*. Springer, Berlin, 465 pp.

De Marsily, G., 1986. *Quantitative Hydrogeology*, Academic Press, London, 440 pp.

Desaulniers, D.E. and Cherry, J.A., 1988. Origin and movement of groundwater and major ions in a thick deposit of Champlain Sea clay near Montreal. *Can. Geotechn. J.* 26, 80–89.

Desaulniers, D.E., Kaufmann, R.S., Cherry, J.A. and Bentley, H.W., 1986. ^{37}Cl-^{35}Cl variations in a diffusion-controlled groundwater system. *Geochim. Cosmochim. Acta* 50, 1757–1764.

DeVault, D., 1943. The theory of chromatography. *J. Am. Chem. Soc.* 65, 532–540.

De Vries, J.J., 1995. Seasonal expansion and contraction of stream networks in shallow groundwater systems. *J. Hydrol.* 170, 15–26.

Dillon, P.J. 1989. An analytical model of contaminant transport from diffuse sources in saturated porous media. *Water Resour. Res.* 25, 1208–1218.

Dogramaci, S.S. and Herczeg, A.L., 2002. Strontium and carbon isotope constraints on carbonate-solution interactions and inter-aquifer mixing in groundwaters of the semi-arid Murray Basin, Australia. *J. Hydrol.* 262, 50–67.

Domenico, P.A. and Schwartz, F.W., 1997. *Physical and chemical hydrogeology*, 2nd ed. Wiley and Sons, New York, 506 pp.

Eggenkamp, H.G.M., 1994. *The geochemistry of chlorine isotopes*. Ph.D. thesis, Utrecht, 150 pp.

Eggenkamp, H.G.M., Middelburg, J.J. and Kreulen, R., 1994. Preferential diffusion of ^{37}Cl in sediments of Kau Bay, Halmahera, Indonesia. *Chem. Geol.* 116, 317–325.

Ekwurzel, B., Schlosser, P., Smethie, W.M., Plummer, L.N., Busenberg, E., Michel, R.L., Weppernig, R. and Stute, M., 1994. Dating of shallow groundwater-comparison of the transient tracers $^{3}H/^{3}He$, chlorofluorocarbons and ^{85}Kr. *Water Resour. Res.* 30, 1693–1708.

Elfeki, A.M.M., Uffink, G.J.M. and Barends, F.B.J., 1999. *Groundwater contaminant transport*. Balkema, Rotterdam, 300 pp.

Engesgaard, P., Jensen, K.H., Molson, J., Frind, E.O. and Olsen, H., 1996. Large-scale dispersion in a sandy aquifer: Simulation of subsurface transport of environmental tritium. *Water Resour. Res.* 32, 3253–3266.

Ernst, L.F. 1973. *Determining the traveltime of groundwater* (in Dutch). ICW-nota 755, Wageningen, 42 pp.

Ernst, L.F. 1978. Drainage of undulating sandy soils with high groundwater tables. *J. Hydrol.* 39, 1–50.

Felmy, A.R. and Weare, J.H., 1991. Calculation of multicomponent ionic diffusion from zero to high concentration: 1. The system Na-K-Ca-Mg-Cl-SO_4-H_2O at 25(C. *Geochim. Cosmochim. Acta* 55, 113–131.

Fetter, C.W., 1994. *Applied hydrogeology*. Macmillan, London, 691 pp.

Foster, S.S.D. and Smith-Carrington, A., 1980. The interpretation of tritium in the Chalk unsaturated zone. *J. Hydrol.* 46, 343–364.

Freeze, R.A. and Cherry, J.A., 1979. *Groundwater*. Prentice-Hall, Englewood Cliffs, 604 pp.

Freeze, R.A. and Witherspoon, P.A., 1966. Theoretical analysis of regional groundwater flow. I. Analytical and numerical solutions to the mathematical model. *Water Resour. Res.* 2, 641–656.

Freyberg, D.L., 1986. A natural gradient experiment on solute transport in a sand aquifer. 2, Spatial moments and the advection and dispersion of nonreactive tracers. *Water Resour. Res.* 22, 2031–2046.

Garabedian, S.P., LeBlanc, D.R., Gelhar, L.W., Celia, M.A., 1991. Large-scale natural gradient tracer test in sand and gravel, Cape Cod, Massachusetts. 2. Analysis of spatial moments for a nonreactive tracer. *Water Resour. Res.* 27, 911–924.

Gelhar, L.W., 1997. Perspectives on field-scale application of stochastic subsurface hydrology. In G. Dagan and S.P. Neuman (eds), *Subsurface flow and transport: a stochastic approach*, 157–176. Cambridge Univ. Press.

Gelhar, L.W. and Axness, C.L., 1983. Three-dimensional stochastic analysis of macrodispersion in aquifers. *Water Resour. Res.* 19, 161–170.

Gelhar, L.W. and Wilson, J.L., 1974. Ground-water quality modeling. *Ground Water* 12, 399–408.

Gelhar, L.W., Mantoglou, A., Welty, C. and Rehfeldt, K.R., 1985. A review of field-scale physical solute transport processes in saturated and unsaturated porous media. EPRI, Palo Alto, CA 94303.

Goltz, M.N. and Roberts, P.V., 1987. Using the method of moments to analyze three-dimensional diffusion-limited solute transport from temporal and spatial perspectives. *Water Resour. Res.* 23, 1575–1585.

Groen, J., Velstra, J. and Meesters, A.G.C.A., 2000. Salinization processes in paleowaters in Surinam. *J. Hydrol.* 234, 1–20.

Harvey, C.F. and Gorelick, S.M., 1995. Temporal moment-generating equations: modeling transport and mass-transfer in heterogeneous aquifers. *Water Resour. Res.* 21, 159–169.

Hendry, M.J. and Wassenaar, L.I., 2000. Controls on the distribution of major ions in pore waters of a thick surficial aquitard. *Water Resour. Res.* 36, 503–513.

Hsieh, P.A., 2001. *TopoDrive and ParticleFlow–Two computer models for simulation and visualization of ground-water flow and transport of fluid particles in two dimensions*. U.S. Geological Survey Open-File Report 01–286, 30 pp.

Hu, Q. and Wang, S.Y., 2003. Aqueous-phase diffusion in unsaturated geologic media: a review. *Crit. Rev. Env. Sci. Technol.* 33, 275–297.

Johnson, R.L., Cherry, J.A. and Pankow, J.F., 1989. Diffusive contaminant transport in natural clay: a field example and implications for clay-lined waste disposal sites. *Env. Sci. Technol.* 23, 340–349.

Kinzelbach, W., 1992. *Numerische Methoden zur Modellierung des Transports von Schadstoffen im Grundwasser*, 2th Auflage. Oldenbourg, München, 343 pp.

Klenk, I.D. and Grathwohl, P., 2002. Transverse vertical dispersion in groundwater and the capillary fringe. *J. Contam. Hydrol.* 58, 111–128.

Kreft, A. and Zuber, A., 1978. On the physical meaning of the dispersion equation and its solutions for different initial and boundary conditions. *Chem. Eng. Sci.* 33, 1471–1480.

Lapidus, L. and Amundson, N.R., 1952. Mathematics of adsorption in beds. VI. The effect of longitudinal diffusion in ion-exchange and chromatographic columns. *J. Phys. Chem.* 56, 984–988.

Lasaga, A.C., 1998. *Kinetic theory in the earth sciences*. Princeton Univ. Press, 811 pp.

Lerman, A., 1979. *Geochemical processes*. Wiley and Sons, New York, 481 pp.

Levenspiel, O., 1972. *Chemical reaction engineering*. Wiley and Sons, New York, 857 pp.

Li, Y.-H. and Gregory, S., 1974. Diffusion of ions in seawater and in deep-sea sediments. *Geochim. Cosmochim. Acta* 38, 703–714.

Lyman, W.J., Reehl, W.F. and Rosenblatt, D.H., 1990. *Handbook of chemical property estimation methods*. Am. Chem. Soc., Washington.

Lyngkilde, J. and Christensen, T.H., 1992. Redox zones of a landfill leachate pollution plume (Vejen, Denmark). *J. Contam. Hydrol.* 10, 273–289.

Mackay, D.M., Freyberg, D.L., Roberts, P.V. and Cherry, J.A., 1986. A natural gradient experiment on solute transport in a sand aquifer. 1. Approach and overview of plume movement. *Water Resour. Res.* 22, 2017–2029.

Maloszewski, P. and Zuber, A., 1985. On the theory of tracer experiments in fissured rocks with a porous matrix. *J. Hydrol.* 79, 333–358.

McNeil, J.D., 1980. *Electrical conductivity of soils and rocks*. Geonics Ltd, Missisauga, 22 pp.

Mercado, A., 1967. The spreading pattern of injected water in a permeability stratified aquifer. IASH Pub. 72, 23–36.

Mercado, A., 1984. A note on micro and macrodispersion. *Ground Water* 22, 790–791.

Molz, F.J., Güven, O. and Melville, J.G., 1983. An examination of scale-dependent dispersion coefficients. *Ground Water* 21, 715–725.

Mills, R. and Harris, K.R., 1976. The effect of isotope substitution on diffusion in liquids. *Chem. Soc. Rev.* 5, 215–231.

Moore, W.J., 1972. *Physical chemistry*, 5th ed. Longman, London, 977 pp.

Nkedi-Kizza, P.S.C., Rao, P.S.C. and Johnson, J.W., 1983. Adsorption of diuron and 2,4,5-T on soil particle-size separates. *J. Env. Qual.* 12, 195–197.

Oster, H., Sonntag, C. and Munnich, K.O., 1996. Groundwater age dating with chlorofluorocarbons. *Water Resour. Res.* 32, 2989–3001.

Patriarche, D., Ledoux, E., Michelot, J.-L., Simon-Coincon, R. and Savoye, S., 2004. Diffusion as the main process for mass transport in very low water content argillites: 2. Fluid flow and mass transport modeling. *Water Resour. Res.* 40, W01517, DOI 10.1029/2003WR002700.

Perkins, T.K. and Johnston, O.C., 1963. A review of diffusion and dispersion in porous media. *Soc. Petrol. Engin. J.*, march 1963, 70–84.

Phillips, F.M., 1994. Environmental tracers for water movement in desert soils of the American Southwest. *Soil Sci. Soc. Am. J.* 58, 15–24.

Pickens, J.F. and Grisak, G.E., 1981. Scale-dependent dispersion in a stratified granular aquifer. *Water Resour. Res.* 17, 1191–1211.

Plumb, O.A. and Whitaker, S., 1988. Dispersion in heterogeneous porous media. I. Local volume averaging and large-scale averaging. *Water Resour. Res.* 24, 913–926.

Rubin, Y. and Bellin, A., 1998. Conditional simulation of geologic media with evolving scales of heterogeneity. In G. Sposito (ed.), *Scale dependence and scale invariance*, 398–420, Cambridge Univ. Press.

Scheidegger, A., Bürgisser, C.S., Borkovec, M., Sticher, H., Meeussen, H. and Van Riemsdijk, W.H., 1994. Convective transport of acids and bases in porous media. *Water Resour. Res.* 30, 2937–2944.

Schlosser, P., Stute, M., Sonntag, C. and Münnich, K.O., 1989. Tritiogenic ^{3}He in shallow groundwater. *Earth Planet. Sci. Lett.* 94, 245–256.

Schröter, J., 1984. Micro- and macrodispersivities in porous aquifers (in German). *Meyniana* 36, 1–34.

Schwartzenbach, R.P., Gschwend, P.M. and Imboden, D.M., 1993. *Environmental organic chemistry*. Wiley and Sons, New York, 681 pp.

Sillén, L.G., 1951. On filtration through a sorbent layer. IV. The ψ condition, a simple approach to the theory of sorption columns. *Arkiv Kemi* 2, 477–498.

Sjöberg, E.L., Georgala, D. and Rickard, D.T., 1983. Origin of interstitial water compositions in postglacial black clays (Northeastern Sweden). *Chem. Geol.* 42, 147–158.

Smith, L., 1981. Spatial variability of flow parameters in a stratified sand. *Mathem. Geol.* 13, 1–21.

Solomon, D.K., Schiff, S.L., Poreda, R.J. and Clarke, W.B., 1993. A validation of the $^{3}H/^{3}He$ method for determining groundwater recharge. *Water Resour. Res.* 29, 2951–2962.

Sudicky, E.A., 1986. A natural gradient experiment on solute transport in a sand aquifer: spatial variability of hydraulic conductivity and its role in the dispersion process. *Water Resour. Res.* 22, 2069–2082.

Sun, M.-Z., 1999. A finite cell method for simulating the mass transport process in porous media. *Water Resour. Res.* 35, 3649–3662.

Tabor, D., 1991. *Gases, liquids and solids*, 3rd ed. Cambridge Univ. Press, 418 pp.

Taylor, S.R. and Howard, K.W.F., 1987. A field study of scale-dependent dispersion in a sandy aquifer. *J. Hydrol.* 90, 11–17.

Toride, N., Leij, F.J. and Van Genuchten, M.T., 1995. *The CXTFIT code for estimating transport parameters from laboratory or field tracer experiments, version 2*. US Salinity Lab. Res. Rep. 137. Riverside, Cal.

Tóth, J., 1962. A theory of groundwater motion in small drainage basins in Central Alberta, Canada. *J. Geophys. Res.* 67, 4375–4387.

Tyrrell, H.J.V. and Harris, K.R., 1984. *Diffusion in liquids*. Butterworths, London, 448 pp.

Uffink, G.J., 1988. Modeling of solute transport with the random walk method. In E. Custodio, A. Gurgui and J.P. Lobo Ferreira (eds), *Groundwater flow and quality modelling*, 247–265, D. Reidel, Dordrecht.

Valocchi, A.J., 1985. Validity of the local equilibrium assumption for modeling sorbing solute transport through homogeneous soils. *Water Resour. Res.* 21, 808–820.

Van Genuchten, M.Th., 1981. Analytical solutions for chemical transport with simultaneous adsorption, zero-order production and first-order decay. *J. Hydrol.* 49, 213–233.

Van Genuchten, M.Th. and Parker, J.C., 1984. Boundary conditions for displacement experiments through short laboratory soil columns. *Soil Sci. Soc. Am. J.* 48, 703–708.

Vogel, J.C., 1967. Investigation of groundwater flow with radiocarbon. Proc. 1966 Symp., 355–369. IAEA, Vienna.

Volker, A., 1961. Source of brackish ground water in Pleistocene formations beneath the Dutch polderland. *Econ. Geol.* 56, 1045–1057.

Yurtsever, Y., 1983. Models for tracer data analysis. In *Guidebook on nuclear techniques in hydrology*, 381–402, IAEA, Vienna.

Zheng, C. and Bennett, G.D., 2002. *Applied contaminant transport modeling*, 2nd ed. Wiley and Sons, New York, 621 pp.

4

Minerals and Water

The solubility of a mineral constrains the maximal concentration of its components in water. For example, the mineral fluorite has the composition CaF_2 and by doing solubility calculations we can predict concentration ranges of Ca^{2+} and F^- in groundwater. In the first part of this chapter we recapitulate the basics of equilibrium chemistry and demonstrate the applicability in hydrogeochemistry.

Minerals present in aquifers are often not pure phases but rather mixtures or solid solutions of different minerals. We will discuss the theory behind solid solutions and analyze the interrelations between the composition of the aqueous solution and the solid solution.

Some minerals react fast upon contact with water. This is particularly the case for the more "soluble" minerals such as gypsum ($CaSO_4 \cdot 2H_2O$), halite (NaCl), fluorite as well as most carbonate minerals, and equilibrium will be attained within a timespan that is short compared to the residence time of groundwater. Other minerals, typically silicates, react so sluggishly that equilibrium is never attained at low temperatures and therefore reaction kinetics have to be considered.

The general strategy of this chapter is to explore first the basic principles in manual calculations. Subsequently follows a demonstration of similar calculations with a geochemical model. Present-day geochemical models are extremely powerful tools to analyze complex natural systems. Throughout this chapter, and in fact throughout the remainder of this book the code PHREEQC (Parkhurst and Appelo, 1999) will be used as an integrated part of the presentation.

4.1 EQUILIBRIA AND THE SOLUBILITY OF MINERALS

Fundamental to any description of equilibria in water is the *law of mass action*. It states that for a reaction of the generalized type:

$$aA + bB \leftrightarrow cC + dD$$

the distribution at equilibrium of the species at the left and right side of the reaction is given by:

$$K = \frac{[C]^c[D]^d}{[A]^a[B]^b} \qquad (4.0)$$

Here K is the equilibrium constant and the bracketed quantities denote *activities* or "effective concentrations". For the present we will equate activities with molal concentrations, but return to the difference between the two in Section 4.2. The law of mass action is applicable to any type of reaction, the dissolution of minerals, the formation of complexes between dissolved species, the dissolution of gases in water, etc. For dissolution of the mineral fluorite (CaF_2) we may write:

$$CaF_2 \leftrightarrow Ca^{2+} + 2F^-$$

At equilibrium the aqueous concentrations obey the solubility product:

$$K_{\text{fluorite}} = [Ca^{2+}][F^-]^2 = 10^{-10.57} \quad \text{at } 25°C \tag{4.1}$$

The solubility product is a direct application of the law of mass action, except that $[CaF_2]$ is omitted in (4.1) since by definition the activity of a pure solid is equal to one. The exact value of K_{fluorite} is somewhat uncertain (Nordstrom and Jenne, 1977). Here we use the value of Handa (1975) in order to be consistent with the following example. Relation (4.1) can be rewritten in logarithmic form as:

$$\log K_{\text{fluorite}} = \log[Ca^{2+}] + 2\log[F^-] = -10.57 \tag{4.2}$$

On a logarithmic plot, the equilibrium condition (4.2) between fluorite and the solution is a straight line, as illustrated in Figure 4.1. All combinations of $[Ca^{2+}]$ and $[F^-]$ that plot below the line are subsaturated with respect to fluorite, while those that plot above the line are supersaturated for fluorite. Together with the fluorite solubility line in Figure 4.1, a number of groundwater analyses from Rajasthan, India are plotted and clearly, the equilibrium with fluorite places an upper limit to the F^- (and Ca^{2+}) concentrations in the groundwater.

Although F^- at low concentration in drinking water has been considered as beneficial in candy bar consuming countries, and in fact is added to drinking water in many water supplies where fluoride is absent in groundwater, it constitutes a health hazard at concentrations above 3 ppm, causing dental fluorosis (tooth mottling) and more seriously, skeletal fluorosis (bone deformation and painful brittle joints in older people). High F^- concentrations in groundwater are found in many places of the world, but notably in Asia and Africa. Fluoride is basically brought into the groundwater by leaching from minerals in the rocks (Handa, 1975; Jacks et al., 1993; Saxena and Ahmed, 2003). Volcanic rocks may have high natural fluor contents, but also the dissolution of fossil shark teeth, containing fluorapatite, has been reported as a source (Zack, 1980). In arid areas evaporation may strongly increase the F^- concentration in water.

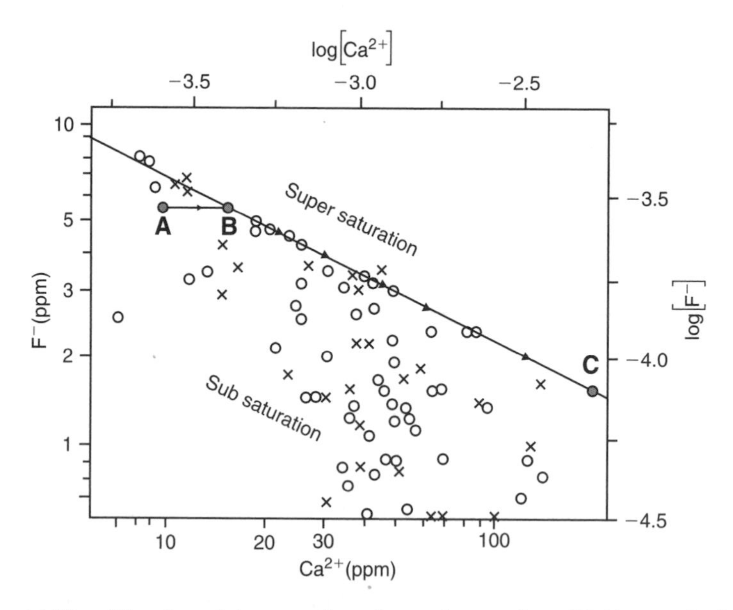

Figure 4.1. The stability of fluorite and the saturation of groundwaters from Sirohi, W. Rajasthan, India (modified from Handa, 1975). The evolution in water chemistry upon addition of gypsum is described by the pathway A, B to C as discussed in the text.

If equilibrium with fluorite imposes a constraint on the groundwater composition then water with a high natural Ca^{2+} concentration will generally contain a low F^- concentration, while groundwater with a low Ca^{2+} concentration will be rich in F^-. This seems substantiated by the field observations of Handa (1975; Figure 4.1). The fluorite equilibrium control on the F^- concentration can be exploited to remove F^- by water treatment. At saturation for fluorite, an increase of $[Ca^{2+}]$ will, according to Equation (4.1), decrease $[F^-]$. To bring about an increase in $[Ca^{2+}]$ it has been proposed (Schuiling, pers. comm.; Jacks et al., 2000) to add gypsum ($CaSO_4 \cdot 2H_2O$). For gypsum dissolution we may write, neglecting the crystal water for simplicity:

$$CaSO_4 \leftrightarrow Ca^{2+} + SO_4^{2-}$$

With the solubility product:

$$K_{gypsum} = [Ca^{2+}][SO_4^{2-}] = 10^{-4.60} \quad \text{at } 25°C \tag{4.3}$$

Note that gypsum is much more soluble than fluorite. The amount of gypsum which has to be added to a water sample with composition A (Figure 4.1), in order to reach equilibrium with fluorite, can be calculated following Example 4.1.

EXAMPLE 4.1. *Gypsum addition to high fluor groundwater*
A groundwater sample contains 10 ppm Ca^{2+} and 5.5 ppm F^-. Is this water saturated with respect to fluorite, and if not, how much gypsum should be added in order to reach saturation for fluorite?

First we recalculate ppm to molal concentrations, which yields $[Ca^{2+}] = 10^{-3.60}$ and $[F^-] = 10^{-3.54}$. A fast answer is obtained by plotting these values directly into Figure 4.1 which results in point A. Alternatively we may calculate the product:

$$[Ca^{2+}][F^-]^2 = [10^{-3.60}][10^{-3.54}]^2 = 10^{-10.68}$$

The value obtained is slightly lower than the solubility product for fluorite (Equation 4.2) and the sample is subsaturated for fluorite.

When gypsum is added, the water composition will change from point A (Figure 4.1) parallel to the Ca-axis until it meets the fluorite saturation line at point B. From the fluorite solubility product (Equation 4.2) we find that:

$$[Ca^{2+}]_B = K_{fluorite} / [F^-]^2 = [10^{-10.57}] / [10^{-3.54}]^2 = 10^{-3.49}$$

Thus to reach equilibrium with fluorite, the amount of gypsum to be added is equal to $[Ca^{2+}]_B - [Ca^{2+}]_A = 10^{-3.49} - 10^{-3.60} = 0.072$ mmol/L.

Once equilibrium with fluorite is reached, the addition of more gypsum will modify the water composition following the fluorite solubility line in Figure 4.1. The change in water chemistry is described quantitatively by Equation (4.2) and results in a decreasing F^- concentration due to the precipitation of fluorite. However, the dissolution of gypsum cannot continue indefinitely since at some point also saturation for gypsum is reached. In Figure 4.1, point C indicates the simultaneous saturation for fluorite and gypsum and its location can be calculated as follows.

When the composition of the solution changes from B to C, the dissolution of gypsum and precipitation of fluorite occur simultaneously. We may define:

x = mol gypsum dissolved per liter water
y = mol fluorite precipitated per liter water

Using these two variables we write the mass balance equations for the concentrations of Ca^{2+} and F^- at point C as:

$$[Ca^{2+}]_C = [Ca^{2+}]_B + x - y \quad (4.4)$$

$$[F^-]_C = [F^-]_B - 2y \quad (4.5)$$

$$[SO_4^{2-}]_C = [SO_4^{2-}]_B + x \quad (4.6)$$

Together with the solubility products for fluorite (Equation 4.1) and gypsum (Equation 4.3), this yields 5 equations with 5 unknowns which can be solved. However, the resulting quadratic equation is rather awkward and a more simple method is to identify in the mass balance equations a parameter which can be neglected. From Figure 4.1 we can surmise that when the water composition changes from B to C, the amount of Ca^{2+} released by gypsum dissolution will be much larger than the amount of Ca^{2+} precipitated as fluorite. We introduce this as an simplifying assumption, and check its validity afterwards:

$$y \ll x$$

which reduces Equation (4.4) to:

$$[Ca^{2+}]_C = [Ca^{2+}]_B + x \quad (4.7)$$

Substitution of this relation, together with Equation (4.6), in the solubility product for gypsum (Equation 4.3) results in:

$$K_{gypsum} = ([Ca^{2+}]_B + x)\,([SO_4^{2-}]_B + x)$$

Rearranging yields:

$$x^2 + ([Ca^{2+}]_B + [SO_4^{2-}]_B)\,x + ([Ca^{2+}]_B \cdot [SO_4^{2-}]_B - K_{gypsum}) = 0$$

And substitution of known values (from Example 4.1):

$$x^2 + 10^{-3.49}x - 10^{-4.60} = 0$$

This is a neat standard equation that solves to:

$$x = 4.84 \times 10^{-3} = 10^{-2.32}$$

Accordingly, 4.8 mmol gypsum/L must dissolve to reach saturation with both gypsum and fluorite. Substituting x in Equation (4.7) produces:

$$[Ca^{2+}]_C = 10^{-2.29}$$

Which again is substituted in the solubility product of fluorite (Equation 4.2):

$$[F^-]_C = (K_{fluorite} / [Ca^{2+}]_C)^{0.5} = 10^{-4.14}$$

This corresponds to a fluoride concentration of 1.4 ppm, and gypsum addition has therefore lowered the fluoride content of the water by 75%. A side effect of the addition of gypsum as water treatment technique is, however, a high sulfate concentration and in our example it amounts to 468 ppm, which exceeds the drinking water limit (Table 1.1) and may cause intestinal problems and diarrhea.

It remains to check the validity of the initial assumption by substitution in Equation (4.5):

$$y = ½([F^-]_B - [F^-]_C) = 1.08\times10^{-4}$$

Since y is about fifty times smaller than x, our initial assumption, $y \ll x$, is quite reasonable. This type of simplification is often very useful in chemical calculations. Any reasonable assumption can be made, as long as it is tested afterwards. Neglection of terms is only allowed in mass balance equations (which are summations), never in mass action equations (which are multiplications or divisions).

At point C there exists simultaneous equilibrium between fluorite and gypsum, and we can write the reaction:

$$SO_4^{2-} + CaF_2 \leftrightarrow CaSO_4 + 2F^- \tag{4.8}$$

The corresponding equilibrium constant is derived by combining the solubility products of fluorite and gypsum:

$$K = \frac{[F^-]^2}{[SO_4^{2-}]} = \frac{[Ca^{2+}][F^-]^2}{[Ca^{2+}][SO_4^{2-}]} = \frac{K_{fluorite}}{K_{gypsum}} = \frac{10^{-10.57}}{10^{-4.60}} = 10^{-5.97} \tag{4.9}$$

Equation (4.9) demonstrates that at simultaneous equilibrium with gypsum and fluorite, the ratio of the squared F^- over SO_4^{2-} concentration becomes invariant. If some other process, for example the oxidation of pyrite (FeS_2), increases the SO_4^{2-} concentration, while equilibrium for both fluorite and gypsum is maintained, then the F^- concentration also increases due to the conversion of fluorite to gypsum (Equation 4.8).

In summary, we have seen in this section how to work with solubility products of minerals and how they may constrain the composition of groundwater. Furthermore, we have calculated a pathway in the evolution of water chemistry as a function of the dissolution and precipitation of two minerals.

4.2 CORRECTIONS FOR SOLUBILITY CALCULATIONS

In the foregoing we have used the molal concentration of a solute to calculate mineral solubility. However, the law of mass action is only valid for the *activity* of ions, which is the measured total concentration corrected for the effects of electrostatic shielding and for the presence of aqueous complexes, as discussed in the next two sections.

4.2.1 *Concentration and activity*

The activity in the law of mass action is a measure for the effective concentration of the species. The effective concentration can be thought of as indicating how, for example, a Ca^{2+} ion would behave when there are no interactions with other ions in solution, i.e. at infinite dilution. In thermodynamics the activity of solutes, gases, components in solid solutions and adsorbed ions are all expressed as a fraction relative to a *standard state* (Table 4.1) and as a fraction, the activity is always dimensionless. For gases the standard state is a pure gas phase at 1 atm. Atmospheric air contains 21% O_2 and the activity, or partial pressure, of O_2 in air is therefore 0.21 and is dimensionless! Similarly, the standard state for a solid is a pure solid, and for an ion exchanger the standard state is an exchanger filled with a single ion only. For aqueous solutes the standard state is defined as an *ideal* solution with solute concentration of 1 mol/kg H_2O = 1 molal. In this context "ideal" means, somewhat cryptically, a 1 M solution behaving like at infinite dilution.

Table 4.1. Conventions for standard states and activities.

	Concentration measure	Symbol	Standard state
Gases	Gas pressure (atm)/1 atm	P_i/P^0	$P^0 = 1$ atm
Solid and liquid mixtures	Mole-fraction	χ	$\chi = 1$
Aqueous solutes	Molality/(1 molal)	m_i/m^0	$m^0 = 1$ mol/kg H_2O
Exchangeable ions	Equivalent- or Mole-fraction	β or β^M	$\beta = 1$, $\beta^M = 1$

The activity is related to the molal concentration by an *activity coefficient* which corrects for non-ideal behavior. For aqueous solutes, the relation is:

$$[i] = \gamma_i \cdot m_i \,/\, m_i^0 \equiv \gamma_i \cdot m_i \tag{4.10}$$

where: $[i]$ is the activity of ion i (dimensionless), γ_i is the activity coefficient (dimensionless), m_i is the molality (mol/kg H_2O), m_i^0 is the standard state, i.e. 1 mol/kg H_2O. The factor $1/m_i^0$ is unity for all species and cancels in the practical enumeration of Equation (4.10) but causes the activity to become dimensionless. Activity coefficients may vary, but if ion i is present at trace concentration, and there are no other ions present, then $\gamma_i \rightarrow 1$. In this book the activity of an ion is denoted by square brackets, but in other texts activity is indicated by braces $\{i\}$, or simply as a_i.

Activity coefficients for solutes are calculated using the Debye-Hückel theory. In this theory, first the *ionic strength*, I, is defined which describes the number of electrical charges in the solution (Atkins and de Paula, 2002):

$$I = \tfrac{1}{2}\Sigma(m_i/m_i^0 \cdot z_i^2) \equiv \tfrac{1}{2}\Sigma\, m_i \cdot z_i^2 \tag{4.11}$$

where z_i is the charge number of ion i, and m_i is the molality of i. Similarly to the definition of activity, the ionic strength becomes dimensionless by division with the standard state m_i^0 i.e. 1 mol/kg H_2O. The ionic strength of freshwater is normally less then 0.02 while seawater has an ionic strength of about 0.7. A highly saline environment like the Dead Sea has an ionic strength of 9.4.

For dilute electrolyte solutions, $I < 0.1$, the Debye-Hückel equation describes the electrostatic interactions as:

$$\log \gamma_i = -\frac{A z_i^2 \sqrt{I}}{1 + B å_i \sqrt{I}} \tag{4.12}$$

where A and B are temperature dependent constants; at 25°C $A = 0.5085$ and $B = 0.3285 \times 10^{10}$/m. Langmuir (1997) lists the dependency of A and B on the temperature. However, in the temperature range of most groundwater, 5–35°C, the variation in A and B remains small.

The empirical ion-size parameter $å_i$ (Table 4.2) is a measure of the effective diameter of the hydrated ion. The value of $å$, multiplied by 10^{10}/m, (roughly) indicates the number of water molecules surrounding the ion. The smaller $å$ is, the closer can oppositely charged ions approach the ion and shield it, thereby lowering its activity coefficient. The symbols in Figure 4.2 depict activity coefficients of various ions as a function of ionic strength as calculated with the Debye-Hückel equation. Activity coefficients are seen to decrease with I, strongest for the divalent ions. At an ionic strength of 0.1, γ is about 0.8 for monovalent ions and 0.4 for divalent ions.

Table 4.2. The ion-size parameter, $\mathring{a}$, in the Debye-Hückel Equation (4.12). Numbers are given in Ångstrom (10^{-10}m) (Garrels and Christ, 1965).

$\mathring{a}/(10^{-10}\text{m})$	Ion
2.5	Rb^+, Cs^+, NH_4^+, Tl^+, Ag^+
3.0	K^+, Cl^-, Br^-, I^-, NO_3^-
3.5	OH^-, F^-, HS^-, BrO_3^-, IO_4^-, MnO_4^-
4.0–4.5	Na^+, HCO_3^-, $H_2PO_4^-$, HSO_3^-, Hg_2^{2+}, SO_4^{2-}, SeO_4^{2-}, CrO_4^{2-}, HPO_4^{2-}, PO_4^{3-}
4.5	Pb^{2+}, CO_3^{2-}, SO_3^{2-}, MoO_4^{2-}
5.0	Sr^{2+}, Ba^{2+}, Ra^{2+}, Cd^{2+}, Hg^{2+}, S^{2-}, WO_4^{2-}
6	Li^+, Ca^{2+}, Cu^{2+}, Zn^{2+}, Sn^{2+}, Mn^{2+}, Fe^{2+}, Ni^{2+}, Co^{2+}
8	Mg^{2+}, Be^{2+}
9	H^+, Al^{3+}, Cr^{3+}, trivalent rare earths
11	Th^{4+}, Zr^{4+}, Ce^{4+}, Sn^{4+}

Various equations have been proposed to derive activity coefficients at ionic strength values higher than 0.1 (Nordstrom and Munoz, 1994; Langmuir, 1997). Truesdell and Jones (1973), and subsequently Parkhurst (1990), fitted a modified version of the Debye-Hückel equation to activity coefficient data for chloride solutions:

$$\log \gamma_i = -\frac{Az_i^2\sqrt{I}}{1 + Ba_i\sqrt{I}} + b_iI \tag{4.13}$$

where A and B are the temperature dependent coefficients from the Debye-Hückel Equation (4.12), and a_i and b_i are ion-specific fit parameters (Table 4.3). Note that a_i differs from the ion-size parameter $\mathring{a}$; for Mg^{2+} $a_i = 5.5\times10^{-10}$m while $\mathring{a} = 8\times10^{-10}$m.

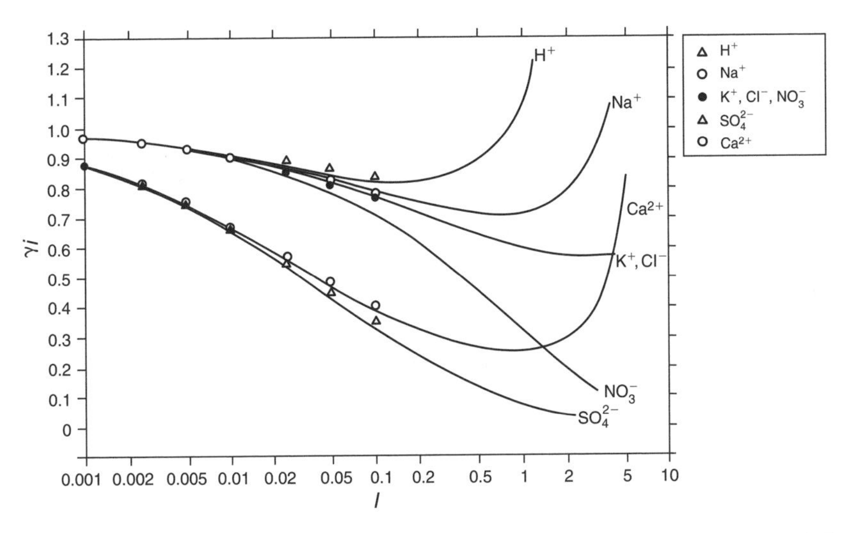

Figure 4.2. Activity coefficients for some common ions as a function of ionic strength. Symbols represent calculations with the Debye-Hückel Equation (4.12). Lines correspond to the Truesdell and Jones Equation (4.13). From Garrels and Christ (1965).

Table 4.3. The individual ion activity coefficient parameters in the Truesdell and Jones (1973) Equation (4.13). Values for other ions can be found in Parkhurst (1990).

Ion	$a/(10^{-10}\,m)$	b
Na^+	4.0	0.075
K^+	3.5	0.015
Mg^{2+}	5.5	0.20
Ca^{2+}	5.0	0.165
Cl^-	3.5	0.015
HCO_3^-	5.4	0.0
CO_3^{2-}	5.4	0.0
SO_4^{2-}	5.0	−0.04

The lines in Figure 4.2 correspond to the Truesdell and Jones equation. At low ionic strength they agree well with the values obtained with the Debye-Hückel equation. At high ionic strength the activity coefficients of cations are seen to increase by the action of the term $(b_i \cdot I)$. The Truesdell and Jones equation is a reasonable approximation up to I values of about 2 in dominantly chloride solutions (Parkhurst, 1990). The Davies equation is another relation often used to calculate activity coefficients and is applicable up to an ionic strength of about 0.5.

$$\log \gamma_i = -Az_i^2\left(\frac{\sqrt{I}}{1+\sqrt{I}} - 0.3I\right) \tag{4.14}$$

where A is the same temperature dependent coefficient as in Equation (4.12).

The net effect of the aqueous activity coefficient is that the solubility of minerals increases with ionic strength up to $I \sim 0.7$. In more saline solutions the solubility of minerals decreases again as activity coefficients increase, an effect known as *salting out*. Salting out already starts at low ionic strength for neutral species, including gases, which have $b_i = 0.1$ (at least in PHREEQC). Another point to note is that the mole fraction of H_2O decreases markedly in a saline solution, which results in a decreased vapor pressure. Garrels and Christ (1965) have given formulas to calculate the activity of H_2O in saline water.

EXAMPLE 4.2. *Calculate ionic strength and ion activity coefficients*

Let's calculate the solubility of fluorite in a solution with 10 mmol NaCl/L. First calculate the ionic strength:

$$I = ½\,[m_{Na^+} \times (1)^2 + m_{Cl^-} \times (-1)^2 + m_{Ca^{2+}} \times (2)^2 + m_{F^-} \times (-1)^2]$$

We expect that only small amounts of fluorite dissolve and neglect $m_{Ca^{2+}}$ and m_{F^-} in the equation. Thus $I = 0.01$.

Now calculate the activity coefficient for Ca^{2+} using (4.14):

$$\log \gamma_{Ca^{2+}} = -0.5085(2)^2\left(\frac{\sqrt{0.01}}{1+\sqrt{0.01}} - 0.3(0.01)\right) = -0.179$$

So that:

$$\gamma_{Ca^{2+}} = 0.66$$

Likewise for F^-

$$\log \gamma_{F^-} = -0.5085(-1)^2\left(\frac{\sqrt{0.01}}{1+\sqrt{0.01}} - 0.3(0.01)\right) = -0.0447$$

which gives:

$$\gamma_{F^-} = 0.90$$

Note that the activity coefficient of the monovalent ion, F^-, is much higher than of the divalent ion, Ca^{2+}. The solubility product of fluorite is given by Equation (4.1):

$$K_{fluorite} = [Ca^{2+}][F^-]^2 = 10^{-10.57} \quad \text{at } 25°C \tag{4.15}$$

Substituting (4.10) in (4.15) produces:

$$\gamma_{Ca^{2+}} m_{Ca^{2+}} \cdot \gamma_{F^-}^2 m_{F^-}^2 = K_{fluorite} \tag{4.16}$$

Rearranging yields:

$$m_{Ca^{2+}} m_{F^-}^2 = \frac{1}{\gamma_{Ca^{2+}} \gamma_{F^-}^2} K_{fluorite} \tag{4.17}$$

And substituting activity coefficients

$$m_{Ca^{2+}} m_{F^-}^2 = \frac{1}{(0.66)(0.90)^2} K_{fluorite} = 1.87 K_{fluorite} \tag{4.18}$$

In distilled water the concentrations of calcium and fluoride in equilibrium with fluorite are so low that the ionic strength approaches zero and the activity coefficients unity. However, in a 10 mmol NaCl/L solution our calculation shows that the solubility of fluorite has increased by almost a factor two. In these calculations, the contribution of Ca^{2+} and F^- concentrations to I were neglected; their inclusion would give an additional (very) slight increase of the solubility of fluorite.

QUESTIONS:

Obtain the concentrations of Ca^{2+} and F^- in 10 mmol NaCl/L? *Hint*: use the mass balance on solutes that dissolve from CaF_2: $m_{F^-} = 2\, m_{Ca^{2+}}$.

ANSWER: $(0.5\, m_{F^-})(m_{F^-})^2 = 1.87 \times 10^{-10.57}$ solves to $m_{F^-} = 0.465$ mmol/L, $m_{Ca^{2+}} = 0.233$ mmol/L

Find the contribution of Ca^{2+} and F^- to the ionic strength in 0.01 M NaCl?

ANSWER: $I = ½[m_{Na^+} \times (1)^2 + m_{Cl^-} \times (-1)^2 + m_{Ca^{2+}} \times (2)^2 + m_{F^-} \times (-1)^2] = 0.0107$

And calculate now the correct γ's and concentrations of Ca^{2+} and F^-?

ANSWER: $\gamma_{Ca^{2+}} = 0.65$, $\gamma_{F^-} = 0.90$, $m_{F^-} = 0.467$ mmol/L, $m_{Ca^{2+}} = 0.233$ mmol/L

How much does the solubility of fluorite increase in 0.1 M NaCl (cf. Figure 4.2)?

ANSWER: roughly $1/(0.4 \times 0.8^2) = 4$ times

4.2.2 *Aqueous complexes*

Ions in aqueous solution may become attached to each other as ion pairs or aqueous complexes. Examples are major cation complexes like $CaSO_4^0$, CaF^+ or $CaOH^+$, but also heavy metals complexes exist like $CdCl^+$, $HgCl_3^-$ or $PbOH^+$ and metal complexes with an organic ligand like oxalate, for example CuC_2O_4. The complexes may be *outer sphere*, with water molecules present in between the constituent ions of the complex, or *inner sphere* when a covalent bond is formed and water molecules that surrounded the ions have been expelled. The inner/outer sphere state is gradual and fluctuates in time (Burgess, 1988).

Complexation will lower the activity of the "free" ion in water, thereby increasing the solubility of minerals and also the mobility of trace metals. Grassi and Netti (2000) found that the Hg concentration in groundwater in a coastal aquifer increased with the Cl^- concentration, apparently due to complexing of Hg^{2+} and Cl^- resulting in the solubilization of Hg from the rock. Complexes may have detoxifying effects, and heavy metal poisoning can be abated by ingesting a strong complexer such as EDTA (ethylene-diamine-tetra-acetate). Complexation finds important application in analytical chemistry and separation technology and in the production of food and medicines.

The majority of complexes consist of a metal cation, surrounded by a number of ligands (Burgess, 1988; Stumm and Morgan, 1996). For example, Al^{3+} in water is surrounded by 6 water molecules which together form the octahedrally shaped complex $Al(OH_2)_6^{3+}$. Depending on the pH, protons are released from the complex and its charge increases:

$$Al(OH_2)_6^{3+} \leftrightarrow AlOH(OH_2)_5^{2+} + H^+$$

This complex is usually written simplified as $AlOH^{2+}$.

The ligand may act as a bridge between two metal ions to give a *binuclear* complex. For example, the chromate anion, CrO_4^{2-}, is present in dilute acid as dichromate, $Cr_2O_7^{2-}$, in the form of two tetrahedra which share a corner oxygen. *Polynuclear* complexes are formed by $Si(OH)_4^0$ or $Al(OH)_3(OH_2)_3$ in polymers that can be a precursor for a solid. If a ligand has various positions to bind the same cation, then it forms so-called *multidentate* complexes. Thus, F^- is a *monodentate* ligand, while oxalate, $C_2O_4^{2-}$, usually is a *bidentate* ligand with two oxygens where the cation is bound. EDTA is a *hexadentate* ligand, complexing both on N and O. Complexes with multidentate ligands are called *chelates*. The chelates can form rings which encapsulate the metal cations completely and these are important carriers of trace elements in biological systems (Buffle, 1990).

The formation of aqueous complexes can be described by equilibria of the type:

$$Ca^{2+} + SO_4^{2-} \leftrightarrow CaSO_4^0 \quad (4.19)$$

The distribution of the species is obtained by applying the law of mass action (4.0):

$$K = \frac{[CaSO_4^0]}{[Ca^{2+}][SO_4^{2-}]} = 10^{2.5} \quad (4.20)$$

The reaction is written as an ion-association reaction and the corresponding mass action constant is therefore termed a *stability constant*. Sometimes equilibria for aqueous complexes are written as dissociation reactions and the *dissociation constant* for the $CaSO_4^0$ complex is obviously the inverse of Equation (4.20). Clearly, these two formulations should not be confused.

4.2.3 *Combined complexes and activity corrections*

Both aqueous complexing and activity coefficients will lower the activity of the free ions, and their combined effect must be calculated. The total amount of Ca^{2+} in solution is given by a mass balance equation comprising the various aqueous complexes of Ca^{2+}, given in molal units:

$$\Sigma Ca = m_{Ca^{2+}} + m_{CaF^+} + m_{CaSO_4^0} + m_{CaOH^+} + \cdots \quad (4.21)$$

Similar mass balance equations can be written for all other substances in solution. In order to calculate the activities of, for example $[Ca^{2+}]$ and $[SO_4^{2-}]$, we have to solve simultaneously a set of mass balance equations, with concentrations in molal units, and a set of mass action equations like (4.20), with concentrations expressed as activities. The calculation proceeds through an iterative procedure.

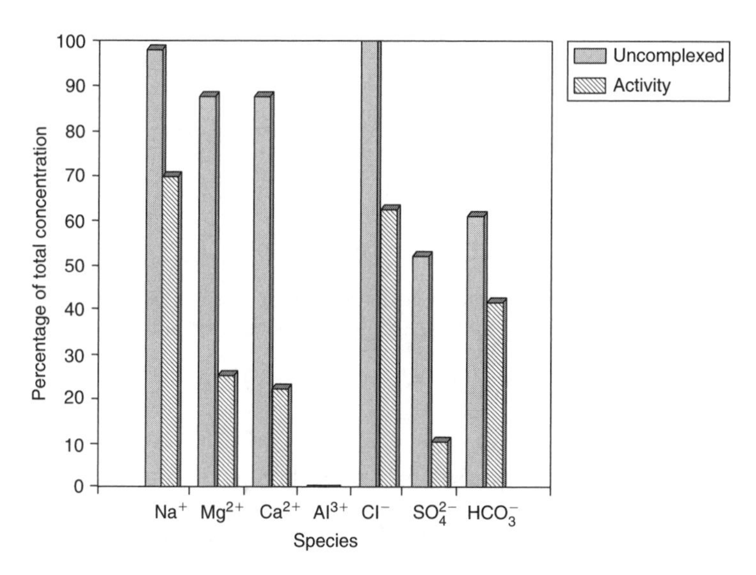

Figure 4.3. The importance of complexing and activity corrections as percentage of total concentrations for 35‰ seawater with a pH of 8.22.

From the analytical data, first the activity coefficients and ion activities are calculated, which are used to calculate activities of complexes. Then the molal concentrations of complexes are obtained which enables us to calculate improved mass balances and thereby a better estimate of the ionic strength. This procedure continues until no further significant improvement is obtained.

The importance of complexing and activity corrections, expressed as percentage of the total concentration, is illustrated for seawater in Figure 4.3. For Na^+ and Cl^- the effect of complexing is small while activity coefficients make the ultimate activity 30–40% lower than the total concentration. For the divalent ions Ca^{2+} and Mg^{2+}, about 10% of the total concentration is complexed, while the activity of Ca^{2+} and Mg^{2+} with the activity coefficient correction goes down to 20–30% of the total concentration. For SO_4^{2-} and HCO_3^-, aqueous complexes constitute 40–50% of the total concentration, and the activity of the free SO_4^{2-} ion is about 10% of the total concentration.

The methods described here work well for electrolyte solutions of up to about seawater ionic strength ($I = 0.7$), or up to $I = 2$ mol/L when HCO_3^- and SO_4^{2-} concentrations are low (Parkhurst, 1990). For more concentrated solutions, as are found in association with evaporites, other methods are available. Particularly so-called Pitzer equations (Pitzer, 1981; Harvie et al., 1982; Monnin and Schott, 1984; Plummer et al., 1988; Millero and Pierrot, 1998) have been successful in describing mineral equilibria in highly concentrated solutions.

EXAMPLE 4.3. *Solubility of gypsum*
The solubility of gypsum in water, considering both aqueous complexes and activity corrections, can be calculated as follows. For equilibrium between gypsum and water we write:

$$CaSO_4 \cdot 2H_2O \leftrightarrow Ca^{2+} + SO_4^{2-} + 2H_2O$$

and the mass action expression is:

$$K = \frac{[Ca^{2+}][SO_4^{2-}][H_2O]^2}{[CaSO_4 \cdot 2H_2O]} = 10^{-4.6}$$

Since the activity of a pure solid like gypsum is unity and also for dilute electrolyte solutions $[H_2O] = 1$, this simplifies to the solubility product:

$$K_{gypsum} = [Ca^{2+}][SO_4^{2-}] = 10^{-4.60} \quad \text{at } 25°C$$

Substituting molal concentrations for activities yields:

$$K_{gypsum} = (\gamma_{Ca^{2+}} \cdot m_{Ca^{2+}})(\gamma_{SO_4^{2-}} \cdot m_{SO_4^{2-}}) = 10^{-4.60}$$

Since equal amounts of Ca^{2+} and SO_4^{2-} are released during gypsum dissolution, $m_{Ca^{2+}} = m_{SO_4^{2-}}$ and according to Equation (4.14), also $\gamma_{Ca^{2+}} = \gamma_{SO_4^{2-}}$, the solubility of gypsum is simplified to

$$(m_{Ca^{2+}})^2 = 10^{-4.60} / (\gamma_{Ca^{2+}})^2$$

or

$$m_{Ca^{2+}} = (5.01 \times 10^{-3}) / \gamma_{Ca^2}$$

$\gamma_{Ca^{2+}}$ can be estimated with the Davies Equation (4.14) and the ionic strength, I:

$$I = ½ \sum m_i \cdot z_i^2 = ½ (m_{Ca^{2+}} \times 4 + m_{SO_4^{2-}} \times 4) = 4m_{Ca^{2+}}$$

The problem is now that $m_{Ca^{2+}}$ depends on $\gamma_{Ca^{2+}}$ which depends on I and again on $m_{Ca^{2+}}$. It may be solved in an iterative procedure. In first approximation $\gamma_{Ca^{2+}}$ is set equal to 1, which enables a first estimate of I and so forth. In tabulated form the calculation proceeds as:

Iteration	$m_{Ca^{2+}}$	I	$\gamma_{Ca^{2+}}$
0			1
1	→5.01×10^{-3}	→0.0201	→0.58
2	→8.71×10^{-3}	→0.0348	→0.50
3	→9.97×10^{-3}	→0.0399	→0.49
4	→10.3×10^{-3}	→0.0413	→0.48
5	→10.4×10^{-3}	→0.0417	→0.48
6	→10.4×10^{-3}		

Note that the application of activity corrections has doubled the calculated solubility of gypsum. However, we also know that the aqueous complex $CaSO_4^0$ is of importance. The stability constant of this complex (Equation 4.20) can be rewritten as:

$$[CaSO_4^0] = 10^{2.5} \times [Ca^{2+}][SO_4^{2-}]$$

Substitution of the solubility product of gypsum gives:

$$[CaSO_4^0] = 10^{2.5}\times10^{-4.60} = 10^{-2.10} = 7.94\times10^{-3}$$

Since for uncharged species the activity coefficient is close to unity, this is also the molal concentration of the complex. Thus, the total solubility of gypsum is $10.4\times10^{-3} + 7.94\times10^{-3} = 18.3\times10^{-3}$ mol/L. Note that 40% of the solubility is accounted for by the complex $CaSO_4^0$, while 30% is due to activity corrections. These results are easily transcribed in grams of gypsum dissolved per liter by multiplication with the molecular weight of gypsum (172.1). The solubility product alone predicts that (5.01×10^{-3}) $(172.1) = 0.86$ gram gypsum can dissolve per liter water, while including complexes and activity corrections leads to a solubility of $(18.3\times10^{-3})(172.1) = 3.15$ g/L.

4.2.4 *Calculation of saturation states*

Once we are able to calculate the activities of free ions in solution, we can also calculate the state of saturation of a groundwater sample with respect to minerals. One way to do this has already been shown in Figure 4.1 where the analytical data, or more correctly the ion activities, are plotted in a stability diagram for fluorite. Another approach is to compare the solubility product K with the analogue product of the activities derived from water analyses. The latter is often termed the *Ion Activity Product* (*IAP*). For example for gypsum:

$$K_{gypsum} = [Ca^{2+}][SO_4^{2-}] \quad \text{(activities at equilibrium)}$$

and

$$IAP_{gypsum} = [Ca^{2+}][SO_4^{2-}] \quad \text{(activities in the water sample)}$$

An example is shown in Figure 4.4. Here the oxidation of pyrite (FeS_2) in the upper part of the profile produces large amounts of sulfate and equilibrium with gypsum appears to place an upper limit to dissolved calcium and sulfate concentrations.

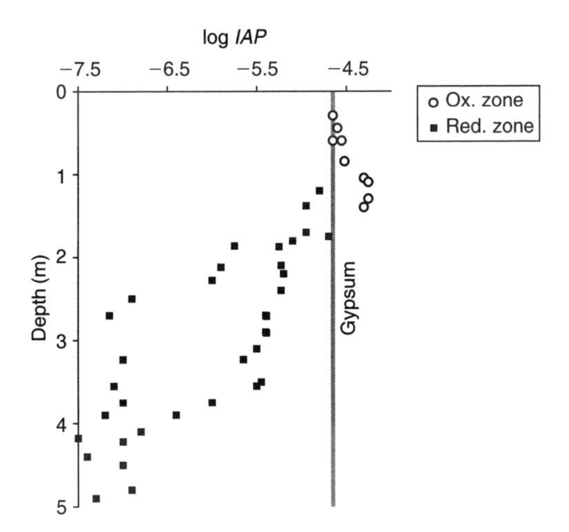

Figure 4.4. The IAP_{gypsum} compared with the solubility product at the site of pyrite oxidation in pore waters of swamp sediments (Postma, 1983).

Saturation conditions may also be expressed as the ratio between *IAP* and *K*, or the *saturation state* Ω:

$$\Omega = IAP/K \tag{4.22}$$

Thus for $\Omega = 1$ there is equilibrium, $\Omega > 1$ indicates supersaturation and $\Omega < 1$ subsaturation. For larger deviations from equilibrium, a logarithmic scale can be useful given by the *saturation index SI*:

$$SI = \log\,(IAP/K) \tag{4.23}$$

For $SI = 0$, there is equilibrium between the mineral and the solution; $SI < 0$ reflects subsaturation, and $SI > 0$ supersaturation. For comparison, it can be useful to normalize Ω and SI to the number of solutes ν in the ion activity product. Thus, $\nu = 2$ for $CaCO_3$ (calcite) and $\nu = 4$ for $CaMg(CO_3)_2$ (dolomite). The saturation state then becomes $\Omega^{1/\nu}$ and the saturation index is SI/ν.

Different types of information can be obtained from saturation data. In few cases an actual control of solutes by equilibrium with a mineral can be demonstrated as clearly as in Figures 4.1 and 4.4. Often there is no equilibrium and then the saturation state merely indicates in which direction the processes may go; for subsaturation dissolution is expected, and supersaturation suggests precipitation.

4.3 MASS ACTION CONSTANTS AND THERMODYNAMICS

Thermodynamics is the science concerned with energy distributions among substances in a system. It offers an impressive framework of formulas which can be derived from a few basic laws. An introduction to thermodynamics can be found elsewhere (Lewis and Randall, 1961; Denbigh, 1971; Nordstrom and Munoz, 1994; Anderson and Crerar, 1993). Here we confine ourselves to some practical aspects and in particular to the calculation of mass action constants and their dependency on temperature from thermodynamic tabulations.

4.3.1 *The calculation of mass action constants*

For the general reaction:

$$a\text{A} + b\text{B} \leftrightarrow c\text{C} + d\text{D}$$

we may write

$$\Delta G_r = \Delta G_r^0 + RT \ln \frac{[\text{C}]^c[\text{D}]^d}{[\text{A}]^a[\text{B}]^b} \tag{4.24}$$

where ΔG_r is the change in Gibbs free energy (kJ/mol) of the reaction, ΔG_r^0 is the standard Gibbs free energy of the reaction and equal to ΔG_r when each product or reactant is present at unit activity (so that the log term becomes zero) at a specified standard state (25°C and 1 atm), $[i]$ denotes the activity of i, R is the gas constant (8.314×10^{-3} kJ/mol/deg), T is the absolute temperature, Kelvin (Kelvin = °C + 273.15).

In older texts, energy is often expressed in kcal/mol which can be converted to kJ/mol by multiplying with 4.184 J/cal (hence $R = 1.987 \times 10^{-3}$ kcal/mol/deg). The prefix Δ is used because energy can be measured only as relative amounts. The direction in which the reaction will proceed is indicated by ΔG_r:

$\Delta G_r > 0$ the reaction proceeds to the left;
$\Delta G_r = 0$ the reaction is at equilibrium;
$\Delta G_r < 0$ the reaction proceeds to the right.

Therefore, in the case of equilibrium, Equation (4.24) reduces to:

$$\Delta G_r^0 = -RT \ln \frac{[\text{C}]^c[\text{D}]^d}{[\text{A}]^a[\text{B}]^b} \tag{4.25}$$

Note that the activity product in the last term is equal to the mass action constant K (Equation 4.0):

$$\Delta G_r^0 = -RT \ln K \tag{4.26}$$

Back substitution of Equation (4.26) in (4.24) results in

$$\Delta G_r = -RT \ln K + RT \ln \frac{[\text{C}]^c[\text{D}]^d}{[\text{A}]^a[\text{B}]^b} \tag{4.27}$$

In Equation (4.27) the distance from equilibrium is expressed in terms of the mass action constant and the solution composition, a formulation analoguous to the saturation index *SI* (Equation 4.23).

Equation (4.26) has the practical application that it allows us to calculate the mass action constant for any reaction from tabulated data of ΔG_f^0 for dissolved substances, minerals, and gases. ΔG_f^0 is the *free energy of formation, i.e.* the energy needed to produce one mole of a substance from pure elements in their most stable form. The latter (and the H^+ ion) have by definition zero values. Tabulations are normally given for 25°C and 1 atm pressure. ΔG_r^0 is calculated from:

$$\Delta G_r^0 = \Sigma \Delta G^0_{f\ products} - \Sigma \Delta G^0_{f\ reactants}$$

A number of compilations of thermodynamic data concerning mineral-water systems are available (Helgeson et al., 1978; Robie et al., 1978; Wagman et al., 1982; Cox et al., 1989; Woods and Garrels, 1987, Nordstrom and Munoz, 1994, etc.). Preferably, consistent sets of data, as presented in the first four references, should be used as otherwise erratic values can be obtained. For example, the listed $\Delta G^0_{f\,calcite}$ could have been calculated from solubility measurements using values of $\Delta G^0_{f CO_3^{2-}}$ and $\Delta G^0_{f Ca^{2+}}$ by the reverse procedure of Example 4.4. When $K_{calcite}$ subsequently is calculated from thermodynamic tables with a different $\Delta G^0_{f\,CO_3^{2-}}$ value, the result will clearly be erroneous. Even when these precautions are taken you will discover that different compilations may give slightly different constants. For example, the value of $K_{calcite}$, calculated in Example 4.4, is slightly higher than the currently accepted value of $10^{-8.48}$ (Plummer and Busenberg, 1982) which includes corrections for aqueous complexes.

EXAMPLE 4.4. *Calculation of solubility products from Gibbs free energy data*
Calculate the solubility product of calcite from the Gibbs free energies of formation at 25°C (Wagman et al., 1982):

$$\Delta G^0_{f CaCO_3} = -1128.8 \text{ kJ/mol}$$
$$\Delta G^0_{f Ca^{2+}} = -553.6 \text{ kJ/mol}$$
$$\Delta G^0_{f CO_3^{2-}} = -527.8 \text{ kJ/mol}$$

ANSWER:
For the reaction $CaCO_3 \leftrightarrow Ca^{2+} + CO_3^{2-}$
we obtain

$$\Delta G_r^0 = \Delta G^0_{f Ca^{2+}} + \Delta G^0_{f CO_3^{2-}} - \Delta G^0_{f CaCO_3}$$
$$\Delta G_r^0 = -553.6 - 527.8 + 1128.8 = 47.4 \text{ kJ/mol}$$
$$\Delta G_r^0 = -RT \ln K$$
$$47.4 = -8.314\times10^{-3} \times 298.15 \times 2.303 \log K = -5.708 \log K$$
$$\log K = 47.4/ -5.708 = -8.30$$

4.3.2 *Calculation of mass action constants at different temperature*

In nature, groundwater is generally not found at the standard conditions of 25°C and 1 atm pressure. While variations in pressure have little effect on the values of the mass action constants, the temperature variations are important. Variations of mass action constants with temperature are usually calculated with the Van't Hoff equation:

$$\frac{\mathrm{d}\ln K}{\mathrm{d}T} = \frac{\Delta H_r}{RT^2} \tag{4.28}$$

where ΔH_r is the *reaction enthalpy*, or the heat lost or gained by the chemical system. For exothermal reactions ΔH_r is negative (the *system* loses energy and heats up) and for endothermal reactions ΔH_r is positive (the system cools). The calculation of ΔH_r and the definition of the standard state is analogous to that of ΔG_r^0. At 25°C, the value of the reaction enthalpy, ΔH_r^0, is calculated from the formation enthalphies, ΔH_f^0, which are listed in thermodynamic tables. Equation (4.28) shows that K increases with temperature for positive ΔH_r^0 (the slope of $\ln K$ with temperature is positive) and decreases with temperature for negative ΔH_r^0. Usually, ΔH_r^0 is constant within the range of a few tenths of degrees and therefore (4.28) can be integrated to give for two temperatures:

$$\log K_{T_1} - \log K_{T_2} = \frac{-\Delta H_r^0}{2.303R}\left(\frac{1}{T_1} - \frac{1}{T_2}\right) \tag{4.29}$$

This equation is generally applicable in groundwater environments.

EXAMPLE 4.5. *Temperature dependency of the solubility product*
Calculate the solubility product of calcite at 10°C from the following enthalphies of formation (Wagman et al., 1982):

$$\Delta H^0_{f\,\mathrm{CaCO_3}} = -1206.9 \text{ kJ/mol}$$
$$\Delta H^0_{f\,\mathrm{Ca^{2+}}} = -542.8 \text{ kJ/mol}$$
$$\Delta H^0_{f\,\mathrm{CO_3^{2-}}} = -677.1 \text{ kJ/mol}$$

ANSWER:
For the reaction $CaCO_3 \leftrightarrow Ca^{2+} + CO_3^{2-}$
we may write,

$$\Delta H_r^0 = -542.8 + (-677.1) - (-1206.9) = -13.0 \text{ kJ/mol}$$

which means that the reaction is exothermal: the system heats up when calcite dissolves. The difference in $\log K$ between 25°C and 10°C, according to Equation (4.29) is:

$$\log K_{25} - \log K_{10} = \frac{-\Delta H_r^0}{2.303\,R}\left(\frac{1}{298.15} - \frac{1}{283.15}\right)$$
$$= \frac{-13.0}{2.303 \times 8.314\times10^{-3}}\left(\frac{1}{298.15} - \frac{1}{283.15}\right)$$
$$= -0.12$$

In EXAMPLE 4.4 it was calculated that $\log K_{\mathrm{calcite}}$ at 25°C is -8.30, so that at 10°C $K_{\mathrm{calcite}} = -8.30 + 0.12 = -8.18$. Thus for an exothermal reaction the solubility increases with decreasing temperature and vice versa for an endothermal reaction.

QUESTION:
Calculate the temperature change when 3 mmol calcite dissolve in 1 L water? (*Hint*: to heat 1 L water from 15 to 16°C needs 1 kcal)
ANSWER: temperature increases by 0.01°C

4.4 EQUILIBRIUM CALCULATIONS WITH PHREEQC

Manual calculations are useful to gain a basic understanding of the underlying principles. Once this basic understanding is reached the calculation of ion activities and saturation states becomes tedious and can be left to the computer. Here we use the code PHREEQC (Parkhurst and Appelo, 1999) which can be downloaded as freeware and has become the standard for doing a variety of hydrogeochemical calculations. Appendix A contains a quick reference guide in the form of a number of *get-going* sheets which explain the basic features of the code. A full description of many alternatives for input and the mathematical backgrounds can be found in the manual of the program by Parkhurst and Appelo (1999). Acquisition and installation of the program is explained in the get-going sheet *Getting started*. PHREEQC was developed for calculating "*real world*" hydrogeochemistry and is a powerful tool for modeling data. It is also instructive to try things out and to gain an understanding how concentrations are affected by chemical processes and transport. One thing is important, however. The results of geochemical computer programs should always be inspected critically since the reactions are often intricate and the results may depend on factors like the quality of the database or the sequence of the calculations.

4.4.1 *Speciation calculations using PHREEQC*

The input file of PHREEQC is organized in KEYWORDS and associated data blocks. To do a speciation calculation, we use the keyword SOLUTION followed by the water composition, and PHREEQC then calculates the ion activities and the saturation states for relevant minerals, cf. Example 4.6.

EXAMPLE 4.6. *Calculate the speciation of a water analysis using PHREEQC*
We construct an input file for the Windows version of PHREEQC. The keyword SOLUTION *m-n* is used to define the composition of the solution and analysis A, used in Example 4.1, is entered (*m* and *n* are integers to indicate that the composition is for a range of solutions). Text following a "#" in the input file is explanatory and is not read by the program. The keyword END is the signal for PHREEQC to start the calculations.

```
SOLUTION 1 Analysis A          # keyword, solution number and name
 -units ppm                    # default units are mmol/kg H2O
 Ca 10
 F 5.5
END                            # keyword
```

Run the file by pressing the calculator icon, or Alt+C, S. Quickly, the program reports "Done" and if Enter is pressed, the Output tab appears where the results of the calculations are recorded. The solution composition is listed as molality and as number of moles of Ca and F. In this case the two are equal since PHREEQC by default assumes that the solution consist of 1 kg H_2O (but it may be defined otherwise in the input file). Then follows the "Description of solution" which gives default values for pH, pe and temperature (they were not entered under keyword SOLUTION), and calculated values for ionic strength, alkalinity and electrical balance. Note that the solution is not well balanced. Next follows the "Distribution of species", first for the water species and then for the elements in alphabetical order. The last section gives the saturation indices. The solution is subsaturated with respect to fluorite, as was calculated in Example 4.1, but the *SI* values are not the same. This is because PHREEQC considers corrections for activity and complexing and also because the log *K* value in the PHREEQC database is slightly different.

```
---------------------------------Solution composition------------------------------

      Elements            Molality            Moles

      Ca                  2.495e-04           2.495e-04
      F                   2.895e-04           2.895e-04

----------------------------------Description of solution--------------------------


   pH                                          =  7.000
   pe                                          =  4.000
   Activity of water                           =  1.000
   Ionic strength                              =  6.427e-04
   Mass of water (kg)                          =  1.000e+00
   Total alkalinity (eq/kg)                    = -4.154e-08
   Total carbon (mol/kg)                       =  0.000e+00
   Total CO2 (mol/kg)                          =  0.000e+00
   Temperature (deg C)                         = 25.000
   Electrical balance (eq)                     =  2.095e-04
   Percent error, 100*(Cat-|An|)/(Cat+|An|) = 26.61
                                  Iterations =  3
   Total H                                     =  1.110124e+02
   Total O                                     =  5.550622e+01


---------------------------------Distribution of species---------------------------

                                                        Log           Log          Log
      Species        Molality          Activity      Molality      Activity       Gamma

      OH-            1.030e-07         1.001e-07      -6.987        -7.000        -0.012
      H+             1.028e-07         1.000e-07      -6.988        -7.000        -0.012
      H2O            5.551e+01         1.000e+00      -0.000        -0.000         0.000
Ca             2.495e-04
      Ca+2           2.489e-04         2.221e-04      -3.604        -3.653        -0.049
      CaF+           5.590e-07         5.431e-07      -6.253        -6.265        -0.012
      CaOH+          3.794e-10         3.687e-10      -9.421        -9.433        -0.012
F              2.895e-04
      F-             2.889e-04         2.807e-04      -3.539        -3.552        -0.012
      CaF+           5.590e-07         5.431e-07      -6.253        -6.265        -0.012
      HF             4.209e-08         4.210e-08      -7.376        -7.376         0.000
      HF2-           4.667e-11         4.534e-11     -10.331       -10.343        -0.012
      H2F2           4.618e-15         4.619e-15     -14.336       -14.335         0.000
H(0)           1.416e-25
      H2             7.078e-26         7.079e-26     -25.150       -25.150         0.000
O(0)           0.000e+00
      O2             0.000e+00         0.000e+00     -42.080       -42.080         0.000


---------------------------------Saturation indices--------------------------------


      Phase            SI          log IAP     log KT

      Fluorite        -0.16        -10.76      -10.60        CaF2
      H2(g)          -21.96        -22.00       -0.04        H2
```

```
H2O(g)          -1.59      -0.00      1.59      H2O
O2(g)           -39.19     44.00      83.19     O2
Portlandite     -12.45     10.35      22.80     Ca(OH)2
```

4.4.2 *The PHREEQC database*

The database of PHREEQC contains the definitions of chemical species, complexes, mineral solubilities etc. and can be viewed by clicking on the *Database* tab. The database is structured by a number of keywords, each followed by chemical definitions and constants needed to do the calculations. The first keyword is SOLUTION_MASTER_SPECIES and defines the elements in solution (Table 4.4).

The first column gives the element names, like Ca, Na and Alkalinity, which are used to enter a concentration under keyword SOLUTION. These names must start with a capital letter. Elements with several redox levels may be repeated with the number of electrons lost, for example Fe, Fe(+2) and Fe(+3). When the redox level of an element is specified in the input file, PHREEQC will maintain this during the initial speciation calculation regardless of the selected pe. The pe $= -\log[e^-]$ and indicates the general redox level of a solution. However, when "total" Fe is input, the distribution over Fe^{2+} and Fe^{3+} will be calculated from the pe. Always keep a watchful eye on the redox species and see if they remain at the expected redox levels during the calculations and consult Chapter 9 for more details on redox calculations.

The second column lists the aqueous species that are used primarily in the speciation calculations. The third column gives the contribution of that species to the alkalinity and requires explanation. Operationally, the alkalinity is defined as the amount of acid that is needed to bring the pH down to about 4.5 (cf. Chapter 5). For most waters, the alkalinity equals $Alk = m_{OH^-} + m_{HCO_3^-} + 2m_{CO_3^{2-}}$, but PHREEQC considers also all the other species that consume protons. Thus, in the output of Example 4.6, the Total Alkalinity is negative, -4.154×10^{-8} (eq/kg), due to the presence of 4.209×10^{-8} M HF. The contribution of F^- (the aqueous master species) to the alkalinity is 0, and therefore HF is considered to be dissociated at pH = 4.5 (the approximate endpoint of an alkalinity titration), producing an equivalent amount of protons which contribute negatively to the alkalinity.

The next two columns give the chemical formula that is used to convert grams into moles when concentrations are entered in mg/L or ppm, and the atomic weight of the element.

Table 4.4. PHREEQC elements in the database.

```
SOLUTION_MASTER_SPECIES
#
#element      species     alk     gfw_formula     element_gfw
#
Ca            Ca+2         0.0    Ca              40.08
Mg            Mg+2         0.0    Mg              24.312
Na            Na+          0.0    Na              22.9898
Fe            Fe+2         0.0    Fe              55.847
Fe(+2)        Fe+2         0.0    Fe
Fe(+3)        Fe+3        -2.0    Fe
S             SO4-2        0.0    SO4             32.064
S(6)          SO4-2        0.0    SO4
S(-2)         HS-          1.0    S
# etc...
```

QUESTIONS:
Find the dominant species of F and Fe(3) at pH 4.5?
ANSWER: The input file `SOLUTION 1; pH 4.5; F 1e-3; Fe(3) 1e-3; END` shows: 95% F^- and 90% $Fe(OH)_2^+$
Why has Fe(3) an alkalinity contribution of -2, cf. Table 4.4?
ANSWER: Fe(3) is dominantly $Fe(OH)_2^+$ at pH 4.5
If you would duly consider the hydroxy complexes of Fe^{3+}, what should be the alkalinity contribution?
ANSWER: 0.9×-2 (for $Fe(OH)_2^+$) + 0.09×-1 (for $FeOH^{2+}$) = -1.89

Under keyword SOLUTION_SPECIES (Table 4.5) the various aqueous complexes are tabulated in the form of an association reaction, with log_k, the association constant. The definition holds for the species right after the equal sign. The first reactions in Table 4.5 are identity reactions that define the primary master species (as entered in the second column of SOLUTION_MASTER_SPECIES) and with log_k = 0. Changes of log_k with temperature are calculated from the reaction enthalphy delta_h (Section 4.3.2) or with an analytical expression. The different options to calculate the activity coefficient from the ionic strength are controlled by the parameter −gamma. For major ions the database PHREEQC.DAT employs the Truesdell-Jones Equation (4.13), the two parameters *a* and *b* follow −gamma. For minor ions the Debye-Hückel Equation (4.12) is used. The ion-size parameter å is given, in Ångstrom, as the first parameter following −gamma, while the second parameter is set to zero. If the line −gamma is absent, then the Davies Equation (4.14) is used. The options are illustrated in Table 4.5. For neutral species, like uncharged complexes and gases, PHREEQC uses log $\gamma = 0.1\, I$.

Table 4.5. Definition of species and complexes in solution, and parameters for activity corrections in the PHREEQC.DAT database.

```
SOLUTION_SPECIES
H+ = H+
          log_k     0.0
          -gamma    9.0       0.0             # Debye-Hückel Equation (4.12)
Ca+2 = Ca+2
          log_k     0.0
          -gamma    5.0       0.1650          # Truesdell-Jones Equation (4.13)
Pb+2 = Pb+2
          log_k     0.0
                                              # Davies Equation (4.14)
Ca+2 + SO4-2 = CaSO4
          log_k     2.3
                                              # uncharged complex
          delta_h   1.650     kcal
# etc...
```

QUESTIONS:
Change the activity coefficient for Pb^{2+} from the Davies to the Debye Hückel equation (find *å* in Table 4.2).
Find *SI* for cerrusite ($PbCO_3$) using the two –gamma's for a solution with C 1, Pb 1e−3, Na 500, Cl 500 mM, pH 7?
ANSWER: *SI* = -1.00 and -1.02 for Davies and Debye-Hückel, respectively (a tiny difference!)

Repeat the calculation for strontianite ($SrCO_3$) for 1 μM Sr^{2+} in 1.5 M NaCl, and compare with the Truesdell-Jones equation?

ANSWER: $SI = -3.61$ (Davies), -4.23 (Debye-Hückel), -4.02 (Truesdell-Jones)

In the PHREEQC database, the keyword PHASES (Table 4.6) lists minerals and gases for which saturation indices are calculated. Note that the equations for minerals are written as dissociation reactions, therefore log_k is for the dissociation reaction. The variation of log_k with temperature is again calculated from the reaction enthalphy (delta_h, section 4.3.2), or from an analytical expression of the type:

$$\log K = A_1 + A_2 \cdot T + A_3 / T + A_4 \cdot \log T + A_5 / T^2 \tag{4.30}$$

where T is temperature in Kelvin and the numbers $A_{1...5}$ are fit parameters. In the input file (Table 4.6), the numbers $A_{1...5}$ follow the indentifier – analytic. If an analytical expression is present, it overrides the reaction enthalpy method.

Table 4.6. Definition of minerals and their solubility in the PHREEQC.DAT database.

```
PHASES
Fluorite
         CaF2 = Ca+2 + 2 F-
         log_k                      -10.6
         delta_h                      4.69     kcal
         -analytic                   66.348    0.0     -4298.2     -25.271
# etc...
```

It is important to emphasize that the results obtained by the speciation calculation never are better than the quality of the analytical input and the constants used. Analytical sources of error have already been discussed in Chapter 1 and particularly errors in pH measurements may affect the results of saturation calculations significantly. For example, the reaction for equilibrium between gibbsite ($Al(OH)_3$) and water is :

$$3H^+ + Al(OH)_3 \leftrightarrow Al^{3+} + 3H_2O$$
$$\log(IAP) = \log[Al^{3+}] + 3pH$$

An error in the pH measurement of about 0.33 unit, which is not uncommon, would affect the log(IAP) by a whole unit! Also for carbonate equilibria (see Chapter 5), errors in the pH may affect the results of saturation calculations seriously and the uncertainty introduced by such errors should be evaluated carefully in each individual case.

The mass action constants which are used in the programs should be internally consistent. For example, the solubility product of gypsum listed in the database is probably derived from experiments during which gypsum has been equilibrated with water. In the calculations to derive K_{gypsum}, a stability constant for the aqueous complex $CaSO_4^0$ has been used. Obviously, the same stability constant should be used in all subsequent calculations involving K_{gypsum}. For major species, good quality databases, like PHREEQC.DAT, are available. However, for trace components the user should maintain a healthy scepticism concerning the quality and internal consistency of the data. In addition to PHREEQC.DAT there are several other databases available in the PHREEQC package and they are listed in Table 4.7.

Table 4.7. Databases distributed with the PHREEQC program.

Database	Features
PHREEQC.DAT	Limited but most consistent database
WATEQ4F.DAT	The PHREEQC.DAT database extended with many heavy metals
MINTEQ.DAT	The database developed for the US EPA MINTEQ program (Allison et al., 1991). Includes some organic compounds
LLNL.DAT	A huge database containing many elements and with a large temperature range. Developed for the program EQ3/6 (Wolery, 1992).

You can change the database with keyword DATABASE (this must be the first keyword in the input file).

The critical use of the available databases and their effect on the calculations is illustrated in Example 4.7.

EXAMPLE 4.7. *Solubility of quartz at 150°C*
Calculate the solubility of Quartz at 150°C, using the three databases PHREEQC.DAT, MINTEQ.DAT and LLNL.DAT in PHREEQC. The log *K*'s and reaction enthalpies are for the reaction:

$$SiO_{2(qu)} + 2H_2O \leftrightarrow H_4SiO_4^0$$

$\log K_{25^\circ C}$	Reaction enthalpy	Database
−3.98	5.99 kcal/mol	PHREEQC.DAT
−4.006	6.22 kcal/mol	MINTEQ.DAT
−3.9993	32.949 kJ/mol	LLNL.DAT

The reaction enthalpy in LLNL.DAT of 32.949 kJ/mol corresponds to 32.949/4.184 = 7.88 kcal/mol and is quite different from the two other databases. We manually calculate the $\log K_{150^\circ C}$ using Equation (4.29), and compare the results with those from PHREEQC using the following input file, run separately with each of the three databases.

```
DATABASE phreeqc.dat      # also with llnl.dat or minteq.dat
SOLUTION 1
  temp 150
  Si 1 Quartz             # the Si concentration is adapted to equilibrium
                          # with quartz
END
```

The answer is:

Van 't Hoff $\log K_{150^\circ C}$	PHREEQC $\log K_{150}$	Database
−2.68	−2.66 (Polynomial)	PHREEQC.DAT
−2.66	−2.66 (Van 't Hoff)	MINTEQ.DAT
−2.29	−2.71 (Polynomial)	LLNL.DAT

Databases LLNL.DAT and PHREEQC.DAT contain a polynomial expression (4.30) for temperature dependency which overrides the reaction enthalphy approach. The log *K* has increased at 150°C, corresponding to the endothermal reaction. The polynomial from PHREEQC.DAT and the Van 't Hoff formula with constant enthalpy used in MINTEQ.DAT yield the same log *K* over this temperature range and are close to the value calculated manually. The polynomial expression from the LLNL.DAT also gives nearly the same value. Apparently the reaction enthalpy listed in LLNL.DAT is incorrect.

4.4.3 *Mineral equilibration*

Besides for calculating the saturation state of a solution with respect to different minerals, PHREEQC can also be used to bring the solution in equilibrium with a specified mineral. This feature is illustrated for the fluorite/gypsum experiment of Figure 4.1, by calculating the water compositions for point B, equilibrium for fluorite by addition of gypsum, and point C, equilibrium with both fluorite and gypsum (Example 4.8).

EXAMPLE 4.8. *Equilibrate a water sample with minerals*
Modify the input file from Example 4.6 and use keyword EQUILIBRIUM_PHASES to obtain equilibrium, first with fluorite by adding gypsum (point B), then with gypsum and fluorite to arrive at point C. Keyword END separates the simulations. The results of a simulation may be stored in temporary memory with keyword SAVE solution *no*, and can be recalled again with USE solution *no* in a later simulation (*no* is a number or a range of numbers *m–n*).

```
SOLUTION 1 Analysis A
 -units ppm
 Ca 10
 F 5.5
END                                       # end first simulation

# Calculate point B
USE solution 1                            # use already stored solution

EQUILIBRIUM_PHASES 1                      # equilibrate with minerals
  Fluorite 0 Gypsum                       # Mineral name, SI, reactant
SAVE solution 2                           # save solution (Point B)
END                                       # end second simulation

# Calculate point C
USE solution 2

EQUILIBRIUM_PHASES 2
  Fluorite 0 0                            # SI = 0, initially 0 moles
  Gypsum                                  # default SI = 0, 10 moles
END
```

Note that equilibrium can be reached by adding a reactant. The reactant can be a phase (defined with keyword PHASES) or a chemical formula (for example Fluorite 0 $CaSO_4$ instead of Fluorite 0 Gypsum in the second simulation). If the reactant is omitted, the mineral itself reacts until the requested *SI* is attained, or until it is exhausted.

An excerpt from the output of the last simulation is copied below, showing that 0.1 mmoles of fluorite have precipitated, while 15.4 mmoles of gypsum have dissolved ('Delta' is positive for fluorite, negative for gypsum). The fluoride concentration has decreased markedly to 0.087 mmol/L × 19 g/mol = 1.6 mg/L, which is almost the same as we calculated by hand. However, the concentration of 15.6 mmol SO_4^{2-}/L × 96 g/mol = 1500 mg SO_4^{2-}/L is much higher than was obtained before and will lead to serious intestinal problems that make the water unfit for consumption.

```
----------------------------------Phase assemblage----------------------------------

                                                    Moles in assemblage
Phase             SI   log IAP    log KT        Initial          Final           Delta

Fluorite        0.00    -10.60    -10.60      0.000e+00      1.011e-04       1.011e-04
Gypsum          0.00     -4.58     -4.58      1.000e+01      9.985e+00      -1.544e-02
```

```
--------------------------------Solution composition--------------------------------

Elements          Molality          Moles

Ca                1.572e-02         1.573e-02
F                 8.731e-05         8.736e-05
S                 1.557e-02         1.558e-02
```

QUESTION:
Use EQUILIBRIUM_PHASES to find the log *K*'s for quartz at 150°C in Example 4.7?

ANSWER: `SOLUTION 1; temp 150; EQUILIBRIUM_PHASES 1; Quarts; END`

4.5 SOLID SOLUTIONS

Up to now the minerals have been considered to be pure phases. However, the analysis of naturally occurring minerals by microanalytical devices, like the microprobe, shows that pure minerals are the exception rather than the rule. Minerals of variable composition may be considered as *solid solutions* of pure end-member minerals. Detrital minerals, originally formed at high temperatures, such as plagioclases, amphiboles, pyroxenes, etc. are solid solutions. Also low temperature phases like carbonates form extensively solid solutions, the best studied example being Mg-calcite which we treat in more detail in Chapter 5. Numerous other cases are known such as replacement of Ca^{2+} by Cd^{2+}, Zn^{2+}, Cu^{2+}, Co^{2+}, Fe^{2+} and Mn^{2+} in calcite (Rimstidt et al., 1998), Ca^{2+} in rhodochrosite ($MnCO_3$), Mn^{2+} in siderite ($FeCO_3$) and Sr^{2+} in aragonite ($CaCO_3$). Additional examples are substitution of F^- for OH^- in apatite ($Ca_5OH(PO_4)_3$), as happens in tooth enamel, and substitution of Al^{3+} in goethite (FeOOH). An example of the latter is illustrated in Figure 4.5.

4.5.1 *Basic theory*

Substitution of Cd^{2+} in $CaCO_3$ can be used as an example of the binary solid solution between otavite ($CdCO_3$) and calcite ($CaCO_3$). Equilibrium between the aqueous solution and the solid solution is defined by two mass action equations which must be fulfilled simultaneously:

$$\underset{calcite}{CaCO_3} \leftrightarrow Ca^{2+} + CO_3^{2-} \qquad K_{cc} = \frac{[Ca^{2+}][CO_3^{2-}]}{[CaCO_3]} = 10^{-8.48} \tag{4.31}$$

and

$$\underset{otavite}{CdCO_3} \leftrightarrow Cd^{2+} + CO_3^{2-} \qquad K_{ot} = \frac{[Cd^{2+}][CO_3^{2-}]}{[CdCO_3]} = 10^{-12.1} \tag{4.32}$$

Here $[CaCO_3]$ and $[CdCO_3]$ are the activities of the two components in the solid phase. For a pure solid phase the activity is equal to one, and usually omitted in the mass action expression.

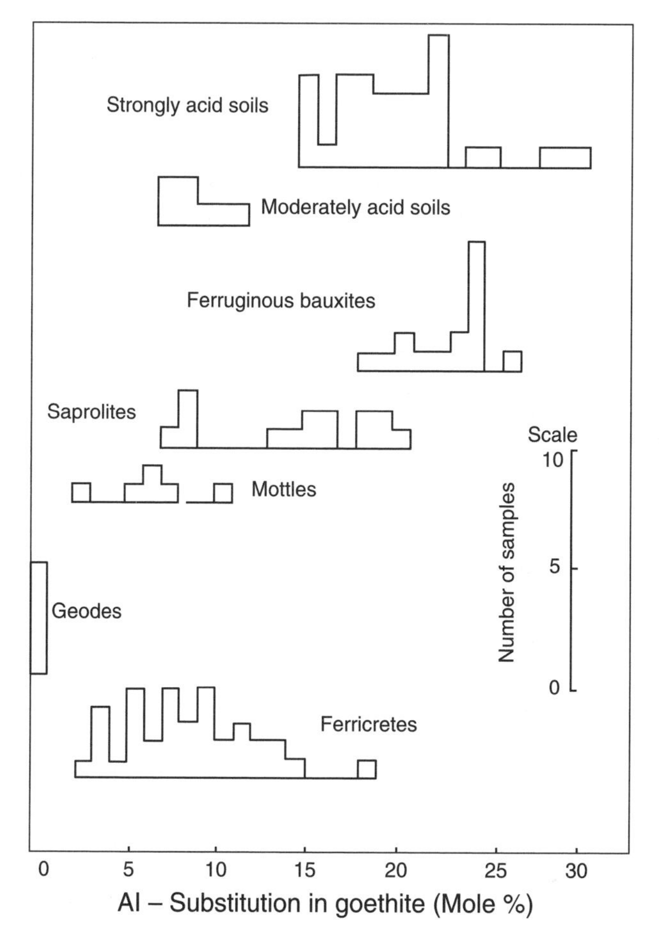

Figure 4.5. The aluminum content of goethites in various soil environments (Fitzpatrick and Schwertmann, 1982).

However for a solid solution the solid phase activity is related to the mole fraction (Table 4.1) by a relation like:

$$[\mathrm{CdCO_3}] = \lambda_{\mathrm{CdCO_3}} \cdot \chi_{\mathrm{CdCO_3}} \tag{4.33}$$

where λ is the activity coefficient that corrects for non-ideal behavior and χ is the mole fraction. In the case of ideal solid solution, λ is equal to one and:

$$[\mathrm{CdCO_3}] = \left(\frac{mol_{\mathrm{Cd}}}{mol_{\mathrm{Ca}} + mol_{\mathrm{Cd}}} \right)_{ss} = \chi$$

and also:

$$[\mathrm{CaCO_3}] = (1 - \chi)$$

Subtraction of reaction (4.32) from (4.31) results in:

$$\mathrm{Cd^{2+}} + \mathrm{CaCO_3} \leftrightarrow \mathrm{CdCO_3} + \mathrm{Ca^{2+}} \tag{4.34}$$

with the mass action expression:

$$K = \frac{[CdCO_3]}{[CaCO_3]} \cdot \frac{[Ca^{2+}]}{[Cd^{2+}]} = \frac{K_{cc}}{K_{ot}} = 10^{3.62} \tag{4.35}$$

What is the effect of calcite and otavite being present as a solid solution instead of separate minerals? Consider the following two situations. In the first situation a solution is in equilibrium with calcite and otavite as separate minerals. Then in Equation (4.35) $[CaCO_3] = 1$ and $[CdCO_3] = 1$ and therefore the ratio $[Ca^{2+}] / [Cd^{2+}]$ in solution is fixed to $10^{3.62} = 4169$. The concentration of Cd^{2+} is much smaller than of Ca^{2+} because the solubility of otavite is much smaller than of calcite.

In the second situation Cd^{2+} is incorporated as an ideal solid solution in the calcite structure, which can be written as $Cd_\chi Ca_{(1-\chi)}CO_3$. Let us take $\chi_{Cd} = [CdCO_3] = 0.01$ and $[CaCO_3] = 0.99$ so that $[CdCO_3] / [CaCO_3] = 0.01 / 0.99 \approx 1 / 100$. Equation (4.35) predicts then that the $[Ca^{2+}] / [Cd^{2+}]$ ratio in solution must be 100 times higher than for separate minerals. If the Ca^{2+} concentration remains the same as before, then the Cd^{2+} concentration must now be a 100 fold lower, cf. Example 4.9. The formation of a solid solution may thus lower the aqueous concentration of the minor component. Therefore, solid solutions are important for scavenging heavy metals from water and for limiting the mobility of heavy metals in the environment (Tesoriero and Pankow, 1996).

EXAMPLE 4.9. *Ideal solid solution of Cd^{2+} in calcite*
Use PHREEQC to calculate the concentrations of Ca^{2+} and Cd^{2+} in equilibrium with the two minerals calcite and otavite, and compare with a solid solution of the two minerals with $\chi_{Cd} = 0.01$. Also use keyword USER_PRINT to print the ratio of the activity of Ca^{2+} and Cd^{2+} in the output file.

ANSWER:

```
SOLUTION 1
 pH        7.5        Calcite    # pH is adjusted to equilibrium with calcite
 C(4)      4.0
 Ca        1.0
 Cd        1 ug/kgw   Otavite    # Cd conc. is adjusted to equilibrium with otavite

SOLID_SOLUTION 1
 CaCdCO3                         # Name of solid solution
 -comp calcite        99         # Ideal solid solution, calcite = 99 mol
 -comp Otavite         1         # otavite 1 mol, x_Cd = 1 / (1+99) = 0.01

USER_PRINT
 -start
 10 print "aCa =", act("Ca+2"), " aCd =", act("Cd+2"), " aCa/aCd =", act("Ca+2")/act("Cd+2")
-end
END
```

gives in the output file:

```
-----------------------------------User print-----------------------------------

aCa = 7.3946e-04    aCd = 1.7732e-07    aCa/aCd = 4.1703e+03 (initial solution)
aCa = 7.3837e-04    aCd = 1.7884e-09    aCa/aCd = 4.1286e+05 (equilibrated solution)
```

First, SOLUTION 1 is defined with a fixed concentration of 1 mmol Ca^{2+}/kg H_2O (the default units) and equilibrated with calcite by adjusting the pH. The initial Cd^{2+} concentration of 1 μg/kg is adapted to equilibrium with otavite. Subsequently, the solution is equilibrated with SOLID_SOLUTION 1 with $\chi_{Cd} = 0.01$.

In PHREEQC, an ideal solid solution can be multicomponent, and have as many components "-comp ..." as relevant. The equilibration will exchange mass among solid and water, but in this example it will not affect the composition of the solid solution, because the number of moles present in the solid is much larger than in the solution. The ratio $[Ca^{2+}]/[Cd^{2+}] = 4.17\times10^3$ in the initial solution equilibrated with the separate minerals (first line in User print), and it increases 100 fold to 4.13×10^5 when equilibrated with the solid solution.

Keyword USER_PRINT calls the BASIC interpreter that has been programmed in PHREEQC and permits to customize printout of model variables. BASIC is a computer language that executes instructions from numbered lines in the order of the numbers. In line 10, act("Ca+2"), gives the activity of the Ca^{2+} species. The BASIC commands are listed in the PHREEQC manual under keyword RATES, and can be used in the various USER_... keywords and for defining rates of kinetic reactions.

QUESTIONS:
If $\chi_{Cd} = 0.001$ in the solid solution, what will $[Ca^{2+}]/[Cd^{2+}]$ become?
ANSWER: 4.13×10^6
Find the ratio m_{Ca}/m_{Cd} in Example 4.9?
ANSWER: 3.95×10^3 and 3.91×10^5, smaller because $CdHCO_3^+$ is a stronger complex than $CaHCO_3^+$: total Cd increases.
Find the ratio m_{Ca}/m_{Cd} in Example 4.9, with 0.1 M Cl^- present.
ANSWER: 6.29×10^2 and 6.22×10^4, even smaller because of strong CdCl complexes

Commonly, solid solutions show deviations from ideal behavior and the activity coefficients λ become a function of the excess free-energy of mixing. The free energy of mixing can be obtained from solubility measurements (Busenberg and Plummer, 1989), from miscibility gaps, variations in fractionation factors or other, non-ideal thermodynamic properties (Glynn, 2000). For a binary solid solution the excess free-energy of mixing, ΔG^E, can be modeled with the Guggenheim series expansion (Glynn, 1991; Guggenheim, 1967):

$$\Delta G^E = (\chi_1\chi_2)\, RT\, (a_0 + a_1(\chi_1 - \chi_2) + a_2(\chi_1 - \chi_2)^2 + \cdots) \quad (4.36)$$

where $a_0, a_1, \ldots$ are empirical coefficients (dimensionless), and χ_1 and χ_2 are the fractions of the two components in the mixture ($\chi_2 = 1 - \chi_1$). When all the coefficients are zero, $\Delta G^E = 0$ and the solid solution is ideal. When a_0 is non-zero, and the other coefficients are zero, the solid solution is classified as regular. When a_0 and a_1 are non-zero, the solid solution is termed subregular. Glynn and Reardon (1990, 1992) observed that a_0 and a_1 are mostly sufficient to relate the activity coefficients to the mole fraction (cf. Glynn, 2000).

The activity coefficients are found from the excess energy with the Gibbs-Duhem relation, and are in the subregular model given by:

$$\ln \lambda_1 = \chi_2^2(a_0 - a_1(4\chi_1 - 1)) \quad (4.37a)$$

and

$$\ln \lambda_2 = \chi_1^2(a_0 + a_1(4\chi_2 - 1)) \quad (4.37b)$$

For the solid solution $(Cd, Ca)CO_3$ Tesoriero and Pankow (1996) found that $a_0 = 1.21$ and $a_1 \approx 0$. If χ_{Cd} is small, like 0.01, $\lambda_{Cd} = \exp((\chi_{Ca})^2 \times 1.21) \approx \exp(0.98 \times 1.21) = 3.27$ and $\lambda_{Ca} = \exp((\chi_{Cd})^2 \times 1.21) \approx \exp(0.0001 \times 1.21) = 1$. This illustrates that the main component in the solid solution behaves close to ideal as long as $\chi > 0.9$. Since $\lambda_{Cd}/\lambda_{Ca} = 3.27$ it follows from

Equation (4.33) that the ratio $[CdCO_3]/[CaCO_3]$ increases compared to ideal solid solution. Equation (4.35) then requires that the ratio $[Ca^{2+}]/[Cd^{2+}]$ decreases (Example 4.10).

EXAMPLE 4.10. *Non-ideal solid solution of Cd^{2+} in calcite*
PHREEQC can model regular and subregular *binary* (two-component) solid solutions. Non-ideal behavior can be defined with the Guggenheim parameters a_0 and a_1 (Equation 4.36), or from various thermodynamic properties. We extend Example 4.9 with a regular solid solution. Simply add the following lines after the END in Example 4.9:

```
USE solution 1
SOLID_SOLUTION 2
 CaCdCO3
 -comp1 Calcite 99              # Non-Ideal solid solution, component 1
 -comp2 Otavite 1               # component 2
 -Gugg_nondim 1.21 0            # a0 = 1.21, a1 = 0, Tesoriero and Pankow, 1996.
END
```

ANSWER:

```
-----------------------------------User print-----------------------------------
aCa = 7.3838e-04          aCd = 5.8541e-09aCa/aCd = 1.2613e+05
```

The ratio $[Ca^{2+}]/[Cd^{2+}]$ is now 1.26×10^5, which is $\exp(1.21 \times 0.99^2) = 3.27$ times smaller than was calculated in Example 4.9.

For solid solution of Sr^{2+} in calcite, Tesoriero and Pankow (1996) obtained $a_0 = 5.7$. The larger value for Sr^{2+} is related to the larger ionic radius of Sr^{2+} (1.18 Å) than of Ca^{2+} (1.0 Å), while Ca^{2+} and Cd^{2+} (0.95 Å) have a nearly similar radius. Although the solubility of $SrCO_3$ is smaller than of $CaCO_3$ and therefore a fractionation factor greater than 1 is expected, Sr^{2+} is rejected from the calcite structure, as is indicated by the very large value of a_0, resulting in a high activity coefficient. However, Tesoriero and Pankow used the solubility constant of strontianite for estimating λ_{SrCO_3} which has an orthorhombic crystal structure, while calcite is rhombohedral and this does influence the thermodynamics of the system (Böttcher, 1997). For most salts, a_0 ranges from 1 to 5 (Glynn, 2000), resulting in activity coefficients in solid solutions that are greater than 1. The activity coefficients can become so high that the solid solution becomes unstable with respect to the end-members and falls apart. Figure 4.6 compares the energy levels of an ideal solid solution, a regular solid solution with $a_0 = 3$, and a mechanical mixture of the two end-members.

For a mechanical mixture, where component 2 is the minor component with mole fraction χ, the energy (kJ/mol) is:

$$\Delta G_{\text{mm}} = \chi\Delta G_{f,\text{comp2}} + (1-\chi)\Delta G_{f,\text{comp1}} \tag{4.38}$$

This is the reference level indicated by the thin dotted line in Figure 4.6. Compared to the reference level, the solid solution has the energy:

$$\Delta G_{\text{ss}} = \Delta G_{\text{mm}} + RT \cdot \chi \cdot \ln[\lambda_2\chi] + RT \cdot (1-\chi) \cdot \ln[\lambda_1(1-\chi)] \tag{4.39}$$

In an ideal solid solution $\lambda_1 = \lambda_2 = 1$, the two log terms in Equation (4.39) are both negative $(0 < \chi < 1)$, and consequently, the energy of an ideal solid solution is always less than of a mechanical mixture (Figure 4.6). Accordingly, an ideal solid solution is more stable than the two separate phases, an effect due to the entropy of mixing.

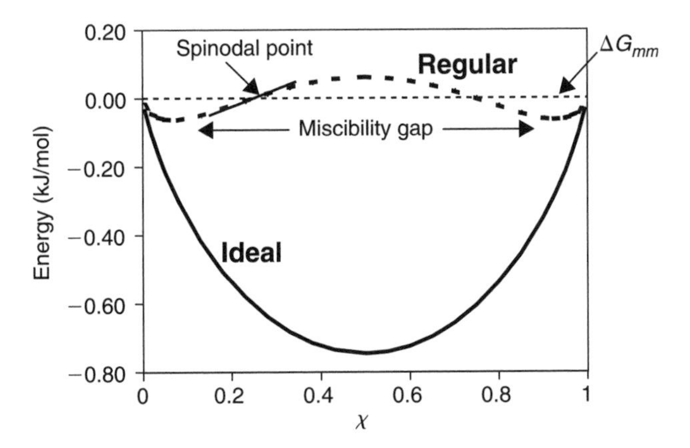

Figure 4.6. Energy of an ideal solid solution and a regular solid solution ($a_0 = 3$), compared with a mechanical mixture (thin dotted line).

The regular solid solution with $a_0 = 3$ has two minima in Figure 4.6 (at $\chi = 0.07$ and $\chi = 0.93$). Thermodynamically, the compositions in between these minima are unstable and result in a *miscibility gap*. The minima are found from:

$$\frac{d\Delta G_{ss}}{d\chi} = \frac{d}{d\chi}(\chi \ln[\lambda_2 \chi] + (1-\chi)\ln[\lambda_1(1-\chi)]) = 0 \tag{4.40}$$

and are also calculated by PHREEQC (Example 4.11). The occurrence of a miscibility gap can be tested with the condition for an inflection point on the solid solution curve ($d^2\Delta G_{ss}/d\chi^2 = 0$). The inflection point is called a *spinodal point* in thermodynamics.

EXAMPLE 4.11. *Miscibility gap in the solid solution*

Increasing a_0 of the solid solution increases also λ, and PHREEQC may calculate a hypothetical miscibility gap for the otavite/calcite solid solution when a_0 is increased to 3. Extend the input file from Example 4.9 with the following lines after the END:

```
USE solution 1
SOLID_SOLUTION 3
 CaCdCO3
 -comp1 Calcite        0
 -comp2 Otavite        0
 -Gugg_nondim          3    0    # hypothetical a0
END
```

The output file gives:

```
-----------------------------------
Description of Solid Solution CaCdCO3
-----------------------------------

                           Temperature:    298.15 kelvin
                    A0 (dimensionless):    3
                    A1 (dimensionless):    0
                           A0 (kJ/mol):    7.43708
                           A1 (kJ/mol):    0
```

```
               Critical mole-fraction of component 2:    0.5
                            Critical temperature:    447.225 kelvin
            Spinodal-gap mole fractions, component 2:    0.211324            0.788675
             Miscibility-gap fractions, component 2:    0.070720             0.92928
-------------------------------------
```

and...

```
------------------------------User print-----------------------------------
aCa = 7.3641e-04            aCd = 8.4148e-08            aCa/aCd = 8.7514e+03
```

The output file repeats the values of a_0 and a_1, and writes them in dimensionalized form by multiplying with RT (= 2.479 kJ/mol at 298 K). The miscibility gap disappears at the *critical* temperature and mole-fraction in the *alyotropic* or *eutectic* point. PHREEQC gives various definitions of this point in T-χ space following Glynn (1991) and Lippmann (1980). We also note that, compared to ideal solid solution $a_0 = 3$ has decreased the ratio $[Ca^{2+}]/[Cd^{2+}]$ by a factor of 47.

4.5.2 *The fractionation factor for solid solutions*

An alternative way to relate the concentration ratios in a solid solution with the ratios in aqueous solution is the *fractionation factor* α which already was introduced for stable isotopes (Equation 2.7). The fractionation factor is often called the distribution coefficient, but in this book we reserve the latter term for the partitioning of a single substance between solid and solution. For the otavite/calcite example, the fractionation factor is:

$$\alpha' = \frac{(mol_{CdCO_3} / mol_{CaCO_3})_{ss}}{(m_{Cd} / m_{Ca})_{aq}} \tag{4.41}$$

When $\alpha' = 1$ both components are taken up in the same ratio as present in the solution. If $\alpha' > 1$, the trace metal is preferentially taken up in the solid, and if $\alpha' < 1$, the trace metal is rejected by the solid. A theoretical fractionation factor can be derived by combining Equations (4.33) and (4.35):

$$\alpha = \frac{(\lambda_{Cd} \cdot \chi_{Cd} / \lambda_{Ca} \cdot \chi_{Ca})_{ss}}{([Cd^{2+}]/[Ca^{2+}])_{aq}} = \frac{K_{cc}}{K_{ot}} \tag{4.42}$$

Hence,

$$\alpha' = \alpha \cdot \frac{\Gamma_{Cd} / \Gamma_{Ca}}{\lambda_{Cd} / \lambda_{Ca}} \tag{4.43}$$

where Γ_i is the correction for aqueous ions, calculated by subtracting complexes from m_i, followed by multiplication with the activity coefficient, for example:

$$[Ca^{2+}] = \Gamma_{Ca}\ m_{Ca} = (m_{Ca} - \Sigma m_{Ca\text{-cplxs}}) \cdot \gamma_{Ca}.$$

A difficulty in defining α' for solutions and minerals in rocks is the often irregular and zoned composition of the solid solutions (Figure 4.7). The zonation reflects changes in the water composition and may be due to a Rayleigh process as was discussed for $\delta^{18}O$ variations in atmospheric vapor during rainout (Chapter 2). Thus, if some calcite/otavite solid solution precipitates, the Cd/Ca ratio in the aqueous solution will change because the two elements precipitate at different rates. We write the differential equation for a Rayleigh process (Equation 2.7):

$$\left(\frac{\mathrm{d}\,mol_{\mathrm{CdCO_3}}}{\mathrm{d}\,mol_{\mathrm{CaCO_3}}}\right)_{ss} = \left(\frac{-\mathrm{d}\,m_{\mathrm{Cd}}}{-\mathrm{d}\,m_{\mathrm{Ca}}}\right)_{aq} = \alpha'\left(\frac{m_{\mathrm{Cd}}}{m_{\mathrm{Ca}}}\right)_{aq} \tag{4.44}$$

which we integrate from composition b to e:

$$\ln\left(\frac{m^e_{\mathrm{Cd}}}{m^b_{\mathrm{Cd}}}\right)_{aq} = \alpha' \cdot \ln\left(\frac{m^e_{\mathrm{Ca}}}{m^b_{\mathrm{Ca}}}\right)_{aq} \tag{4.45}$$

From the change of the aqueous molar ratios in time, the fractionation factor can be calculated. Since $(\mathrm{d}\,mol_{\mathrm{Cd}}/\mathrm{d}\,mol_{\mathrm{Ca}})_{ss} = (-\mathrm{d}\,m_{\mathrm{Cd}}/-\mathrm{d}\,m_{\mathrm{Ca}})_{aq}$, the aqueous molar ratios are conserved in the solid solution, which can be analyzed in a mineral with the electron microprobe. If we replace the ratio of the aqueous concentrations with those in the solid solution at two points in space, Equation (4.45) describes a solid solution with an onion shell structure as depicted in Figure 4.7. The equation is known as the Doerner-Hoskins relation.

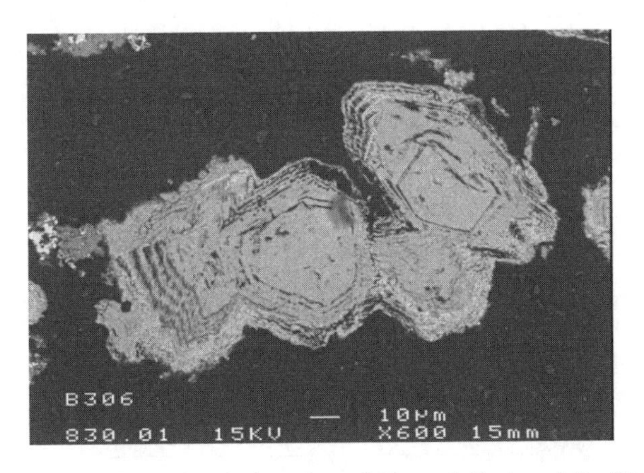

Figure 4.7. Solid solutions showing chemical zoning of Mg and Ca in calcite (Beets and Immerhauser, pers. comm.).

4.5.3 *Kinetic effects on the fractionation factor*

The incorporation of elements in different proportions in a solid solution can easily lead to kinetic effects (Rimstidt et al., 1998). During precipitation, the solutes have to diffuse through a stagnant water layer adjacent to the mineral surface and then further, on the surface, towards the crystal growth site. This situation is illustrated in Figure 4.8 and the otavite/calcite solid solution is again used as example. The diffusional flux of solute i through the stagnant water layer is:

$$F_i = -D_i(m_i^{aq} - m_i^0)\,/\,\delta \tag{4.46}$$

where D_i is the diffusion coefficient for i, m_i^{aq} is the concentration in the free solution, m_i^0 the concentration at the surface, and δ is the thickness of the stagnant layer. The diffusion coefficient will be nearly the same for HCO_3^-, Ca^{2+}, Cd^{2+} and H^+ (Section 3.5.1). The flux to the surface will then depend on the concentration difference ($m_i^{aq} - m_i^0$) and reflect the rate of precipitation of the elements. The concentration gradient will be equal but opposite in sign for HCO_3^- and H^+, and for HCO_3^- equal to the sum of Ca^{2+} and Cd^{2+}. The apparent molar fractionation factor is now given by:

$$\alpha'^a = \frac{F_{Cd} / F_{Ca}}{m_{Cd} / m_{Ca}} = \frac{(m_{Cd}^{aq} - m_{Cd}^0) / (m_{Ca}^{aq} - m_{Ca}^0)}{m_{Cd}^{aq} / m_{Ca}^{aq}} \tag{4.47}$$

At high supersaturation, m_i^0 is small compared to m_i^{aq}, and Equation (4.47) predicts that the fractionation factor α'^a will approach 1. Furthermore, the minimal surface concentration of the trace element limits the degree of supersaturation that gives the molar fractionation factor. If $m_{Cd}^0 = 0$, and $\alpha' = 4169$ (the equilibrium fractionation factor for the otavite/calcite solid solution) then Equation (4.47) can be reworked to give $m_{Ca}^{aq} / m_{Ca}^0 = 1.0002$. Any larger ratio of m_{Ca}^{aq} / m_{Ca}^0 increases the flux of Ca^{2+} relative to Cd^{2+} and thus reduces α' to an apparent α'^a. Since the ratio m_{Ca}^{aq}/ m_{Ca}^0 is indicative for the degree of supersaturation, it is evident that the equilibrium fractionation factor only can be measured in a system that is very close to equilibrium, or when the precipitation takes place very slowly. In consequence, most experimentally measured fractionation factors are kinetically controlled (Rimstidt et al., 1998).

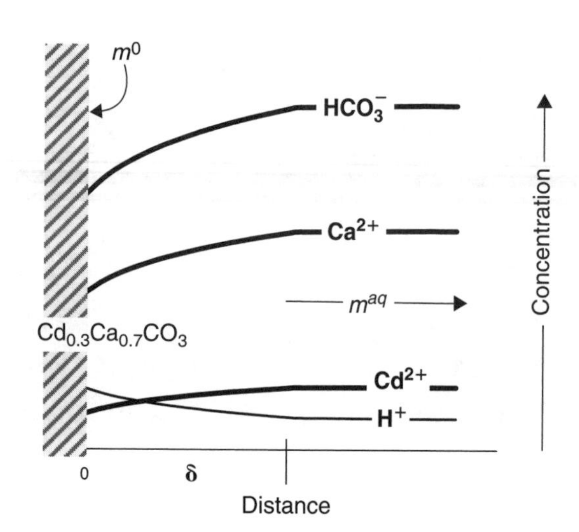

Figure 4.8. Schematic concentration gradients at the surface of an otavite/calcite solid solution.

The effect of the degree of supersaturation on the apparent fractionation factor can be calculated by assuming that the concentrations in the stagnant water layer adjacent to the mineral surface (m_i^0) are controlled by the molar (equilibrium) fractionation factor (Riehl et al., 1960; McIntyre, 1963). In that case,

$$m_{Cd}^0 = m_{Ca}^0 \cdot (F_{Cd} / F_{Ca}) / \alpha' \tag{4.48}$$

Furthermore, let's assume that the saturation ratio is given by $\Omega = m_{Ca}^{aq} / m_{Ca}^0$. This equation, in combination with (4.48), can be used to eliminate m_{Cd}^0 and m_{Ca}^0 from

$$F_{Cd} / F_{Ca} = (m_{Cd}^{aq} - m_{Cd}^0) / (m_{Ca}^{aq} - m_{Ca}^0) \tag{4.49}$$

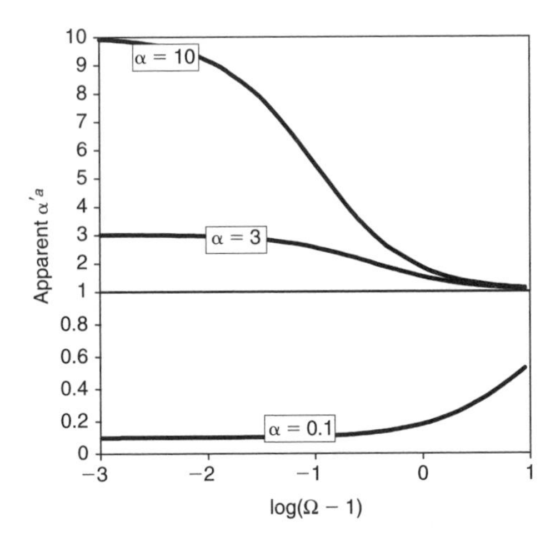

Figure 4.9. The apparent fractionation factors for solid solutions as function of a modified saturation index ($\log \Omega - 1$) for molar fractionation factors of 10, 3 and 0.1.

and to rewrite it as:

$$\frac{F_{\mathrm{Cd}} \,/\, F_{\mathrm{Ca}}}{m_{\mathrm{Cd}}^{aq} \,/\, m_{\mathrm{Ca}}^{aq}} = \frac{\alpha' \cdot \Omega}{(\Omega - 1) \cdot \alpha' + 1} = \alpha'^{a} \tag{4.50}$$

Figure 4.9 shows the apparent fractionation factor as a function of the modified saturation index ($\log(\Omega - 1)$) for molar fractionation factors $\alpha' = 10$, 3 and 0.1, calculated according to Equation (4.50). Clearly, if $\Omega = 1$ (and $\log(\Omega - 1) = -\infty$), then $\alpha'^{a} = \alpha'$, and the apparent and the molar fractionation factors are the same. On the other hand, if the solution is highly supersaturated, $\Omega \gg 1$, then $\alpha'^{a} \rightarrow 1$, as we also noted before.

The lines in Figure 4.9 can be compared with the experimental data from Tesoriero and Pankow (1996) for strontium and cadmium solid solutions in calcite (Figure 4.10).

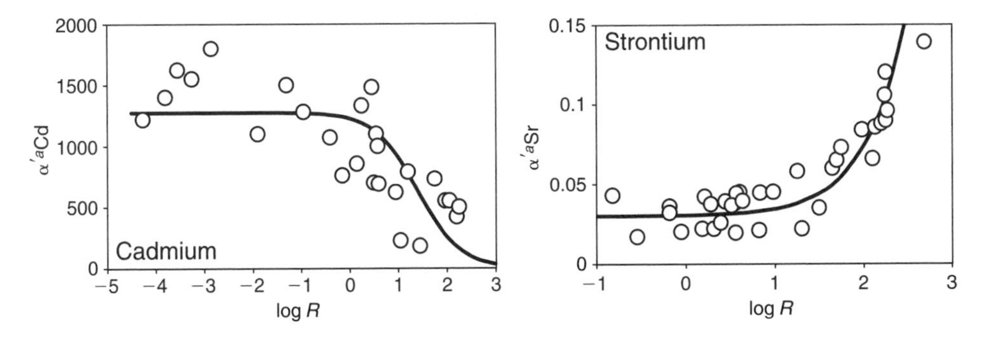

Figure 4.10. Apparent fractionation factor (α'^{a}) for solid solutions of strontium ($\alpha'^{a} < 1$) and cadmium ($\alpha'^{a} > 1$) in calcite as function of the precipitation rate. R is the precipitation rate in nmol/mg/min. Approximately 900 mg of seed calcite was added to 400 mL water in the reaction vessel; a rate of 1 nmol/mg/min ($\log R = 0$) corresponds to 0.135 mmol $CaCO_3$/L/hr, or 5.1×10^{-8} mol/s/m^2 seed crystal surface area, which is low, also for natural conditions (cf. Chapter 5). Lines are calculated with Equation (4.50), using $R = 10^{4.5}(\Omega - 1)$ for Cd^{2+}, and $R = 10^{1.8}(\Omega - 1)$ for Sr^{2+}; experimental points are from Tesoriero and Pankow (1996).

The apparent fractionation factors are plotted as function of $\log R$, the rate of carbonate precipitation in the experiment. The experimental data for Cd show the expected trend with a high apparent fractionation factor at low precipitation rates and a decline at higher rates. For Sr, the apparent fractionation factor is less than 1 and nearly constant when the precipitation rate is smaller than 10 nmol/mg/min, but it increases towards 1 when the rate increases.

The important conclusion to be drawn is that at high precipitation rates the ions will be incorporated in the solid in the ratio they are present in the solution, while at lower precipitation rates the thermodynamic effects may become increasingly more important. The effect of precipitation rate will be stronger when the fractionation factor is larger. Overall, the effects render the derivation of thermodynamic properties of solid solutions from laboratory or field data rather doubtful (Curti, 1999; Rimstidt et al., 1998; Schwartz and Ploethner, 1999).

QUESTION:
Find the miscibility gap for calcite/strontianite solid solution with PHREEQC, -comp 2 Strontianite; -Gugg_non 5.7 0.

ANSWER: Miscibility-gap fractions, component 2: 0.0035 0.9965, i.e. barely miscible

4.6 KINETICS OF GEOCHEMICAL PROCESSES

4.6.1 *Kinetics and equilibrium*

If the mineral halite (NaCl) dissolves in distilled water, and the concentrations are followed as a function of time, you may find a curve as shown in Figure 4.11. The $[Na^+]$ increases with time until equilibrium between water and mineral is attained at time t_2. From t_2 onwards, $[Na^+]$ becomes independent of time and is determined by the equilibrium constraint:

$$K_{halite} = [Na^+][Cl^-]$$

K_{halite} can be measured directly in the laboratory or calculated from thermodynamic tables. As we have seen, also the dependency of K_{halite} on temperature can be calculated. In many cases, the equilibrium concept gives a satisfactory explanation for the observed groundwater chemistry. Due to its firm fundament and ease in calculations, it should always be the first approach to any problem.

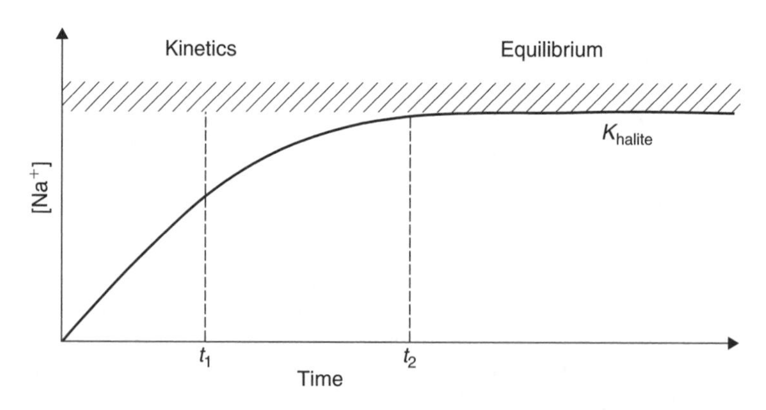

Figure 4.11. The realms of kinetics and equilibrium, illustrated in a dissolution experiment with halite.

However, some chemical phenomena in aquifers are insufficiently explained by equilibrium chemistry. We already noted that the composition of precipitating solid solutions becomes kinetically controlled when the fractionation factor for the minor component is high. Many silicate minerals, like feldspars, amphiboles, pyroxenes, etc. are thermodynamically unstable at the earth surface (groundwaters/surface waters are strongly subsaturated for these minerals). Still they persist for thousands or tens of thousands of years. Other examples are the observed subsaturation for calcite in some aquifers as well as the survival of aragonite in aquifers for thousands of years. There are two fundamental causes for the lack of equilibrium. The first is the tectonic activity that transports minerals from the environments where they were formed and are stable, often deep down in the earth at high pressures and temperatures, to environments at the earth's surface where they are unstable. The second is photosynthesis which uses solar energy to convert CO_2 into thermodynamically unstable organic matter; its subsequent kinetic decomposition affects the equilibria of redox sensitive elements such as iron, oxygen and nitrogen. Biological activity also disturbs carbonate equilibria through the production or consumption of CO_2 and through biomineralization. To understand, and to describe quantitatively such deviation from equilibrium requires the use of reaction kinetics.

In the example of halite dissolution (Figure 4.11), the first part of the reaction, before equilibrium with halite is attained, should clearly be studied from a kinetic point of view. Qualitatively, one can expect that the rate of dissolution of halite depends on factors such as grain size (small crystals have a larger surface area than large ones per unit weight), the amount of stirring, temperature, and the distance from equilibrium. The link to equilibrium chemistry is to predict in which direction the reaction will go. At time t_1, equilibrium chemistry predicts that dissolution will take place rather than precipitation. It does not tell us, however, whether it will take 10 seconds or 10 million years before equilibrium is attained.

Our understanding of reaction kinetics is at a much lower level than of equilibrium chemistry although empirical information is rapidly accumulating (e.g. Sparks, 1989; Hochella and White, 1990; Schwartzenbach et al., 1993; White and Brantley, 1995; Lasaga, 1998). There is no unifying theory, and different notations are used to describe even the same problem. The approach to a kinetic problem is usually divided into two steps:

1. to describe quantitatively the rate data measured in the laboratory or in the field.
2. to interpret the quantitative description of the rate data in terms of mechanisms.

It is often found that different mechanisms fit equally well to the same rate data. Therefore, kinetic data derived from solution chemistry may at best be demonstrated as consistent with a given mechanism. Surface chemical techniques like AFM, SIMS, XPS, on the other hand, may elucidate the mechanisms at the atomic scale (Stipp, 1994).

4.6.2 *Chemical reactions and rates*

Consider a simple reaction where compound A is converted to compound B by the reaction:

$$A \rightarrow B$$

This reaction can be followed by recording the concentration of A as a function of time as shown in Figure 4.12. The *reaction rate* is the change of A with time. Thus at time x_1, the rate can be determined from the slope of the tangent, c_{x1} / t_{x1}, at this point. For the whole curve, this is equal to:

$$\text{rate} = -\mathrm{d}c_{\mathrm{A}} / \mathrm{d}t$$

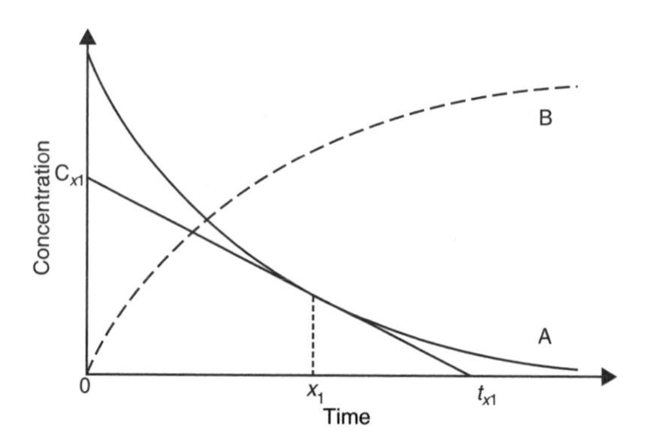

Figure 4.12. Derivation of rates from concentration/time data.

The units of the rate depend on the definition of the reaction and could for example be mol/L/s. For a decrease in concentration of a reactant, the rate is given a negative sign (the slope of the tangent is negative), while for the increase of a product, the rate is positive (the slope of the tangent is positive). Thus:

$$\text{rate} = -\mathrm{d}c_\mathrm{A} / \mathrm{d}t = \mathrm{d}c_\mathrm{B} / \mathrm{d}t$$

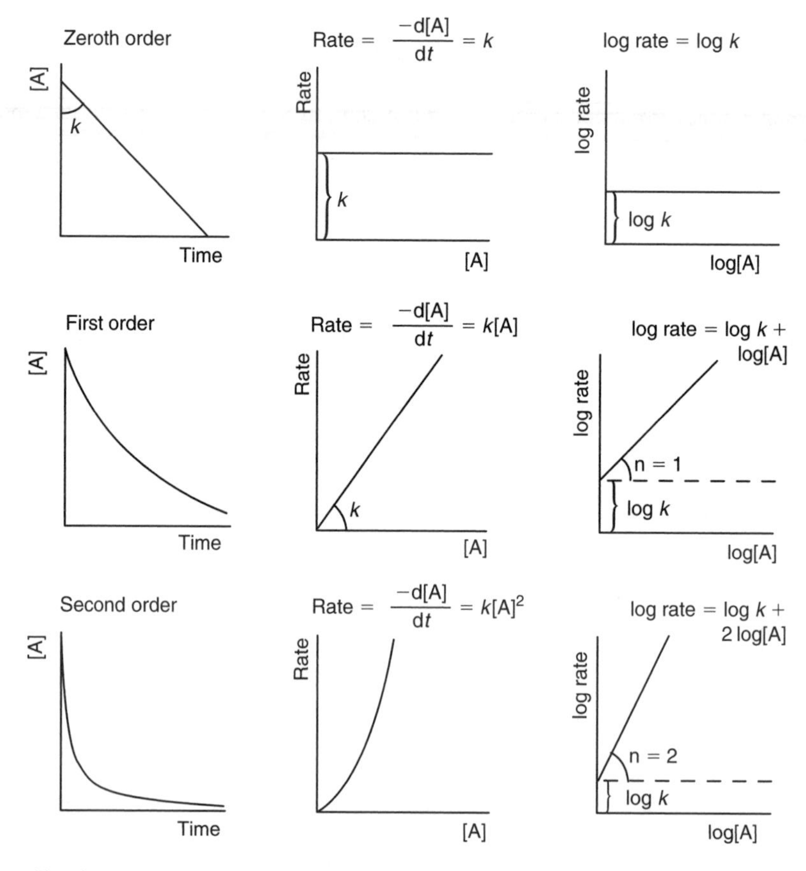

Figure 4.13. Simple rate laws and the differential method for the reaction A → B.

For the more complicated reaction

$$A + 2B \rightarrow 3C$$

we find

$$\text{rate} = dc_C / dt = -3\,dc_A / dt = -3/2\,dc_B / dt$$

This shows the importance of defining the reaction rate explicitly. The manner in which the reaction rate varies with the concentrations of reacting substances is referred to as the *reaction order*. For example, if it is found experimentally that the reaction rate is proportional to the αth power of the concentration of reactant A and to the βth power of reactant B etc., then

$$\text{rate} = k \cdot c_A^{\alpha} \cdot c_B^{\beta} \cdots \tag{4.51}$$

Such a reaction is said to be αth order with respect to [A] and βth order for [B]. The *overall order n* of this reaction is simply

$$n = \alpha + \beta + \cdots$$

The coefficient k in (4.51) is the *rate constant*, also known as the *specific rate*, and is numerically equal to the reaction rate when all reactants are present at unit concentrations. Since the rate has fixed units (for example mol/L/s), the units of the rate constant depend on the overall order of the reaction. In general for an n-order reaction, the units of k are $(\text{mol/L})^{1-n}$/s. The effect of the reaction orders on relations between concentration and time, and concentration and rate is illustrated in Figure 4.13.

For zeroth order kinetics, the rate is independent of the reactant concentration and this implies that other factors or reactants control the rate. An example is shown in Figure 4.14 for bacterial reduction of sulfate by organic matter, which shows zeroth order behavior as long as $m_{SO_4^{2-}}$ is higher than 2 mmol/L. To find the concentration of SO_4^{2-} at a given time, for example $t = 15$ days, we integrate the rate:

$$dm_{SO_4^{2-}} / dt = -k = -0.92 \text{ mmol/L/day} \tag{4.52}$$

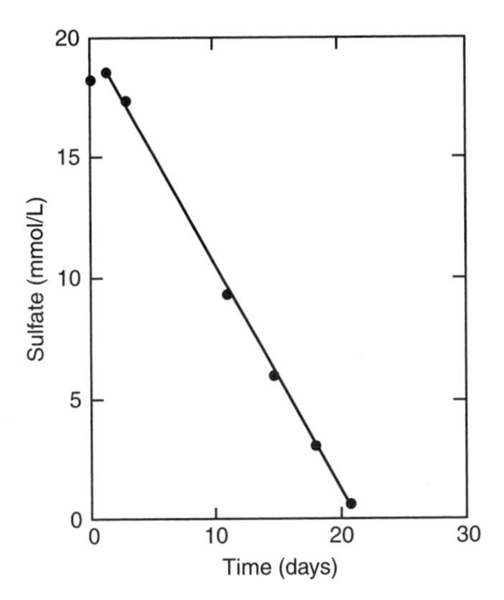

Figure 4.14. Dissolved sulfate versus time for marine sediment incubated in the laboratory (Berner, 1981). The straight line indicates a zeroth order dependence on the sulfate concentration.

from $m_{SO_4^{2-}} = 20$ mmol/L at $t = 0$ to the concentration $m_{SO_4^{2-}}$ at $t = 15$ days. Thus:

$$\int_{20}^{m_{SO_4^{2-}}} dm_{SO_4^{2-}} = -0.92 \int_0^{15} dt$$

which gives $m_{SO_4^{2-}} - 20 = -0.92 \times (15 - 0)$, or $m_{SO_4^{2-}} = 20 - 13.8 = 6.2$ mmol/L (after 15 days).

However, for times longer than 21.74 days, Equation (4.52) predicts a negative concentration of SO_4^{2-}, which is impossible of course. Actually, for concentrations smaller than 2 mM SO_4^{2-} the rate slows down and becomes first order with respect to SO_4^{2-} (Boudreau and Westrich, 1984; Section 10.4.1). As shown in Figure 4.13, a first order rate decreases proportionally as the concentration goes down and reaches zero when the concentration becomes zero.

First order rate laws are common and comprise, for example, all radioactive decay reactions (Equation 3.20). Example 4.12 illustrates the effect of first and second order rate laws for the oxidation of Fe^{2+}. The rate is first order with respect to both Fe^{2+} and P_{O_2}, implicating that doubling the concentrations increases the rate with a factor of two. The rate is second order for $[OH^-]$ and the rate increases fourfold when the OH^- activity is doubled. It is thus clear that increasing the pH is a much more effective way to accelerate Fe(2) oxidation than increasing the P_{O_2}.

EXAMPLE 4.12. *Oxidation of Fe(2)*
For oxidation of Fe^{2+} in aqueous solutions at 20°C and 1 atm., the following rate law has been reported by Stumm and Morgan (1996):

$$\text{rate} = -d\, m_{Fe^{2+}} / dt = k \cdot m_{Fe^{2+}} \cdot [OH^-]^2 \cdot P_{O_2}$$

where $k = 8 \times 10^{13}$ /min/atm.

Thus the rate equation is formulated as a decrease in $m_{Fe^{2+}}$ and is first order for $m_{Fe^{2+}}$ and P_{O_2}, and second order for $[OH^-]$. What is the rate of Fe^{2+} oxidation if $m_{Fe^{2+}} = 1$ mM and $P_{O_2} = 0.2$ atm, for pH values 5 and 7?

ANSWER:

$$pH = 5; \quad K_w = [H^+][OH^-] = 10^{-14} \rightarrow [OH^-] = 10^{-9}$$

$$-d\, m_{Fe^{2+}} / dt = 8 \times 10^{13} \times 10^{-3} \times (10^{-9})^2 \times 0.2 = 1.6 \times 10^{-8} \text{ mmol/L/min}$$
$$= .001 \text{ mmol/L/hour}$$

pH = 7:

$$-d\, m_{Fe^{2+}} / dt = 8 \times 10^{13} \times 10^{-3} \times (10^{-7})^2 \times 0.2 = 2.0 \times 10^{-4} \text{ mol/L/min}$$
$$= 9.4 \text{ mmol/L/hour}$$

Note the very large difference in rates at pH 5 and 7, due to the second order rate dependence on $[OH^-]$.

The time needed to reach a given concentration is found by integrating the rate equation, which is possible for the rate of Fe(2) oxidation (Example 4.12), when pH and P_{O_2} are kept constant. For example, the time needed to halve the Fe^{2+} concentration from 1 to 0.5 mmol/L, at pH = 5 and $P_{O_2} = 0.2$ atm:

$$-d\, m_{Fe^{2+}} / dt = 8 \times 10^{13} \times (10^{-9})^2 \times 0.2 \times m_{Fe^{2+}} = 1.6 \times 10^{-5}\, m_{Fe^{2+}} \quad (4.53)$$

which we integrate from $m_{Fe^{2+}} = 1$ mM at $t = 0$, to $m_{Fe^{2+}} = 0.5$ mM at $t = t_{½}$:

$$\int_{1e-3}^{0.5e-3} \frac{dm_{Fe^{2+}}}{m_{Fe^{2+}}} = -1.6\times10^{-5}\int_{0}^{t_{½}} dt$$

It gives $\ln(0.5\times10^{-3}/10^{-3}) = -1.6\times10^{-5} \times (t_{½} - 0)$. Thus, $t_{½} = \ln(1/2)/-1.6\times10^{-5} = 43321$ s or 12 hr.

In first order rate laws, the time needed to halve the reactant concentration is known as the *half-life* of the reaction, which is independent of the initial concentration. In a more loose sense, the half-life concept is also used for other types of reaction rates, but then the value will depend on the initial reactant concentration.

QUESTIONS:
Find the time to lower the Fe^{2+} concentration from 0.5 to 0.25 mM, pH = 5, $P_{O_2} = 0.2$ atm.
ANSWER: 12 hr
Find the half-life of Fe^{2+} oxidation at pH = 7 and $P_{O_2} = 0.2$ atm?
ANSWER: 4.3 seconds

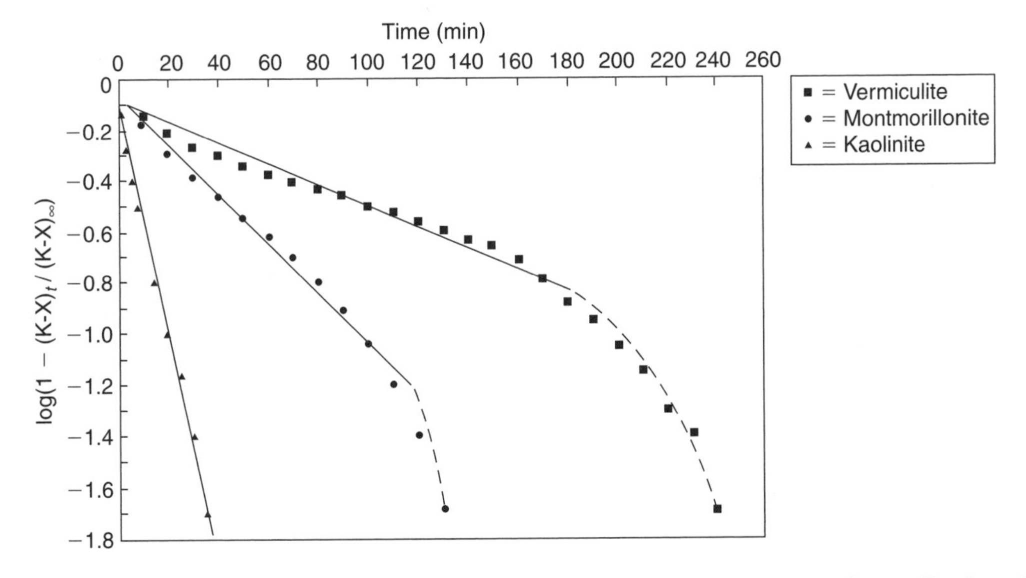

Figure 4.15. Kinetic exchange of K^+ on three clay minerals, plotted according to a first order rate (Sparks and Jardine, 1984; Sparks, 1989).

Another example of a first order reaction is shown in Figure 4.15 for sorption of K^+ on three clay minerals. It depicts the result of the exchange reaction:

$$K^+ + ½Ca\text{-}X_2 \leftrightarrow K\text{-}X + ½Ca^{2+} \tag{4.54}$$

where X is the exchange site on the mineral, with a charge of X^-. Thus, K-X is the exchangeable K^+, which is expressed as a fraction of the total amount of exchange sites. $K\text{-}X_\infty$ is the fraction at infinite time and equals 1 (exchange theory is discussed in Chapter 6). Sparks and Jardine (1984) postulated the rate equation:

$$\mathrm{d\,(K\text{-}X)} / \mathrm{d}t = -k\ \{\mathrm{(K\text{-}X)} - \mathrm{(K\text{-}X)}_\infty\} \tag{4.55}$$

which is first order with respect to $\{\mathrm{(K\text{-}X)} - \mathrm{(K\text{-}X)}_\infty\}$. Again, to find the concentration (K-X) at a given time $t = t_e$, we integrate the equation from (K-X) = 0 at $t = 0$. First collect the terms:

$$\int_0^{\mathrm{(K\text{-}X)}_{t_e}} \frac{\mathrm{d(K\text{-}X)}}{\mathrm{(K\text{-}X)} - \mathrm{(K\text{-}X)}_\infty} = -k \int_0^{t_e} \mathrm{d}t$$

To solve the integral, we subsitute $z = \{\mathrm{(K\text{-}X)} - \mathrm{(K\text{-}X)}_\infty\}$. Then, $\mathrm{d}z/\mathrm{d\,(K\text{-}X)} = 1$, and $\mathrm{d(K\text{-}X)} = \mathrm{d}z$. The integration boundaries change to $z_{t=0} = \{\mathrm{(K\text{-}X)} - \mathrm{(K\text{-}X)}_\infty\} = -\mathrm{(K\text{-}X)}_\infty$ and to $z_{t=t_e} = \{\mathrm{(K\text{-}X)} - \mathrm{(K\text{-}X)}_\infty\} = \{\mathrm{(K\text{-}X)}_{t_e} - \mathrm{(K\text{-}X)}_\infty\}$. Thus:

$$\int_{-\mathrm{(K\text{-}X)}_\infty}^{\{\mathrm{(K\text{-}X)}_{t_e} - \mathrm{(K\text{-}X)}_\infty\}} \frac{\mathrm{d}z}{z} = -k \int_0^{t_e} \mathrm{d}t$$

which solves to

$$\ln\left(\frac{\{\mathrm{(K\text{-}X)}_{t_e} - \mathrm{(K\text{-}X)}_\infty\}}{-\mathrm{(K\text{-}X)}_\infty}\right) = -k(t_e - 0)$$

or

$$\log(1 - \mathrm{(K\text{-}X)}_{t_e} / \mathrm{(K\text{-}X)}_\infty) = -k\ t_e/2.303 \tag{4.56}$$

The form of this equation is typical for an integrated first order rate in having the logarithm of the concentration as linear function of time.

Experimental results were plotted according to Equation (4.56) in Figure 4.15. The slope for the clay mineral kaolinite is larger than for montmorillonite and vermiculite, indicating that k is largest for kaolinite, and therefore the reaction is quickest for this mineral. At longer times, the slope increases for montmorillonite and vermiculite, which suggests that the first order reaction mechanism invoked by Equation (4.55) is only applicable for the initial part of the reaction.

QUESTION:
Find k for K^+ exchange on kaolinite and montmorillonite in the first 100 min

ANSWER: $k = 0.115$/min for kaol; $k = 0.0242$/min for montm

In the examples presented so far, the reaction orders have been whole numbers. However, there is no a priori reason why they could not be fractional numbers and these are frequently reported. As the first guess for the rate equation one often takes the product of reactants, just as in the law of mass action. For K^+-Ca^{2+} exchange (Reaction 4.54) the forward rate could be written as:

$$r_{fw} = k_{fw} \cdot [\mathrm{K^+}]\ [\mathrm{Ca\text{-}X_2}]^{1/2} \tag{4.57}$$

and the backward rate as:

$$r_{bw} = k_{bw} \cdot [\mathrm{Ca^{2+}}]^{1/2} [\mathrm{K\text{-}X}] \tag{4.58}$$

Both rate equations are half-order with respect to the activity of Ca^{2+} as solute and exchangeable species. But in contrast to the easy integration of the first order rate law that led to Equation (4.56), the chemically more attractive half-order rate formulation needs numerical integration.

If the overall reaction is split-up in a forward and a backward rate equation, the rate constants are interconnected since at equilibrium the overall rate is zero and the two rates must cancel:

$$R = 0 = r_{fw} - r_{bw} = k_{fw} \cdot [K^+][Ca\text{-}X_2]^{1/2} - k_{bw} \cdot [Ca^{2+}]^{1/2}[K\text{-}X] \tag{4.59}$$

or:

$$\frac{k_{fw}}{k_{bw}} = \frac{[Ca^{2+}]^{1/2}\,[K\text{-}X]}{[K^+]\,[Ca\text{-}X_2]} = K \tag{4.60}$$

where K is the equilibrium constant of the reaction. This links the forward and the backward rate constants as:

$$k_{fw} = k_{bw} \cdot K \tag{4.61}$$

In deriving relation (4.61) we used the principle of *detailed balancing*.

4.6.3 *Temperature dependency of reaction rates*

The rate of most reactions is highly influenced by the temperature. According to the Arrhenius equation the rate constant changes with temperature as:

$$k = A \cdot \exp\left(\frac{-E_a}{RT}\right) \tag{4.62}$$

where A is the *pre-exponential factor* (same units as k) and E_a is the *activation energy* (kJ/mol), R the gas constant and T absolute temperature. Taking the logarithm of Equation (4.62) gives:

$$\ln k = \ln A - E_a / RT \tag{4.63}$$

A plot of log k versus $1/T$ therefore gives a straight line with a slope of

$$\frac{d \log k}{d(1/T)} = \frac{-E_a}{2.303\, R} \tag{4.64}$$

Table 4.8. Typical values (and ranges) of activation energies for chemical and transport processes, and relative increase of the rate coefficient when temperature increases from 10 to 20°C according to the Arrhenius equation.

Process	E_a (kJ/mol)	$k_{20°C}/k_{10°C}$
Photochemical reactions	15 (10–30)[a]	1.24
Diffusion in water	20 (Equation 3.42)	1.34
Dissolution/Precipitation	60 (20–80)[b]	2.4
Biochemical reactions	63 (50–75)	2.5
Solid state diffusion	500 (200–650)[b]	1400

[a]Schwarzenbach et al., 1993. [b]Lasaga, 1998.

Since $d(T^{-1}) / dT = -T^{-2}$, we can rewrite Equation (4.64) as:

$$\frac{d \log k}{dT} = \frac{-E_a}{2.303\, RT^2} \qquad (4.65)$$

Equation (4.65) has the same form as the Van 't Hoff equation (4.28), which related the temperature dependency of equilibrium constants to the reaction enthalpy. In *transition state theory*, the activation energy is linked to the energy absorbed by the reactants when an *activated complex* is formed in an intermediate step from which the products of the reaction are generated (Lasaga, 1998).

Activation energies differ markedly for various types of chemical reactions and transport processes and the temperature dependency of a set of experimental rate data may indicate the probable reaction mechanism (Table 4.8). For example, the relatively small temperature effect on diffusion in water suggests that the temperature dependency of a transport controlled dissolution reaction is much less than of a surface controlled reaction. The activation energy is very high for diffusion in solids which implies a huge effect for a small temperature increase from 10 to 20°C. However, the diffusion coefficient at this temperature is so small, that the actual movement of the atoms in the solid at 20°C is still negligible.

4.6.4 *Mechanisms of dissolution and crystallization*

The dissolution or growth of minerals proceeds through a chain of processes, including transport of solutes between the bulk solution and the mineral surface, the ad- and desorption of solutes at the surface, hydration and dehydration of ions and surface migration. These processes must take place simultaneously but some may proceed at a faster rate than others. Since the overall rate will depend on the rate of the slowest process, it is of importance to identify the *rate limiting* process or mechanism.

There is an essential distinction between an overall rate control by transport of solutes from the bulk solution to the mineral surface, and by the reaction occurring at the mineral surface (Nielsen, 1986). The difference is illustrated in Figure 4.16 for a dissolving mineral. If transport of ions to and from the mineral surface is rate limiting, equilibrium will be approached at the mineral surface and a concentration gradient develops from the mineral surface to the bulk solution. On the other hand when a surface reaction is rate limiting, any ion released from the crystal lattice will be carried away quickly and no concentration gradient develops. Also intermediate cases of surface and transport controlled reactions can be found.

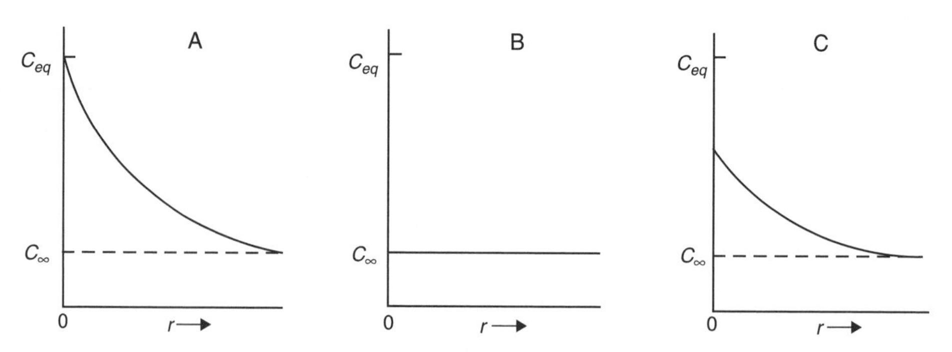

Figure 4.16. Schematic concentration profiles from the mineral surface to the bulk of solution, for different rate controlling processes during dissolution. C_∞ is the concentration in the bulk solution and C_{eq} the concentration in equilibrium with the mineral. A. Transport control, B. Surface reaction control, C. Mixed control (Berner, 1978).

How to decide whether a mineral dissolution reaction is transport or surface reaction controlled? In a dissolution experiment the stirring rate can be changed. If the reaction is transport controlled, it affects the concentration gradients at the mineral surface and the rate changes. The reverse is not always true, since very small crystals (<about 5 μm) are carried along with the whirling water, so that the convection at the mineral surface remains the same (Nielsen, 1986). A more elegant way is to calculate whether the reaction is slower or faster than molecular diffusion (Example 4.13). Since molecular diffusion is the slowest form of transport possible, slower rates are necessarily controlled by surface reaction.

EXAMPLE 4.13. *Dissolution of hydroxyapatite; transport or surface reaction controlled?*
Christoffersen and Christoffersen (1979) studied the dissolution kinetics of hydroxyapatite (HAP) using suspensions of fine crystals. The overall dissolution rate is expressed as $R = -dn/dt$, where n is the amount of undissolved HAP. For steady state diffusion through a sphere circumscribing the crystal we may use the equation:

$$R = 4\pi D_{app}\ N\ r^2\ (c_{eq} - c)$$

where D_{app} is the apparent diffusion coefficient, N the number of crystals, r the radius of the crystals, c_{eq} the concentration at equilibrium and c the concentration in the bulk solution. Substituting measured values of R, N, r, c_{eq} and c, yields a value of $D_{app} \approx 10^{-9}\ cm^2/s$. Diffusion coefficients in aqueous solution are in the order $10^{-5}\ cm^2/s$ (Section 3.5.1). Since the apparent diffusion coefficient D_{app}, is four orders of magnitude lower, one must conclude that dissolution of HAP is much slower than predicted by molecular diffusion. Therefore, the reaction is surface reaction controlled.

The solubility of minerals appears to be related to the dissolution mechanism (Berner, 1980). The dissolution of sparingly soluble minerals is generally controlled by surface processes whereas the dissolution of soluble minerals predominantly is controlled by transport processes (Table 4.9). Silicate minerals, like feldspars, pyroxenes and amphiboles are also insoluble and their dissolution is controlled by surface reactions while most salt minerals are soluble and their dissolution is transport controlled.

Table 4.9. Rate controlling dissolution mechanism and solubility for various substances (Berner, 1980).

Substance	Solubility (mol/L)	Dissolution rate control
$Ca_5(PO_4)_3OH$	$2 \cdot 10^{-8}$	Surface-reaction
$KAlSi_3O_8$	$3 \cdot 10^{-7}$	Surface-reaction
$NaAlSi_3O_8$	$6 \cdot 10^{-7}$	Surface-reaction
$BaSO_4$	$1 \cdot 10^{-5}$	Surface-reaction
$AgCl$	$1 \cdot 10^{-5}$	Transport
$SrCO_3$	$3 \cdot 10^{-5}$	Surface-reaction
$CaCO_3$	$6 \cdot 10^{-5}$	Surface-reaction
Ag_2CrO_4	$1 \cdot 10^{-4}$	Surface-reaction
$PbSO_4$	$1 \cdot 10^{-4}$	Mixed
$Ba(IO_3)_2$	$8 \cdot 10^{-4}$	Transport
$SrSO_4$	$9 \cdot 10^{-4}$	Surface-reaction
Opaline SiO_4	$2 \cdot 10^{-3}$	Surface reaction
$CaSO_4 \cdot 2H_2O$	$5 \cdot 10^{-3}$	Transport
$Na_2SO_4 \cdot 10H_2O$	$2 \cdot 10^{-1}$	Transport
$MgSO_4 \cdot 7H_2O$	$3 \cdot 10^{0}$	Transport
$Na_2CO_3 \cdot 10H_2O$	$3 \cdot 10^{0}$	Transport
KCl	$4 \cdot 10^{0}$	Transport
$NaCl$	$5 \cdot 10^{0}$	Transport
$MgCl_2 \cdot 6H_2O$	$5 \cdot 10^{0}$	Transport

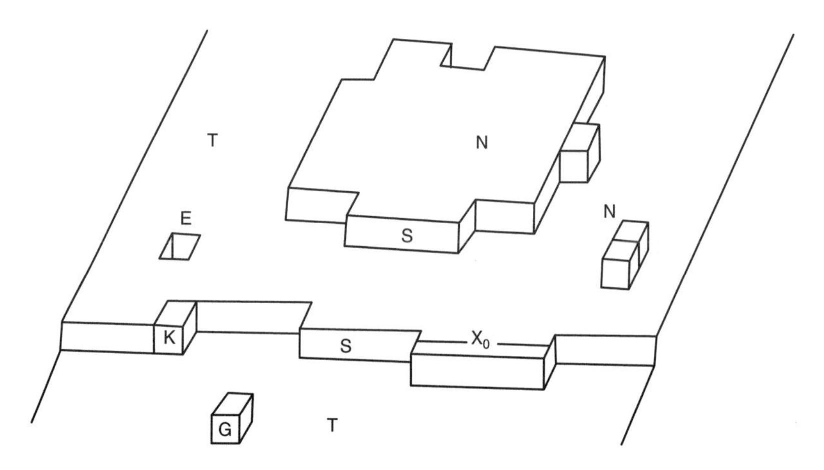

Figure 4.17. The schematic surface of a simple cubic crystal. Energetically favorable sites include kinks (K), steps (S), clusters (N) and etch pits (E) (Zhang and Nancollas, 1990).

During a surface controlled reaction, dissolution or growth only takes place at energetically favorable sites on the mineral surface. A schematic picture of a mineral surface (Figure 4.17) identifies the favorable sites as irregularities on the crystal face. Clearly, ions or atoms which are not surrounded by other crystal units on three sides have a lower binding energy and are the first to be released upon dissolution. Also crystal growth will take place preferentially at those sites where bonds in several directions can be established. Selective dissolution will yield beautiful etch patterns on mineral surfaces that can be observed by SEM on a microscopic level. Such etch patterns can be used as an indicator for surface reaction dissolution control (Berner, 1978), and may be used to establish rates of dissolution (Gratz et al., 1990).

Various models have been derived to describe crystal growth or dissolution mechanisms at reactive surface sites. These include spiral growth or dissolution, polynuclear growth/dissolution, etc., which are discussed in more detail in texts like Nielsen (1964, 1984), Zhang and Nancollas (1990) and Lasaga (1998).

4.6.5 *Rate laws for mineral dissolution and precipitation*

A general rate law for the change in solute concentration due to mineral dissolution/precipitation reactions can be written as:

$$R = k\frac{A_0}{V}\left(\frac{m}{m_0}\right)^n g(C) \qquad (4.66)$$

where R is the overall reaction rate (mol/L/s), k is the specific rate (mol/m^2/s), A_0 is the initial surface area of the solid (m^2), V is the volume of solution (m^3), m_0 is the initial moles of solid, and m is the moles of solid at a given time. The overall rate depends first of all on the specific rate and the ratio of surface area A_0 over the solution volume V. Next, the factor $(m/m_0)^n$ accounts for changes in surface area, or more precise, of reactive surface sites, during dissolution. These changes may be due to changes in crystal size during dissolution/precipitation, changes in the size distribution of the crystal population, but also factors like selective dissolution and aging of the solid may be important. For a monodisperse population of uniformly dissolving or growing spheres and cubes, $n = 2/3$, because m is proportional to the volume, or r^3 (here r is the radius of the sphere or the side of the cube) while the surface area is proportional to r^2. For polydisperse crystal populations, the finest crystals will dissolve selectively and the size distribution is modified. Therefore, n becomes a function of the initial grain size distribution and for log-normal size distributions values of up to $n = 3.4$ are possible (Dixon and Hendrix, 1993). Also the disintegration of the crystals during dissolution may modify n (Larsen and Postma, 2001). Finally, g(C) is a function that comprises the effects of the solution composition on the rate, like the pH, the distance from equilibrium, and the effects of catalysis and inhibition (Lasaga, 1998).

Dissolution of minerals

Let us use the dissolution of quartz (SiO_2) as example. The reaction is

$$SiO_2 + 2H_2O \leftrightarrow H_4SiO_4^0$$

In an early study, Van Lier et al. (1960) measured a zeroth order rate for dissolution of quartz,

$$r_{quartz} = 10^{-13.5}\ \text{mol/m}^2\text{/s} = 10^{-6}\ \text{mol/m}^2\text{/yr.} \tag{4.67}$$

Let's calculate the concentration of $H_4SiO_4^0$ in soil water (Example 4.14).

EXAMPLE 4.14. *Dissolution rate of quartz*

Estimate the concentration of $H_4SiO_4^0$ in water in a quartz soil after one year, $\rho_{quartz} = 2.65\ \text{g/cm}^3$, $\varepsilon_w = 0.3$, grain-size = 0.1 mm.

ANSWER:

First calculate the surface to volume factor A/V: determine the number of spherical grains in 1 kg soil and the surface area of a single grain.

volume of 1 grain = $\frac{4}{3}\pi r^3 = \frac{4}{3} \times 3.14 \times (0.05\times10^{-3})^3 = 5.24\times10^{-13}\,\text{m}^3$

volume of 1 kg quartz grains = $1\,\text{kg} / (2650\,\text{kg/m}^3) = 3.77\times10^{-4}\,\text{m}^3$

number of grains in 1 kg = $3.77\times10^{-4}/5.24\times10^{-13} = 7.21\times10^8$

surface of 1 grain = $4\pi r^2 = 4 \times 3.14 \times (0.05\times10^{-3})^2 = 3.14\times10^{-8}\,\text{m}^2$

$\rightarrow$ surface area of 1 kg soil, $A = 7.21\times10^8 \times 3.14\times10^{-8} = 22.6\,\text{m}^2\text{/kg}$

The volume of water in contact with 1 kg soil, $V = 3.77\times10^{-4}\,\text{m}^3 \cdot \varepsilon_w/(1 - \varepsilon_w)$

$\rightarrow V = 3.77\times10^{-4} \times 0.3/(0.7) = 1.62\times10^{-4}\,\text{m}^3$

The increase in concentration is $R_{qu} = dm_{H_4SiO_4^0}/dt = r_{qu} \cdot A/V = 10^{-6} \times 22.6/(0.162) = 0.14\,\text{mmol/L/yr.}$

The solubility of quartz is about 0.1 mM $H_4SiO_4^0$, and for the quartz soil in Example 4.14, Equation (4.67) predicts that this concentration should be reached in less than one year. However, a zero order rate equation is perhaps not appropriate since the rate must slow down as equilibrium is approached. Rimstidt and Barnes (1980) therefore proposed a rate equation that includes the saturation state $\Omega = IAP/K$:

$$r_{qu} = k_{qu} \cdot (1 - IAP/K) \tag{4.68}$$

where $k_{qu} = 10^{-13.7}\,\text{mol/m}^2\text{/s}$ at 25°C. The rate as defined by Equation (4.68) is positive at subsaturation, zero at equilibrium, and negative at supersaturation which intuitively appears correct. The rate equation can be integrated numerically with PHREEQC as shown in Example 4.15.

EXAMPLE 4.15. *Kinetic dissolution of quartz with PHREEQC*
The rate equation is defined under keyword RATES using Equation (4.68) as part of Equation (4.66). The keyword KINETICS invokes the rate for the case of the quartz soil of Example 4.14.

```
RATES
 Quartz                                  # Rate name
  -start
  10 A0 = parm(1)
  20 V = parm(2)
  30 rate = 10^-13.7 * (1 - SR("Quartz")) * A0/V * (m/m0)^0.67
  40 moles = rate * time
  50 save moles                          # time and save must be in the rate definition
-end


KINETICS 1
 Quartz
  -formula SiO2
  -m0 103                                # initial quartz, mol/L
  -parms 22.6 0.162                      # A0 in m2, V in dm3
  -step 1.5768e8 in 15                   # 5 years in seconds, in 15 steps
INCREMENTAL_REACTIONS true
SOLUTION 1                               # distilled water
USER_GRAPH
  -axis_titles Years "mmol Si / l"
  -axis_scale x_axis 0 5
  -start
  10 graph_x total_time/3.1536e7                 # time in years on x-axis
  20 graph_y 1000 * tot("Si")                    # Si conc. on y-axis
  -end
END
```

RATES starts by defining the name 'Quartz'. Then, the rate equation is written in a BASIC program between '-start' and '-end'. The BASIC commands are given in the PHREEQC manual under keyword RATES. The initial surface area 'A0' in line 10 and the solution volume 'V' in line 20 are defined as the variables parm(1) and parm(2). KINETICS starts by calling Quartz and then provides values for these variables in the rate equation.

The keyword 'INCREMENTAL_REACTIONS true' signals PHREEQC to do the calculations incrementally for each timestep, adding each time to the results of the previous calculation; 'false' would mean that the calculation starts from zero for each timestep, defined with '-step' in KINETICS. Running the file produces the plot shown in Figure 4.18.

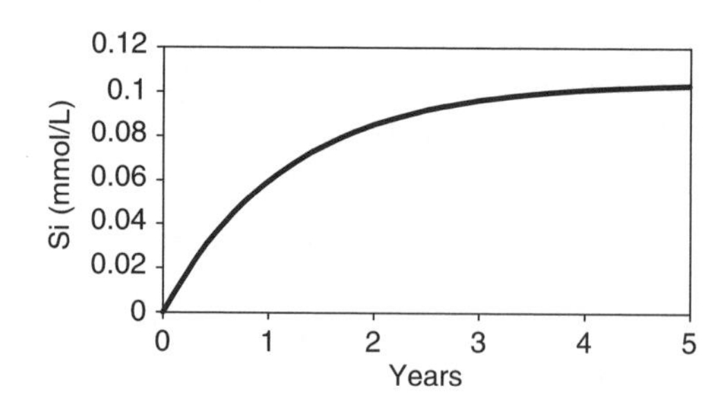

Figure 4.18. Kinetic dissolution of quartz based on Equation (4.68) for a quartz soil.

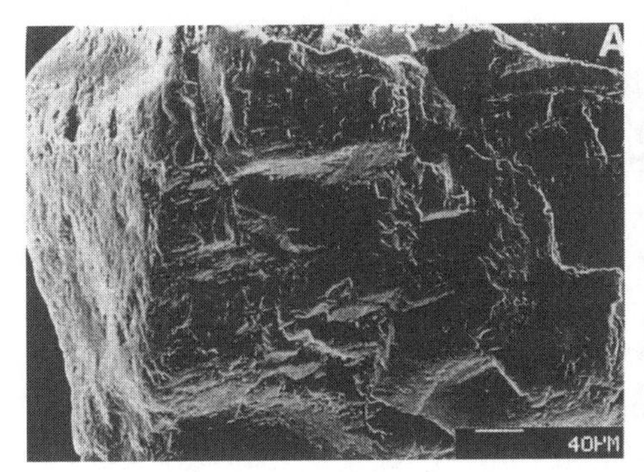

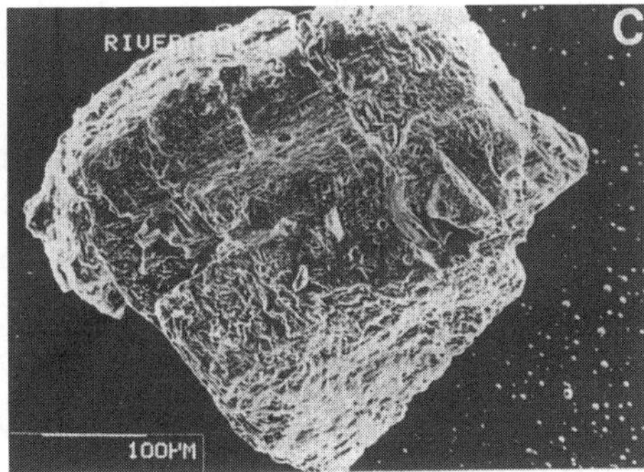

Figure 4.19. Scanning electron microscopy pictures of mineral grains showing etch pits and surface irregularities. A is from a 10 ka old soil, C stems from a 250 ka old profile (White, 1995).

The calculated Si release in the quartz soil defined in Example 4.14 is shown in Figure 4.18. The initial rise of the $H_4SiO_4^0$ concentration slows down when the equilibrium concentration is approached, which takes about 4 years. The term $(1 - IAP/K)$ has a small influence when $IAP/K < 0.1$ and in that domain the kinetics are linear in time, but the term gains importance as the quartz solubility is approached. This seems to be a general feature for the dissolution of silicate minerals (Nagy and Lasaga, 1992).

The calculated concentration trend not necessarily complies with field observations since the conditions in a real soil are much more complex. First, other minerals such as clay minerals and feldspars are present in the soil and may dissolve more quickly than quartz, impeding quartz dissolution as soon as its solubility has been reached. Second, the surface area of quartz was calculated based on a smooth, geometric surface of an average grain; in reality, the grains have pits and irregularities that increase the surface area, as illustrated in Figure 4.19.

In laboratory studies, the surface area is usually measured with the BET method which is based on gas adsorption (Atkins and de Paula, 2002). Figure 4.20 shows a comparison of BET surface areas with surface areas estimated from geometric calculations. The BET surface area of freshly crushed grains is approximately 7 times higher than the geometric estimate, due to sub-microscopic surface roughness and crevices (Sverdrup and Warfvinge, 1988; Anbeek, 1992; White, 1995).

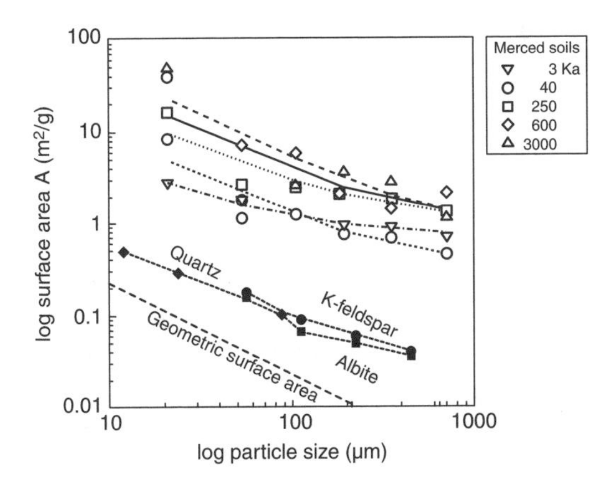

Figure 4.20. Surface areas as a function of particle diameter for a) Geometric areas calculated for smooth spheres (dashed line), b) BET surface areas for freshly crushed grains of quartz, K-feldspar and albite (solid symbols) and c) BET surface areas of soil grains, cleaned for clay and Fe-oxides, of different age (White, 1995).

The BET surfaces of mineral grains isolated from soils are significantly higher and tend to increase with age due to further development of dissolution etch pits.

Also the solution composition may influence the rate of quartz dissolution. Both the dissolution and precipitation kinetics of quartz may increase by up to 35-fold when the concentration of alkalis in solution increases (Dove and Rimstidt, 1993), and probably similar effects can be connected with other electrolytes which modify the surface properties of quartz. The sub-processes involved are:

background dissolution:	$k_{\text{qu},\,bck}$
proton promoted dissolution:	$k_{\text{H}}[\equiv\text{OH}_2^+]$
hydroxyl assisted dissolution:	$k_{\text{OH}}\ [\equiv\text{O}^-]$
ligand promoted dissolution:	$k_L\ [\equiv\text{O}L]$
dissolution inhibition:	$-k_{IB}\ [\equiv\text{O}IB]$

where $[\equiv\text{O}i]$ indicates the activity of a species adsorbed on the surface (dimensionless), which may be found by a sorption model, or in first approximation be related to the activity of the driving solute species $[i]$, and k_i is the rate constant for the sub-process. Since the sorbed species are activities (dimensionless), the k_i's all have the dimension mol/m^2/s. The rate constant in Equation (4.68) is summarizing all the sub-processes of (4.69):

$$k_{\text{qu}} = ff(k_{\text{qu},bck} + \sum k_i\ [\equiv\text{O}i]) \tag{4.70}$$

Here ff is a factor that relates the field and laboratory dissolution rates. Judged from the greater surface roughness of field grains compared with fresh, crushed grains (Figure 4.20), this factor is expected to be higher than 1, but in reality it is found to be less than 1, in the order of 0.1 to 0.01. Apparently, the grains used in laboratory studies have unstable surfaces and sites which dissolve relatively quickly on the timescale of the laboratory experiment, while these highly reactive sites have disappeared over time in the field.

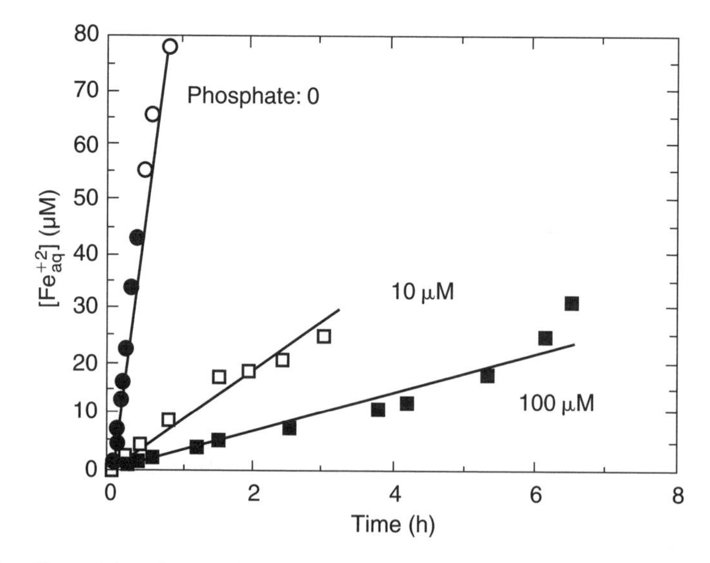

Figure 4.21. The effect of phosphate on the reductive dissolution of hematite (Fe_2O_3, 0.017 g/L) H_2S (partial pressure of $H_2S = 10^{-2}$ atm.) (Biber et al., 1994).

Possibly the field rates are also impeded by the presence of inhibitors which are substances that reduce the rate of reaction (Ganor and Lasaga, 1998). For quartz dissolution, and also other silicate minerals, aluminum is a well known inhibitor, just as phosphate inhibits fluorite and calcite dissolution even at micromolar concentrations (Morse and Berner, 1979; Christoffersen et al., 1988). Inhibitors are able to block active sites on the mineral surface; an example is shown in Figure 4.21 for the release of Fe^{2+} during reductive dissolution of hematite (Fe_2O_3) by H_2S. In the absence of phosphate there is a steep immediate release of Fe^{2+} over time. However in the presence of a phosphate concentration as low as 10 μM, hematite dissolution is strongly inhibited. It is the recognition that dissolution and growth only takes place at specific sites on the crystal surface that explains why substances even at very low concentrations may have an inhibiting effect on the dissolution or precipitation rate.

Crystal formation

Crystal formation is a two step process: the first step is the formation of a crystal embryo by nucleation, and the second step is crystal growth. The distinction is based on the fact that very small crystals are dominated by edges and corners, and since atoms located at such positions are not charge compensated, very small crystals are less stable than larger crystals. In terms of energy this can be described as:

$$\Delta G_{\text{n}} = \Delta G_{\text{bulk}} + \Delta G_{\text{interf}}$$

where ΔG_{n} is the free energy of formation of a crystal as a function of the number of atoms n in the crystal, ΔG_{bulk} the bulk free energy identical to the one used in Section 4.3.1, and ΔG_{interf} the interfacial free energy which reflects the excess free energy of very small crystals and is determined by the crystal-water surface tension. A direct result of this equation is that very small crystals are more soluble than larger crystals. However, already at crystal sizes larger than 2 μm, the effect of the interfacial free energy becomes negligible.

The change in ΔG_{n} with increasing crystal size is shown in Figure 4.22. During the growth of the crystal embryo an energy barrier has to be passed in order to allow further growth. The crystal size corresponding to this energy barrier is known as the *critical nucleus* and its size is dependent on the saturation state Ω. For increasing Ω, both the energy barrier and the size of the critical nucleus decreases. Once the size of the critical nucleus has been passed, the crystal enters the domain of crystal growth.

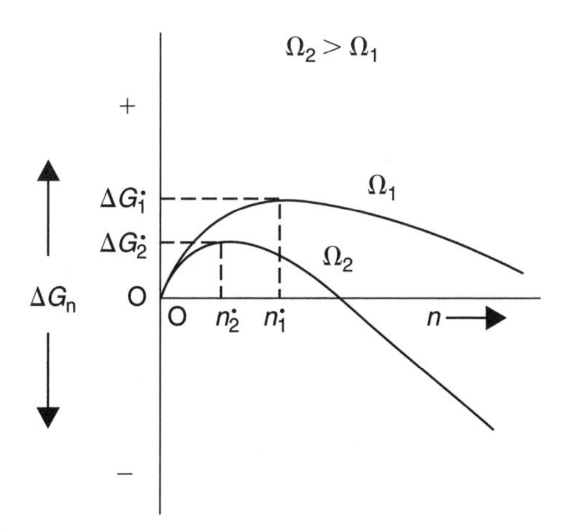

Figure 4.22. Free energy of a single crystal as a function of the number of atoms (n) in the crystal (Nielsen, 1964; Berner, 1980).

Qualitatively these relations may explain why a high degree of supersaturation yield poor crystals. In this case the nucleation threshold is insignificant which allows many crystal nuclei to form, but few to grow larger. Typical natural examples are Fe-oxyhydroxides and Al-hydroxides. On the other hand, at low supersaturation, crystal growth will dominate over nucleation and result in few but larger crystals. The energy barrier also explains why minerals in aquifers rarely form by spontaneous nucleus formation from solution (*homogeneous nucleation*), but preferentially do so on pre-existing surfaces (*heterogeneous nucleation*) which decrease the energy barrier considerably.

The energy barrier decreases with the crystal-water surface energy, which in turn tends to decrease with increasing solubility of the mineral (Nielsen, 1986). This effect may explain the precipitation in sequential order of lesser and lesser soluble minerals constituted of the same elements, known as the Ostwald step rule (Steefel and Van Cappellen, 1990). First, the mineral with the lowest surface energy precipitates, as it needs less supersaturation. This mineral may reduce the surface energy term of the less soluble mineral by acting as template for nucleation and decreasing the surface tension with water. Steefel and Van Cappellen (1990) explained the presence of halloysite in soils, by defining the interfacial energy of the less soluble kaolinite to be so high that the homogeneous nucleation rate would be very small, and Lasaga (1998) demonstrates why quartz is sluggish to precipitate compared with amorphous silica.

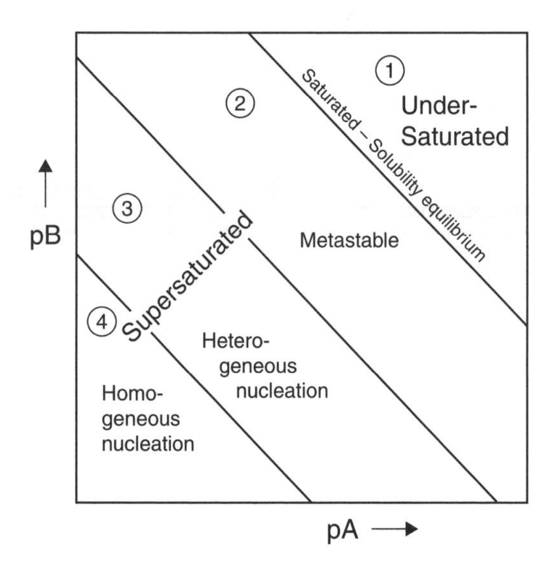

Figure 4.23. Regimes of crystal growth for ionic substances A and B that may form mineral AB. $pA = -\log[A]$ and $pB = -\log[B]$. (Nielsen and Toft, 1984).

The different regimes of crystal formation are illustrated schematically in Figure 4.23 for the activities of ionic substances A and B that may form a mineral AB. In subsaturated solutions, region (1), only dissolution may take place. The line separating region (1) and (2) corresponds to the solubility product. In region (2), existing crystals may grow but no nucleation can take place. With increasing supersaturation, region (3), crystal growth can be accompanied by nucleation on existing surfaces of other minerals etc. And finally in region (4) the degree of supersaturation has become high enough to allow homogeneous nucleation.

An illustration of crystal growth is shown for fluorite in Figure 4.24. The rate of growth is plotted versus the distance from equilibrium and since the straight line describing the data points has a slope of 3, the rate is proportional to $(\Omega - 1)^3$.

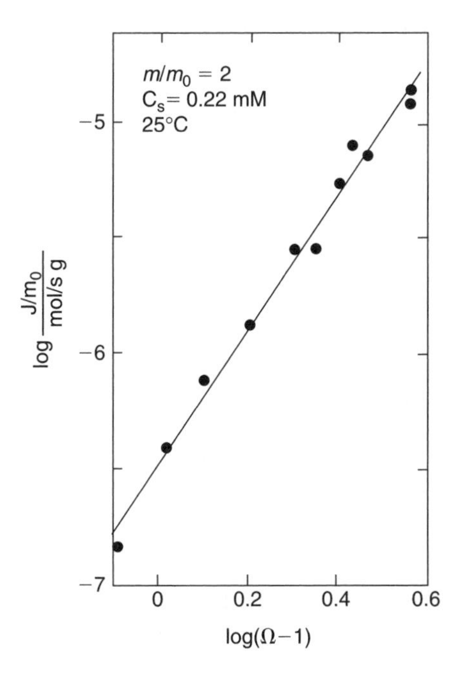

Figure 4.24. The rate (J) of fluorite crystal growth as a function of the distance from equilibrium ($\Omega = IAP/K$) (Christoffersen et al., 1988).

PROBLEMS

4.1. The following table lists the composition of some fluoride rich waters from A. Maarum, Denmark, B. Rajasthan, India, C. Lake Abiata, Ethiopia.

Sample:		A	B	C
pH		7.8	7.3	9.62
Na^+	(mM)	19.1	47.9	194
K^+	(mM)	0.358	0.150	4.91
Mg^{2+}	(mM)	1.19	0.785	0.023
Ca^{2+}	(mM)	1.05	0.675	0.042
Cl^-	(mM)	5.67	17.4	53.9
HCO_3^-	(mM)	17.8	14.8	138
SO_4^{2-}	(mM)	0.00	5.20	0.15
NO_3^-	(mM)	0.032	7.80	–
F^-	(mM)	0.089	0.356	6.28

a. Calculate the saturation state of these waters for fluorite without making any corrections for ionic strength and complexes.
b. Do the same but correct for the ionic strength.
c. Do the same with PHREEQC.

4.2. The mineral *villiaumite* (NaF) has been proposed as a possible solubility control on dissolved fluoride. While the solubility of fluorite can be found in any geochemistry book, this is not the case for villiaumite.

Fortunately, Wagman et al. (1982) list the following thermodynamic data for villiaumite and associated species.

	ΔG_f^0	ΔH_f^0
Villiaumite	−543.5 kJ/mol	−573.6 kJ/mol
Na^+	−261.9 kJ/mol	−240.1 kJ/mol
F^-	−278.8 kJ/mol	−332.6 kJ/mol

a. Calculate $K_{villiaumite}$ at 25°C
b. Calculate $K_{villiaumite}$ at 10°C
c. Check the saturation state of the water samples in Problem 4.1 for villiaumite and consider whether this mineral could limit dissolved fluoride concentrations.

4.3. One could consider whether it is possible to decrease the fluoride content of the Rajasthan water in 4.1 by addition of gypsum.
   a. Calculate the saturation state of the Rajasthan water for gypsum. Would it increase or decrease when complex $CaSO_4^0$ is included. (The complete calculation considering both ionic strength and complexes is to cumbersome to carry out by hand calculations.)
   b. To what level could the fluoride concentration be reduced by gypsum addition if you neglect the effects of ionic strength and complexes. Would these additional corrections increase or decrease the molar fluoride concentration?
   c. Check your last conclusion by doing the calculation with PHREEQC.

4.4. It has been argued in the literature that ion exchange of Ca^{2+} replacing adsorbed Na^+ may increase fluoride concentrations in groundwater. Inspect the water analyses of 4.1 to find evidence that supports such a mechanism. Which additional condition is required?

4.5. Calculate the solubility (mg/L) of atmospheric O_2 in water at 5°C, 15°C and 25°C, check with PHREEQC.

	$O_{2(g)}$	$O_{2(aq)}$
ΔG_f^0	0	16.5 kJ/mol
ΔH_f^0	0	−10.0 kJ/mol

4.6. Redo Example 4.9 and the questions using smithsonite ($ZnCO_3$) and calcite

4.7. Redraw Figure 4.16 for mineral precipitation.

4.8. Pyrite forms in sediments mostly by a two step process: First metastable FeS is formed, which is then transformed to FeS_2 by the overall reaction:

$$FeS + S^0 \rightarrow FeS_2$$

Two reaction mechanisms have been proposed: 1) A solid state reaction between FeS and S^0. 2) Dissolution of both FeS and S^0 and subsequent reaction between Fe^{2+} and polysulfides. (Polysulfides are S_n–S^{2-} compounds which form by the reaction between H_2S and S^0). Experimental data of Rickard (1975) are listed below. Rates are given as $d(FeS_2)/dt$.

1) P_{H_2S} variable (pH = 7, FeS surf. area = $1.6 \times 10^5\,cm^2$, S^0 surf. area = $1.4 \times 10^3\,cm^2$)

P_{H_2S} (atm)	Rate (mol/L/s)
1.00	8.3×10^{-7}
0.50	4.8×10^{-7}

0.25	2.3×10^{-7}
0.10	9.5×10^{-8}
0.05	4.9×10^{-8}

2) S^0-surf. area variable (pH = 7, P_{H_2S} = 1 atm, FeS-surf. area = 1.6×10^5 cm^2).

S^0-surf.area (cm^2)	Rate (mol/L/s)
1.4×10^3	8.3×10^{-7}
7.0×10^2	4.8×10^{-7}
3.5×10^2	1.9×10^{-7}
1.4×10^2	5.4×10^{-8}

3) FeS-surf. area variable (pH = 7, P_{H_2S} = 1 atm, S^0-surf. area = 1.4×10^3 cm^2).

FeS-surf.area (cm^2)	Rate (mol/L/s)
2.5×10^5	2.0×10^{-6}
1.6×10^5	8.3×10^{-7}
8.2×10^4	1.5×10^{-7}
4.1×10^4	4.8×10^{-8}

4) pH variable (P_{H_2S} = 1 atm, S^0-surf. area = 1.4×10^3 cm^2, FeS-surf. area = 1.6×10^5 cm^2).

pH	Rate (mol/L/s)
7.0	8.3×10^{-7}
7.5	7.0×10^{-7}
8.0	6.4×10^{-7}

a. Determine the reaction order for P_{H_2S}, pH, FeS-surface area and S°-surface area.
b. Construct a rate equation for the reaction.
c. Which of the two mechanisms is supported by the experimental data.

4.9. Compare CO_2 emission by activity of man, with natural escape from soil in the Netherlands. Use: 0.1 m thick humic soil (40% *OC*, ρ_b = 1.4 g/cm^3), aerobic respiration rate $dC/dt = -k\,C$, k = 0.025/yr. Surface of the Netherlands is 3×10^{10} m^2, 16×10^6 inhabitants emit 10 ton/yr/person.

4.10. A reaction flask is filled with harbor sludge and 20 mM Na_2SO_4 solution. Use PHREEQC to model the reduction of SO_4^{2-} with the rate for Organic_C given under RATES in PHREEQC.DAT. The flask contains 2.68 M *OC* which is 8×10^4 times more reactive than the terrestric organic matter of the Organic_C rate (After data from Westrich and Berner, 1984).
  a. Calculate the first order decay rate for *OC* in the experiment.

REFERENCES

Allison, J.D., Brown, D.S. and Novo-Gradac, K.J., 1991. *MINTEQA2, a geochemical assessment data base and test cases for environmental systems*. Report EPA/600/3-91/-21. US EPA, Athens, GA.

Anbeek, C., 1992. Surface roughness of minerals and implications for dissolution studies. *Geochim. Cosmochim. Acta* 56, 1461–1469.

Anderson, G.M. and Crerar, D.A., 1993. *Thermodynamics in geochemistry, the equilibrium model.* Oxford Univ. Press, 588 pp.

Atkins, P.W. and de Paula, J. 2002. *Atkins' Physical chemistry*, 7th ed. Oxford Univ. Press, 1149 pp.

Berner, R.A., 1978. Rate control of mineral dissolution under earth surface conditions. *Am. J. Sci.* 278, 1235–1252.

Berner, R.A., 1980. *Early diagenesis – a theoretical approach*. Princeton Univ. Press, 241 pp.

Berner, R.A., 1981. Authigenic mineral formation resulting from organic matter decomposition in modern sediments. *Fortschr. Miner.* 59, 117–135.

Biber, M.V., Dos Santos Afonso, M. and Stumm, W., 1994. The coordination chemistry of weathering: IV. Inhibition of the dissolution of oxide minerals. *Geochim. Cosmochim. Acta* 58, 1999–2010.

Böttcher, M.E., 1997. Comment on "Solid solution partitioning of Sr^{2+}, Ba^{2+}, and Cd^{2+} to calcite". *Geochim. Cosmochim. Acta* 61, 661–662.

Boudreau, B.P. and Westrich, J.T., 1984. The dependence of bacterial sulfate reduction on sulfate concentration in marine sediments. *Geochim. Cosmochim. Acta* 48, 2503–2516.

Buffle, J.D., 1990. *Complexation reactions in aquatic systems.* Ellis Horwood, Chichester, 692 pp.

Burgess, J., 1988. *Ions in solution.* Ellis Horwood, Chichester, 191 pp.

Busenberg, E. and Plummer, L.N., 1989. Thermodynamics of magnesian calcite solid-solutions at 25°C and 1 atm. total pressure. *Geochim. Cosmochim. Acta* 53, 1189–1208.

Christoffersen, J., Christoffersen, M.R., Kibalczyc, W. and Perdok, W.G., 1988, Kinetics of dissolution and growth of calcium fluoride and effects of phosphate. *Acta Odontol Scand.* 46, 325–336.

Christoffersen, J. and Christoffersen, M.R., 1979. Kinetics of dissolution of calcium hydroxy apatite. *J. Crystal Growth* 47, 671–679.

Cox, J.D., Wagman, D.D. and Medvedev, V.A., 1989. *CODATA key values for thermodynamics.* Hemisphere Publ. Corp., Washington, D.C.

Curti, E., 1999. Coprecipitation of radionuclides with calcite: estimation of partition coefficients based on a review of laboratory investigations and geochemical data. *Appl. Geochem.* 14, 433–455.

Denbigh, K., 1971. *The principles of chemical equilibrium.* 3rd ed. Cambridge Univ. Press, 496 pp.

Dixon, D.G. and Hendrix, J.L., 1993. Theoretical basis for variable order assumption in the kinetics of leaching of discrete grains. *AIChE J.* 39, 904–907.

Dove, P. and Rimstidt, J.D., 1993. Silica-water interactions. *Rev. Mineral.* 29, 259–308.

Fitzpatrick, R.W. and Schwertmann, U., 1982. Al-substituted goethite – an indicator of pedogenic and other weathering environments in South Africa. *Geoderma* 27, 335–347.

Ganor, J. and Lasaga, A.C., 1998. Simple mechanistic models for inhibition of a dissolution reaction. *Geochim. Cosmochim. Acta* 62, 1295–1306.

Garrels, R.M. and Christ, C.L., 1965. *Solutions, minerals, and equilibria.* Harper and Row, New York, 450 pp.

Glynn, P.D., 1991. MBSSAS: A code for the computation of Margules parameters and equilibrium relations in binary solid-solution aqueous-solution systems. *Comp. Geosc.* 17, 907–920.

Glynn, P.D., 2000. Solid-solution solubilitites and thermodynamics: sulfates, carbonates and halides. *Rev. Mineral.* 40, 481–511.

Glynn, P.D. and Reardon, E.J., 1990. Solid-solution aqueous-solution equilibria: thermodynamic theory and representation. *Am. J. Sci.* 278, 164–201.

Glynn, P.D. and Reardon, E.J., 1992. Reply to a comment by Königsberger, E. and Gämsjager, H. *Am. J. Sci.* 292, 215–225.

Grassi, S. and Netti, R., 2000. Sea water intrusion and mercury pollution of some coastal aquifers in the province of Grosseto (Southern Tuscany-Italy). *J. Hydrol.* 237, 198–211.

Gratz, A.I., Bird, P. and Quiro, G.B., 1990. Dissolution of quartz in aqueous basic solution. *Geochim. Cosmochim. Acta* 54, 2911–2922.

Guggenheim, E.A., 1967. *Thermodynamics.* Elsevier, Amsterdam, 390 pp.

Handa, B.K., 1975. Geochemistry and genesis of fluoride-containing ground waters in India. *Ground Water* 13, 275–281.

Harvie, C.E., Eugster, H.P. and Weare, J.H., 1982. Mineral equilibria in the six-component seawater system, Na-K-Mg-Ca-SO_4-Cl-H_2O at 25°C. II: Compositions of the saturated solutions *Geochim. Cosmochim. Acta* 46, 1603–1618.

Helgeson, H.C., Delany, J.M., Nesbitt, W.H. and Bird, D.K., 1978. Summary and critique of the thermodynamic properties of rock forming minerals. *Am. J. Sci.*, 278A, 1–229.

Hochella, M.F. and White, A.F. (eds), 1990. *Mineral-water interface geochemistry. Rev. Mineral.* 23, 603 pp.

Jacks, G., Rajagopalan, K., Alveteg, T. and Jönsson, M., 1993. Genesis of high-F groundwaters. Southern India. *Appl. Geochem. Suppl. Issue* 2, 241–244.

Jacks, G., Bhattacharya, P. and Singh, K.P., 2000. High-fluoride groundwaters in India. In P.L. Bjerg, P. Engesgaard and Th.D. Krom (eds), *Groundwater 2000*, 193–194, Balkema, Rotterdam.

Langmuir, D., 1997. *Aqueous environmental geochemistry.* Prentice-Hall, Englewood Cliffs, 600 pp.

Larsen, O. and Postma, D., 2001. Kinetics of bulk dissolution of lepidocrocite, ferrihydrite, and goethite. *Geochim. Cosmochim. Acta* 65, 1367–1379.

Lasaga, A.C., 1998. *Kinetic theory in the earth schiences*. Princeton Univ. Press, 811 pp.

Lewis, G.N. and Randall, M., 1961. *Thermodynamics*. 2nd ed. McGraw-Hill, New York, 723 pp.

Lippmann, F., 1980. Phase diagrams depicting aqueous solubility of binary mineral systems. *N. Jb. Min. Abh.* 139, 1–25.

McIntire, W.L., 1963. Trace element partition coefficients – a review of theory and applications to geology. *Geochim. Cosmochim. Acta* 27, 1209–1264.

Millero, F.J. and Pierrot, D., 1998. A chemical equilibrium model for natural waters. *Aq. Geochem.* 4, 153–199.

Monnin, C. and Schott, J., 1984. Determination of the solubility products of sodium carbonate minerals and an application to trona deposition in Lake Magadi (Kenya). *Geochim. Cosmochim. Acta* 48, 571–581.

Morse, J.W. and Berner, R.A., 1979. Chemistry of calcium carbonate in the deep oceans. In E.A. Jenne (ed.), *Chemical Modelling in Aqueous Systems*. ACS Symp. Ser. 43, 499–535.

Nagy, K.L. and Lasaga, A.C., 1992. Dissolution and precipitation kinetics of gibbsite at 80°C and pH 3: The dependence on solution saturation state. *Geochim. Cosmochim Acta* 56, 3093–3111

Nielsen, A.E., 1964. *Kinetics of precipitation*. Pergamon Press, Oxford, 151 pp.

Nielsen, A.E., 1984. Electrolyte crystal growth mechanisms. *J. Crystal Growth* 67, 289–310.

Nielsen, A.E., 1986. Mechanisms and rate laws in electrolyte crystal growth. In J.A. Davis and K.F. Hayes (eds), *Geochemical Processes at Mineral Surfaces*. ACS Symp. Ser. 323, 600–614.

Nielsen, A.E. and Toft, J.M., 1984. Electrolyte crystal growth kinetics. *J. Crystal Growth* 67, 278–288.

Nordstrom, D.K. and Jenne, E.A., 1977. Fluorite solubility equilibria in selected geothermal waters. *Geochim. Cosmochim. Acta* 41, 175–188.

Nordstrom, D.K. and Munoz, J.L., 1994. *Geochemical thermodynamics*, 2nd ed. Blackwell, Oxford, 493 pp.

Parkhurst, D.L., 1990. Ion association models and mean activity coefficients of various salts. In D.C. Melchior and R.L. Basset (eds), *Chemical modeling of aqueous systems II*. ACS Symp. Ser. 416, 30–43.

Parkhurst, D.L. and Appelo, C.A.J., 1999. *User's guide to PHREEQC (version 2)-a computer program for speciation, batch-reaction, one-dimesnional transport, and inverse geochemical calculations*. U.S. Geol. Surv. Water Resour. Inv. Rep. 99–4259, 312 pp.

Pitzer, K.S., 1981. Characteristics of very concentrated aqueous solutions. In F.E. Wickman and D.T. Rickard (eds), *Chemistry and geochemistry of solutions at high temperatures and pressures*. *Phys. Chem. Earth* 3 and 4, 249–272.

Plummer, L.N. and Busenberg, E., 1982. The solubilities of calcite, aragonite and vaterite in CO_2-H_2O solutions between 0 and 90°C, and an evaluation of the aqueous model for the system $CaCO_3$-CO_2-H_2O. *Geochim. Cosmochim. Acta* 46, 1011–1040.

Plummer, L.N., Parkhurst, D.L., Fleming, G.W. and Dunkle, S.A., 1988. *A computer program incorporating Pitzer's equations for calculation of geochemical reactions in brines*. U.S. Geol. Surv. Water Resour. Inv. Rep. 88–4153.

Postma, D., 1983. Pyrite and siderite oxidation in swamp sediment. *J. Soil Sci.* 34, 163–182.

Rickard, D.T., 1975. Kinetics and mechanism of pyrite formation at low temperatures. *Am. J. Sci.* 275, 636–652.

Riehl, N., Sizmann, R. and Hidalgo, P., 1960. Zur Deutung der Verteilung kleinster Fremdsubstanzmengen zwischen einem wachsenden Kristall und der Lösung. *Z. Phys. Chem.* 25, 351–359.

Rimstidt, J.D. and Barnes, H.L., 1980. The kinetics of silica water reactions. *Geochim. Cosmochim. Acta* 44, 1683–1699.

Rimstidt, J.D., Balog, A. and Webb, J., 1998. Distribution of trace elements between carbonate minerals and aqueous solutions. *Geochim. Cosmochim. Acta* 62, 1851–1863.

Robie, R.A., Hemingway, B.S. and Fisher, J.R., 1978. *Thermodynamic properties of minerals and related substances at 298.15 K and 1 bar (105 Pascals) pressure and at higher temperatures*. U.S. Geol. Surv. Bull. 1452, 456 pp.

Saxena, V.K. and Ahmed, S., 2003. Inferring the chemical parameters for the dissolution of fluoride in groundwater. *Env. Geol.* 43, 731–736.

Schwartz, M. O. and Ploethner, D., 1999. Removal of heavy metals from mine water by carbonate precipitation in the Grootfontein-Omatako canal, Namibia. *Env. Geol.* 39, 1117–1126.

Schwarzenbach, R.P., Gschwend, P.M. and Imboden, D.M., 1993. *Environmental organic chemistry*. Wiley and Sons, New York, 681 pp.

Sparks, D.L., 1989. *Kinetics of soil chemical processes*. Academic Press, San Diego, 210 pp.

Sparks, D.L. and Jardine, P.M., 1984. Comparison of kinetic equations to describe K-Ca exchange in pure and in mixed systems. *Soil Sci.* 138, 115–122.

Steefel, C.I. and Van Cappellen, P., 1990. A new approach to modeling water-rock interactions: the role of precursors, nucleation, and Ostwald ripening. *Geochim. Cosmochim. Acta* 54, 2657–2677.

Stipp, S.L., 1994. Understanding interface processes and their role in the mobility of contaminants in the geosphere: The use of surface sensitive techniques. *Eclogae geol. Helv.* 87, 335–355.

Stumm, W. and Morgan, J.J., 1996. *Aquatic Chemistry.* 3rd ed. Wiley and Sons, New York, 1022 pp.

Sverdrup, H. and Warfvinge, P., 1988. Weathering of primary silicate minerals in the natural soil environment in relation to a chemical weathering model. *Water Air Soil Poll.* 38, 387–408.

Tesoriero, A.J. and Pankow, J.F., 1996. Solid solution partitioning of Sr^{2+}, Ba^{2+}, and Cd^{2+} to calcite. *Geochim. Cosmochim. Acta* 60, 1053–1063.

Truesdell, A.H. and Jones, B.F., 1973. *Wateq, a computer program for calculating chemical equilibria of natural waters.* U.S. Geol. Surv. NTIS PB-220 464, 73 pp.

Van Lier, J.A., De Bruyn, P.L. and Overbeek, J.T.G., 1960. The solubility of quartz. *J. Phys. Chem.* 64, 1675–1682.

Wagman, D.D., Evans, W.H., Parker, V.B., Schumm, R.H., Halow, I., Bailey, S.M., Churney, K.L. and Nuttall, R.L., 1982. *The NBS tables of chemical thermodynamic properties: Selected values for inorganic and C1 and C2 organic substances in SI units. J. Phys. Chem. Ref. Data Suppl. 2*, 392 pp.

Westrich, J.T. and Berner, R.A., 1984. The role of sedimentary organic matter in bacterial sulfate reduction: the G model tested. *Limnol. Ocanogr.* 29, 236–249.

White, A.F., 1995. Chemical weathering rates of silicate minerals in soils. *Rev. Mineral.* 31, 407–461.

White, A.F. and Brantley, S.L. (eds), 1995. Chemical weathering rates of silicate minerals. *Rev. Mineral.* 31, 583 pp.

Wolery, T.J., 1992. *EQ3/6, A solftware package for geochemical modeling of aqueous systems.* UCRL-MA-110662 Pts I-III, Lawrence Livermore Nat. Lab., Cal.

Woods, T.L. and Garrels, R.M., 1987. *Thermodynamic values at low temperature for natural inorganic materials.* Oxford Univ. Press, New York, 242 pp.

Zack, A.L., 1980. *Geochemistry of fluoride in the Black Creek Aquifer system of Horry and Georgetown counties, South Carolina-and its physiological implications.* U.S. Geol. Surv. Water-Supply Paper 2067, 40 pp.

Zhang, J.W. and Nancollas, G.H., 1990. Mechanisms of growth and dissolution of sparingly soluble salts. *Rev. Mineral.* 23, 365–396.

5

Carbonates and Carbon Dioxide

Limestones and dolomites often form productive aquifers with favorable conditions for groundwater abstraction. The main minerals in these rocks are Ca- and Mg-carbonates which dissolve easily in groundwater and give the water its 'hard' character. Usually, the carbonate rock consists of recrystallized biological material with a high porosity and a low permeability. Groundwater flow is then restricted to more permeable fracture zones and karst channels produced by carbonate dissolution. Prolonged dissolution over periods of thousands of years may result in the development of karst landscapes and fantastic features in caves. Apart from the carbonate rocks that consist exclusively of carbonate minerals, sands or sandstones, and marls and clays may contain carbonate minerals as accessory minerals or as cement around the more inert grains.

On a world-wide basis, the effect of carbonate dissolution on water compositions is quite conspicuous. Figure 5.1 compares the average river water composition from different continents by the relation between total dissolved solids (*TDS*) and the concentrations of individual ions.

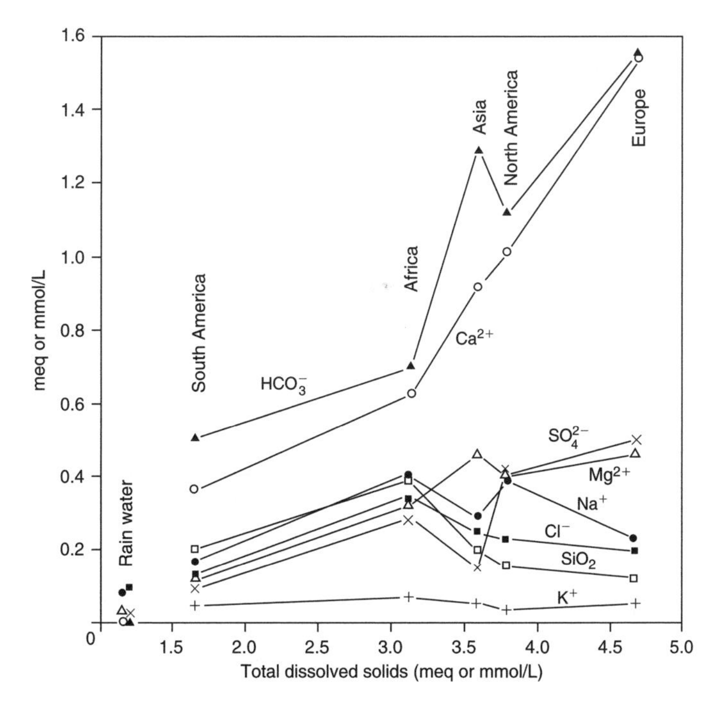

Figure 5.1. The relation between total dissolved solids and concentrations of individual ions in average river water compositions from different continents (Modified from Garrels and Mackenzie, 1971; mmol/L for SiO_2).

The high *TDS* values are mainly related to increases of Ca^{2+} and HCO_3^-, which predominantly are due to carbonate dissolution. The abundance of carbonate rocks in Europe is clearly reflected in the river water composition.

This chapter presents aspects of the mineralogy, the aqueous chemistry and the isotope chemistry required to appreciate the role of carbonate dissolution as regulator of natural water compositions.

5.1 CARBONATE MINERALS

Some carbonate minerals and their solubility products are listed in Table 5.1. Carbonate minerals crystallize with either a trigonal or an orthorhombic crystal structure, depending on the ionic radius of the cation. The smaller ions form trigonal minerals, while the larger ones form orthorhombic minerals. The Ca^{2+} ion is intermediate with the trigonal mineral calcite, being the main component of older limestones, and the orthorhombic aragonite, which is common in recent carbonate sediments and the mineral forming pearls and mother of pearl. Rock-forming carbonate minerals are calcite, dolomite and aragonite. The other carbonate minerals, like siderite or rhodochrosite, occur only in small amounts if present at all. Still, they may exercise an important control on dissolved concentrations of Fe^{2+} and Mn^{2+} (Al et al., 2000; Jurjovec et al., 2002).

Solid solutions of different carbonate minerals are common (Reeder, 1983; Chapter 4). Solid solution may occur among minerals that belong to the same crystal group, within the limits determined by the degree of deformation allowed in the crystal lattice. For example, Mg atoms may substitute for Ca in calcite, and Sr for Ca in aragonite, but Sr will not replace much of Ca in calcite because the ionic radii are too different. Solid solutions between calcite and magnesite are called Mg-calcites. Usually, they are subdivided into low Mg-calcites with <5 mol% and high Mg-calcites with 5–30 mol% Mg. The composition of Mg-calcites in recent marine sediments, mostly from tropical to subtropical areas, is displayed in Figure 5.2, showing a composition with 10–15 mol% $MgCO_3$. The biological uptake of Mg in the shells of marine organisms is temperature dependent (Morse and Mackenzie, 1990). Carbonate shells sampled along the coast of Holland where the summer seawater temperature is about 18°C, contain less than 1% Mg.

Another important rock-forming carbonate mineral is dolomite with a structure displayed in Figure 5.3. Dolomite shows a high degree of ordering with Ca and Mg atoms located in separate layers in the crystal structure. This is different from the structure of high Mg-calcite in which Ca and Mg atoms are distributed more or less at random in the Ca layer (Figure 5.3). A 50 mol% Mg-calcite is therefore not the same as dolomite. Thermodynamically, dolomite should be treated as a pure phase while for Mg-calcites the rules of solid solutions apply.

Table 5.1. Mineralogy and solubility of some carbonates. Log *K*'s from Nordstrom et al., 1990; cation radii from Dowty, 1999 except for Sr.

Trigonal	Formula	$-\log K$	Cation radius (Å)	Orthorhombic	Formula	$-\log K$	Cation radius (Å)
Calcite	$CaCO_3$	8.48	1.12	Aragonite	$CaCO_3$	8.34	1.12
Magnesite	$MgCO_3$	8.24	0.72	Strontianite	$SrCO_3$	9.27	1.18
Rhodochrosite	$MnCO_3$	11.13	0.84	Witherite	$BaCO_3$	8.56	1.42
Siderite	$FeCO_3$	10.89	0.74	Cerussite	$PbCO_3$	13.1	1.18
Smithsonite	$ZnCO_3$	10.01	0.83				
Otavite	$CdCO_3$	13.74	0.99				
Dolomite	$CaMg(CO_3)_2$	17.09					

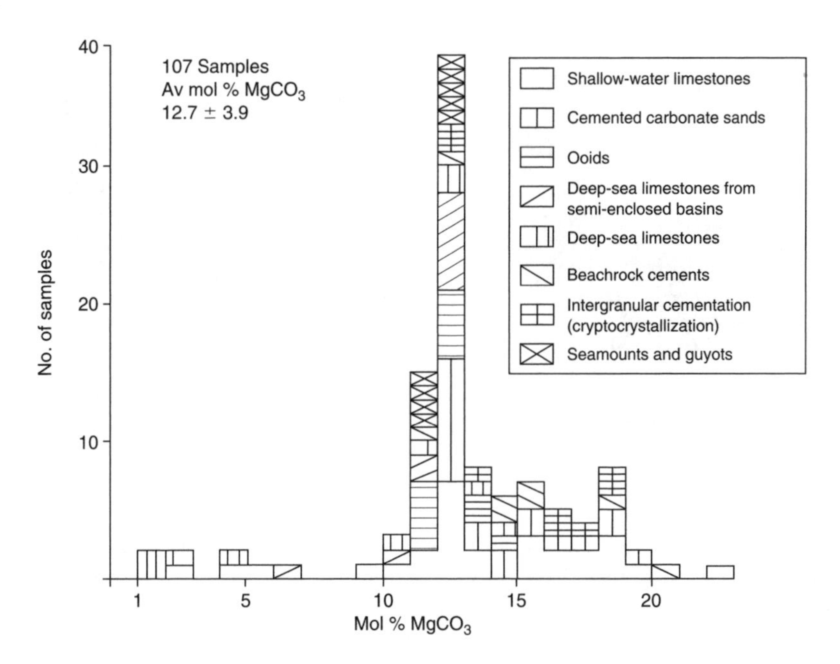

Figure 5.2. The composition of Mg-calcite cements precipitated from seawater (Mucci 1987).

On the other hand, dolomite does show solid solution like most other minerals. The Ca/Mg ratio may vary, and substitution of Fe and Mn is common. The highly ordered structure of dolomite complicates its precipitation from aqueous solutions at low temperatures and the formation of dolomite is still a matter of lively debate (Hardie, 1987; Budd, 1997; Van Lith et al., 2002; Deelman, 2003).

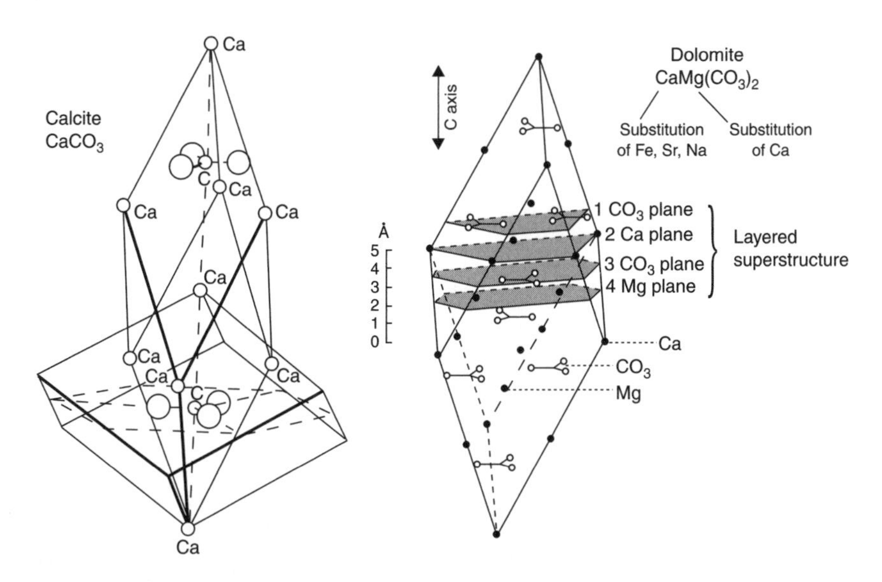

Figure 5.3. The structures of calcite and dolomite. Note the separation of Ca and Mg atoms in different layers in the dolomite (Morrow, 1982; Deer, Howie and Zussman, 1966). The structure of minerals can be displayed easily with the computer program Atoms (Dowty, 1999; www.shapesoftware.com).

Figure 5.4. Scanning electron microscope photographs of recent chalk ooze (left) consisting skeletal remains of organisms and (right) partly recrystallized chalk ooze where inorganic calcite crystals become dominant. Picture size is 3×3 μm (Fabricius, 2003).

Carbonate rocks are with few exceptions, like travertine, biogenic in origin. They consist of the skeletal remains of different organisms as is illustrated in Figure 5.4 showing a subrecent chalk ooze. Older carbonate rocks still bear traces of their organic origin but generally are extensively recrystallized (cf. Reeder, 1983). Recrystallization is also reflected by the observation that recent carbonate sediments generally consist of aragonite and high Mg-calcite while older carbonate rocks (Pre-Pleistocene ($>10^6$ yr)) contain the more stable calcite and dolomite (Hanshaw and Back, 1979).

5.2 DISSOLVED CARBONATE EQUILIBRIA

If we put small calcite crystals in a beaker with distilled water, cover the beaker to exclude CO_2 from air, and return the next day to analyze the solution, we would expect from the dissolution reaction:

$$CaCO_3 \rightarrow Ca^{2+} + CO_3^{2-} \tag{5.1}$$

that:

$$[Ca^{2+}] = [CO_3^{2-}]$$

and in combination with the solubility product (Table 5.1)

$$K_{calcite} = [Ca^{2+}][CO_3^{2-}] = [Ca^{2+}]^2 = 10^{-8.48}$$

Thus, the predicted calcium concentration is

$$[Ca^{2+}] = 10^{-4.25} = 0.06 \text{ mmol/L}.$$

For comparison, Table 5.2 shows the composition of two groundwaters in a carbonate aquifer. The Ca^{2+} concentration in the groundwater of the Alpujarras aquifer (Table 5.2) is 63 and 24 mg/L, corresponding to 1.6 and 0.60 mmol/L respectively. Both values are many times higher than the value of 0.06 mmol/L estimated above and the chemistry involved is clearly more complex.

Table 5.2. Groundwater composition in the Triassic Alpujarras aquifer, southern Spain (Cardenal et al., 1994). Units are mg/L. Alkalinity is listed as HCO_3^- in mg/L.

Sample	Temp°C	pH	Na^+	Mg^{2+}	Ca^{2+}	Cl^-	*Alkalinity*	SO_4^{2-}
2	18.3	7.23	9	20	63	16	269	24
44	18.5	7.91	10	20	24	13	133	20

The higher Ca^{2+} concentration in the groundwater is due to the reaction of calcite with carbon dioxide derived from respiration or oxidation of organic matter. Carbon dioxide reacts with H_2O to form carbonic acid (H_2CO_3). The acid provides protons (H^+) which associate with the carbonate ion (CO_3^{2-}) from calcite to form bicarbonate (HCO_3^-). This is similar to complexation of Ca^{2+} and SO_4^{2-} ions leading to an increase of the solubility of gypsum (Example 4.3). The overall reaction between carbon dioxide and $CaCO_3$ is:

$$CO_{2(g)} + H_2O + CaCO_3 \rightarrow Ca^{2+} + 2HCO_3^- \tag{5.2}$$

This reaction is fundamental for understanding the behavior of $CaCO_3$ dissolution and precipitation in nature. An increase of CO_2 results in dissolution of $CaCO_3$. Removal of CO_2 causes $CaCO_3$ to precipitate. An example is CO_2 degassing from carbonate water in springs and stream rapids where $CaCO_3$ precipites in the form of travertine. Reaction (5.2) also demonstrates the direct coupling between the biological cycle and carbonate mineral-reactions. Degradation of organic matter, or respiration, produces CO_2 and enhances the dissolution of carbonates in soils, while photosynthesis consumes CO_2 and results, for example, in $CaCO_3$ precipitation in lakes.

5.2.1 *The carbonic acid system*

When CO_2 dissolves in water, gaseous $CO_{2(g)}$ becomes aqueous $CO_{2(aq)}$, and some of this associates with water molecules to form carbonic acid, H_2CO_3:

$$CO_{2(g)} \rightarrow CO_{2(aq)} \tag{5.3}$$

and subsequently:

$$CO_{2(aq)} + H_2O \rightarrow H_2CO_3 \tag{5.4}$$

Actually $CO_{2(aq)}$ is at 25°C about 600 times more abundant than H_2CO_3, but to facilitate calculations a convention is adopted in which the two species are summed up as $H_2CO_3^*$. The overall reaction becomes then:

$$CO_{2(g)} + H_2O \rightarrow H_2CO_3^* \tag{5.5}$$

where $H_2CO_3^* = CO_{2(aq)} + H_2CO_3$.

During dissociation, carbonic acid stepwise releases two protons. The concentrations of the dissolved carbonate species therefore depend on the pH of the solution. The reactions and mass action constants that are needed to calculate the composition of the solution are given in Table 5.3. The constants tabulated are approximate values, adequate for manual calculations.

Table 5.3. Equilibria in the carbonic acid system with approximate equilibrium constants at 25°C. More precise values can be found in PHREEQC.DAT.

$H_2O \leftrightarrow H^+ + OH^-$	$K_W = [H^+][OH^-] = 10^{-14.0}$	
$CO_{2(g)} + H_2O \leftrightarrow H_2CO_3^*$	$K_H = [H_2CO_3^*] / [P_{CO_2}] = 10^{-1.5}$	(5.5)
$H_2CO_3^* \leftrightarrow H^+ + HCO_3^-$	$K_1 = [H^+][HCO_3^-] / [H_2CO_3^*] = 10^{-6.3}$	(5.6)
$HCO_3^- \leftrightarrow H^+ + CO_3^{2-}$	$K_2 = [H^+][CO_3^{2-}] / [HCO_3^-] = 10^{-10.3}$	(5.7)

Equation (5.6) can be rewritten as

$$\frac{[HCO_3^-]}{[H_2CO_3^*]} = \frac{10^{-6.3}}{10^{-pH}} \tag{5.6a}$$

At a pH of 6.3, the activities of HCO_3^- and H_2CO_3 are equal. With pH $>$ 6.3, HCO_3^- is the predominant species and at pH $<$ 6.3 there is more $H_2CO_3^*$. The same relation for the CO_3^{2-} / HCO_3^- couple shows that the two species have equal activity at pH = 10.3. At a pH $>$ 10.3, CO_3^{2-} becomes the predominant species while HCO_3^- is more abundant at pH $<$ 10.3. Figure 5.5 summarizes how the different aqueous carbonate species vary with pH.

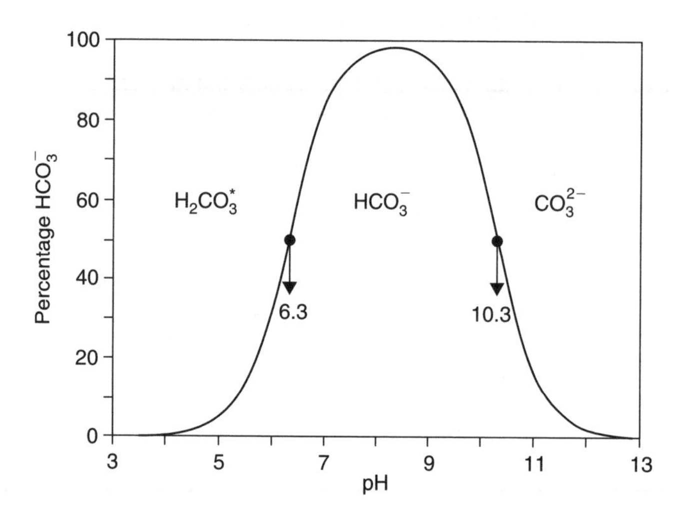

Figure 5.5. Percentage of HCO_3^- of total dissolved carbonate as function of pH.

The distribution of dissolved carbonate species can be calculated for two idealized cases. In the first case the CO_2 pressure is known and constant, a system that is termed *open* with respect to carbon dioxide gas. In the second case, total inorganic carbon (*TIC*) is known and constant and the system is called *closed* with respect to CO_2.

The system *open* with respect to CO_2 gas considers the $[P_{CO_2}]$ as constant. Using a log-transformation of the mass action Equation (5.5) we obtain:

$$\log[H_2CO_3^*] = \log[P_{CO_2}] - 1.5 \tag{5.8}$$

indicating that also $[H_2CO_3^*]$ is constant and independent of the pH. The activity of HCO_3^- as a function of carbonic acid activity and pH is derived from Equation (5.6):

$$\log[HCO_3^-] = -6.3 + \log[H_2CO_3^*] + pH \tag{5.9}$$

To calculate the activity of CO_3^{2-} as a function of pH at constant $[H_2CO_3^*]$ we combine Reactions (5.6) and (5.7). Adding two reactions, means multiplication of the two mass action constants, and subtraction requires similarly that the two be divided.

$$\begin{array}{lll} H_2CO_3^* \leftrightarrow H^+ + HCO_3^-; & & K_1 = 10^{-6.3} \\ HCO_3^- \leftrightarrow H^+ + CO_3^{2-}; & & K_2 = 10^{-10.3} \\ \hline & + & \qquad\qquad \Pi \\ H_2CO_3^* \leftrightarrow 2H^+ + CO_3^{2-}; & & K = 10^{-16.6} \end{array} \tag{5.10}$$

And the law of mass action is, after log-transformation:

$$\log[CO_3^{2-}] = -16.6 + \log[H_2CO_3^*] + 2pH \tag{5.11}$$

When the gas pressure of CO_2, or the activity $[P_{CO_2}]$, is known, the activities of all dissolved carbonate species can be calculated accordingly. Figure 5.6 shows a *pH-log[activity]* diagram in which the activities of the CO_2-species are given for a constant gas pressure of CO_2 of 0.01 atm. The bold line indicates the sum of the dissolved carbonate species (ΣCO_2 or *TIC*). The $\log(\Sigma CO_2)$ increases 1:1 with pH for $\log K_1 < pH < \log K_2$, and 2:1 for $pH > \log K_2$.

The sum of the dissolved carbonate species is given by the mass balance equation:

$$TIC = m_{H_2CO_3^*} + m_{HCO_3^-} + m_{CO_3^{2-}} \tag{5.12}$$

The mass of a species is expressed in molal concentration while the mass *action* equations in Table 5.3 contain activities. For simplicity we equate activity and molal concentration, and then the concentrations can be calculated and added up as shown in Example 5.1.

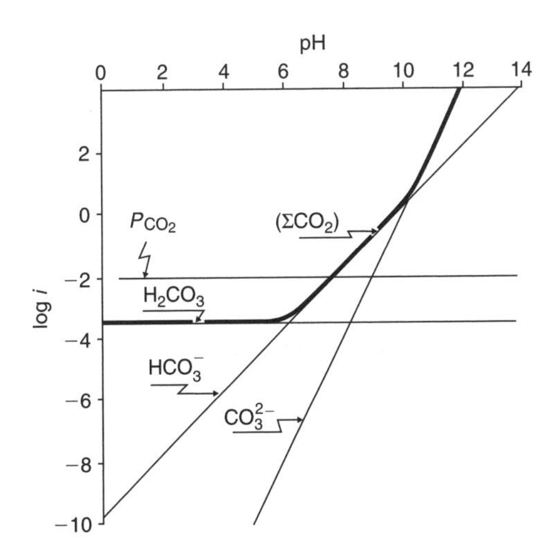

Figure 5.6. The activity of CO_2 species in water as function of pH, at a constant CO_2 pressure of 0.01 atm. The bold line indicates the sum of the carbonate species.

EXAMPLE 5.1. *Calculate TIC in water at pH = 7 and pH = 10, when CO_2 pressure is 0.01 atm*

ANSWER:
We calculate the concentrations of the individual species, and add them up:

	pH = 7	pH = 10
$[H_2CO_3^*] = 10^{-1.5}\,[P_{CO_2}]$	$10^{-3.5}$	$10^{-3.5}$
$[HCO_3^-] = 10^{-6.3}\,[H_2CO_3^*]/[H^+]$	$10^{-2.8}$	$10^{0.2}$
$[CO_3^{2-}] = 10^{-10.3}\,[HCO_3^-]/[H^+]$	$10^{-6.1}$	$10^{-0.1}$
$\Sigma CO_2 =$	$10^{-2.72}$	$10^{0.38}$

In a system *closed* with respect to CO_2, the sum of the dissolved carbonate species is considered constant. The distribution of the dissolved carbonate species, as a function of pH, can be calculated by substituting the mass action expressions (5.6) and (5.7) in (5.12). The results are shown in Figure 5.7 and the line indicating *TIC* (or ΣCO_2) is now parallel to the pH axis. Again we note that carbonic acid is dominant at pH < 6.3 while CO_3^{2-} becomes dominant at pH > 10.3, and in between these pH values HCO_3^- is the major dissolved carbonate species in water. At pH > 6.3, the activity of carbonic acid, and the P_{CO_2} decreases one log-unit with each unit of pH increase; above pH = 10.3, it decreases two log-units with each unit pH increase. At pH's in the near vicinity of the *equivalence* pH (where two species have equal activity) a light rounding of the curves indicates that two species have a significant contribution to the *TIC*.

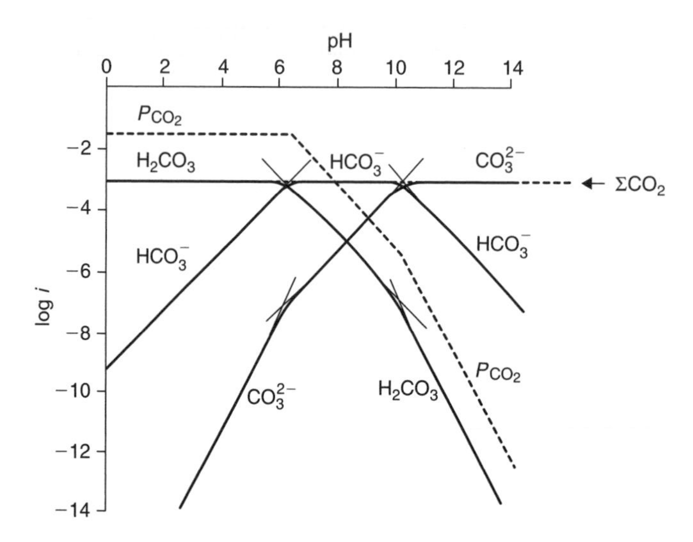

Figure 5.7. The activity of CO_2 species in water at constant total inorganic carbon (*TIC* = 1 mmol/L).

The pH-log[activity] diagram is illustrative for depicting effects of pH on the relative concentrations of solute species. The actual concentration of an individual species as a function of pH can be found quickly by using the *speciation factor* α. For example for HCO_3^-:

$$m_{HCO_3^-} = \alpha \cdot \Sigma CO_2 \tag{5.13}$$

where α gives the fraction of *TIC* made up by HCO_3^-. The speciation factor $\alpha = m_{HCO_3^-} / \Sigma CO_2$, or:

$$\alpha^{-1} = \frac{\sum CO_2}{m_{HCO_3^-}} = \frac{m_{H_2CO_3}}{m_{HCO_3^-}} + \frac{m_{HCO_3^-}}{m_{HCO_3^-}} + \frac{m_{CO_3^{2-}}}{m_{HCO_3^-}}$$

$$= \frac{[H_2CO_3^*]\gamma_1}{[HCO_3^-]\gamma_0} + \frac{[HCO_3^-]}{[HCO_3^-]} + \frac{[CO_3^{2-}]\gamma_1}{[HCO_3^-]\gamma_2}$$

$$= \frac{[H^+]\gamma_1}{K_1\gamma_0} + 1 + \frac{K_2\gamma_1}{[H^+]\gamma_2}$$

where $\gamma_{0,1,2}$ are the activity coefficients for the neutral molecule and the singly and doubly charged ion, respectively. This equation was used to calculate the distribution of HCO_3^- as function of pH shown in Figure 5.5.

QUESTIONS:
Using Equation (5.7), give the pH where HCO_3^- and CO_3^{2-} have equal activity
ANSWER: 10.3.
Give the ratio $[CO_3^{2-}] / [HCO_3^-]$ at pH = 9.3 and 11.3?
ANSWER: 1:10 and 10:1.
Estimate the acidity constant of H_2CO_3 (which is about 1/600 part of $H_2CO_3^*$, cf. Kern, 1960)?
ANSWER: $600 \times 10^{-6.3} = 10^{-3.5}$

5.2.2 *Determining the carbonate speciation in groundwater*

This section describes how the speciation of dissolved carbonate is determined for specific groundwater samples. The four equations (5.5), (5.6), (5.7) and (5.12) together define the carbonate species. The number of variables in the equations amounts to six. Therefore the determination of any two variables fixes the remaining variables in the aqueous carbonate system. In usual practice one of the following pairs of variables are analyzed: pH and alkalinity, *TIC* and alkalinity, P_{CO_2} and alkalinity, or P_{CO_2} and pH. For each of these couples PHREEQC calculates the same carbonate species concentrations, as shown in Example 5.2.

EXAMPLE 5.2. *Calculation of the aqueous carbonate system with PHREEQC*
Given the aqueous carbonate composition (temperature is 25°C):

pH = 7.23; *Alk* = 4.41 meq/L; *TIC* = 4.959 mmol/L; $\log P_{CO_2} = -1.789$

The following input solutions all produce identical results (note that meq is not a valid input format, mmol is used for *Alk*):

```
SOLUTION 1                    # pH and Alkalinity...
  pH 7.23
  Alkalinity 4.41
END

SOLUTION 2                    # TIC and Alkalinity...
  C(4) 4.959
  Alkalinity 4.41
END
```

```
SOLUTION 3                    # PCO2 and Alkalinity...
  pH 7 CO2(g) -1.789          # pH is adapted to equilibrium with [P_CO2] = 10^-1.789
  Alkalinity 4.41
END

SOLUTION 4                    # pH and PCO2...
  pH 7.23
  C(4) 1 CO2(g) -1.789        # Total C(4) is adapted to equilibrium with
                              # [P_CO2] = 10^-1.789
END
```

Enter these input files in PHREEQC and inspect the results.

In most cases the carbonate speciation is determined by measuring pH and alkalinity. This is best done in the field immediately after sample retrieval, to avoid degassing of CO_2. The alkalinity of a water sample is equal to the number of equivalents of all dissociated weak acids. Phosphoric acid and other weak acids may contribute to some extent, and in leachates from waste site various organic acids may be important. However, normally only the carbonate ions are of quantitative importance for the measured alkalinity. It is then equal to:

$$Alk = m_{HCO_3^-} + 2\, m_{CO_3^{2-}} \tag{5.14}$$

At pH values below 8.3, less than 1% of the carbonic acid is present as CO_3^{2-}, and then only the first term is of importance.

The alkalinity is often determined by titration with a HCl or H_2SO_4 solution of known normality towards an endpoint pH of about 4.5, as indicated by a color reaction. The most accurate way to determine alkalinity is the so-called *Gran titration* (Stumm and Morgan, 1996). The principle of the method is that past the point where all HCO_3^- has been converted to H_2CO_3 (at a pH of about 4.3), the concentration of H^+ increases linearly with the amount of H^+ added. In practice, the volume of HCl added is plotted against the Gran function

$$F = (V + V_0) \cdot 10^{-\text{pH}} \tag{5.15}$$

where V is volume of acid added, and V_0 is start volume of the sample. The equivalence point is obtained by backwards extrapolation from the linear part of the curve as shown in Figure 5.8.

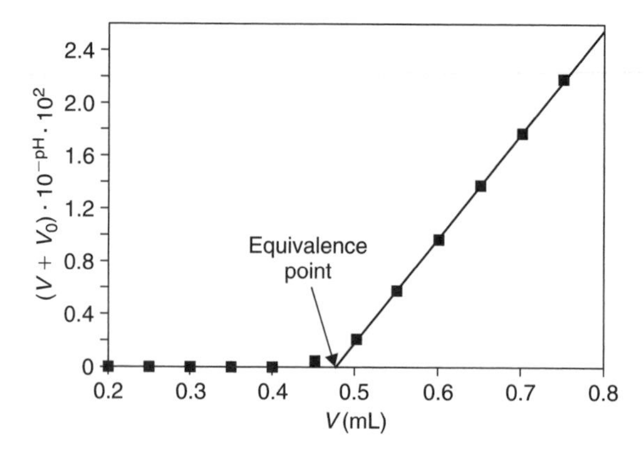

Figure 5.8. Illustration of the Gran plot method for determining alkalinity by titration with acid.

The advantages of this method are several. First, the equivalent point need not be determined accurately during the titration as such, second, the equivalent point is determined by linear regression on a number of pH readings, and third, calibration errors of the pH meter are not important since only changes in $[H^+]$ are used. Once the pH and alkalinity of the water sample are known, the activity of CO_3^{2-}, the P_{CO_2} and the saturation index for carbonate minerals can be calculated as is illustrated in Example 5.3.

EXAMPLE 5.3. *Manual calculation of carbonate speciation and* $SI_{calcite}$
For water sample 2 of the Alpujarras aquifer (Table 5.2) calculate the $SI_{calcite}$ and P_{CO_2} at 25°C:

$pH = 7.23$; $Alk = 269\,mg\,HCO_3^-/L = 10^{-2.36} mol/L$; $Ca^{2+} = 63\,mg/L = 10^{-2.80} mol/L$

ANSWER:
Using the equations from Table 5.3:

$$\log[CO_3^{2-}] = \log K_2 + pH + \log[HCO_3^-] = -10.3 + 7.23 + (-2.36) = -5.43$$

$$\rightarrow SI_{calcite} = \log([Ca^{2+}][CO_3^{2-}]) - \log K_{cc} = -2.80 + (-5.43) - (-8.48) = 0.25$$

$$\log[H_2CO_3^*] = -\log K_1 - pH + \log[HCO_3^-] = 6.3 - 7.23 + (-2.36) = -3.29$$

$$\rightarrow \log [P_{CO_2}] = -\log K_H + \log[H_2CO_3^*] = 1.5 - 3.29 = -1.79$$

The more accurate calculation which includes activity coefficients, complexing and correction of mass action constants to in situ temperature is carried out by PHREEQC for both samples of the Alpujarras aquifer (Table 5.2) in Example 5.4.

EXAMPLE 5.4. *PHREEQC calculation of carbonate speciation and* $SI_{calcite}$
Speciate the groundwater analyses of the Alpujarras aquifer listed in Table 5.2.

ANSWER:
For sample 2, the PHREEQC input file is

```
SOLUTION 1 # Cardenal et al 94 sample 2
  pH 7.23
  temp 18.3
  units mg/L
  Na 9
  Mg 20
  Ca 63
  Cl 16
  Alkalinity 269 as HCO3
  S(6) 24
END
```

and similar for sample 44. In the output file we find:

Sample	*TIC* (mmol/L)	$\log[P_{CO_2}]$	$\log[CO_3^{2-}]$	$\log[Ca^{2+}]$	$SI_{calcite}$
2	4.98	−1.85	−5.566	−2.979	−0.1
44	2.22	−2.83	−5.185	−3.364	−0.1

The PHREEQC calculations indicate that both samples are very slightly subsaturated for calcite even though their water compositions actually are quite different. The manual calculation for sample 2 shows slight supersaturation, the difference resulting from the effects of temperature, activity coefficients and complexing.

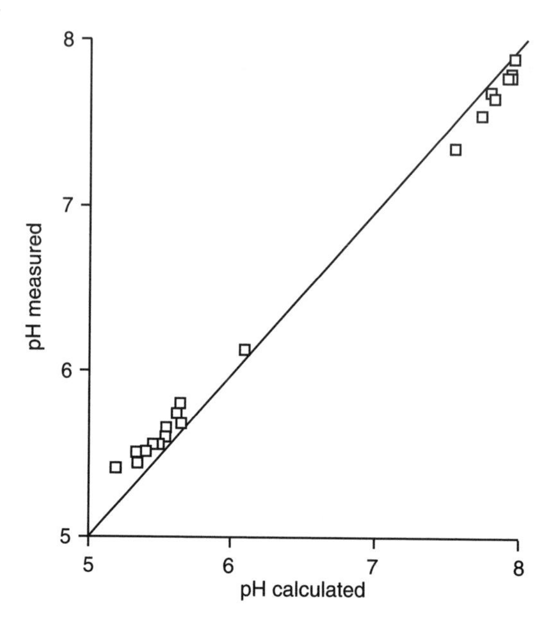

Figure 5.9. A comparison of pH calculated with PHREEQC from *TIC*/alkalinity measurements, with pH values measured in the field.

A reliable estimate of $[CO_3^{2-}]$ requires a good measurement of pH and alkalinity. Sometimes the pH measurement of field samples is not reliable due to temperature differences of electrode, buffers and sample, degassing of CO_2, sluggish or faulty electrodes, etc. As seen above, errors in measured pH directly influence the calculation of the CO_3^{2-} activity from alkalinity. As an alternative the *TIC* (cf. Table 1.6) and alkalinity can be measured, and the pH and the aqueous carbonate speciation is then calculated using PHREEQC (cf. Example 5.2). A comparison of measured field pH values and pH values calculated with PHREEQC according to this method is shown in Figure 5.9. Generally there is good agreement between measured and calculated pH which strengthens the confidence in both measurements and theory. However, at low pH, where pH buffering by alkalinity is poorest, the measured pH values tend to be slightly higher, perhaps due to some degassing.

QUESTION:
Make a PHREEQC input file to calculate the speciation for sample 44, Table 5.2.
Find $SI_{calcite}$ when the samples' temperature is 25°C?
ANSWER: −0.01

5.3 CARBON DIOXIDE IN SOILS

The CO_2 concentration in the atmosphere is low, 0.03 vol%, which is equivalent to $P_{CO_2} = 3\times10^{-4}$ of 1 atm $= 10^{-3.5}$ atm. The CO_2 pressure of groundwater is easily one or two orders of magnitude higher, primarily as result of uptake of carbon dioxide during the infiltration of rainwater through the soil. In the soil, CO_2 is generated by root respiration and decay of labile organic material (Hanson et al., 2000):

$$CH_2O + O_2 \rightarrow CO_2 + H_2O \tag{5.16}$$

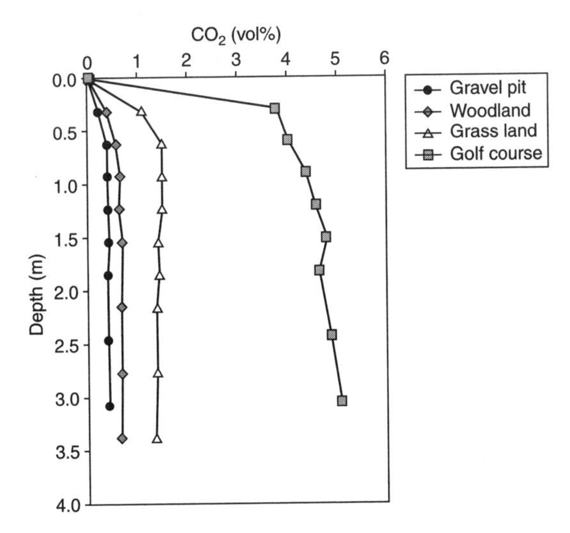

Figure 5.10. Concentration of CO_2 gas in sandy unsaturated zone with different land use. The data were collected in the growth season, July 1992. The concentration at 0 m depth corresponds to the atmospheric CO_2 concentration. Modified from Lee 1997.

The mean residence time of organic carbon in soils is short, 10–100 years for a temperate climate soil (Russell, 1973; Keller and Bacon, 1998). Most of the CO_2 that is produced in the soil escapes to the atmosphere, where it has a residence time of about 10 years. The short residence time means that permutations in agricultural practice may have important effects on the CO_2 content of the atmosphere, much like the burning of fossil carbon, which adds to the complexity of assessing the global CO_2 budget and the effects of CO_2 on climatic change.

Figure 5.10 shows the concentration of CO_2 gas in unsaturated sandy soils below different vegetations. The gravel pit has a sparse vegetation cover and the CO_2 concentration is low but still nearly an order of magnitude higher than in the atmosphere. The woodland consists of mixed hardwood and conifer trees and has a higher CO_2 concentration. Below uncultivated grass lands the CO_2 concentration may be 1 vol% and higher. Finally, below the cultivated grass fields of the golf course CO_2 reaches as much as 5 vol% CO_2, probably an effect of high density growth of hybrid grasses in combination with intense irrigation and fertilization. In all the profiles, a steep gradient of the CO_2 distribution towards the surface (0 m depth) indicates a loss of CO_2 to the atmosphere. Globally this loss of CO_2 was estimated to be 77 Gt C/yr (Raich and Potter, 1995).

Seasonal variations of CO_2 in soil air are apparent in Figure 5.11. During winter the biological CO_2 production stops and the CO_2 gradient indicates that groundwater is degassing to the atmosphere. In summer the CO_2 production is high and the CO_2 distribution suggests both upward and downward diffusion. The CO_2 concentrations in the atmosphere also fluctuate seasonally, concentrations are lowest during the late summer (September on the northern hemisphere, and March on the southern hemisphere) mainly as result of the photosynthesis of land plants (Bolin et al., 1979).

Due to higher biological activity, the CO_2 concentration in soil air increases with temperature (Hamada and Tanaka, 2001; Yoshimura et al., 2001; Drake, 1983). However, temperature cannot be the only controlling variable for biological productivity, also the availability of water and the soil conditions must play a role. Accordingly, it has been suggested to use evapotranspiration as a measure of biological production and hence of the average CO_2 pressure in the soil. Apart from biological activity, there are also physical feedbacks between the soil CO_2 content and evapotranspiration. When soils dry up, CO_2 diffusion out of the soil increases and the CO_2 concentration goes down.

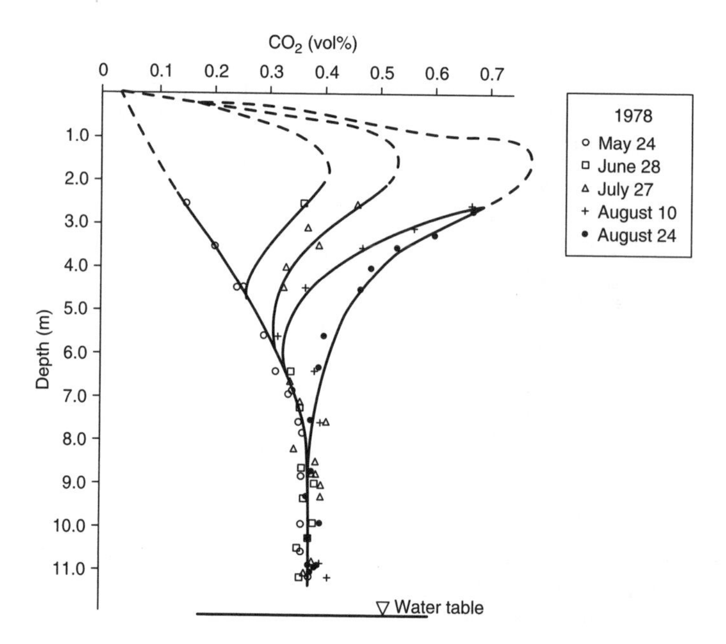

Figure 5.11. Seasonal fluctuations of soil CO_2 pressures in an unsaturated sand (Reardon et al., 1979).

Figure 5.12 shows the relation between evapotranspiration and soil P_{CO_2} (Brook et al., 1983). Cool continental climates like Nahanni and Saskatchewan, Canada, have a low evapotranspiration and P_{CO_2}, while warm and humid places like Puerto Rico and Thailand are found at the other end of the scale. The relationship has been used to produce a world map of CO_2 pressure based on actual evapotranspiration (Figure 5.13), showing the highest P_{CO_2} values in the tropics.

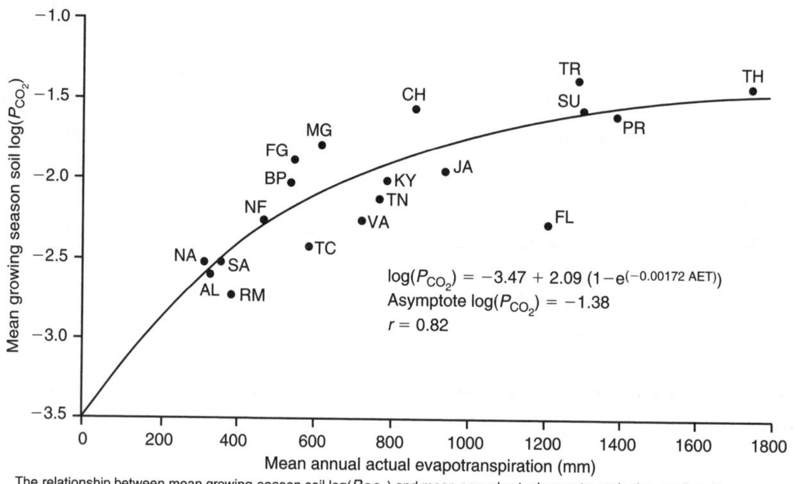

Figure 5.12. Relationship between soil CO_2 pressure and actual evapotranspiration (Brook et al., 1983).

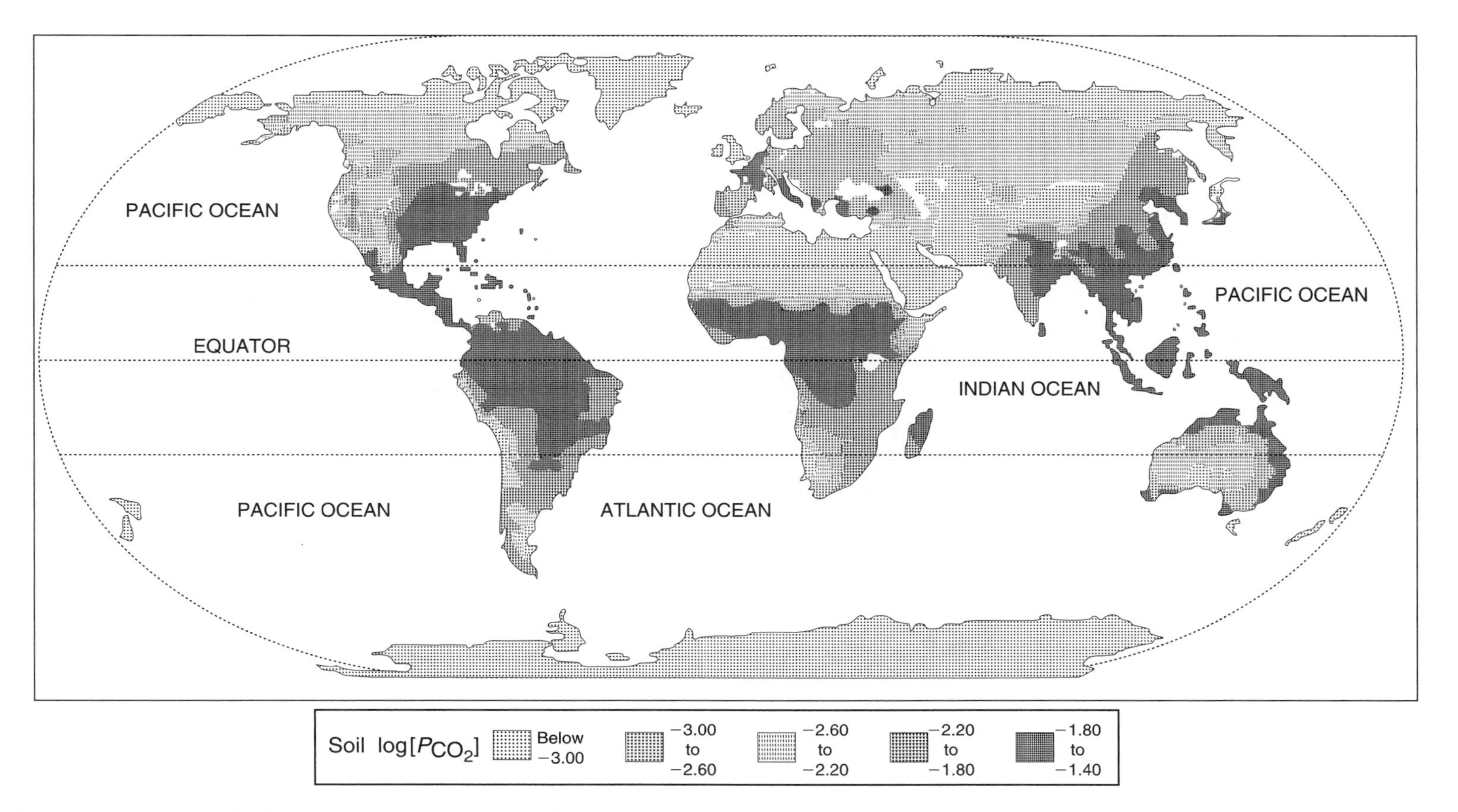

Figure 5.13. Map of soil CO_2 pressure produced with the relationship shown in Figure 5.12 (Brook et al., 1983).

The data have also been used to estimate the global flux of dissolved carbonate moving downward into groundwater, by combining the world P_{CO_2} map with maps of recharge and soil pH (Kessler and Harvey, 2001). Using P_{CO_2} and pH, *TIC* is calculated and multiplied with recharge. The resulting estimate amounts to 0.2 Gt C/yr, which is much smaller than the upward loss of CO_2 to the atmosphere of 77 Gt C/yr, but lies in the range of the 0.3 Gt C/yr carried as *TIC* by rivers to the oceans (Suchet and Probst, 1995). For comparison, the total human emission of CO_2 is about 10 Gt C/yr, which augments the natural CO_2 flow into the atmosphere by about one seventh.

Soil respiration is not the only source of CO_2 entering groundwater. Also the oxidation of organic carbon present in sediments (Keller and Bacon, 1998; see also Chapter 9) or of *DOC* carried with the groundwater (Alberic and Lepiller, 1998) is an important source of *TIC*. Furthermore, CO_2 derived from deep sources, which include degassing of magma or thermal metamorphosis of oceanic carbonate rock, may migrate upward along fracture zones and contribute to the groundwater CO_2 input (Etiope, 1999; Yoshimura et al., 2001; Chiodini et al., 1998).

Figures 5.10 and 5.11 indicate the range of CO_2 content in soil gas to lie in between the atmospheric value of 0.03 vol% ($P_{CO_2} = 10^{-3.5}$) and 3 vol% ($P_{CO_2} = 10^{-1.5}$). Let us approximate the soil water as pure water in which CO_2 dissolves, and calculate the composition of this hypothetical "soil moisture" in equilibrium with a constant CO_2 pressure. When CO_2 dissolves in water the aqueous species are H_2CO_3, HCO_3^- and CO_3^{2-} and the activity ratios among the species are given by the mass action equations in Table 5.3. The electroneutrality condition is:

$$m_{H^+} = m_{HCO_3^-} + 2m_{CO_3^{2-}} + m_{OH^-} \tag{5.17}$$

Dissolution of CO_2 in water forms carbonic acid and we may surmise that the pH will be slightly acidic. In first approximation we therefore neglect $m_{CO_3^{2-}}$ and m_{OH^-}. This reduces the electroneutrality equation to:

$$m_{H^+} = m_{HCO_3^-} \tag{5.18}$$

We can express $[HCO_3^-]$ as function of $[H^+]$ and $[P_{CO_2}]$ with Equations (5.5) and (5.6) to find:

$$[HCO_3^-] = 10^{-7.8}\ [P_{CO_2}]/[H^+] \tag{5.19}$$

By combining this relation with Equation (5.18), and neglecting the differences between activity and molality, we obtain the concentration of H^+:

$$m_{H^+} = \sqrt{10^{-7.8}[P_{CO_2}]} \tag{5.20}$$

Table 5.4. Calculated CO_2 species in soil water and rainwater.

$[P_{CO_2}]$	$10^{-1.5}$	$10^{-3.5}$	
H^+	$10^{-4.6}$	$10^{-5.6}$	mol/L
pH	4.6	5.6	
H_2CO_3	$10^{-3.0}$	$10^{-5.0}$	mol/L
HCO_3^-	$10^{-4.6}$	$10^{-5.6}$	mol/L
CO_3^{2-}	$10^{-10.3}$	$10^{-10.3}$	mol/L
OH^-	$10^{-9.4}$	$10^{-8.4}$	mol/L

Back substitution in the equations of Table 5.3 gives the concentrations entered in Table 5.4. The P_{CO_2} of the atmosphere is $10^{-3.5}$ atm and the pH of unpolluted rainwater should therefore be 5.6 (compare with the values in Table 2.2). A P_{CO_2} of $10^{-1.5}$ atm is near the upper limit for respiration and organic matter degradation, which means that the pH of soil water should not become lower than 4.6 (actual soil waters can be more acidic due to acid rain and dissolved organic acids, cf. Chapter 8).

Furthermore, note that our approximation to neglect $m_{CO_3^{2-}}$ and m_{OH^-} in the electroneutrality equation is acceptable. PHREEQC could easily calculate the pH of a solution in contact with a constant $[P_{CO_2}]$, without approximations, following Example 5.2.

QUESTIONS:

Estimate the CO_2 pressure in soil below grassland, Figure 5.10?

ANSWER: CO_2 = 1.1 vol% = $10^{-1.96}$ atm

Check the results of the manual calculation in Table 5.4 with PHREEQC (temp = 25°C).

ANSWER: `SOLUTION; EQUILIBRIUM_PHASES; CO2(g) -1.5; END` gives pH = 4.66

5.4 CALCITE SOLUBILITY AND P_{CO_2}

Intuitively we expect that a higher P_{CO_2} and the associated increase of dissolved carbonic acid allows more calcite to dissolve (Equation 5.2). More formally we may combine the dissociation reaction for calcite with the reactions for aqueous carbonate:

$$\begin{array}{llr} CaCO_3 \leftrightarrow Ca^{2+} + CO_3^{2-} & & K_{cc} = 10^{-8.5} \\ CO_3^{2-} + H^+ \leftrightarrow HCO_3^- & & K_2^{-1} = 10^{10.3} \\ H_2CO_3^* \leftrightarrow H^+ + HCO_3^- & & K_1 = 10^{-6.3} \\ CO_{2(g)} + H_2O \leftrightarrow H_2CO_3^* & + & K_H = 10^{-1.5} \quad \Pi \\ \hline CO_{2(g)} + H_2O + CaCO_3 \leftrightarrow Ca^{2+} + 2HCO_3^- & & K = 10^{-6.0} \end{array}$$

The corresponding mass action equation is:

$$\frac{[Ca^{2+}][HCO_3^-]^2}{[P_{CO_2}]} = 10^{-6.0} \tag{5.21}$$

For a system containing only calcite, CO_2 and H_2O, the electroneutrality equation in the solution is:

$$2m_{Ca^{2+}} + m_{H^+} = m_{HCO_3^-} + m_{OH^-} + 2m_{CO_3^{2-}} \tag{5.22}$$

Assume the pH of the soil water to be less than 8.3, then $m_{CO_3^{2-}}$ and m_{OH^-} can be neglected in (5.22). Furthermore, assume that m_{H^+} is negligible in comparison to $m_{Ca^{2+}}$ and the electroneutrality is reduced to:

$$2m_{Ca^{2+}} = m_{HCO_3^-} \tag{5.23}$$

Combining with Equation (5.21), and assuming that activity equals molality yields:

$$m_{Ca^{2-}} = \sqrt[3]{10^{-6.0}[P_{CO_2}]/4} \tag{5.24}$$

Table 5.5. Solution speciation in equilibrium with a given P_{CO_2} and calcite.

$[P_{CO_2}]$	$10^{-1.5}$	$10^{-3.5}$	
Ca^{2+}	2.0	0.44	mmol/L
CO_3^{2-}	1.64×10^{-3}	7.6×10^{-3}	mmol/L
HCO_3^-	4.0	0.87	mmol/L
H^+	1.24×10^{-4}	5.7×10^{-6}	mmol/L
pH	6.9	8.2	

Now we have obtained a simple relationship between the Ca^{2+} concentration and the CO_2 pressure for equilibrium with calcite in pure H_2O. Once $m_{Ca^{2+}}$ ($\approx [Ca^{2+}]$) is known, $[CO_3^{2-}]$ can be calculated from the solubility of calcite; the amount of HCO_3^- can be calculated with Equation (5.23), and the other species follow from the equations provided in Table 5.3. The results of such calculations have been entered in Table 5.5 for CO_2 pressures of $10^{-1.5}$ and $10^{-3.5}$ atm. Note that concentrations of H^+, CO_3^{2-} and OH^- are small compared to concentrations of HCO_3^- and Ca^{2+}, and were rightly neglected.

Equation (5.24) predicts that the Ca^{2+} concentration increases with the cube root of the CO_2 pressure as shown in Figure 5.14. The relation has an important consequence. Mixing of two waters which are both at equilibrium with calcite but which have a different CO_2 pressure (A and B in Figure 5.14), produces a water composition on an almost straight line that connects A and B.

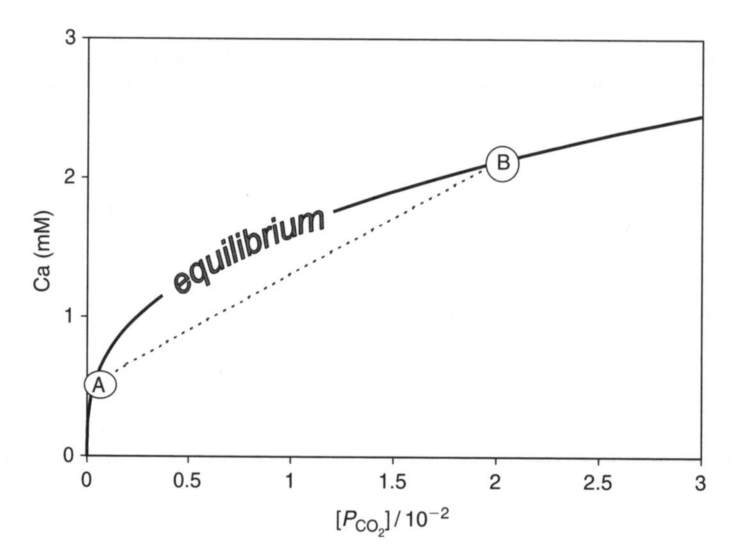

Figure 5.14. The solubility of calcite in H_2O as a function of P_{CO_2}. The curved line indicates equilibrium with calcite according to Equation (5.24). Mixtures of waters A and B are positioned on the (nearly) straight line connecting A and B.

The straight line lies below the equilibrium curve, indicating that the Ca^{2+} concentration is smaller than is required for equilibrium. In other words, a mixture of two waters which are both saturated for calcite, becomes subsaturated for calcite. This effect may cause calcite dissolution in carbonate aquifers and has been called *Mischungskorrosion* in the German literature (Bögli, 1978). It is probably the dominant process in cave formation in karst, and even plays a role when seawater (that is supersaturated for calcite) is mixed with fresh groundwater in equilibrium with calcite, but with a higher CO_2 pressure (Section 5.5.2). Example 5.5 shows how to calculate the effects of mixing with PHREEQC.

EXAMPLE 5.5. *Calculate mixing effects on calcite saturation state with PHREEQC*
The input file to calculate and plot Figure 5.14 is listed below. In the first simulation, the equilibrium line is calculated. CO_2 is added stepwise to distilled water with keyword REACTION, while equilibrium with calcite is maintained. In the second and third simulation the two solutions B and A are defined. In the fourth simulation, keyword MIX is used to mix the two solutions to illustrate that the mixing line is slightly curved. Note how in USER_GRAPH the special function sr ("CO2 (g)") is used. It gives the Saturation Ratio of CO_2(g), or the partial pressure of this gas with respect to the standard state of 1 atm.

```
SOLUTION 0                          # Calculate the equilibrium line...
EQUILIBRIUM_PHASES 1
 Calcite
REACTION 1
 CO2 1; 3.5e-3 in 30
USER_GRAPH; -head CO2/vol% Ca/mM
 -start; 10 graph_x 100 * sr("CO2(g)"); 20 graph_y 1e3 * tot("Ca") ; -end
END

SOLUTION 2                          # point B in Figure 5.14...
 pH 7 charge; Ca 1 Calcite; C(4) 1 CO2(g) -1.7
USER_GRAPH; -connect_simulations
END

SOLUTION 1                          # point A in fig 5.14...
 pH 7 charge; Ca 1 Calcite; C(4) 1 CO2(g) -3.5
END

MIX
 1 0.97; 2 0.03
END
```

QUESTIONS:
Why is the mixing line in Figure 5.14 slightly curved near point A?

ANSWER: $H_2CO_3^*$ does not behave conservatively upon mixing (when pH changes)

Calculate $SI_{calcite}$ when solution A and B (Figure 5.14) are mixed 1:1, and how much calcite dissolves in the mixture?

ANSWER: $SI_{cc} = -0.27$, 0.12 mM calcite dissolves

5.4.1 *Calcite dissolution in systems open and closed for CO_2 gas*

Figure 5.15 sketches two situations for calcite dissolution in the field. In both cases root respiration provides a continuous source of CO_2. In the case at left, $CaCO_3$ is present above the water table, and it will dissolve in contact with the CO_2 charged water. The CO_2 that is consumed is resupplied by root respiration, and the system is open with respect to CO_2. In the calculations this is simulated by keeping $[P_{CO_2}]$ constant. The open system extends through most of the unsaturated zone due to rapid gas exchange through the gas phase (Laursen, 1991). The water composition can be calculated for a given $[P_{CO_2}]$ with Equation (5.24) (see also Table 5.5).

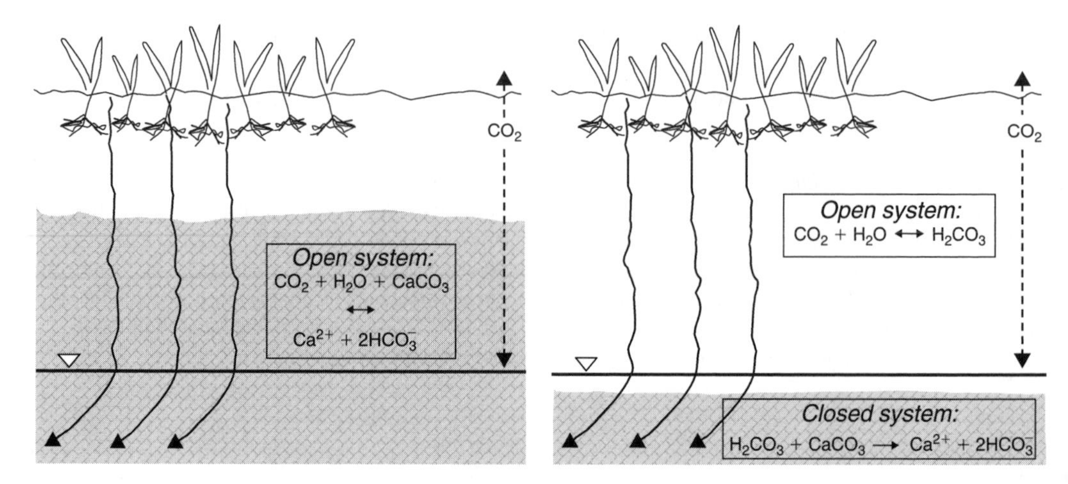

Figure 5.15. Calcite dissolution in systems open (left) and closed (right) with respect to CO_2. In the open system calcite dissolution proceeds in contact with CO_2 production. During closed system dissolution CO_2 production and calcite dissolution are spatially separated.

On the situation at right (Figure 5.15), calcite is absent in the unsaturated zone. The CO_2 charged water first encounters a calcite layer below the groundwater table. When calcite dissolves there, the CO_2 consumed is not replenished, because diffusion of CO_2 through water is slow. The system is *closed* with respect to CO_2 gas and Equation (5.24) is not applicable. However, the water composition can be calculated assuming that all the initial CO_2 is used up by dissolution of calcite:

$$CO_2 + H_2O + CaCO_3 \rightarrow Ca^{2+} + 2HCO_3^- \qquad (5.25)$$

The amount of Ca^{2+} released equals the initial amount of CO_2 (=$(\Sigma CO_2)_{root}$):

$$m_{Ca^{2+}} = (\Sigma CO_2)_{root} \qquad (5.26a)$$

while the amount of HCO_3^- is twice as much:

$$m_{HCO_3^-} = 2(\Sigma CO_2)_{root} \qquad (5.26b)$$

Here it is assumed that pH <8.3, so that CO_3^{2-} is unimportant. For a given initial $[P_{CO_2}]$ the concentration of $H_2CO_3^*$ can be calculated from Henry's law (Equation 5.5) which is equal to $(\Sigma CO_2)_{root}$. The calculation of the other species in a hand-calculation follows the same route as in the open case. The analogue calculations with PHREEQC are shown in Example 5.6.

EXAMPLE 5.6. *PHREEQC calculation of open and closed system calcite dissolution*
In open system dissolution, water is equilibrated with both calcite and CO_2 gas at the desired pressure. With closed system dissolution, water is first equilibrated with CO_2 gas. This solution is SAVEd, and USEd in the next simulation where it is equilibrated with calcite. This stepwise calculation closely follows the sequence in the field where the water first is charged with CO_2 at a given pressure in the soil, and then contacts calcite below the groundwater table where the CO_2 pressure will decrease as calcite dissolves.

```
SOLUTION 1                    # open system dissolution of CO2 with calcite...
EQUILIBRIUM_PHASES 1
 CO2(g) -1.5; Calcite 0.0
END

SOLUTION 1                    # open system dissolution of CO2 in the
                              # unsaturated soil...
EQUILIBRIUM_PHASES 1
 CO2(g) -1.5
SAVE solution 2
END

USE solution           2 # equilibrate with calcite below the groundwater
                         # table...
EQUILIBRIUM_PHASES 2
 Calcite
END
```

Table 5.6 summarizes the effect of CO_2 pressure and calcite dissolution on the water composition for $[P_{CO_2}]$ ranging from $10^{-1.5}$ in a productive soil to $10^{-3.5}$ in a desert sand. The resulting pH varies from 7.0–8.3 for open system calcite dissolution, and from 7.6–10.1 for closed system calcite dissolution. With the same initial $[P_{CO_2}]$, calcite dissolution in an open system yields higher Ca^{2+} and HCO_3^- concentrations than if the system is closed with respect to CO_2 gas.

Table 5.6. Summary of possible concentrations in water in which calcite dissolves at 15°C (PHREEQC calculations).

		Open system, constant $[P_{CO_2}]$		Closed system, with known initial $[P_{CO_2}]$		
$[P_{CO_2}]$	Initial	$10^{-1.5}$	$10^{-3.5}$	$10^{-1.5}$	$10^{-3.5}$	
	Final	$10^{-1.5}$	$10^{-3.5}$	$10^{-2.5}$	$10^{-6.4}$	
pH		6.98	8.29	7.62	10.06	
Ca^{2+}		2.98	0.58	1.32	0.12	mmol/L
Alk		5.96	1.16	2.65	0.24	mmol/L
EC		600	120	265	25	μS/cm

QUESTION:
Check the concentrations in Table 5.6

5.4.2 *Two field examples*

Figure 5.16 shows the dissolution of carbonate in unsaturated sand that contains calcite and dolomite at depths greater than 1.7 m. The water table at the site is located at 12 m depth. As soon as infiltrating water reaches the level where calcite is present, the concentrations of Ca^{2+} and HCO_3^-, and the pH increase, and the saturation index for calcite approaches equilibrium. The P_{CO_2} in the profile remains constant which is typical for open system dissolution.

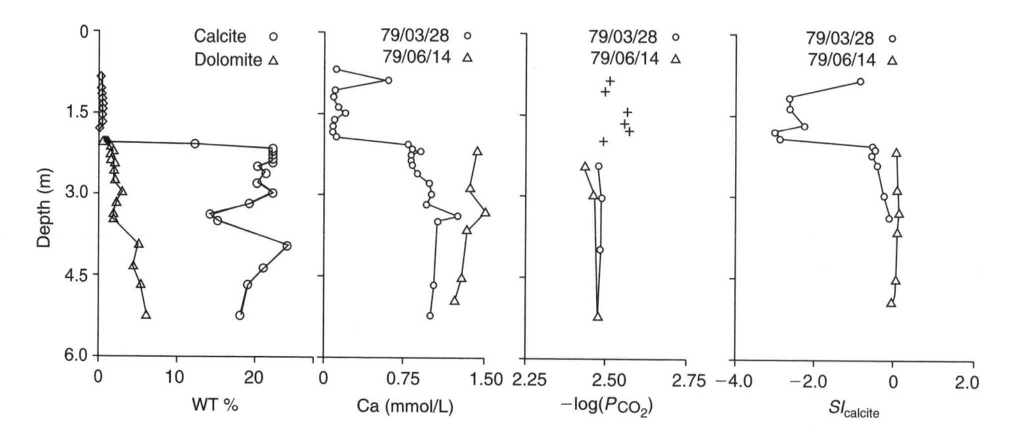

Figure 5.16. Calcite and dolomite distribution and calcite dissolution at constant CO_2 pressure in unsaturated sand (open system), at Trout Creek, Canada, (Reardon et al., 1980). P_{CO_2} values given by crosses were calculated from pH and alkalinity while the open symbols are direct measurements of soil gas.

The depth distributions of the two carbonates calcite and dolomite are different in the profile. For calcite, the transition in the sediment concentration is very sharp, from zero in the upper part to 20 wt% at 1.7 m. This suggests that calcite dissolves rapidly until saturation is reached, whereafter dissolution stops as indicated by the sudden jump to $SI \approx 0$. The dissolution front moves slowly downward as calcite is depleted. On the other hand, the concentration of dolomite increases gradually with depth beyond 1.7 m, apparently because dolomite dissolves slower and needs more time before saturation is attained. The kinetic dissolution of dolomite causes supersaturation for calcite which will, in principle, precipitate (cf. Problem 5.15). The result is that dolomite is transformed into calcite, in other words, the sediment is *dedolomitized.*

Figure 5.17 illustrates calcite dissolution in groundwater in the Rømø aquifer (Jakobsen and Postma, 1999). The upper 4 m of the saturated zone is free of $CaCO_3$, the pH of groundwater is around 5, the $H_2CO_3^*$ concentration is about 1 mM, and the groundwater is strongly subsaturated for calcite.

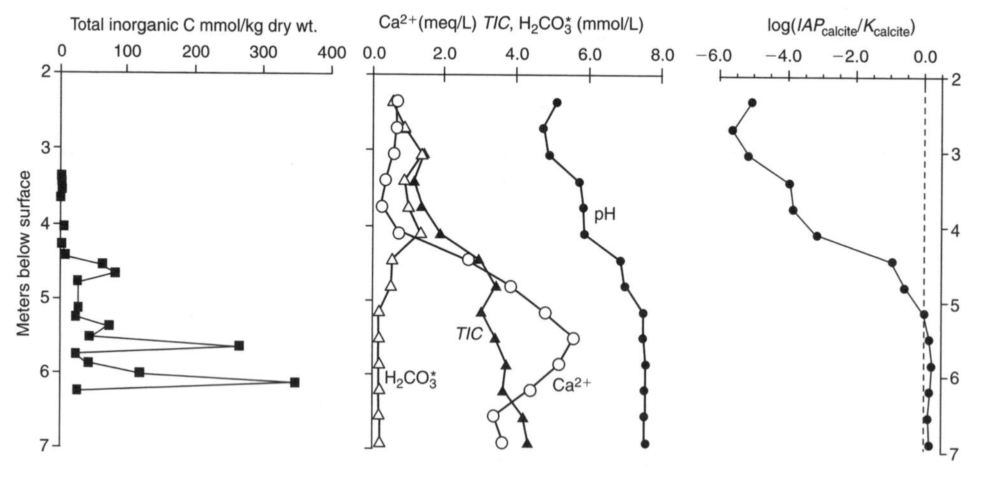

Figure 5.17. Calcite dissolution under conditions closed with respect to CO_2 in the Rømø aquifer, Denmark (Jakobsen and Postma, 1999).

Below 4.5 m depth, $CaCO_3$ is present and dissolves. In a distinct dissolution front, pH, Ca^{2+} and *TIC* increase, while $H_2CO_3^*$ decreases until the groundwater reaches saturation with calcite at 5 m depth. In this case, calcite dissolves spatially separated from the point where CO_2 is generated in the soil, a situation that conforms to calcite dissolution in a system closed with respect to CO_2 gas. The important difference with the situation at Trout creek (Figure 5.16) is that $H_2CO_3^*$, and hence $[P_{CO_2}]$, is decreasing when calcite dissolves.

EXAMPLE 5.7. *Propagation of the calcite dissolution front in the Rømø aquifer*
Calculate the time needed to decalcify the upper 4 m of the Rømø aquifer. The pristine sediment contains 50 mmol $CaCO_3$/kg and has $\rho_b = 1.86\,g/cm^3$, $\varepsilon = 0.3$. The downward flow velocity $v_{H_2O} = 0.8$ m/yr.

ANSWER:
The downward rate of the $CaCO_3$ front is calculated with the retardation equation:

$$v_{CaCO_3} = v_{H_2O} \;/\; R = v_{H_2O} \;/\; (1 + \Delta q/\Delta c)$$

where Δq and Δc are the change in $CaCO_3$ and in solute *TIC* across the dissolution front. $\Delta q = (50\,mmol/kg) \times (1.86\,kg/L) / 0.3 = 310$ mmol/L pore water. $\Delta c = 1.5$ mmol/L (for *TIC*, Figure 5.17; note that Ca^{2+} increases more). Hence,

$$v_{CaCO_3} = 0.8 \;/\; (1 + 207) = 0.004 \text{ m/yr}$$

Accordingly, the present day profile has developed during the last 4/0.004 = 1000 years. While this value is only approximate since considerable uncertainty is involved, particularly in Δq, given the young age of the sediment (<1500 yrs old), it is not unreasonable.

The concentration of Ca^{2+} is higher in the Rømø aquifer (Figure 5.17, closed system) than at the Canadian site (Figure 5.16, open system), which is the opposite of the expected trend (Table 5.6). Primarily, the reason lies in a different P_{CO_2} in the unsaturated zone of the two studied plots. Within a single area, the concentration contrast is expected to follow the theory more closely. Open *vs* closed system dissolution was invoked to explain the different compositions in the Triassic Alpujarras aquifer (Table 5.2 and Example 5.4) and was used to identify the environment where Ca^{2+} entered the water and to discern infiltration areas (Langmuir, 1971; Pitman, 1978; Hoogendoorn, 1983; Problem 5.23). In addition, ^{13}C can be used to distinguish chemical processes in the source area (Section 5.8).

QUESTIONS:
Estimate the CO_2 pressure in the unsaturated zone above the Rømø aquifer? (*Hint*: look at the $H_2CO_3^*$ concentration)

ANSWER: $P_{CO_2} \approx 10^{-1.5}$ atm (compare with the Canadian site: $10^{-2.5}$ atm).

Compare concentrations in Figure 5.17 with values in Table 5.6?

ANSWER: concentrations in Figure 5.17 are somewhat higher, but temperature in the field is lower (10°C).

5.5 CARBONATE ROCK AQUIFERS

There are two modes of groundwater transport through a limestone aquifer, matrix or diffuse flow and conduit flow. In the case of matrix flow, groundwater is transported through the rock pores or through a possibly dense network of micro-fissures in a manner not unlike Darcy flow. Conduit flow occurs along larger fissures and openings that result from dissolution of carbonate rock and is much quicker, and may be even turbulent. The two flow types interchange water continuously as illustrated in Figure 5.18. Due to fast conduit flow, pollutants may arrive earlier than with plug-flow which considers displacement of the whole resident mass of matrix and conduit water together. The peak concentrations in conduits are slightly diminished by exchange with the matrix, while slow bleeding from a polluted matrix yields a continuous source of contaminants, albeit at a possibly low concentration. Atkinson (1975) derived from hydrograph-analysis and other parameters in the Mendip Hills, England, that conduit flow probably accounts for between 60% and 80% of the transmission of water in the aquifer, although conduits contain only 1/29 of all the phreatic groundwater when the aquifer is fully recharged.

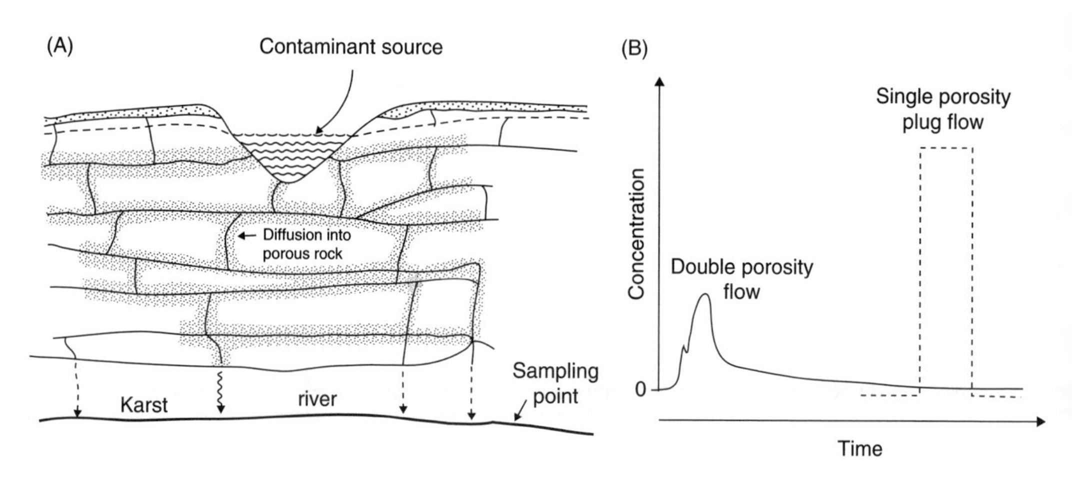

Figure 5.18. Dual porosity character of limestones (A) results in two flow modes with typical effects on solute (pollutant) transport and arrival times of a pulse injection (B).

The flowpath has an important effect on the solute concentrations in the upper, phreatic zone where dissolution is going on. Figure 5.19 shows different flowpaths in carbonate karst in New Zealand, together with mean Ca^{2+} and Mg^{2+} concentrations (Gunn, 1981). The concentrations are lowest in overland flow and throughflow, where contact-time and -surface are small. Water in the subcutaneous zone has the highest concentrations, and this water seems to feed the larger karst pipes (termed shafts in Figure 5.19). The presence of lower concentrations in smaller fissures seems illogical, but perhaps there is a rapid flow component involved which brings water rapidly downwards in these particular fissures.

Seasonal variations in water qualities of carbonate rock springs have been related to the type of flow mechanism in limestone aquifers. Shuster and White (1972) indicated that conduit flow springs have large variations in concentrations, temperature and discharge over the year, whereas diffuse flow springs have in all aspects a more stable regime (Figure 5.20). Martin and Dean (2001) found that during low flow the concentrations generally increased due to a greater matrix contribution.

The dissolution of carbonate along conduits may result in spectacular features such as sinkholes (cenotes) caves, rivers disappearing underground and a karstic landscape may develop over time. Conduit flow is often rapid, giving sometimes even turbulent flow with flow rates of up to 150 m/h (Alberic and Lepiller, 1998). Because of their highly dynamic character, karst systems are vulnerable towards pollution (Katz et al., 1998). Solution features are also visible on the surface of all barren limestones.

Figure 5.21 shows the typical solution flutes (sometimes denoted with the German term "Rillenkarst") that are often observed. Surface solution phenomena as well as cave formation are likewise, though more seldom, found in dolomite rock (Zötl, 1974; Ford and Williams, 1989; Figure 5.21).

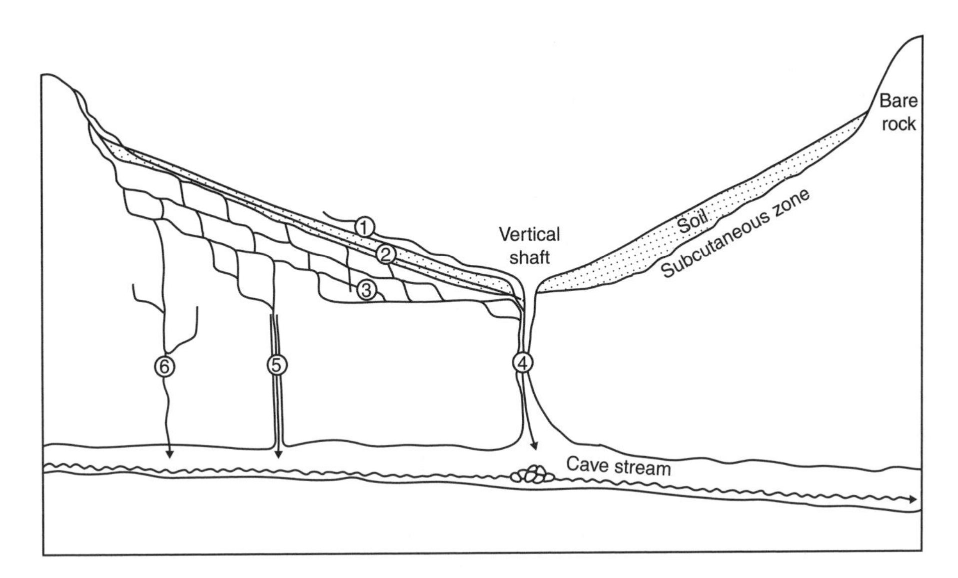

Flow component (key below)	Number of observations	Mean	Standard deviation	Coefficient of variation (%)	Range
(a) Calcium ion concentrations (mg/L)					
1	6	7.5	2.49	33.2	4.2–10.4
2	162	18.7	5.04	26.9	8.0–31.7
3	154	46.5	6.75	14.5	34.8–61.4
4	61	46.8	6.66	14.2	30.0–56.4
5	186	30.6	17.86	58.3	10.4–61.4
6	1714	36.1	6.72	18.6	20.6–59.0
(b) Magnesium ion concentrations (mg/L)					
1	6	0.6	0.308	47.4	0.2–1.0
2	154	1.1	0.270	23.9	0.5–1.8
3	139	1.3	0.212	16.2	0.8–1.5
4	58	1.3	0.224	16.9	0.7–1.8
5	183	1.0	0.294	30.4	0.5–1.5
6	1754	1.5	0.427	28.0	0.7–3.0
(c) Temperature (°C)					
2	78	10.3	2.43		4.9–14.7
3	95	10.7	0.58		9.4–12.2
4	34	10.0	1.33		8.3–12.8
5	105	11.1	1.53		8.1–13.4

Key to flow components: 1. Overland flow; 2. Throughflow; 3. Subcutaneous flow; 4. Shaft flow; 5. Vadose flow. 6. Vadose seepage.

Figure 5.19. Types of flow in New Zealand karst, and properties of flow components (Gunn, 1981).

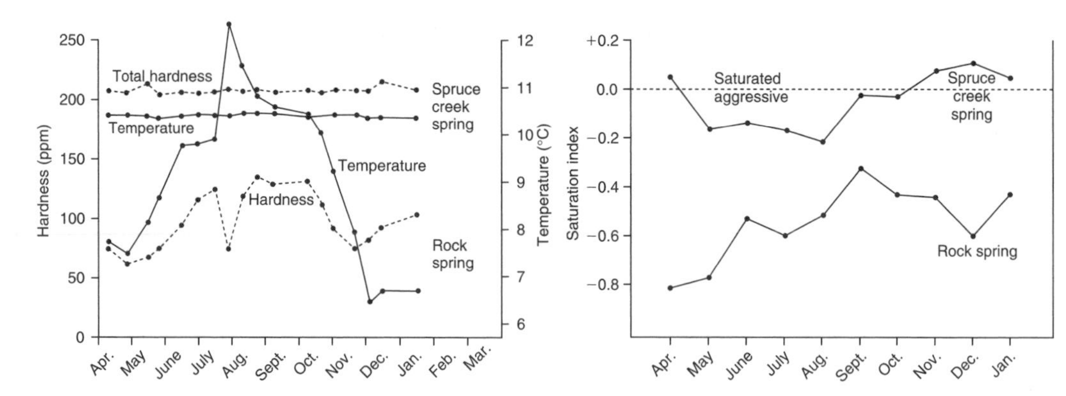

Figure 5.20. (Left) Comparison of seasonal variation of hardness and temperature for a typical diffuse-flow spring (Spruce Creek spring) and a typical conduit-flow spring (Rock spring). (Right) Seasonal variation in saturation index for the two types of springs (Shuster and White, 1972).

Further contributing to dissolution along conduits in a karst system is the variability of water compositions, illustrated in Figure 5.19. Mixing of waters with different compositions will produce waters subsaturated for calcite (Figure 5.14), allowing renewed dissolution in the subsurface. Finally, surface water that is transported rapidly into a karst system may contain reactive organic carbon. Decomposition of the organics down in the karst channels causes CO_2 production and allows calcite dissolution (Batoit et al., 2000; Alberic and Lepiller, 1998).

Figure 5.21. (Left) Rillenkarst on Limestone in the N.-Italian Alps. (Right) Karstic developments of shafts in the Dolomites (N.-Italy) (Appelo, priv. coll.).

5.5.1 *Dolomite and dedolomitization*

Aquifers of dolomite rock behave in many ways similar to limestones. For the dissolution of dolomite we can write:

$$2CO_{2(g)} + 2H_2O + CaMg(CO_3)_2 \leftrightarrow Ca^{2+} + Mg^{2+} + 4HCO_3^- \tag{5.27}$$

Groundwater in a monomineralic dolomite aquifer should ideally contain equal amounts of Ca^{2+} and Mg^{2+} while the bicarbonate concentration should be four times the Ca^{2+} and Mg^{2+} concentrations. Models for dissolution in systems open or closed with respect to CO_2 gas apply also to dolomite. The law of mass action for Reaction (5.27) is (cf. Equation 5.21):

$$\frac{[Ca^{2+}][Mg^{2+}][HCO_3^-]^4}{[P_{CO_2}]^2} = K_{dol}\ (K_H K_1 / K_2)^2 \tag{5.28}$$

The electroneutrality equation $2m_{Ca^{2+}} + 2m_{Mg^{2+}} = m_{HCO_3^-}$ and the mass balance $m_{Ca^{2+}} = m_{Mg^{2+}}$ combine to:

$$4m_{Ca^{2+}} = m_{HCO_3^-} \tag{5.29}$$

which yields together with Equation (5.28):

$$m_{Ca^{2+}} = \sqrt[6]{\frac{1}{256} K_{dol} \left([P_{CO_2}] K_H \frac{K_1}{K_2} \right)^2} \tag{5.30}$$

where $K_{dol} = 10^{-16.7}$ at 10°C.

EXAMPLE 5.8. *Dissolution of dolomite in the Italian Dolomites*
The Dolomites in N.-Italy form isolated, barren plateaus at levels up to 3000 m. Scree slopes which are vegetated at lower altitudes, extend from the plateaus. The timberline is at around 2000 m. The situation, depicted in Figure 5.22, gives low concentrations in water infiltrated on the plateaus (flowline A), and higher concentrations in water which infiltrates in the forests on the slopes (flowline B).

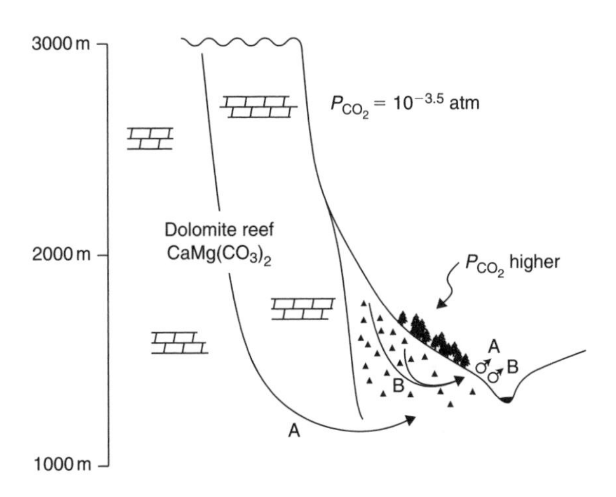

Figure 5.22. Schematic cross section through a dolomite plateau.

The difference suggests rapid dissolution of dolomite to near saturation levels, dictated by the CO_2 pressure of the infiltration area.

When discharge of the springs increases in response to rain, it is often observed that the concentrations in the discharge increase as well. Apparently it is primarily the contribution of runoff from the scree slopes that increases in spring discharge. The corresponding flowline goes through highly permeable scree slopes which can give very rapid response to rainfall, while at the same time the soil slopes have a higher CO_2 pressure than the plateaus at 3000 m.

For a CO_2 pressure at the plateau of $10^{-3.5}$ atm, similar to the atmosphere, and for the forested scree $P_{CO_2} = 10^{-2}$ atm, the calculated concentrations in flowline A and B are as given in the table, in close correspondance to observed concentrations.

Flowline	A	B
$Ca^{2+} \equiv Mg^{2+}$	0.33	1.04 mmol/L
HCO_3^-	1.31	4.16 mmol/L
pH	8.42	7.42
EC	131	416 μS/cm

For closed system dissolution of dolomite identical arguments provide us with:

$$HCO_3^- = 2(\Sigma CO_2)_{root}; \quad m_{Ca^{2+}} = m_{Mg^{2+}} = \frac{1}{4} m_{HCO_3}; \text{ etc.}$$

Aquifer rocks often contain both calcite and dolomite either in separate layers or as disseminated crystals. Figure 5.23 shows the saturation state of groundwater in Paleozoic rock with respect to both calcite and dolomite. The spring waters generally have a short residence time or consist of mixed waters and are subsaturated, while the well waters, representing a longer residence time, approach saturation.

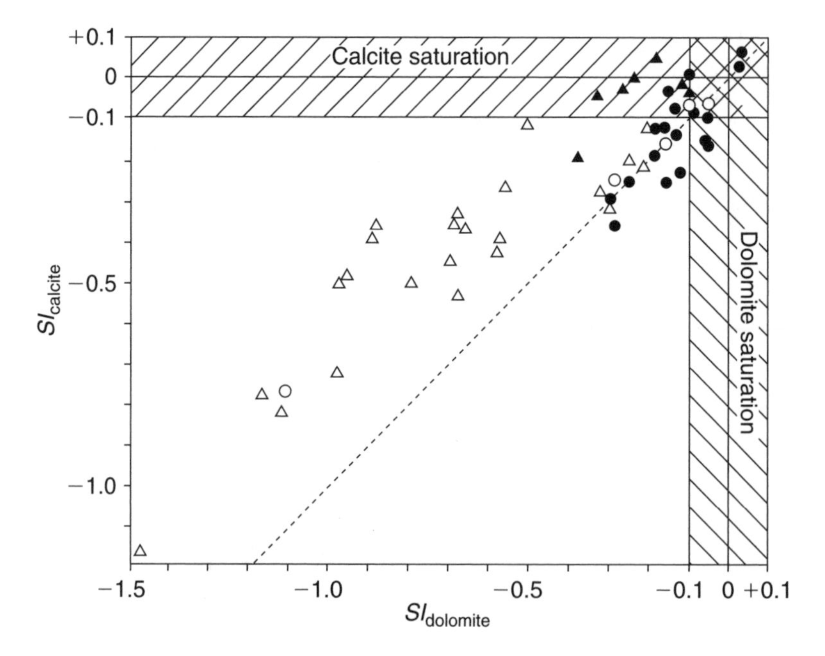

Figure 5.23. The saturation state of groundwater in paleozoic limestones with respect to calcite and dolomite. Open symbols are spring waters and filled symbols are groundwaters (Langmuir, 1971).

The point where the groundwater is in equilibrium with both calcite and dolomite has special chemical properties. For his situation we can write the reaction:

$$Ca^{2+} + CaMg(CO_3)_2 \leftrightarrow 2CaCO_3 + Mg^{2+} \tag{5.31}$$

The corresponding mass action equation is:

$$K = [Mg^{2+}]/[Ca^{2+}] = K_{dol} / (K_{cc})^2 = 10^{-17.09} / (10^{-8.48})^2 = 0.8 \tag{5.32}$$

at 25°C. In other words, in case of equilibrium with both calcite and dolomite, the $[Mg^{2+}]/[Ca^{2+}]$ ratio becomes invariant (at constant temperature and pressure).

When simultaneous equilibrium with calcite and dolomite is combined with the dissolution of gypsum or anhydrite, the process of dedolomitization may take place. For dissolution of anhydrite or gypsum we write:

$$CaSO_4 \rightarrow Ca^{2+} + SO_4^{2-} \tag{5.33}$$

The increasing Ca^{2+} concentration due to gypsum dissolution causes calcite to precipitate, which is similar to the effect of gypsum dissolution on fluorite precipitation discussed in Section 4.1. The CO_3^{2-} concentration decreases as calcite precipitates, and this provokes the dissolution of dolomite and an increase of the Mg^{2+}-concentration. When Mg^{2+} increases, Ca^{2+} must increase as well according to relation (5.32). The net result is that the dissolution of gypsum induces the transformation of dolomite to calcite in the rock and produces waters with increased Mg^{2+}, Ca^{2+} and SO_4^{2-} concentrations. Dedolomitization may occur in aquifers containing dolostones (dolomite-type carbonates) associated with gypsiferous layers (Back et al., 1983; Plummer et al., 1990; Saunders and Toran, 1994; Cardenal et al., 1994; Sacks et al., 1995; Capaccioni et al., 2001; López-Chicano et al., 2001). Also groundwater pumping may draw sulfate rich waters from gypsum layers into dolomite rock and thereby cause dedolomitization in a well field (López-Chicano et al., 2001).

The process of dedolomitization is illustrated for the Madison aquifer, USA, in Figures 5.24 and 5.25 (Plummer et al., 1990). The sulfate concentration was used to monitor the extent of gypsum dissolution. The groundwater is at saturation or slightly supersaturated for calcite throughout the aquifer (Figure 5.24), regardless of the sulfate concentrations. The data scatter for saturation with respect to dolomite is somewhat larger, but still within a SI variation of ± 0.5 from equilibrium. (The calculation of SI_{dol} involves 4 solutes, increasing the effect of analytical errors by a factor of 2 compared to SI_{cc}). For gypsum, the saturation state ranges from strongly subsaturated to equilibrium. When equilibrium with both calcite and dolomite is maintained, the ratio of $[Mg^{2+}]/[Ca^{2+}]$ must remain about 0.8. Since the two ions behave similarly with regard to complexation and activity coefficient corrections, the ratio of total concentrations of the two ions should also remain about 0.8. The total mass transfer is thus given by the reaction:

$$1.8CaSO_4 + 0.8CaMg(CO_3)_2 \rightarrow 1.6CaCO_3 + Ca^{2+} + 0.8Mg^{2+} + 1.8SO_4^{2-} \tag{5.34}$$

For each mole of dolomite that dissolves, two moles of calcite are precipitated. The reaction scheme predicts that the SO_4^{2-} concentration must increase with both the concentration of Mg^{2+} and Ca^{2+}, as occurs in the Madison aquifer (Figure 5.25). The ideal stoichiometry according to Reaction (5.34) is indicated by lines, labelled with 1.8 as the coefficient for SO_4^{2-}. The suggested trend is closely followed by Ca^{2+}, but not by Mg^{2+}. However, Figure 5.24 shows the groundwater to be supersaturated with respect to calcite, which influences the $[Mg^{2+}]/[Ca^{2+}]$ ratio. If we take into account that the waters on the average have $SI_{cc} = 0.2$ and $SI_{dol} = 0$, the ratio becomes:

$$[Mg^{2+}] / [Ca^{2+}] = 10^{-17.09} / (10^{-8.48+0.2})^2 = 0.3$$

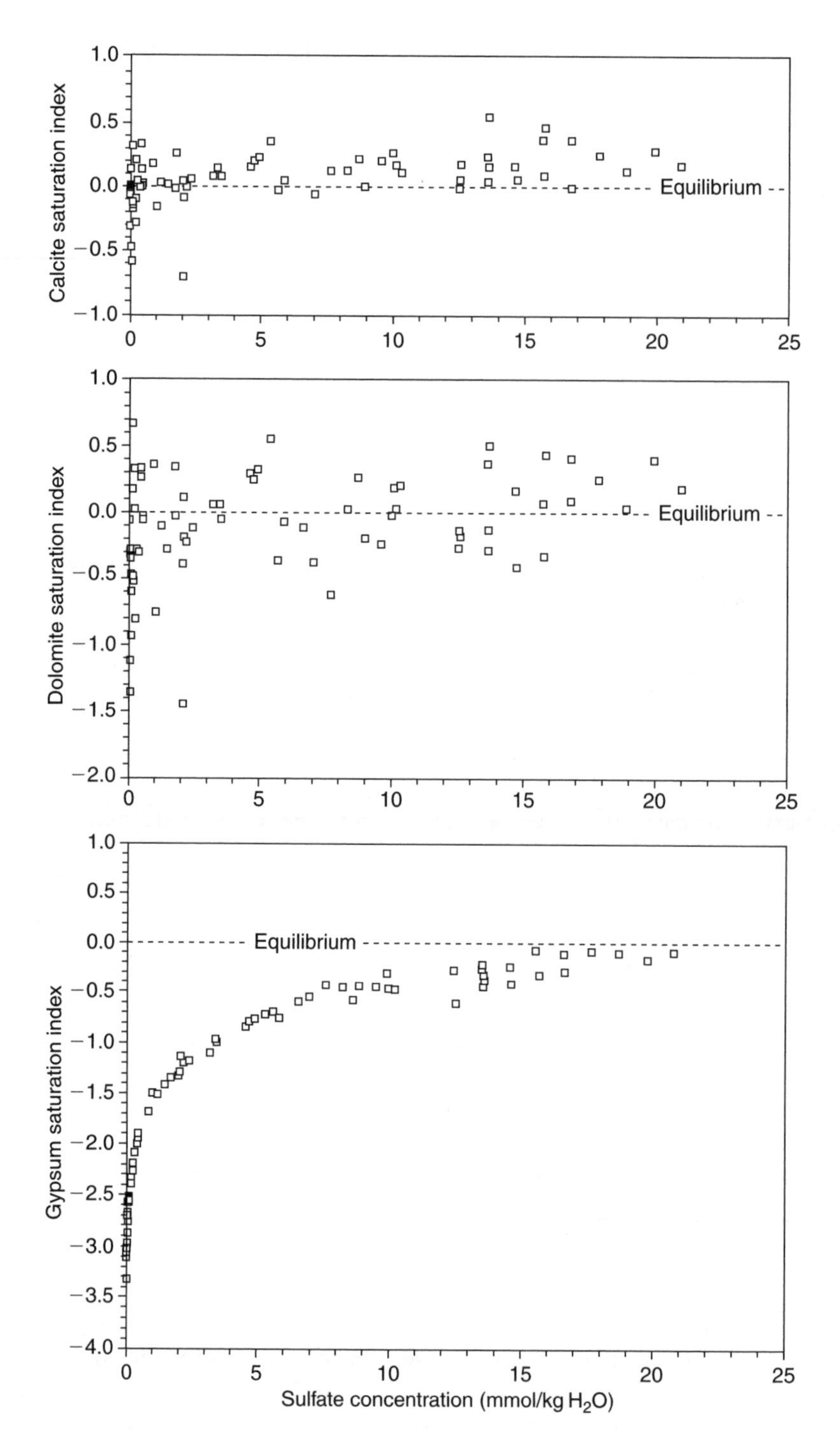

Figure 5.24. Calcite, dolomite and gypsum saturation indices as a function of total dissolved sulfate content for wells and springs in the Madison aquifer (Plummer et al., 1990).

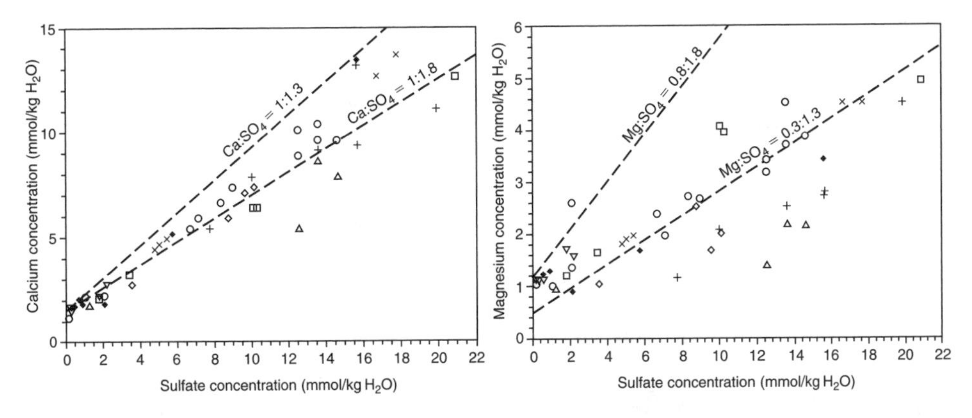

Figure 5.25. Dissolved calcium and magnesium concentrations as a function of dissolved sulfate content for waters from the Madison aquifer. Lines refer to the mass transfer predicted by Equations (5.34) and (5.35). (Modified from Plummer et al., 1990).

The overall mass transfer now changes to

$$1.3CaSO_4 + 0.3CaMg(CO_3)_2 \rightarrow 0.6CaCO_3 + Ca^{2+} + 0.3Mg^{2+} + 1.3SO_4^{2-} \qquad (5.35)$$

Lines according to this stoichiometry have also been drawn in Figure 5.25, and are labelled with 1.3 as the coefficient for SO_4^{2-}. The relation presents an upper boundary for the increase of Ca^{2+} with SO_4^{2-}, while the increase of Mg^{2+} with SO_4^{2-} is fitted quite well.

Apparently the mass transfer coefficients are determined by the failure of calcite to attain complete equilibrium during precipitation. This may be caused by precipitation inhibitors present in the groundwater. Plummer et al. (1990) found that the average apparent rates remain very low in this aquifer: 0.59 μmol/L/yr for calcite precipitation, 0.25 μmol/L/yr for dolomite dissolution and 0.95 μmol/L/yr for gypsum dissolution. These rates have no mechanistic kinetic significance, but depend on the haphazard encounter of water with gypsum, which dissolves immediately.

QUESTION:
Calculate the dedolomitization reaction with PHREEQC, plot Ca^{2+} and Mg^{2+} *vs* SO_4^{2-} concentrations for SI_{cc} = 0 and 0.2

ANSWER:
```
SOLUTION 1; pH 7 charge; C 2
EQUILIBRIUM_PHASES 1; Calcite 0.; Dolomite
REACTION; CaSO4; 20e-3 in 20; END
```

5.5.2 *Pleistocene carbonate aquifers*

Recent marine carbonate sediments found in tropical areas like the Caribbean and the South Pacific consist of aragonite and high Mg-calcite of biogenic origin, such as shells, skeletal debris, carbonate mud, etc. A minor part is played by abiotic precipitation at high supersaturation which also gives aragonite and high Mg-calcite (Given and Wilkinson, 1985). Since these sediments are formed in the marine environment we first inspect their behavior in contact with seawater. Using PHREEQC we can calculate the saturation states for surface ocean seawater at 25°C:

$$SI_{calcite} = 0.74 \quad SI_{aragonite} = 0.60 \quad SI_{dolomite} = 2.37.$$

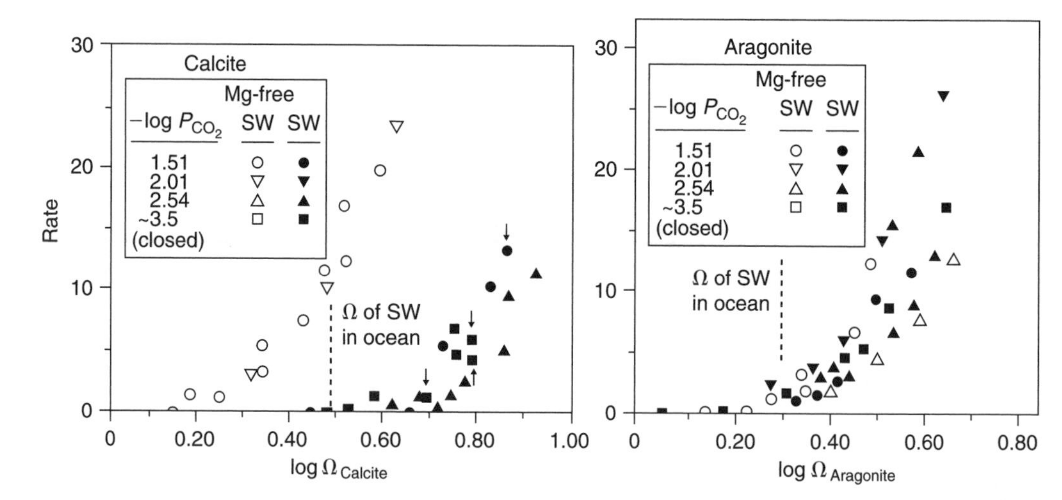

Figure 5.26. The rate of precipitation of calcite and aragonite *vs* the degree of saturation in artificial seawater with or without Mg. $\log \Omega = SI = \log IAP/K$, or the saturation index (Berner, 1975).

Surface seawater is apparently strongly supersaturated with respect to calcite, aragonite, and dolomite. These minerals should therefore precipitate extensively from seawater, but this does not happen because of kinetic inhibition. Calcite and aragonite precipitate only slowly and locally while dolomite does not seem to precipitate at all in the surface layers of the ocean.

The lack of precipitation is due to *inhibitors* present in seawater. Seawater contains about five times as much Mg^{2+} as Ca^{2+} on a molar base. Berner (1975) showed in a series of precipitation experiments with calcite and aragonite, that the rate of calcite precipitation in the presence of Mg^{2+} at seawater concentration becomes negligible (Figure 5.26). Apparently, Mg^{2+} disturbs the crystal growth of calcite. An alternative explanation is that incorporation of Mg^{2+} enhances the solubility and thereby inhibits calcite growth (Davis et al., 2000). The Mg^{2+}-ion can become incorporated in the structure when supersaturation becomes very high and the resulting calcites may contain up to 30% $MgCO_3$ (Figure 5.2). In contrast, aragonite, having another crystal structure is not affected by the presence of Mg^{2+}.

Organic acids (Berner et al., 1978) and phosphates (Walter and Hanor, 1979; Mucci, 1986) are also well known inhibitors and may also inhibit the precipitation of aragonite. The overall conclusion is, that in contact with seawater the most stable minerals, low Mg-calcite and dolomite, cannot form because of kinetic inhibition. Instead, metastable minerals like aragonite and Mg-calcite are formed.

What happens to these minerals when they enter the freshwater environment? Sea level variations, related to glaciations and deglaciations, may amount to several hundreds of meters and can lift marine sediments into the freshwater zone, for example forming the coastal aquifers on the Yucatan peninsula in Mexico. In contact with fresh water the precipitation of calcite is no longer inhibited and a giant recrystallization process starts whereby aragonite and high-magnesium calcite dissolve and calcite precipitates.

The transition of aragonite into calcite is predictable from the solubility products (Table 5.1). The reaction:

$$CaCO_3 \leftrightarrow Ca^{2+} + CO_3^{2-}$$

has

$$K_{calcite} = [Ca^{2+}][CO_3^{2-}] = 10^{-8.48} = 3.3 \times 10^{-9} \quad (25°C)$$

or

$$K_{\text{aragonite}} = [\text{Ca}^{2+}][\text{CO}_3^{2-}] = 10^{-8.34} = 4.6\times10^{-9} \quad (25°\text{C})$$

Water in equilibrium with aragonite is therefore supersaturated for calcite so that calcite precipitates and aragonite dissolves. The reaction is a two step process with sequential dissolution and precipitation:

$$\text{CaCO}_{3\ \text{aragonite}} \rightarrow \text{Ca}^{2+} + \text{CO}_3^{2-} \rightarrow \text{CaCO}_{3\ \text{calcite}} \tag{5.36}$$

For high Mg-calcite the situation is more complex because it is a solid solution. Figure 5.27 shows what happens when a high Mg-calcite is placed in distilled water. Initially both the Mg^{2+} and Ca^{2+} concentrations increase, corresponding to congruent dissolution. However, after some time, the Ca concentration starts to fall. Apparently a new Mg-calcite with a lower $MgCO_3$ content precipitates, and the process can be formalized as:

$$\begin{aligned}\text{Ca}_{1-x}\text{Mg}_x\text{CO}_3 \rightarrow\ & a \cdot \text{Ca}_{1-y}\text{Mg}_y\text{CO}_3 + (x - ay) \cdot \text{Mg}^{2+} + \\ & + (1 - x - a + ay) \cdot \text{Ca}^{2+} + (1 - a) \cdot \text{CO}_3^{2-}\end{aligned} \tag{5.37}$$

where $y < x$.

In this fashion, a row of Mg-calcites with slowly decreasing $MgCO_3$ contents will precipitate, and the water composition is enriched in Mg^{2+}. The process of dissolution whereby another, less soluble solid precipitates, is termed *incongruent dissolution*.

Figure 5.27 shows that it is quite impossible to measure the solubility of a Mg-calcite with a specific composition from such dissolution experiments since the solid phase composition changes during the experiment. This inability to reach equilibrium in experiments with Mg-calcite, and which probably also is valid in many natural settings, led Thorstenson and Plummer (1977) to reevaluate the equilibria criteria for two-component solid solutions.

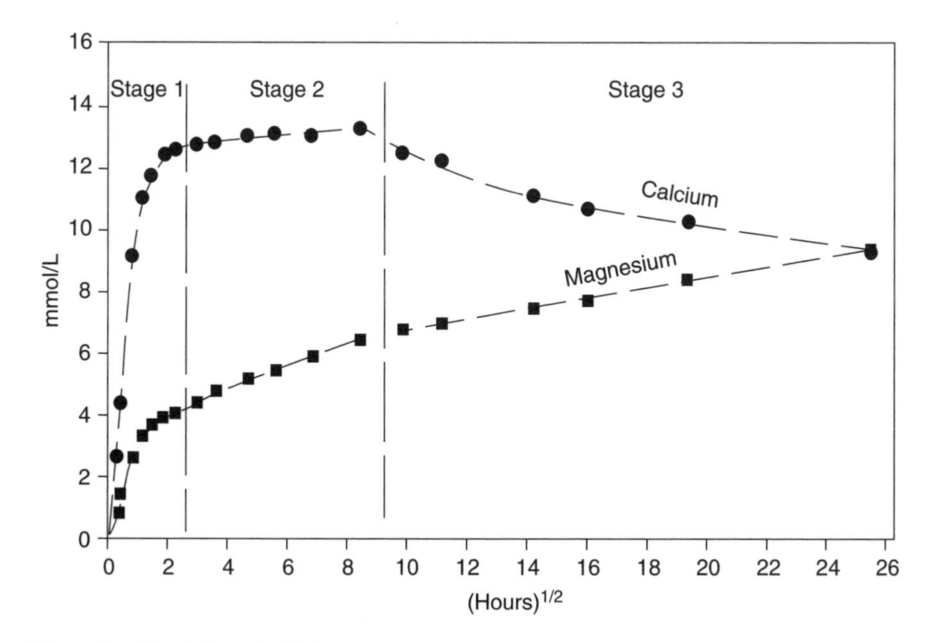

Figure 5.27. The dissolution of a high Mg-calcite in distilled water (Plummer and Mackenzie, 1974).

They proposed the concept of *stoichiometric saturation*, which defines equilibrium between an aqueous solution and a homogeneous solid solution of *fixed* composition. For Mg-calcite the stoichiometric solubility expression becomes:

$$Ca_{1-x}Mg_xCO_3 \rightarrow (1-x)Ca^{2+} + x\,Mg^{2+} + CO_3^{2-} \tag{5.38}$$

for which the stoichiometric solubility product is:

$$K_{(x)} = [Mg^{2+}]^x \cdot [Ca^{2+}]^{1-x} \cdot [CO_3^{2-}] \tag{5.39}$$

For stoichiometric saturation as expressed by Equation (5.39), the $[Mg^{2+}]/[Ca^{2+}]$ ratio in solution is decoupled from the $MgCO_3$ content of the solid, which means that saturation with any Mg-calcite can be attained (in principle) in a solution with a very low Mg^{2+} concentration. This differs from solid solution behavior where the $[Mg^{2+}]/[Ca^{2+}]$ ratio of the solution determines the composition of the solid and vice versa (Equation 4.35). The stoichiometric saturation concept has been applied with some success to the Mg-calcite system (Busenberg and Plummer, 1989) and the strontianite-aragonite system (Plummer and Busenberg, 1987).

Criteria for attaining stoichiometric saturation are, that dissolution proceeds congruent, as in stage 1 of Figure 5.27, that a constant *IAP* (corresponding to Equation 5.39) is maintained over a considerable range of time, and that the solution composition does not change when the crystals are placed in a solution with the same $IAP = K_{(x)}$. Busenberg and Plummer (1989) prevented reprecipitation of lower Mg-calcite, as happens during stage 3 in Figure 5.27, by adding phosphate to their experiments which strongly inhibits calcite precipitation. The stability of Mg-calcite, obtained with the stoichiometric saturation concept, is shown in Figure 5.28. It may be noted that high Mg-calcite, as present in recent marine carbonates, is indeed unstable relative to low Mg-calcite. Furthermore, calcite with a few mol% $MgCO_3$ is more stable than pure calcite.

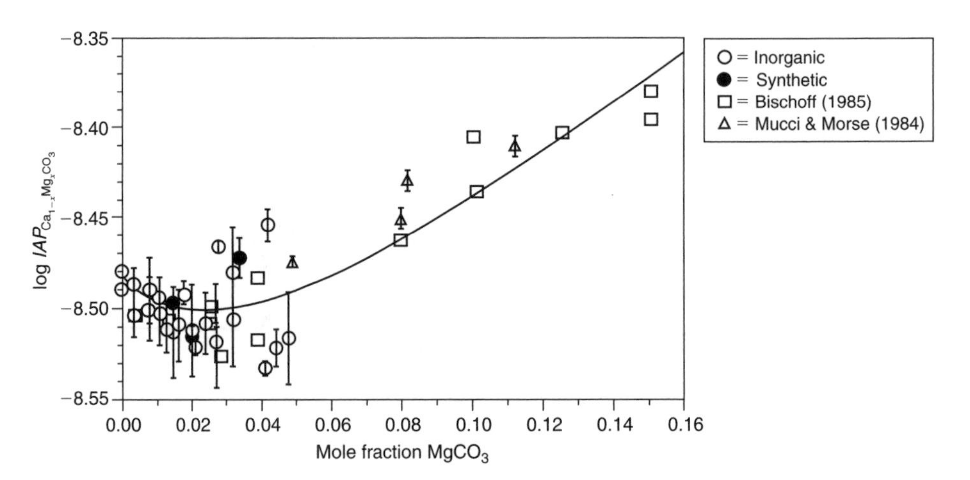

Figure 5.28. The stoichiometric solubility of well crystallized Mg-calcite From Busenberg and Plummer (1989).

The recrystallization process can be traced in the present-day groundwater chemistry of the Bermuda aquifer. The local sediments consist of calcarenites (beach deposits, dunes, etc.). The sediments are oldest in the central part of the island and become younger towards the marginal parts. Recent sediments contain about 35% aragonite and 50% high Mg-calcite with 14 mol% magnesite on average.

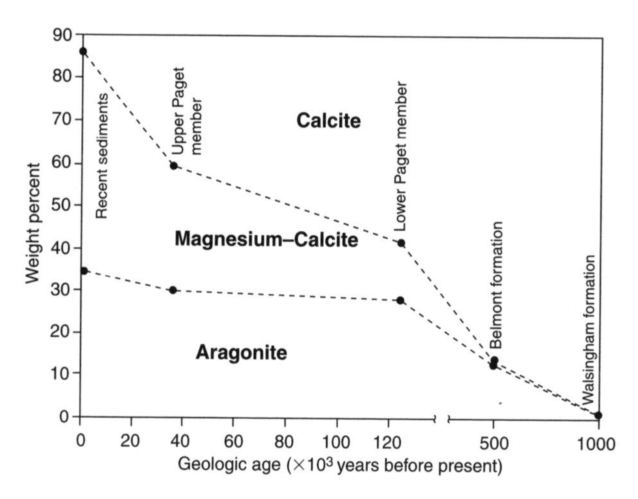

Figure 5.29. Relative composition by weight percent of Bermuda limestones as a function of geological age (Plummer et al., 1976).

At the other end of the scale, 10^6 yr old limestones consist exclusively of low Mg-calcite (Figure 5.29). In the central parts of the island, groundwater is near equilibrium with calcite but subsaturated for aragonite (Figure 5.30). In the marginal parts, the groundwaters are close to equilibrium with aragonite (which dissolves), but supersaturated for calcite (which will precipitate). An additional indication for aragonite dissolution is the high strontium concentration in the marginal groundwaters (Plummer et al., 1976). Small amounts of strontium are easily incorporated in aragonite but not in calcite (cf. Table 5.1), and Sr^{2+} will be enriched in groundwater when aragonite recrystallizes to calcite.

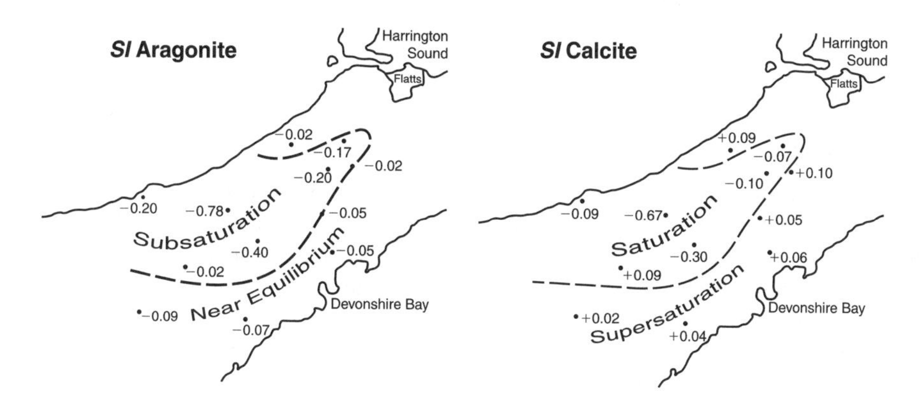

Figure 5.30. Saturation indexes of Bermuda groundwater with respect to calcite and aragonite (Plummer et al., 1976).

Figure 5.31 presents an overall summary of the processes in coastal carbonate aquifers, indicating both dissolution ("corrosion") and precipitation. Three corrosion processes are discerned, of which mixing corrosion is probably the most important for cave formation. Mixing corrosion takes place when waters with different P_{CO_2} mix (Figure 5.14) and applies for mixing seawater and groundwater (Wigley and Plummer, 1976; Back et al., 1986; Sanford and Konikow, 1989; Problem 5.16).

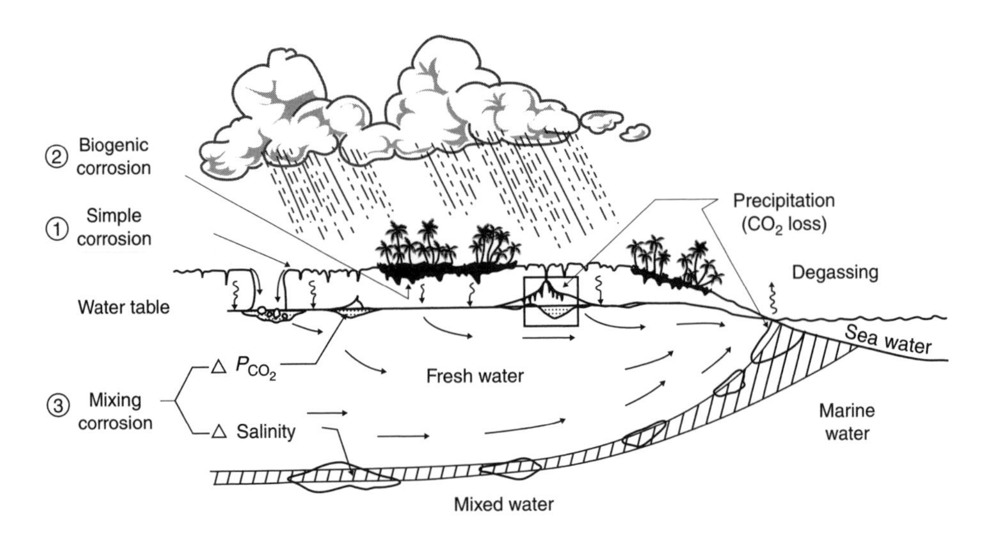

Figure 5.31. Dissolution features in a coastal carbonate aquifer (James and Choquette, 1984).

However, the mixing effects become often more complicated when other geochemical processes participate. Particularly redox processes may affect the dissolved carbonate composition and change both the mixing endmember and the saturation state of the mixed waters (Stoessel et al., 1993; Whitaker and Smart, 1997; Andersen, 2001).

QUESTION:
Estimate the difference in log K for calcite and aragonite from Figure 5.30?

ANSWER: $\log K_{cc} = \log K_{arg} - 0.11$ (varies a bit from 0.10 to 0.12)

5.6 KINETICS OF CARBONATE REACTIONS

5.6.1 *Dissolution*

The general pattern of calcite dissolution kinetics as a function of pH and CO_2 pressure is shown in Figure 5.32 (Plummer et al., 1978). Three regions are discerned as a function of pH. The first region is confined to pH values below 3.5, where the rate is proportional to $[H^+]$. At these low pH values a strong dependence on the stirring rate is found which indicates that transport of H^+ to the calcite surface is rate controlling. At higher pH, the rate becomes independent of pH but instead dependent on P_{CO_2}. Here, the rate appears to be controlled by both transport and surface reaction (Rickard and Sjöberg, 1983). The third region indicates a sharp drop of the dissolution rate when the pH of saturation is approached, depending on the specific CO_2 pressure.

Different approaches have been used to model these general results (Morse and Arvidson, 2002). Sjöberg (1978), Rickard and Sjöberg (1983) and Morse (1983) described calcite dissolution kinetics in seawater by empirical rate expressions of the type

$$R = k(A / V)(1 - \Omega)^n \tag{5.40}$$

where A is the calcite surface area, V is the volume of the solution, Ω is the saturation state IAP/K, and k and n are coefficients which depend on the composition of the solution, and are obtained by fitting with observed rates.

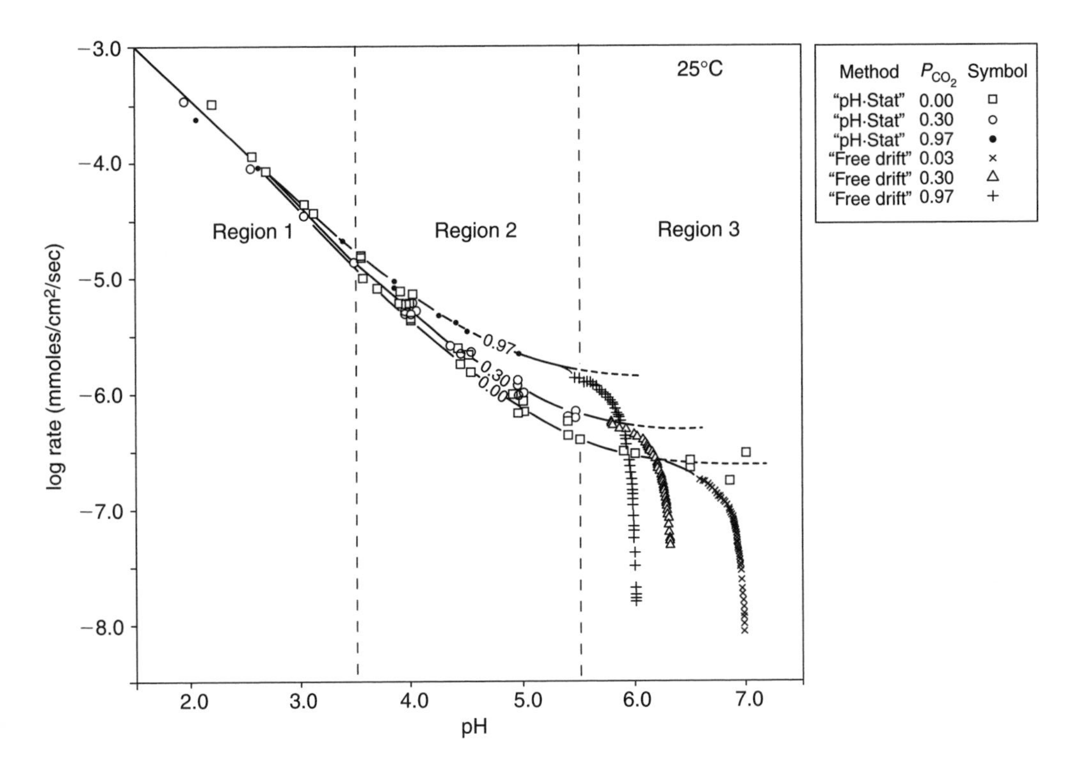

Figure 5.32. Dissolution rates of calcite as a function of pH and P_{CO_2} (Plummer et al., 1978).

Plummer, Wigley and Parkhurst (1978, denoted further as "PWP") developed a rate model for calcite dissolution based on three dissolution reactions:

$$CaCO_3 + H^+ \rightarrow Ca^{2+} + HCO_3^- \tag{5.41a}$$

$$CaCO_3 + H_2CO_3^* \rightarrow Ca^{2+} + 2HCO_3^- \tag{5.41b}$$

$$CaCO_3 + H_2O \rightarrow Ca^{2+} + HCO_3^- + OH^- \tag{5.41c}$$

The first reaction reflects the dominating process of proton attack at pH < 3.5, and the second reaction incorporates the effect of $H_2CO_3^*$ at higher pH. At still higher pH, above 7, the third reaction becomes important, which reflects simple hydrolysis of calcite. Finally the backward precipitation reaction was added:

$$Ca^{2+} + HCO_3^- \rightarrow CaCO_3 + H^+ \tag{5.42}$$

These reactions form the basis for the rate equation which covers both dissolution and precipitation:

$$r = \underbrace{k_1[H^+] + k_2[H_2CO_3] + k_3[H_2O]}_{r_{fw}} - \underbrace{k_4[Ca^{2+}][HCO_3^-]}_{r_b} \tag{5.43}$$

where r is the specific rate of calcite dissolution (mmol/cm²/s), separated in a forward rate r_{fw} and a backward rate r_b, and $k_1, \ldots, k_4$ are the rate constants. The forward rate constants have been fitted to the experimental data as function of temperature (T, in K):

$$\begin{aligned} \log k_1 &= 0.198 - 444 / T \\ \log k_2 &= 2.84 - 2177 / T \\ \log k_3 &= -5.86 - 317 / T \qquad \text{for } T \leq 298 \\ \log k_4 &= -1.1 - 1737 / T \qquad \text{for } T > 298 \end{aligned} \tag{5.44}$$

The backward rate was derived from the principle of microscopic reversibility of the individual forward reactions as a function of $[H^+]_{sf}$ and $[H_2CO_3^*]_{sf}$, which are activities on the crystal surface. It was assumed that saturation exists at the crystal surface during the entire dissolution process, so that $[H^+]_{sf}$ and $[H_2CO_3^*]_{sf}$ are given by their saturation values at the bulk solution $[P_{CO_2}]$. This rather complicated reasoning was necessary, since the backward reaction was *observed* to depend on $[Ca^{2+}]$ and $[HCO_3^-]$ only, i.e. giving only the backward reaction of the first dissolution reaction. Strict microscopic reversibility would require a likewise dependence of the backward reaction on $[OH^-]$ from the third term, and a quadratic dependence on $[HCO_3^-]$ from the second term (Chou et al., 1989). Anyhow, Plummer et al. (1978) argued that as long as $[H^+]_{sf}$ and $[H_2CO_3^*]_{sf}$ remain constant during the dissolution process, the value of k_4 remains constant as well. This allows a straightforward calculation of k_4 since at saturation the net rate $r = r_{fw} - r_b = 0$. Constant $[H_2CO_3^*]_{sf}$ is true when CO_2 pressure is constant (open system dissolution). For example, at constant $[P_{CO_2}] = 10^{-1.5}$ atm, the water composition at equilibrium with calcite was calculated to be (Table 5.5): $[H^+] = 10^{-6.9}$; $[H_2CO_3^*] = 10^{-3.0}$; $[Ca^{2+}] = 10^{-2.7}$, and $[HCO_3^-] = 2[Ca^{2+}]$. Hence we find, at 10°C (= 283 K):

$$\begin{aligned} k_1 \cdot [H^+] &= 10^{-1.4} \cdot 10^{-6.9} &&= 10^{-8.3} \\ k_2 \cdot [H_2CO_3^*] &= 10^{-4.9} \cdot 10^{-3.0} &&= 10^{-7.9} \\ k_3 \cdot [H_2O] &= 10^{-7.0} \cdot 1 &&= 10^{-7.0} \\ r_{fw} & &&= 10^{-6.9} \end{aligned}$$

And with $r_b = k_4 \cdot 2 \cdot [Ca^{2+}]^2 = r_{fw}$, we obtain $k_4 = 10^{-6.9}/(2 \cdot (10^{-2.7}))^2 = 10^{-1.8}$. We furthermore define $k_4' = 2 \cdot k_4 = 10^{-1.5}$.

The contribution from the first term, $k_1[H^+]$, rapidly diminishes as pH increases. The term can be neglected when pH is more than 6: the forward rate then becomes a constant which increases slightly with P_{CO_2}. The overall rate which describes calcite dissolution in a system with a constant CO_2-pressure of $10^{-1.5}$ atm is therefore simplified to:

$$r = \frac{dm_{Ca}}{dt} = r_{fw} - k_4' m_{Ca}^2 = 10^{-6.9} - 10^{-1.5} m_{Ca}^2 \tag{5.45}$$

Equation (5.45) can be integrated to provide the Ca^{2+}-concentration as a function of time:

$$\int_0^{m_{Ca_t}} \frac{dm_{Ca}}{(r_{fw} - k_4' m_{Ca}^2)} = \int_0^{t} dt \tag{5.46}$$

where m_{Ca_t} is the Ca^{2+} concentration reached at time t. The solution for Equation (5.46) is:

$$\frac{1}{2\sqrt{r_{fw} k_4'}} \ln \left[\frac{\sqrt{r_{fw}} + \sqrt{k_4'} \cdot m_{Ca_t}}{\sqrt{r_{fw}} - \sqrt{k_4'} \cdot m_{Ca_t}} \right] = t \tag{5.47}$$

Inserting values for r_{fw} and k'_4, and for $A/V = 1\ \text{cm}^2/\text{cm}^3$, the timespace is obtained to reach a concentration in solution of, say, 95% of the value at saturation. (The time to reach 100% saturation is infinite). In the case of $[P_{CO_2}] = 10^{-1.5}$, for which we calculated $m_{Ca^{2+}} = 0.002$ mol/L at saturation, it is:

$$t_{.95} = 30\,000\ \text{s} = 8.3\ \text{hr} \tag{5.48}$$

Let's compare this result with a PHREEQC calculation, first integrating Equation (5.45) by defining it as rate under keyword RATES, and then also apply the parent Equation (5.43) which is available in PHREEQC.DAT (Example 5.9).

EXAMPLE 5.9. *Kinetic dissolution of calcite calculated with PHREEQC, comparing simplified and parent "PWP" rates*

The simplified rate of calcite dissolution is defined as "Calcit2" in 2 BASIC lines under keyword RATES. Line 10 contains the formula (Equation 5.45) in which $m_{Ca^{2+}}$ is obtained with the function tot ("Ca"). Line 20 multiplies with 1/cm to convert mmol/cm² to mol/dm³ (PHREEQC adds moles and the solution contains 1 dm³ by default). The functions "save" and "time" in line 20 signal the start of the integration (cf. Chapter 4). The rate is invoked with keyword KINETICS where the chemical formula is entered for Calcit2, the initial amount of 1 mol reactant and the timespace of 30 000 s. SOLUTION 1 is pure water with a $[P_{CO_2}] = 10^{-1.5}$, and the CO_2 pressure is kept constant with EQUILIBRIUM_PHASES 1. Finally, keyword USER_GRAPH defines the plot.

The next simulation invokes the rate for Calcite which is defined in PHREEQC.DAT. The rate (look in the database) requires the surface to volume ratio (A/V in 1/dm) and the exponent for surface change (n in $(m/m_0)^n$) which are defined as '-parms 10 0.67' under KINETICS. Thus, $A/V = 10/\text{dm} = 1/\text{cm}$, equal to the ratio in the previous rate, and the exponent $n = 0.67$.

Running the file, and inspecting the results of the simple rate in the grid, shows that the Ca^{2+} concentration after 30 000 s is 1.91 mM according to the simple rate, which is indeed 95% of the estimated saturation concentration of 2.0 mM.

```
DATABASE phreeqc.dat           # contains the PWP calcite rate
RATES
  Calcit2                      # simplified rate cf. Equation (5.45)...
  -start
  10 rate = 10^-6.91 - 10^-1.52 * (tot("Ca"))^2  # mmol/cm2/s
  20 save 1 * rate * time      # integrate in moles/L
  -end
SOLUTION 1
  temp 10; pH 6 charge; C 1 CO2(g) -1.5
EQUILIBRIUM_PHASES 1
  CO2(g) -1.5

KINETICS 1
  Calcit2; formula CaCO3; -m0 1; -step 30000 in 20
INCREMENTAL_REACTIONS


USER_GRAPH
  -head time Ca pH; -axis_titles "Time / 1000s" "Ca / mM" "pH"
  -start
  10 graph_x total_time / 1e3; 20 graph_y tot("Ca") * 1e3; 30 graph_sy -la("H+")
  -end
END
                             # Simulation 2, with the parent PWP rate...
USE solution 1; USE equilibrium_phases 1
KINETICS 1
  Calcite; -m0 1; -parms 10 0.67; -step 30000 in 20
END
```

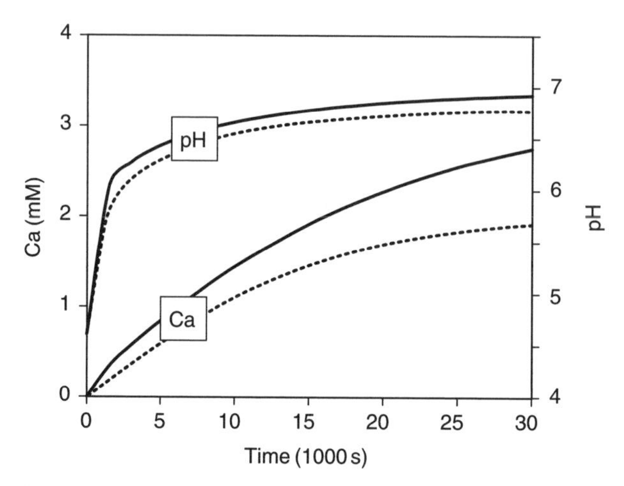

Figure 5.33. pH and Ca increase with kinetic dissolution of calcite. Full line indicates the PWP rate, dotted line is for the simplified rate.

The plot in Figure 5.33 shows for the simple formula a smaller final Ca concentration, since complexes and activity coefficients and the increase of the calcite solubility at 10°C were all neglected. Also, pH and Ca^{2+} increase slower than with the PWP calcite rate because the proton effect on the rate is neglected. Similar differences will be found with all other rate equations such as Equation (5.40) that do not account for specific reactants.

QUESTIONS:

Calculate the concentrations for 25°C instead of 10°C, comment on the change in shape of the Ca/time curves.

ANSWER: the simple formula yields the same curve. In the calcite rate the proton contribution is enhanced (cf. temperature dependence of k_1), the solubility is reduced.

Calculate the concentrations with time for $[P_{CO_2}] = 10^{-3.5}$. Is the reaction quicker?

ANSWER: saturation is attained about 4.6 times faster.

Compare the solubility of calcite at $10^{-3.5}$ and $10^{-1.5}$ $[P_{CO_2}]$?

ANSWER: About $(10^2)^{1/3} = 4.6$ times smaller at $[P_{CO_2}] = 10^{-3.5}$ (hence, saturation is reached 4.6 times faster).

Plot the log(rate) vs pH, and compare the two rate formulas. (*Hint*: in calcit2 add line 12 if rate > 0 then put (log10(rate), 1). In USER_GRAPH, graph_x -la("H1") and graph_y get (1). For the calcite rate, copy the rate definition from PHREEQC.DAT into the input file, add the line: 122 if moles > 0 then put (log10(moles), 1).

ANSWER: The simple rate is independent of pH < 6, in contrast to the calcite rate.

Plot log (rate) vs pH, starting at pH $= 3$ (*Hint*: define pH 3 in solution 1, add Cl 1 charge).

Consider the effect of changing the exponent on the saturation ratio in the PWP calcite rate from ⅔ to 1 (the exponent of 1 is given by strict microscopic reversibility).

ANSWER: with an exponent of 1 the log(rate)/pH curve turns more sharply to saturation.

We may apply the calculated saturation time to a real world situation, for example a shaft in limestone along which water flows in small films (Weyl, 1958; Dreybrodt, 1981; Gabrovšek and

Dreybrodt, 2000), or a small film of water which covers calcite fragments in a soil. The reaction rate is in units mmol/cm^2/s and must be multiplied by the available surface area A (cm^2) of calcite, and divided by the solution volume V (cm^3) to obtain the actual rate in mol/L/s. Take a waterfilm of 0.05 cm thickness, which gives $A/V = 20$/cm, and t_{95} becomes 25 minutes only. The distance covered during that timespan can be calculated for a specific hydrological situation. For a soil with a net precipitation of 0.3 m/yr and 10% water filled porosity, the percolation velocity is 3 m/yr. During 25 minutes a flowpath of only 0.2 mm is covered, sufficient to reach concentrations corresponding to 95% of calcite saturation values.

The average flow velocity in a shaft can be calculated from:

$$v_{H_2O} = \frac{\delta^2 \rho g}{3\eta} \frac{dh}{dz} \tag{5.49}$$

where δ is waterfilm thickness (m), ρ is density of water (1000 kg/m^3), g is the acceleration in the earth's gravitation field (10 m/s^2), η is viscosity of water (1.4×10^{-3} Pa · s at 10°C), and dh/dz is the hydraulic gradient (m H_2O/m). The average flow velocity in a waterfilm of 0.5 mm, with a hydraulic gradient of 1, is $v = 0.6$ m/s (or about 2 km/hr). The distance, covered in 25 minutes is accordingly about 890 m.

Clearly, the dissolution kinetics of calcite are fast enough to reach equilibrium in a soil containing calcite fragments, whereas the flow of water over barren rock may be so fast that the water reaches the inner realms of the rock while still undersaturated, and thereby contributing to karst formation. However, the dominant process that explains cave formation in pure carbonate rocks is undoubtedly the mixing of waters with different CO_2 pressures discussed before.

The dissolution kinetics of dolomite are much slower than for calcite. Busenberg and Plummer (1982) interpreted dolomite dissolution experiments with a rate expression similar to the one for calcite. However, the resulting equation predicts negative rates while the solution is still strongly subsaturated ($SI < -0.5$), and allows no extrapolation from the highly subsaturated pH-stat experiments to the "free-drift" conditions which apply in nature. Other experiments at low CO_2 pressure indicate that the dissolution rate can be appreciable, but still 10–20 times slower than for calcite (James, 1981; Appelo et al., 1984; Schulz, 1988; Chou et al., 1989; Pokrovsky and Schott, 2001). Even equilibrium may be reached, permitting the derivation of the dolomite solubility product from the solution composition (Sherman and Barak, 2000). Appelo et al. (1984) related the dissolution rate to the natural logarithm of the saturation ratio:

$$r = dm_{Ca} / dt = -k_d \cdot \ln(IAP / K_{dol}) \tag{5.50}$$

It is possible to integrate this equation analytically, but it is easier to let PHREEQC do the work. For the conditions used previously for calcite (10°C, $[P_{CO_2}] = 10^{-1.5}$, $A/V = 1$/cm) and $k_d = 1.2 \times 10^{-10}$ mmol/cm^2/s, the time to reach 95% saturation is $t_{95} = 1.4 \times 10^7$ s (160 days; Problem 5.18). This is about 450 times longer than required for calcite.

In contrast, field observations indicate that karst phenomena are identical for both calcite and dolomite rocks, with similar dissolved concentrations in springs and streams. How can these field observations be reconciled with the difference in laboratory rates? The main reason is that dolomite rock weathers (or rather, falls apart) into a grainy, friable material which almost consists of individual crystals. This material gives the magnificent scree slopes which are renowned in the Dolomites of N.-Italy and alps all over the world. The grainy texture gives a high ratio of surface area over water volume and increases the overall dissolution rate. Karst rivers draining dolomite rocks may in fact show higher and more steady concentrations during runoff peaks than rivers draining calcitic terrains (Zötl, 1974).

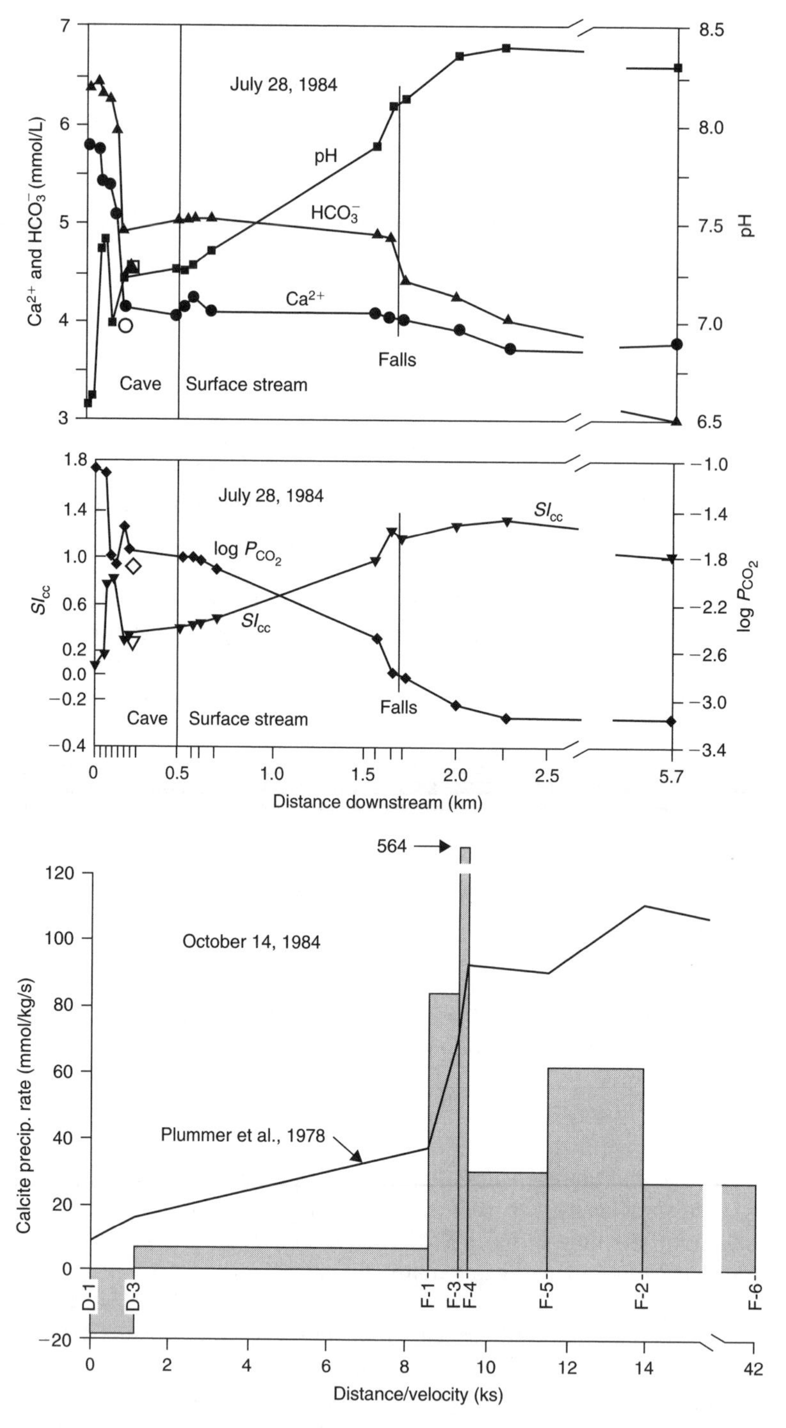

Figure 5.34. Calcite precipitation along Falling Spring Creek. (Upper 2 graphs) Change of concentrations along the reach. (Lower graph) Observed mass which is deposited, compared with predicted mass according to the Plummer et al. (1978) rate equation. From: Herman and Lorah, 1988.

QUESTIONS:
The solubility of dolomite for $[P_{CO_2}] = 10^{-3.5}$ is 5.2 times smaller than for $[P_{CO_2}] = 10^{-1.5}$ (calculated with PHREEQC). Estimate the time to reach 95% saturation for $[P_{CO_2}] = 10^{-3.5}$.

ANSWER: (160 / 5.2) = 31 days. (check with PHREEQC, Problem 5.18!)

Estimate the solubility decrease of dolomite, when $[P_{CO_2}]$ decreases 100-fold.

ANSWER: $100^{1/3}$ = 4.6 times smaller solubility.

5.6.2 *Precipitation*

The calcite dissolution rate model of Plummer et al. (1978) contains a backward term, which has been used successfully to describe precipitation of calcite (Plummer et al., 1979; Reddy et al., 1981; Busenberg and Plummer, 1986). An interesting study by Herman and Lorah (1988) compared the mass of calcite deposited by a karst stream with the amounts predicted by the Plummer et al. (1978) rate equation. The actual deposited mass was calculated from decline of concentrations at sampling points along the stream, as shown in Figure 5.34. The theoretical mass was obtained by multiplying the rate, calculated for the streamwater composition, with the wetted perimeter of the streambed. The comparison is shown in Figure 5.34, and we may note some discrepancies; the predicted amounts are lower where a waterfall increases outgassing of CO_2 and precipitation, and are higher in the other stretches of the stream. However, agreement is generally within a factor of 2–3.

One problem is to estimate the correct reactive surface area. It is clear that irregularities in the deposited travertine present a much larger microscopic surface than is given by the wetted perimeter of the stream. Experiments where small blocks of calcite were placed in a calcite depositing stream and the weight increase of the block was measured (Dreybrodt et al., 1992; Zaihua et al., 1995) yielded much less precipitation than predicted from the laboratory rate equation, even though the initial surface area of these blocks rapidly increased as they became covered by a myriad of small calcite crystals. The slower rates in streams may also be due to inhibition by foreign ions such as Mg^{2+} (Zaihua et al., 1995) or organic matter (Plummer et al., 2000). In relatively simple systems where the limestone is pure, the field data agree better with predictions from the laboratory rate equation (Baker and Smart, 1995).

Organic acids and phosphates inhibit the precipitation of carbonate in fresh water environments (Inskeep and Bloom, 1986). The inhibitory action already starts at low concentrations of phosphates ($<$1.0 μmol/L) and organic acids ($<$10 μmol/L, Reddy, 1977; Lebron and Suarez, 1996). The inhibition of calcite precipitation by the high Mg^{2+} content of seawater has already been mentioned. Generally, divalent cations that form solid carbonates tend to inhibit the growth rate of calcite increasingly more as the solubility of their solid carbonate form decreases (Terjesen et al., 1961; Gutjahr et al., 1996). Figure 5.35 shows that the growth rate is halved by a concentration of Mg^{2+} of $10^{-3.8}$ M (0.16 mM) and of Fe^{2+} of $10^{-7.3}$ M (0.05×10^{-3} mM) while the pK's of their carbonates are 8.24 and 10.89, respectively (Table 5.1). Brown and Glynn (2003) suspended sacks with calcite and dolomite in wellscreens in acid groundwater and noted dissolution rates that were 1000 times smaller than theoretically expected, which may be related to the high concentrations of heavy metals in the acid mine drainage that was the source of the groundwater.

QUESTION:
Estimate the inhibitory action of Cd^{2+} on calcite precipitation? (*Hint*: find $K_{otavite}$ in Table 5.1)

ANSWER: Figure 5.35 suggests very strong inhibition for $pK_{otavite} = 13.7$.

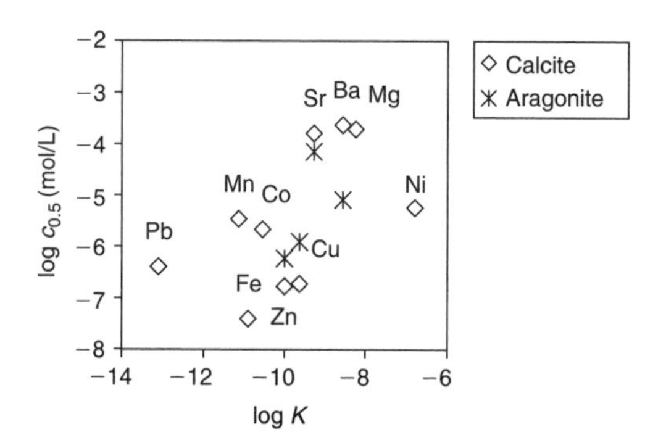

Figure 5.35. Concentrations of divalent cations by which the calcite and aragonite precipitation rates are halved as a function of the solubility product of the metal-carbonate (Modified from Gutjahr et al., 1996).

5.7 CARBON ISOTOPES

Carbon occurs in nature with two stable isotopes, ^{12}C and ^{13}C (Table 2.4). The $\delta^{13}C$ of dissolved carbon species in groundwater depends on the $\delta^{13}C$ signature of the dissolving C-source and the fractionation among the carbonate species in the solution (cf. Table 5.7). Atmospheric $CO_{2(g)}$ has a $\delta^{13}C \approx -7‰$. When this $CO_{2(g)}$ dissolves in water, ^{13}C is slightly depleted, resulting in a $\delta^{13}C \approx -8‰$ for $CO_{2(aq)}$. However, subsequent hydration of $CO_{2(aq)}$ favors the heavier isotope and produces a $\delta^{13}C \approx 2‰$ in HCO_3^-. The overall $\delta^{13}C$ of the dissolved carbonate depends therefore on the relative proportions of $CO_{2(aq)}$, HCO_3^- and CO_3^{2-}.

Table 5.7. Enrichment of ^{13}C in carbonate species with respect to CO_2(gas), and $\delta^{13}C$ in species in equilibrium with a given $\delta^{13}C$ of CO_2(gas).

Species i	‰ enrichment $\varepsilon^{13}C_{i\,/\,CO2(g)}$	$\delta^{13}C$ of species	
		$\delta^{13}C_{CO2(g)} = -7‰$	$\delta^{13}C_{CO2(g)} = -23‰$
$CO_{2(aq)}$	−1	−8	−24
HCO_3^-	9	2	−14
CO_3^{2-}	8	1	−15
$CaCO_3$	11	4	−12

QUESTIONS:

What is the enrichment of ^{13}C in the reaction $HCO_3^- \rightarrow CO_3^{2-} + H^+$?

ANSWER: $\varepsilon^{13}C_{CO_3^{2-}/HCO_3^-} = \varepsilon^{13}C_{CO_3^{2-}/CO_2(g)} - \varepsilon^{13}C_{HCO_3^-/CO_2(g)} \approx -1‰$

Water in equilibrium with calcite at $[P_{CO_2}] = 10^{-1.5}$ and $\delta^{13}C_{CO_2(g)} = -23‰$ has pH = 7, $H_2CO_3^* = 1.1$ mM, $HCO_3^- = 5.0$ mM and $CO_3^{2-} = 8$ μM. Estimate $\delta^{13}C$ of the solution?

ANSWER: $\delta^{13}C = ((-24 \times 1.1) + (-14 \times 5)) / 6.1 = -16‰$

Estimate (roughly) $\delta^{13}C$ of water in equilibrium with calcite and $[P_{CO_2}] < 10^{-1.5}$?

ANSWER: the proportion of $H_2CO_3^*$ decreases, hence $\delta^{13}C > -16‰$

The majority of terrestrial plants (C3 plants) uses CO_2 for photosynthesis. During diffusion of atmospheric CO_2 ($\delta^{13}C \approx -7‰$) through the stomata and with enzymatic conversion of CO_2 to CH_2O in the chloroplast, ^{13}C is depleted by about 20‰ to give $\delta^{13}C \approx -27‰$ for carbon in the C3 plant tissue. Some arid zone vegetations (C4 plants) are able to close off the stomata to reduce transpiration and incorporate HCO_3^- from the cell fluid, which is isotopically heavier and results in a heavier $\delta^{13}C \approx -13‰$ of C4 plant material (Farquhar et al., 1989; Vogel, 1993). The name C4 plant refers to the 4-carbon molecule *malate* that forms in the cells by reaction of HCO_3^- with pyruvate and transports the CO_2 to the sites where conversion to CH_2O takes place (cf. Berg et al., 2002).

When the plants and roots decay in the soil, CO_2-gas with $\delta^{13}C \approx -27‰$ from C3 plants diffuses upward to the atmosphere. The diffusion coefficient of $^{12}CO_{2(g)}$ is 4.4‰ higher than of the heavier $^{13}CO_{2(g)}$ (Chapter 3). To obtain fluxes into the atmosphere which balance the production of the two isotopes, the concentration gradient of $^{13}CO_{2(g)}$ must be larger by about 4.4‰, which increases the $\delta^{13}C$ of soil CO_2 gas to $\approx -23‰$ in the temperate climate (Dörr and Münnich, 1980; Cerling, 1984). In very poor soils with a CO_2 pressure lower than $10^{-2.7}$ atm, $\delta^{13}C$ of soil gas increases with the reciprocal of the CO_2 gas pressure to $-7‰$, the value of the atmosphere (Van der Kemp et al., 2000; Amundson et al., 1998).

During open system dissolution of calcite, the isotopes in CO_2 gas determine the $\delta^{13}C$ of water because the CO_2 gas/water exchange is quick (Chapter 10). Thus, water in equilibrium with calcite and $[P_{CO_2}] = 10^{-1.5}$ will have $\delta^{13}C \approx -16‰$ (Question below Table 5.7). In the case of closed system calcite dissolution, the *TIC* is derived in about equal proportions from CO_2 gas ($\delta^{13}C \approx -24‰$ in $H_2CO_3^*$ which is mostly $CO_{2(aq)}$) and from $CaCO_3$ ($\delta^{13}C \approx 0‰$ in marine carbonate, similar to the PDB standard), which results in $\delta^{13}C \approx -12‰$. We can calculate the change of $\delta^{13}C$ of *TIC* during dissolution precisely with PHREEQC (Example 5.10).

EXAMPLE 5.10. *PHREEQC calculation of $\delta^{13}C$ during calcite dissolution*

For the case of open system dissolution, the principle is to calculate the isotopic composition of the carbonate species with respect to CO_2 gas. Summing them up in proportion to their contribution to *TIC* yields $\delta^{13}C$ of the solution. The $\delta^{13}C$ of the carbonate species are calculated with respect to the constant $\delta^{13}C$ of CO_2 gas using enrichment formulas selected by Clark and Fritz (1997). The concentrations of the carbonate species and of *TIC* (which all change during calcite dissolution) are obtained from PHREEQC (lines 40–70 in USER_GRAPH in the input file). Dissolution is modeled using keyword KINETICS.

In the case of closed system dissolution, the isotopic composition of the initial (calcite free) solution is calculated as in the open system. The mass of $\delta^{13}C \times TIC_0$ (d13 ms in line 90 in USER_GRAPH) is stored in memory location 1 (line "92 put(d13 ms, 1)"). Adding $\delta^{13}C$ from calcite to the inital mass, and dividing by *TIC*, gives $\delta^{13}C$ of the solution. The calculation of initial $\delta^{13}C$ is bypassed during stepwise addition of the reaction (line "2 if step_no > 0 then goto 100").

The figure produced by USER_GRAPH (Figure 5.36) shows that closed systems have higher $\delta^{13}C$ values than open systems (this is also true when the final Ca^{2+} concentrations are the same, cf. Question). The $\delta^{13}C$ value will increase when initial $[P_{CO_2}]$ is smaller (cf. Question).

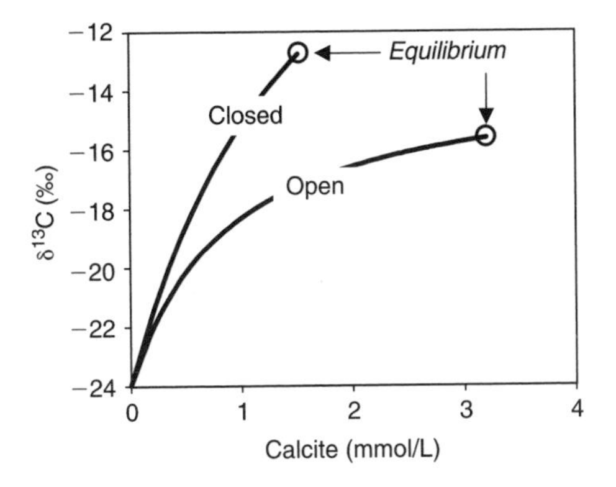

Figure 5.36. $\delta^{13}C$ of water when calcite dissolves in open or closed conditions with respect to CO_2 gas.

```
DATABASE phreeqc.dat
                         # Find d13C. Open system dissolution of calcite...
PRINT; -reset false
SOLUTION 1
  -temp 10
EQUILIBRIUM_PHASES 1; CO2(g) -1.5
KINETICS 1
  Calcite; -m0 1; -parms 200 0.67; -steps 3600 in 100
INCREMENTAL_REACTIONS true

USER_GRAPH
  -head cc open_d13C
                         # Find d13C in solute species 1 = CO2aq, 2 = HCO3-,3 = CO3-2
  -start
  2 if step_no = 0 then goto 120
# Define enrichments (eps) relative to g = CO2_gas from Clark and Fritz, 1997, p. 121 ...
  4 T = TK
  10 eps_1_g = 0.19 - 373 / T
  20 eps_2_g = -24.1 + 9.552e3 / T
  30 eps_3_g = -3.4 + 0.87e6 / T^2
                         # Find carbonate species...

  40 co2 = mol("CO2")
  50 hco3 = mol("HCO3-") + mol("CaHCO3+")
  60 co3 = mol("CO3-2") + mol("CaCO3")
  70 tic = co2 + hco3 + co3
                         # and d13 of the species...
  80 d13g = -23
  90 d13tot = (co2 * (eps_1_g + d13g) + hco3 * (eps_2_g + d13g) +\
             co3 * (eps_3_g + d13g)) / tic
                         # chart...
  100 graph_x tot("Ca") * 1e3
  110 graph_y d13tot
  120 end
-end
END


                         # simulation 2. Find d13C. Closed system dissolution...
SOLUTION 2
  -temp 10; pH 6 charge; C 1 CO2(g) -1.5
KINETICS 2
  Calcite; -m0 1; -parms 200 0.67; -steps 1200 in 30
USER_GRAPH
  -head cc closed_d13C
  -start
  2 if step_no > 0 then goto 100
# Copy lines 4 .. 80 from the open system above ...
# 4 ...
# ...
# 80 ...
                         # Calculate δ13C x TIC of the initial solution...
  90 d13ms = co2 * (eps_1_g + d13g) + hco3 * (eps_2_g + d13g) +
          \co3 * (eps_3_g + d13g)
                         # and store in memory 1...
  92 put(d13ms, 1)
  94 goto 120
                         # chart...
```

```
100 graph_x tot("Ca") * 1e3
                    # d13C from calcite (d13C = 0) and from initial
                      solution (memory 1)...
110 graph_y (tot("Ca") * 0+get(1)) / tot("C(4)")
120 end
-end
END
```

QUESTIONS:

Find $\delta^{13}C$ for closed system dissolution of calcite that gives gives the same Ca^{2+} concentration as for constant $[P_{CO_2}] = 10^{-1.5}$ (*Hint*: use initial $[P_{CO_2}] = 10^{-1.05}$ in solution 2).

ANSWER: $m_{Ca^{2+}} = 3.2$ mM; $\delta^{13}C = -15.6‰$ (open for CO_2 gas); $\delta^{13}C = -14.5‰$ (closed for CO_2 gas)

Find $\delta^{13}C$ for open and closed system dissolution of calcite when initial $[P_{CO_2}] = 10^{-2.5}$, $10^{-3.5}$?

ANSWER: -13.99 and $-10.80‰$ for $[P_{CO_2}] = 10^{-2.5}$ and -13.54 and $-3.36‰$ for $[P_{CO_2}] = 10^{-3.5}$.

5.7.1 *Carbon-13 trends in aquifers*

Figure 5.37 shows a plot of $\delta^{13}C$ of various karst water types (pools, fast drip, and stalactite and film waters) sampled in the Soreq cave in Israel (Bar-Matthews et al., 1996). The $\delta^{13}C$ values range from -15 to $-5‰$ and are hyperbolically related to *TIC* (grey outline in Figure 5.37). When calcite precipitates, *TIC* decreases and $\delta^{13}C$ of the water increases. This is illustrated by the lines in Figure 5.37 which connect the compositions of three fast-dripping waters with the pool waters in which they fall and from which calcite precipitates. The reaction is:

$$Ca^{2+} + 2HCO_3^- \rightarrow CaCO_3 + CO_{2(g)} + H_2O \tag{5.51}$$

The ^{13}C enrichment is largely due to the escape of isotopically light CO_2 that enriches *TIC* by 9‰ (Table 5.7), whereas the precipitation of calcite will lighten $\delta^{13}C$ of *TIC* by only about 1‰. Because calcite and CO_2(g) are produced in equal proportion, the water is enriched by $(9 + (-1)) / 2 = 4‰$ in a Rayleigh process that generates the hyperbolic relation between *TIC* and $\delta^{13}C$ (Bar-Matthews et al., 1996).

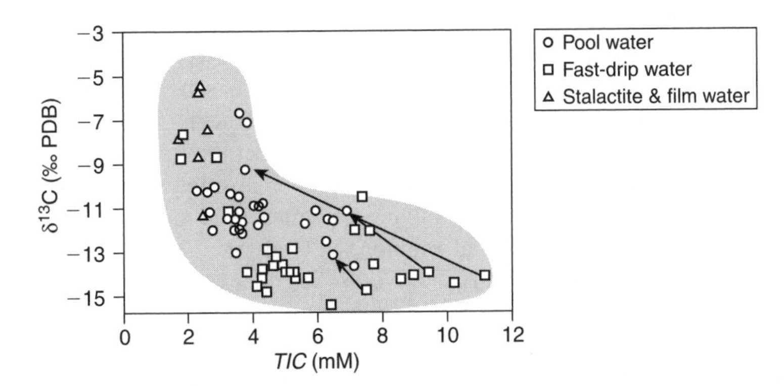

Figure 5.37. Plot of $\delta^{13}C$ *vs* *TIC* of waters from the Soreq cave, Israel. The arrows connect pool water compositions with their fast-drip source (Bar-Matthews et al., 1996).

The process is modeled with PHREEQC in Chapter 11. The enrichment of ^{13}C in calcite depends on the precipitation rate (Turner, 1982), in the same way as the composition of solid solutions is a function of kinetics (Chapter 4). Bar-Matthews et al. (1996) found an enrichment near -1‰, but for slowly precipitating calcite the enrichment may become as low as -3‰.

Groundwaters, which have equilibrated with calcite and are closed with respect to gas exchange, are expected to maintain a stable chemical and isotopic composition during flow through the aquifer. Nevertheless, groundwaters in many aquifers show a trend of increasing $\delta^{13}C$ with increasing age along the flowpath (Bath et al., 1979; Chapelle and Knobel, 1985; Plummer et al., 1990; Aravena et al., 1995; Kloppmann et al. 1998). The increase may be the result of renewed dissolution of carbonates, due to dedolomitization (Edmunds et al., 1982; Plummer et al., 1990, cf. Section 5.5.1) or loss of Ca^{2+} from solution into the cation exchanger (Chapelle and Knobel, 1985; Chapter 6). Methanogenesis, whereby CO_2 is converted into CH_4 by bacteria, will also increase $\delta^{13}C$ of the remaining CO_2 (Aravena et al., 1995) and is discussed in Chapter 9.

As an example of dedolomitization, let's consider the Triassic Sherwood sandstone aquifer in the English East Midlands (Figure 5.38). The aquifer is about 200 m thick and consists of quartz with some detrital and authigenic feldspars, and 1–4% detrital dolomite and calcite. Figure 5.38 presents concentrations plotted as a function of the water temperature, which increases downdip in the aquifer and is used as a proxy for the residence time of the groundwater (Edmunds et al., 1982; Smedley and Edmunds, 2002). Modern water has high Cl^- and SO_4^{2-} concentrations compared with older Holocene water in which Cl^- is generally less than 30 mg/L. In older water, SO_4^{2-} increases from 10 to 100 mg/L as result of gypsum dissolution. The Mg^{2+}/Ca^{2+} ratio and the HCO_3^- concentration remain more or less constant which suggests that dedolomitization takes place (Section 5.5.1). The evolution in groundwater composition has been attributed to 1) congruent dissolution of dolomite, 2) dedolomitization and 3) reduction which introduces ferrous iron and depletes oxygen. The molar Mg^{2+}/Ca^{2+} ratio tends to decrease somewhat with increasing temperature, which is expected when the groundwater remains in equilibrium with calcite and dolomite. The Mg^{2+}/Ca^{2+} ratio is smaller than predicted for pure dolomite, perhaps because ankerite ($CaFe(CO_3)_2$) is also present. Finally, $\delta^{13}C$ increases from -13‰ to -10‰ in conjunction with the increase of SO_4^{2-}.

The evolution of $\delta^{13}C$ during dedolomitization can be calculated from the mass balance in which calcite precipitates and dolomite dissolves:

$$\delta^{13}C = \frac{\delta^{13}C_i \times TIC_i + \delta^{13}C_{dol} \times 2 \times D_{dol} + (\delta^{13}C + \varepsilon_{cc/sol}) \times P_{cc}}{TIC_i + D_{dol} - P_{cc}} \tag{5.52}$$

where D indicates moles of dissolution, P is moles of precipitation, and subscripts i, dol and cc stand for the initial solution composition, dolomite and calcite, respectively. The $\delta^{13}C$ of the solution is also present on the right of the equal sign where it determines the isotopic composition of the calcite precipitate. Therefore, the equation must be integrated in small steps from 0 to the amount of precipitate for a given amount of gypsum. However, if we neglect the fractionation in the precipitate (assume $\varepsilon_{cc/sol} = 0$), we can directly calculate the isotopic composition of the solution from the mass balance of the original solution and the dissolving dolomite. If as much carbon from dolomite is added as was present as *TIC* in the initial solution, $\delta^{13}C$ of the solution will become -7.5‰, if $\delta^{13}C_i = -13$‰ and $\delta^{13}C_{dol} = -2$‰. Thus, the solution becomes heavier by dissolution of dolomite, and the calcite precipitate (which has the $\delta^{13}C$ of the solution) is lighter than the dolomite.

This behavior was confirmed in simple dissolution experiments with Sherwood aquifer rock samples in acid (Figure 5.39, Edmunds et al., 1982). Initially, the acid dissolves calcite with a low $\delta^{13}C$, followed by the slower reaction of dolomite with a higher $\delta^{13}C$. Thus, $\delta^{13}C$ of the solution increases over time, as illustrated in Figure 5.39. In the experiment, $\delta^{18}O$ of CO_2 increases together with $\delta^{13}C$, indicating that fresh water carbonate (with the low ^{18}O of fresh water) dissolves at the onset of the experiment.

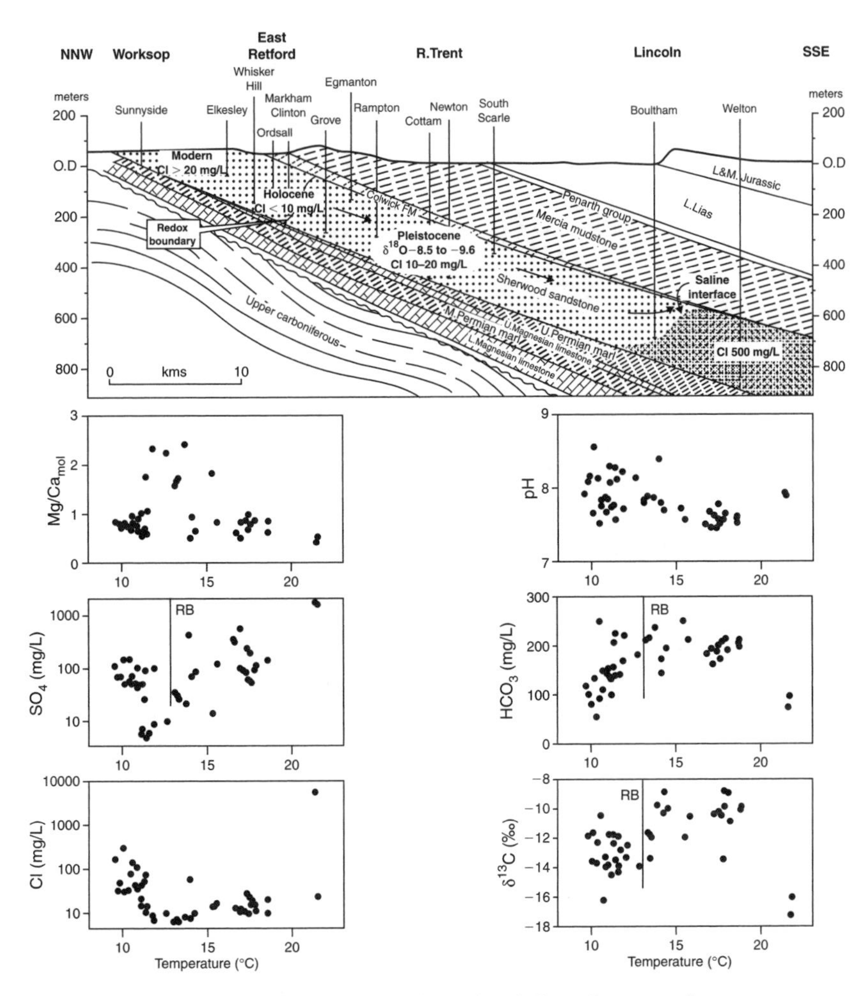

Figure 5.38. Cross section over the Sherwood aquifer and variation of waterquality parameters as a function of the water temperature, which is used as a proxy for residence time. RB indicates the redox boundary where water becomes depleted of oxygen (Smedley and Edmunds, 2002).

The $\delta^{18}O$ in Figure 5.39 is given relative to the PDB standard (which is a marine carbonate with about the same $\delta^{18}O_{PDB}$ as the dolomite in the aquifer).

The calculation of isotopic compositions which includes fractionation into precipitates can be done accurately with the KINETICS module of PHREEQC as shown in Example 5.11.

EXAMPLE 5.11. *PHREEQC calculation of ^{13}C evolution during dedolomitization*
Calculate $\delta^{13}C$ when 1 mmol gypsum/L is added to a solution with 3.3 mM *TIC* with $\delta^{13}C = -13$‰, in equilibrium with calcite and dolomite.

ANSWER:
We use the kinetic integrator of PHREEQC, which automatically adjusts the steps in the integration process to attain the required accuracy of the answer. The reactions are programmed for 1 second overall time since we only want the net result. Thus, line 10 in the rate for gypsum in the input file below will add 1 mM gypsum in total.

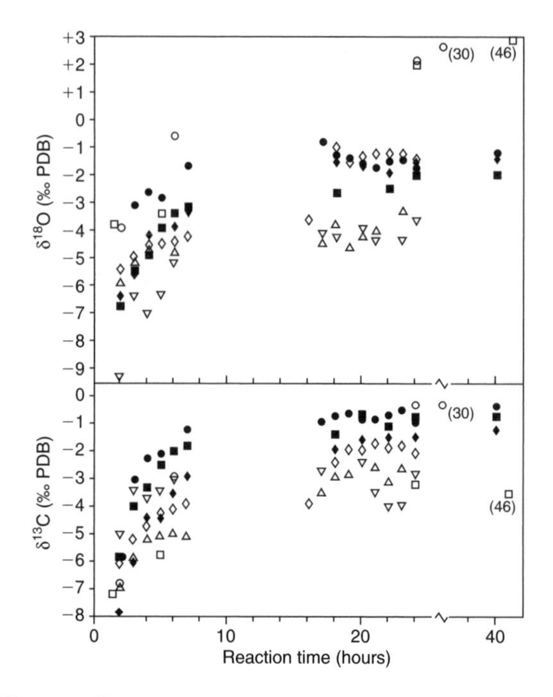

Figure 5.39. Trends in $\delta^{13}C$ and $\delta^{18}O$ in CO_2 released during dissolution experiments with Sherwood aquifer samples. The increase with time indicates the transient dissolution of fresh water carbonate followed by marine dolomite (Edmunds et al., 1982).

The rates for calcite and dolomite provide reaction amounts in small steps. The reaction amounts are stored in memory and used by the rate "Carbon13" which keeps track of the concentration of ^{13}C (in mol/L). Note that the rates for calcite and dolomite are so high (0.1 M/s) that equilibrium is always attained when gypsum is added.

In the rate "Carbon13" we use isotope fractionation factors rather than enrichments since the actual concentration of the individual isotopes is calculated. With the speciation formula (Equation 5.13) the fraction of HCO_3^- of *TIC* is obtained (this is ^{12}C), and similarly, the fraction of $H^{13}CO_3^-$ of the total moles of ^{13}C in solution (the total is given by "m" in line 90). The ratio of the two, and the fractionation factor for calcite and solution, determine the loss of ^{13}C. The gain in ^{13}C from dolomite dissolution is found in line 120, where the reaction amount of dolomite is multiplied with the moles of ^{13}C in the dolomite and with 2 (the formula is $CaMg(CO_3)_2$).

As initial solution we define a water with 3.3 mM *TIC* (195 mg HCO_3^-/L) and 0.1 mM SO_4^{2-} in equilibrium with calcite and dolomite.

```
        # Find 13C when gypsum is added to calcite/dolomite equilibrated solution
RATES
Gypsum                         # Add 1 mmol gypsum in 1 s
  -start
  10 save 1e-3 * time
  -end

Calcite                        # rate is for 13C integration only. . .
  -start
  10 moles = 1e-1 * (1 - sr("Calcite")) * time
  20 put(moles, 1)             # . . .for C13 kinetics
  30 save moles
  -end
```

```
Dolomite                        # rate is only for 13C integration. . .
  -start
  10 moles = 1e-1 * (1 - sr("Dolomite")) * time
  20 put(moles, 2)       # . . .for C13 kinetics
  30 save moles
  -end

Carbon13                        # Distribute 13C among solute species
                                # 1 = CO2aq, 2 = HCO3-,3 = CO3-2, and in calcite.
  -start
                                # Define alpha's from Clark and Fritz, 1997, p. 121 ...
  4 T = TK
  10 aa_1_2 = exp(24.29e-3 - 9.925 / T)
  20 aa_3_2 = exp(20.7e-3 - 9.552 / T + 870 / T^2)
  30 aa_cc_2 = exp(26.561e-3 - 17.2183 / T + 2988.0 / T^2)
                                # Find H12CO3- and H13CO3-...
  50 aH = act("H+")
  60 K1 = aH * act("HCO3-") / act("CO2") / act("H2O")
  70 K2 = aH * act("CO3-2") / act("HCO3-")
  80 m12C = (tot("C(4)")) / (aH/K1 + 1 + K2/aH)
  90 m13C = m / (aa_1_2 * aH/K1 + 1 + aa_3_2 * K2/aH)
                                # and the ratio 13C/12C in HCO3-. . .
  100 R2 = m13C / m12C
                                # Fractionate 13C into calcite . . .
  110 d13C_cc = aa_cc_2 * -get(1) * R2
                                # Add 13C from dolomite with d13C = -2 permil . . .
  120 d13C_dol = get(2) * 2 * 0.0112145
                                # Integrate. . .
  130 save d13C_cc - d13C_dol
  -end
KINETICS 1
  Gypsum;        Calcite;        Dolomite
  Carbon13; -formula C 0; -m0 0.0366E-3 # d13C = -13 for TIC = 3.3e-3
  -step 1                       # only 1 step in 1 second
SOLUTION 1
  -temp 15; pH 7 charge; S(6) 0.1; Ca 1 Calcite; Mg 1 Dolomite; C(4) 3.3
USER_PRINT
  -start
  10 Rst = 0.011237
  20 print 'd13C of solution :', (kin("Carbon13") / tot("C(4)") / Rst - 1) * 1e3
  -end
END
```

The results are presented in Table 5.8. Note that $\delta^{13}C$ increases from -13‰ to -11‰, that the alkalinity remains equal, and that pH decreases in agreement with the observed water qualities (Figure 5.38, consider the unpolluted samples to the right of the RB). Adding 10 mM gypsum will increase $\delta^{13}C$ to -6.2‰. In the aquifer, however, concentrations above 1000 mg SO_4^{2-}/L are accompanied by $\delta^{13}C \approx -17$‰ (Figure 5.38). This low value is indicative of methanogenesis.

Table 5.8. $\delta^{13}C$ and water composition during dedolomitization of Sherwood aquifer water. Concentrations in mmol/L, pH units and ‰ for $\delta^{13}C$.

Water	SO_4^{2-}	$\delta^{13}C$	*TIC*	*Alk*	pH	Mg^{2+}	Ca^{2+}	cc	dol
Initial	0.1	−13	3.30	3.19	7.76	0.85	0.84	0	0
+ reaction	1.1	−11.1	3.26	3.11	7.64	1.32	1.33	−0.98	0.47

QUESTIONS:
Give an estimate of $\delta^{13}C$ using only the reaction amounts listed in Table 5.8 (i.e. neglect fractionation in the precipitate)

ANSWER: $(-13 \times 3.3 + (-2) \times (0.47 \times 2)) / (3.3 + (0.47 \times 2)) = -10.6‰$

Why is the solution, calculated with PHREEQC, lighter?

ANSWER: the solution loses relatively more ^{13}C than ^{12}C to calcite.

Calculate $\delta^{13}C$ with PHREEQC when 5 mM gypsum enters the solution?

ANSWER: by hand: $-6.6‰$, with PHREEQC: $-7.3‰$

Program the calculation for total time 3×10^9 s (100 yrs, gives about 1.5 km flow distance in the field)

ANSWER: in all rates, redefine "time" to "time/3e9", in KINETICS use -step 3e9

5.7.2 ^{14}C *and groundwater age*

The travel time of groundwater can be calculated from the decay of radioactive carbon, ^{14}C according to (cf. Chapter 4):

$$dc / dt = -\lambda c \tag{5.53}$$

which solves for ^{14}C to:

$$t_{^{14}C} = \ln(^{14}A / {}^{14}A_0) / -1.21\times10^{-4} = 8267 \ln(^{14}A_0 / {}^{14}A) \text{ (yr)} \tag{5.54}$$

where $\lambda = 1.21\times10^{-4}$ is the decay constant (yr^{-1}), ^{14}A is the activity of ^{14}C in *TIC* in the downstream sample (expressed in % modern carbon, pmc) and $^{14}A_0$ is the activity in the upstream sample. The difficulty is to define the $^{14}A_0$ and the chemical reactions which may have modified ^{14}A during flow (Kalin, 2000).

Usually, $^{14}A_0$ must be found for soil water in the recharge area. When soil water equilibrates with CO_2 from the atmosphere and young organic matter, it obtains $^{14}A = 100$ pmc (since nuclear bomb testing in the atmosphere this value has augmented, but also over the Pleistocene and Holocene the values have fluctuated). When "dead" carbon from old carbonates dissolves in the unsaturated soil, ^{14}A of soil water still remains 100 pmc because of CO_2 exchange with soil gas. Therefore, for open system dissolution of carbonate, $^{14}A_0 = 100$ pmc. If the carbonate dissolves below the water table, gas exchange is absent and *TIC* is diluted by carbon from the old carbonate. We calculated for closed system dissolution that half of the *TIC* comes from $H_2CO_3^*$ from the soil, and half from the solid carbonate. Accordingly, $^{14}A_0 = 50$ pmc for closed system carbonate dissolution (50 pmc is correct for $P_{CO_2,soil} > 10^{-2}$ atm, otherwise $^{14}A_0$ will be lower, Problem 5.20). Empirical evidence so far suggests that some intermediate value in between 50 and 100 pmc should be used (Clark and Fritz, 1997).

In the Triassic Sherwood aquifer (Figure 5.38), the groundwater samples from the recharge zone which contain tritium, have ^{14}A ranging from 40 to 80 pmc. The pH of the groundwater of around 8 indicates closed system carbonate dissolution with an initial $[P_{CO_2}] > 10^{-2}$, and therefore, a value of 50 pmc for ^{14}A is expected. The low value of 40 pmc may indicate mixing of old and young water in the long screen of the production well. On the other hand, the high value of 80 pmc suggests a mixture of waters with open and closed system dissolution. Using the average of $^{14}A_0 = 60$ pmc for water in the recharge zone of the Sherwood aquifer, we can calculate the groundwater age, corrected for dilution by the dedolomitization reaction (Example 5.12).

EXAMPLE 5.12. *Groundwater age from ^{14}C*
Calculate the age of water in the Sherwood aquifer from well Ompton 2 which has 50 pmc, 2.9 mmol *TIC*/L, $\delta^{13}C = -12.7‰$ and 5 mg SO_4^{2-}/L. Also for well Gainsborough 2 with 4.8 pmc, 4.3 mmol *TIC*/L, $\delta^{13}C = -10.4‰$ and 201 mg SO_4^{2-}/L. Assume $^{14}A_0 = 60$ pmc. ($\delta^{13}C$ from Smedley, pers. comm., ^{14}A from Bath et al., 1979).

ANSWER:
Water from Ompton 2 with 5 mg SO_4^{2-} shows no dedolomitization, hence $t = 8267 \ln(60/50) = 1507$ yr.
In water from Gainsborough 2 with 201 mg SO_4^{2-}, 0.92 mM dolomite has dissolved, diluting *TIC* to $4.3 + 2 \times 0.92 = 6.14$ mM (calculated according to Example 5.11). Thus, ^{14}C is diluted to $(^{14}A_0)_r = 60 \times (4.3/6.14) = 42.02$ pmc (the subscript *r* indicates reaction corrected). The age of the water is $t = 8267 \ln(42.02/4.8) = 17{,}935$ yr. Model $\delta^{13}C = -10.4‰$, equals observed.

In Example 5.12, we have neglected the fractionation of ^{14}C into precipitating calcite. We have seen already that ^{13}C enrichment will change the $\delta^{13}C$ of the solution by about $-0.5‰$ (Questions with Example 5.11). ^{14}C fractionates 2.3 times more than ^{13}C and changes by $-1.3‰$, which gives $(^{14}A_0)_r = 42.02 - 0.13 = 41.89$ pmc. The age then becomes 17,909 instead of 17,935 yr, which is a negligible difference.

The uncertainties of translating the ^{14}C age into real time comprise dilution and fractionation effects associated with other reactions such as ion exchange and oxidation-reduction which are discussed in later chapters. The computer code NETPATH (Plummer et al., 1994) is useful for calculating the corrections. Often, the reactions can be traced to concentration changes of specific ions, and $^{14}A_0$ can be corrected as was done for the dedolomitization reaction with SO_4^{2-} (Plummer et al., 1990, 1994; Parkhurst, 1997; Van der Kemp et al., 2000). For common reactions such as the conversion of aragonite into calcite and of high-Mg calcite into low-Mg calcite, the correction can be quantified with $\delta^{13}C$. Figure 5.40 shows the relation of ^{14}C and $\delta^{13}C$ in Chalk groundwaters from the Paris and N.-German Basins, and Figure 5.41 gives the Mg^{2+}/Ca^{2+} ratio of these waters as a function of $\delta^{13}C$ (Kloppmann et al., 1998).

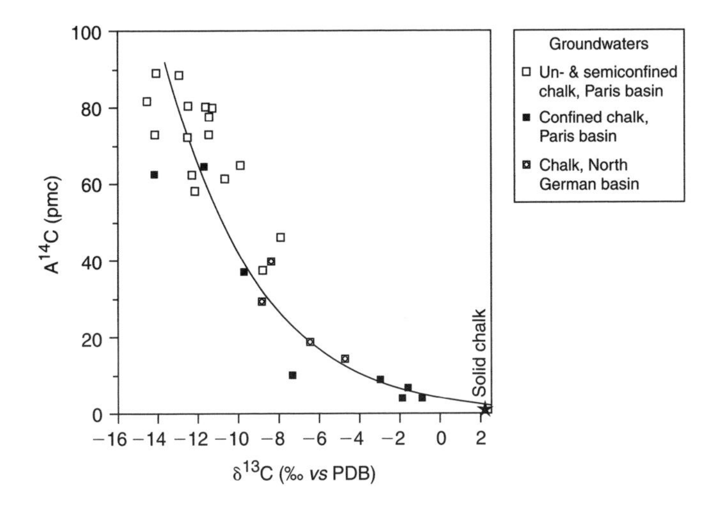

Figure 5.40. Plot of ^{14}C *vs* $\delta^{13}C$ in Chalk groundwaters from the Paris and N.-German Basin (Kloppmann et al., 1998).

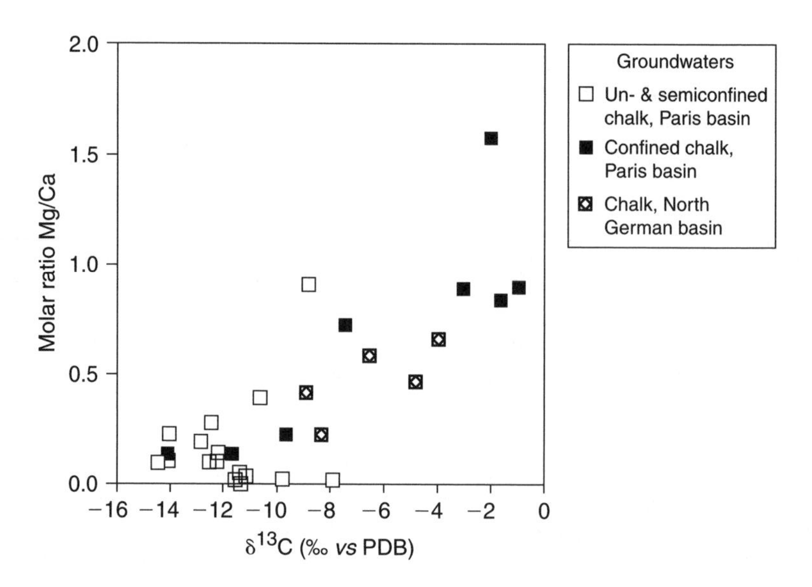

Figure 5.41. The Mg/Ca ratio increases with $\delta^{13}C$ in the groundwaters from the Paris and N.-German Basin (Kloppmann et al., 1998).

The exponential correlation of ^{14}C and $\delta^{13}C$ suggests that the decrease of ^{14}C by radioactive decay is enhanced by the dissolution of dead carbon from Chalk that has $\delta^{13}C$ of 2.2‰. The increase of the Mg^{2+} / Ca^{2+} ratio with $\delta^{13}C$ indicates that the chemical reaction involves the incongruent dissolution of high-Mg calcite. Because the latter reaction is a gliding one that produces calcites of intermediate compositions, it is still uncertain what the contribution is of these intermediate products on the water composition, and whether these are really completely "dead". The process can also be lumped into a first-order kinetic recrystallization factor that increases λ (Gonfiantini and Zuppi, 2003). In the case of the Paris Basin, λ increases from 1.21×10^{-4} to 2.95×10^{-4}/yr, which reduces the "conventional" ^{14}C age by a factor of 2.4.

5.7.3 *Retardation by sorption and stagnant zone diffusion*

Sometimes, unexpected low values of ^{14}A are encountered in groundwater which give a much higher age than follows from the flow velocity calculated with groundwater models, even after the various chemical corrections have been applied. The discrepancy may be due to adsorption of carbonate, which retards the flowtime of carbon and hence increases the decay of ^{14}C, but is not visible in changes of concentrations of other solutes or of ^{13}C if the latter has attained sorption equilibrium (Mozeto et al., 1984; Garnier, 1985; Garnier et al., 1985; Maloszewski and Zuber, 1991). With sorption, the age of the water becomes (Chapter 3):

$$t_{\mathrm{H_2O}} = t_{^{14}\mathrm{C}} \;/\; R_c \tag{5.55}$$

where R_c is the retardation, $R_c = 1 + \mathrm{d}q/\mathrm{d}c$, q is sorbed concentration expressed in the units of solute concentrations, *e.g.* mol/L. The retardation can be included in a "working" decay constant which becomes:

$$(\lambda_{^{14}\mathrm{C}})_R = R_c \, \lambda_{^{14}\mathrm{C}} \tag{5.56}$$

Actually, it is difficult to discern sorption on carbonate minerals from recrystallization. Mozeto et al. (1984) found that sorption of ^{13}C was reversible on aged, crushed calcite, but irreversible on freshly crushed calcite which dissolved partly and incorporated the isotope in the reprecipitated crystals. Garnier (1985) used a mixture of aragonite and calcite to measure the retardation of ^{14}C in column experiments and noted clogging due to calcite precipitation which apparently resulted from recrystallization of aragonite. One molecular layer of the calcite surface was sufficient to explain the amount of reversible sorption in the experiments of Mozeto. Labotka et al. (2000) measured carbon diffusion in calcite in between 600 and 800°C and noted very small diffusion lengths, in agreement with unaltering carbon isotope ratios in calcite in metamorphic rocks. Although uncertain because it grossly surpasses the experimental temperature range, their data extrapolate to $D_{^{13}C} = 6 \times 10^{-42}\,m^2/s$ at 25°C in calcite. This gives $\sigma = \sqrt{(2Dt)} = 2 \times 10^{-15}\,m$ in 10,000 yrs, which is even less than the single layer of carbonate ions that exchanged isotopes in the experiments of Mozeto et al. Thus, sorption on calcite is limited to the outer surface layer, but probably with notable effects (cf. Example 5.13). In addition, sorption of carbonate takes place onto iron- and aluminum-oxyhydroxides, which gives a retardation of about 2 in the case of the earlier discussed Sherwood (Bunter) sandstone aquifer (of which the grains are red-stained by iron-oxyhydroxides, cf. Problem 7.2). This sorption reduces the ages calculated in Example 5.12 by a factor of 2.

EXAMPLE 5.13. *Estimate exchangeable carbonate on Chalk and the retardation of* ^{14}C
The rhombohedral face of a calcite crystal contains 1 molecule CO_3/21 Å^2 (Reeder, 1983). Estimate the outer layer of carbonate ions of calcite (mol/L pore water) for Chalk with 10 μm calcite crystals and 30% porosity, and find the retardation of ^{14}C when $TIC = 5\,mM$.

ANSWER:
One liter water contacts $1 \times (1 - 0.7) / 0.3 = 2.33\,L$ sediment (solids only). There are $2.33 / (10^{-4})^3 = 2.33 \times 10^{12}$ crystals with $1399\,m^2$ total surface area. Thus, the Chalk has $1399\,m^2 / (21 \times 10^{-20}\,m^2$ per molecule) / (6.022×10^{20} molecules per mmol) = 10.9 mmol/L exchangeable carbonate ions (q). Assume linear sorption and negligible fractionation for ^{14}C, and find $R = 1 + dq/\,dc = 1 + q/\,c = 1 + 10.9 / 5 = 3$.

Example 5.13 shows that the retardation of ^{14}C by exchange with carbonate groups on the surface of calcite in Chalk may be 3. The same retardation applies for ^{13}C, and therefore the exchange reaction becomes "invisible" for the usual ^{14}C correction methods after water with a given ^{13}C has flushed the porosity of the Chalk more than three times. The observed trend in ^{13}C in the Paris Basin Chalk was attributed to recrystallization of high-Mg calcite which gives a retardation for ^{14}C of 2.4. However, if the system has been flushed more than 3 times, the exchange reaction should be added, which augments the overall retardation to $R = 2.4 + 2 = 4.4$. The uncorrected ^{14}C ages would then be a factor 4.4 too high.

Another retardation mechanism that may go unnoticed in ^{14}C-age determinations is diffusion into the matrix of limestone or in the low permeability layers of a sedimentary sequence (Neretnieks, 1981; Sudicky and Frind, 1981; Maloszewski and Zuber, 1991; Sanford, 1997). Consider the situation of a regular alternation of sand and clay layers or of conduits in a marly limestone, etc., depicted in Figure 5.42. Neglecting dispersion in the mobile zone, the steady state transport equation in the zone with mobile water (x-direction) is (Chapter 3):

$$\frac{\partial c_m}{\partial t} = 0 = -v_{H_2O} \frac{\partial c_m}{\partial x} - \lambda c_m - Q_z \tag{5.57}$$

where c_m is the concentration in the mobile water (mol/m^3) and Q_z is the mass transfer into the stagnant zone (mol/m^3/s). The mass transfer equals the diffusive flux into the stagnant zone ($= \varepsilon_{im} F_z$),

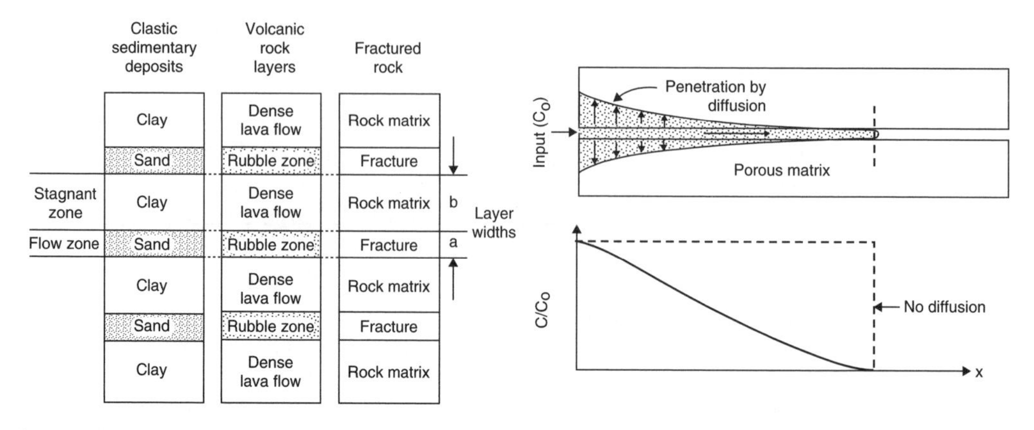

Figure 5.42. A regular alternation of mobile and mobile zones to model diffusion into the matrix of limestones, volcanic rocks, or sand/clay alterations (Sanford, 1997), and concentration profile in the mobile zone, with and without diffusion into the immobile zone.

counted twice for upper and lower contact, and normalized to the amount of water in the mobile zone ($= \varepsilon_m a$), or:

$$Q_z = \frac{2\varepsilon_{im}}{\varepsilon_m a}\left(-D_e \frac{\partial c_{im}}{\partial z}\right)_{z=0} \tag{5.58}$$

where a is the width of the mobile zone (m).

The concentration gradient in Equation (5.58) is obtained by first solving the steady state diffusion equation in the stagnant zone:

$$\frac{\partial c_{im}}{\partial t} = 0 = D_e \frac{\partial^2 c_{im}}{\partial z^2} - \lambda c_{im} \tag{5.59}$$

This is an ordinary differential equation ($t = \infty$):

$$\frac{\mathrm{d}^2 c_{im}}{\mathrm{d}z^2} = \frac{\lambda}{D_e} c_{im} = l^2 c_{im} \tag{5.60}$$

with the general solution:

$$c_{im} = k_1 \mathrm{e}^{lz} + k_2 \mathrm{e}^{-lz} \tag{5.61}$$

The two constants k_1 and k_2 are obtained from the boundary conditions. At the contact of the mobile and immobile zone, $c_{im,z=0} = c_m = k_1 + k_2$. Midway in the immobile zone of width b, the concentration gradient is zero, $(\partial c_{im}/\partial z)_{z=b/2} = 0 = k_1 l e^{lb/2} - k_2 l e^{-lb/2}$. Together in (5.61) this yields:

$$c_{im} = c_m \frac{e^{l(\frac{b}{2}-z)} + e^{-l(\frac{b}{2}-z)}}{e^{l\frac{b}{2}} + e^{-l\frac{b}{2}}} = c_m \frac{\cosh\left[l\left(\frac{b}{2} - z\right)\right]}{\cosh\left[l\frac{b}{2}\right]} \tag{5.62}$$

We can now find the concentration gradient at $z = 0$:

$$\frac{dc_{im}}{dz} = -c_m \, l \frac{e^{l\frac{b}{2}} - e^{-l\frac{b}{2}}}{e^{l\frac{b}{2}} + e^{-l\frac{b}{2}}} = -c_m \, l \tanh\left(l\frac{b}{2}\right) = -lwc_m \quad (5.63)$$

When Equations (5.57), (5.58) and (5.63) are combined, a simple, ordinary differential equation (valid for $t = \infty$) is obtained:

$$v_{H_2O}\frac{dc_m}{dx} = -\lambda\left(1 + \frac{2\varepsilon_{im} w}{\varepsilon_m al}\right)c_m = -\lambda R_{im} c_m \quad (5.64)$$

which gives the concentration profile in the mobile zone:

$$c_m(x, \infty) = c_0 e^{-\lambda R_{im}\frac{x}{v}} = c_0 e^{-\lambda R_{im} t} \quad (5.65)$$

where c_0 is the concentration (*e.g.* $^{14}A_0$) that enters at $x = 0$.

Equations (5.64) and (5.65) state that the decay constant of ^{14}C is multiplied by a retardation factor which depends on the interaction of mobile and immobile water. When the immobile zone is fully penetrated by ^{14}C from the conduits, we can sense already that

$$R_{im} = (1 + (b\varepsilon_{im})/(a\varepsilon_m)),$$

i.e. the retardation is given by the ratio of the total and the mobile pore volume. Thus, R_{im} can be quite high, depending on the volume of water in the matrix and in the conduits of the limestone (Example 5.14).

EXAMPLE 5.14. *Estimate the retardation of* ^{14}C *by matrix diffusion*
Calculate R_{im} in Equation (5.64) given conduit width $a = 1$ mm, fully filled with water $\varepsilon_m = 1$, and matrix block size $b = 0.3$ m with porosity $\varepsilon_{im} = 0.3$. The diffusion coefficient in the matrix $D_e = 0.3\times10^{-9}$ m^2/s.

ANSWER:
The various constants solve to:

$l = (\lambda/D_e)^{0.5} = (3.84\times10^{-12}/3\times10^{-10})^{0.5} = 0.113$/m (note that λ is per second)

$b/2 = 0.15$ m

$\exp(l \times b/2) = 1.0171 = A$, $\exp(-l \times b/2) = 0.9832 = B$

$w = \tanh(l \times b/2) = (A - B)/(A + B) = 0.01696$

$R_{im} = 1 + (2 \times \varepsilon_{im} \times w/(\varepsilon_m \times a \times l)) = 90.99$

(which compares with $(R_{im})_{max} = (1 + (b\varepsilon_{im})/(a\varepsilon_m)) = 91$).

QUESTIONS:
Find the retardation by sorption of carbonate when calcite crystals in Chalk are 20 μm?
ANSWER: $1 + 5.4/5 = 2$.
Find R_{im} and $(R_{im})_{max}$ for $b = 10$ m instead of 0.3 m in Example 5.14?
ANSWER: $R_{im} = 2717$ and $(R_{im})_{max} = 3001$.

Give the decay constant for ^{14}C which includes mobile/immobile exchange and linear sorption, $q/c = 2$?

ANSWER: We assumed steady state ($t = \infty$), hence the contribution by sorption $dq_{im}/dc_{im} \cdot dc_{im}/dt = 0$, sorption does not add to the retardation at $t = \infty$. Transient profiles must be calculated numerically with PHREEQC using option stagnant in keyword TRANSPORT (Chapter 6).

PROBLEMS

Laboratory exercise: dissolution of calcite in water and carbon dioxide

To simulate the dissolution process of limestone, 5 g calcite crystals are added to 200 mL distilled water in an erlenmeyer flask that is bubbled with 99% N_2 and 1% CO_2 gas. After one day, determine the solution pH (pH-meter) and, after filtration over 0.45 μm, Ca^{2+}, HCO_3^- and CO_2 (titrations).

Find:

pH = ... Ca^{2+} = ... mmol/L HCO_3^- = ... mmol/L CO_2 = ... mmol/L

From analyzed CO_2 we calculate P_{CO_2} = ... atm

From CO_2/ HCO_3^- equilibrium follows pH = ... , and $[CO_3^{2-}]$ = ... The saturation index for calcite, SI_{cc} = ... (without activity coefficients), and SI_{cc} = ... (with activity coefficients). (*Hint*: Note how to use activity coefficients when calculating $[CO_3^{2-}]$).

From *measured* pH, $[CO_3^{2-}]$ = ... The saturation index for calcite, SI_{cc} = ... (without activity coefficients).

The calcite crystals consist of rhomboeders. Assume them to be cubes with a side of 50 μm with density 2.72 g/cm^3. Calculate the time to reach 95% saturation with PHREEQC.

What would be the solution composition, pH, Ca^{2+}, CO_3^{2-}, HCO_3^-, if we used 100 mM NaCl instead of distilled water (*Hint*: Derive Equation (5.24) once again for 0.1 M NaCl while correcting for activity coefficients).

pH = ... Ca^{2+} = ... mmol/L CO_3^{2-} = ... mmol/L HCO_3^- = ... mmol/L

5.1. Calculate *TIC* in Example 5.1 for pH = 7, when the solution contains 10 mmol NaCl/L. Na^+ balances the alkalinity. Correct for activity-coefficients. Why is the correction problematic when pH = 10?

5.2. Calculate pH and $[CO_3^{2-}]$ in a sample with *TIC* of 1.7, and $m_{HCO_3^-}$ of 1.3 mmol/L. (*Hint*: consider as a first estimate that $TIC = m_{H_2CO_3^*} + m_{HCO_3^-}$, i.e. pH < 8.5). Same question for the case that *TIC* = 1.3 and *Alk* = 1.7 mmol/L

5.3. Calculate the activity and concentration of CO_3^{2-} at pH = 10.5 and pH = 6.3, when *TIC* = 2.5 mmol/L at both pH's.

5.4. Calculate pH and CO_2 species in pure water at a P_{CO_2} of 10^{-2} atm.

5.5. Calculate Ca^{2+} concentration when the CO_2 pressure is 0.01 atm and calcite dissolves till saturation.

5.6. What CO_2 pressure is associated with a Ca^{2+} concentration of 2 mmol/L, assuming equilibrium with calcite, and absence of other reactions?

5.7. Soil-water in equilibrium with calcite has pH = 7.6. Calculate the CO_2 pressure? (*Hint*: find HCO_3^- as function of P_{CO_2} with Equations (5.23) and (5.24); same for CO_3^{2-} with Equation (5.24) and calcite equilibrium).

5.8. Calculate the time to reach 95% saturation with calcite under open conditions, initial $[P_{CO_2}] = 10^{-3.5}$. Calculate the distance covered in this time in a limestone fracture of width $\delta = 0.05$ cm.

5.9. Groundwater-samples from the Veluwe (Netherlands) are listed below (concentrations in mmol/L).
a) Calculate P_{CO_2}. b) Which water was in contact with calcite? Did dissolution occur in an open or in a closed system? c) Do you see effects of contamination?

	1.	2.	3.	4.
pH	4.55	7.54	7.86	6.24
Na^+	0.16	0.41	0.59	0.81
K^+	0.03	0.02	0.05	0.10
Mg^{2+}	0.03	0.20	0.39	0.26
Ca^{2+}	0.05	1.55	2.62	1.43
Cl^-	0.17	0.32	0.51	1.11
HCO_3^-	0.07	3.57	2.75	0.76
SO_4^{2-}	0.22	0.01	1.86	0.55
NO_3^-	0.17	<.002	0.008	1.47
Fe	<.001	0.02	0.02	<.001
Al	0.14	<.001	0.008	<.001

5.10. Pouhon is the local name for a spring with Fe^{2+} and CO_2 rich water in Spa, Belgium. An analysis is given below (concentration in mmol/L):

pH	?	Cl^-	.54
Na^+	1.0	*Alk*	6.38
K^+	.15	SO_4^{2-}	.43
Mg^{2+}	.54	Fe	.34
Ca^{2+}	.77	CO_2	66.0

a) Calculate pH from HCO_3^-/CO_2-equilibrium. b) What would be pH of Pouhon-water, if P_{CO_2} decreases to atmospheric level. c) Check the analysis of the Pouhon-water. Alkalinity and CO_2 were titrated at the spring, Fe was analysed in the laboratory, in a non-acidified sample. Explain the error in the charge balance. d) What would be the change in alkalinity, if all Fe precipitates? e) CO_2 in Pouhon-water might originate from Eifel-volcanism, and is called "juvenile". What is "juvenile"?

5.11. Write reactions which take place when the following substances are added to a water sample. Does Alkalinity increase, remain constant, or decrease?

1. CO_2	4. $MnCl_2$	7. Na_2CO_3
2. H_2SO_4	5. $MnCl_2 + O_2$	8. $NaHCO_3$
3. $FeCl_3$	6. NaOH	9. Na_2SO_4

5.12. An analysis of water from a playa (evaporating lake) gave as results:

$pH = 10.0 \qquad SiO_2 = 60\,mg/L$

What is the contribution of Si to titrated Alkalinity?
Data: Alkalinity $= HCO_3^- + 2CO_3^{2-} + OH^- +$ base – acid.
$H_4SiO_4^0 \leftrightarrow H_3SiO_4^- + H^+$; $\log K = -9.1$

5.13. Calculate the denudation rate of limestone in mm/yr when P_{CO_2} is 10^{-2} atm, precipitation surplus is 300 mm/yr. Density of the limestone is 2.0 g/cm^3. *Hint*: calculate the Ca^{2+} concentration in water, multiply with the water flux.

5.14. Calculate the decalcification rate for the unsaturated sand in Figure 5.16. P = 0.1 m/yr, temp = 10°C, ε= 0.3, ρ_b = 1.8 g/cm^3.

5.15. Model the profile of Figure 5.16 with PHREEQC. Take decalcification as equilibrium reaction and dedolomitization as kinetic reaction. The pristine sediment contains per L pore water (ε_w = 0.1) 0.18 mol calcite and 0.04 mol dolomite (200 times less than in the field). Use 10 cells of 0.1 m each. Discuss the calculated profiles of dolomite, calcite and of Mg^{2+} and Ca^{2+} after 25, 50 and 75 yrs.

5.16. With PHREEQC, calculate *SI* for calcite and aragonite of a mixture (0–100% in 10 steps) of Caribbean seawater with local fresh groundwater (temp. 25°C). Discuss what happens with the solid carbonates when up to 50% seawater mixes with fresh water, when 50–60% seawater contacts aragonite, and when the mixture contains >60% seawater. Use the composition (mg/L, except pH):

	pH	Na^+	K^+	Mg^{2+}	Ca^{2+}	Cl^-	HCO_3^-	SO_4^{2-}
Seawater	8.18	10000	423	1180	410	17700	180	2610
Groundwater	7.10	14	3.3	7.1	80	34	253	1.8

5.17. From the PWP calcite dissolution rate (Equation 5.43) and the Ca^{2+} / P_{CO_2} relation (Equation 5.24) find that the overall rate will be $r = r_{fw}\,(1 - (\Omega_{cc})^{2/3})$.

5.18. Program the dolomite dissolution rate (Equation 5.50) in PHREEQC and calculate the time to reach 95% saturation with A/V = 1/cm for $[P_{CO_2}] = 10^{-1.5}$ and $10^{-3.5}$.

5.19. Construct a PHREEQC rate for the kinetics of CO_2 hydration. Reaction 1:

$$CO_2 + H_2O \rightarrow H_2CO_3$$

with $R_{fw} = k_1\,[CO_2]$ and $R_{bw} = k_{-1}\,[H_2CO_3]$.
$H_2CO_3 / H_2CO_3^* = H_2CO_3 / (CO_2 + H_2CO_3) = 1/(630 + 1)$.
$k_1 = 10^{-3}\exp(34.69 - 9252/T)$/s (Welch et al., 1969), T is temperature in Kelvin.

Reaction 2:

$$CO_2 + OH^- \rightarrow HCO_3^-$$

with $R_{fw} = k_2\,[CO_2][OH^-]$ and $R_{bw} = k_{-2}\,[HCO_3^-]$. $k_{-2} = 10^{-6}\exp(48.08 - 12720/T)$/s (Welch et al., 1969).
Hint: introduce a new SOLUTION_MASTER_SPECIES Cunh for unhydrated CO_2. Let Cunh react to give $H_2CO_3^*$ according to:

$$d\,[\text{Cunh}]/dt = -k_1[\text{Cunh}] + k_{-1}[H_2CO_3] - k_2[\text{Cunh}]\,[OH^-] + k_{-2}[HCO_3^-]$$

a. Calculate the half-life for the reaction when $m_{\text{Cunh},\,0}$ = 1 mM at 10° and 25°C.
b. What is $[P_{CO_2}]$ when 1 mM Cunh has hydrated to equilibrium at 25°C?
c. Construct a rate which adds Cunh to the solution and hydrates Cunh until $[P_{CO_2}] = 10^{-1.5}$.
d. Include the kinetics of CO_2 hydration in the PWP rate model of calcite dissolution.

5.20. Calculate $^{14}A_0$ of water in which calcite dissolves closed with respect to CO_2 for initial $[P_{CO_2}] = 10^{-1.5}$, $10^{-2.5}$ and $10^{-3.5}$.

5.21.

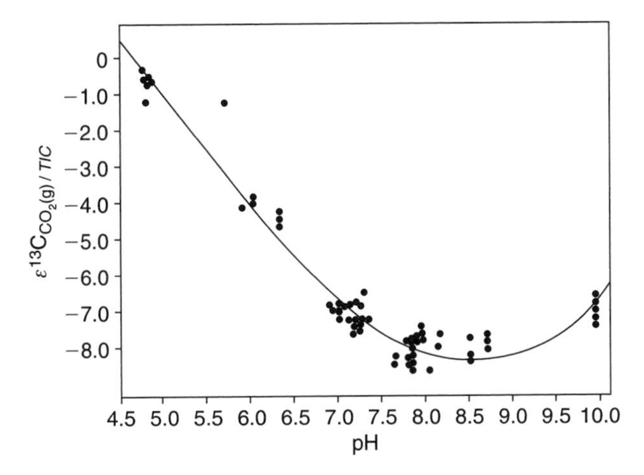

Turner (1982) analyzed enrichment of ^{13}C of $CO_2(g)$ with respect to *TIC* in water ($\varepsilon^{13}C_{CO_2(g)/TIC}$) at 25°C as a function of pH. Calculate the trend with PHREEQC, use Example 5.10, or in Example 5.11, $\alpha_{CO_2(g)/HCO_3^-} = \exp(24.1e\text{–}3 - 9.552/T)$.

5.22. Model the kinetic dissolution experiment with which we started Section 5.1. After how much time can we expect $\Omega = 0.98$ if we put 5 g calcite in 200 mL water (cf. the laboratory exercise)?

5.23. The cross section through an alluvial landscape shows groundwater flowlines. The CO_2 pressure in the water and the Ca^{2+} concentration depend on the presence of calcite in the rooted soil. Indicate the groundwater composition (Ca^{2+} concentration and P_{CO_2}) in boreholes A and B. *Hint*: Start with the groundwater flowlines; draw the flowline for water that infiltrates at the boundary of the calcite/calcite-free zone. Then start with borehole B, using concentrations from Table 5.6 for initial $P_{CO_2} = 10^{-1.5}$. Follow the water quality from borehole B along the flowlines.

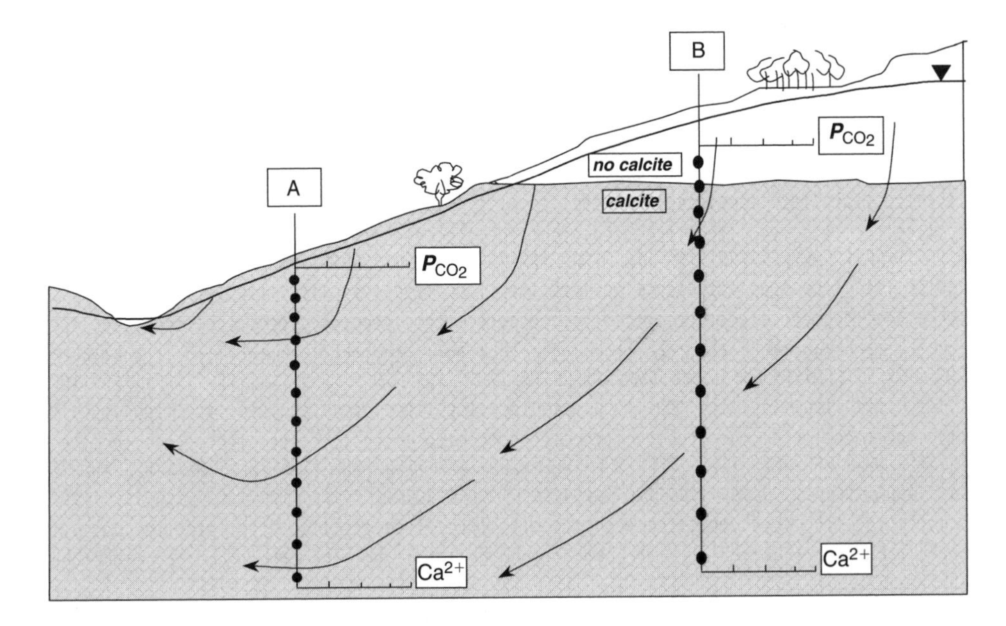

REFERENCES

Al, T.A., Martin, C.J. and Blowes, D.W., 2000. Carbonate–mineral/water interactions in sulfide-rich mine tailings. *Geochim. Cosmochim. Acta* 64, 3933–3948.

Alberic, P. and Lepiller, M., 1998. Oxydation de la matière organique dans un système hydrologique karstique alimenté par des pertes fluviales (Loiret, France), *Water Res.* 32, 2051–2064.

Amundson, R., Stern, L. Baisden, T. and Wang, Y., 1998. The isotopic composition of soil and soil-respired CO_2. *Geoderma* 82, 83–114.

Andersen, M.S., 2001. Geochemical processes at a seawater–freshwater interface. Ph.D. thesis, Lyngby, 145 pp.

Appelo, C.A.J., Beekman, H.E. and Oosterbaan, A.W.A., 1984. Hydrochemistry of springs from dolomite reefs in the southern Alps of Northern Italy. IAHS Pub. 150, 125–138.

Aravena, R., Wassenaar, L.I. and Plummer, L.N., 1995. Estimating ^{14}C groundwater ages in a methanogenic aquifer. *Water Resour. Res.* 31, 2307–2317.

Atkinson, T.C., 1975. Diffuse flow and conduit flow in limestone terrain in the Mendip Hills, Somerset, England. IAH Congress, Huntsville, Alabama.

Back, W., Hanshaw, B.B., Plummer, L.N., Rahn, P.H., Rightmire, C.T. and Rubin, M., 1983. Process and rate of dedolomitisation: mass transfer and ^{14}C dating in a regional carbonate aquifer. *Geol. Soc. Am. Bull.* 94, 1415–1429.

Back, W., Hanshaw, B.B., Herman, J.S., Van Driel, J.N., 1986. Differential dissolution of a Pleistocene reef in the ground-water mixing zone of coastal Yucatan, Mexico. *Geology* 14, 137–140.

Baker, A. and Smart, P.L., 1995. Recent flowstone growth rates: field measurements in comparison to theoretical predictions. *Chem. Geol.* 122, 121–128.

Bar-Matthews, M., Ayalon, A., Matthews, A., Sass, E. and Halicz, L., 1996. Carbon and oxygen isotope study of the active water-carbonate system in a karstic Mediterranean cave: implications for paleoclimate research in semiarid regions. *Geochim. Cosmochim. Acta* 60, 337–347.

Bath, A.H., Edmunds, W.M. and Andrews, J.N., 1979. Palaeoclimatic trends deduced from the hydrochemistry of a Triassic sandstone aquifer, United Kingdom. In *Isotope Hydrology II*, IAEA-SM-228, 545–568, IAEA, Vienna.

Batoit, C., Emblanch, C., Blavoux, B., Simler, R. and Daniel, M., 2000. Organic matter in karstic aquifers: a potential tracer in the carbon cycle. A small scale laboratory model approach. IAHS Pub. 262, 459–463.

Berg, J.M., Tymoczko, J.L. and Stryer, L., 2002. *Biochemistry*, 5th ed. Freeman, New York, 974 pp.

Berner, R.A., 1975. The role of magnesium in the crystal growth of calcite and aragonite from sea water. *Geochim. Cosmochim. Acta* 39, 489–504.

Berner, R.A., Westrich, J.T., Graber, R., Smith, J. and Martens, C.S., 1978. Inhibition of aragonite precipitation from supersaturated seawater: a laboratory and field study. *Am. J. Sci.* 278, 816–837.

Bögli, A., 1978. *Karsthydrographie und physische Speläologie*. Springer, Berlin, 292 pp.

Bolin, B., Degens, E.T., Duvigneaud, P. and Kempe, S., 1979. The global biogeochemical carbon cycle. In B. Bolin et al. (eds), *The global carbon cycle, SCOPE 13*, 1–56, Wiley and Sons, New York.

Brook, G.A., Folkoff, M.E. and Box, E.O., 1983. A world model of soil carbon dioxide. *Earth Surf. Proc.* 8, 79–88.

Brown, J.G. and Glynn, P.D., 2003. Kinetic dissolution of carbonates and Mn oxides in acidic water: measurement of in situ field rates and reactive transport modeling. *Appl. Geochem.* 18, 1225–1239.

Budd, D.A., 1997. Cenozoic dolomites of carbonate islands: their attributes and origin. *Earth Sci. Rev.* 42, 1–47.

Busenberg, E. and Plummer, L.N., 1982. The kinetics of dissolution of dolomite in CO_2-H_2O systems at 1.5 to 65°C and 0 to 1 atm PCO_2, *Am. J. Sci.* 282, 45–78.

Busenberg, E. and Plummer, L.N., 1986. A comparative study of the dissolution and crystal growth kinetics of calcite and aragonite. *U.S. Geol. Surv. Bull.* 1578, 139–168.

Busenberg, E. and Plummer, L.N., 1989. Thermodynamics of magnesium calcite solid-solutions at 25° and 1 atm total pressure. *Geochim. Cosmochim. Acta* 53, 1189–1208.

Capaccioni, B., Didero, M., Paletta, C. and Salvadori, P., 2001. Hydrogeochemistry of groundwaters from carbonate formations with basal gypsiferous layers: an example from Mt Catria-Mt Nerone ridge (Northern Appenines, Italy). *J. Hydrol.* 253, 14–26.

Cardenal, J., Benavente, J. and Cruz-Sanjulián, J.J., 1994. Chemical evolution of groundwater in Triasic gypsum-bearing carbonate aquifers (Las Alpujarras, southern Spain). *J. Hydrol.* 161, 3–30.

Cerling, T.E., 1984. The stable isotopic composition of modern soil carbonate and its relationship to climate. *Earth Planet. Sci. Lett.* 71, 229–240.

Chapelle, F.H. and Knobel, L.L., 1986. Stable carbon isotopes of HCO_3^- in the Aquia aquifer, Maryland: evidence for an isotopically heavy source of CO_2. *Ground Water* 23, 592–599.

Chiodini, G., Cioni, R., Guidi, M. Raco, B. and Marini, L., 1998. Soil CO_2 flux measurements in volcanic and geothermal areas. *Appl. Geochem.* 13, 543–552.

Chou, L., Garrels, R.M. and Wollast, R., 1989. Comparative study of the kinetics and mechanisms of dissolution of carbonate minerals. *Chem. Geol.* 78, 269–282.

Clark, I.D. and Fritz, P., 1997. *Environmental isotopes in hydrogeology*. CRC Press, Boca Raton, 328 pp.

Davis, K.J., Dove, P.M. and De Yoreo, J.J., 2000. The role of Mg^{2+} as an impurity in calcite growth. *Science* 290, 1134–1137.

Deelman, J.C., 2003. Low-temperature formation of dolomite and magnesite.http://users.skynet.be/infolib/dolomite/bookprospectus.htm

Deer, W.A., Howie, R.A. and Zussman, J., 1966. *An introduction to the rock forming minerals*. Longman, London, 528 pp.

Dörr, H. and Münnich, K.O., 1980. Carbon-14 and carbon-13 in soil CO_2. *Radiocarbon* 22, 909–918.

Dowty, E., 1999. ATOMS for Windows and Macintosh, Vs 5. www.shapesoftware.com.

Drake, J.J., 1983. The effects of geomorphology and seasonality on the chemistry of carbonate groundwater. *J. Hydrol.* 61, 223–236.

Dreybrodt, W., Buhmann, D., Michaelis, J. and Usdowski, E., 1992. Geochemically controlled calcite precipitation by CO_2 outgassing: field measurements of precipiation rates to theroretical predictions. *Chem. Geol.* 97, 287–296.

Dreybrodt, W., 1981. Kinetics of the dissolution of calcite and its applications to karstification. *Chem. Geol.* 31, 245–269.

Edmunds, W.M., Bath, A.H. and Miles, D.L., 1982. Hydrochemical evolution of the East Midlands Triassic sandstone aquifer, England. *Geochim. Cosmochim. Acta* 46, 2069–2081.

Etiope, G., 1999. Subsoil CO_2 and CH_4 and their advective transfer from faulted grassland to the atmosphere. *J. Geophys. Res.* 104, 16,889–16,894.

Fabricius, I.L., 2003. How burial diagenesis of chalk sediments controls sonic velocity and porosity. *AAPG Bull.* 87, 1755–1778.

Farquhar, G.D., Ehleringer, J.R. and Hubick, K.T., 1989. Carbon isotope discrimination and photosynthesis. *Ann. Rev. Plant Physiol. Plant Biol.* 40, 503–537.

Ford, D.C. and Williams, P.W., 1989. *Karst geomorphology and hydrology*. Unwin Hyman, Winchester, MA, 601 pp.

Gabrovšek, F. and Dreybrodt, W., 2000. Role of mixing corrosion in calcite-agressive H_2O-CO_2-$CaCO_3$ solutions in the early evolution of karst aquifers in limestone. *Water Resour. Res.* 36, 1179–1188.

Garnier, J.-M., 1985. Retardation of dissolved radiocarbon through a carbonated matrix. *Geochim. Cosmochim. Acta* 49, 683–693.

Garnier, J.-M., Crampon, N., Préaux, C., Porel, G. and Vreulx, M., 1985. Traçage par ^{13}C, ^2H, I^- et uranine dans la nappe de la craie Sénonienne en écoulement radial convergent (Béthune, France). *J. Hydrol.* 78, 379–392.

Garrels, R.M. and Mackenzie, F.T., 1971. *Evolution of sedimentary rocks*. Norton, New York, 397 pp.

Given, R. and Wilkinson, B.H., 1985. Kinetic control of morphology, composition and mineralogy of abiotic sedimentary carbonates. *J. Sed. Petrol.* 55, 109–119.

Gonfiantini, R. and Zuppi, G.M., 2003. Carbon isotope exchange rate of DIC in karst groundwater. *Chem. Geol.* 197, 319–336.

Gunn, J., 1981. Hydrological processes in karst depressions. *Z. Geomorph. N.F.* 25, 313–331.

Gutjahr A., Dabringhaus H. and Lacmann R., 1996. Studies of the growth and dissolution kinetics of the $CaCO_3$ polymorphs calcite and aragonite. II. The influence of divalent cation additives on the growth and dissolution rates. *J. Cryst. Growth* 158, 310–315.

Hamada, Y. and Tanaka, T., 2001. Dynamics of carbon dioxide in soil profiles based on long-term field observation. *Hydrol. Proc.* 15, 1829–1845.

Hanshaw, B.B. and Back, W., 1979. Major geochemical processes in the evolution of carbonate aquifer systems. *J. Hydrol.* 43, 287–312.

Hanson, P.J., Edwards, N.T., Garten, C.T. and Andrews, J.A., 2000. Separating root and soil microbial contributions to soil respiration: A review of methods and observations. *Biogeochem.* 48, 115–146.
Hardie, L.A., 1987. Dolomitization: a critical view of some current views. *J. Sed. Petrol.* 57, 166–183.
Herman, J.S. and Lorah, M. M., 1988. Calcite precipitation rates in the field: Measurement and prediction for a travertine—depositing stream. *Geochim. Cosmochim. Acta* 52, 2347–2355.
Hoogendoorn, J.H., 1983. *Hydrochemistry of Eastern Netherlands* (in Dutch). DGV-TNO, Delft.
Inskeep, W.P. and Bloom, P.R., 1986. Kinetics of calcite precipitation in the presence of water-soluble organic ligands. *Soil Sci. Soc. Am. J.* 50, 1167–1172.
Jakobsen R. and Postma D., 1999. Redox zoning, rates of sulfate reduction and interactions with Fe-reduction and methanogenesis in a shallow sandy aquifer, Rømø, Denmark. *Geochim. Cosmochim. Acta* 63, 137–151.
James, A.N., 1981. Solution parameters of carbonate rocks. *Bull. Int. Ass. Eng. Geol.* 24, 19–25.
James, N.P. and Choquette, P.W., 1984. Limestones – the meteoric diagenetic environment. *Geosci. Can.* 11, 161–194.
Jurjovec, J., Ptacek, C.J. and Blowes, D.W., 2002. Acid neutralization mechanisms and metal release in mine tailings: A laboratory column experiment. *Geochim. Cosmochim. Acta* 66, 1511–1526.
Kalin, R.M., 2000. Radiocarbon dating of groundwater systems. In P.G. Cook and A.L. Herczeg (eds), *Environmental tracers in subsurface hydrology*, Chapter 4. Kluwer, Boston.
Keller, C.K. and Bacon, D.H. 1998. Soil respiration and georespiration distinguished by transport analyses of vadose CO_2, $^{13}CO_2$, and $^{14}CO_2$. *Global Biogeochem. Cycles* 12, 361–372.
Kessler, T.B. and Harvey, C.F., 2001. The global flux of carbon dioxide into groundwater. *Geophys. Res. Lett.* 28, 279–282.
Kern, D.M., 1960. The hydration of carbon dioxide. *J. Chem. Educ.* 14–23.
Kloppmann, W., Dever, L. and Edmunds, W.M., 1998. Residence time of Chalk groundwaters in the Paris Basin and the North German Basin: a geochemical approach. *Appl. Geochem.* 13, 593–606.
Labotka, T.C., Cole, D.R. and Riciputi, L.R., 2000. Diffusion of C and O in calcite at 100 MPa. *Am. Mineral.* 85, 488–494.
Langmuir, 1971. The geochemistry of some carbonate ground waters in central Pennsylvania, *Geochim. Cosmochim. Acta* 35, 1023–1045.
Laursen, S., 1991. On gaseous diffusion of CO_2 in the unsaturated zone. *J. Hydrol.* 122, 61–69.
Lebron, I. and Suarez, D.L., 1996. Calcite nucleation and precipitation kinetics as affected by dissolved organic matter at 25°C and pH > 7.5. *Geochim. Cosmochim. Acta* 60, 2765–2776.
Lee, R.W., 1997. Effects of carbon dioxide variations in the unsaturated zone on water chemistry in a glacial-outwash aquifer. *Appl. Geochem.* 12, 347–366.
López-Chicano, M., Bouamama, M., Vallejos, A. and Pulido-Bosch, A., 2001. Factors which determine the hydrogeochemical behaviour of karstic springs. A case study from the Betic Cordilleras, Spain. *Appl. Geochem.* 16, 1179–1192.
Martin, J.B. and Dean, R.W., 2001. Exchange of water between conduits and matrix in the Floridan aquifer. *Chem. Geol.* 179, 145–165.
Maoszewski, P. and Zuber, A., 1991. Influence of matrix diffusion and exchange reactions on radiocarbon ages in fissured carbonate aquifers. *Water Resour. Res.* 27, 1937–1945.
Morrow, D.W., 1982. The chemistry of dolomitization and dolomite precipitation. *Geosci. Can.* 9, 5–13.
Morse, J.W., 1983. The kinetics of calcium carbonate dissolution and precipitation. In R.J. Reeder (ed.), *Rev. Mineral.* 11, 227–264.
Morse, J.W. and Arvidson, R.S., 2002. The dissolution kinetics of major sedimentary carbonate minerals. *Earth Sci. Rev.* 58, 51–84.
Morse, J.W. and Mackenzie, F.T., 1990. *Geochemistry of sedimentary carbonates*. Elsevier, Amsterdam, 707 pp.
Mozeto, A.A., Fritz, P. and Reardon, E.J., 1984. Experimental observations on carbon isotope exchange in carbonate–water systems. *Geochim. Cosmochim. Acta* 48, 495–504.
Mucci, A., 1986. Growth kinetics and composition of magnesian calcite overgrowths precipitated from seawater: quantitative influence of orthophosphate ions. *Geochim. Cosmochim. Acta* 50, 217–233.
Mucci, A., 1987. Influence of temperature on the composition of magnesian calcite overgrowths precipitated from seawater. *Geochim. Cosmochim. Acta* 51, 1977–1984.
Neretnieks, I., 1981. Age dating of groundwater in fissured rock: influence of water volume in micropores. *Water Resour. Res.* 17, 421–422.

Nordstrom, D.K., Plummer, L.N., Langmuir, D., Busenberg, E., May, H.M., Jones, B.F and Parkhurst, D.L., 1990. Revised chemical equilibrium data for major water–mineral reactions and their limitations. In D.C Melchior and R.L Bassett (eds), *Chemical modeling of aqueous systems II*, ACS Symp Ser 416, 398–413.

Parkhurst, D.L., 1997. Geochemical mole-balancing with uncertain data. *Water Resour. Res.* 33, 1957–1970.

Pitman, J.I., 1978. Carbonate chemistry of groundwater from chalk, Givendale, East Yorkshire. *Geochim. Cosmochim. Acta* 42, 1885–1897.

Plummer, L. N. and Busenberg, E., 1987. Thermodynamics of aragonite strontianite solid-solutions: Results from stoichiometric solubility at 25° and 76°C. *Geochim. Cosmochim. Acta* 51, 1393–1411.

Plummer, L.N. and MacKenzie, F.T., 1974. Predicting mineral solubility from rate data: application to the dissolution of magnesian calcites. *Am. J. Sci.* 61–83.

Plummer, L.N., Vacher, H.L., Mackenzie, F.T., Bricker, O.P. and Land, L.S., 1976. Hydrogeochemistry of Bermuda: a case history of groundwater diagenesis of biocalcarenites. *Geol. Soc. Am. Bull.* 87, 1301–1316.

Plummer, L.N., Wigley, T.M.L. and Parkhurst, D.L., 1978. The kinetics of calcite dissolution in CO_2–water systems at 5° to 60°C and 0.0 to 1.0 atm CO_2. *Am. J. Sci.* 278, 179–216.

Plummer, L.N., Parkhurst, D.L. and Wigley, T.M.L., 1979. Critical review of the kinetics of calcite dissolution and precipitation. In E.A. Jenne (ed.), *Chemical modeling in aqueous systems*, ACS Symp. Ser. 93, 537–573.

Plummer, L.N., Busby, J.F., Lee, R.W. and Hanshaw, B.B., 1990. Geochemical modeling in the Madison aquifer in parts of Montana, Wyoming and South Dakota. *Water Resour. Res.* 26, 1981–2014.

Plummer, L.N., Prestemon, E.C. and Parkhurst, D.L., 1994. *An interactive code (NETPATH) for modeling net geochemical reactions along a flow path, 2.0.* U.S. Geol. Surv. Water Resour. Inv. Rep. 94-4169, 130 pp.

Plummer, L.N., Busenberg, E. and Riggs, A.C., 2000. In-situ growth of calcite at Devils Hole, Nevada: comparison of field and laboratory rates to a 500,000 year record of near-equilibrium calcite growth. *Aq. Geochem.* 6, 257–274.

Pokrovsky, O.S. and Schott, J., 2001. Kinetics and mechanism of dolomite dissolution in neutral to alkaline solutions revisited. *Am. J. Sci.* 301, 597–626.

Raich, J.W. and Potter, C.S., 1995. Global patterns of carbon dioxide emissions from soils. *Global Biogeochem. Cycles* 9, 23–36.

Reardon, E.J., Allison, G.B. and Fritz, P., 1979. Seasonal chemical and isotopic variations of soil CO_2 at Trout Creek, Ontario. *J. Hydrol.* 43, 355–371.

Reardon, E.J., Mozeto, A.A. and Fritz, P., 1980. Recharge in northern clime calcareous sandy soils: soil water chemical and carbon-14 evolution. *Geochim. Cosmochim. Acta* 44, 1723–1735.

Reddy, M.M., 1977. Crystallization of calcium carbonate in the presence of trace concentrations of phosphorous containing anions. *J. Cryst. Growth* 41, 287–295.

Reddy, M.M., Plummer, L.N. and Busenberg, E., 1981. Crystal growth of calcite from calcium bicarbonate solutions at constant P_{CO_2} and 25°C: a test of a calcite dissolution model. *Geochim. Cosmochim. Acta* 45, 1281–1289.

Reeder, R.J., 1983. Crystal chemistry of the rhombohedral carbonates. *Rev. Mineral.* 11, 1–47.

Rickard, D. and Sjöberg, E.L., 1983. Mixed kinetic control of calcite dissolution rates. *Am. J. Sci.* 283, p. 815–830.

Russell, E.W., 1973. *Soil conditions and plant growth*. Longman, London, 849 pp.

Sacks, L.A., Herman, J.S. and Kauffman, S.J., 1995. Controls on high sulfate concentrations in the Upper Floridan aquifer in southwest Florida. *Water Resour. Res.* 31, 2541–2551.

Sanford, W.E., 1997. Correcting for diffusion in Carbon-14 dating of ground water. *Ground Water* 35, 357–361.

Sanford, W. and Konikow, L.F., 1989. Simulation of calcite dissolution and porosity changes in saltwater mixing zones in coastal aquifers. *Water Resour. Res.* 25, 655–667.

Saunders, J.A. and Toran, L.E., 1994. Evidence for dedolomitization and mixing in Paleozoic carbonates near Oak Ridge, Tennessee. *Ground Water* 32, 207–214.

Sherman, L.A. and Barak, P., 2000. Solubility and dissolution kinetics of dolomite in Ca-Mg-HCO_3/CO_3 solutions at 25 degrees C and 0.1 MPa carbon dioxide. *Soil Sci. Soc. Am. J.* 64, 1959–1968.

Schulz, H.D., 1988. Labormessung der Sättigungslänge als Mass für die Lösungskinetik von Karbonaten im Grundwasser. *Geochim. Cosmochim. Acta* 52, 2651–2657.

Shuster, E.T. and White, W.B., 1972. Source areas and climatic effects in carbonate groundwaters determined by saturation indices and carbon dioxide pressures. *Water Resour. Res.* 8, 1067–1073.

Sjöberg, E.L., 1978. Kinetics and mechanism of calcite dissolution in aqueous solutions at low temperatures. *Stockholm Contrib. Geol.* XXXII, 1, 1–92.

Smedley, P.L. and Edmunds, W.M., 2002. Redox patterns and trace-element behavior in the East Midlands Triassic sandstone aquifer, U.K. *Ground Water* 40, 44–58.

Stoessell, R.K., Moore, Y.H. and Coke, J.G., 1993. The occurrence and effect of sulfate reduction and sulfide oxidation on coastal limestone dissolution in Yucatan cenotes. *Ground Water* 31, 566–575.

Stumm, W. and Morgan, J.J., 1996. *Aquatic chemistry*, 3rd ed. Wiley and Sons, New York, 1022 pp.

Suchet, P.A. and Probst, J.L. 1995. A global model for present-day atmospheric/soil CO_2 consumption by chemical erosion of continental rocks. *Tellus* 47B, 273–280.

Sudicky, E. and Frind, E.O., 1981. Carbon 14 dating of groundwater in confined aquifers: implications of aquitard diffusion. *Water Resour. Res.* 17, 1060–1064.

Terjesen, S.G., Erga, O., Thorsen, G. and Ve, A., 1961. Phase boundary processes as rate determining steps in reactions between solids and liquids. *Chem. Eng. Sci.* 74, 277–288.

Thorstenson, D.C. and Plummer, L.N., 1977. Equilibrium criteria for two-component solids reacting with fixed composition in an aqueous phase – example: the magnesian calcites. *Am. J. Sci.* 277, 1203–1223.

Turner, J.V., 1982. Kinetic fractionation of carbon-13 during calcium carbonate precipitation. *Geochim. Cosmochim. Acta* 46, 1183–1191.

Van der Kemp, W., Appelo, C.A.J. and Walraevens, K., 2000. Inverse chemical modeling and radiocarbon dating of palaeogroundwaters: the Tertiary Ledo-Paniselian aquifer in Flanders, Belgium. *Water Resour. Res.* 36, 1277–1287.

Van Lith, Y., Vasconcelos, C., Warthmann, R., Martins, J.C.F. and McKenzie, J.A., 2002. Bacterial sulfate reduction and salinity: two controls on dolomite precipitation in Lagoa Vermelha and Brejo do Espinho (Brazil). *Hydrobiol.* 485, 35–49.

Vogel, J.C. 1993. Variability of carbon isotope fractionation during photosynthesis. In J.R. Ehleringer et al. (eds), *Stable isotopes and plant carbon-water relations*, 29–46, Academic Press, London.

Walter, L.M. and Hanor, J.S., 1979. Effect of orthophosphate on the dissolution kinetics of biogenic magnesian calcites. *Geochim. Cosmochim. Acta* 43, 1377–1385.

Welch, M.J., Lifton, J.F. and Seck, J.A., 1969. Tracer studies with radioactive Oxygen-15 exchange between carbon dioxide and water. *J. Phys. Chem.* 73, 3351–3356.

Weyl, P.K., 1958. Solution kinetics of calcite. *J. Geol.* 66, 163–176.

Whitaker, F. and Smart, P.L. 1997. Groundwater circulation and geochemistry of a karstified bank-marginal fracture system, South Andros Island, Bahamas. *J. Hydrol.* 197, 293–315.

Wigley, T.M.L. and Plummer, L.N., 1976. Mixing of carbonate waters. *Geochim. Cosmochim. Acta* 40, 989–995.

Yoshimura, K., Nakao, S., Noto, M., Inokura, Y., Urata, K., Chen, M. and Lin, P.-W., 2001. Geochemical and stable isotope studies on natural water in the Taroko Gorge karst area, Taiwan-chemical weathering of carbonate rocks by deep source CO_2 and sulfuric acid. *Chem. Geol.* 177, 415–430.

Zaihua, L., Svensson, U., Dreybrodt, W., Daoxian, Y. and Buhmann, D., 1995. Hydrodynamic control of inorganic calcite precipitation in Huanglong Ravine, China: field measurements and theoretical prediction of deposition rates. *Geochim. Cosmochim. Acta* 59, 3087–3097.

Zötl, J.G., 1974. *Karsthydrogeologie*. Springer, Berlin, 291 pp.

6

Ion Exchange

Soils and aquifers contain materials like clay minerals, organic matter and metal oxy-hydroxides which can sorb chemicals. The various processes indicated by the general term *sorption* are illustrated in Figure 6.1. The term *adsorption* refers to the adherence of a chemical to the surface of the solid, *absorption* suggests that the chemical is taken up into the solid, and *exchange* involves replacement of one chemical for another one at the solid surface.

It is sometimes difficult to separate sorption and other types of reactions involving solids such as precipitation and dissolution (Sposito, 1984). A major difference is that sorption depends on the presence of a preexisting solid surface, in contrast to, for example, precipitation. Sorption and ion exchange have become important topics for hydrologists since these processes regulate the transport of pollutant chemicals in aquifers and soils. In this chapter we focus on ion exchange while the following chapter will treat sorption processes for heavy metals.

Under steady-state chemical conditions, the composition of a cation exchanger will be in equilibrium with the resident groundwater. When the water composition changes as a result of pollution or acidification, or due to a moving salt/fresh water interface, the cation exchanger readjusts its composition to the new groundwater concentrations. In this way the exchanger acts as a *temporary buffer* which may completely alter the concentrations in water through a process known as *ion-chromatography*.

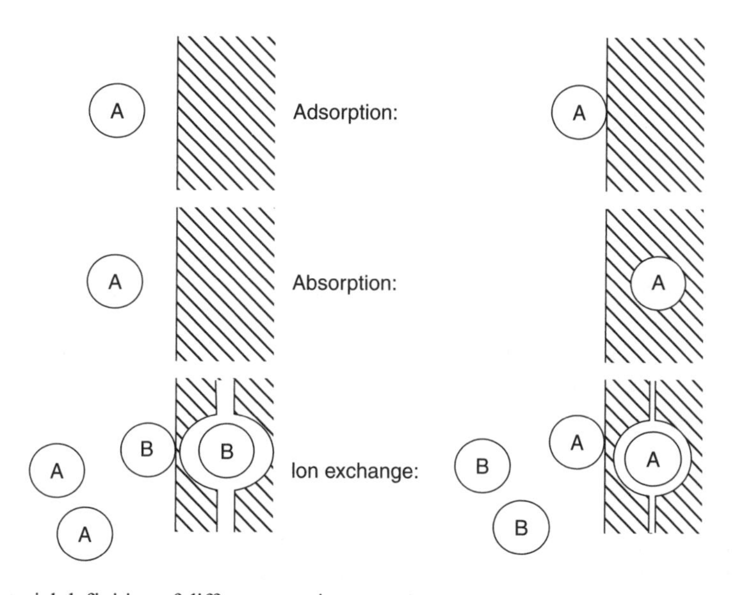

Figure 6.1. Pictorial definition of different sorption processes.

A hydrologist who tries to decipher hydrogeological conditions may use the hydrochemical patterns which are the result of cation exchange. As aquifers become more exploited and polluted, the water quality along flowlines will change, but due to cation exchange, the water composition may still reflect aspects of former water quality. This chapter demonstrates the principles of ion exchange in aquifers and soils. We start with examples of cation exchange as indicators of salinization or desalinization of aquifers. Then we treat exchanger materials, exchange equations, and give examples of salt water intrusion and chromatographic patterns in aquifers. We also show how exchange and chromatography can be modeled with PHREEQC.

6.1. CATION EXCHANGE AT THE SALT/FRESH WATER INTERFACE

The composition of fresh groundwater in coastal areas is often dominated by Ca^{2+} and HCO_3^- ions which result from calcite dissolution. The cation exchanger is then also dominated by adsorbed Ca^{2+}. In seawater, Na^+ and Cl^- are the dominant ions, and sediment in contact with seawater will have mostly Na^+ on the exchanger. When seawater intrudes into a coastal fresh water aquifer, an exchange of cations takes place:

$$Na^+ + ½Ca\text{-}X_2 \rightarrow Na\text{-}X + ½Ca^{2+} \tag{6.1}$$

where X indicates the soil exchanger. In the reaction Na^+ is taken up by the exchanger, while Ca^{2+} is released. Since the dominant anion Cl^- remains the same, the water quality changes from a NaCl to a $CaCl_2$-type of water. The reverse process takes place during freshening, i.e. when fresh water, with Ca^{2+} and HCO_3^- ions, flushes a salt water aquifer:

$$½Ca^{2+} + Na\text{-}X \rightarrow ½Ca\text{-}X_2 + Na^+ \tag{6.2}$$

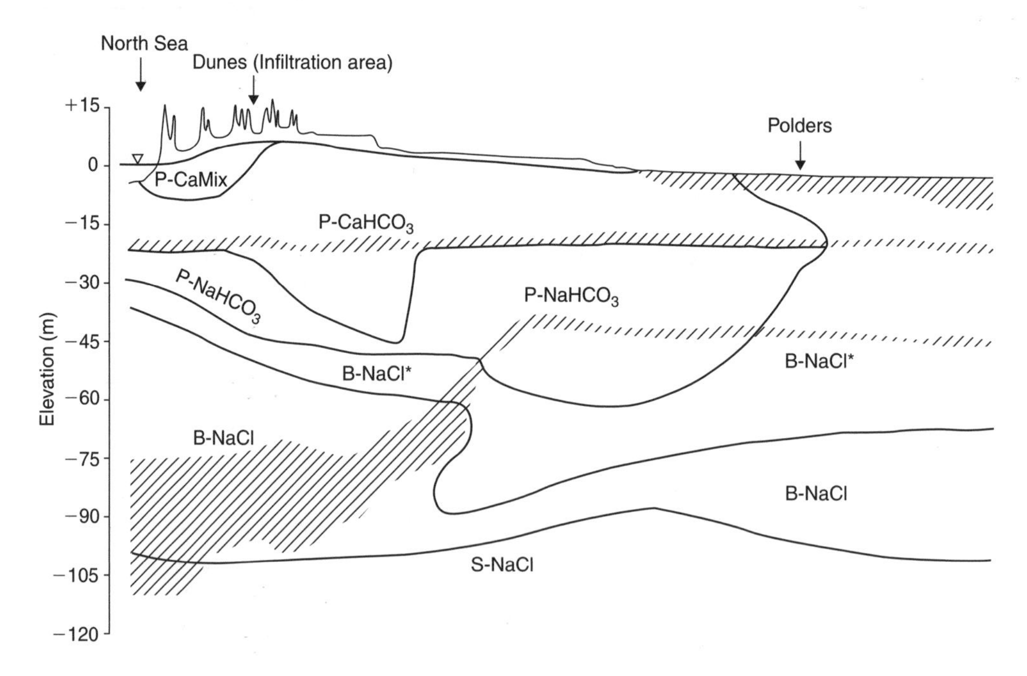

Figure 6.2. Section through the coastal dunes near Castricum (Netherlands), showing refreshening of a brackish water aquifer. The prefixes P, B and S before the water types indicate fresh, brackish and salt water. Clay layers are indicated by hatched lines (Stuyfzand, 1985).

The sediment now adsorbs Ca^{2+} while Na^+ is released and a $NaHCO_3$-type water results. In this way the water composition can indicate whether upconing of seawater occurs, or conversely whether fresh water is flushing salt water from the aquifer.

An example is given in Figure 6.2, showing a section through the Dutch coastal dunes (Stuyfzand, 1985). Groundwater has been pumped from the area for drinking water purposes for a long time. An alarming rate of upconing of salt water was observed several decades ago, which threatened to salinize all the pumping wells. To relieve the problem, fresh water from the river Rhine was allowed to infiltrate into the dunes and then recovered from shallow wells. Deliberately, more water was infiltrated than abstracted in order to flush the fresh-salt water boundary backwards. This strategy has been successful, as indicated by the 'freshened' $NaHCO_3$-type water now present in the middle part of the aquifer (Figure 6.2). The quality pattern shown in Figure 6.2 was constructed based on a detailed network of sampling wells and allows for an interpretation of flowlines in the aquifer. When Na^+ has been flushed from the exchanger, the $Ca(HCO_3)_2$-type fresh water returns. This water type will appear sooner when flushing is more rapid, i.e. along the more rapid flowlines, or where the cation exchange capacity is smaller.

An example of salt water intrusion in the Nile delta, Egypt, is shown in Figure 6.3. The Stiff diagrams (see also Chapter 1) summarize the groundwater composition and indicate the intrusion of salt water in boreholes 121 and 129, as demonstrated by the relative excess of Ca^{2+} compared to seawater (shown in the top of Figure 6.3). Fresh groundwater in the Nile delta originates almost exclusively from Nile water, which has a $Ca(HCO_3)_2$-type composition (shown in Figure 6.3 west of Cairo).

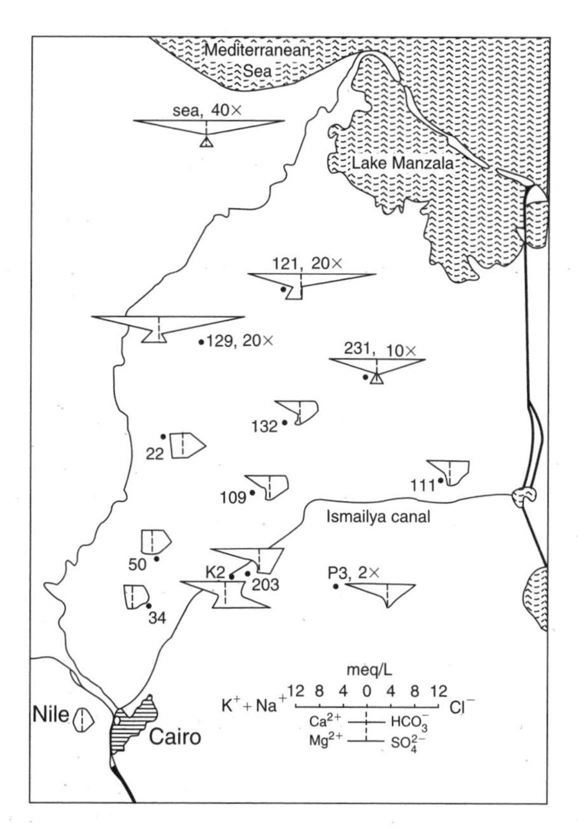

Figure 6.3. Stiff diagrams of groundwater in the Nile delta, Egypt, indicating salt water upconing in borehole 121 and 129, and Ca^{2+}/Na^+-exchange near the Ismailya canal.

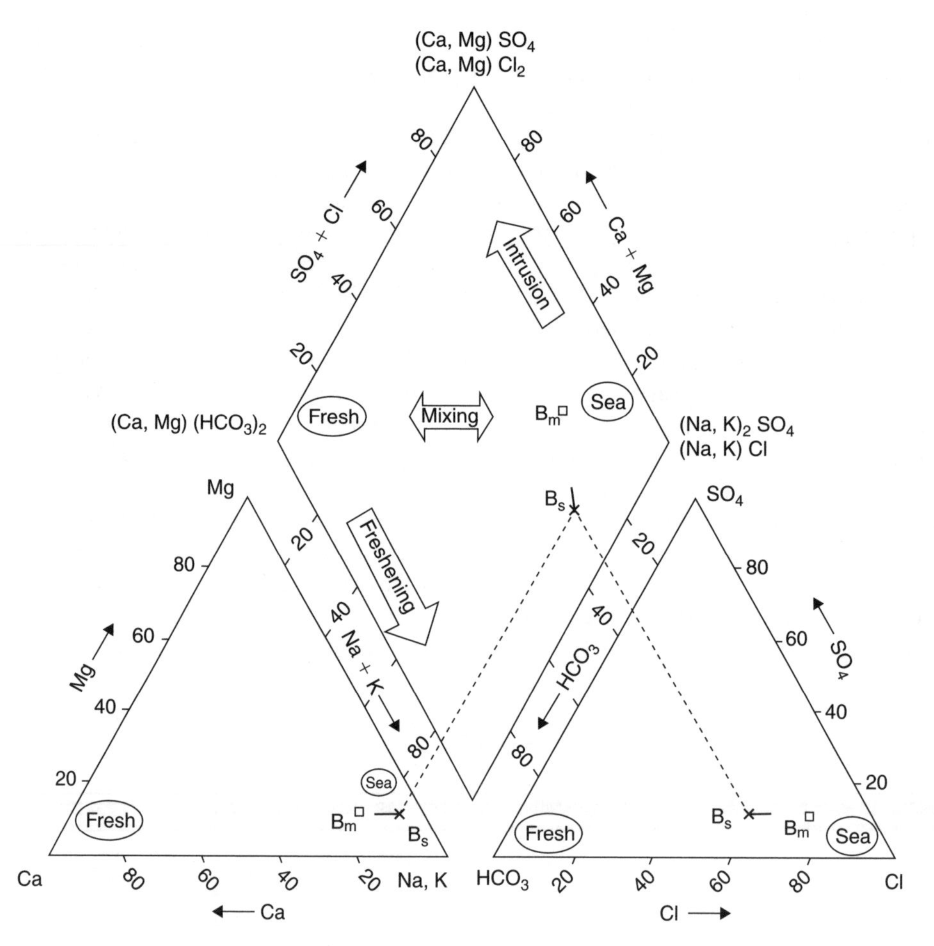

Figure 6.4. Piper plot showing average compositions of fresh water and seawater, and $NaHCO_3$-sample (B_s) from Table 6.2. The tail at B_s points to the calculated composition of a conservative mixture B_m.

Groundwater along the Ismailya canal, which runs from Caïro to the Red Sea, shows increasing Cl^- concentrations as a result of evapotranspiration of Nile water used for irrigation. The Na^+ concentration increases even more than Cl^-, while Ca^{2+} relatively decreases, thus suggesting cation-exchange of Ca^{2+} for Na^+ in the soil. The sodium originates from Na-containing loess that is blown in from the desert and settles in the wetter area of the delta, as was observed in the Negev desert (Nativ et al., 1983).

The effects of cation exchange are particularly evident when analyses are plotted in a Piper diagram (see also Chapter 1), as shown in Figure 6.4. The average compositions of fresh water and seawater are shown and a straight line between the two indicates water compositions due to conservative mixing. If groundwater samples in a coastal area show a surplus of Ca^{2+}, compared to the conservative mixture, then this can imply seawater intrusion, while a surplus of Na^+ may indicate freshening of the aquifer. Typical groundwaters showing these trends are given in Table 6.1.

The chemical reactions occurring during fresh/salt water displacement can be deduced more precisely by calculating the expected composition based on conservative mixing of salt water and fresh water, and then comparing the result with the measured concentrations in the water sample.

Table 6.1. Groundwaters from the Netherlands showing "only-mixing" water composition, and a composition influenced by cation exchange ('$CaCl_2$' and '$NaHCO_3$'-types). Concentrations in mmol/L.

	Only mixing		$CaCl_2$ water		$NaHCO_3$ water	
pH	7.5	7.2	6.91	6.6	8.7	8.3
Na^+	24.53	54.33	341	124	40	2.0
K^+	0.82	1.41	2.8	2.4		
Mg^{2+}	2.9	7.1	27.9	30.7	2.8	1.1
Ca^{2+}	3.0	7.9	39.6	47.2	0.7	0.65
Cl^-	27.2	70.8	440	271	25.6	1.4
HCO_3^-	9.2	15.3	7.0	3.8	14.4	4.0
SO_4^{2-}	0.07	0	18.8	4.7	2.7	0.2

The concentration of an ion *i*, by conservative mixing of seawater and fresh water is:

$$m_{i,\mathrm{mix}} = f_{\mathrm{sea}} \cdot m_{i,\mathrm{sea}} + (1 - f_{\mathrm{sea}}) \cdot m_{i,\mathrm{fresh}} \qquad (6.3)$$

where m_i is the concentration of *i* (mmol/L), f_{sea} the fraction of seawater in the mixed water, and subscripts $_{\mathrm{mix}}$, $_{\mathrm{sea}}$, and $_{\mathrm{fresh}}$ indicate a conservative mixture, seawater and fresh water, respectively. Any change in the concentration $m_{i,\,\mathrm{react}}$ due to reactions then becomes simply:

$$m_{i,\mathrm{react}} = m_{i,\mathrm{sample}} - m_{i,\mathrm{mix}} \qquad (6.4)$$

where $m_{i,\mathrm{sample}}$ is the measured concentration in the sample. The fraction of seawater, f_{sea}, is normally calculated using the Cl^- concentration of the sample. Chloride is assumed to be a conservative parameter, just as it was used in Chapter 2 to find the sea salt contribution in rainwater. The Cl^- based fraction of seawater is:

$$f_{\mathrm{sea}} = \frac{m_{\mathrm{Cl^-,sample}} - m_{\mathrm{Cl^-,fresh}}}{m_{\mathrm{Cl^-,sea}} - m_{\mathrm{Cl^-,fresh}}} \qquad (6.5)$$

If seawater is the only source of Cl^-, then $m_{\mathrm{Cl^-,fresh}} = 0$, which simplifies Equation (6.5) to:

$$f_{\mathrm{sea}} = m_{\mathrm{Cl^-,sample}} / 566 \qquad (6.6)$$

where the Cl^- concentration is expressed in mmol/L, and 566 mmol/L is the Cl^- concentration of 35‰ (grams of salt per kilogram) seawater. While the salinity of seawater may vary, the ratios among the ions remain constant for all major ions and can be used to calculate the composition of conservative mixtures.

Table 6.2 illustrates the results of such a calculation for some analyses in Table 6.1. The relative increases of Na^+ in $NaHCO_3$-water, and of Ca^{2+} in $CaCl_2$-water are evident. Another conspicuous change is the increase of HCO_3^- in the $NaHCO_3$-type water. This relative HCO_3^- increase is commonly found, and may easily amount to 10 mmol/L or more. When Ca^{2+} exchanges for Na^+ (Reaction 6.2) the water becomes undersaturated for calcite and dissolution results (Back, 1966; Chapelle, 1983). Also, the pH of $NaHCO_3$ water often becomes high, above 8, due to calcite dissolution. $CaCl_2$ type of water on the other hand, often has a low pH, less than 7. This may be the result of calcite *precipitation*, driven by the increase of Ca^{2+} caused by cation exchange (Reaction 6.1). However, other reactions may also take place. In Table 6.2 the $CaCl_2$ water shows a deficit in sulfate.

Table 6.2. Recalculated water analyses of Table 6.1, showing actual reaction amounts. Values in mmol/L.

	Seawater	$CaCl_2$ water			$NaHCO_3$ water			Fresh water
		Sample	Mix	React	Sample	Mix	React	
Na^+	485.0	341	377	−36	40	22	18	0
K^+	10.6	2.8	8.2	−5.4				0
Mg^{2+}	55.1	27.9	42.8	−14.9	2.8	2.5	0.3	0
Ca^{2+}	10.7	39.6	9.0	30.6	0.7	3.3	−2.6	3.0
Cl^-	566	440	440	–	25.6	25.6	–	0
HCO_3^-	2.4	7.0	3.2	3.8	14.4	5.8	8.6	6.0
SO_4^{2-}	29.3	18.8	22.8	−4.0	2.7	1.3	1.4	0
% seawater	100	78			5			0
Symbol in Figure 6.4:					B_s	B_m		

Sample: given water composition from Table 6.1. Mix: calculated water composition based on conservative mixing. React: sample – mix, showing effects of cation exchange and other reactions.

Seawater has a high content of sulfate and when it intrudes an anoxic coastal aquifer this may result in sulfate reduction (Andersen, 2001):

$$SO_4^{2-} + 2CH_2O \rightarrow H_2S + 2HCO_3^- \tag{6.7}$$

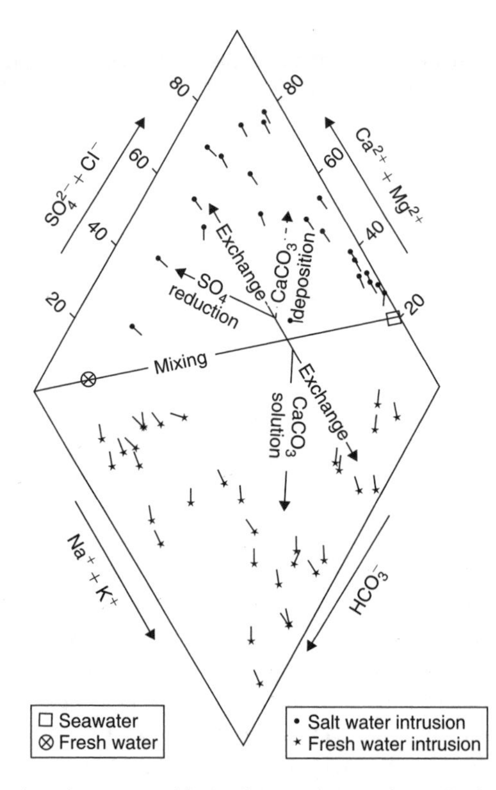

Figure 6.5. Piper plot showing the composition of groundwater from Zeeland and Western Brabant in the Netherlands, displaying cation-exchange. Each symbol has a tail that points towards the position on the mixing line based on Cl^- concentration of the groundwater sample.

Such additional reactions can be visualized in a Piper-diagram by adding a tail going from the sample point towards the corresponding position on the conservative mixing line (download the program from www.xs4all.nl/~appt).

The $NaHCO_3$-water of Table 6.2 has been entered in Figure 6.4, as point B_s (the actual sample) and point B_m (the calculated mixture). When only cation exchange takes place, the tail is parallel to the cation axis of the Piper diamond; when other reactions occur this is not the case (Figure 6.5). The dissolution of $CaCO_3$ brings the mixture towards the $Ca^{2+} + HCO_3^-$ corner, the reduction of SO_4^{2-} gives a shift parallel to the anion-axis, deposition of $CaCO_3$ leads towards the Na^+ and $(SO_4^{2-} + Cl^-)$ corner. Figure 6.5 shows groundwater analyses from Zeeland and Western Brabant, the Netherlands, where seawater inundations have been common in the past centuries. During freshening of the aquifer (Figure 6.5), the relative increase of HCO_3^- indicates calcite dissolution, while during salt water intrusion not much $CaCO_3$ precipitation appears to take place. It is possible that $CaCO_3$-deposition is masked in the Piper diagram by SO_4^{2-} reduction, since Reaction (6.7) lowers the percentage of $(SO_4^{2-} + Cl^-)$ and produces HCO_3^-.

6.2 ADSORBENTS IN SOILS AND AQUIFERS

While in principle all solid surfaces in soils and aquifers can act as adsorbers, solid phases with a large specific surface area will adsorb most, and the adsorption capacity therefore depends on the grain size. Solids with a large specific surface area reside in the clay fraction (<2 μm), but coarser grains in a sediment are often coated with organic matter and iron oxyhydroxides. The adsorption capacity is therefore linked to the clay content (fraction <2 μm), clay minerals, organic matter (% C), and oxide or hydroxide content. The current convention is to express the *CEC*, or *Cation Exchange Capacity* of a soil in meq/kg, replacing the unit meq/100 g. In the US soil science literature, $mmol_c$ is often used to indicate "mmol *charge*" for meq.

An empirical formula which relates the *CEC* to the percentages of clay (<2 μm) and organic carbon at near neutral pH is (e.g. Breeuwsma et al., 1986):

$$CEC\ (\text{meq/kg}) = 7 \cdot (\%\ \text{clay}) + 35 \cdot (\%\ \text{C}) \tag{6.8}$$

Table 6.3 gives the *CEC* of common soil constituents. Clay minerals show a wide range in *CEC* depending upon mineral structure, structural substitutions and the specific surface of the mineral accessible to water. For iron oxides and organic matter, the *CEC* is also a function of pH since the surface oxygens behave as amphoteric acids. Oxyhydroxides may even acquire a positive charge below pH 8 (discussed in Chapter 7).

Table 6.3. Cation exchange capacities of common soil and sediment materials.

	CEC, meq/kg
Kaolinite	30–150
Halloysite	50–100
Montmorillonite	800–1200
Vermiculite	1000–2000
Glauconite	50–400
Illite	200–500
Chlorite	100–400
Allophane	up to 1000
Goethite and hematite:	up to 1000 (pH > 8.3, discussed in Chapter 7)
Organic matter (C)	1500–4000 (at pH = 8, discussed in Chapter 7)
or, accounting for pH-dependence:	510 × pH – 590 = *CEC* per kg organic carbon (Scheffer and Schachtschabel, 2002).

EXAMPLE 6.1. *Recalculate CEC (meq/kg soil) to concentration (meq/L pore water)*
The cation exchange capacity (*CEC*) of a sediment is expressed in units of meq per kg of dry soil. Express a *CEC* of 10 meq/kg of a sediment with porosity $\varepsilon = 0.2$, in meq/L pore water. The sediment consists mainly of quartz grains with specific weight of 2.65 g/cm^3.

ANSWER:
Since quartz has a specific weight of 2.65 g/cm^3, 1 kg of dry sediment contains $1000/2.65 = 377$ mL "grains". The amount of pore water in 1 kg sediment can be calculated and *CEC* expressed in units meq/L pore water (Example 3.7): Total sediment volume is $377/(1 - \varepsilon) = 472$ mL; pore water $= 472 \times \varepsilon = 94$ mL. $CEC = 10$ meq/kg $= 10/94$ meq/mL $= 106$ meq/L pore water.

$$\text{bulk density } \rho_b = 1000 / 472 = 2.12 \text{ g/cm}^3.$$

Generally $$CEC \times \rho_b / \varepsilon = \text{meq/kg} \times \text{kg/L} = \text{meq/L}$$

QUESTION:
Express $CEC = 15$ meq/kg in terms of pore water concentration. $\varepsilon = 0.3$, $\rho_b = 1.86$ g/cm^3.
ANSWER: 93 meq/L

Example 6.1 shows how a *CEC* of 10 meq/kg, a reasonable value for a sandy aquifer, corresponds to 106 meq/L of pore water. Fresh groundwater normally contains less than 10 meq/L cations, ten times less than the amount of cations located on the exchanger. The example serves to illustrate that the amount of adsorbed cations is often very large in comparison to the amount of cations in solution.

6.2.1 *Clay minerals*

The basic coordination units for clay minerals are tetrahedra and octahedra with oxygens forming the corners, and a cation residing in the centre (Figure 6.6). In tetrahedral coordination the central cation is surrounded by four oxygens, in octahedral coordination the cation is surrounded by six oxygens. The oxygens in the coordination units are envisaged as spheres with radii of about 1.4 Å that should just touch each other. Tetrahedra therefore contain the smaller metal-ions such as Si^{4+} or Al^{3+}, and octahedra the larger metal-ions such as Al^{3+}, Mg^{2+}, Fe^{2+}, Mn^{2+}. (Still larger metal-ions such as Ca^{2+}, Na^+, K^+, have even larger coordination polyhedra). Within the clay minerals, tetrahedra and octahedra form layers which share oxygens. The way of stacking of these layers determines the type of clay mineral. In *kaolinite* the layers of tetrahedra (⊤) and octahedra (○) show the repetition: ⊢○ ⊢○ ⊢○ ⊢○. The chemical formula of kaolinite is:

$$[Al_2]^{vi}[Si_2]^{iv}O_5(OH)_4 \tag{6.9}$$

where vi and iv indicate six- and four-fold coordination by oxygen. Stacking of a tetrahedral layer of approximately 3 Å with an octahedral layer of 4 Å gives a repetition of the ⊢○ structure every 7 Å. This is the characteristic *c*-axis spacing (or 001 reflection) observed in X-ray analysis of kaolinite. The outer oxygens of the octahedra of one layer share protons with the tetrahedra from the next layer.

In *mica* or soil clay minerals such as *smectite* or *montmorillonite* the octahedra are sandwiched between two layers of tetrahedra, and the structure repeats as: ⊢○⊣ ⊢○⊣ ⊢○⊣. The ⊢○⊣ structure is about 10 Å thick, and the cohesion of the ⊢○⊣ layers is mediated by *interlayer cations*. Another group of clay minerals, of which *chlorite* is an example, has its interlayers filled with an octahedral layer. These structures can be indicated as repetitions of ⊢○⊣ ○ ⊢○⊣ ○ (of 14 Å each). All these minerals are found in the finest fraction of soils and aquifers.

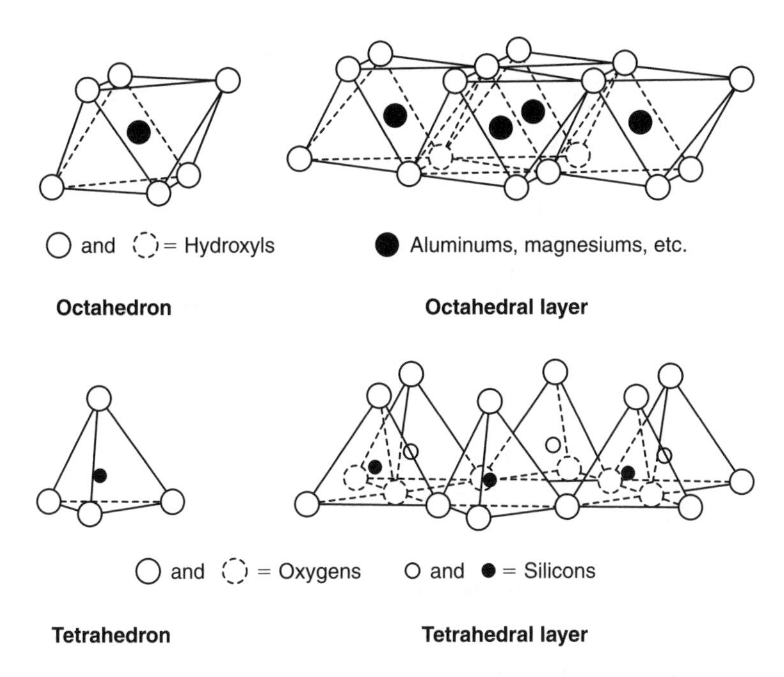

Figure 6.6. A perspective drawing of tetrahedra and octahedra as coordinating units in minerals (Grim, 1968).

Clay minerals can have a deficit of positive charge which arises from substitutions of cations in the structure of the crystal. The substitution of Si^{4+} for Al^{3+} reduces the positive charge by one in an otherwise not much affected structure. The possible substitutions can be understood most easily with Pauling's rules for the buildup of crystal structures (Pauling, 1960). The first rule entails that in filling the *sites* of cations in the framework of oxygens, the length of the cation-oxygen bond is the most important. This length is even more important than the charge on the cation. Charge imbalance can, if needed, be compensated in other units some distance away from the point of imbalance (although Pauling's second rule states that the most stable structure is obtained when charges are balanced nearby).

A charge imbalance is mostly limited to the mica-type clay minerals. The charge imbalance can be calculated from the structural formula, starting with the mica *muscovite*:

$$K^{xii}[Si_3Al]^{iv}[Al_2]^{vi}O_{10}(OH)_2 \tag{6.10}$$

The 10 oxygens and 2 hydroxyls give 22 negative charges; four tetrahedral metal sites, and two octahedral metal sites must be filled in this formula. When the charge of the cations adds up to 21, one single charge is left for the interlayer cation, which is K^+ in muscovite. This K^+ is actually in twelve-coordination with oxygens from two adjacent tetrahedral layers. When Si^{4+} takes a larger share in the tetrahedral cation sites, the interlayer charge is diminished. An example is

$$K^{xii}_{0.7}[Si_{3.3}Al_{0.7}]^{iv}[Al_2]^{vi}O_{10}(OH)_2 \tag{6.11}$$

which is the typical formula for the clay mineral *illite*. A further decrease of the interlayer charge is possible, but reduces the cohesion of the tetrahedral layers around the interlayer cation. The interlayer cation may then become hydrated, and the ⊢O⊣ layers separate as more and more layers of water are incorporated in the structure.

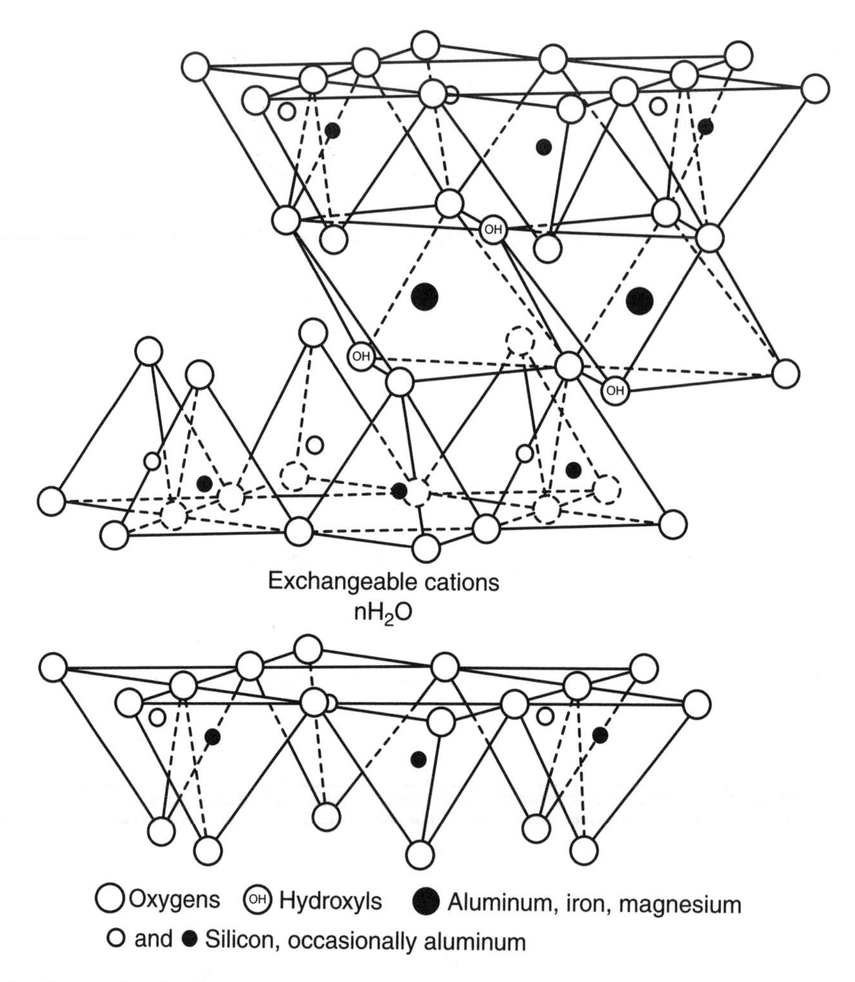

Figure 6.7. Perspective drawing of the structure of the clay mineral smectite. (Grim, 1968).

Clay minerals with a mica structure, which expand to 18 Å with Mg-ions as interlayer cations, are called *smectites* (Figure 6.7). The hydrated cations in the interlayer space can move freely into solution to be exchanged for other ions. Smectites (or *montmorillonites*) typically have an interlayer charge that ranges from 0.6 to 0.3.

Substitutions also occur in the octahedral layer, for example in *vermiculite*:

$$K^{xii}_{0.9}[Si_{3.8}Al_{0.2}]^{iv}[Al_{1.3}Mg_{0.7}]^{vi}O_{10}(OH)_2 \tag{6.12}$$

The low tetrahedral charge of 0.2 is here surpassed by an octahedral charge of 0.7, and the full interlayer charge is 0.9. Vermiculites are able to swell to 14 Å with strongly hydrated cations such as Mg^{2+}, despite the large interlayer charge, because the charge imbalance is predominantly located in the octahedral layer, i.e. at a relatively large distance from the oxygens in the basal tetrahedral plane to which the interlayer cations are bonded. However, most soil clay minerals with such a high charge will tend to collapse when ions of suitable size enter the interlayer space, for example K^+, NH_4^+ or ions of similar size and low hydration number. A collapsed structure can only exchange through solid state diffusion, which is 3–4 orders of magnitude slower than diffusion in solution.

EXAMPLE 6.2. *Structural charge of smectite*
A smectite has the chemical formula $Na_{0.6}[Si_{7.6}Al_{0.4}]^{iv}[Al_{3.8}Mg_{0.2}]^{vi}O_{20}(OH)_4$ for one unit cell. Calculate the amount of exchangeable cations per gram of mineral, and calculate the charge density on the basal plane in μeq/m². The size of the unit cell is $a \times b \times c = 5.2 \times 9 \times 10$ Å³, and $a \times b = 5.2 \times 9 = 46.8$ Å² is the size of the basal plane where the interlayer cations reside.

ANSWER:
The gram-formula weight of the smectite is 733.5 g/mol. One kg contains 1.36 moles of the unit cell formula, hence the exchange capacity is $0.6 \times 1.36 \times 1000 = 818$ meq/kg. The charge of 0.6 is divided over the two basal planes of a completely separated ⊢O⊣ layer, or $0.3\,q_e$ per 46.8 Å². This is $6.4 \times 10^{17}\,q_e/\text{m}^2$ (q_e is the elementary charge). Multiply with the elementary charge 1.6×10^{-19} C, and obtain 0.1 C/m²; or divide by Avogadro's number $N_a = 6 \times 10^{23}$/mole, and obtain the charge density as 1.07 μeq/m².

6.3 EXCHANGE EQUATIONS

For ion exchange of Na^+ for K^+ we write the reaction:

$$Na^+ + \text{K-X} \leftrightarrow \text{Na-X} + K^+ \tag{6.13}$$

and the distribution of species is given by the law of mass action:

$$K_{\text{Na}\backslash\text{K}} = \frac{[\text{Na-X}]\,[K^+]}{[\text{K-X}]\,[Na^+]} \tag{6.14}$$

The ions in the subscript below the equilibrium constant are written in the order in which they appear as *solute* ions in the reaction. (Note that the convention for isotope exchange equations is to start with the product of the reaction, cf. Chapter 2).

Square brackets in the equations denote activities. The Debye-Hückel theory offers a straightforward model to relate concentrations and activities in water (Chapter 4). However, for adsorbed cations there is no unifying theory to calculate activity coefficients and different conventions are in use. Since the convention has a bearing on the results of exchange calculations, we must discuss this matter.

The standard state, i.e. where the activity of the exchangeable ion is equal to 1, is in all cases an exchanger which is totally occupied by the same cations (Table 4.1). The activity of each exchangeable ion is expressed as a fraction of the total, either as molar fraction (as for solutes, Table 4.1), or as equivalent fraction. Furthermore, the total number can be based on the number of exchange sites, or on the number of exchangeable cations.

For ion I^{i+} the equivalent fraction β_I is calculated as:

$$\beta_I = \frac{\text{meq } I\text{-X}_i \text{ per kg sediment}}{CEC} = \frac{\text{meq}_{I\text{-X}_i}}{\sum_{I,J,K,\ldots} \text{meq}_{I\text{-X}_i}} \tag{6.15}$$

where $I, J, K, \ldots$ are the exchangeable cations, with charges i, j, k.

A molar fraction β_I^M is likewise obtained from:

$$\beta_I^M = \frac{\text{mmol } I\text{-X}_i \text{ per kg sediment}}{TEC} = \frac{(\text{meq}_{I\text{-X}_i})/i}{\sum_{I,J,K,\ldots} (\text{meq}_{I\text{-X}_i})/i} \tag{6.16}$$

where *TEC* denotes total exchangeable cations, in mmol/kg sediment. The use of fractions always gives $\Sigma\beta = 1$. When the activity is calculated with respect to the number of exchangeable cations this is indicated as $[I\text{-X}_i]$, and with respect to the number of exchange sites as $[I_{1/i}\text{-X}]$.

For homovalent exchange it makes no difference what convention is used, but for heterovalent exchange the effect is quite notable. For the exchange of Na^+ for Ca^{2+}, using the number of exchangeable cations convention:

$$Na^+ + ½Ca\text{-}X_2 \leftrightarrow Na\text{-}X + ½Ca^{2+} \tag{6.17}$$

with

$$K_{Na\backslash Ca} = \frac{[Na\text{-}X][Ca^{2+}]^{0.5}}{[Ca\text{-}X_2]^{0.5}[Na^+]} = \frac{\beta_{Na}[Ca^{2+}]^{0.5}}{\beta_{Ca}^{0.5}[Na^+]} \tag{6.18}$$

When the equivalent fraction of the exchangeable cations is used in Equation (6.18) it conforms to the *Gaines-Thomas* convention, after Gaines and Thomas (1953) who were among the first to give a rigorous definition of a thermodynamic standard state of exchangeable cations. The use of molar fractions in Equation (6.18) would follow the *Vanselow* convention (Vanselow, 1932).

If, on the other hand, the activities of the adsorbed ions are expressed as a fraction of the number of exchange sites (X^-), then Reaction (6.17) becomes:

$$Na^+ + Ca_{0.5}\text{-}X \leftrightarrow Na\text{-}X + ½Ca^{2+} \tag{6.17a}$$

with

$$K_{Na\backslash Ca} = \frac{[Na\text{-}X]\,[Ca^{2+}]^{0.5}}{[Ca_{0.5}\text{-}X]\,[Na^+]} = \frac{\beta_{Na}[Ca^{2+}]^{0.5}}{\beta_{Ca}[Na^+]} \tag{6.18a}$$

Equation (6.17a) agrees with the *Gapon* convention (Gapon, 1933). In this case the molar and equivalent fractions become identical since both are based on a single exchange site with charge -1.

The differences in Na\Ca exchange behavior resulting from the three different conventions are shown in Figure 6.8. Both the exchangeable Na^+ and the solute Na^+ are expressed as fractions. In the calculations $K_{Na\backslash Ca}$ was set to 0.5 for all three equations, which means that $[Ca\text{-}X_2]$ or $[Ca_{0.5}\text{-}X]$ is twice the [Na-X] when both ions have an activity of 1 in solution. Thus, Ca^{2+} is the preferred, or selected, ion relative to Na^+. As is apparent in Figure 6.8, the total solute concentration affects the selectivity of the exchanger for Ca^{2+} in that the higher charged ion is preferred more strongly when the total solute concentration decreases. When a clay suspension in a solution containing Na^+ and Ca^{2+} is diluted, the concentration of adsorbed Ca^{2+} must increase while the solute Ca^{2+} concentration decreases. This effect is a consequence of the exponent that is used in the mass action equation.

Figure 6.8 shows only small differences among the exchange conventions at low total solute concentrations, and one might consider the practical aspects of calculation when making a choice. The *CEC* of a sediment is mostly constant, whereas *TEC* of a heterovalent system varies with the relative amounts of cations with different charge that neutralize the constant *CEC*. In most situations the activities of exchangeable cations are therefore calculated more conveniently (the argument of Gaines and Thomas, 1953) as equivalent fractions with respect to a fixed *CEC*. The Gapon convention is popular among soil scientists. It facilitates the calculation of exchangeable cations in heterovalent systems, and forms the basis for calculating the amount of exchangeable Na^+ from the Sodium Adsorption Ratio (*SAR*, the ratio of Na^+ over $Ca^{2+} + Mg^{2+}$ in water), which is an important parameter for estimating irrigation water-quality (Section 6.7). However, the equation does not perform well when several heterovalent cations are present, which suggests that the Gaines-Thomas or the Vanselow convention should be preferred (Bolt, 1982; Evangelou and Phillips, 1988). In this book the Gaines-Thomas convention is generally used.

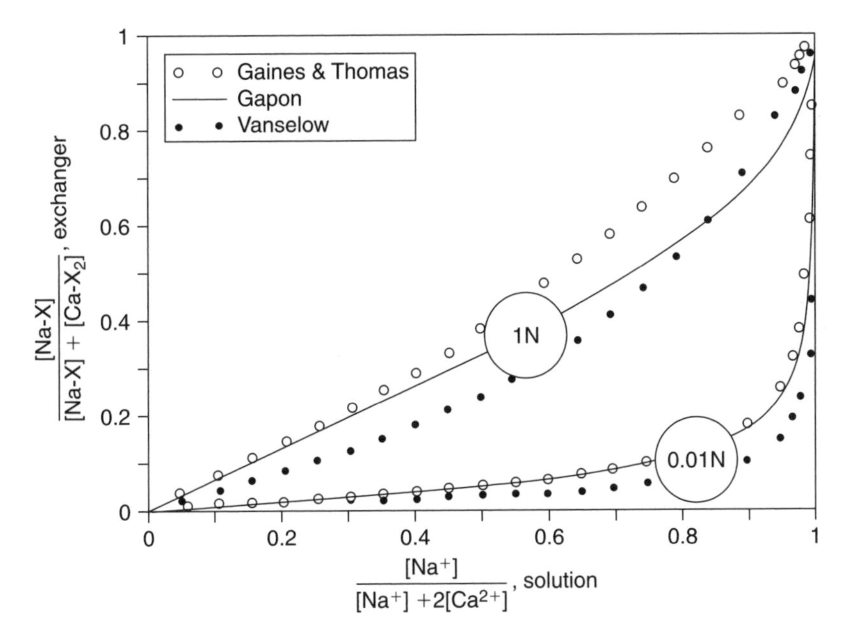

Figure 6.8. The Na/Ca exchanger composition calculated from solute concentrations using three different conventions. $K_{\mathrm{Na\backslash Ca}} = 0.5$ and at two solution normalities, 1 N and 0.01 N.

EXAMPLE 6.3. *Exchange coefficients derived from different conventions as a function of solution normality*
Wiklander (1955) determined the molar Ca/K-ratios on several clay minerals and a synthetic resin when in equilibrium with a solution that contains equivalent amounts of Ca^{2+} and K^+ in Cl^--solutions of different normality. His results are presented below:

	CEC (meq/kg)	0.1 N	0.01 N	0.001 N	10^{-4}N
Kaolinite	23	–	1.8	5.0	11.1
Illite	162	1.1	3.4	8.1	12.3
Smectite	810	1.5	(10.0)	22.1	38.8
Resin	2500	3.27	10.8	36.0	89.9

Calculate the equivalent and molar fraction of Ca^{2+} and K^+ on the clay, and also the exchange coefficients according to the conventions of Gaines-Thomas, Vanselow, and Gapon. Assume that solute concentrations are equal to activities.

ANSWER:
A 10-fold dilution of the solution gives an increase in the $\sqrt{([Ca^{2+}])}/[K^+]$ ratio of $\sqrt{10} = 3.2$ in Equation (6.18). For each 10-fold dilution the β_{Ca}/β_K-ratio should therefore increase by about a factor of 3.2. The resin (a synthetic ion-exchanger) behaves quite nicely according to this relation. The behavior of the clay minerals deviates somewhat although the same trend is followed. We are therefore confident that it is meaningful to apply the mass action equation and to calculate the exchange coefficient. The calculation is carried out only for smectite, leaving the calculation for the other clay minerals as an exercise.

From the measured molar ratio on the clay, we calculate the molar fractions in combination with:

$$\beta^M_{Ca} + \beta^M_K = 1,$$

which for example for 0.1 N gives $\beta^M_{Ca} = 1.5\beta^M_K$, and $\beta^M_K = 1 / 2.5 = 0.40$, $\beta^M_{Ca} = 0.60$.

The equivalent fractions are obtained with the help of Equation (6.16):

$$\beta_K^M = \frac{meq_{K\text{-}X}}{meq_{K\text{-}X} + (meq_{Ca\text{-}X_2})/2} = \frac{(meq_{K\text{-}X})/CEC}{(meq_{K\text{-}X})/CEC + (meq_{Ca\text{-}X_2})/(2 \cdot CEC)} = \frac{\beta_K}{\beta_K + \beta_{Ca}/2}$$

while also in this case

$$\beta_{Ca} + \beta_K = 1$$

which gives $\beta_K = \beta_K^M/(\beta_K^M + 2\,\beta_{Ca}^M)$.
For 0.1 N we obtain $\beta_K = 0.25$, and $\beta_{Ca} = 0.75$.

The molal concentrations in solution are for the 0.1 N solution $m_{K^+} = 0.05$ and $m_{Ca^{2+}} = 0.025$ mol/kg H_2O, taken equal to activities. The exchange coefficients for the different conventions are

Gaines & Thomas

$$K_{Ca\backslash K} = \frac{\beta_{Ca}^{0.5}}{\beta_K}\,\frac{[K^+]}{[Ca^{2+}]^{0.5}}$$

Vanselow

$$K_{Ca\backslash K}^V = \frac{(\beta_{Ca}^M)^{0.5}}{\beta_K^M}\,\frac{[K^+]}{[Ca^{2+}]^{0.5}}$$

Gapon

$$K_{Ca\backslash K}^G = \frac{\beta_{Ca}}{\beta_K}\,\frac{[K^+]}{[Ca^{2+}]^{0.5}}$$

Thus, the exchange coefficient for Ca/K sorption on smectite is:

	0.1 N	0.01 N	0.001 N	10^{-4}N
$K_{Ca\backslash K}$	1.10	(2.05)	1.41	0.78
$K_{Ca\backslash K}^V$	0.61	(1.05)	0.71	0.39
$K_{Ca\backslash K}^G$	0.95	(2.00)	1.40	0.78

The Ca\K selectivity of smectite varies somewhat with total solute concentrations, and none of the conventions performs decisively better than the others. In the calculations we assumed that activity coefficients are equal to 1 for both exchangeable and solute cations, but this may be incorrect. However, if we apply activity corrections only to the solute ions, the variation increases.

QUESTION:
Calculate $K_{Ca\backslash K}$ by applying solute activity corrections for $I = 0.1$ and 10^{-4}N?
ANSWER: $K_{Ca\backslash K} = 1.39$ and 0.79 for 0.1 and 10^{-4}N, respectively.

6.3.1 *Values for exchange coefficients*

Table 6.4 lists exchange coefficients for various ions following the Gaines-Thomas convention. The table provides coefficients rather than constants since the values depend upon the type of exchanger present in the soil, and also on the water composition (Example 6.3). This is due to non-ideal behavior of the exchanger, and because the activity coefficients are assumed to be equal for the solute and the exchangeable ion. The given ranges represent many measurements from different soils and for different clay minerals.

The exchange coefficients in Table 6.4 are relative to Na^+. Exchange coefficients among other cation-pairs are obtained by combining two reactions. For example, the exchange reaction for Al^{3+} and Ca^{2+} is obtained from:

$$Na^+ + \tfrac{1}{2}Ca\text{-}X_2 \leftrightarrow Na\text{-}X + \tfrac{1}{2}Ca^{2+}, \qquad K_{Na\backslash Ca} = 0.4$$

and

$$Na^+ + \tfrac{1}{3}Al\text{-}X_3 \leftrightarrow Na\text{-}X + \tfrac{1}{3}Al^{3+}, \qquad K_{Na\backslash Al} = 0.7$$

Table 6.4. Ion exchange coefficients relative to Na^+ following the Gaines-Thomas convention. Based partly on a compilation by Bruggenwert and Kamphorst, 1982.

Equation: $$Na^+ + 1/i \cdot I\text{-}X_i \leftrightarrow Na\text{-}X + 1/i \cdot I^{i+}$$

with

$$K_{Na\backslash I} = \frac{[Na\text{-}X]\,[I^{i+}]^{1/i}}{[I\text{-}X_i]^{1/i}[Na^+]} = \frac{\beta_{Na}[I^{i+}]^{1/i}}{\beta_I^{1/i}[Na^+]}$$

Ion I^+	$K_{Na\backslash I}$	Ion I^{2+}	$K_{Na\backslash I}$	Ion I^{3+}	$K_{Na\backslash I}$
Li^+	1.2 (0.95–1.2)	Mg^{2+}	0.50 (0.4–0.6)	Al^{3+}	0.7 (0.5–0.9)
K^+	0.20 (0.15–0.25)	Ca^{2+}	0.40 (0.3–0.6)	Fe^{3+}	?
NH_4^+	0.25 (0.2–0.3)	Sr^{2+}	0.35 (0.3–0.6)		
Rb^+	0.10	Ba^{2+}	0.35 (0.2–0.5)		
Cs^+	0.08	Mn^{2+}	0.55		
		Fe^{2+}	0.6		
		Co^{2+}	0.6		
		Ni^{2+}	0.5		
		Cu^{2+}	0.5		
		Zn^{2+}	0.4 (0.3–0.6)		
		Cd^{2+}	0.4 (0.3–0.6)		
		Pb^{2+}	0.3		

Subtracting the two reactions, and dividing the two exchange coefficients gives:

$$\tfrac{1}{3}Al^{3+} + \tfrac{1}{2}Ca\text{-}X_2 \leftrightarrow \tfrac{1}{3}Al\text{-}X_3 + \tfrac{1}{2}Ca^{2+}, \qquad K_{Al\backslash Ca} = 0.6$$

In case of homovalent exchange, the coefficients for the Gapon and Vanselow conventions are identical to the Gaines-Thomas values. For heterovalent exchange it is possible to derive the coefficients for the binary case (two cations only, Problem 6.9).

For sediments and soils, the selectivity of cations generally follows the lyotropic series. Cations with the same charge are more strongly held when their hydration number is smaller, i.e. when the hydration shell of water molecules is smaller (cf. Section 6.6.1). A similar sequence exists for synthetic, strongly acid ion-exchange resins (Helfferich, 1959; Rieman and Walton, 1970). In contrast, weakly acid resins exhibit the reverse selectivity in the high pH range when the resin is fully deprotonated. This may indicate that strong acid cation exchangers generally are the most important in the natural environment.

6.3.2 *Calculation of exchanger composition*

The composition of the exchanger can be calculated by combining the mass action expressions with the mass balance for the sum of the exchangeable cations. Analogous to Reactions (6.17) and (6.18) we can write the general reaction:

$$1/i \cdot I^{i+} + 1/j \cdot J\text{-}X_j \leftrightarrow 1/i \cdot I\text{-}X_i + 1/j \cdot J^{j+}$$

which gives:

$$\beta_J = \frac{\beta_I^{j/i} \cdot K_{J\backslash I}^{j} \cdot [J^{j+}]}{[I^{i+}]^{j/i}} \qquad (6.19)$$

and also β_K, β_L, ..., all expressed as a function of β_I. All fractions are introduced in the mass balance:

$$\beta_I + \beta_J + \beta_K + \cdot\cdot = 1 \tag{6.20}$$

to give an equation with one unknown, β_I. When only monovalent and divalent ions are present, a quadratic equation results that can be solved (Example 6.4).

EXAMPLE 6.4. *Calculate the cation exchange complex in equilibrium with groundwater*
Calculate the exchangeable cations in dune sand with a *CEC* of 10 meq/kg, in equilibrium with dunewater, $Na^+ = 1.1$, $Mg^{2+} = 0.48$, $Ca^{2+} = 1.9$ mmol/L. Assume that activity is numerically equal to concentration in mol/L. Use the exchange coefficients in Table 6.4.

ANSWER:
First express the exchangeable fractions of divalent cations as a fraction of β_{Na}. Thus:

$$\beta_{Mg} = \beta_{Na}^2 \cdot \frac{[Mg^{2+}]}{K_{Na\backslash Mg}^2 \cdot [Na^+]^2}$$

and

$$\beta_{Ca} = \beta_{Na}^2 \cdot \frac{[Ca^{2+}]}{K_{Na\backslash Ca}^2 \cdot [Na^+]^2}$$

are introduced into

$$\beta_{Mg} + \beta_{Ca} + \beta_{Na} = 1$$

Therefore:

$$\beta_{Na}^2 \cdot \left(\frac{[Mg^{2+}]}{K_{Na\backslash Mg}^2 \cdot [Na^+]^2} + \frac{[Ca^{2+}]}{K_{Na\backslash Ca}^2 \cdot [Na^+]^2} \right) + \beta_{Na} - 1 = 0$$

Substitutions of the coefficients from Table 6.4, and concentrations for solute activities yields:

$$\beta_{Na}^2 \cdot \{0.48\times10^{-3} / (0.5\times1.1\times10^{-3})^2 + 1.9\times10^{-3} / (0.4\times1.1\times10^{-3})^2\} + \beta_{Na} - 1 = 0$$

This quadratic equation is solved to give $\beta_{Na} = 0.00932$, and by back-substitution, $\beta_{Mg} = 0.138$ and $\beta_{Ca} = 0.853$. With the factor $\rho_b / \varepsilon = 6$ kg/L we can recalculate the concentrations from meq/kg to meq/L pore water (cf. Example 6.1). Thus, the exchangeable cations amount to $meq_{Na\text{-}X} = \beta_{Na} \cdot CEC \cdot 6 =$ 0.5 meq/L, $meq_{Mg\text{-}X_2} = 8.3$ meq/L, and $meq_{Ca\text{-}X_2} = 51.2$ meq/L.

The use of the Vanselow convention gives mole fractions. These can be recalculated to equivalent fractions with (cf. Problem 6.14)

$$\beta_I = \frac{\beta_I^M \cdot i}{\beta_I^M \cdot i + \beta_J^M \cdot j + \beta_K^M \cdot k + \cdots} \tag{6.21}$$

The Gapon convention makes it particularly easy to calculate exchangeable fractions. All fractions are linearly related according to:

$$\beta_J = \frac{\beta_I \cdot K_{J\backslash I}^G \cdot [J^{j+}]^{1/j}}{[I^{i+}]^{1/i}}$$

When all fractions are entered in the sum $\Sigma\beta = 1$, some rearrangement gives:

$$\beta_I = \frac{[I^{i+}]^{1/i}}{\sum_{J=I,J,K,\ldots} K^G_{J\backslash I} \cdot [J^{j+}]^{1/j}} \tag{6.22}$$

This last equation is also known as the *multicomponent Langmuir* equation.

6.3.3 *Calculation of exchanger composition with PHREEQC*

In the database of PHREEQC, cation exchange is defined according to the Gaines-Thomas convention. The exchanger X is specified under EXCHANGE_MASTER_SPECIES, followed by exchange half-reactions under keyword EXCHANGE_SPECIES. The exchange master species must be defined as the first species with log $K = 0$, similarly as for solution species. An excerpt of the default database is:

```
EXCHANGE_MASTER_SPECIES
        X        X-
EXCHANGE_SPECIES
        X-   = X-
        log_k      0.0

        Na+ + X- = NaX
        log_k      0.0
        -gamma     4.0   0.075
        Ca+2 + 2X- = CaX2
        log_k      0.8
        -gamma     5.0   0.165
```

PHREEQC handles exchange reactions by splitting them into half reactions. For example for $Ca^{2+}\backslash Na^+$ exchange we may rewrite the reaction in Table 6.4 as:

$$Ca^{2+} + 2Na\text{-}X \leftrightarrow Ca\text{-}X_2 + 2Na^+ \tag{6.23}$$

with $\log K = \log(1 / K^2_{Na\backslash Ca}) = \log(1/0.4^2) = 0.8$.
One half reaction needs to be defined as the point of reference, which is:

$$Na^+ + X^- \leftrightarrow Na\text{-}X \qquad \log K = 0.0 \tag{6.24}$$

If we add Reaction (6.24) twice to Reaction (6.23) (also the log K's are added), we obtain:

$$Ca^{2+} + 2X^- \leftrightarrow Ca\text{-}X_2 \qquad \log K = 0.8 \tag{6.25}$$

which is identical to the equation in the database (see above).

PHREEQC uses the same activity coefficient for the exchangeable species as for the aqueous species. This procedure is based on observations of Na^+/Ca^{2+} exchange in saline soils. Here, the use of activity coefficients for the solute ions only leads to an overestimate of Na-X (or, more generally, of the monovalent ions compared to the divalent ions, cf. Example 6.3). For the calculation of the exchangeable fractions, the free ion concentrations are used, equal to the total (analytical) concentrations *minus* the aqueous complexes.

EXAMPLE 6.5. *Calculate the exchanger composition in contact with groundwater, using PHREEQC*
Repeat the hand calculation of Example 6.4 with PHREEQC. The input file is:

```
SOLUTION 1
 Na 1.1
 Mg 0.48
 Ca 1.9

EXCHANGE 1
 X 0.06                   # moles
 -equilibrate 1

PRINT
 -reset false
 -totals true
 -exchange true
END
```

Keyword EXCHANGE defines $X^- = 0.06$ mol to be in equilibrium with solution 1. Since the solution has 1 kg H_2O by default, the *CEC* is 60 meq/L, in accordance with Example 6.4. The keyword PRINT is used to reduce the lengthy output. The first line "-reset false" indicates that all printout is to be suppressed, while "-totals true" and "-exchange true" will print these specific variables. In the output file we find:

```
----------------------Exchange composition-------------------------------------

X              6.000e-02 mol

                                                      Equivalent
      Species      Moles           Equivalents        Fraction          Log Gamma

      CaX2         2.564e-02       5.127e-02          8.545e-01         -0.132
      MgX2         4.086e-03       8.172e-03          1.362e-01         -0.130
      NaX          5.572e-04       5.572e-04          9.287e-03         -0.033
```

The results correspond to Example 6.4 within roundoff errors of the logarithms of the *K* values from Table 6.4.

Other exchange conventions can be implemented also in PHREEQC, as shown by Appelo and Parkhurst (cf. www.xs4all.nl/~appt). For example, the Rothmund-Kornfeld equation has been used recently in several studies (Bond, 1995; Voegelin et al., 2000; Bloom and Mansell, 2001). The equation contains an empirical exponent for adapting the activity-ratio of the solute ions and can simulate the sigmoidal shape often observed in binary isotherms (which reflects that exchangeable cations are preferred more when present in small concentrations since the sites which favor the cation are filled first).

Thus, for $Ca^{2+}\backslash K^+$ exchange:

$$Ca^{2+} + 2K\text{-}X \leftrightarrow Ca\text{-}X_2 + 2K^+$$

the law of mass action for Brucedale subsoil was (Bloom and Mansell, 2001):

$$\frac{[Ca\text{-}X_2]}{[K\text{-}X]^2}\left(\frac{[K^+]^2}{[Ca^{2+}]}\right)^{0.618} = 10^{-0.281} \tag{6.26}$$

Example 6.6 illustrates how this equation is introduced in PHREEQC.

EXAMPLE 6.6. *Calculate the exchanger composition with PHREEQC, using the Rothmund-Kornfeld equation*
For Brucedale subsoil, the exchange species CaX_2 is defined as:

$$0.618\text{Ca+2} + 2\text{KX} = \text{CaX2} + 1.236\text{K+}$$

which has the coefficients of mass action Equation (6.26). However, the stoichiometric coefficients for Ca^{2+} and K^+ are incorrect and the equation would generate an error message of PHREEQC, unless "-no_check" and "mole_balance" are used. These options allow exponents in the mass action law that differ from the stoichiometric coefficients in the reaction equation.

The input file further defines a solution and exchanger which contain K^+ only. Subsequently, by using the keyword REACTION, Ca^{2+} is added and K^+ is removed and the isotherm is plotted that covers the range of β_K from 0–1.

```
EXCHANGE_SPECIES
 0.618Ca+2 + 2KX = CaX2 + 1.236K+ ; -log_k - 0.281
 -no_check
 -mole_balance CaX2
 -gamma 5.0 0.165
SOLUTION 1
 K 100                  # 100mmol K/kg H2O
EXCHANGE 1
 KX 0.417               # 417 mmol KX, total K in the system = 517 mmol

REACTION
 K -1 Ca 0.5            # Remove 517 mmol K+, replace with 258.5 mmol Ca2+
 0.517 in 20 steps
USER_GRAPH              # Plot isotherm...
 -head alpha_K beta_K
 -axis_titles "Solute K / (K + 2Ca)" "Exchangeable K / (K + 2Ca)"
 -axis_scale x_axis 0 1 0.2 0.1; -axis_scale y_axis 0 1 0.2 0.1
 -start
 10 graph_x tot("K") / 0.1
 20 graph_y mol("KX") / 0.417
 -end
END
```

A plot of the K^+/Ca^{2+} isotherm is obtained with USER_GRAPH and compared with measured data in Figure 6.9. The sigmoidal shape of the isotherm indicates that the cation with a relatively low concentration is favored by the exchanger.

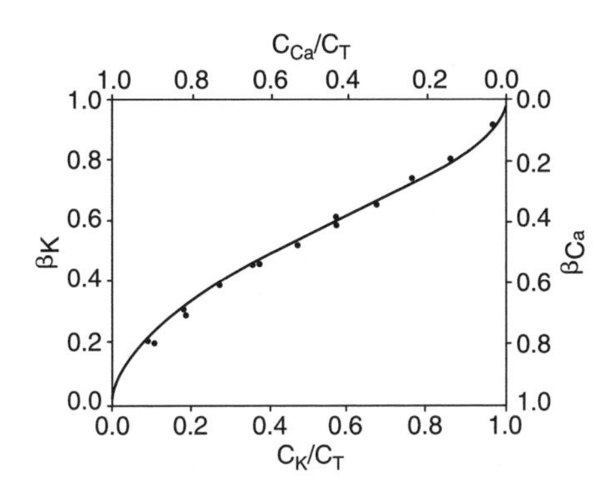

Figure 6.9. The K/Ca isotherm according to the Rothmund-Kornfeld equation (Bloom and Mansell, 2001).

Bond (1995) and Bond and Verburg (1997) have shown how results from the Rothmund-Kornfeld equation can be interpreted in terms of activity coefficients. They obtained activity coefficients for the exchangeable cations which varied from 0.2 to 1 in Ca^{2+}-K^+ systems, and from 1 to 3 in Ca^{2+}-Na^+ systems. One remaining problem in their study was that several minerals were present in the soil, with every mineral having various cation exchange sites, and each site exhibiting probably different cation preferences. Moreover, Bond and coworkers found that the exchange coefficients and exponents in the Rothmund-Kornfeld formulas changed with ionic strength, leading to activity coefficients which vary with salinity. Unfortunately, these variations are poorly understood although models derived from solid or liquid solutions have been applied (Elprince et al., 1980; Liu et al., 2004). Thus, the Rothmund-Kornfeld expression may be helpful for detailed description of accurate laboratory experiments where everything is well known and fixed, but it may not altogether be applicable in field situations where solution compositions and salinities vary markedly.

6.3.4 *Determination of exchangeable cations*

Most methods start with the removal of solute cations by centrifuging a field-moist sample, or by displacing the pore solution with alcohol or a fluid immiscible with water (Figure 6.10). Subsequently, the sorbed cations are exchanged with an alien cation, while trying to avoid the dissolution of minerals that can yield false cation contributions. Alternatively, corrections can be made by analyzing the anion released by the dissolving mineral (alkalinity in the case of carbonates, sulfate in case gypsum is present).

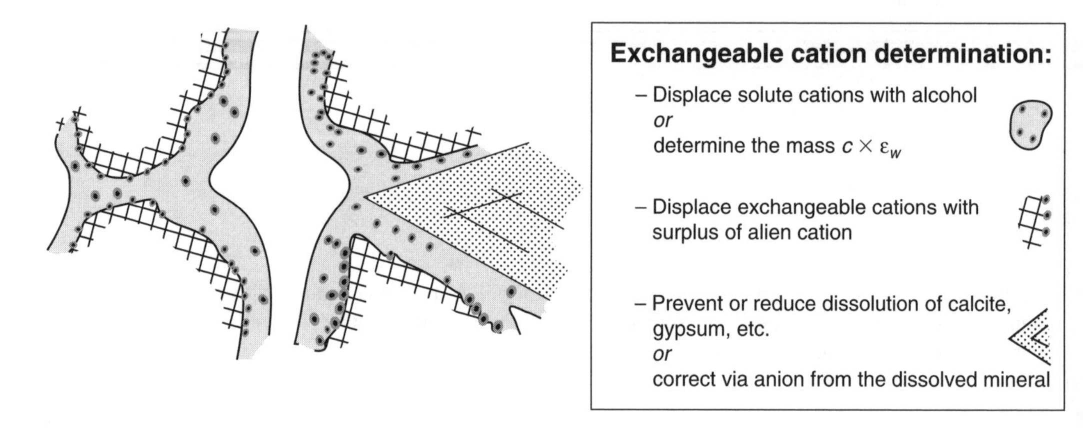

Figure 6.10. Procedure for determining exchangeable cations in soils and sediments.

Many cations have been advocated as the displacing cation, but usually each one yields a different result since the cations have varying displacing power and induce various side reactions. A popular method is displacement with 1 M NH_4-acetate in which case a single extraction is sufficient to remove all the exchangeable cations, while not much calcite dissolves if the pH is adjusted to 8.2 (Rhoades, 1982). However, even a small amount of calcite that may dissolve could contribute substantially to "exchangeable" Ca^{2+} when the exchange capacity is low. Figure 6.11 shows on the left the computed amounts of calcite that dissolve as a function of the pH of the acetate solution and the *CEC* ($\equiv$X). On the right, Figure 6.11 shows the contribution of calcite dissolution to "exchangeable Ca". For example when 20 mL NH_4-acetate solution is added to 5 g soil with a *CEC* of 10 meq/kg, then X = 2.5 mM.

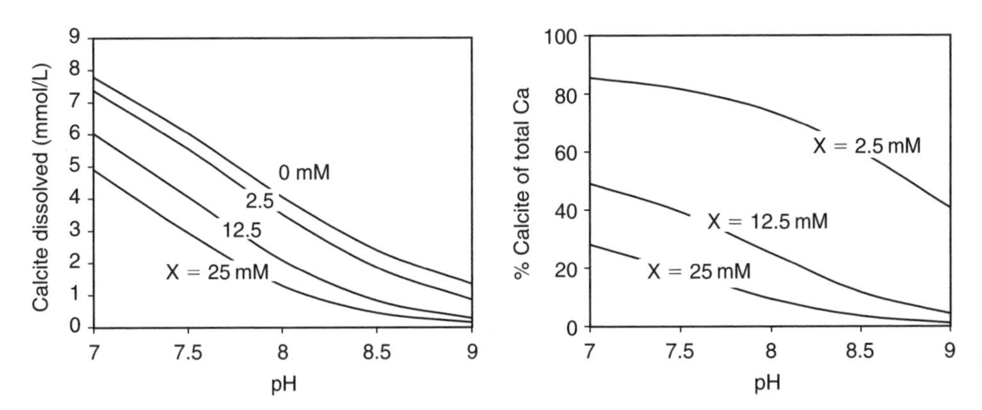

Figure 6.11. The calculated effect of calcite dissolution on determination of Ca-X_2 using 1 M NH_4-acetate. Left: the amount of dissolved calcite as a function of pH and X (≡*CEC*). Right: the contribution of calcite dissolution to "exchangeable Ca". Calculations included the Ca-acetate complex with an association constant of $10^{1.18}$ in Martell and Smith (1977).

According to Figure 6.11, 1.9 mM calcite may dissolve at pH = 8.5 and X = 2.5 mM, which amounts to $(1.9/(1.9 + 1.25)) \times 100\% = 60\%$ of the total Ca^{2+} in the extract (the sum of calcite and Ca-X_2). The Ca-acetate aqueous complex contributes significantly to calcite dissolution and the Ca^{2+} that originates from calcite cannot be corrected, since an alkalinity titration of the 1 M acetate solution is useless.

In contrast to the results of batch experiments, the exchange constants for the major cations derived from column experiments and field injections are more uniform. To obtain constants from batch experiments that are comparable to flow experiments requires the use of a cation that is strongly hydrated and only weakly displaces protons, for example Na^+ (Zuur, 1938; Van der Molen, 1958) or Li^+ (Hurz, 2001; Reardon et al., 1983). The recommended procedure is to carry out 2–3 subsequent extractions with 1 M NaCl, followed by an alkalinity titration to correct for dissolving calcite. Exchangeable Na^+ can be determined separately, in an extraction with 1 M NH_4Cl.

Example 6.7 simulates the extraction of exchangeable cations in our dune sand (Example 6.4) with PHREEQC. The keyword MIX is used to work in absolute units. MIX takes fractions of the actual amount of solution (you may compare the amounts of water that PHREEQC calculated with the weight of the solutions in the real-life analytical procedure). The extraction must be repeated once since not all Ca-X_2 is displaced in the first step. The last simulation mixes the two extractions and the total concentrations are calculated in much the same way as would be done when a laboratory experiment is analyzed. Note that in this sample with a low *CEC*, the low concentrations can be analyzed only if analytically pure NaCl is used in the laboratory extraction.

EXAMPLE 6.7. *Simulate the analytical measurement of exchangeable cations*
Place 5 g (dry) soil in a centrifuge tube and extract twice with 20 mL 1 M NaCl. We neglect the contribution from the original pore water, but do consider that 2.5 mL of the extractant remains in the centrifuge tube after decanting. As a result, in the 2nd extraction a fraction 2.5 / 20 = 0.125 of the first extraction is added to the second one.

```
# Define exchanger and solutions...
SOLUTION 1                          # Pore water
 Na 1.1; Mg 0.48; Ca 1.9
EXCHANGE 1                          # 5g soil with CEC = 10meq/kg, or 0.05mmol X-
 -equilibrate 1
 X 0.05e-3                          # moles
```

```
SOLUTION 2; Na 1e3; Cl 1e3                # Extractant solution
END

USE exchange 1                            # ...put 5g soil in centrifuge tube
MIX                                       # ...add 20g 1M NaCl
 2 20e-3
SAVE exchange 1; SAVE solution 3          # centrifuge and decant
END

USE exchange 1                            # repeat extraction, 2nd time
MIX                                       # 20g new extractant, 2.5g of the old extractant
 2 20e-3
 3 0.125                                  # 2.5 / 20 part of the old extractant
SAVE solution 4
END

MIX                                       # combine the two centrifuged solutions...
 3 0.875                                  # 17.5 / 20 part decanted after centrifuging
 4 0.8888                                 # 20 / 22.5 idem
USER_PRINT
 -start
 10 print "CaX2 = ", (tot("Ca") * tot("water") * 4e5), "meq/kg"
 20 print "MgX2 = ", (tot("Mg") * tot("water") * 4e5), "meq/kg"
 -end
END
```

The last simulation combines the two decanted solutions and prints the exchangeable cations. In line 10 of USER_PRINT, tot("Ca") gives the molality of total Ca^{2+}, tot("water") provides the kg of water and the product of the two, the moles of Ca^{2+}. When multiplied with 1000 (g/kg)/(5 g soil) $\times$ 2000 (meq/mol) = 4×10^5, it yields meq/kg soil. The result is:

```
--------------------------------------------------User print----------------------------------------------

CaX2 = 8.4129e+00 meq/kg
MgX2 = 1.3418e+00 meq/kg
```

which is nearly equal to *CEC* $\times$ the exchangeable fraction, calculated in Example 6.4 (there is still some trace of Ca^{2+} and Mg^{2+} left on the exchanger in the centrifuge tube). Exchangeable Na^+ must be determined by means of a separate extraction with NH_4Cl.

QUESTION:
Perform the extraction with 1 N NH_4Cl. Look in PHREEQC.DAT how NH_4^+ is defined (a special case, because oxidation to N_2 and NO_3^- is to be avoided). Also assume that 5 g soil now contains 1% pore water.

6.4 CHROMATOGRAPHY OF CATION EXCHANGE

The principles of ion exchange are routinely used in the laboratory in *ion-chromatography*, a method to separate and analyze ions from a complex mixture. The chemical analyst uses a column packed with beads of an ion-exchanger. The pores within and between the beads are filled with fluid, and the total amount of fluid in the column is termed the *column pore volume*, or just the pore volume (V_0, m^3). The column is percolated with different solutions, and as a result of cation exchange a concentration pattern develops at the end of the column. Strongly selected cations will displace other ions from the exchanger and be transported at a relatively low velocity. In *displacement chromatography*, the cation in the injected solution is sorbed more strongly than the resident cations, which are flushed in an ordered manner from the column. The ion with the lowest selectivity is leached first, then the next favored, etc., and finally the cation arrives that is injected in the column. If, on the other hand, the cation in the incoming solution has a lower affinity for the exchanger, the process is called *elution chromatography*.

In this case, the concentrations at the outlet of the column change more gradually. Sometimes the total ion concentration of the injected fluid is different from the resident solution. This will induce a *salinity front*, or an *anion-jump* when one pore volume has been displaced from the column.

6.4.1 *Field examples of freshening*

Displacement chromatography applies to field situations where fresh water, containing mainly Ca^{2+}, displaces seawater dominated by Na^+ and Mg^{2+}. Ca^{2+} is favored by natural ion exchangers over Mg^{2+} and Na^+. It is possible to obtain some idea of the timescales involved for flow and cation exchange, if we simplify the fresh water to contain only Ca^{2+} and HCO_3^-. Example 6.8 shows how such a calculation is performed.

EXAMPLE 6.8. *Flushing of an exchange complex*
Cultivated fields on islands in the Dutch Wadden Sea often show brackish patches in an otherwise fresh groundwater area. The brackish water is a remnant of transgressions during the last century. They occur where clay layers at the surface obstruct the downward percolation of fresh water. Estimate the time needed for refreshening and flushing of the exchange complex for the situation depicted in Figure 6.12. The *CEC* of the sediment is 10 meq/kg, bulk density $\rho_b = 2.0\,g/cm^3$ and porosity $\varepsilon = 0.30$. The salt water has the composition of seawater, the fresh water is pure Ca^{2+}- and HCO_3^--water in equilibrium with $[P_{CO_2}] = 0.01$ atm.

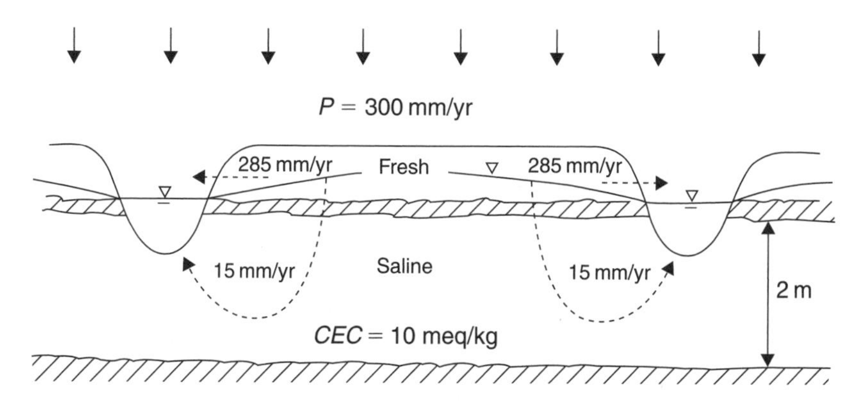

Figure 6.12. Flushing of salt water below a clay layer: The 2 m. thick aquifer has a porosity ε=0.3 and therefore contains 0.6 m water. The precipitation surplus is 300 mm/yr, of which 285 mm/yr flows directly into the ditches, while 15 mm/yr percolates downward into the aquifer.

ANSWER:
First we calculate the number of years needed to flush the pore volume with a rainfall surplus of 15 mm/yr. The seawater chloride is flushed after one pore volume, and the concentrations drop to the fresh HCO_3^--level. Then, the exchange complex is flushed. Initially the fresh water equilibrates with the original seawater exchange complex. This continues until all the Na^+ has been flushed. Next comes Mg^{2+}, which in turn is displaced by Ca^{2+}. The flushing time is in all cases the ratio of the amount of exchangeable cation over the amount of fresh water Ca^{2+}, multiplied with the flushing time for one pore volume.

The aquifer contains initially 0.6 m salt water, which is displaced in 40 years with 15 mm/yr. The amount of exchangeable cations in the sediment can be calculated on a pore water basis: 2 g sediment has a volume of 1 cm^3, and has 0.02 meq exchangeable cations, and 0.3 mL water. A sediment column of $2 \times 1 \times 1\,m^3$ contains $2 \times 2 \times 10^6 = 4 \times 10^6$ g sediment, with 4×10^4 meq exchangeable cations, and 0.6 m^3 water.

Percolation gives $0.015 \times 1 \times 1 \times 1\,m^3 = 0.015\,m^3/yr$. The fraction of exchangeable Na^+ in equilibrium with seawater is calculated as in Example 6.4:

$$\beta_{Na}^2 \cdot \{[Ca^{2+}] / (K_{Na\backslash Ca} \cdot [Na^+])^2 + [Mg^{2+}] / (K_{Na\backslash Mg} \cdot [Na^+])^2\} + \beta_{Na} - 1 = 0$$

Assume that activity equals molal concentration i.e. $[Na^+] = 0.485$, $[Mg^{2+}] = 0.055$ and $[Ca^{2+}] = 0.011$, and use the exchange constants from Table 6.4, $K_{Na\backslash Ca} = 0.4$, and $K_{Na\backslash Mg} = 0.5$, to find $\beta_{Na} = 0.583$, and subsequently $\beta_{Mg} = 0.318$, and $\beta_{Ca} = 0.099$. The sediment column contains $0.583 \times 4 \times 10^4 = 23316$ meq Na-X.

Fresh water in equilibrium with calcite at $P_{CO_2} = 0.01$ atm contains: $Ca^{2+} = 10^{-2.1} \times [P_{CO_2}]^{1/3} = 1.7$ mmol/L (cf. Chapter 5). The percolating volume thus brings $0.015 \times 1700 \times 2 = 51$ meq Ca^{2+} each year into the soil. The time needed to flush all Na-X is then $23316 / 51 = 457$ years. Similarly we find the time needed for flushing of Mg^{2+}, using the assumption that $Mg\text{-}X_2$ flushing first starts after all exchangeable Na-X has been removed, i.e. after 249 years (this time for Mg removal is only approximate). The results are summarized below.

	Years needed for removal
Salt (high Cl^-)	40
Na (poor soil structure)	457
Mg (hydrological tracer)	249
Total:	746

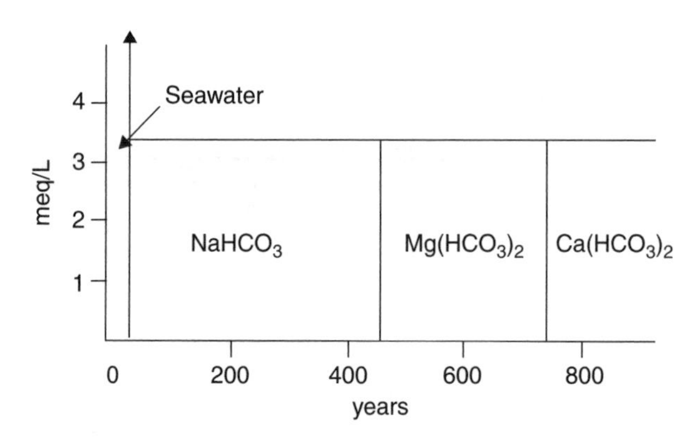

Figure 6.13. Water compositions at the end of a flowline as seawater is displaced by fresh water.

Figure 6.13 shows how the water composition changes over time at the end of a flowline. We will discuss the bad soil structure associated with $NaHCO_3$ type water in later sections of this Chapter, but note already that gypsum is used to accelerate the reclamation of saline soils. Gypsum dissolution will increase the Ca^{2+} concentration in the percolating water and thereby speed up flushing of Na^+ (Šimůnek and Suarez, 1997). In these (simple) calculations, the effects of dispersion and geological heterogeneity were neglected. These tend to accelerate the initial breakthrough of Ca^{2+}, but slow down the complete removal of Na^+ and Mg^{2+}.

QUESTIONS:

What are the flushing times for Na^+ and Mg^{2+} if the *CEC* were twice higher?

ANSWER: twice longer

What are the flushing times when the Ca^{2+} concentration is 10 times higher because gypsum is added to the soil?

ANSWER: 10 times shorter, 45.7 year for $NaHCO_3$ type water.

Figure 6.13 shows how the water composition changes with time at the end of a flowline as calculated in Example 6.8. Alternatively the flushing times can be calculated using the retardation formula for sharp fronts (cf. Example 3.6):

$$R = 1 + \Delta q \,/\, \Delta c \tag{3.33}$$

which for cation exchange becomes:

$$R = 1 + CEC\ \Delta\beta \,/\, \Delta c \tag{6.27}$$

Since the *CEC* is constant, only the fraction β of the exchangeable ions varies. At the front where Na^+ is being flushed, the initial β_{Na} is 0.583 and with $CEC = 66.7$ meq/L pore water, this corresponds to $q_{Na} = 0.583 \times 66.7 = 38.9$ meq/L. After the front, solute Na^+ concentration must balance $m_{HCO_3^-} = 2 \times m_{Ca^{2+}}$, therefore $c_{Na} = 2 \times 1.7 = 3.4$ meq/L. Once all Na^+ has been flushed from the column, $q_{Na} = 0$ and $c_{Na} = 0$. Hence:

$$R_1 = 1 + \frac{(q_{Na})_1 - (q_{Na})_2}{(c_{Na})_1 - (c_{Na})_2} = 1 + \frac{38.9 - 0}{3.4 - 0} = 1 + 11.4 = 12.4 \tag{6.28}$$

The Na^+-front is 12.4 times retarded compared to the conservative front, which corresponds to $12.4 \times 40 = 496$ years from the onset of freshening.

The second front where Mg^{2+} is flushed is obtained similarly, and also more accurately than in Example 6.8. The injected Ca^{2+} has not yet arrived at the end of the flowline (it exchanges with Mg^{2+}), so that Ca^{2+} remains unchanged, i.e. $\beta_{Ca} = 0.099$. After Na^+ has been flushed the remainder is Mg-X_2 and $\beta_{Mg} = 1 - 0.099 = 0.901$ with $q_{Mg} = 60.1$ meq/L. Given the exchangeable fractions of Ca^{2+} and Mg^{2+}, the solution composition must obey:

$$\frac{\beta_{Mg}[Ca^{2+}]}{\beta_{Ca}[Mg^{2+}]} = \left(\frac{0.4}{0.5}\right)^2$$

or $[Ca^{2+}] = 0.0703\ [Mg^{2+}]$. Since the sum of the two cations is 3.4 meq/L,

$$c_{Ca} + c_{Mg} = (0.0703 + 1)\ c_{Mg} = 3.4.$$

Hence $c_{Mg} = 3.18$ meq/L.

The Mg^{2+}-front arrives with a retardation (R_2) of:

$$R_2 = 1 + \frac{(q_{Mg})_2 - (q_{Mg})_3}{(c_{Mg})_2 - (c_{Mg})_3} = 1 + \frac{60.1 - 0}{3.18 - 0} = 1 + 18.9 = 19.9 \tag{6.29}$$

which amounts to $19.9 \times 40 = 796$ years. This is longer than calculated in Example 6.8 since the elution of Ca^{2+} together with Mg^{2+} is now duly accounted for (for an exact answer, see Appelo et al., 1993).

We have calculated the retardation relative to flushing one pore volume through the column. If the flowline is twice as long, twice as much time is needed for all the transitions and fronts. In other words, we have found the *characteristic velocity* $v = dx \,/\, dt$ of a front. This is illustrated in a time / distance diagram (Figure 6.14), where the transitions appear as straight lines.

The freshening chromatographic pattern calculated in Example 6.8 has been observed in the injection experiment of Valocchi et al. (1981). Fresh water from a sewage plant with tertiary treatment (i.e. giving near drinking water quality) was injected in a brackish-water aquifer, shown in Figure 6.15. The native brackish water contained large amounts of Na^+ and Mg^{2+}, and the fresh water had Ca^{2+} as the major cation; the cation exchange capacity of the silty aquifer sediment was as high as 100 meq/kg.

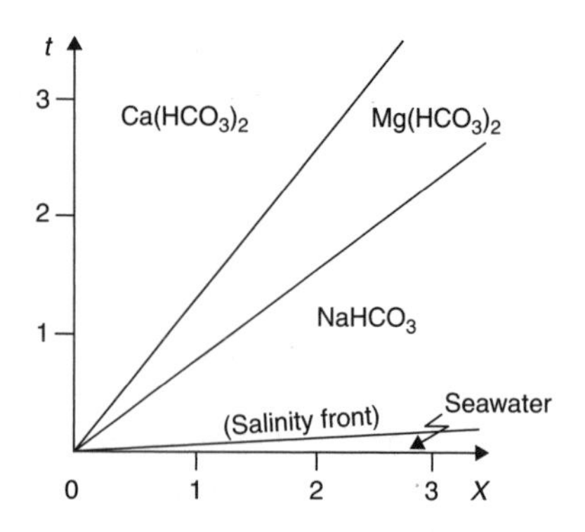

Figure 6.14. Front propagation and characteristic velocity $v = dx / dt$ as calculated in Example 6.8.

Figure 6.16 shows the Ca^{2+} and Mg^{2+} concentrations in an observation well 16 m downstream from the point of injection. The initial Ca^{2+} and Mg^{2+} concentrations are those in the undisturbed aquifer. After injection of about 200 m^3, the concentrations start to decrease and at 500 m^3, the fresh water has replaced all of the brackish water. The Ca^{2+} (and Mg^{2+}) concentrations become now lower than in the injected water, and remain low until about 5000 m^3 have been injected. This stage corresponds to the dilution step after the salinity front where Ca^{2+} (and Mg^{2+}) are exchanged for Na^+. When exchangeable Na^+ becomes exhausted, the Ca^{2+} and Mg^{2+} concentrations increase again. Ca^{2+} increases towards the concentration in the injection water. However, Mg^{2+} substantially surpasses the injected concentration because exchangeable Mg^{2+} is displaced. The water changes quality variations were modeled with a numerical model using exchange constants obtained from batch experiments with the same aquifer material, obviously with very good results (Figure 6.16).

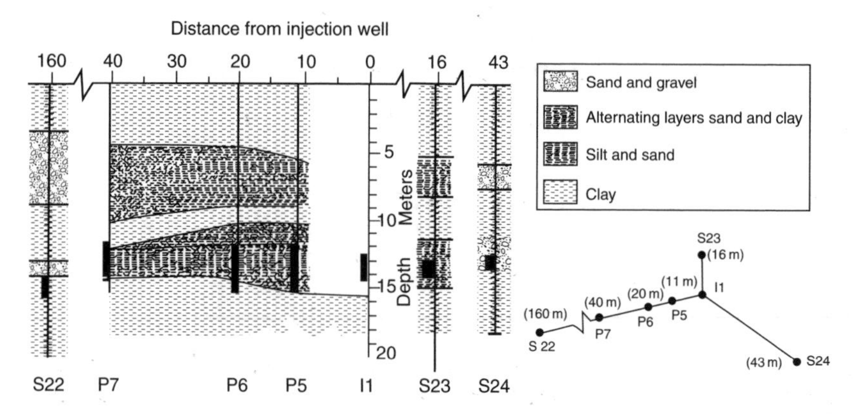

Figure 6.15. Aquifer outline for the Valocchi experiment, showing the fresh water injection well I1 and the observation wells (Valocchi et al., 1981. Copyright by the Am. Geophys. Union).

The injection experiment by Valocchi et al. (1981) displayed the cation exchange phenomena especially clearly since the *CEC* was high and the conditions in the aquifer before injection were uniform. This is not always true for field studies where freshening and salinization may have occurred repeatedly in the same aquifer. A reversal of flow does not restore the original quality conditions since the cation exchange processes are non-linear. The chromatographic pattern may then become blurred.

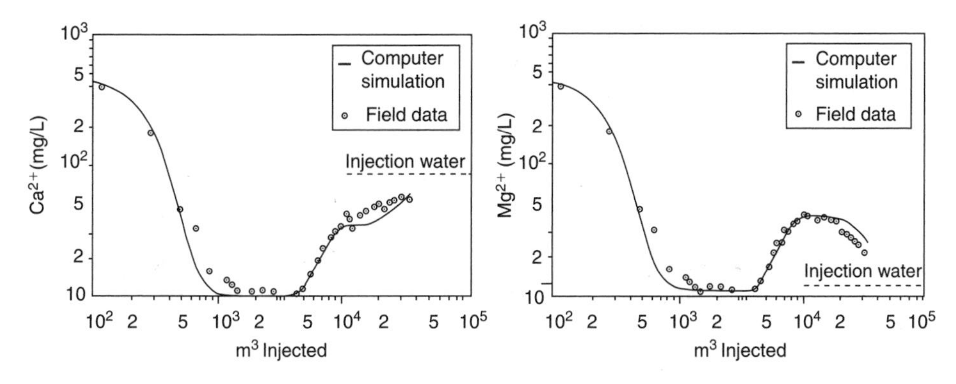

Figure 6.16. Breakthrough data for Ca^{2+} and Mg^{2+} in boring S23 of Figure 6.15 during injection of fresh water into a brackish aquifer. The drawn lines represent simulated water qualities (Valocchi et al., 1981. Copyright by the Am. Geophys. Union).

However, freshening quality patterns have been observed in studies by Lawrence et al. (1976), Chapelle and Knobel (1983), Stuyfzand (1985, 1993, cf. Figure 6.2), Beekman (1991), Walraevens and Cardenal, (1994), and in a diffusion profile by Manzano et al. (1992) and Xu et al. (1999).

The freshening chromatographic pattern was observed in exemplary fashion in the Aquia aquifer (Chapelle and Knobel, 1983; Appelo, 1994). This aquifer consists of a Paleocene, glauconite rich sand, confined by thick clay layers (Figure 6.17). The vertical exaggeration in the figure is misleading; the aquifer is 70 m thick and extends over 90 km. On a laboratory scale, this corresponds to a hair-thin chromatographic column. The area was uplifted and emerged in the Pleistocene. As a result, fresh $CaHCO_3$-water infiltrated and displaced the saline pore water. While Cl^- has been flushed almost entirely from the aquifer, cation exchange still reveals the previously more saline conditions (Figure 6.18). The increased alkalinity in the downstream end is due to calcite dissolution as Ca^{2+} exchanges for Na^+, and $NaHCO_3$ type of water is formed. Behind the $NaHCO_3$ type water follows first a $KHCO_3$, then a $MgHCO_3$ type of water and finally the fresh $CaHCO_3$ infiltration water quality.

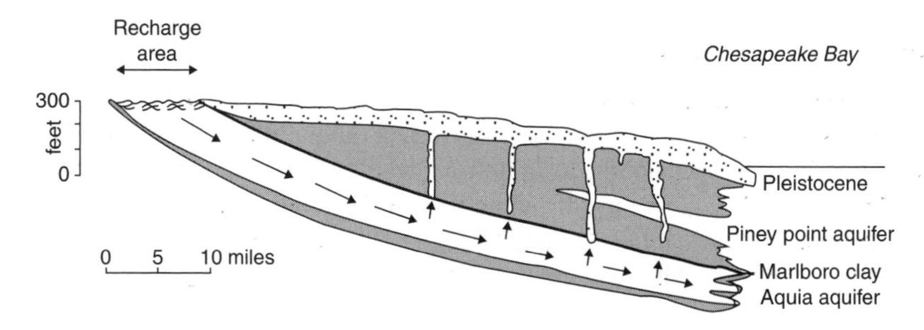

Figure 6.17. Cross section showing the outline of the Aquia aquifer, Maryland, USA (Appelo, 1994).

Chromatographic patterns allow a time-frame to be associated with the succession of various water qualities. The calculated profiles in Figure 6.18 were obtained when the first 50 km of the aquifer had been flushed 8 times. Using ^{14}C dating, Purdy et al. (1992) found an age of 12 ka for water at a distance of 50 km from the recharge area. The chromatographic pattern required therefore $8 \times 12{,}000 = 96{,}000$ years to become established under present day flow conditions.

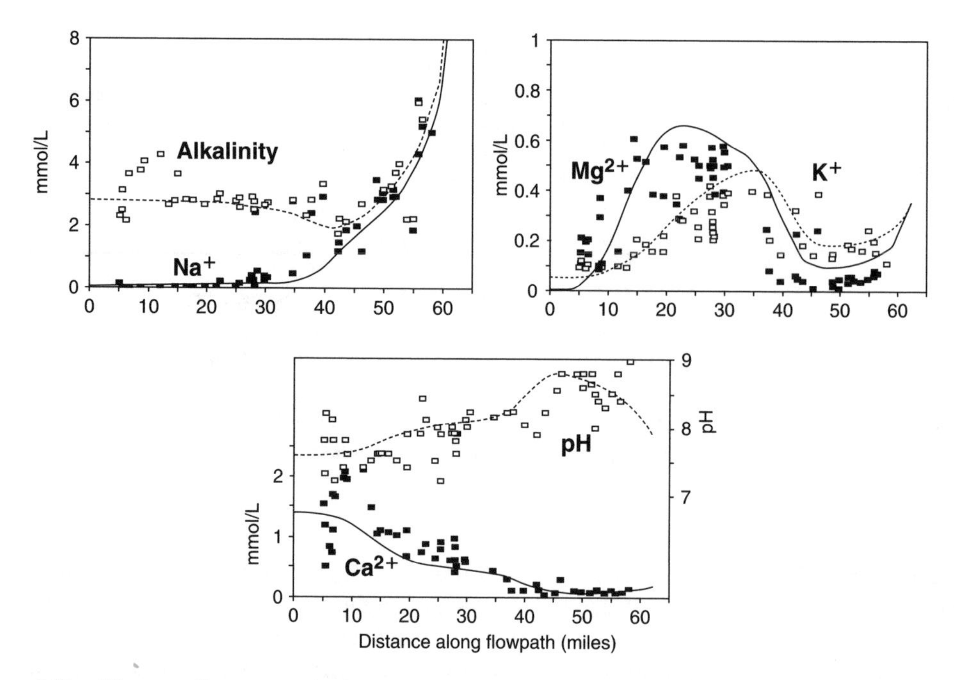

Figure 6.18. Water quality patterns in the Aquia aquifer; symbols indicate observed concentrations, lines are modeled (Appelo, 1994).

Compared to the duration of the Pleistocene this seems a short timespan. Probably flow in the aquifer was enhanced when the confining clay layers were sliced up by glacier tongues from Pleistocene icesheets, thus creating an upward outlet for water from the aquifer. An almost identical water quality pattern has been found in the Ledopaniselian aquifer at the Dutch / Belgian border (Walraevens and Cardenal, 1994). Also in this case, flow times were established based on ^{14}C measurements corrected for the dissolution of calcite (Van der Kemp et al., 2000).

6.4.2 *Salinity effects in cation exchange*

Both in the Valocchi experiment (Figure 6.16) and in the Aquia aquifer (Figure 6.18) the salinity jumped when fresh water displaced the salt water. The anions traveled unhindered through the aquifer and determined the speed of the salinity front, while the injected cations exchanged with the sediment and were retarded. Consequently, directly after the salinity front, the water attained equilibrium with the original seawater exchanger. Example 6.9 calculates the composition of $NaHCO_3$ water which has lost Ca^{2+}, and has equilibrated with the seawater exchange complex.

EXAMPLE 6.9. *The water composition after passage of a salinity front during freshening*
Calculate the water quality when fresh water with 3.4 meq anions/L displaces seawater (Example 6.8)

ANSWER:
The exchangeable fractions in equilibrium with seawater are $\beta_{Na} = 0.583$, $\beta_{Mg} = 0.318$, $\beta_{Ca} = 0.099$. The anion concentration is 3.4 meq HCO_3^-/L, which limits the total cations to 3.4 meq/L. Hence:

$$m_{Na^+} + 2\,m_{Mg^{2+}} + 2\,m_{Ca^{2+}} = 0.0034 \text{ eq/L} \tag{6.30}$$

We furthermore have from exchange equilibrium

$$[Ca^{2+}] = \frac{\beta_{Ca} \cdot (K_{Na\backslash Ca} \cdot [Na^+])^2}{\beta_{Na}^2} \tag{6.31}$$

and

$$[Mg^{2+}] = \frac{\beta_{Mg} \cdot (K_{Na\backslash Mg} \cdot [Na^+])^2}{\beta_{Na}^2} \tag{6.32}$$

Again, we set $[I^{i+}] = m_{I^{i+}}$, and obtain on combining Equations (6.30), (6.31) and (6.32):

$$m_{Na^+}^2 \cdot \left(\frac{2\,\beta_{Ca} K_{Na\backslash Ca}^2}{\beta_{Na}^2} + \frac{2\,\beta_{Mg} K_{Na\backslash Mg}^2}{\beta_{Na}^2} \right) + m_{Na} - 0.0034 = 0$$

This quadratic equation can be readily solved with the fractions β_I in Example 6.8, and with $K_{Na\backslash Ca} = 0.4$, and $K_{Na\backslash Mg} = 0.5$, to give $m_{Na^+} = 3.39 \times 10^{-3}$ mol/L. Further back-substitution in Equations (6.31) and (6.32) gives $m_{Mg^{2+}} = 2 \times 10^{-6}$, and $m_{Ca^{2+}} = 0.5 \times 10^{-6}$ mol/L. This shows that our assumption that Na^+ is the only cation of importance after the salinity jump, was indeed justified.

During dilution, divalent ions are preferentially adsorbed in comparison to monovalent Na^+ (Figure 6.8). When fresh water displaces saltwater, dilution takes place and Na^+ is therefore desorbed from the exchanger (Example 6.9). For the exchange reaction of Ca^{2+} with Na^+:

$$\tfrac{1}{2}Ca^{2+} + Na\text{-}X \leftrightarrow \tfrac{1}{2}Ca\text{-}X_2 + Na^+$$

the law of mass action yields:

$$\frac{[Na^+]}{\sqrt{[Ca^{2+}]}} = K_{Ca\backslash Na} \frac{\sqrt{[Ca\text{-}X_2]}}{[Na\text{-}X]}$$

If Na^+ is diluted 10 times, then Ca^{2+} must be diluted 100 times to maintain equilibrium with the same exchangeable activities [Na-X] and [Ca-X2]. Similarly for Al^{3+} / Na^+ exchange, if Na^+ is diluted 10 times, Al^{3+} must be diluted 1000 times. For ions with the same charge the dilution is the same. The dilution factor f can be used to calculate (more easily than in Example 6.9) the composition behind a salinity front from the solute concentrations ahead of the front:

$$\frac{(m_{Na^+})_0}{f} + \frac{2(m_{Mg^{2+}} + m_{Ca^{2+}})_0}{f^2} = \Sigma\,(i \cdot m_{I^{i+}})_1 \tag{6.33}$$

where the subscripts 0 and 1 indicate concentrations before and after the salinity front, and f is the dilution factor. For Example 6.9 this yields:

$$\frac{485}{f} + \frac{2 \times (55 + 11)}{f^2} = 3.4 \tag{6.34}$$

which is a quadratic equation that can be solved to yield $f = 142.9$ (hence, $m_{Na^+} = 3.39$ mmol/L, as before).

QUESTION:
Calculate the concentrations of Ca^{2+} and Mg^{2+} (mg/L) after the salinity jump in the injection experiment of Valocchi et al. (Figure 6.16). The original concentrations in the aquifer were: $(m_{Na^+})_0 = 86.5, (m_{Mg^{2+}})_0 = 18.2, (m_{Ca^{2+}})_0 = 11.1$ mmol/L. Anions in the injected water sum up to $-\Sigma(i \cdot m_I^{i-})_1 = 14.7$ meq/L.
ANSWER: $f = 6.5$, $Ca^{2+} = 10.5$ mg/L, $Mg^{2+} = 10.5$ mg/L, the same as in Figure 6.16.

The salinity effect is particularly clear when the concentrations decrease behind a salinity front since the exchange complex will control the water composition longer when water is more dilute. But, sometimes the effects are unmistakable when the salinity increases. Ceazan et al. (1989) injected NH_4Br solution in an aquifer with a relatively low total anion concentration of 0.5 meq/L.

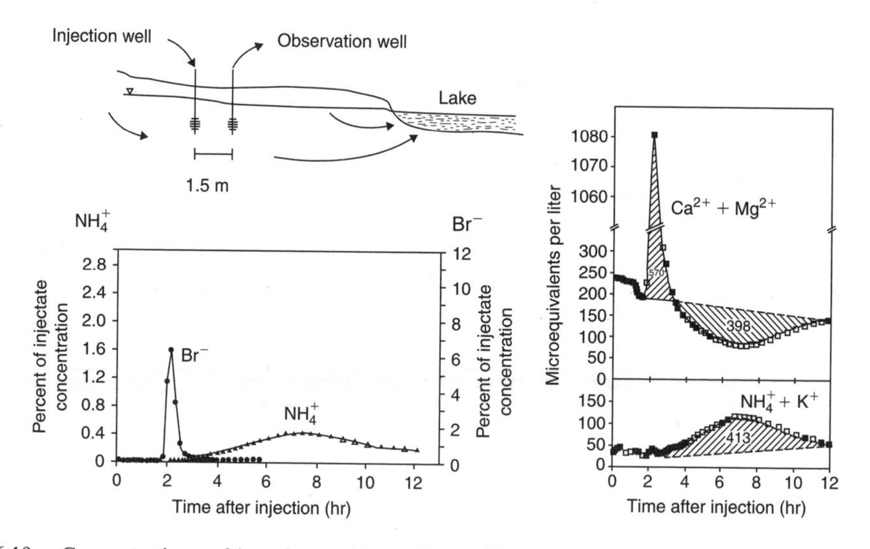

Figure 6.19. Concentrations of ions in an observation well 1.5 m downstream of the point of injection of NH_4Br (Ceazan et al., 1989).

The passage of the injected fluid was monitored in an observation well located 1.5 m downstream from the injection well (Figure 6.19). The arrival of Br^- after 2 hours was accompanied by an increase in the total anion concentration to 1.8 meq/L. The NH_4^+ concentration did not follow Br^- since NH_4^+ was adsorbed upstream. Instead, the divalent cations Ca^{2+} and Mg^{2+} showed a marked increase (Figure 6.19). The ratio $(c_{Ca^{2+}+Mg^{2+}}) / c_{Na^+}$ was 0.67 (meq/meq) in the native water (Table 6.5) and changed to 1.54 at the peak of the Br^- concentration. This is again due to the salinity effect, and, using Equation (6.28) we can calculate that $f = 0.43$, $c_{Na^+} = 0.7$ meq/L and $c_{Ca^{2+}+Mg^{2+}} = 1.08$ meq/L in the peak. These numbers are in excellent agreement with the observations of Ceazan et al. (Figure 6.19).

Table 6.5. Concentrations (meq/L) in the injection experiment of Ceazan et al., 1989.

Time (h)	Br^-	Σanions	Σc^+	Σc^{2+}	c^{2+}/c^+ ratio
1.4	0	0.50	0.3	0.2	0.67
2.08	1.28	1.78	0.7	1.08	1.54

QUESTION:
The Cd^{2+} concentration in resident groundwater in the experiment of Ceazan is estimated to be 1 μg/L. Calculate the concentration at the Br^- peak.

ANSWER: $1 / f^2 = 1 / 0.43^2 = 5.4$ μg/L.

6.4.3 *Quality patterns with salinization*

The development of a chromatographic pattern depends on the ratio of exchangeable and solute concentration, which gives the velocity of a front (Equation 6.27). When the ratio is small, because the *CEC* of the aquifer is low or the solute concentrations are high, the succession of the fronts will be rapid. In case of seawater intrusion, the high concentrations accelerate the transitions of the water types and reduce their lateral extent in the aquifer. This may be the reason why a chromatographic sequence of salinization has not been observed at the field scale, even though $CaCl_2$-water type is an ubiquitous indicator of salt water upconing in coastal areas (Section 6.1). Information on the sequence of compositions with salt water intrusions must therefore be obtained from column or small-scale field experiments (Appelo et al., 1990; Beekman, 1991; Bjerg et al., 1993; Van Breukelen et al., 1998; Gomis-Yagnes et al., 2000; Andersen, 2001), or from models as discussed in the next section.

Beekman and Appelo (1990) performed column experiments, displacing fresh water with seawater diluted 1:1 with distilled water (Figure 6.20). When about 75 mL had been leached from the column, the brackish water had displaced the fresh water and both Ca^{2+} and Mg^{2+} increased because of the exchange with injected Na^+ and the salinity effect, where cations must balance the increased Cl^-. The amount of adsorbed Ca^{2+} was small and therefore Ca^{2+} rapidly dropped again towards the Ca^{2+} concentration in the brackish water. The Mg^{2+} concentration showed a small decrease after the initial salinity jump and then an increase towards the brackish water Mg^{2+} content. The high Ca^{2+} concentrations resulting from the salinity jump, may, in combination with high SO_4^{2-} concentrations, lead to gypsum precipitation (Gomis et al., 2000).

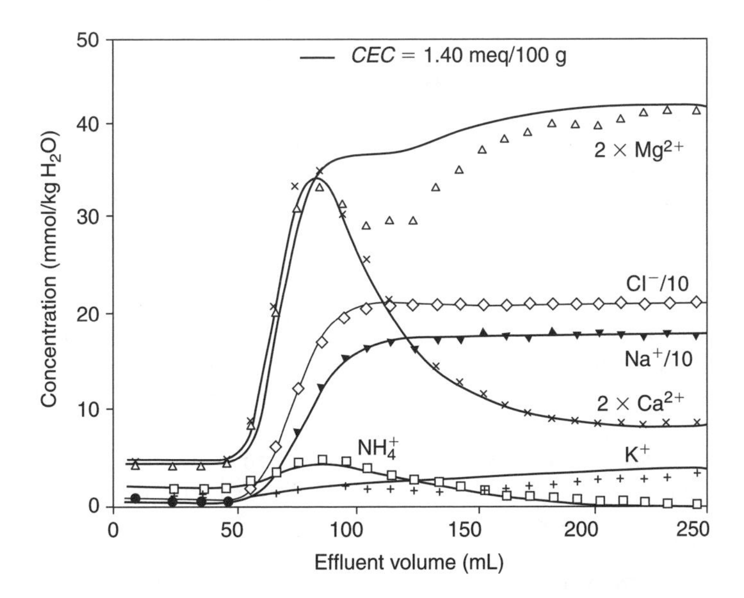

Figure 6.20. Column experiment with once diluted seawater displacing fresh water (Beekman and Appelo, 1990).

The simulated lines in Figure 6.20 were obtained with the Gapon exchange model, in which Ca^{2+} / Mg^{2+} exchange is written as:

$$0.5Mg^{2+} + Ca_{0.5}X \leftrightarrow Mg_{0.5}X + 0.5Ca^{2+} \tag{6.35}$$

which has the mass action equation:

$$\frac{[Ca_{0.5}\text{-}X][Mg^{2+}]^{0.5}}{[Mg_{0.5}\text{-}X][Ca^{2+}]^{0.5}} = K_{Ca\backslash Mg} \tag{6.36}$$

The square roots for the solute activities in this equation induce the particular effect of favoring sorption of small concentrations of Ca^{2+} or Mg^{2+}. Gomis et al. (1996) have shown later that the Mg^{2+} concentration in the experiment is modeled more accurately when all species are given an equal exponent of 1:

$$\frac{[Ca_{0.5}\text{-}X][Mg^{2+}]}{[Mg_{0.5}\text{-}X][Ca^{2+}]} = K_{Ca\backslash Mg} \tag{6.37}$$

(cf. Problem 6.15).

6.4.4 *Fronts and chromatographic sequences*

Chromatographic sequences in heterovalent, multicomponent systems are difficult to calculate, and models that discretize the Advection-Reaction-Dispersion equation may have problems because small concentration variations disappear by numerical dispersion. Van Veldhuizen et al. (1998) developed a model that integrates the matrix of the retardations $\{dq / dc\}$, calculates the position of the fronts and is dispersion-free. However, the program solves only for uniform initial conditions and one input solution. For such systems a few rules can be formulated (Helfferich and Klein, 1970):

- In a system with n ions, a chromatographic sequence develops with n constant compositions ("plateaux") from the initial to the final composition.
- The initial solution composition may change in response to a salinity change.
- Transitions between the plateaux may be sharp (a shock), or smooth (a wave).
- The $n - 1$ transitions involve the sequential reaction of two ions. The first to react are the ions with the lowest affinity; they are not much retarded because dq / dc is small.
- The transition is a shock if the ion with the higher affinity of the two increases in concentration, and a wave if the concentration decreases. All ions with still higher affinity will similarly increase or decrease in concentration.
- The ion with the lower affinity of the two reacts counter to the higher affinity ion (decreases in concentration in a shock, increases in a wave). All ions with still lower affinity react likewise.
- A concentration increase in the injected solution (compared to the initial resident solution) of the ion with the higher affinity of the two gives a shock, whereas concentration decrease in the injected solution normally produces a wave.

Table 6.6 gives for a system with 6 ions, Li^+, Na^+, K^+, Mg^{2+}, Ca^{2+} and Al^{3+}, the initial concentration in a column and the composition of the injected solution. The ions are listed in the order of increasing affinity for the exchanger.

Table 6.6. Initial concentrations (mol/L) in a model column and the composition of the injected solution. $CEC = 0.1$ eq/L. The resulting chromatographic sequence is illustrated in Figure 6.21.

	Li^+		Na^+		K^+		Mg^{2+}		Ca^{2+}		Al^{3+}
Initial	0.015		0.005		0.011		0.004		0.008		0.0015
Injected	0.008		0.015		0.004		0.01		0.006		0.003
Transition		1, shock		2, wave		3, shock		4, wave		5, shock	

The chromatographic sequence after a transport distance of 9 m is shown in Figure 6.21. Six plateaux are visible where the concentrations remain constant, while five transitions exist where all the concentrations change. At transition 1, Na^+ displaces Li^+ and, because the concentration of Na^+ is higher in the injected solution than initially in the column this leads to a sharp front. Next K^+ exchanges for Na^+ (transition 2), and since the injected concentration of K^+ is lower than in the initial column solution, this produces a wave. In this fashion, the rules stated above will become clear when the traces of the individual ions are followed in Figure 6.21. Basically it is the retardation equation which controls the velocity of the fronts, depending on the relative masses of the adsorbed and solute ions. If the change in sorbed mass is relatively high, then more time is necessary for transporting it via a relatively small change in solute concentration.

Manual calculation of the position of sharp fronts was already demonstrated in Section 6.4.1 and is elaborated in Appelo (1994b) and Appelo et al. (1993). Broadening fronts in ion exchange can be solved with a modification of the retardation Equation (6.27):

$$R = 1 + \frac{CEC \; d\beta}{dc} = 1 + \frac{CEC}{A_0} \frac{d\beta}{d\alpha} \tag{6.38}$$

Here $dc = A_0 d\alpha$ where A_0 is the total ion concentration (meq/L) and α the equivalent fraction in solution. By integrating Equation 6.38 it is possible to define the isotherm when the dispersivity in the column is small and the flow velocity is low (DeVault, 1943; Bürgisser et al., 1993; Appelo, 1996, cf. Example 6.10).

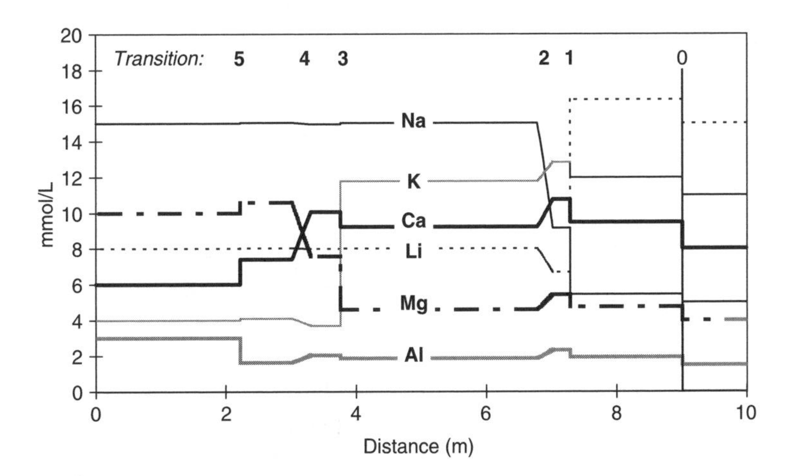

Figure 6.21. A chromatographic sequence along a flowline for a system with heterovalent cations, specified in Table 6.6. The conservative front has traveled 9 m. The transitions were calculated with the model MIE of Van Veldhuizen et al., 1998.

EXAMPLE 6.10. *Flushing of* K^+ from a column

Van Eijkeren and Loch (1984) packed a laboratory column with cation exchange resin-beads. The column was equilibrated with 0.3 meq/L KCl, and then flushed with 0.3 meq/L NaCl. The column length $L = 6.9$ cm; water flux = 56.7 cm/day; dispersion coefficient $D_L = 11.3$ cm²/day. Experimental data of the fraction of K^+ in the effluent, $\alpha_{K^+} = m_{K^+} / (m_{Na^+} + m_{K^+})$, are shown in Figure 6.22. Let us model these data with the retardation Equation (6.38).

A small dispersion coefficient yields a variance $\sigma^2 = 2Dt$ of the concentrations around the midpoint breakthrough. The time needed to flush one pore volume $t = 6.9/56.7 = 0.12$ days. Hence $\sigma = \sqrt{(2Dt)} = 1.67$ cm. Or, scaled to column length, $\sigma = 1.67 / L = 1.67 / 6.9 = 0.24$. The breakthrough curve for a conservative element can now be constructed following Example 3.13, with $c / c_i = 0.84$ at $V = (1 - 0.24)V_0$; $c / c_i = 0.5$ at $V = V_0$; $c/c_i = 0.16$ at $V = (1 + 0.24)V_0$; etc.; V = volume injected in the column. The conservative breakthrough is plotted in Figure 6.22 (line $V_{0,\alpha}$).

The next step is to obtain the *CEC* of the column by integrating the area between the conservative breakthrough and the K^+-experimental points. This gives $CEC = 5.66\ A_0$, where A_0 is the total ion concentration in the effluent (=0.3 meq/L). The exchange equation for K^+ / Na^+ is reworked with $[K^+] = \alpha_{K^+}A_0$ and $[Na^+] = (1 - \alpha_{K^+})A_0$ to:

$$\beta_K = \frac{K_{K\backslash Na} \cdot \alpha_K}{1 - \alpha_K + K_{K\backslash Na} \cdot \alpha_K}$$

and the derivative is:

$$\frac{d\beta_K}{d\alpha_K} = \frac{K_{K\backslash Na}}{\left(1 - \alpha_K + K_{K\backslash Na} \cdot \alpha_K\right)^2}$$

From Equation (6.38), $R - 1 = CEC / A_0 \cdot (d\beta / d\alpha) = V^*$, where V^* is the number of pore volumes that an exchanged element arrives later than the same fraction of a conservative ion. V^* can be read for a few fractions from the experimental data to solve for $K_{K\backslash Na}$. The average is $K_{K\backslash Na} = 2.7$. All variables have now been estimated and the elution curve can be constructed as shown in Figure 6.22. (Rather than multiplying $V_{0,\alpha}$ with R, we added $R - 1$ to $V_{0,\alpha}$, since dispersion for K^+ is mostly determined by the exchange isotherm).

Van Eijkeren and Loch modeled the experimental data with a more intricate model, incorporating mobile and immobile fractions. They used a lower $K_{K\backslash Na} = 1.7$ obtained from batch experiments. It is interesting that a similar discrepancy between *K*'s from batch experiments and a flow experiment was found by Rainwater et al. (1987), who used essentially the same theory as given here. Rainwater et al. modeled a laboratory sand box with injection and withdrawal wells.

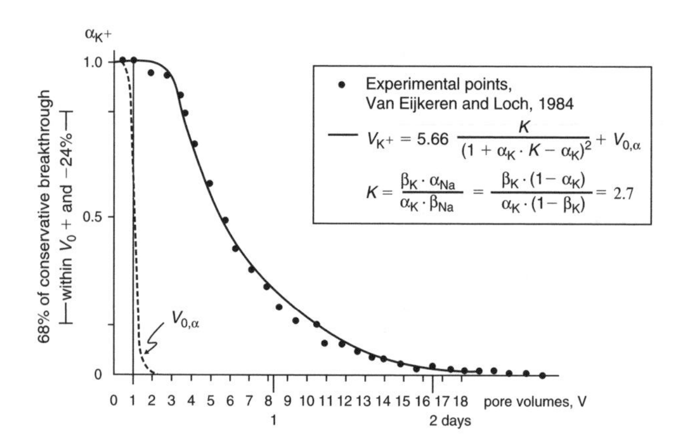

Figure 6.22. The exchange of K^+ by Na^+ in a column packed with cation exchange resin-beads. Dots: relative K^+ concentrations from the experiment (Van Eijkeren and Loch, 1984). Line: modeled results.

QUESTION:
Find $d\beta / d\alpha$ for $\alpha_K = 0.5$ and $\alpha_K = 0.1$, and calculate the retardation for these fractions.
ANSWER: for $\alpha_K = 0.5$: $d\beta / d\alpha = 0.79$, $R = 1 + 5.66 \times 0.79 = 5.47$. For $\alpha_K = 0.1$: $d\beta / d\alpha = 1.97$, $R = 12.2$.

6.4.5 *Modeling chromatographic sequences with PHREEQC*

PHREEQC calculates ion chromatographic sequences via the *ARD* Equation (3.66). The three components of this equation, *Advection, Reaction and Dispersion*, are calculated sequentially for a 1D column or flowline. Figure 6.23 illustrates the discretization of the flowline in a number of cells. One timestep, or a *shift* using PHREEQC terminology, moves the mobile cell contents into the next cell. Subsequently, the reactions between the immobile entities (exchangers, minerals, etc.) and the solution are calculated. Then for each timestep, dispersion is calculated by mixing the contents of adjacent cells. Once more this is followed by calculating reactions between the mobile and immobile entities. During the next timestep, everything is repeated. The mathematical details of the model are discussed in Chapter 11; here we use the model to investigate the combined effects of transport and ion-exchange.

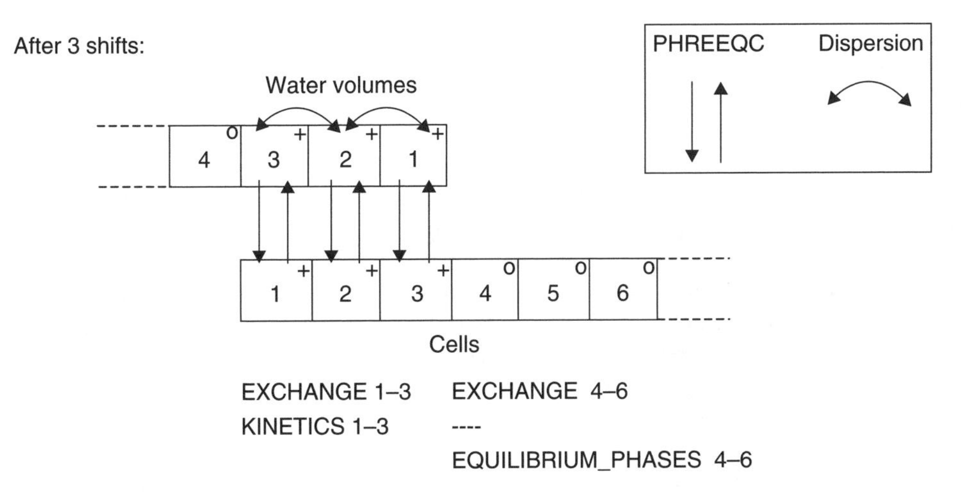

Figure 6.23. PHREEQC in a 1D flowline.

Consider an 8 cm long laboratory column, filled with very coarse sand, $CEC = 1.1$ meq/L pore water. The initial solution is 1 mM $NaNO_3$, and the column is flushed with 0.6 mM $CaCl_2$ solution. The pore water flow velocity is 3.17×10^{-6} m/s (100 m/yr), the dispersivity α is 0.2 mm. First, we investigate whether the reaction front will be broadening or sharpening for these conditions. The following input file provides the isotherm, plotted in Figure 6.24.

```
SOLUTION 1
 Na 1
EXCHANGE 1
 X 1.1e-3; -equil 1
REACTION 1
 Na -1 Ca 0.5; 2.1e-3 in 20 steps
```

```
USER_GRAPH
 -head a_Ca b_Ca
 -start
 10 graph_x 2*tot("Ca")/1e-3
 20 graph_y 2*mol("CaX2")/1.1e-3
 -end
END
```

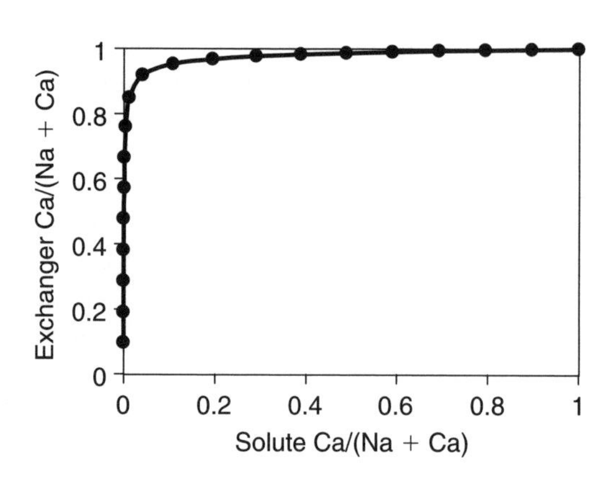

Figure 6.24. Ca/Na exchange isotherm for $CEC = 1.1$ meq/L.

The isotherm is extremely curved, with a steep slope (high retardation $R = 1 + dq / dc$) for small Ca^{2+}-concentrations and a small slope (low retardation) for high Ca^{2+} concentrations. Therefore the front in the column experiment will be self-sharpening.

What results do we expect from the column experiment? The Cl^- front arrives at the outlet after 1 pore volume (*PV*) (Figure 6.25A). The total anion concentration then increases from 1 to 1.2 meq/L and Na^+ must increase as well to 1.2 meq/L. The Ca^{2+} ion arrives later since it is taken up by the exchanger. After one *PV*, all the Na^+ in the effluent comes from the exchanger, until it is exhausted after $\Delta q / \Delta c = 1.1 / 1.2 = 0.917\,PV$. The retardation of the Na^+ front is therefore 1.917. At the same time Ca^{2+} appears at the column outlet.

QUESTIONS:
What is the retardation of the Na^+ / Ca^{2+} front if the $CEC = 2.4$ meq/L?
ANSWER: $1 + 2.4 / 1.2 = 3\,PV$
What is the retardation if the $CEC = 2.2$ meq/L and 1.2 mM $CaCl_2$ is injected?
ANSWER: $1 + 2.2 / (2 \times 1.2) = 1.917\,PV$

The same calculations can be done with PHREEQC using the input file:

```
# Define column...                # right hand: change required for 40 cell column
SOLUTION 1-10                     # SOLUTION 1-40
 Na 1; N(5) 1
EXCHANGE 1-10                     # EXCHANGE 1-40
 X 1.1e-3; -equilibrate 1
END
```

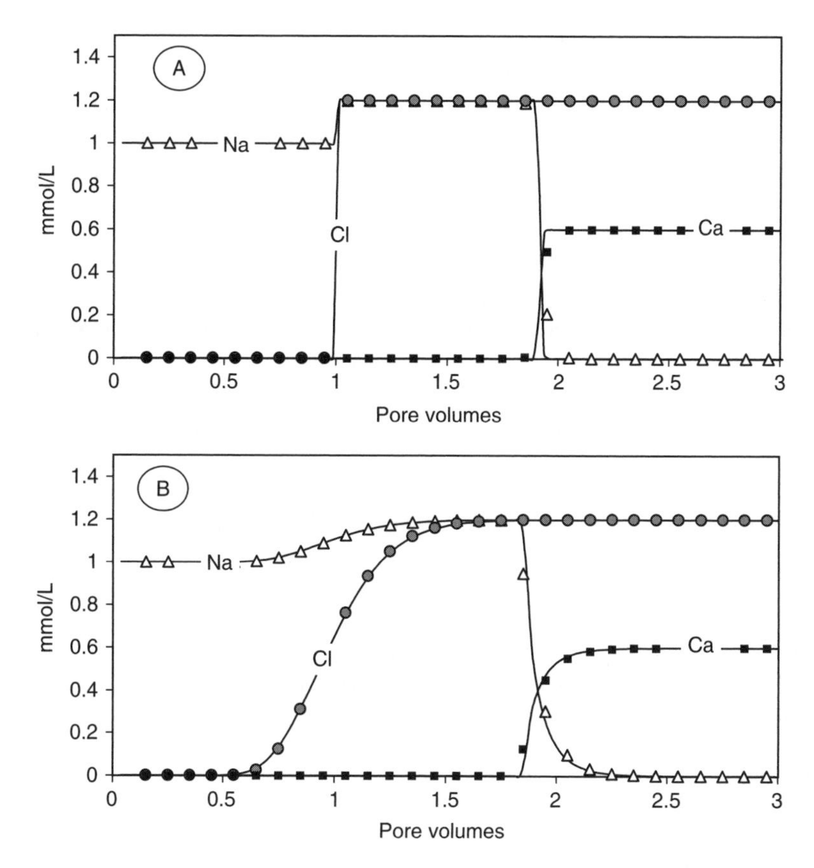

Figure 6.25. A column containing 1 mM $NaNO_3$ solution and 1.1 mM exchangeable Na-X is eluted with a 0.6 mM $CaCl_2$ solution; A: without dispersion, B: with dispersion. Symbols are for a 10 cell model, lines for a 40 cell model.

```
# Define injected solution...
SOLUTION 0
 Ca 0.6; Cl 1.2

TRANSPORT
 -cells 10                              # 40
 -length 8e-3                           # 2e-3 meter
 -shifts 30                             # 120
 -time_step 2523                        # 631 seconds
 -punch 10                              # 40 cell number that is graphed
 -flow_direction forward
 -boundary_conditions flux flux
 -diffusion_coefficient 0.0e-9          # effective D_e, m2/s
 -dispersivity 0                        # meter

USER_GRAPH
 -headings PV Na Cl Ca
 -plot_concentration_vs time
 -start
 10 graph_x (step_no + 0.5) / cell_no
 20 graph_y tot("Na")*1e3, tot("Cl")*1e3, tot("Ca")*1e3
 -end
END
```

In the PHREEQC file, SOLUTION 1–10 and EXCHANGE 1–10 define solution and exchanger for 10 cells in the column. SOLUTION 0 is the injected solution. Keyword TRANSPORT defines the column discretization and the transport steps. The column consists of 10 cells, each with length $\Delta x = 8$ mm. Solution 0 is transported 30 times ("shifts") to the next cell. A total of 30 shifts for 10 cells entails that 3 pore volumes are injected. The time step for one shift is $\Delta t = 2523$ seconds and is the same for all cells. The pore water flow velocity is $v = \Delta x / \Delta t = 8 \times 10^{-3}/2523 = 3.17 \times 10^{-6}$ m/s. To obtain the concentrations in the effluent of the column, the last cell 10 is punched, "-punch 10". The boundary conditions for the column end are of the "flux" type where a given mass enters the column per unit time, without a diffusive contribution (cf. Section 3.6.1). This boundary condition applies to laboratory columns where the in- and outlet tubings have a much smaller diameter than the column and the concentration gradient at the end of the column is therefore zero. The dispersion coefficient, $D_L = D_e + \alpha v$, consists of the *effective diffusion coefficient*, D_e, and the product of flow velocity and *dispersivity*, α, and both are zero in the input file.

PHREEQC calculates concentrations midway in each cell (cell-centered). Since the midpoint concentration arrives half a shift later at the cell boundary, the pore volume eluted (or injected) is the number of (shifts + 0.5) divided by the number of cells in the column:

$$PV = (\text{step\_no} + 0.5)/\text{cell\_no}$$

Here "step_no" is a special BASIC word for shift number and "cell_no" stands for the cell number that is processed. In this case only the last cell is graphed ("punch 10" in TRANSPORT).

The results of this input file are displayed in Figure 6.25A and comply with our manual predictions. Also PHREEQC calculates Ca^{2+} breakthrough at 1.917 *PV*. The best check on the behavior of a numerical model is a comparison with an analytical solution of the problem. Another, easier, but approximate check is to refine the grid in the model and see if the results remain the same. We can change the number of cells to 40 (as indicated on the right side in the input file), with a cell length of 2 mm and 120 shifts, and now punch cell 40. The results of the 40 cell model are identical to the 10 cell model, and clearly, the model behaves well. When we include a dispersivity of 2×10^{-3} m in the input file, Cl^- breakthrough becomes sigmoidal (Figure 6.25B), but Ca^{2+} breakthrough is not much affected because the sharpening front counteracts dispersion.

Next, we add 0.2 mM KNO_3 to the column solution of 1 mM $NaNO_3$. With $K_{Na\backslash K} = 0.2$ in the database, the concentrations of Na-X and K-X become the same. Again, we expect Cl^- to arrive after one *PV* at the end of the column, while Ca^{2+} is retarded because it displaces Na^+ and K^+. In the meantime, the effluent maintains a Na^+ / K^+ ratio of 5 in equilibrium with the initial exchanger, while on the exchanger the Na^+ / K^+ ratio is 1. Therefore, the exchanger is emptied of Na^+ before K^+. The elution of Na^+ ends with the retardation $R_1 = 1 + (0.55 - 0) / (1 - 0) = 1.55$ *PV*. Next the K^+ concentration increases to 1.2 mM to compensate the anion charge. When K^+ is the only cation in solution, K^+ is also the only ion on the exchanger and K-X increases to the exchange capacity of 1.1 mM. K^+ is depleted with the retardation $R_2 = 1 + (1.1 - 0) / (1.2 - 0) = 1.917$ *PV*. Run the file, after having added graphic output for K^+ on line 20 of USER_GRAPH, and observe the result (Figure 6.26).

On average, the modeled front positions agree well with the calculated retardations. After the depletion of Na^+, K^+ increases but does not attain 1.2 mM because dispersion smears the front and K-X is exhausted beforehand. The K^+ / Ca^{2+} front is steep compared to the Na^+ / K^+ transition since the slope of the Ca^{2+} / K^+ isotherm for the final concentration is much smaller than of K^+ / Na^+. When comparing the 10 and 40 cell model results, the modeled Na^+ / K^+ transition is seen to be more affected by coarse discretization than the K^+ / Ca^{2+} front. A coarse discretization may lead to *numerical dispersion*, particularly for an *indifferent* front, but this can be checked by refining the grid. A sharpening front will counteract numerical dispersion, while a broadening front will overrule the numerical dispersion.

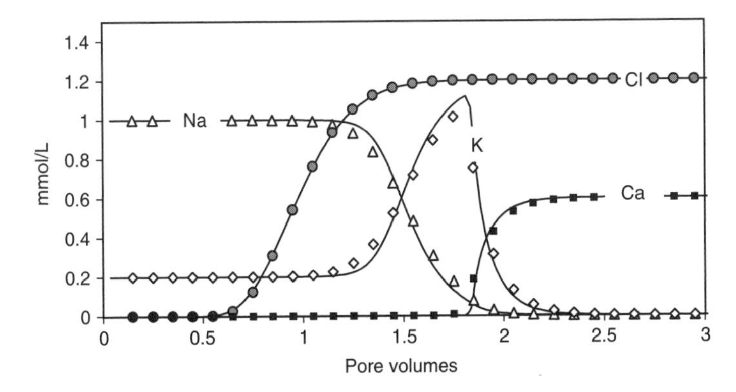

Figure 6.26. A column with 1 mM $NaNO_3$ and 0.2 mM KNO_3 in solution, and 0.55 mM Na-X and 0.55 mM K-X is eluted with a 0.6 mM $CaCl_2$ solution. Points are for a model with 10 cells, lines for 40 cells.

QUESTION:
Rerun the file with 1.198 mM Na^+ and 1 μM Mg^{2+} instead of K^+. In this case the amounts of Na-X and $Mg\text{-}X_2$ are approximately equal. Discuss the shape of the Mg^{2+} / Na^+ and Ca^{2+} / Mg^{2+} fronts.

EXAMPLE 6.11. *PHREEQC model for Valocchi's field injection experiment*
Table 6.7 gives the water compositions of Valocchi's experiment discussed earlier (Figure 6.16). Model the data for well S23 at 16 m from the injection well. The pore volume $V_0 = 295\,m^3$ (a conservative tracer arrives at S23 when 295 m³ have been injected), the dispersivity is 1 m.

Table 6.7. Water compositions in the fresh water injection experiment (Valocchi et al., 1981). $K_{Na\backslash i}$ is for the reaction $1/i \cdot I\text{-}X_i + Na^+ \leftrightarrow 1/i \cdot I^{i+} + Na\text{-}X$. Concentrations in mmol/L.

	Na^+	Mg^{2+}	Ca^{2+}	Cl^-	X^-
Native	86.5	18.2	11.1	160	750
Injected	9.4	0.5	2.13	14.7	
$K_{Na\backslash i}$	1.0	0.54	0.41		

ANSWER:
In the input file we first define the initial aquifer with exchangeable cations. Since the exchange coefficients used by Valocchi are slightly different from those in Table 6.4, the log K values are redefined under EXCHANGE_SPECIES following Equations (6.23–6.25). Next, the injected solution and the transport parameters are defined. We use three cells, with well S23 located at the end of the third cell, but add a fourth cell to mimick an "infinite" aquifer. Each cell has the same volume, and because we have radial flow, the lengths decrease with the square root of distance. The first cell has the length:

$$\text{length(l)} = \text{total\_length} \,/\, \sqrt{n_{tot}} \tag{6.39}$$

and successive cells have

$$\text{length}(n) = \text{length(l)} \cdot (\sqrt{n} - \sqrt{(n-1)}) \tag{6.40}$$

In our case, length(1) $= 16 / \sqrt{3} = 9.24\,\text{m}$.

```
SOLUTION 1-4 Well S23                                    # define aquifer...
 Na 86.5; Mg 18.2; Ca 11.2
 Cl 160.0
EXCHANGE_SPECIES
 Mg+2 + 2X- = MgX2; -log_k 0.54; -gamma 5.5 0.2
 Ca+2 + 2X- = CaX2; -log_k 0.78; -gamma 5.0 0.165
EXCHANGE 1-4
 X 0.7500; -equilibrate 1
PRINT; -reset false                                      # ... note to reduce printout
END

SOLUTION 0                                               # define injected solution and xpt...
 Na 9.4; Mg 0.5; Ca 2.13
 Cl 14.66
TRANSPORT
 -cells    4
 -shifts  400
 -length 9.24 3.83 2.94 2.48
 -dispersivity 1.0
 -punch    3
USER_GRAPH
 -headings Injected Na Cl Mg Ca
 -axis_scale x_axis 100 100000 auto auto true            # use logarithmic axes
 -axis_scale y_axis 0.1 160 auto auto true
 -plot_concentration_vs time
 -start
 10 V0 = 295                                             # m3 from injection well to S23
 20 graph_x V0 * (step_no + 0.5)/cell_no
 30 graph_y 1e3 * tot("Na"), 1e3 * tot("Cl"), 1e3 * tot("Mg"), 1e3 * tot("Ca")
 -end
END
```

The model results are shown in Figure 6.27 and match the concentrations observed by Valocchi et al. well. Note that the length of the first three cells of the model sum to 16 m. The role of the fourth cell is to allow dispersion beyond this point, as happens in the aquifer. When the grid is refined twice, the fronts sharpen only slightly which indicates that numerical dispersion is negligible. However, now two extra cells are needed to describe dispersion beyond the model path correctly, the total length of the additional cells should at least sum up to the dispersivity.

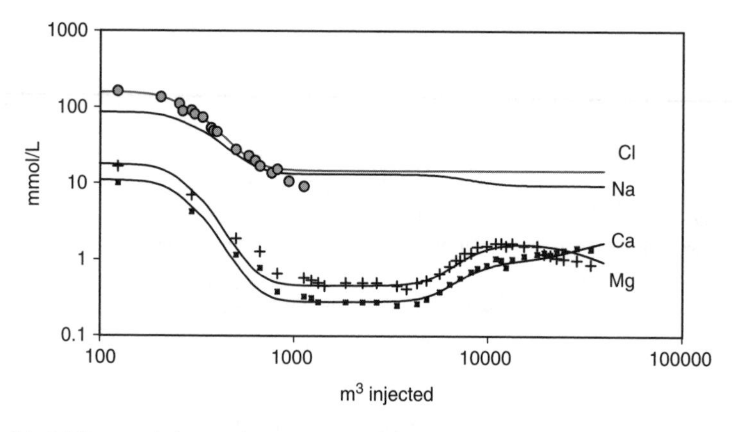

Figure 6.27. Model lines and observed concentrations for the Valocchi case.

QUESTIONS:
Refine the grid twice.
Include equilibrium with Calcite and a CO_2 pressure of 10^{-2} atm in the model; make sure to adapt pH and alkalinity in solution 0 and the aquifer solutions so that the Ca^{2+} concentration remains equal.
Invert the quality pattern, *i.e.* inject brackish water in a fresh aquifer.

Lastly in this section, we compare a model of seawater intrusion by advective transport and by diffusion. The following PHREEQC input file models 20 years advective transport of a twice diluted seawater into a fresh aquifer with a *CEC* of 0.1 eq/L, and calcite and CO_2 as mineral buffers.

```
SOLUTION 1-30                                          # Fresh pore water and sediment...
 pH 7.27
 Na 0.485; Mg 0.8; Ca 2.0
 Cl 0.566; C(4) 3.5
EQUILIBRIUM_PHASES 1-30
 Calcite; CO2(g) -2.0
SAVE solution 1-30
PRINT; -reset false
END                                                    # Note the END (see text)
EXCHANGE 1-30
 X 0.1; -equilibrate 1
END

SOLUTION 0                                             # 1/3 seawater enters the aquifer...
 pH 8.3
 Na 162; Mg 18.4; Ca 3.6
 Cl 189; Alkalinity 2.4
TRANSPORT                                              # 20 yr flow, 5 m/yr, in 150 m...
-cells 30;                 -length 5
-time_step 3.15e7
-flow_direction forward; -shifts 20
-dispersivity 1;           -punch_frequency 20
USER_GRAPH
 -heading dist Na Mg*10 Ca Cl Alk*10
 -init false                                           # -init is shorthand for initial_solutions
 -plot_concentration_vs x
 -start
 10 graph_x dist
 20 graph_y tot("Na")*1e3, tot("Mg")*1e4, tot("Ca")*1e3,\
            tot("Cl")*1e3, Alk*1e4                     # Note line continuation with \
 -end
END
```

First, the pore water is brought to equilibrium with calcite and a $[P_{CO_2}] = 10^{-2}$. The solution is saved and following "END", the next simulation equilibrates the exchanger with the solution. The order of the calculations is important here. If END were left out, then the solution would first equilibrate with the exchanger, after which the mineral reactions would be calculated. The exchanger would then provide buffering. CO_2 is included to balance the pH decrease during calcite precipitation. Normally, there are proton sources and sinks by sorption on mineral surfaces and organic matter, which can be included in the model with SURFACE, as will be discussed in Chapter 7.

The profile for seawater intrusion in Figure 6.28 displays overall the same pattern as Figure 6.20. As seawater intrudes, Na^+ is taken up by the exchanger, releasing Ca^{2+} and Mg^{2+} which display maxima due to the salinity effect. Calcite precipitation is a secondary result and reflected by a decrease in the alkalinity.

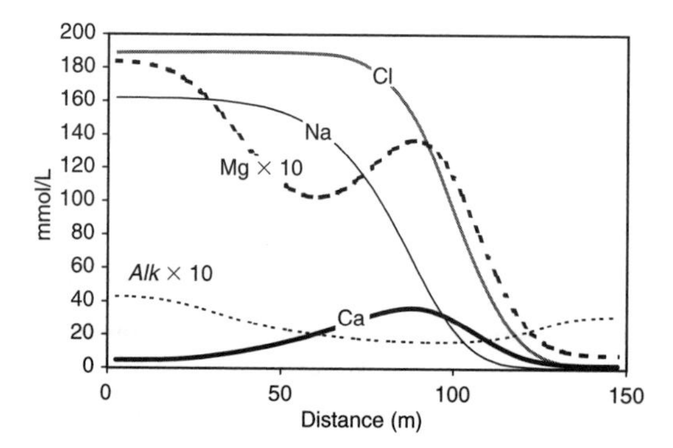

Figure 6.28. PHREEQC simulation of salt water intrusion by advective transport into a fresh water aquifer.

To model diffusion with PHREEQC, only the TRANSPORT keyword needs to be adapted in the previous file.

```
TRANSPORT
 -cells 30;                          -length 0.2
 -time_step 3.15e9;
 -flow_direction diffusion;          -shifts 1
 -boundary_conditions constant closed
 -diffusion_coefficient 0.3e-9
```

The model is now for 100 years diffusion into 6 m sediment. The "flow_direction" is diffusion, and the boundary conditions are "constant" for constant concentrations at the first cell of the column and "closed" at the last cell. Shifts define now the number of time steps that are calculated for diffusion (here $1 \times 3.15 \times 10^9$ give 100 years total time). Figure 6.29 shows the results of the modified file, with concentration patterns that follow more the trace of Cl^- than in the previous case. For example, the Mg^{2+} peak is completely absent.

When seawater diffuses into a stagnant fresh groundwater, part of the exchanged ions from the sediment diffuse away at the same rate as the displacing ions arrive. The difference in diffusion flux of a conservative element and an exchangeable element is partially determined by the *CEC*. When the *CEC* is not too high, the effect of cation exchange on the water composition is modest and less conspicuous than when seawater intrudes the aquifer by flow. In case of diffusive transport the resulting water compositions will resemble those of a simple mixture of fresh and salt water.

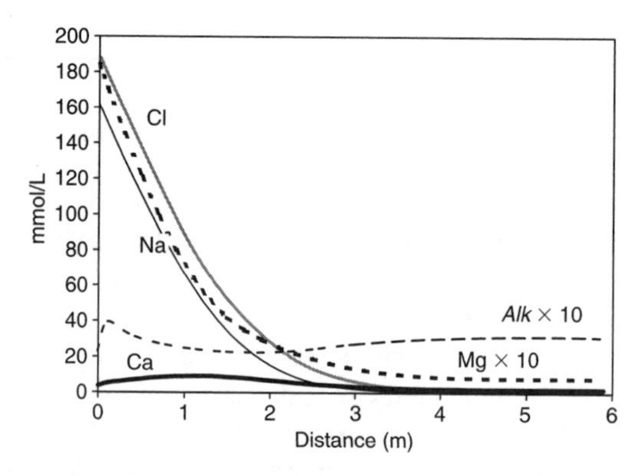

Figure 6.29. PHREEQC simulation of salt water intrusion by diffusion (initial concentrations are identical to Figure 6.28).

6.5 PHYSICAL NON-EQUILIBRIUM

Flow conditions in a porous medium are often inhomogeneous with water flowing in certain parts of the medium and being stagnant within particle aggregates, cemented clusters or organic clods. A diffusive exchange of ions will take place between the stagnant zones and the mobile flow region. Figure 6.30 visualizes the situation. First, there is *film diffusion* through a boundary layer of stagnant water that surrounds the particles. Next *intraparticle diffusion* may occur inside particle aggregates which is a much slower process. Lastly, there are the chemical reaction rates which vary from instantaneous for ion exchange to slow for silicate mineral reactions.

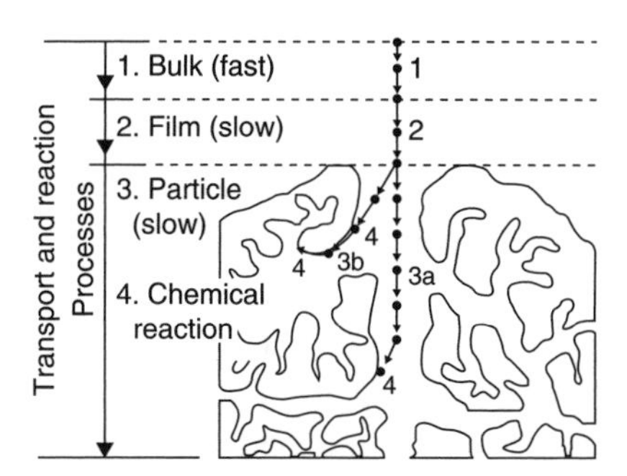

Figure 6.30. Rate determining processes for reactions in a porous medium (Weber and Smith, 1987).

When diffusion into the stagnant zones is unable to equalize the concentration gradients brought about by advective flow, the system is in physical non-equilibrium. An estimate of the pore size that can be homogenized by diffusion is obtained by using the formula (Section 3.5.2):

$$\sigma^2 = 2D_e t \tag{3.60}$$

where σ^2 is the variance of the diffusion (Gaussian) curve, D_e is the effective diffusion coefficient and t is time. The distance covered by diffusion can be compared with the distance traveled by advective flow during the same period of time, as illustrated in Figure 6.31. The lines are drawn for the equality $\sigma = x$, where $\sigma = \sqrt{2D_e t}$ (diffusion) and $x = vt$ (advective transport).

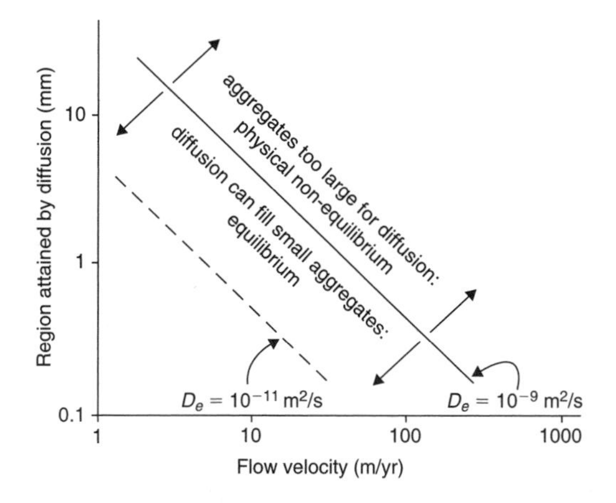

Figure 6.31. Approximate size of stagnant zones that can be homogenized by diffusion as a function of the pore water flow velocity for two diffusion coefficients.

For example, when $D_e = 10^{-9}$ m^2/s, $\sigma = 1$ mm enclosing 68% of the chemical, is reached in $t = (10^{-3})^2/2 \times 10^{-9} = 500$ s. The distance of 1 mm is covered by advective flow in the same time when $v = x / t = 10^{-3}/500 = 2 \times 10^{-6}$ m/s, or 63 m/yr. Clearly, the size of a stagnant zone that will be homogenized with a normal groundwater flow velocity of around 20–200 m/yr is small. For a homogeneous sand the size of the stagnant zone corresponds to the pore size, which is about equal to the particle size below which 10% of the particles exist (Perkins and Johnston, 1963). The size of stagnant zones is also about equal to the dispersivity and an idea of flow velocity required for full equilibrium in columns can be obtained from the dispersion of the breakthrough curve of a conservative tracer. Figure 6.31 shows that column dispersivities of a few mm require flow velocities below about 50 m/yr. Field dispersivities of several meters, on the other hand, indicate stagnant zones that will never attain full equilibrium when the concentrations vary in water that flows around them.

A *mobile fraction* of the porosity can be defined as

$$\Phi_m = \varepsilon_m / \varepsilon_w \tag{6.39}$$

where ε_w is the total water filled pore volume and ε_m is the mobile part.

Figure 6.32 shows how the fraction of mobile water may affect the shape of tritium breakthrough curves. In the upper graph the aggregates are smaller than 2 mm and for the applied flow velocity, all of the pore water is mobile. The lower graph shows results for a column packed with aggregates up to 6 mm.

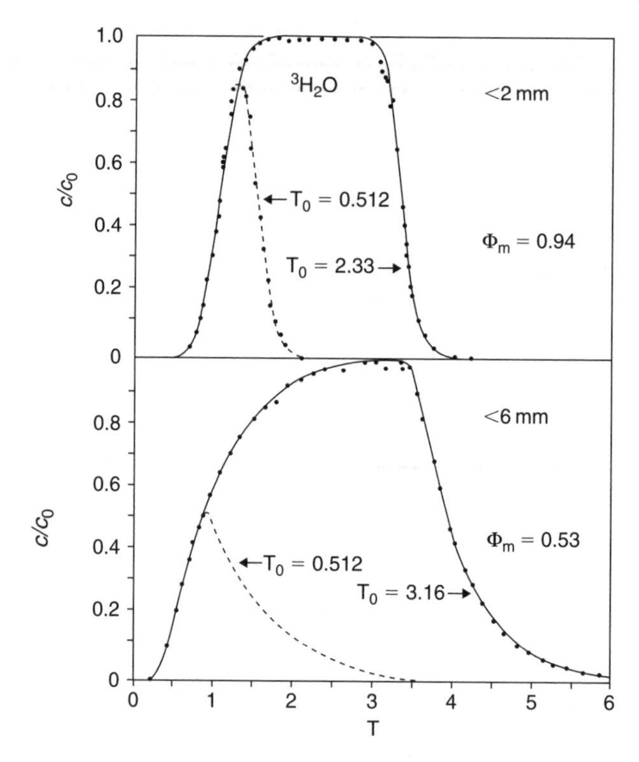

Figure 6.32. Calculated and observed effluent curves for tritium movement through a clay loam for small aggregates <2 mm (upper) and larger aggregates <6 mm (lower graph). T = number of pore volumes; T_0 is injected volume of tritiated water or the change to unlabeled water (Van Genuchten and Cleary, 1982).

In this case the fraction of mobile water Φ_m is 0.53, as obtained by curve-fitting the concentrations in the column effluent. The tailing of the effluent concentrations in the lower graph is due to diffusion into or out of the dead-end pores. The breakthrough of tritium also occurs earlier (Figure 6.32) because the velocity of the liquid in the mobile region increases by a factor $(\Phi_m)^{-1}$ when the volume of the stagnant zone is inaccessible.

6.5.1 *Modeling stagnant zones*

We assume that the mobile and immobile zones consist of two homogeneous boxes which exchange mass depending upon the concentration difference:

$$\mathrm{d}M_{im} \,/\, \mathrm{d}t = V_{im} R_{im} \, \mathrm{d}c_{im} \,/\, \mathrm{d}t = \alpha\,(c_m - c_{im}) \tag{6.40}$$

where the subscript m indicates mobile and im indicates immobile area. M_{im} are moles of chemical in the immobile zone, V_{im} is the volume of the immobile part (m^3), R_{im} is the retardation ($R_{im} = 1 + (\mathrm{d}q \,/\, \mathrm{d}c)_{im}$), c_m is the concentration in the mobile area, and c_{im} the concentration in the immobile part (both in mol/L), and α is the exchange factor or mass transfer coefficient (s^{-1}). The concentration c_{im} is an average for the entire stagnant zone, even though in reality the concentration can be expected to vary with distance from the interface.

A mass increase in the immobile zone is balanced by mass decrease in the mobile zone and *vice versa*:

$$\Delta M_{im} = V_{im} R_{im} \cdot \Delta c_{im} = -\Delta M_m = -V_m R_m \cdot \Delta c_m \tag{6.41}$$

Thus, the concentration changes depend on the relative size of the boxes (V_m, V_{im}), the concentration changes in the solid part (Δq_m, Δq_{im} in the retardation equation), and the exchange factor α. Instead of the volumes V_m and V_{im}, the porosities ε_m and ε_{im} as fractions of the total volume may be used in the formulas.

We can integrate Equation (6.40) with the initial concentrations $c_m = c_{m0}$ and $c_{im} = c_{im0}$ at $t = 0$, and the mass balance (6.41) to obtain:

$$c_{im} = \beta_{im} f \cdot c_{m0} + (1 - \beta_{im} f) \cdot c_{im0} \tag{6.42a}$$

where

$$\beta_{im} = R_m \varepsilon_m \,/\, (R_m \varepsilon_m + R_{im} \varepsilon_{im}) \tag{6.43a}$$

and

$$f = 1 - \exp\left(\frac{-\alpha t}{\beta_{im} R_{im} \varepsilon_{im}} \right) \tag{6.44}$$

The concentration in the mobile box is found similarly:

$$c_m = \beta_m f \cdot c_{im0} + (1 - \beta_m f) \cdot c_{m0} \tag{6.42b}$$

with

$$\beta_m = R_{im} \varepsilon_{im} \,/\, (R_m \varepsilon_m + R_{im} \varepsilon_{im}) \tag{6.43b}$$

Since β is constant and f is also constant for a given simulation time, Equation (6.42) describes a mixing process in which a fraction of the mobile cell mixes into the immobile cell and *vice versa*. For large α and t (or large distance), the exponential term in (6.44) becomes negligible, $f \rightarrow 1$, and mixing is complete. In that case, the solute concentrations in the two boxes become the same, as you may check.

The exchange factor α is a function of the contact surface of the two boxes and can be related to the shape of the immobile zone (Van Genuchten, 1985):

$$\alpha = \frac{D_e \varepsilon_{im}}{(a f_{s\rightarrow 1})^2} \tag{6.45}$$

where D_e is the diffusion coefficient in the stagnant zone, a is the size of the immobile area (an effective diffusion length), and $f_{s\rightarrow 1}$ is a shape factor. The shape factors given by Van Genuchten (1985) are listed in the PHREEQC manual (Table 1, p. 53). For example for spheres, a is the radius of the sphere, and $f_{s\rightarrow 1} = 0.21$.

The effects of stagnant zones can be calculated with PHREEQC, using the option "-stagnant" in keyword TRANSPORT. A two-box model can be simulated by adding a stagnant cell to each mobile cell in the column. Let us calculate again the column in which 1.1 mM Na-X is exchanged by injecting 0.6 mM $CaCl_2$ solution, but now we increase the flow velocity 50 times to 1.59×10^{-4} m/s (5 km/yr). We subdivide the porosity in a mobile ($\varepsilon_m = 0.2$) and a stagnant part ($\varepsilon_{im} = 0.1$). The stagnant part consists of exchanger beads with a diameter $2a = 2$ mm (equal to the dispersivity) and a diffusion coefficient $D_e = 0.3\times10^{-9}$ m²/s. With the shape factor for spheres, $f_{s\rightarrow 1} = 0.21$, the exchange factor $\alpha = 6.8\times10^{-4}$/s. We compare the outflow profiles when the exchange sites are distributed homogeneously over the mobile and immobile cells and when the exchanger is confined to the immobile parts of the column.

```
SOLUTION 1-81                    # mobile 1-40, immobile 42-81
Na 1
EXCHANGE 1-81                    # Distribute exchanger homogeneously...
X 1.1e-3; -equilibrate 1
                                 # Distribute X heterogeneously, ε_im = 0.1 ε_m = 0.2
                                 # All X in mobile, X_m = X * (ε_im + ε_m)/ε_m...
#EXCHANGE 1-40
# X 1.65e-3; -equil 1
                                 # All X in immobile, X_im = X * (ε_im + ε_m)/ε_im ...
#EXCHANGE 42-81
# X 3.3e-3; -equil 1
END

SOLUTION 0; Ca 0.6; Cl 1.2
TRANSPORT
 -cells 40; -length 2e-3; -shifts 176
 -time_step 12.6                 # note smaller timestep: higher flow velocity
 -punch 40
                                 # 1 stagnant layer, α = 6.8e-4/s, ε_m = 0.2, ε_im = 0.1
 -stagnant 1 6.8e-4 0.2 0.1
 -disp 0.002
USER_GRAPH
 -head PV Na Cl Ca
 -plot t
 -start
 10 graph_x (step_no + 0.5) / (cell_no * 1.5)
 20 graph_y tot("Na")*1e3, tot("Cl")*1e3, tot("Ca")*1e3
 -end
PRINT
 -reset false
END
```

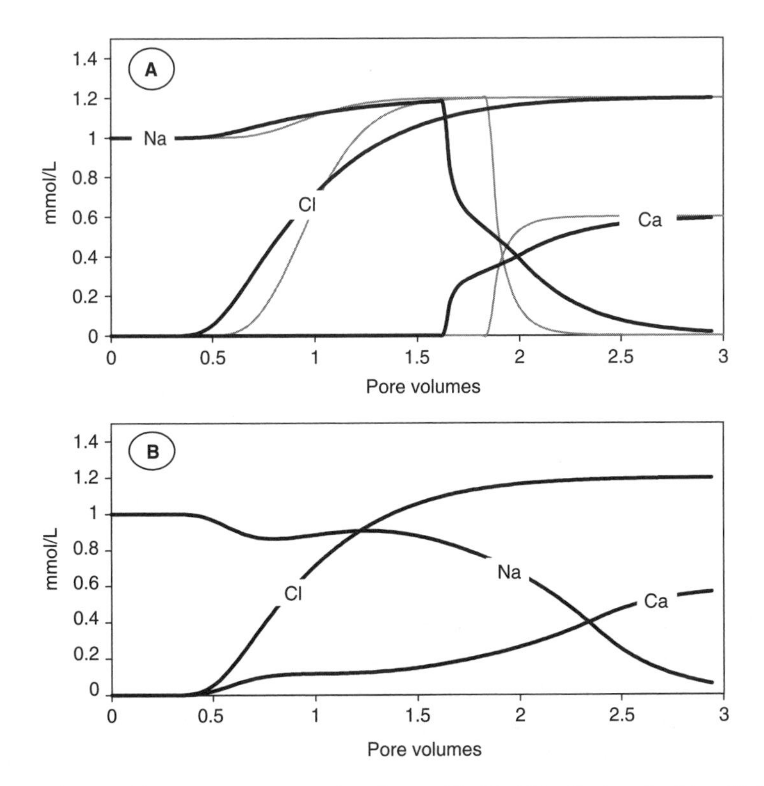

Figure 6.33. A column with stagnant zones and exchangeable Na-X is displaced with a $CaCl_2$ solution. In 6.33A, the thin lines are the concentrations for a column without stagnant zones (from Figure 6.25B), the thick lines are for Na-X distributed homogeneously over the mobile and immobile zones. In 6.33B, the exchange sites are confined to the immobile zone.

In the input file, the mobile cells are numbered 1–40, and the immobile cells are 42–81. Immobile cell 42 mixes with cell 1, cell 43 mixes with cell 2, etc. The exchanger can be assigned to any of these cells and corrected for the volumetric contribution of the cell (stagnant or mobile) to maintain equal total exchange capacity in the column (cf. the input file). Only the solutions of the mobile cells are displaced, so that the number of shifts must be increased to obtain displacement of the total pore volume of mobile plus immobile water. For this same reason, the formula for the x-axis in the plot is modified since the total pore volume is 1.5 times larger than the mobile volume.

Run the file, and compare the results shown in Figure 6.33A with the column without stagnant zones (Figure 6.25B). The Cl^- curve has become more disperse because the ion enters the stagnant zones which initially are free of Cl^-. The Ca^{2+} / Na^+ front arrives earlier because only ⅔ of the total cation exchange capacity is readily accessible. Next we confine the exchange sites to the stagnant zone only (Figure 6.33B). The cations follow now a more gradual curve. A small part of Ca^{2+} arrives together with Cl^- at the end of the column because exchange with the stagnant zone is too slow to take away all the Ca^{2+} from the mobile water.

QUESTIONS:
Will the pattern for Ca^{2+}/Na^+ be different when exchange is limited to the mobile zone?
What happens with the Cl^- concentrations if the flow velocity is doubled to 10 km/yr? And how are the Ca^{2+}/Na^+ concentrations affected? (Run the files)

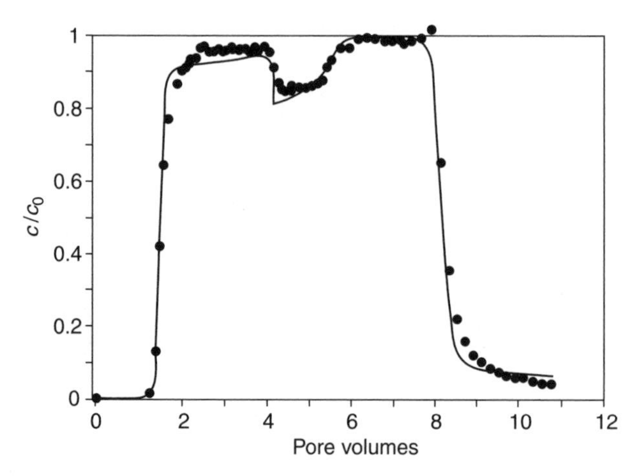

Figure 6.34. Effects of non-equilibrium during flow-interruption. The flow was interrupted after 4 pore volumes (Brusseau et al., 1989).

Dispersion, and chemical and physical non-equilibrium may produce similar effluent curves, and therefore experiments can often be simulated equally well with different models (Nkedi-Kizza et al. 1984). For example, the Cl^- breakthrough in Figure 6.33 for the column with stagnant zones, can be matched almost perfectly by assuming a homogeneous column and a dispersivity of 8 mm (instead of 2 mm). Similarly in Figure 6.32, the tritium curve in the column with stagnant aggregates of 6 mm, can be simulated by increasing the dispersivity to 6 mm. On the other hand, the effluent curves for exchanging or sorbing species are more unique and less easily fitted by changing the dispersivity.

The effect of non-equilibrium in column experiments or aquifer clean-up schemes can be assessed by stop-flow or *flow-interruption* (Brusseau et al., 1989; Koch and Flühler, 1993). In Figure 6.34 the flow was stopped after 4 pore volumes. When flow was reactivated, the concentration had decreased because of diffusion into the immobile zone. Similarly during decreasing concentrations, when the chemical seems leached out, a temporary flow stop may cause an increase in concentration when the flow is restarted because of diffusion out of the immobile zone. During aquifer clean-up an increase in concentration after pumping is temporarily halted may be due to this effect (Bahr, 1989).

QUESTIONS:

Consider the effect of dispersion on Cl^- breakthrough. Make the column of Figure 6.33 homogeneous and change the dispersivity to 8 mm.

Estimate the equilibration time for the column of Figure 6.33. (*Hint*: in Equation 6.44, make αt / $(\beta_{im} R_{im} \varepsilon_{im}) > 4.6$)

ANSWER: $t > 406$ s.

Calculate the effect of flow-interruption for the experiment depicted in Figure 6.33B. Stop the flow for the estimated equilibration time after 1.5 *PV*'s have been injected, then resume the flow. repeat the experiment, but allow for a 10 times longer equilibration time. Is the estimated equilibration time correct?

6.6 THE GOUY-CHAPMAN THEORY OF THE DOUBLE LAYER

We have considered macroscopic equations that relate concentrations in solution to those on the exchanger surface. But if we want to describe what occurs mechanistically when a charged surface attracts or repels ions from a solution, we must go down to the molecular level (Moore, 1972; Feynman et al., 1989). In our case, we could consider the cations and anions as point charges which can move in an electric field. A bulk solution in which the number of cations and anions are equal, has zero electric potential. A negatively charged surface has a negative potential that attracts positively charged ions and repels negative ones. However, diffusion will smear out concentration-steps, and cause the transitions to become more gradual. The net effect (Figure 6.35) is a gradual change in concentrations in the so-called Gouy-Chapman *diffuse double layer* (*DDL*) which surrounds the charged surfaces in a solution.

The electric potential will not change abruptly at the solid-solution interface, but gradually lessen as distance from the interface increases. The relative concentrations in the electric field are given by Boltzmann's law:

$$m_{i,a} \,/\, m_{i,b} = \exp\left(-z_i F \psi_a \,/\, RT\right) \,/\, \exp\left(-z_i F \psi_b \,/\, RT\right) \tag{6.46}$$

where m_i is the concentration of ion i (kmol/m^3; formally the *activity* should be used) with charge number z_i, F is Faraday's constant (C/mol), ψ is potential (Volt), R is the gas constant (J/mol/K), and T absolute temperature (K). Subscripts a and b indicate the position in the electric field.

In the electrically neutral solution at infinite distance from the surface, the potential $\psi_\infty = 0$. Therefore:

$$m_i \,/\, m_{i,\infty} = \exp\left(-z_i F \psi \,/\, RT\right) \tag{6.47}$$

Any concentration difference between anions and cations will create a charge density ρ (C/m^3):

$$\rho = F\, \Sigma\, z_i\, m_i = F\, \Sigma\, z_i\, m_{i,\infty} \exp\left(-z_i F \psi \,/\, RT\right) \tag{6.48}$$

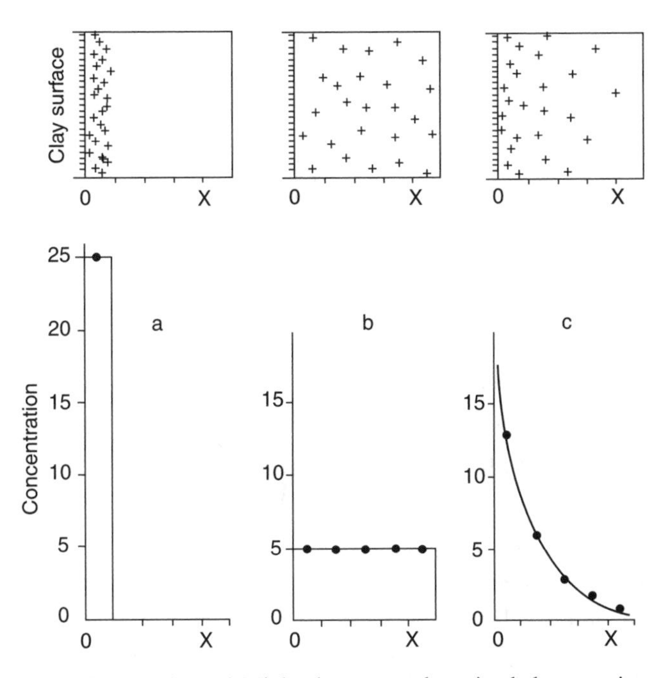

Figure 6.35. Distribution of counterions. a) Minimal energy and maximal electrostatic attraction; b) maximal concentration distribution and entropy; c) compromise: the distribution in a diffuse double layer (Bolt and Bruggenwert, 1978).

Let us calculate, for a planar surface, how the concentrations change in the x-direction. The electric field strength changes with the charge density as:

$$dE / dx = \rho_x / \varepsilon \tag{6.49}$$

where ε is the dielectric constant of water, 6.95×10^{-10} F/m at 25°C.
A stronger electric field strength conforms to a steeper potential gradient:

$$E = -d\psi / dx \tag{6.50}$$

Combining the three equations gives the Poisson-Boltzmann equation:

$$d^2\psi / dx^2 = -(F / \varepsilon) \Sigma z_i \, m_{i,\infty} \exp(-z_i F\psi / RT) \tag{6.51}$$

This equation can be integrated once (first multiply both sides with $2(d\psi / dx)$, then integrate over x):

$$(d\psi / dx)^2 = 2(RT / \varepsilon) \Sigma m_{i,\infty} \exp\left(-z_i F\psi / RT\right) + constant \tag{6.52}$$

In the free solution, $d\psi / dx = 0$ and $\psi = 0$, and the constant is:

$$constant = -2(\mathrm{RT} / \varepsilon) \Sigma m_{i,\infty} \tag{6.53}$$

Therefore, the *fundamental* double layer equation for a planar surface is:

$$(d\psi / dx)^2 = 2(RT / \varepsilon) \Sigma m_{i,\infty} \left\{\exp\left(-z_i F\psi / RT\right) - 1\right\} \tag{6.54}$$

The next step is to integrate once more for simplified conditions, or to use a numerical integrator. For a simple mono-monovalent solution with $z_i = \pm 1$, the double layer equation is:

$$(d\psi / dx)^2 = 2(RT / \varepsilon) \, m_{i,\infty} \left\{\exp\left(-F\psi / RT\right) + \exp\left(F\psi / RT\right) - 2\right\} \tag{6.55}$$

or

$$d\psi / dx = \sqrt{(2(RT / \varepsilon) \, m_{i,\infty}) \left\{2\cosh(F\psi / RT) - 2\right\}} \tag{6.56}$$

or

$$d\psi / dx = \sqrt{(8(RT / \varepsilon) \, m_{i,\infty})} \sinh(F\psi / 2RT) \tag{6.57}$$

With $\psi = \psi_0$ at $x = 0$, the latter integrates to:

$$\psi = \frac{1}{b} \ln\left(\tanh\left(\frac{-bax}{2} + \tfrac{1}{2} \ln\left(-\frac{\exp(b\psi_0) + 1}{\exp(b\psi_0) - 1}\right)\right)\right) \tag{6.58}$$

where $b = F / (2RT)$ and $a = \sqrt{8(RT / \varepsilon) \cdot m_{i,\infty}}$.

The potential can now be calculated for any x, and subsequently the concentrations of the ions in the double layer follow from the Boltzmann equation. An example is provided in Figure 6.36. At the negatively charged surface, the cations are *counter-ions* and attracted to the surface, while the anions are *co-ions* and repelled from the surface. Accordingly, the cations increase in concentration towards the surface, while the anions decrease in concentration. The surface charge is balanced both by the counter-ion excess, and by the co-ion deficit or co-ion exclusion.

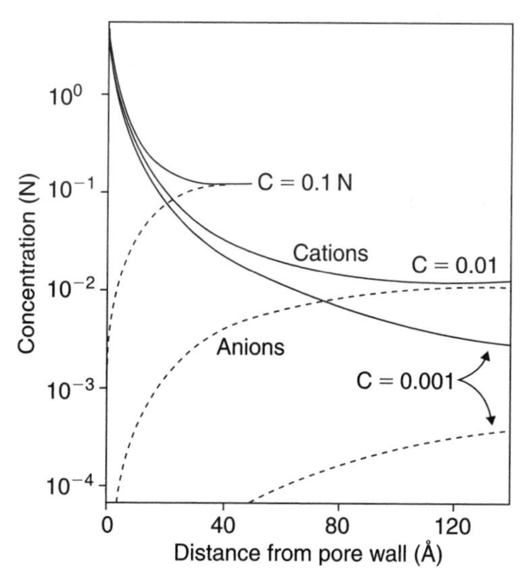

Figure 6.36. Distribution of cations (full lines) and anions (dotted lines) in the double layer on a negatively charged surface at 3 concentration levels in the free solution (Nielsen et al., 1986. Copyright by the Am. Geophys. Union).

The potential at the surface can also be calculated simply for a mono-monovalent solution. The charge of the double layer, σ_{DL}, is equal and opposite to the charge of the surface:

$$\sigma_{DL} = F\sigma_D / A_s = -F\sigma / A_s, \tag{6.59}$$

where σ_{DL} is in C/m^2 and σ_D and σ indicate the charge in moles of the double layer and on the surface, respectively.

The double layer charge equals the integrated charge density:

$$\sigma_{DL} = \int_0^{\infty} \rho \, dx = -\varepsilon \int_0^{\infty} \frac{d^2\psi}{dx^2} \, dx \tag{6.60}$$

After integration we obtain:

$$\sigma_{DL} = + \ \varepsilon(d\psi / dx)_{x=0} \tag{6.61}$$

With (6.57) and $\psi = \psi_0$ at $x = 0$ we obtain the Gouy-Chapman double layer equation:

$$\sigma_{DL} = \sqrt{(8(RT / \varepsilon)\, m_{i,\infty})} \ \sinh(F\psi_0 / 2RT) \tag{6.62}$$

The above derivation is for a mono-monovalent solution, but conventionally, the equation is applied also for an equivalent ionic strength I:

$$I = \tfrac{1}{2} \Sigma m_i z_i^2 \tag{6.63}$$

Thus,

$$\sigma_{DL} = 0.1174 \ I^{0.5} \ \sinh(F\psi_0 / 2RT) \tag{6.64}$$

EXAMPLE 6.12. *Calculate the surface potential for* 1.07×10^{-6} *eq/m*^{2} *surface charge on montmorillonite* in a 0.1 M NaCl and in a 0.025 M $MgSO_4$ solution.
Use the identities $\sinh(x) = (e^x - e^{-x})/2$ and $\text{arcsinh}(y) = \ln(y + (y^2 + 1)^{1/2})$.

ANSWER:
For both solutions, $I = 0.1$ mol/L. Thus, $F\ \sigma = 96485\times1.07\times10^{-6} = 0.103$ C/m^2. Divide by $0.117 \times (0.1)^{0.5}$ and find 2.78. Hence $F\psi_0 / 2RT = \ln(2.78 + (2.78^2 + 1)^{0.5}) = 1.75$, and $\psi_0 = 89.7$ mV.
However, for a di-divalent solution, the correct equation is (note the disappearance of the factor 2):

$$\sigma_{DL} = 0.117\ m_{i,\infty}^{0.5} \sinh(F\psi_0 / RT) \tag{6.65}$$

For the $MgSO_4$ solution we divide 0.103 C/m^2 by $0.117 \times (0.025)^{0.5}$, and find 5.56. Hence, $F\psi_0 / RT = 2.42$, and $\psi_0 = 62.1$ mV.

For small arguments, $\exp(\pm x) \approx (1 \pm x)$. Therefore, when $\psi < 25$ mV, $\sinh(F\psi / 2RT) \approx F\psi / 2RT$, and we obtain the simple formula (at 25°C):

$$\sigma_{DL} = 2.29\ I^{0.5}\ \psi_0 \tag{6.66}$$

Already at an earlier stage, the approximation for small ψ can be applied to the Poisson-Boltzmann Equation (6.52) for a homovalent solution:

$$d^2\psi / dx^2 = 2(F^2/\varepsilon RT)\ z_i^2\ m_{i,\infty}\ \psi \tag{6.67}$$

which can be simplified using

$$\kappa^2 = 2(F^2/\varepsilon RT)\ z_i^2\ m_{i,\infty} \tag{6.68}$$

to:

$$d^2\psi / dx^2 = \kappa^2\psi \tag{6.69}$$

This equation is much easier to integrate, and yields (check by differentiating!):

$$\psi = \psi_0 \exp(-\kappa x). \tag{6.70}$$

In essence the potential decays exponentially from the charged surface, and so will the extent of a diffuse double layer, given the exponential concentration changes of cations and anions.

The units of κ^2 are m^{-2}, as can be verified from Equation (6.68). The parameter κ^{-1} is termed the Debye length; this parameter also appears in the Debye-Hückel theory for calculating ion activity coefficients. The ionic strength can be incorporated in the Debye length to give:

$$\kappa^{-1} = \sqrt{\frac{\varepsilon RT}{2(N_a q_e)^2} \frac{1}{I}} \tag{6.71}$$

With $\varepsilon = 6.95\times10^{-10}$ F/m, $R = 8.314$ J/K/mol, $T = 298$ K, $N_a = 6.022\times10^{23}$/mol, $q_e = 1.6\times10^{-19}$C, and I expressed in mol/dm^3, Equation (6.71) becomes:

$$\kappa^{-1} = 3.09/\sqrt{I} \quad (\text{Å}) \tag{6.72}$$

It is customary to consider the point where $\psi = \psi_0/e$, i.e. where $x = \kappa^{-1}$, as the *thickness* of the *DDL*. The influence of the ionic strength on the thickness of the *DDL* is shown in Figure 6.37.

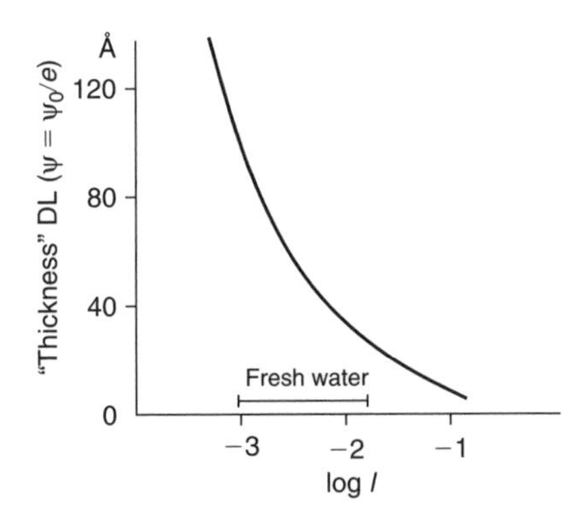

Figure 6.37. The extension of the double layer ("thickness") as a function of the ionic strength *I*.

6.6.1. *Numerical integration of the double layer equations*

The double layer equations have been used to estimate selectivities among mono- and divalent ions (Eriksson, 1952; Bolt, 1967). Equation (6.55) can also be integrated numerically, which allows to include the closest approach distance of the ions to the surface (Neal and Cooper, 1983). Ions with a large hydration shell (given by the ion-size parameter in the Debye-Hückel equation, Table 4.2) cannot come close to the surface and will thus exhibit a relatively small selectivity. In the following we use the solver for ordinary differential equations (ode's) provided in Matlab. The instructions for Matlab are written in the two socalled "*m*" files given in Table 6.8. The calculation is for a system with Na^+, Ca^{2+} and Cl^-. In the first *m*-file some basic definitions of the physical constants, Debye-Hückel parameters and concentrations in the free solution are provided. Subsequently, the solver is called with:

$$[\text{psi, x}] = \text{ode45}(\text{“x\_psidot”}, \psi_0, \psi_x, x_0, \textit{tolerance});$$

Matlab will give $\psi(x)$ in the form of two vectors from ψ_0 at x_0 up to ψ_∞ at x. In the *m*-file in Table 6.8, ψ_0 is -0.1 V, x_0 is 3.5 Å (which is the Debye-Hückel å parameter for Cl^-, the smallest of the three ions), ψ_∞ is -1×10^{-11}, and x is obtained from the integration.

The statement "x_psidot" in ode45 invokes the *m*-file with the differential equation. In this file, the sum of the ion-concentrations is calculated (i.e. the right hand side of Equation (6.55)) and returned in inverted form, $(dx / d\psi)$. Thus, the integration is along ψ, and the *tolerance* of 1×10^{-17} applies to x (m). We *could* integrate along x, going piecewise from the smallest to the largest Debye-Hückel value and account explicitly for the singularities at the closest ion-approach distances. However, we do not know exactly the distance where the potential is so small that it can be considered zero ($\approx -1\times10^{-11}$) which may cause the integration to fluctuate if it is continued too far along x. The Runge-Kutta procedure of ode45 is quite stable and accounts for the singularities by a step-size decrease at those points.

After integration of the double layer equation, the program calculates the concentrations as a function of x with the Boltzmann equation, and plots them in a graph similar to Figure 6.38. The example calculation is for a solution with 8 mM Na^+, 1 mM Ca^{2+}, and 10 mM Cl^-. Several things can be noted in Figure 6.38. First, the cations increase very rapidly towards the surface. For example, the Ca^{2+} concentration at 5 Å from the surface (1.5 Å from the point where the potential is -0.1 V) is about 1.6 mol/L. Second, the amount of Ca^{2+} in the double layer is much larger than of Na^+, despite the equivalent ratio of 1 : 4 in solution. Third, the concentration of Na^+ descends more quickly when Ca^{2+} enters the double layer, as can be seen in the break of the Na^+ curve at 5 Å.

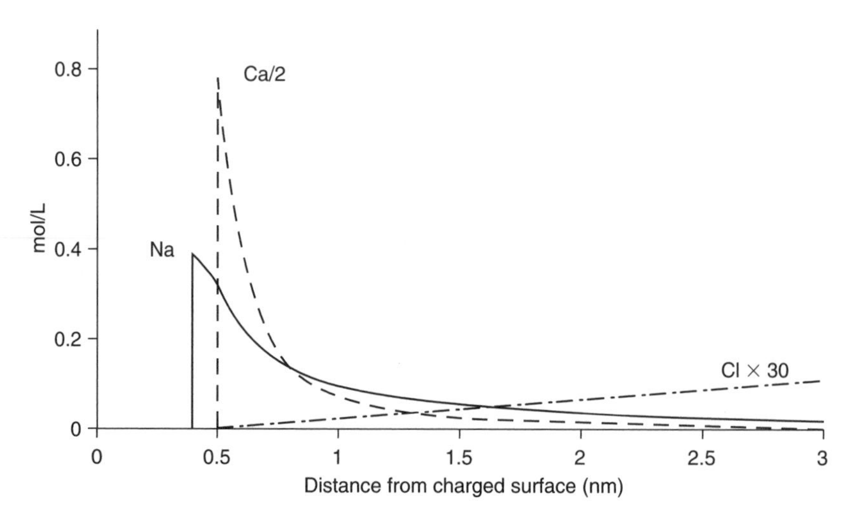

Figure 6.38. Concentrations of Na^+, Ca^{2+} and Cl^- in a double layer in which the approach to the charged surface is limited by the Debye-Hückel å parameter.

The high proportion of Ca^{2+} is due to the quadratic weight of charge in the Boltzmann term (exp(z)). The increased descent of the Na^+ concentration is related to a more rapid decrease in the charge density when the divalent Ca^{2+} ion is also present.

Table 6.8. Matlab program to calculate concentrations in the double layer at a charged surface, with varying closest approach distance for different ions. Note: the % sign indicates a comment.

```
------------------------------------ file: dl.m ------------------------------------
function dl % (% = Comment) integrates double layer

% Uses : x_psidot.m

global Na Ca Cl Na_0 Ca_0 Cl_0 dha_Na dha_Ca dha_Cl;
global RT F epw x        psi;

%          J/kmol       C/kmol        e_w          e_0
RT = 2477.57e3; F = 96485.e3; epw = 78.5 * 8.854e-12;

% Closest distance to surface = Debye-Huckel A (m) ...
 dha_Na = 4.0e-10; dha_Ca = 5.0e-10; dha_Cl = 3.5e-10;

% conc's in kmol/m3 at x = inf, psi = 0 ...
Na_0 = 0.008; Ca_0 = 0.001; Cl_0 = 0.01;

% Call Matlab ODE solver ...
%                              psi_0     psi_end     x_0    tol
 [psi, x] = ode45('x_psidot', -0.100, -1.e-11, dha_Cl, 1e-17);

% Calculate c = c_0 * exp(-zFpsi/RT) ...
lx = length(x); Na = zeros(lx,1); Ca = zeros(lx,1); Cl = zeros(lx,1);
d = exp(-F*psi/RT);
Na = Na_0 * d; Ca = Ca_0 * d.*d; Cl = Cl_0 ./ d;
% c = 0 for x < DH A param ...
for i = 1:lx, if x(i) < 5.1e-10
 if x(i) < dha_Na, Na(i) = 0; end; if x(i) < dha_Ca, Ca(i) = 0; end;
 if x(i) < dha_Cl, Cl(i) = 0; end, else break, end, end;
```

Table 6.8. (*Contd*)

```
figure(1);
clf; hold on; axis([0, 30e-10, 0, 1]);
plot(x, 30*Cl, 'y:'); plot(x, Na, 'm-'); plot(x, 0.5*Ca, 'c- -');
```

```
-------------------------------- file:x_psidot.m   --------------------------------
function ans = x_psidot(psi, x);

% derivative for ode ...
% dx/d(psi) = (2RT/epw * SUM c_i*[exp(-z_i F psi / RT) - 1])^-.5

global Na_0 Ca_0 Cl_0 dha_Na dha_Ca dha_Cl RT F epw;

d = exp(-F*psi/RT);

if x >= dha_Na,  d1 = Na_0 * (d - 1)   ; else d1 = 0; end;
if x >= dha_Ca,  d2 = Ca_0 * (d*d - 1); else d2 = 0; end;
if x >= dha_Cl,  d3 = Cl_0 * (1/d - 1); else d3 = 0; end;
dall = d1 + d2 + d3; if dall == 0,  dall = 1e-21; end;

ans = (2 * RT / epw * abs(dall))^(-0.5);
```

The program is easily extended to more ions. Figure 6.39 shows a multicomponent example with Na^+, K^+, Mg^{2+} and Ca^{2+} for the solution composition in Table 6.9. As before, the figure shows peaks for the various ions at the closest approach distance. The traces for $Ca^{2+}/5$ and Mg^{2+} in this figure are equal since the concentration of Ca^{2+} in the free solution is 5 times higher than of Mg^{2+}. Otherwise, the relations among the approach distances, head-end potentials and the relative enrichment of an ion in the double layer are not simple to generalize. However, some feeling of the diffuse double layer's behavior can be gained by calculating exchange constants $K_{Na\backslash I}$ and comparing with empirical cation exchange constants noted in Table 6.4.

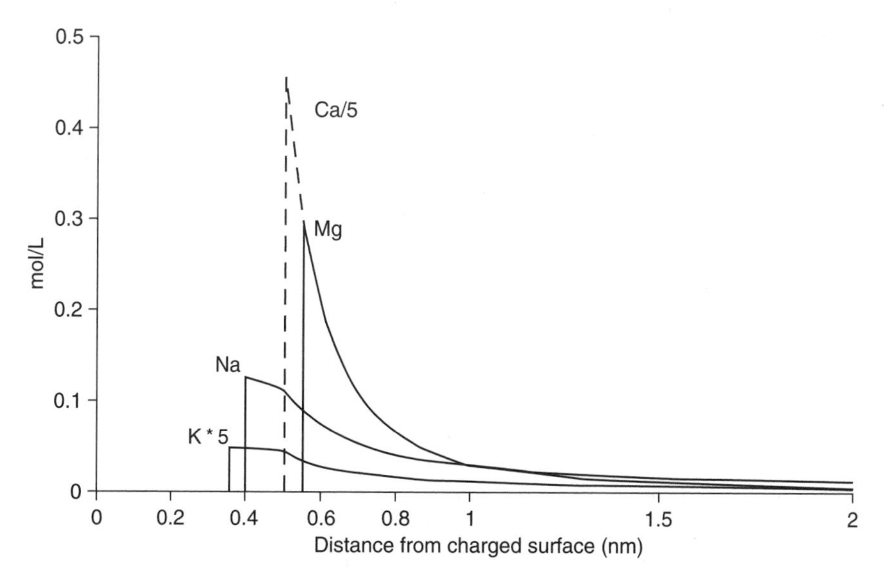

Figure 6.39. Concentrations of Na^+, K^+, Mg^{2+} and Ca^{2+} in the double layer, with variable closest approach to the surface for the different ions.

Table 6.9. Apparent cation exchange selectivities with respect to Na^+ from double layer accumulation at a surface with charge $-0.1\ C/m^2$ (cf. Example 6.1). The 0.01 M Cl^- solution contains (in mmol/L): Na^+ 3.7; K^+ 0.3; Mg^{2+} 0.5; Ca^{2+} 2.5. The 0.1 M Cl^- solution is 10 times more concentrated. Debye-Hückel å parameters are 4, 3.5, 5, 5.5 and 3.5 Å for Na^+, K^+, Mg^{2+}, Ca^{2+} and Cl^-, respectively.

Cl^- (mol/L)	ψ_0 (V)	$K_{Na\backslash K}$	$K_{Na\backslash Mg}$	$K_{Na\backslash Ca}$	σ_{Cl} (%)
0.01	−0.091	0.91	1.13	0.98	3.5
0.1	−0.063	0.87	0.82	0.71	10.6
Table 6.4:		0.2	0.5	0.4	

The integrated amounts in the double layer give *I*-X, which can be used to calculate selectivities according to:

$$Na^+ + 1/i \cdot I\text{-}X_i \leftrightarrow Na\text{-}X + 1/i \cdot I^{i+}; \quad K_{Na\backslash I}$$

Results for cation concentrations which are near to the (present day) average Rhine water in Holland, and for a 10 times more concentrated water, are given in Table 6.9.

The *K* values in Table 6.9 are all close to 1.0, both for K^+ and the divalent cations. When concentrations increase in the free solution, the *K* value for the divalent cations decreases, indicating more enrichment for the divalent ions. The decrease in *K* is less than was calculated by Bolt (1982) with approximative analytical solutions. The relative constancy of the double layer *K* indicates that exchange formulas can be used to estimate double layer contents. Furthermore, the observed exchange constants (from Table 6.4) are significantly smaller than the estimated double layer counterparts. In other words, the greater *observed* selectivity for ions other than Na^+ is not captured by the double layer theory, at least not with the standard Debye-Hückel parameters. For example, to obtain the observed $K_{Na\backslash K} = 0.2$ with double layer theory, the approach distance for K^+ must be diminished to zero and ψ_0 must be lowered to -0.2 V. For the divalent ions, the observed selectivity can be obtained by diminishing the approach distance to 0 and 0.5 Å for Ca^{2+} and Mg^{2+}, respectively. This suggests that inner-sphere bonds are contributing to ion exchange of divalent ions.

In Table 6.9 the charge compensation by anion exclusion is given as a percentage of the total charge of $-0.1\ C/m^2$, amounting to 3 and 10% for the 0.01 and 0.1 M Cl^- solutions, respectively. Actually, most of the structural charge of clay minerals such as montmorillonite is balanced in the interlayer region, which probably contains only water and cations (Karaborni et al., 1996).

The double layer theory was developed here for a planar surface, which permits an easy first integration step. All other shapes, even spheres, require integration of the second order Poisson-Boltzmann equation, for example with a central differences algorithm as is used for calculating diffusion (cf. Chapter 11). Alternatively, for spherical shapes, the polynomial approximations derived by Bartschat et al. (1992) can be used.

6.6.2 *Practical aspects of double layer theory*

The thickness of the double layer (Figure 6.37) indicates how far a negative potential from a clay mineral surface extends into solution. Clay particles are repelled from each other when the negative potentials from opposite clay particles overlap. A suspension of clay particles becomes stable when the double layer is so thick that the clay particles repel each other: the colloidal solution remains then *peptized*. A thin double layer on the other hand, may induce *flocculation*, the suspension is separated in a clear supernatant solution and a flocculated clay. The double layer thickness increases with decreasing ionic strength and with decreasing charge of the counterions. If a $CaSO_4$ solution is replaced by a NaCl solution of equal molality, the ionic strength decreases by a factor of 4, and the Debye length increases by 2.

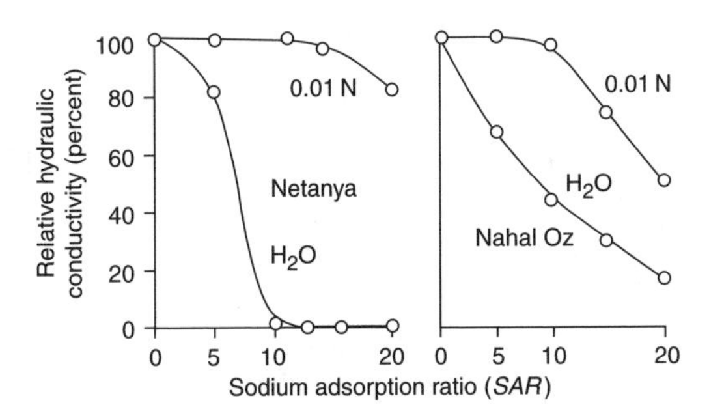

Figure 6.40. Hydraulic conductivity of a sandy loam (Netanya) and a silty loam (Nahal Oz) soil as a function of the *SAR* and the concentration of Nacl in the leaching solutions. From Shainberg and Oster, 1978.

Swelling of clay in the soil may therefore be associated with low ionic strength and monovalent cations in the soil solution. Swelling of clay may induce transport of clay particles, leading to a clogging of pores, and to a decrease of the hydraulic conductivity (Quirk and Schofield, 1955; Suarez et al., 1984). Figure 6.40 illustrates the decrease of the hydraulic conductivity as a function of *SAR*, i.e. the $Na^+/\sqrt{Ca^{2+}}$ ratio in solution, and of total concentration (normality).

Swelling has even more outspoken effects when fresh water displaces seawater in an aquifer. Initially, in seawater, the concentrations are so high that the clay remains flocculated, irrespective of the high Na^+ concentration in seawater. Displacement and dilution by fresh water will decrease the ionic strength and increase the $Na^+/\sqrt{Ca^{2+}}$ ratio in solution. Both factors promote the movement of clay particles. Goldenberg (1985) used dune sand mixed with smectite clay and found a marked decrease in hydraulic conductivity when seawater was replaced by fresh water. In later experiments he could actually observe the migration of clay particles (Goldenberg and Mandel, 1988). Experiments by Seaman et al. (1995), Fauré et al. (1996) and Grolimund et al. (1996) showed similar results. Transport of clay particles also influences other properties of an aquifer, such as porosity and dispersivity (Mehnert and Jennings, 1985; Beekman and Appelo, 1990), and can lead to clogging of wells when fresh water is injected into brackish water aquifers (Brown and Signor, 1974). Also, transport with colloids may facilitate the transport of hazardous chemicals, radioactive elements and strongly sorbing organic chemicals, which would otherwise be immobile because of their large tendency for sorption and consequent high retardation (Van der Lee et al., 1992; Puls and Powell, 1992; Saiers and Hornberger, 1996). For describing colloid mobility, filtration theory is often used (Yao et al., 1971) which considers solution composition effects only implicitly (see Kretzschmar et al., 1999, for a review).

The extent of the double layer determines whether pores are electrostatically "free". Pores which are so small that the double layers overlap are completely filled with a negative potential. Anions are then unable to pass the pores. Passage of cations is hindered as well since the outflowing solution must remain neutral. A consequence is that only water molecules and uncharged species (H_4SiO_4, uncharged complexes) can pass the pore. This process is termed *hyperfiltration* or *membrane filtration* (Fritz, 1986). Figure 6.37 shows that the double layer extends to about 0.01 μm when the ionic strength is 0.001 mol/L. Some effect may therefore be expected in clay soils (particles and pores $<<1$ μm), or compacted clays. The process is self-regulating since the double layer will shrink as concentrations and ionic strength increase when a residual brine forms, which allows the passage of more and more solute ions. The double layer approaches the size of a single layer of ions if concentrations are higher than 1 mol/L, and hyperfiltration would be reduced in that case. The process of membrane filtration has nevertheless been invoked for explaining anomalous ion-ratios in deep brines (Kharaka and Berry, 1973; Demir, 1988) and calcite precipitation (Fritz and Eady, 1985). A warning with sampling of turbid waters is also appropriate here. Hyperfiltration may take place when water samples are filtered over membrane-filters (pore size 0.45 μm, or sometimes even 0.1 μm) with too high pressure and against a too large resistance of the filtered solids.

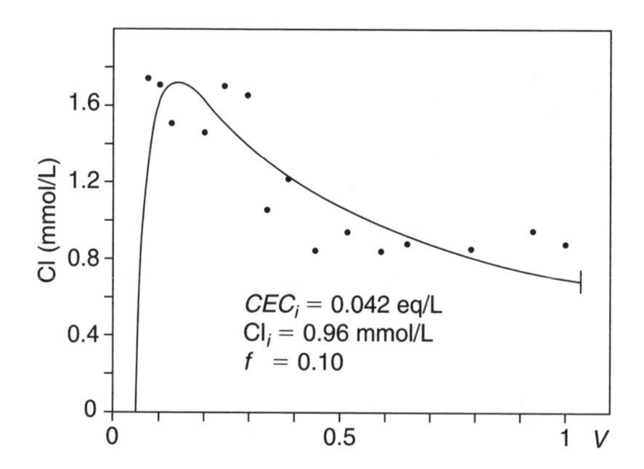

Figure 6.41. Concentrations in pressure filtrate from a Na-illite suspension. Data points from Bolt (1961), line calculated by Appelo (1977). V is the fraction of water remaining in the suspension, CEC_i is the initial (CEC), Cl_i is the intial Cl^- concentration in the suspension, f is the ratio of activity coefficients inside the suspension and in the outer solution.

A related phenomenon takes place when water is squeezed from compacting sediments or clay suspensions. The pressure filtrate may then have *higher* ion concentrations than the original pore water (Bolt, 1961), and, consequently, the concentrations in the remaining pore water will decrease when water is pressed out the suspension (Von Engelhardt and Gaida, 1963). These effects are difficult to trace in stratigraphic sequences in the field since other water-rock interaction processes may be more dominating. However, the effects are of importance when sediment samples are centrifuged or squeezed to obtain interstitial water for chemical analyses (De Lange, 1984; Saager et al., 1990).

The pressure filtration effects are related to compaction of the diffuse double layer, and can be quantitatively described with the Donnan theory of suspensions (Appelo, 1977). Two examples are shown in Figures 6.41 and 6.42. Figure 6.41 gives the composition of water squeezed from a Na-illite suspension as a function of V, the relative volume of water that remains in the suspension.

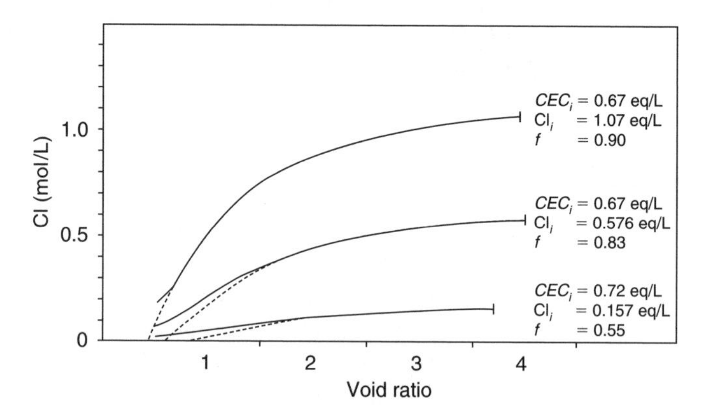

Figure 6.42. Concentrations in pore water of a compacted clay suspension. Solid line indicates data from Von Engelhardt and Gaida (1963), dotted line (shown where deviating from the measured data) is calculated by Appelo (1977). Void ratio is the volume of water divided by the volume of solids. For explanation of other symbols see the caption with Figure 6.41.

Initially the salt concentration is 0.96 mmol Na^+ / L, while the clay content amounts to 42 meq *CEC*/L. The figure shows that the salt concentration in the pressure filtrate increases sharply when about half of the water has been pressed from the suspension. Another example, in Figure 6.42, shows the concentrations in the pore solutions of a clay as a function of the void ratio, defined as the volume of water divided by the volume of solids. Three different salt solutions in the initial pore water, varying from 0.157 to 1.07 eq/L, show marked decreases in concentration when water is squeezed from the clay suspension. The effect is quite notable due to the high *CEC* of 0.67 or 0.72 eq/L in the initial suspension.

Another aspect of the relative anion-deficit in the *DDL* is *anion-exclusion* (Bresler, 1973). Whenever pores are so small that double layers overlap, or when waterfilms become, in unsaturated soils, thinner than the *DDL*, the anions are hesitant to enter pores with a negative potential. The path of anions is then confined to only a part of the porespace, which is easily measured by means of column experiments. An anion, affected by exclusion, will show an early breakthrough compared to water flow (found, for example, with tritium); the faster flow may easily be as much as 10% in soils (Bond et al., 1982; James and Rubin, 1986), or even 50% in compacted clay (Demir, 1988). A field experiment by Gvirtzmann and Gorelick (1991) showed similar effects.

We have seen (Figure 6.36) that the concentration of the anions will decrease towards a negatively charged surface. The relative deficit of anions (or more generally, of the *co-ions* which have the same valence as the adsorbing solid) balances part of the charge of the solid, and is thus rightly termed *negative adsorption* (Van Olphen, 1977). Negative adsorption increases with higher salt concentrations on solids with a fixed charge, such as clay minerals. A higher anion deficit means that less cations are adsorbed, and the *CEC* (defined from cations alone) may be expected to decrease when solute concentrations increase. This effect may lead to an about 10% lower *apparent CEC* of soils in a 0.1N Cl^- solution (Table 6.9). Since other adsorption mechanisms may play a role, such as competition with protons, or the adsorption of complexes (Sposito et al. 1983; Griffioen and Appelo, 1993), the net effect is very difficult to detect experimentally in a sediment or soil which generally consists of a mixture of clay minerals, organic matter and oxides which all exchange somewhat differently.

6.7 IRRIGATION WATER QUALITY

Soils containing a high concentration of Na^+ in the soil solution are notorious for having a poor soil structure. The permeability is reduced (Figure 6.40), and heavy agricultural machinery is not supported. The effects are related to a high degree of swelling of clays caused by extended double layers surrounding the clay particles. Diffuse double layers increase in thickness when monovalent ions make up a high proportion of the exchangeable cations. Problems with soil structure arise when Na^+ forms about 15% of the exchangeable cations, and the ionic strength is less than about 0.015 (i.e. fresh water). With Na^+, Mg^{2+}, and Ca^{2+} as the dominant cations in natural water, the fraction of exchangeable Na^+ is a function of the ratio of Na^+ over the square root of Ca^{2+} and Mg^{2+} concentrations. The fraction of Na^+ can be calculated with the Gaines-Thomas convention (Example 6.4):

$$\beta_{Na}^2 \cdot \{[Ca^{2+}] / (K_{Na\backslash Ca} \cdot [Na^+])^2 + [Mg^{2+}] / (K_{Na\backslash Mg} \cdot [Na^+])^2\} + \beta_{Na} - 1 = 0$$

However, in irrigation studies the Gapon convention is more popular, and offers an easier relationship between exchanger-fraction and solute-concentrations. From the reaction:

$$Na^+ + Ca_{0.5}\text{-}X \leftrightarrow Na\text{-}X + \tfrac{1}{2}Ca^{2+} \qquad (6.17a)$$

we have:

$$K^G_{\text{Na}\backslash\text{Ca}} = \frac{[\text{Na-X}][\text{Ca}^{2+}]^{0.5}}{[\text{Ca}_{0.5}\text{-X}][\text{Na}^+]} = \frac{\beta_{\text{Na}}[\text{Ca}^{2+}]^{0.5}}{\beta_{\text{Ca}}[\text{Na}^+]} \tag{6.18a}$$

or

$$\frac{\beta_{\text{Na}}}{\beta_{\text{Ca}}} = K^G_{\text{Na}\backslash\text{Ca}} \cdot \frac{[\text{Na}^+]}{[\text{Ca}^{2+}]^{0.5}} \tag{6.73}$$

The value of the Gapon constant $K^G_{\text{Na}\backslash\text{Ca}} = 0.5$. A further simplification is to assume that Mg^{2+} and Ca^{2+} are equally selected by the soil, i.e. $K^G_{\text{Na}\backslash\text{Ca}} = K^G_{\text{Na}\backslash\text{Mg}}$, and to add up the Ca^{2+} and Mg^{2+} concentrations. Since furthermore $\beta_{\text{Mg+Ca}} = 1 - \beta_{\text{Na}}$ in a soil with only Na^+, Mg^{2+} and Ca^{2+}, we obtain:

$$\frac{\beta_{\text{Na}}}{1 - \beta_{\text{Na}}} = K^G_{\text{Na}\backslash\text{Ca}} \cdot \frac{[\text{Na}^+]}{\sqrt{[\text{Ca}^{2+}] + [\text{Mg}^{2+}]}} \tag{6.74}$$

where the quotient $\beta_{\text{Na}}/(1 - \beta_{\text{Na}})$ is termed the *Exchangeable Sodium Ratio* (*ESR*). The *ESR* is, according to Equation (6.74), a function of the activity of $[Na^+]$ divided by the square root of the sum of Ca^{2+} and Mg^{2+} activities.

EXAMPLE 6.13. *Calculation of the Exchangeable Sodium Ratio (ESR).*
Calculate the ratio of $\text{Na}/\sqrt{\text{Ca}}$ in water where the limit of 15% Na in the exchange complex is reached.

ANSWER:
Equation (6.74) gives:

$$\frac{0.15}{1 - 0.15} = 0.5 \cdot \frac{[\text{Na}^+]}{\sqrt{[\text{Ca}^{2+}]}}$$

and the limiting ratio $\text{Na}/\sqrt{\text{Ca}}$ is 0.35 $(\text{moles/L})^{.5}$ (assuming concentrations equal to activities).

Soil scientists often use the *sodium adsorption ratio* (*SAR*) of water as a measure of *ESR* to estimate the suitability of irrigation water. In the formula for *SAR*, concentrations in mmol/L are used, and the Mg^{2+} concentration is added to the Ca^{2+} concentration. With this convention Equation (6.74) becomes:

$$\frac{\beta_{\text{Na}}}{1 - \beta_{\text{Na}}} = 0.5 \cdot \frac{\sqrt{1000}}{1000} \cdot \frac{m_{\text{Na}}}{\sqrt{m_{\text{Ca}} + m_{\text{Mg}}}} \tag{6.74a}$$

where m_i is now the concentration of i in mmol/L. In this equation

$$SAR = \frac{m_{\text{Na}}}{\sqrt{m_{\text{Ca}} + m_{\text{Mg}}}}$$

so that Equation (6.74a) may be written as:

$$ESR = 0.0158 \cdot SAR \tag{6.75}$$

The critical $ESR = 0.15$, is reached when $SAR = 10$. The equation as obtained from exchange theory can be compared with the empirical relationship (US Soil Salinity Lab, 1954):

$$ESR = -0.013 + 0.015 \cdot SAR \tag{6.76}$$

which is the relationship found for a number of soils (Figure 6.43). The figure also includes a plot of ESR calculated with the Gaines-Thomas convention, with $K_{\text{Na\backslash Ca}} = 0.37$. The value of K^G has been found to decrease when soil organic matter increases (Gheyi and Van Bladel, 1976; Curtin et al., 1995), indicating specific binding (complexation) of Ca^{2+} to organic matter.

The quality of irrigation water is not only determined by the SAR in water at the time of application. The ratio of Na^+ over the square root of $Ca^{2+} + Mg^{2+}$ concentrations of water may change by evapotranspiration and chemical reactions in the soil. Most important are the concentrating effect of evapotranspiration and the precipitation of calcite, which both increase SAR. Evapotranspiration has this effect because of the square root of the Ca^{2+} concentration in the formula for SAR, while the precipitation of calcite simply decreases the Ca^{2+} concentration in water. An adjusted SAR can be calculated in which calcite precipitation is taken into account (Ayers and Westcot, 1985; Example 6.14).

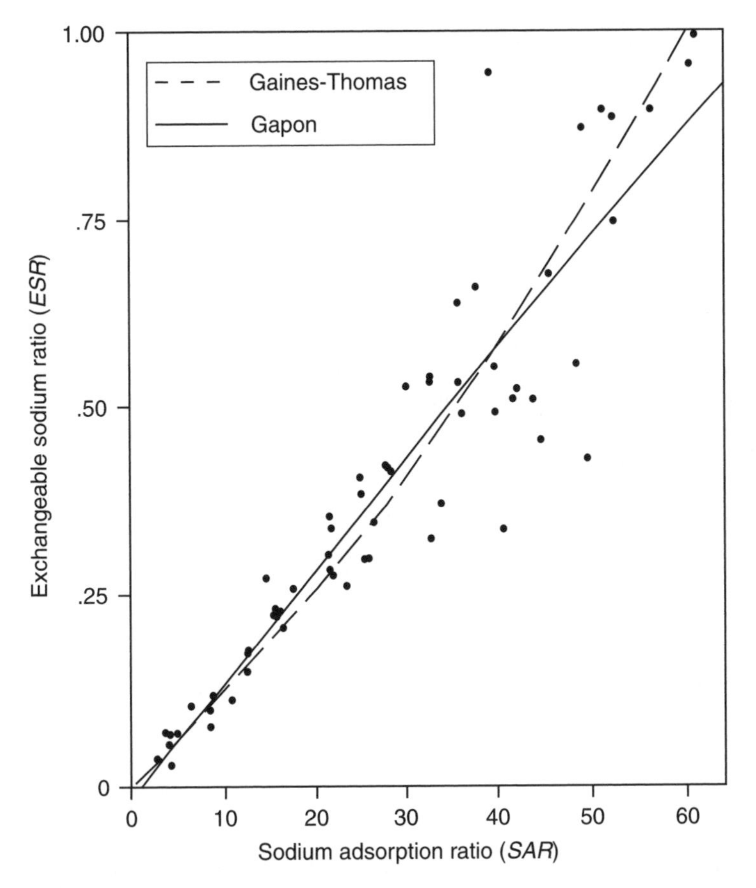

Figure 6.43. Relationship between SAR and ESR in a number of soils; full line gives the empirical relationship (agrees with the Gapon convention), dotted line gives Gaines-Thomas convention with $K_{\text{Na\backslash Ca}} = 0.37$. Adapted from US Soil Salinity Lab (1954).

EXAMPLE 6.14. *Calculation of SAR adjusted for calcite precipitation*
An analysis of groundwater provided the results:

pH	Na^+	K^+	Mg^{2+}	Ca^{2+}	Cl^-	HCO_3^-	SO_4^{2-}	
8.5	62	0.1	9.0	81	168	183	9.0	mg/L

- Is the analysis reliable?
- Is the water quality suitable for irrigation when irrigation return flow is 10%?

ANSWER:
Recalculating the analysis in meq/L gives:

Na^+	K^+	Mg^{2+}	Ca^{2+}	$\Sigma+$	Cl^-	HCO_3^-	SO_4^{2-}	$\Sigma-$
2.7	0.002	0.74	4.05	7.49	−4.74	−3.00	−0.19	−7.93

The difference is −0.44, or 3% of Σ(cations–anions), and the analysis is correct. *SAR* of the water is 1.7, and it offers excellent irrigation quality. However, concentrations increase upon evapotranspiration of 90% of the water, and this gives, by itself, an increase of *SAR*. Moreover, one can expect that calcite precipitates, so that the Ca^{2+} concentration might decrease despite the increase in concentration due to evapotranspiration. The Ca^{2+} concentration in equilibrium with calcite can be calculated with the equations derived in Chapter 5. It is easiest to assume a constant CO_2-pressure:

$$Ca^{2+} + 2HCO_3^- \leftrightarrow CaCO_3 + CO_2 + H_2O; \quad K = 10^6$$

With a CO_2-pressure of 0.01 atm, then at equilibrium with calcite:

$$[Ca^{2+}][HCO_3^-]^2 = 10^{-8} \tag{6.77}$$

Note that HCO_3^- as used here, is the actually measured concentration which may have resulted from other reactions than calcite dissolution: we cannot assume $HCO_3^- = 2Ca^{2+}$, as was did in Chapter 5.

Neglecting activity coefficients and complexes allows to use concentrations instead of activities. When the cations are 10 times concentrated through evapotranspiration, the concentration of Ca^{2+} increases to 20.2 mmol/L and of HCO_3^- to 30 mmol/L. or

$$(0.02)(0.03)^2 = 10^{-4.74}$$

An amount Δ will precipitate, which can be calculated with Equation (6.77):

$$(0.02 - \Delta)(0.03 - 2\Delta)^2 = 10^{-8}$$

Trial and error gives Δ = 0.0143 mol/L, and the final concentrations are:

Na^+	Mg^{2+}	Ca^{2+}	HCO_3^-	
27	3.7	5.9	1.4	mmol/L

In the resulting water $SAR = 27/\sqrt{(3.7 + 5.9)} = 8.7$, which is still on the safe side.
We note once more, that water types with more equivalents Ca^{2+} than HCO_3^- offer relatively safe irrigation water qualities from the *SAR* point of view.

QUESTION:
Repeat the calculation with PHREEQC. Evaporate water with keyword REACTION.
ANSWER: Δ = 13.7 mmol Calcite/L, SAR = 8.5.

PROBLEMS

Laboratory simulation of the exchange reaction:
Add a $CaCl_2$ solution to an exchanger in Na^+ form. Ca^{2+} displaces part of the Na^+ from the exchanger sites and hence both the Ca^{2+} and Na^+ concentrations change. We calculate the equilibrium constant $K_{Na\backslash Ca}$. The resin DOWEX 50 with an exchange capacity CEC = 5 meq/g is used as a substitute for the soil.

You receive:

- 300 mL 2.5 mM $CaCl_2$ solution. Determine Ca^{2+}, Cl^- (titrations)
- 200 mg Na-X (DOWEX 50) in a 500 mL erlenmeyer flask

Add 200 mL of the $CaCl_2$ solution to the exchanger in the erlenmeyer flask. Stir for 5 minutes, and analyze the solution for Ca^{2+} and Cl^- (titrate), and Na^+ (flame-photometer). Estimate the Na^+ concentration beforehand, and determine the correct dilution factor to be in the analysis range of the flamephotometer.

Analyze:
in the $CaCl_2$ solution:
Ca^{2+} = ... mmol/L $\quad$ Cl^- = ... mmol/L
after reaction with Na-X:
Ca^{2+} = ... mmol/L $\quad$ Na^+ = ... mmol/L $\quad$ $\Sigma+$ = ... meq/L (*correct?*)

Calculate:
$Ca\text{-}X_2$ = ... meq/L $\quad$ Na-X = ... meq/L
β_{Ca} = ... $\quad$ β_{Na} = ... $\quad$ $K_{Na\backslash Ca}$ = ...

The distribution-coefficient is for Na^+ .... L/kg, and for Ca^{2+} .... L/kg.
(Use the *analyzed* concentrations, when calculating the distribution coefficients).

6.1. Calculate the structural charge per gram of illite, $K_{0.7}[Si_{3.4}Al_{0.6}]^{iv}[Al_{1.9}Fe^{2+}_{0.1}]^{vi}O_{10}(OH)_2$, as well as the charge density. The ⊢O⊣ layers are stacked 20 units deep; the unit cell size is as of smectite, one unit cell contains 24 oxygens.

6.2. Also for vermiculite, $Mg_{0.4}[Si_{3.8}Al_{0.2}]^{iv}[Mg_{0.6}Al_{1.4}]^{vi}O_{10}(OH)_2$; unit cell size is as of smectite, the layers are completely separated.

6.3. Repeat the calculation of Example 6.4 with all concentrations in water 10 times higher (note how the relative proportion of exchangeable Na-X changes). Do the same for seawater using the concentrations listed in Table 6.2.

6.4. Determine the exchange coefficient $K_{Na\backslash Cd}$ on smectite from the following data of Garcia-Miragaya and Page (1976). The amount of smectite is 2.05 meq/L.

$NaClO_4$ (N)	Cd_{total} (μM)	adsCd (%)
0.03	0.40	91.1
	0.80	87.8

6.5. In the dunes of the island Ameland two boreholes (A and B) have been drilled. Water from three filters (1–3), and the nearby sea (4) has the following composition (concentrations in mmol/L):

	1	2	3	4
pH	6.7	6.8	7.0	8.2
Na^+	0.86	18.4	112	485
K^+	0.04	0.1	3.9	10.6
Mg^{2+}	0.09	0.45	10	55.1
Ca^{2+}	3.0	0.9	31.3	10.7
Cl^-	1.0	3.0	200	566
HCO_3^-	6.1	18.0	0.3	2.4
SO_4^{2-}	0.02	–	–	29.3

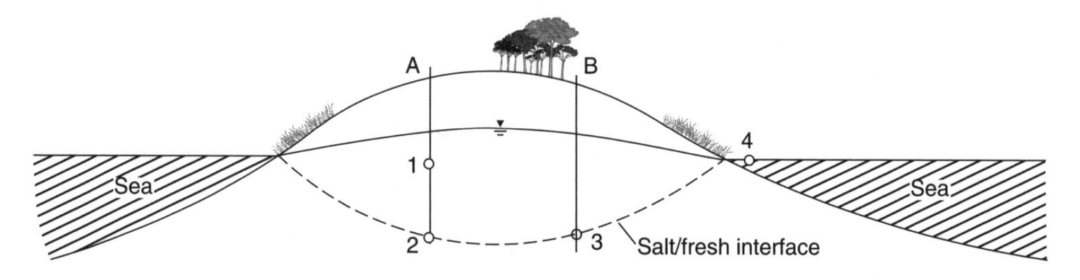

a. Explain the composition of these water samples.
b. Is the salt / fresh-water interface moving upward or downward at A, B?
c. Calculate the exchangeable fractions β_{Na}, β_K, β_{Mg} and β_{Ca} on the sediment with the Gaines-Thomas convention for sample 1 and 4 (assume activity = molality).
d. Recalculate the fractions of exchangeable cations in meq/L pore water, given $CEC = 10$ meq/kg; $\rho_b = 1.8$ g/cm^3; $\varepsilon_w = 0.3$.

6.6. In the southern part of the Nile-delta the composition of the groundwater is determined by:
– composition of Nile-water, used as irrigation-water
– processes during infiltration of Nile-water in the soil
– processes in the soil
a. What are the processes during infiltration? Data (concentrations in mmol/L):

	Na^+	K^+	Mg^{2+}	Ca^{2+}	HCO_3^-	Cl^-	SO_4^{2-}	pH
Nile-water	0.5	0.1	0.4	0.7	2.2	0.5	0.1	
Groundwater	8.0	0.1	0.2	0.7	2.7	4.5	1.0	8.1

b. In the northern part of the delta, near the sea, water is pumped with a composition of:

Na^+	K^+	Mg^{2+}	Ca^{2+}	HCO_3^-	Cl^-	SO_4^{2-}	
180	2.5	35	15	7.5	270	1.5	(borehole 121)

– What processes influenced this composition?
– If more water is pumped up, will water become more brackish or fresher?
c. Calculate TDS (Total Dissolved Solids) of the analysis in b.

6.7. Native groundwater in the injection test of Valocchi et al. (1981) had the composition $Na^+ = 86.5$, $Mg^{2+} = 18.2$ and $Ca^{2+} = 11.1$ mmol/L (Figures 6.16, 6.27). Injected water has 14.66 meq Cl^-/L.

Selectivity coefficients were (Gaines and Thomas convention, activity = molal concentration) $K_{Na\backslash Mg} = 0.54$ and $K_{Na\backslash Ca} = 0.41$. Sediment *CEC* = 750 meq/L pore water.
Calculate

a. the composition of the exchange complex,
b. the composition of water after the salinity jump, in equilibrium with the original exchange complex (compare with Figure 6.16),
c. the exchange complex in equilibrium with injected water (cf. Example 6.11),
d. the number of porevolumes that the exchange complex can maintain the high Na^+ concentration found under b) (When the outcome is multiplied with $V_0 = 295\,m^3$, the porevolume for observation well S23, the rise of Mg^{2+} must be found: compare with Figure 6.16).

6.8. The fresh groundwater resident in the column shown in Figure 6.20 has $Na^+ = 5.6$, $Mg^{2+} = 1.9$, and $Ca^{2+} = 2.0$ mmol/L. Injected is water with 235 meq anions/L. The CEC of the sediment is 60 meq/L pore water. $K_{Na\backslash Mg} = 0.55$, $K_{Na\backslash Ca} = 0.35$ (Gaines and Thomas convention, solute activities = molal concentrations).
Calculate

a. the composition of the exchange complex,
b. the composition of water after the salinity jump, in equilibrium with the original exchange complex,
c. the number of porevolumes that the exchange complex can maintain the high Ca^{2+}concentration found under b).

6.9. Derive equations for heterovalent exchange of divalent ions and Na^+ which relate the Gaines-Thomas *K* to the Vanselow *K* and the Gapon *K*.

6.10. Derive the equation for the thickness of the double layer, Equation (6.71).

6.11. Calculate the potential at a smectite surface in a solution of 0.01 M NaCl, and estimate the concentration of Na^+ and Cl^- at 0, 10, 50 and 100Å distance from the smectite surface. The smectite has a surface charge of 0.09 C/m^2.

6.12. Calculate *SAR* in the irrigation return flow water of Example 6.14 without taking calcite precipitation into account. Compare with *SAR* of the irrigation water.

6.13. Calculate the exchange coefficients for Ca / K exchange on illite and ion-exchange resin from the data of Wiklander (1955), as given in Example 6.3.

6.14. Derive a general formula to calculate equivalent fractions from molar fractions, and *vice versa*. *Hint:* use $\beta_I \cdot CEC = i \cdot \beta_I \cdot TEC$, and find two relations for *TEC* / *CEC* in which only β^M's or β's appear.

6.15. Calculate a Ca^{2+}/Mg^{2+} isotherm with PHREEQC, using Gapon's convention:

$$0.5\,Ca^{2+} + Mg_{½}\text{-}X \leftrightarrow Ca_{½}\text{-}X + 0.5Mg^{2+}$$

and also for

$$Ca^{2+} + Mg_{½}\text{-}X \leftrightarrow Ca_{½}\text{-}X + Mg^{2+}$$

Derive K^G's from Table 6.4. Use "-no_check" and "-mole_balance" for the 2nd case. Consult Figure 6.24 for obtaining an isotherm.

6.16. Make a plot of the K / Na exchange-isotherm (β_K *vs* α_K) with PHREEQC for $K_{K/Na} = 2.7$, used in Example 6.10.

6.17. Calculate the retardation of Na^+ in the aquifer below the Vejen waste site, Figure 3.21. $m_{Ca} = 1$ mM, m_{Na} varies from 1 μM to 10 mM, *CEC* = 50 meq/L, $K_{Ca\backslash Na} = 2.85$.

REFERENCES

Andersen, M.S., 2001. *Geochemical processes at a seawater-freshwater interface*. Ph.D. thesis, Lyngby, 145 pp.

Appelo, C.A.J., 1977. Chemistry of water expelled from compacting clay layers: a model based on Donnan equilibrium. *Chem. Geol.* 19, 91–98.

Appelo, C.A.J., 1994. Cation and proton exchange, pH variations, and carbonate reactions in a freshening aquifer. *Water Resour. Res.* 30, 2793–2805.

Appelo, C.A.J., 1994b. Some calculations on multicomponent transport with cation exchange in aquifers. *Ground Water* 32, 968–975.

Appelo, C.A.J., 1996. Multicomponent ion exchange and chromatography in natural systems. In P.C. Lichtner, C.I. Steefel and E.H. Oelkers (eds), *Rev. Mineral.* 34, 193–227.

Appelo, C.A.J., Willemsen, A., Beekman, H.E. and Griffioen, J., 1990. Geochemical calculations and observations on salt water intrusions. II. Validation of a geochemical model with laboratory experiments. *J. Hydrol.* 120, 225–250.

Appelo, C.A.J., Hendriks, J.A. and Van Veldhuizen, M., 1993. Flushing factors and a sharp front solution for solute transport with multicomponent ion exchange. *J. Hydrol.* 146, 89–113.

Ayers, R.S. and Westcot, D.W., 1985. *Water quality for agriculture*. FAO Irr. Drain. Pap. 29, 174 pp.

Back, W., 1966. *Hydrochemical facies and groundwater flow patterns in northern part of Atlantic coastal plain*. U.S. Geol. Surv. Prof. Paper 498-A, 42 pp.

Bahr, J.M., 1989. Analysis of nonequilibrium desorption of volatile organics during field test of aquifer decontamination. *J. Contam. Hydrol.* 4, 205–222.

Bartschat, B.M., Cabaniss, S.E. and Morel, F.M.M., 1992. Oligoelectrolyte model for cation binding by humic substances. *Env. Sci. Technol.* 26, 264–294.

Beekman, H.E., 1991. *Ion chromatography of fresh and salt water intrusions*. Ph.D. thesis, Free University, Amsterdam, 198 pp.

Beekman, H.E. and Appelo, C.A.J., 1990. Ion chromatography of fresh- and salt-water displacement: laboratory experiments and multicomponent transport modelling. *J. Contam. Hydrol.* 7, 21–37.

Bjerg, P.L., Ammentorp, H.C. and Christensen, T.H., 1993. Model simulations of a field experiment on cation exchange-affected multicomponent solute transport in a sandy aquifer. *J. Contam. Hydrol.* 12, 291–311.

Bloom, S.A. and Mansell, R.S., 2001. An algorithm for generating cation exchange isotherms from binary selectivity coefficients. *Soil Sci. Soc. Am. J.* 65, 1426–1429.

Bolt, G.H., 1961. The pressure filtrate of colloidal suspensions, II. experimental data on homoionic clays. *Kolloid-Z.* 175, 144–150.

Bolt, G.H., 1967. Cation exchange equations used in soil science – a review. *Neth. J. Agric. Sci.* 15, 81–103.

Bolt, G.H. (ed.), 1982. *Soil chemistry, B. Physico-chemical models*. Elsevier, Amsterdam, 527 pp.

Bolt, G.H. and Bruggenwert, M.G.M. (eds), 1978. *Soil chemistry, A. Basic elements*. Elsevier, Amsterdam, 281 pp.

Bond, W.J., 1995. On the Rothmund-Kornfeld description of cation exchange. *Soil Sci. Soc. Am. J.* 59, 436–443.

Bond, W.J. and Verburg, K., 1998. Comparison of methods for predicting ternary exchange from binary isotherms. *Soil Sci. Soc. Am. J.* 61, 444–454.

Bond, W.J., Gardiner, B.N. and Smiles, D.E., 1982. Constant flux absorption of a tritiated calcium chloride solution by a clay soil with anion exclusion. *Soil Sci. Soc. Am. J.* 46, 1133–1137.

Breeuwsma, A., Wösten, J.H.M., Vleeshouwer, J.J., Van Slobbe, A.M. and Bouma, J. 1986. Derivation of land qualities to assess environmental problems from soil surveys. *Soil Sci. Soc. Am. J.* 50, 186–190.

Bresler, E., 1973. Anion exclusions and coupling effects in nonsteady transport through unsaturated soils: I. Theory. *Soil Sci. Soc. Am. Proc.* 37, 663–669.

Brown, R.F. and Signor, D.C., 1974. Artificial recharge – state of the art. *Ground Water* 12, 152–160.

Bruggenwert, M.G.M. and Kamphorst, A., 1982. Survey of experimental information on cation exchange in soil systems. In G.H. Bolt, 1982, *op. cit.*, 141–203.

Brusseau, M.L., Rao, P.S.C., Jessup, R.E. and Davidson, J.M., 1989. Flow interruption: a method for investigating sorption nonequilibrium. *J. Contam. Hydrol.* 4, 223–240.

Bürgisser, C., Černík, M., Borkovec, M. and Sticher, H., 1993. Determination of nonlinear adsorption isotherms from column experiments: an alternative to batch studies. *Env. Sci. Technol.* 27, 943–948.

Ceazan, M.L., Thurman, E.M. and Smith, R.L., 1989. Retardation of ammonium and potassium transport through a contaminated sand and gravel aquifer: the role of cation exchange. *Env. Sci. Technol.* 23, 1402–1408.

Chapelle, F.H., 1983. Groundwater geochemistry and calcite cementation of the Aquia aquifer in Southern Maryland. *Water Resour. Res.* 19, 545–558.

Chapelle, F.H. and Knobel, L.L., 1983. Aqueous geochemistry and the exchangeable cation composition of glauconite in the Aquia aquifer, Maryland. *Ground Water* 21, 343–352.

Curtin, D., Selles, F. and Steppuhn, H., 1995. Sodium-calcium exchange selectivity as influenced by soil properties and method of determination. *Soil Sci.* 159, 176–184.

De Lange, G.J., 1984. Shipboard pressure-filtration system for interstitial water extraction. Med. Rijks Geol. Dienst (Haarlem, the Netherlands) 38, 209–214.

Demir, I., 1988. The interrelation of hydraulic and electrical conductivities, streaming potential and salt fitration during the flow of chloride brines through a smectite layer at elevated pressures. *J. Hydrol.* 98, 31–52.

DeVault, D., 1943. The theory of chromatography. *J. Am. Chem. Soc.* 65, 532–540.

Elprince, A.M., Vanselow, A.P. and Sposito, G., 1980. Heterovalent, ternary cation exchange equilibria: NH_4^+-Ba^{2+}-La^{3+} exchange on montmorillonite. *Soil Sci. Soc. Am. J.* 44, 964–969.

Eriksson, E., 1952. Cation-exchange equilibria on clay minerals. *Soil Sci.* 74, 103–113.

Evangelou, V.P. and R.E. Phillips, 1988. Comparison between the Gapon and Vanselow exchange selectivity coefficients. *Soil Sci. Soc. Am. J.* 52, 379–382.

Fauré, M.-H., Sardin, M. and Vitorge, P., 1996. Transport of clay particles and radioelements in a salinity gradient: experiments and simulations. *J. Contam. Hydrol.* 21, 255–267.

Feynman, R., Leighton, R.B. and Sands, M.L., 1989. *The Feynman lectures on physics*, vol. 2. California Inst. Technol.

Fritz, S.J., 1986. Ideality of clay membranes in osmotic processes: a review. *Clays Clay Min.* 34, 214–223.

Fritz, S.J. and Eady, C.D., 1985. Hyperfiltration-induced precpitation of calcite. *Geochim. Cosmochim. Acta* 49, 761–768.

Gaines, G.L. and Thomas, H.C., 1953. Adsorption studies on clay minerals. II. A formulation of the thermodynamics of exchange adsorption. *J. Chem. Phys.* 21, 714–718.

Gapon, E.N., 1933. Theory of exchange adsorption V. *J. Gen. Chem. (USSR)* 3, 667–669, (*Chem. Abstr.* 28, 4516, 1934)

Garcia-Miragaya, J. and Page, A.L., 1976. Influence of ionic strength and inorganic complex formation on the sorption of trace amounts of Cd by montmorillonite. *Soil Sci. Soc. Am. J.* 40, 658–663.

Gheyi, H.R. and Van Bladel, R., 1976. Calcium-sodium exchange in some calcareous soils and a montmorillonite clay as compared with predictions based on double layer threory. *Geoderma* 16, 159–169.

Goldenberg, L.C., 1985. Decrease of hydraulic conductivity in sand at the interface between seawater and dilute clay suspensions. *J. Hydrol.* 78, 183–199.

Goldenberg, L.C. and Mandel, S., 1988. Some processes in the sea-water/fresh-water interface, as influenced by the presence of gases. *Natuurwet. Tijdschr.* 70, 288–299, (SWIM Conf. Ghent, Belgium).

Gomis, V., Boluda, N. and Ruiz, F., 1996. Application of a model for simulating transport of reactive multispecies components to the study of the hydrochemistry of salt water intrusions. *J. Contam. Hydrol.* 22, 67–81.

Gomis-Yagues, V., Boluda-Botella, N. and Ruiz-Bevia, F. 2000. Gypsum precipitation/dissolution as an explanation of the decrease of sulphate concentration during seawater intrusion. *J. Hydrol.* 228, 48–55.

Griffioen, J. and Appelo, C.A.J., 1993. Adsorption of calcium and its complexes by two sediments in calcium-hydrogen-chlorine-carbon dioxide systems. *Soil Sci. Soc. Am. J.* 57, 716–722.

Grim, R.E., 1968. *Clay mineralogy*. McGraw-Hill, New York, 596 pp.

Grolimund, D., Borkovec, M., Barnettler, M. and Sticher, H., 1996. Colloid-facilitated transport of strongly sorbing contaminants in natural porous media. *Env. Sci. Technol.* 30, 3118–3123.

Gvirtzman, H. and Gorelick, S.M., 1991. Dispersion and advection in unsaturated porous media enhanced by anion exclusion. *Nature* 352, 793–795.

Helfferich, F., 1959. *Ionenaustauscher Band 1*. VCH, Weinheim, 520 pp.

Helfferich, F. and Klein, G., 1970. *Multicomponent chromatography*. M. Dekker, New York, 419 pp.

Hurz, G., 2001. Lithium chloride solution as an extraction agent for soils. *J. Plant Nutr. Soil Sci.* 164, 71–75.

James, R.V. and Rubin, J., 1986. Transport of chloride ion in a water-unsaturated soil exhibiting anion-exclusion. *Soil Sci. Soc. Am. J.* 50, 1142–1149.

Karaborni, S., Smit, B., Heidug, W., Urai, J. and Van Oort, E., 1996. The swelling of clays: molecular simulations of the hydration of montmorillonite. *Science* 271, 1102–1104.

Kharaka, Y.K. and Berry, F.A.F., 1973. Simultaneous flow of water and solutes through geological membranes. I. Experimental investigation. *Geochim. Cosmochim. Acta* 37, 2577–2603.

Koch, S. and Flühler, H., 1993. Non-reactive solute transport with micropore diffusion in aggregated porous media determined by a flow-interruption method. *J. Contam. Hydrol.* 14, 39–54.

Kretzschmar, R., Borkovec, M., Grolimund, D. and Elimelech, M., 1999. Mobile subsurface colloids and their role in contaminant transport. *Adv. Agron.* 66, 121–193.

Lawrence, A.R., Lloyd, J.W. and Marsh, J.M., 1976. Hydrochemistry and groundwater mixing in part of the Lincolnshire Limestone aquifer, England. *Ground Water* 14, 12–20.

Liu, C., Zachara, J.M. and Smith, S.C., 2004. A cation exchange model to describe Cs^+ sorption at high ionic strength in subsurface sediments at Hanford site, USA. *J. Contam. Hydrol.* 68, 217–238.

Manzano, M., Custodio, E. and Carrera, J., 1992. Fresh and salt water in the Llobregat delta aquitard. *Proc. 12th Saltwater Intrusion Meeting*, 207–238, CIMNE, Barcelona.

Martell, A.E. and Smith, R.M., 1977. *Critical stability constants, vol. 3*. Plenum Press, New York.

Mehnert, E. and Jennings, A.A., 1985. The effect of salinity-dependent hydraulic conductivity on saltwater intrusion episodes. *J. Hydrol.* 80, 283–297.

Moore, W.J., 1972. *Physical chemistry*, 5th ed. Longman, London, 977 pp.

Nativ, R., Issar, I. and Rutledge, J., 1983. Chemical composition of rainwater and floodwaters in the Negev desert, Israel. *J. Hydrol.* 62, 201–223.

Neal, C. and Cooper, D.M., 1983. Extended version of Gouy-Chapman electrostatic theory as applied to the exchange behavior of clay in natural waters. *Clays Clay Miner.* 31, 367–376.

Nielsen, D.R., Van Genuchten, M.Th. and Biggar, J.W., 1986. Water flow and solute transport processes in the saturated zone. *Water Resour. Res.* 22, 89S-108S.

Nkedi-Kizza, P., Biggar, J.W., Selim, H.M., Van Genuchten, M.Th., Wierenga, P.J., Davidson, J.M. and Nielsen, D.R., 1984. On the equivalence of two conceptual models for describing ion exchange during transport through an aggregated oxisol. *Water Resour. Res.* 20, 1123–1130.

Pauling, L., 1960. *The nature of the chemical bond*. Cornell Univ. Press, 644 pp.

Perkins, T.K. and Johnston, O.C., 1963. A review of diffusion and dispersion in porous media. *Soc. Petrol. Engin. J.*, march 1963, 70–84.

Puls, R.W. and Powell, R.M., 1992. Transport of inorganic colloids through natural aquifer material: implications for contaminant transport. *Env. Sci. Technol.* 26, 614–621.

Purdy, C.B., Burr, G.S., Rubin, M., Helz, G.R. and Mignerey, A.C., 1992. Dissolved organic and inorganic ^{14}C concentrations and ages for coastal plain aquifers in southern Maryland. *Radiocarbon* 34, 654–663.

Quirk, J.P. and Schofield, R.K., 1955. The effect of electrolyte concentration on soil permeability. *J. Soil Sci.* 6, 163–178.

Rainwater, K.A., Wise, W.R. and Charbeneau, R.J., 1987. Parameter estimation through groundwater tracer tests. *Water Resour. Res.* 23, 1901–1910.

Reardon, E.J., Dance, J.T. and Lolcama, J.L., 1983. Field determination of cation exchange properties of calcareous sand. *Ground Water* 21, 421–428.

Rhoades, J.D., 1982. Cation exchange capacity. In A.L. Page et al. (eds), *Methods of soil analysis*, Agron. Monogr. 9, 149–157. Soil Sci. Soc. Am., Madison.

Rieman, W. and Walton, H.F., 1970. *Ion exchange in analytical chemistry*. Pergamon Press, Oxford, 295 pp.

Saager, P.M., Sweerts, J.-P. and Ellermeijer, H.J., 1990. A simple pore-water sampler for coarse, sandy sediments of low porosity. *Limnol. Oceanogr.* 35, 747–751.

Saiers, J.E. and Hornberger, G.M., 1996. The role of colloidal kaolinite in the transport of cesium through laboratory sand columns. *Water Resour. Res.* 32, 33–41.

Scheffer, F. and Schachtschabel, P., 2002. *Lehrbuch der Bodenkunde,* 15th Aufl. Elsevier, Amsterdam, 607 pp.

Seaman, J.C., Bertsch, P.M. and Miller, W.P., 1995. Chemical controls on colloid generation and transport in a sandy aquifer. *Env. Sci. Technol.* 29, 1808–1815.

Shainberg, I. and Oster, J.D., 1978. *Quality of irrigation water*. Int. Irr. Inf. Cent. Bet Dagan, Israel.

Šimůnek, J. and Suarez, D.L., 1997. Sodic soil reclamation using multicomponent transport modeling. *J. Irr. Drain. Eng.* 123, 367–376.

Sposito, G., 1984. *The surface chemistry of soils*. Oxford Univ. Press, New York, 234 pp.

Sposito, G., Holtzclaw, K.M., Charlet, L., Jouany, C. and Page, A.L., 1983. Na-Ca and Na-Mg exchange in Wyoming bentonite in ClO_4^- and Cl^- background ionic media. *Soil Sci. Soc. Am. J.* 47, 51–56.

Stuyfzand, P.J., 1985. Hydrochemistry and hydrology of the dune area between Egmond and Wijk aan Zee (in Dutch). KIWA, SWE, 85.012, Nieuwegein.

Stuyfzand, P.J., 1993. *Hydrochemistry and hydrology of the coastal dune area of the Western Netherlands*. Ph.D. Thesis, Free University, Amsterdam, 366 pp.

Suarez, D.L., Rhoades, J.D., Lavado, R. and Grieve, C.M., 1984. Effect of pH on saturated hydraulic conductivity and soil dispersion. *Soil Sci. Soc. Am. J.* 48, 50–55.

US Soil Salinity Lab. Staff, 1954. *Diagnosis and improvement of saline and alkali soils. USDA Handb.* 60, US Gov. Print. Office, Washington DC.

Valocchi, A.J., Street, R.L. and Roberts, P.V., 1981. Transport of ion-exchanging solutes in groundwater: chromatographic theory and field simulation. *Water Resour. Res.* 17, 1517–1527.

Van Breukelen, B.M., Appelo, C.A.J. and Olsthoorn, T.N., 1998. Hydrogeochemical transport modelling of 24 years of Rhine water infiltration in the dunes of the Amsterdam water supply. *J. Hydrol.* 209, 281–296.

Van der Kemp, W., Appelo, C.A.J., Walraevens, K., 2000. Inverse chemical modeling and radiocarbon dating of palaeogroundwaters: the Tertiary Ledo-Paniselian aquifer in Flanders, Belgium. *Water Resour. Res.* 36, 1277–1287.

Van der Lee, J., Ledoux, E. and Demarsily, G., 1992. Modeling of colloidal uranium transport in a fractured medium. *J. Hydrol.* 139, 135–158.

Van der Molen, W.H., 1958. *The exchangeable cations in soils flooded with seawater*. Staatsdrukkerij, Den Haag, 167 pp.

Van Eijkeren, J.C.H., annd Loch, J.P.G., 1984. Transport of cationic solutes in sorbing porous media. *Water Resour. Res.* 20, 714–718.

Van Genuchten, M. Th., 1985. A general approach for modeling solute transport in structured soils. IAH Mem. 17, 513–525.

Van Genuchten, M. Th. and Cleary, R.W., 1982. Movement of solutes in soil: computer-simulated and laboratory results. In G.H. Bolt, 1982, *op. cit.*, 349–386.

Van Olphen, H., 1977. *An introduction to clay colloid chemistry*, 2nd ed. Wiley and Sons, New York, 318 pp.

Van Veldhuizen, M., Hendriks, J.A. and Appelo, C.A.J., 1998. Numerical computation in heterovalent chromatography. *Appl. Numer. Mathem.* 28, 69–89.

Vanselow, A.P., 1932. Equilibria of the base-exchange reactions of bentonites, permutites, soil colloids and zeolites. *Soil Sci.* 33, 95–113.

Voegelin, A., Vulava, V.M., Kuhnen, F. and Kretzschmar, R., 2000. Multicomponent transport of major cations predicted from binary adsorption experiments, *J. Contam. Hydrol.* 46, 319–338.

Von Engelhardt, W. and Gaida, K.H., 1963. Concentration changes of pore solutions during the compaction of clay sediments. *J. Sed. Petrol.* 33, 919–930.

Walraevens, K. and Cardenal, J., 1994. Aquifer recharge and exchangeable cations in a Tertiary clay layer (Bartonian clay, Flanders-Belgium). *Min. Mag.* 58A, 955–956.

Weber, W.J. and Smith, E.H., 1987. Simulation and design models for adsorption processes. *Env. Sci. Technol.* 21, 1040–1050.

Wiklander, L., 1955. Cation and anion exchange phenomena. In F.E. Bear (ed.), *Chemistry of the soil*, Chapter 4. Wiley and Sons, New York.

Xu, T., Samper, J., Ayora, C., Manzano, M. and Custodio, E. 1999. Modeling of non-isothermal multi-component reactive transport in field scale porous media flow systems. *J. Hydrol.* 214, 144–164.

Yao, K.M., Habibian, M.T. and O'Melia, C.R., 1971. Water and waste water filtration: concepts and applications. *Env. Sci. Technol.* 5, 1105–1112.

Zuur, A.J., 1938. Trans. 2nd Comm. and Alkali-Subcomm., Int. Congr. Soil Sci., Helsinki, B 66–67.

7

Sorption of Trace Metals

Heavy metals like Cd, Pb and Ni, and also As, are strongly poisonous and therefore the highest admissible concentration for these substances is very low (Table 1.1). Enhanced heavy metal concentrations constitute a serious problem for water supplies around the world. Once released to groundwater, the mobility of trace metals is to a large extent controlled by sorption processes. Part of trace metal sorption is due to ion exchange and can be estimated with the formulas from Chapter 6. However, most sorption is connected with specific binding of heavy metals to the variable charge surfaces of oxides and organic matter.

Variable charge solids sorb ions from solution without releasing other ions in equivalent proportion. Their surface charge can be positive or negative depending on the pH and the solution composition. Variable charge solids are important in regulating the mobility of both positively charged heavy metals such as Pb^{2+} and Cd^{2+}, and of oxyanions such as $HAsO_4^{2-}$ and $H_2PO_4^-$. Modeling of sorption on variable charge solids is more complex than ion exchange, and generally requires computer programs for an exact answer. Essentially, the surfaces of oxides and hydroxides gain a pH dependent charge due to sorption of protons and other ions from solution. The surface charge creates a potential difference between the solution and the surface. This potential will influence the approach of ions towards the surface, and thus couples back to the charge development on the surface. However, some ions are bound so strongly that they even remain fixed on a surface that has the same charge and repels the ion electrostatically.

7.1 THE ORIGIN AND OCCURRENCE OF HEAVY METALS IN GROUNDWATER

Groundwater is often heavily polluted near mines of sulfide minerals like pyrite (FeS_2), galena (PbS), sphalerite (ZnS) or arsenopyrite ($FeAsS$). Waste materials, containing residual sulfide minerals, are piled up in tailings and may become oxidized, producing sulfuric acid:

$$FeS_2 + \tfrac{15}{4}O_2 + \tfrac{7}{2}H_2O \rightarrow Fe(OH)_3 + 2SO_4^{2-} + 4H^+ \qquad (7.1)$$

Concomitantly, large amounts of heavy metals are released which remain soluble in the *acid mine drainage* (Stollenwerk, 1994; Appleyard and Blowes, 1994). Figure 7.1 shows the pore water composition of a mine tailing. In the upper part, the pH is less than 2.5, and the water contains over 20,000 mg Zn/L and over 100 mg Cu/L. Downward the pH increases to above 4 and the Cu concentration drops to <1 mg/L, while Zn remains high and only decreases in the saturated part of the tailings. Apparently, the pH has a major influence on the mobility of the heavy metals, but the effect is different for the various metals. Acid mine drainage is globally a major environmental issue. There are thousands and thousands of metal sulfide mines, many of them abandoned, that continuously release low pH and heavy-metal loaded water into the environment.

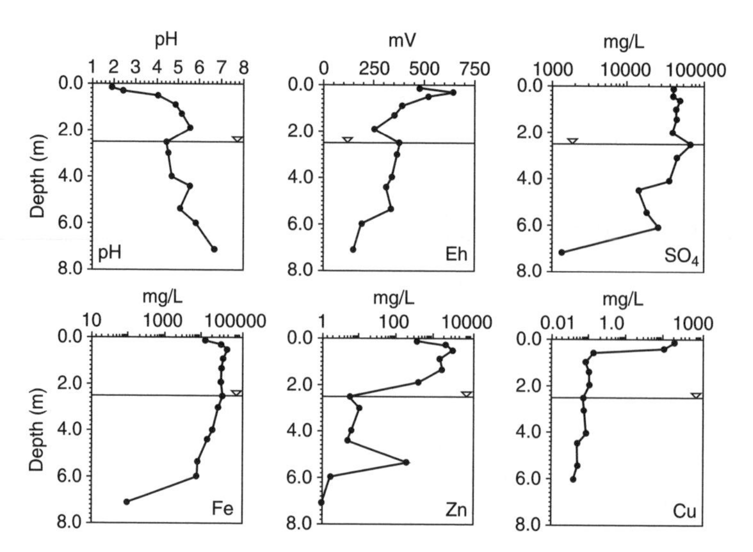

Figure 7.1. Pore water geochemical profiles for pH, Eh and selected elements in a mine tailings impoundment. Note the logarithmic scale for SO_4, Fe, Zn and Cu. ∇ indicates the water table at time of sampling (Appleyard and Blowes, 1994).

Lowering of the water table by pumping may give access of atmospheric oxygen to sulfides which were once submerged and protected against oxidation (Kinniburgh et al., 1994; Larsen and Postma, 1997). Figure 7.2 shows the Ni^{2+} distribution in the zone of drawdown in the Beder aquifer (Larsen and Postma, 1997). The high nickel concentration is accompanied by a high sulfate concentration which suggests that pyrite oxidation is taking place. Nickel has not moved far in the aquifer since pumping started, apparently because sorption on iron oxyhydroxide and manganese oxides retards its movement.

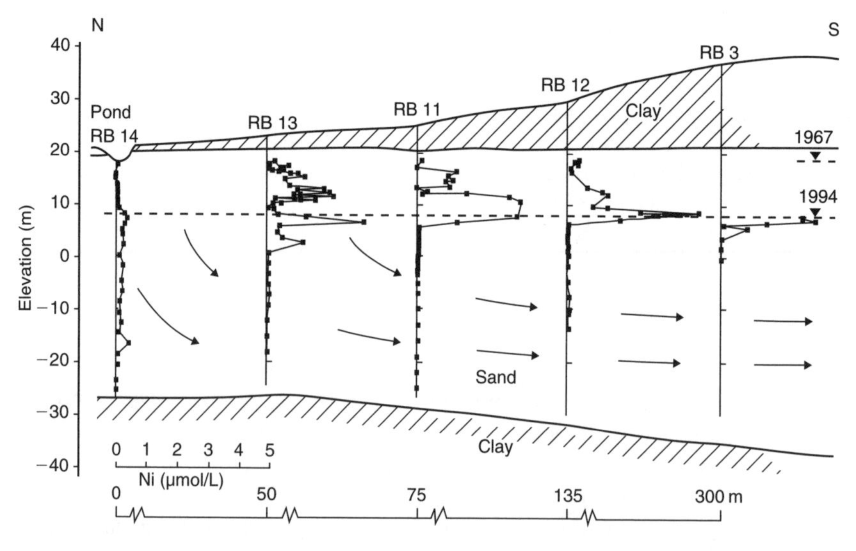

Figure 7.2. Distribution of dissolved nickel in the unsaturated and saturated zone of the Beder aquifer, Denmark. The water table at the time of sampling in 1994 and the pre-production water table in 1967 is indicated (Larsen and Postma, 1997).

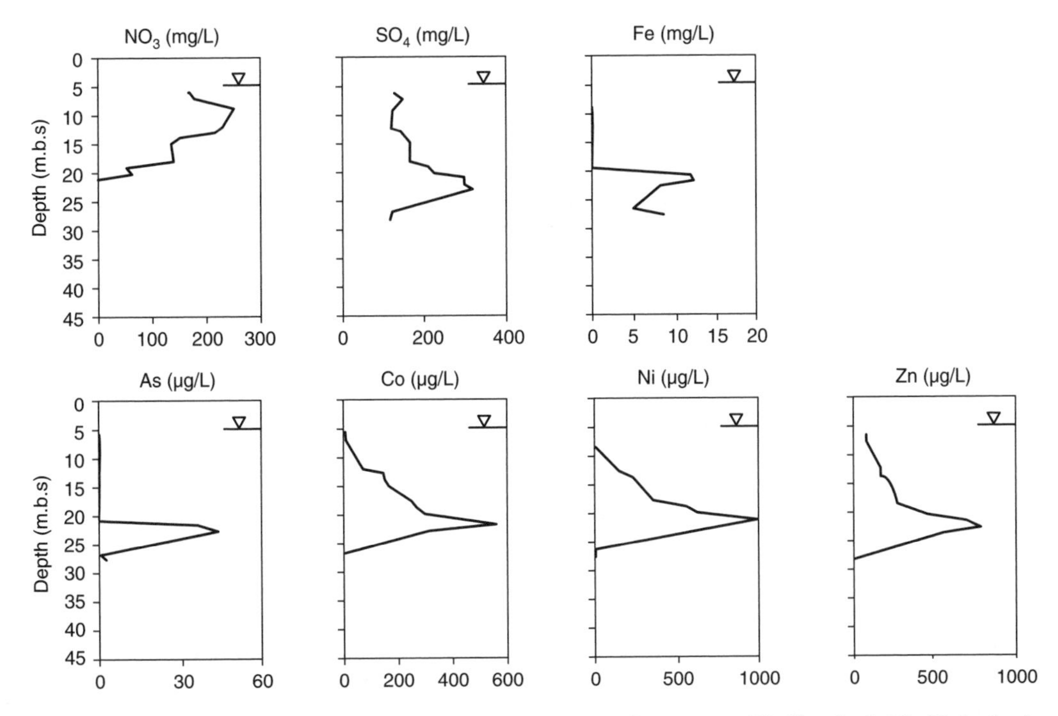

Figure 7.3. Groundwater composition in a drinking water protection area near Vierlingsbeek, The Netherlands. Pyrite oxidation is coupled to nitrate reduction and leading to the release of trace metals to the groundwater (Van Beek and Van der Jagt, 1996).

However, the SO_4^{2-} concentration has already increased in the pumping wells, because SO_4^{2-} is more mobile and not sorbed at the groundwater pH of 7.

Another oxidant for pyrite is nitrate, leached from increased agricultural application of fertilizer and manure. The reaction is in this case:

$$FeS_2 + 3NO_3^- + 2H_2O \rightarrow Fe(OH)_3 + 1.5N_2 + 2SO_4^{2-} + H^+ \qquad (7.2)$$

and is further discussed in Chapter 9. Figure 7.3 shows the groundwater composition of a drinking water protection area at Vierlingsbeek. From 15 to 20 m below surface there is a decrease of 150 mg NO_3^-/L accompanied by an increase of 155 mg SO_4^{2-}/L, in accordance with Reaction (7.2). Together with sulfate, trace components As, Co, Ni and Zn are released and reach concentrations many times higher than permissible in drinking water. Again, the pumping wells in the area showed an increase of sulfate, but not of the trace metals. The trace metals are sorbed and thus retarded with respect to groundwater flow, although eventually they will also arrive at the production well.

The chemical weathering of detrital silicates like amphiboles and biotite, and of clay minerals and oxyhydroxides may also release trace metals (Edmunds et al., 1992). Atmospheric input can be important as well. At circum-neutral pH, the heavy metals remain fixed by sorption or solid equilibria, but when water turns acid they can be mobilized. Figure 7.4 shows the water chemistry in a shallow aquifer consisting of unreactive siliceous sand. The upper groundwater has become acid mainly due to acid rain, and has a pH of 4.5 and increased concentrations of heavy metals like Ni, Co and Be (Kjøller et al., 2004). Further down, the pH increases to around 6.5 and the concentrations of heavy metals are much reduced.

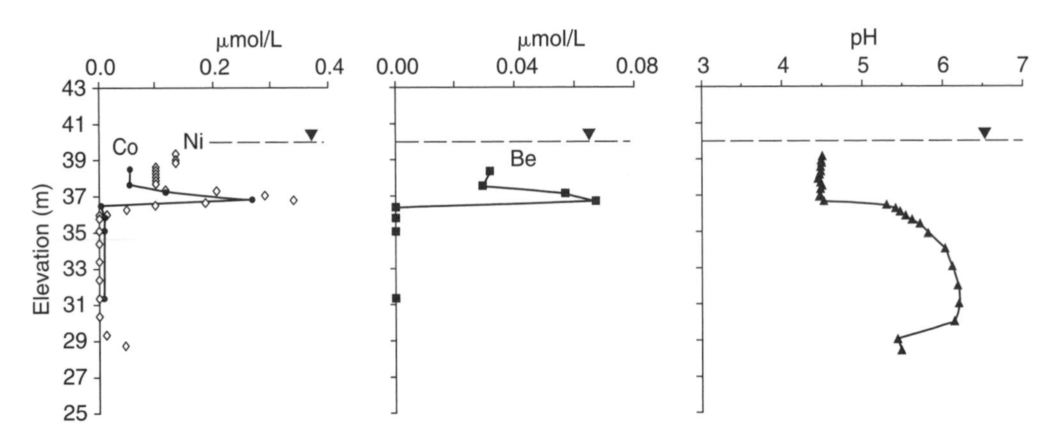

Figure 7.4. The mobilization of trace metals in an aquifer acidified by acid rain (Kjøller et al., 2004).

The contamination of groundwater by arsenic is a serious problem in Bangladesh and West Bengal (Smedley and Kinniburgh, 2002) and Vietnam (Berg et al., 2001). Figure 7.5 shows the arsenic distribution in well waters of Bangladesh. In the upper 100 m, arsenic is as high as 1000 μg/L, while the highest admissible WHO concentration is 10 μg/L (Table 1.1). Arsenic may substitute for sulfide in pyrite and the oxidation of pyrite can be one of the causes for the high arsenic contents of groundwater (Smedley and Kinniburgh, 2002). Another hypothesis is that arsenic adsorbed on Fe-oxide particles had been transported by the Ganges and other large rivers down to the delta. Here the sediment was deposited together with organic material. Subsequent development of anoxic conditions caused the Fe-oxides to become reduced, resulting in the release of the arsenic to the groundwater (Ravenscroft et al., 2001). Recently, Appelo et al. (2002) have proposed that the displacement of adsorbed arsenic by dissolved carbonate could be the cause of the mobilization of arsenic (cf. Chapter 11).

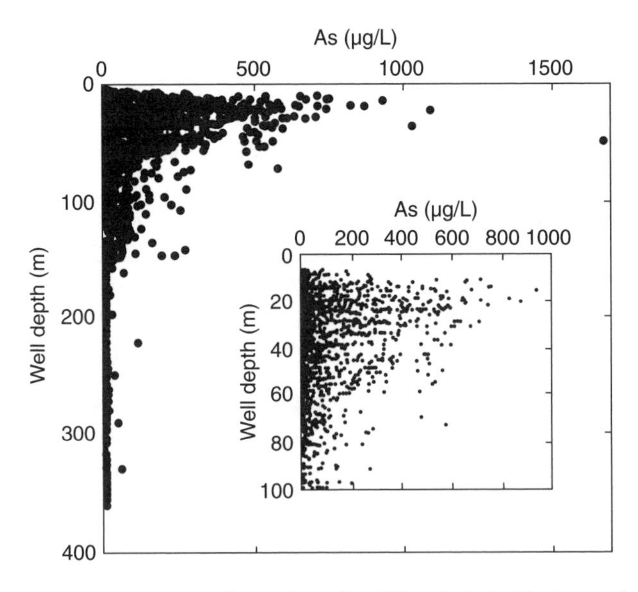

Figure 7.5. Arsenic concentration of groundwater in wells of Bangladesh. The insert shows that most arsenic is found in the uppermost aquifer (Kinniburgh and Smedley, 2001).

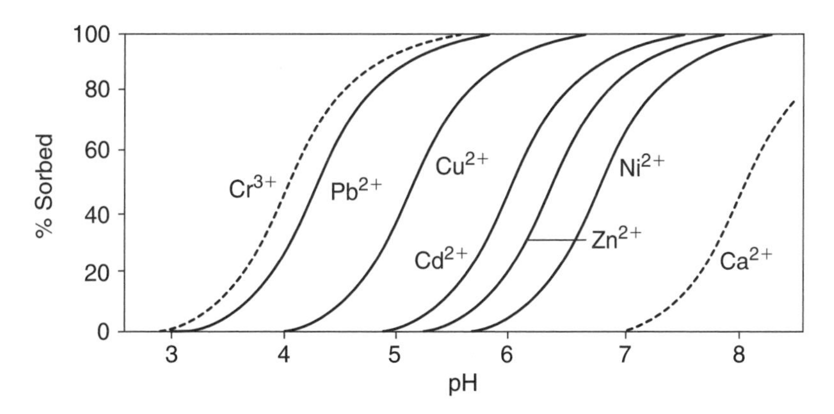

Figure 7.6. Adsorption of heavy metals on the surface of ferrihydrite as a function of pH (Stumm, 1992, based on the compilation of Dzombak and Morel, 1990).

In Chapter 6 the capacity of sediment particles to sorb major ions like Na^+ and Ca^{2+} was given by the cation exchange capacity, the *CEC*. The *CEC* was a constant related to the clay mineral and organic carbon content. The exchange concept works well for major cations in general and for trace metals when the pH is constant, but trace metals behave differently when pH varies. Figure 7.6 shows the sorption of heavy metals on amorphous iron oxyhydroxide (ferrihydrite or hydrous ferric oxide) as a function of pH. All metals show zero sorption at low pH and increased sorption as pH increases. Apparently, H^+ ions compete with the heavy metals for the sorption sites. However, the pH where 50% of the total amount of metal is sorbed varies for different metal ions. One of the challenges of hydrogeochemistry is to quantify this sorption behavior, and to assess its influence on the mobility of heavy metals and to predict arrival times of increasing concentrations of these elements.

7.2 SORPTION ISOTHERMS AND DISTRIBUTION COEFFICIENTS

The relation between sorbed and dissolved solute concentrations at a fixed temperature is called a *sorption isotherm*. Two equations are often employed to describe the relation, the Langmuir and the Freundlich isotherm. The Freundlich isotherm has the form:

$$s_I = K_F \cdot c_I^n \tag{7.4}$$

where s_I is the sorbed concentration (mol/kg, μg/g, etc.), c_I is the solute concentration of chemical I (mol/L, μg/mL, the same chemical mass units as for s), and K_F and n are adjustable coefficients. Usually, n is smaller than 1, so that the increase of sorbed concentration lessens as the solute concentration increases. For fitting of experimental data, the Freundlich equation can be linearized by a log transform to:

$$\log s_I = \log K_F + n \log c_I \tag{7.5}$$

which enables the constants to be derived by linear regression. Alternatively, a non-linear regression can be performed directly on Equation (7.4), with the advantage that the errors can be weighted as desired (Kinniburgh, 1986). The procedure is easy using a spreadsheet program (Example 7.1).

EXAMPLE 7.1. *Freundlich sorption isotherm for Cd^{2+} on loamy sand*
Christensen (1984) determined sorption of Cd^{2+} on loamy sand in 1 mM $CaCl_2$, pH = 6.0, as tabulated in Table 7.1. Find the coefficients of the Freundlich sorption isotherm with logarithmic weighting of the data $(\log s_{est} - \log s_{obs})^2$ and compare with absolute weighting $(s_{est} - s_{obs})^2$, where s is sorbed concentration in μg/g soil. Use a non-linear solver, for example from Excel.

Table 7.1. Dissolved and sorbed concentrations of Cd^{2+} on loamy sand.

Dissolved c_{Cd} (μg/L)	Sorbed s_{Cd} (μg/g)
3.1	0.86
6.1	1.12
5.9	1.71
7.1	1.68
8.1	2.03
9.9	2.46
12.3	2.85
13.0	3.36
13.6	3.22
16.0	3.25
19.1	3.56
24.1	3.76
25.8	4.17
27.6	4.58
33.2	4.82
36.4	5.19

ANSWER:
The data are entered in columns **A2..17** and **B2..17** of the spreadsheet. Guesses for K_F and n are typed in **E2** and **D2**, respectively. Calculate the estimated sorbed concentration in **C2** using the formula **+E$2 * A2 ^ D$2** and copy the formula into **C3..17** for the other datapoints. Calculate in column **D2..17** the squared difference of the values in **B** and **C**, **+(C2 − B2)^2** etc. for absolute weighting of the errors, or the log of the numbers, **+(log(C2) − log(B2))^2** etc. for logarithmic weighting, and sum the column in **D19**.

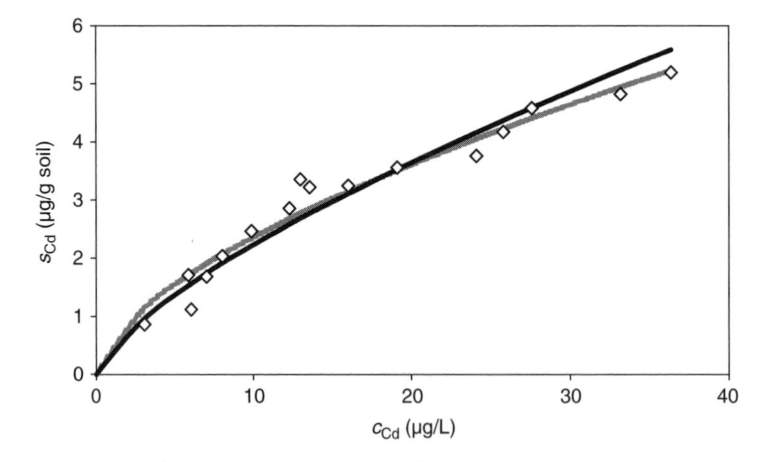

Figure 7.7. Freundlich isotherms for Cd^{2+} sorbed on loamy sand. Data points from Christensen (1984), grey and black lines are best-fits using absolute and logarithmic weighting of errors, respectively.

This sum is to be minimized by adjusting the values of K_F and n in **E2** and **D2**, for example in Excel. Use Tools, Solver and follow the instructions. The result will be,

with absolute weights: $s_{Cd} = 0.568 \cdot (c_{Cd})^{0.617}$
with logarithmic weights: $s_{Cd} = 0.431 \cdot (c_{Cd})^{0.713}$

The isotherms are slightly different (Figure 7.7). In general, a logarithmic or relative weighting may be preferable as it weights more highly the small values in the fit which may be the most accurately determined and the most important for extrapolation.

QUESTION:
Derive the Freundlich parameters with relative weighting of the errors (**+((C2 − B2) / C2)^2**, etc.).
ANSWER: $s_{Cd} = 0.468 \cdot (c_{Cd})^{0.690}$

With the Freundlich equation, sorption extends infinitely as concentrations increase, which is unrealistic since the number of sorption sites can be expected to be limited. Also, it is generally observed that the *distribution coefficient*, i.e. the ratio $K'_d = s_I / c_I$, becomes a constant when concentrations are small, but in the Freundlich equation the value of

$$K'_d = K_F \cdot c^{n-1} \tag{7.6}$$

increases indefinitely as concentrations decrease (for $n < 1$).

The Langmuir equation has a better theoretical background and can be derived from the law of mass action for a sorption reaction:

$$s + I \leftrightarrow s_I; \qquad K_{s_I} = \frac{[s_I]}{[s][I]} \tag{7.7}$$

With the mass balance for sorption sites:

$$s + s_I = s_{tot} \tag{7.8}$$

it gives:

$$[s_I] = \frac{[s_{tot}]K_{s_I}[I]}{1 + K_{s_I}[I]} \tag{7.9}$$

where s_{tot} is the total concentration of sorption sites. The activity of the sorbed species can be expressed as a fractional surface coverage by dividing by s_{tot}. Langmuir used a kinetic approach to derive the equation for gas adsorption on a surface and obtained:

$$s_I = \frac{s_{max}\, c_I}{K_L + c_I} \tag{7.10}$$

It can be checked that the equations are identical if $s_{tot} = s_{max}$ and $K_L = 1 / K_{s_I}$.

The Langmuir equation shows that the sorbed concentration s_I increases linearly with solute concentration c_I, if $c_I << K_L$. When the concentration of I is very high and $c_I >> K_L$, the surface becomes saturated, and $s_I = s_{tot}$.

In the linear part of the curve, the ratio of sorbed and solute concentrations is constant:

$$K'_d = s_I / c_I = s_{tot} / K_L \tag{7.11}$$

If s_I is expressed in mass/kg solid and c_I in mass/L water, K'_d has the dimensions L/kg. For the experiment with loamy sand and Cd^{2+} from Example 7.1, a Langmuir equation can be fitted (using logarithmic weighting of the errors) with $s_{tot} = 9.50\ \mu g/g$ soil and $K_L = 30.9\ \mu g/L$. Hence, $K'_d = 307$ L/kg. It is of interest to extrapolate these numbers to field conditions, assuming $\rho_b/\varepsilon_w = 6$ kg/L. The retardation $R = 1 + K'_d \cdot \rho_b/\varepsilon_w = 1845$ will slow down the movement of Cd^{2+} appreciably, but the sorption maximum of 0.084 mmol/kg is fairly small and must be considered when calculating the mass that is retained by the soil.

7.2.1 *Distribution coefficients from ion exchange*

We can use ion exchange equations to calculate *a priori* values for distribution coefficients of heavy metals. For a heavy metal I^{i+} in exchange equilibrium with Ca^{2+}, we have:

$$1/i\, I^{i+} + \tfrac{1}{2}\text{Ca-X}_2 \leftrightarrow 1/i\, I\text{-X}_i + \tfrac{1}{2}\text{Ca}^{2+} \tag{7.12}$$

with

$$K_{I\backslash \text{Ca}} = \frac{[\text{I-X}_i]^{1/i}\, [\text{Ca}^{2+}]^{0.5}}{[\text{Ca-X}_2]^{0.5}\, [I^{i+}]^{1/i}} \tag{7.13}$$

from which we obtain:

$$\frac{[I\text{-X}_i]}{[I^{i+}]} = \frac{\beta_I}{[I^{i+}]} = \left(K_{I\backslash \text{Ca}} \frac{\beta_{\text{Ca}}^{0.5}}{[\text{Ca}^{2+}]^{0.5}} \right)^i \tag{7.14}$$

The equivalent fraction β_I is multiplied by the *CEC* to give the exchangeable I-X$_i$ in meq/kg; the activity $[I^{i+}]$ is multiplied by the charge i and 1000 to give (numerically) the aqueous concentration in meq/L. Thus, we get the distribution coefficient in L/kg:

$$K'_d = \frac{CEC}{1000\, i} \left(K_{I\backslash \text{Ca}} \frac{\beta_{\text{Ca}}^{0.5}}{[\text{Ca}^{2+}]^{0.5}} \right)^i \tag{7.15}$$

The K'_d can subsequently be made non-dimensional, by multiplying by the sediment bulk density ρ_b (kg/L) and dividing by the porosity ε_w.

In fresh water β_{Ca} approaches 1 (cf. Example 6.4) and the Ca^{2+}-concentration can be estimated for a given CO_2-pressure and equilibrium with calcite (Chapter 5). With an estimate of the *CEC*, the distribution coefficient is obtained (Examples 7.2 and 7.3).

EXAMPLE 7.2. *Estimate the distribution coefficient of Cd^{2+}*

The loamy sand of Example 7.1 contained 6.2% clay (<2 μm) and 0.35% organic carbon. Estimate the distribution coefficient in the experiment with $m_{Ca^{2+}} = 1$ mM, for Cd^{2+} at trace quantities, and express the distribution coefficient in dimensionless units assuming $\varepsilon_w = 0.3$.

ANSWER:

Estimate the *CEC* with Equation (6.8): $CEC = 7 \times 6.2 + 35 \times 0.35 = 55.7$ meq/kg (Christensen (1984) analyzed 75 meq/kg). The exchange coefficient $K_{Cd\backslash Ca} = K_{Na\backslash Ca}/K_{Na\backslash Cd} = 1$ (from Table 6.4). With Equation (7.15) the distribution coefficient is: $K'_d = 55.7\ /\ 2000 \times (1 \times 1\ /\ \sqrt{10^{-3}})^2 = 27.8$ L/kg.

Bulk density $\rho_b = 2.65 \times (1 - \varepsilon) = 1.86$ kg/L (cf. Example 6.1). Hence the dimensionless distribution coefficient is $27.8 \times \rho_b / \varepsilon = 172$.

QUESTIONS:
What is the effect of doubling the Ca^{2+} concentration?
ANSWER: K_d halves
What is the effect of doubling the *CEC*?
ANSWER: K_d doubles
Calculate K_d with PHREEQC
ANSWER: $K_d = 19$ L/kg, note the effect of activity corrections for CdX_2, Problem 7.20.

The distribution coefficient, estimated in Example 7.2, is about a factor 10 smaller than calculated with the Langmuir fit ($K_d' = 307$ L/kg). Also, the maximal sorption capacity for Cd^{2+} would be the $CEC / 2 = 27.8$ mmol/kg which is 330 times higher than was estimated with the Langmuir equation ($s_{tot} = 9.50\ \mu g/g = 0.085$ mmol/kg). Apparently, the soil contains a small amount of sorbent that binds Cd^{2+} very strongly in a process not captured by the ion exchange formulae. Christensen (1984) and many others, e.g Boekhold et al. (1991), noted a strong pH dependency of the distribution coefficient of Cd^{2+} in soils that may be related to sorption on variable charge surfaces discussed later in this chapter.

Major elements are normally neglected in the calculation of the distribution coefficient from experimental data, but they can be incorporated in the multicomponent ion exchange formulae from Chapter 6. If the charge of the trace element and major ions is equal, a simple linear equation is obtained from Equation (6.22). For example for Sr^{2+}, as a trace element, with respect to the major ions Mg^{2+} and Ca^{2+}:

$$\beta_{Sr} = \frac{[Sr^{2+}]}{[Sr^{2+}] + K^2_{Ca\backslash Sr} \cdot [Ca^{2+}] + K^2_{Mg\backslash Sr} \cdot [Mg^{2+}]} \qquad (7.16)$$

EXAMPLE 7.3. *Distribution coefficient for* Sr^{2+}

Let us consider the applicability of the multicomponent relation (7.16) for data provided by Johnston et al. (1985) who carefully determined Sr^{2+} distribution coefficients for sediments near a radioactive waste site. The distribution coefficients were obtained for sands and tills at varying Sr^{2+} concentrations in groundwater and synthetic solutions. A typical result is shown in Figure 7.8 for two samples of weathered till. Synthetic groundwater composition, and *CEC* of the two samples are given in Table 7.2.

Table 7.2. Water compositions and sediment *CEC* in an experiment for determining the Sr^{2+} distribution coefficient (Johnston et al., 1985). Aqueous concentrations in mmol/L.

Sample	*CEC* (meq/kg)	Ca^{2+}	Mg^{2+}	Na^+	K^+	Sr^{2+}
A4	53	.75	.80	.09	.08	10^{-8}–50
A8	34	.75	.33	.3	.43	10^{-8}–50

The groundwater used for sample A4 contains mainly Ca^{2+} and Mg^{2+} in solution, which allows the experimental results to be modeled as a cation exchange reaction with respect to only the divalent ions (Mg^{2+} and Ca^{2+}). The distribution coefficient is in our exchange model:

$$K_d' = \frac{CEC}{2000} \frac{\beta_{Sr}}{m_{Sr^{2+}}}$$

where m_{Sr} is molality of Sr^{2+} (mol/kg H_2O). By combining the reactions listed in Table 6.4 we obtain the exchange coefficients $K_{Ca\backslash Sr^{2+}} = 0.77$, and $K_{Mg\backslash Sr^{2+}} = 0.49$. We furthermore assume $[I] = m_I$ and find in combination with Equation (7.16):

$$K'_d = \frac{CEC/2000}{c_{Sr^{2+}}/87600 + 0.77 \cdot m_{Ca^{2+}} + 0.49 \cdot m_{Mg^{2+}}} \quad (7.17)$$

where c_{Sr} is concentration of Sr^{2+} in mg/L (we use mg/L to allow for a direct comparison with experimental data from Johnston et al., 1985). The conditions for sample A4 thus lead to:

$$K'_d(A4) = 0.0265/(c_{Sr}/87600 + 9.7\times10^{-4}),$$

and for sample A8:

$$K'_d(A8) = 0.017/(c_{Sr}/87600 + 7.4\times10^{-3}).$$

Lines corresponding to these equations are plotted in Figure 7.8, and are in good agreement with experimental values.

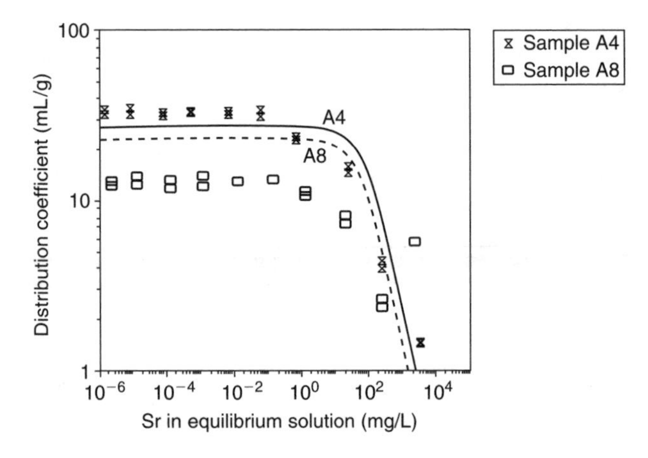

Figure 7.8. Strontium distribution curves for two soil samples. Experimental data from Johnston et al., 1985. Lines are estimated here.

EXAMPLE 7.4. *Exchange coefficient of Cd^{2+} vs Na^+ on montmorillonite*
Garcia-Miragaya and Page (1976) determined adsorption percentages of Cd^{2+} from $NaClO_4$ solutions of different normality on 2.05 meq/L montmorillonite. A selection of their data is presented in the Table below.

$NaClO_4$ (N)	Cd_{total} (μM)	adsCd (%)	$NaClO_4$ (N)	Cd_{total} (μM)	adsCd (%)
0.01	0.40	97.1	0.05	0.40	73.6
	0.80	96.3		0.80	69.2

Determine the exchange coefficient $K_{Na/Cd}$ for the Gaines and Thomas convention.

ANSWER:
Inspection of the data shows that the percentage of Cd^{2+} adsorbed (adsCd) appears to be independent of the amount of Cd^{2+} added to the solution. In other words, the ratio of sorbed Cd^{2+} over solute Cd^{2+} is

independent of Cd_{total}, and the distribution coefficient is a constant. The concentration of Na^+ has a marked effect, however, and we expect that cation exchange is operative. The reaction is:

$$Na^+ + 0.5Cd\text{-}X_2 \leftrightarrow Na\text{-}X + 0.5Cd^{2+}$$

with

$$K_{Na\backslash Cd} = \frac{\beta_{Na}}{[Na^+]} \frac{[Cd^{2+}]^{0.5}}{\beta_{Cd}^{0.5}} \tag{7.18}$$

We have X = exchange capacity of the montmorillonite, 2.05×10^{-3} M,
$[Na^+] = m_{Na^+}$ = solution normality N,
$\beta_{Cd} \cdot X/2 = Cd_{total} \cdot adsCd/100$, or $\beta_{Cd} = Cd_{total} \cdot adsCd/(50\,X)$,
$[Cd^{2+}] = m_{Cd^{2+}} = Cd_{total}(1 - adsCd/100)$,
and assume $\beta_{Na} = 1$. This gives in (7.18):

$$K_{Na\backslash Cd} = \frac{1}{N} \frac{(1 - adsCd\,/\,100)}{adsCd\,/\,50X}$$

and we obtain: for N = 0.01: $K_{Na\backslash Cd} = 0.55$ and 0.63, and for $N = 0.05$: $K_{Na\backslash Cd} = 0.38$ and 0.43. Compare with Table 6.4.

The distribution coefficient formulae generally work quite well for metal ions such as Ni^{2+}, Zn^{2+}, Cd^{2+}, Sr^{2+} and alkaline metals on montmorillonites (Garcia-Miragaya and Page, 1976; Shiao et al., 1979; Baeyens and Bradbury, 1997), but deviations occur. For example, in Figure 7.9, distribution coefficients are plotted for various metal ions at trace concentration in exchange with Na^+ on montmorillonite. We replace $(\beta_{Ca}/[Ca^{2+}])^{0.5}$ with $(\beta_{Na}/[Na^+])$ in Equation (7.15) and expect to find, in a plot of log K'_d versus log m_{Na^+}, a slope of -2 for Sr^{2+} and Co^{2+}, and of -1 for Cs^+.

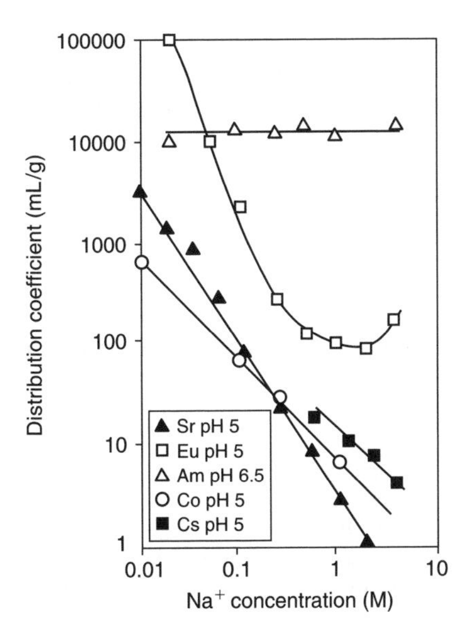

Figure 7.9. Distribution coefficients of various radionuclides as a function of the Na^+ concentration, on a Na^+ smectite (Shiao et al., 1979).

However, the experimental data in Figure 7.9 indicate that the distribution coefficient for Co^{2+} has a slope of -1, similar to Cs^+, while americium is apparently so strongly sorbed that an increase of the Na^+ concentration has no effect at all. The slope for Eu^{3+} is -3 at low concentrations of Na^+, but displays an unexpected minimum at higher concentrations. Clearly, the high Na^+ concentrations are beyond the usually encountered range in fresh ground water, but the anomalies are important for assessing the behavior of radionuclides at waste sites. We have already calculated that in a natural soil, specific sorbers may be present which bind heavy metals very strongly (Examples 7.1, 7.2). Likely candidates for these solids are the edges of clay minerals, oxyhydroxides and organic matter that we will discuss in the following sections.

7.3 VARIABLE CHARGE SURFACES

Goethite (α-FeOOH) is a ubiquitous mineral in soils and aquifers and is probably the most often used mineral for studying sorption behavior of trace metals because it can be easily synthesized in the laboratory. The primary structural unit is an octahedron of 6 oxygens which coordinates to the central Fe^{3+} ion (Figure 6.6 displays an octahedron). In the crystal structure of goethite, the octahedra share oxygens to form rows of two octahedra as illustrated in Figure 7.10. The rows are interconnected at the corners, and give the impression of being surrounded by open channels which are, however, occupied by the protons. On the outside of the crystal, the channels become "grooves" in the (010) and ($0\bar{1}0$) faces. These can act as a trap for ions and may be responsible for the distinct sorption behavior of goethite (Russell et al., 1974).

The tendency for oxygens to bind or to loose protons can be calculated with Pauling's bond valence concept and is related to the number of bonds to Fe^{3+} atoms (Hiemstra et al., 1989; Venema et al., 1996a). One Fe^{3+} ion donates $+3 / 6 = 0.5$ bond valence to each of the 6 surrounding oxygens. An oxygen connected to 3 Fe^{3+} has a formal charge of $-2 + 3 \times 0.5 = -0.5$. It needs half a proton for charge balance and thus, will share a proton with another oxygen. All the internal oxygens in the goethite structure are triply coordinated to iron, but at the surface of the crystal, doubly or singly coordinated oxygens exist which are especially greedy for charge compensation by heavy metal ions. The singly coordinated oxygens are indicated with "A" in Figure 7.10, the hydroxyls as "B", and the doubly coordinated oxygens as "C". The singly coordinated oxygens are present on all the faces which parallel the c-axis, but are most abundant on the $\{010\}$ form.

The reactivity of goethite thus depends on the crystal morphology, which will change depending on the growth stage of the crystal. The planes of the $\{110\}$ form grow rapidly and are found on small crystals in the laboratory and in soils (Cornell and Schwertmann, 2003). However, quickly developing faces fade away during extended growth and the two planes of the $\{010\}$ form, which possess the deep groove, become dominant on larger, natural crystals. Most experimental binding constants apply to the surface properties of small, needle-like crystals, but it can be expected that the (010) and ($0\bar{1}0$) faces of larger goethite crystals provide more sites per given surface area and will also assert a stronger influence on binding.

7.3.1 *Titration curves with suspended oxide particles*

The surfaces of Fe-oxides, clays, other minerals or organics may become protonated or deprotonated and obtain different surface properties as a function of pH. In many ways surfaces behave similarly to amphoteric solutes as illustrated in the following. First we titrate a volume of H_2O with NaOH or HCl. Electroneutrality requires that:

$$m_{H^+} + m_{Na^+} = m_{OH^-} + m_{Cl^-} \tag{7.19}$$

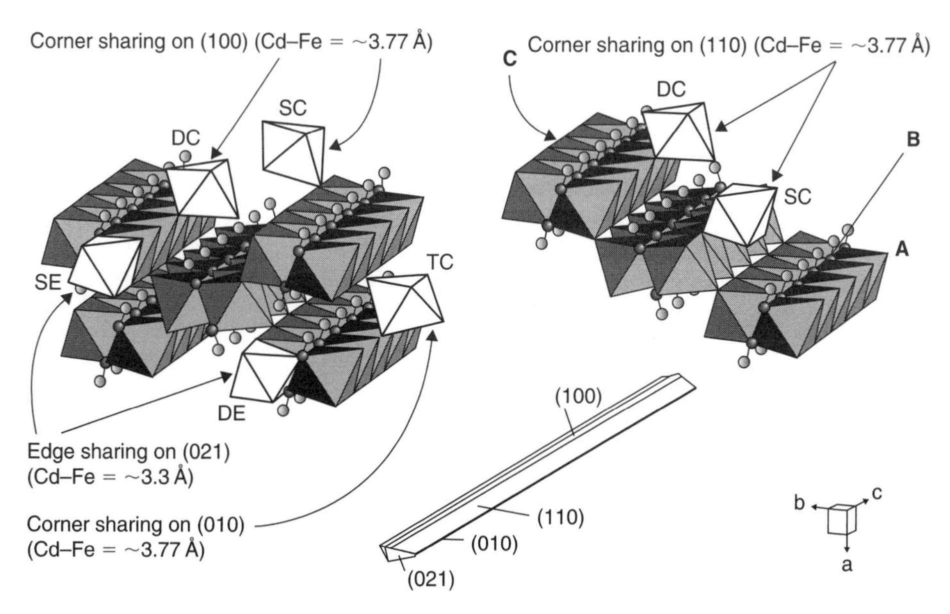

Figure 7.10. Atoms in the structure of goethite and, at the same orientation, morphology of the lath shaped crystals. Cadmium adsorbs mainly in the form of a corner sharing octahedron with "A" type oxygens (see text). The sorption coordination units are indicated by SC (single corner), DC (double corner), TC (triple corner), SE (single edge) (Randall et al., 1999).

From the H_2O dissociation constant we derive $m_{OH^-} = 10^{-14}/m_{H^+}$, neglecting the difference between activity and molal concentration. The relation between $[H^+]$ or m_{H^+} and added base m_{Na^+} is then:

$$m_{Na^+} = 10^{-14} / m_{H^+} - m_{H^+} + m_{Cl^-} \tag{7.20}$$

and we calculate the titration curve of pure water (Figure 7.11).

Next we add an amphoteric solute AH, i.e. a substance which can act both as acid and as base, following the reactions:

$$AH_2^+ \leftrightarrow AH + H^+; \quad K_{a1} \tag{7.21}$$

and

$$AH \leftrightarrow A^- + H^+; \quad K_{a2} \tag{7.22}$$

where K_{a1} and K_{a2} are the acid dissociation constants. The species AH_2^+ and A^- are included in the electroneutrality equation, which becomes:

$$m_{Na^+} = 10^{-14} / m_{H^+} - m_{H^+} + m_{Cl^-} + m_{A^-} - m_{AH_2^+} \tag{7.23}$$

where the species AH_2^+ and A^- are obtained as function of $[H^+]$ with the speciation formula (Equation 5.13). Again, we calculate the titration curve (Figure 7.11).

The titration curve for pure water (Figure 7.11) shows no buffering and the pH is a simple logarithmic function of the amount of NaOH or HCl added. The amphoteric acid titration curve shows two steps, corresponding to Reactions (7.21) and (7.22). AH_2^+ is the dominant species below pH 7.29, AH is predominant between pH 7.29 and 8.93 and A^- above pH 8.93.

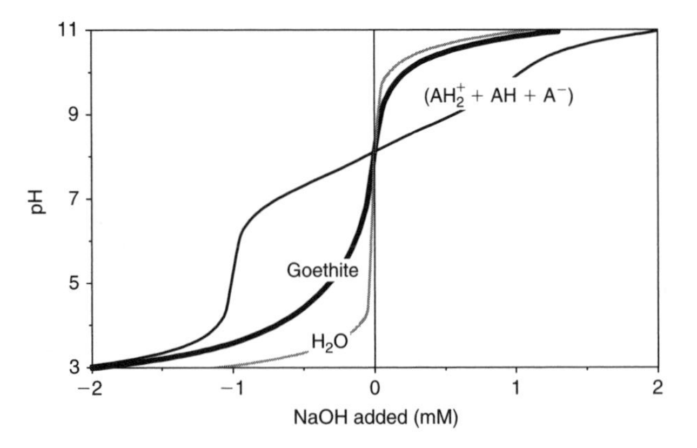

Figure 7.11. Titration-curves, showing the change of pH with addition or subtraction of NaOH (subtraction means addition of HCl) to pure H_2O, 1 mM AH, and a suspension of goethite. The amphoteric acid AH and goethite have both $\log K_{a1} = -7.29$ and $\log K_{a2} = -8.93$ as intrinsic acidity constants. Background electrolyte is 0.01 N NaCl.

The protonation/deprotonation reactions at the surface of goethite are:

$$\equiv FeOH_2^+ \leftrightarrow \equiv FeOH + H^+; \quad K'_{a1} \tag{7.24}$$

and

$$\equiv FeOH \leftrightarrow \equiv FeO^- + H^+; \quad K'_{a2} \tag{7.25}$$

where K'_{a1} and K'_{a2} are apparent equilibrium constants. For comparison, the titration-curve of goethite is shown in Figure 7.11 using the same values of K'_{a1} and K'_{a2} as for AH and an equivalent amount of acid neutralizing capacity. The titration curve of goethite shows a more gradual variation without the stepwise changes of AH.

In solution, the individual AH molecule is not affected by dissociation of other AH molecules. On the surface of goethite, however, the adsorption sites are positioned closely together. If one group has lost a H^+, it will be more difficult to desorb a H^+ from the neighbor groups since the increased negative charge holds the remaining H^+ more strongly. When part of the $\equiv FeOH$ surface sites has dissociated into $\equiv FeO^-$ and H^+ due to increasing pH, the apparent dissociation constant K'_{a2} for the remaining $\equiv FeOH$ will decrease. Likewise for the association of $\equiv FeOH$ with H^+, as more $\equiv FeOH_2^+$ has formed in response to decreasing pH, the more difficult is the attachment of additional H^+, and the apparent K'_{a1} will increase. The gradual change in the goethite titration curve is therefore due to the development of electrostatic charge on the goethite surface. When studying the association of ions with mineral surfaces one needs to consider both the chemical binding of the ion at the surface and the charge development on the surface (Dzombak and Morel, 1990).

7.3.2 *Surface charge and point of zero charge, PZC*

The surface charge of the solid is the sum of structural deficits, unbalanced bonds at the crystal surface and charge generated by the adsorbed ions:

$$\sigma = \sigma_0 + \sigma_i = \sigma_0 + \sigma_H + \sigma_M + \sigma_A \tag{7.26}$$

σ is the specific charge of the mineral (eq/kg) and the sum of σ_0, the structural charge due to crystal structure substitutions, and σ_i which is due to the various complexed ions as specified below.

σ_H is the charge due to protons:

$$\sigma_H = \{\equiv SOH_2^+\} - \{\equiv SO^-\} \tag{7.27}$$

For cation sorption the surface charge contribution is given by σ_M:

$$\sigma_M = (m-1)\{\equiv SOM^{(m-1)+}\} \tag{7.28}$$

From sorption of anion A^{a-}, the surface charge is:

$$\sigma_A = -(a+1)\{\equiv SA^{(a+1)}\} \tag{7.29}$$

The anions also comprise oxyanions such as PO_4^{3-}.

The sign of the surface charge can be measured using the *electrophoretic mobility* of suspended particles in an electric field. An example is shown in Figure 7.12 with rutile (TiO_2) particles placed in solutions of different composition (Fuerstenau et al., 1981). At the lowest $Ca(NO_3)_2$ concentration (Figure 7.12a) it is protonation that determines the surface charge. At low pH the rutile surface is positively charged and the particles move towards the negative electrode, at high pH the surface is negatively charged and the particles migrate to the positive electrode. The point where the surface charge is zero in these measurements, is called the iso-electric point (*IEP*). For the rutile used, it was pH 6.5.

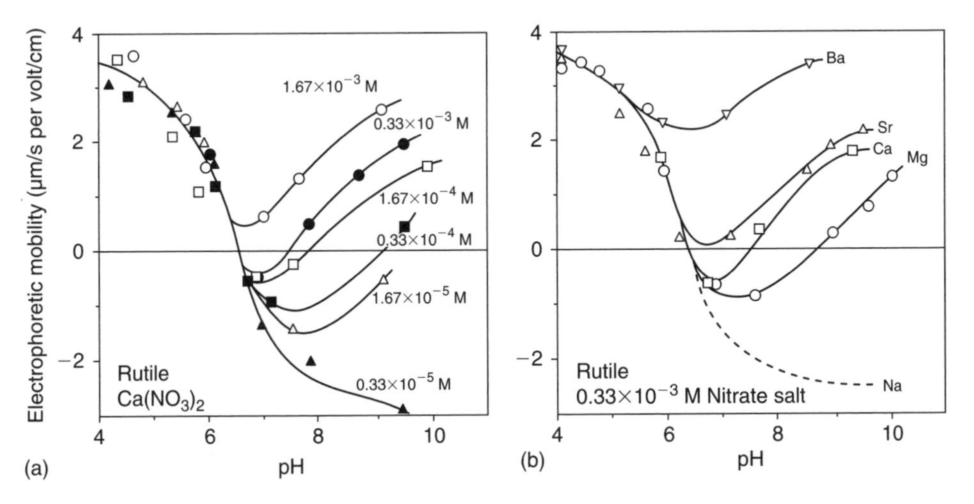

Figure 7.12. Dependence of the electrophoretic mobility of rutile particles on pH and electrolyte concentration of NO_3-salts of alkaline-earth ions. (a) Mobility at various concentrations of $Ca(NO_3)_2$. (b) Mobility at an electrolyte concentration of 0.33 mM of alkaline-earths. From Fuerstenau et al. (1981).

As the $Ca(NO_3)_2$ concentration increases, more Ca^{2+} will adsorb on the rutile surface and diminish the negative surface charge. As the result, the electrophoretic mobility decreases and it can even become reversed in the sense that the particles move towards the negative electrode. Thus, simple alkaline-earth ions, which are normally considered as inert background-electrolytes, can reverse the surface potential of oxides (Parks, 1990). In the low pH range the electrophoretic mobility of the rutile particles was not affected by the $Ca(NO_3)_2$ concentration and clearly nitrate does not change the surface charge of the rutile particles.

Figure 7.12b compares the ability of different ions to change the surface charge of the rutile particles. At the one end, the Na^+ ion seems to have little influence on the surface charge while at the other end, Ba^{2+} changes the surface charge to strongly positive even at the pH corresponding to the iso-electric point. Apparently, ions have different abilities to displace protons from the surface. Alkaline-earth ions like Ca^{2+} and Mg^{2+} act mainly as background electrolytes as long as the protonated surface is positive or neutral and it is only when their concentration becomes high that they may become potential determining ions. Heavy metals such as Cu^{2+}, Pb^{2+}, Cd^{2+}, but also Ba^{2+}, on the other hand, have a very strong ability to displace protons from the surface and that is the reason why heavy metal adsorption is strongly pH dependent (Figure 7.6).

The charge of the solid is compensated in the *diffuse double layer* (Section 6.6):

$$\sigma + \sigma_D = 0, \tag{7.30}$$

where σ_D is the charge of the double layer (eq/kg).

The distribution of ions at the particle surface is visualized in Figure 7.13. Alkali metals, including Na^+, and acid anions such as Cl^- keep a shell of hydration water and bind via outer-sphere complexes to the surface. Protons, heavy metals, and most oxyanions are found closer to the surface where they bind via inner-sphere complexes to the structural oxygens. In Figure 7.13b a Cu^{2+} ion has displaced a H^+ on a structural oxygen atom. Note also the inner-sphere, bidentate surface-complex of phosphate. The binding mechanism and the environment around the molecules can be deduced from electron resonance spectra, using various means of excitation (Hawthorne, 1988; Atkins and de Paula, 2002). Lastly, at some distance away from the surface, the diffuse layer extends into the solution, with counter-ions at higher and co-ions at lower concentration than in the free solution (Section 6.6).

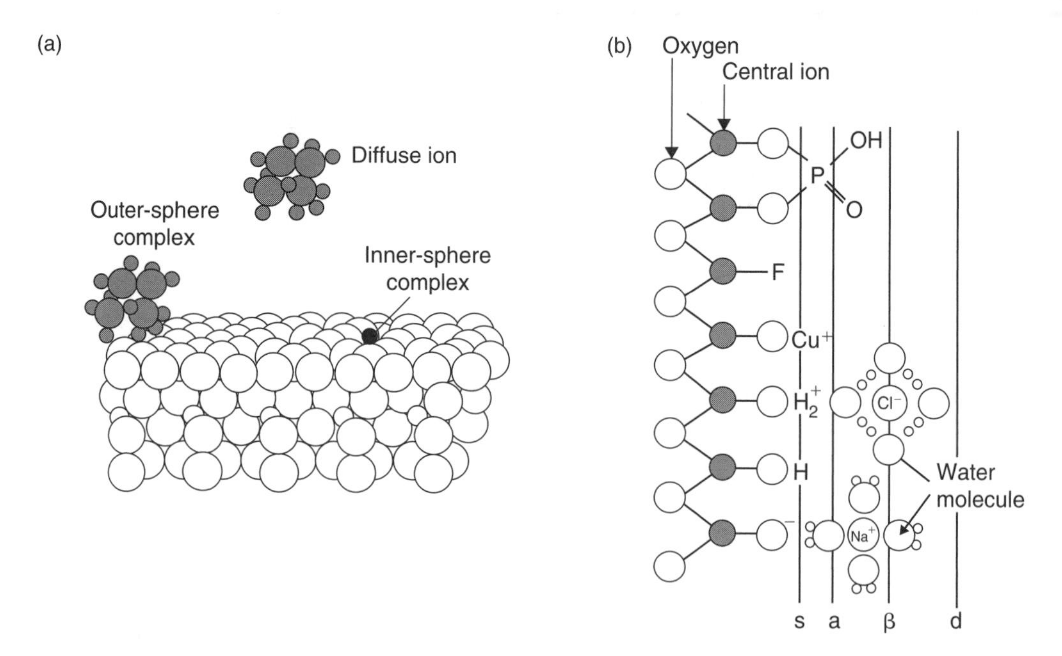

Figure 7.13. Sposito's visualization of a charged surface showing inner- and outer-sphere bonding and ions in the diffuse double layer. Figure 7.13b shows planes associated with the various types of bonding: "s" for surface hydroxyl groups, "a" for inner-sphere complexes, "β" for outer-sphere complexes and "d" for ions in the diffuse layer (Stumm, 1992).

At a given pH, the proton charge will compensate all other charge, and $\sigma = 0$. This pH is called the point of zero charge, *PZC*, or pH_{PZC}:

$$\sigma = 0 \text{ at } pH_{PZC}. \tag{7.31}$$

At the *PZC*, the surface has also zero potential, $\psi_0 = 0$, and therefore *PZC* equals the iso-electric point *IEP* when specific adsorption is absent. More precise definitions of *PZC* exist which account for the various processes that determine the charge (Sposito, 1984). The *Point of Zero Net Proton Charge* recognizes that the mineral may have a permanent structural charge, but at the pH_{PZNPC} the contributions of surface protonation and deprotonation reactions to the total charge are balanced:

$$\sigma_H = 0 \text{ at } pH_{PZNPC}. \tag{7.32}$$

The *Pristine Point of Zero Charge*, pH_{PPZC}, is the *PZC* for a mineral without structural charge, and without specific adsorption of other ions than protons:

$$\sigma = \sigma_H = 0 \text{ at } pH_{PPZC}. \tag{7.33}$$

Lastly, the *Point of Zero Salt Effect*, pH_{PZSE}, indicates the intersection point of titration curves at different ionic strengths:

$$d\sigma \, / \, dI = 0 \text{ at } pH_{PZSE}. \tag{7.34}$$

Table 7.3 lists *PZC*'s for a number of minerals. Iron oxides have *PZC*'s ranging from 8.5 to 9.3 and will at most groundwater pH values be neutral or weakly positive. Birnessite, δ-MnO_2, has a *PZC* of 2.2 and is negatively charged at the pH of most groundwaters. Minerals have a general capacity for anion exchange (in the double layer) when the pH is below the *PZC*, and a cation exchange capacity when the pH is above *PZC*. The capacity for exchange depends on the difference between the *PZC* and the pH of the solution. Figure 7.14 shows adsorption of Na^+ and Cl^- on the oxides SnO_2 and ZrO_2 as a function of pH. The *PZC* of SnO_2 is found at pH = 4.8, and is slightly lower than the *PZC* of ZrO_2 at 6.8. When the pH is higher than the *PZC* the oxide surface adsorbs Na^+ from solution, when the pH is below the *PZC*, Cl^- is adsorbed. Both ions are sorbed in small, equal quantities at the *PZC* of the oxide.

Table 7.3. The Point of Zero Charge, pH_{PZC}, of clays and common soil oxides and hydroxides (Parks, 1967; Stumm and Morgan, 1996; Davis and Kent, 1990; Venema et al., 1996a).

	pH_{PZC}
Kaolinite	4.6 (Parks)
Montmorillonite	<2.5 (Parks)
Corundum, α-Al_2O_3	9.1 (Stumm and Morgan)
γ-Al_2O_3	8.5 (Stumm and Morgan)
alpha-$Al(OH)_3$	5.0 (Stumm and Morgan)
Hematite, α-Fe_2O_3	8.5 (Davis and Kent)
Goethite, α-FeOOH	9.3 (Venema et al.)
$Fe(OH)_3$	8.5 (Stumm and Morgan)
Birnessite, δ-MnO_2	2.2 (Davis and Kent)
Rutile, TiO_2	5.8 (Davis and Kent)
Quartz, SiO_2	2.9 (Davis and Kent)
Calcite, $CaCO_3$	9.5 (Parks)
Hydroxyapatite, $Ca_5OH(PO_4)_3$	7.6 (Davis and Kent)

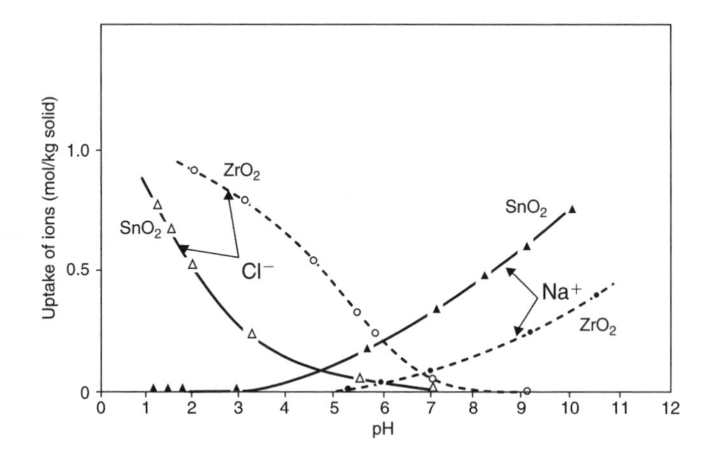

Figure 7.14. pH-dependent sorption on SnO_2 and ZrO_2: Cl^- is adsorbed at low pH, Na^+ at high pH (Kraus and Phillips, 1956).

7.3.3 *Sorption edges*

The pH dependency of metal sorption on oxides can be depicted in a plot of sorbed fraction versus the pH. Figure 7.15 shows adsorption curves for three divalent metals on rutile. The pH where 50% of the total metal in the suspension is sorbed, is called the pH_{50} or sorption edge. In Figure 7.15 the pH_{50} is at pH = 5.5 for Cd^{2+}, at 3.7 for Cu^{2+}, and at pH = 2.0 for Pb^{2+}. The sorption edge of all these cations moves to a lower pH when the concentration of rutile particles increases.

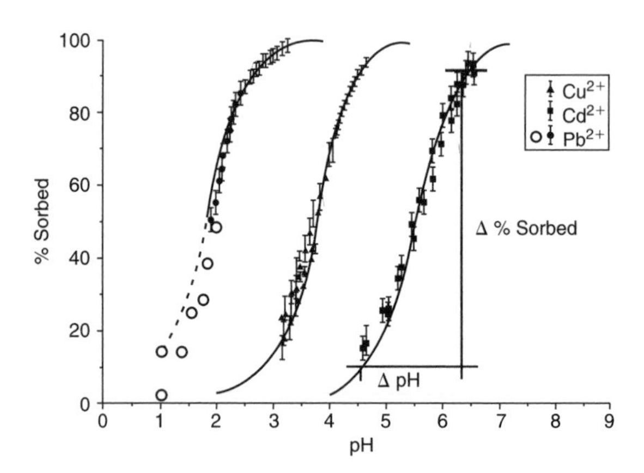

Figure 7.15. Adsorption of some divalent metal ions on TiO_2 (rutile). Comparison of pH-values at 10 and at 90% adsorption allows the reaction mechanism to be determined. Modified from Schindler and Stumm (1987).

Rutile has its *PZC* at ca. pH = 5.8 and in contrast to the preceding example (Figure 7.14), the heavy metals are sorbing on the rutile surface at a pH below the *PZC*, i.e. in the pH range where the oxide surface has a positive charge. The reaction where a heavy metal (M^{2+}) competes with a proton for the surface site is:

$$\equiv SOH + M^{m+} \leftrightarrow \equiv SOM^{(m-1)+} + H^+; \qquad K_{SOM} \tag{7.35}$$

This is called a *monodentate ligand* formation. The heavy metal with a divalent charge may also bind to two surface sites, and the reaction becomes:

$$2\equiv\text{SOH} + M^{2+} \leftrightarrow (\equiv\text{SO})_2M + 2\text{H}^+; \qquad K_{(\equiv\text{SO})_2M} \tag{7.36}$$

The association of a metal ion with two surface sites is often termed *bidentate ligand* formation. However, the binding of the metal ion to two oxygens at the mineral surface is not an actual bidentate bond. Because the oxygen atoms at the mineral surface are positioned closely together, they act as one unit in the law of mass action and Reaction (7.36) is normally written as:

$$(\equiv\text{SOH})_2 + M^{2+} \leftrightarrow (\equiv\text{SO})_2M + 2\text{H}^+; \qquad K_{(\equiv\text{SO})_2M} \tag{7.37}$$

If we calculate the ratio $\bar{M}/M$, where $\bar{M}$ is the adsorbed and M the solute concentration, for a high and a low metal sorption ratio, e.g. of 10 / 1 and of 1 / 10, we find for Equation (7.35):

$$\frac{[\text{H}^+]_{high\ sorption}}{[\text{H}^+]_{low\ sorption}} = \frac{1}{100} \cdot \frac{[\equiv\text{SOH}]_{high\ sorption}}{[\equiv\text{SOH}]_{low\ sorption}} \tag{7.35a}$$

and for Equation (7.36):

$$\frac{[\text{H}^+]_{high\ sorption}}{[\text{H}^+]_{low\ sorption}} = \frac{1}{10} \cdot \frac{[\equiv\text{SOH}]_{high\ sorption}}{[\equiv\text{SOH}]_{low\ sorption}} \tag{7.36a}$$

If $\bar{M}$ is present in trace quantities only, $[\equiv\text{SOH}]_{high\ sorption} = [\equiv\text{SOH}]_{low\ sorption}$, and the increase of the adsorption ratio from 1 / 10 to 10 / 1 requires a pH-increase of either 2 units (for Equation 7.35) or of 1 unit (for Equation 7.36). This pH-increase corresponds to an adsorption increase from 9.1 to 90.9%. Inspection of Figure 7.15 shows that ΔpH is 2.0 for Cd^{2+}, and about 1.6 for Cu^{2+} and Pb^{2+}. It suggests that Equation (7.35) is appropriate for Cd^{2+} on rutile, while a combination of mono- and bidentate bonds could be surmised for Cu^{2+} and Pb^{2+}.

However, the situation is more complicated for Cu^{2+} and Pb^{2+}. The sorption edge for these metals is located at a low pH where $[\equiv\text{SOH}_2^+]$ is the dominant surface species. When the pH increases, $[\equiv\text{SOH}]$ will increase and the assumption $[\equiv\text{SOH}]_{high\ sorption} = [\equiv\text{SOH}]_{low\ sorption}$ is no longer valid. Instead, in Equations (7.35a) and (7.36a), the ratio $[\equiv\text{SOH}]_{high\ sorption} / [\equiv\text{SOH}]_{low\ sorption}$ increases and therefore ΔpH becomes smaller. According to (7.35), sorption of a divalent metal ion will increase the positive surface charge and surface potential which will also favor $[\equiv\text{SOH}]$ as compared to $[\equiv\text{SOH}_2^+]$. In conclusion, a steep sorption edge can result from, either bidentate (7.36a) versus monodentate (7.35a) binding near the *PZC* where $[\equiv\text{SOH}]$ is the dominant surface species, or at a low pH because of competition between $[\equiv\text{SOH}]$ and $[\equiv\text{SOH}_2^+]$.

Sorption of arsenate and arsenite

Oxyanions will be attracted to positively charged surfaces and repelled from negative surfaces and so will exhibit high sorption at low pH and low sorption at high pH. We may use arsenic as an example. Arsenic is found in groundwaters as arsenate, As(5), (Smedley et al., 2002) and arsenite, As(3) (Mukherjee et al., 2000). Both As(5) and As(3) form protolytes which may release protons stepwise in a similar way to carbonic acid. For As(5), the first dissociation constant of H_3AsO_4, is $\log K_1 = -2.24$ and this species is therefore rarely important. The second dissociation reaction is:

$$\text{H}_2\text{AsO}_4^- \leftrightarrow \text{HAsO}_4^{2-} + \text{H}^+; \qquad \log K = -6.9 \tag{7.38}$$

At pH < 6.9, $H_2AsO_4^-$ is the dominant form and at pH > 6.9 $HAsO_4^{2-}$ predominates. These two As(5) species will be most abundant in groundwater.

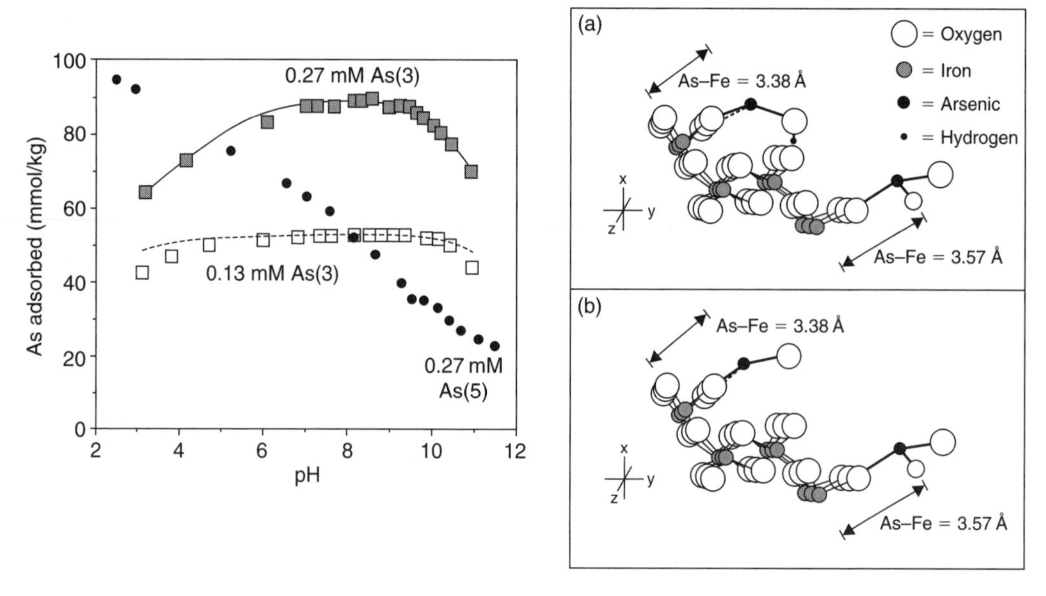

Figure 7.16. Left; pH dependent sorption of As(3) and As(5) on goethite (2.5 g/L). Right; section through goethite along the (001) plane showing the position of adsorbed As(3) on the (110) plane (Manning et al., 1998).

For As(3) the first dissociation reaction is:

$$H_3AsO_3 \leftrightarrow H_2AsO_3^- + H^+; \qquad \log K = -9.2 \tag{7.38a}$$

Therefore at pH < 9.2, the uncharged H_3AsO_3 is the predominant species.

Figure 7.16 shows the sorption of As(3) and As(5) to goethite. For As(5) there is a steady decrease in sorption as pH increases corresponding to the decreasing positive surface charge and increasing importance of the $HAsO_4^{2-}$ species. For As(3) there is a sorption maximum at pH 4–9.2 with sorption decreasing both at higher and lower pH. Apparently, H_3AsO_3 is outcompeted from the surface by protons when the pH is less than 4. Essentially all of the arsenite is sorbed on goethite in the range $7 < \text{pH} < 10$ from the 0.133 mM As(3) solution, corresponding to 53 mmol/kg adsorbed As (Figure 7.16). In the 0.267 mM As(3) solution, up to 84% of the total concentration of 107 mmol/kg is adsorbed. At near neutral pH the adsorption of uncharged As(3) is larger than of negatively charged As(5) (but this depends on the concentration of As, cf. Problem 7.6). The right side of Figure 7.16 shows a section through goethite along the (001) plane with the double row of Fe-centered octahedra. The As atoms adsorb on the (110) plane, sharing oxygens with the Fe^{3+} octahedra in the mineral structure.

Sorption on Mn-oxide, clay and calcite

Manganese oxides are highy effective in sorbing heavy metals and the low *PZC* of birnessite (Table 7.3) indicates that the surface will be negatively charged under most conditions. Sorption of various trace metals on the manganese oxide birnessite, δ-MnO_2, is illustrated in Figure 7.17 and shows strong sorption even at a low pH. For example, Pb^{2+} is 100% sorbed at pH 5. Sorption of transition metal ions on birnessite is only partially reversible (Murray, 1975; McKenzie, 1980). The structure of birnessite allows an easy uptake of metal ions in the octahedral layer when the atomic radii fit. Such behavior is transitionary between adsorption processes and solid solution behavior. However, for alkaline-earths and Mn^{2+}, adsorption is reversible. The sorption site on the birnessite surface is centered on 3 oxygens surrounding a vacancy in the octahedral layer with a charge of -2. The surface complexation model based on this diprotic site contains the following reactions (Appelo and Postma, 1999):

$$\equiv MnO_3H_2 + H^+ \leftrightarrow \equiv MnO_3H_3^+ \quad (7.39)$$

$$\equiv MnO_3H_2 \leftrightarrow \equiv MnO_3H^- + H^+ \quad (7.39a)$$

$$\equiv MnO_3H^- \leftrightarrow \equiv MnO_3^{2-} + H^+ \quad (7.39b)$$

$$\equiv MnO_3^{2-} + M^{m+} \leftrightarrow \equiv MnO_3M^{(m-2)+} \quad (7.39c)$$

$$\equiv MnO_3H^- + M^{m+} \leftrightarrow \equiv MnO_3HM^{(m-1)+} \quad (7.39d)$$

The metals M^{m+} bind via inner-sphere complexes to the three oxygens. The association reactions (7.39c and d) are similar to complexation of metal ions to diprotic acids such as H_2CO_3.

The diprotic sorption site has a distinctive broadening effect on the sorption edge. While for most oxyhydroxides an increase in the adsorption ratio from 1 / 10 to 10 / 1 was covered in 1–2 pH units (Figures 7.6 and 7.15), this needs 5 pH units or more for birnessite (Figure 7.17). As the pH increases, more and more negative charge accumulates on the surface and desorption of additional H^+ becomes increasingly more difficult. Because the diprotic site is divalent, the potential effect increases more with pH than for other oxides.

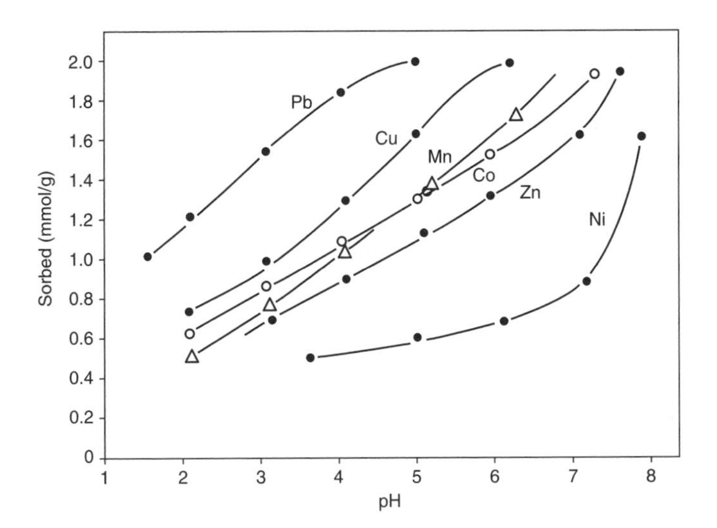

Figure 7.17. pH dependent sorption of trace metals to birnessite (δ-MnO_2). Sorption is expressed as mmol trace metal per gram of birnessite. The total amount of trace metal added is 2 mmol trace metal per 2 g of birnessite (McKenzie, 1980).

The oxygen groups on surfaces of clay minerals may be protonated and deprotonated in a similar way and display pH dependent sorption in addition to the permanent negative charge which accounts for most of their cation exchange properties (Chapter 6). Figure 7.18 shows sorption of a trace concentration of Ni^{2+} on the surface of montmorillonite. The background electrolyte in the experiments is 0.33 mM $Ca(NO_3)_2$ and since the affinity of Ca^{2+} and Ni^{2+} for ion exchange sites is almost the same (Table 6.4), Ca^{2+} fills most of the constant charge exchange sites.

Nickel is only present in trace amounts and shows pH dependent sorption somewhat like on oxides. However, at pH < 5 about 10% of total Ni^{2+} remains sorbed to the constant charge exchange sites instead of going down to zero as on ferrihydrite (cf. Problem 7.19). Accordingly, clay minerals show additive sorption behavior towards heavy metals, featuring both pH dependent specific adsorption and pH independent ion exchange (Bayens and Bradbury, 1997; Kraepiel et al., 1999). Note in Figure 7.18 that sorption diminishes at pH > 10 due to complex formation in solution.

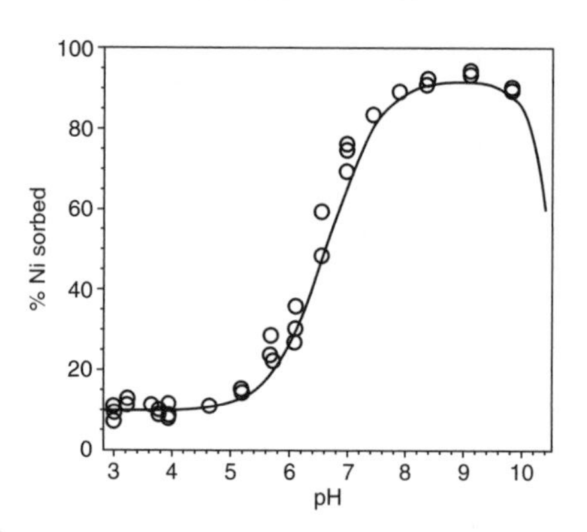

Figure 7.18. Sorption of trace amounts of nickel on montmorillonite. The background electrolyte concentration is 3×10^{-3}M $Ca(NO_3)_2$ and the total Ni^{2+} concentration around 10^{-8}M (Bradbury and Baeyens, 1999).

Also calcite which is ubiquitous in aquifers, may be important as a sorbent. When trace metal concentrations are very low, sorption on calcite may govern the aqueous concentration, while at higher concentrations solid solutions may become controlling. Zachara et al. (1991) determined sorption of various cations on calcite as function of pH (Figure 7.19). A 10^{-7}M solution of the divalent cation was added after the calcite had equilibrated at a given pH, and the loss from solution was recorded as percent adsorbed. Zachara et al. (1991) interpreted the sorption behavior as the result of competition between Ca^{2+} and the foreign ion for surface sites. Sorption increases with pH (Figure 7.19), because the Ca^{2+} concentration decreases towards higher pH to maintain calcite equilibrium.

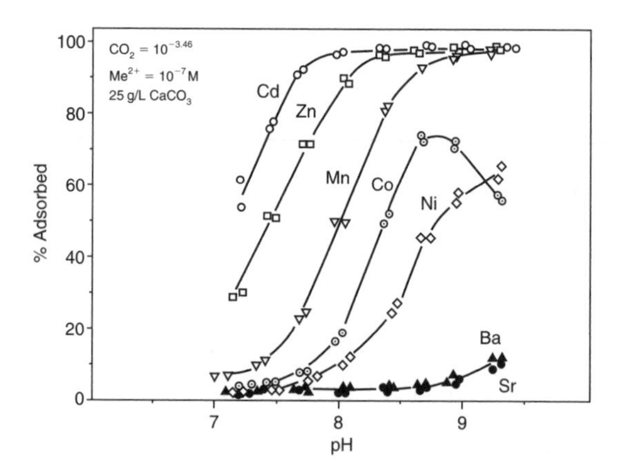

Figure 7.19. Trace metal sorption on calcite in a system brought in equilibrium with calcite at varying pH (Zachara et al., 1991).

At a given pH, sorption decreases in the order $Cd^{2+} > Zn^{2+} > Mn^{2+} > Co^{2+} > Ni^{2+} > Ba^{2+}$, Sr^{2+} (Figure 7.19). Since sorption of an alien cation on the crystal sites impedes the dissolution and growth of calcite, this order of binding strength is expected to reflect the inhibitory drive of the cations. Comparison with Figure 5.35, which illustrates the growth inhibition effect of heavy metals on calcite and aragonite, indeed gives the same sequence: $Zn^{2+} > (Co^{2+} \geqslant Mn^{2+} \geqslant Ni^{2+}) > Sr^{2+}$, Ba^{2+}.

Sorption of protons and other species on calcite and siderite has also been measured with a rapid titration technique that prevents calcite dissolution by Charlet et al. (1990) and subsequently on the much slower dissolving magnesite by Pokrovsky et al. (1999). These data were modeled with a surface complexation model while accounting for electrostatic effects (Van Cappellen et al., 1993). However, the results indicate that electrostatic effects are minor on the surface of carbonates.

7.3.4 *Sorption, absorption, and coprecipitation*

Before continuing with the modeling of sorption, we will have a brief look at the complexities of the natural processes. Sorption comprises a whole suite of reactions ranging from adsorption to solid solution formation. There is a continuum going from one stage to the other and the processes may be sequential. Often there is an initial fast adsorption step followed by a slow step where the adsorbed species are incorporated into the crystal structure to form a solid solution. This is for example observed for Cd^{2+} on the calcite surface (Davis et al., 1987; Stipp et al., 1992). For Mn-oxide it was already mentioned that adsorbed heavy metal ions can diffuse into octahedral positions and build a solid solution. Experimentally these phenomena are noted by the irreversible adsorption of ions on mineral surfaces.

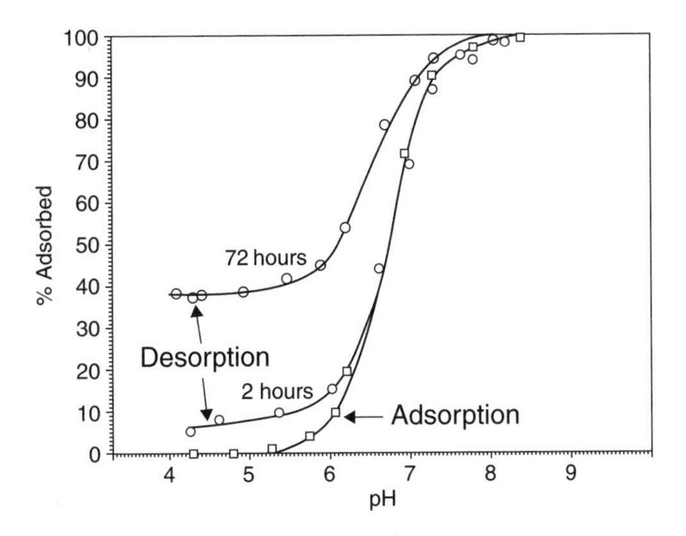

Figure 7.20. Adsorption/desorption of 10^{-5} M Zn^{2+} on 10^{-3} M Fe-ferrihydrite, $I = 0.1$. Desorption was initiated after holding the system at pH 9 for respectively 2 and 72 hours (modified from Schultz et al., 1987).

Figure 7.20 shows experimental results for adsorption and desorption of Zn^{2+} on ferrihydrite. Desorption was initiated by lowering the pH of suspensions that were aged at pH 9. A short aging period of two hours produced a desorption curve almost similar to the sorption edge (Figure 7.20). However, when the aging period was increased to 72 hours, only about 60% of the Zn^{2+} could be desorbed and the remainder stayed bound in the ferrihydrite structure. Similar time-dependent sorption phenomena have been noted by e.g. McKenzie (1980), Brümmer et al. (1988), Barrow et al. (1989), and Gerth et al. (1993).

Fuller et al. (1993) investigated both the adsorption of arsenate on ferrihydrite, and coprecipitation of arsenate in ferrihydrite. After 24 hours of reaction approximately twice as much arsenate was removed from solution by precipitation as compared to adsorption, while the As / Fe molar ratio in the precipitate became as high as 0.2. During adsorption a continued slow uptake of As(5) was observed. However, during coprecipitation some of the arsenate was released over time due to recrystallization of the ferrihydrite.

Irreversible sorption phenomena are difficult to deal with in field and modeling studies. Usually, an emphasis is given to reversible surface complexation models, as is done in the remainder of this chapter, but it should be noted that we are dealing with relatively low concentrations of trace metals. In general, these models are not as definitive and robust as the model for cation exchange of major ions discussed in Chapter 6.

7.4 SURFACE COMPLEXATION

Sorption involves two effects; a chemical bond between the ion and the surface atoms, and an electrostatic effect that depends on the surface charge. These yield two terms in the equation for the Gibbs free energy of *surface complexation*:

$$\Delta G_{des} = \Delta G_{chem} + \Delta G_{coul} \tag{7.40}$$

The Coulombic term (here for a desorption reaction) reflects the electrical work required to move ions away from a charged surface. ΔG_{coul} corresponds to the difference in energy of the state where one mole of an ion resides at the surface with the potential ψ_0, and the state where the ion resides in the bulk of the solution with potential $\psi = 0$ (Volt). For desorption it is:

$$\Delta G_{coul} = \Delta G_{\psi=0} - \Delta G_{\psi=\psi 0} = zF \cdot (0 - \psi_0) = -zF\psi_0 \tag{7.41}$$

where z is the charge of the ion, and F is the Faraday constant (96,485 C/mol).

The Gibbs free energy is related to the mass action constant by the equation (Chapter 4):

$$RT \ln K_a = -\Delta G_{des} \tag{7.42}$$

Substitution in Equation (7.40) yields:

$$RT \ln K_a = -\Delta G_{chem} - \Delta G_{coul} = RT \ln K_{int} + zF\psi_0 \tag{7.43}$$

or:

$$\log K_a = \log K_{int} + \frac{zF\psi_0}{RT \ln 10} \tag{7.44}$$

Here ΔG_{des} results in an *apparent* dissociation constant K_a that can be measured experimentally. ΔG_{chem} gives K_{int}, the *intrinsic* dissociation constant describing the chemical binding. While K_{int} is a constant, K_a must vary with the charge of the surface.
For example for dissociation of a surface proton:

$$\equiv SOH_2^+ \leftrightarrow \equiv SOH + H^+ \tag{7.45}$$

the apparent dissociation constant is:

$$K_{a1} = \frac{[\equiv\text{SOH}][\text{H}^+]}{[\equiv\text{SOH}_2^+]} \tag{7.46}$$

The symbol K_{a1} is used to indicate the first of two possible deprotonation steps of the surface. The apparent constant for dissociation of $\equiv$SOH into $\equiv SO^-$ is denoted as K_{a2}. It is difficult to relate the concentration of a surface species to its activity. As discussed by Sposito (1983, 1984), the surface complex has an activity of 1 when fully covering the surface, but in a charge-free environment. Thus, surface activity includes the Coulombic term, but this term is not included in the standard state.

The surface potential ψ_0 can be calculated using the *constant capacitance model* (Schindler and Stumm, 1987). This model considers the charged surface to be balanced by a parallel layer of counter ions by analogy with a parallel plate condenser. It also assumes a linear relation between surface charge and surface potential:

$$\sigma = \kappa\varepsilon\, \psi_0 \tag{7.47}$$

where σ is the specific surface charge in Coulombs/m^2, and $\kappa\varepsilon$ is the specific capacitance in Farads/m^2. The form of Equation (7.47) can be derived from the simplified double layer equation for small potentials:

$$\sigma_{DL} = 2.29\, I^{0.5}\, \psi_0 \tag{6.66}$$

The specific capacitance $\kappa\varepsilon$ depends both on the solution composition and the properties of the surface. If only the surface species $\equiv SOH_2^+$ is present, the surface charge is:

$$\sigma = F\left\{\equiv\text{SOH}_2^+\right\} / A_s \tag{7.48}$$

where A_s is the specific surface in m^2/kg, and $\{\equiv SOH_2^+\}$ is expressed in moles per kg adsorbing solid. In the constant capacitance model, the surface potential is (Equation 7.47):

$$\psi_0 = \frac{F\left\{\equiv\text{SOH}_2^+\right\}}{\kappa\varepsilon\, A_s} \tag{7.49}$$

We can combine Equations (7.49) and (7.44), and obtain for the first dissociation constant, with $z = 1$ for H^+:

$$\log K_{a1} = \log K_{\text{int}} + \frac{F^2}{\kappa\varepsilon\, A_s\, RT \ln 10} \cdot \left\{\equiv\text{SOH}_2^+\right\} \tag{7.50}$$

Part of the factor before $\{\equiv SOH_2^+\}$ is customarily lumped together as:

$$\alpha = \frac{F^2}{\kappa\varepsilon\, A_s\, RT}$$

so that we write Equation (7.50) more simply as:

$$\log K_{a1} = \log K_{\text{int}} + \frac{\alpha}{\ln 10} \cdot \left\{\equiv\text{SOH}_2^+\right\} \tag{7.51}$$

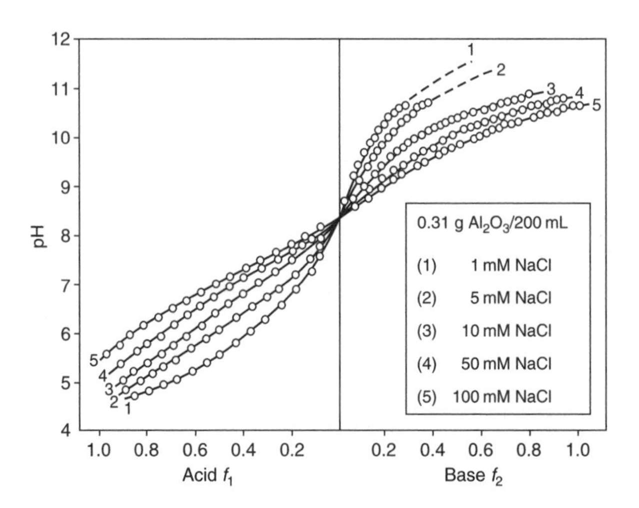

Figure 7.21. Effect of different concentrations of background electrolyte on titration curves of γ-Al_2O_3. f is equivalent fraction of titrant added. Modified from Stumm and Morgan, 1981 (*Aquatic chemistry*, 2nd ed.).

and similarly for K_{a2}:

$$\log K_{a2} = \log K_{\text{int}} - \frac{\alpha_2}{\ln 10} \cdot \{\equiv \text{SO}^-\} \tag{7.52}$$

The factor α (and α_2) can be obtained from the slope of a plot of the surface charge (i.e. of $\equiv SOH_2^+$, or $\equiv SO^-$) *vs* the apparent $\log K_a$, and the intercept provides an estimate of K_{int} (cf. Stumm and Morgan, 1996). The factors α and α_2 should be identical for K_{a1} and K_{a2}, as follows from the derivation, but this is often not the case (Schindler and Stumm, 1987).

The factor α (and α_2) must be determined for each background electrolyte concentration. The constant capacitance model does not describe the effects of changes in concentration of the background electrolyte. A background electrolyte such as NaCl has the effect of equalizing the slope of the titration curve over a large pH-interval, as illustrated in Figure 7.21. This equalization indicates that surface hydroxyls dissociate more readily in the basic branch, while protons associate easier in the acid limb. In other words, the effect of the electrostatic term decreases with increasing concentration of NaCl. This shows (cf. Equations 7.47 and 6.66) that the surface capacitance ($\kappa\varepsilon$) increases as the concentration of the background electrolyte increases.

EXAMPLE 7.5. *Calculate the specific capacitance (F/kg) of a γ-Al_2O_3 suspension* for $I = 0.1$ M NaCl from the equation

$$\text{p}K_a = 7.2 - 14.5\{\equiv \text{AlOH}_2^+\}.$$

Find the ratio $\{\equiv \text{AlOH}\} / \{\equiv \text{AlOH}_2^+\}$ at pH = 7 and 6 for this ionic strength. The total number of sites is 0.03 mol/kg Al_2O_3. Also estimate the specific capacitance for $I = 0.001$ M NaCl from the titration plot shown in Figure 7.21.

ANSWER:
The capacitance $\kappa\varepsilon A_s$ is contained in the slope of pK_a *vs* $\{\equiv \text{AlOH}_2^+\}$ (Equation 7.50). In this case,

$$\kappa\varepsilon\, A_s = F^2/(14.5RT \ln 10) = 1.1 \times 10^5 \text{ F/kg}.$$

From Equation (7.51) we have:

$$\log K_{a1} = -\log\left(\{\equiv AlOH\}/\{\equiv AlOH_2^+\}\right) + pH = 7.2 - 14.5\{\equiv AlOH_2^+\} \qquad (7.53)$$

At pH = 7, we can neglect $\{\equiv AlO^-\}$, and therefore $\{\equiv AlOH_2^+\} + \{\equiv AlOH\} = 0.03$ mol/kg. Set $\{\equiv AlOH_2^+\} = x$, and fill in Equation (7.53):

$$-\log((0.03 - x)/x) = 0.2 - 14.5x,$$

which can be solved by trial and error to give $x = \{\equiv AlOH_2^+\} = 0.0148$ mol/kg. Thus, the ratio $\{\equiv AlOH\}$ / $\{\equiv AlOH_2^+\} = 0.0148/0.0152 = 1$. (Without electrostatic contribution, their ratio would be 0.63).

Similarly, at pH = 6, $\{\equiv AlOH_2^+\} = 0.026$ mol/kg, and the ratio $\{\equiv AlOH\}$ / $\{\equiv AlOH_2^+\} = 0.15$. (Without electrostatics 0.064). Dividing the calculated $\{\equiv AlOH_2^+\}$ by 0.03 mol/kg yields the fraction of sites occupied by protons, 0.5 and 0.87 for pH 7 and 6, respectively, which can also be found from curve 5 in Figure 7.21.

For $I = 0.001$ M we read from Figure 7.21, at pH = 7, $\{\equiv AlOH_2^+\} = 0.15 \times 0.03 = 0.0045$ mol/kg solid. Hence $\{\equiv AlOH\}$ / $\{\equiv AlOH_2^+\} = 5.67$, and from Equation (7.51) we obtain $\alpha / \ln 10 = 212$. Similarly for pH = 6, $\{\equiv AlOH_2^+\} = 0.35 \times 0.03 = 0.01$ mol/kg solid, and $\alpha / \ln 10 = 140$. We take the average, and find the capacitance $\kappa\varepsilon A_s = F^2 / (176\ RT\ \ln 10) = 9.3\times10^3$ F/kg. Thus, the capacitance has decreased by about a factor of 10 for a 100 times dilution of the background concentration.

QUESTIONS:
Why is the ratio $\{\equiv AlOH\}/\{\equiv AlOH_2^+\}$ smaller when electrostatics are excluded?
ANSWER: the positive surface potential repels positive species.
Is the ratio also smaller (without electrostatics) when pH > 8.3?
ANSWER: no, the surface carries a negative charge, so the ratio will be larger.

Example 7.5. shows that the capacitance of Al_2O_3 increases with the square root of the ionic strength, in excellent agreement with the simplified double layer equation, Equation (6.66). For larger potentials, $\psi > 25$ mV, the full double layer equation should be used:

$$\sigma_{DL} = 0.1174\ I^{0.5}\ \sinh(F\psi_0/2RT) \qquad (6.64)$$

which also relates charge and potential approximately by the square root of the ionic strength. This relation is used in various computer models for calculating surface complexation.

EXAMPLE 7.6. *Calculate the surface potential for* 2.6×10^{-7} meq/m² on γ-Al_2O_3 in 0.1 M NaCl solution. Use the identities $\sinh(x) = (e^x - e^{-x}) / 2$ and $\text{arcsinh}(y) = \ln(y + (y^2 + 1)^{½})$.

ANSWER:
We have $F\sigma = 96485 \times 2.6\times10^{-7} = 0.0251$ C/m². Divide by $0.1174 \times (0.1)^{0.5}$ and find 0.676. Hence $F\psi_0 / 2RT = \ln(0.676 + (0.676^2 + 1)^{0.5}) = 0.633$, and $\psi_0 = 32.5$ mV.

QUESTION:
Calculate the potential using the simplified double layer equation?
ANSWER: $\psi_0 = 34.7$ mV.

Experimentally, the surface characteristics are determined by titration at different concentrations of a background electrolyte (Kinniburgh et al., 1995). The procedure is to start at a low electrolyte concentration, titrate forward (and backward if desired) to a given endpoint, and then add more electrolyte. Adding background electrolyte at the acid endpoint will increase pH, while at the basic endpoint it will decrease pH (cf. Figure 7.21). The amount of titrant which must be added to return to the endpoint pH before salt was added, gives the starting point for the next titration in the graph. The titrations will intersect at the point where the surface has zero charge, as in Figure 7.21. Formally, this *PZC* is the point of zero salt effect, *PZSE*.

If a single intersection point is not obtained, other processes are operative. For example, sorption of CO_2 from the laboratory atmosphere is notorious for giving different forward and backward titration curves and CO_2 must be removed by purging the suspension for extended time with nitrogen gas under acid conditions (Evans et al., 1979). Sorption of CO_2 is also known to lower the *PZSE* of goethite from about 9 to 8.5 while the *IEP* is not affected (Zeltner and Anderson, 1988), whereas for hematite both *PZSE* and *IEP* decrease from 9.3 to 8.5 (Lenhart and Honeyman, 1999). Modeling of CO_2 sorption experiments indicates that the neutral species is dominant on the surface (Appelo et al., 2002). Thus, the *IEP* of goethite is not much affected, while the *PZC* takes into account the buffering by sorbed CO_2 during the titration.

Figure 7.21 shows how adding a salt, such as KCl, will shift the pH of the suspension towards the *PZC*. Usually, the pH of a soil in 1 M KCl solution is lower than the pH in distilled water, which indicates that *PZC* is lower than the pH of soil water. This is typical for most soils in temperate climates with clay minerals such as illite and smectite which have a $PZC < 3$. However, tropical soils with iron oxide as the main charging mineral often have a higher pH in 1 M KCl, which we can relate now to the higher *PZC* of the oxides that are present in the soil.

7.4.1 *Surface complexation models*

Two types of models are commonly used for surface complexation: the two-layer model and the triple layer model. In the first type, the diffuse double layer starts immediately at the charged surface, and the surface potential is directly connected to the surface charge via the capacitance (Equation 7.47). The capacitance is user-definable in the constant capacitance model, and defined by the ionic strength in the double layer model via the Gouy-Chapman relation (Equation 6.64). In the second type, three different layers with different capacitances are assumed, one layer at the surface called the Stern layer, another layer starting at a closest approach distance (for hydrated ions, but the distance is assumed equal for all), and one more corresponding with the diffuse double layer. Three capacitances must be defined for the three layers, but only the one for the Stern layer is usually adjusted (Davis et al., 1978; Davis and Kent, 1990). The capacitance for the intermediate layer is usually fixed at 0.2 F/m (Davis and Kent, 1990), but varied by others (Hiemstra and Van Riemsdijk, 1991). The capacitance of the diffuse double layer is obtained from the Gouy-Chapman relation (Equation 6.64). Figure 7.22 shows the essential features of the two types of models.

Surface binding can be purely electrical, as for ions in the diffuse double layer, or it can involve a combination of complexation and electrical work. However, both model types consider only complexed ions and completely ignore the composition of the diffuse double layer. In the two-layer models, the potential in the Boltzmann term is identical for all the ions which complex to the surface. In the triple layer model, ions can be assigned to bind directly to the surface (the zero plane), or be separated from it by a hydration shell (the beta plane). This gives a choice of two potentials for the modeling (Davis and Kent, 1990). In addition, the charge of the surface-bound ion can also be distributed over the zero plane and the β plane which influences the distribution of potential because the capacitances of the layers are different (Hiemstra and Van Riemsdijk, 1996).

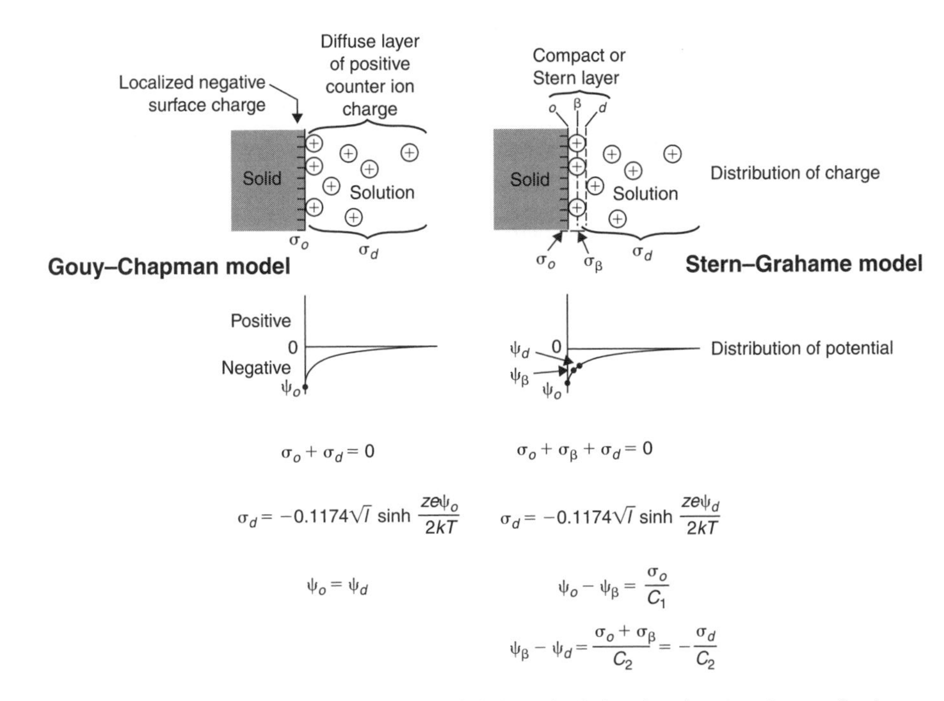

Figure 7.22. The two-layer and the triple layer models for calculating the electrostatic contribution to surface complexation. In two-layer model, the diffuse double layer starts at the charged surface while in the triple layer model, the diffuse double layer start two layers away from the surface.

Most *chemical* observations such as sorption-edges, proton-metal release ratios and ionic strength effects can be very well described by all the models (Westall and Hohl, 1980; Stumm, 1992; cf. also Venema et al., 1996b). The triple layer model has more options to account for mechanistic details of the sorption process (Hiemstra and Van Riemsdijk, 1996) and probably also for the physical behavior of the particles in an electric field (Davis and Kent, 1990). However, Dzombak and Morel (1990) have derived a comprehensive database for sorption on ferrihydrite with the double layer model which fits many observations. Thus, the double layer model has become the de-facto standard for heavy metal sorption modeling in natural environments, while the greater flexibility of the triple layer model can be helpful for modeling detailed laboratory experiments (Venema et al., 1996b) and for incorporating the charge distributions of oxyanions (Rietra et al., 1999). The double layer model is included in PHREEQC, and the triple layer model can be computed with MINTEQ (Allison et al., 1991; a Windows version can be found at www.lwr.kth.se/english/OurSoftware/Vminteq/index.htm).

EXAMPLE 7.7. *Calculate the equivalent ionic strength for a constant capacity model*
An arsenic sorption experiment on goethite reported by Manning and Goldberg, 1996, in 0.1 M NaCl, was modeled with the constant capacity relation: $\sigma_{DL} = 1.06\,\psi$. What is the ionic strength?

ANSWER:
The theory is in Section 6.6. For small potentials <25 mV, we use Equation (6.66), $\sigma_{DL} = 2.29\sqrt{I}\,\psi = 1.06\,\psi$, which gives $\sqrt{I} = 1.06 / 2.29 = 0.463$, and $I = 0.21$ mol/L. For larger potentials we use the double layer Equation (6.64), $\sigma_{DL} = 0.1174\sqrt{I}\sinh(F\psi / 2RT) = 1.06\,\psi$ which is solved by trial and error for I. For small ψ, we find again $I = 0.21$ mol/L. For higher ψ, I varies: $I = 0.1$ mol/L for $\psi = 80$ mV, and $I = 0.069$ mol/L for $\psi = 150$ mV. Manning and Goldberg's (1996) results can be well modeled with $I = 0.2$ mol/L.

Sorption site densities

For calculating the surface charge, the sorption density must be derived from sorption data (mol/kg) and surface area of the solid (m^2/kg), giving the specific number of sorption sites (sites/nm^2, mol/m^2, or mol/mol metal-oxide). The number of sorption sites can also be estimated geometrically from the crystal structure and morphology, but it is usually determined from acid-base titration for protons, or from sorption plateaux for specific elements (Davis and Kent, 1990). The number of sorption sites may be different for various elements.

Table 7.4. Measured sorption densities for various elements on goethite, after Davis and Kent, 1990.

Ion	H^+	F^-	SeO_3^{2-}	PO_4^{3-}	Pb^{2+}
Sites/nm^2	2.6–4.0	5.2–7.3	1.5	0.8	2.6–7

In Table 7.4, F^- has a higher sorption density on goethite than any of the other elements because it exchanges with structural oxygens besides being sorbed on the positive goethite surface (Davis and Kent, 1990). In principle, it is possible to calculate sorption for various sites, each with its own sorption density and set of competing ions, but most models consider no more than two types of sites. The number of sites per nm^2 can be recalculated to moles per liter, as shown in Example 7.8.

EXAMPLE 7.8. *Recalculate* sites/nm^2 to mol/L for a suspension with 0.2 mol goethite/L (=17.8 g/L). The goethite has 2 sites/nm^2 and a surface area of 40 m^2/g.

ANSWER:
First recalculate sites/nm^2 to mol/m^2. There are 2 sites/nm^2 = 2×10^{18} nm^2/m^2/(N_a = 6.022×10^{23} sites/mol) = 3.32 μmol/m^2. Multiply by the surface area of goethite: 3.32 × 40 m^2/g = 133 μmol/g × 88.85 g/mol goethite = 11.8 mmol/mol goethite. Find the site concentration in mol/L: 11.8 × 0.2 mol/L = 2.36 mmol/L.

7.4.2 *The ferrihydrite ($Fe(OH)_3$) database*

Dzombak and Morel (1990) have obtained a coherent database for surface complexation on hydrous ferric oxide (Hfo, or ferrihydrite) by fitting the results of numerous laboratory experiments. This database is also available in PHREEQC. Dzombak and Morel's model defines complexation reactions for two sites on hydrous ferric oxide (Table 7.5), a strong site Hfo_sOH and a weak site Hfo_wOH, while the Gouy-Chapman double layer equation is used for defining the surface potential as function of surface charge and ionic strength.

The database included in PHREEQC is obviously not complete, but constants for other elements can be estimated with *linear free energy relations* (*LFER*). The *LFER* relations listed in Table 7.5 are based on the regression of optimized surface complexation *K*'s with the K_{MOH} for the first hydrolysis constant of the metal in water, or the second dissociation constant of the acid anion. For example, the constant for surface complexation of carbonate:

$$\text{Hfo\_wOH} + CO_3^{2-} + H^+ = \text{Hfo\_wCO}_3^- + H_2O \qquad (7.54)$$

is estimated using:

$$HCO_3^- = H^+ + CO_3^{2-}; \quad \log K_{a2} = -10.33 \qquad (7.54a)$$

For reaction (7.54) the *LFER* relation (Table 7.5) gives $\log K_2 = 6.384 - 0.724 \times (-10.33) = 13.86$. However, this value is much higher than the $\log K_2 = 12.56$, which was optimized using measurements of CO_2 sorption on ferrihydrite (Appelo et al., 2002).

Table 7.5. Properties of Hfo (ferrihydrite) in Dzombak and Morel's (1990) database, and sorption-reaction equations for various ions. *LFER* stands for linear free energy relations and are used to estimate unknown values by analogy to aquesous complexation reactions.

Weak sites	Strong sites	Surface area		Mol · wt	*PZC*
Hfo_w, mol/mol Fe	Hfo_s, mol/mol Fe	m^2/g	m^2/mol	g/mol	pH
0.2	0.005	600	5.33×10^4	88.85	8.11

...... Reactions ......

Reaction	Constant
Protons on strong sites:	
$Hfo\_sOH_2^+ = Hfo\_sOH + H^+$	$\log K_{a1}$ −7.29
$Hfo\_sOH = Hfo\_sO^- + H^+$	$\log K_{a2}$ −8.93
Protons on weak sites:	
$Hfo\_wOH_2^+ = Hfo\_wOH + H^+$	$\log K_{a1}$ −7.29
$Hfo\_wOH = Hfo\_wO^- + H^+$	$\log K_{a2}$ −8.93
Transition metals:	
$Hfo\_sOH + M^{m+} = Hfo\_sOM^{(m-1)+} + H^+$	K_1
LFER:	$\log K_1 = -4.374 + 1.166 \cdot \log K_{MOH}$ [1]
$Hfo\_wOH + M^{m+} = Hfo\_wOM^{(m-1)+} + H^+$	K_2
LFER:	$\log K_2 = -7.893 + 1.299 \cdot \log K_{MOH}$
Alkaline earth cations:	
$Hfo\_sOH + M^{2+} = Hfo\_sOHM^{2+}$	K_1
$Hfo\_wOH + M^{2+} = Hfo\_wOM^+ + H^+$	K_2
$Hfo\_wOH + M^{2+} + H_2O = Hfo\_wOMOH + 2H^+$	K_3
Di- and monovalent anions:	
$Hfo\_wOH + A^{a-} + H^+ = Hfo\_wA^{(a+1)-} + H_2O$	K_2
LFER:	$\log K_2 = 6.384 - 0.724 \cdot \log K_{a2}$ [2]
$Hfo\_wOH + A^{a-} = Hfo\_wOHA^{a-}$	K_3
LFER:	$\log K_3 = -0.485 - 0.668 \cdot \log K_{a2}$

[1] K_{MOH} is for: $M^{m+} + OH^- = MOH^{(m-1)+}$. [2] K_{a2} is for: $HA^{(a-1)-} = H^+ + A^{a-}$.

In the optimization an uncharged complex appeared to be the dominant species, which in Dzombak and Morel's database is reserved for the trivalent anions only. The overall sorption of CO_2 is therefore overestimated by the *LFER* as illustrated by a comparison of the optimized and *LFER*-derived sorption envelopes (Figure 7.23).

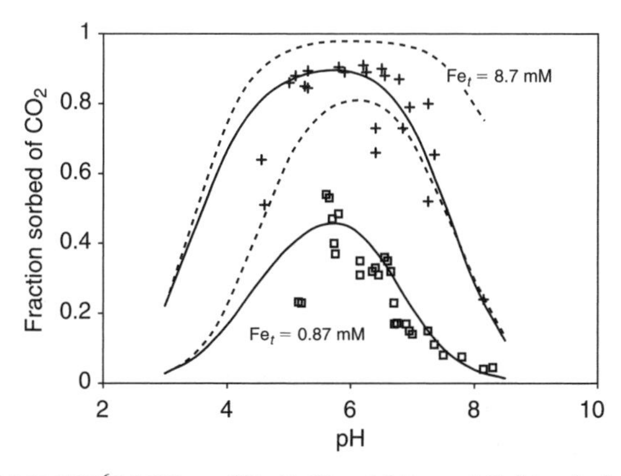

Figure 7.23. Sorption of 4.6×10^{-6} M CO_2 on Hfo (0.87 and 8.7 mmol Fe/L) calculated with the double layer model. Full lines are for optimized complexation constants and species, dotted lines are for $Hfo\_wCO_3^-$ with a complexation constant estimated using *LFER*. Datapoints from Zachara et al., 1987.

Let us return to Examples 7.1 and 7.2 dealing with Cd^{2+} sorption on sand. Can we explain the measured value of the distribution coefficient $K_d = 307$ L/kg as due to sorption on iron oxyhydroxide in the sand?

EXAMPLE 7.9. *Sorption of Cd^{2+} to iron oxyhydroxide in loamy sand*
To measure the content of iron oxyhydroxides, Christensen (1984) extracted the sand with dithionite, and found 2790 ppm Fe. Derive the K_d (L/kg) for Cd^{2+}.

ANSWER:
The iron oxyhydroxides in the soil probably consist of goethite, which is more crystalline than ferrihydrite and has an about 10 times smaller surface area (for goethite prepared in the laboratory). We assume therefore 10 times fewer sites as compared to ferrihydrite, i.e. 0.02 mol weak sites/mol Fe, and 5×10^{-4} mol strong sites/mol Fe, and also a 10 times lower specific surface area (60 m^2/g) and grams of "Ferrihydrite" in the PHREEQC input file.

```
SOLUTION 1
 pH 6.0; pe 14 O2(g) -0.68
 Ca 1; Cl 2; Cd 1e-6

SURFACE 1
# 2790 ppm Fe / 55.85 = 50 mmol/kg * 89 = 4.45 g "Ferrihydrite"/kg
 Hfo_w 1e-3 60 4.45    # 1e-3 mol weak sites, 60 m2/g specif. surf., 4.45 g ferrihydrite
 Hfo_s 0.025e-3         # 0.025e-3 mol strong sites
 -equil 1

USER_PRINT
 -start
 10 print "K_d (L/kg) = ", (mol("Hfo_wOCd+")+mol("Hfo_sOCd+"))/ tot("Cd")
 -end
END
```

which gives in the output file:

```
------------------------------------------------------User print------------------------------------------

K_d (L/kg) = 4.2411e-01
```

QUESTIONS:
What is the K_d when there are 10 times more surface sites?
ANSWER: 10 times higher
What is the K_d at pH = 7, 8 and 9?
ANSWER: $K_d = 23.5$ at pH = 7, $K_d = 151$ at pH = 8, $K_d = 864$ at pH = 9

The value of $K_d = 0.42$ L/kg calculated for sorption of Cd^{2+} onto iron oxyhydroxides in the soil (Example 7.9) is much lower than the measured value of 307 L/kg. Therefore, iron oxyhydroxides contribute very little to Cd^{2+} sorption on sand at pH = 6. This corroborates experiments of Zachara et al. (1992) who determined sorption on soil separates and noted that the oxides may even block the access of Cd^{2+} to sorption sites on clay minerals. In near neutral or slightly acidic water, sorption to iron oxyhydroxides is probably mostly important for Pb^{2+} (Lion et al., 1982) and Cu^{2+} (inspect the PHREEQC database), and for oxyanions such as As and Se.

7.4.3 *Diffuse double layer concentrations in surface complexation models*

Neither the double layer model, nor the triple layer model considers the composition of the diffuse double layer. The surface charge is simply balanced by ions in solution, whereas in reality the double layer is attached to the surface for charge balance. When a model run separates the solution and the solid, without considering the double layer, then both will carry a net charge, which is of course impossible. Let us imagine an experiment where 89 mg neutral ferrihydrite is weighed in 3 centrifuge tubes. Next, 20 ml 0.01 M HCl is added to the first tube. After equilibration, the tube is centrifuged, and the fluid is poured into the next tube containing ferrihydrite, and so on. What is then the Cl^- concentration in the third tube? The PHREEQC input file for this experiment is given in Table 7.6. Note that SURFACE 1 is defined with the "-diffuse_layer" option, for calculating the composition of the diffuse double layer.

The results in Table 7.6 indicate that the Cl^- concentration decreases from 10 to 2.16 mmol/L in the first tube and to 0.009 mmol/L in the third tube. The Hfo has become protonated by the HCl solution and acquired a positive charge. The charge is balanced by an excess of anions in the diffuse layer that remain with the surface when solution and solid are separated, and causes a decrease in the Cl^- concentration. When the diffuse layer calculation is switched off (the default option in PHREEQC), the Cl^- concentration remains 10 mmol/L. The positive charge of the surface is now counterbalanced by a negative charge balance in solution. The loss of protons is higher than when diffuse_layer option is considered and the pH is calculated (wrongly!) to increase to 8.1 in the third tube.

The diffuse_layer option in PHREEQC is computer intensive and usually the results are insensitive to its setting, for example, because the pH is determined by other equilibria. Therefore, modeling of variable charge surfaces may often proceed with the double layer option off, but it should be ascertained whether the effects are indeed minor for the problem at hand (cf. Problems 7.4 and 7.5).

Table 7.6. PHREEQC input file for modeling the reaction of Hfo with HCl in 3 centrifuge tubes, and calculation results.

```
# Put 89 mg neutral Hfo in 3 centrifuge tubes...
SURFACE 1-3
Hfo_wOH 2e-4 600 0.089 # 0.2 mmol weak sites, 600 m2/g specif. surf., 89 mg Hfo
-diffuse_layer
END

# Take 1st tube, add 20 g of 10 mM HCl, react and decant in tube 2...
USE surface 1
SOLUTION 1; pH 2.0 charge; Cl 10; water 0.02
SAVE solution 2
END

# 2nd tube, react and decant in tube 3...
USE surface 2; USE solution 2
SAVE solution 3
END

# 3rd tube...
USE surface 3; USE solution 3
END
```

The output gives:	Initial solution	tube 1	2	3
Cl^-	10.0	2.16	0.17	0.009 mmol/L
pH	2.04	2.69	3.78	5.04

7.5 COMPLEXATION TO HUMIC ACIDS

Organic material is commonly extracted from soils and sediments with 0.5 M NaOH. When the resulting, dark-colored solution is brought to pH = 1 with HCl, part of the organic matter precipitates. The fraction which precipitates is called *humic acid*, while the fraction that remains soluble at pH = 1 is called *fulvic acid*. The organic extracts contain macromolecules with 60–10,000 atoms of C, H, O, N and S and other elements, and molecular weights ranging from 500 to 1500 for fulvic acid, and from 1000 to 85,000 for humic acid. Average elemental compositions of soil extracted organic matter are given in Table 7.7.

Table 7.7. Average elemental composition of humic and fulvic acid extracted from soil. Data in Schnitzer and Khan (1978), recalculated to 100% for the given elements.

	C (mol %)	H (mol %)	O (mol %)	N (mol %)	S (mol %)	Ash (g/kg)
Humic	40.2	38.8	19.9	1.0	0.08	30
Fulvic	35.1	39.5	24.7	0.6	0.1	15

In soil water the concentration of dissolved organic carbon (*DOC*) ranges from 0.1 to 3 mM, in groundwater from 0.01 to 1 mM, and in rivers draining swamps the concentration may be as high as 5 mM. Small size organic acids and saccharides make up about 75% of the *DOC* in groundwater and only about 25% consists of fulvic and humic acid (Thurman, 1985; Routh et al., 2001). However, the contribution of humic acid may increase to over 90% in high pH, high *DOC* (>10 mM C) groundwater (Artinger et al., 1999).

The charge on humic and fulvic acids can be determined by titrating with acid or base and, as shown in Figure 7.24, it is strongly pH dependent. The charge usually also depends on the ionic strength and an increase in ionic strength, by adding neutral salt, will tend to decrease the pH.

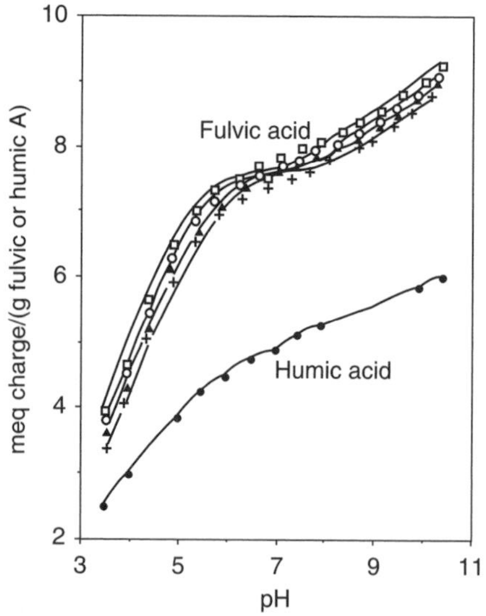

Figure 7.24. Titration curves for fulvic acid extracted from groundwater in □ 0.1, ▲ 0.03, + 0.01 and ○ 0.003 M NaCl (upper curves) and for average humic acid (lower curve) in ● 0.1 M NaCl. Lines modeled according to Tipping and Hurley's model (Higgo et al., 1993).

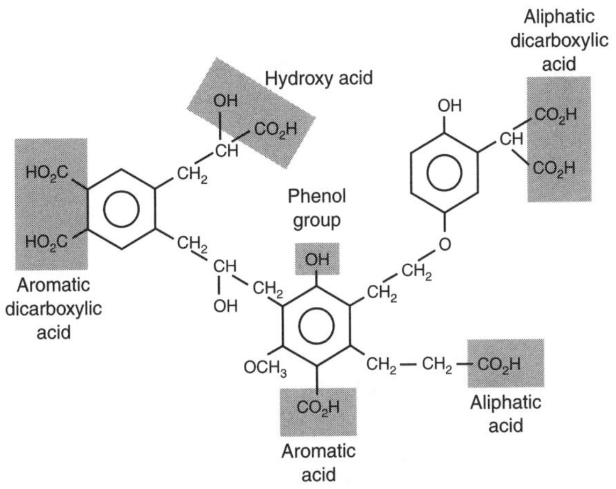

Figure 7.25. Various types of carboxylic acids and phenolic acid on humic acids (Thurman, 1985).

However, the ionic strength effects are irregular and within the measurement error in Figure 7.24. The fulvic acid shown in Figure 7.24 has a total charge of -11.4 meq/g, while the humic acid has a total charge of -7.1 meq/g. Not even at pH $= 11$ are all sites deprotonated, and likewise, at pH $= 3$ not all sites are protonated.

The charge buildup is attributed to two main groups of acids, carboxylic (R-COOH groups) with $pK_a < 5$, and phenolic with $pK_a > 8$. The acidity of the functional groups is determined by their position within the humic molecule, as illustrated in Figure 7.25. For example, the (linear) aliphatic acid has a pK_a of 4.8, while the two carboxylic groups on the aromatic ring have pK_a's of 2.9 and 4.4 (Thurman, 1985). One may expect a more or less Gaussian distribution of the acidity constants, akin to the variation for simple organic acids shown in Figure 7.26. Furthermore, the apparent acidity constants will change as charge builds up on the molecule, just as for oxide surfaces. Therefore, the approach used for modeling the deprotonation and complexation behavior of humics is similar to that for variable charge oxide surfaces, in that a general electrostatic contribution is combined with an intrinsic surface complexation K for the various ions. However, humic molecules have small, spherical or cylindrical shapes without a fixed structure which complicates the electrostatic calculations, and the variation in the acidic groups and their orientation within the molecule is much greater than for an oxide, for which at least some structural order exists. Example 7.10 shows how to estimate the size and charge density for humic particles.

EXAMPLE 7.10. *Estimate the spherical surface area of a fulvic acid* of 1000 g/mol, density 1 g/cm^3. Also estimate the surface charge density for fulvic acid at pH $= 7$ (-7.5 meq/g according to Figure 7.24).

ANSWER:
The volume of 1 molecule of fulvic acid is 1 cm^3/1 mmol/6.022×10^{20} (molecules/mmol) $\times$ 10^{24} (Å/cm)^3 = 1661 Å^3. The radius of the molecule is 7.3 Å, and the surface area is 4.08×10^{10} cm^2/mol fulvic acid. The calculated radius agrees well with physical determinations (Aiken and Malcolm, 1987; Buffle, 1990; Clapp and Hayes, 1999) and estimates from electrostatic calculations (De Wit et al., 1993a). The molecule expands with decreasing ionic strengths below 0.1 and with increasing charge or pH (Avena et al., 1999).

The humic acid has a charge of -7.5 eq/kg at pH $= 7$, hence the charge density is -7.5×96485 C/mol/4.08×10^{10} $= -17.7$ μC/cm^2. Note that the charge density is high compared to oxide surfaces, and that it will further increase 1:1 with the molecular weight for spherical particles if charge increases 1:1 with the molecular weight (Bartschat et al., 1992).

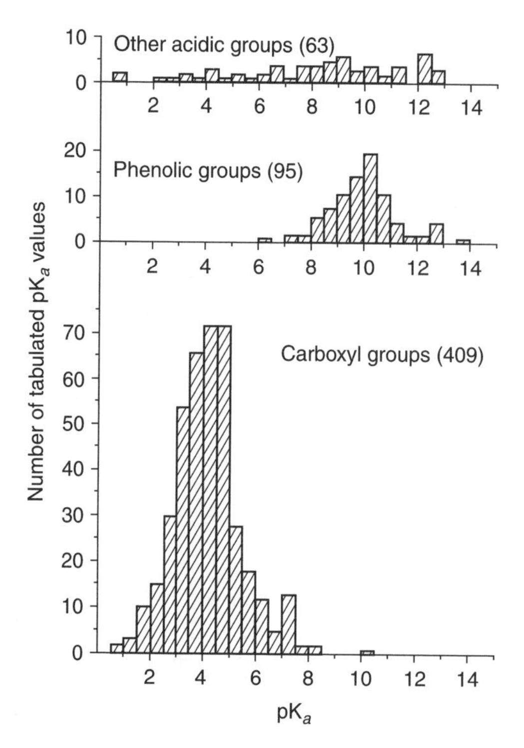

Figure 7.26. The acidity constants of simple organic acids show an approximately Gaussian distribution (Perdue et al., 1984).

Compared to oxide minerals, the ionic strength has a relatively small effect on the pH titration curves of humics, but a greater bearing on sorption of metals at a fixed pH (Bartschat et al., 1992). Basically, the small effect on titration curves is due to the wide range of acidity constants in humics: when the salt concentration increases at a given pH_x, the proton dissociation increases (according to electrostatic theory) for the group with $pK_a \approx pH_x$, but the protons are taken up by the next, more acid group. The large effect on metal sorption indicates that fewer sites are responsible for metal sorption (K's are more restricted), or that ion exchange with the cation from the background electrolyte is important. Electrostatics by itself, does not offer a sufficient correction to obtain the smooth and gradual pH titration curves if only one type of site is available with two pK_a's as on oxide surfaces. The necessary larger variation in intrinsic K's is accounted for by taking various discrete sites (Tipping and Hurley, 1992) or by assuming a continuous distribution of K's (Perdue and Lytle, 1983; De Wit et al., 1993b; Kinniburgh et al., 1999). The behavior of humics is accordingly dominated by site heterogeneity rather than electrostatics. The electrostatic contribution is bypassed altogether by some authors and proton release is modeled as an ion exchange process of H^+ *versus* the cation of the background electrolyte (Westall et al., 1995).

7.5.1 *The ion association model*

When present, the electrolyte effect can be explained as due to an ion association of the background electrolyte with a charged fulvic acid ($\equiv F^-$). For example, Na^+ with $\equiv F^-$ gives the reaction:

$$\equiv FNa \leftrightarrow Na^+ + \equiv F^-; \qquad K_{\equiv FNa} \tag{7.55}$$

with

$$K_{\equiv \text{FNa}} = \frac{[\text{Na}^+][\equiv\text{F}^-]}{[\equiv\text{FNa}]} = \frac{[\text{Na}^+]\,\beta_{\equiv\text{F}^-}}{\beta_{\equiv\text{FNa}}} \tag{7.56}$$

where $\beta_{\equiv F..}$ is the fraction of the surface species on fulvic acid. The apparent acid dissociation constant is calculated from experimental data:

$$K_{a2} = \frac{[\text{H}^+]\cdot f}{(1-f)} = \frac{[\text{H}^+]\cdot f}{\beta_{\equiv\text{FH}}} \tag{7.57}$$

Here, f is the fraction which has dissociated, as found from a titration experiment, and is equal to:

$$f = \beta_{\equiv\text{F}^-} + \beta_{\equiv\text{FNa}} \tag{7.58}$$

Combining with Equation (7.56) gives:

$$f = \beta_{\equiv\text{F}^-} \cdot (1 + [\text{Na}^+] / \text{K}_{\equiv\text{FNa}}) \tag{7.59}$$

We introduce this relation in Equation (7.57). The quotient $[\text{H}^+] \cdot \beta_{\equiv\text{F}^-} / \beta_{\equiv\text{FH}}$ has an apparent constant defined by Equation (7.52) and therefore:

$$\log K_{a2} = \log K_{int} - \frac{\alpha_2}{\ln 10} \cdot \{\equiv\text{F}^-\} + \log\left(1 + \frac{[\text{Na}^+]}{K_{\equiv\text{FNa}}}\right) \tag{7.60}$$

This equation shows that K_{a2} *decreases* with increasing dissociation of the surface groups (increasing pH), and *increases* when the salt concentration increases. If the dissociation constant $K_{\equiv\text{FNa}}$ is small, the increase of K_{a2} would be about ten-fold for every ten-fold increase in concentration. This is indeed observed in titrations of polymer solutions (Marinsky, 1987). Titrations of humic and fulvic acid follow a similar trend (Figure 7.27).

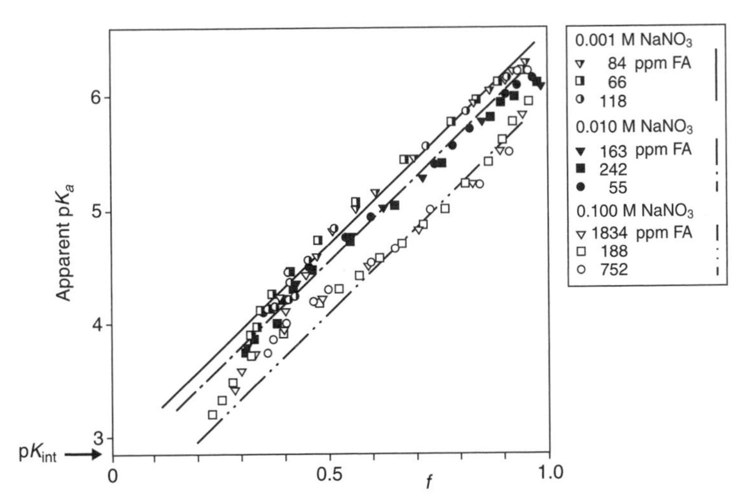

Figure 7.27. The variation of pK_a of fulvic acid with degree of dissociation at three different ionic strength levels. The ionic medium is $NaNO_3$. Lines are drawn according to the constant capacitance/ion association model. The fulvic acid is from Armadale Bh soil horizon. Experimental data after Marinsky (1987).

Assume that the electrostatic term for a fulvic acid may be approximated by:

$$\frac{\alpha_2}{\ln 10} \cdot \beta_{\equiv F^-} = \alpha_f \cdot f \tag{7.61}$$

so that Equation (7.60) can be rewritten as (with $-\log K = \mathrm{p}K$):

$$\mathrm{p}K_a = \mathrm{p}K_{\mathrm{int}} + \alpha_f \cdot f - \log\left(1 + \frac{[\mathrm{Na}^+]}{K_{\equiv \mathrm{FNa}}}\right) \tag{7.62}$$

where the constants for the fulvic acid $\equiv$FH follow the notation used earlier for a solid surface $\equiv$SOH. Figure 7.27 shows an approximately linear increase of $\mathrm{p}K_a$ with f, and also a decrease as the concentration of $NaNO_3$ increases. The value of $K_{\equiv \mathrm{FNa}}$ can be estimated from two points in the plot, using:

$$K_{\equiv \mathrm{FNa}} + m_{\mathrm{Na},1} = 10^{pK_{a,1} - pK_{a,2}} \cdot (K_{\equiv \mathrm{FNa}} + m_{\mathrm{Na},2}) \tag{7.63}$$

The lines in Figure 7.27 were drawn with $\mathrm{p}K_{\mathrm{int}} = 2.8$, $\alpha_f = 3.7$, and $K_{\equiv \mathrm{FNa}} = 0.03$.

7.5.2 *Tipping and Hurley's discrete site model "WHAM"*

Tipping and Hurley (1992) and Tipping (1998) have formulated the model WHAM for complexing of heavy metals by humic acid. The acronym WHAM stands for "Windermere humic acid model", for Lake Windermere in the English Lake District. In the model, the carboxylic ("HA") and the phenolic ("HB") acid groups are each assigned 4 sites. Each site in turn is given a charge $n_{HA}(i)$ or $n_{HB}(i)$ (meq/g), a $\mathrm{p}K_a(i)$ for proton dissociation and a $\mathrm{p}K_{M\mathrm{HA}}$ or $\mathrm{p}K_{M\mathrm{HB}}$ for metal dissociation. The parameters for the A and B sites are determined with the algorithm:

```
for i = 1 to 4:
  nHA(i)  = (1-fpr) x nHA/4;                    {...fixed charge on each site}
  pKa(i)  = pKHA + (2i - 5)/6 x ΔpKHA;                          {...varying pKa}
  pKMHA(i) = pKMHA;                      {...fixed metal complexation constant}
end.
```

and similar for HB.

The values for the various parameters are given in Table 7.8. Tipping and Hurley suggest that some of the sites are so close together that they form bidentate bonds, controlled by a proximity factor f_{pr}. These bidentate sites have two $\mathrm{p}K_a$'s equal to the $\mathrm{p}K_a$ of the constituent monoprotic sites, and a $\mathrm{p}K_{M\mathrm{H2AB}}$ equal to the sum of the two individual $\mathrm{p}K_M$'s (($\mathrm{p}K_{M\mathrm{HA}} + \mathrm{p}K_{M\mathrm{HA}}$) or ($\mathrm{p}K_{M\mathrm{HA}} + \mathrm{p}K_{M\mathrm{HB}}$)).

Table 7.8. Basic parameters for Tipping and Hurley's (1992) model for complexation of protons and metals to average humic acid.

	n_{HA} [a]	n_{HB} [a]	f_{pr}	$\mathrm{p}K_{\mathrm{HA}}$	$\mathrm{p}K_{\mathrm{HB}}$	$\Delta\mathrm{p}K_{\mathrm{HA}}$	$\Delta\mathrm{p}K_{\mathrm{HB}}$	$-P$
	−4.73	$n_{\mathrm{HA}}/2$	0.4	3.26	9.64	3.34	5.52	103
Metals	Mg^{2+}	Ca^{2+}	Ni^{2+}	Cu^{2+}	Zn^{2+}	Cd^{2+}	Pb^{2+}	
$\mathrm{p}K_{M\mathrm{HA}}$	3.3[d]	3.2[d]	1.4	0.63	1.7[c]	1.52	0.81	
$\mathrm{p}K_{M\mathrm{HB}}$	7.1[b]	7.0[b]	4.5[b]	3.75	4.9[b]	5.57	3.04	

[a] n_{H} in meq/g humic acid. [b] Estimated with the relation: $\mathrm{p}K_{M\mathrm{HB}} = 2.57 + 1.38\mathrm{p}K_{M\mathrm{HA}}$.
[c] Christensen and Christensen, 1999. [d] Lofts and Tipping, 2000.

(Note that Table 2 of Tipping and Hurley, 1992, incorrectly assigns (pK_{MHA} + pK_{MHB}) to *all* sites, Tipping, pers. comm.). Tipping and Hurley (1992) estimate a proximity factor $f_{pr} = 0.4$ which allocates 40% of the total charge to 12 bidentate bonds. Perhaps, this can be seen more clearly by inspecting the PHREEQC input file shown in Table 7.9.

The empirical electrostatic term

Tipping and Hurley (1992) mimicked the electrostatic Boltzmann factor, $\exp(-zF\psi/RT)$, by an empirical relation (for $z = 1$):

$$\exp(-F\psi_0/RT) = \exp(2P\log(I)\,Z) = \gamma_{TH}, \qquad (7.64)$$

where P is an adjustable coefficient ($P = -103$ for average humic acid), and Z is the charge on the humic acid, eq/g. The factor γ_{TH} in Equation (7.64) varies with I and Z, and you may check that with Z varying from -3 to -6 meq/g, for $I = 0.3$, γ_{TH} will change from 0.72 to 0.52 (low to high charge Z). For $I = 0.003$, γ_{TH} changes from 0.21 to 0.044 (low to high charge). At higher charge γ_{TH} becomes smaller, and the cations are more attracted to the humic acid. Also, at a lower ionic strength, γ_{TH} decreases, because the capacitance around the charged particles decreases. This increases the negative potential for a given charge ($\psi = \sigma/\kappa\varepsilon$), and draws cations stronger to the humic acid.

The double layer theory for a planar surface also links charge and ionic strength with ψ_0, and we may try to connect formula (7.64) with the double layer Equation (6.64). The relation between potential and charge for a monovalent electrolyte according to double layer theory is (cf. Examples 6.12 and 7.6):

$$F\psi_0/2RT = \ln(y + (y^2 + 1)^{1/2})$$

where $y = \sigma_H/(0.1174\sqrt{I})$, and $\sigma_H = ZF/A$ is the specific charge on the humic particle (C/m^2). Thus, the equivalent double layer $\gamma_{DL} = \exp(-F\psi_0/RT) = (y + (y^2 + 1)^{1/2})^{-2}$. For the desired connection of $\gamma_{TH} = \gamma_{DL}$, we may now adapt A (in Example 7.7 we linked the constant capacity model with the double layer model by adapting I, since that parameter was then the unique variable for the double layer model). With a least squares fit we obtain:

$$A = 159300 - 220800/I^{0.09} + 91260/I^{0.18} \qquad (7.65)$$

The equation implies an increase of the specific surface as the ionic strength decreases. Thus, the calculated charge density becomes smaller at lower ionic strength, which reduces the Boltzmann factor for humic acid. This is in agreement with the smaller effect of ionic strength on a spherical surface (of humic acid) than on a planar surface (of ferrihydrite).

The Donnan term

The humic molecule carries a negative charge which radiates out to form a double layer, in which cations are enriched in addition to the complex bound amounts. Tipping and Hurley (1992) and others (Kinniburgh et al., 1999; Kraepiel et al., 1999) calculate the cations in the double layer with a so-called Donnan model. In the Donnan model a single, averaged potential is assumed. The cations are enriched in the Donnan double layer according to:

$$m_{i,DL} = m_{i,\infty}\exp(-z_i F\psi_{DL}/RT) \qquad (7.66)$$

where m_{DL} is the concentration in the Donnan phase, where the potential is ψ_{DL}. The ratio of, for example, Na^+ and Ca^{2+} in the two phases is:

$$m_{Na,DL}/m_{Na,\infty} = (m_{Ca,DL}/m_{Ca,\infty})^{1/2} \qquad (7.67)$$

and the ions sum up to counterbalance the negative charge of the humic acid.

Table 7.9. A PHREEQC data file for calculating metal complexation to humic acid according to Tipping and Hurley (1992).

```
# download database T&H.DAT from www.xs4all.nl/~appt
DATABASE T&H.DAT
SURFACE_SPECIES
# Cd (Example in fig. 10 of Tipping and Hurley, 1992)
  H_aH + Cd+2 = H_aCd+ + H+; log_k -1.500
  H_bH + Cd+2 = H_bCd+ + H+; log_k -1.500
  H_cH + Cd+2 = H_cCd+ + H+; log_k -1.500
  H_dH + Cd+2 = H_dCd+ + H+; log_k -1.500

  H_eH + Cd+2 = H_eCd+ + H+; log_k -4.640
  H_fH + Cd+2 = H_fCd+ + H+; log_k -4.640
  H_gH + Cd+2 = H_gCd+ + H+; log_k -4.640
  H_hH + Cd+2 = H_hCd+ + H+; log_k -4.640

  H_abH2 + Cd+2 = H_abCd + 2H+; log_k -3.000
  H_adH2 + Cd+2 = H_adCd + 2H+; log_k -3.000
  H_afH2 + Cd+2 = H_afCd + 2H+; log_k -6.140
  H_ahH2 + Cd+2 = H_ahCd + 2H+; log_k -6.140
  H_bcH2 + Cd+2 = H_bcCd + 2H+; log_k -3.000
  H_beH2 + Cd+2 = H_beCd + 2H+; log_k -6.140
  H_bgH2 + Cd+2 = H_bgCd + 2H+; log_k -6.140
  H_cdH2 + Cd+2 = H_cdCd + 2H+; log_k -3.000
  H_cfH2 + Cd+2 = H_cfCd + 2H+; log_k -6.140
  H_chH2 + Cd+2 = H_chCd + 2H+; log_k -6.140
  H_deH2 + Cd+2 = H_deCd + 2H+; log_k -6.140
  H_dgH2 + Cd+2 = H_dgCd + 2H+; log_k -6.140
SURFACE 1
# For Psi vs I (= ionic strength) dependence, adapt specific surface A in PHRC:
# A = 159300 - 220800/(I)^0.09 + 91260/(I)^0.18
# Example: A = 34170 m2/g for I = 0.01 mol/L
# 1 g humic acid
  H_a 7.10e-4 34.17e3 1
  H_b 7.10e-4; H_c 7.10e-4; H_d 7.10e-4

  H_e 3.55e-4; H_f 3.55e-4; H_g 3.55e-4; H_h 3.55e-4

  H_ab 1.18e-4; H_ad 1.18e-4; H_af 1.18e-4; H_ah 1.18e-4
  H_bc 1.18e-4; H_be 1.18e-4; H_bg 1.18e-4; H_cd 1.18e-4
  H_cf 1.18e-4; H_ch 1.18e-4; H_de 1.18e-4; H_dg 1.18e-4
  -equil 1
  -diffuse_layer                          # diffuse layer calculations included
SOLUTION 1
  pH 7.0
  Cd 1e-6 Fix_cd -9.0
  Na 10; N(5) 2 charge
PHASES
  Fix_cd; Cd+2 = Cd+2; -log_k 0.0
SELECTED_OUTPUT
  -file fig_10.prn; -reset false
  -act Cd+2
  -mol H_aCd+ H_bCd+ H_cCd+ H_dCd+ H_eCd+ H_fCd+ H_gCd+ H_hCd+\
       H_abCd H_adCd H_afCd H_ahCd H_bcCd H_beCd H_bgCd H_cdCd H_cfCd H_chCd H_deCd
H_dgCd
USE solution none
END
```

Equation (7.67) has the form of an exchange formula, with $K_{(7.67)} = 1$. However, when the different standard state of exchangeable ions is taken into account, it appears that $K_{(7.67)} = K_{Ca\backslash Na}\,(2\,CEC)^{1/2}$, where *CEC* is the charge of the humic acid in eq/L. In Section 6.6.1 we calculated that the enrichment of cations in a double layer follows exchange theory with $K_{I\backslash Na} \approx 1$, and the simple Donnan model may not be correct. Generally, the Donnan model underestimates the contribution of the higher-charged ions (Problem 7.22).

An example calculation

A PHREEQC input file for Tipping and Hurley's model is given in Table 7.9. The 4 carboxylic sites are defined as H_a, H_b, H_c and H_d, and the 4 phenolic sites are H_e..H_h. The bidentate sites are entered separately with letters noting the corresponding monoprotic bonds (H_ah is the combination of sites H_a and H_h). The input file is for calculating the distribution of Cd^{2+} over the various sites of average humic acid and the double layer and uses the database T&H.DAT that can be downloaded from www.xs4all.nl/~appt. Diffuse layer calculations are included, which diminishes the surface potential and reduces sorption of Cd^{2+}. Output in file "fig_10.prn" closely matches the results depicted in Figure 10 in Tipping and Hurley (1992).

Again, we return to Example 7.1, since we still want to know whether the organic carbon content can explain the $K_d = 307$ L/kg for sorption of Cd^{2+} on sand.

EXAMPLE 7.11. *Sorption of Cd^{2+} to organic matter in loamy sand*

Christensen (1984) analyzed 0.7% OM (Organic Matter, ≈0.35% OC) in the sand. Find the K_d (L/kg) for Cd^{2+}.

ANSWER:

Let's make an input file for the actual experiment, 1 g soil or 7 mg OM reacts with 50 mL solution with a trace concentration of Cd^{2+}. The data for Tipping and Hurley's model can be downloaded from www.xs4all.nl/~appt and will not be repeated here, only the actual calculation of the distribution coefficient is presented:

```
# Cd sorption on OC in Christensen's xpt
DATABASE T&H.DAT # from www.xs4all.nl/~appt
PRINT; -reset false; -user_print true

SURFACE 1
# 1 g loamy sand = 0.007 g HA, multiply site conc's for 1 g HA with 7e-3
# Specific surface = 46.5e3 m2/g for 1 mM CaCl2 (I = 0.003 mol/L)
  H_a   4.97e-6  46.5e3  0.007
  H_b   4.97e-6; H_c 4.97e-6; H_d 4.97e-6

  H_e   2.483e-6; H_f 2.483e-6; H_g 2.483e-6; H_h 2.483e-6

  H_ab 8.278e-7; H_ad 8.278e-7; H_af 8.278e-7; H_ah 8.278e-7
  H_bc 8.278e-7; H_be 8.278e-7; H_bg 8.278e-7; H_cd 8.278e-7
  H_cf 8.278e-7; H_ch 8.278e-7; H_de 8.278e-7; H_dg 8.278e-7
  -equil 1
  -diffuse_layer

SOLUTION 1
  pH 6.0
  Ca 1; Cd 1e-6; Cl 2 charge
  -water 0.05

USER_PRINT
  -start
  10 H_Cd = mol("H_aCd+") + mol("H_bCd+") + mol("H_cCd+") + mol("H_dCd+")
  20 H_Cd = H_Cd + mol("H_eCd+") + mol("H_fCd+") + mol("H_gCd+") + mol("H_hCd+")
```

```
  30 H_Cd = H_Cd + mol("H_abCd") + mol("H_adCd") + mol("H_afCd") + mol("H_ahCd")
  40 H_Cd = H_Cd + mol("H_bcCd") + mol("H_beCd") + mol("H_bgCd") + mol("H_cdCd")
  50 H_Cd = H_Cd + mol("H_cfCd") + mol("H_chCd") + mol("H_deCd") + mol("H_dgCd")
  60 print " ug Cd/g soil                  =",   H_Cd * 112.4e6 * 0.05,\
           " ug Cd/L                       =", tot("Cd") * 112.4e6,\
           " Kd (L/kg)                     =", H_Cd / tot("Cd") * 50
 70 print  " ug Cd/g soil in DL             =", edl("Cd") * 112.4e6, chr$(13)
 80 dl_Ca = edl("Ca")*2 - edl("water")*tot("Ca") * 2
 90 dl_Cl = edl("Cl")   - edl("water")*tot("Cl")
100 print "Excess eq Ca in DL =", dl_Ca
110 print "Excess eq Cl in DL =",       dl_Cl, "Ca - Cl charge in DL =", dl_Ca-dl_Cl
120 print "Surface charge         =", edl("Charge")
-end
END
```

In keyword USER_PRINT the concentrations of surface complexes are summed up in mol/kg H_2O and converted to μg Cd/g by multiplying with the atomic weight in μg/mol, and with 0.05 (=kg H_2O/g soil). The distribution coefficient is likewise found by multiplying the quotient of complexed and solute concentrations by 50 kg H_2O/kg soil. The concentration of Cd^{2+} in the double layer and the charge due to excess Ca^{2+} and Cl^- in the double layer are also printed. These quantities are given in moles by PHREEQC, not in molality, and they are not multiplied with 0.05. The result is:

```
---------------------------------------------User print---------------------------------------------------------------------

  ug Cd/g soil          = 1.5080e-01
  ug Cd/L = 1.1240e-01 Kd (L/kg) = 1.3416e+03
  ug Cd/g soil in DL    = 1.5122e-03

Excess eq Ca in DL      = 2.2492e-05
Excess eq Cl in DL      = -3.2998e-06 Ca - Cl charge in DL = 2.5792e-05
Surface charge          = -2.5796e-05
```

The calculated distribution coefficient $K_d = 1342$ L/kg is about 5 times larger than determined. However, we assumed that all the organic matter is humic acid whereas a large part probably consists of more unreactive lignin. The contribution of the double layer to the overall sorbed Cd^{2+} is about 1%. The total (possible) charge of the OM is 0.007 g OM × 7.095 meq/g HA = 5×10^{-5} eq. Approximately half the sites are dissociated at pH = 6 (Surface charge = -2.58×10^{-5} mol). About 90% of this charge is compensated by an excess Ca^{2+} and 10% by a deficit of Cl^-. The double layer content for Cd^{2+} has been neglected which is reasonable given the small amounts Cd^{2+}. On the other hand, the double layer contains 5 times more Ca^{2+} than is complex-bound and clearly the complex constants for a weakly bound cation are difficult to obtain if only traditional wet-chemical analyses are used.

QUESTIONS:

Obtain a printout of total complexed Ca^{2+}?

ANSWER: complexed mol Ca/g soil = 2.8367×10^{-6}

Calculate K_d for pH = 7.0 and 8.0?

ANSWER: K_d = 1749 L/kg at pH = 7, K_d = 2663 L/kg at pH = 8.

Calculate K_d when I increases to 0.01 mol/L by adding NaCl (use Equation (7.65) to adapt the specific surface A)?

ANSWER: K_d = 888 L/kg

Calculate K_d when I increases to 0.01 mol/L by adding $CaCl_2$?

ANSWER: K_d = 384 L/kg

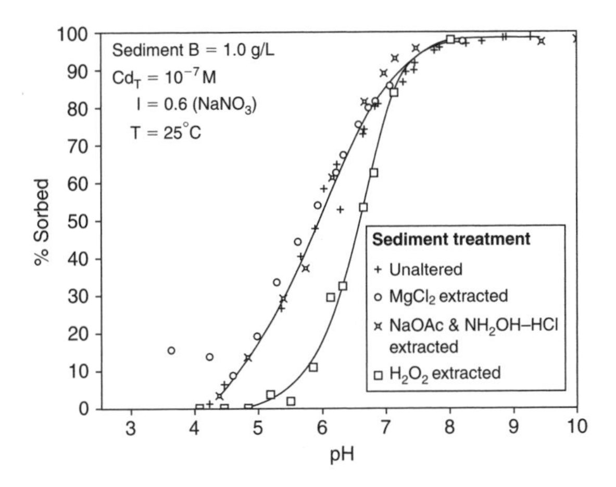

Figure 7.28. Cadmium adsorption onto sequentially extracted estuarine salt marsh sediment as a function of pH (Lion et al., 1982).

Clearly, sorption of Cd^{2+} to the loamy sand of Christensen (1984) is likely to be located mainly in the organic fraction. This corroborates the experiments of Lion et al. (1982) who extracted oxides and organic matter from an estuarine sediment and determined sorption edges for Cd^{2+} and Pb^{2+} on the residues and also on the intact sediment (Figure 7.28). Treatment with acetate and hydroxylamine/hydrochloride extractant showed no effects for Cd^{2+} (this extract removes manganese oxides and mainly amorphous iron oxyhydroxides), but the removal of organic carbon with H_2O_2 shifted the sorption edge from pH = 5.9 to pH = 6.5, indicating that a major sorbent had been eliminated. For Pb^{2+}, however, the sorption edge shifted 0.7 unit higher pH when the oxides were removed, and remained at that pH when organic carbon subsequently was extracted.

We can calculate the contributions of the various fractions in the sand to the sorption isotherm of Example 7.1 and compare the total with the observed distribution (Figure 7.29). The amount of organic carbon was reduced by 7 (the amounts of 'H_' in Example 7.11 were divided by 7) to obtain the observed distribution coefficient for small concentrations of Cd^{2+}.

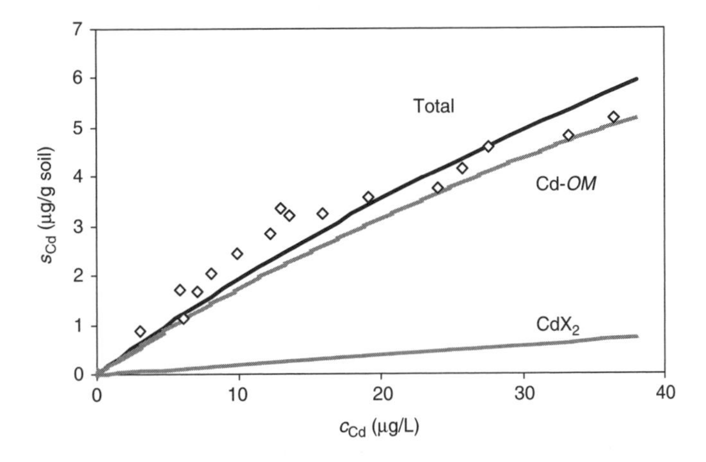

Figure 7.29. Cd^{2+} sorption on sandy loam, modeled as the sum of sorption on iron oxyhydroxide (negligible and omitted from the graph), ion exchange (Cd-X_2) and complexation on organic matter (Cd-*OM*).

The amount sorbed on iron oxyhydroxide is negligible, exchangeable Cd-X_2 on clay contributes about 10%, while organic matter provides the dominant contribution to the total isotherm. Note that the modeled sorption isotherm is less curved than the data, indicating that the number of strongly sorbing sites in the model is too small. Similar comparisons of a deterministic model for heavy metal sorption have been presented by Cowan et al. (1992) and Lofts and Tipping (2000). Generally, modeling seems able to describe various data to within a factor of ten.

7.5.3 *Distribution models*

In Section 7.3.3 we found that the stoichiometry of the exchange reaction determined the steepness of the sorption edge. To generalize this finding we write the sorption Reaction (7.7) as:

$$s + q \cdot I \leftrightarrow s_I; \quad K_{s_I} \tag{7.68}$$

Here q is the stoichiometric reaction coefficient. Charges have been left out for simplicity, but you may think of sorption of neutral molecules, gases or organics. Most of this theory was originally developed for gas-adsorption, and subsequently applied to sorption of solute ions (Van Riemsdijk et al., 1986). The mass action equation for Reaction (7.68) yields a modified form of the Langmuir equation:

$$[s_I] = \frac{[s_{tot}]\, K_{s_I}\, [I]^q}{1 + K_{s_I}\, [I]^q} \tag{7.69}$$

This equation can be rewritten as:

$$\theta = \frac{K_{s_I}\, [I]^q}{1 + K_{s_I}\, [I]^q} \tag{7.70}$$

where $\theta = s_I / s_{tot}$ is the fraction of surface sites covered by I. When q equals one, then Equation (7.70) is identical to the Langmuir equation (7.9). If $[I]$ is small, Equation (7.70) approaches the Freundlich equation (7.4).

When θ is plotted *vs* log $[I]$, a sigmoidal curve results, which broadens as q decreases. Sips (1948) has shown that this curve closely approaches an integrated probability curve when the log K's of the sorption sites are normally distributed around an average log K_{av} and with a standard deviation σ:

$$\theta_i = \frac{1}{\sigma\sqrt{2\pi}} \int_{-\infty}^{\log K_i} \exp\left(-\left(\frac{\log K - \log K_{av}}{\sigma\sqrt{2}}\right)^2\right) \mathrm{d} \log K \tag{7.71}$$

where θ_i are all the sites with binding strength up to log K_i. The relation among the Sips parameters and those characterizing the normal distribution is:

$$\log K_{s_I} = q \log K_{av} \tag{7.72a}$$

and

$$q = 1/(\sigma\sqrt{2}) \tag{7.72b}$$

Figure 7.30 shows sorption of protons as a function of pH in case $\log K_{av} = -4$, for two values of $\sigma = 1/\sqrt{2}$ and $\sigma = 1.7$ (corresponding to $q = 1$ and $q = 0.416$, respectively). There is a close correspondence but the Sips curve is slightly broader in the tails.

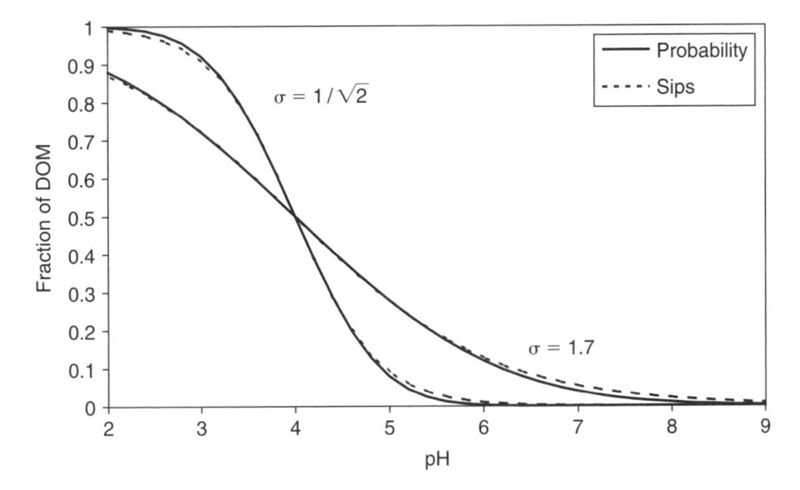

Figure 7.30. Sorption of protons on a surface with probability-distributed binding strength. Comparison of the Sips curve with a probability curve, $\log K_{av} = -4$, $q = 1$ and 0.416.

Equation (7.70) can be generalized for a multicomponent system in the form of the so-called Henderson-Hasselbalch (HH) equation:

$$\theta = \frac{K_{s_I}[I]^q}{1 + \sum_I (K_{s_I}[I]^q)} \tag{7.73}$$

However, the deviation with respect to the normal distribution increases because of tailing of the distribution. Van Riemsdijk et al. (1986) therefore derived what is called the multicomponent Langmuir-Freundlich (LF) equation:

$$\theta_I = \frac{K_{s_I}[I]}{\sum_I K_{s_I}[I]} \cdot \frac{\left(\sum_I K_{s_I}[I]\right)^p}{1 + \left(\sum_I K_{s_I}[I]\right)^p} \tag{7.74}$$

where p indicates the heterogeneity among the sites, with the same function as q has for the individual ions. The integration of the normal distribution of log K's (Equation 7.71) is an option in the geochemical model MINTEQ (Allison et al., 1991) for calculating sorption of heavy metals to humic acids. The comparison of the Sips distribution (Equation 7.72) and MINTEQ is excellent for a single component (similar to Figure 7.30). In multicomponent solutions, the different tailings of the distributions give notable effects. Figure 7.31 compares competitive sorption of protons and Ca^{2+} on humic acid, at variable pH and $[Ca^{2+}] = 0.001$, calculated by MINTEQ and the HH (7.73) and LF (7.74) approximations. The LF equation approximates the Gaussian distribution quite well, while the HH curves are too broad.

The multicomponent Sips' equations are used in the non-ideal competitive adsorption (NICA) model developed by Koopal and coworkers (Koopal et al., 1994; Kinniburgh et al., 1998, 1999):

$$\theta_I = \frac{q_I K_{s_I}[I]^{q_I}}{\sum_I (K_{s_I}[I]^{q_I})} \cdot \frac{\left\{\sum_I (K_{s_I}[I]^{q_I})\right\}^p}{1 + \left\{\sum_I (K_{s_I}[I]^{q_I})\right\}^p} \tag{7.75}$$

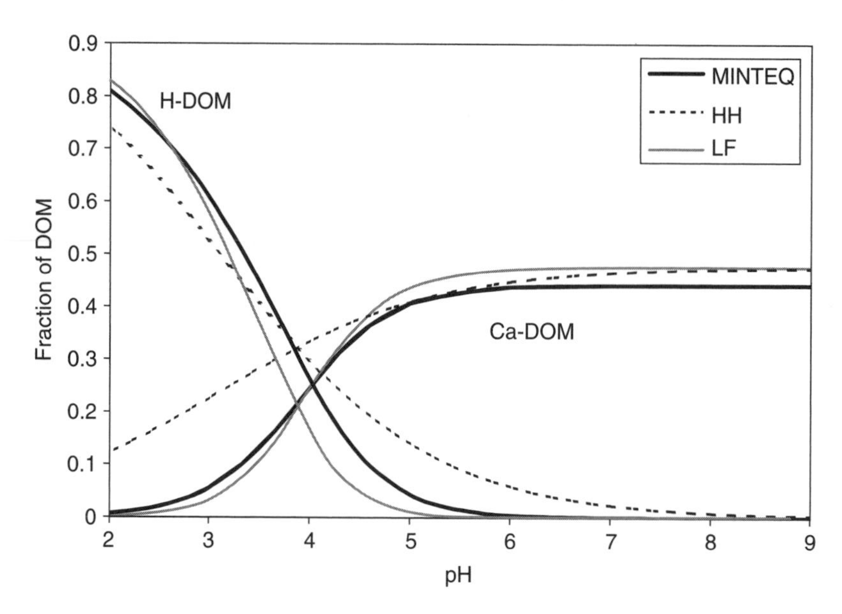

Figure 7.31. Comparison of two-component sorption of protons and Ca^{2+} by three equations: MINTEQ is a numerical integration of the normal distribution of log *K*'s; HH and LF correspond to Equations (7.73) and (7.74), respectively. Data: $\log K_{m,\mathrm{H}} = -3.78$, $\log K_{m,\mathrm{Ca}} = -2.9$, $\sigma = 1.7$ for both, $[Ca^{2+}] = 0.001$.

The NICA model considers two types of sites (carboxylic and phenolic, similar to the Tipping and Hurley model), each with a parameter p which reflects the heterogeneity within the sites. Basically, sorption of protons is determined by titrations at various ionic strength. These data are corrected to a master curve with an electrostatic model for spherical or cylindrical particles (De Wit et al., 1993a, b). For the master curve, values of q_{H} and p are derived by curve fitting (Buffle, 1990; Rusch et al., 1997). Parameters for heavy metal sorption are likewise determined from sorption data at a fixed pH (Benedetti et al., 1995). More recently, the parameters are estimated with non-linear estimation programs (Milne et al., 2001, 2003).

7.5.4 *Humic acids as carriers of trace elements*

Most of the experimental work for deriving sorption parameters has been done with an extracted and subsequently purified humic acid. But how applicable are the obtained parameters for groundwater *DOC*? Christensen and Christensen (1999) equilibrated solutions with groundwater *DOC*, heavy metals and cation exchange resins or aquifer sands. *DOC* was derived from a waste site percolate and added in variable concentrations. Equilibrium will be achieved between the resin and the free metal in solution. When increased *DOC* complexes more metal, the concentration of free metal decreases. The total solute concentration of Ni^{2+}, Zn^{2+} and Cd^{2+} increases with *DOC* as a result of increasing complexation in solution, and desorption from the resin or aquifer materials (Figure 7.32).

Also shown in Figure 7.32 are model lines for the concentration ratio, given by Tipping and Hurley's 'WHAM' model, and by MINTEQ. Clearly, the modeled lines are not exactly matching the observations. The WHAM model predicts too much complexation for sample L2, and too little for sample L1. The MINTEQ model tends to underestimate complexation to *DOC*, but still matches the data within a factor of 3. Thus, at least an idea of the importance of humic acids as carriers of heavy metals can be derived with the models. In the experiments of Figure 7.32, the solubility is enhanced by a factor of 2 with about 100 mg C/L for Cd^{2+} and Zn^{2+}, and with 40 mg C/L for Ni^{2+}.

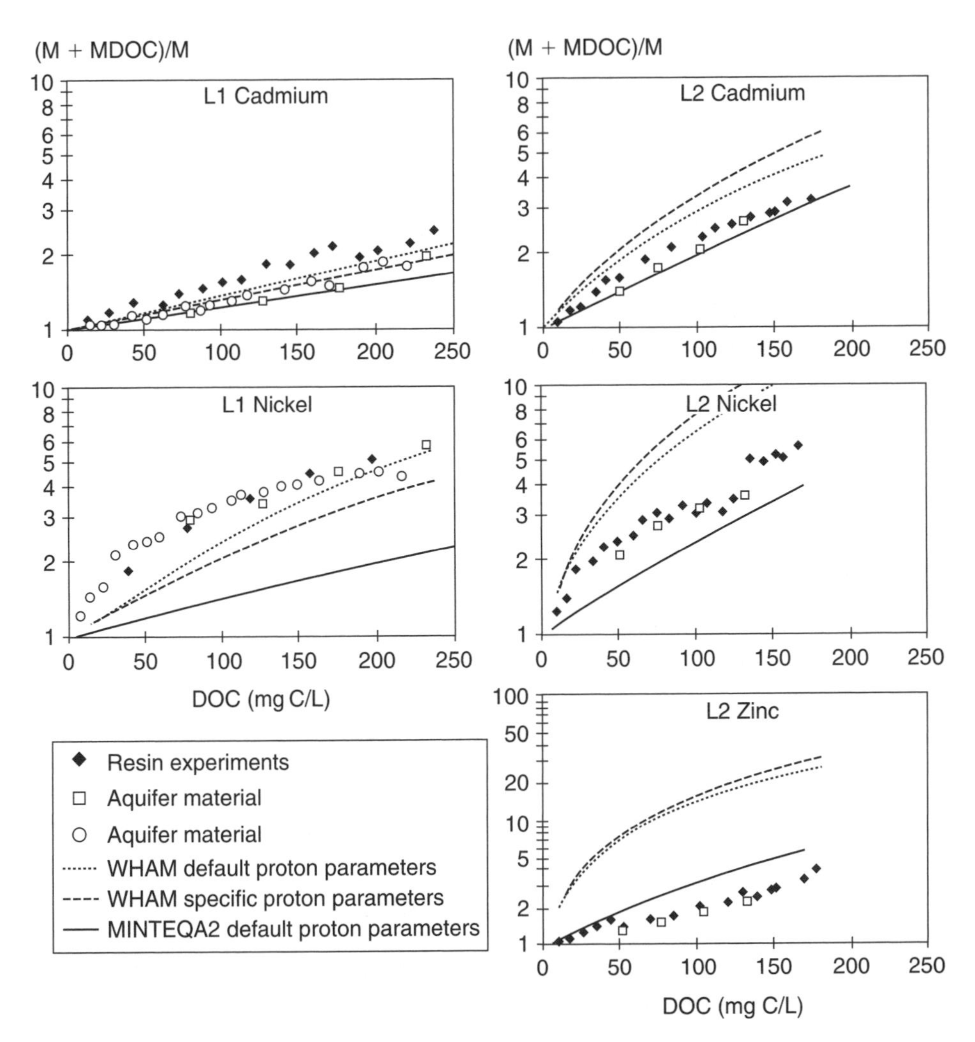

Figure 7.32. Solute concentration of Ni^{2+}, Zn^{2+} and Cd^{2+} as function of *DOC* in water in equilibrium with exchange resin or aquifer materials, relative to water without *DOC* (*M* is free, *MDOC* is complexed to *DOC*) (Christensen and Christensen, 1999). L1 and L2 are two different water samples, L1 with a higher Ca^{2+} concentration.

These *DOC* concentrations are much higher than encountered in average groundwater. Therefore, enhanced transport of heavy metals by dissolved humic acids in groundwater takes place only in dark, tea colored waters and it seems that it may be ignored in most other cases.

QUESTION:
Obtain the distribution coefficient K_d (L/kg) for Cd^{2+} on OC in water L2 for 70 mg *DOC*/L from the experiment in Figure 7.32. Water L2 has pH = 6.6, Ca^{2+} = 1.3 mM, I = 0.013 mM. Calculate K_d as in Example 7.11.

ANSWER: $K_d \approx 10^4$ L/kg in the experiment. The WHAM model estimates: $K_d = 1.9 \times 10^4$ L/kg

7.6 KINETICS OF SURFACE COMPLEXATION

Sorption of trace metals on humic acid or oxides proceeds quickly, with half-lives of minutes or less (Bunzl and Schimmack, 1991; Hachiya et al., 1984). However, longer reaction times can enhance sorption and shift the sorption edge to lower pH (Gerth et al., 1993). Desorption generally takes more time than adsorption, and the time needed for desorption increases further when the duration of the adsorption experiment is longer (Section 7.3.4). This suggests that several mechanisms operate in the sorption process, for example sorption on weak and strong sites on oxides, with slower kinetics for the strong sites. Another explanation may be that metals slowly diffuse into the oxide structure during adsorption, and that outward migration during desorption is slower because inward diffusion continues as long as the internal structure has not been saturated. We can try to model a kinetics experiment, and establish the various processes by following the time dependent variation in the species concentrations.

The pressure jump technique is used to study rapid kinetics (Sparks, 1989). A high pressure of about 100 atm is applied to the sample and suddenly released. The return to normal pressure within less than 100 ms is accompanied by changes in the solution composition that can be recorded with an accurate conductivity meter. Hayes and Leckie (1986) have applied this technique to measure the sorption kinetics of lead on goethite and used the triple layer model to interpret the data. Equilibrium sorption of lead followed the reaction:

$$\equiv\text{FeOH} + \text{Pb}^{2+} \leftrightarrow \equiv\text{FeOPb}^{+} + \text{H}^{+} \qquad (7.76)$$

The rate data were modeled with the forward rate:

$$r_f = k_1\, m_{\equiv\text{FeOH}}\, [\text{Pb}^{2+}]_s \qquad (7.77)$$

where r_f is the forward rate in mol/L/s, k_1 the rate constant (1/s), m_i the concentration in mol/L, and $[\text{Pb}^{2+}]_s = [\text{Pb}^{2+}]\exp(-2F\psi_0/RT)$ is the (solute) activity of Pb^{2+} at the surface. The backward reaction was likewise:

$$r_b = k_{-1}\, m_{\equiv\text{FeOPb}^+}\, [\text{H}^+]_s \qquad (7.78)$$

The rate constants are linked by detailed balancing. At equilibrium the net rate is zero, $r_f = r_b$, and therefore $k_1 / k_{-1} = K_{\text{int}}\exp(F\psi_0/RT)$, where K_{int} is the intrinsic equilibrium constant for the sorption reaction, Equation (7.76). Hayes and Leckie (1986) found $k_1 \approx 2.5\times10^5$/s at 100 atm pressure. The value of k_1 decreased when pressure was increased, suggesting that stronger sorption sites gained importance at greater pressure. A fast and a slow relaxation was also observed for sorption of Cu^{2+} on goethite (Grossl et al., 1994).

The kinetic reaction can be linearized for small perturbations which permits an easy evaluation of the effect of various solution parameters on the rate (Hayes and Leckie, 1986; Zhang and Sparks, 1990). For larger changes, e.g. modeling sorption from zero metal concentration onwards, the linearization is incorrect. However, the full rate equations can be included in PHREEQC, thus permitting a modeling approach which was hitherto not possible. The sorption edge of Pb^{2+} can be reasonably modeled with the strong and weak sites of Hfo, but log K must be adapted for the strong sites (Figure 7.33). Note in Figure 7.33 that the model appears to shift the sorption edge too much for changes in ionic strength.

For modeling kinetic sorption, we need the surface potential ψ_0. This can be obtained by calculating sorption on a dummy surface with a very small concentration. The complexation model is not electrically neutral and requires a double layer calculation for charge compensation. However, the surface is positively charged at all times in this experiment, and we can, more simply, neutralize the surface reaction with NO_3^-. Part of the input file for PHREEQC which calculates kinetic sorption of Pb to ferrihydrite is shown in Table 7.10 (download the complete file from www.xs4all.nl/~appt).

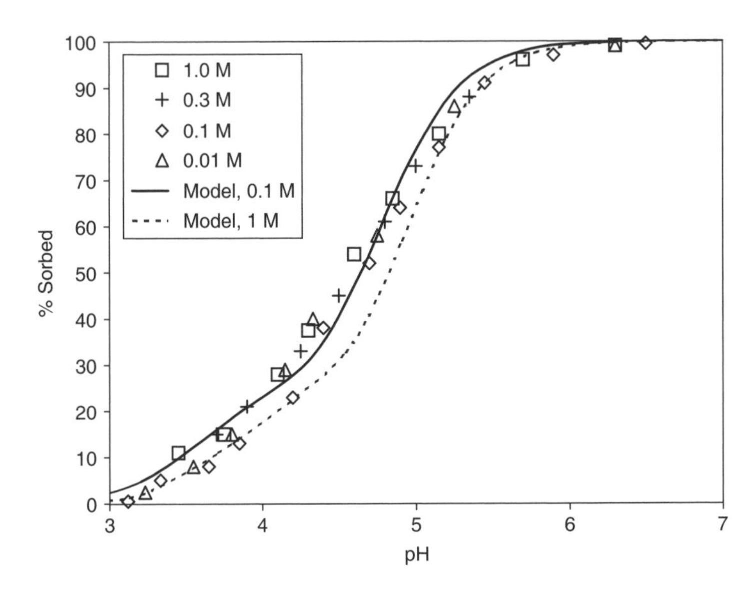

Figure 7.33. Sorption edge for Pb^{2+} on goethite. Data from Hayes and Leckie (1986). Double layer model with log *K* changed to 3.7 for the strong sites.

The file starts with the definition of the basic setup of the experiment. Imagine a titrator which maintains the pH at 5.0 in a beaker with 10 mM NaCl solution and some ferrihydrite. We will explore time-dependent sorption as 2 mM $Pb(NO_3)_2$ is added to the beaker and use RATES and KINETICS to define the reactions and *not* do so with keyword SURFACE. However, we do need the surface potential of ferrihydrite which we obtain using the SURFACE keyword. The concentration of surface sites is given an insignificant concentration, say 10^{-8} smaller than actual. The potential at the surface is independent of the concentration of ferrihydrite. Furthermore, when a surface species is multiplied by 10^8, it will give the sorbed *equilibrium* concentration.

The rates are calculated according to Equations (7.77) and (7.78) for all the surface species. A few specific points may bev noted when studying the input file in Table 7.10:

- The Boltzmann factor is used in various rate equations, and is therefore stored in temporary memory ("put(Boltzm, 99)") in the first rate, for later recall with "Boltzm = get(99)".
- Line 20 in the first rate defines the forward rate for the species. Note that "kin("kin_surfs_oh")" provides the moles of the Hfo_sOH species.
- Line 30 defines the backward rate. The moles of the species is given by *m*, and is divided by K_{int} of the reaction ($K_{int} = 10^{3.7}$ for sorption of Hfo_sOPb^+).
- Line 40 calculates the moles of reaction by multiplying by k_1 = parm(1) and by "time". The rate equations have very different rates and the fastest reactions may need very small timesteps for reducing oscillations. These "stiff" equations should be solved with the ordinary differential equation solver implemented in PHREEQC-2.9.
- Line 50 stores the moles of reaction in temporary memory ("put(moles, 1)") for calculating mass and charge balance for the sorption sites, and line 60 "saves" the reaction, finishing the rate definition.
- The rate "kin_surfs_oh" is for species "Hfo_sOH". This species is defined by mass balance on the strongly sorbing surface sites. The same applies to the corresponding species on the weak sites.
- The rate "kin_surf_ntr" sums the charge of all the kinetics reactions, and balances it by adding or subtracting NO_3^-. In this way, pH effects which result from balancing the electrical neutrality in

Table 7.10. Partial PHREEQC input file for calculating kinetic sorption of Pb to weak and strong sites on ferrihydrite.

```
# Rates for calculating kinetic sorption of Pb
# This is a partial file. Download complete file sk_pb_t.ode from www.xs4all.nl/~appt

PHASES; Fix_ph; H+ = H+; -log_k 0.0
EQUILIBRIUM_PHASES 1; Fix_ph -5.0 HNO3

SOLUTION 0-2
pH 5.0; Na 10; N(5) 10 charge

# dummy surface for calculating ψ and limiting conc's...
SURFACE 1
 Hfo_w 18.14e-11 600 7.981e-8          # ...site conc'n * 1e-8
 Hfo_s 0.45e-11
 -equil 0
                                       # Rate equations...
RATES                                  # Strong surface sites...
 kin_surfs_Pb
-start
 10 Boltzm = exp(-38.924 * edl("Psi"))
 12 put(Boltzm, 99)
 20 r_f = kin("kin_surfs_oh") * act("Pb+2") * (Boltzm*Boltzm)
 30 r_b = m / (10^3) * act("H+") * Boltzm
 40 moles = parm(1) * (r_f - r_b) * time
 50 put(moles, 1); 60 save -moles
-end

# Similar rates are defined for the other species on strong surface sites
# kin_surfs_h, kin_surfs_o

# kin_surfs_OH species is obtained by mass balance on sites...
 kin_surfs_oh
-start
 10 r = get(1) + get(2) - get(3)
 20 moles = m - r
 30 save m - moles
-end

# And also for all the species on the weak sites...
# kin_surfw_Pb, kin_surfw_h, kin_surfw_o, kin_surfw_oh

# charge balance, sum the reactions...
 kin_surf_ntr
-start
  5 s = 0
 10 for i = 1 to 6
 20 if exists(i) then s = s + get(i)
 30 next
 40 save -s
-end
END

KINETICS 1
kin_surfs_Pb;  -formula Pb 1. H -1.; -m0        0.0; -parm 2.5e5
# also other species on strong sites...
kin_surfs_oh;  -formula OH 0.      ; -m0   2.880e-4
```

(Contd)

Table 7.10. (*Contd*)

```
# also other species on weak sites...
kin_surf_ntr; -formula NO3 1.          ; -m0          1.0
-step 1 3 11 25 40 80 100 240 780 2000 4000 10000 20000 49120
-cvode true
                                                          # Adsorption in pH-stat...
USE equilibrium_phases 1
SOLUTION 1
  pH 5.0; Na 10; N(5) 10 charge; Pb 2.0                   # Add Pb to solution
USE surface 1
# etc. for print
```

the solution are avoided. For a negative surface, or for humic acids, one would counterbalance with cations from solution, as in the Donnan approach discussed earlier. It is possible to calculate neutralization by a mixture of ions with the exchange formulae given in Chapter 6, "put" the amounts for the different ions in storage, and then "get" the amount in a rate which contains the reactant formula for the specific ion only.

- The initial amount ("m0") (only exemplified for a few species in Table 7.10) of each kinetic reactant equals the initial surface composition. It can be obtained by multiplying the equilibrium surface species concentration from the first simulation by 10^8.

This sorption model does consider weak and strong sites, and illustrates an interesting feature noted by Morel and Hering (1993) for the sorption of aqueous complexes with different complexation strengths. Sorption on the weaker sites is usually faster because more sites are present. This results in temporary over-loading of the weak sites. The overshoot retards sorption on the strong sites since it must take place via redistribution and desorption from the weak sites.

Figure 7.34 shows adsorption of Pb^{2+} on the weak sites to go through a maximum at about $10^{2.7} = 500$ s, while sorption on the strong sites increases steadily with a half-life of $10^{3.4}$ s ≈ 45 min. Without weak sites, the half-life is shorter by a factor of 2.5. The half-life decreases for high concentrations of Pb^{2+} as is illustrated in Figure 7.35, which compares sorption isotherms for 45 minutes kinetics and at equilibrium. Equilibrium sorption on the weak sites is very nearly linear for the concentration range from 0–2 mM Pb^{2+}, and 45 minutes are sufficient to reach near equilibrium for the weak sites.

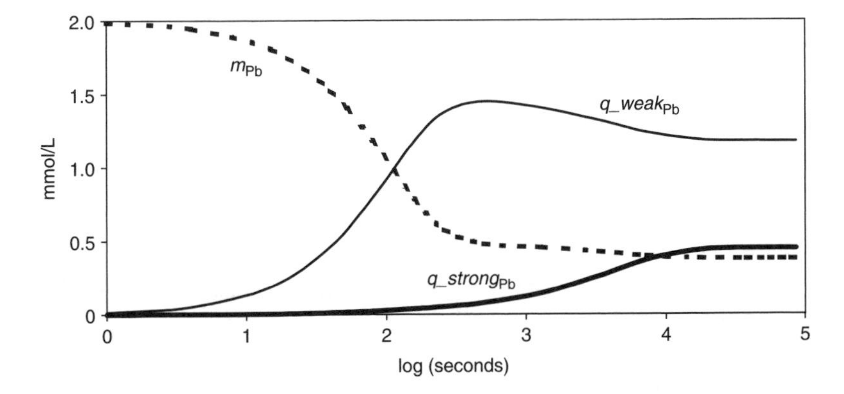

Figure 7.34. Adsorption of Pb^{2+} on strong and weak sites of ferrihydrite, modeled as a kinetic process. Note that the redistribution of sorbed complexes continues after $10^{2.7} = 500$ s, even when the solute concentration of Pb^{2+} is already near the final value.

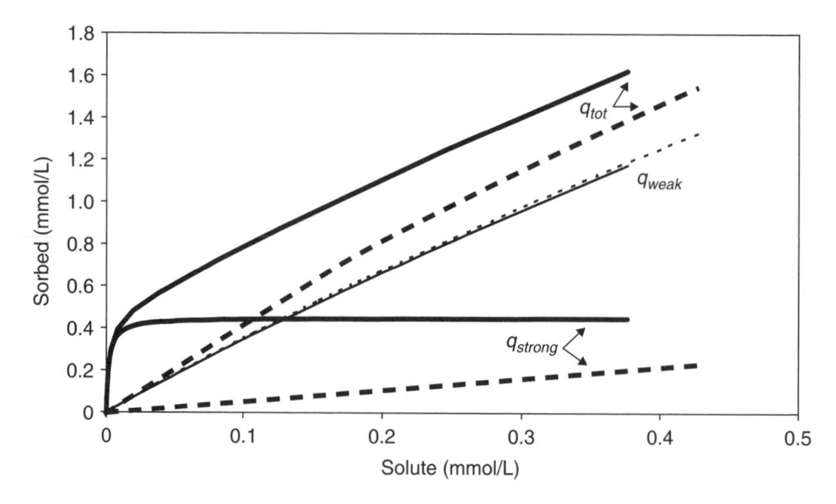

Figure 7.35. Adsorption isotherms of Pb^{2+} at 45 min. (kinetics, dotted lines) and at equilibrium (continuous lines), weak and strong sites of ferrihydrite.

The initial part of the overall equilibrium sorption isotherm is curved because the strong sites are filled up almost completely with Pb^{2+} (the concentration of the weak sites is 18 mM, the concentration of strong sites is 0.45 mM). The kinetic sorption isotherm for strong sites lies below the equilibrium isotherm, but approaches it at higher concentrations. This suggests that the experimental determination of sorption isotherms may require more equilibration time for low than for high concentrations.

When adsorption is stopped after 45 minutes, by replacing the fluid with a NaCl solution without Pb^{2+}, the redistribution of sorbed complexes is not yet complete. The concentration of Pb on the strong sites therefore continues to increase during the overall desorption process. The solute concentration may even become higher due to redistribution among the species than in the final equilibrium solution, as is shown in Figure 7.36. When adsorption is extended over 1 day, the system approaches equilibrium. Desorption occurs then almost exclusively from the weak sites.

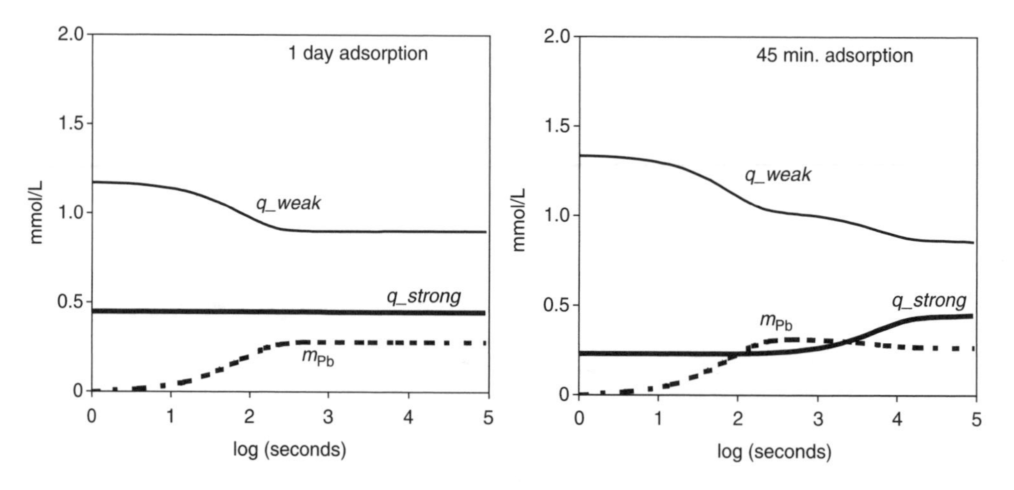

Figure 7.36. Time curves for desorption of Pb^{2+} from ferrihydrite after 45 minutes and 1 day of adsorption. Note that Pb^{2+} adsorption on the strong sites increases over short times even though there is overall desorption.

Desorption is modeled with PHREEQC by defining a new KINETICS block with initial concentrations m_0 equal to the concentrations m at the end of the adsorption step, as shown in the complete, downloaded input file.

Lastly, we may note that the final solute equilibrium for the 45 minutes experiment is slightly less than for one day adsorption time, due to the kinetic readjustment on the strong and the weak sites. The observation that desorption is less when the adsorption time has been in the order of several days, therefore cannot be explained by kinetic rearrangements among strong and weak sorption sites. Rather, it must be related to diffusion of the metal ions into the mineral structure.

7.6.1 *Extrapolation of adsorption kinetics for other metal ions*

Surface complexation of heavy metals on oxides involves an inner-sphere reaction where the water molecules from the hydration shell of the ion are removed and replaced by direct coordination to the surface oxygens of the solid. Generally, the rate for the water exchange rate is 4–5 orders of magnitude greater than for the inner-sphere adsorption reaction (Wehrli et al., 1990; Figure 7.37) and can be used for estimating adsorption rates. For example, the rate for water exchange on Zn^{2+} is $10^{7.5}$/s (Figure 7.37). This is 500 times slower than for Pb^{2+}, and accordingly, we estimate the half-life for Zn^{2+} sorption on the weak sites to be around 500 × 90 s (½ day), while for sorption on the strong sites it is estimated to be 500 × 45 min (15 days).

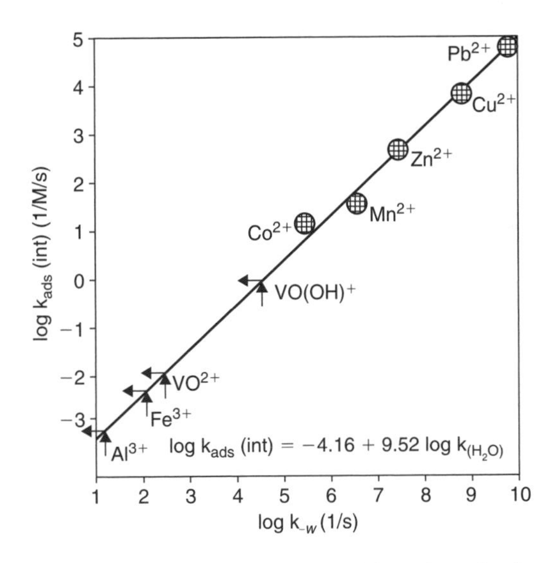

Figure 7.37. Rate constants for adsorption of metal ions on oxide surfaces (k_{ads}) are related to exchange rates of water in the hydration shell (k_{-w}) (Stumm, 1992).

7.7 FIELD APPLICATIONS

Surface complexation theory is useful to obtain a conceptual understanding of heavy metal sorption behavior on mineral surfaces. However, the theory has been developed from experimental data using a single mineral, ferrihydrite or goethite, and the application to natural sediments poses complications. Grains in aquifer sediment are normally coated by mixtures of iron oxyhydroxides, clay minerals and organics. Often the coating consists of sandwiches of organics and clays with predominantly negative surface charge, and iron oxyhydroxides with predominantly positive surface charges. It would be quite unlikely that the Dzombak and Morel (1990) database for ferrihydrite is completely adequate for a natural sediment. There are, however, some environmental settings where it is reasonable to apply the Dzombak and Morel (1990) database. For example as a model substance for heavy metal behavior in acid mine drainage, or for an iron oxide filter bed in a drinking water plant. As an illustration, the concentration changes in a reduced groundwater which contains heavy metals, and is oxidized in a drinking water plant, can be calculated as shown in Example 7.12.

EXAMPLE 7.12. *Calculate heavy metal removal from groundwater with Fe^{2+} oxidation during aeration and filtration*

The groundwater composition is (mg/L):

pH	Ca^{2+}	*Alk*	Cl^-	Fe^{2+}	As	Cd^{2+}	Ni^{2+}
6.5	100	120	100	5	10×10^{-3}	10×10^{-3}	30×10^{-3}

ANSWER:

We make an input file for PHREEQC, define the groundwater (SOLUTION) and calculate the composition after oxidation (EQUILIBRIUM_PHASES) with surface complexation (SURFACE). The surface complexes of CO_2 on Hfo are copied from the database (SURFACE_SPECIES). NaOH or HCl is added to increase or decrease pH (REACTION). Output is directly in μg/L for the heavy metals (USER_ PRINT). The file must be run with the wateq4f.dat database that includes the elements As and Ni.

```
DATABASE wateq4f.dat
SOLUTION 1
 pH 6.5; -units mg/l; Ca 100; Alkalinity 120; Cl 100; Fe(2) 5
 As 10e-3; Cd 10e-3; Ni 30e-3

PHASES; Ferrihydrite; Fe(OH)3 + 3H+ = Fe+3 + 3H2O; -log_k 2.0
EQUILIBRIUM_PHASES 1
 O2(g) -0.68                              # P_O2 = 0.21 atm (log 0.21 = -0.68)
 Ferrihydrite 0.0 0.0                     # SI = 0.0, initial amount = 0.0
# CO2(g) -3.0                             # For bubbling out CO2

SURFACE 1
 Hfo_w Ferrihydrite 0.2 5.33e4  # weak sites, 0.2 mol/mol, surface area 5.33e4 m2/mol
 Hfo_s Ferrihydrite 0.5e-2                # strong sites 2.5% of weak sites
 -equil 1
SURFACE_SPECIES
 Hfo_wOH + CO3-2 + H+  = Hfo_wCO3- + H2O; log_k 12.56
 Hfo_wOH + CO3-2 + 2H+ = Hfo_wHCO3 + H2O; log_k 20.62

#REACTION 1
# NaOH 1; 1.e-3                           # Increase pH for increased metal removal
# HCl 1; 2.e-3                            # Reduce alkalinity and bubble out CO2

USER_PRINT
 -start
 10 print '           pH          As_tot           Cd          Ni            '
 20 print -la("H+"), tot("As") * 74.92e6, tot("Cd") * 112.4e6, tot("Ni") * 58.71e6
 -end
END
```

Results are given in Table 7.11 for five cases, 1) oxidation and sorption onto the precipitated Hfo; 2) oxidation and sorption, with CO_2 partially removed; 3) with 1 mmol NaOH/L added to increase pH for possibly increased metal removal; 4) with 2 mmol/L HCl added to decrease carbonate complexation for nickel by bubbling out CO_2, and 5) the effect of removing CO_2 surface complexes in case 1.

Table 7.11. Effects of operational conditions on heavy metal concentrations in groundwater after aeration and precipitation of $Fe(OH)_3$. Concentrations in μg/L.

	pH	As_{tot}	Cd^{2+}	Ni^{2+}
Groundwater	6.50	10	10	30
Oxidized	6.42	1.3	9.6	29.5
Aerated, $P_{CO_2} = 10^{-3}$	8.10	0.13	4.5	28.7
NaOH added	8.25	0.14	3.8	29.1
HCl added	7.11	0.01	8.2	25.9
Oxidized, no CO_2 cplx	6.42	0.003	9.6	29.5

The results of Example 7.12 indicate that aeration, with precipitation of ferrihydrite and sorption on its surface, is removing most As, but not much of Cd^{2+} and Ni^{2+}. During aeration, part of the CO_2 will be degassing, which is modeled by imposing CO_2 equilibrium with a partial pressure of 10^{-3} atm. The pH increases to 8.10 due to the loss of CO_2, and more than half of the Cd^{2+} is now sorbed. Increasing the pH further will enhance sorption of the heavy metal cations, as is shown by adding 1 mmol NaOH/L. Adding more NaOH will increase total CO_2 when the CO_2 pressure is fixed. This leads to carbonate complexes in solution which reduce the free metal ion activity and thereby diminish sorption. Apparently, sorption of Ni^{2+} on Hfo is not very strong. Lastly, a comparison of case 1 and 5, with and without CO_2 complexes, shows that sorption of As is much reduced by competition with CO_2.

In the case of acid mine drainage, streamwaters often have a low pH and contain high amounts of dissolved iron and heavy metals. When base is added to the streamwater, iron oxyhydroxides are expected to precipitate, removing by adsorption and coprecipitation some of the dissolved heavy metals. Figure 7.38 shows the copper content of an acid mine drainage stream.

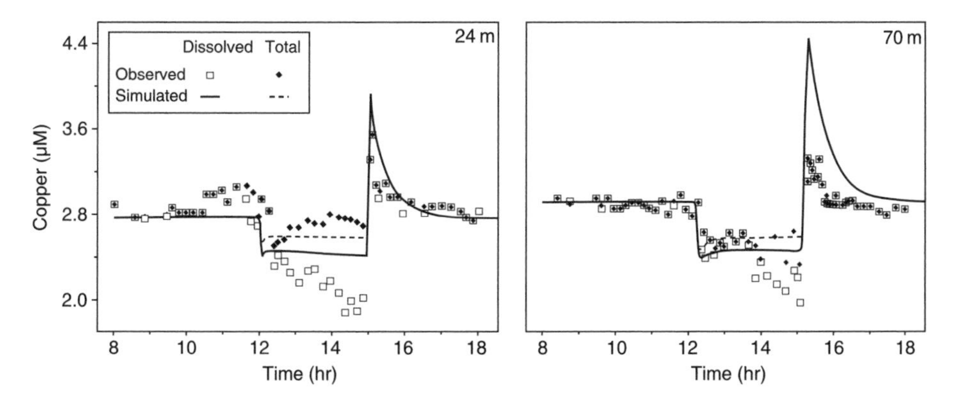

Figure 7.38. The variation in copper concentration in streamwater, during an experiment where a pulse of Na_2CO_3 was added to an acid mine drainage stream at Leadville, Colorado, USA. Shown is the copper concentration over time at 24 and 70 m downstream from the point of Na_2CO_3 injection. Closed symbols give the total Cu concentration and open symbols the Cu concentration in samples passed through 0.1 μm filters. The solid lines are model predictions based on the Dzombak and Morel (1990) surface complexation model for ferrihydrite (Runkel et al., 1999).

The upstream pH value is about 3.4, but in the pulse, where a strong Na_2CO_3 solution was added, the pH increased to above 5.6. The high pH water arrived at 12.00 hr at the sampling point at 24 m and the dissolved copper concentration decreased markedly (open symbols), while the total (unfiltered) concentration did not change much. This was attributed to increased sorption of Cu^{2+} on the surface of iron oxyhydroxide, suspended in water and sedimented on the streambed. After the high pH spike had passed, the Cu^{2+} concentration increased temporarily above the background concentration due to desorption from the iron oxyhydroxide in the streambed. The lines give the model results based on a surface complexation model using the ferrihydrite database of Dzombak and Morel (1990) and generally match the field data well. Recently, Swedlund and Webster (2001) found that in sulfate-rich water, such as acid mine drainage, a ternary sulfate complex, like $\equiv FeOHCuSO_4$, also needs to be taken into consideration.

As mentioned above, sandy aquifer sediments will have other sorption properties than ferrihydrite because the surface coatings consist of mixtures with clay. Wang et al. (1997) measured trace metal sorption on a wide range of natural sediments and their results for copper are shown in Figure 7.39. The results show a wide variation in the sorption properties of the various sediments. The pH where 50% is adsorbed varies from about 4.7 to 6.7. A surface complexation model was fitted to the data with variable complexation constants. These constants were found to be proportional to the first hydrolysis constant of the metal ion in water and also related to the relative amounts of organic carbon and various oxide minerals in the sediment. Possibly, the variation can be explained by assigning each sediment the properties of its individual sorbing components in the correct proportion, as we did for Cd^{2+} in a sand (Figure 7.29).

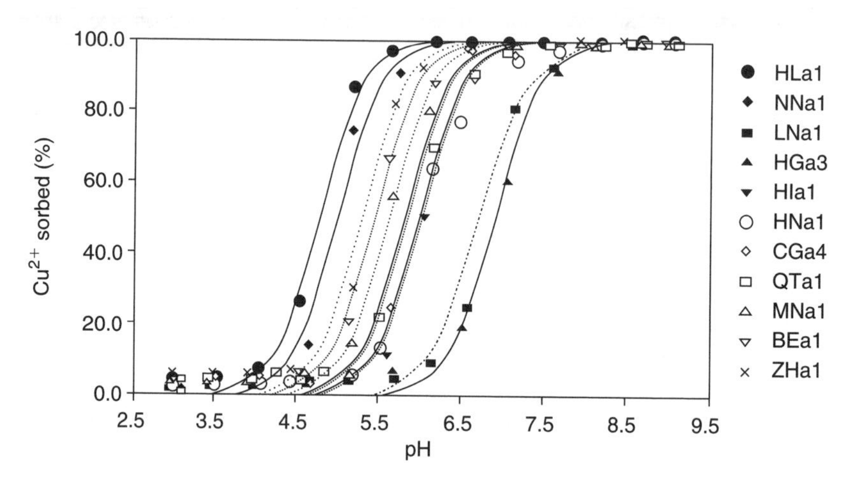

Figure 7.39. pH-dependent sorption of copper on a range of natural sediments (Wang et al., 1997).

The spatial heterogeneity of the adsorbers in aquifer sediments deserves also attention. Mn-oxides are extremely effective scavengers of heavy metals and often observed in distinct layers which may become enriched in trace elements. Figure 7.40 shows the speciation of Ni^{2+} in a sediment core with separate bands of Mn-oxide and Fe-oxide. The two plots to the left in Figure 7.40 show results of sediment extractions using a mild reductant, hydroxylamine (HA). Just below 2 m depth, HA extracts a large amount of Mn from Mn-oxide and Ni^{2+} is clearly associated with the Mn-oxide.

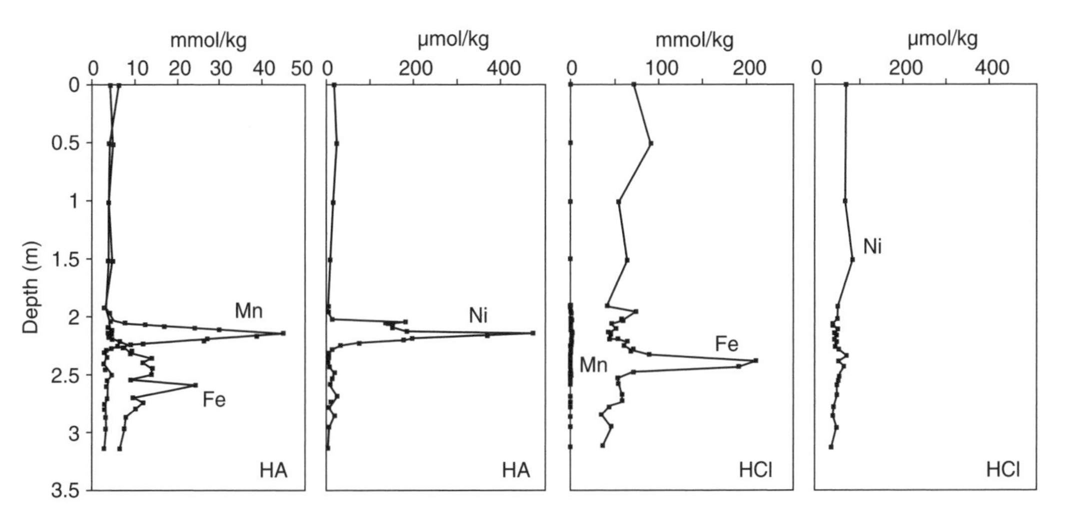

Figure 7.40. Sequential extraction of Mn, Fe and Ni^{2+} from a sandy sediment with a distinct Mn-oxide band over a more diffuse Fe-oxide band. HA: extraction with hydroxylamine, HCl: with 6 N HCl (Larsen and Postma, 1997).

Extraction with HCl releases a lot of Fe from the sediment, corresponding to the Fe-oxide band, but there is little Ni^{2+} associated with this layer. If the Mn-oxides become reduced, for example by reaction with Fe^{2+} (Postma and Appelo, 2000), large amounts of Ni^{2+} may be mobilized (Stollenwerk, 1994; Larsen and Postma, 1997).

PROBLEMS

7.1. Find the specific capacitance (F/kg) of γ-Al_2O_3 at pH = 5 in 0.001 M NaCl from Figure 7.21.

7.2. Find the ratio $\{\equiv AlOH\} / \{\equiv AlOH_2^+\}$ for γ-Al_2O_3 at pH = 5.4 in 0.1 M NaCl, using Equation 7.53. There are 0.03 mol sites/kg. Is this correct in Figure 7.21?

7.3. The plot of pK_{app} *vs* f for a humic acid at 3 different concentrations of $Al(NO)_3$ is presented in Figure 7.41. Estimate pK_{int}, α / ln 10, and $K_{h\text{-}Na}$. Assume that Al^{3+} forms the complex:

$$\equiv hOAl^{2+} \leftrightarrow Al^{3+} + \equiv hO^-; \quad K_{h\text{-}Al}$$

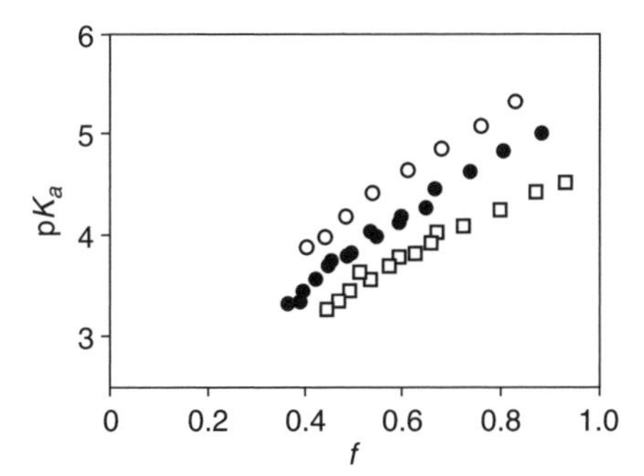

Figure 7.41. Apparent acid dissociation constants of humic acid plotted against degree of dissociation at concentrations of 0.001, 0.01, and 0.1 M of background electrolyte. From Tipping et al. (1988).

7.4. Seaman et al. (1996) measured the retardation of Br^- on a quartz sediment coated with iron oxides as a function of the Br^- concentration, see table below. The sediment contained 3.6% Fe, $\rho_b = 1.8\,g/cm^3$, $\varepsilon = 0.3$.

Use PHREEQC to find the retardation for Br^- in the double layer of Hfo at pH = 5, 8.5 mM weak sites, 600 m^2/g, 3.8 g Hfo, for these KBr concentrations. Explain the trend with concentration. How much of the analyzed iron is active as Hfo?

mM KBr	Retardation
1	2.15
10	1.35
100	1.08

7.5. Model a column experiment of Br^- retardation, Figure 7.42. Experiment D has the same iron oxyhydroxide parameters as Problem 7.4. Column length 10 cm. Initial solution: pH = 5, 4 mM K^+ and Cl^- balances charge in solution. Influent solution: 1 mM KBr, pH = 5, K^+ balances charge. Find the dispersivity from Experiment A.

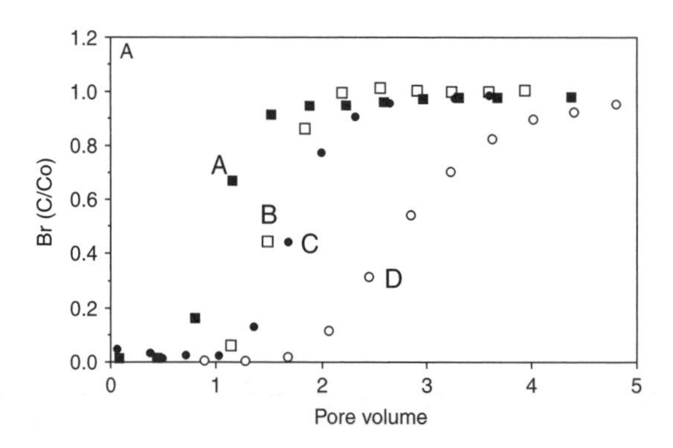

Figure 7.42. Bromide breakthrough in column experiments with quartz sands coated with variable amounts of iron oxide (Seaman et al., 1996).

7.6. Compare sorption of As(3) and As(5) on ferrihydrite at concentrations of 0.267 mM and 0.001 mM, both in 0.1 M NaCl solution with pH = 7. For comparison with Figure 7.16: assume that goethite has 10 times less sorption sites and surface area than ferrihydrite, but with the same chemical properties. Explain the relative increase of As(5) sorption (arsenate anion) at the low concentration.

7.7. Wehrli et al. (1990) have shown that the rate constants for surface complexation are related to the water exchange rate of the metal ion (cf. Figure 7.37). If the exchange rate for Pb^{2+} sorption on ferrihydrite decreases from 2.5×10^5/s to 2.5×10^3/s, how will the half-life of the sorption change?

7.8. The water exchange rate for Zn^{2+} is approximately 500 times smaller than for Pb^{2+} (cf. Figure 7.37). Modify the PHREEQC input file for kinetic sorption of Pb^{2+} (Table 7.10) to calculate the kinetics of Zn^{2+} sorption on ferrihydrite.

7.9. Determine the half-life for sorption of 0.1 mM Pb^{2+} on the strong sites of ferrihydrite with the PHREEQC input file given in Table 7.10.

7.10. Calculate the equilibrium sorbed concentrations of Pb^{2+} on Hfo_w = 18.14×10^{-3}M and Hfo_sOPb = 0.45×10^{-11}M, 7.981 g Hfo. Solution composition: fixed at 2 mM $PbCl_2$ in 10 mM $NaNO_3$ solution, pH = 5.

7.11. Modify the PHREEQC input file from Table 7.10 and compare equilibrium and 45 min. kinetic sorption isotherms for weak and strong sites over the range 0–2.0 mM Pb. *Hint*: Set INCREMENTAL_REACTIONS false, and use keyword REACTION to add $PbCl_2$.

7.12. Modify the PHREEQC input file in Table 7.10 to calculate desorption after 45 min. adsorption. *Hint*: the initial masses m_0 of kinetic reactants must be changed for desorption.

7.13. Check that a decrease of the rate constant by a factor 100 will increase the reaction half-life by 100. *Hint:* change parm(1) in the PHREEQC input file.

7.14. Find the Langmuir parameters for sorption of Cd^{2+} on loamy sand, data in Example 7.1. Define the concentration range where the Langmuir isotherm is linear to within 10%. Also the concentration where $q_I = 0.75\ q_{max}$.

7.15. Calculate the distribution coefficient of 0.2 μM Zn^{2+} for a sediment with CEC = 7 meq/kg, $\rho_b = 2.1$ g/cm^3, $\varepsilon_w = 0.2$, $m_{Ca} = 2$ mmol/L, using PHREEQC. What is the effect of b) doubling the Zn concentration, c) doubling the Ca^{2+} concentration, d) doubling the CEC, e) changing the pH from 7 to 5?

7.16. Calculate the distribution coefficient of Zn^{2+} for a solution with 88 mg ferrihydrite/L at pH = 7, $m_{Mg^{2+}} = 2$ mmol/L, using PHREEQC. What is the effect of b) doubling the Zn concentration, c) doubling the Mg^{2+} concentration, d) doubling the amount of ferrihydrite, e) changing the pH from 7 to 5?

7.17. Discuss the effects of doubling the Mg^{2+} concentration in Problem 7.16 in terms of surface potential changes due to increased sorption of Mg^{2+} and increased ionic strength. Compare with the effect of Ca^{2+}, when Mg^{2+} is replaced by Ca^{2+}.

7.18. Use PHREEQC to study sorption isotherms for Pb^{2+}, Zn^{2+} and Ca^{2+} on ferrihydrite, pH range 3–9, Hfo_w 20×10^{-3} 600 8.8, Hfo_s 0.5×10^{-3}, 0.1 M $NaNO_3$, 2 mM $M(NO_3)_2$. What is the effect of halving the concentration of Hfo; decreasing the concentration of $NaNO_3$ to 0.01 M; decreasing the concentration of the metal ion to 0.2 mM.

7.19. Estimate the percentage Ni^{2+} at trace (10^{-8}M) concentration sorbed on smectite in 0.33 mM Ca^{2+} solution, 1 g smectite/L, CEC = 8.7 meq/g. Compare with Figure 7.18, pH < 5. Note to include NiX_2 (cf. Table 6.4) and HX with log_k 4.5.

b. Include 8.8 g ferrihydrite, vary log_k for sorption of Ni^{2+} on the strong sites?

7.20. In Example 7.2, include the activity correction for CdX_2 (cf. Section 6.3.3) and recalculate K'_d with PHREEQC.

7.21. Estimate the retardation of ^{14}C due to sorption of HCO_3^- on ferrihydrite in the Sherwood sandstone aquifer (cf. Chapter 5). Smedley and Edmunds (2002) found 400 ppm Fe in 0.2 M oxalate extractions, equivalent to 42 mM ferrihydrite/L pore water. *Hint*: Use water in equilibrium with calcite and dolomite at $[P_{CO_2}] = 10^{-2.5}$, and equilibrate with the surface.

7.22. Calculate the counterions which balance the charge of 1 g HA (humic acid)/L at pH 7, 1 mM NaCl, 1 mM $CaCl_2$ according to the Donnan model (*Hint*: use Example 6.4) and according to the double layer composition (*Hint*: cf. Example 7.11). Calculate $K_{Ca\backslash Na}$ for the double layer composition.

7.23. Lets make a pill with $FeSO_4$ and $NaClO_4$ as purificants for water with a harmful content of As and F^-. $NaClO_4$ oxidizes bacteria and also Fe^{2+}, which precipitates as ferrihydrite and sorbs As and F^-. This reaction is acid, and if we add calcite, the Ca^{2+} concentration will increase and CaF_2 may precipitate. Determine the composition of the pill to bring the As and F^- concentrations down to below the drinking water limits in a water with:
pH 7.67 and Na^+ 2.93, Ca^{2+} 0.59, Alk 3.85, F^- 0.26 and As 1.3×10^{-3} mmol/L (4.9 mg F^-/L and 97 μg As/L).

REFERENCES

Aiken, G.R. and Malcolm, R.L., 1987. Molecular weight of aquatic fulvic acids by vapor pressure osmometry. *Geochim. Cosmochim. Acta* 51, 2177–2184.

Allison, J.D., Brown, D.S. and Novo-Gradac, K.J., 1991. *MINTEQA2, a geochemical assessment data base and test cases for environmental systems.* Report EPA/600/3-91/-21. US EPA, Athens, GA. http://www.epa.gov/ceampubl/mmedia/minteq/index.htm

Appelo, C.A.J. and Postma, D., 1999. A consistent model for surface complexation on birnessite (d-MnO_2) and its application to a column experiment. *Geochim. Cosmochim. Acta* 63, 3039–3048.

Appelo, C.A.J., Van der Weiden, M.J.J., Tournassat, C. and Charlet, L., 2002. Surface complexation of ferrous iron and carbonate on ferrihydrite and the mobilization of arsenic: *Env. Sci. Technol.* 36, 3096–3103.

Appleyard, E.C. and Blowes, D.W., 1994. Applications of mass-balance calculations to weathered sulfide mine tailings. In C.N. Alpers and D.W. Blowes (eds), *Environmental geochemistry of sulfide oxidation*. ACS Symp. Ser. 550, 516–534.

Artinger, R., Buckau, G., Geyer, S., Fritz, P., Wolf, M. and Kim, J.I., 1999. Characterization of groundwater humic substances: influences of sedimentary organic carbon. *Appl. Geochem.* 15, 97–116.

Atkins, P.W. and de Paula, J, 2002. *Atkins' Physical Chemistry*, 7th ed. Oxford Univ. Press, 1149 pp.

Avena, M.J., Vermeer, A.W.P. and Koopal, L.K., 1999. Volume and structure of humic acids studied by viscometry pH and electrolyte concentration effects. *Coll. Surf. A* 151, 213–224.

Baeyens, B. and Bradbury, M.H., 1997. A mechanistic description of Ni- and Zn sorption on Na-montmorillonite. *J. Contam. Hydrol.* 27, 199–248.

Barrow, N.J., Gerth, J. and Brümmer, G.W., 1989. Reaction—kinetics of the adsorption and desorption of nickel, zinc and cadmium by goethite. 2. Modeling the extent and rate of reaction. *J. Soil Sci.* 40, 437–450.

Bartschat, B.M., Cabaniss, S.E. and Morel, F.M.M., 1992. Oligoelectrolyte model for cation binding by humic substances. *Env. Sci. Technol.* 26, 264–294.

Benedetti, M.F., Milne, C.J., Kinniburgh, D.G., Van Riemsdijk, W.H. and Koopal, L.K., 1995. Metal ion binding to humic substances: application of the non-ideal competitve adsorption model. *Env. Sci. Technol.* 29, 446–457.

Berg, M., Tran, H.C., Nguyen, T.C., Pham, H.V., Schertenleib, R. and Giger, W., 2001. Arsenic contamination of ground and drinking water in Vietnam: a human health threat. *Env. Sci. Technol.* 35, 2621–2626.

Boekhold, A.E., Van der Zee, S.E.A.T.M. and De Haan, F.A.M., 1991. Spatial patterns of cadmium contents related to soil heterogeneity. *Water Air Soil Poll.* 57, 479–488.

Bradbury, M.H. and Baeyens, B., 1999. Modelling the sorption of Zn and Ni on Ca-montmorillonite. *Geochim. Cosmochim. Acta* 63, 325–336.

Brümmer, G.W., Gerth, J., Tiller, K.G., 1988. Reaction kinetics of the adsorption and desorption of nickel, zinc and cadmium by goethite. I. Adsorption and diffusion of metals. *J. Soil Sci.* 39, 37–51.

Buffle, J. 1990. Complexation reactions in aquatic systems: an analytical approach. Ellis Horwood, Chichester, 692 pp.

Bunzl, K. and Schimmack, W., 1991. Kinetics of ion sorption on humic substances. In D.L. Sparks and D.L. Suarez (eds), *Rates of soil chemical processes*. Soil Sci. Soc. Am. Spec. Pub. 27, 119–134.

Charlet, L., Wersin, P. and Stumm, W., 1990. Surface charge of $MnCO_3$ and $FeCO_3$. *Geochim. Cosmochim. Acta* 54, 2329–2336.

Christensen, T.H., 1984. Cadmium soil sorption at low concentrations: 1. Effect of time, cadmium load, pH, and calcium. *Water Air Soil Poll.* 21, 105–114.

Christensen, J.B. and Christensen, T.H., 1999. Complexation of Cd, Ni, and Zn by DOC in polluted groundwater: a comparison of approaches using resin exchange, aquifer material sorption, and computer speciation models (WHAM and MINTEQA2). *Env. Sci. Technol.* 33, 3857–3863.

Clapp, C.E. and Hayes, M.H.B., 1999. Sizes and shapes of humic substances. *Soil Sci.* 164, 777–789.

Cornell, R.M. and Schwertmann, U., 2003. *The Iron Oxides*, 2nd ed. Wiley-VCH, Weinheim, 664 pp.

Cowan, C.E., Zachara, J.M., Smith, S.C. and Resch, C.T., 1992. Individual sorbent contributions to cadmium sorption on ultisols of mixed mineralogy. *Soil Sci. Soc. Am. J.* 56, 1084–1094.

Davis, J.A. and Kent, D.B., 1990. Surface complexation modeling in aqueous geochemistry. *Rev. Mineral.* 23, 177–260.

Davis, J.A., James, R.O. and Leckie, J.O., 1978. Surface ionization and complexation at the oxide/water interface. 1. Computation of electrical double layer properties in simple electrolytes. *J. Coll. Interf. Sci.* 63, 480–499.

Davis, J.A., Fuller, C.C. and Cook, A.D., 1987. A model for trace metal sorption processes at the calcite surface: Adsorption of Cd^{2+} and subsequent solid solution formation. *Geochim. Cosmochim. Acta* 51, 1477–1490.

De Wit, J.C.M., Van Riemsdijk, W.H. and Koopal, L.K., 1993a. Proton binding to humic substances. 1. Electrostatic effects. *Env. Sci. Technol.* 27, 2005–2014.

De Wit, J.C.M., Van Riemsdijk, W.H. and Koopal, L.K., 1993b. Proton binding to humic substances. 2. Chemical heterogeneity and adsorption models. *Env. Sci. Technol.* 27, 2015–2022.

Dzombak, D.A. and Morel, F.M.M., 1990. *Surface complexation modeling: hydrous ferric oxide*. Wiley and Sons, New York, 393 pp.

Edmunds, W.M., Kinniburgh, D.G. and Moss, P.D., 1992. Trace metals in interstitial waters from sandstones: acidic inputs to shallow groundwaters. *Env. Poll.* 77, 129–141.

Evans, T.D., Leal, J.R. and Arnold, P.W., 1979. The interfacial electrochemistry of goethite (a-FeOOH), especially the effect of CO_2 contamination. *J. Electroanal. Chem.* 105, 161–167.

Fuerstenau, D.W., Manmohan, D. and Raghavan, S., 1981, The adsorption of alkaline-earth metal ions at the rutile/aqueous interface. In P.H. Tewari (ed.), *Adsorption from aqueous solutions*, 93–117, Plenum Press, New York.

Fuller, C.C., Davis, J.A. and Waychunas, G.A., 1993. Surface chemistry of ferrihydrite: Pt. 2. Kinetics of arsenate adsorption and coprecipitation. *Geochim. Cosmochim. Acta* 57, 2271–2282.

Garcia-Miragaya, J. and Page, A.L., 1976. Influence of ionic strength and inorganic complex formation on the sorption of trace amounts of Cd by montmorillonite. *Soil Sci. Soc. Am. J.* 40, 658–663.

Gerth, J., Brümmer, G.W. and Tiller, K.G., 1993. Retention of Ni, Zn and Cd by Si-associated goethite. *Z. Pflanzenern. Bodenk.* 156, 123–129.

Grossl, P.R., Sparks, D.L. and Ainsworth, C.C., 1994. Rapid kinetics of Cu(II) adsorption/desorption on goethite. *Env. Sci. Technol.* 28, 1322–1429.

Hachiya, K., Sasaki, M., Saruta, Y., Mikami, N. and Yasunaga, T., 1984. Static and kinetic studies of adsorption-desorption of metal ions on the ?-Al_2O_3 surface. *J. Phys. Chem.* 88, 23–27, 27–31.

Hawthorne, F.C. (ed.) 1988. Spectroscopic methods in mineralogy and geology. *Rev. Mineral.* 18, 798 pp.

Hayes, K.F. and Leckie, J.O., 1986. Mechanism of lead ion adsorption at the goethite-water interface. ACS Symp. Ser. 323, 114–141.

Hiemstra, T. and Van Riemsdijk, W.H., 1991. Physical-chemical interpretation of primary charging behavior of metal (hydr)oxides. *Coll. Surf.* 59, 7–25.

Hiemstra, T. and Van Riemsdijk, W.H., 1996. A surface structural approach to ion adsorption: the charge distribution (CD) model. *J. Coll. Interf. Sci.* 179, 488–508.

Hiemstra, T., Van Riemsdijk, W.H. and Bolt, G.H., 1989. Multisite proton adsorption modeling at the solid/solution interface of (hydr)oxides: a new approach, 1. Model description and evaluation. *J. Coll. Interf. Sci.* 133, 91–104.

Higgo, J.J.W., Kinniburgh, D., Smith, B. and Tipping, E., 1993. Complexation of Co^{2+}, Ni^{2+}, $UO_2{}^{2+}$ and Ca^{2+} by humic substances in groundwaters. *Radiochim. Acta* 61, 91–103.

Johnston, H.M., Gillham, R.W. and Cherry, J.A., 1985. Distribution coefficients for strontium and cesium at a storage area for low-level radioactive waste. *Can. Geotechn. J.* 22, 6–16.

Kinniburgh, D.G., 1986. General purpose adsorption isotherms. *Env. Sci. Technol.* 20, 195–204.

Kinniburgh, D.G. and Smedley, P.L. (eds), 2001. *Arsenic contamination of groundwater in Bangladesh*. BGS Techn. Rep. WC/00/19, British Geological Survey.

Kinniburgh, D.G., Gale, I.N., Smedley, P.L., Darling W.G., West, J.M., Kimblin, R.T., Parker, A., Rae, J.E., Aldous, P.J. and O'Shea, M.J., 1994. The effect of historic abstraction of groundwater from the London basin aquifers on groundwater quality. *Appl. Geochem.* 9, 175–195.

Kinniburgh, D.G., Milne, C.J. and Venema, P., 1995. Design and construction of a personal-computer-based automatic titrator. *Soil Sci. Soc. Am. J.* 59, 417–422.

Kinniburgh, D.G., Van Riemsdijk, W.H., Koopal, L.K. and Benedetti, M.F., 1998. Ion binding to humic substances. In E.A. Jenne (ed.) *Adsorption of metals by geomedia*, 483–520, Academic Press, San Diego.

Kinniburgh, D.G., Van Riemsdijk, W.H., Koopal, L.K., Borkovec, M., Benedetti, M.F. and Avena, M.J., 1999. Ion binding to natural organic matter: competition, heterogeneity, stoichiometry and thermodynamic consistency. *Coll. Surf. A* 151, 147–166.

Kjøller, C., Postma, D. and Larsen, F., 2004. Groundwater acidification and the mobilization of trace metals in a sandy aquifer. *Env. Sci. Technol.* 38, 2829–2835.

Koopal, L.K., Van Riemsdijk, W.H., De Wit, J.C.M. and Benedetti, M.F., 1994. Analytical isotherm equations for multicomponent adsorption to heterogeneous surfaces. *J. Coll. Interf. Sci.* 166, 51–60.

Kraepiel, A.M.L., Keller, K. and Morel, F.M.M., 1999. A model for metal adsorption on montmorillonite. *J. Coll. Interf. Sci.* 210, 43–54.

Kraus, K.A. and Phillips, H.O., 1956. *J. Am. Chem. Soc.* 78, 249.

Larsen, F. and Postma, D., 1997. Nickel mobilization in a groundwater well field: release by pyrite oxidation and desorption from manganese oxides. *Env. Sci. Technol.* 31, 2589–2595.

Lenhart, J.J. and Honeyman, B.D., 1999. Uranium(VI) sorption to hematite in the presence of humic acid. *Geochim. Cosmochim. Acta* 63, 2891–2901.

Lion, L.W., Altmann, R.S. and Leckie, J.O., 1982. Trace metal adsorption characteristics of estuarine particulate matter: evaluation of Fe/Mn oxide and organic surface coatings. *Env. Sci. Technol.* 16, 660–666.

Lofts, S. and Tipping, E., 2000. Solid-solution metal partitioning in the Humber rivers: application of WHAM and SCAMP. *Sci. Total Env.* 251, 381–399.

Manning, B.A., Fendorf, S.E. and Goldberg, S., 1998. Surface structures and stability of arsenic (III) on goethite: spectroscopic evidence for inner-sphere complexes. *Env. Sci. Technol.* 32, 2383–2388.

Manning, B.A. and Goldberg, S., 1996. Modeling competitive adsorption of arsenite and molybdate on oxide minerals. *Soil Sci. Soc. Am. J.* 60, 121–131.

Marinsky, J.A., 1987, A two-phase model for the interpretation of proton and metal ion interaction with charged polyelectrolyte gels and their linear analogs. In W. Stumm (ed.), *Aquatic surface chemistry*, 49–81, Wiley and Sons, New York.

McKenzie, R., 1980. The adsorption of lead and other heavy metals on oxides of Mn and Fe. *Aust. J. Soil Res.* 18, 61–73.

Milne, C.J., Kinniburgh, D.G. and Tipping, E., 2001. Generic NICA-Donnan model parameters for proton binding by humic substances. *Env. Sci. Technol.* 35, 2049–2059.

Milne, C.J., Kinniburgh, D.G., Van Riemsdijk, W.H. and Tipping, E., 2003. Generic NICA-Donnan model parameters for metal-ion binding by humic substances. *Env. Sci. Technol.* 37, 958–971.

Morel, F.M.M. and Hering, J.G., 1993. *Principles and applications of aquatic chemistry*. Wiley and Sons, New York, 374 pp.

Mukherjee, M., Sahu, S.J., Jana, J., Bhattacharya, R., Chatterjee, D., De Dadal, S.S., Bhattacharya, P. and Jacks, G., 2000. The governing geochemical processes responsible for mobilisation of arsenic in sedimentary aquifer of Bengal Delta Plain. In P.L. Bjerg, P. Engesgaard and Th.D. Krom (eds), *Groundwater 2000*, 201–202, Balkema, Rotterdam.

Murray, J.W., 1975. The interaction of metal ions at the manganese dioxide-solution interface. *Geochim. Cosmochim. Acta* 39, 505–519.

Parks, G.A., 1967. Surface chemistry of oxides in aqueous systems. In W. Stumm (ed.), *Equilibrium concepts in aqueous systems*. Adv. Chem. Ser. 67, 121–160.

Parks, G.A., 1990. Surface energy and adsorption at mineral-water interfaces: an introduction. *Rev. Mineral.* 23, 133–175.

Perdue, E.M. and Lytle, C.R., 1983. Distribution model for binding of protons and metal ions by humic substances. *Env. Sci. Technol.* 17, 654–660.

Perdue, E.M., Reuter, J.H. and Parrish, R.S., 1984. A statistical model of proton binding by humus. *Geochim. Cosmochim. Acta* 48, 1257–1263.

Pokrovsky, O.S., Schott, J. and Thomas, F., 1999. Processes at the magnesium-bearing carbonates/solution interface. 1. A surface speciation model for magnesite. *Geochim. Cosmochim. Acta* 63, 863–880.

Postma, D. and Appelo, C.A.J., 2000. Reduction of Mn-oxides by ferrous iron in a flow system: Column experiment and reactive transport modeling. *Geochim. Cosmochim. Acta* 64, 1237–1247.

Randall, S.R., Sherman, D.M., Ragnarsdottir, K.V. and Collins, C.R., 1999. The mechanism of cadmium surface complexation on iron oxyhydroxide minerals. *Geochim. Cosmochim. Acta* 63, 2971–2987.

Ravenscroft, P., McArthur, J.M. and Hoque, B.A., 2001. Geochemical and palaeohydrological controls on pollution of groundwater by arsenic. In W.R. Chappell et al. (eds), *Arsenic exposure and health effects IV*, Elsevier, Amsterdam.

Rietra, R.P.J.J., Hiemstra, T. and Van Riemsdijk, W.H., 1999. The relationship between molecular structure and ion adsorption on variable charge minerals. *Geochim. Cosmochim. Acta* 63, 3009–3015.

Routh, J., Grossman, E.L., Murphy, E.M. and Benner, R., 2001. Characterization and origin of dissolved organic carbon in Yegua ground water in Brazos County, Texas. *Ground Water* 39, 760–767.

Runkel, R.L., Kimball, B.A., McKnight, D.M. and Bencala, K.E., 1999. Reactive solute transport in streams: a surface complexation approach for trace metal sorption. *Water Resour. Res.* 35, 3829–3840.

Rusch, U., Borkovec, M., Daicic, J. and Van Riemsdijk, W.H., 1997. Interpretation of competitive adsorption isotherms in terms of affinity distributions. *J. Coll. Interf. Sci.* 191, 247–255.

Russell, J.D., Parfitt, R.L., Fraser, A.R. and Farmer, V.C., 1974. Surface—structures of gibbsite, goethite and phosphated goethite. *Nature* 248, 220–221.

Schindler, P.W. and Stumm, W., 1987. The surface chemistry of oxides, hydroxides and oxide minerals. In W. Stumm (ed.), *Aquatic surface chemistry*, 83–110, Wiley and Sons, New York.

Schnitzer, M. and Kahn, S.U., 1978. *Soil organic matter*. Elsevier, Amsterdam, 319 pp.

Schultz, M.F., Benjamin, M.M. and Ferguson, J.F., 1987. Adsorption and desorption of metals on ferrihydrite: Reversibility of reaction and sorption properties of the regenerated solid. *Env. Sci. Technol.* 21, 663–669.

Seaman, J.C., Bertsch, P.M., Korom, S.F. and Miller, W.P., 1996. Physicochemical controls on nonconservative anion migration in coarse-textured alluvial sediments. *Ground Water* 34, 778–783.

Shiao, S.Y., Rafferty, P., Meyer, R.E. and Rogers, W.J., 1979. Ion exchange equilibriums between montmorillonite and solutions of moderate to high ionic strength. ACS Symp. Ser. 100, 297–324.

Sips, R., 1948. On the structure of a catalyst surface. *J. Chem. Phys.* 16, 490–495.

Smedley, P.L. and Edmunds, W.M., 2002. Redox patterns and trace-element behavior in the East Midlands Triassic sandstone aquifer, U.K. *Ground Water* 40, 44–58.

Smedley, P.L. and Kinniburgh, D.G., 2002. A review of the source, behaviour and distribution of arsenic in natural waters. *Appl. Geochem.* 17, 517–568.

Smedley, P.L., Nicolli, H.B., Macdonald, D.M.J., Barros, A.J. and Tullio, J.O., 2002. Hydrogeochemistry of arsenic in groundwater from La Pampa, Argentina. *Appl. Geochem.* 17, 259–284.

Sparks, D.L., 1989. *Kinetics of soil chemical processes*. Academic Press, San Diego, 210 pp.

Sposito, G., 1983. On the surface complexation model of the oxide-aqueous solution interface. *J. Coll. Interf. Sci.* 91, 329–340.

Sposito, G., 1984. *The surface chemistry of soils*. Oxford Univ. Press, New York, 234 pp.

Stipp, S.L.S., Hochella, M.F., Parks, G.A. and Leckie, J.O., 1992. Cd^{2+} uptake by calcite, solid-state diffusion, and the formation of solid-solution: Interface processes observed with near-surface sensitive techniques (XPS, LEED and AES). *Geochim. Cosmochim. Acta* 56, 1941–1954.

Stollenwerk, K.G., 1994. Geochemical interactions between constituents in acidic groundwater and alluvium in an aquifer near Globe, Arizona. *Appl. Geochem.* 9, 353–369.

Stumm, W., 1992. *Chemistry of the solid-water interface*. Wiley and Sons, 428 pp.

Stumm, W. and Morgan, J.J., 1996. *Aquatic chemistry*, 3rd ed. Wiley and Sons, New York, 1022 pp.

Swedlund P.J. and Webster J.G., 2001. Cu and Zn ternary surface complex formation with SO_4 on ferrihydrite and schwertmannite. *Appl. Geochem.* 16, 503–512.

Thurman, E.M., 1985. *Organic geochemistry of natural waters*. Nijhoff. Dordrecht, 497 pp.

Tipping, E., 1998. Humic ion-binding model VI: An improved description of the interactions of protons and metal ions with humic substances. *Aq. Geochem.* 4, 3–48.

Tipping, E. and Hurley, M.A., 1992. A unifying model of cation binding by humic substances. *Geochim. Cosmochim. Acta* 56, 3627–3641.

Tipping, E., Backes, C.A. and Hurley, M.A., 1988. The complexation of protons, aluminium and calcium by aquatic humic substances: a model incorporating binding-site heterogeneity and macroionic effects. *Water Res.* 22, 597–611.

Van Beek, C.G.E.M. and Van der Jagt, H., 1996. Mobilization and speciation of trace elements in groundwater. Int. Workshop "Natural origin of inorganic micropollutants", Vienna.

Van Cappellen, P., Charlet, L., Stumm, W. and Wersin, P., 1993. A surface complexation model of carbonate mineral-aqueous solution interface. *Geochim. Cosmochim. Acta* 57, 3505–3518.

Van Riemsdijk, W.H., Bolt, G.H., Koopal, L.K. and Blaakmeer, J., 1986. Electrolyte adsorption on heterogeneous surfaces: adsorption models. *J. Coll. Interf. Sci.* 109, 219–228.

Venema, P., Hiemstra, T. and Van Riemsdijk, W.H., 1996a. Comparison of different site binding models for cation sorption: description of pH dependency, salt concentration and cation-proton exchange. *J. Coll. Interf. Sci.* 181, 45–59.

Venema, P., Hiemstra, T. and Van Riemsdijk, W.H., 1996b. Multi site adsorption of cadmium on goethite. *J. Coll. Interf. Sci.* 183, 515–527.

Wang, F.Y., Chen, J.S. and Forsling, W., 1997. Modeling sorption of trace metals on natural sediments by surface complexation model. *Env. Sci. Technol.* 31, 448–453.

Wehrli, B., Ibric, S. and Stumm, W., 1990. Adsorption kinetics of vanadyl(IV) and chromium(III) to aluminum oxide: evidence for a two-step mechanism. *Coll. Surf.* 51, 77–88.

Westall, J.C. and Hohl, H. 1980. A comparison of electrostatic models for the oxide/solution interface. *Adv. Coll. Interf. Sci.* 12, 265–294.

Westall, J.C., Jones, J.D., Turner, G.D. and Zachara, J.M., 1995. Models for association of metal ions with heterogeneous environmental sorbents. 1. Complexation of Co(II) by Leonardite humic acid as a function of pH and $NaClO_4$ concentration. *Env. Sci. Technol.* 29, 951–959.

Zachara, J.M., Girvin, D.C., Schmidt, R.L. and Resch, C.T., 1987. Chromate adsorption on amorphous iron oxy-hydroxide in the presence of major groundwater ions. *Env. Sci. Technol.* 21, 589–594.

Zachara, J.M., Cowan, C.E. and Resch, C.T., 1991. Sorption of divalent metals on calcite. *Geochim. Cosmochim. Acta* 55, 1549–1562.

Zachara, J.M., Smith, S.C., Resch, C.T. and Cowan, C.E., 1992. Cadmium sorption to soil separates containing layer silicates and iron and aluminum oxides. *Soil Sci. Soc. Am. J.* 56, 1074–1084.

Zeltner, W.A. and Anderson, M.A., 1988. Surface-charge development at the goethite aqueous-solution interface-effects of CO_2 adsorption. *Langmuir* 4, 469–474.

Zhang, P.C. and Sparks, D.L., 1990. Kinetics and mechanisms of sulfate adsorption/desorption on goethite using pressure-jump relaxation. *Soil Sci. Soc. Am. J.* 54, 1266–1273.

8

Silicate Weathering

Weathering of silicate minerals is a slow process and the resulting changes in water chemistry will be gradual and less conspicuous than in carbonate aquifers (Chapter 5). Still, weathering of silicate minerals is estimated to contribute about 45% of the total dissolved load of the world's rivers (Stumm and Wollast, 1990). In the global cycle of CO_2, the weathering of silicate minerals acts as an important CO_2 sink. Furthermore, silicate weathering is the most important pH-buffering mechanism in sediments without carbonate minerals. Because the rate of silicate dissolution is slow, aquifers in silicate rock are vulnerable towards acidification. Also, forest deterioration due to acid rain has focused attention on silicate weathering as pH-buffer.

8.1 WEATHERING PROCESSES

Traditionally, weathering of detrital silicate minerals has been studied in soils. Soils may be exposed to chemical weathering over thousands of years and the variation in mineralogical composition which develops as a function of depth and time displays even very slow degradation and transformation processes of minerals.

Figure 8.1 shows the mineralogical variation in a soil formed on a granodiorite. Cosmogenic ^{10}Be in quartz revealed that weathering has been going on here for at least 90,000 years (Schroeder et al., 2001).

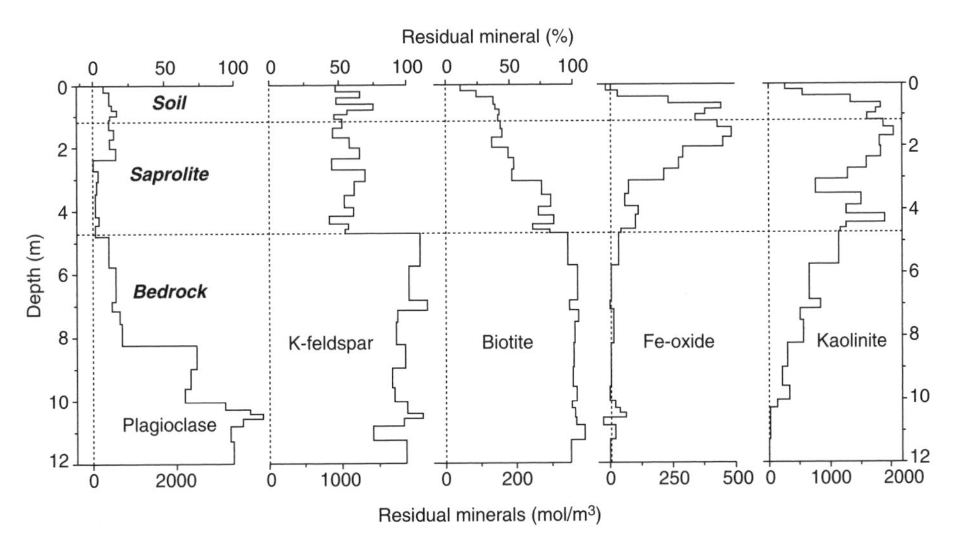

Figure 8.1. Mineralogy of a soil developed on granodiorite (White et al., 2001). Weathered rock residue is also called saprolite.

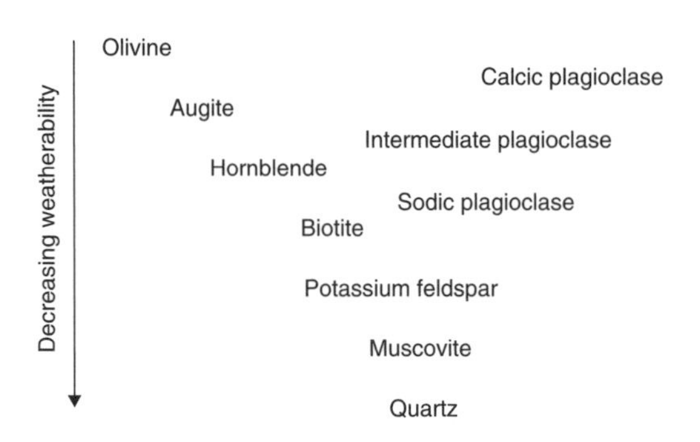

Figure 8.2. The Goldich weathering sequence based on observations of the sequence of their disappearance in soils (Goldich, 1938).

The parent granodiorite rock consists of plagioclase (32%), quartz (28%), K-feldspar (21%), biotite (13%), muscovite (7%) and small amounts of hornblende (<2%). In Figure 8.1, the pristine bedrock composition is found at the greatest depth. Already in the uppermost part of the bedrock the plagioclase disappears. Biotite and K-feldspar are apparently more resistant to weathering but decrease in the saprolite (partially weathered rock). The sequential disappearance of different silicate minerals reflects the overall kinetic control on the distribution of primary silicate minerals during weathering. This was recognized as early as 1938 by Goldich who used field observations to compile the weathering sequence shown in Figure 8.2. Olivine and Ca-plagioclase are listed as the most easily weatherable minerals, and quartz as the mineral most resistant to weathering. The Goldich sequence is the reverse of the Bowen reaction series which predicts the sequence of minerals precipitating during cooling from basaltic magma. The minerals that formed at the highest temperature in the earth (olivine) are the most unstable under weathering conditions at the surface of the earth.

The second important observation to be made from Figure 8.1 is the formation of secondary minerals like clays, here kaolinite, and Fe-oxides during the weathering process. These are the insoluble remnants which form during *incongruent dissolution* of primary silicate minerals. Incongruent dissolution strictly means that the ratio of elements appearing in the solution differs from the ratio in the dissolving mineral. In silicate weathering studies the term incongruent dissolution is commonly extended to include the effect of secondary precipitates. Weathering reactions for some common primary minerals are listed in Table 8.1, where the clay mineral kaolinite is used as an example of a weathering product.

Table 8.1. Weathering reactions for different silicate minerals to the clay mineral kaolinite.

Sodic Plagioclase (Albite)		*Kaolinite*
$2Na(AlSi_3)O_8 + 2H^+ + 9H_2O$	$\rightarrow$	$Al_2Si_2O_5(OH)_4 + 2Na^+ + 4H_4SiO_4$
Calcic Plagioclase (Anorthite)		
$Ca(Al_2Si_2)O_8 + 2H^+ + H_2O$	$\rightarrow$	$Al_2Si_2O_5(OH)_4 + Ca^{2+}$
K-feldspar (Microcline)		
$2K(AlSi_3)O_8 + 2H^+ + 9H_2O$	$\rightarrow$	$Al_2Si_2O_5(OH)_4 + 2K^+ + 4H_4SiO_4$
Pyroxene (Augite)		
$(Mg_{0.7}CaAl_{0.3})(Al_{0.3}Si_{1.7})O_6 + 3.4H^+ + 1.1H_2O$	$\rightarrow$	$0.3Al_2Si_2O_5(OH)_4 + Ca^{2+} + 0.7Mg^{2+} + 1.1H_4SiO_4$
Mica (Biotite)		
$2K(Mg_2Fe)(AlSi_3)O_{10}(OH)_2 + 10H^+ + 0.5O_2 + 7H_2O$	$\rightarrow$	$Al_2Si_2O_5(OH)_4 + 2K^+ + 4Mg^{2+} + 2Fe(OH)_3 + 4H_4SiO_4$
$CO_2 + H_2O$	$\rightarrow$	$H^+ + HCO_3^-$

The formation of secondary products is due to the insolubility of Al-compounds. The reactions in Table 8.1 are therefore written so that aluminum remains conserved in the solid phase. Generally, these reactions can be derived by following steps 1 to 4.

(1) balance all the elements except H and O,
(2) balance O by adding H_2O,
(3) balance H by adding H^+ and
(4) check the charge balance of the resulting equation (for redox reactions consult Chapter 9).

The effect of silicate weathering on the water chemistry is primarily the addition of cations and silica. Nearly all silicate weathering reactions consume acid and increase the pH. Under unpolluted conditions, carbonic acid and organic acids are the most important sources of protons. As indicated by the last equation in Table 8.1, bicarbonate will be produced during weathering of silicates. Finally, iron that is present in silicate minerals, like biotite or hornblende, may form Fe-oxide as an insoluble weathering product (Figure 8.1).

Weathering reactions will also take place in silicate rocks or sandstones percolated by groundwater, although the mineralogical transformations here often are more difficult to detect than in soils. Some examples of groundwater compositions which may result from weathering of silicate minerals are shown in Figure 8.3. The high silica content indicates active degradation of silicate minerals. The highest concentration is found in volcanic rocks (rhyolite and basalt) which contain more reactive material than rock-types like mica-schists or granite. Sodium contributes in all the 4 groundwaters significantly to the cations (K^+ which has been added to Na^+ in Figure 8.3 is in most cases a minor constituent) and its presence is not balanced by chloride as expected when seawater was the source of Na^+. Sodium is mainly derived from weathering of Na-feldspar like albite or any member of the plagioclase solid solution series between albite and anorthite (Ca-feldspar). Ca^{2+} is released both from weathering of plagioclase and of hornblendes (also called amphiboles) and pyroxenes. As suggested by Table 8.1, the increase in cation concentration is accompanied by an increase in bicarbonate and it appears that in some cases even carbonate precipitation can be the result of silicate weathering.

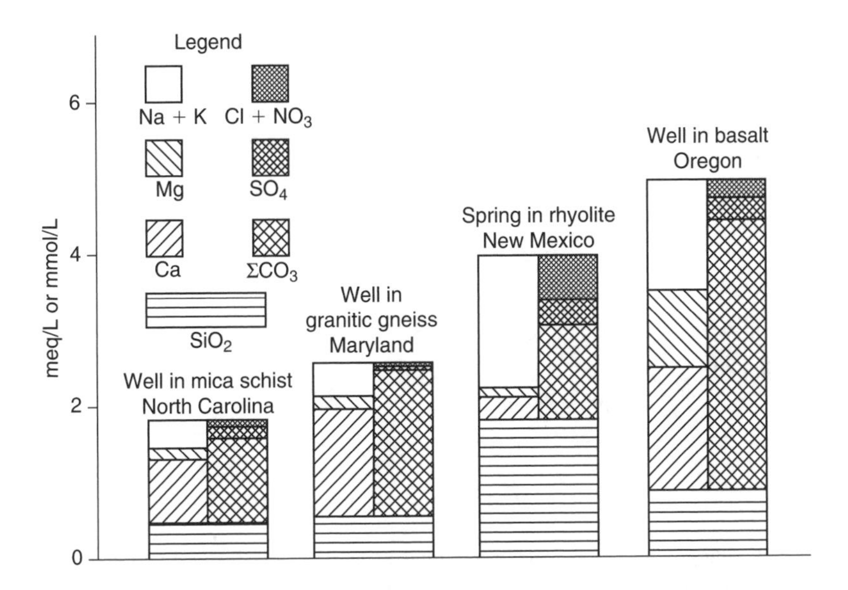

Figure 8.3. Examples of groundwater compositions in igneous and metamorphic rock. Dissolved silica is expressed as mmol/L while charged ions are displayed in meq/L (modified from Hem (1985)).

Weathering of primary silicates results in the formation of clay minerals like montmorillonite and kaolinite, and of gibbsite. Using albite as an example, its transformation to different weathering products is described by the following equations. For montmorillonite the presence of Mg^{2+}, for example leached from pyroxenes, amphiboles or biotite, is assumed:

$$\underset{\textit{albite}}{3\text{Na(AlSi}_3\text{)O}_8} + \text{Mg}^{2+} + 4\text{H}_2\text{O} \rightarrow \underset{\textit{montmorillonite}}{2\text{Na}_{0.5}(\text{Al}_{1.5}\text{Mg}_{0.5})\text{Si}_4\text{O}_{10}(\text{OH})_2} + 2\text{Na}^+ + \text{H}_4\text{SiO}_4 \quad (8.1)$$

$$\underset{\textit{albite}}{2\text{Na(AlSi}_3\text{)O}_8} + 2\text{H}^+ + 9\text{H}_2\text{O} \rightarrow \underset{\textit{kaolinite}}{\text{Al}_2\text{Si}_2\text{O}_5(\text{OH})_4} + 2\text{Na}^+ + 4\text{H}_4\text{SiO}_4 \quad (8.2)$$

$$\underset{\textit{albite}}{\text{Na(AlSi}_3\text{)O}_8} + \text{H}^+ + 7\text{H}_2\text{O} \rightarrow \underset{\textit{gibbsite}}{\text{Al(OH)}_3} + \text{Na}^+ + 3\text{H}_4\text{SiO}_4 \quad (8.3)$$

The alteration of albite to montmorillonite consumes no acid, but with kaolinite and gibbsite as weathering products, increasing amounts of protons are consumed. Furthermore, when albite alters to montmorillonite, 89% of the Si is preserved in the weathering product, decreasing to 33% for weathering to kaolinite and to 0% for gibbsite. The sequence of weathering products, going from montmorillonite over kaolinite to gibbsite, reflects increasing intensity of leaching, removing silica and cations from the rock.

The hydrological conditions in combination with the rate of mineral weathering determine the nature of the weathering product. Montmorillonite is preferentially formed in relatively dry climates, where the rate of flushing of the soil is low. Its formation is further enhanced when rapidly dissolving material such as volcanic rock is available. Gibbsite, on the other hand, typically forms in tropical areas with intense rainfall and well drained conditions. Under such conditions, gibbsite and other Al-hydroxides may form a thick weathering residue, *bauxite*, that constitutes the most important Al-ore.

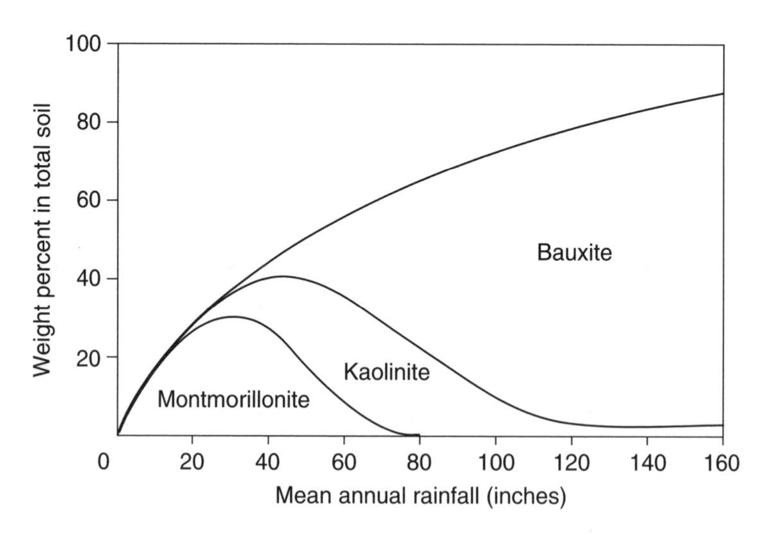

Figure 8.4. Weathering products on volcanic rocks on the island of Hawaii as a function of mean annual rainfall. Bauxite corresponds to Al-hydroxides and the contents of different minerals are plotted cumulatively as weight percent of the total soil (Berner, 1971).

A classic example of the effect of precipitation and leaching intensity on the composition of weathering residues is found in the soils on Hawaii where bauxite, kaolinite and montmorillonite form differently with altitude on the volcanic slopes (Figure 8.4). Montmorillonite is found in areas with low rainfall and therefore with a longer residence time of water in the soil, so that the solute concentrations become higher. At the other end of the scale, bauxite forms in high-rainfall areas where leaching is intense and the residence time of water in the soil is short, so that dissolved ion concentrations are low. Figure 8.4 also shows the amounts of weathering product to increase with the amount of rainfall.

Authigenic clay mineral formation due to the alteration of primary silicates is studied extensively in sedimentary petrology, since the type of clay mineral may seriously affect the permeability of the rock. Kaolinite forms neat booklets which have only a moderate effect on the permeability, but illite forms hairy aggregates that clog pore throats and reduce the permeability very significantly. This feature is well recognized in oil exploration and is illustrated in Figure 8.5. Sandstones cemented with illite are shown to have a permeability that is at least an order of magnitude lower than sandstones with the same porosity, but cemented with kaolinite.

Clay minerals that originate from silicate weathering remain fairly stable during erosion and transport, and are deposited as coatings on sand grains, often closely intermingled with iron oxides. These coatings are important for the ion exchange properties of the aquifer. In some cases the clay particles are washed down from the upper soil layers deeper into the profile (Ryan and Gschwend, 1992). The total amount of clay that is present in sand deposits is normally less than a few percent and consists in most cases of a mixture of various clay minerals. An example of the relative distribution of clay minerals in a sandy deposit is shown in Figure 8.6. Illite, montmorillonite and kaolinite are the dominant clay minerals.

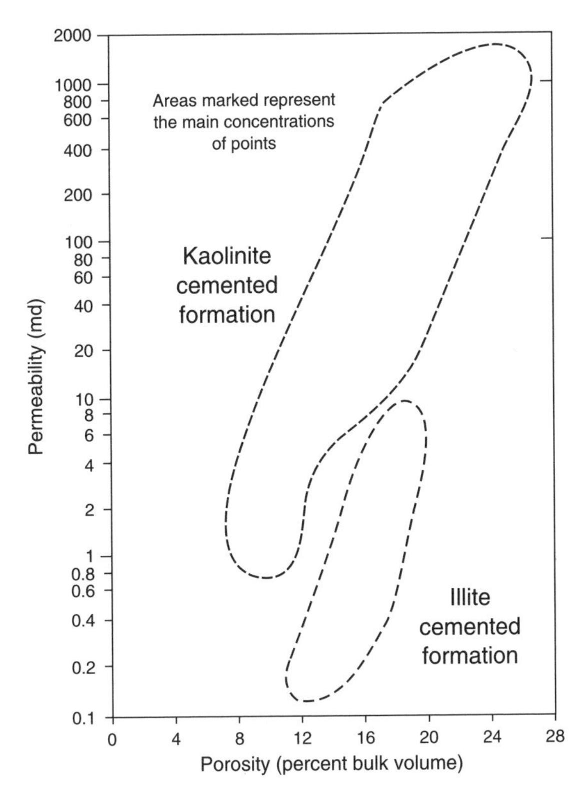

Figure 8.5. The effect of type of authigenic clay on the permeability of a sandstone (Blatt et al., 1980).

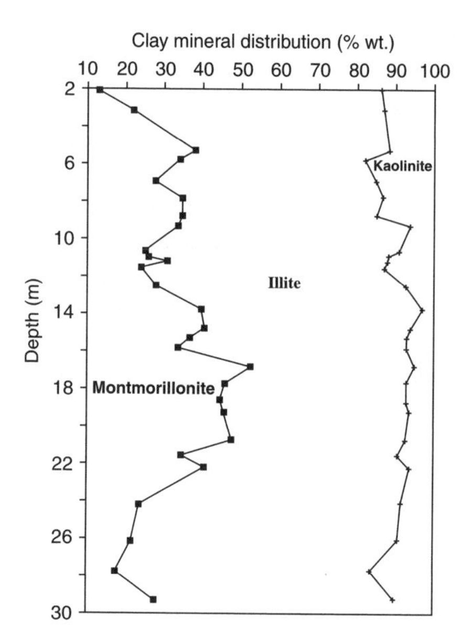

Figure 8.6. The relative distribution of clay minerals in the unsaturated zone of a quaternary sand deposit (Matthess et al., 1992).

In the upper part of the sequence the montmorillonite content tends to increase with depth and perhaps montmorillonite is being transformed into illite. An alternative interpretation would be that the sequence reflects differences in the source areas of the sediments. Sometimes, the source areas of the clay minerals can be recognized by their $\delta^{18}O$ composition which is a function of the $\delta^{18}O$ of the soil water (Savin and Hsieh, 1998; Salomons and Mook, 1987).

QUESTION:
Write balanced reactions for dissolution or weathering to kaolinite for: *Olivine*, Mg_2SiO_4; *Olivine (Hortonolite)*, $(FeMg)SiO_4$; *Intermediate Plagioclase (Oligoclase)*, $(Na_{0.8}Ca_{0.2})(Al_{1.2}Si_{2.8})O_8$; *Hornblende or Amphibole (Glaucophane)*, $Na_2(Mg_3Al_2)Si_8O_{22}(OH)_2$, *Mica (Muscovite)*, $KAl_2(AlSi_3)O_{10}(OH)_2$. Also with gibbsite as weathering product.

ANSWER: (for oligoclase only)

$$(Na_{0.8}Ca_{0.2})(Al_{1.2}Si_{2.8})O_8 + 3.8H_2O + 1.2H^+ \rightarrow$$

$$0.8Na^+ + 0.2Ca^{2+} + 0.6Al_2Si_2O_5(OH)_4 + 1.6H_4SiO_4$$

and

$$(Na_{0.8}Ca_{0.2})(Al_{1.2}Si_{2.8})O_8 + 6.8H_2O + 1.2H^+ \rightarrow$$

$$0.8Na^+ + 0.2Ca^{2+} + 1.2Al(OH)_3 + 2.8H_4SiO_4$$

8.2 THE STABILITY OF WEATHERING PRODUCTS

Gibbsite or Al-hydroxide is the most extreme product of silicate weathering and this mineral is used as a starting point for our discussion of the stability of weathering products. The stability is described by the reaction:

$$Al(OH)_{3\ gibbsite} + 3H^+ \leftrightarrow Al^{3+} + 3H_2O \tag{8.4}$$

with the mass action equation

$$K_{gibbsite} = [Al^{3+}] / [H^+]^3 \approx 10^{10} \tag{8.5}$$

Equation (8.5) indicates that the activity of aluminum in water $[Al^{3+}]$, will be strongly dependent on the pH. Dissolved aluminum has, however, a distinct tendency to form hydroxy complexes which may increase the solubility of gibbsite significantly. The stability of the most important Al-hydroxy complexes is listed in Table 8.2.

Table 8.2. The stability of the predominant dissolved Al-hydroxy complexes at 25°C, taken from PHREEQC.DAT.

Reaction	log *K*
$Al^{3+} + H_2O \leftrightarrow Al(OH)^{2+} + H^+$	−5.09
$Al^{3+} + 2H_2O \leftrightarrow Al(OH)_2^+ + 2H^+$	−10.1
$Al^{3+} + 4H_2O \leftrightarrow Al(OH)_4^- + 4H^+$	−22.7

The total amount of aluminum in solution consists of both complexed and uncomplexed aluminum and is described by the mass balance equation:

$$\Sigma Al = m_{Al^{3+}} + m_{AlOH^{2+}} + m_{Al(OH)_2^+} + m_{Al(OH)_4^-} \tag{8.6}$$

In some cases also complexes between Al^{3+} and fluoride, sulfate and organic matter have to be considered (Driscoll, 1980).

The set of equations consisting of the mass action equations for the complexing reactions in Table 8.2, the equation for gibbsite stability (8.5) and the mass balance equation (8.6) can be solved by the methods given in the Section 4.2, to describe the solubility of total dissolved Al^{3+} as a function of pH (Figure 8.7).

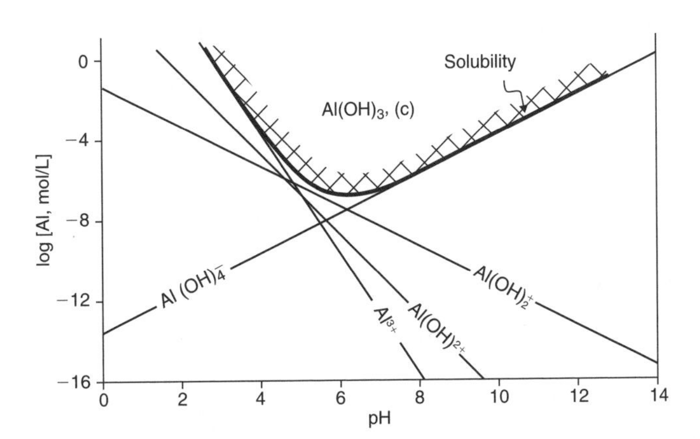

Figure 8.7. The solubility of total dissolved aluminum in equilibrium with gibbsite ($Al(OH)_3$) and aqueous Al-hydroxy complexes.

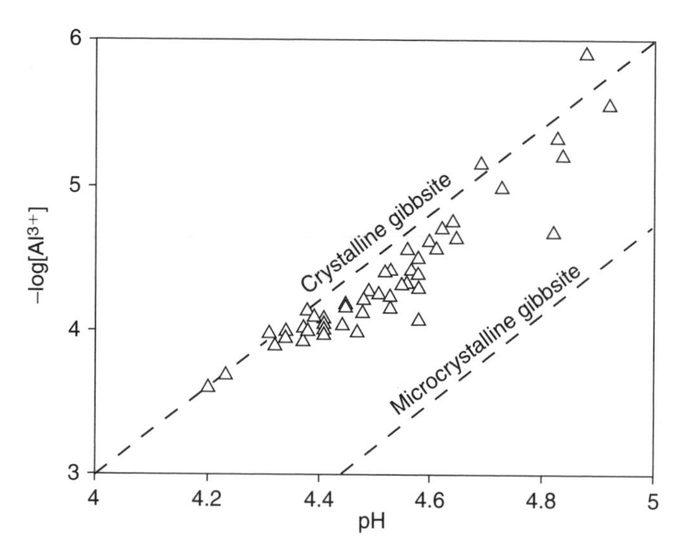

Figure 8.8. The saturation state of shallow groundwater for gibbsite. Activity calculations have been carried out with PHREEQC at the *in situ* temperature of 10°C. The lines indicate saturation for gibbsite according to Equation (8.5). At 10°C log *K* is 9.00 for crystalline gibbsite and 10.30 for microcrystalline gibbsite (Hansen and Postma, 1995).

Below pH 4.5 the aluminum solubility increases and the Al^{3+} concentration may by far exceed the drinking water limit of 0.2 mg/L. In the near-neutral pH range, the total dissolved Al^{3+} concentration becomes as low as 1 μmol/L. Above pH 7 the aluminate complex dominates dissolved aluminum and the total solubility increases again.

Figure 8.8 shows the pH and $[Al^{3+}]$ activity of some acid groundwaters. Also shown are the stability lines for crystalline gibbsite or microcrystalline gibbsite, both corresponding to Equation (8.5). Clearly the data covariate with the stability lines indicating a solubility control for Al^{3+} in groundwater by a gibbsite-type mineral. The use of two stability lines in Figure 8.8 illustrates that the gibbsite phase in a natural sediment is not well defined. Inspection of the PHREEQC databases shows the log *K* of Reaction (8.5) to vary from 8.11 for crystalline gibbsite to 10.8 for amorphous $Al(OH)_3$, that is two orders of magnitude difference in solubility! The microcrystalline gibbsite in Figure 8.8 corresponds to log $K = 9.0$ and is intermediate.

Next after gibbsite, the stability of kaolinite is considered and the dissociation reaction is:

$$Al_2Si_2O_5(OH)_4 + 6H^+ \leftrightarrow 2Al^{3+} + 2H_4SiO_4 + H_2O \tag{8.7}$$

In logarithmic form the mass action equation is:

$$\log K = 2\log[Al^{3+}] + 2\log[H_4SiO_4] + 6pH = 7.4 \tag{8.8}$$

Equation (8.8) can be rewritten as

$$\log[Al^{3+}] + 3pH = (7.4 - 2\log[H_4SiO_4]) / 2 \tag{8.9}$$

The $\log[H_4SiO_4]$ of groundwater is mostly in the range from -4 to -3.3 and varies much less than $\log[Al^{3+}]$ or pH. If $\log[H_4SiO_4]$ is approximately constant, Equation (8.9) becomes identical to Equation (8.5). Substitution of a realistic value of -4 for $\log[H_4SiO_4]$ reduces the right hand part of

Equation (8.9) to 7.7 which is close to the value of 8.11 for crystalline gibbsite. Accordingly, it is difficult to identify the actual controlling mineral from saturation calculations.

Additional uncertainty concerns the slow dissolution kinetics of clay minerals which makes it questionable whether true equilibrium is ever attained. This problem is encountered in the interpretation of field data, as well as in the determination of clay mineral stability by equilibration with water in the laboratory. May et al. (1986) found it necessary to equilibrate kaolinite with water for 1237 days in order to approach equilibrium, while smectite (montmorillonite) did not attain equilibrium in that period. Despite these uncertainties, there is little doubt that the dissolved aluminum concentration in groundwater is controlled by the solubility of weathering products like gibbsite or kaolinite.

QUESTION:
Find the pH where the solubility of gibbsite surpasses the drinking water limit (0.2 mg Al/L)?
ANSWER: `SOLUTION; Al 0.2 mg/kgw; pH 4 Gibbsite; END`. Here pH is adjusted to equilibrium with gibbsite. Note: for `pH 8 Gibbsite` PHREEQC gives a different answer, why?

8.3 INCONGRUENT DISSOLUTION OF PRIMARY SILICATES

Primary silicate minerals comprise feldspars, amphiboles, pyroxenes, micas, etc, that are present in igneous and metamorphic rocks and as detrital minerals in sand and sandstones. The mineralogical composition of a quaternary sand deposit is illustrated in Figure 8.9 and shows that quartz is by far the dominant component of the sand. However, more easily weatherable minerals, like plagioclase, K-feldspar and mica are also present in significant amounts.

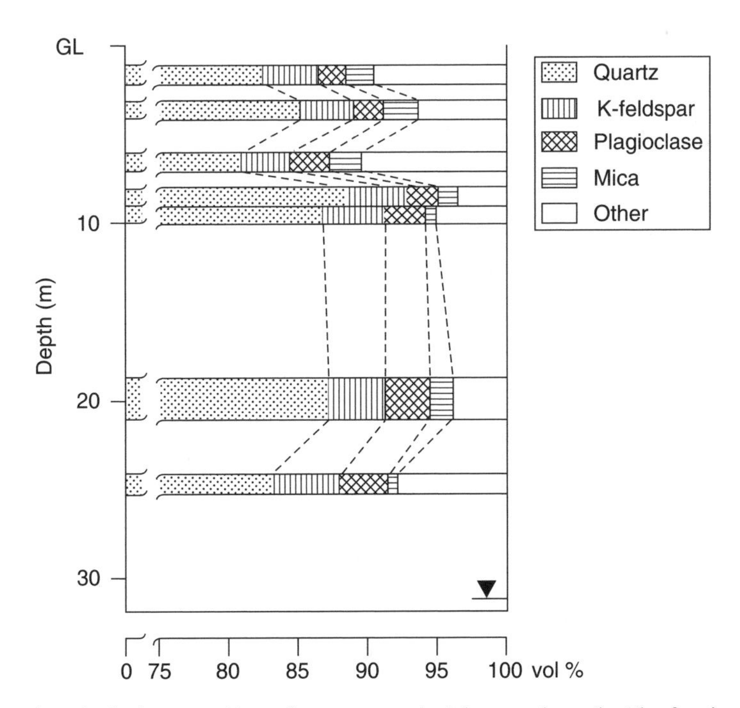

Figure 8.9. The mineralogical composition of quaternary glacial outwash sands. The fraction "other" comprises heavy minerals, organic matter, etc. (Ohse et al., 1984).

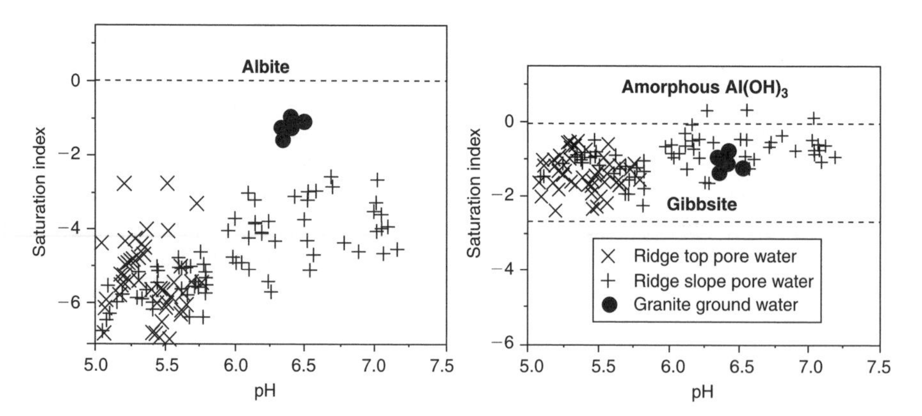

Figure 8.10. The saturation index (log (*IAP/K*) of groundwater for albite and $Al(OH)_3$ in an area with granite weathering (White et al., 2001).

The stability of these minerals in a groundwater system can be evaluated by calculating the saturation state of the groundwater for a given mineral. For example for albite, we write the dissociation reaction:

$$\underset{albite}{NaAlSi_3O_8} + 4H^+ + 4H_2O \rightarrow Na^+ + Al^{3+} + 3H_4SiO_4 \tag{8.10}$$

As for other primary silicate weathering reactions (Table 8.1), the single arrow in the equation is used to indicate that the reaction is irreversible at temperatures less than 50°C (Stéfansson and Arnorsson, 2000). The mass action expression for Reaction (8.10) is

$$\log K = \log[Na^+] + \log[Al^{3+}] + 3\log[H_4SiO_4] + 4pH \tag{8.11}$$

The saturation index for albite is shown for groundwater in a granitic area in Figure 8.10 and the groundwater is found to be subsaturated for albite by orders of magnitude. Since albite is present in the rock, this indicates that the kinetics of albite dissolution are slow. The saturation state for gibbsite in the same waters is also included and shows approximate equilibrium with a phase ranging between gibbsite and amorphous $Al(OH)_3$, similar as in Figure 8.8. This pattern demonstrates the general behavior of primary silicate minerals during weathering. The dissolution of the primary silicate is slow and kinetically controlled while the precipitation of the secondary weathering product is faster and approaches equilibrium (Helgeson et al., 1969). The dissolution kinetics of the primary silicate are therefore overall rate limiting in the weathering process.

Although the saturation state approach outlined above is useful, it also has several drawbacks. First of all, the Al^{3+} concentration in groundwater of near neutral pH is often very low and cumbersome to analyze accurately. Second, clusters of Al-hydroxy complexes can be present in the water which may not be removed completely by filtering over 0.45 or even 0.1 μm filters, and which are unaccounted for in the speciation model. Third, even in the simple case of albite (Equation 8.11) already 4 variables affect the saturation state, and the stability of albite is therefore difficult to display in a diagram that includes all the relevant parameters. To circumvent these problems, silicate stability diagrams have been devised which assume that all Al^{3+} is preserved in the weathering product. Such a diagram for Ca-silicates is shown in Figure 8.11. It contains stability fields for the Ca-feldspar anorthite and its possible weathering products, gibbsite, kaolinite and Ca-montmorillonite expressed as a function of $\log([Ca^{2+}]/[H^+]^2)$ and $\log[H_4SiO_4]$.

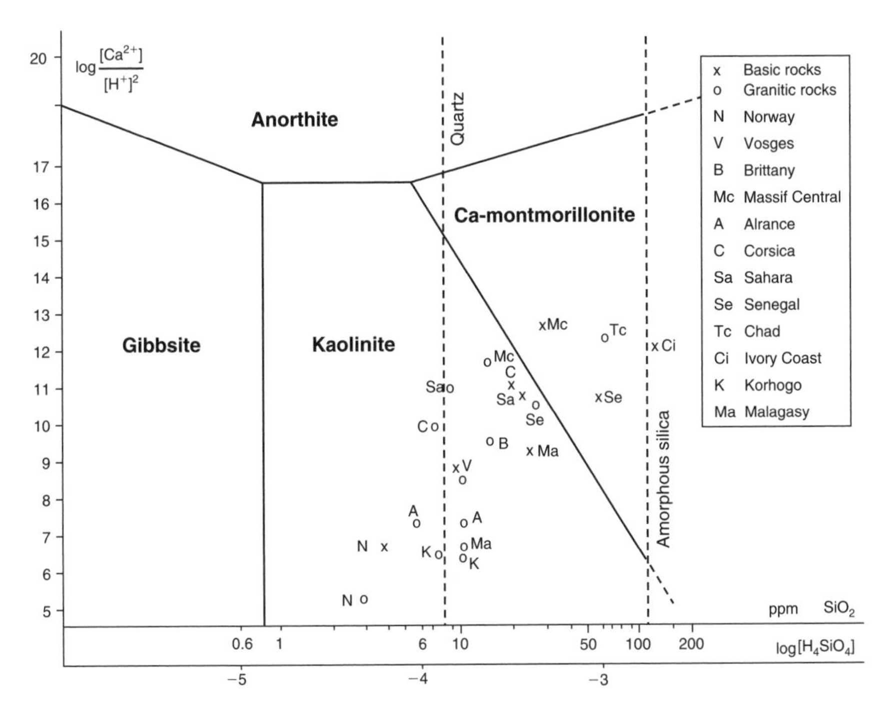

Figure 8.11. The stability of anorthite and its possible weathering products gibbsite, kaolinite and Ca-montmorillonite. Also included are the compositions of selected surface water samples from areas with crystalline rocks (Tardy, 1971).

The peculiar choice of the y-axis parameter can be understood by considering the reaction between anorthite and gibbsite:

$$\underset{anorthite}{CaAl_2Si_2O_8} + 2H^+ + 6H_2O \rightarrow \underset{gibbsite}{2Al(OH)_3} + Ca^{2+} + 2H_4SiO_4 \tag{8.12}$$

In this reaction all the Al^{3+} released from anorthite is preserved in gibbsite. The relative stability between the two minerals is therefore controlled by dissolved silica, Ca^{2+} and pH. The mass action equation of Reaction (8.12) is:

$$\log K = \log[Ca^{2+}] + 2\log[H_4SiO_4] - 2\log[H^+] = 6.78 \tag{8.13}$$

Rearranging Equation (8.13) yields:

$$\log K = \log([Ca^{2+}]/[H^+])^2 + 2\log[H_4SiO_4] = 6.78 \tag{8.14}$$

Equation (8.14) is plotted as a straight line with slope -2 in the stability diagram (Figure 8.11) and the four variables have been reduced to two. In general, the expression on the y-axis has the form $\log([\text{cation}^{n+}]/[H^+]^n)$ and reflects that the charge of released cations must be balanced by consumption of H^+. Similarly, the equilibrium between Ca-montmorillonite and kaolinite is described by the reaction:

$$2H^+ + \underset{Ca-montmorillonite}{3Ca_{.33}[Si_{7.33}Al_{.67}][Al_4]O_{20}(OH)_4} + 23H_2O \leftrightarrow \underset{kaolinite}{7Al_2Si_2O_5(OH)_4} + 8H_4SiO_4 + Ca^{2+} \tag{8.15}$$

and

$$\log K = \log([Ca^{2+}] / [H^+]^2) + 8 \log[H_4SiO_4] = -15.7 \tag{8.16}$$

The boundary between kaolinite and gibbsite is described by the reaction:

$$\underset{kaolinite}{Al_2Si_2O_5(OH)_4} + 5H_2O \leftrightarrow \underset{gibbsite}{2Al(OH)_3} + 2H_4SiO_4 \tag{8.17}$$

and

$$\log K = 2 \log[H_4SiO_4] = -9.8 \tag{8.18}$$

Accordingly, a H_4SiO_4 activity of $10^{-4.9}$ indicates equilibrium between kaolinite and gibbsite and this phase boundary results in a straight line parallel to the y-axis in the stability diagram (Figure 8.11). Finally in the reaction of anorthite forming kaolinite, all silica is preserved in the solid phase (Table 8.1) and the boundary therefore plots parallel to the $\log[H_4SiO_4]$ axis.

QUESTION:
Write the mass action equation for the reaction of anorthite forming kaolinite. Find the mass action constant by combining Equations (8.18) and (8.13).

ANSWER: $\log K = 6.78 - (-9.8) = 16.58$.

The discussed silicate stability diagrams contain the implicit assumption that Al^{3+} is present in the water, and in equilibrium with the depicted phases. However, a water sample that plots in the kaolinite field in Figure 8.11 may well be subsaturated for kaolinite because of a low $[Al^{3+}]$. A better statement would be that according to Figure 8.11, kaolinite is more likely to be stable than for example gibbsite.

The composition of surface waters from different crystalline massifs in Europe and Africa plot dominantly in the field of kaolinite (Figure 8.11). Basic rocks (basalts etc.) are more reactive than granitic rocks and yield higher total dissolved solid concentrations (Figure 8.3). Accordingly, waters from basic rocks plot in general closer to the stability field of montmorillonite than those derived from granitic rocks. The leaching intensity also affects the water composition. Therefore, water from a basic rock area in rainy Norway plots way down in the kaolinite field, while a sample from a granitic area in arid Chad plots in the montmorillonite field.

Figure 8.11 considers only Ca-feldspar while most rocks contain several feldspars. Stability diagrams for sodium and potassium feldspars can be constructed in a similar way, but this is still a simplification, since feldspars in rocks often are solid solutions rather than pure end-member minerals. For example, plagioclases consist of a series of solid solutions of albite and anorthite. Also the variation in the composition of clay minerals such as montmorillonite is large. The stability and dissolution/precipitation behavior of solid solutions and of pure end-member minerals will affect the stability boundaries in Figure 8.11 considerably and the conclusions drawn from these diagrams must be used with caution. The second point of concern is again the slow reaction kinetics of silicate minerals and it seems questionable whether true equilibrium is ever attained. Tardy (1971) reports that many waters which plot in the kaolinite stability field were sampled from springs where montmorillonite is in the soils of the watershed.

EXAMPLE 8.1. *Incongruent dissolution of K-feldspar*

Irreversible dissolution of K-feldspar (microcline) as primary mineral, and the precipitation of secondary weathering products at partial equilibrium is illustrated in the following PHREEQC input file.

```
SOLUTION 1
 pH 4.8 charge; C(4) 1 CO2(g) -1.5
END

USE solution 1
EQUILIBRIUM_PHASES 1
 Gibbsite 0 0
 Kaolinite -1. 0
PRINT; -reset false
REACTION 1
 KAlSi3O8 1; 10 umol in 20

USER_GRAPH
 -head Si K-feldspar Gibbsite Kaolinite
 -axis_titles Si umol/L
 -start
 10 graph_x tot("Si") * 1e6
 20 graph_y tot("K") * 1e6, equi("Gibbsite") * 1e6, equi ("Kaolinite") * 1e6
 -end
END
```

The initial solution is soil water containing carbonic acid. Equilibrium is specified for the weathering products gibbsite and kaolinite with initial amounts of zero, implying that precipitation will start once saturation is reached. The *SI* for kaolinite is set to -1.0 to comply with the relatively more stable kaolinite used by Tardy in Figures 8.11 and 8.13. Irreversible dissolution of K-feldspar is simulated with REACTION, the amount dissolved is given by the K^+ concentration (tot("K") in USER_GRAPH).

Figure 8.12 shows the dissolution and precipitation reactions. K-feldspar dissolves stoichiometrically until 1.6 μmol have reacted and the solution reaches saturation with gibbsite. Subsequently, gibbsite precipitates while the concentrations of K^+ and of Si increase. When saturation with kaolinite is reached, gibbsite is converted to kaolinite by the reverse of Reaction (8.17). The reaction consumes all the Si coming from feldspar dissolution and $[H_4SiO_4]$ remains fixed until all gibbsite has been converted to kaolinite. Once gibbsite has disappeared, the silica concentration increases again. The solution remains, throughout, strongly subsaturated for microcline.

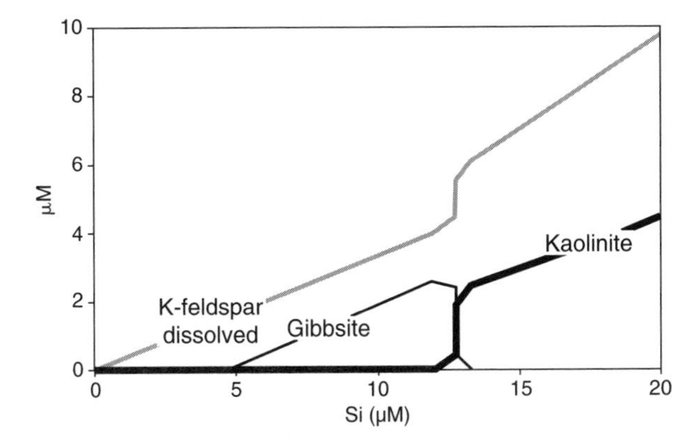

Figure 8.12. Incongruent dissolution of K-feldspar resulting in sequential precipitation of gibbsite and kaolinite, as modeled by PHREEQC.

The water composition pathway is also shown in Figure 8.13, the potassium sister-diagram of Figure 8.11. Initially the water composition plots at the lower left in the gibbsite stability field. As microcline dissolves the water moves to the right until the kaolinite boundary. Here gibbsite is recrystallized to kaolinite and then the water moves towards the microcline field. The pathway in water composition evolution is reasonably consistent with the plotted water compositions from different localities and rock types.

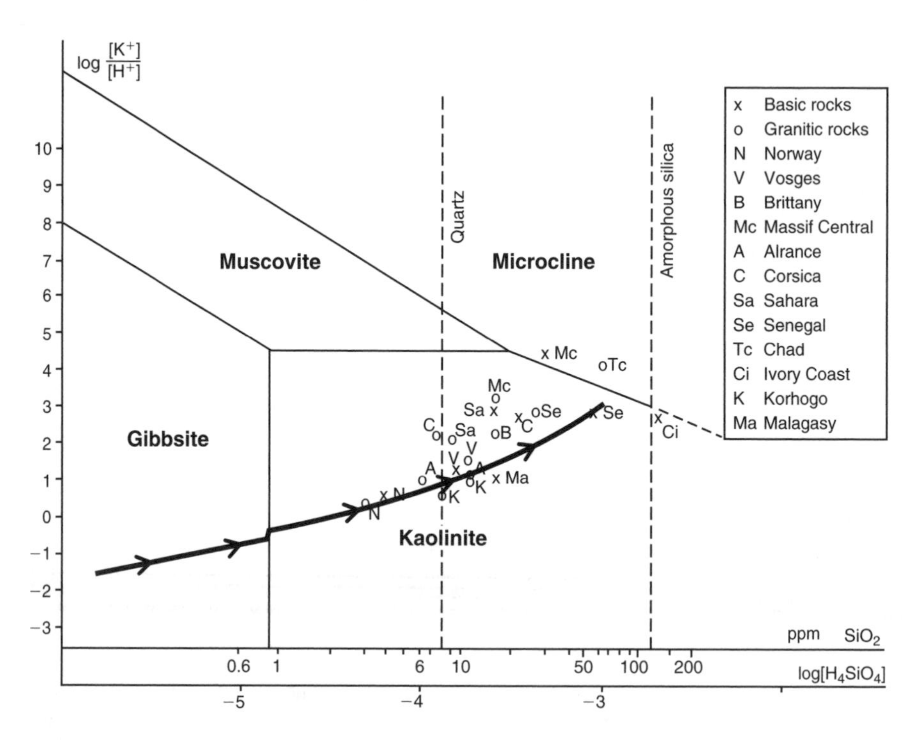

Figure 8.13. The stability diagram for K-feldspar microcline and its weathering products (modified from Tardy, 1971). Superimposed is the pathway modeled with PHREEQC for incongruent dissolution of microcline using an initial solution of H_2O in contact with $[P_{CO_2}] = 10^{-1.5}$. To model the full pathway, the input file in Example 8.1 was extended to dissolve a total of 0.5 mM $KAlSi_3O_8$.

QUESTIONS:

Plot the concentrations of Figure 8.12 logarithmically with PHREEQC (Note to use log10, cf. RATES in the PHREEQC manual).

Plot the reaction path with log10($[K^+]$ / $[H^+]$) and log10$[H_4SiO_4]$ in Figure 8.13.

Find the names of K-feldspars in the PHREEQC databases?

ANSWER: Microcline, K-feldspar, Adularia, Sanidine

Continue dissolution of K-feldspar until saturation in the PHREEQC log10 plot. Discuss the (fixed) end-point state.

Change the CO_2 pressure of initial water to $10^{-2.5}$. Why does the line plot at a higher $[K^+]$ / $[H^+]$ than before?

Figures 8.11 and 8.13 include the stability lines for quartz and amorphous silica. For both substances, the solubility is described by the reaction:

$$SiO_{2(s)} + 2H_2O \leftrightarrow H_4SiO_4 \tag{8.19}$$

H_4SiO_4 remains undissociated at pH values below 9, and the stability of the $SiO_{2(s)}$ phases is determined by the solubility product of Reaction (8.19):

$$K = [H_4SiO_4] \tag{8.20}$$

For quartz, $K = 10^{-3.98}$ at 25°C is generally used (see PHREEQC.DAT) but a recent study (Rimstidt, 1997) has suggested a higher solubility product of $10^{-3.74}$. Quartz has extremely sluggish reaction kinetics and solutions grossly supersaturated for quartz are common. The slow precipitation of quartz allows the formation of less stable forms of $SiO_{2(s)}$ like amorphous $SiO_{2(s)}$ also called *opal*, found in marine sediments as diatom tests, *chalcedony* (cryptocrystalline quartz), *cristobalite*, *tridymite* and *opal-CT* (disordered cristobalite-tridymite) (Williams et al., 1985). The most soluble phase is amorphous $SiO_{2(s)}$ which has a solubility product of about $10^{-2.7}$ (25°C) and places the upper constraint on the dissolved silica concentration.

The water compositions in Figures 8.11 and 8.13 range from subsaturated for quartz to close to equilibrium with amorphous silica. The silica has probably been released to the water by weathering of silicates like feldspars and because the precipitation of quartz apparently cannot keep pace with the silica release, the water becomes supersaturated. In contact with slowly weathered granites (Figure 8.11 and Schulz and White, 1999) and low reactive sandy aquifers (Kjøller, 2001; Ohse et al., 1983) the silica concentration may approach quartz saturation quite closely. Volcanic rocks on the other hand often contain volcanic glass that readily dissolves and silica concentrations may become much higher (Figure 8.11 and Miretzky et al., 2001).

QUESTION:
Will the Al^{3+} concentration change with pH in the kaolinite stability field of Figure 8.11 at fixed $[H_4SiO_4] = 10^{-4}$?

ANSWER: Yes, it follows the pH / ΣAl trend of Figure 8.8.

8.4 THE MASS BALANCE APPROACH TO WEATHERING

The problems encountered when treating silicate-water reactions as equilibrium systems have stimulated the exploration of alternative approaches. One of the best alternatives is the use of mass balance calculations, which relate changes in water chemistry to the dissolution or precipitation of minerals and basically have the character of bookkeeping. For reactions between mineral and water we can write the general reaction:

$$\text{reactant phase} \rightarrow \text{weathering residue} + \text{dissolved ions} \tag{8.21}$$

For the congruent dissolution of calcite, in water containing carbonic acid, the dissolution reaction is:

$$H_2CO_3 + CaCO_3 \rightarrow Ca^{2+} + 2HCO_3^- \tag{8.22}$$

Water that percolates through a soil with calcite should accordingly be enriched with two moles of HCO_3^- for each mole of Ca^{2+}. For the incongruent dissolution of albite to kaolinite:

$$\underset{albite}{2NaAlSi_3O_8} + 2H_2CO_3 + 9H_2O \rightarrow \underset{kaolinite}{Al_2Si_2O_5(OH)_4} + 2Na^+ + 2HCO_3^- + 4H_4SiO_4 \tag{8.23}$$

the release of H_4SiO_4 is twice the release of HCO_3^- or of Na^+.

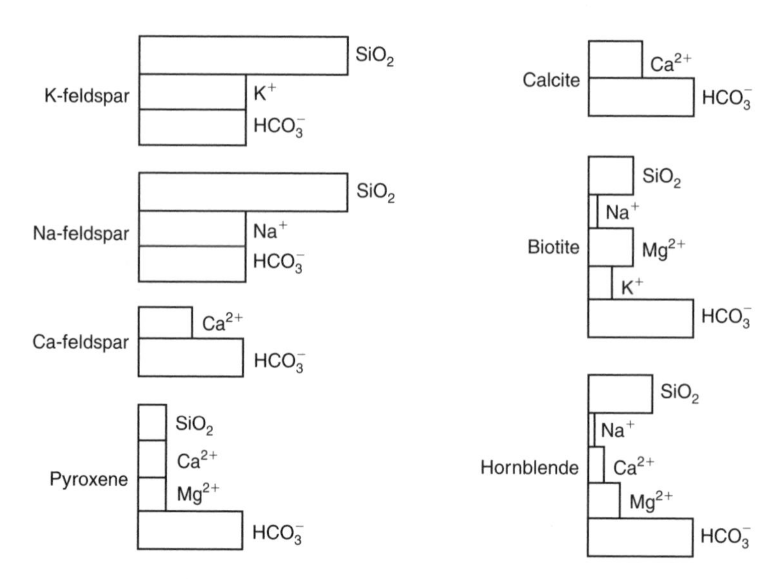

Figure 8.14. The composition of waters resulting from alteration of different silicate minerals to kaolinite in the presence of carbonic acid, according to the reactions listed in Table 8.1 (Garrels and Mackenzie, 1971).

Exactly the same can be done for the other reactions listed in Table 8.1. The results are displayed graphically in Figure 8.14 and show the expected water quality for each reaction. Note that a given water composition is not necessarily unique for a specific mineral; the dissolution of calcite (Reaction 8.22) and the weathering of anorthite to kaolinite (Table 8.1) yield the same relative concentrations of Ca^{2+} and HCO_3^- (Figure 8.14).

Rocks consist normally of mixtures of minerals and in some cases it is possible to reconstruct the contributions of different weathering reactions from concentration changes along a flowpath.

EXAMPLE 8.2. *Mass balance and the water chemistry of the Sierra Nevada (USA)*
In a classic paper, Garrels and Mackenzie (1967) related the groundwater chemistry in a granitic area, Sierra Nevada (USA), to silicate weathering reactions by the use of mass balance calculations. The approach used by Garrels and Mackenzie was stepwise substraction of different weathering reactions. The chemistry of snow and ephemeral spring waters is listed in Table 8.3. The concentrations in snow were subtracted from the spring water, to obtain the contribution by rock weathering.

Table 8.3. The composition of snow and ephemeral spring water (Garrels and Mackenzie, 1967). Concentrations in mmol/L.

	Snow	Spring	Rock weathering
Na^+	0.024	0.134	0.110
Ca^{2+}	0.01	0.078	0.068
Mg^{2+}	0.007	0.029	0.022
K^+	0.008	0.028	0.020
HCO_3^-	0.018	0.328	0.310
Si	0.003	0.273	0.270
pH		6.2	

The granites of the Sierra Nevada contain the primary minerals plagioclase, biotite, quartz and K-feldspar and the weathering product kaolinite. Quartz was not considered because its reactivity is so low. The intuitive expectation is that weathering of the primary minerals to kaolinite (Table 8.1) may explain the spring water composition. The average composition of the minerals found in the area are listed in Table 8.4.

Table 8.4. The composition of some minerals in the granitic rocks in the Sierra Nevada.

Plagioclase	$Na_{0.62}Ca_{0.38}Al_{1.38}Si_{2.62}O_8$
Biotite	$KMg_3AlSi_3O_{10}(OH)_2$
K-Feldspar	$KAlSi_3O_8$
Kaolinite	$Al_2Si_2O_5(OH)_4$

The calculation proceeds as follows. First all Na^+ is attributed to weathering of plagioclase. This requires the alteration of 0.110 / 0.62 = 0.177 mmol/L plagioclase to kaolinite:

$$.177Na_{0.62}Ca_{0.38}Al_{1.38}Si_{2.62}O_8 + .246CO_2 + .367H_2O \rightarrow$$

$$.122\ Al_2Si_2O_5(OH)_4 + .11Na^+ + .068Ca^{2+} + .246HCO_3^- + .22SiO_2$$

Subtracting the contribution of plagioclase weathering (Table 8.5) from the water chemistry shows that it also accounts for all Ca^{2+}, as well as most of the HCO_3^- and SiO_2. The next step is to attribute all Mg^{2+} to biotite weathering and 0.022 / 3 = 0.0073 mmol/L biotite is altered to kaolinite:

$$.0073KMg_3AlSi_3O_{10}(OH)_2 + .051CO_2 + .026H_2O \rightarrow$$

$$.0037Al_2Si_2O_5(OH)_4 + .0073K^+ + .022Mg^{2+} + .051HCO_3^- + .015SiO_2$$

This removes most of the HCO_3^- as well as some of the K^+. The final step is to attribute the remainder of K^+ to K-feldspar weathering and accordingly 0.013 mmol/L K-feldspar must alter to kaolinite:

$$.013KAlSi_3O_3 + .013CO_2 + .0195H_2O \rightarrow$$

$$.0065Al_2Si_2O_5(OH)_4 + .013K^+ + .013HCO_3^- + .026SiO_2$$

What is left is only a minor amount of SiO_2 and the calculation shows that the three weathering reactions may explain the water chemistry surprisingly well. According to this reaction scheme, the dominant reaction is the transformation of plagioclase into kaolinite under the consumption of carbonic acid, while minor amounts of biotite and K-feldspar dissolve.

Table 8.5. The contributions of weathering various silicates to the composition of ephemeral spring waters in the Sierra Nevada (mmol/L).

	Rock weathering	Minus plagioclase	Minus biotite	Minus K-feldspar
Na^+	0.110	0.000		
Ca^{2+}	0.068	0.000		
Mg^{2+}	0.022	0.022	0.000	
K^+	0.020	0.020	0.013	0.000
HCO_3^-	0.310	0.064	0.013	0.000
Si	0.270	0.050	0.035	0.009

In a mass balance calculation a set of linear equations, that describe the components involved, is solved and for this purpose dedicated computer programs like NETPATH (Plummer et al., 1991) are available. The problems that can be solved comprise water composition evolution in terms of mineral dissolution and precipitation and simple mixing of end-member waters. The approach is also termed inverse modeling: from the final result, the chemical reactions are unraveled by backward calculus. NETPATH handles mass balances including redox reactions, isotopic mole balances, isotopic fractionation etc. PHREEQC also contains an option for mass balance calculation or inverse modeling and it can adjust concentrations in the optimization process to achieve charge and mass balance. In addition, PHREEQC employs an electron balance equation for redox processes, and an alkalinity-balance equation. In Example 8.3 the mass balance calculation is illustrated for Sierra Nevada spring water.

EXAMPLE 8.3. *Mass balance calculation of mineral weathering as source for solutes in ephemeral spring waters in the Sierra Nevada using PHREEQC*

The overall mass balance equation is:

$$\text{initial solution} + \text{reactant phases} \rightarrow \text{final solution} + \text{product phases} \tag{8.24}$$

and the mass balance equations on elements is:

$$\Delta m_{T,k} = m_{T,k(final)} - m_{T,k(initial)} = \sum_{p=1}^{P} \alpha_p b_{p,k} \tag{8.25}$$

For each element $k = 1$ to J.

Here $\Delta m_{T,k}$ is the change in total molality of the kth element between final and initial solution, while P is the number of total reactant and product phases, α_p is the mass transfer of the pth phase and $b_{p,k}$ denotes the stoichiometric coefficient of the kth element in the pth phase. For solving the set of linear equations, the number of unknowns (independent elements) should equal the number of equations. For Equation (8.25), this means that the number of independent elements equals the number of phases. The mass balance equation for the various elements becomes:

$\Delta m_{T,\text{Na}} = 0.134 - 0.024 = 0.62\alpha_{\text{plagioclase}}$

$\Delta m_{T,\text{Ca}} = 0.78 - 0.01 = 0.38\alpha_{\text{plagioclase}}$ (Note that Na^+ and Ca^{2+} are dependent by the stoichiometric ratio in plagioclase)

$\Delta m_{T,\text{Mg}} = 0.029 - 0.007 = 3\alpha_{\text{biotite}}$

$\Delta m_{T,K} = 0.028 - 0.008 = 1\alpha_{\text{biotite}} + 1\alpha_{\text{K-feldspar}}$

$\Delta m_{T,\text{Si}} = 0.273 - 0.003 = 2.62\alpha_{\text{plagioclase}} + 3\alpha_{\text{biotite}} + 3\alpha_{\text{K-feldspar}} + 2\alpha_{\text{kaolinite}}$

$\Delta m_{T,\text{C}} = 0.328 - 0.018 = 1\alpha_{CO_2\text{-gas}}$

We could also have chosen to balance on $\Delta m_{T,\text{Al}} = 0$ with the appropriate coefficients instead of on Si, when doing the calculation by hand. The PHREEQC input file is:

```
DATABASE wateq4f.dat                  # used for phlogopite (biotite)

SOLUTION_SPREAD
Number   pH    Na      Ca      Mg      K       Alkalinity   Si      Cl     S(6)
1        5.6   0.024   0.01    0.007   0.008   0.018        0.003   .014   .01
2        6.2   0.134   0.078   0.029   0.028   0.328        0.273   .014   .01

INVERSE_MODELING
 -solutions 1 2                       # find solution 2 from 1
 -uncertainty 0.05                    # default is 0.05 or 5%
 -phases
  CO2(g)
```

```
 Kaolinite force prec
 Plagioclase force dis
 Phlogopite                    # iron-free biotite
 Adularia
 -balances
  Cl; S(6) -0.7e-5

PHASES
Plagioclase
Na0.62Ca0.38Al1.38Si2.62O8 + 5.52H+ + 2.48H2O = .62Na+ + .38Ca+2 + 1.38Al+3 + 2.62H4SiO4
log_k 0.0
END
```

Solutions number 1 and 2 correspond to the snow and ephemeral springs respectively in Table 8.3. The pH of snow water was not given in Garrels and Mackenzie (1967) but is estimated to be 5.6, corresponding to equilibrium with the atmospheric P_{CO_2} (Chapter 5). The keyword INVERSE_MODELING controls the mass balance calculations. **-solutions 1 2** indicates that solution 2 is to be balanced from the initial solution 1, using the constraints:

-uncertainty 0.05 specifies that 5% may be added to or subtracted from all the concentrations to achieve charge balance in the solutions ($E.B. = 0$) and to obtain mass balance with the phase transfer;
-phases lists the minerals whose mass transfer must be considered to explain the evolution of solution 1 to 2. Following the mineral names, the option "dis" indicates dissolution only, "prec" precipitation only, and "force" that the mineral should be included in every model considered. **-phases** reads mineral data from the database, and since biotite is not listed, we look for another one. Phlogopite is in wateq4f.dat, and is the name for iron-free biotite. The specific plagioclase (andesine) is not present in any database and is defined separately under the keyword PHASES;
-balances lists the elements not in **-phases**, in this case, Cl^- and SO_4^{2-}. Solution 1 has a large charge imbalance of 11%, and can be balanced only when the allowed uncertainty for all the elements is increased to 0.12, or when Cl^- or SO_4^{2-} is given a higher uncertainty. Here, an absolute uncertainty of -0.7×10^{-5} (negative number) is defined for SO_4^{2-}, which is just the amount of SO_4^{2-} needed for charge-balance in solution 1.

PHREEQC finds one model and a selection of the output file showing the mineral mass transfer is:

```
Phase mole transfers:
                      CO2(g)      6.473e-04       CO2
                   Kaolinite     -1.330e-04       Al2Si2O5(OH)4
                 Plagioclase      1.783e-04       Na0.62Ca0.38Al1.38Si2.62O8
                  Phlogopite      7.333e-06       KMg3AlSi3O10(OH)2
                    Adularia      1.267e-05       KAlSi3O8
```

Positive values indicates dissolution and a negative values precipitation. Comparison with the manual calculations in Example 8.2 shows excellent agreement, although more CO2(g) is consumed because the PHREEQC mass balance includes H_2CO_3. The most important reaction is still the alteration of plagioclase to kaolinite, while only small amounts of biotite and K-feldspar dissolve.

The concentration of solutes in rain and snow by evapotranspiration was not considered in the mass balance calculations of Examples 8.2 and 8.3. It appears unimportant in the example since the Cl^- concentrations in the snow and spring water are the same. Actually, average Cl^- concentrations were found to be lower in the springwaters than in the precipitation, which suggested sorption of Cl^- in the soil (cf. Chapter 2). Evapotranspiration can be included in manual calculations (Example 8.2) by increasing all the concentrations in the initial water by a factor that includes evapotranspiration and dry deposition (Chapter 2). In PHREEQC calculations, the phase "Water" can be added under -phases (it needs the definition of Water; $H_2O = H_2O$; $-$log_k 0 with keyword PHASES).

Mass balance calculations are a useful approach for identifying possible reactions that may explain differences in water chemistry along a flow path, not only in the case of silicate weathering, but also for carbonates, redox reactions, etc. Often they form the first step in elucidating geochemical processes before attempting the more complex approach of forward modeling. One should always keep the following limitations of mass balances in mind:

(1) The solution of a mass balance calculation is not necessarily unique. Different choices of phases may lead to equally consistent reaction schemes.
(2) There are no thermodynamic constraints on mass balance calculations. The mass balance calculation may predict impossible reactions, like precipitation of gibbsite even though the water is subsaturated.
(3) Mass balance calculations do not consider what is kinetically consistent. They may predict that quartz is dissolving, but not plagioclase, although this kinetically would be unreasonable.
(4) Mass balance calculations assume steady state; water samples along a flow path are usually taken at the same time and the differences in water chemistry are assumed to be due to reactions with minerals and not to temporal variations in the composition of water entering the system.
(5) Mass balances assume a homogeneous reaction between the points of analysis. This condition is hard to test in aquifer systems and certainly questionable in soil systems unless mass balances are carried out for the different soil horizons.

In other words, use your geochemical common sense in the interpretation of mass balance calculations. In the Sierra Nevada example, one could plot the water chemistry of the spring in the silicate stability diagram (Figure 8.11) and check whether kaolinite is stable relative to gibbsite (do this yourself, neglecting differences between activities and molar concentrations). Also the preferential dissolution of plagioclase compared to K-feldspar as observed in Examples 8.2 and 8.3 appears kinetically reasonable. Note that the rate of mineral dissolution or precipitation can be derived from a mass balance calculation when the travel distance and the groundwater flow rate are known.

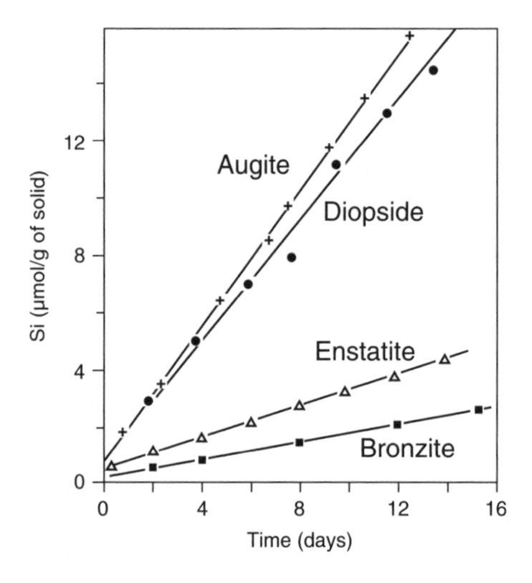

Figure 8.15. The release of silica to the solution over time for pyroxenes and amphiboles at pH 6. Experiments with enstatite, augite and diopside were carried out at 50°C while bronzite was dissolved at 20°C and $P_{O_2} = 0$ atm (Schott and Berner, 1985).

8.5 KINETICS OF SILICATE WEATHERING

The dissolution kinetics of silicate minerals have been assessed both in laboratory experiments and in the field. In laboratory studies, mineral grains are dissolved in solutions of various composition (White and Brantley, 1995; Sverdrup, 1990) while monitoring the rate of dissolution from the Si release (Figure 8.15). This type of results allows to quantity the dissolution kinetics of various minerals and the change in crystal size during dissolution. Table 8.6 shows the calculated lifetime of 1 mm crystals at pH 5 for various minerals and indicates a very large range. For example, Ca-feldspar (anorthite) dissolves about 700 times faster than K-feldspar. More recent data suggest, however, a smaller difference between the dissolution rates of K-feldspar and albite (Blum and Stillings, 1995). Qualitatively, the lifetimes calculated from laboratory experiments are in good agreement with the Goldich weathering sequence (Figure 8.2). The long dissolution times are also in agreement with the high age of the Panola granodiorite weathering profile shown in Figure 8.1.

Table 8.6. The mean lifetime in years of 1 mm crystals of various minerals calculated from laboratory dissolution studies at 25°C and pH 5 (Lasaga, 1984).

Mineral	Lifetime
Quartz	34,000,000
Muscovite	2,700,000
Forsterite	600,000
K-feldspar	520,000
Albite	80,000
Enstatite	8,800
Diopside	6,800
Nepheline	211
Anorthite	112

Apart from the mineralogy, other factors like solution composition may influence the dissolution rate. Equation (8.26) contains the various parameters used in rate equations by Sverdrup and Warfvinge (1995).

$$r = k_{H^+} \frac{[H^+]^n}{f_H} + k_{H_2O} \frac{1}{f_{H_2O}} + k_{OH^-} \frac{[OH^-]^o}{f_{OH}} + k_{CO_2} \frac{[P_{CO_2}]^{0.6}}{f_{CO_2}} + k_{org} \frac{[R^-]^{0.5}}{f_{org}} \quad (8.26)$$

where r is the reaction rate in mol/m^2/s, k_i are the rate coefficients for the solutes that influence the rate (mol/m^2/s), $[i]$ indicate the solute activities of H^+, OH^-, free organic radicals R^- and the partial pressure of CO_2, n and o are the apparent reaction orders, and f_i are inhibition factors. The first term on the right hand describes the effect of protons on the rate, the second term the rate contribution due to hydrolysis, the third term entails the contribution of OH^-, the fourth gives the influence of CO_2 and the last term the effect of organic acids.

The influence of pH on the dissolution kinetics of several silicate minerals is shown in Figure 8.16. In all cases, the rate shows a distinct minimum at neutral pH and the rate increases both towards lower and higher pH. However, the absolute rates and the rate dependency on the pH vary significantly. For example, in the near neutral pH range the dissolution rate of hornblende is two orders of magnitude higher than of albite.

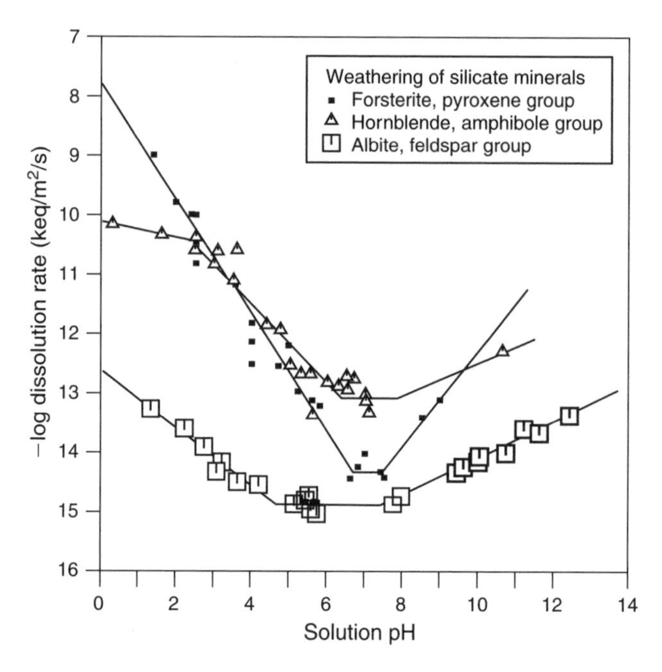

Figure 8.16. Experimental dissolution rates for different silicate minerals at 25°C (Sverdrup and Warfvinge, 1988; Chou and Wollast, 1985).

The inhibition factor f_H in Equation (8.26) indicates that proton-promoted dissolution is inhibited by solution elements included in the term f_H. The inhibition factors have the form:

$$f_H = \left[1 + \frac{[BC]}{Lim_{BC,H}}\right]^{x_{BC}} \left[1 + \frac{[Al^{3+}]}{Lim_{Al,H}}\right]^{x_{Al}} \tag{8.26a}$$

$$f_{H_2O} = \left[1 + \frac{[BC]}{Lim_{BC,H_2O}}\right]^{z_{BC}} \left[1 + \frac{[Al^{3+}]}{Lim_{Al,H_2O}}\right]^{z_{Al}} \tag{8.26b}$$

$$f_{org} = \left[1 + \frac{[R^-]}{Lim_{org}}\right]^{0.5} \tag{8.26c}$$

$$f_{OH} = f_{CO_2} = 1 \tag{8.26d}$$

where *Lim* is the limiting activity below which the inhibitive effect of the specific solute starts to become negligible. [*BC*] indicates the sum of the activities of the base cations Na^+, K^+, Mg^{2+} and Ca^{2+}. The exponents x_i and z_i are empirical.

Equations (8.26a and b) indicate that dissolved aluminum may influence the dissolution rate. The effect of aluminum on the dissolution rate of albite is shown in Figure 8.17 and apparently, already low concentrations of dissolved aluminum inhibit the dissolution of albite. Since the Al-concentration under natural conditions also is strongly pH dependent (Figure 8.7), the two rate-controlling parameters interact in a complex way. The first term in Equation (8.26a) indicates that the presence of base cations in solution may also slow down silicate dissolution. The fourth and fifth term of Equation (8.26) show that

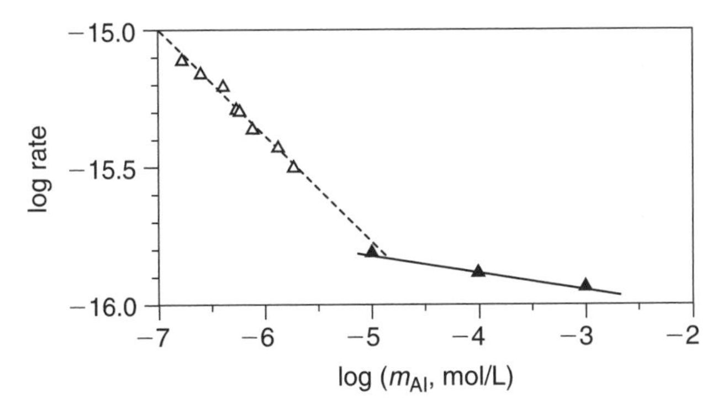

Figure 8.17. The effect of dissolved Al concentrations on the dissolution rate of albite at pH 3 (Chou and Wollast, 1985).

both a high CO_2 and a high content of free organic acid radicals may accelerate the dissolution rate (Sverdrup, 1990; Sverdrup and Warfvinge, 1995). The role of CO_2 is partly to act as a provider of protons, but particularly in the near neutral and basic pH range, the carbonate ion may accelerate silicate weathering directly (Berg and Banwart, 2000). The coefficients for calculating approximate rates with Equation (8.26) were obtained from Sverdrup and Warfvinge (1995) and Sverdrup (1990) and are listed in Tables 8.7 and 8.8. An application for K-feldspar dissolution is presented in Example 8.4.

Table 8.7. Coefficients for calculating approximate dissolution rates of silicate minerals in soils for use with Equation (8.26). Coefficients k_i are in mol/m²/s and given as $pk_i = -\log k_i$ for 8°C, coefficients Lim_i are $\times 10^6$. Arrhenius factors for recalculating pk_i to other temperatures are in Table 8.8. $k_{CO_2} \approx 10^{-13}$ mol/m²/s, $k_{org} \approx 10^{-12}$ mol/m²/s, $Lim_{org} = 5 \times 10^{-6}$.

Mineral	pk_H	n_H	Lim_{Al}	x_{Al}	Lim_{BC}	x_{BC}	pk_{H_2O}	z_{Al}	z_{BC}	pk_{OH}	o_{OH}
K-feldspar, $K(AlSi_3)O_8$	11.7	0.5	4	0.4	500	0.15	14.5	0.14	0.15	13.1	0.3
Albite, $Na(AlSi_3)O_8$	11.5	0.5	4	0.4	500	0.2	13.7	0.14	0.15	11.8	0.3
Anorthite, $Ca(Al_2Si_2)O_8$	6.9	1.0	4	0.4	500	0.25	13.2	0.14	0.25	12.0	0.25
Pyroxene, $Mg_{0.9}Ca_{0.5}Fe_{0.6}Si_2O_6$	9.9	0.7	500	0.2	200	0.3	15.1	0.1	0.3	8.6	0.5
Hornblende, $NaCa_2Mg_4(Fe)(Al_2Si_6)O_{22}(OH)_2$	11.4	0.7	30	0.4	200	0.3	14.0	0.3	0.3	11.2	0.3
Epidote, $Ca_2(Al_2Fe)Si_3O_{12}OH$	11.4	0.5	500	0.3	200	0.2	15.3	0.2	0.2		
Biotite, $K(Mg_2Fe)(AlSi_3)O_{10}(OH)_2$	12.6	0.6	10	0.3	500	0.2	14.5	0.2	0.2		
Muscovite, $KAl_2(AlSi_3)O_{10}(OH)_2$	12.2	0.5	4	0.4	500	0.1	14.5	0.2	0.1	12.7	0.3
Chlorite, $(Mg_8Fe_2Fe_3Al_2)(Al_3Si_5)O_{20}(OH)_{16}$	12.5	0.7	50	0.2	200	0.2	14.9	0.1	0.1	13.0	0.3
Apatite, $Ca_5(PO_4)_3OH$	10.8	0.7			300	0.4	13.8		0.2		

Table 8.8. Temperature factors (E_a/ (2.303 R)) in the Arrhenius equation (4.63) for calculating silicate weathering rates k_i at another temperature (T K) than 8°C (271 K):

$$pk_{i,T} = -\log k_{i,T} = pk_{i,271} + \frac{E_a}{2.303\,R}\left[\frac{1}{T} - \frac{1}{271}\right] \tag{8.27}$$

	pk_H	pk_{H2O}	pk_{OH}	pk_{CO2}	pk_{org}
E_a / (2.303 R)	3500	2000	2500	2000	3000

EXAMPLE 8.4. *Dissolution kinetics of K-feldspar calculated by PHREEQC*

We calculate the dissolution kinetics of K-feldspar in the soil from Example 4.14 and consider the various influences on the rate given by Equation (8.26). The soil contains 10% K-feldspar as 0.1 mm spheres, $A_0 = 2.26\,m^2/kg$, $\varepsilon_w = 0.3$ hence $V_{H_2O} = 0.162$ L/kg (cf. Example 4.14). The PHREEQC input file is:

```
DATABASE phreeqc.dat                              # for K-feldspar
SOLUTION 1
 temp 10; pH 4; Cl 0.1 charge
EQUILIBRIUM_PHASES 1
 Gibbsite 0 0
# CO2(g) -2.5
KINETICS 1
 K-feldspar
 -m0 2.16                                         # 10% K-fsp, mol/L pore water
 -parms 2.26 0.162                                # A0 in m2, V in L
 -steps 4.7e7 in 40                               # 1.5 years
INCREMENTAL_REACTIONS true

RATES
 K-feldspar
 -start
                    # specific rates from Table 8.7 in mol/m2/s
                    # parm(1) = A in m2, parm(2) = V in L (recalc's sp. rate to mol/kgw)
 1 A0 = parm(1);  2 V = parm(2)
                    # find activities of inhibiting ions. . .
 3 a_Al = act("Al+3"); 4 BC = act("Na+") + act("K+") + act("Mg+2") + act("Ca+2")
                    # temp corrected with the Arrhenius eqn, Table 8.8
                    # the difference in temperature, TK gives solution temp in Kelvin...
 10 dif_T = 1/TK - 1/271
                      # rate by H+...
 20 pk_H = 11.7 + 3500 * dif_T
 22 rate_H = 10^-pk_H * act("H+")^0.5 / ((1 + a_Al / 4e-6)^0.4 * (1 + BC / 5e-4)^0.15)
                    # rate by hydrolysis...
 30 pk_w = 14.5 + 2000 * dif_T
 32 rate_w = 10^-pk_w / ((1 + a_Al / 4e-6)^0.14 * (1 + BC / 5e-4)^0.15)
                    # rate by OH-...
 40 pk_OH = 13.1 + 2500 * dif_T
 42 rate_OH = 10^-pk_OH
                    # rate by CO2...
 50 pk_CO2 = 13.0 + 2000 * dif_T
 52 rate_CO2 = 10^-pk_CO2 * (10^SI("CO2(g)"))^0.6
                    # Sum the rate contributions...
 60 rate = rate_H + rate_w + rate_OH + rate_CO2
                    # normalize to mol/kgw, correct for m/m0 and the approach to equi...
 70 rate = rate * A0 / V * (m/m0)^0.67 * (1 - SR("K-feldspar"))
                    # integrate...
 80 moles = rate * time
 90 save moles
 -end

USER_GRAPH
 -head year Si
 -axis_titles Years "mmol Si/L"
 -start
 10 graph_x Total_time / (365 * 24 * 3600)
 20 graph_y tot("Si") * 1e3
 -end
END
```

The initial solution has a pH of 4 and temperature is 10°C. From the previous discussion, incongruent dissolution of K-feldspar is expected and therefore the input file allows gibbsite to precipitate once the solution becomes supersaturated. The result of this calculation is shown in Figure 8.18 as case a), indicated by the bold line. Dissolved Si, monitoring K-feldspar dissolution, increases almost linearly until equilibrium for K-feldspar is reached and then remains constant.

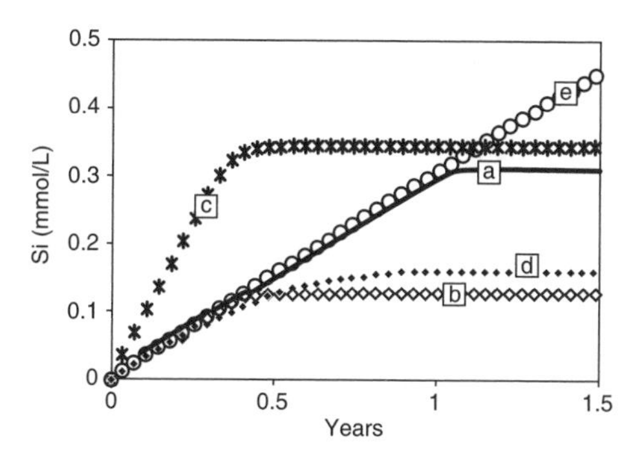

Figure 8.18. PHREEQC calculation of K-feldspar dissolution kinetics using Equation (8.26). The conditions are varied as follows: (**a**) Equilibrium for gibbsite imposed, initial pH = 4 and 10°C. (**b**) as case (a) but without equilibrium for gibbsite, (**c**) as case (a) but using a temperature of 25°C, (**d**) as case (a) but with an initial pH of 5, (**e**) as case (a) but with $P_{CO_2} = 10^{-2.5}$ atm imposed.

Small modifications in the input file allow the exploration of the effects of the different parameters in Equation (8.26). Only one parameter is changed each time while the others remain the same as in case a). First the equilibrium constraint for gibbsite was removed, b), and as the result less than half the K-feldspar dissolves to reach equilibrium. Without gibbsite precipitation, both Al^{3+} and OH^- remain in solution and saturation with K-feldspar is reached more rapidly. In c) the temperature is increased from 10 to 25°C and as the result the dissolution rate of K-feldspar more than doubles, illustrating the importance of the temperature on weathering rates. Also, the solubility of K-feldspar increases. In d) the initial pH was increased from 4 to 5 which decreased the amount dissolved at equilibrium by about 50%. Finally, Equation (8.26) predicts that CO_2 stimulates the dissolution rate and therefore in e) a P_{CO_2} of $10^{-2.5}$ atm was included. CO_2 has two different effects. First the CO_2 increases the dissolution rate slightly and second, hydroxyl ions react with H_2CO_3 and form HCO_3^- allowing more K-feldspar to dissolve before saturation is reached.

QUESTIONS:
Include dissolution of quartz (Example 4.15), discuss the effect on the Si concentration with time?
Let quartz precipitate 10 times slower than it dissolves?
The quartz rate is for 25°C, include the temperature effect, the Arrhenius activation energy is $E_a = 90$ kJ/mol?
Compare the dissolution rates of K-feldspar and quartz at 100°C?
Alkalis *increase* the quartz dissolution rate by $(1 + 1.5\ m_{Na})$, m_{Na} in mmol/L (Dove and Rimstidt, 1993). Include the alkali effect assuming that K^+ reacts similar to Na^+?

ANSWER: add the rate for quartze and modify KINETICS

```
RATES
 Quartz
 -start
 1 A0 = parm(1);   2 V = parm(2)
 10 dif_T = 1/TK - 1/298
 20 rate = 10^-(13.7 + 4700*dif_T)  *  (1 - SR("Quartz"))\
            * A0/V *  (m/m0)^0.67 *  (1 + 1500*tot("K"))
 30 if SR("Quartz") > 1 then rate = rate * 0.1
 40 moles = rate * time
 50 save moles
 -end
```

8.6 FIELD WEATHERING RATES

Silicate weathering rates have been studied extensively in soils (Sverdrup, 1990; Rosén, 1991; White and Brantley, 1995). However, agreement between field and laboratory dissolution rates is frequently poor and subject to continuous debate (White and Brantley, 1995; Langan et al., 2001; White et al., 2001).

The kinetic dissolution model PROFILE (Sverdrup and Warfvinge, 1988, 1993; Sverdrup, 1990) uses laboratory dissolution rate data to estimate the field rate of silicate weathering. The distribution of minerals in the soil must be determined and the mineral surface areas estimated. Then, using the kinetic rate laws determined in the laboratory, the model calculates the dissolution of different minerals considering the effects of the soil solution composition, available surface areas and the temperature. The total weathering rate is given in terms of amount of acid neutralized by base cation release as:

$$ANR_{eq} = \sum_{i=1}^{horizons} z_i \theta_i \sum_{j=1}^{minerals} r_j A_{w_{ij}} n_{BCj} \tag{8.28}$$

where ANR_{eq} is the acid neutralization rate for the soil (eq/m^2/yr), z_i is thickness of layer i (m), θ_i is amount of water in the layer (kg H_2O/m^3 soil), r_j is in mol/m^2/yr, $A_{w_{ij}}$ the total exposed surface area (m^2/kg H_2O) and $n_{BC,j}$ are the equivalents of base cations released per mol ($Na^+ + K^+ + 2(Ca^{2+} + Mg^{2+})$ per formula unit).

The model has been applied to a number of watersheds (Sverdrup and Warfvinge, 1993; Hodson et al., 1997). Results for the Gårdsjön watershed, which is located on a granitic bedrock with shallow podzolic soils are shown in Figure 8.19. Generally, the weathering rate increases down through the upper 0.65 m of the profile. Plagioclase, hornblende, epidote and apatite are the main contributors to weathering. Hornblende and epidote constitute at most 2% of the mineral content of the soil, while the content of microcline and plagioclase ranges between 12 and 16%. The important role of even minor amounts of hornblende and epidote is obviously due to their fast dissolution kinetics. This type of model clearly requires detailed information on soil mineralogy, the dissolution kinetics for the various minerals, and particularly, the exposed reactive surface areas of the minerals (Hodson et al., 1997).

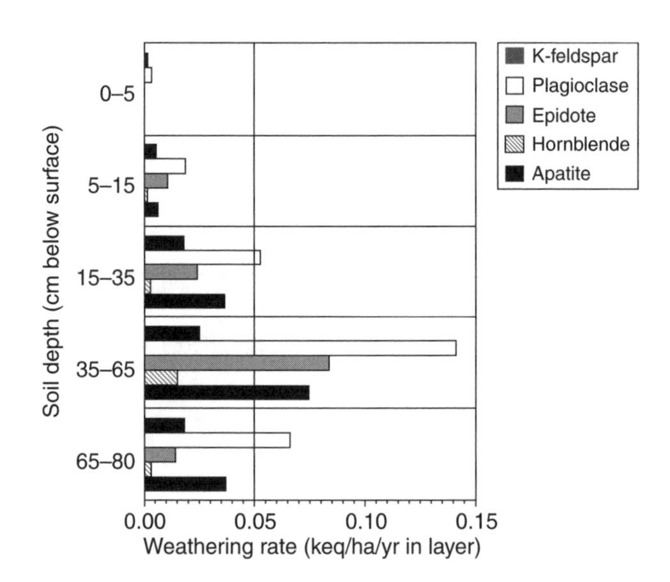

Figure. 8.19. Contribution of different minerals to weathering in four soil horizons in the granitic watershed of Gårdsjön, Sweden (Sverdrup and Warfvinge, 1993).

An alternative way to estimate field weathering rates is to consider the change in mineralogical composition through a soil sequence which has developed over a known period of time (Bain et al., 1991; Olsson and Melkerud, 1991; Sverdrup, 1990; White et al., 1996). The disappearance of minerals is calculated relative to an inert tracer such as quartz or zircon. The historical acid neutralization rate is obtained by summing the relative loss over depth:

$$ANR_{eq} = \left[\frac{1}{\Delta t}\right] \sum_i z_i \sum_j \left[\frac{x_{S,qu}}{x_{R,qu}} x_{R,j} - x_{S,j}\right] \frac{w_{S,j}}{MW_j} n_{BC,j} \tag{8.29}$$

where Δt is the time interval of soil formation, z_i is the thickness of the soil layer (m), x indicates the weight fraction in the parent rock (subscript R) or soil (subscript S) of quartz (qu) or mineral j, $w_{S,j}$ is the amount of mineral j in the soil layer (g/m^3), MW_j is the molecular weight of mineral j (g/mol) and $n_{BC,j}$ gives again the equivalent base cations per formula unit.

Figure 8.20 shows the weathering rates calculated from the mineral distributions in granitic sands found in alluvial terraces of different age in the Merced chronosequence (White et al., 1996).

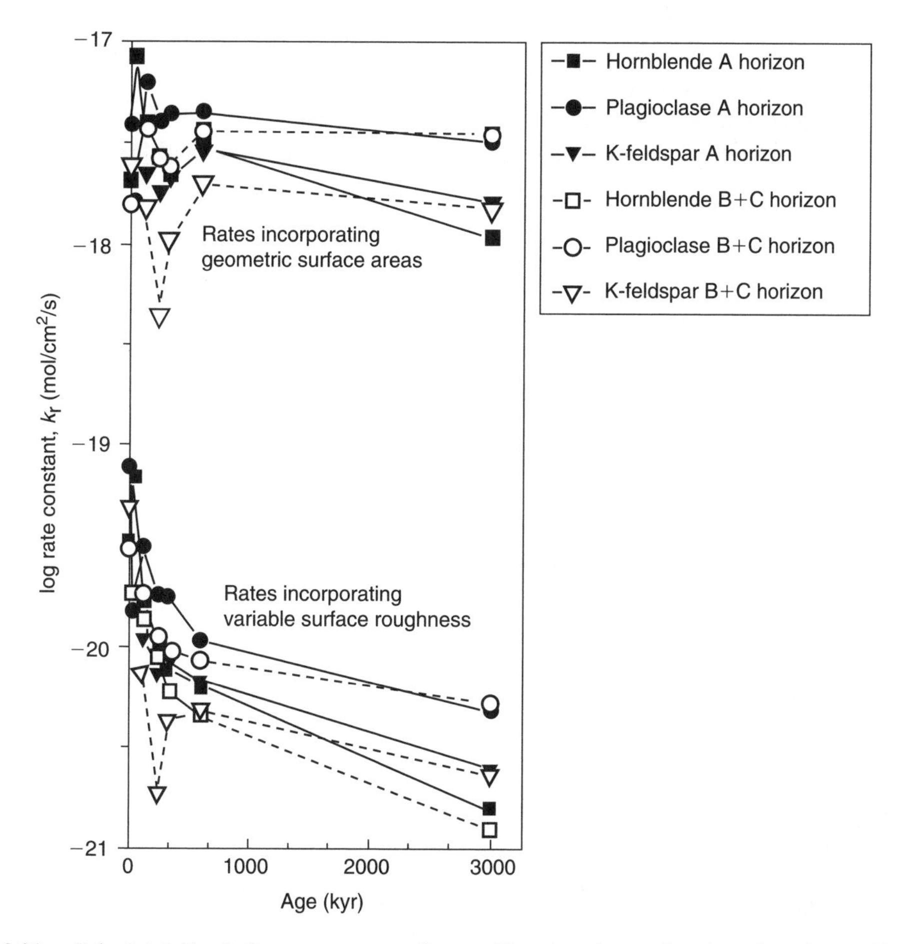

Figure 8.20. Calculated dissolution rate constants for specific minerals as a function of age in granitic sands of the Merced chronosequence. Rates are given both for geometric surface areas and for variable surface roughness based on the BET surface area (White et al., 1996).

Rates are expressed both as a function of the geometric surface area and relative to the actual surface area based on the BET surface area (cf. Chapter 4). Naturally, the BET based rates are much lower than those based on the geometric surface area. Furthermore, the weathering rates for individual minerals decrease with soil age, fresh mineral grains dissolve faster than old mineral grains (Bain et al., 1991). Another uncertainty is that the weathering rate calculated from mineralogical changes in the soil profile provides an average weathering rate over a long period of time, during which the conditions controlling the weathering rate, like the pH of the soil solution, may well have varied.

The third approach towards obtaining field weathering rates is the mass balance as already introduced in Section 8.4 for a groundwater flow path. To obtain the rate of weathering for a whole watershed the general mass balance equation for base cations (Ca, Mg, K, Na) becomes:

$$\begin{aligned}[\text{flux from weathering}] = {} & [\text{flux with outflow}] + [\text{flux in exchange pool}] \\ & + [\text{flux in biomass}] - [\text{flux from atmosphere}]\end{aligned} \tag{8.30}$$

In a steady state situation both the changes in the exchange and biomass will must be zero, and the weathering rate is then obtained as the difference between solute outflow and atmospheric input. Generally, young systems where the soil develops on freshly exposed rock have a high contribution of solutes from weathering, while in strongly weathered areas the atmospheric input becomes increasingly important (Chadwick et al., 1999).

Table 8.9 shows the mass balance for the Hubbard brook catchment area, which we already introduced in Chapter 2, to highlight some of the problems in obtaining weathering rates from mass balances.

Table 8.9. Yearly element balances in the Hubbard Brook area, New Hampshire, USA (Likens and Bormann, 1995).

Component	Chemical element							
	Ca	Mg	Na	K	N	S	P	Cl
	Standing stock (kg/ha)							
Aboveground biomass	383	36	1.6	155	351	42	34	*
Belowground biomass	101	13	3.8	63	181	17	53	*
Forest floor	372	38	3.6	66	1256	124	78	*
	Annual flux (kg/ha.yr)							
Bulk precipitation input	2.2	0.6	1.6	0.9	6.5	12.7	0.04	6.2
Gaseous or aerosol input	*	*	*	*	14.2	6.1	*	?
Weathering release	21.1	3.5	5.8	7.1	0	0.8	?	*
Streamwater output								
Dissolved substances	13.7	3.1	7.2	1.9	3.9	17.6	0.01	4.6
Particulate matter	0.2	0.2	0.2	0.5	0.1	<0.1	0.01	*
Vegetation uptake	62.2	9.3	34.8	64.3	79.6[a]	24.5[a]	8.9	*
Litter fall	40.7	5.9	0.1	18.3	54.2	5.8	4.0	*
Root litter	3.2	0.5	0.01	2.1	6.2	0.6	1.7	*
Throughfall and stemflow	6.7	2.0	0.3	30.1	9.3	21.0	0.7	4.4
Root exudates	3.5	0.2	34.2	8.0	0.9	1.9	0.2	1.8
Net mineralization	42.4	6.1	0.1	20.1	69.6	5.7	?	?
Aboveground biomass accretion	5.4	0.4	0.03	4.3	4.8	0.8	0.9	*
Belowground biomass accretion	2.7	0.3	0.12	1.5	4.2	0.4	1.4	*
Forest floor accretion	1.4	0.2	0.02	0.3	7.7	0.8	0.5	*

*Small, unmeasured, [a]Root uptake.

First of all, the weathering rate is very small compared to the amount of elements fixed in the biomass. Also, the rate of annual element cycling by the biomass is much larger than the release by weathering. Accordingly, small changes in total biomass, which can be difficult to measure, may yield significant apparent variations in the weathering rate.

The same problem applies to the pool of exchangeable cations, which, under the stress of acid input, typically is not in steady state. The change in composition of the exchange pool is also difficult to analyze and the weathering rates obtained by the mass balance approach may include the depletion of the pool of exchangeable cations. The buffering effects become clear when weathering rates are calculated for different seasons in which discharge varies (Drever and Clow, 1995). Usually, the concentrations in surface water are not simply diluted as discharge increases, but show the hydrological response of the watershed in which water qualities vary with depth in the soil profile, and most often also laterally, over the area (Richards and Kump, 2003).

For Ca^{2+}, some of the problems of estimating the weathering rate from mass balance calculations can be solved using the strontium isotope ratio $^{87}Sr\,/\,^{86}Sr$ (Åberg et al., 1989; Bailey et al., 1996; Clow et al., 1997; Bullen et al., 1997; Stewart et al., 2001). ^{87}Sr is a decay product from ^{87}Rb, a process with a half-life of 48.8 billion year. Due to similar geochemical properties, Sr^{2+} can be considered as an analogue to Ca^{2+} for studying Ca^{2+} behavior in catchments. The $^{87}Sr\,/\,^{86}Sr$ ratio in atmospheric deposition differs from that in bedrock so that the contribution by weathering can be estimated from the difference between the $^{87}Sr\,/\,^{86}Sr$ ratio in deposition and in runoff. This approach assumes isotopic equilibration between soil solution, exchange complex and runoff. The weathering rate is calculated from the mass balance:

$$R_{Ca} = P_{Ca} \frac{q - a}{s - q} \tag{8.31}$$

Where R_{Ca} and P_{Ca} are the Ca^{2+} weathering and deposition rate, respectively, a is the $^{87}Sr\,/\,^{86}Sr$ ratio in the deposition, q the $^{87}Sr\,/\,^{86}Sr$ ratio in the runoff and s the $^{87}Sr\,/\,^{86}Sr$ ratio in the mineral matrix. The combination of weathering rates obtained using the Sr-isotope method, with those estimated from traditional mass balances also permits to separate of the weathering rate from the depletion of the cation exchange complex (Åberg et al., 1989; Bailey et al., 1996). The method assumes isotopic equilibrium between the soil solution and the exchanger which may not always be attained (Bullen et al., 1997). Furthermore, the $^{87}Sr\,/\,^{86}Sr$ ratio of different minerals in the bedrock will vary as it depends on their original Rb / Sr ratio. The weathering rate based on the whole rock $^{87}Sr\,/\,^{86}Sr$ ratio therefore assumes the congruent release of Sr. However, since different minerals weather at different rates, and also may dissolve incongruently, caution is warranted (Bullen et al., 1997; Brantley et al., 1998). Preferably, the strontium isotopic behavior of the various minerals in the rock should be evaluated.

QUESTIONS:

Estimate the contributions of K-feldspar, plagioclase, epidote and apatite to ANR_{eq} in Gårdsjön (Figure 8.19).

ANSWER: K-fsp 0.09, plag 0.28, epi 0.135, ap 0.16 keq/ha/yr. Total: 0.67 keq/ha/yr.

Estimate the background acid input (keq/ha/yr) by 600 mm rain in Sweden (Table 2.2; note to convert NH_4^+ to NO_3^-)

ANSWER: 0.087 keq/ha/yr

What causes the weathering of the silicates in Gårdsjön?

ANSWER: CO_2 produced in the soil

Estimate the acid deposition with 800 mm rain in Beek (NL) in ’78–’83 and in ’97–’98 (Table 2.2)?

ANSWER: 2.2 and 1.2 keq/ha/yr

Estimate ANR_{eq} of a 1 m thick sandy soil with 90% quartz and 10% albite, $\varepsilon_w = 0.1$, pH = 4.5 (Use Table 8.7, neglect inhibition)?

ANSWER: 0.05 keq/ha/yr

8.7 ACID GROUNDWATER

Acid groundwater is found at an increasing number of places throughout the industrialized world (e.g. Appelo et al., 1982; Eriksson, 1981; Hultberg and Wenblad, 1980; Böttcher et al., 1985; Moss and Edmunds, 1992; Hansen and Postma, 1995; de Caritat, 1995; Hinderer and Einsele, 1997; Donovan, 1997). The threat of acidification towards the groundwater resource is illustrated in Figure 8.21, showing a strong decrease of pH in the youngest groundwater of a carbonate-free sandy aquifer in the Netherlands.

Several processes may cause the acidification of groundwater. First, there is natural acidification through CO_2 production and root respiration in the soil by the overall reaction:

$$CH_2O + O_2 \rightarrow H_2O + CO_2 \quad (8.32)$$

As was pointed out in Chapter 5, the lower pH limit from CO_2 production in soil is around 4.6, in the absence of buffering processes within the aquifer.

A second potential source of acidification is the excessive use of ammonia and manure as fertilizers. Nitrification of ammonia is a major acidifying process in the soil:

$$NH_4^+ + 2O_2 \rightarrow NO_3^- + 2H^+ + H_2O \quad (8.33)$$

If, subsequently, the nitrate is removed by denitrification, the proton production due to (8.33) is balanced again by the HCO_3^- production of denitrification (Chapter 9):

$$5CH_2O + 4NO_3^- \rightarrow 2N_2 + 4HCO_3^- + CO_2 + 3H_2O \quad (8.34)$$

However, the ubiquitous presence of nitrate in aquifers shows that this is not the case and nitrification of ammonia must be considered as an important acidifying process. The net effect of nitrification on the pH of recharge water is largely determined by the amount of lime which is applied together with fertilizers to the soil.

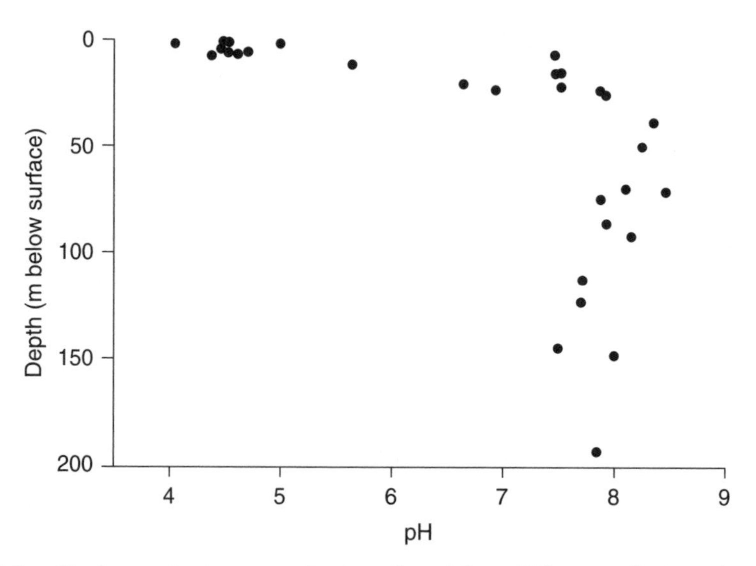

Figure 8.21. The pH of groundwater versus depth, collected from different wells in sandy aquifers of the Veluwe area, the Netherlands (Appelo et al., 1982).

Table 8.10. The composition of 1938 rain and 1980 acid rain and the effect of evapotranspiration in the Netherlands (Appelo, 1985). Concentrations are in mmol/L. 3 × gives values calculated by PHREEQC using an evapotranspiration factor of 3 and $[P_{CO_2}] = 10^{-3.5}$, 2) Na + K.

	1938 rain	3×	1980 rain	3×
pH	5.4	5.62	4.52	3.54
Na^+	0.07[2)]	0.21[2)]	0.073	0.219
K^+			0.004	0.012
Mg^{2+}	0.03	0.09	0.009	0.027
Ca^{2+}	0.04	0.12	0.014	0.042
NH_4^+	0.03	–	0.13	–
Cl^-	0.09	0.27	0.085	0.255
Alk	0.06	0.002	–	–
SO_4^{2-}	0.045	0.135	0.071	0.213
NO_3^-	–	0.09	0.061	0.573

A third major acidifying process is the oxidation of pyrite (FeS_2). Pyrite is found, at least in small quantities, in most reduced sediments and lowering the groundwater table may cause the oxidation of pyrite. This process is discussed in detail in Chapter 9. For our present purpose it is summarized by the overall reaction:

$$2FeS_2 + {}^{15}\!/\!_2O_2 + 5H_2O \leftrightarrow 2FeOOH + 4SO_4^{2-} + 8H^+ \qquad (8.35)$$

Pyrite oxidation is one of the most strongly acid-producing reactions found in nature.

Among the possible causes for acid groundwater, acid rain is the one that impacts the largest tracts of land and its detrimental effect on ecosystems of northern Europe and North America is well documented (Wright and Henriksen, 1978; Likens and Bormann, 1995; Ulrich et al., 1979; Drabløs and Tollan, 1980; Kirchner, 1992). Acid rain is, to an increasing extent, the cause of groundwater acidification (de Caritat, 1995; Hansen and Postma, 1995; Hinderer and Einsele, 1997; Kjøller et al., 2004). The acid rain problem originates from fossil fuel combustion, which produces nitrous oxides and SO_2, that subsequently are oxidized in the atmosphere and precipitate as dilute sulfuric and nitric acid solutions (Berner and Berner, 1996, see also Chapter 2).

Table 8.10 compares the composition of rainwater before heavy industrialization in 1938 and acid rain around 1980 in the Netherlands. A decrease in pH of almost one unit is evident and associated with significant increases of sulfate and the nitrogen compounds. Evapotranspiration further concentrates the solutes and also changes the pH of the solution. In the 1938 rainwater the pH increases from 5.4 to 5.6 upon concentration. However, the pH of 1980 rain changes from an initial value of 4.52 to 3.54 after concentration. Most of this pH drop is due to oxidation of ammonia, present at a high concentration in 1980 rainwater, to nitrate (Reaction 8.33).

8.7.1 *Buffering processes in aquifers*

The acid water that is introduced in soils and aquifer systems may be neutralized by reactions with soil and aquifer materials. To predict the extent of neutralization, the first assessment is to consider the relation between the geology and the sensitivity for acidification. Obviously, rocks which contain carbonate minerals are unlikely to develop acid groundwater because of the fast dissolution kinetics of carbonates (Chapter 5). In a regional survey of groundwater in the UK, Edmunds and Kinniburgh (1986) pointed out that groundwaters low in alkalinity are most vulnerable to acidification. In fact a decreasing alkalinity over time is a good warning for groundwater acidification in the future (Hultberg and Wenblad, 1980).

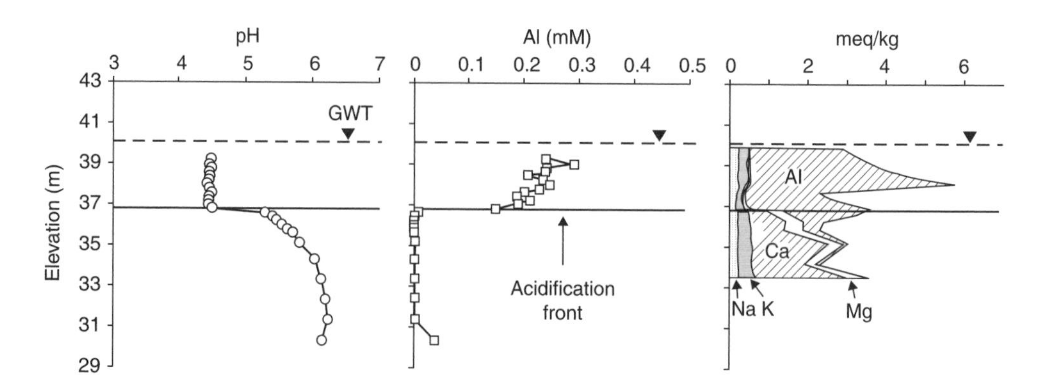

Figure 8.22. Groundwater acidification in the saturated zone of a sandy aquifer below a coniferous forest in Denmark. Displayed are groundwater pH, dissolved aluminum and the exchanger composition (Kjøller et al., 2004).

As carbonate-free rocks, Edmunds and Kinniburgh (1986) listed granites, acid igneous rock and clean quartz sandstones, which contain mainly slowly dissolving silicate minerals, as the most susceptible to acidification, while basic and ultrabasic rocks, that are dominated by more quickly dissolving silicates, were found at the other end of the scale. According to Sverdrup (1990) the presence of a few percent of pyroxene, hornblende or biotite should suffice to prevent acidification when the acid load is less than 0.5 keq/ha/yr.

Figure 8.22 shows groundwater acidification in a low reactive sandy aquifer below a coniferous forest. The groundwater in the upper part of the profile has a pH of 4.5, and suddenly increases at a depth to about 5.5–6.0 in what appears as a distinct acidification front. The acidification front is situated at 3 m below the water table. Since the downward pore water velocity is 1–1.4 m/yr, and acid rain has been infiltrating the soil for decades, buffering must take place within the aquifer. The upper groundwater with the low pH, also has a high Al^{3+} concentration and this suggest that buffering is related to the dissolution of Al^{3+} containing minerals. According to Table 8.10 the pH of the acid rain that enters the soil is close to 3 while the pH in the uppermost groundwater is slightly above 4 (Figure 8.22). The high Al^{3+} concentration indicates that most mineral dissolution and buffering already takes place in the unsaturated soil. The minerals being dissolved could be primary silicates, clay minerals, $Al(OH)_3$, or any combination of these.

EXAMPLE 8.5. *Acid groundwater formation and gibbsite buffering*

In the Veluwe (the Netherlands), acid groundwater develops from acid rain. Groundwater is potentially even more acid than rain because acidity is concentrated by evapotranspiration and NH_4^+ is nitrified. The pH increases again when gibbsite dissolves or when NO_3^- is denitrified. This example shows how to calculate buffering and dissolution reactions when acid rain becomes concentrated groundwater.

1. Rainwater from 1980 (Table 8.10) is concentrated 3 times by evapotranspiration and becomes groundwater. All NH_4^+ is oxidized to NO_3^-. Calculate the composition.
2. $Al(OH)_3$ (gibbsite) dissolves in the soil. Calculate pH and Al^{3+}-concentration.
3. Nitrate is denitrified in peat layers in the aquifer. Calculate pH and Al^{3+}-concentration.

ANSWER:

We use PHREEQC, enter the rain as SOLUTION 0 with the sum of NH_4^+ and NO_3^- as NO_3^- (=N(5)) and obtain charge balance on pH. From this water H_2O is removed ('evapotranspirated') with keyword REACTION and the result is saved as SOLUTION 1. In the next simulation equilibrium with gibbsite is imposed. With USER_PRINT we obtain the pH and the Al^{3+} concentration in mg/L for easy comparison with the drinking water limit of 0.2 mg Al/L. In the third simulation, we add CH_2O to the previous solution for denitrification (5/4 times the moles of NO_3^- according to Reaction (8.34)).

```
SOLUTION 0        # Rainwater at Deelen .(1980)
 temp 10
 pH 4.5 charge
 -units umol/L
 Na 73; K 4;    Mg 9;    Ca 14
 Cl 85; S(6) 71; N(5) 191
REACTION 0
 H2O -1; 37.005      # remove 37/55.5 = 2/3 of the water
SAVE solution 1
END

USE SOLUTION 1
EQUILIBRIUM_PHASES 1
 Gibbsite
USER_PRINT
 -start
 10 print 'pH = `, -la("H+"), '.     mg Al/L = `, tot("Al") * 26981
 -end
SAVE solution 2
END

USE SOLUTION 2; USE equilibrium_phases 1
REACTION 2
 CH2O 1.25; 191e-6
END
```

The result is:

Solution number	pH	mg Al/L	mM NO_3^-
1, evapotranspired rain	3.54	–	0.573
2, soil water	4.00	7.2	0.573
3, groundwater + peat	4.17	2.3	0.0

Clearly, the Al concentration is above the drinking water limit.

QUESTIONS:

The concentration of SO_4^{2-} in this rainwater has decreased to 21 μM in 2000. Calculate the pH and Al concentration in soil- and groundwater?

ANSWER: No.1: pH = 3.25; No.2: pH = 4.04, 4.5 mg Al/L; No.3: pH = 4.7, 0.07 mg Al/L.

Discuss why the SO_4^{2-} decrease lowers the pH when NH_4^+ is oxidized. (*Hint*: consider the effect of HSO_4^-)

In a dry year the acid rain was concentrated 5 times. Calculate the concentrations?

ANSWER: No.2: pH = 3.9, 15 mg Al/L.

Example 8.5 suggests that 0.27 mmol Al/L was mobilized in the Dutch rain of the 1980's. In Figure 8.22 the Al^{3+}-concentration in groundwater is around 0.25 mmol/L showing that the same conditions apply to Denmark. As found elsewhere in Denmark (Figure 8.8) the activity of dissolved Al^{3+} and the pH are in near equilibrium with microcrystalline gibbsite. Figure 8.22 shows the sediment exchanger to be filled nearly exclusively with Al^{3+} in the upper part, while in the lower part Ca^{2+} is the predominant adsorbed cation with subordinate amounts of Mg^{2+}.

At exactly the same depth where the Al^{3+} concentration in groundwater is decreasing, also exchangeable Al^{3+} decreases (Figure 8.22). Apparently, the acid water loses Al^{3+} by ion exchange (Dahmke et al., 1986; Hansen and Postma, 1995; Kjøller et al., 2004):

$$\tfrac{1}{3}Al^{3+} + \tfrac{1}{2}Ca\text{-}X_2 \leftrightarrow \tfrac{1}{3}Al\text{-}X_3 + \tfrac{1}{2}Ca^{2+} \quad (8.36)$$

Since gibbsite is present, the adsorption of Al^{3+} will cause dissolution of gibbsite:

$$Al(OH)_{3\ gibbsite} + 3H^+ \leftrightarrow Al^{3+} + 3H_2O \quad (8.4)$$

The result is an increase in pH, controlled by the two simultaneous equilibria.

The progression of the acidification front is restricted by the exchange of Al^{3+}. We can apply the retardation equation on the changes in solid and dissolved Al over the front,

$$R = 1 + \Delta q_{Al} / \Delta m_{Al} \quad (8.37)$$

Most of the change in solid Δq_{Al} is located in exchangeable Al-X_3. Since the exchanger above the front is filled with Al, and below the front adsorbed Al is negligible, $\Delta q_{Al} \approx \Delta m_{CEC}$ and equal to 3 meq/kg = 6.17 mmolAl/L. Δm_{Al} is about 0.25 mmolAl/L and accordingly, the retardation $R = 1 + 6.17 / 0.25 = 25.7$. Since the downward rate of water transport is 1.0–1.4 m/y, the current depth of the acidification front at 3 m below the water table, corresponds to 77–55 years of acidification, which seems a reasonable estimate considering the history of industrialization.

EXAMPLE 8.6. *Modeling acidification with PHREEQC*

The file below models the downward migration of acid water into an aquifer in equilibrium with gibbsite and an exchanger. The $SI_{gibbsite}$ of 0.38 corresponds to field observations and the exchanger capacity of 18.5 meq/L is equivalent to measured values of around 3 meq/kg. The pre-acid water quality is unknown (it has long been flushed), but is estimated by analogy with the Dutch rainwater trend (Table 8.10) to have half the SO_4^{2-} and, 0.2 times the NO_3^- concentration of the acid groundwater. Furthermore, the Ca^{2+} / Mg^{2+} ratio was probably 2.5, the same as in water below the acidity front, and C(4) was set to equilibrium with $[P_{CO_2}] = 10^{-2}$. See also Kjøller et al. (2004).

```
DATABASE phreeqc.dat
SOLUTION 1-20
 pH 6 charge;  temp 8
 Na 1.0;      K 0.048;      Mg 0.07;      Ca 0.17
 Cl 1.0;      S(6) 0.16;    N(5) 0.02;    C(4) 0.6      CO2(g) -2
EQUILIBRIUM_PHASES 1-20
 Gibbsite 0.38 1.0
SAVE solution 1-20
END

EXCHANGE 1-20
 X 18.5e-3;     -equil 1
END

SOLUTION 0
 pH 4.5 charge;  temp 8
 Na 1.0;      K 0.048;      Mg 0.068;     Ca 0.02;      Al 0.2 Gibbsite 0.38
 Cl 1.0;      S(6) 0.37;    N(5) 0.1;     C(4) 0.3
END

TRANSPORT
 -cells 20; -lengths 0.25
 -shifts 330;   -diffusion_coefficient 0
```

```
 -punch_frequency 110
 -time_step 7.9e6
PRINT
 -reset false
SELECTED_OUTPUT
 -file Grind.csv
 -totals Al Ca Mg Na K
 -molalities AlX3 AlOHX2 NaX CaX2 MgX2 KX
 -equilibrium_phases    Gibbsite
END
```

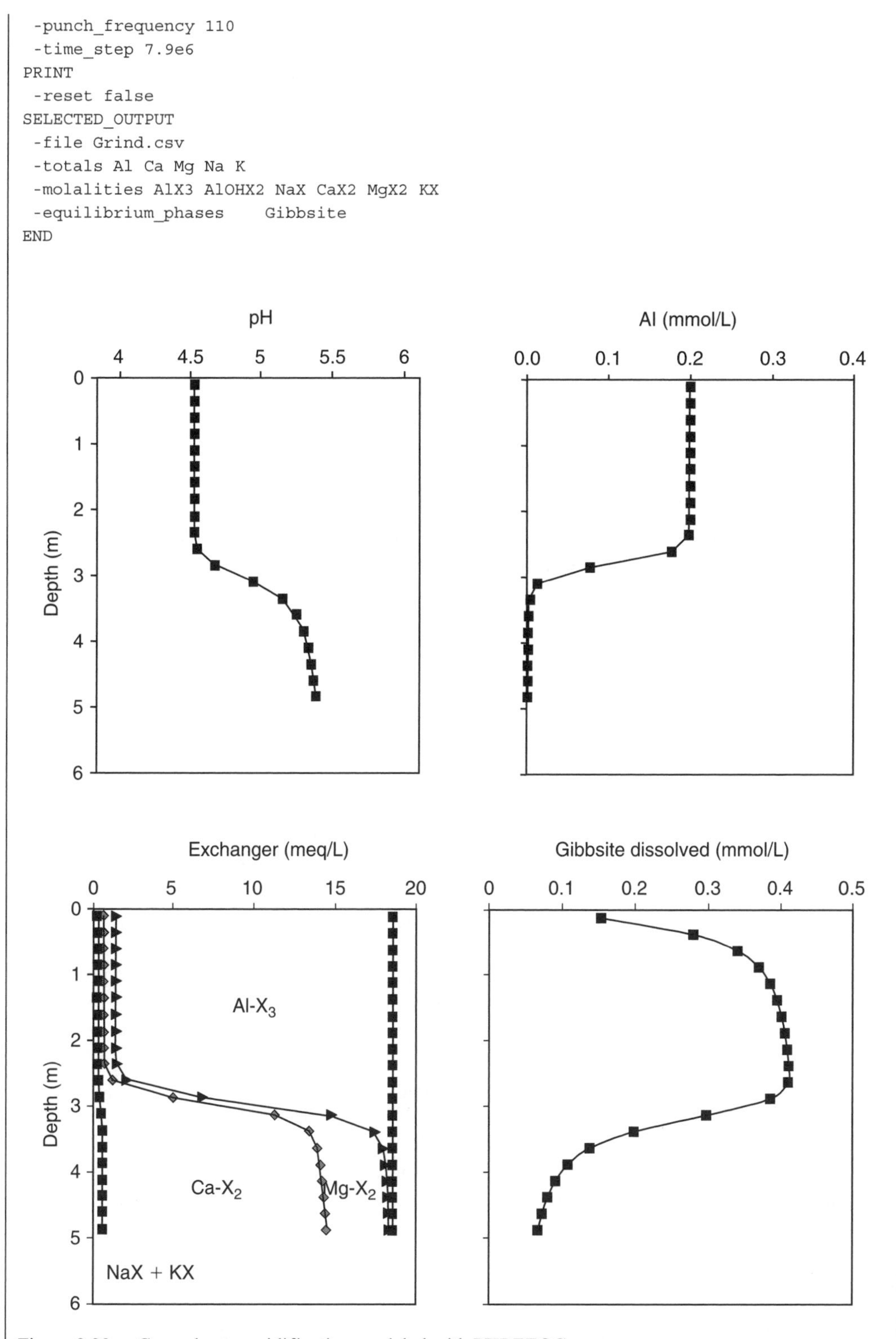

Figure 8.23. Groundwater acidification modeled with PHREEQC.

The vertical pore water velocity at Grindsted is 1.0–1.4 m/yr and with a cell length of 0.25 m the 330 shifts in the modeled sequence correspond to 83–59 years. The results are presented in Figure 8.23. Acid water with 0.2 mM Al^{3+} is transported downward. When it reaches the acidification front, Al^{3+} is sorbed on the exchanger, mostly as $Al\text{-}X_3$ and a small part as $AlOH\text{-}X_2$. The decrease of Al^{3+} and the ensuing dissolution of gibbsite raises the pH. The acidification front therefore reflects the depth where the exchanger is filled with Al^{3+}. The retardation is:

$$R = 1 + \frac{\Delta m_{Al\text{-}X_3} + \Delta m_{AlOH\text{-}X_2} + \Delta m_{gibbsite}}{\Delta m_{Al^{3+}}} = 1 + \frac{(0 - 5.58) + (0 - 0.15) + (1 - 0.6)}{(0 - 0.2)} = 27.6 \quad (8.38)$$

Note that the change in gibbsite concentration is positive and reduces the total change Δq_{Al} (the change $\Delta m_{gibbsite}$ is approximate because some gibbsite dissolves beyond the front due to changes in the solution composition). Again, for a vertical pore water velocity of 1 m/yr it would take 84 years to move the acidification front 3 m downward. This simple model is adequate for explaining the field data in Figure 8.22. Silicate weathering is not included in the model, and has only a slight effect on reducing the acidity (cf. Question).

QUESTIONS:
Inspect the calculated compositions before and after the acid front and note the differences?
ANSWER: anions have the same concentration, Ca^{2+} has diminished and is replaced by Al^{3+}.
Include the kinetic dissolution of 10% K-feldspar in the model and note the effect on the Al^{3+} concentration?
ANSWER: The concentration of Al^{3+} decreases with 0.01 mM/m in the acid zone
Estimate the time needed to reduce the Al^{3+} concentration from 0.2 mmol/L to below the drinking water limit (0.0074 mmol Al/L) by the feldspar buffering reaction?
ANSWER: $(0.2 - 0.0074) / 0.01 = 19.3\,m \approx 20$ year

In addition to progression of the front under uniform conditions as in Example 8.6, major variations in the input of Na^+, for example derived from sea salt deposition or by concentration by evapotranspiration in dry years, will be accompanied by 1000-fold larger changes in Al^{3+} by the salinity jump effect (Section 6.4.2). Displacement of Al^{3+} from the exchanger will then cause gibbsite precipitation and vice versa (Moss and Edmunds, 1992; Hansen and Postma, 1995; de Caritat, 1995).

The exchange complex forms a storage medium for acidity when it neutralizes the acid input. If a remediation scheme is planned for deacidification, for example by applying lime to the soil (Warfvinge, 1988) the processes occurring during acidification are reversed: the addition of base induces the precipitation of gibbsite, and thereby the Al^{3+} activity in solution is lowered, causing the desorption of Al^{3+} from the exchange complex and exchange by Ca^{2+}. The amount of base applied must therefore be sufficient to precipitate all adsorbed Al^{3+} as gibbsite before the pH can increase to near-neutral values. The high retardation of the acidification front through the aquifer implies a very long lag period before acid water appears in production wells. Late discovery of aquifer acidification entails the storage of large amounts of acidity on the sediment, which make it increasingly more difficult to remediate the problem.

PROBLEMS

8.1. Which of the following minerals do you expect in a thoroughly weathered sediment: olivine, K-feldspar, quartz, gibbsite and smectite.

8.2. The pH and aluminum speciation (in molality) of a groundwater sample is shown below.

pH	Al^{3+}	$AlOH^{2+}$	$Al(OH)_2^+$
4.53	6.0×10^{-5}	7.7×10^{-6}	6.0×10^{-7}

What is the pH of the water entering the reservoir when gibbsite dissolution has been the only buffering process? Neglect differences between activity and molal concentrations.

8.3. Derive the relationship between pH and $[SO_4^{2-}]$ at simultaneous equilibrium between gibbsite and jurbanite ($AlOHSO_4 = Al^{3+} + OH^- + SO_4^{2-}$; $K = 10^{-17.8}$).

8.4. Listed below are the compositions of ephemeral and perennial springs in the Sierra Nevada (Garrels and Mackenzie, 1967). Perennial springs are considered to represent a longer residence time in the reservoir and the difference between perennial and ephemeral spring is considered to be due to rock weathering. Concentrations are given as mmol/L.

	Perennial spring	Ephemeral spring	Rock weathering
Na	0.259	0.134	0.125
Ca	0.260	0.078	0.182
Mg	0.071	0.029	0.042
K	0.040	0.028	0.012
HCO_3	0.895	0.328	0.567
Si	0.410	0.273	0.137
SO_4	0.025	0.010	0.015
Cl	0.030	0.014	0.016
pH	6.8	6.2	

The minerals presumed to be present are:

Halite	$NaCl$
Gypsum	$CaSO_4$
Plagioclase	$Na_{0.62}Ca_{0.38}Al_{1.38}Si_{2.62}O_8$
Biotite	$KMg_3AlSi_3O_{10}(OH)_2$
Kaolinite	$Al_2Si_2O_5(OH)_4$
Ca-smectite	$Ca_{0.17}Al_{2.33}Si_{3.67}O_{10}(OH)_2$
Calcite	$CaCO_3$

(a) Perform a mass balance calculation to explain the change in composition in terms of dissolution and precipitation of the minerals listed below. *Hints*: Use the change in Al^{3+} concentration (which is O). Assign all Cl^- and SO_4^{2-} to respectively halite and gypsum. Do not include K^+ in the mass balance.

(b) Plot the water composition of the perennial spring into the stability diagram Figure 8.11 and evaluate whether the chosen weathering products are reasonable.

(c) Are the calculated mineral transfers kinetically consistent?

8.5. Aqua de Penha is a mineral water from a granitic area in Portugal with the composition (mmol/L, except pH) pH 5.77; Na 0.343; K 0.015; Mg 0.036; Ca 0.06; Cl 0.24; Alkalinity 0.292; SO_4^{2-} 0.007; Si 0.293. Calculate how much Na^+ comes from rain and how much from weathering of albite? (Assume that the rain has the Na^+/Cl^- ratio of seawater). Perform an inverse model calculation balancing Aqua da Penha water from seawater and distilled water, with the phases albite; biotite; K-feldspar; anorthite; gibbsite; kaolinite and $CO_2(g)$. Estimate the composition of the plagioclase $(Na_xCa_{1-x})(Al_{1+x}Si_{3-x})O8$ in the rock.

8.5. In Example 8.5, include equilibrium with atmospheric oxygen in evapotranspired rain- and soil water, and calculate the reaction of peat sequentially, first with O_2, then with NO_3^-.

REFERENCES

Åberg, G., Jacks, G. and Hamilton, P.J., 1989. Weathering rates and $^{87}Sr/^{86}Sr$ ratios; an isotopic approach. *J. Hydrol.* 109, 65–78.

Appelo, C.A.J., 1985. CAC, computer aided chemistry, or the evaluation of groundwater quality with a geochemical computer model (in Dutch). *H_2O* 26, 557–562.

Appelo, C.A.J., Krajenbrink, G.J.W., Van Ree, C.C.D.F. and Vasak, L., 1982. *Controls on groundwater quality in the NW Veluwe catchment* (in Dutch). Soil Protection Series 11, Staatsuitgeverij, Den Haag, 140 pp.

Bailey, S.W., Hornbeck, J.W., Driscoll, C.T. and Gaudette H.E., 1996. Calcium inputs and transport in a base-poor forest ecosystem as interpreted by Sr isotopes. *Water Resour. Res.* 32, 707–719.

Bain, D.C., Mellor, A., Robertson, M.S.E. and Buckland, S.T., 1991. Variations in weathering processes and rates with time in a chronosequence of soils from Glen Feshie, Scotland. In O. Selinus (ed.), Proc. 2nd Int. Symp. Env. Geochemistry, Uppsala.

Berg, A. and Banwart, S.A., 2000. Carbon dioxide mediated dissolution of Ca-feldspar: implications for silicate weathering. *Chem. Geol.* 163, 25–42.

Berner, R.A., 1971. *Principles of chemical sedimentology.* McGraw-Hill New York, 240 pp.

Berner, E.K. and Berner, R.A., 1996. *Global Environment.* Prentice-Hall, Englewood Cliffs, 376 pp.

Blatt, H., Middleton, G. and Murray, R., 1980. *Origin of sedimentary rocks*, 2nd ed. Prentice-Hall, Englewood Cliffs, 782 pp.

Blum, A.E. and Stillings, L.L., 1995. Feldspar dissolution kinetics. In A.F. White and S.L. Brantley (eds), *Chemical weathering rates of silicate minerals. Rev. Mineral.* 31, 291–351.

Böttcher J., Strebel, O. and Duynisveld, H.M., 1985. Vertikale Stoffkonzentrationsprofile im Grundwasser eines Lockergesteins-Aquifers und deren Interpretation (Beispiel Fuhrberger Feld). *Z. dt. geol. Ges.* 136, 543–552.

Brantley, S.L., Chesley, J.T. and Stillings, L.L., 1998. Isotopic ratios and release rates of strontium measured from weathering feldspars. *Geochim. Cosmochim. Acta* 62, 1493–1500.

Bullen, T.D., White, A.F., Blum, A.E., Harder, J.W. and Schulz, M.S., 1997. Chemical weathering of a soil chronosequence on granitoid alluvium: II. Mineralogical and isotopic constraints on the behavior of strontium. *Geochim. Cosmochim. Acta* 61, 291–306.

Chadwick, O.A., Derry, L.A., Vitousek, P.M., Huebert, B.J. and Hedin, L.O., 1999. Changing sources of nutrients during four million years of ecosystem development. *Nature* 397, 491–496.

Chou, L. and Wollast, R., 1985. Steady-state kinetics and dissolution mechanisms of albite. *Am. J. Sci.* 285, 963–993.

Clow, D.W., Mast, M.A., Bullen, T.D. and Turk, J.T., 1997. Strontium 87 – strontium 86 as a tracer for mineral weathering reactions and calcium sources in an alpine/subalpine watershed, Loch Vale, Colorado. *Water Resour. Res.* 33, 1335–1351.

de Caritat, P., 1995. Intensifying groundwater acidification at Birkenes, southern Norway. *J. Hydrol.* 170, 47–62.

Dahmke, A., Matthess, G., Pekdeger, A., Schenk, D. and Schulz, H.D., 1986. Near-surface geochemical processes in Quaternary sediments. *J. Geol. Soc.* 143, 667–672.

Donovan, J.J., Frysinger, K.W. and Maher, T.P., 1997. Geochemical response of acid groundwater to neutralization by alkaline recharge. *Aq. Geochem.* 2, 227–253.

Dove, P. and Rimstidt, J.D., 1993. Silica-water interactions. *Rev. Mineral.* 29, 259–308.

Drabløs, D. and Tollan, A. (eds), 1980. *Ecological impact of acid precipitation.* Proc. Int. Conf. Sandefjord, SNSF Project, Oslo, Norway.

Drever, J.I. and Clow, D.W., 1995. Weathering rates in catchments. In A.F. White and S.L. Brantley (eds), *Chemical weathering rates of silicate minerals. Rev. Mineral.* 31, 463–483.

Driscoll, C.T., 1980. Aqueous speciation of aluminum in the Adirondack region of New York State, USA. In D. Drabløs and A. Tollan, *op. cit.*, 214–215.

Edmunds, W.M. and Kinniburgh, D.G., 1986. The susceptibility of UK groundwaters to acidic deposition. *J. Geol. Soc.* 143, 707–720.

Eriksson, E., 1981. Aluminum in groundwater possible solution equilibria. *Nordic Hydrol.* 12, 43–50.

Garrels, R.M. and Mackenzie, F.T., 1967. Origin of the chemical compositions of some springs and lakes. In W. Stumm (ed.), *Equilibrium concepts in natural water systems.* Adv. Chem. Series 67, 222–242.

Garrels, R.M. and Mackenzie, F.T., 1971. *Evolution of Sedimentary Rocks.* Norton, New York, 397 pp.

Goldich, S.S., 1938. A study in rock-weathering. *J. Geol.* 46, 17–58.

Hansen, B.K. and Postma, D., 1995. Acidification, buffering, salt effects in the unsaturated zone of a sandy aquifer, Klosterhede, Denmark, *Water Resour. Res.* 31, 2795–2809.

Helgeson, H.C., Garrels, R.M. and MacKenzie, F.T., 1969. Evaluation of irreversible reactions in geochemical processes involving minerals and aqueous solutions. II. Applications. *Geochim. Cosmochim. Acta* 33, 455–481.
Hem, J.D., 1985. *Study and Interpretation of the chemical characteristics of natural water*, 3rd ed. U.S. Geol. Surv. Water Supply Paper 2254, 264 pp.
Hinderer, M. and Einsele, G., 1997. Groundwater acidification in Triassic sandstones: prediction with MAGIC modeling. *Geol. Rundsch.* 86, 372–388.
Hodson, M.E., Langan, S.J. and Wilson, M.J., 1997. A critical evaluation of the use of the PROFILE model in calculating mineral weathering rates. *Water Air Soil Poll.* 98, 79–104.
Hultberg, H. and Wenblad, A., 1980. Acid groundwater in southwestern Sweden. In D. Drabløs and A. Tollan, *op. cit.*, 220–221.
Kirchner, J.W., 1992. Heterogeneous geochemistry of catchment acidification. *Geochim. Cosmochim. Acta* 56, 2311–2327.
Kjøller, C., 2001. Nickel mobilization in response to groundwater acidification, Ph.D. thesis, Lyngby, 140 pp.
Kjøller, C., Postma, D. and Larsen, F., 2004. Groundwater acidification and the mobilization of trace metals in a sandy aquifer. *Env. Sci. Technol.*, 38, 2829–2835.
Langan, S., Hodson, M., Bain, D., Hornung, M., Reynolds, B., Hall, J. and Johnston, L., 2001. The role of mineral weathering rate determinations in generating uncertainties in the calculation of critical loads of acidity and their exceedance. *Water Soil Air Poll. Focus* 1, 299–312.
Lasaga, A.C., 1984. Chemical kinetics of water–rock interactions. *J. Geophys. Res.* 89, 4009–4025.
Likens, G.E. and Bormann, F.H., 1995. *Biogeochemistry of a forested ecosystem.* 2nd ed. Springer, New York, 159 pp.
Matthess, G., Petersen, A., Schenk, D. and Dahmke, A., 1992. Field studies on the kinetics of silicate minerals/water interaction. In G. Matthess, F.H. Frimmel, P. Hirsch, H.D. Schulz and E. Usdowski (eds), *Progress in hydrogeochemistry*, 298–307, Springer, Berlin.
May, H.M., Kinniburgh, D.G., Helmke, P.A. and Jackson, M.L., 1986. Aqueous dissolution, solubilities and thermodynamic stabilities of common aluminosilicate clay minerals: kaolinite and smectites. *Geochim. Cosmochim. Acta* 50, 1667–1677.
Miretzky, P., Conzonno, V. and Cirelli, A.F., 2001. Geochemical processes controlling silica concentrations in groundwaters of the Salado River drainage basin, Argentina. *J. Geochem. Expl.* 73, 155–166.
Moss, P.D. and Edmunds, W.M., 1992. Processes controlling acid attenuation in the unsaturated zone of a Triassic sandstone aquifer (U.K.) in the absence of carbonate minerals. *Appl. Geochem.* 7, 573–583.
Ohse, W., Matthess, G. and Pekdeger, A., 1983. Gleichgewichts- und Ungleichgewichtsbeziehungen zwischen Porenwassern und Sedimentgesteinen im Verwitterungsbereich. *Z. dt. geol. Ges.* 134, 345–361.
Ohse, W., Matthess, G., Pekdeger, A. and Schulz, H.D., 1984. Interaction water–silicate minerals in the unsaturated zone controlled by thermodynamic disequilibria. IASH Pub. 150, 31–40.
Olsson, M. and Melkerud, P.A., 1991. Weathering rates in mafic soil mineral material. In K. Rosén (ed.), *Chemical weathering under field conditions.* Rep. Forest Ec. Forest Soil 63, 63–78, Swedish Univ. Agri. Sci.
Plummer, L.N., Prestemon, E.C. and Parkhurst, D.L., 1991. *An interactive code (NETHPATH) for modeling net geochemical reactions along a flow path.* U.S. Geol. Surv. Water Resour. Inv., 91-4078, 100 pp.
Rosén, K. (ed.), 1991. *Chemical weathering under field conditions.* Rep. Forest Ec. Forest Soil 63, Swedish Univ. Agri. Sci., 185 pp.
Richards, P.L. and Kump, L.R., 2003. Soil pore-water distributions and the temperature feedback of weathering in soils. *Geochim. Cosmochim. Acta* 67, 3803–3815.
Rimstidt, J.D., 1997. Quartz solubility at low temperatures. *Geochim. Cosmochim. Acta* 61, 2553–2558.
Ryan, J.N. and Gschwend, P.M., 1992. Effect of iron diagenesis on the transport of colloidal clay in an unconfined aquifer. *Geochim. Cosmochim. Acta*. 56, 1507–1521.
Salomons, W. and Mook, W.G., 1987. Natural tracers for sediment transport studies. *Cont. Shelf Res.* 7, 1333–1343.
Savin, S.M. and Hsieh, J.C.C., 1998. The hydrogen and oxygen isotope geochemistry of pedogenic clay minerals: principles and theoretical background. *Geoderma* 82, 227–253.
Schott, J. and Berner, R.A., 1985. Dissolution mechanisms of pyroxenes and olivines during weathering. In J.I. Drever (ed.), *The chemistry of weathering*, 35–53, D. Reidel, Dordrecht.
Schulz, M.S. and White, A.F., 1999. Chemical weathering in a tropical watershed, Luquillo Mountains, Puerto Rico III: Quartz dissolution rates. *Geochim. Cosmochim. Acta* 63, 337–350.
Schroeder, P.A., Melear, N.D., Bierman, P., Kashgarian, M. and Caffee, M.W., 2001. Apparent gibbsite growth ages for regolith in the Georgia piedmont. *Geochim. Cosmochim. Acta* 65, 381–386.

Stefansson, A. and Arnorsson, S., 2000. Feldspar saturation state in natural waters. *Geochim. Cosmochim. Acta* 64, 2567–2584.

Stewart, B.W., Capo, R.C. and Chadwick, O.A., 2001. Effects of rainfall on weathering rate, base cation provenance, and Sr isotope composition of Hawaiian soils. *Geochim. Cosmochim. Acta* 65, 1087–1099.

Stumm, W. and Wollast, R., 1990. Coordination chemistry of weathering. *Rev. Geophys.* 28, 53–69.

Sverdrup, H.U., 1990. *The kinetics of base cation release due to chemical weathering*. Lund Univ. Press, 246 pp.

Sverdrup, H.U. and Warfvinge, P., 1988. Weathering of primary silicate minerals in the natural soil environment in relation to a chemical weathering model. *Water Air Soil Poll.* 38, 387–408.

Sverdrup, H.U. and Warfvinge, P., 1993. Calculating field weathering rates using a mechanistic geochemical model PROFILE. *Appl. Geochem.* 8, 273–283.

Sverdrup, H.U. and Warfvinge, P., 1995. Estimating field weathering rates using laboratory kinetics. In A.F. White and S.L. Brantley (eds), *Chemical weathering rates of silicate minerals. Rev. Mineral.* 31, 485–541.

Tardy, Y., 1971. Characterization of the principal weathering types by the geochemistry of waters from some European and African crystalline massifs. *Chem. Geol.* 7, 253–271.

Ulrich, B., Mayer, R. and Khanna, P.K., 1979. *Deposition von Luftverunreinigungen und ihre Auswirkungen in Waldökosystemen im Solling*. Sauerländer, Frankfurt a.M., 291 pp.

Warfvinge, P., 1988. *Modeling acidification mitigation in watersheds*. Ph D. Thesis, Lund, 180 pp.

White, A.F. and Brantley, S.L. (eds), 1995. *Chemical weathering rates of silicate minerals. Rev. Mineral.* 31, 583 pp.

White, A.F., Blum, A.E., Schulz, M.S., Bullen, T.D., Harder, J.W. and Peterson, M.L., 1996. Chemical weathering rates of a soil chronosequence on granitic alluvium: 1. Quantification of mineralogical and surface area changes and calculation of primary silicate reaction rates. *Geochim. Cosmochim. Acta* 60, 2533–2550.

White, A.F., Bullen, T.D., Schulz, M.S., Blum, A.E., Huntington, T.G. and Peters, N.E., 2001. Differential rates of feldspar weathering in granitic regoliths. *Geochim. Cosmochim. Acta* 65, 847–869.

Williams, L.A., Parks, G.A. and Crerar, D.A., 1985. Silica Diagenesis, I. Solubility Controls. *J. Sed. Petrol.* 55, 301–311.

Wright, R.F. and Henriksen, A., 1978. Chemistry of small Norwegian lakes, with special reference to acid precipitation. *Limnol. Oceanogr.* 23, 487–498.

9

Redox Processes

Reduction and oxidation processes exert an important control on the natural concentrations of O_2, Fe^{2+}, SO_4^{2-}, H_2S, CH_4, etc. in groundwater. They also determine the fate of pollutants like nitrate leaching from agricultural fields, contaminants leaching from landfill sites, industrial spills, or heavy metals in acid mine drainage. Redox reactions occur through electron transfer from one atom to another and the order in which they proceed can be predicted from standard equilibrium thermodynamics. However, the electron transfer is often very slow and may only proceed at significant rates when mediated by bacterial catalysis. An example is the reduction of sulfate by organic matter which occurs both in aquifers and in marine sediments. The reaction is immeasurably slow abiotically, but microbes like *Desulfovibrio sp.* produce enzymes that catalyze the process and the reaction proceeds rapidly in natural environments.

Redox processes in groundwater typically occur through the addition of an oxidant, like O_2 or NO_3^- to an aquifer containing a reductant. However, the addition of a reductant, such as dissolved organic matter (*DOC*) that leaches from soils or landfills can also be important. In the following, we first treat some basic redox theory, and subsequently use the principles to discuss the redox processes in aquifers.

9.1 BASIC THEORY

As an example of a redox process, consider the reaction between Fe^{2+} and Mn^{4+} as it would take place in acid solution:

$$2Fe^{2+} + MnO_2 + 4H^+ \leftrightarrow 2Fe^{+} + Mn^{2} + 2H_2O \tag{9.1}$$

In this reaction two electrons are transferred from Fe(2) to reduce Mn(4) in MnO_2. Ferrous iron acts as the *reductant* and reduces Mn(4), while MnO_2 can be called the *oxidant* that oxidizes Fe^{2+}. An alternative terminology is to call Fe^{2+} the *electron donor*, and Mn(4) the *electron acceptor*. Reaction (9.1) also demonstrates that redox reactions may have a significant pH effect.

Electrons exchange between atoms and do not exist in a "free" state in solution (Figure 9.1). Accordingly, electrons do not appear in a balanced redox reaction such as (9.1). However, when deriving a complete redox reaction it is convenient to consider first the half-reactions:

$$Fe^{2+} \leftrightarrow Fe^{3+} + e^- \tag{9.2}$$

and

$$Mn^{2+} + 2H_2O \leftrightarrow MnO_2 + 4H^+ + 2e^- \tag{9.3}$$

Figure 9.1. Electrons can only be exchanged (Nordstrom and Munoz, 1994).

Subtracting Reaction (9.2) twice from (9.3) cancels the electrons and produces Reaction (9.1). With the guidelines listed in Table 9.1, any redox reaction can be balanced.

Table 9.1. Guide to balancing redox equations.

1. For each redox couple, write the oxidized and reduced species in an equation and balance the amount of element.
2. Balance the number of oxygen atoms by adding H_2O.
3. Balance the number of hydrogen atoms by adding H^+.
4. Balance electroneutrality by adding electrons.
5. Subtract the two half-reactions, canceling electrons to obtain the complete redox reaction.

For example, in Reaction (9.1) we could replace pyrolusite (MnO_2) by birnessite ($MnO_{1.9}$). Birnessite contains small and variable amounts of Mn^{2+} or Mn^{3+} instead of Mn^{4+} (the framework of oxygens remains intact with all the oxygens having valence O(−2)):

Step from Table 9.1

1. $MnO_{1.9} = Mn^{2+}$
2. $MnO_{1.9} = Mn^{2+} + 1.9H_2O$
3. $MnO_{1.9} + 3.8H^+ = Mn^{2+} + 1.9H_2O$
4. $MnO_{1.9} + 3.8H^+ + 1.8e^- = Mn^{2+} + 1.9H_2O$
5. $MnO_{1.9} + 3.8H^+ + 1.8Fe^{2+} \leftrightarrow Mn^{2+} + 1.9H_2O + 1.8Fe^{3+}$ (9.1b)

QUESTION:
Balance the following redox reactions:
$NO_3^- + Fe^{2+} \leftrightarrow N_2 + FeOOH$
$SO_4^{2-} + C \leftrightarrow CO_2 + HS^-$
$C_2HCl_3 + Fe \leftrightarrow C_2H_3Cl + Cl^- + Fe^{2+}$ (trichloroethene reacts with iron to vinylchloride)

ANSWER: $NO_3^- + 5Fe^{2+} + 7H_2O \leftrightarrow ½N_2 + 5FeOOH + 9H^+$; $SO_4^{2-} + 2C + H^+ \leftrightarrow 2CO_2 + HS^-$; $C_2HCl_3 + 2Fe + 2H^+ \leftrightarrow C_2H_3Cl + 2Cl^- + 2Fe^{2+}$

Reactions (9.1) to (9.3) can be written in their general form as

$$bB_{red} + cC_{ox} \rightarrow dD_{ox} + gG_{red} \tag{9.4}$$

and half-reactions

$$bB_{red} \rightarrow dD_{ox} + ne^- \tag{9.5}$$

$$gG_{red} \rightarrow cC_{ox} + ne^- \tag{9.6}$$

In terms of Gibbs free energy (Section 4.3.1) we may write for Reaction (9.4)

$$\Delta G_r = \Delta G_r^0 + RT \ln \frac{[D_{ox}]^d [G_{red}]^g}{[B_{red}]^b [C_{ox}]^c}$$

The Gibbs free energy of a reaction can be related to the voltage developed by a redox reaction in an electrochemical cell by the relation

$$\Delta G = nFE \tag{9.7}$$

E is the potential (emf) in Volts, F the Faraday's constant (96.42 kJ/Volt gram equivalent) and n the number of electrons transferred in the reaction. Substitution of (9.7) into (4.24) produces the *Nernst* equation:

$$E = E^o + \frac{RT}{nF} \ln \frac{[D_{ox}]^d [G_{red}]^g}{[B_{red}]^b [C_{ox}]^c} \tag{9.8}$$

Here E^0 is the standard potential (Volt) where all substances are present at unit activity at 25°C and 1 atm, similar to ΔG_r^0. As before, R is the gas constant (8.314×10^{-3} kJ/deg/mol), and T the absolute temperature. For oxidation of H_2 the half-reaction is:

$$H_2 \leftrightarrow 2H^+ + 2e^- \tag{9.9}$$

By definition Reaction (9.9) has $\Delta G_r^0 = 0$, at 25°C and 1 atm, and according to (9.7) $E^0 = 0$ Volt. Substituting (9.9) for half-reaction (9.6) in (9.8) gives:

$$E = E^0 + \frac{RT}{nF} \ln \frac{[D_{ox}]^d P_{H_2}}{[B_{red}]^b [H^+]^2} \tag{9.10}$$

The setup for such a redox cell is illustrated in Figure 9.2. The left hand part is the standard hydrogen electrode, consisting of a Pt-electrode over which H_2 gas is bubbled in a solution of pH = 0, so that standard state conditions are fulfilled. In the right hand part of the cell, an inert Pt electrode is placed in a solution containing Fe^{2+} and Fe^{3+}, corresponding to half-reaction (9.2). The two electrodes are connected to a voltmeter and the electrical circuit is closed by a salt bridge. Under these conditions, both P_{H_2} and $[H^+]$ are equal to one in (9.10). When Fe^{2+} and Fe^{3+} are present in solution at unit activity, the voltmeter will register the E^0 of (9.2) but at other activity ratios of Fe^{3+} and Fe^{2+}, different E values are measured.

Since both P_{H_2} and $[H^+]$ are always equal to one in this setup, they are usually omitted from Equation (9.10) and instead indicated by adding the postscript h from hydrogen to E:

$$Eh = E^0 + \frac{RT}{nF} \ln \frac{[D_{ox}]^d}{[B_{red}]^b} \tag{9.11}$$

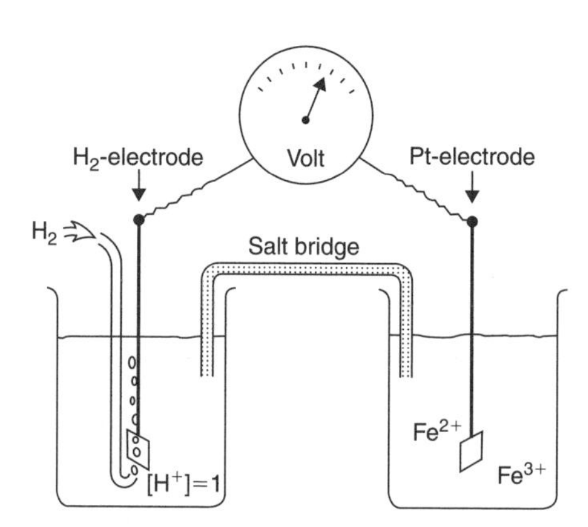

Figure 9.2. A schematic drawing of a redox cell.

The *Eh* and E^0 of a half-reaction are thus defined as potentials relative to the standard state H_2/H^+ reaction. The value of the standard potential E^0 of a half-reaction indicates the tendency to release or to accept electrons. Standard potentials for a few reactions are listed in Table 9.2 and such a list is useful to identify possible reactions. There is some confusion concerning sign conventions of redox reactions. In many texts half-reactions are written as reduction reactions instead of oxidation reactions which are used here. The advantage of writing oxidation reactions is that the oxidized species appears in the numerator of the mass action equation, similar to the Nernst equation. When reduction reactions are written instead, a minus sign is added in Equation (9.7). With both conventions reducing agents have a more negative E^0 and oxidizing agents a more positive E^0.

If we apply the Nernst equation to Reaction (9.2), using E^0 from Table 9.2, we obtain:

$$\begin{aligned} Eh &= E^0 + \frac{RT}{1F} \ln \frac{[Fe^{3+}]}{[Fe^{2+}]} \\ &= 0.77 + \frac{(8.314 \times 10^{-3})(298.15)(2.303)}{96.42} \log \frac{[Fe^{3+}]}{[Fe^{2+}]} \\ &= 0.77 + 0.059 \log \frac{[Fe^{3+}]}{[Fe^{2+}]} \end{aligned} \tag{9.12}$$

Table 9.2. Standard potentials for a few reactions at 25°C, 1 atm.

Reaction	E^0, Volt
$Fe(s) \leftrightarrow Fe^{2+} + 2e^-$	−0.44
$Cr^{2+} \leftrightarrow Cr^{3+} + e^-$	−0.41
$H_2 \leftrightarrow 2H^+ + 2e^-$	0.00
$Cu^+ \leftrightarrow Cu^{2+} + e^-$	+0.16
$S^{2-} + 4H_2O \leftrightarrow SO_4^{2-} + 8H^+ + 8e^-$	+0.16
$As(s) + 3H_2O \leftrightarrow H_3AsO_{3(aq)} + 3H^+ + 3e^-$	+0.25
$Cu(s) \leftrightarrow Cu^+ + e^-$	+0.52
$H_3AsO_{3(aq)} + H_2O \leftrightarrow H_3AsO_{4(aq)} + 2H^+ + 2e^-$	+0.56
$Fe^{2+} \leftrightarrow Fe^{3+} + e^-$	+0.77
$Fe^{2+} + 3H_2O \leftrightarrow Fe(OH)_3 + 3H^+ + e^-$	+0.98
$2H_2O \leftrightarrow O_{2(g)} + 4H^+ + 4e^-$	+1.23
$Mn^{2+} + 2H_2O \leftrightarrow MnO_2 + 4H^+ + 2e^-$	+1.23

where the factor 2.303 converts natural to base ten logarithms. Similarly for Reaction (9.3):

$$\begin{aligned} Eh &= E^0 + \frac{RT}{2F}\ln\frac{[H^+]^4}{[Mn^{2+}]} \\ &= 1.23 + \frac{(8.314\times10^{-3})(298.15)(2.303)}{2\times 96.42}\log\frac{[H^+]^4}{[Mn^{2+}]} \\ &= 1.23 + 0.03\log\frac{[H^+]^4}{[Mn^{2+}]} \end{aligned} \tag{9.13}$$

Equation (9.13) contains neither MnO_2 nor H_2O, since they both have unit activity. At equilibrium between the two half-reactions, the *Eh* for both reactions should be the same. In other words, for a given *Eh* the distribution of all redox equilibria is fixed (Example 9.1).

EXAMPLE 9.1. *Calculation of redox speciation with the Nernst equation*
A water sample contains $[Fe^{2+}] = 10^{-4.95}$ and $[Fe^{3+}] = 10^{-2.29}$ with a pH = 3.5 at 25°C. What would be $[Mn^{2+}]$ if this water sample were in equilibrium with sediment containing MnO_2?

ANSWER:
First rewrite Equation (9.12) as

$$Eh = 0.77 + 0.059(\log[Fe^{3+}] - \log[Fe^{2+}])$$

Substitute iron activities

$$Eh = 0.77 + 0.059(-2.29 - (-4.95)) = 0.927 \text{ Volt.}$$

Next rewrite Equation (9.13) as

$$Eh = 1.23 + 0.03(-4\,\text{pH} - \log[Mn^{2+}])$$

$$\log[Mn^{2+}] = (1/0.03)(1.23 - Eh - 0.12\,\text{pH})$$

substitute the given pH and the *Eh* calculated from the Fe^{3+}/Fe^{2+} couple

$$\begin{aligned} \log[Mn^{2+}] &= 33.33(1.23 - 0.927 - 0.42) = -3.90 \\ &\rightarrow [Mn^{2+}] = 10^{-3.90} \end{aligned}$$

Thus, for any lower $[Fe^{3+}]/[Fe^{2+}]$ ratio, or lower pH value, the $[Mn^{2+}]$ will increase and vice versa. The example illustrates that once the *Eh* is evaluated from one redox couple, the distribution of all other redox couples is fixed.

In order to obtain the E^0 for Reaction (9.1), we simply subtract the E^0 of (9.13) from the E^0 of (9.12) (without multiplying the E^0 with the number of electrons transferred since E^0 already is expressed as potential per electron according to Equation (9.8)).

$$E^0 = 0.77 - 1.23 = -0.46 \text{ Volt.}$$

The negative voltage indicates that the reaction should proceed spontaneously to the right when all activities are equal to one. For redox reactions like (9.1), we return to the general form of the Nernst Equation (9.8):

$$\begin{aligned} E &= E^0 + \frac{RT}{2F}\ln\frac{[Fe^{3+}]^2[Mn^{2+}]}{[Fe^{2+}]^2[H^+]^4} \\ &= -0.46 + 0.03\log\frac{[Fe^{3+}]^2[Mn^{2+}]}{[Fe^{2+}]^2[H^+]^4} \end{aligned} \tag{9.14}$$

In this fashion, the Nernst equation can be used to express the distribution of species in any redox reaction at equilibrium.

As indicated by Equations (9.7) and (9.8), redox potentials can be related to Gibbs free energies. The practical significance of this relation is that it allows us to calculate the standard potential of any redox reaction directly from thermodynamic tables (see Section 4.3.1) since we may rewrite Equation (9.7) for standard conditions as:

$$\Delta G_r^0 = nFE^0 \tag{9.15}$$

EXAMPLE 9.2. *Calculate* E^0 *from* ΔG_r^0
For Reaction (9.2)

$$Fe^{2+} \leftrightarrow Fe^{3+} + e^-$$

$$\Delta G^0_{f\ Fe^{2+}} = -78.9 \text{ kJ/mol;}$$

$$\Delta G^0_{f\ Fe^{3+}} = -4.7 \text{ kJ/mol;}$$

$$\Delta G^0_f\ e^- = 0.0 \text{ kJ/mol (by definition).}$$

$$\Delta G_r^0 = \Delta G^0_{f\ Fe^{3+}} + \Delta G^0_{f\ e-} + \Delta G^0_{f\ Fe^{2+}} = -4.7 + 0 - (-78.9) = 74.2 \text{ kJ/mol}$$

Using (9.15) $\quad E^0 = \Delta G_r^0 / nF = 74.2 / 96.45 = 0.77 \text{ Volt}$

QUESTIONS:
Calculate the concentration of Fe^{2+} in equilibrium with solid iron at $Eh = 0$ and -0.3 V?
ANSWER: Using Table 9.2, 8×10^{14} and 5.6×10^{4} mol Fe^{2+}/L for $Eh = 0$ and -0.3 V (huge concentration!)
And of Cu^{2+} in contact with solid copper at $Eh = 0$?
ANSWER: $E^0 = 0.52 + 0.16$ V; $Cu^{2+} = 8.9\times10^{-24}$ mol/L (very low concentration)
Which metal is more stable in water, Cu(s) or Fe(s)?
ANSWER: Cu
Calculate ΔG_f^0 of Cu^+ from Table 9.2?
ANSWER: -50.15 kJ/mol

9.1.1 *The significance of redox measurements*

Theoretically, the *Eh* determines the distribution of all redox equilibria in a similar way as the pH expresses the distribution of all acid-base equilibria. In contrast to pH, unfortunately *Eh* cannot be measured unambiguously in most natural waters.

Eh-measurements are performed using an inert Pt-electrode against a standard electrode of a known potential (see Grenthe et al., 1992 and Christensen et al., 2000 for detailed procedures). What we really want to know is the potential relative to the standard hydrogen electrode. However, since the latter is rather impractical to carry around in the field, a reference electrode of known potential is used and measured potentials are corrected accordingly

$$Eh = E_{\text{meas.}} + E_{\text{ref}} \tag{9.16}$$

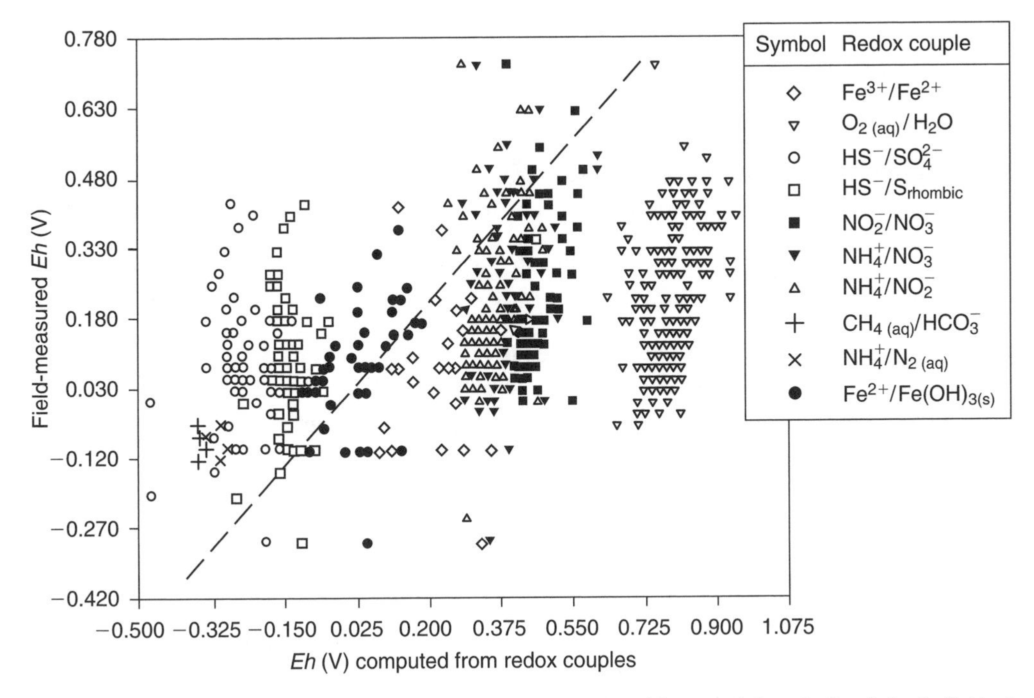

Figure 9.3. Comparison of groundwater field *Eh* measurements with potentials calculated for individual redox species (Lindberg and Runnels, 1984, Science, 225, 925–927 Copyright 1985 by the AAAS).

For example, the potential of a calomel reference electrode ($KCl_{(sat)}$, $Hg_2Cl_{2(s)}|Hg_{(l)}$) is $E_{ref} = 244.4$ mV at 25°C.

Although waters from oxidized environments generally yield higher *Eh* values than those from reduced environments, it has proven very difficult to obtain a meaningful quantitative interpretation in the sense of the Nernst equation. This is illustrated in Figure 9.3 were *Eh* values measured in the field are compared with those calculated with the Nernst equation from the analytic data for several half-reactions. The results show disturbingly large variations between the two sets of data. For example, for the important reaction which relates the O_2 content to the *Eh*,

$$2H_2O \leftrightarrow O_{2(g)} + 4H^+ + 4e^- \tag{9.17}$$

the Pt-electrode is apparently unaffected by the O_2 concentration. The results for the other half-reactions are not much better.

There are two reasons for the large discrepancies: lack of equilibrium between different redox couples in the same water sample (Lindberg and Runnels, 1984) and analytical difficulties in measuring with the Pt-electrode (Stumm and Morgan, 1996). The latter include lack of electroactivity at the Pt surface, like for O_2, mixing potentials, and poisoning of the electrode. An example of poisoning is the precipitation of FeOOH on the Pt-electrode which occurs when the electrode is immersed in an anoxic, Fe^{2+}-rich, sample because of O_2 adsorbed on the electrode surface (Doyle, 1968). Others have used graphite electrodes instead but also these have their limitations (Walton-Day et al., 1990; Grenthe et al., 1992).

Probably, *Eh* measurements are applicable in acid mine waters, where high concentrations of Fe^{2+} and Fe^{3+} appear to control the electrode response (Nordstrom et al., 1979). Encouraging results of *Eh* measurements in the iron system at higher pH have also been obtained (Macalady et al., 1990;

Grenthe et al., 1992). Still, the overall conclusion remains that *Eh*-measurements only should be interpreted quantitatively when there is a rigorous control on what is really being measured.

9.1.2 *Redox reactions and the pe concept*

An alternative theoretical treatment of redox reactions simplifies the algebra considerably. In this approach the law of mass action is used in redox half-reactions. For example, for Reaction (9.2):

$$K = \frac{[Fe^{3+}][e^-]}{[Fe^{2+}]} = 10^{-13.05} \tag{9.18}$$

where K is the equilibrium constant. In contrast to the Nernst equation, the electron activity appears explicitly in the activity product. The electron activity should not be interpreted in terms of a concentration of electrons, since electrons are only exchanged, but rather as the tendency to release or accept electrons. In analogy to pH, one may define the parameter pe:

$$\mathrm{pe} = -\log[e^-] \tag{9.19}$$

Just as for *Eh*, high positive values of pe indicate oxidizing conditions and low negative values reducing conditions. Rewriting (9.18) in logarithmic form yields the equation:

$$\log K = \log[Fe^{3+}] - \mathrm{pe} - \log[Fe^{2+}] = -13.05 \tag{9.20}$$

In a similar fashion we may write for Reaction (9.3)

$$\log K = -4\mathrm{pH} - 2\mathrm{pe} - \log[Mn^{2+}] = -41.52 \tag{9.21}$$

Equations (9.20) and (9.21) can be compared with the corresponding Nernst Equations (9.12) and (9.13). The calculation of redox speciation, using the pe concept is demonstrated in Example 9.3. It is a direct analogue to Example 9.1 where the Nernst approach was used.

EXAMPLE 9.3. *Calculation of redox speciation using the pe concept*
A water sample contains $[Fe^{2+}] = 10^{-4.95}$ and $[Fe^{3+}] = 10^{-2.29}$ with a pH = 3.5 at 25°C. What would be $[Mn^{2+}]$ if this water sample were in equilibrium with sediment containing MnO_2?

ANSWER:
First pe is calculated from Equation (9.20) which is then substituted into Equation (9.21). Rewriting (9.20) yields:

$$\mathrm{pe} = -\log K + \log[Fe^{3+}] - \log[Fe^{2+}]$$

Substituting values gives:

$$\mathrm{pe} = -(-13.05) + (-2.29) - (-4.95) = 15.71$$

Rewriting (9.21) yields:

$$\log[Mn^{2+}] = -\log K - 4\mathrm{pH} - 2\mathrm{pe}$$

Substituting $\log K$, pH and the pe calculated above yields

$$\log[Mn^{2+}] = -(-41.52) - 4 \times 3.5 - 2 \times 15.71 = -3.90$$

Which is the same result as obtained in Example 9.1.

The values of K can be calculated using thermodynamic tables just as any other mass action constant (Example 9.4).

EXAMPLE 9.4. *Calculation of K from thermodynamic data*
This example is an analogue to Example 9.2 where the Nernst equation was used. The mass action constant is related to ΔG_r^0 by Equation (4.26)

$$\Delta G_r^0 = -\mathrm{RT} \ln K$$

At 25°C this is equal to

$$\Delta G_r^0 = -5.701 \log K$$

Following Example 9.2 the ΔG_r^0 of the Fe^{2+} / Fe^{3+} reaction is:

$$\Delta G_r^0 = \Delta G_{f\,\mathrm{Fe}^{3+}}^0 + \Delta G_{fe-}^0 - \Delta G_{f\,\mathrm{Fe}^{2+}}^0 = -4.7 + 0 - (-78.9) = 74.2 \text{ kJ/mol}$$

Note again that $\Delta G_{f\,\mathrm{e}^-}^0 = 0.0$ kJ/mol by definition. Substitution results in

$$\log K = 74.2 / -5.701 = -13.02$$

There must of course be a simple relationship between Eh and pe, which can be found by combining Equations (4.26), (9.7) and (9.8) written for a half-reaction. It yields:

$$Eh = \frac{2.303\ RT}{F}\,\mathrm{pe} \tag{9.22}$$

At 25°C this is equal to

$$Eh = 0.059\mathrm{pe}\ (\mathrm{Volt}) \tag{9.23}$$

Both the Nernst equation and the pe-concept are commonly used in the literature. The advantage of the pe concept is that the algebra of redox reactions becomes identical to other mass action expressions, allowing the same algorithm to be used in computer programs. The disadvantage of the use of pe is, that it is a non-measurable quantity. For the sake of keeping calculations as simple as possible, we will in this book mainly use the pe concept.

QUESTION:
What are the pe values corresponding to $Eh = 0.59$ and -0.118 Volt, and what is the activity of the electrons?

ANSWER: pe = 10 and −2, $[e^-] = 10^{-10}$ and 10^2

9.2 REDOX DIAGRAMS

The number of dissolved species and mineral phases for which the redox conditions affect the stability can be overwhelming. Furthermore many redox reactions, like for Example (9.3), are strongly influenced by pH. In order to retain an overview in such complicated systems, redox diagrams are useful to display the stability of both dissolved species and minerals as a function of pe (or *Eh*) and pH. The prime force of the diagrams is that possible stable phases and species can be identified at a glance. Garrels and Christ (1965) and Brookins (1988) have presented redox diagrams for many systems serving that purpose, although more detailed calculations are often needed to confirm critical aspects. The construction of redox diagrams, and some of their limitations are discussed below. Detailed guidelines for construction of redox diagrams can be found in Garrels and Christ (1965), Stumm and Morgan (1996) and Drever (1997).

9.2.1 *Stability of water*

Although H_2O is not intuitively considered a redox sensitive substance, it may take part in the following redox reactions

$$H_2O \leftrightarrow 2e^- + 2H^+ + \tfrac{1}{2}O_{2(g)} \qquad (9.24)$$

where O^{2-} in water is oxidized, and

$$\tfrac{1}{2}H_2 \leftrightarrow H^+ + e^- \qquad (9.25)$$

where H^+ is being reduced. Very strong oxidants, that drive Reaction (9.24) to the right, cannot persist in natural environments because they react with H_2O. Likewise, very strong reductants will reduce H_2O. Accordingly, the stability of water sets limits to the possible redox conditions in natural environments. These can be quantified with the mass action equations for Reactions (9.24) and (9.25). For Reaction (9.24) it is:

$$\log K = \tfrac{1}{2}\log[P_{O_2}] - 2\text{pe} - 2\text{pH} = -41.55 \qquad (9.26)$$

Substituting the atmospheric concentration of oxygen, $[P_{O_2}] = 0.2$, leads to

$$\text{pe} = 20.60 - \text{pH} \qquad (9.27)$$

Similarly for Equation (9.25):

$$\log K = -\text{pH} - \text{pe} - \tfrac{1}{2}\log[P_{H_2}] \qquad (9.28)$$

Here log K is zero by definition. To assign a value to $[P_{H_2}]$ is more arbitrary, but the upper limit at the earth's surface must be a value of one, which reduces (9.28) to

$$\text{pe} = -\text{pH} \qquad (9.29)$$

Relationships (9.27) and (9.29) can be plotted in a pe–pH diagram (Figure 9.4) to delineate the range of redox conditions to be expected in natural environments. Together with the upper and lower stability limits of water, ranges of pe/pH conditions encountered in natural environments are shown. Some caution is warranted since these ranges are based on *Eh* measurements in nature (Section 9.1.1). Groundwater environments are seen to cover a broad range from oxidizing to reducing environments.

QUESTION:
If a more realistic value for P_{H_2} of 10^{-5} is used in Equation (9.28), how would this displace the H_2O/H_2 line in Figure 9.4?

ANSWER: $\text{pe} = -\text{pH} - \tfrac{1}{2}\log[H^+]$, $\text{pe} = -\text{pH} + 2.5$

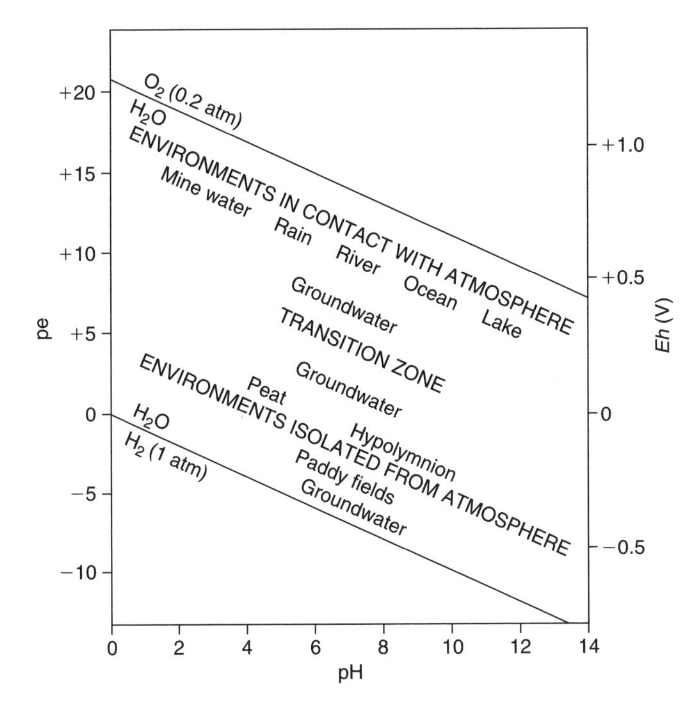

Figure 9.4. The stability of water the ranges of pe- and pH-conditions in natural environments (Modified from Garrels and Christ, 1965).

9.2.2 *The stability of dissolved species and gases: Arsenic*

Arsenic will be used to illustrate the construction of redox diagrams for dissolved species. As already discussed in Chapter 1, a high arsenic concentration in groundwater is a threat towards the health of millions of people in West Bengal and Bangladesh.

Arsenic is found in groundwaters as arsenate, As(5) (Smedley et al., 1998) and arsenite, As(3) (Mukherjee et al., 2000). We discuss here how to plot the stability fields of the dissolved As species in a redox diagram. Both As(5) and As(3) form protolytes, which may release protons stepwise in the same way as carbonic acid. The mass action constants for the reactions among the selected species can be found in the WATEQ4F.DAT database distributed with PHREEQC, and are given in Table 9.3.

Table 9.3. Log *K*'s for reactions among (selected) As species.

	Reaction	log *K*	
As(5)	$H_2AsO_4^- \leftrightarrow HAsO_4^{2-} + H^+$	−6.76	(9.30)
As(3)	$H_3AsO_3 \leftrightarrow H_2AsO_3^- + H^+$	−9.23	(9.31)
As(5)/As(3)	$H_3AsO_3 + H_2O \leftrightarrow H_2AsO_4^- + 3H^+ + 2e^-$	−21.14	(9.32)

For the second dissociation of H_3AsO_4 (Reaction (9.30) in Table 9.3) the mass action equation is:

$$\log[HAsO_4^{2-}] - pH - \log[H_2AsO_4^-] = -6.76 \qquad (9.33)$$

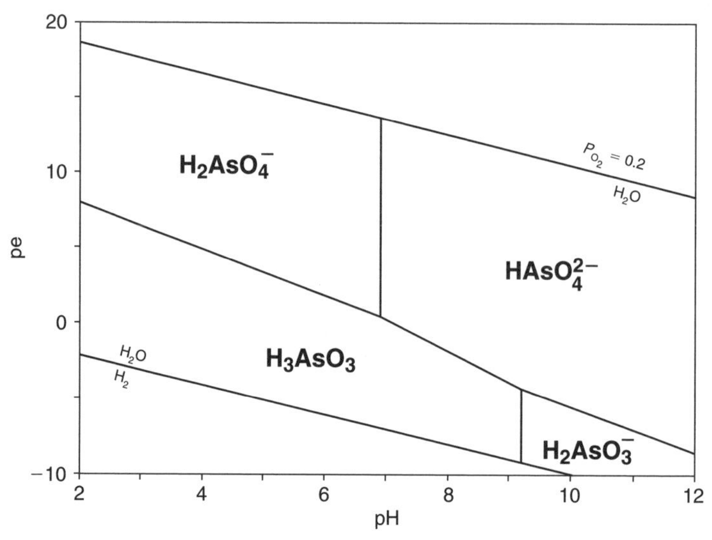

Figure 9.5. A partial pe–pH stability diagram for dissolved As species. Boundaries indicate equal activities of both species.

In case of equal activity of $H_2AsO_4^-$ and $HAsO_4^{2-}$, Equation (9.33) reduces to pH = 6.76; at lower pH the species $H_2AsO_4^-$ is the dominant form and at higher pH, $HAsO_4^{2-}$ predominates. The stability fields of the two species are separated by a vertical line at pH 6.76 in a pe–pH diagram (Figure 9.5), since the boundary is independent of the pe. It is important to realize that in the stability field of $H_2AsO_4^-$, $HAsO_4^{2-}$ is also present albeit, at a lower concentration.

In the same way the first dissociation reaction of H_3AsO_3 gives:

$$\log[H_2AsO_3^-] - pH - \log[H_3AsO_3] = -9.23 \tag{9.34}$$

Again a vertical line indicating equal activities of both species at pH 9.23 is shown in our pe–pH diagram (Figure 9.5).

The boundary between the stability fields of the As(5)-species $H_2AsO_4^-$ and the As(3)-species H_3AsO_3 is determined from Reaction (9.32) in Table 9.3 and results in:

$$\log[H_2AsO_4^-] - 3pH - 2pe - \log[H_3AsO_3] = -21.14 \tag{9.35}$$

which for equal activities of both species reduces to

$$2pe = -3pH + 21.14$$

This boundary depends on both pe and pH and is also plotted in Figure 9.5.

The boundaries between $HAsO_4^{2-}$ and H_3AsO_3, and between $HAsO_4^{2-}$ and $H_2AsO_3^-$ are found by combining reactions from Table 9.3. For the $HAsO_4^{2-}$ / H_3AsO_3 boundary, we add (9.30) to (9.32) and obtain:

$$H_3AsO_3 + H_2O \leftrightarrow HAsO_4^{2-} + 4H^+ + 2e^- \tag{9.36}$$

and

$$\begin{aligned} &\log[HAsO_4^{2-}] - 4pH - 2pe - \log[H_3AsO_3] = -27.9 \\ \text{or} \quad &2pe = -4pH + 27.9 \end{aligned} \tag{9.37}$$

And similar for the $HAsO_4^{2-}/H_2AsO_3$ boundary (Figure 9.5):

$$2pe = -3pH + 18.67$$

> QUESTIONS:
> At what pe and pH have the 3 species $H_2AsO_4^-$, $HAsO_4^{2-}$ and H_3AsO_3 equal activity?
> ANSWER: pH = 6.76, pe = 0.43
> Find the pH where $HAsO_4^{2-}$ and AsO_4^{3-} have equal activity from the WATEQ4F database?
> ANSWER: pH = 11.6
> Find log K for Reaction (9.32) from MINTEQ.DAT?
> ANSWER: −21.68 (note the difference with WATEQ4F.DAT!)

Calculating redox speciation with PHREEQC

We check the answer to the first question above with PHREEQC, using the input file:

```
DATABASE wateq4f.dat
SOLUTION 1
 pH 6.76; pe 0.43
 As 1e-3 # 75ug As/L
END
```

The speciation of As(3) and As(5) is here calculated from the input pe. Since there is only a very small amount of As in distilled water, the activity coefficients will be close to unity and the output shows indeed almost equal concentrations of the three species:

```
        Species                  Molality     Activity
As(3)                3.329e-07                              (←this is total As(3) in mol/L)
        H3AsO3                   3.318e-07    3.318e-07
As(5)                6.671e-07
        HAsO4-2                  3.349e-07    3.333e-07
        H2AsO4-                  3.322e-07    3.318e-07
```

Alternatively, we can input the concentrations of As(3) and As(5), together with a default pe (=4):

```
SOLUTION 1
 pH 6.76; pe 4
 As(3) 0.333e-3; As(5) 0.667e-3
REACTION
END
```

In this case the input As speciation is maintained and PHREEQC calculates the pe corresponding to the As(3) / As(5) redox couple:

```
        Redox couple          pe       Eh(volts)
        As(3)/As(5)        0.4299        0.0254
```

The pe of 4 is used in the initial equilibrium calculation for all other redox couples, e.g. of H_2O/O_2. Next, the keyword REACTION signals to do an equilibrium calculation in which the pe is adapted to achieve overall equilibrium. In this case, only the As species are of importance, and the pe becomes 0.43 "adjusted to redox equilibrium" in agreement with the initial concentrations of As(5) and As(3).

> QUESTION:
> At what pe and pH have the 3 species $HAsO_4^{2-}$, H_3AsO_3 and $H_2AsO_3^-$ equal activity?
> ANSWER: pH = 9.23, pe = −4.51
> Check the answer with PHREEQC?

Nitrogen

Nitrogen is an important component in the biogeochemical cycle, it is found in organic matter, in dissolved species and as various gases. In contrast, minerals containing N are generally very soluble and therefore rare in nature (but adsorption of NH_4^+ can be highly significant). Increasing fertilizer usage and spreading of manure with subsequent leaching of nitrogen from the soil has raised an interest in the fate of nitrogen in aquifers. Section 9.5 will deal with this subject.

Nitrogen is found in nature in valences ranging from +5 in NO_3^-, to −3 in NH_4^+, and a reduction series can be written as:

$$NO_3^- \rightarrow NO_2^- \rightarrow N_{2(g)} \rightarrow NH_4^+$$

Intermediates between NO_2^- and $N_{2(g)}$, such as $NO_{(g)}$ and $N_2O_{(g)}$ are known to occur in aquifers, although seldom in significant amounts. Bacteria play an essential role in catalyzing the reactions of nitrogen in nature, but the stability relationships among the nitrogen species will guide us in what can be expected in various environments. From PHREEQC.DAT we obtain the constants listed in Table 9.4.

Table 9.4. Log *K*'s for redox reactions among N species.

	Reaction	log *K*	
N(5)/N(3)	$NO_2^- + H_2O \leftrightarrow NO_3^- + 2H^+ + 2e^-$	−28.57	(9.38)
N(5)/N(0)	$\frac{1}{2}N_2 + 3H_2O \leftrightarrow NO_3^- + 6H^+ + 5e^-$	−103.54	(9.39)
N(0)/N(−3)	$NH_4^+ \leftrightarrow \frac{1}{2}N_2 + 4H^+ + 3e^-$	−15.54	(9.40)

First, we calculate the boundaries of N_2 with NO_3^- as the most oxidized species and with NH_4^+ as the most reduced species. For NO_3^- and N_2 the mass action equation of Reaction (9.39) is:

$$\log K = \log[NO_3^-] - 6\text{pH} - 5\text{pe} - \tfrac{1}{2}\log[P_{N_2}] = -103.54 \quad (9.41)$$

And for N_2 and NH_4^+ the mass action equation of Reaction (9.40):

$$\log K = \tfrac{1}{2}\log[P_{N_2}] - 4\text{pH} - 3\text{pe} - \log[NH_4^+] = -15.54 \quad (9.42)$$

In order to plot these relations in a redox diagram, we have to substitute values for the activities of NO_3^-, N_2, NH_4^+. The pressure of N_2 in the atmosphere is 0.77 atm and is a reasonable value to chose. The choice for $[NO_3^-]$ and $[NH_4^+]$ is more arbitrary, but an activity of 10^{-3} for NO_3^- is commonly found in polluted groundwater and is used here. For consistency, we may then choose the same activity for NH_4^+ even though it is usually much lower in groundwater. Substitution of these values simplifies Equations (9.41) and (9.42) to:

$$6\text{pH} + 5\text{pe} = 101.0 \text{ for the } NO_3^- / N_2 \text{ equilibrium} \quad (9.43)$$

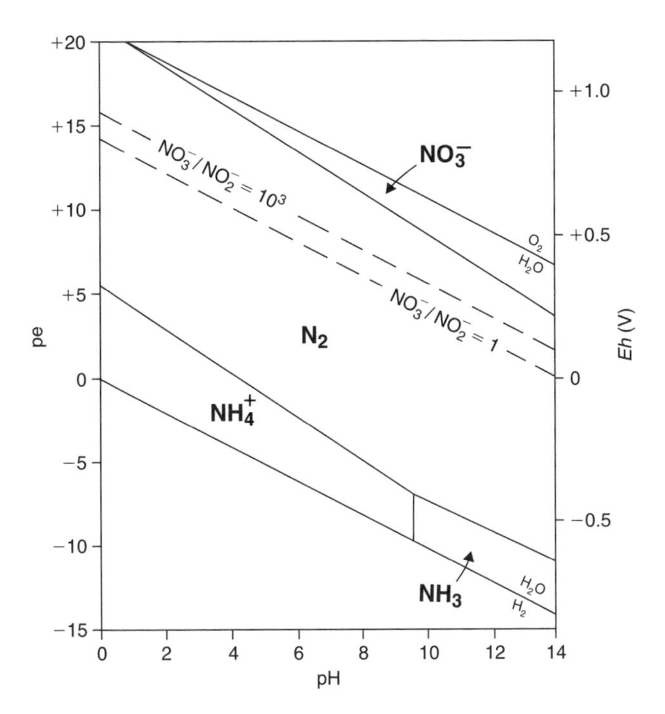

Figure 9.6. pe–pH diagram for the nitrogen system at 25°C. The diagram is drawn for $[P_{N_2}] = 0.77$ and 10^{-3} as activity of dissolved species unless otherwise specified. Metastable boundaries are indicated by dashed lines.

and

$$4\text{pH} + 3\text{pe} = 18.48 \text{ for the } N_2 / NH_4^+ \text{ equilibrium} \tag{9.44}$$

These equations are displayed in Figure 9.6. The results show a large stability field for N_2. NO_3^- is only stable near the upper stability limit of water and NH_4^+ first becomes stable near the lower stability limit of water. Above pH 9.25, NH_4^+ deprotonates to $NH_{3(aq)}$ and the pe-pH slope for the N_2/NH_3 boundary changes to −1.

In order to evaluate the stability of NO_2^-, which is intermediate in the redox sequence between NO_3^- and N_2, we consider Reaction (9.38) from Table 9.4:

$$\log K = \log[NO_3^-] - 2\text{pH} - 2\text{pe} - \log[NO_2^-] = -28.57 \tag{9.45}$$

Equation (9.45) can be drawn by assuming equal activities for NO_3^- and NO_2^-. The resulting line plots in Figure 9.6 in the N_2 stability field. Since N_2 is more reduced than NO_2^-, nitrite is unstable relative to N_2 with the chosen activities. One might argue that the assumption of equal activities of NO_3^- and NO_2^- is unreasonable, since the concentration of NO_2^- in groundwater normally is much lower than of NO_3^-. Therefore, we replot Equation (9.45) with the activity of NO_3^- a thousand times larger than of NO_2^-. The NO_3^-/NO_2^- boundary now moves upward to a higher pe but NO_2^- remains unstable relative to N_2. Only when the NO_3^-/NO_2^- ratio is increased to above 10^7, would NO_2^- attain its own stability field. But in that case NO_2^- becomes unmeasurable low, which leads us to conclude that the presence of NO_2^- in groundwater is kinetically controlled as an intermediate product in the reduction of NO_3^- to N_2. In the same way it can be shown that intermediates like $NO_{(g)}$ and $N_2O_{(g)}$ are unstable. In general, the example illustrates how unstable boundaries are identified during the construction of a redox diagram.

Figure 9.6 shows that at the P_{O_2} of the earth's atmosphere, N_2-gas is thermodynamically unstable relative to NO_3^-. The atmosphere consists of 77% N_2 and 21% O_2, and if N_2 should react with O_2 to form nitrate, the atmosphere would become depleted of oxygen. The associated proton production would acidify ocean waters to around 1.7 when mineral buffering reactions are disregarded. Fortunately, equilibrium thermodynamics do not predict a feasible reaction in this particular case.

EXAMPLE 9.5. *Oxidation of the atmosphere's N_2 content to nitrate*
Thermodynamically, N_2 is unstable relative to nitrate at the P_{O_2} of the atmosphere. What would be the pH of ocean water if all the O_2 in the atmosphere was consumed by oxidation of N_2 to NO_3^-? The O_2 content of the atmosphere is 3.7×10^{19} mol. The oceans contain 13.7×10^{23} g H_2O with an alkalinity of 2.3 meq/L.

ANSWER:
For oxidation of N_2 to NO_3^- we may write

$$N_2 + 2.5O_2 + H_2O \rightarrow 2NO_3^- + 2H^+$$

Accordingly, the 3.7×10^{19} moles of O_2 in the atmosphere could produce 3.0×10^{19} mol H^+, or 21.6 mmol H^+ per kg seawater. Subtracting the alkalinity content of seawater would leave us with 19.3 mmol H^+/L, corresponding to a pH of 1.7. Dissolution of carbonate and silicate minerals would in the long run probably buffer most of this acidity, but it would still leave us with an atmosphere deprived of oxygen.

QUESTIONS:
How would the NO_3^-/N_2 line move when $[NO_3^-] = 10^{-2}$ instead of 10^{-3}?
ANSWER: upward by ⅕ pe unit
Evaluate the fate of NH_4^+, commonly present in rainwater (Table 2.2)?
ANSWER: it should oxidize to NO_3^-

Non-equilibrium redox in PHREEQC
The lack of reaction among nitrogen species may conflict with the equilibrium calculations of PHREEQC. For example, you can dissolve the chemical $Fe(NO_3)_2$ and prepare a 1 mM solution in an anoxic glove box. Under anoxic conditions the concentrations of Fe^{2+} and NO_3^- remain unchanged. However, the simultaneous presence of Fe^{2+} and NO_3^- is not a stable combination (compare Figure 9.6 with 9.8) and if we calculate the equilibrium composition with PHREEQC, the result is that almost all Fe^{2+} oxidizes to Fe^{3+}, while 0.2 mM NO_3^- reduces to N_2. The reaction can be blocked in PHREEQC by decreasing the stability of NO_2^- and N_2 relative to NO_3^-:

```
SOLUTION_SPECIES
NO3- + 2H+ + 2e- = NO2- + H2O;      log_k -28.570   # +28.57 in database
2NO3- + 12H+ + 10e- = N2 + 6H2O;    log_k -207.08   # +207.08
SOLUTION 1
 Fe(2) 1
 N(5) 2
REACTION
END
```

Of course this also disables all redox reactions involving nitrate in further calculations.

When nitrate becomes reduced in aquifers the predominant reaction product is N_2. Further reduction of N_2 to NH_3 does not occur. In the PHREEQC database it is therefore useful to decouple NH_3 from the rest of the nitrogen system. In the database under SOLUTION_MASTER_SPECIES, N(-3) has been deleted and a substitute "Amm" species is defined:

```
SOLUTION_MASTER_SPECIES
#N(-3)          NH4+            0.0             14.0067
Amm             AmmH+           0.0             AmmH   17.0
```

The molecular weight of Amm is the same as for NH_3, and the default input species to recalculate input from mg/L to molar units is AmmH ($= NH_4^+$). In the database, all the SOLUTION_SPECIES, EXCHANGE_SPECIES, etc. which contain NH_3 have been redefined as Amm.

Sulfur and carbon

Also redox reactions involving sulfur and carbon play crucial roles in biogeochemical cycling of organic matter. The redox diagrams for these elements are shown in Figure 9.7. The stable valences of dissolved sulfur are sulfate with S(6) and hydrogen sulfide with S(−2). At near neutral pH both H_2S and HS^- can be found, but never S^{2-} because of the very small second dissociation constant of H_2S, $\log K = -13.9$. Elemental sulfur S(0) is shown as an intermediate wedge at the low pH side. This is a solid and the methods to delineate solids in redox diagrams will be presented in the next section.

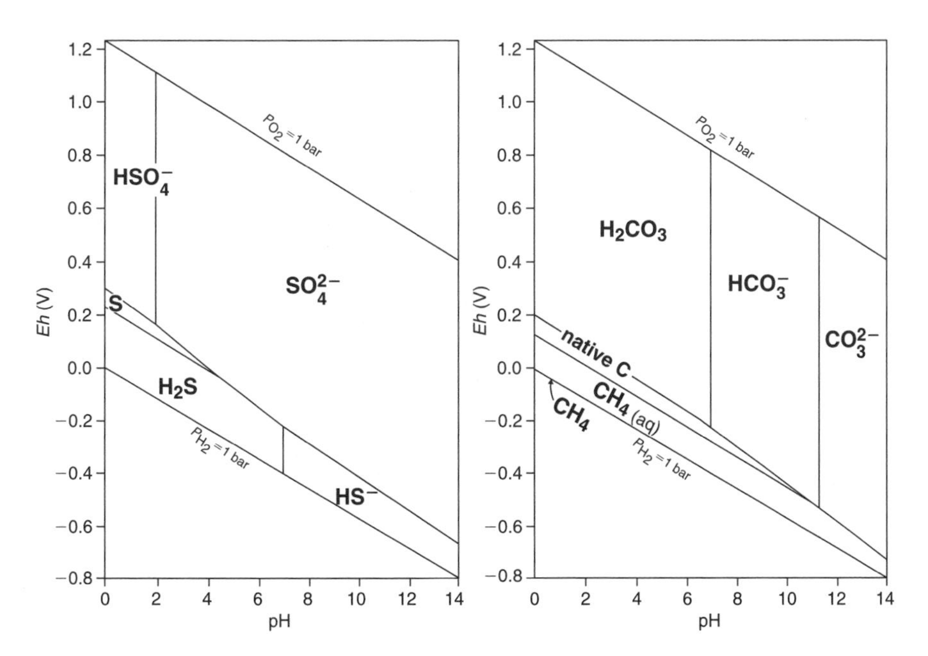

Figure 9.7. Redox diagrams for sulfur and carbon at 25°C. Total activity of dissolved species is 10^{-3}. Native C indicates dissolved CH_2O. S indicates elemental sulfur (Brookins, 1988).

The redox diagram of carbon displays C(4) at high *Eh* in the form of the pH dependent carbonate species. Under highly reducing conditions methane, C(−4), becomes stable. The intermediate wedge of "native C" is given for aqueous CH_2O, but is indicative for the stability of carbon compounds with C(0) in general.

Comparison of redox diagrams immediately delineates the thermodynamically stable species of different elements. For example, at a high pe (or *Eh*), O_2, NO_3^-, SO_4^{2-} and CO_3^{2-} are stable according to Figures 9.4, 9.6, and 9.7.

QUESTION:
Find the stable species of C, N and S in highly reducing conditions in water?
ANSWER: CH_4, NH_4^+ and H_2S

9.2.3 *The stability of minerals in redox diagrams: iron*

The construction of redox diagrams containing minerals is analogous to those for dissolved species, although the number of assumptions tends to increase. The construction of stability fields for minerals will be illustrated with a diagram for the iron system (Figure 9.8), considering the aqueous species and solids listed in Table 9.5.

Table 9.5. Log *K*'s for reactions among Fe species (Figure 9.8).

	Reaction	log *K*	
$FeOH^{2+}$	$Fe^{3+} + H_2O \leftrightarrow FeOH^{2+} + H^+$	−2.4	(9.46)
Fe(3)/ Fe(2)	$Fe^{2+} \leftrightarrow Fe^{3+} + e^-$	−13.05	(9.20)
Ferrihydrite	$Fe(OH)_{3(s)} + 3H^+ \leftrightarrow Fe^{3+} + 3H_2O$	3.7	(9.47)
Siderite	$FeCO_3 \leftrightarrow Fe^{2+} + CO_3^{2-}$	−10.45	(9.48)
$Fe(OH)_{2(s)}$	$Fe(OH)_{2(s)} + 2H^+ \leftrightarrow Fe^{2+} + 2H_2O$	13.9	(9.49)

Stability lines corresponding to Reactions (9.46) and (9.20) can be plotted in Figure 9.8 by again assuming equal activities for the dissolved species Fe^{2+}, Fe^{3+} and $Fe(OH)^{2+}$.

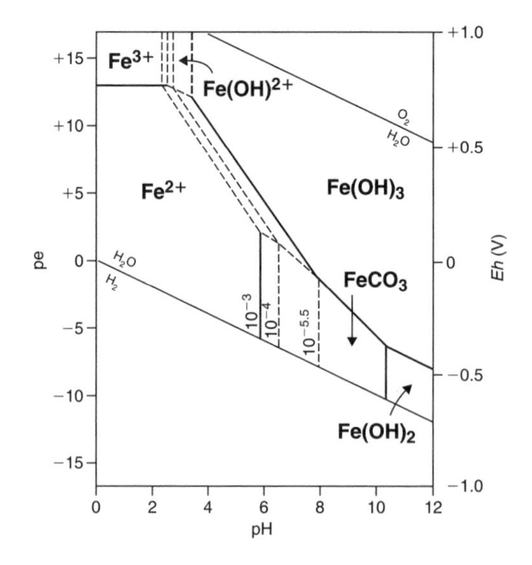

Figure 9.8. Stability relations in the system, Fe-H_2O-CO_2 at 25°C. *TIC* = $10^{-2.5}$. Solid/solution boundaries are specified for different [Fe^{2+}]. Heavy lines indicate "realistic" boundaries that correspond to usual field conditions.

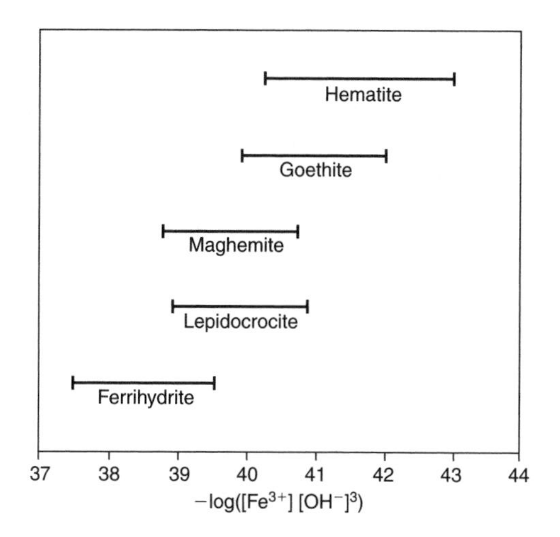

Figure 9.9. The stability ranges of common Fe-oxyhydroxides, hematite (α-Fe_2O_3), goethite (α-FeOOH), maghemite (γ-Fe_2O_3), lepidocrocite (γ-FeOOH) and amorphous $Fe(OH)_3$ or ferrihydrite ($5Fe_2O_3 \cdot 9H_2O$). The activity products are given for 25°C and 1 atm (modified after Langmuir, 1969, using data tabulated by Cornell and Schwertmann, 2003).

To depict the stability field of iron oxide we need to select one among several minerals. Some common iron oxide minerals comprise hematite (α-Fe_2O_3), goethite (α-FeOOH), maghemite (γ-Fe_2O_3), lepidocrocite (γ-FeOOH) and freshly precipitated hydrous ferric oxide or ferrihydrite ($5Fe_2O_3 \cdot 9\,H_2O$) (Cornell and Schwertmann, 2003). For the dissociation of lepidocrocite and goethite we may write:

$$FeOOH + H_2O \;\rightarrow\; Fe^{3+} + 3OH^- \tag{9.50}$$

for hematite and maghemite:

$$\tfrac{1}{2}Fe_2O_3 + \tfrac{3}{2}H_2O \;\rightarrow\; Fe^{3+} + 3OH^- \tag{9.51}$$

and for ferrihydrite:

$$Fe(OH)_3 \;\leftrightarrow\; Fe^{3+} + 3OH^- \tag{9.52}$$

In all three cases the mass action equation is

$$K = [Fe^{3+}][OH^-]^3 \tag{9.53}$$

Figure 9.9 compares the solubility of the various iron oxides and demonstrates a range in stability of about six orders of magnitude.

QUESTIONS:

Find log K for ferrihydrite in the PHREEQC databases?

ANSWER: PHREEQC.DAT, MINTEQ.DAT, WATEQ4F.DAT: −37.11; LLNL.DAT: −36.35

Find log K for goethite in the PHREEQC databases?

ANSWER: PHREEQC.DAT, WATEQ4F.DAT: −43; MINTEQ.DAT: −41.5; LLNL.DAT: −41.47

The most unstable form is freshly precipitated hydrous ferric oxide ($Fe(OH)_3$) while the most stable forms are hematite and goethite. Even the more crystalline minerals goethite and hematite show a range in solubility which partly is due to experimental uncertainty, but also covers a real variation due to different crystallinity, crystal size, solid solution, etc. In aquifers a mixture of different iron oxides is commonly found. Since only the least soluble iron oxide is stable, the presence of other Fe-oxides reflects the slow formation kinetics of the more stable phases. For Figure 9.8 we have used amorphous iron oxyhydroxide, $Fe(OH)_3$, with $K = [Fe^{3+}][OH^-]^3 = 10^{-38.3}$.

Substituting the dissociation reaction of H_2O yields Reaction (9.47) in Table 9.5, and the mass action equation in logarithmic form:

$$\log[Fe^{3+}] + 3pH = 3.7 \tag{9.54}$$

To display Equation (9.54) in the redox diagram, we select a small value for $[Fe^{3+}]$ (for example 10^{-6}) to indicate a low concentration under most field conditions. Actually, the situation at hand is more complex, since the aqueous complex $Fe(OH)^{2+}$ is also present in significant amounts. For total ferric iron in solution the mass balance is:

$$\Sigma Fe(3) = m_{Fe^{3+}} + m_{FeOH^{2+}} \tag{9.55}$$

The total concentration of dissolved ferric iron in equilibrium with $Fe(OH)_3$ is found by substituting the mass action equations for Reactions (9.46) and (9.47) in Equation (9.55). We disregard the difference between activity and molar concentration and obtain:

$$\Sigma Fe(3) = 10^{3.7}\,[H^+]^2\,([H^+] + 10^{-2.4}) \tag{9.56}$$

This equation was solved for $\Sigma Fe(3)$ equal to 10^{-3}, 10^{-4} and $10^{-5.5}$ M, and the resulting boundaries are displayed in Figure 9.8. Note that the aqueous complex $Fe(OH)^{2+}$ doubles the solubility of $Fe(OH)_3$ at $pH = 2.4$.

The boundary between the Fe^{2+} and $Fe(OH)_3$ is obtained by subtracting Reaction (9.47) from (9.20) and writing the mass action equation:

$$\log K = -3pH - pe - \log[Fe^{2+}] = -16.8 \tag{9.57}$$

Usually a low Fe^{2+} concentration is chosen to draw a stability line in contact with Fe-oxide and a boundary for $[Fe^{2+}] = 10^{-5.5}$ is marked by a full line in Figure 9.8. Note the strong pH dependency of this boundary. The size of the stability field of Fe-oxide will vary with the stability of the selected Fe-oxide. In Figure 9.8 we used the most soluble form, ferrihydrite, but for Figure 9.10 (discussed later) the least soluble Fe-oxide (hematite) was taken and the Fe-oxide stability field becomes much larger.

For siderite ($FeCO_3$, Reaction (9.48) in Table 9.5) we need to define the carbonate speciation. Assume *TIC* to be constant, and obtain CO_3^{2-} as a function of pH with the speciation factor $\alpha = m_{CO_3^{2-}}/TIC$, where (cf. Equation 5.13):

$$\alpha^{-1} = \frac{[H^+]^2}{K_1K_2} + \frac{[H^+]}{K_2} + 1 \tag{9.58}$$

Substituting in Equation (9.48) and filling in constants yields:

$$TIC = \frac{10^{-10.45}}{[Fe^{2+}]}\left[\frac{[H^+]^2}{10^{-16.6}} + \frac{[H^+]}{10^{-10.3}} + 1\right] \tag{9.59}$$

This equation has been solved for $[H^+]$, with $TIC = 10^{-2.5}$M and $[Fe^{2+}]$ activities of 10^{-3}, 10^{-4} and $10^{-5.5}$. The size of the siderite stability field changes with $[Fe^{2+}]$ and for a boundary concentration of 10^{-6} the siderite stability field would disappear altogether. In this case the choice of a low equilibrium $[Fe^{2+}]$ is not reasonable since siderite in nature is found in environments high in $[Fe^{2+}]$ (Postma, 1981, 1982). The boundaries at $[Fe^{2+}] = 10^{-3}$ and 10^{-4} are therefore the most realistic ones.

The boundary between $FeCO_3$ and $Fe(OH)_3$ is found graphically as the intersection of the $Fe^{2+}/FeCO_3$ and $Fe^{2+}/Fe(OH)_3$ boundaries, or more formally by combining Reactions (9.20), (9.47) and (9.48):

$$FeCO_3 + 3H_2O \leftrightarrow Fe(OH)_3 + CO_3^{2-} + 3H^+ + e^- \quad (9.60)$$

where the CO_3^{2-} activity should again be related to *TIC* and pH. The resulting boundary between $FeCO_3$ and $Fe(OH)_3$ is also shown in Figure 9.8.

QUESTIONS:
Find log K for Reaction (9.60)?
ANSWER: −27.2
Find $[H^+]$ and $[CO_3^{2-}]$ for $TIC = 10^{-2.5}$ M and $[Fe^{2+}] = 10^{-3}$ in equilibrium with siderite?
ANSWER: 1.27×10^{-6} (pH = 5.90), $[CO_3^{2-}] = 3.55\times10^{-8}$
In addition to the previous question, find the pe for equilibrium with ferrihydrite?
ANSWER: 2.05
Calculate the pe/pH for these conditions with PHREEQC? The solubility products of siderite and ferrihydrite used in the manual calculation differ from those in PHREEQC.DAT. Saturation indices following the mineral names are used to apply the manual calculation constants in PHREEQC.
ANSWER:
```
SOLUTION 1; C(4) 3.16; Fe 1; pH 7 Siderite 0.44;
pe 4 Fe(OH)3(a) -1.19;END
```
gives pH = 6.02, pe = 1.78 (Why are the results still different?)
Find the pH for the boundary between siderite and $Fe(OH)_{2(s)}$ in Figure 9.8?
ANSWER: 10.87

In summary, boundary lines in redox diagrams have different meanings: between dissolved species they usually indicate equal activities, between solids and dissolved species they indicate equilibrium for a specified concentration, and if other components, such as dissolved carbonate or sulfur species, are involved their concentrations need to be stipulated. Therefore the specifications in small print below redox diagrams should be read carefully.

Iron and manganese

Compare, for example, Figure 9.8 with the often reproduced diagram for the iron system in Figure 9.10. The more stable Fe-oxide hematite (Figure 9.10) has a much larger stability field than $Fe(OH)_3$ (Figure 9.8), while magnetite (Fe_3O_4), which was not considered in Figure 9.8, replaces $Fe(OH)_2$. Figure 9.10 includes sulfur in the system and a stability field for pyrite appears within that of siderite. Inspection of the figure caption shows the diagram to be drawn for an unrealistically high *TIC* content of 1 M, while ΣS is very low. These values were chosen in order to display both pyrite and siderite stability fields together in the same diagram. With a more realistic (smaller) *TIC*, the siderite field would simply disappear. Clearly the diagram is not suitable to evaluate the relative stability of pyrite and siderite.

The great value of redox diagrams is to obtain a quick overview of complex geochemical systems, as illustrated in a comparison of the diagrams for Fe and Mn (Figure 9.10). The presence of MnO_2 in sediments indicates strongly oxidizing conditions. In contrast hematite is stable over a much broader pe range.

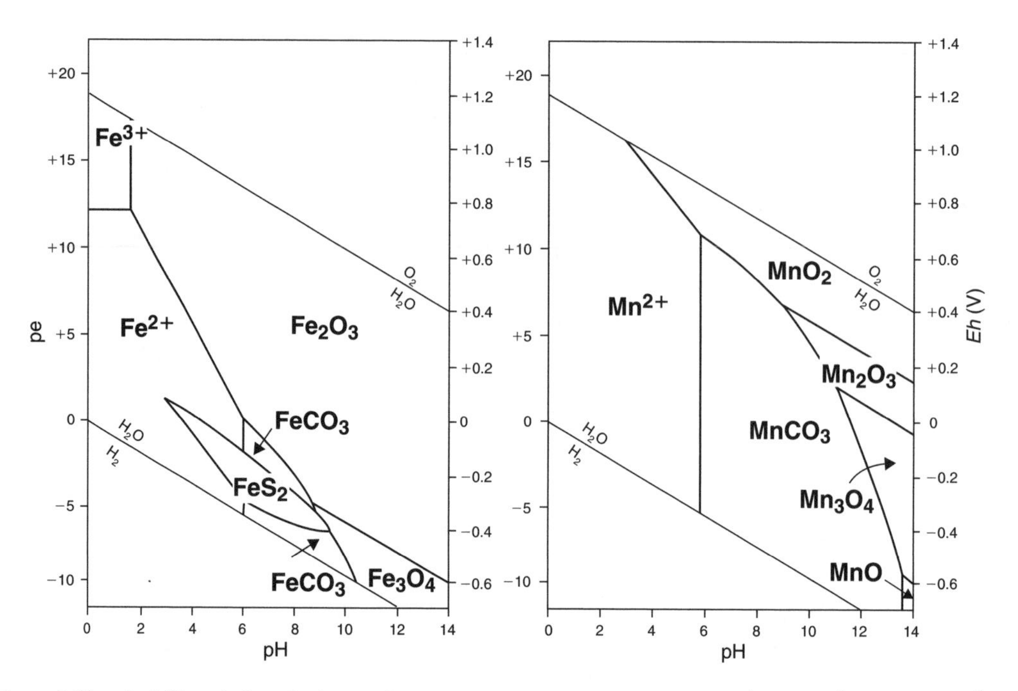

Figure 9.10. Stability relations for iron and manganese at 25°, both assuming that $\Sigma S = 10^{-6}$ and $TIC = 10^{0}$ M. Solid-solution boundaries are drawn for $[Fe^{2+}] = 10^{-6}$ (Modified from Krauskopf, 1979).

Dissolved Mn^{2+} is stable over a wide range in contact with hematite, while reversely Fe^{2+} is unstable in contact with MnO_2. Furthermore, rhodochrosite ($MnCO_3$) is stable over a wide pe range, while the presence of siderite ($FeCO_3$) indicates strongly reducing conditions. Finally, Mn-sulfide does not appear in the diagram, because it is much more soluble than Fe-sulfide, and therefore Mn-sulfide is extremely rare in recent environments. These important conclusions can be derived immediately from pe / pH diagrams, and alternatively would require many hours of calculations.

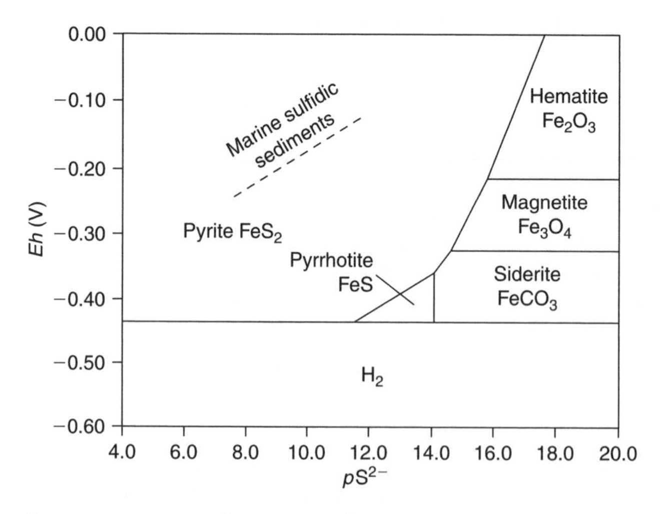

Figure 9.11. Eh–pS^{2-} diagram where $pS^{2-} = -\log[S^{2-}]$, pH = 7.37, $P_{CO_2} = 10^{-2.4}$, 25°C (Berner, 1971).

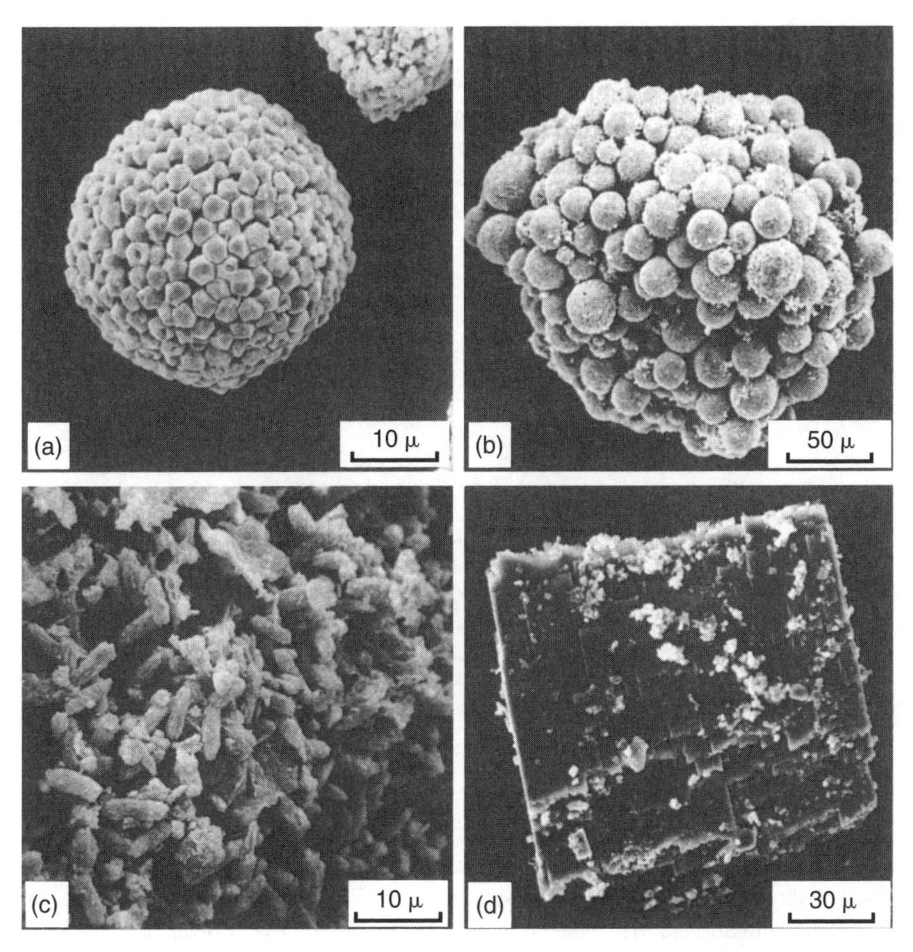

Figure 9.12. SEM micrographs of some common authigenic Fe(2) minerals. (a) Framboidal pyrite which are aggregates of pyrite crystals, (b) polyframboidal pyrite, (c) siderite and (d) a crystal fragment of vivianite. From Postma (1982).

In some cases the stability relations among redox species are better visualized in diagrams using variables other than pe and pH. For example, the relative stability of pyrite and siderite is relatively unaffected by pe and pH. The decisive factor is the insolubility of iron sulfide. It is only once dissolved sulfide becomes exhausted, that the Fe^{2+} concentration can increase sufficiently to stabilize siderite. This point will become clear when in the redox diagram of iron, the pH is replaced by pS^{2-} as controlling variable (Figure 9.11). Siderite first becomes stable at extremely low dissolved sulfide concentrations. In practice this implies an anoxic environment where sulfate is absent or has become exhausted. Therefore, siderite may be found in freshwater environments low in sulfate, like swamps (Postma, 1977, 1981, 1982) and lakes (Anthony, 1977) and is here often accompanied by the ferrous phosphate mineral vivianite ($Fe_3(PO_4)_2 \cdot 8\,H_2O$). Ferrous-iron rich groundwater is often supersaturated for siderite (Margaritz and Luzier, 1985; Morin and Cherry, 1986; Ptacek and Blowes, 1994; Jakobsen and Postma, 1999) indicating slow precipitation kinetics. Siderite is a common constituent of sedimentary rocks either in finely dispersed form or as concretions. Furthermore, siderite may form solid solutions with calcite and dolomite. Some SEM micrographs of siderite and other authigenic iron minerals are shown in Figure 9.12.

9.3 SEQUENCES OF REDOX REACTIONS AND REDOX ZONING

Figure 9.13 shows the evolution in groundwater composition along the flow path in the Middendorf aquifer. The groundwater is already anoxic near the zone of recharge but nitrate and sulfate are still present as electron acceptors. Downstream, first nitrate is reduced, and next the reduction of Fe-oxides leads to an increase of the Fe^{2+} concentration. Subsequently sulfate is reduced and the precipitation of iron sulfide causes a decrease in the Fe^{2+} concentration. Finally, methane appears in the groundwater.

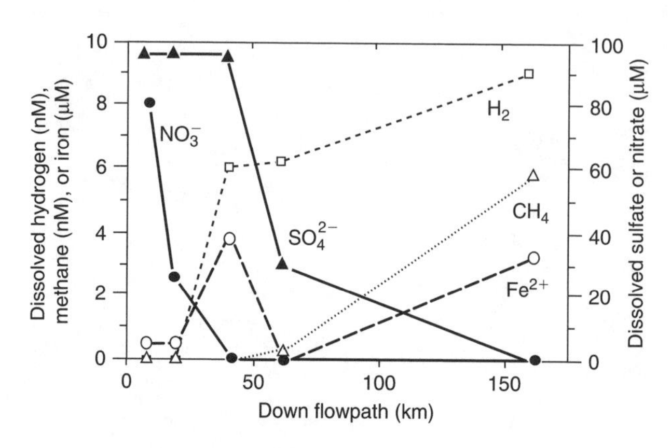

Figure 9.13. Redox zoning along the flow path in the Cretaceous Middendorf aquifer, South Carolina, USA (Lovley and Goodwin, 1988).

The Middendorf aquifer is a good example of a regional aquifer, with long groundwater residence times, where the slow degradation of sedimentary organic material generates sequential changes in the water chemistry. On a much smaller scale Figure 9.14 displays the water chemistry in a Pleistocene phreatic aquifer near Bocholt, Germany. In the profile, the residence time of the groundwater increases with depth as discussed in Chapters 3 and 11. Going downwards in the borehole, first O_2 disappears, then NO_3^-, and then the concentration of SO_4^{2-} decreases, similar to what is observed with increasing flow distance in the Middendorf aquifer.

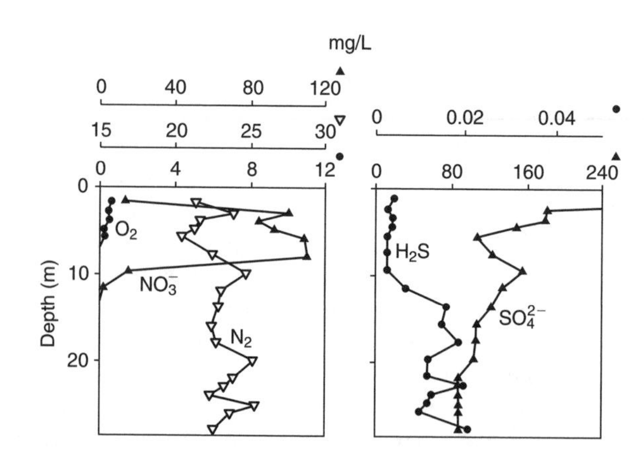

Figure 9.14. Redox zoning in a sandy Pleistocene aquifer, Bocholt, Germany. The borehole (DFG6) is located in the upstream part of the aquifer with coherent flowlines (Leuchs, 1988).

Apparently, as water contacts reductants in the subsoil, it loses its oxidants in a sequence that follows the pe from high to low and the changes in water chemistry can be predicted by the redox diagrams. In the upper part of the Bocholt aquifer the groundwater age increases with 1 yr/m depth and it takes about 5 years before O_2 is fully depleted. In the Middendorf aquifer hundreds of years are involved. Taken together, the two aquifers show that the sequence of redox reactions can be predicted from equilibrium thermodynamics, but the reaction rates are rather variable.

The sequence of predominant redox half reactions that can be predicted from the redox diagrams is summarized in Figure 9.15 for pH 7. The upper part lists the reduction reactions going from O_2 reduction, NO_3 reduction and reduction of Mn-oxides that occur at a high pe, to the reduction of Fe-oxides, sulfate reduction and methanogenesis taking place at a lower pe. The lower half of Figure 9.15 lists in the same way the oxidation reactions, with the oxidation of organic matter having the lowest pe. A reduction reaction will proceed with any oxidation reaction that is located (origin of the arrow) at a lower pe. For example, the reduction of sulfate can be combined with the oxidation of organic matter, but not with the oxidation of Fe(2).

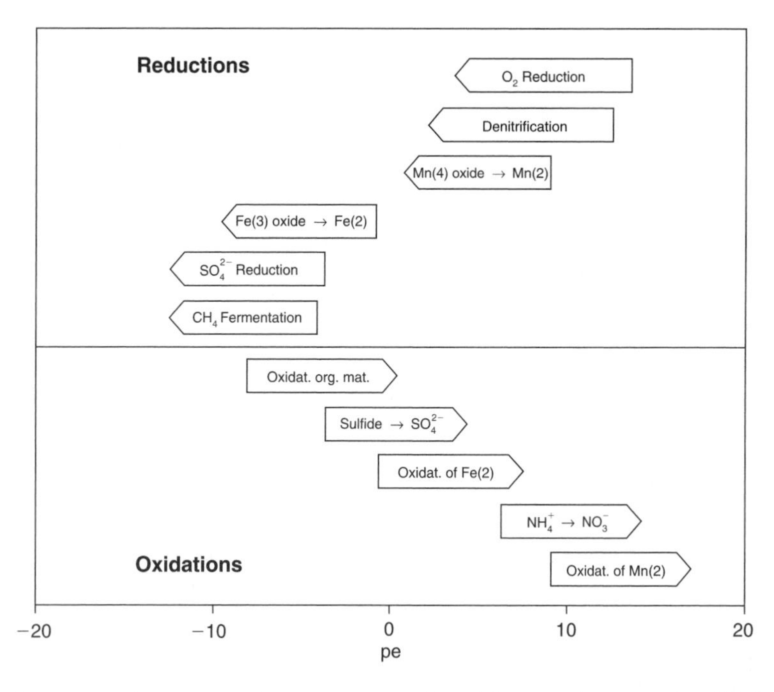

Figure 9.15. Sequences of important redox processes at pH = 7 in natural systems (modified and corrected after Stumm and Morgan, 1996).

Taking organic matter as the driving reductant, the water chemistry may change as illustrated schematically in Figure 9.16. For some electron acceptors (O_2, NO_3^-, SO_4^{2-}) it is the disappearance of a reactant while in other cases (Mn^{2+}, Fe^{2+}, H_2S and CH_4) it is the appearance of a reaction product that is notable in the groundwater composition. In the region dominated by sulfate reduction the Fe^{2+} concentration may decrease as the result of precipitation of iron sulfides. Redox environments are often characterized by the dominant ongoing redox process as indicated by the water chemistry (Champ et al., 1979; Berner, 1981). Berner (1981) (Figure 9.16) distinguishes between *oxic* and *anoxic* environments, i.e. whether they contain measurable amounts of dissolved O_2 ($\geqslant 10^{-6}$M).

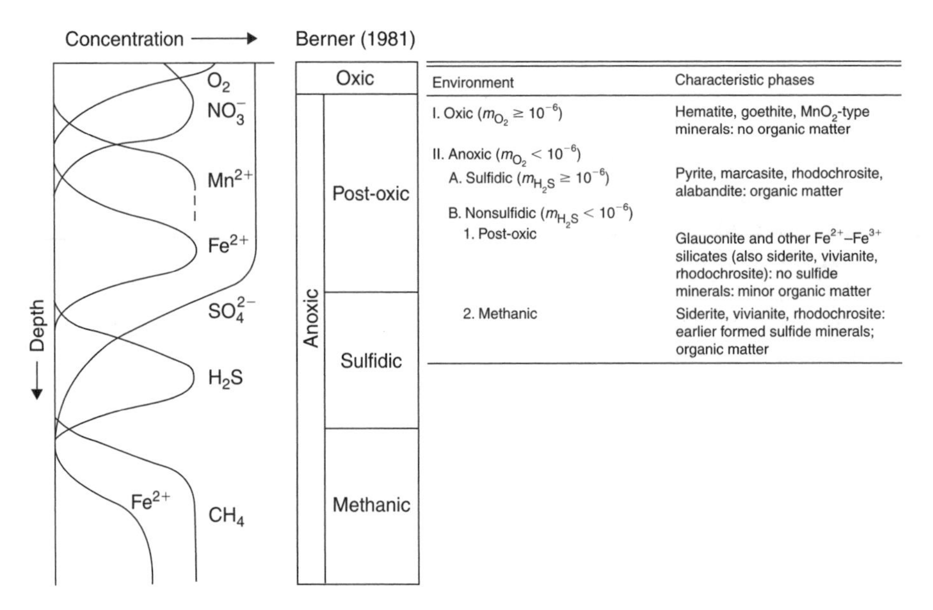

Figure 9.16. The sequence of reduction processes as displayed in groundwater chemistry. At right is Berner's (1981) classification of redox environments together with solids expected to form in each zone.

The increase in nitrate in the oxic zone is due to oxidation of ammonia released from oxidizing organic matter. The anoxic environments are subdivided in *post-oxic*, dominated by the reduction of nitrate, Mn-oxide and Fe-oxide, *sulfidic*, where sulfate reduction occurs and finally the *methanic* zone. A further subdivision of the post-oxic zone into a *nitric* ($m_{NO_3^-} \geqslant 10^{-6}$M, $m_{Fe^{2+}} \leqslant 10^{-6}$M) and a *ferrous* zone ($m_{Fe^{2+}} \geqslant 10^{-6}$M, $m_{NO_3^-} \leqslant 10^{-6}$M) can be useful in groundwater environments. Other classifications of groundwater redox zoning are given by Bjerg et al. (1995) and Chapelle et al. (1995). Not all zones are necessarily visible in a reduction sequence and in many cases the groundwater never passes the post-oxic state. In other cases, the sulfidic zone may seem to follow the oxic zone directly, but the post-oxic zone is probably always present on a microscopic scale. Figure 9.16 also shows some minerals which could be stable in the different redox zones. Their occurrence can be derived from the redox diagrams on the preceding pages. To characterize the aquifer redox state from the water composition is certainly a useful approach but some caution is warranted. For example, the enrichment of groundwater with Fe^{2+} may be caused by partial oxidation of pyrite (see Section 9.4.2) a process totally different from the reduction of iron oxides by organic matter. Reaction products like methane can be transported over long distances through the aquifer and the mere presence of methane in groundwater does not implicate ongoing methanogenesis at the site of sampling.

EXAMPLE 9.6. *Calculation of redox zonation with PHREEQC*
The sequence of redox reactions can be modeled by adding organic carbon stepwise through an irreversible reaction to a mixture of water and sediment that contains oxidants. A PHREEQC input file might look like:

```
SOLUTION 1
 pH       6.0
 Na       1.236;   K        0.041;   Mg        0.115;   Ca              0.067
 Cl       1.467;   N(5)     0.058;   S(6)      0.085;   Alkalinity      0.26
 O(0)     0.124
```

```
EQUILIBRIUM_PHASES 1
 Goethite 0 2.5e-3
 FeS(ppt) 0 0
 Pyrolusite 0 4e-5

REACTION 1
 C; 0.572E-3 in 26 steps
INCREMENTAL_REACTIONS true

USER_GRAPH
 headings C O2 NO3 Mn Fe SO4 S(-2) C(-4)
 axis_titles "Carbon added, mmol/L" "Concentration, mol/L"
 -start
 10 graph_x step_no * 0.572/26
 20 graph_y tot("O(0)")/2, tot("N(5)"), tot("Mn(2)"), tot("Fe(2)"),
 \tot("S(6)"), tot("S(-2)"), tot("C(-4)")
 -end
END
```

The results are plotted in Figure 9.17 and show the characteristic stepwise development in redox zones. Hydrogen sulfide appears only at low concentrations in the water because of excess iron and the precipitation of FeS.

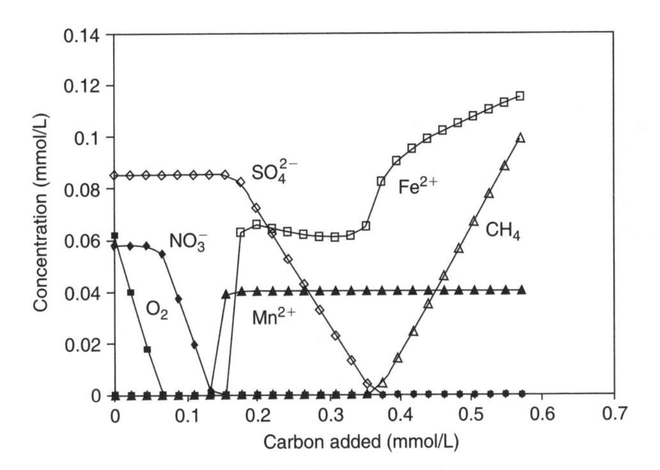

Figure 9.17. Development of redox zones modeled by PHREEQC using stepwise addition of carbon to an oxidized sediment.

QUESTIONS:
Plot pe and pH in Figure 9.17 on the secondary y-axis?
What is the effect of defining pe 16 in SOLUTION 1?

Redox zoning may develop in an opposite way downstream of landfills (see Christensen et al., 2000 for a review). Figure 9.18 displays the extent of different redox zones in the plume of the Vejen landfill, Denmark. The different redox zones are delineated based on the water composition, similar to the classification given in Figure 9.16. Directly below the landfill the conditions are methanogenic, but they change downstream to sulfate reducing, iron and manganese reducing, then to nitrate reducing and about 400 m downstream of the landfill the groundwater is oxic (this is the pristine groundwater). In contrast to the natural situation of infiltrating rainwater, the landfill plume becomes less reduced downstream and electron acceptors must be provided to oxidize the organics in the plume. The most important electron acceptors are the Fe-oxides present in the sediment (Heron and Christensen, 1995) through which the leachate passes, and the iron reducing zone also has the largest extent in Figure 9.18. Dissolved electron acceptors, like sulfate, nitrate and oxygen can only be introduced into the organic-rich leachate plume through mixing, dispersion and diffusion. Therefore, a sulfate reducing zone is not logically generated downstream of a methanogenic zone. In this case, local sulfate rich sources within the landfill, like gypsum waste, are emitting sulfate enriched groundwater which is mixed with the organic leachate. Likewise, a nitrate reducing zone can only result from mixing of organic leachate with nitrate-bearing groundwater. The oxidation of a large amount of organics in the leachate plume may generate abundant CO_2 which can cause the dissolution of carbonates. Similar plumes of organic contaminants are often found in aquifers below cities in developing countries due to infiltration of unsewered urban wastewaters (Lawrence et al., 2000).

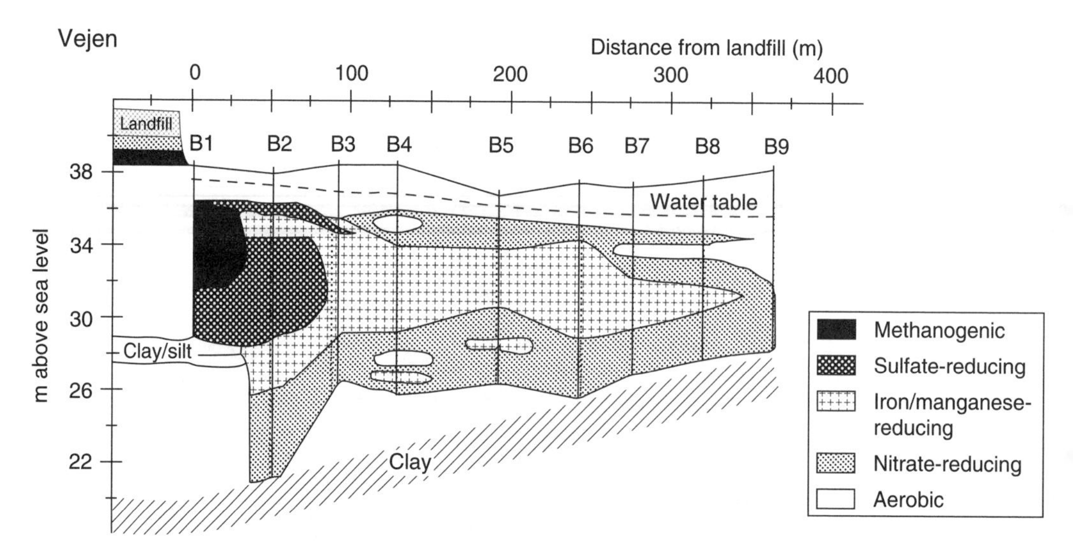

Figure 9.18. Redox zones in a sandy aquifer downstream the Vejen Landfill, Denmark (Lyngkilde and Christensen, 1992).

9.3.1 *Decomposition of organic matter*

Organic matter in aquifers is mostly of sedimentary origin and consists of the remains of plants in terrestrial sediments, supplemented by plankton and animal residues in the case of marine and lacustrine sediments. The organic carbon content is commonly lowest in sand deposits and chalk but higher in clays and silts. Generally, the degradation of sedimentary organic matter is very slow. Table 9.6 provides the rate of organic carbon degradation of some unpolluted aquifers and shows rates to vary from a few micromoles to more than millimoles per liter of groundwater per year. The older aquifer systems generally contain organic matter that reacts more slowly although the reactivity of organic matter will depend on the diagenetic history.

Table 9.6 also lists the inverse of the first order reaction rate constant, derived by arbitrarily assuming that all the aquifers contain 12‰ organic C ($\approx$ 6 mol C/L groundwater), i.e. $(k)^{-1}$ in the equation $dC/dt = -kC$. With this organic C content, k^{-1} is approximately equal to the age of the sediment, illustrating in a general sense how organic matter becomes less reactive as it grows older (Middelburg, 1989; Boudreau, 1997). Middelburg (1989) found a relation that can be reworked to:

$$k = 0.16 / t^{0.95} \tag{9.61}$$

where t is the age of the organic matter. The organic matter reactivity is ultimately the engine that drives most redox reactions and the difference in organic matter reactivity of more than five orders of magnitude is the key towards understanding differences in redox behavior among aquifers.

Table 9.6. Rates of organic matter degradation in aquifers. Modified from Jakobsen and Postma (1994).

Aquifer	Rate (mmol C/L/yr)	Sediment age	k^{-1} (yr)	Reference
Rømø, Denmark	$1.0–90\times10^{-1}$	Holocene	1.3×10^{3}	Jakobsen and Postma (1994)
Tuse Næs, Denmark	$3.0–7.4\times10^{-1}$	Pleistocene	1.2×10^{4}	Jakobsen and Postma (1994)
Bocholt, Germany	4.6×10^{-2}	Pleistocene	1.3×10^{5}	Leuchs (1988)
Führberg, Germany	2.8×10^{-2}	Pleistocene	2.1×10^{5}	Böttcher et al. (1989)
Sturgeon Falls, Canada	$2.0–3.2\times10^{-2}$	Pleistocene	2.4×10^{5}	Robertson et al. (1989)
Fox Hills, USA	4×10^{-4}	Cretaceous	1.5×10^{7}	Thorstenson et al. (1979)
Florida, USA	2×10^{-4}	Tertiary	3×10^{7}	Plummer (1977)
Black Creek, USA	$1.4–30\times10^{-5}$	Cretaceous	10^{8}	Chapelle and McMahon (1991)

The most abundant source of sedimentary organic matter is plant material, consisting mainly of lignin and polysaccharides. Since lignin is largely recalcitrant, the degradation of carbohydrates becomes the most important process and it is mediated by different groups of microorganisms (Conrad, 1999). In an overall sense, this is what happens. As the first step, fermenting bacteria excrete enzymes that hydrolyze the polysaccharides and break them down to alcohols, fatty acids and H_2. During step two, other bacteria degrade the alcohols and long chained fatty acids further to acetic acid (CH_3COOH), formic acid ($HCOOH$), H_2 and CO_2. In the final step, the fermentative intermediates acetate, formate and H_2 are oxidized by oxidants or so-called terminal electron accepting processes (TEAP's). For the microbial reduction of Fe-oxide, Mn-oxide and sulfate, and for methanogenesis, the intermediate products of the fermentation process are necessary. For the reduction of oxygen and nitrate the pathway is slightly different since the microorganisms mediating these processes may metabolize the products of step one directly.

Figure 9.19 illustrates the overall pathway of organic matter decomposition, combining step one and two in the production of H_2, acetate and formate, and indicating step three with the competition of different oxidants for the same pool of substrate. In a kinetic process consisting of several successive reaction steps, the slowest step will become overall rate limiting. The rate limiting step can be identified from the behavior of the pool of reaction intermediates: if the first step is rate limiting then the concentration of intermediates remains low since everything produced is quickly consumed by step two. Conversely when step two is rate limiting, the intermediates are accumulating.

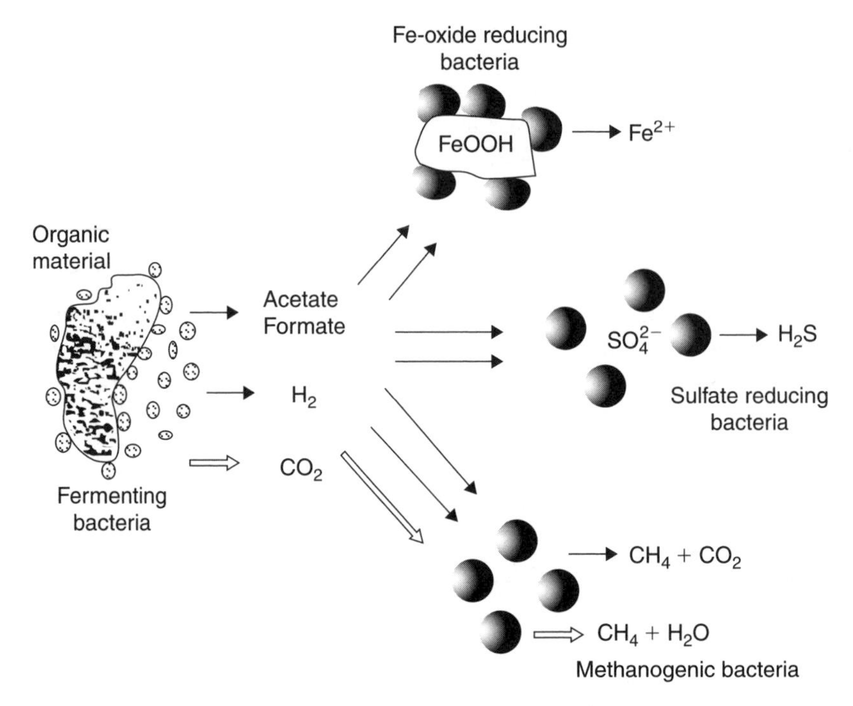

Figure 9.19. Schematic pathway of organic matter decomposition under anaerobic conditions (Courtesy Rasmus Jakobsen).

Figure 9.20 shows the redox chemistry and the distribution of acetate, formate and H_2 in the anoxic aquifer at Rømø, Denmark (Hansen et al., 2001; Jakobsen and Postma, 1999). H_2 is present at a concentration level of a few nanomoles per liter, and acetate and formate are at a micromolar level. Since the oxidants are turned over on a scale of mmol/L, as reflected by the increases in Fe^{2+} and CH_4 and the decrease in SO_4^{2-}, the first (fermenting) step must be rate limiting. The residence time for H_2, acetate and formate can be calculated from rates and pool sizes and ranges from minutes to hours, indicating a highly dynamic behavior of the intermediates. If the fermentive step is rate limiting and the overall process therefore kinetically controlled why can the sequence of the different redox reaction be predicted from equilibrium thermodynamics? As a first approximation Postma and Jakobsen (1996) proposed a partial equilibrium model where organic matter fermentation provides the overall kinetic control, while the subsequent oxidation step approaches equilibrium. The order of the redox processes is in this model derived from equilibrium calculation using, for example, H_2 as electron donor (Table 9.7).

On a microbial level, the different microbial communities mediating the oxidation step are competing for the energy available from the oxidation of H_2, acetate or formate. For each microbial community there is an energy threshold in the range 3–25 kJ/mol H_2, corresponding to the energy needed to sustain microbial activity. This energy threshold is the reason why H_2 concentrations measured in the field (Figures 9.13 and 9.20) are about an order of magnitude higher than calculated for the reaction at equilibrium (cf. Question). The microbial communities utilize the available energy, bringing it down to the minimum threshold where they can operate (Hoehler, 1998; Conrad, 1999; Jakobsen and Postma, 1999). Reactions with a high energy gain (Table 9.7) are therefore able to bring down the H_2 concentration to a lower level than reactions with a lower energy gain. As the result, the reactions requiring a higher H_2 concentration become inhibited because they cannot pass their energy threshold.

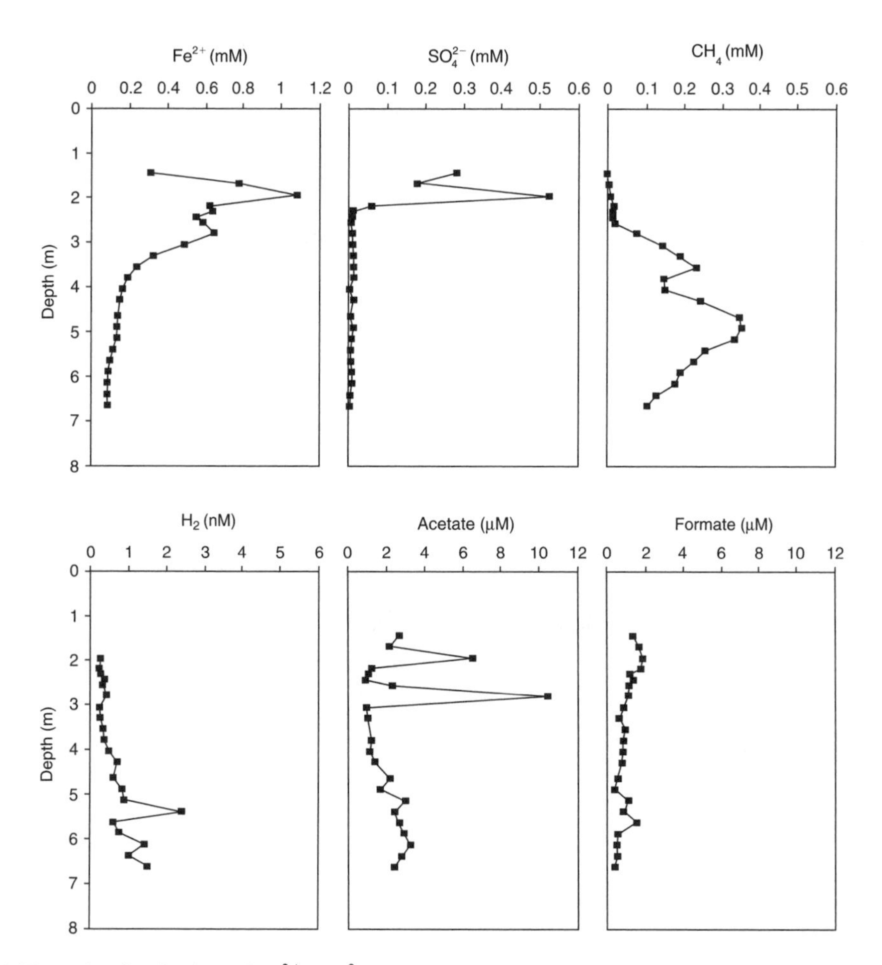

Figure 9.20. The distribution of Fe^{2+}, SO_4^{2-} and CH_4 and of fermentative intermediates H_2, acetate and formate in the Rømø aquifer (Hansen et al., 2001).

For example, microbes reducing $Fe(OH)_3$ may, due to the higher energy yield of the process (Table 9.7), lower the H_2 concentration, and thereby the energy gain, to below the level where a sulfate reducer can be active. The result is that different electron accepting processes proceed in the same order as predicted for equilibrium even though in reality they never attain equilibrium because of the energy thresholds.

Table 9.7. Energy gains of reactions between H_2 and electron acceptors under standard conditions at pH 7 (Lovley and Goodwin, 1988).

Reaction	Released energy (kJ/mol H_2)
$2H_2 + O_2 \rightarrow 2H_2O$	−237
$5H_2 + 2NO_3^- + 2H^+ \rightarrow N_2 + 6H_2O$	−224
$H_2 + 2Fe(OH)_3 + 4H^+ \rightarrow 2Fe^{2+} + 6H_2O$	−50
$4H_2 + SO_4^{2-} + H^+ \rightarrow HS^- + 4H_2O$	−38
$4H_2 + H^+ + HCO_3^- \rightarrow CH_4 + 3H_2O$	−34

As hinted above, the energy gain of the various redox reactions is related to the H_2 concentration. In earlier studies (Lovley and Goodwin, 1988; Chapelle and Lovley, 1992; Vroblesky and Chapelle, 1994; Chapelle et al., 1995) it was therefore proposed to use the H_2 level as an indicator for an ongoing electron accepting process. It has been observed indeed, that H_2 concentrations increase as conditions become more reducing (Figures 9.13 and 9.20). However, the concentrations of H_2 will be influenced by the activities of the other reactants in the redox reaction, notably the concentration of SO_4^{2-} and the type of iron-oxyhydroxide available. Furthermore, the redox reaction in which H_2 is consumed is endothermal and the equilibrium constant decreases with temperature (Hoehler et al., 1998; Jakobsen et al., 1998; Problem 9.9).

QUESTIONS:

Use the reaction $\frac{1}{2}H_{2(aq)} \leftrightarrow H^+ + e^-$; $\log K = 1.57$, to estimate the concentration of H_2 in nM, at pH 7, in equilibrium with 0.3 mM O_2/H_2O; 0.3 mM SO_4^{2-}/1 μM HS^-; 5 mM HCO_3^-/0.3 mM CH_4 (*Hint*: Find $\log K$ values from PHREEQC.DAT)

ANSWER: $H_2 = 3.8\times10^{-36}$, 0.04, 0.4nM for O_2/H_2O, SO_4^{2-}/HS^-, HCO_3^-/CH_4, respectively.

Compare the estimated concentrations of H_2 with observations in Figure 9.20.

ANSWER: equilibrium concentrations are about 10 times smaller than observed.

Plot the concentration of H_2 during the reaction in the graph of Example 9.6?

9.4 OXYGEN CONSUMPTION

The atmosphere has a P_{O_2} of 0.21 atm, and using Henry's law the calculated dissolved oxygen content is 0.27 mmol/L (8.6 mg/L) at 25°C, increasing to 0.40 mmol/L (13 mg/L) at 5°C. Water saturated with oxygen by contact with the atmosphere is continuously infiltrating through soils into aquifers and O_2 is therefore an important oxidant in aquifer systems.

EXAMPLE 9.7. *Applying Henry's law to oxygen dissolution*

$$K = [O_{2(aq)}]/[P_{O_2}] = 10^{-2.898}$$

and $\log K = -7.5001 + 7.8981\times10^{-3} \times T + 2.0027\times10^5/T^2$

The PHREEQC input file to calculate the O_2 concentration at different temperature(s) (beware, some databases may contain an incorrect K; compare with the values above):

```
SOLUTION 1
EQUILIBRIUM_PHASES
 O2(g)   -0.678   #log 0.21
REACTION_TEMPERATURE
 25 10 in 2
END
```

Figure 9.21 shows the O_2 distribution in a sandy aquifer. The unsaturated zone is 15 m thick and the upper 12 m of the saturated zone displays a nearly constant O_2 content. Since the average annual temperature is near 8°C, the water is in equilibrium with the O_2 content of the atmosphere (Example 9.7). Any O_2 consumption that may have occurred in the soil zone is apparently resupplied by gaseous O_2 transport through the permeable sandy soil and there is no significant reduction of O_2 in the upper 12 m of the saturated zone. Below 27 m depth O_2 is rapidly consumed, because the water enters a layer containing reduced substances.

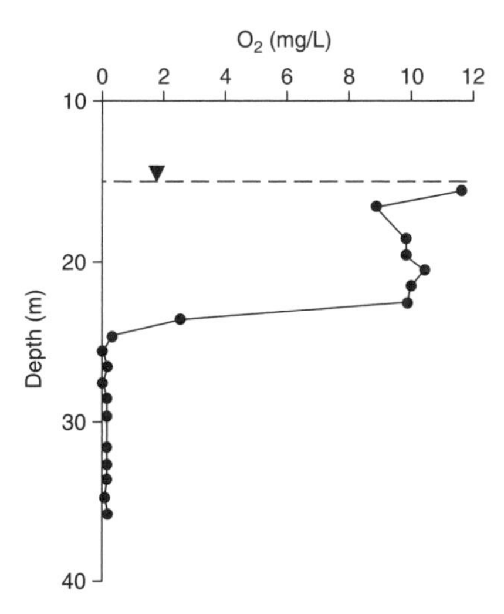

Figure 9.21. The O_2 distribution in a sandy aquifer (Postma et al., 1991).

In the absence of such reducing substances, O_2 saturated groundwater may travel a long way through aquifers. Winograd and Robertson (1982) reported groundwaters more than 10,000 years old that traveled up to 80 kilometers from their point of recharge and were still rich in oxygen.

Figure 9.22 illustrates how the thickness of the unsaturated zone influences the O_2 content of the groundwater in the saturated zone. At Alliston the water table is located 4 m below the surface and the travel time of water through the unsaturated zone is almost a year. During this time, *DOC* may react with O_2 which is resupplied by gas diffusion from the atmosphere.

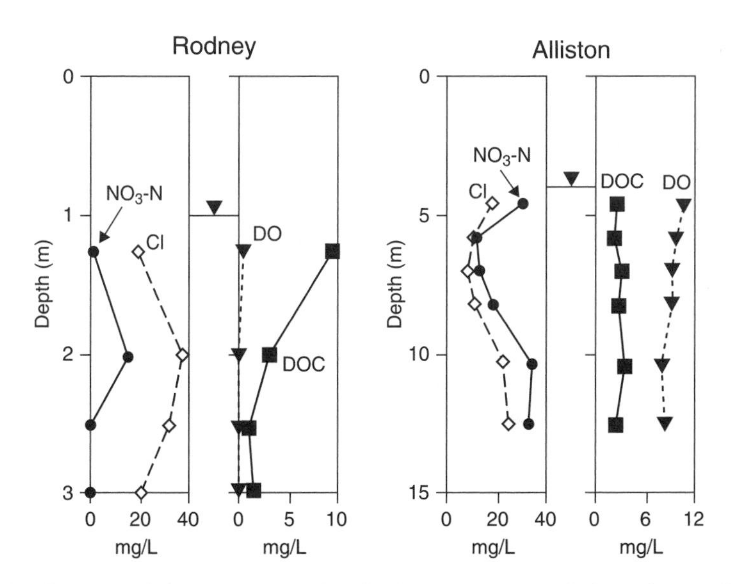

Figure 9.22. Dissolved organic carbon (DOC), dissolved oxygen (DO) and nitrate in a sandy aquifer at two locations in a sandy aquifer; Rodney (left) with a groundwater table at 1 m below the surface and at Alliston (right) with a water table at 4 m depth (Modified from Starr and Gillham, 1993).

As the result, most *DOC* is degraded and the groundwater remains oxic. At the Rodney site, the water table is at 1 m below the surface and, due to the short residence time, the *DOC* leaching from the soil survives transport through the unsaturated zone. Therefore the *DOC* content in the top of the saturated zone is high and the groundwater anoxic. Clearly, the oxygen concentration of groundwater is influenced by the residence time of *DOC* in the unsaturated zone in the recharge area.

When the water table is high and the soils are less permeable, the O_2 concentration of groundwater may also become depleted due to reduced diffusion of O_2 through the soil, in combination with the O_2 consumption by organic matter oxidation. Gas diffusion in the soil pores depends on the porosity of the soil and the fraction of the porosity that is water filled. The effective diffusion coefficient can be estimated with one of the Millington-Quirk relations (Jin and Jury, 1996):

$$D_{e,a} = D_a \varepsilon_g^2 / \varepsilon^{0.67} \tag{9.62}$$

where D_a is the diffusion coefficient in "free" air ($D_a \approx 10^{-5}\, m^2/s$), and ε_g and ε are the fractions of gas-filled and total porosity, respectively. In a completely dry soil with $\varepsilon_g = \varepsilon = 0.4$, the effective diffusion coefficient becomes:

$$D_{e,a} = D_a\, \varepsilon^{1.33} = 0.3 \times 10^{-5}\ m^2/s$$

If the same soil is partly saturated (for example $\varepsilon_w = 0.32$ and $\varepsilon_g = \varepsilon - \varepsilon_w = 0.08$), $D_{e,a}$ diminishes to:

$$D_{e,a} = D_a \varepsilon_g^2 / \varepsilon^{0.67} = 0.01 \times 10^{-5}\ m^2/s$$

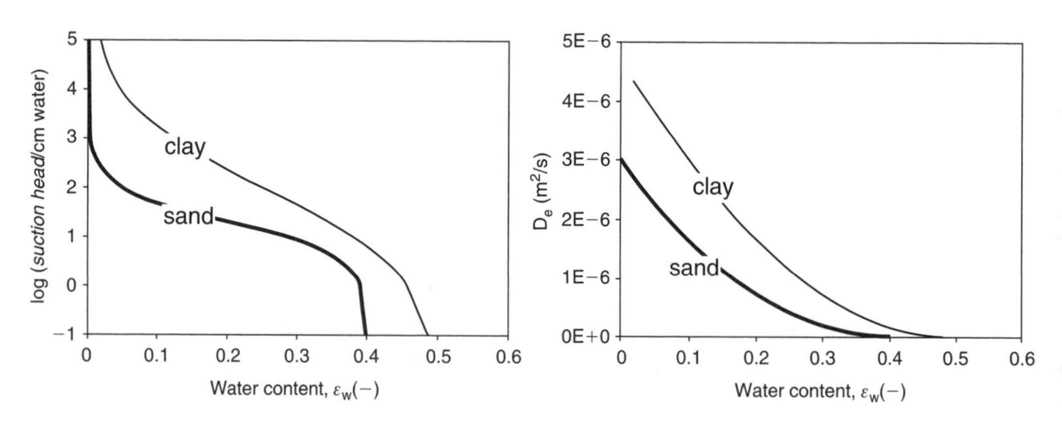

Figure 9.23. (Left) pF-curves which relate soil suction (log h, h in cm water) and water content ε_w for a sandy soil and a clay. At equilibrium, h equals height above the water table. (Right) Estimated effective diffusion coefficient ($D_{e,a}$ m^2/s) as function of the water filled porosity (ε_w). ($D_a \approx 10^{-5}$ m^2/s)

The relation between $D_{e,a}$ and ε_w as predicted by Equation (9.62) is shown in Figure 9.23 (right) for a clay soil with a high porosity and a sand with a lower porosity. With an increasing water content, more air-filled pores become isolated, and gas transport is hindered as it must diffuse via water with a diffusion coefficient that is approximately 10^4 times lower than in air (cf. Chapter 3). We can estimate the yearly flux of oxygen in a soil and compare it with the carbon productivity (Example 9.8).

EXAMPLE 9.8. *Compare oxygen flux and carbon productivity in a soil*
A loamy soil, with the water table at 1 mbs, has $\varepsilon = 0.4$, and average $\varepsilon_g = 0.15$ in the upper 0.9 m. Below that depth, ε_g is smaller than 0.05 and O_2 diffusion is negligible in the gas phase. Estimate the yearly flux of oxygen into the soil.

ANSWER:
Assume a zero O_2 concentration at 0.9 m depth, and a linear gradient. Following Fick's first law the O_2 flux is:

$$F = -\varepsilon_g D_{e,a} \frac{c_{air} - 0}{0 - 0.9}$$

with $\varepsilon_g = 0.15$, $\varepsilon = 0.4$
$D_{e,a} = 0.15^2 / 0.4^{0.67} \times 10^{-5}$ (m²/s) $\times$ 3.15×10^7 (s/yr) = 13 m²/yr
$c_{air} = P$ (atm) / RT (atm · L/mol) = 0.21 (atm) / 24 (atm · L/mol) $\times 10^3$ (L/m³) = 8.75 mol/m³
we find $F_{O_2} = 19$ mol/m²/yr.
This flux can oxidize 19 mol CH_2O/m²/yr, or 5.7 ton CH_2O/ha/yr, which amounts to the productivity of a rich soil (Russell, 1973).

QUESTION:
Show, using the steady state diffusion equation (Equation 5.59) that a linear concentration gradient for oxygen is valid when organic carbon degrades with a first order rate coefficient of less than 1/yr (*Hint*: Find k_1 and k_2 in Equation (5.61) using c_{O_2} at 0 and 0.9 m)

Example 9.8 suggests that soils with a deep water table will not accumulate organic matter because O_2 transport from the atmosphere is sufficient to oxidize all the carbon produced. The O_2 gradient in a wet soil will be steep (e.g. if all O_2 is consumed at 0.09 m depth, the gradient is 10 times higher) but since the effective gas-diffusion coefficient is much lower, the flux of O_2 is smaller than in a dry soil and organic matter may accumulate to generate peaty soils.

As Equation (9.62) shows, the diffusion coefficient of a soil depends strongly on its water content, which in turn depends on the soil type and weather conditions. The distribution of soil moisture over depth can be predicted from the so-called *pF-curve* which relates the water content of the soil to soil suction, and may be defined as:

$$\varepsilon_w = \frac{\varepsilon}{1 + (\alpha\, h)^n} \qquad (9.63)$$

where h is soil suction in cm water, and α and n are Van Genuchten fitting parameters. In a soil profile at equilibrium, the soil suction h equals the height above the water table. The parameters α and n depend on the characteristics of the soil. For a sandy soil $\alpha = 0.05$ and $n = 1.2$ may be found, while in a clay soil $\alpha = 0.01$ and $n = 0.5$ can apply. Figure 9.23 (left) shows how with increasing depth, the water content increases which in turn decreases the diffusion coefficient of O_2 and thereby the downward flux of O_2.

QUESTION:
Estimate the diffusional oxygen flux when the soil in Example 9.8 is fully water saturated?
ANSWER: $F = -\varepsilon_w D_e \,\mathrm{grad}(m_{O_2})$ (cf. Chapter 3). Take a gradient $\Delta m_{O_2} / \Delta z = 0.3 / 0.9$ mol/m², and find $F = 1.7$ mmol O_2/m²/yr

9.4.1 *Pyrite oxidation*

The oxidation of pyrite and other metal-sulfide minerals by oxygen has a large environmental impact and plays a key role in acid mine drainage and the formation of acid sulfate soils resulting from drainage of lowlands. It acts as a source of sulfate and iron in groundwater, and of heavy metals in general in the environment. The oxidation of pyrite has been studied for almost a century but some aspects remain unclarified (Lowson, 1982; Nordstrom, 1982; Alpers and Blowes, 1994; Evangelou and Zhang, 1995; Herbert, 1999; Rimstidt and Vaughan, 2003). The overall process is described by the reaction:

$$FeS_2 + {}^{15}\!/_4O_2 + {}^{7}\!/_2H_2O \rightarrow Fe(OH)_3 + 2SO_4^{2-} + 4H^+ \qquad (9.64)$$

It illustrates the strong generation of acid by pyrite oxidation. In extreme cases this may result in negative pH values, corresponding to a concentrated sulfuric acid solution (Nordstrom et al., 2000). The full oxidation process involves both the oxidation of the disulfide S_2^{2-} and of the Fe^{2+}.

The initial step is the oxidation of the disulfide to sulfate by O_2:

$$FeS_2 + {}^{7}\!/_2O_2 + H_2O \rightarrow Fe^{2+} + 2SO_4^{2-} + 2H^+ \qquad (9.65)$$

Subsequently Fe^{2+} is oxidized by oxygen to Fe^{3+}:

$$Fe^{2+} + {}^{1}\!/_4O_2 + H^+ \rightarrow Fe^{3+} + {}^{1}\!/_2H_2O \qquad (9.66)$$

The oxidation of disulfide proceeds at a lower redox potential (Figures 9.7, 9.8) than Fe^{2+} oxidation. Incomplete pyrite oxidation, due to an insufficient supply of electron acceptor, therefore results in a solution enriched in Fe^{2+} and SO_4^{2-} following Reaction (9.65). Unless pH is extremely low, Fe^{3+} will precipitate according to:

$$Fe^{3+} + 3H_2O \rightarrow Fe(OH)_3 + 3H^+ \qquad (9.67)$$

Equation (9.67) is highly pH dependent and generates three quarters of the acidity in the overall Reaction (9.64).

9.4.2 *Kinetics of pyrite oxidation*

Laboratory experiments show the oxidation of pyrite by O_2 to be a slow process, a conclusion contrary to the dramatic results of pyrite oxidation in the field. This discrepancy has induced extensive research in the kinetics of pyrite oxidation (Williamson and Rimstidt, 1994; Rimstidt and Vaughan, 2003). In order to oxidize an S_2^{2-} group of pyrite to sulfate, 14 electrons have to be transferred and therefore complicated reaction kinetics are to be expected. The different reactions involved in the oxidation of pyrite are summarized in Figure 9.24.

Initially FeS_2 reacts with O_2 following Reaction (9.65), either through a direct reaction ((a) in Figure 9.24)) or through dissolution followed by oxidation (a′), but in both cases the rates remain low. The second pathway of pyrite oxidation is by reaction with Fe^{3+}:

$$FeS_2 + 14Fe^{3+} + 8H_2O \rightarrow 15Fe^{2+} + 2SO_4^{2-} + 16H^+ \qquad (9.68)$$

The reaction between pyrite and Fe^{3+} (Figure 9.24(c)) is fast and yields a low pH. The produced Fe^{2+} may become oxidized by O_2 to Fe^{3+} (Reaction 9.66, Figure 9.24(b)). The kinetics of Fe^{2+} oxidation have already been discussed in Chapter 4 and are slow at low pH and increase steeply with increasing pH. On the other hand, only at low pH does Fe^{3+} remain in solution since it otherwise precipitates as $Fe(OH)_3$ (Equation 9.67). In a purely inorganic system the rate of Fe^{2+} oxidation (b) therefore rapidly becomes rate limiting as the pH decreases. However, in a natural setting iron oxidizing bacteria may accelerate process (b) by orders of magnitude (Kirby and Elder Brady, 1998).

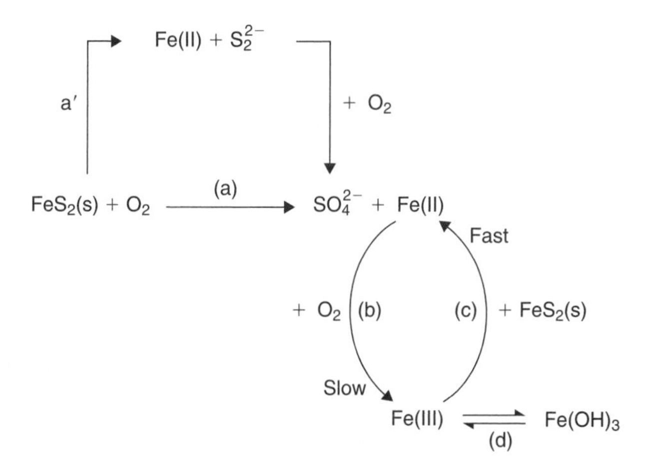

Figure 9.24. Reaction pathways in the oxidation of pyrite (Stumm and Morgan, 1996).

Bacterial catalysis therefore enables rapid pyrite oxidation at low pH (pH < 4) through reactions (b) and (c). Towards higher pH, pyrite oxidation by O_2 becomes dominant because the Fe^{3+} pathway is inhibited by the low solubility of $Fe(OH)_3$ (reaction (d)) and Equation (9.67)) which keeps the Fe^{3+} concentration very low.

Figure 9.25 shows experimental results for both pathways. At high pH, oxidation by oxygen is the dominant pathway but the process is slow. Thiosulfate appears as an intermediate oxidation product between sulfide and sulfate. The specific rate of pyrite oxidation by O_2 is described by (Williamson and Rimstidt, 1994):

$$r = 10^{-8.19}\ m_{O_2}^{0.5}\ m_{H^+}^{-0.11}\ (\text{mol/m}^2/\text{ s}) \tag{9.69}$$

The rate has a square root dependency on the oxygen concentration, indicating a large effect at low O_2 concentrations while at higher O_2 concentration the effect is small. This corresponds to the pyrite surface becoming saturated with O_2 (Nicholson et al., 1988). The effect of pH on the rate is very small.

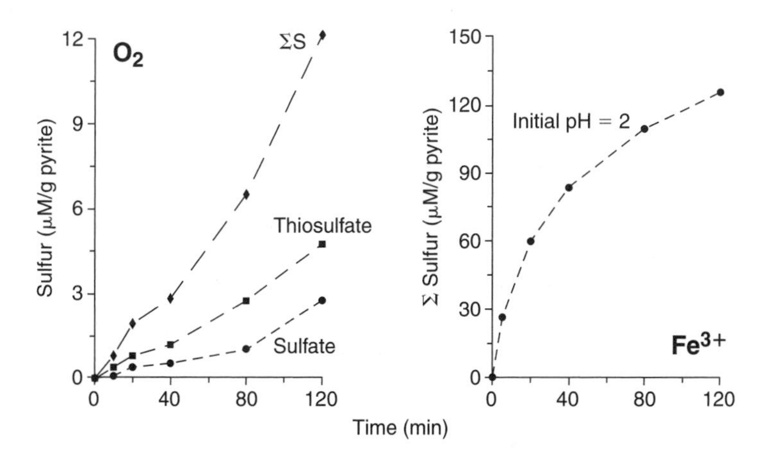

Figure 9.25. Pyrite oxidation by an oxygen saturated solution at pH 9 (left), and by ferric iron at pH 2 (right). Note the differences in scale on the y-axis implying that oxidation by Fe^{3+} is much faster (Moses et al., 1987).

The rate of pyrite oxidation by Fe^{3+} is many times faster (Figure 9.25) and in the presence of O_2 it is described by (Williamson and Rimstidt, 1994):

$$r = 10^{-6.07}\ m_{Fe^{3+}}^{0.93}\ m_{Fe^{2+}}^{-0.40} \tag{9.70}$$

and when O_2 is absent by:

$$r = 10^{-8.58}\ m_{Fe^{3+}}^{0.3}\ m_{Fe^{2+}}^{-0.47}\ m_{H^+}^{-0.32} \tag{9.71}$$

Pyrite is a semiconductor where sulfide atoms are oxidized at an anodic site, releasing electrons that are transported through the crystal to a cathodic Fe(II) site, where they are acquired by the aqueous oxidant. The overall rate determining step is the reaction at the cathodic site (Rimstidt and Vaughan, 2003). Fe^{2+}, and H^+ compete with Fe^{3+} for sorption on the cathodic sites on the pyrite, and therefore have a negative effect on the rate, as is expressed in the rate equations. Other cations can be expected to compete with Fe^{3+} as well, but the effects have not been quantified yet (Evangelou and Zhang, 1994).

The rates as calculated from Equations (9.70) and (9.71) will increase infinitely as the Fe^{2+} and H^+ concentrations decrease, whereas the competitive effect should become negligible towards low concentrations. To avoid this artefact, numerical models contain an inhibition factor which approaches 1 when the concentration of the inhibiting ion becomes smaller than a limiting concentration:

$$(1 + m_{Fe^{2+}}/m_{Lim})^n \tag{9.72}$$

where m_{Lim} is a small limiting concentration and n an empirical coefficient. The data compilation by Williamson and Rimstidt (1994) was used to obtain coefficients of modified rate equations that include an inhibition factor, using a non-linear estimation algorithm (cf. Example 7.1), which gave an optimized value of $m_{Lim} = 1.3\ \mu M$. Using instead a round number of $m_{Lim} = 1\ \mu M$, the rate equation for the case with oxygen present becomes:

$$r = 6.3\times10^{-4}\ m_{Fe^{3+}}^{0.92}\ (1 + m_{Fe^{2+}}/10^{-6})^{-0.43} \tag{9.70a}$$

and without oxygen:

$$r = 1.9\times10^{-6}\ m_{Fe^{3+}}^{0.28}\ (1 + m_{Fe^{2+}}/10^{-6})^{-0.52}\ m_{H^+}^{-0.3} \tag{9.71a}$$

with r in mol/m^2/s. The maximum pH in the data is 2.5 which limits the applicability of the rates to the pH range 0.5–3.0. However, the rate will go to zero anyway for pH > 3.0 because the Fe^{3+} concentration becomes limited by the solubility of iron oxyhydroxide.

All the rates given by Equations (9.69, 9.70a and 9.71a) must be multiplied with the surface area of pyrite in m^2/L to obtain the reaction in mol/L/s. Experimentally, the reactive surface area is found to be smaller than geometric surface area, and furthermore, to decrease in the course of the reaction as result of armoring by precipitates of $Fe(OH)_3$ and iron-sulfates (Wiese et al., 1987; Nicholson et al., 1990). Observed rates in tailings and waste dumps may therefore be smaller than estimated from the laboratory rates (cf. Questions). Higher rates are also found because microbes enhance the reactions in the field (Herbert, 1999).

QUESTIONS:

Estimate the maximal yearly rate of oxidation of framboidal pyrite in the presence of oxygen and calcite. Assume 3 μm framboids, $\rho_{pyr} = 5$ g/cm^3.

ANSWER: pH $= 7$, $A = 0.4$ m^2/g, $m_{O_2} = 3\times10^{-4}$ M yield 0.007 mol FeS_2/(g pyrite)/yr.

What is the rate at pH = 2, oxygen present?

ANSWER: From solubility of $Fe(OH)_3$: $m_{Fe^{3+}} = 10^{-40}/(10^{-12})^3 = 10^{-4}M$ (hence $Fe^{2+} = 10^{-10}M$) gives 1.7 mol FeS_2/g/yr.

And with initial conditions as previous, but no resupply of oxygen?

ANSWER: From electron balance, $Fe(OH)_3 : FeS_2 = 14 : 1$, hence $m_{Fe^{2+}} / m_{SO_4^{2-}} = 15/2$. From electrical balance, $2m_{Fe^{2+}} = 2m_{SO_4^{2-}} + 0.01$. Together, $m_{SO_4^{2-}} = 0.77$mM, $FeS_2 = 0.38$mM. (you may want to check with PHREEQC?).

9.4.3 *Oxygen transport and pyrite oxidation*

Apart from the kinetics of pyrite oxidation, the transport of oxygen towards pyrite may become the limiting factor for pyrite oxidation in the field. Similar to the different oxidation rates of organic carbon in wet and dry soils, pyrite oxidation also depends strongly on moisture conditions. When pyrite is situated below the water table advective transport of dissolved O_2 is the only mode of O_2 transport. Since air saturated groundwater contains about 0.33 mM O_2 the maximum increases are $(4/7) \times 0.33 = 0.19$ mM SO_4^{2-} and $(2/7) \times 0.33 = 0.09$ mM Fe^{2+} for incomplete pyrite oxidation (Reaction 9.65) or $(8/15) \times 0.33 = 0.18$ mM SO_4^{2-} in the case of complete pyrite oxidation (Reaction 9.64). This situation is illustrated at the left in Figure 9.26 where the upper part of the saturated zone in a sandy aquifer has a constant O_2 content that matches air saturation, while the increase in SO_4^{2-} and Fe^{2+}, at the depth where O_2 disappears corresponds to the ones predicted by Reaction (9.65).

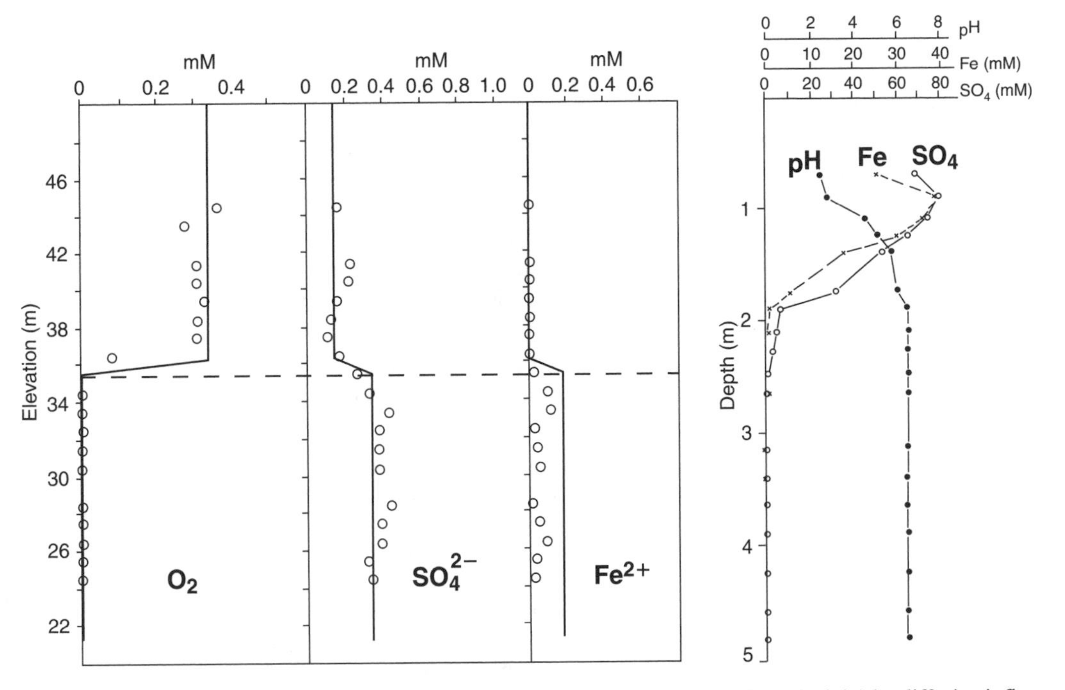

Figure 9.26. Pyrite oxidation by oxygen supplied by purely advective flow (left) and (right) by diffusive influx under stagnant water conditions. In the first case (Postma et al., 1991) groundwater, air-saturated with O_2, is transported through a pyritic layer. In the diagram at right, drained pyritic swamp sediments are oxidized by gaseous O_2 diffusion through the unsaturated zone into the pyritic layer (Postma, 1983). Note the huge difference in the amounts of pyrite oxidation and the associated effects.

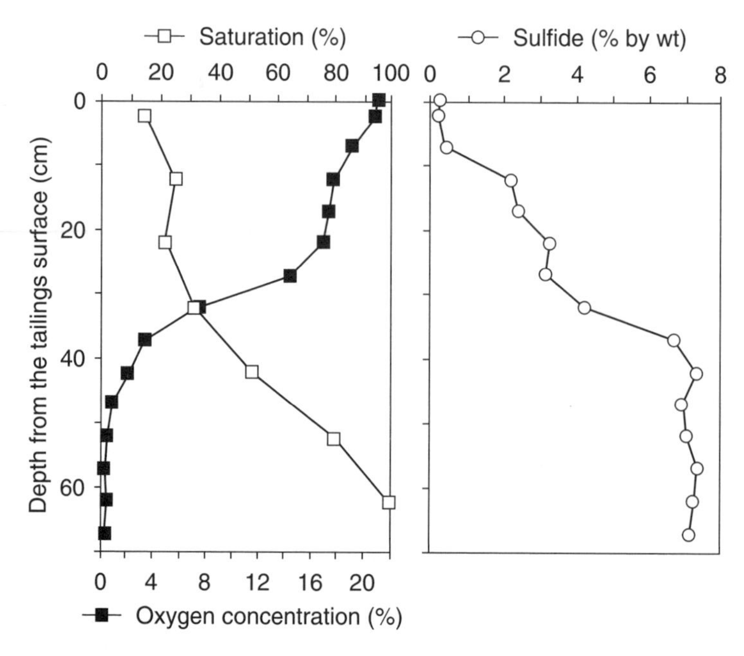

Figure 9.27. Sulfide oxidation in mine tailings: (left) O_2 concentration in the gas phase and the water saturation percentage. (right) sulfide distribution in the solid phase (Elberling and Nicholson, 1996).

A completely different water chemistry results when the pyrite is located above the groundwater table. Gaseous diffusion may now convey a much larger flux of oxygen and produce dramatic changes in water chemistry. This situation causes the severe effects of acid mine drainage and critically deteriorates soils which are drained for agricultural ue and contain pyrite. The resulting sulfate and iron concentrations (Figure 9.26, right) are huge and the pH decreases to values close to 2.

Figure 9.27 illustrates the O_2 transport through a mine tailing deposit. The upper 60 cm of the tailing are unsaturated and the O_2 content in the soil gas is already close to zero at 40 cm below the surface. Oxygen is transported downward by diffusion. The right hand graph shows the distribution of the sulfide in the solid phase which mirrors the O_2 gas distribution. If downward diffusion of O_2 was the sole process to limit the rate of sulfide oxidation, then a sharp front is expected in the distribution of sulfide in the sediment. The more gradual increase in sulfide distribution over depth indicates that O_2 transfer over the air-water interface and reaction kinetics are both of importance. Therefore the O_2 profile is different from what is expected for simple diffusion with linear retardation (cf. Question with Example 9.8). The shape suggests that O_2 is mostly consumed at 30 cm while above that depth pyrite has become unreactive or shielded by reaction products.

QUESTIONS:

Estimate the yearly flux of O_2 in the mine tailings of Figure 9.27, cf. Example 9.8?

ANSWER: take $\varepsilon = 0.3$, then $\varepsilon_g = 0.3 \times (100 - 20) / 100 = 0.24$. $\Delta c_{O_2} / \Delta z = ((0.21 - 0.17) / 24 \times 1000) / -0.2 = -8.3$ mol/m^4. The flux is 81 mol O_2/m^2/yr.

Estimate the amount of sulfide oxidation in the tailings?

ANSWER: $81 \times 4 / 15 = 22$ mol FeS_2/m^2/yr

Estimate sulfide oxidation if the tailings are wetted by sprinklers to 90% saturation?

ANSWER: 0.04 mol FeS_2/m^2/yr

Transport of gaseous O_2 occurs not only by simple diffusion. When O_2 is used up from soil air, the total gas pressure decreases by 20% resulting in downward advective transport of air into the soil. This advective gas transport brings in nitrogen, which builds up pressure in the soil and tends to diffuse back into the air. A multicomponent diffusion process takes place which is further complicated by temperature and atmospheric pressure variations (Thorstenson and Pollock, 1989; Massmann and Farrier, 1992; Elberling et al., 1998).

The drop in gas pressure also depends on accompanying buffering reactions. If the proton production by pyrite oxidation (Reaction 9.64) causes calcite dissolution according to:

$$CaCO_3 + 2H^+ \rightarrow Ca^{2+} + CO_2 + H_2O \qquad (9.73)$$

then the net volume loss of the two reactions would be close to 10%. Wisotzky (1994) and Van Berk and Wisotzky (1995) estimated that the contributions of O_2 transport by diffusion and advection are about the same. In the field barometric pressure changes and the action of wind may also contribute to advective transport of O_2 in the unsaturated zone.

EXAMPLE 9.9. *Modeling gas loss during pyrite oxidation with oxygen*

Unsaturated aquifer sediment containing pyrite was incubated in gas-impermeable polymer bags containing a gas mixture with 8% O_2 and 92% N_2 (Andersen et al., 2001). The sediment contains $CaCO_3$ that functions as pH buffer. The gas composition in contact with the incubated sediment is followed over time as shown in Figure 9.28.

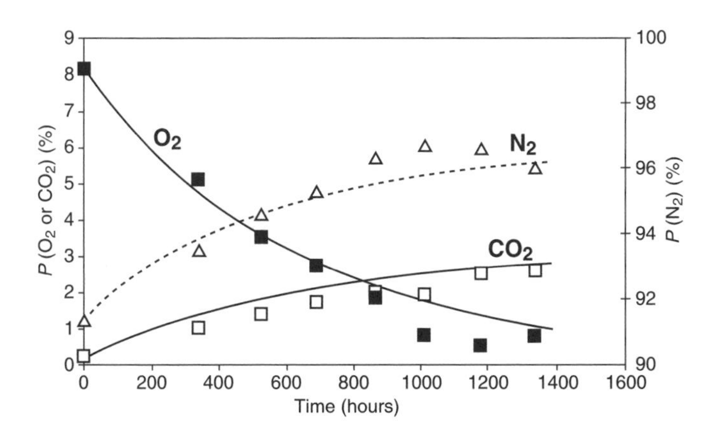

Figure 9.28. Development in gas phase composition during the oxidation of pyrite in unsaturated aquifer sediment incubated in gas-impermeable polymer bags. The initial gas volume was 21.3 ml, the gas/water ratio 3:1, the sediment weight 82.79 g and the FeS_2 content 1.59 mmol/kg. Symbols indicate experimental data and lines are modelled with PHREEQC.

Since N_2 is a conservative gas, and is quite insoluble in water, its increase indicates a loss of total volume given by

$$\Delta V_{loss} = V_0 \left(1 - P_{N_{2(0)}} / P_{N_{2(t)}}\right) \text{ (mL)} \qquad (9.74)$$

Here (0) and (t) refer to times zero and t. The change in N_2 gas pressure in Figure 9.28 indicates a total loss of 1.1 mL, corresponding to 5% of the initial gas volume.The evolution of the gas phase composition during pyrite oxidation was modeled using PHREEQC and the input file is shown below.

```
RATES
 Pyrite                                             # rates from Equations (9.69), (9.70a)
                                                    # and (9.71a)
 -start
  1 A = 15e3 * m0    # initial surface area in m2 for 0.01 um crystals
 10 if SI("Pyrite") >0 then goto 100               # step out when supersaturated...
 20 fH = mol("H+")
 30 fFe2 = (1 + tot("Fe(2)")/1e-6)
 40 if mol("O2") < 1e-6 then goto 80
 50 rO2 = 10^-8.19 * mol("O2") * fH^-0.11        # ...rate with oxygen
 60 rO2_Fe3 = 6.3e-4 * tot("Fe(3)")^0.92 * fFe2^-0.43 # rate O2 + Fe3+
 70 goto 90
 80 rem                                            # rate with Fe3+ without oxygen,
                                                   # and for pH < 3...
 81 rFe3 = 1.9e-6 * tot("Fe(3)")^0.28 * fFe2^-0.52 * fH^-0.3
 90 rate = A * (m/m0)^0.67 * (rO2 + rO2_Fe3 + rFe3) * (1 - SR("Pyrite"))
 100 save rate * time
 -end

SOLUTION_SPECIES                                   # make N2 the only N species...
 2NO3- + 12H+ + 10e- = N2 + 6H2O; log_k 500        # 207.080 in the database

SOLUTION 1
 -water 0.0069239
 -temp 20
 pH 7 charge; pe 14 O2(g) -1.0878
 Ca 1 Calcite; C 1 CO2(g) -2.6021
 Fe 1e-3 Goethite 2; N 1.3  N2(g) -0.0382

EQUILIBRIUM_PHASES 1
 Goethite 2; Calcite 0; Gypsum 0 0

KINETICS 1
 Pyrite; -m0 1.32e-4; -step 0 5e5 5e5 5e5 5e5 5e5 5e5 5e5 5e5 5e5 5e5
INCREMENTAL_REACTIONS true

GAS_PHASE 1
 -fixed_pressure                                   # default 1 atm
 -volume 0.02127; -temp 20.0
 CO2(g) 0.0025; O2(g) 0.0817; N2(g) 0.9157

USER_GRAPH
 -head time CO2 O2 N2
 -axis_titles "Time/hour" "P_O2 and P_CO2/atm" "P_N2/atm"
 -start
 10 graph_x total_time/3600
 20 graph_y 10^si("CO2(g)"), 10^si("O2(g)"); 30 graph_sy 10^si("N2(g)")
 -end
END
```

The rate of pyrite oxidation is defined according to Equations (9.69)–(9.71a). In the experiment only the rate with O_2 gives a reaction, but the other terms were added for illustration. The overall rate is multiplied with (1 − SR("Pyrite")) to let it zero out towards equilibrium (line 90). N_2 must behave like a conservative gas during the simulation and is made the only N-species by increasing its stability relative to NO_3^-. (Note that $N^{(-3)}$ should also be minimized if defined in the database). The SOLUTION specifies the actual amount of water in the incubation bag, and EQUILIBRIUM_PHASES and KINETICS define the solids. GAS_PHASE states the volume and composition of the gas, and lastly, USER_GRAPH plots the data.

Pyrite oxidation is simulated as kinetic reaction, fitting the rate to the experimental data by adapting the specific surface (which appears to be rather high, line 1 in the Pyrite rate). The reaction consumes O_2 (Reaction 9.64). The modeled gas concentrations are given as solid lines in Figure 9.28 and show good correspondence with the measured compositions. Note that the amount of CO_2 that appears in the gas phase is smaller than predicted by the sum of Reactions (9.64) and (9.73) because CO_2 is more soluble in water than N_2 and O_2; its dissolution explains part of the loss of volume.

QUESTION:
Find the specific surface of 4 wt% pyrite (Figure 9.27) necessary to consume 22 mol $O_2/m^2/yr$ from 25 to 35 cm depth? *Hint*: use the rate of PHREEQC and a gas phase that contains 22 mol O_2

ANSWER: 0.005 m^2/mol pyrite (valid for 3 cm crystals). Note the increase in rate for pH < 3, when the rate with Fe^{3+} and O_2 takes the lead. Also note that allowing for precipitation of JarositeH gives sulfate concentrations close to field observations (Figures 9.26, 9.29; 7.1)

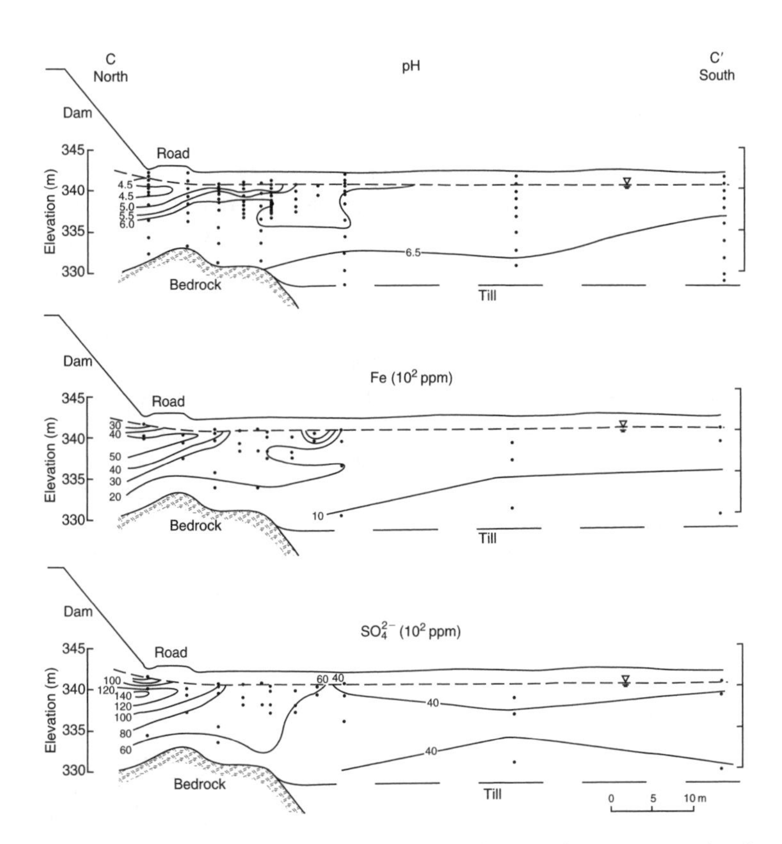

Figure 9.29. A plume of contaminants, including iron sulfate and low pH emanating from sulfidic mine tailings into a sandy aquifer. The tailings are located just north of the dam. (Modified from Dubrovsky et al., 1984).

Figure 9.29 shows drainage water from sulfide mine tailings entering an aquifer. A plume of contaminants spreads through a sandy aquifer from sulfide mine tailings located at the left side of the dam. In its center the plume contains more than 11,000 ppm sulfate, more than 5000 ppm dissolved Fe, and the pH is less than 4.5. Sulfide ores contain, apart from pyrite, a range of sulfide minerals like sphalerite (ZnS), chalcopyrite ($CuFeS_2$) and arsenopyrite ($FeAsS$). These minerals provide an ample source of heavy metals for groundwater contamination. The low pH values causes large increases in dissolved Al (Chapter 8) and a variety of secondary minerals may precipitate under such extreme pyrite oxidation conditions. These include gypsum, ferric hydroxysulfates like jarosite ($KFe_3(SO_4)_3 \cdot 9H_2O$ and several others (van Breemen, 1976; Nordstrom, 1982; Postma, 1983). Generally these precipitates are soluble and will be leached out by infiltration, except for FeOOH. The plume displayed in Figure 9.29 was found to migrate at a much slower rate than the groundwater flow velocity, and neutralization and precipitation of sulfate and ferrous iron must occur at the margins of the plume. Morin and Cherry (1986) found equilibrium with gypsum throughout the plume and saturation to supersaturation for siderite. They proposed that calcite, which is present in small amounts in the sediments, dissolves. The resulting high Ca^{2+}, Fe^{2+} and dissolved carbonate concentrations apparently generate the simultaneous precipitation of gypsum and siderite, the latter possibly as solid solutions with calcite. Carbonate reactions were also observed and evaluated in the Pinal Creek aquifer (Glynn and Brown, 1996).

9.5 NITRATE REDUCTION

Nitrate pollution of groundwater is an increasing problem in all European and North American countries (e.g. Strebel et al., 1989; Korom 1992; Spalding and Exner, 1993; Feast et al., 1998; Tesoriero et al., 2000) and poses a major threat to drinking water supplies based on groundwater. The admissible nitrate concentration in drinking water (see Chapter 1) is 50 mg NO_3^-/L (corresponding to 11 mg NO_3^--N/L or 0.8 mmol/L) and the recommended level is less than 25 mg/L NO_3. A high nitrate concentration in drinking water is believed to be a health hazard because it may cause methaemoglobinaemia in human infants, a potentially fatal syndrome in which oxygen transport in the bloodstream is impaired.

The groundwater nitrate content is derived from various point and non-point sources, including cattle feed lots, septic tanks, sewage discharge and the oxidation of organically bound nitrogen in soils. However, the main cause for the increasing nitrate concentration in shallow groundwater is, without doubt, the excessive application of fertilizers and manure in agriculture since the early sixties (cf. Example 2.3). The relation between land-use and nitrate pollution of aquifers is illustrated in Figure 9.30.

Little nitrate is leached from the forest and heath areas, but downstream from the arable land, plumes of nitrate containing waters spread through the aquifer. The question is to what extent nitrate is transported as a conservative substance through the aquifer, and how geochemical processes within the aquifer may attenuate the nitrate concentration.

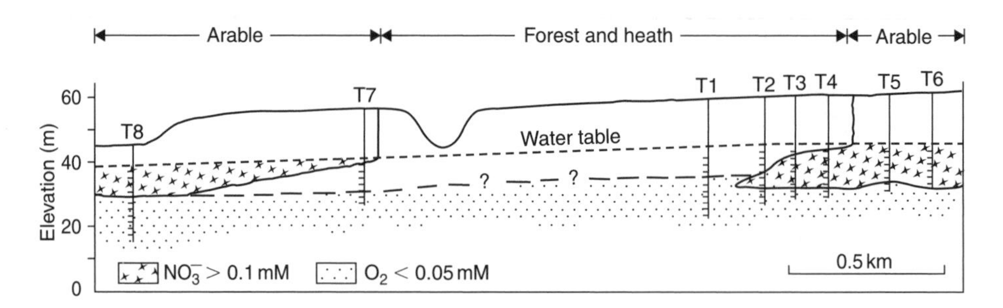

Figure 9.30. Nitrate pollution plumes emanating from agricultural fields into an unconfined sandy aquifer (Rabis Creek, Denmark). The groundwater flows from right to left. Numbers T1 through T8 refer to locations of multilevel samplers on which the plume distribution is based (Postma et al., 1991).

9.5.1 *Nitrate reduction by organic matter oxidation*

Nitrate forms neither insoluble minerals that could precipitate, nor is it adsorbed significantly. Therefore the only way for *in situ* nitrate removal from groundwater is by reduction:

$$2NO_3^- + 12H^+ + 10e^- \rightarrow N_2 + 6H_2O \quad (9.75)$$

The redox diagram for nitrogen (Figure 9.6) indicates that nitrate is only stable under highly oxidizing conditions while N_2 is the stable form at intermediate pe. Ammonia is stable only under highly reduced conditions.

In microbiology (Krumbein, 1983; Zehnder, 1988) the two important overall reactions for N-cycling are *denitrification* and *nitrification* (Figure 9.31). Denitrification is the term used for the microbial reduction of nitrate to N_2 by organic carbon. The overall reaction comprises a transfer of five electrons per N-atom and proceeds through a complicated pathway with several metastable intermediates (see also Section 9.2.2 on nitrogen):

$$NO^-_{3(aq)} \rightarrow NO^-_{2(aq)} \rightarrow NO_{(enzyme\ complex)} \rightarrow N_2O_{(gas)} \rightarrow N_{2(gas)} \quad (9.76)$$

Intermediates like NO_2^- and N_2O are often found at trace levels in natural environments and are then used as evidence for ongoing denitrification. Nevertheless, N_2 is always the predominant reaction product. Denitrification is not a reversible reaction; fortunately there are no bacteria which are able to live on the energy available from oxidizing N_2 to NO_3^- (Example 9.5). Dissimilatory nitrate reduction to NH_4^+ is possible in groundwater systems (Smith et al., 1991a) but normally plays a subordinate role. During nitrification, bacteria oxidize amines from organic matter to nitrite and nitrate and this process is of most importance in the soil zone.

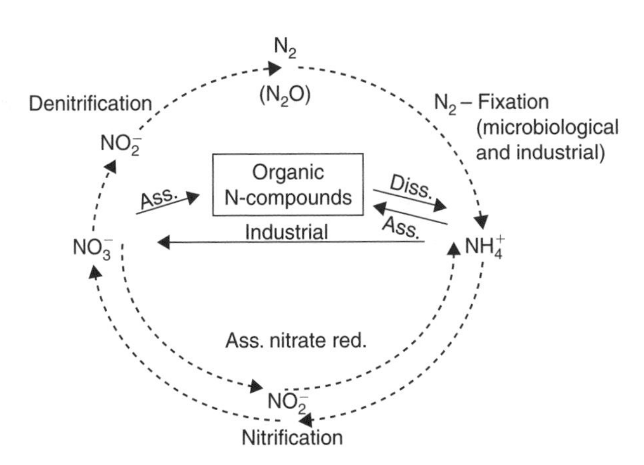

Figure 9.31. Pathways in the redox reactions of nitrogen. Ass. is assimilation into, diss. is dissimilation from organic matter.

The reduction of nitrate by organic matter is well documented for soils and marine sediments and is also important in aquifers (Smith and Duff, 1988; Bradley et al., 1992; Korom, 1992; Starr and Gillham, 1993; Smith et al., 1996; Bragan et al., 1997). The overall reaction stoichiometry is:

$$5CH_2O + 4NO_3^- \rightarrow 2N_2 + 4HCO_3^- + H_2CO_3 + 2H_2O \quad (9.77)$$

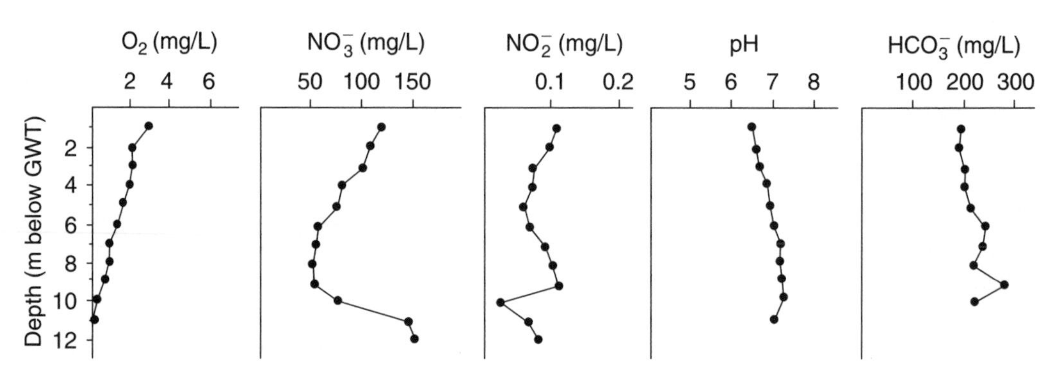

Figure 9.32. Nitrate reduction by organic matter oxidation. Average concentrations in the saturated zone at the Mussum waterworks, Germany. Modified from Obermann (1982).

Figure 9.32 illustrates the reduction of nitrate by oxidation of organic matter in a sandy aquifer. At the groundwater table the O_2 concentration is already depleted to less than half of the value at air saturation. As predicted thermodynamically, O_2 is preferentially reduced compared to nitrate but the presence of small amounts of nitrite confirms that denitrification takes place as well. The oxidation of organic matter is reflected by increases in HCO_3^- and pH, resulting from the increase in HCO_3^-/H_2CO_3 ratio, predicted by Equation (9.77). As demonstrated in Example 9.10, 60% of the decrease in nitrate (and O_2) between the water table and 8 m depth can be explained by the increase in total dissolved inorganic carbon, which strongly suggests organic matter to be the predominant electron donor.

EXAMPLE 9.10. *Construct a redox balance for nitrate reduction by organic matter oxidation*
Based on the changes in concentrations between the water chemistry at the top of the saturated zone (Figure 9.32) and those at 8 m depth, the following redox balance can be constructed (modified from Obermann, 1982). All concentrations are in mmol/L.

	Water table	8 m Depth	Difference
NO_3^-	2.18	0.81	−1.37
Ca^{2+}	3.37	3.24	−0.13
HCO_3^-	3.11	4.42	+1.31
TIC	3.75	4.74	+0.99
O_2	0.11	0.03	−0.08

If the decrease in nitrate is caused by oxidation of organic matter according to Reaction (9.77), it should be balanced by an increase in total dissolved inorganic carbon (*TIC*) of $5/4 \times 1.37 = 1.71$ mmol/L. The observed ΔTIC amounts to 0.99. Additional processes which affect *TIC* may be $CaCO_3$ precipitation induced by increasing pH and HCO_3^- during nitrate reduction, and also oxidation of organic matter by O_2 that produces an equivalent amount of *TIC*. Correcting for these reactions yields $\Delta TIC = 0.99 + 0.13 - 0.08 = 1.04$ mmol/L. In other words, $1.04/1.71 \times 100\% = 61\%$ of the decrease in nitrate can be explained by organic matter oxidation while the remainder probably is due to variations in nitrate input. Sulfate concentrations in the profile are constant so that pyrite oxidation (see next section) is of no importance.

QUESTIONS:

What is the pH effect of Reaction (9.77) when the pH of the NO_3^- water is 6.3, 6.9, or 7.3?

ANSWER: At pH = 6.3, $[HCO_3^-]$ equals $[H_2CO_3]$, thus pH increases. At pH = 6.9, ¼$[HCO_3^-] = [H_2CO_3]$ and Reaction (9.77) is pH neutral. At pH = 7.3, $[HCO_3^-] = 10\ [H_2CO_3]$, thus pH decreases.

Is it likely that the pH increase with depth in Figure 9.32 is caused by denitrification?

ANSWER: Yes, down to 6 m depth, the pH tends to the pH-neutral value of 6.9.

Use PHREEQC to estimate the pH at the water table and 8 m depth from HCO_3^- and *TIC* in the table (temp = 10°C)?

ANSWER: pH = 7.13 at w.t., pH = 7.57 at −8 m (Probably, *TIC* was higher in the groundwater, CO_2 escaped before analysis).

The gradual decreases of oxygen and nitrate with depth (Figure 9.32) indicates that O_2 and NO_3^- reduction rates are slow compared to the downward rate of water transport. The reactivity of organic matter is probably the overall rate controlling factor.

More precise evidence for the rate of denitrification in aquifer sediment can be obtained with the acetylene block technique. In the presence of acetylene the bacterial reduction of N_2O to N_2 is inhibited. An increase in N_2O is much easier to measure than an increase in N_2 because the background level of N_2 is so high. In the technique, acetylene is added to sediment incubations and the denitrifying activity is quantified as the rate of accumulation of N_2O. Rates of denitrification obtained by the acetylene block technique are in the range of 0.08 to 85 mmol/L/yr (Starr and Gillham, 1993; Smith et al., 1996; Bragan et al., 1997). This broad range again reflects the huge variability in the reactivity of organic carbon in aquifers. However, the highest values suggest that the conditions during sediment incubation differ from those *in situ* in the aquifer and therefore, yield higher rates (cf. Table 9.6; Question below Figure 9.33). Figure 9.33 shows the rate of denitrification versus the organic content of the sediment for incubations in the presence of acetylene. The good correlation of the N_2O production rate with the organic content suggest organic carbon to be the controlling parameter limiting the rate of denitrification.

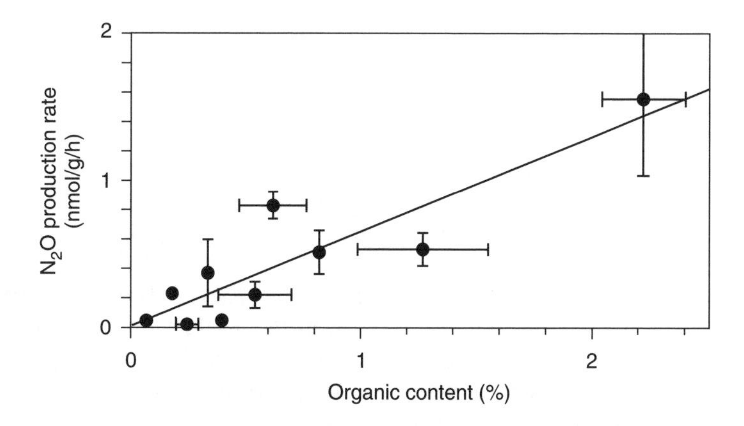

Figure 9.33. The rate of denitrification versus sediment organic carbon content. Results from sediment incubations using the acetylene block technique where N_2O accumulation gives the rate of denitrification (Bradley et al., 1992).

QUESTIONS:
Estimate from Figure 9.33 the denitrification rate in mmol C/L/yr for 1% carbon.
ANSWER: 0.7×10^{-9} (mol/g/hr) $\times$ 8760 (hr/yr) $\times$ 1000 (g/kg) $\times$ 6 (L/kg) = 37 mmol/L/yr.
Estimate the age of the organic carbon used by Bradley et al. (1992)?
ANSWER: k = (0.037 mol C/L/yr)/(5 mol C/L) = 0.00736/yr. Using Equation (9.61), t = 26 yr.

Another method to identify nitrate reduction is the N_2/Ar method (Wilson et al., 1994; Feast et al., 1998; Blicher-Mathiesen et al., 1998). The reduction of nitrate produces N_2 gas and since N_2 is virtually inert, denitrification should result in excess dissolved N_2 in the groundwater. The background concentration of N_2 in groundwater depends on the temperature at the time of infiltration and the entrapment of air bubbles during recharge. In order to correct for variations in background N_2 concentration, the amount of excess N_2 is usually evaluated from the N_2/Ar ratio, where argon is used as tracer. Also nitrogen stable isotopes, $^{14}N/^{15}N$, have been used both to identify the process of denitrification and the source of nitrate in groundwater (Feast et al., 1998; Wilson et al., 1994; Böttcher et al., 1990).

9.5.2 *Nitrate reduction by pyrite and ferrous iron*

Organic carbon is not the only electron donor available for the reduction of nitrate in aquifers. Also the reduction of NO_3^- by reduced groundwater components, such as Fe^{2+}, H_2S and CH_4, is thermodynamically favored. However, the nitrate concentration introduced into the aquifer by fertilizers usually by far exceeds the reducing capacity of these dissolved species and the required electron donor must be found in the solid phase. Besides organic matter, pyrite (FeS_2) is the other important solid phase electron donor for nitrate reduction. The energy yield of NO_3^- reduction with organic matter is larger than with pyrite (Figure 9.15) and, thermodynamically, nitrate reduction by organic matter should occur before the reduction by pyrite. However, the relative sequence of these two reactions is also strongly affected by the reaction kinetics.

Nitrate reduction coupled with pyrite oxidation in aquifers has been reported widely (Kölle et al., 1983; Strebel et al., 1985; Van Beek et al., 1988, 1989, Postma et al., 1991; Robertson et al., 1996; Tesoriero et al., 2000). The process involves the oxidation of both sulfur and Fe(2):

$$5FeS_2 + 14NO_3^- + 4H^+ \rightarrow 7N_2 + 5Fe^{2+} + 10SO_4^{2-} + 2H_2O \tag{9.78}$$

and

$$5\,Fe^{2+} + NO_3^- + 7\,H_2O \rightarrow 5\,FeOOH + \tfrac{1}{2}N_2 + 9\,H^+ \tag{9.79}$$

The energy gain for sulfide oxidation is larger than for Fe(2) oxidation (Figure 9.15). In the presence of excess pyrite, Fe^{2+} will therefore remain unoxidized. Inorganic oxidation of pyrite by nitrate does not seem possible (Postma, unpublished results) and bacterial catalysis is therefore required. *Thiobacillus denitrificans* is able to oxidize sulfur in pyrite (Kölle et al., 1987) while Fe^{2+} may be oxidized with nitrate by *Gallionella ferruginea* (Gouy et al., 1984) and *Escherichia coli* (Brons et al., 1991).

Nitrate reduction by pyrite oxidation is illustrated in Figure 9.34 for the sandy Rabis Creek aquifer. The groundwater in the upper part is derived from forest and heath (Figure 9.30) and is free of nitrate, while the water in the lower part is derived from arable land and contains nitrate. The O_2 content is near constant and close to air saturation in the oxidized zone. At the redoxcline oxygen and nitrate disappear simultaneously, and therefore the reduction process must be fast compared to the downward water transport rate of about 0.75 m/yr. The oxidation of pyrite is reflected by increases in sulfate and Fe^{2+}, in agreement with the pyrite distribution in the sediment.

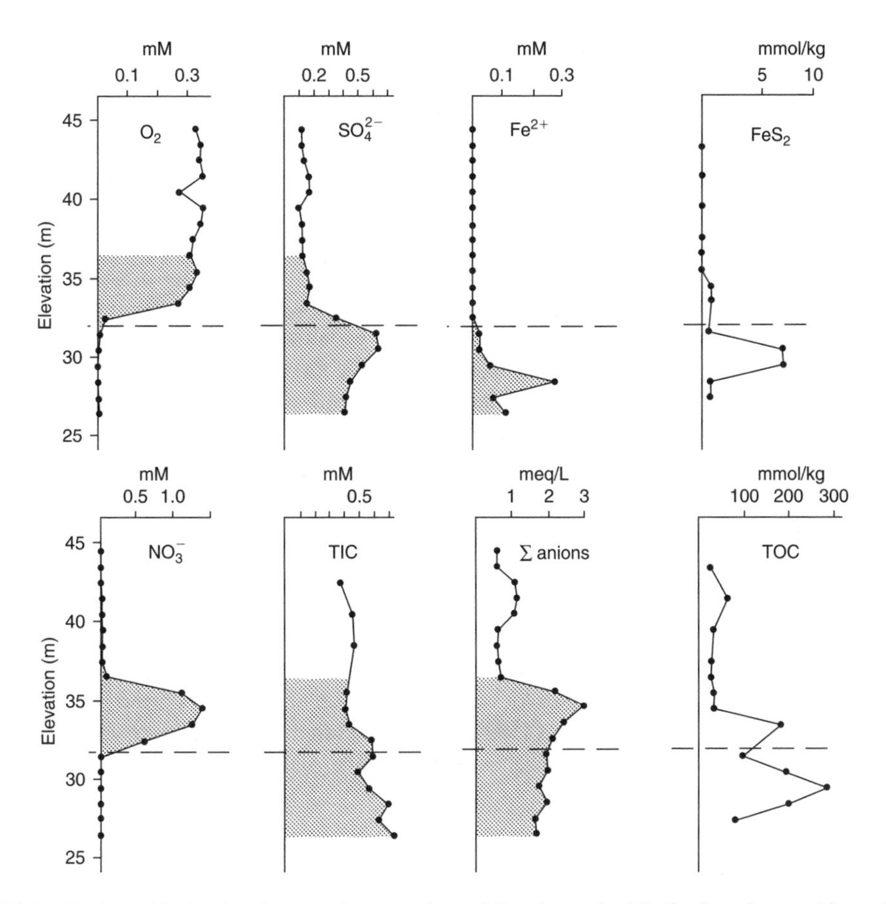

Figure 9.34. Pyrite oxidation by nitrate and oxygen in multilevel sampler T2 (for location see Figure 9.30) in the saturated zone of the Rabis Creek aquifer, Denmark. The dashed line indicates the depth where O_2 disappears and the shaded parts indicate nitrate contaminated water derived from arable land which is overlain by nitrate free water from a forested area (Modified from Postma et al., 1991).

However, organic carbon (*TOC*) is also present in the sediment, and in fact at a higher concentration than pyrite. The slight increase in *TIC* over the redoxcline could indicate the oxidation of organic carbon. The relative contributions of organic matter and pyrite oxidation can be estimated from an electron balance across the redoxcline (Postma et al., 1991; Tesoriero et al., 2000). The processes expected are listed in Table 9.8 and the contribution of each half reaction in the system is estimated by multiplying concentrations with the number of electrons transferred.

Table 9.8. Electron equivalents for dissolved redox species.

Reaction		Electron equivalents
$NO_3^- \rightarrow \frac{1}{2}N_2$	$+5e^-$	$5 \cdot m_{NO_3^-}$
$O_2 \rightarrow 2O^{2-}$	$+4e^-$	$4 \cdot m_{O_2}$
$CH_2O \rightarrow CO_2$	$-4e^-$	$4 \cdot TIC$
$S_{FeS_2} \rightarrow SO_4^{2-}$	$-7e^-$	$7 \cdot m_{SO_4^{2-}}$
$Fe_{FeS_2} \rightarrow FeOOH$	$-1e^-$	$0.5 \cdot (m_{SO_4^{2-}} - 2m_{Fe^{2+}})$

The electron equivalents obtained in this way are only valid for the specified half reactions and carbonates should not dissolve or precipitate, a condition which is fulfilled in the Rabis aquifer. As shown in Example 9.10, carbonate dissolution/precipitation can to some extent be corrected for by taking Ca^{2+} concentrations into account.

Electron equivalents are plotted cumulatively for two boreholes in Figure 9.35. T5 displays nitrate containing groundwater derived from agricultural fields. The disappearance of nitrate over depth is balanced by the oxidation of pyrite-S(−1) as the main electron donor. The contributions of both the oxidation of organic carbon and of pyrite-Fe(2) are minor. The organic matter present in these sediments consists of reworked Miocene lignite fragments, and apparently has a very low reactivity.

Borehole T2 shows water from the forested area, free of nitrate, on top of water from arable land containing nitrate. Compared to the input of only O_2, the contamination with nitrate increases the load of electron acceptor on the aquifer by a factor of five. But even then, the nitrate/pyrite interface moves downward at a rate of less than 2 cm/yr in the Rabis aquifer (Postma et al., 1991) and even slower in other systems (Robertson et al., 1996). Apparently the pyrite content of aquifer sediment may attenuate groundwater nitrate over long periods of time but not indefinitely.

Fe(2) in sediment is another possible electron donor for nitrate reduction, according to the overall reaction:

$$10Fe^{2+} + 2NO_3^- + 14H_2O \rightarrow 10FeOOH + N_2 + 18H^+ \tag{9.80}$$

In sediments, Fe(2) is present in clay minerals, detrital silicates like amphiboles, pyroxenes and biotite and in magnetite and its precursor green rust. Ernsten et al. (1998) reported Fe(2) in clay minerals to be able to reduce nitrate. Postma (1990) found that amphiboles and pyroxenes could reduce nitrate at low rates, but only in the presence of secondary reaction products, including FeOOH.

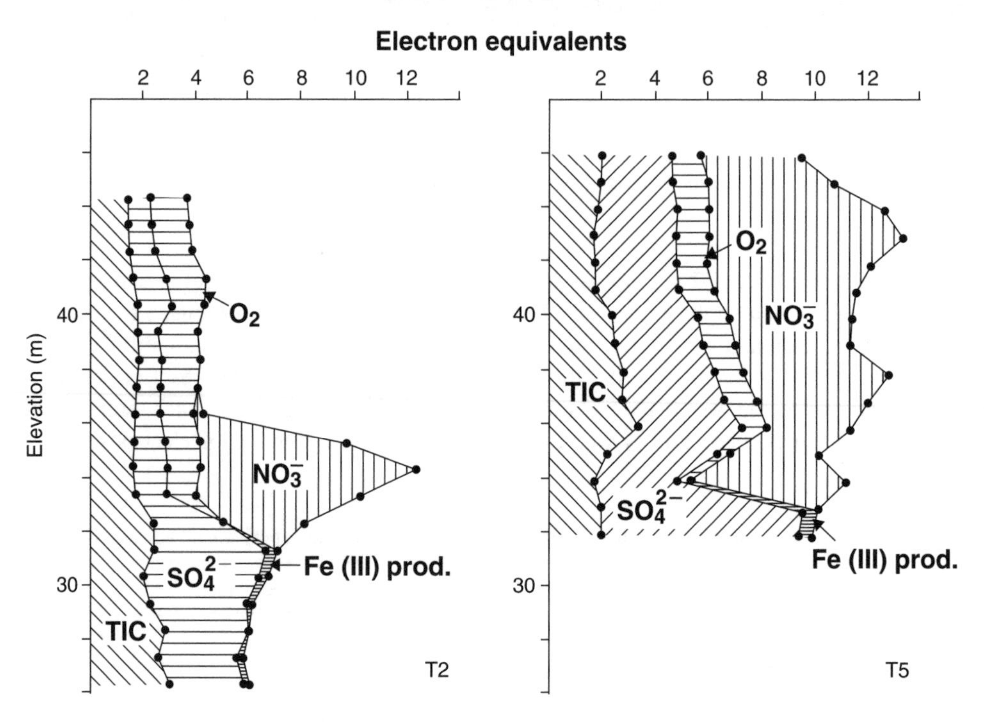

Figure 9.35. Cumulative distribution of electron equivalents in Rabis Creek multisampler T2 and T5. Electron equivalents are defined in Table 9.8 (Postma et al., 1991).

Hansen et al. (1996) found that green rusts (mixed Fe(2,3)oxides) readily reduced nitrate but the reaction product was ammonia! Probably the reduction of nitrate by the oxidation of structural Fe(2) in minerals is most important in fine grained sediments with slow groundwater transport rates.

QUESTIONS:

Estimate the reactivity of Miocene (10^7 yr) organic carbon in mmol C/L/yr using Equation (9.61).

ANSWER: $k \times C = 0.16 / (10^7)^{0.95}$ (/yr) $\times 100$ (mmol C/kg) $\times 6$ (kg/L) $= 2 \times 10^{-5}$ mmol C/L/yr.

Estimate the time for reducing 1 mM NO_3^- by reaction with Miocene OC?

ANSWER: 58,000 yr

9.6 IRON REDUCTION AND SOURCES OF IRON IN GROUNDWATER

Ferrous iron is a common constituent of anoxic groundwater. Its origin may be the partial oxidation of pyrite, the dissolution of Fe(2) containing minerals, or reductive dissolution of iron oxides. Redox diagrams (Figures 9.8, 9.10) suggest Fe^{2+} to be the dominant form of dissolved iron in the pH range of most groundwaters (5–8), since under these conditions Fe^{3+} is insoluble. During drinking water production, anoxic groundwater containing Fe^{2+} is aerated and Fe-oxyhydroxides precipitate. Although iron in drinking water is not poisonous and perhaps even beneficial, Fe-oxyhydroxides may clog distribution systems, and stain clothing and sanitary installations, and iron is therefore removed during water treatment. In hand-pumped wells, slimy deposits from Fe^{2+}-oxidizing bacteria can also be a problem. Most of the following discussion applies for manganese as well. However, Mn-oxides become reduced at a higher pe than Fe-oxides (Figure 9.10) and manganese is much less abundant than iron in aquifers.

9.6.1 *Iron in aquifer sediments*

The distribution of iron in a sandy aquifer sediment is shown in Figure 9.36. The light fraction (a) contains quartz and feldspars and here iron is present as coatings of iron oxides covering the grains. Fraction (b) contains primary Fe(2)-bearing silicates such as amphiboles, pyroxenes or biotite. Finally fraction (c) contains iron containing oxide minerals comprising magnetite and ilmenite.

Under anoxic conditions, the dissolution of Fe(2)-bearing silicates, including amphiboles and pyroxenes, and magnetite may release Fe^{2+} to the groundwater. Because the dissolution rates of such minerals generally are very low, the resulting Fe^{2+} concentration will also be low. Postma and Brockenhuus Schack (1987) found amphiboles and pyroxenes in a sandy aquifer with distinct dissolution features and a Fe^{2+} concentration of 10–13 μM. Under oxic conditions the Fe^{2+} released during dissolution precipitates as an iron oxyhydroxide coating that inhibits further dissolution (Schott and Berner, 1983; Ryan and Gschwend, 1992; White, 1990). Common iron oxides and oxyhydroxides found in sediments are ferrihydrite, goethite, lepidocrocite and hematite (Figure 9.9, Cornell and Schwertmann, 2003). For example, the color of red sandstones is due to thin hematite coatings covering all sediment grains. In areas of groundwater discharge, Fe^{2+}-rich groundwater comes into contact with atmospheric oxygen and the precipitation of iron oxyhydroxides forms deposits in streams beds and bog iron ores.

Iron speciation in aquifer sediments can be determined by solid state methods such as Mössbauer spectroscopy and X-ray diffraction and/or physical separation as shown in Figure 9.36. However, in most cases iron speciation in sediments is determined using wet chemical extraction methods.

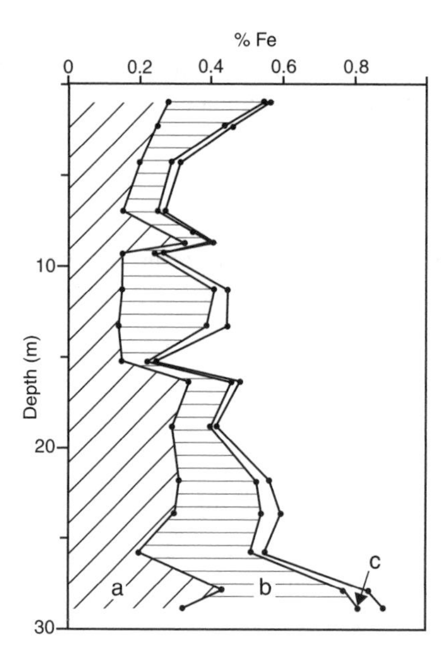

Figure 9.36. Cumulative iron distribution (wt%) in an oxidized sandy aquifer based on physical separation methods: (a) light fraction with quartz and feldspar, (b) weak magnetic fraction containing pyroxenes and amphiboles, (c) strong magnetic fraction with magnetite and ilmenite (Postma and Brockenhuus-Schack, 1987).

These methods are empirically defined and basically non-mineral specific. Popular methods for extracting poorly crystalline material, mainly ferrihydrite, comprise extraction with ammonium oxalate (Schwertmann, 1964) or 0.5 M HCl (Heron and Christensen, 1995). Less reactive iron oxides, including hematite and goethite, can be extracted with dithionite-citrate-bicarbonate (Mehra and Jackson, 1960), and Ti-citrate-EDTA-bicarbonate (Ryan and Gschwend, 1991). A method to quantify the reactivity of iron oxides in sediments is given by Postma (1993).

9.6.2 *Reductive dissolution of iron oxides*

The reduction of iron oxides by organic matter is of major importance for the evolution of water quality in pristine aquifers (Jakobsen and Postma, 1999). Also, iron oxides often form an important redox buffer that may limit the spreading of organic pollutants (Tucillo et al., 1999; Heron and Christensen, 1995). The overall reaction for the reduction of iron oxide by organic carbon is:

$$CH_2O + 4FeOOH + 7H^+ \rightarrow 4Fe^{2+} + HCO_3^- + 6H_2O \qquad (9.81)$$

The process entails a strong increase in the Fe^{2+} concentration and a large proton consumption. Microbial catalysis is important for the reduction of iron oxides by organic matter and the microbes mediating the reaction must be in direct contact with the iron oxide surface, or excrete complex formers and reductants that enhance the dissolution (Thamdrup, 2000). Humics have been proposed to function as electron carriers between iron reducing bacteria and the iron oxide (Nevin and Lovley, 2000).

For oxidants like O_2 and nitrate, the overall kinetic control by organic matter fermentation has been emphasized in Section 9.3.1. When iron is the oxidant, the reactivity of the iron oxide present in the sediment may exert additional kinetic control. In a sediment containing highly reactive organic carbon the reactivity of iron oxide is likely to become rate limiting, while in the presence of poorly reactive organic matter the fermentation step becomes rate limiting. Accordingly, Albrechtsen et al. (1995) found iron oxide reactivity to limit the rate of iron reduction in the proximal part of a landfill plume while organic matter limited the rate in the more distal part.

The reaction kinetics of inorganic reductive dissolution of iron oxides may be described by the general rate equation for crystal dissolution (Postma, 1993):

$$R = k\frac{A_0}{V}\left[\frac{m}{m_0}\right]^n g(c) \tag{4.66}$$

where R is the overall rate (mol/L/s), A_0 the initial surface area (m^2), V the solution volume (L), k the specific rate constant (mol/m^2/s), m_0 the initial mass of crystals (mol), m the mass of undissolved crystals at time t (mol), n a constant, and g(c) contains terms representing the influence of the solution composition.

In initial rate experiments a negligible amount of mineral matter is dissolved and $(m/m_0)^n$ remains constant (=1). Such experiments are used to find the specific rate $r = k \cdot g(c)$ and to study the effect of solution composition (Zinder et al., 1986; Banwart et al., 1989; Hering and Stumm, 1990 and others). Dissolution of iron oxides by the attack of protons proceeds slowly and the rate depends on the pH of the solution. For goethite, the experiments of Zinder et al. (1986) yield:

$$r = 10^{-11}\,[H^+]^{0.45}\ (\text{mol/m}^2/\text{s}) \tag{9.82}$$

The fractional exponent for $[H^+]$ may be related to formation of the protonated surface complex, e.g. $Hfo\_wOH_2^+$, but the relation between the rate and this complex is not simply linear (Stumm, 1992; Problem 9.8).

Organic compounds that form a ligand with iron at the surface of the oxide enhance the dissolution rate considerably (Figure 9.37). Small amounts of Fe^{2+} in solution accelerate the process even more and apparently the oxalate-Fe^{2+} complex acts as a catalytic electron shuttle in which Fe^{2+} is regenerated after dissociation of an iron atom from the crystal lattice (Sulzberger et al., 1989). Also a reductant like ascorbate facilitates dissolution by electron exchange with surface Fe(3) through an inner sphere complex. The reduction rate is found to increase with the ascorbate concentration until surface saturation with ascorbate is reached (Banwart et al., 1989). The combination of a ligand and a reductant gave the highest rate of hematite dissolution (Figure 9.37). In these laboratory experiments relatively simple synthetic organic compounds were used but natural organic compounds, such as phenols, tannic acid and cysteine are also known to reduce Fe-oxides (LaKind and Stone, 1989; Lovley et al., 1991). Finally, hydrogen sulfide is probably the most powerful reductant of iron oxides (Yao and Millero, 1996).

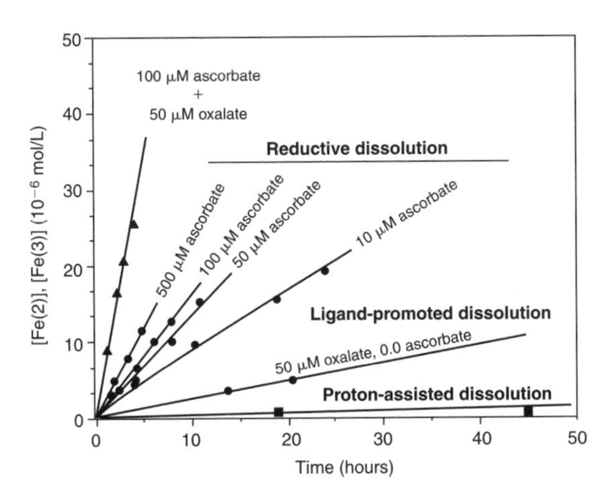

Figure 9.37. Dissolution of hematite at pH 3, by proton assisted dissolution, ligand (oxalate) promoted dissolution, reductant (ascorbate) promoted dissolution and combined ligand and reductant promoted dissolution (Banwart et al., 1989).

The effect of the iron oxide reactivity on the reduction rate can be evaluated by using a high ascorbate concentration. When the surface of the Fe-oxide is saturated with ascorbate, the rate of reductive dissolution is no longer affected by small variations in the ascorbic acid concentration (Banwart et al., 1989; Postma, 1993). Figure 9.38 shows the results of iron oxide dissolution in 10 mM ascorbate. While ferrihydrite is completely dissolved within two hours, the dissolution of a poorly crystalline goethite takes about five days. The solid lines correspond to data fits of the integrated form of Equation (4.66) (Larsen and Postma, 2001). The effect of iron oxide reactivity is contained in the specific rate $r = k \cdot g(c)$, with $g(c)$ being constant, and the exponent n. The specific rate r varies from 7.6×10^{-4}/s for ferrihydrite to 5.4×10^{-6}/s for the poorly crystalline goethite (Larsen and Postma, 2001), or more than two orders of magnitude. The exponent n (Section 4.6.4) was found to be close to 1 for goethite and lepidocrocite and to vary between 1 and 2.3 for ferrihydrite and indicates a strong decrease in the rate of dissolution as the iron oxide dissolves (Figure 9.38). The sum these two factors suggest a variation in the rate of iron oxide dissolution of at least three orders of magnitude, depending on the initial mineralogy and the extent of dissolution.

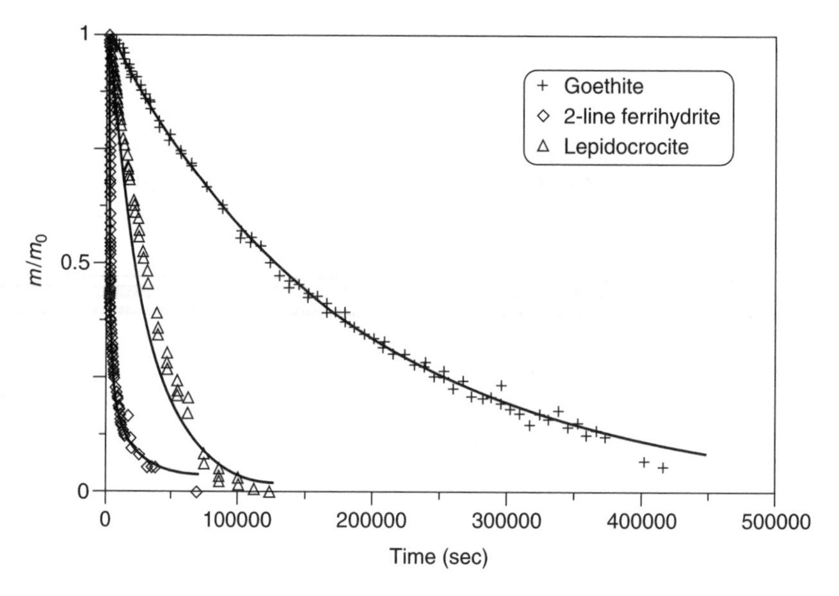

Figure 9.38. The rate of dissolution of iron oxides in 10 mM ascorbic acid at pH 3. m/m_0 is the remaining fraction of mineral mass (Modified from Larsen and Postma, 2001).

The reduction of iron oxide by organic pollutants has been studied extensively at an oil spill due to a pipeline rupture near Bemidji, Minnesota (Lovley et al., 1989). The spill released soluble BTEX (Benzene, Toluene, Ethylbenzene, Xylene) compounds to the groundwater, resulting in a reducing contaminant plume (Baedecker et al., 1993; Tucillo et al., 1999; Cozarelli et al., 2001). The BTEX reduced iron oxides in the aquifer and created also methanogenic conditions according to the reactions:

$$C_6H_6 + 30FeOOH + 54H^+ \rightarrow 6HCO_3^- + 30Fe^{2+} + 42H_2O \tag{9.83}$$

and

$$\tfrac{8}{3}C_6H_6 + 18\,H_2O \rightarrow 10\,CH_4 + 6\,HCO_3^- + 6\,H^+ \tag{9.84}$$

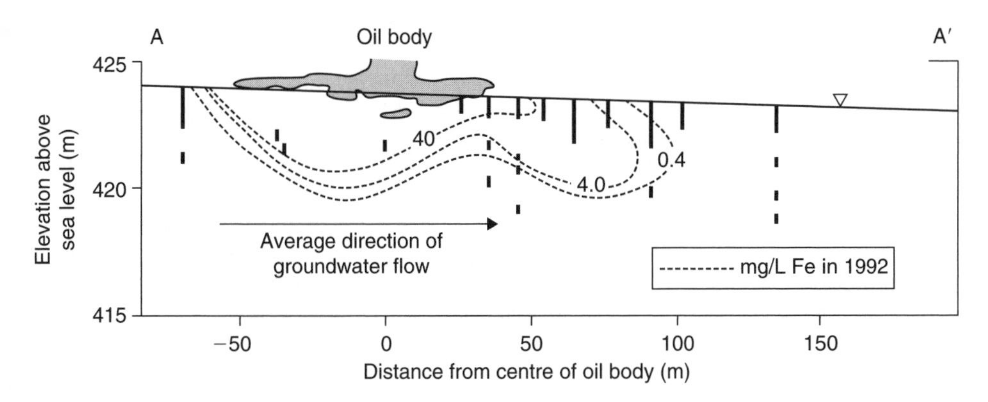

Figure 9.39. Dissolved ferrous iron (Fe^{2+}) contours in groundwater below the Bemidji oil spill. The screens of observation wells are shown as bars (Cozzarelli et al., 2001).

In the previously aerobic aquifer, the Fe^{2+} concentration reached more than 40 mg/L in 1992, 13 years after the accident (Figure 9.39). The increase in concentrations over time, directly below the floating oil, is shown in Figure 9.40. The concentrations of Mn^{2+} and Fe^{2+} went up first, but three years later the CH_4 concentration had increased markedly while Fe^{2+} stabilized at about 1 mM. After 13 years about 30% of HCl extractable iron-oxide in the sediment below the oil had reacted (8 mmol Fe(3)/kg), and methanogenic conditions appeared to extend further in the aquifer than iron-reduction. This sequence indicates a gradual depletion of the pool of easily reducible iron-oxides.

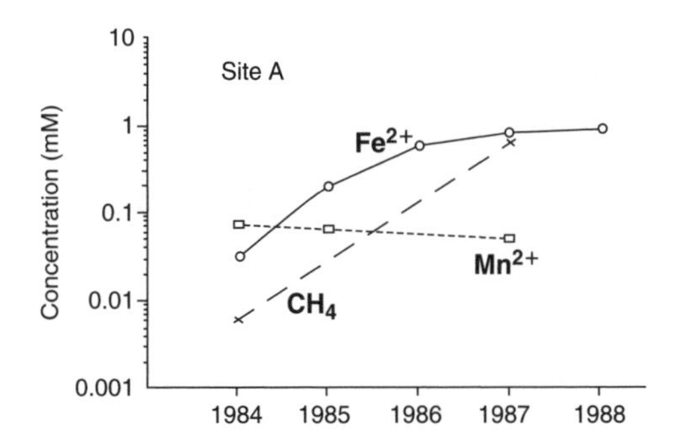

Figure 9.40. The changing concentrations of Mn^{2+}, Fe^{2+} and CH_4 with time in groundwater directly below the Bemidji oil spill (Baedecker et al., 1993).

Reaction (9.83) will result in the precipitation of siderite with the net effect of increasing the pH while lowering the CO_2 pressure:

$$C_6H_6 + 30FeOOH + 24H^+ + 24HCO_3^- \rightarrow 30FeCO_3 + 42H_2O \qquad (9.85)$$

To maintain the observed, neutral pH, Reaction (9.85) must be balanced by the acidifying Reaction (9.84). As the pool of reactive iron oxides becomes depleted, Reaction (9.85) becomes less important resulting in an increase of the CO_2 pressure and a decrease of pH. In good agreement, the field data show that, as the increase in Fe^{2+} levels off, there is a slight decrease of pH and an increase of the CO_2 pressure.

EXAMPLE 9.11. *Reaction of benzene with iron-oxide and via methanogenesis*
Benzene polluting an aquifer may be degraded by iron-reducing and methane generating bacteria. When the most reactive iron oxides have been removed by reduction, the importance of Fe(3) reduction decreases and that of methanogenesis increases. In the following PHREEQC input file, benzene degradation occurs at a constant rate and is controlled by REACTION. The concentration of CH_4 is limited to 1 mM, by using $SI\,CH_{4(g)} = -0.3$, i.e. excess CH_4 is degassing. The dissolution of goethite over a period of 13 years is described by a kinetic rate law (RATES) similar to Equation (4.66) and its effect is monitored in the lower part of Figure 9.41. Most of the released Fe^{2+} is reprecipitated as siderite. The reduction of iron oxide is proton consuming (Reaction 9.48) and causes the initial increase in pH in Figure 9.41. As the reduction of iron oxides diminishes, methanogenesis takes over, Reaction (9.85) is impeded compared to Reaction (9.84), and the pH must decrease.

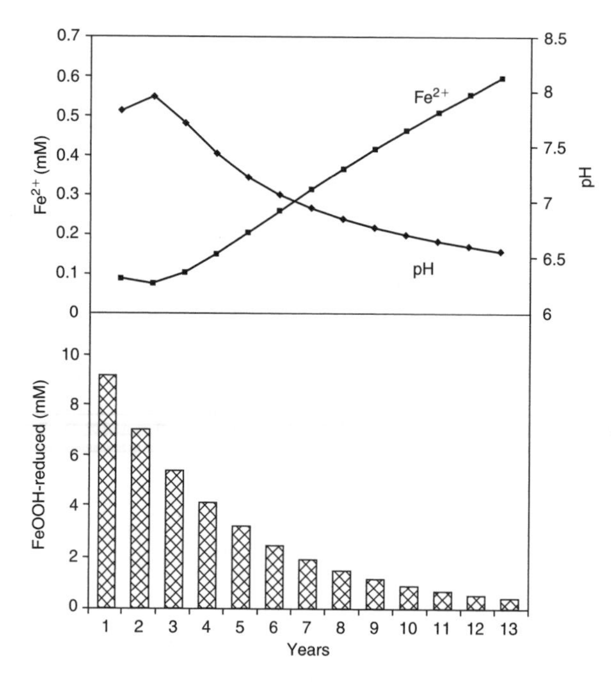

Figure 9.41. The degradation of benzene by initially iron reduction (lower graph) followed by methanogenesis. As the latter process becomes dominant the pH decreases. The Fe^{2+} concentration is controlled by equilibrium with siderite and will in consequence increase.

```
# Bemidji oil spill... react C6H6 with goethite and/or ferment
# kinetically controlled dissolution of goethite
# Limit CH4 to 1e-3 by degassing, Fe+2 precipitates in siderite
RATES
 Goethite
 -start
 10 moles = 1e-10 * (m/m0)        # ...empirical rate
 20 save moles * time
 -end

KINETICS 1
 Goethite; -m0 0.012
 -steps 4.1e8 in 13 steps
INCREMENTAL_REACTIONS true
```

```
SOLUTION 1
 pH 7 charge; -temp 10
 Ca 1 Calcite; C 1 CO2(g) -2.

REACTION 1; C6H6 1; 10.0e-3 in 13

EQUILIBRIUM_PHASES 1
 Calcite; CH4(g) -0.3 0; Siderite 1.2 0     # Limit CH4 to 1 mM
                                            # SI_Siderite = 1.2

USER_GRAPH
 -headings yr Fe+2 go pH
 -start
 10 graph_x total_time / (3600 * 24 * 365)
 20 graph_y tot (*Fe(2)*)*1e4, kin("Goethite")*1e3
 30 graph_sy -la("H+")
 -end
END
```

EXAMPLE 9.12. *pH buffering by pyrite and kinetically dissolving iron oxides*
In pH–pe diagrams, the pyrite/iron-oxyhydroxide equilibrium is located at near-neutral pH (Figure 9.10). This suggests that acid mine drainage leaking into an aquifer may be buffered to neutral pH by the reaction with these minerals. Pyrite oxidation consumes ferric iron (Fe^{3+}), causing the dissolution of iron oxides which again raises the pH. The following PHREEQC file calculates the reaction when both goethite and ferrihydrite are present. We use the rate constant from Equation (9.82) for goethite and a 100 times larger rate for ferrihydrite (cf. Figure 9.38)

```
RATES
Ferrihydrite
 -start
 1 A0 = 54000 * m0                          # initial surface area, 600m2/g
 2 SS = (1 - SR("Goethite")/1e3)            # 1000 * more soluble than goethite
 10 if SS < 0 then goto 30                  # dissolve only
 20 moles = A0 * (m/m0)^2.3 * 10^-9 * act("H+")^0.45 * SS
 30 save moles * time
 -end

Goethite
 -start
 1 A0 = 5400 * m0                           # initial surface area, 60m2/g
 2 SS = (1 - SR("Goethite"))                # SI = 0 for goethite
 10 if SS < 0 then goto 30
 20 moles = A0 * (m/m0) * 10^-11 * act("H+")^0.45 * SS
 30 save moles * time
 -end

SOLUTION 1
 pH 1.5; pe 10 O2(g) -0.01
 Fe 1. Goethite 1
 S 1. charge

EQUILIBRIUM_PHASES 1; Pyrite

KINETICS 1
 Ferrihydrite; -m0 0.025; -form FeOOH
 Goethite; -m0 0.1
 -step 0 2e6 2e6 5e6 5e6 5e6 5e6 5e6 5e6 5e6 5e6 5e6 5e6 5e6 5e6 3e7
INCREMENTAL_REACTIONS true
```

```
USER_GRAPH
 -head time Fh Go pH SI_go
 -start
 10 graph_x total_time/3.15e7
 20 graph_y kin("Ferrihydrite"), kin("Goethite")
 30 graph_sy -la("H+"), SI("Goethite")
 -end
END
```

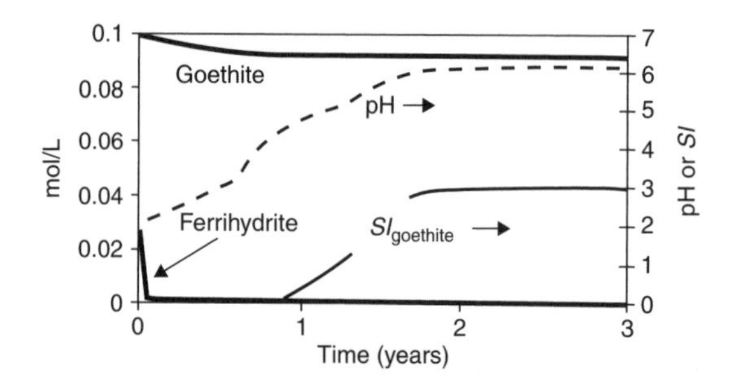

Figure 9.42. pH buffering by dissolution of iron-oxides in acid mine water. The dissolution of goethite stops when $SI_{goethite} > 0$ and, because of the higher solubility (Figure 9.9) for ferrihydrite when $SI_{goethite} > 3$.

QUESTION:
Explain what happens when m_0 of ferrihydrite is increased to 0.05?

9.7 SULFATE REDUCTION AND IRON SULFIDE FORMATION

Sulfate reduction by organic matter is catalyzed by various strains of bacteria according to the overall reaction:

$$2CH_2O + SO_4^{2-} \rightarrow 2HCO_3^- + H_2S \tag{9.86}$$

Here CH_2O is used as a simplified representation of organic matter. H_2S gives the foul smell of rotting eggs and is also highly toxic. For both reasons H_2S is undesirable in drinking water. Sulfate reduction is a common process in anoxic aquifers (Böttcher et al., 1989; Robertson et al., 1989; Chapelle and McMahon, 1991; Stoessell et al., 1993; Jakobsen and Postma, 1999). The importance of sulfate reduction in the aquifer depends on both the availability of reactive organic matter and the supply of sulfate. Freshwaters are generally low in sulfate. However, important local sources of sulfate can be the dissolution of gypsum, the oxidation of pyrite, and mixing of fresh water with seawater. Also acid rain, sea spray, fertilizers and waste deposits may be sources of sulfate. Enhanced circulation around production wells may stimulate sulfate reducing bacteria to grow, clogging the well screens with iron sulfides (Van Beek and Van der Kooij, 1982). Sulfate reduction is also an important electron accepting process in landfill plumes (Christensen et al., 2000) and can accompany BTEX pollution (Chapelle et al., 1996; Wisotzky and Eckert, 1997).

Ongoing sulfate reduction is often manifested by the presence of H_2S in groundwater (Figure 9.14). However, the produced H_2S may react with Fe-oxides present in the sediment and form iron sulfide minerals (see Section 9.7.1). If excess Fe-oxide is available, this may consume even all H_2S (Jakobsen and Postma, 1999). Particularly in aquifers where mixing of freshwater and seawater occurs, the Cl^- / SO_4^{2-}

ratio may indicate sulfate reduction. Seawater has a fixed Cl^-/SO_4^{2-} molar ratio of 19 so that significantly higher values in combination with an increased Cl concentration, point to sulfate reduction.

Stable sulfur isotopes are also a suitable indicator since during sulfate reduction, ^{32}S is preferentially consumed compared to ^{34}S. The $\delta^{34}S$ value in rain is less than +10‰ and a higher $\delta^{34}S$ in combination with decreasing sulfate concentrations (Figure 9.43) must indicate sulfate reduction. Rates of sulfate reduction in sediments can be measured directly by using a radiotracer method (Jørgensen, 1978).

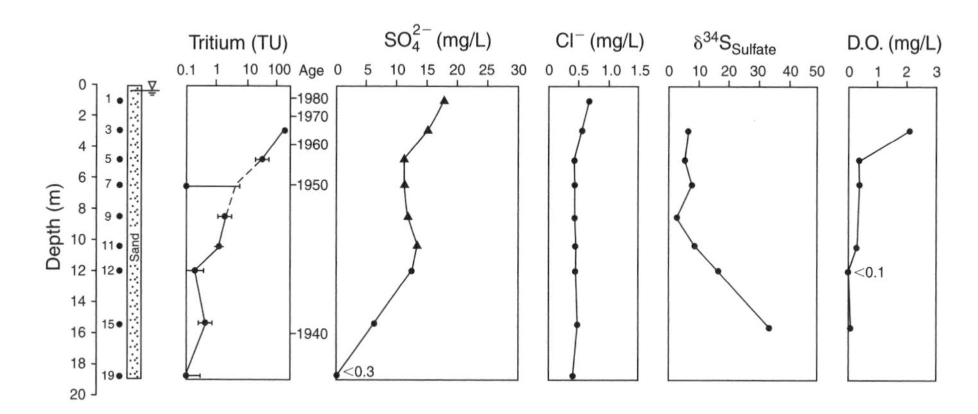

Figure 9.43. Sulfate reduction as indicated by $\delta^{34}S$ values in a sandy aquifer (Robertson et al., 1989).

Ludvigsen et al. (1998) spiked sediment incubations with $^{35}SO_4$ to identify sulfate reduction, using aquifer sediment downstream a landfill. Jakobsen and Postma (1994, 1999) and Cozzarelli et al. (2000) determined *in situ* rates of sulfate reduction by injecting a trace amount of $^{35}SO_4^{2-}$ in a sediment core which was incubated for 24 hours in a borehole, followed by determination of the fraction of tracer that was reduced. Some results are displayed in Figure 9.44 and show considerable variation. However, a good correlation exists between the depth range where sulfate is depleted and the occurrence of high rates of sulfate reduction. At some depths apparently highly reactive organic matter is present and rates as high as several mM/yr are measured while at other depth ranges there is little or no activity.

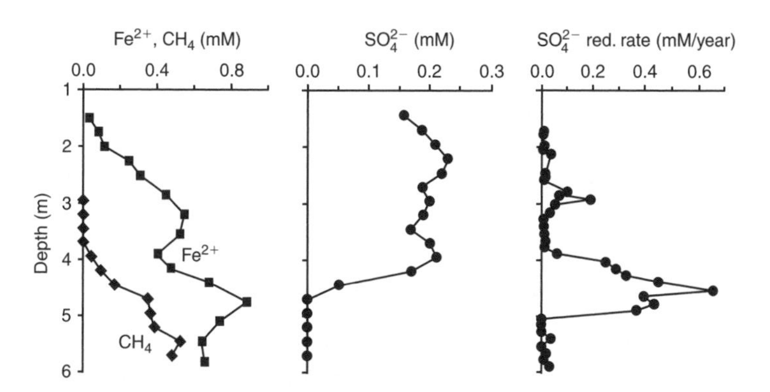

Figure 9.44. Depth distribution of ferrous iron, methane and sulfate and rates of sulfate reduction measured by the radiotracer method in the sandy aquifer at Rømø, Denmark. The H_2S concentration is below the detection limit (Jakobsen and Postma, 1994).

Sulfate reducing bacteria have a size range of 0.5–1 μm and it has been suggested that they are too big to pass the pore necks of fine grained sediments (Fredrickson et al., 1997). Fermenting organisms, in contrast, are smaller and a spatial separation of the two processes can result. This is illustrated in Figure 9.45, where fermentation occurs in low permeable clayey aquitard layers, producing formate and acetate that diffuse into sandy aquifer layers where the dissolved organics are consumed by sulfate reduction.

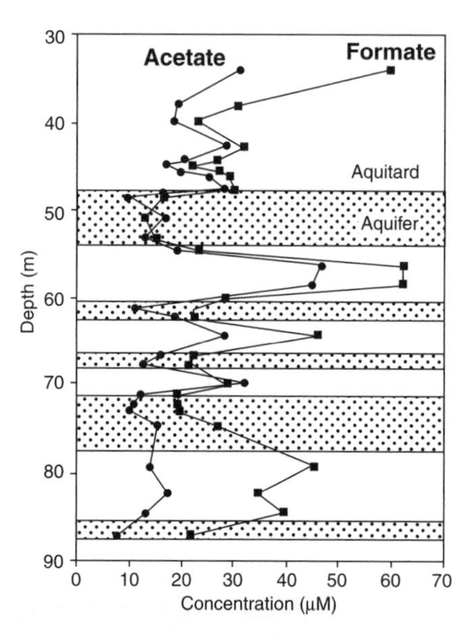

Figure 9.45. Concentrations of dissolved formate and acetate in pore waters of aquitard and aquifer sediments, indicating diffusion of dissolved organics into aquifer layers and consumption by sulfate reduction (McMahon and Chapelle, 1991).

Although pH–pe diagrams (Figures 9.7 and 9.8) suggest that iron reduction takes place at a higher pe than sulfate reduction, the two reactions may proceed concurrently when the available iron-oxide has low solubility (Postma and Jakobsen, 1996). An equilibrium reaction for the two redox couples can be written as:

$$8Fe^{2+} + SO_4^{2-} + 20H_2O \leftrightarrow 8Fe(OH)_3 + HS^- + 15H^+ \qquad (9.87)$$

The reaction will be displaced to the right when sulfate reduction is favorable, and to the left when $Fe(OH)_3$ reduction gives more energy. However, the reaction depends also on the activities of the other reactants and we wish to eliminate some variables to simplify. Sulfate reducing environments are commonly found to be near equilibrium with FeS (Wersin et al., 1991; Perry and Pedersen, 1993) and we can write:

$$H^+ + FeS \leftrightarrow Fe^{2+} + HS^- \qquad (9.88)$$

Reactions (9.87) and (9.88) can be combined in two ways, eliminating Fe^{2+} or HS^-. Both options may be valid for natural conditions, since due to the insolubility of FeS, Fe^{2+} and dissolved sulfide normally are mutually exclusive in anoxic environments. Eliminating HS^-, the reactions combine to:

$$9Fe^{2+} + SO_4^{2-} + 20H_2O \leftrightarrow 8Fe(OH)_3 + FeS + 16H^+ \qquad (9.89)$$

with

$$K_{(9.89)} = \frac{[H^+]^{16}}{[Fe^{2+}]^9[SO_4^{2-}]} \tag{9.90}$$

The logarithmic form of Equation (9.90) is displayed for different Fe-oxides in Figure 9.46, using pH and $\log[Fe^{2+}]$ as variables and at a fixed $\log[SO_4^{2-}] = -3$. For each iron oxyhydroxide, the solid line indicates equilibrium. On the right side of the equilibrium line sulfate reduction is favored ($[Fe^{2+}]$ or pH is higher than at the assumed equilibrium condition and the reaction tends to proceed to the right). Similarly, on the left side of the line Fe(3) reduction will take place. In the presence of amorphous $Fe(OH)_3$, iron reduction is favored under most aquatic conditions ($[Fe^{2+}] < 10^{-3}$ and pH < 8.5). However, if goethite or hematite are present, sulfate reduction is often more favorable. Figure 9.46 demonstrates the major influence of pH on the relative stability of iron and sulfate reduction. In contrast, the effect of $[SO_4^{2-}]$ variations is small as indicated by the dotted lines for $Fe(OH)_3$.

Figure 9.46 also illustrates that Fe(3) reduction and sulfate reduction thermodynamically may proceed simultaneously when the most unstable Fe-oxides are absent or have dissolved completely. Data from natural environments are included in Figure 9.46 where iron and sulfate reduction were observed to proceed concomitantly. All environments are able to reduce the least stable Fe-oxides, $Fe(OH)_3$ and lepidocrocite, but compared to goethite and hematite, sulfate reduction is more favorable except in case (5).

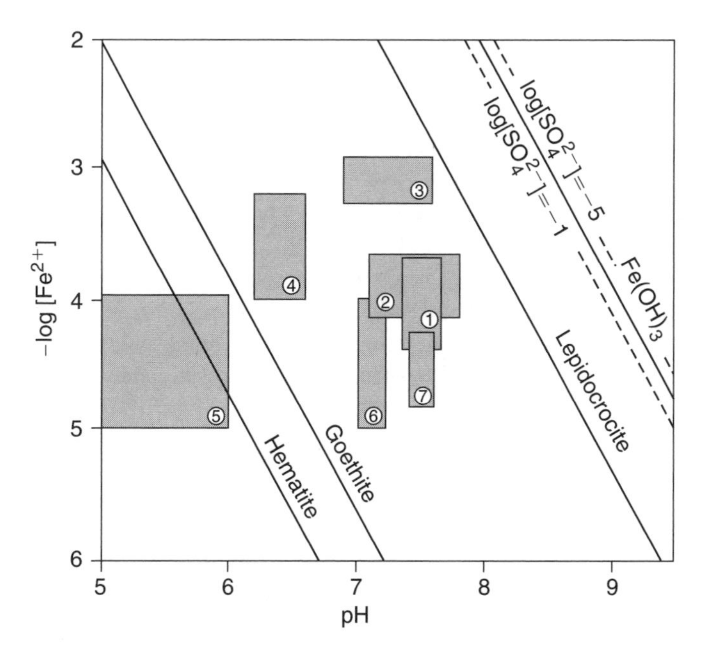

Figure 9.46. Stability relations for simultaneous equilibrium between Fe(3) and sulfate reduction at equilibrium with FeS in an Fe^{2+}-rich environment. The solid lines correspond to Equation (9.89) at $\log[SO_4^{2-}] = -3$ for different Fe-oxides. Shaded areas reflect field data with the simultaneous occurrence of Fe(3) and sulfate reduction taken from: 1. Canfield et al. (1993, and pers. comm.), marine sediment, core S4 > 1 cm depth and S6 > 3 cm depth, 2. Leuchs (1988), aquifer, core DGF7 > 2.5 m depth, 3. Jakobsen and Postma (1999) core 6, 4.45–6 m depth, 4. White et al. (1989), lake sediment, 17 m site, 5. Marnette et al. (1993), lake Kliplo sediment, 2–10 cm, 6. Wersin et al. (1991), lake sediment, 0–10 cm, 7. Simpkins and Parkin (1993), aquifer, Nest 1/4, 6–18 m depth. (Postma and Jakobsen, 1996).

QUESTIONS:
For an enrichment $\varepsilon^{34}S_{H_2S/SO_4^{2-}} = -27‰$, estimate $\delta^{34}S$ when the SO_4^{2-} concentration decreases from 13 mg/L ($\delta^{34}S = 5‰$) to 5 mg/L by bacterial reduction (*Hint*: use the Rayleigh formula)
ANSWER: $\delta^{34}S = 30.8‰$, cf. Figure 9.43, 15 m depth
From Figure 9.46, estimate log $K_{(9.89)}$ with hematite and ferrihydrite?
ANSWER: −51 (hematite) and −104 (ferrihydrite)
What is difference in solubility of hematite and ferrihydrite in Figure 9.46?
ANSWER: log $K_{ferrihydrite}$ = log $K_{hematite}$ + 6.7 (cf. Figure 9.9)

9.7.1 *The formation of iron sulfides*

Pyrite (cubic FeS_2) and its polymorph marcasite (orthorhombic FeS_2) are the two most abundant iron sulfide minerals in ancient sedimentary rocks. Pyrite ($\Delta G_f^0 = -160.2$ kJ/mol) is more stable than marcasite ($\Delta G_f^0 = -158.4$ kJ/mol) and pyrite is also the form that is normally encountered in recent environments. In ancient deposits pyrite is particularly abundant in fine grained sediments, but it may also be present in sand-sized deposits. It occurs as disseminated single crystals, globular aggregates of crystals with a size of up to several hundreds microns (Love and Amschutz, 1966) called framboids (Figure 9.12), and as concretions. Marcasite is less stable, its formation is poorly understood, and it is often found as concretions in carbonate rocks. The overall pathway of pyrite formation is in most cases initiated by the reaction:

$$2FeOOH + 3HS^- \rightarrow 2FeS + S^\circ + H_2O + 3OH^- \tag{9.91}$$

This reaction comprises the reduction of iron oxide by H_2S which generally is a fast process (Canfield et al., 1992). While part of the sulfide reduces Fe(3) and produces S°, the remainder of dissolved sulfide precipitates as FeS. FeS is also called acid volatile sulfide (AVS), since in contrast to pyrite it readily dissolves in HCl, and this property forms the base of the analytical assay. FeS stains the sediment black and consist of amorphous FeS and extremely fine grained minerals like mackinawite ($Fe_{1-x}S$) and greigite (Fe_3S_4). FeS is less stable than pyrite but precipitates rapidly and is formed therefore due to the more sluggish precipitation kinetics of pyrite. The transformation of FeS to FeS_2 is an oxidation process and the overall reaction can be written as:

$$FeS + S^\circ \rightarrow FeS_2 \tag{9.92}$$

The process probably consists of a continuous sulfurization of FeS (Schoonen and Barnes, 1991) and proceeds presumably through a dissolution-precipitation pathway (Wang and Morse, 1996). The oxidation of FeS may also proceed through the reduction of protons according to the reaction (Drobner et al., 1990; Rickard, 1997).

$$FeS + H_2S \rightarrow FeS_2 + H_2 \tag{9.93}$$

This reaction enables continuous pyrite formation in a completely anoxic environment. In contrast, the consumption of S° by Reaction (9.92) is twice its production, relative to FeS, by Reaction (9.91) and an external oxidant is therefore required to sustain continuous pyrite formation. In some cases, where dissolved sulfide concentrations remain very low, fast direct formation of pyrite seems also possible (Howarth, 1979; Howarth and Jørgensen, 1984).

QUESTION:
Find the H_2 concentration where Reaction (9.93) proceeds if $H_2S = 10^{-6}$M? (Use K values from PHREEQC.DAT).

ANSWER: log K(9.93) = 4.43. $H_{2(aq)} < 0.027$ M.

9.8 THE FORMATION OF METHANE

The final stage in the reductive sequence is methane formation. Methane is a common constituent of anoxic groundwater (Barker and Fritz, 1981; Leuchs, 1988; Grossman et al., 1989; Simkins and Parkin, 1993; Aravena et al., 1995; Zhang et al., 1998). Its presence can be a problem for water supplies due to its potentially explosive nature (Aravena et al., 1995). Methane is also a common constituent of groundwater pollution plumes downstream of landfills containing organic waste (Adrian et al., 1994; Albrechtsen et al., 1999). Finally, methane is the second most important atmospheric greenhouse gas. The release of methane from wetlands (Avery and Martens, 1999) and landfills (Chanton and Liptay, 2000) contributes significantly to the increase of atmospheric methane (Dlugokencky et al., 1998).

The origin of methane may either be *bacterial*, produced by microbial activity, or *thermogenic*. The latter takes place non-biologically at several kilometers depth and is often related to oil formation. The produced methane may, however, migrate upwards into aquifers through fracture zones etc. Distinction between the two types of methane is usually made by comparison of stable carbon and hydrogen isotopes (Clark and Fritz, 1997). Thermogenic methane has $\delta^{13}C$ values in the range -50 to -20‰, while bacterial methane has $\delta^{13}C$ values of less than -50‰ (Whiticar, 1999). Biogenic methane from landfill sites is reported to have intermediate $\delta^{13}C$ values. Biogenic methane forms through a series of complex biogenic reactions (Conrad, 1999; Whiticar, 1999; Vogels et al., 1988). In general, the predominant processes of methane formation are the reduction of CO_2 by free hydrogen:

$$CO_2 + 4H_2 \rightarrow 2H_2O + CH_4 \tag{9.94}$$

and the fermentation of acetate

$$CH_3COOH \rightarrow CH_4 + CO_2 \tag{9.95}$$

In the latter reaction the acetate methyl group is transferred directly to CH_4. Both Reactions (9.94) and (9.95) are based on competitive substrate reactions, i.e. methanogenic bacteria have to compete with sulfate reducing and iron reducing bacteria for the same pools of hydrogen and acetate (Figure 9.19). There are also non-competitive substrates, such as methylated amines which are fermented directly to CH_4 by methanogens (Whiticar, 1999). The relative importance of non-competitive substrates in natural environments has not yet been clarified.

The pathway of methane formation in sediments has been determined by radiotracer methods. The sediment is incubated with trace amounts of either $^{14}CO_2$ or $^{14}CH_3COOH$ and the production rate of $^{14}CH_4$ is determined. Both pathways of CH_4 production are found to proceed simultaneously in sediments and their relative importance may vary from close to 100% acetate fermentation (Phelps and Zeikus, 1984) to close to 100% CO_2 reduction (Lansdown et al., 1992). Generally, acetate fermentation is the primary pathway in lake sediments (Phelps and Zeikus, 1984; Kuivila et al., 1989) while CO_2 reduction tends to dominate in marine sediments (Crill and Martens, 1986; Hoehler et al., 1994). Sulfate is abundant in seawater, and sulfate reduction is therefore an important precursor for methanogenesis in marine sediments and at the same time accounts for most acetate metabolism. Below the zone of sulfate reduction, CO_2 reduction will become the dominant pathway of methanogenesis. In a typical freshwater lake sediment there is little sulfate and therefore more reactive organic carbon is available for methanogenesis. Acetate fermentation may then be the dominant pathway (Whiticar, 1999).

The deuterium content of methane is another, albeit rather complicated key for deriving the dominant pathway of methane formation in sediments (Whiticar et al., 1986; Clark and Fritz, 1997; Whiticar, 1999). In the interpretation of Whiticar, CO_2 reduction produces δD_{CH_4} ranges from -250 to -150‰ while fermentation of acetate produces generally lower values. However, Sugimoto and Wada (1995) found much lower δD_{CH_4} values for CO_2 reduction in freshwater sediments. Recently Waldron et al. (1999) showed that the δD_{CH_4} in low sulfate freshwater environments may be independent of the methanogenic pathway but strongly influenced by the δD_{H_2O}.

Figure 9.47 shows the depth distribution of methane in the Rømø aquifer. Below 2 m depth, methane is present in the groundwater and radiotracers were used to determine the source of CH_4. CO_2 reduction was found to be the dominant pathway of methane formation. In the depth range 4–5.5 m the rate of CO_2 reduction is particularly high with values up to 3.5 mM/yr. However, acetate fermentation also occurs throughout the profile and in depth intervals where the rate of CO_2 reduction is low, acetate fermentation may contribute with up to 50% of the total methane production rate.

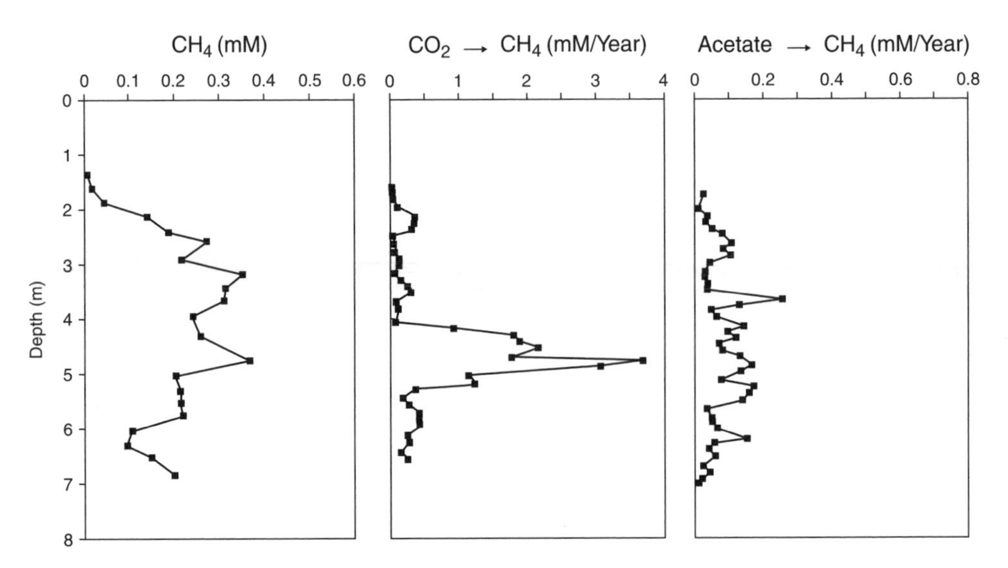

Figure 9.47. The distribution of methane and rates and pathways of methane formation in the Rømø aquifer. Rates of methanogenesis were measured by injecting $^{14}HCO_3^-$ or $^{14}CH_3COOH$ in sediment cores and measuring the production of $^{14}CH_4$ after incubation (Hansen et al., 2001).

The rate of methane formation is rather variable over depth but this variation is not reflected in the methane distribution over depth. Once formed, methane is quite stable in anoxic aquifers and it may be transported further downstream without showing any reaction.

However, when methane enters an aerobic environment, by degassing into soil or a landfill cover, or by mixing in an aquifer, bacteria will oxidize methane in the presence of O_2. Chanton and Liptay (2000) found that 10–60% of methane emitted from landfills is reoxidized in the overlying cover. Smith et al. (1991b) studied methane oxidation by O_2 in groundwater using a natural gradient approach and described the results using Michaelis-Menten kinetics. Anaerobic oxidation of methane by sulfate has in marine sediments been studied by Iversen and Jørgensen (1985) and Hoehler et al. (1994). However, in aquifer sediments the process was found to be insignificant (Hansen, 1998).

PROBLEMS

9.1 Groundwater samples from wells in the Nadia district of West Bengal show the following composition (Mukherjee et al., 2000). Concentrations are in micrograms per liter.

Sample	pH	*Eh*/mV	As(3)	As(5)
S_1	6.90	−152	84.7	88.3

a. How do these samples compare with drinking water limits.
b. Calculate the *Eh* from dissolved arsenic.
c. Compare with measured *Eh* values and discuss differences.

9.2 a. List, using Figure 9.10, those Fe and Mn minerals which are stable together. Do the same for combinations of dissolved species and minerals.
b. Calculate and plot in Figure 9.8 the redox boundary between Fe^{2+} and hematite.
c. Drainwater from the Haunstrup browncoal mine, in Denmark, showed the following variations over time:

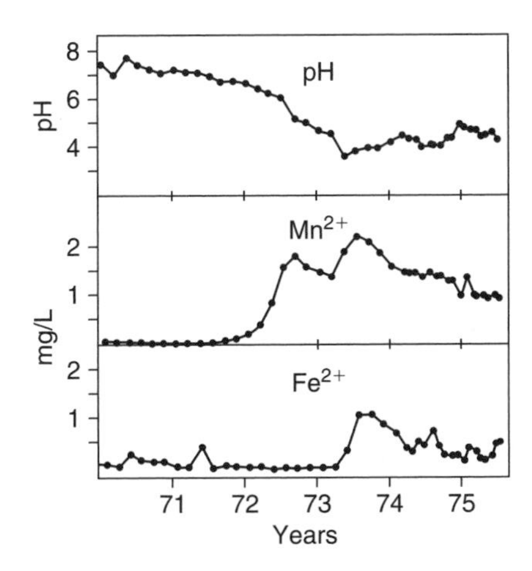

Find two explanations, in terms of pe/pH variations why Mn^{2+} is released earlier than Fe^{2+}.
d. An analysis of groundwater gave the following concentrations:

pH	NO_3^-	NO_2^-
7.16	110	2.1 (mg/L)

Calculate the amount of Mn^{2+} in water, when the aquifer contains MnO_2.

9.3 Construct a pe–pH diagram for Pb^{2+} with the following data:

1) $PbO_2 + 4H^+ + 2e^- \leftrightarrow 2H_2O + Pb^{2+}$ $\log K_1 = 49.2$
2) $Pb^{2+} + 2e^- \leftrightarrow Pb$ $\log K_2 = -4.26$
3) $PbO + 2H^+ \leftrightarrow Pb^{2+} + H_2O$ $\log K_3 = 12.7$
4) $Pb^{2+} + 3H_2O \leftrightarrow Pb(OH)_3^- + 3H^+$ $\log K_4 = -28.1$

Use activity levels of Pb^{2+} and $Pb(OH)_3^-$ 10^{-6} at the solid/solution boundaries.

9.4 a. Calculate the amount of $MnCO_3$ (rhodochrosite) that can dissolve in water, if $P_{CO_2} = 10^{-1.5}$

$$MnCO_3 \leftrightarrow Mn^{2+} + CO_3^{2-}; \quad K = 10^{-9.3}$$

b. Explain the terms "congruent/incongruent" dissolution of a mineral.
c. Which type of rhodochrosite dissolution is expected in a well-aerated soil: congruent dissolution or incongruent dissolution? Which mineral(s) will form? What is the effect on the P_{CO_2}?

9.5 Write a balanced equation for oxidation of organic material (CH_2O) by $KMnO_4$ (i.e. determination of *COD*)

9.6 Draw a pe/pH-diagram for sulfur-species. Consider $S_{(s)}$, $H_2S_{(aq)}$, HS^-, SO_4^{2-}. Activity of a species in solution is 10 mmol/L.

1) $H_2S_{(aq)} \leftrightarrow HS^- + H^+$ $\qquad K = 10^{-7}$
2) $H_2S_{(aq)} + 4H_2O \leftrightarrow SO_4^{2-} + 10H^+ + 8e^-$ $\qquad K = 10^{-40.6}$
3) $S_{(s)} + 4H_2O \leftrightarrow SO_4^{2-} + 8H^+ + 6e^-$ $\qquad K = 10^{-36.2}$

If S precipitates from spring-water, what can be concluded about pH and H_2S-content?

9.7 Zinder et al. (1986) measured the rate of dissolution of goethite as function of pH:

pH	r, mol/m^2/s
1.5	8.72×10^{-9}
2	4.46×10^{-9}
3	1.02×10^{-9}
3.5	8.91×10^{-9}
3.5	1.56×10^{-9}
3.7	2.34×10^{-9}

a. Determine a rate law that includes the pH dependency (cf. Example 7.1, Equation (9.82))
b. Find a rate law that includes the effect of the protonated surface complex, Hfo_wOH_2^+. (*Hint*: regress on the calculated fraction of the surface complex).

9.8 A threshold concentration of H_2 is needed to trigger acetogenesis, methanogenesis and sulfate reduction, in energy terms amounting to -2 to -4 kJ per transferred electron. By how much must H_2 exceed the equilibrium concentration to correspond to a energy threshold of -3 kJ/e?

9.9 The enthalpy for the reaction $4H_2 + HCO_3^- + H^+ \leftrightarrow CH_4 + 3H_2O$ is $\Delta H_r^0 = -237.8$ kJ/mol. Estimate the change in H_2 concentration when temperature decreases from 25 to 10°C in methanogenic aquifers?

9.10. The free energy of the reaction $CH_3COO^- + 4H_2O \leftrightarrow 4H_2 + 2HCO_3^- + H^+$ is $\Delta G_r^0 = 144.29$ kJ/mol. Plot the concentration of acetate in Example 9.6?

REFERENCES

Adrian, N.R., Robinson, J.A. and Suflita, J.M., 1994. Spatial variability in biodegradation rates as evidenced by methane production from an aquifer. *Appl. Env. Microbiol.* 60, 3632–3639.

Albrechtsen, H.-J., Heron, G. and Christensen, T.H., 1995. Limiting factors for microbial Fe(III)-reduction in a landfill leachate polluted aquifer (Vejen, Denmark). *FEMS Microbiol. Ecol.* 16, 233–248.

Albrechtsen, H.-J., Bjerg, P.L., Ludvigsen, L., Rügge, K. and Christensen, T.H., 1999. An anaerobic field injection experiment in a landfill leachate plume, Grindsted, Denmark. 2. Deduction of anaerobic methanogenic, sulfate- and Fe(III)-reducing redox conditions. *Water Resour. Res.* 35, 1247–1256.

Alpers, C.N. and Blowes, D.W. (eds), 1994. *Environmental geochemistry of sulfide oxidation*. ACS Symp. Ser. 550, 681 pp.

Andersen, M.S., Larsen, F. and Postma, D., 2001. Pyrite oxidation in unsaturated aquifer sediments; reaction stoichiometry and rate of oxidation. *Env. Sci. Technol.* 35, 4074–4079.

Anthony, R.S., 1977. Iron-rich rhythmically laminated sediments in Lake of the Clouds, northeastern Minnesota. *Limnol. Oceanogr.* 22, 45–54.

Aravena, R., Wassenaar, L.I. and Barker, J.F., 1995. Distribution and isotopic characterization of methane in a confined aquifer in southern Ontario. *Can. J. Hydrol.* 173, 51–70.

Avery Jr, G.B. and Martens, C.S. 1999. Controls on the stable isotopic composition of biogenic methane produced in a tidal freshwater estuarine sediment. *Geochim. Cosmochim. Acta* 63, 1075–1082.

Baedecker, M.J., Cozzarelli, I.M., Siegel, D.I., Bennett, P.C. and Eganhouse, R.P., 1993. Crude oil in a shallow sand and gravel aquifer. *Appl. Geochem.* 8, 569–586.

Banwart, S., Davies, S. and Stumm, W., 1989. The role of oxalate in accelerating the reductive dissolution of hematite (α-Fe_2O_3) by ascorbate. *Coll. Surf.* 39, 303–309.

Barker, J.F. and Fritz, P., 1981. The occurrence and origin of methane in some groundwater flow systems. *Can. J. Earth Sci.* 18, 1802–1816.

Berner, R.A., 1971. *Principles of chemical sedimentology*. McGraw-Hill, New York, 240 pp.

Berner, R.A., 1981. A new geochemical classification of sedimentary environments. *J. Sed. Petrol.* 51, 359–365.

Bjerg, P.L., Rügge, K., Pedersen, J.K. and Christensen, T.H., 1995. Distribution of redox-sensitive groundwater quality parameters downgradient of a landfill. *Env. Sci. Technol.* 29, 1387–1394.

Blicher-Mathiesen, G., McCarty, G.W. and Nielsen, L.P., 1998. Denitrification and degassing in groundwater estimated from dissolved dinitrogen and argon. *J. Hydrol.* 208, 16–24.

Böttcher, J., Strebel, O. and Duynisveld, W.H.M., 1989. Kinetik und Modellierung gekoppelter Stoffumsetzungen im Grundwasser eines Lockergesteins-Aquifers. *Geol. Jb. C* 51, 3–40.

Böttcher, J., Strebel, O., Voerkelius, S. and Schmidt, H.L., 1990. Using isotope fractionation of nitrate-nitrogen and nitrate-oxygen for evaluation of microbial denitrification in a sandy aquifer. *J. Hydrol.* 114, 413–424.

Boudreau, B.P., 1997. *Diagenetic models and their implementation.* Springer, Berlin, 414 pp.

Bradley, P.M., Fernandez, M. and Chapelle, F.H., 1992. Carbon limitation of denitrification rates in an anaerobic groundwater system. *Env. Sci. Technol.* 26, 2377–2381.

Bragan, R.J., Starr, J.L. and Parkin, T.B., 1997. Shallow groundwater denitrification rate measurement by acetylene block. *J. Env. Qual.* 26, 1531–1538.

Brons, H.J., Hagen, W.R. and Zehnder, A.J.B., 1991. Ferrous iron dependent nitric oxide production in nitrate reducing cultures of *Escheridia coli*. *Arch. Microbiol.* 155, 341–347.

Brookins, D.G., 1988. *Eh-pH diagrams for geochemistry*. Springer, Berlin, 176 pp.

Canfield, D.E., Raiswell, R. and Bottrell, S., 1992. The reactivity of sedimentary iron minerals toward sulfide. *Am. J. Sci.* 292, 659–683.

Canfield, D.E., Thamdrup, B. and Hansen, J.W., 1993. The anaerobic degradation of organic matter in Danish coastal sediments: Iron reduction, manganese reduction, and sulfate reduction. *Geochim. Cosmochim. Acta* 57, 3867–3883.

Champ, D.R., Gulens, J. and Jackson, R.E., 1979. Oxidation-reduction sequences in ground water flow systems. *Can. J. Earth Sci.* 16, 12–23.

Chanton, J. and Liptay, K., 2000. Seasonal variation in methane oxidation in a landfill cover soil as determined by an in situ stable isotope technique. *Global Biogeochem. Cycles* 14, 51–60.

Chapelle, F.H. and Lovley, D.R., 1992. Competitive exclusion of sulfate reduction by Fe(III)-reducing bacteria: A mechanism for producing discrete zones of high-iron ground water. *Ground Water* 30, 29–36.

Chapelle, F.H. and McMahon, P.B., 1991. Geochemistry of dissolved inorganic carbon in a coastal plain aquifer. 1. Sulfate from confining beds as an oxidant in microbial CO_2 production. *J. Hydrol.* 127, 85–108.

Chapelle, F.H., McMahon, P.B., Dubrovsky, N.M., Fujii, R.F., Oaksford, E.T. and Vroblesky, D.A., 1995. Deducing the distribution of terminal electron-accepting processes in hydrologically diverse groundwater systems. *Water Resour. Res.* 31, 359–371.

Chapelle, F.H., Bradley, P.M., Lovley, D.R. and Vroblesky, D.A., 1996. Measuring rates of biodegradation in a contaminated aquifer using field and laboratory methods. *Ground Water* 34, 691–698.

Christensen, T.H., Bjerg, P., Banwart, S.A. Jakobsen, R., Heron, G. and Albrechtsen, H.J., 2000. Characterization of redox conditions in groundwater contaminant plumes. *J. Contam. Hydrol.* 45, 165–241.

Clark, I.D. and Fritz, P., 1997. *Environmental isotopes in hydrogeology*. CRC Press, Boca Raton, 328 pp.

Conrad, R., 1999. Contribution of hydrogen to methane production and control of hydrogen concentrations in methanogenic soils and sediments. *FEMS Microbiol. Ecol.* 28, 193–202.

Cornell, R.M. and Schwertmann, U., 2003. *The Iron Oxides*, 2nd ed. Wiley-VCH, Weinheim, 664 pp.

Cozarelli, I.M., Bekins, B.A., Baedecker, M.J., Aiken, G.R., Eganhouse, R.P. and Tucillo, M.E., 2001. Progression of natural attenuation processes at a crude-oil spill site: 1. Geochemical evolution of the plume. *J. Contam. Hydrol.* 53, 369–385.

Cozzarelli, I.M., Suflita, J.M., Ulrich, G.A., Harris, S.H., Scholl, M.A., Schlottmann, J.L. and Christenson, S., 2000. Geochemical and microbiological methods for evaluating anaerobic processes in an aquifer contaminated by landfill leachate. *Env. Sci. Technol.* 34, 4025–4033.

Crill, P.M. and Martens, C.S., 1986. Methane production from bicarbonate and acetate in an anoxic marine sediment. *Geochim. Cosmochim. Acta* 50, 2089–2097.

Dlugokencky, E.J., Masarie, K.A., Lang, P.M. and Tans, P.P., 1998. Continuing decline in the growth rate of the atmospheric methane burden. *Nature* 393, 447–450.

Doyle, R.W., 1968. The origin of the ferrous ion-ferric oxide Nernst potential in environments containing ferrous iron. *Am. J. Sci.* 266, 840–859.

Drever, J.I., 1997. *The geochemistry of natural waters*. 3rd ed. Prentice Hall, Englewood-Cliffs, 436 pp.

Drobner, E., Huber, H., Wächtershauser, G., Rose, D. and Stetter, K.O., 1990. Pyrite formation linked with hydrogen evolution under anaerobic conditions. *Nature* 346, 742–744.

Dubrovsky, N.M., Morin, K.A., Cherry, J.A. and Smyth, D.J.A., 1984. Uranium tailings acidification and subsurface contaminant migration in a sand aquifer. *Water Poll. Res. J. Can.* 19, 55–89.

Elberling, B. and Nicholson, R.V., 1996. Field determination of sulphide oxidation rates in mine tailings. *Water Resour. Res.* 32, 1773–1784.

Elberling, B., Larsen, F., Christensen, S. and Postma, D., 1998. Gas transport in a confined unsaturated zone during atmospheric pressure cycles. *Water Resour. Res.* 34, 2855–2862.

Ernsten, V., Gates, W.P. and Stucki, J.W., 1998. Microbial reduction of structural iron in clay – A renewable source of reduction capacity. *J. Env. Qual.* 27, 761–766.

Evangelou, V.P. and Zhang, Y.L., 1995. A review: Pyrite oxidation mechanisms and acid mine drainage prevention. *Crit. Rev. Env. Sci. Technol.* 25, 141–199.

Fredrickson, J.K., McKinley, J.P., Bjornstad, B.N., Long, P.E., Ringelberg, D.B., White, D.C., Krumholz, L.R., Suflita, J.M., Colwell, F.S., Lehman, R.M. and Phelps, T.J., 1997. Pore-size constraints on the activity and survival of subsurface bacteria in a late Cretaceous shale-sandstone sequence, northwestern New Mexico. *Geomicrobiol. J.* 14, 183–202.

Feast, N.A., Hiscock, K.M., Dennis, P.F. and Andrews, J.N., 1998. Nitrogen isotope hydrochemistry and denitrification within the chalk aquifer system of north Norfolk, UK. *J. Hydrol.* 211, 233–252.

Garrels, R.M. and Christ, C.L., 1965. *Solutions, Minerals, and Equilibria.* Harper and Row, New York, 450 pp.

Glynn, P. and Brown, J., 1996. Reactive transport modeling of acidic metal-contaminated ground water at a site with sparse spatial information. *Rev. Mineral.* 34, 377–438.

Gouy, J., Bergé, P. and Labroue, L., 1984. *Gallionella ferruginea*, facteur de dénitrification dans les eaux pauvre en matière organique. *C. R. Acad. Sc. Paris* 298, 153–156.

Grenthe, I., Stumm, W., Laaksuharju, M., Nilson, A.-C. and Wikberg, P., 1992. Redox potentials and redox reactions in deep groundwater systems. *Chem. Geol.* 98, 131–150.

Grossman, E.L., Coffman, B.K., Fritz, S.J. and Wada, H., 1989. Bacterial production of methane and its influence on ground-water chemistry in east-central Texas aquifer. *Geology* 17, 495–499.

Hansen, H.C., Koch, C.B., Krogh, H.N., Borggaard, O.K. and Sørensen, J., 1996. Abiotic nitrate reduction to ammonium: key role of green rust. *Env. Sci. Technol.* 30, 2053–2056.

Hansen, L.K., 1998. Biogeochemistry of methane in a shallow sandy aquifer. Ph.D. thesis, Lyngby, 111 pp.

Hansen, L.K., Jakobsen, R. and Postma, D., 2001. Methanogenesis in a shallow sandy aquifer, Rømø, Denmark, *Geochim. Cosmochim. Acta* 65, 2925–2935.

Herbert, R.B., 1999. Sulfide oxidation in mine waste deposits – A review with emphasis on dysoxic weathering. MIMI 1999:1, 45 pp. (www.mimi.kiruna.se).

Hering, J.G. and Stumm, W., 1990. Oxidation and reductive dissolution of minerals. In M.F. Hochella and A.F. White (eds), *Mineral water interface geochemistry*. *Rev. Mineral.* 23, 427–465.

Heron, G. and Christensen, T.H., 1995. Impact of sediment-bound iron on redox buffering in a landfill leachate polluted aquifer (Vejen, Denmark). *Env. Sci. Technol.* 29, 187–192.

Hoehler, T.M., Alperin, M.J., Albert, D.B. and Martens, C.S., 1994. Field and laboratory studies of methane oxidation in an anoxic marine sediment. *Global Biogeochem. Cycles* 8, 451–463.

Hoehler, T.M., Alperin, M.J., Albert, D.B. and Martens, C.S., 1998. Thermodynamic control on H_2 concentrations in anoxic sediments. *Geochim. Cosmochim. Acta* 62, 1745–1756.

Howarth, R.W., 1979. Pyrite: Its rapid formation in a salt marsh and its importance in ecosystem metabolism. *Science* 203, 49–51.

Howarth, R.W. and Jørgensen, B.B., 1984. Formation of ^{35}S-labelled elemental sulfur and pyrite in coastal marine sediments (Limfjorden and Kysing Fjord, Denmark) during short-term $^{35}SO_4^{2-}$ reduction measurements. *Geochim. Cosmochim. Acta* 48, 1807–1818.

Iversen, N. and Jørgensen, B.B., 1985. Anaerobic methane oxidation rates at the sulfate-methane transition in marine sediments from Kattegat and Skagerrak (Denmark). *Limnol. Oceanogr.* 30, 944–955.

Jakobsen, R. and Postma, D., 1994. In situ rates of sulfate reduction in an aquifer (Rømø, Denmark) and implications for the reactivity of organic matter. *Geology* 22, 1103–1106.

Jakobsen, R. and Postma, D., 1999. Redox zoning, rates of sulfate reduction and interactions with Fe-reduction and methanogenesis in a shallow sandy aquifer, Rømø, Denmark. *Geochim. Cosmochim. Acta* 63, 137–151.

Jakobsen, R., Albrechtsen, H.J., Rasmussen, M., Bay, H., Bjerg, P.L. and Christensen, T.H., 1998. H_2 concentrations in a landfill leachate plume (Grindsted, Denmark): In situ energetics of terminal electron acceptor processes. *Env. Sci. Technol.* 32, 2142–2148.

Jin, Y. en Jury, W.A., 1996. Characterizing the dependence of gas diffusion coefficient on soil properties. *Soil Sci. Soc. Am. J.* 60, 66–71.

Jørgensen, B.B., 1978. A comparison of methods for the quantification of bacterial sulfate reduction in coastal marine sediments. I. Measurement with radiotracer techniques. *Geomicrobiol. J.* 1, 11–27.

Kirby, C.S. and Elder Brady, J.A., 1998. Field determination of Fe^{2+} oxidation in acid mine drainage using a continuously-stirred tank reactor. *Appl. Geochem.* 13, 509–520.

Korom, S.F., 1992. Natural denitrification in the saturated zone: a review. *Water Resour. Res.* 28, 1657–1668.

Kölle, W., Strebel, O. and Böttcher, J., 1987. Reduced sulfur compounds in sandy aquifers and their interactions with groundwater. Int. Symp. Groundwater Monitoring, Dresden, 12 pp.

Kölle, W., Werner, P., Strebel, O. and Böttcher, J., 1983. Denitrifikation in einem reduzierenden Grundwasserleiter. *Vom Wasser* 61, 125–147.

Krauskopf, K.B., 1979. *Introduction to Geochemistry*. McGraw-Hill, New York, 617 pp.

Krumbein, W.E. (ed.), 1983. *Microbial Geochemistry*. Blackwell, Oxford, 330 pp.

Kuivila, K.M., Murray, J.W., Devol, A.H. and Novelli, P.C., 1989. Methane production, sulfate reduction and competition for substrates in the sediments of Lake Washington. *Geochim. Cosmochim. Acta* 53, 409–416.

LaKind, J.S. and Stone, A.T., 1989. Reductive dissolution of goethite by phenolic reductants. *Geochim. Cosmochim. Acta* 53, 961–971.

Langmuir, D., 1969. The Gibbs free energies of substances in the system Fe-O_2-H_2O-CO_2 at 25 °C. U.S. Geol. Surv. Prof. Paper, 650-B, B180–B184.

Lansdown, J.M., Quay, P.D. and King, S.L., 1992. CH_4 production via CO_2 reduction in a temperate bog: A source of ^{13}C-depleted CH_4. *Geochim. Cosmochim. Acta* 56, 3493–3503.

Larsen, O. and Postma, D., 2001. Kinetics of reductive bulk dissolution of lepidocrocite, ferrihydrite and goethite. *Geochim. Cosmochim. Acta* 65, 1367–1379.

Lawrence, A.R., Gooddy, D.C., Kanatharana, P., Meeslip, W. and Ramnarong, P., 2000. Groundwater evolution beneath Hat Yai, a rapidly developing city in Thailand. *Hydrogeol. J.* 8, 564–575

Leuchs, W., 1988. *Vorkommen, Abfolge und Auswirkungen anoxischer redoxreaktionen in einem pleistozanen Porengrundwasserleiter*. Bes. Mitt. dt. Gewässerk. Jb. 52, 106 pp.

Lindberg, R.D. and Runnels, D.D., 1984. Ground water redox reactions: An analysis of equilibrium state applied to Eh measurements and geochemical modeling. *Science* 225, 925–927.

Love, L.G. and Amstutz, G.C., 1966. Review of microscopic pyrite from the Devonian Chattanooga Shale and Rammelsberg Banderz. *Fortschr. Miner.* 43, 273–309.

Lovley, D.R. and Goodwin, S., 1988. Hydrogen concentrations as an indicator of the predominant terminal electron-accepting reactions in aquatic sediments. *Geochim. Cosmochim. Acta* 52, 2993–3003.

Lovley, D.R., Baedecker, M.J., Lonergan, D.J., Cozzarelli, I.M., Phillips, E.J.P. and Siegel, D.I., 1989. Oxidation of aromatic compounds coupled to microbial iron reduction. *Nature* 339, 297–299.

Lovley, D.R., Phillips, E.J.P. and Lonergan, D.J., 1991. Enzymatic versus nonenzymatic mechanisms for Fe(III) reduction in aquatic sediment. *Env. Sci. Technol.* 25, 1062–1067.

Lowson, R.T., 1982. Aqueous oxidation of pyrite by molecular oxygen. *Chem. Rev.* 82, 461–497.

Ludvigsen, L., Albrechtsen, H.-J., Heron, G., Bjerg, P.L. and Christensen, T.H., 1998. Anaerobic microbial redox processes in a landfill leachate contaminated aquifer (Grindsted, Denmark). *J. Contam. Hydrol.* 4, 1–26.

Lyngkilde, J. and Christensen, T.H., 1992. Redox zones of a landfill leachate pollution plume (Vejen, Denmark). *J. Contam. Hydrol.* 10, 273–289.

Macalady, D.L., Langmuir, D., Grundl, T. and Elzerman, A., 1990. Use of model-generated Fe^{3+} ion activities to compute Eh and ferric oxyhydroxide solubilities in anaerobic systems. In D.C. Melchior and R.L. Basset (eds), *Chemical modelling of aqueous systems II*. ACS Symp. Ser. 416, 350–367.

Magaritz, M. and Luzier, J.E., 1985. Water-rock interactions and seawater-freshwater mixing effects in the coastal dunes aquifer, Coos Bay, Oregon. *Geochim. Cosmochim. Acta* 49, 2515–2525.

Marnette, E.C.L., Van Breemen, N., Hordijk, K.A. and Cappenberg, T.E., 1993. Pyrite formation in two freshwater systems in the Netherlands. *Geochim. Cosmochim. Acta* 57, 4165–4177.

Massmann, J. and Farrier, D.F., 1992. Effects of atmospheric pressure on gas transport in the vadose zone. *Water Resour. Res.* 28, 777–791.

McMahon, P.B. and Chapelle, F.H., 1991. Microbial production of organic acids in aquitard sediments and its role in aquifer geochemistry. *Nature* 349, 233–235.

Mehra, O.P. and Jackson, M.L., 1960. Iron oxide removal from soils and clays by a dithionite-citrate sytem buffered with sodium bicarbonate. *Clays Clay Min.* 5, 317–327.

Middelburg, J.J., 1989. A simple rate model for organic matter decomposition in marine sediments. *Geochim. Cosmochim. Acta* 53, 1577–1581.

Morin, K.A. and Cherry, J.A., 1986. Trace amounts of siderite near a uranium-tailings impoundment, Elliot Lake, Ontario, Canada, and its implication in controlling contaminant migration in a sand aquifer. *Chem. Geol.* 56, 117–134.

Moses, C.O., Nordstrom, D.K., Herman, J.S. and Mills, A.L., 1987. Aqueous pyrite oxidation by dissolved oxygen and ferric iron. *Geochim. Cosmochim. Acta* 51, 1561–1572.

Mukherjee, M., Sahu, S.J., Roy, D., Jana., J., Bhattacharya, R., Chatterjee, D., De Dalal, S.S., Bhattacharya, P. and Jacks, G., 2000. The governing geochemical processes responsible for mobilisation of arsenic in sedimentary aquifer of Bengal Delta Plain. In P.L. Bjerg., P. Engesgaard and Th.D. Krom (eds), *Groundwater 2000*, 201–202, Balkema, Rotterdam.

Nevin, K.P. and Lovley, D.R., 2000. Potential for nonenzymatic reduction of Fe(III) via electron shuttling in subsurface sediments. *Env. Sci. Technol.* 34, 2472–2478.

Nicholson, R.V., Gillham, R.W. and Reardon, E.J., 1988. Pyrite oxidation in carbonate-buffered solution: 1. Experimental kinetics. *Geochim. Cosmochim. Acta* 52, 1077–1085.

Nicholson, R.V., Gillham, R.W. and Reardon, E.J., 1990. Pyrite oxidation in carbonate-buffered solution: 2. Rate control by oxide coatings. *Geochim. Cosmochim. Acta* 54, 395–402.

Nordstrom, D.K., 1982. Aqueous pyrite oxidation and the consequent formation of secondary iron minerals. In D.K. Nordstrom (ed.), *Acid Sulphate Weathering*. Soil Sci. Soc. Am. Spec. Pub. 10, 38–56.

Nordstrom, D.K. and Munoz, J.L., 1994. *Geochemical thermodynamics,* 2nd ed. Blackwell, Oxford, 493 pp.

Nordstrom, D.K., Jenne, E.A. and Ball, J.W., 1979. Redox equilibria of iron in acid mine waters. In E.A. Jenne (ed.), *Chemical Modelling in Aqueous Systems*. ACS Symp. Ser. 93, 49–79.

Nordstrom, D.K., Alpers, C.N., Ptacek, C.J. and Blowes, D.W., 2000. Negative pH and extremely acidic mine waters from Iron Mountain, California. *Env. Sci. Technol.* 34, 254–258.

Obermann, P., 1982. *Hydrochemische/hydromechanische Untersuchungen zum Stoffgehalt von Grundwasser bei landwirtschaftlicher Nutzung*. Bes. Mitt. dt. Gewässerk. Jb. 42.

Perry, K.A. and Pedersen, T.F., 1993. Sulphur speciation and pyrite formation in meromictic ex-fjords. *Geochim. Cosmochim. Acta* 57, 4405–4418.

Phelps, T.J. and Zeikus, J.G., 1984. Influence of pH on terminal carbon metabolism in anoxic sediments from a mildly acidic lake. *Appl. Env. Microbiol.* 48, 1088–1095.
Plummer, L.N., 1977. Defining reactions and mass transfer in part of the Floridan aquifer. *Water Resour. Res.* 13, 801–812.
Postma, D., 1977. The occurrence and chemical composition of recent Fe-rich mixed carbonates in a river bog. *J. Sed. Petrol.* 47, 1089–1098.
Postma, D., 1981. Formation of siderite and vivianite and the pore-water composition of a recent bog sediment in Denmark. *Chem. Geol.* 31, 255–244.
Postma, D., 1982. Pyrite and siderite formation in brackish and freshwater swamp sediments. *Am. J. Sci.* 282, 1151–1183.
Postma, D., 1983. Pyrite and siderite oxidation in swamp sediments. *J. Soil Sci.* 34, 163–182.
Postma, D., 1990. Kinetics of nitrate reduction by detrital Fe(II)-silicates. *Geochim. Cosmochim. Acta* 54, 903–908.
Postma, D., 1993. The reactivity of iron oxides in sediments: A kinetic approach. *Geochim. Cosmochim. Acta* 57, 5027–5034.
Postma, D. and Brockenhuus-Schack, B.S., 1987. Diagenesis of iron in proglacial sand deposits of late- and post-Weichselian age. *J. Sed. Petrol.* 57, 1040–1053.
Postma, D. and Jakobsen, R., 1996. Redox zonation: Equilibrium constraints on the Fe(III)/SO_4-reduction interface. *Geochim. Cosmochim. Acta.* 60, 3169–3175.
Postma, D., Boesen, C., Kristiansen, H. and Larsen, F., 1991. Nitrate reduction in an unconfined sandy aquifer: Water chemistry, reduction processes, and geochemical modeling. *Water Resour. Res.* 27, 2027–2045.
Ptacek, C.J. and Blowes, D.W., 1994. Influence of siderite on the pore-water chemistry of inactive mine-tailings impoundments. In C.N. Alpers and D.W. Blowes (eds), *Environmental Geochemistry of sulfide oxidation.* ACS Symp. Ser. 550, 172–189.
Rickard, D., 1997. Kinetics of pyrite formation by the H_2S oxidation of iron(II)monosulfide in aqueous solutions between 25 and 125°C: The rate equation. *Geochim. Cosmochim. Acta* 61, 115–134.
Rimstidt, J.D. and Vaughan, D.J., 2003. Pyrite oxidation: A state-of-the-art assessment of the reaction mechanism. *Geochim. Cosmochim. Acta* 67, 873–880.
Robertson, W.D., Cherry, J.A. and Schiff, S.L., 1989. Atmospheric sulfur deposition 1950–1985 inferred from sulfate in groundwater. *Water Resour. Res.* 25, 1111–1123.
Robertson, W.D., Russel, B.M. and Cherry, J.A., 1996. Attenuation of nitrate in aquitard sediments of southern Ontario, *J. Hydrol.* 180, 267–281.
Russell, E.W., 1973. Soil conditions and plant growth. Longman, London, 849 pp.
Ryan, J.N. and Gschwend, P.M., 1991. Extraction of iron oxides from sediments using reductive dissolution by titanium(III). *Clays Clay Min.* 39, 509–518.
Ryan, J.N. and Gschwend, P.M., 1992. Effect of iron diagenesis on the transport of collodial clay in an unconfined sand aquifer. *Geochim. Cosmochim. Acta* 56, 1507–1521.
Schoonen, M.A.A. and Barnes, H.L., 1991. Reactions forming pyrite and marcasite from solution: II Via FeS precursor below 100°C. *Geochim. Cosmochim. Acta* 55, 1505–1514.
Schott, J. and Berner, R.A., 1983. X-ray photoelectron studies of the mechanism of iron silicate dissolution during weathering. *Geochim. Cosmochim. Acta* 47, 2233–2240.
Schwertmann, U., 1964. Differenzierung der Eisenoxide des Bodens durch Extraktion mit Ammoniumoxalatlösung. *Z. Pflanzenern. Bodenk.* 105, 194–202.
Simpkins, W.W. and Parkin, T.B., 1993. Hydrogeology and redox geochemistry of CH_4 in a late Wisconsinan till and loess sequence in central Iowa. *Water Resour. Res.* 29, 3643–3657.
Smedley, P.L., Nicolli, H.B., Barros, A.J. and Tullio, J.O., 1998. Origin and mobility of arsenic in groundwater from the Pampean Plain. In G.B. Arehart and J.R. Hulston (eds), *Proc. 9th Water Rock Interaction Symp.*, 275–278, Balkema, Rotterdam.
Smith, R.L. and Duff, J.H., 1988. Denitrification in a sand and gravel aquifer. *Appl. Env. Microbiol.* 54, 1071–1078.
Smith, R.L., Howes, B.L. and Duff, J.H., 1991a. Denitrification in nitrate-contaminated groundwater: occurrence in steep vertical geochemical gradients. *Geochim. Cosmochim. Acta* 55, 1815–1825.
Smith, R.L., Howes, B.L. and Garabedian, S.P., 1991b. In situ measurement of methane oxidation in groundwater by using natural-gradient tracer tests. *Appl. Env. Microbiol.* 57, 1997–2004.

Smith, R.L., Garabedian, S.P. and Brooks, M. H., 1996. Comparison of denitrification activity measurements in groundwater using cores and natural-gradient tracer tests. *Env. Sci. Technol.* 30, 3448–3456.

Spalding, R.F. and Exner, M.E., 1993. Occurrence of nitrate in groundwater – a review. *J. Env. Qual.* 22, 392–402.

Starr, R.C. and Gillham, R.W., 1993. Denitrification and organic carbon availability in two aquifers. *Ground Water* 31, 934–947.

Stoessell, R.K., Moore, Y.H. and Coke, J.G., 1993. The occurrence and effect of sulfate reduction and sulfide oxidation on coastal limestone dissolution in Yucatan cenotes. *Ground Water* 31, 566–575.

Strebel, O., Böttcher, J. and Kölle, W., 1985. Stoffbilanzen im Grundwasser eines Einzugsgebietes als Hilfsmittel bei Klärung und Prognose von Grundwasserqualitätsproblemen (Beispiel Fuhrberger Feld). *Z. dt. geol. Ges.* 136, 533–541.

Strebel, O., Duynisveld, W.H.M. and Böttcher, J., 1989. Nitrate pollution of groundwater in western Europe. *Agr. Ecosys. Env.* 26, 189–214.

Stumm, W., 1992. *Chemistry of the solid-water interface*. Wiley and Sons, New York, 428 pp.

Stumm, W. and Morgan, J.J., 1996. *Aquatic chemistry*, 3rd ed. Wiley and Sons, New York, 1022 pp.

Sugimoto, A. and Wada, E., 1995. Hydrogen isotopic composition of bacterial methane: CO_2/H_2 reduction and acetate fermentation. *Geochim. Cosmochim. Acta* 59, 1329–1337.

Sulzberger, B., Suter, D., Siffert, C., Banwart, S. and Stumm, W., 1989. Dissolution of Fe(III)-(hydro)oxides in natural waters: laboratory assessment on the kinetics controlled by surface coordination. *Mar. Chem.* 28, 127–144.

Tesoriero, A.J., Liebscher, H. and Cox, S.E., 2000. Mechanism and rate of denitrification in an agricultural watershed: Electron and mass balance along groundwater flow paths. *Water Resour. Res.* 36, 1545–1559.

Thamdrup, B., 2000. Bacterial manganese and iron reduction in aquatic sediments. In B. Schink (ed.), *Adv. Microbial Ecol.* 16, 41–84.

Thorstenson, D.C. and Pollock, D.W., 1989. Gas transport in unsaturated zones: Multicomponent systems and the adequacy of Fick's laws. *Water Resour. Res.* 25, 477–507.

Thorstenson, D.C., Fisher, D.W. and Croft, M.G., 1979. The geochemistry of the Fox Hill-basal Hell Creek aquifer in the southwestern North Dakota and northwestern South Dakota. *Water Resour. Res.* 15, 1479–1498.

Tuccillo, M.E., Cozzarelli, I.M. and Herman, J.S., 1999. Iron reduction in the sediments of a hydrocarbon-contaminated aquifer. *Appl. Geochem.* 14, 655–667.

Van Beek, C.G.E.M. and Van der Kooij, D., 1982. Sulfate-reducing bacteria in ground water from clogging and nonclogging shallow wells in the Netherlands river region. *Ground Water*, 20, 298–302.

Van Beek, C.G.E.M., Boukes, H., Van Rijsbergen, D. and Straatman, R., 1988. The threat of the Netherlands waterworks by nitrate in the abstracted groundwater, as demonstrated on the well field Vierlingsbeek. *Water Supply* 6, 313–318.

Van Beek, C.G.E.M., Hettinga, F.A.M. and Straatman, R., 1989. The effects of manure spreading and acid deposition upon groundwater quality in Vierlingsbeek, the Netherlands. IAHS Pub. 185, 155–162.

Van Berk, W. and Wisotzky, F., 1995. Sulfide oxidation in brown coal overburden and chemical modeling of reactions in aquifers influenced by sulfide oxidation. *Env. Geol.* 26, 192–196.

Van Breemen, N., 1976. *Genesis and solution chemistry of acid sulfate soils in Thailand*. Agric. Res. Rep., Wageningen, 263 pp.

Vogels, G.D., Keltjens, J.T. and Van der Drift, C., 1988. Biochemistry of methane production. In A.J.B. Zehnder (ed.), *op. cit.*, 707–770.

Vroblesky, D.A. and Chapelle, F.H., 1994. Temporal and spatial changes of terminal electron-accepting processes in a petroleum hydrocarbon-contaminated aquifer and the significance of contaminant biodegradation. *Water Resour. Res.* 30, 1561–1570.

Waldron, S., Landsdown, J.M., Scott, E.M., Fallick, A.E. and Hall, A.J., 1999. The global influence of the hydrogen isotope composition of water on that of bacteriogenic methane from shallow freshwater environments. *Geochim. Cosmochim. Acta* 63, 2237–2245.

Walton-Day, K., Macalady, D.L., Brooks, M.H. and Tate, V.T., 1990. Field methods for measurement of ground water redox chemical parameters. *Ground Water Mon. Rev.* 10, 81–89.

Wang, Q. and Morse, J.W., 1996. Pyrite formation under conditions approximating those in anoxic sediments: I. Pathway and morphology. *Mar. Chem.* 52, 99–121.

Wersin, P., Höhener, P., Giovanoli, R. and Stumm, W., 1991. Early diagenetic influences on iron transformations in a freshwater lake sediment. *Chem. Geol.* 90, 233–252.

White, A.F., 1990. Heterogeneous electrochemical reactions associated with oxidation of ferrous oxide and silicate surfaces. *Rev. Mineral.* 23, 467–509.

White, J.R., Gubala, C.P., Fry, B., Owen, J. and Mitchell, M.J., 1989. Sediment biogeochemistry of iron and sulfur in an acidic lake. *Geochim. Cosmochim. Acta* 53, 2547–2559.

Whiticar, M.J., 1999. Carbon and hydrogen isotope systematics of bacterial formation and oxidation of methane. *Chem. Geol.* 161, 291–314.

Whiticar, M.J., Faber, E. and Schoell, M., 1986. Biogenic methane formation in marine and freshwater environments: CO_2 reduction vs. acetate fermentation-isotope evidence. *Geochim. Cosmochim. Acta* 50, 693–709.

Wiese, R.G., Powell, M.A. and Fyfe, W.S., 1987. Spontaneous formation of hydrated iron sulfates on laboratory samples of pyrite- and marcasite-bearing coals. *Chem. Geol.* 63, 29–38.

Williamson, M.A. and Rimstidt, J.D., 1994. The kinetics and electrochemical rate-determining step of aqueous pyrite oxidation. *Geochim. Cosmochim. Acta* 58, 5443–5454.

Wilson, G.B., Andrews, J.N. and Bath, A.H., 1994. The nitrogen isotope composition of groundwater nitrates from the East Midlands Triassic Sandstone aquifer, England. *J. Hydrol.* 157, 35–46.

Winograd, I.J. and Robertson, F.N., 1982. Deep oxygenated ground water: Anomaly or common occurrence. *Science* 216, 1227–1230.

Wisotzky, F., 1994. *Untersuchungen zur Pyritoxidation in Sedimenten des Rheinischen Braunkohlenreviers und deren Auswirkungen af die Chemie des Grundwassers*. Bes. Mitt. dt. Gewässerk. Jb. 58, 153 pp.

Wisotzky, F. and Eckert, P., 1997. Sulfat-dominierter BTEX-Abbau im Grundwasser eines ehemaligen Gaswerksstandortes. *Grundwasser* 2, 11–20.

Yao, W. and Millero, F.J., 1996. Oxidation of hydrogen sulfide by hydrous Fe(III) oxides in seawater. *Mar. Chem.* 52, 1–16.

Zehnder, A.J.B. (ed.), 1988. *Biology of anaerobic microorganisms*. Wiley and Sons, New York, 872 pp.

Zhang, C., Grossman, E.L. and Ammerman, J.W., 1998. Factors influencing methane distribution in Texas ground water. *Ground Water* 36, 58–66.

Zinder, B., Furrer, G. and Stumm, W., 1986. The coordination chemistry of weathering: II Dissolution of Fe(III) oxides. *Geochim. Cosmochim. Acta* 50, 1861–1869.

10

Pollution by Organic Chemicals

We cannot think of a modern society without manufactured organic chemicals. Their production and application is now regulated and spilling in the environment happens, at least in principle, only by accident. However, a few decades ago dumping of industrial organic chemicals at a waste site was common practice, storage tanks with gasoline and petrol were usually leaking into the underlying soil, and agricultural chemicals were spread over the land without considering that groundwater might be polluted by their application and rendered undrinkable. Figure 10.1 shows the number of wells in Denmark that were closed because the concentrations of nitrates, pesticides or other chemicals exceeded drinking water limits. The large increase since 1993 of wells abandoned due to pesticides is highly disturbing.

The pollutants of former practice and of recent accidents may still reside in the subsoil and will not have traveled far, given that the flow velocity of groundwater is usually limited to a few meters per year. To describe and model the hazards of transport of organic chemicals into and with groundwater, we consider in this chapter (Figure 10.2; Tiktak et al., 2000) volatilization as a mechanism that reduces the mass that remains in the soil (but it may pollute the atmosphere and cause a larger spreading), leaching into groundwater, sorption that retards the movement and lastly, transformation and decay by chemical and microbiological reactions.

10.1 GAS–WATER EXCHANGE

The volatilization of organic chemicals is important because these pollutants can form individual, neutral gas molecules, which contrasts with most elements that dissolve as ions in water.

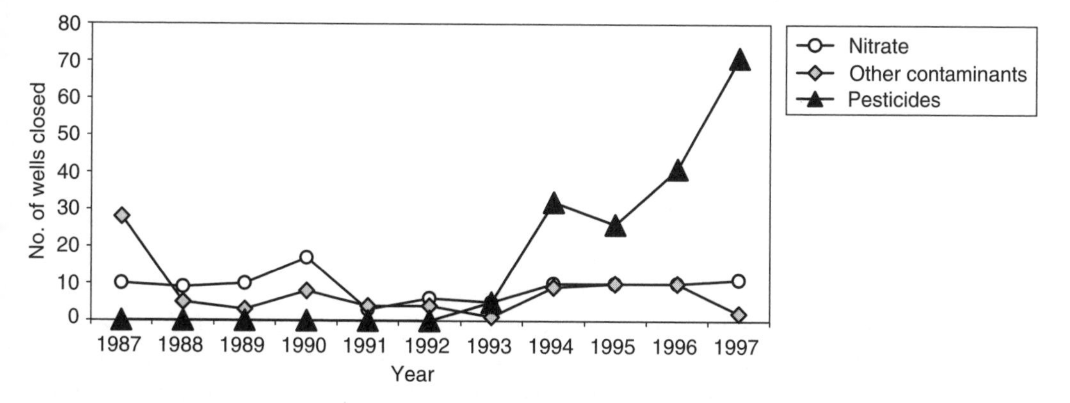

Figure 10.1. Abandoned groundwater wells in Denmark due to excess of pesticides, nitrate, or other chemicals (Miljøprojekt, 1998).

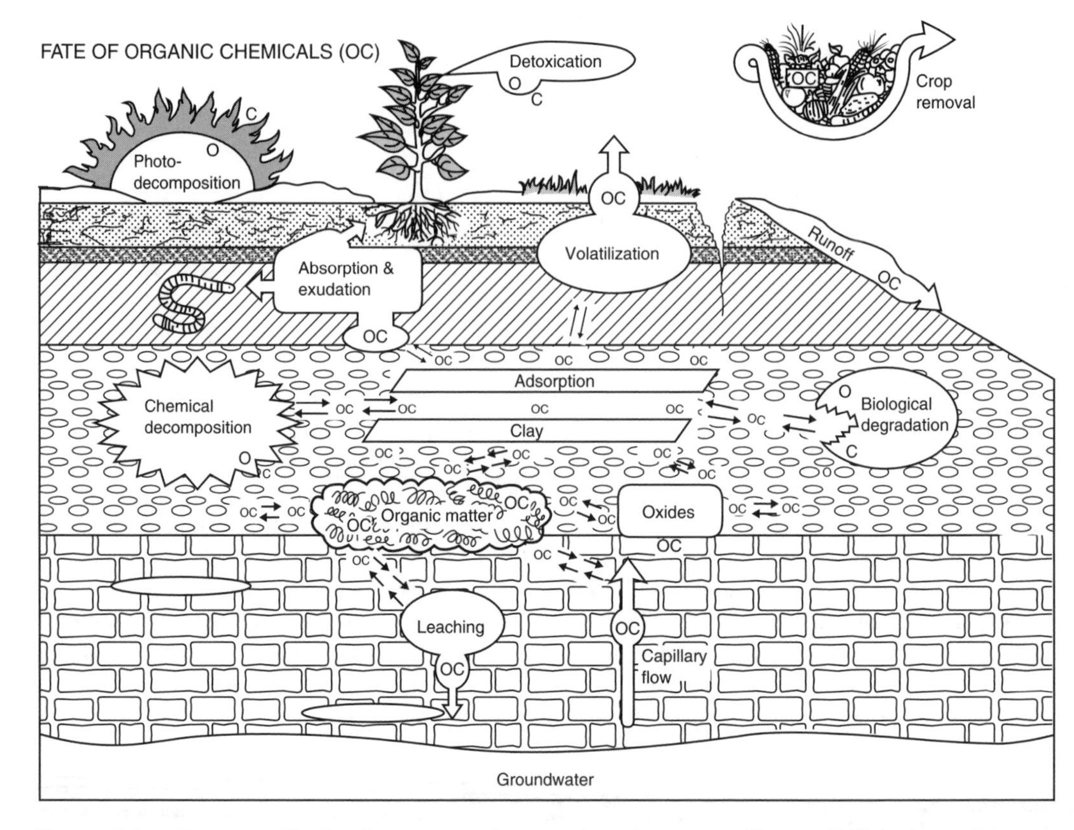

Figure 10.2. Processes affecting the spread and attenuation of organic pollutants (OC) in the environment (Weber and Miller, 1989).

Volatilization is a function of the difference between the actual gas pressure of a substance and the gas pressure in equilibrium with its concentration in water. The equilibrium concentrations are related by the law of mass action (Henry's law for the link between gas pressure and solute concentration).

For example, for benzene (C_6H_6):

$$C_6H_{6(aq)} \leftrightarrow C_6H_{6(g)} \tag{10.1}$$

with

$$K_H = \frac{P_{C_6H_6}}{m_{C_6H_6\,(aq)}} = 5.5 \tag{10.2}$$

where K_H has the units atm · L/mol (K in the law of mass action would be non-dimensional). Often, the gas concentration is expressed in ppmv (parts per million by volume, equal to 10^{-6} atm for a total pressure of 1 atm). For example, in urban air the benzene concentration may be 0.01 ppmv, which yields according to Equation (10.2) an equilibrium concentration of 1.8 nM or 0.14 μg benzene/L in the ponds of the city.

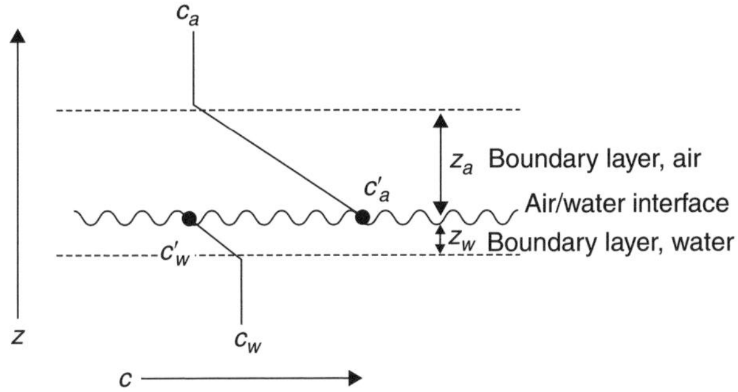

Figure 10.3. Diffusion of a chemical through two boundary layers at the gas–water interface. At the interface itself, the concentrations in air, c'_a, and in water, c'_w, are in equilibrium: $K'_H = c'_a / c'_w$.

The gas–water exchange rate can be calculated with diffusion models (Danckwerts, 1970; Schwarzenbach et al., 1993). The major difficulty of modeling the process is to characterize the interface where the mass transfer takes place. In the two-film model of Liss and Slater (1974), the diffusive flux is calculated for two boundary layers at the air / water interface, cf. Figure 10.3.

In this model, the concentration in air (c_a) is expressed in mol/L air, i.e. the pressure of the chemical (in atm) is divided by the gas constant ($R = 0.08206$ atm · L/mol/K) and the temperature (K) ($RT = 24$ atm · L/mol at 20°C). At the interface, the gas and water concentration are considered to be in equilibrium,

$$K'_H = \frac{K_H}{RT} = \frac{c'_{\text{air}}}{c'_w} \tag{10.3}$$

where K'_H has the dimension (L water) / (L air), i.e. indicates how many liters of water contain the same mass of chemical as 1 liter air at equilibrium. For the model, we eliminate the concentrations at the interface as follows.

The flux through the water boundary-layer is given by the concentration gradient multiplied with the diffusion coefficient:

$$F_w = -D_f \frac{c'_w - c_w}{z_w} \quad (\text{mol/m}^2/\text{s}) \tag{10.4}$$

where z_w is the thickness of the (stagnant) boundary layer of water. Likewise, the flux through the air boundary-layer is:

$$F_a = -D_a \frac{c_{\text{a}} - c'_{\text{a}}}{z_a} \tag{10.5}$$

At steady state, the two fluxes are equal ($F_w = F_a = F$). By introducing the interface equilibrium (Equation 10.3), the concentrations at the interface can be expressed in the (known) bulk concentrations and eliminated from the combination of Equations (10.4) and (10.5). Thus, c'_w in the water film is:

$$c'_w = \frac{(D_a / z_a) c_a + (D_f / z_w) c_w}{K'_H D_a / z_a + D_f / z_f} \tag{10.6}$$

This equation is inserted in the flux through the water layer (Equation 10.4) to obtain:

$$F_w = \frac{1}{z_w / D_f + z_a / (D_a K'_H)} \left(c_w - \frac{c_a}{K'_H} \right) \tag{10.7}$$

The term $(c_w - c_a / K'_H) \equiv (c_w - c_{w\,(\text{in equilibrium with air})})$ is the concentration difference that drives the mass transport, while the denominator in (10.7) represents the resistance to transport.

Similarly, by eliminating the concentration c'_a in the flux through the air boundary-layer (Equation 10.5) we find:

$$F_a = \frac{1}{z_w K'_H / D_f + z_a / D_a} \left(c_w K'_H - c_a \right) \tag{10.8}$$

In Equation (10.8) it is the concentration difference in the air boundary-layer that drives the mass transport, but the same flux is obtained as with Equation (10.7).

The two terms in the denominator of Equations (10.7) and (10.8) represent the resistances against diffusive transport in the water and air films (resistances may be added, as in an electrical circuit). The inverse of this resistance has the dimension m/s, and indicates a velocity. Thus, Equation (10.7) has two velocity terms, for the water layer $v_w = D_f / z_w$, and for the air boundary layer $v_a = D_a / (z_a \; K'_H)$, in which the diffusion coefficients are pretty well fixed ($D_f \approx 10^{-9}\,m^2/s$ and $D_a \approx 10^{-5}\,m^2/s$), but where the thickness of the diffusion layers needs to be defined. To do so, let us calculate the evaporation of water with these formulas. For water molecules in the water film, the resistance for transport will be near-zero since the layer itself consists of water. In other words, the transfer velocity is infinite, $v_w = D_f / z_w = \infty$, or $z_w = 0$. Thus, evaporation is determined solely by the transfer velocity v_a through the air film. The evaporation of water can be measured with an evaporation pan, and, depending on the wind speed, the incoming radiation, the temperature and the moisture content of the air, it ranges from 0.6–3 m/yr at various locations on earth. We can now estimate the thickness of the air boundary-layer z_a to range from 1–5 mm, with an average of about 3 mm for small open waterbodies in the temperate climate.

EXAMPLES

Find the transfer velocity in the air film, $z_a = 3\,mm$, $D_a = 10^{-5}\,m^2/s$.

ANSWER: $v_a = 3.3 \times 10^{-3}\,m/s$.

Estimate for this v_a the yearly water evaporation rate E in air with a relative humidity of 50% at 25°C, $K'_H = 2.3 \times 10^{-5}$ L water/L air.

ANSWER: $E = F_a / c_{w,H_2O} = v_a(K'_H - K'_H / 2) = 3.3 \times 10^{-3} \times (2.3 \times 10^{-5} - 1.65 \times 10^{-5}) = 3.8 \times 10^{-8}\,m/s = 1.2\,m/yr$.

Estimate the theoretical evaporation rate when limited by resistance in the water layer, $z_w = 20\,\mu m$ for the oceans.

ANSWER: $E_{theor} = v_w = D_f / z_w = 1734\,m/yr$.

Calculate the thermodynamic K for the reaction $H_2O_{(l)} \leftrightarrow H_2O_{(g)}$.

ANSWER: $K = (2.3 \times 10^{-5}$ L water/L air) $\times$ (24 atm·L air/mol) $\times$ (55.6 mol/L water) = 0.03(−). Compare with PHREEQC.DAT where the reaction is written as $H_2O_{(g)} \leftrightarrow H_2O_{(l)}$.

Calculating a water evaporation rate from the resistance in the water film does not make sense, as can be seen from the example calculation above. However, the calculation does show that the flux is not only determined by the concentration gradient, but also by the concentration in the medium. In the case of water, the air concentration is small, as expressed by the small Henry constant of 2.3×10^{-5} L water/L air. On the other hand, for sparingly soluble gases the concentration in the water film is much less than in the air film, for example for O_2, $K'_H = 33$ L water/L air. Transport is now limited by the water film, and again, the thickness of the boundary layer can be determined experimentally. It is found to depend strongly on the wind speed in the form (Schwarzenbach et al., 1993):

$$v_{wO_2} = 4\times10^{-7}(v_{10})^2 + 4\times10^{-6}(\text{m/s}) \quad (10.9)$$

where v_{10} is the wind speed in m/s at 10 m above the water surface. The quadratic dependence on windspeed is related to the fact that the stress of the wind on the water surface, and hence the force tending to reduce the liquid phase resistance, is proportional to the square of the friction velocity (Liss, 1973; but it may be cubic, Wanninkhof and McGillis, 1999). With $D_{f,O_2} = 1.1\times10^{-9}$ m^2/s, z_w is in the range of 20–200 μm (note that mass transfer over the ocean/air interface ($z_w \approx 20$ μm) is relatively large because of wind and waves). It is assumed that the transfer velocity for other chemicals varies in proportion to the ratio of the diffusion coefficients (i.e. the film thickness is the same for all). Multiplying the velocity with the surface area permits to derive the total mass exchange, e.g. of freon that enters the oceans (Example 10.1).

EXAMPLE 10.1. *Estimate the flux of freon-11 (CCl_3F) into the sea (Liss and Slater, 1974)*
The concentration of freon in the marine atmosphere was 5×10^{-5} ppmv in 1973, and in surface seawater 7.6×10^{-9} cm^3 freon/L was found. $K'_H = 5$ L water/L air. The surface area of the oceans is 3.6×10^{14} m^2. Find the flux in or out of the oceans relative to oxygen with $v_{wO_2} = 5.5\times10^{-5}$ (m/s), and compare with the yearly production of about 3×10^{11} g/yr (in 1973).

ANSWER:
First express 'ppmv' in concentration per L air, then calculate with Henry's constant the equilibrium concentration per L water and compare with the analyzed concentration.

The atmospheric concentration of 5×10^{-5} ppmv = $5\times10^{-5} \times (10^{-6}$ L/ppmv$) \times (10^3$ cm^3/L$)$ = 5×10^{-8} cm^3 CCl_3F/L air. Water in equilibrium would contain 1×10^{-8} cm^3 CCl_3F/L water which is higher than analyzed. Thus, the flux is into the ocean.

For the given K'_H, the transfer velocity is limited by the water film and can be calculated relative to oxygen. The ratio of the diffusion coefficients, $D_{f,CCl_3F}/D_{f,O_2} = (32\times(138+18)/(138\times(32+18))^{1/2} = 0.85$ (cf. Equation 3.60).
The flux is:

$$F = 0.85\, v_{wO_2}\, \Delta c = 0.85\times5.5\times10^{-5}\,\text{m/s}\times(10^{-8} - 0.76\times10^{-8}\,\text{dm}^3/\text{m}^3)$$
$$= 1.12\times10^{-13}\,\text{dm}^3/\text{m}^2/\text{s}.$$

Multiply with the surface area of the oceans and s/yr, and find a total transfer of 1.29×10^9 dm^3 freon/yr. 1 dm^3 gas is $1/RT = 0.0409$ mol, meaning that we had a flow of 7.2×10^9 g freon or 2.4% of the world production into the oceans in 1973.

QUESTION:
The freon concentration in surface sea water was 3.5 pM in 1993 (Yeats and Measures, 1998); find the concentration in air from Figure 3.9 and estimate the flux.

ANSWER: $(3.5\times10^{-12}$ mol/L$) \times (24{,}400$ cm^3 air/mol$) = 8.5\times10^{-8}$ cm^3 freon/L. From Figure 3.9, freon in air was 3×10^{-4} ppmv in 1993, and the equilibrium concentration in seawater should be 6×10^{-8} cm^3 freon/L. The flux would be out of the oceans, but the depth profiles in the sea indicate a downward flux: apparently the data do not agree, or are insufficiently accurate.

QUESTION:
For which value of K'_H are the transfer velocities in the water and the air film about equal? Assume $z_a = 3$ mm, $z_w = 0.25$ mm.

ANSWER: $K'_H = (z_a / D_a) / (z_w / D_f) = 2 \times 10^{-3}$ L water/L air.

It is not very likely that the film thickness is the same for all substances that pass the air-water interface. Another model considers the distance covered by chemicals by diffusion, $\sigma = \sqrt{(2Dt)}$, and assumes that the mass transfer is related by the square root of the diffusion coefficients of the chemicals (Danckwerts, 1970). For the case of freon in Example 10.1, it would mean that the transfer velocity of freon is $\sqrt{0.85} = 0.92$ less than of oxygen, which is not really much different. In experimental studies, the relationship lies mostly in between a square root and a linear one (Schwarzenbach et al., 1993).

For chemicals that form complexes in water, the water transfer rate may be increased by transport of the complexes. For example for CO_2 exchange, the concentration difference in the water film is:

$$\Delta c = (m_{CO_2} - m'_{CO_2}) + (m_{HCO_3^-} - m'_{HCO_3^-}) + (m_{CO_3^{2-}} - m'_{CO_3^{2-}}) \tag{10.10}$$

At equilibrium, $m_{HCO_3^-} = m_{CO_2} \times K_1 / [H^+]$ and $m_{CO_3^{2-}} = m_{CO_2} \times K_1K_2 / [H^+]^2$ for the bulk solution, and similar for the interface. Hence, Equation (10.10) becomes:

$$\Delta c = (1 + K_1 / [H^+] + K_1K_2 / [H^+]^2)(m_{CO_2} - m'_{CO_2}) = (\alpha^{-1})(m_{CO_2} - m'_{CO_2}) \tag{10.11}$$

where α is the speciation factor (Equation 5.13).

Thus, at pH $= 6.3$ and 8.3, the transfer rate would increase by factors of 2 and 102, respectively. However, the observed pH dependency of the CO_2 transfer is much smaller, which indicates that the conversion among the species and to CO_2 (which only can pass the interface) is relatively slow. Notably, the dehydration of HCO_3^- is rate limiting.

By comparing *characteristic times*, the rates of mass transfer over the interface and of hydration of CO_2 can be compared. For transport through the water layer with $z_w = 100$ μm, the residence time is $\tau = z_w / v_w = 10^{-4} / 5.5 \times 10^{-5} = 1.8$ s. The dehydration reaction of HCO_3^- has a half-life of approximately 100 s at pH $= 8.3$ (Problem 10.4), much longer than the residence time. Thus, HCO_3^- gives only a very small contribution to CO_2 transfer.

10.1.1 *Evaporation of a pure organic liquid*

The evaporation of a pure organic liquid can be estimated by analogy with the evaporation of water assuming that transport is controlled by the transfer velocity through the air film, $v_a = D_a / z_a = 3.3 \times 10^{-3}$ m/s. We neglect differences in diffusion coefficients (cf. Example 10.1), which permits to estimate the evaporation from the concentration difference over the air film:

$$E = -3.3 \times 10^{-3}(c_a - c'_a) \tag{10.12}$$

where c'_a is the concentration at the vapor pressure of the substance, and c_a may be taken 0 (all the chemical is rapidly dispersed). For example, the vapor pressure for benzene, $P^0 = 0.126$ atm. Then, $c'_a = 0.126 / (24.4 \text{ atm} \cdot \text{L/mol}) \times (78.1 \text{ g/mol}) \times (1000 \text{ L/m}^3) = 409 \text{ g/m}^3$.

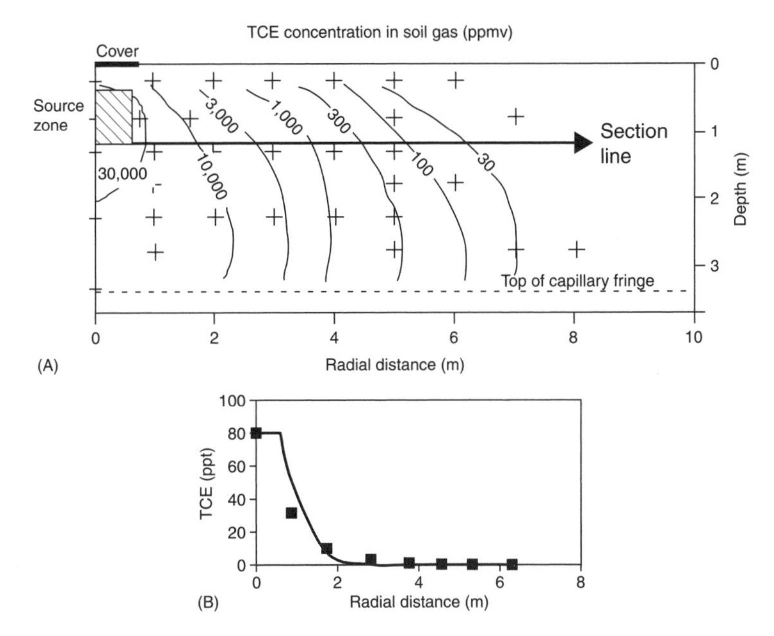

Figure 10.4. TCE concentrations in soil gas, 18 days after 60 L TCE were placed in the source zone. Figure 10.4B compares the observed concentration profile and the diffusion model, Equation (10.14). Data from Conant et al., 1996.

The estimated evaporation is $E = 3.3 \times 10^{-3} \times 409 \times (86{,}400\ \text{s/day}) = 117\ \text{kg/m}^2/\text{day}$, or, with the density of benzene $\rho = 0.88$ kg/L, 0.13 m/day. Thus, we may express evaporation of substance X relative to water at a given location as:

$$E_X = E_{H_2O} \frac{P_X^0}{P_{H_2O}^0 (1 - r_h)} \frac{MW_X}{18} \frac{1}{\rho_X} \tag{10.13}$$

where r_h is the relative humidity of air at the site.

By taking $c_a = 0$, we assume that all the chemical is removed rapidly by transport through the air, but what happens in a soil, where the air is stagnant? Figure 10.4 shows concentrations of TCE vapor in the unsaturated zone in an experiment where a volume of pure liquid TCE was mixed with sand and placed in the soil. The observed vapor concentration profile after 18 days shows the typical exponential decay of concentrations that indicates diffusion.

An approximate solution of the diffusion equation in radial direction is (cf. Equation 3.58):

$$c = c_i + (c_0 - c_i)\,\mathrm{erfc}\left(\frac{(r - r_0)}{\sqrt{\dfrac{4}{1 + \dfrac{\sqrt{2}}{r}} D_{e,a} \dfrac{t}{R}}} \right) \tag{10.14}$$

where r_0 is the radius of the cylinder with TCE, R is the retardation through sorption on the solid, and the factor $(1 + \sqrt{2}\,/\,r)$ corrects for radial instead of linear diffusion.

Gas diffusion in the soil pores will depend on the pore volume of the soil and the water content, and the effective diffusion coefficient can be estimated with the Millington-Quirk relations (Jin and Jury, 1996; Equation 9.62):

$$D_{e,a} = D_a \varepsilon_g^2 / \varepsilon^{0.6} \tag{10.15}$$

where ε_g and ε are the gas-filled and the total pore volume, respectively. For the sandy soil in the experiment of Figure 10.4, $\varepsilon_g = 0.3$ and $\varepsilon = 0.4$, which yields:

$$D_{e,a} = 8.1\times10^{-6}(0.3^2 / 0.4^{0.67}) = 1.34\times10^{-6}\ \text{m}^2/\text{s}.$$

After 18 days, with retardation $R = 4.8$ and $r_0 = 0.6$ m (Conant et al., 1996), and this diffusion coefficient, the calculated concentrations are compared with the experimental data in Figure 10.4B.

In applying Equation (10.14) to the experiment of Figure 10.4 we neglected evaporation from the soil surface into the air and advective vapor transport of the dense TCE molecules down to the capillary fringe (note that a cover layer was placed above the TCE in the experiment to reduce evaporation). Furthermore, TCE dissolves into the (saturated) groundwater (Jellali et al., 2003). Nevertheless, it is clear that our simple equation provides a good approximation for describing the TCE vapor concentration in the soil.

Gas diffusion in a soil (Equation 10.14) is much reduced when gas content decreases to less than about 0.1 (Troeh et al., 1982). Apparently, gas then resides in isolated, unconnected pockets, and only is transported by diffusion via water. This phenomenon of a *residual* occupation of the soil pores expresses itself in the persistence of organic chemicals in the soil upon water flooding.

QUESTIONS:
The vapor pressure of TCE is 0.08 atm at 20°C. Estimate the concentration of TCE in ppmv?
ANSWER: 80,000 ppmv
Henry's constant for TCE at 20°C is $K_H = 7.2$ atm · L/mol. Calculate the retardation for TCE vapor, when it dissolves into soil water, $\varepsilon_g = 0.3$, $\varepsilon_w = 0.1$?
ANSWER: $K'_H = 0.3\ L_{water}/L_{air}$. Hence, $R = 1 + K'_H\ \varepsilon_g / \varepsilon_w = 1.9$. (Note that the retardation given by Conant et al. includes sorption of TCE by organic carbon; its contribution will be calculated in Section 10.3).

10.2 TRANSPORT OF PURE ORGANIC LIQUIDS THROUGH SOIL

Organic chemicals spilled in excess of their aqueous solubility may form a separate non-aqueous phase liquid, a NAPL. The liquid may be denser than water, a DNAPL, and it will sink through the groundwater to larger depths where it may be halted by impermeable layers to accumulate in depressions in the aquifer (Figure 10.5). The liquid can be lighter than water, a LNAPL, and it will then float on the groundwater and be confined to the capillary zone from where it evaporates and also dissolves in percolating soil water. An intermediate form of transport is exhibited by highly viscous tar products which have fingered from above and form tar blobs in the aquifer. Examples of DNAPLs are degreasing solvents such as trichloroethylene (C_2HCl_3, "tri" or "TCE") and tetrachloroethylene (C_2Cl_4, "per" or "PCE") with densities of 1.46 and 1.63 g/cm^3 (Pankow and Cherry, 1996). The most common LNAPL is gasoline, and tar products are often found below coke-gas factories.

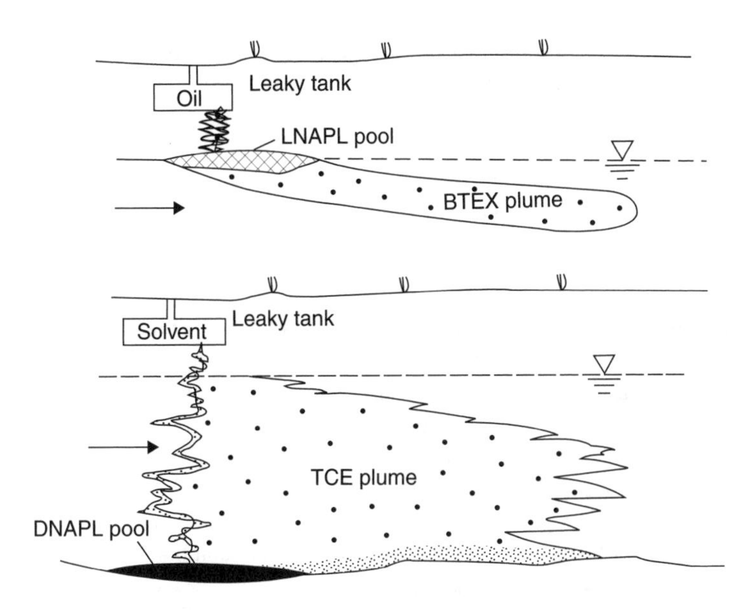

Figure 10.5. Examples of spreading in groundwater of a dense non-aqueous phase liquid (DNAPL, TCE), and a light one (LNAPL, gasoline) (Pankow and Cherry, 1996).

If we would take samples from a borehole through the zone affected by the DNAPL spill of Figure 10.5 and analyze it for water and TCE content, a profile like in Figure 10.6 may be the result. In the upper, unsaturated part of the profile the water content follows the pF-curve that relates soil water suction and water content (cf. Figure 9.22). TCE evaporates quickly from the unsaturated zone when the leakage has been stopped and is already gone in Figure 10.6. Below the water table, the proportions of TCE and water follow an erratic pattern that depends on whether a TCE tongue has passed the location and whether the TCE has been displaced again by the groundwater flow. Down in the profile a pool of TCE has accumulated above the aquiclude.

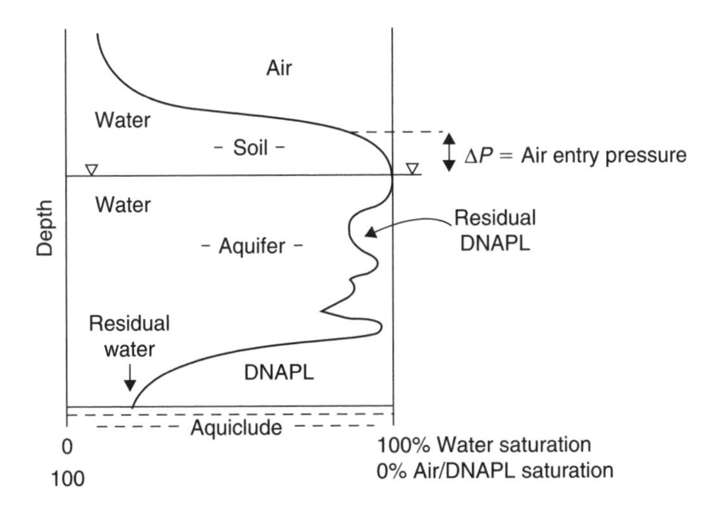

Figure 10.6. Relative saturations of water, air and TCE in a profile through a TCE polluted, phreatic aquifer.

Quite conspicuous in the concentration profile of Figure 10.6 is, that the TCE never attains full saturation, there is always some water left in the pores at the location of the DNAPL pool. Likewise for TCE, when it has entered the pores for once, a residual part remains entrapped that is very difficult to remove with water.

When water and air, or water and TCE reside both in the pores of the soil, the interface between the fluids is curved as result of the pressure difference (Figure 10.7):

$$\Delta P = 2\sigma / r \tag{10.16}$$

where P is pressure (N/m^2), σ is the interfacial tension (N/m) and r is the pore radius (m). In Figure 10.7, the convex side of the interface points downward, indicating that the pressure in air is higher than in water. The capillary pressure results from differences in molecular properties, cohesion within the fluid and strong adhesion of water to the solid capillary.

For the water/air interface, $\sigma = 0.073$ N/m, and we can calculate for a soil that pores with $r = 10\,\mu$m will be filled with water above the water table up to $2 \times 0.073 / (10^{-5}\text{m} \times 1000\,\text{kg/m}^3 \times 9.8\,\text{m/s}^2) = 1.5$ m (in fine sandy soils this may correspond to 10% water saturation). For water / DNAPL's, $\sigma \approx 0.035$ N/m, and similarly, we can calculate that 0.48 m of a DNAPL, with density 1.5 g/cm^3, can displace water from 10 μm pores.

The important thing is to realize that transport of water and DNAPL is coupled to the pore space occupied by each fluid. Initially, when DNAPL reaches the water table, a minimum pressure is necessary to enable further flow through the water saturated soil. This is similar to the air entry pressure (Figure 10.6), which is the required minimal pressure before air can start to flow through a water-saturated soil. It corresponds to the pressure needed to create a network of interconnected pores from which water has been removed, corresponding to a reduction of the water content to about 0.8–0.95 of full water saturation. At this stage the air or NAPL enters the largest pores, with the largest relative permeability. Further displacement of water requires a higher pressure, which can be achieved by the DNAPL when its downward flow is halted by an aquitard with a finer pore space.

The simplest approximation for calculating transport is the Green and Ampt model (Weaver et al., 1994; Hussein et al., 2002):

$$v_{\text{NAPL}} = K_{\text{NAPL}}(h + d - H_f) / d \tag{10.17}$$

where v is the specific discharge (m/day), K the hydraulic conductivity (m/day), h the pressure head of the NAPL where it enters the aquifer (m water), d is the depth of the front (m) and H_f is the NAPL entry pressure (m water). The equation states that as the infiltration depth (d) increases, the pressure head difference ($(h + d - H_f) / d$) decreases, and we recognize the Darcy equation.

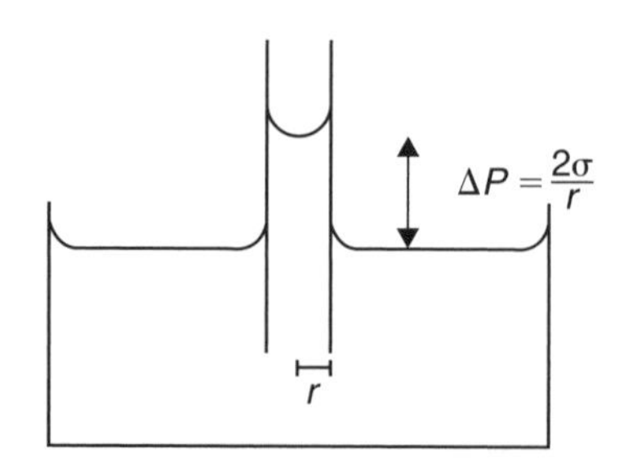

Figure 10.7. Water adheres to the walls of a glass capillary and rises in it because it is at lower pressure than in the free fluid.

In the Green and Ampt approach we assume that the NAPL filled porosity ε_{NAPL} is constant, and we can integrate Equation (10.17) with $v = \varepsilon_{NAPL} \mathrm{d}d / \mathrm{d}t$ from $d = 0$ at time $t = 0$:

$$t = \frac{\varepsilon_{NAPL}}{K_{NAPL}} \left(d - (h - H_f) \ln \left(\frac{h + d - H_f}{h - H_f} \right) \right) \quad (10.18)$$

The permeability is a function of the pores occupied, and of the viscosity of the liquid. Assuming that the viscosities of the NAPL and of water are approximately equal and $K_{NAPL} = 2$ m/day for $\varepsilon_{NAPL} = 0.3$, and that $H_f = 0.1$ m and $h = 1$ m, we can calculate that it takes only 2.6 days to reach 20 m depth. Introductions to more complete calculations of DNAPL transport have been given by Bedient et al. (1994) and McWhorter and Kueper (1996).

QUESTIONS:

Discuss the effects of h on ε_{NAPL} and K_{NAPL}, and on t.

ANSWER: Smaller h will reduce ε_{NAPL} but we expect that K_{NAPL} reduces more. Anyhow, t increases.

Compare rates of TCE evaporation and its infiltration into a sandy soil?

ANSWER: Equation (10.13) gives $E \approx 0.08$ m/day into the air, Equation (10.17) gives $v \approx 2-10$ m/day into the soil. Clearly, leakage into the soil is much larger than evaporation.

10.3 SORPTION OF ORGANIC CHEMICALS

Many organic pollutants are *hydrophobic*, which indicates that these substances have a low affinity for solution in water (a polar liquid), and prefer solution in a-polar liquids. These pollutants are readily taken up in organic matter of sediments. The tendency to become absorbed (i.e. the distribution coefficient for these organic chemicals) is related to the distribution coefficient of the chemical between water and an a-polar liquid like octanol. The latter is termed a *partition constant* or *extraction coefficient* by analytical chemists; the soil sorption process, by analogy, is also envisaged as a partition process, where the hydrophobic pollutant partitions itself between water and the soil organic matter (Chiou et al., 1979, 1987; Karickhoff, 1984). Such solution of the hydrophobic compound into organic matter is appropriately termed *ab*sorption, rather than *ad*sorption.

The distribution coefficient between water and octanol is obtained in a *separatory funnel* as shown in Figure 10.8. The organic chemical is introduced in a funnel containing water and octanol.

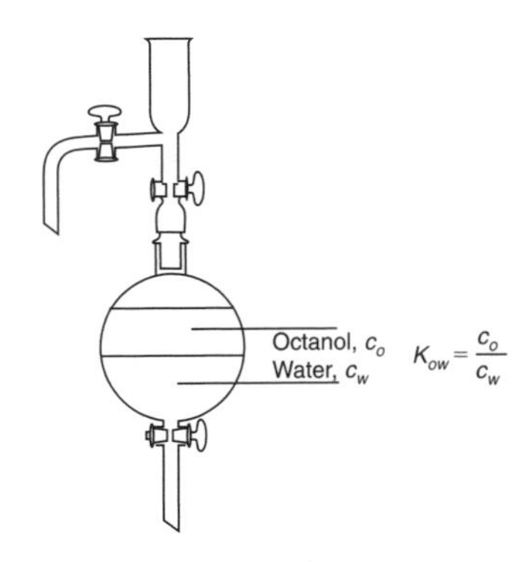

Figure 10.8. Separatory funnel used to obtain octanol/water distribution coefficients.

The funnel is shaken and the two phases are separately collected. Analysis of the concentrations in the water- and octanol-phases gives c_w and c_o respectively, from which the distribution coefficient

$$K_{ow} = c_o / c_w \tag{10.19}$$

is readily obtained. The distribution coefficient is highly correlated with the distribution coefficient between organic carbon and water, K_{oc} (Karickhoff, 1981; Schwarzenbach and Westall, 1985).

Karickhoff (1981) suggests

$$\log K_{oc} = \log K_{ow} - 0.35 \tag{10.20}$$

for the chemicals listed in Table 10.1. For groups of these chemicals, Schwarzenbach and Westall (1985) propose a linear regression of the form:

$$\log K_{oc} = \mathbf{a} \log K_{ow} + \mathbf{b} \tag{10.21}$$

Values of **a** and **b** are given in Table 10.2. More values of octanol/water distribution coefficients and other related data can be found in Verschueren (2001), and especially Lyman et al. (1990) and Schwarzenbach et al. (1993) are recommended for guidelines for estimating properties of organic pollutants.

Table 10.1. Partition coefficients for octanol–water (K_{ow}) and organic carbon–water (K_{oc}) (Karickhoff, 1981).

Compound	log K_{ow}	log K_{oc}
Hydrocarbons and chlorinated hydrocarbons		
3-methyl cholanthrene	6.42	6.09
Dibenz[a,h]anthracene	6.50	6.22
7,12-dimethylbenz[a]anthracene	5.98	5.35
Tetracene	5.90	5.81
9-methylanthracene	5.07	4.71
Pyrene	5.18	4.83
Phenanthrene	4.57	4.08
Anthracene	4.54	4.20
Naphthalene	3.36	2.94
Benzene	2.11	1.78
1,2-dichloroethane	1.45	1.51
1,1,2,2-tetrachloroethane	2.39	1.90
1,1,1-trichloroethane	2.47	2.25
Tetrachloroethylene	2.53	2.56
γ HCH (lindane)	3.72	3.30
α HCH	3.81	3.30
β HCH	3.80	3.30
1,2-dichlorobenzene	3.39	2.54
pp'DDT	6.19	5.38
Methoxychlor	5.08	4.90
22',44',66'PCB	6.34	6.08
22',44',55'PCB	6.72	5.62
Chloro-s-triazines		
Atrazine	2.33	2.33
Propazine	2.94	2.56
Simazine	2.16	2.13
Trietazine	3.35	2.74
Ipazine	3.94	3.22
Cyanazine	2.24	2.26
Carbamates		
Carbaryl	2.81	2.36
Carboturan	2.07	1.46
Chlorpropham	3.06	2.77
Organophosphates		
Malathion	2.89	3.25
Parathion	3.81	3.68
Methylparathion	3.32	3.71
Chlorpyrifos	3.31	4.13
Phenyl ureas		
Diuron	1.97	2.60
Fenuron	1.00	1.43
Linuron	2.19	2.91
Monolinuron	1.60	2.30
Monuron	1.46	2.00
Fluometuron	1.34	2.24
Miscellaneous compounds		
13Hdibenzo[a,i]carbazole	6.40	6.02
2,2'biquinoline	4.31	4.02
Dibenzothiophene	4.38	4.05
Acetophenone	1.59	1.54
Bromacil	2.02	1.86
Terbacil	1.89	1.71

Table 10.2. Estimation of K_{oc} from K_{ow} by the expression log K_{oc} = **a** · log K_{ow} + **b** (Schwarzenbach and Westall, 1985).

Regression coefficient				
a	**b**	Correlation coefficient	Number of compounds	Type of chemical
0.544	1.337	0.74	45	Agricultural chemicals
1.00	−0.21	1.00	10	Polycyclic aromatic hydrocarbons
0.937	−0.006	0.95	19	Triazines, nitroanilines
1.029	−0.18	0.91	13	Herbicides, insecticides
1.00	−0.317	0.98	13	Heterocyclic aromatic compounds
0.72	0.49	0.95	13	Chlorinated hydrocarbons alkylbenzenes
0.52	0.64	0.84	30	Substituted phenyl ureas and alkyl-N-phenyl carbamates

The log K_{oc} refers to partitioning between water and a 100% organic carbon phase; the actual distribution coefficient for the soil or sediment is then obtained as

$$K'_d = K_{oc} \cdot f_{oc} \tag{10.22}$$

where f_{oc} is the fraction of organic carbon. This relationship holds when $f_{oc} > 0.001$, otherwise sorption on non-organic solids can become relatively important (although giving only low K_d's, Karickhoff, 1984; Curtis et al., 1986; Madsen et al., 2000). The solubility of the (solid) organic compound in water is another parameter which has been suggested in order to estimate the organic carbon partitioning coefficient. It is clearly related to the octanol/water distribution coefficients as shown in Figure 10.9.

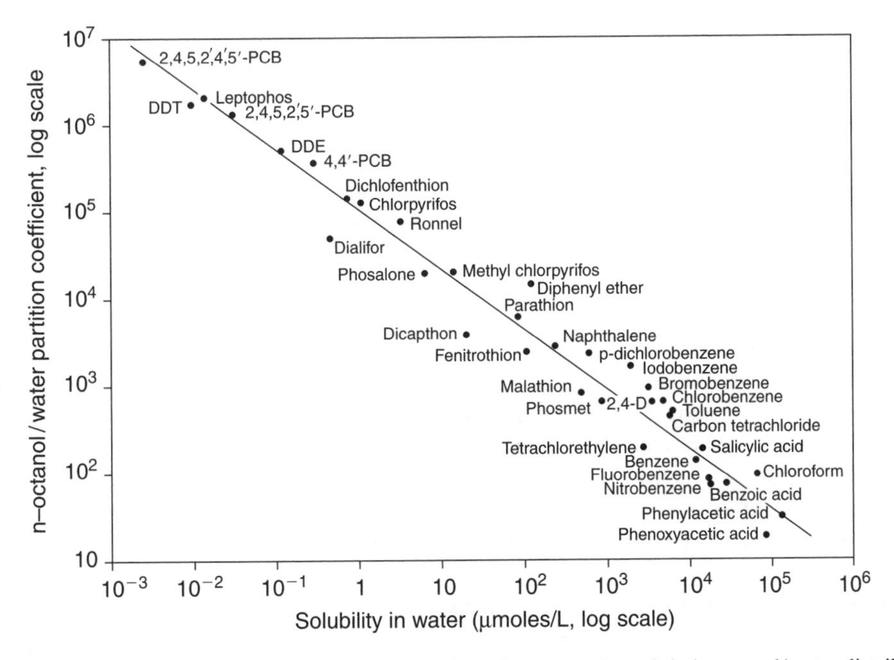

Figure 10.9. Relationship between solubility of a number of compounds and their octanol/water distribution (or partition) coefficient (Chiou et al., 1979).

EXAMPLE 10.2. *Retardation of Lindane and PCB*
Calculate the retardation of Lindane and of PCB with respect to groundwater flow in a sediment with 0.3% organic carbon.

ANSWER:
From Table 10.1, find the distribution coefficient K_{oc} for lindane, which is $10^{3.3}$, and obtain with Equation (10.22):

$\log K_d' = 3.3 + \log 0.003 = 0.8$

$K_d' = 6.3\,\text{mL/g}$, and $K_d = 6.3 \cdot \rho_b / \varepsilon_w \approx 38$.

Similarly for PCB:

$\log K_d' = 5.8 + \log 0.003 = 3.3$

$K_d' \approx 2000\,\text{mL/g}$, and $K_d = 2000 \cdot \rho_b / \varepsilon_w \approx 12{,}000$.

Lindane (γ-HCH) would have a velocity which is about 40 times less than the water velocity, whereas PCB would move imperceptibly slowly at 12,000 times less than the water velocity.

QUESTION:
Estimate the retardation (R) for TCE in the volatilization experiment (Figure 10.4). $K_{oc} = 125$ L/kg. Organic carbon is 0.02% on average in the sediment, and 2% in the organic-rich layer.

ANSWER: We noted already that the ratio gaseous TCE / dissolved TCE = 1 / 0.9. Express sorbed TCE with respect to dissolved TCE, $\rho_b = 1.8\,\text{g/cm}^3$, $\varepsilon_w = 0.1$: $q_{\text{TCE}} = (125\,\text{L/kg OC}) \times (2 \times 10^{-4}\ \text{kg OC/kg}) \times (1.8\,\text{kg/L}) / (0.1\ (-)) \times (0.9\,\text{mmol TCE/L}) = 0.40\,\text{mmol TCE/L}$ for the sediment, and 40 mmol TCE/L for the organic-rich layer. It gives $R = 1 + (0.9 + 0.4)/1 = 2.3$ and $1 + (0.9 + 40)/1 = 42$ for gas transport in the two layers, respectively. The fitted $R = 4.8$ points to the importance of the organic layer in retarding transport (although a smaller diffusion coefficient would have a similar effect).

When sediment contains more than 0.1% organic carbon, the adsorption of nonionic organic chemicals is wholly attributed to organic carbon. One might expect that the deposition environment has an influence on organic matter content, e.g. organic matter will be low in sediments deposited in glacial times, and higher during interglacial periods. However, the scarce data available on sandy sediments in the Netherlands show no definite trend in this respect (Figure 10.10). On the other hand, the figure does illustrate that organic matter content is so low in the common aquifers that our K_d approximation method may not be valid. Since sorption to mineral surfaces is also small, K_d''s smaller than 1 L/kg are expected.

Use of a single K_{oc} assumes that all natural organic matter can be lumped together. However, differences of an order of magnitude have been observed among different humic and fulvic acids (Garbarini and Lion, 1986, Chiou et al., 1987). Grathwohl (1990) investigated organic matter from sediments with different ages, and a different degree of maturing or ripening, ranging from recent organic matter to precursors of oil. He found that a higher H/O-ratio in the organic matter gives an increase of the distribution coefficient for sorption of hydrophobic organic chemicals (Figure 10.11). This effect is associated with lower polarity and higher hydrophobicity of organic matter as the H/O ratio increases. Also, diffusion in coaled organic matter is slower than in soft humic acid, which may lead to curved isotherms in short-term experiments and a decreasing curvature of the isotherm when time progresses (Weber et al., 2001).

The H / O ratio of young organic matter is much more uniform than was used for the experiments in Figure 10.11. In recent soils, K_{oc} tends to be constant to within 0.3 log K_{oc} units for a given substance (Schwarzenbach et al., 1993). Nevertheless, trends may exist even within this small range.

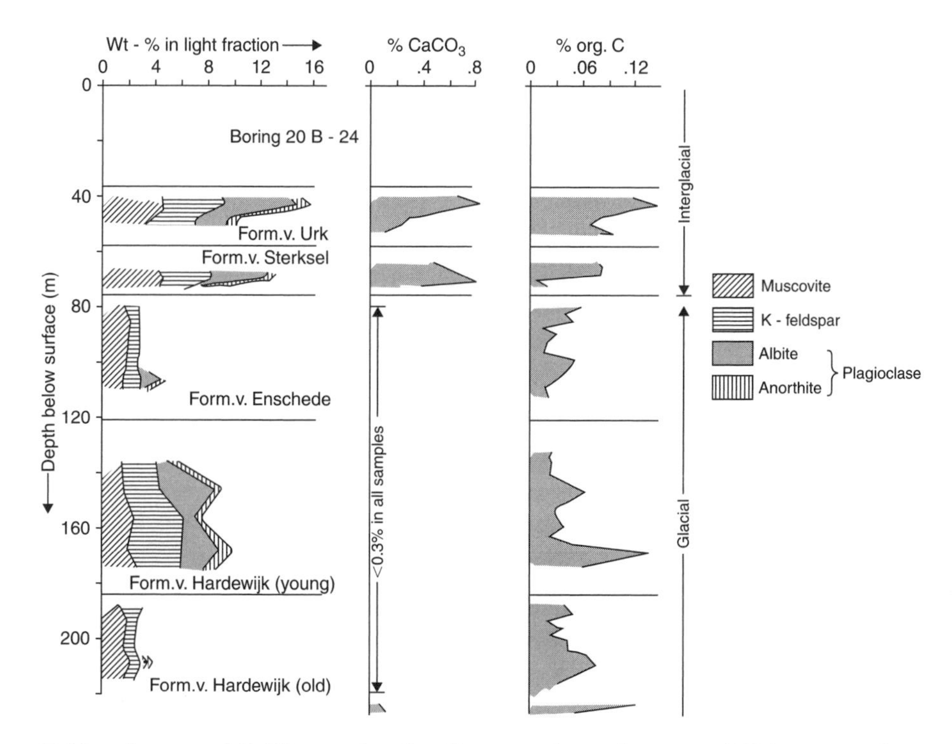

Figure 10.10. Contents of $CaCO_3$, weatherable silicates, and organic matter of Pleistocene sediments in a borehole in The Netherlands.

Kile et al. (1999) and Ahmad et al. (2001) investigated a multitude of soils and sediments and found that K_{oc} increased with the aromaticity of organic carbon. However, the overall variation was small; it can be calculated from their data that $\log K_{oc} = 1.88 \pm 0.12$ for tetrachloromethane (CCl_4), $\log K_{oc} = 2.58 \pm 0.13$ for carbaryl, and $K_{oc} = 3.82 \pm 0.15$ for phosalone (both are pesticides, cf. Table 10.1 for carbaryl).

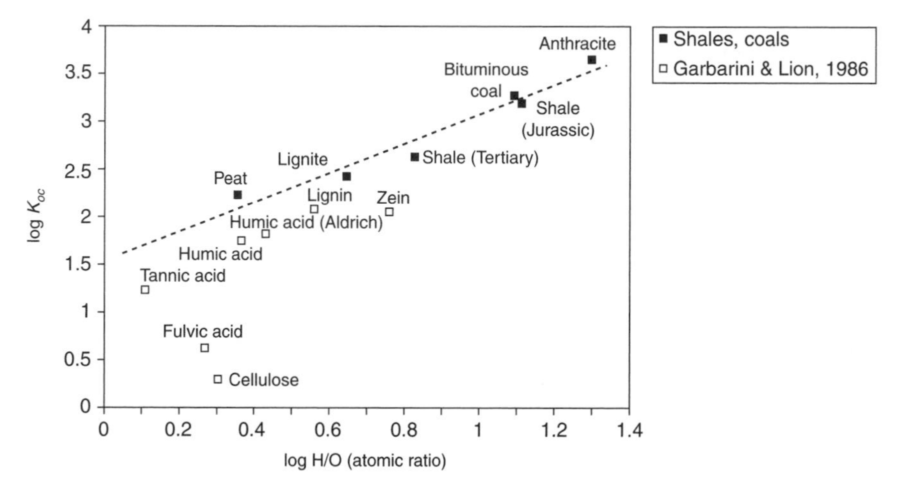

Figure 10.11. The distribution coefficient K_{oc} of trichloroethylene (TCE) increases with increasing H/O ratio (Grathwohl, 1990).

Also, with increasing pH the surface charge of organic matter will increase, rendering it more polar and perhaps less penetrable for a-polar organic chemicals (Kan and Tomson, 1990). Again, the observed reduction of log K_{oc} was minor, in this case for naphthalene decreasing from 2.9 to 2.7 when pH increased from 5 to 9.

10.3.1 *Sorption of charged organic molecules*

If the organic chemical develops a negative charge, sorption to organic carbon and sediments in general is much reduced (Haderlein and Schwarzenbach, 1993; Broholm et al., 2001). The herbicide 2-methyl-4,6-dinitrophenol ($C_6H_2CH_3(NO_2)_2OH$, abbreviated DNOC-OH) dissociates according to

$$\text{DNOC-OH} = \text{DNOC-O}^- + \text{H}^+; \quad \log K_a = -4.31 \tag{10.23}$$

The fraction DNOC-OH of total DNOC is (cf. Equation 5.13):

$$\text{DNOC-OH} = \text{DNOC}/(1 + K_a/[\text{H}^+]) \tag{10.24}$$

In case only the neutral species is sorbed, the distribution coefficient becomes:

$$K_d' = K_{oc} \cdot f_{oc}/(1 + K_a/[\text{H}^+]) \tag{10.25}$$

For pH $\ll$ pK_a, the denominator of Equation (10.25) is 1, and the distribution coefficient is equal to the organic carbon sorption coefficient. When pH = pK_a, the distribution coefficient is just half of this value, and it will further decrease when pH increases and the negative species becomes dominant. The effect is illustrated in Figure 10.12 for DNOC (Broholm et al., 2001).

Actually, the sediment in Figure 10.12 had a low organic carbon content of 0.02%. With log K_{ow} = 2.12 for DNOC, the estimated K_d' = 0.026, much smaller than observed (Figure 10.12). The difference was attributed to sorption on mineral surfaces, but iron-oxyhydroxides which are positively charged in the pH-range of this study would sorb the anionic form rather than the neutral form (Clausen and Fabricius, 2001).

If the chemical develops a positive charge, sorption to the dominantly negatively charged soil particles is enhanced below the pK_a. We can use PHREEQC to model the sorption of the neutral fraction as surface complexation reaction, and of the charged form as ion exchange as shown in Example 10.3.

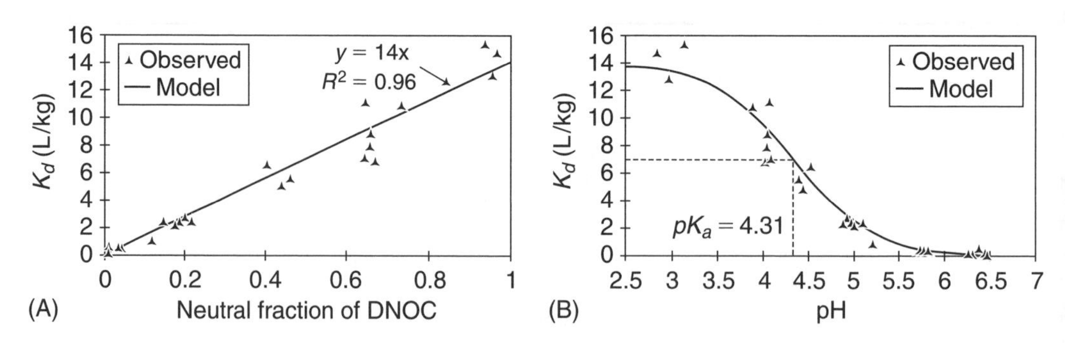

Figure 10.12. pH dependent sorption of DNOC-OH on sediment (Broholm et al., 2001).

EXAMPLE 10.3. *Calculate sorption and ion exchange of Quinoline with PHREEQC*
Quinoline becomes protonated for pH's smaller than $pK_a = 4.94$. The neutral form has $K_{oc} = 100$ L/kg. Zachara et al. (1986) measured K'_d of quinoline for a soil horizon with $f_{oc} = 0.0024$, $CEC = 84$ meq/kg, in 10 mM $CaCl_2$ solution at pH 4.2 and 7.5. Estimate the K'_d for these pH's with PHREEQC.

ANSWER:
We make an input file where quinoline is defined as aqueous species that can associate with H^+. Sorption of the neutral form to organic matter is modeled with SURFACE, for the positive species EXCHANGE is used. The $\log K$ of the surface species is defined in a special way to simulate the simple distribution reaction.

```
SOLUTION_MASTER_SPECIES                                  # define quinoline...
 Quin Quin 0 129 129
SOLUTION_SPECIES
 Quin = Quin; -log_k 0
 Quin + H+ = QuinH+; -log_k 4.94

SURFACE_MASTER_SPECIES                                   # define absorption into OC..
 Org_c Org_c
SURFACE_SPECIES
 Org_c = Org_c; -log_k 0
 Org_c + Quin = Org_cQuin; -log_k -98.0                  # divide K_oc = 100 by 10^100

EXCHANGE_SPECIES
 X- + QuinH+ = XQuinH; -log_k 0                          # equal to K_NaX
END

SOLUTION 1; pH 4.2; Ca 10; Cl 20; Quin 1e-3              # the experimental solutions
SOLUTION 2; pH 7.5; Ca 10; Cl 20; Quin 1e-3
END

USER_PRINT
 -start
 10 K_d_OC = mol("Org_cQuin") / tot("Quin")
 20 K_d_X = mol("XQuinH") / tot("Quin")
 30 print "pH = ",-la("H+"), ". K_d(OC) =",K_d_OC, ". K_d(X) =",K_d_X,\
        ". K_d_tot (L/kg) =", K_d_OC + K_d_X
 -end

USE solution 1
SURFACE 1; Org_c 0.0024e100 1 1; -equil 1                # sites of Org_c times 10^100
EXCHANGE 1; X 0.084; -equil 1
END
USE solution 2
SURFACE 2; Org_c 0.0024e100 1 1; -equil 2
EXCHANGE 2; X 0.084; -equil 2
END
```

The names of the chemical must be defined starting with a capital letter followed by lower case letters in keyword SOLUTION_MASTER_SPECIES. The species in the second column (here "Quin") must also be defined in an identity reaction with log_k = 0 under SOLUTION_SPECIES. The protonated form of quinoline is defined as well.

In the K_d approach, organic carbon is a phase which does not enter in the reaction equation and it is also not affected by the reaction. To translate the K_d formula into a formal chemical reaction that is valid in PHREEQC terms, we rewrite the distribution a bit. For the surface complexation reaction

$$\text{Org\_c} + \text{Quin} = \text{Org\_cQuin},$$

the mass action equation is

$$\frac{[\text{Org\_cQuin}]}{[\text{Quin}][\text{Org\_c}]} = K_{\text{PHREEQC}} \quad \text{or} \quad \frac{[\text{Org\_cQuin}]}{[\text{Quin}]} = ([\text{Org\_c}]K_{\text{PHREEQC}}) = K_{oc}$$

A constant distribution coefficient requires that the product ([Org_c] $\cdot K_{\text{PHREEQC}}$) $= K_{oc}$ is constant. By imposing a huge concentration of surface sites $m_{\text{Org\_c}}$, [Org_c] will not change when a relatively small fraction associates to form [Org_cQuin]. The value for the association constant of the SURFACE_SPECIES then becomes $K_{\text{PHREEQC}} = K_{oc} / m_{\text{Org\_c}}$, or $\log K_{\text{PHREEQC}} = \log K_{oc} - \log m_{\text{Org\_c}}$. In the example, $m_{\text{Org\_c}}$ is multiplied with 10^{100}, and K is accordingly divided by 10^{100} ($\log K_{\text{PHREEQC}} = \log(10^2 / 10^{100}) = -98$). Since the surface species is neutral, charge effects are absent and we do not care about the surface area of the organic carbon and simply make it 1.

The exchange reaction is defined in the usual manner (cf. Chapter 6), we estimate that K for a singly charged species is identical to the one for Na^+.

The result is:

```
------------------------------------User print-----------------------------------
pH = 4.2000e+00 . K_d(OC) = 3.2175e-02 . K_d(X) = 2.4637e-01. K_d_tot (L/kg) = 2.7855e-01
pH = 7.5000e+00 . K_d(OC) = 2.4088e-01 . K_d(X) = 9.2441e-04. K_d_tot (L/kg) = 2.4180e-01
```

Zachara et al. (1986) found $K'_d = 3.5$ and 0.75 L/kg for pH $= 4.2$ and 7.5. Clearly, at low pH where the ion exchange reaction underestimates observed sorption, we are fairly far off with our model. Schwarzenbach et al. (1993) modeled the same data, but they used $K = 20$ for the exchange reaction.

QUESTION:
Find K'_d when $K = 20$ for the exchange reaction?
ANSWER: $K'_d = 5$ L/kg.

10.3.2 *Sorption in stagnant zones*

In chapter 6 we discussed the effects of stagnant layers on transport of solutes retarded by ion exchange. Peat layers, buried soils, and in general clayey sediments with a high organic carbon content are particularly effective absorbers which diminish the spreading of organic pollutants. These layers have a relatively low permeability and water flow through them is restricted. Typically for situations where permeable and stagnant zones alternate, the shape of the breakthrough curve displays an early advent of the chemical and a slow bleeding upon flushing (Section 6.5.1). A column experiment with naphthalene shows the effects (Figure 10.13).

The experiment is to be modeled with PHREEQC (Problem 10.3), but it is instructive to obtain the general picture first. The soil contains natural organic carbon which sorbs naphthalene ($K_{oc} = 302$ L/kg, $f_{oc} = 0.014$) and retards its flow velocity by 8.5. However, the high flow velocity in the experiment incited disequilibrium for 1/3 of the sorption sites. In the breakthrough curve, this disequilibrium is expressed by an early arrival of $c_{0.5}$ (appears at $PV = 5.3$ instead of 8.5) and tailing towards $c_{1.0}$.

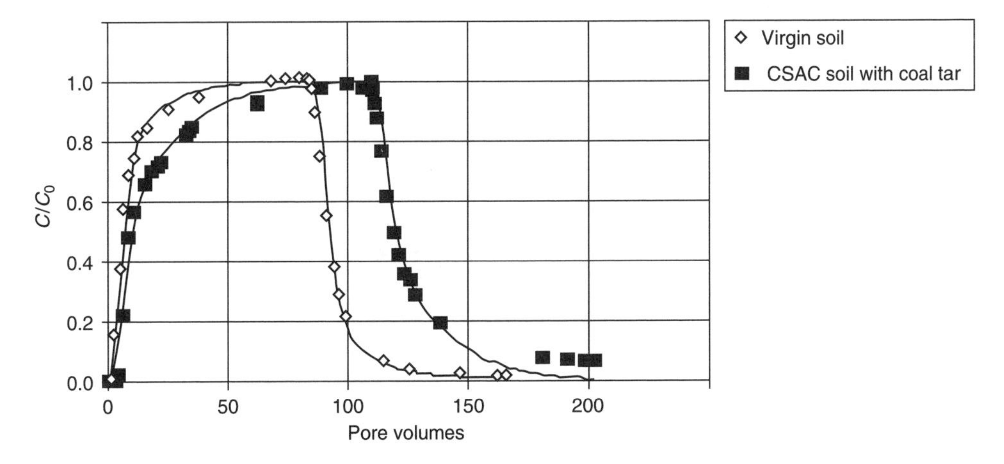

Figure 10.13. Breakthrough curves of naphthalene in column experiments with soil with natural organic carbon ("virgin soil") and with the same soil amended with 4 g coaltar/kg soil (Bayard et al., 2000).

When coaltar was added to the soil (4 g/kg with $f_{oc} = 0.38$, $K_{oc} = 2026$ L/kg, note the much higher K_{oc}), the breakthrough of $c_{0.5}$ of napthalene was retarded only slightly more (to $PV = 5.8$). However, the overall breakthrough occurred much later because sorption to the coaltar was mostly kinetic. The same happened in the elution stage, but note carefully that the model, though excellent for sorption, is not as good for desorption where the release rate is overestimated (the complete elution of naphthalene occurs too quickly in the model). This is because the first order exchange model assumes the same rate for uptake and release, whereas the physical process involves a continuing uptake in the inner realms of the immobile zone, even when the outer layers of the stagnant part release the chemical already to the mobile zone.

The first order exchange process between two boxes is calculated with the formula (Section 6.5.2):

$$\frac{dM_{im}}{dt} = V_{im}R_{im}\frac{dc_{im}}{dt} = \alpha(c_m - c_{im}) \tag{6.40}$$

where subscripts m and im indicate mobile and immobile, and α is the exchange factor (1/s). The exchange factor can be related to the physical properties of the stagnant zone:

$$\alpha = \frac{D_e \varepsilon_{im}}{(af_{s\to 1})^2} \tag{6.45}$$

where $D_{e,a}$ is the effective diffusion coefficient in the stagnant region, a is the size (m), and $f_{s\to 1}$ is a conversion factor which depends on the shape of the stagnant body.

Let us calculate transport of CCl_4 ("tetra") in a heterogeneous aquifer in which clay lenses retard transport. The aquifer is a regular sequence of sandy and clayey layers, of 0.66 m and 0.33 m thickness, respectively, and with equal porosity of 0.36, sketched in Figure 10.14. Flow velocity in the sandy layer is 50 m/year and zero in the clay layer. A 5 year pulse of 1 mg CCl_4/L is followed by 10 years flushing of the chemical. The retardation is zero in the sand, and estimated to be 4 in the clay, given that the clay contains 0.5% OC, $K_{oc} = 100$ L/kg and $\rho_b / \varepsilon = 6$ kg/L. The effective diffusion coefficient is $D_{e,a} = 0.33\times10^{-10}$ m²/s in the clay. We compare transport of Cl^- as a conservative tracer and of CCl_4, and consider the effect of the clay lenses. The shape factor $f_{s\to 1}$ is 0.53 for plane sheets.

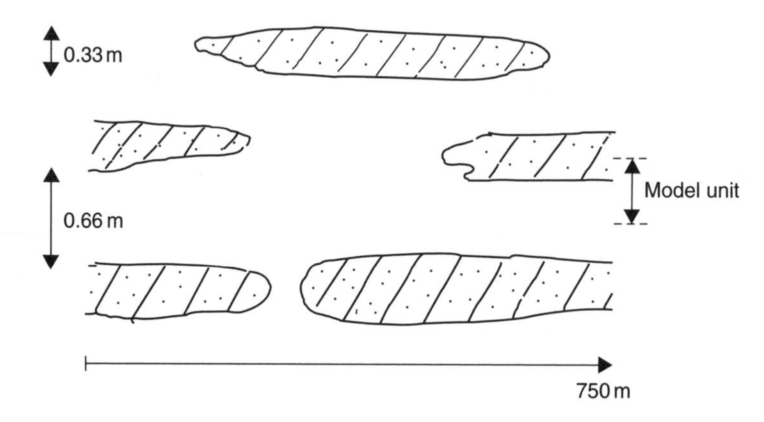

Figure 10.14. Small section of an aquifer with uniform sand and clay layer alternations.

The regular alternation of clay and sand means that the concentrations are symmetric around planes midway of the layers. Therefore, the configuration can be modeled as a unit of 0.33 m sand in contact with 0.165 m clay. The exchange factor $\alpha = 0.33\times10^{-10} \times 0.36/(0.165 \times 0.53)^2 = 1.55\times10^{-9}$/s. Table 10.3 lists the input file for PHREEQC.

Table 10.3. Input file for sorption of CCl_4 in a dual porosity aquifer.

```
SOLUTION_MASTER_SPECIES; Tetra Tetra 0.0 Tetra 1.0
SOLUTION_SPECIES; Tetra = Tetra; log_k 0.0
SURFACE_MASTER_SPECIES; Sor Sor
SURFACE_SPECIES
 Sor = Sor; log_k 0.0
 Sor + Tetra = SorTetra; log_k -99.52                     # log(Kf = 3) - log(m_Sor)

SOLUTION 1-201                                            # The sand and clay aquifer

SURFACE 102-201                           # Sorption in stagnant cells (clay layer)
 Sor 1e100 1 1
END

SOLUTION 0; Cl 1; Tetra 1                                 # The polluted water
PRINT; -reset false
TRANSPORT; -cells 100; -shifts 25 1; -timest 0.73e7       # 5 years pollution
 -length 10; -disp 4.999
 -stagnant 1 1.55e-9 0.24 0.12      # 1 stagnant layer, alpha, mobile por, immobile por
END

SOLUTION 0                                                # Clean water enters
TRANSPORT; -shifts 50 1                                   # 10 years cleaning
 -punch_fr 50                                             # punch/graph only 50th shift
 -punch 1-100                                             # punchout mobile flow
# -punch 102-201                                          # punchout stagnant area
USER_GRAPH
 -heading Dist CCl4 Cl
 -start
 10 graph_x Dist
 20 graph_y tot("Tetra") * 1e3, tot("Cl") * 1e3
 -end
END
```

Note that under keyword TRANSPORT, the option stagnant is invoked with the porosities of the mobile and immobile parts defined as fractions of the total volume. However, in Equation (6.45) the porosity corrects for the surface area accessible to diffusion, and it is a fraction of the stagnant volume there.

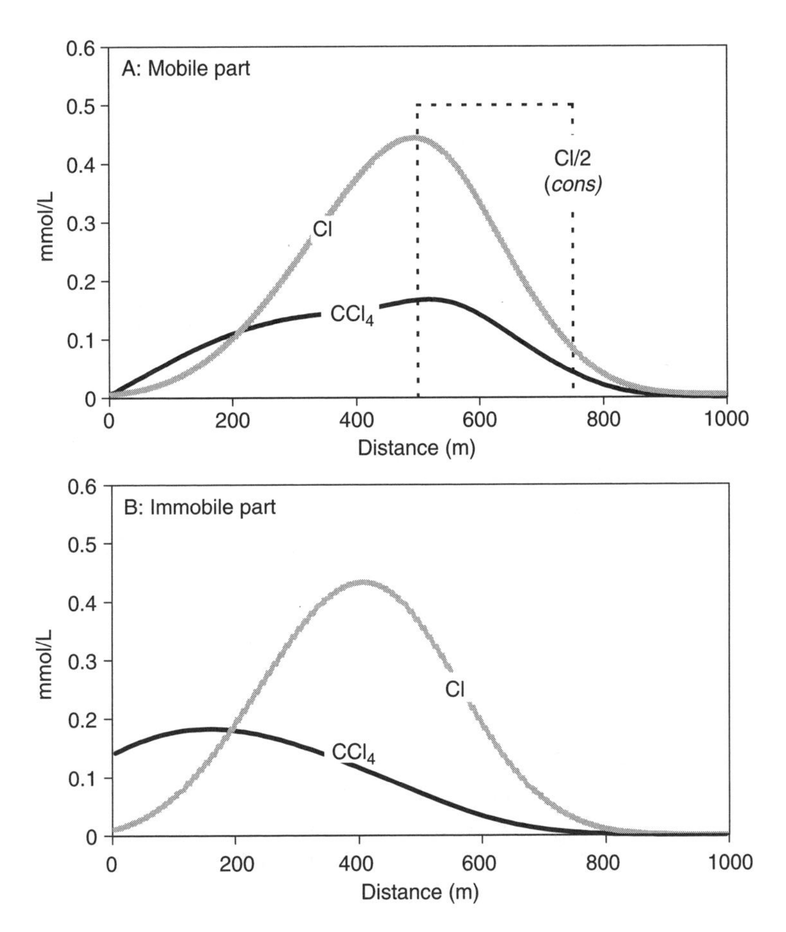

Figure 10.15. Concentration profiles of Cl^- and CCl_4 in the mobile part (A) and the stagnant part (B) of a dual porosity aquifer after 5 years pollution and 10 years flushing.

The concentration profiles are plotted in Figure 10.15. The thin dotted line in Figure 10.15A indicates the expected "conservative" concentration of Cl^- (divided by 2 for scaling to the y-axis) in the absence of dispersion and without clay layers. In this case, the forward front of the pulse is located at 750 m and the backward end is at 500 m. In the model situation, Cl^- shows rounding due to dispersion and retardation as part of the concentration moves into the immobile zone. Furthermore, the pulse has become slightly asymmetric, with a broader backward than forward limb. The concentration of CCl_4 has decreased more than of Cl^- because the chemical is sorbed in the clay. During flushing, the clay releases CCl_4 and the concentration remains higher than of Cl^- for an extended time period. Finally, note that the mass of chemical in the system is given by the sum of the concentrations (mol/L) times the volume fraction of water in the system, to which the sorbed amounts must be added as well.

QUESTIONS:
Run the PHREEQC file (download from www.xs4all.nl/~appt) to obtain a chart of transport without dispersion and without stagnant zones. The concentration peak is located between .... and .... m.
Next, include dispersion and run the file (do not forget to fix the chart to allow for an easy intercomparison).

Stagnant cells are numbered as:

$$cell\_no + 1 + n \times end\_cell\_no$$

where *cell_no* is the cell with mobile water to which the stagnant cell is connected, n is the number of the stagnant layer, and *end_cell_no* is the last cell of the flowtube.

Include the stagnant zones, but display only the mobile zone.
Run the file again to display the concentrations in the stagnant zone. Why is the concentration of CCl_4 smaller than of Cl^-?
Run the file to display the sorbed concentrations of CCl_4 (plot mol ("SorTetra")*1e3).
Investigate the effect of increasing the effective diffusion coefficient in the stagnant region to 10^{-10} m²/s.

10.3.3 *Release from stagnant zones and blobs*

An obnoxious aspect is the continuous bleeding of chemicals from stagnant zones or pools. In the case of NAPL's this may go up to their aqueous solubility with subsequent dilution by dispersal with groundwater flow. Usually, the release of BTEX compounds (benzene, toluene, ethylbenzene and xylene) from pools and tar blobs is modeled with a linear driving force expression that is similar to the first order exchange relation (Miller et al., 1990, Powers et al., 1994):

$$M_x = k_{rr} \, A \, (c_{x,s} - c_x) \tag{10.26}$$

where M_x is the mass-flow of chemical x (mol/L/s), k_{rr} is the specific release rate ($m^{-2}\ s^{-1}$), A is the interfacial area among the blob and water (m^2), $c_{x,s}$ is the concentration of a BTEX compound in equilibrium with the blob (mol/L), and c_x is the actual concentration of the compound in the groundwater. To define the changing interface, it may be assumed that the blobs consist of spheres, with a radius a (m) related to the mass of BTEX in the pool, m_{BTEX} in mol/L:

$$a \sim (m_{BTEX})^{1/3} \tag{10.27}$$

The surface area of the spheres is:

$$A \sim a^2 \sim (m_{BTEX})^{2/3} \tag{10.28}$$

which gives in Equation (10.26):

$$M_x = k_{orr} \, (m_{BTEX} / m_{BTEX,0})^{2/3} (c_{x,s} - c_x) \tag{10.29}$$

where $m_{BTEX,0}$ is the pool at the start (mol/L), and k_{orr} is the overall release rate from the pool ($\equiv k_{rr} A$). The value of the exponent, ⅔, is the same as used in defining the kinetic dissolution rate for ideal spheres or cubes (Chapter 4), but we have noted already that for minerals usually a higher value is found because the surface characteristics change during dissolution. It is also interesting to derive the value for BTEX pools from some observations.

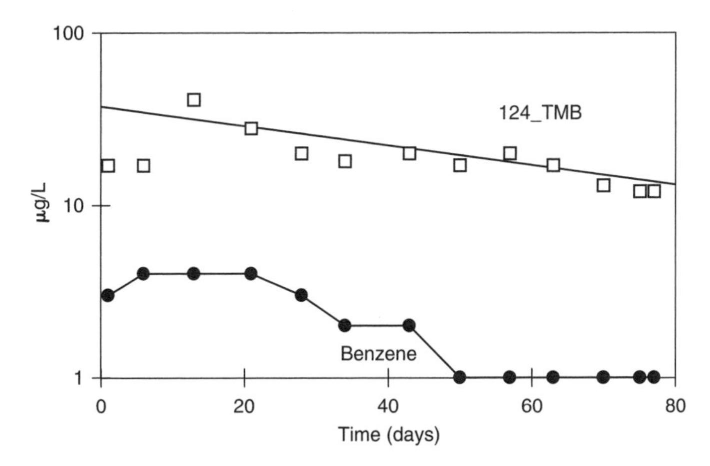

Figure 10.16. Concentrations of BTEX compounds during the remediation of the aquifer below a former benzene factory. TMB is trimethylbenzene (Eckert and Appelo, 2002).

The flux of chemical x will become independent of the concentration difference when the actual concentration c_x is much smaller than the saturated concentration, $c_{x,s}$. We expect then, that the rate becomes:

$$M_x \sim -\mathrm{d}m_{\mathrm{BTEX}} / \mathrm{d}t = k_{orr} \, (m_{\mathrm{BTEX}} / m_{\mathrm{BTEX},0})^f \cdot c_{x,s}$$

where f is the constant to be derived. When $f = 1$, the pool decreases logarithmically with time, and consequently, the BTEX concentrations in water also decrease logarithmically in time. When $f = ⅔$, the pool will decrease with the cubed power of time, and the concentrations decrease more quickly than logarithmic.

Both concentration patterns were observed in a BTEX remediation scheme where groundwater with KNO_3 was injected to enhance the oxidation capacity (Figure 10.16). In an observation well the concentration of 1,2,4-trimethylbenzene decreased logarithmically, while the concentration of benzene decreased more rapidly as a result of biodegradation. The exponent of ⅔ was found applicable in column experiments with toluene and benzene (Geller and Hunt, 1993).

Pools with mixtures of chemicals

The value of $c_{x,s}$ in Equation (10.26) is the aqueous solubility of the pure compound, corrected for the activity in the pool of BTEX. These pools behave thermodynamically nearly ideal, so that the activity is given by the mole fraction in the mixture (Feenstra and Guiguer, 1996; Eberhardt and Gratwohl, 2002). For example, for trichloroethane, TCA, in a mixture of j chemicals:

$$[\mathrm{TCA}] = \chi_{\mathrm{TCA}} = \frac{m_{\mathrm{TCA}}}{\sum_{i=1}^{j} m_i} \tag{10.31}$$

and the saturated concentration is less than the solubility of the pure compound:

$$c_{\mathrm{CTA}} = S_{\mathrm{TCA}} \, \chi_{\mathrm{TCA}} \tag{10.32}$$

where S_{TCA} is the solubility of TCA (in water $S_{\mathrm{TCA}} \approx 1300\,\mathrm{mg/L}$).

We can estimate the fractions in the pool from concentrations in water as shown in Example 10.4.

EXAMPLE 10.4. *Estimate the composition of a DNAPL pool in an aquifer*
The solubilities of 1,1,1-trichloroethane (TCA) and tetrachloroethylene (PCE) in water are 1300 and 200 mg/L, respectively. Observed concentrations in groundwater are 2.9 and 1.7 mg/L. Estimate the mole fractions of TCA and PCE in the pool assuming that only TCA and PCE are present and that the low concentrations are the result of dilution.

ANSWER:
For TCA we have: $2.9\,\text{mg/L} = S_{TCA}\, \chi_{TCA} / f$
and for PCE: $1.7\,\text{mg/L} = S_{PCE}\, \chi_{PCE} / f = S_{PCE}\,(1 - \chi_{TCA}) / f$

where f is the dilution factor.
We have two equations with two unknowns, which can be solved to:

$f = 93.2$, $\chi_{TCA} = 0.21$ and $\chi_{PCE} = 0.79$.

The dilution factor of 93.2 indicates that polluted water, saturated with the chlorinated compounds in contact with the pool, is mixed with uncontaminated groundwater in the ratio 1:92.2. When more compounds are present, another equation is added which permits to estimate the unknown mole fraction.

QUESTION:
Estimate the mole fractions in the pool when also 4 mg 1,1,2-trichloroethane/L is present? This more polar compound has a higher solubility of 4400 mg/L. *Hint*: start finding $f = 1 / \Sigma(c_i / S_i)$.
ANSWER: $f = 85.9$, $\chi_{PCE} = 0.73$, $\chi_{TCA} = 0.19$, $\chi_{1,1,2\text{-}TCA} = 0.08$.

As a result of different solubilities, the composition of a pool of NAPL's will change in time because the most soluble components disappear most rapidly. Likewise, the ratios of the aqueous concentrations of the various compounds will change with time. For example, we can imagine an experiment where TCA and PCE are mixed in equal molar amounts in a beaker, whereafter a number of times a volume of water is added, and removed again after equilibration. Initially, TCA will have the highest concentration of the two chemicals in water. However, since more TCA dissolves from the pool than PCE, the mole fraction of TCA decreases. The lower activity of TCA gives lower aqueous concentrations in successive extractions until the pool is emptied of TCA. Since the mixture of organic liquids is thermodynamically ideal (mole fraction = activity), we can model the experiment with PHREEQC, using keyword SOLID_SOLUTIONS (Example 10.5).

EXAMPLE 10.5. *Model the extraction of a DNAPL pool with PHREEQC*
Mix 0.8 mmol TCA (106.7 mg) and 0.8 mmol PCE (132.6 mg) in a separatory funnel and add 100 mL water. Shake the funnel and separate the aqueous solution. Repeat the extraction with water 9 times. Find the aqueous concentrations with PHREEQC.

ANSWER:
The PHREEQC input file to model the experiment is:

```
SOLUTION_MASTER_SPECIES                 # Define the substances...
 Tca Tca 0 Tca 133.4                    # 1,1,1-trichloroethane C2H3Cl3, TCA
 Pce Pce 0 Pce 165.8                    # Tetrachloroethene C2Cl4, PCE
```

```
SOLUTION_SPECIES
 Tca = Tca; log_k 0
 Pce = Pce; log_k 0
                                        # Define the solubility...

PHASES
 Tca_lq; Tca = Tca; log_k -2.01
 Pce_lq; Pce = Pce; log_k -2.92
                                        # weigh the chemicals in the funnel...

SOLID_SOLUTIONS 1
 Pool
 -comp Tca_lq 0.0008
 -comp Pce_lq 0.0008
                                        # take 0.1 L pure water...

SOLUTION 1
 -water 0.1
END
                                        # mix water and chemicals...

USE solution 1
USE solid_solutions 1
SAVE solid_solutions 1                  # the pool of chemicals after dissolution

USER_GRAPH                              # plot the composition...
 -connect_simulations
 -headings Extr_no TCA PCE
 -start
 10 graph_x sim_no - 1                  # Extraction No. on x-axis
 20 graph_y tot("Tca") * 133.4e3, tot("Pce") * 165.8e3
 -end
END
                                        # 2nd extraction..

USE solid_solutions 1; USE solution 1; SAVE solid_solutions 1
END
                                        # 3rd extraction..
USE solid_solutions 1; USE solution 1; SAVE solid_solutions 1
END
# etc... 6 times more
```

The solubility of the pure chemical is defined with keyword PHASES. The name of the phase is given, followed by the reaction and the equilibrium constant, here the solubility of the liquid chemical in mol/L. The name is used again when the mixture of organic liquids is defined with keyword SOLID_SOLUTIONS.

After the preparations (keyword END, which ends the first simulation), we start the experiment with USE solution 1, USE solid_solutions 1. The composition of the mixture of organic liquids after extraction is saved with SAVE solid_solutions 1. The extraction is repeated 9 times, each time the original solution 1 is used (it keeps the initial composition if not saved) and the varying solid_solution (changing with each SAVE). Run the file, and inspect the chart (Figure 10.17).

The concentrations in the first extraction are close to half of the solubility of the pure substance, in line with the initially equal molar concentrations in the liquid mixture. In the subsequent extractions, the concentration of TCA decreases, while the concentration of PCE increases. When the concentration of TCA is very small the concentration of PCE approaches the solubility of pure PCE (200 mg/L). In the 8th extraction the organic liquid dissolves completely, the available amount of PCE is insufficient to reach the solubility limit.

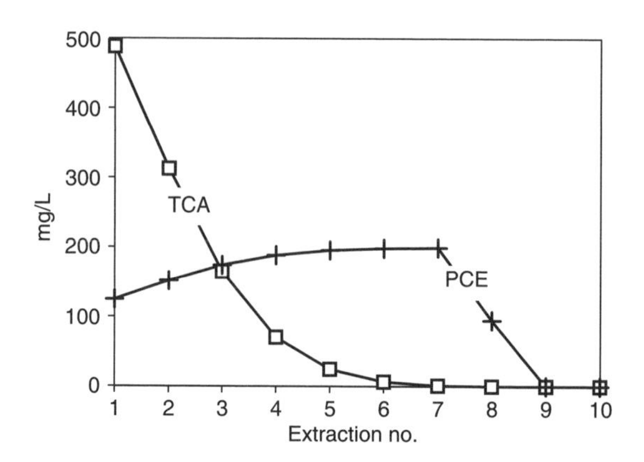

Figure 10.17. The varying aqueous concentrations of TCA and PCE in the extraction experiment of a liquid mixture.

QUESTIONS:
Which chemical is more soluble, TCA or PCE?
ANSWER: TCA
From the input file, find the solubilities of TCA and PCE in mmol/L.
ANSWER: 9.7 mM TCA, 1.2 mM PCE
What would be the initial aqueous concentrations if the size of the pool were very large?
ANSWER: 4.85 mM TCA (650 mg/L), 0.6 mM PCE (100 mg/L)
Plot the fractions of TCA and PCE in the organic liquid on the secondary Y-axis. *Hint*: use the special BASIC function s_s("Tca_lq") which gives the moles of TCA, and similar for PCE (note that the *name* of the component is to be used and not the chemical formula).
Add 1.2 mmol of 1,1,2-TCA to the initial organic liquid and redo the extractions. The solubility in water is 4.4 g/L. What do you expect?
ANSWER: 1,1,2-TCA is extracted first, followed by a peak of 1,1,1-TCA, and lastly of PCE.

Figure 10.18 shows an example where PCE and TCA concentrations are declining during the remediation of a plume (Feenstra and Guiguer, 1996). The ratio PCE/TCA in an observation well increases with time, in agreement with the lower solubility of PCE compared to TCA which is the first to be dissolved completely. We expect that the dilution factor (f) will increase with time as the pool dissolves and is removed, as appears to be the trend in Figure 10.18, but there are irregularities.

Actually, many problems arise when dilution factors and pool sizes are estimated. In the field, the pools may be large and diffusion within the blob may limit the release rate of the compounds. Moreover, sorption in aquifers will be different for the chemicals, which leads to chromatographic separation. Both effects will change the concentration ratios of the components in water from the calculated ones in equilibrium with an ideal pool.

Figure 10.19 shows results from an experiment where an equimolar mixture of benzene and toluene was injected in the center of a column and subsequently eluted with water (Geller and Hunt, 1993). As expected, benzene, with a solubility of 1780 mg/L, was eluted quicker than toluene which has a solubility of 515 mg/L. The concentrations in the effluent did not reach the theoretical solubilities because of the high water flow velocity, and it can be calculated (cf. Question) that the dilution factor was 9 initially.

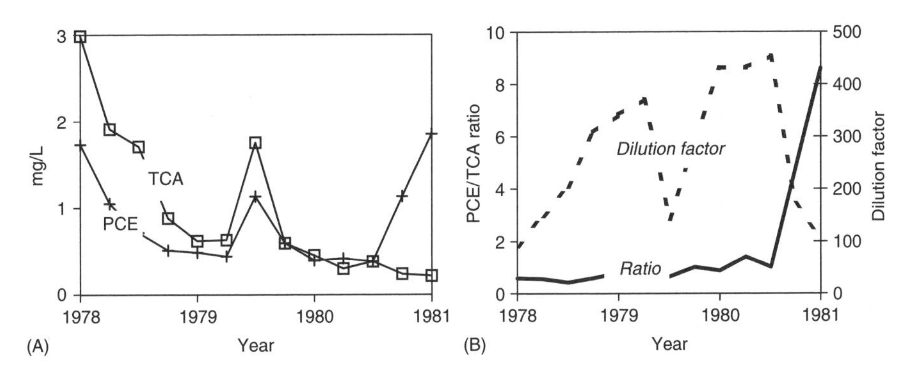

Figure 10.18. Concentrations of PCE and TCA in an observation well during remediation (A), and PCE/TCA concentration ratio and water/(NAPL saturated water) dilution factor (B). (Concentrations from Feenstra and Guiguer, 1996).

Thus, the NAPL formed a stagnant zone that exchanged relatively slowly with water that flowed around it. In the course of the experiment, the dissolution rate increased because the contact surface of the blobs and water increased. Hence, the dilution factor decreased. Furthermore, the dissolution took place in heterogeneous fashion along the column. The NAPL in the first part of the column, already emptied of more soluble benzene, released toluene at a higher concentration than was dissolved from the mixture of NAPL's downstream. Besides these details (and of course, dispersion in the column), the overall elution pattern is very similar to the batch extraction experiment that we calculated before in Example 10.5.

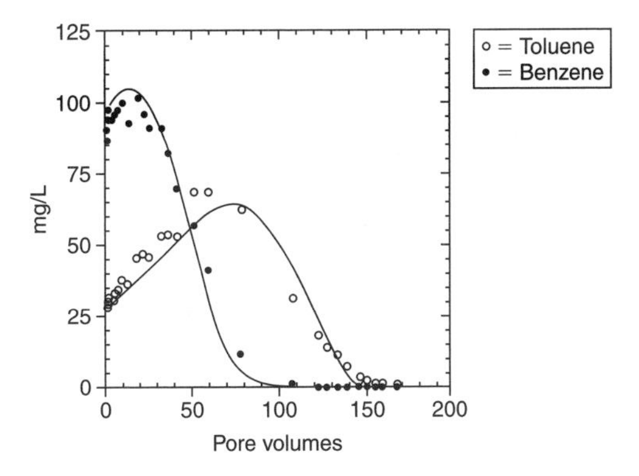

Figure 10.19. Concentrations of toluene and benzene in the effluent of a column with an initially equimolar amount of both chemicals in the column (Geller and Hunt, 1993).

QUESTIONS:

Give the theoretical (equilibrium) concentrations in the initial phase of Geller and Hunt's experiment?

ANSWER: $c_{\text{tol}} = S_{\text{tol}} \times 0.5 = 207\,\text{mg/L}$, $c_{\text{ben}} = S_{\text{ben}} \times 0.5 = 890\,\text{mg/L}$.

Find the dilution factor and χ_{toluene} in Geller and Hunt's experiment at $PV = 1$ and 50?

ANSWER: $PV = 1$: $c_{\text{tol}} = 30\,\text{mg/L}$, $c_{\text{ben}} = 80\,\text{mg/L}$. $f = 9.7$, $\chi_{\text{tol}} = 0.56$.

$PV = 50$: $c_{\text{tol}} = 57\,\text{mg/L}$, $c_{\text{ben}} = 68\,\text{mg/L}$. $f = 6.1$, $\chi_{\text{tol}} = 0.80$.

10.4 TRANSFORMATION REACTIONS OF ORGANIC CHEMICALS

Organic chemicals are transforming via manifold pathways in the environment. The reactions may involve the *substitution* or *elimination* of atoms or groups at the carbon framework of the organic chemical and are usually accompanied by *oxidation* or *reduction*. Take methyl iodide, a chemical that you may have used while preparing microscopic thin slides. It is also produced by marine algae, and is probably one source of the distinct odor of the coastal environment. With OH^-, it transforms to methanol:

$$OH^- + CH_3I \rightarrow CH_3OH + I^- \qquad (10.33)$$

The reaction follows the kinetic rate:

$$\frac{dm_{CH_3I}}{dt} = -k_{OH} m_{OH^-} m_{CH_3I} \qquad (10.34)$$

where $k_{OH} = 10^{-4.2}$/s/(mol/L) (Schwarzenbach et al., 1993, Figure 12.5). Two species figure in this *bimolecular* rate. We can calculate the half-life of methyl iodide with this reaction in seawater with pH $= 8.3$, $m_{OH^-} = 2.6\times10^{-6}$ mol/L:

$$\ln(0.5) = -10^{-4.2}\ m_{OH^-}\ t_{½}, \text{ or } t_{½} = 4.2\times10^9\,\text{s} = 134 \text{ years.} \qquad (10.35)$$

However, the reaction with water is quicker:

$$H_2O + CH_3I \rightarrow CH_3OH + H^+ + I^- \qquad (10.36)$$

which has the same bimolecular rate mechanism, with $k_{H_2O} = 10^{-8.9}$/s/(mol/L). The half-life for this reaction is:

$$\ln(0.5) = -10^{-8.9}\ m_{H_2O}\ t_{½}, \text{ or } t_{½} = 0.3 \text{ years.} \qquad (10.37)$$

(since $m_{H_2O} = 55.6$ mol/L).

An even faster reaction transforms methyl iodide into methyl chloride:

$$Cl^- + CH_3I \rightarrow CH_3Cl + I^- \qquad (10.38)$$

with $k_{Cl} = 10^{-5.5}$/s/(mol/L). With this reaction the half-life of methyl iodide in seawater is only 0.012 years, and the toxic methyl chloride is produced. Much of this substance escapes to the stratosphere, where it is an ozone depleting gas.

QUESTION:
What is the half-life of methyl iodide in fresh water?

ANSWER: $t_{½} = 0.3$ yr, m_{Cl^-} is too small for Reaction (10.38) to be effective.

All the methyl halogenides are thermodynamically unstable with respect to CO_2 and the halogenide ion, but they persist in nature because the reactions are kinetically controlled. The reaction rates for anion substitutions have been determined in the laboratory or were deduced from field observations, and were linked in the *nucleophilicity index* for anions. The nucleophilicity is the logarithm of the reaction rate for an anion, relative to H_2O. In the above examples of methyl iodide reactions, chloride has a nucleophilicity index of $(-5.5 - (-8.9)) = 3.4$. Since the concentration of water is constant at $m_{H_2O} = 55.6$ mol/L, a concentration of chloride of $55.6/10^{3.4} = 22$ mmol/L suffices for an equal reaction rate for chloride and water. The nucleophilicity index can be used to estimate the kinetic reaction rates of similar chemicals (Example 10.6).

EXAMPLE 10.6. *Estimate the hazard of groundwater pollution by methyl bromide*
Methyl bromide is used as fumigant (disinfectant gas) in greenhouse gardening. The bimolecular rate constant of the hydrolysis reaction with water is $10^{-8.4}$/s/(mol/L) (three times faster than for methyl iodide). The most important anion in the soil water, HCO_3^-, has a concentration of 3 mM, and its nucleophilicity index is 3.8. Estimate the hazard of groundwater pollution.

ANSWER:
The rate constant for hydrolysis by water is $10^{-8.4}\, m_{H_2O} = 10^{-8.4}$/s/(mol/L) × 55.6 mol/L = $10^{-6.65}$/s. The rate for substitution by HCO_3^- is $10^{-8.4+3.8} = 10^{-7.12}$/s, slightly smaller than for water. Hence, the half-life is approximately 100 days. In five half-lives, or 1½ year, the concentration will have diminished to 3%, and with the usual travel time of groundwater of a few m/yr, the hazards of methyl bromide pollution are estimated to be small. The Br^- concentration in groundwater may increase, however.

Nucleophilicity expresses the ease with which the carbon nucleus is approached and the substitution reaction accomplished. If access is facilitated, the reaction will speed up. One way is to alter the shape and orientation of the organic molecule by sorbing it onto a solid surface, thus to divulge the nucleus where the reaction takes place. Enzymes in the cells of microbes play a similar role in adapting the orientation of the organic molecule, rendering it more susceptible to the reactions. The half-life of methyl bromide in experiments with unamended soil samples was about 100 hrs (Gan et al., 1998), which is 20 times faster than calculated for the pure hydrolysis reaction in Example 10.6, and low concentrations of methyl bromide could be degraded in the order of minutes by soil microbes (Hines et al., 1998).

Another important type of reaction involves the *elimination* of a molecule or group, removing it altogether, and installing a double bond instead. For example, 1,1,2,2-tetrachloroethane may convert to trichloroethylene by reaction with OH^-:

$$Cl_2HC\text{-}CHCl_2 + OH^- \rightarrow Cl_2C{=}CHCl + H_2O + Cl^- \qquad (10.39)$$

which is again a bimolecular reaction with $k_{OH} = 2$/s/(mol/L). At pH = 7, it gives a half-life of 40 days for 1,1,2,2-TCA.

Trichloroethylene can be *reduced* with H_2 or with Fe(0) (Gillham and O'Hannesin, 1994; Arnold and Roberts, 2000) to dichloroethylene:

$$Cl_2C{=}CHCl + H_2 \rightarrow ClHC{=}CHCl + H^+ + Cl^- \qquad (10.40)$$

and quickly on to vinyl chloride:

$$ClHC{=}CHCl + H_2 \rightarrow ClHC{=}CH_2 + H^+ + Cl^- \qquad (10.41)$$

which again may be reduced to the more innocent molecules ethylene ($H_2C{=}CH_2$) and ethane ($H_3C\text{-}CH_3$). The reaction with Fe(0) involves the reduction of H_2O to produce H_2 and Fe^{2+}:

$$Fe(0) + 2H_2O \rightarrow Fe^{2+} + H_2 + 2OH^- \qquad (10.42)$$

where the resulting hydrogen enhances the reduction of the chlorinated compounds, perhaps assisted by the pH-increase of the reaction. When Fe(0) is introduced in aquifers in *reactive iron barriers*, the pH increases to 10–11, and various precipitates and sorption complexes remove Ca^{2+} and Mg^{2+} (Mayer et al., 2001; Yabusaki et al., 2001).

The highly toxic and mobile vinyl chloride accumulates in aquifers when reducing agents such as pyrite or an iron barrier wall are absent. However, as a fairly reduced substance, it is quite susceptible to *oxidation*:

$$ClHC{=}CH_2 + 2.5\,O_2 \rightarrow 2\,CO_2 + H_2O + H^+ + Cl^- \tag{10.43}$$

Generally, highly reduced BTEX compounds (benzene, toluene, ethylbenzene and xylene) are more easily oxidized than reduced. They may be utilized as substrate by various microbes, including denitrifiers, Fe(3) and SO_4^{2-} reducers and fermentation bacteria (Baedecker et al., 1993; Chapelle, 2001). The oxidation/reduction reactions are catalyzed in sorbed form and accelerated by microbes as discussed extensively in the next section.

10.4.1 *Monod biotransformation kinetics*

The concentrations of phenol and biomass during methanogenic degradation in a laboratory experiment are shown in Figure 10.20 (Bekins et al., 1998). The phenol concentration decreases linearly with time (characteristic for a zeroth order degradation rate), but the degradation seems to decline at very small concentrations.

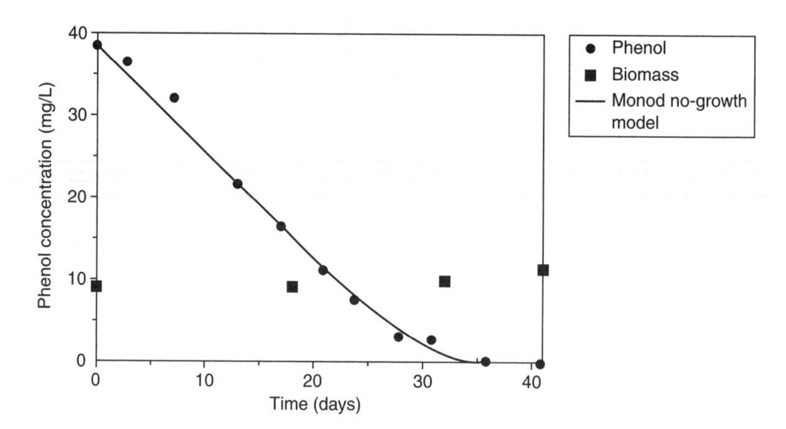

Figure 10.20. Phenol and biomass concentrations (both in mg/L) during methanogenic degradation of phenol in a laboratory experiment (Bekins et al., 1998).

QUESTION:
Write a balanced reaction for the methanogenic degradation of phenol (C_6H_5OH) (cf. Table 9.1)?

ANSWER: $C_6H_5OH + 4H_2O \rightarrow 3.5CH_4 + 2.5CO_2$.

The breakdown was modeled with the Monod kinetic rate equation:

$$\frac{dS}{dt} = R_{Monod} = -k_{max}\frac{S}{k_{1/2} + S} \tag{10.44}$$

where S is the concentration of phenol (mg/L), t is time (s), k_{max} is the maximal rate (mg/L/s), and $k_{1/2}$ is the half-saturation constant (mg/L). The degradation rate is just half of the maximal rate when the concentration of phenol (=S) equals the half-saturation constant.

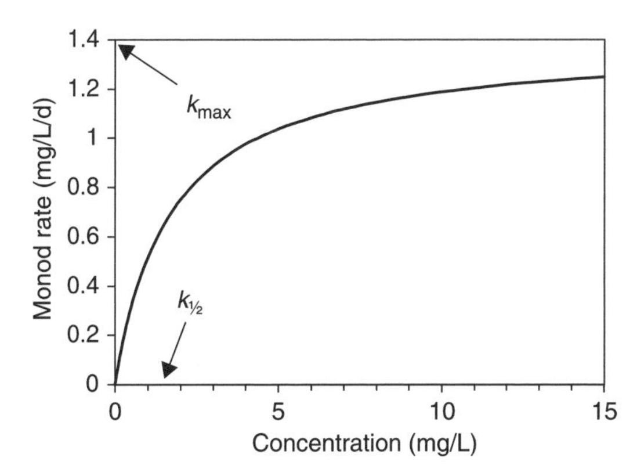

Figure 10.21. The rate of a Monod reaction depends on the concentration of the reactant. The values $k_{½} =$ 1.7 mg/L and $k_{max} = 1.39$ mg/L/d were found in the phenol experiment (Figure 10.20).

The Monod rate depends on the concentration of S and on the value of $k_{½}$ as illustrated in Figure 10.21. When the concentration of S is small, $S < k_{½}$, the rate increases with S. At higher concentrations, when $S > k_{½}$, the rate increases less and becomes more and more independent of S. The Monod equation has been developed for modeling bacterial metabolism and S stands for the substrate or the nutrient utilized by the microbes. At low substrate concentrations the bacteria consume substrate in proportion with the concentration, but at high concentrations the consumption is restricted to what can be utilized by the biomass, for example because an enzyme or another nutrient is limiting the reaction.

Thus, when S can be neglected compared to $k_{½}$ in the denominator of Equation (10.44), it becomes:

$$\frac{dS}{dt} = -\frac{k_{max}}{k_{½}} S \tag{10.44a}$$

The rate is now first order with respect to S.

On the other hand, when $k_{½}$ is small compared to S, Equation (10.44) simplifies to:

$$\frac{dS}{dt} = -k_{max} \tag{10.44b}$$

The rate is now fixed to the maximal rate. It is zero order with respect to S, and therefore, the concentration of S drops linearly with time as in the initial stage of the phenol experiment. The linear decrease of the SO_4^{2-} concentration in a marine sediment shown in Figure 4.14 is also typical for this domain.

EXAMPLE 10.7. *PHREEQC model of phenol degradation*

In their experiment of biodegradation of phenol, Bekins et al., found $k_{max} = 1.39$ mg phenol/L/day and $k_{½} = 1.7$ mg phenol/L. Model the data with PHREEQC.

We make an input file in which phenol is defined as species and the Monod rate is programmed.

```
SOLUTION_MASTER_SPECIES; Phenol Phenol 0 94 94
SOLUTION_SPECIES; Phenol = Phenol; -log_k 0

RATES
 S_degradation
# dS/dt = -k_max * S / (k_half + S).    k_max is maximal growth rate, mol_phenol/L/s.
```

```
# S is nutrient conc, mol phenol/L.      k_half is half saturation conc., mol_phenol/L.
 -start
  1 k_max = parm(1); 2 k_half = parm(2)     # parms defined in KINETICS
 10 S = tot("Phenol")
 20 if < 1e-9 then goto 50                  # quit at low conc.
 30 rate = - k_max * S /(k_half + S)
 40 dS = rate * time
 50 save dS
 -end

SOLUTION 1; -units mg/L; Phenol 40.0

KINETICS 1
 S_degradation
 -formula Phenol 1
 -m0 0
 -parms 1.71e-10 1.81e-5                    # k_max in mol phenol/L/s, k_half in mol/L
 -steps 3.3e6 in 20                         # 40 days
INCREMENTAL_REACTIONS true

USER_GRAPH
 -headings time phenol
 -axis_titles "Time / days" "mg phenol/ L"
 -start
 10 graph_x total_time / 86400
 20 graph_y tot("Phenol") * 94e3
 -end
END
```

Running this file will give the model line of Figure 10.20. The small program that defines the Monod rate starts with assigning the values of parm(1) and parm(2) to k_{max} and $k_{½}$ in lines 1 and 2. These parameters are defined under keyword KINETICS. Next, in line 10, the concentration of phenol is obtained in mol/L (=g/L) with the special function tot("..."). If the concentration is below a limit of 10^{-9} mol/L, we are done and leave the degradation calculation with dS undefined and thus equal to zero (goto line 50). Otherwise, line 30 is processed, where the rate is written just like Equation (10.44). The rate is multiplied with "time" to give the reaction amount dS in line 40. In the last line, dS is passed on to PHREEQC with "save".

The Michaelis-Menten approach

The functioning of the Monod equation can be understood in terms of enzymes in the microbial cell which bind the chemical and facilitate the reaction. Take the enzyme *E*, which associates with *S* to form *ES*, and assume that the reaction rate depends completely on the concentration of *ES*:

$$dS / dt = -k\, ES \tag{10.45}$$

We rewrite *ES* in terms of the total enzyme concentration and the substrate concentration *S*. Thus, the association reaction

$$E + S \leftrightarrow ES; \quad K_{ES} = \frac{[ES]}{[E][S]} \tag{10.46}$$

and a total enzyme concentration $E_t = E + ES$ give:

$$E_t = \frac{ES}{K_{ES} \cdot S} + ES = ES\left(\frac{1}{K_{ES} \cdot S} + 1\right) \tag{10.47}$$

or

$$ES = \frac{E_t \cdot K_{ES} \cdot S}{1 + K_{ES} \cdot S} = \frac{E_t \cdot S}{(K_{ES})^{-1} + S} \tag{10.48}$$

This is the Michaelis-Menten equation. (Note that it is similar to the Langmuir isotherm, cf. Equation (7.9), also note that we have left out the activity signs).

On combining Equations (10.45) and (10.48), we obtain:

$$\frac{dS}{dt} = \frac{-k \cdot E_t \cdot S}{(K_{ES})^{-1} + S} \tag{10.49}$$

Returning now to the Monod equation, we can see that the half-saturation constant, $k_{½}$, equals the inverse of the association constant, K_{ES}, in the Michaelis-Menten equation. Since the association has the specific function to promote a reaction, K_{ES} will be on the high side in the range of association constants, say 10^5 (cf. PHREEQC.DAT). The value of $k_{½}$ is therefore expected to be in the μmol/L range. (Note that we assumed that the complex *ES* forms instantaneously at equilibrium. If not, $k_{½}$ may be larger than $(K_{ES})^{-1}$, cf. Berg. et al., 2002).

Biomass effects

The Michaelis-Menten approach shows that the rate depends on the total enzyme concentration, E_t. Since the enzymes are located in bacterial cells or produced by them, the rate depends on the number of microbes, and the latter may grow as substrate *S* is consumed. Thus, in the Monod equation, the rate of substrate consumption also depends on the number of bacteria present. Actually, in Equation (10.44) the bacterial mass concentration was included in k_{max}:

$$k_{max} = \mu_{max} \, B / Y \tag{10.50}$$

where μ_{max} is the specific bacterial growth rate (s^{-1}), *B* is the biomass (mol $CH_{1.4}ON_{0.2}$/L), and *Y* is the yield factor. The biomass formula is for dry bacterial cells according to Rittmann and McCarty (2001).

The yield factor indicates the fraction of substrate that is converted to biomass. Usually, *Y* is given in (g biomass) / (g substrate) irrespective of the chemical composition of *S* (Madigan et al., 2003). For a more direct link between reaction energetics and cell growth, Rittmann and McCarty (2001) suggest to express *Y* in (electron-equivalents needed for cell growth)/(electron-equivalents from *S*-degradation). For the same reason, and often resulting in simpler arithmetics, VanBriesen (2002) uses (mol $CH_{1.4}O_{0.4}N_{0.2}$/mol C in *S*).

For organic matter and methanogenic conditions, *Y* is small, about 0.05 (mol $CH_{1.4}O_{0.4}N_{0.2}$/mol C in *S*) as discussed later. Now, one bacterial cell contains about 10^{-14} mol carbon (Balkwill et al., 1988), and thus, a yield factor of 0.05 means that degradation of 40 mg phenol/L will have increased the bacterial population with:

$$(0.04 \text{ g phenol}) \times (0.05 \text{ mol C in biomass / mol C in phenol}) \times (6 \text{ mol C / mol phenol}) / (94 \text{ g/mol phenol}) / (10^{-14} \text{mol C/bacterium}) = 1.3 \times 10^{10} \text{ bacteria.}$$

The bacterial cells consist of $CH_{1.4}O_{0.4}N_{0.2}$ and are equivalent to

$$1.3 \times 10^{10} \times 10^{-14} \times (22.6 \text{ g/mol } CH_{1.4}O_{0.4}N_{0.2}) = 2.9 \text{ mg.}$$

which is about the increase shown in Figure 10.20 (squares). In the phenol degradation experiment, k_{max} was assumed constant and fitted together with $k_{½}$ on the experimental data.

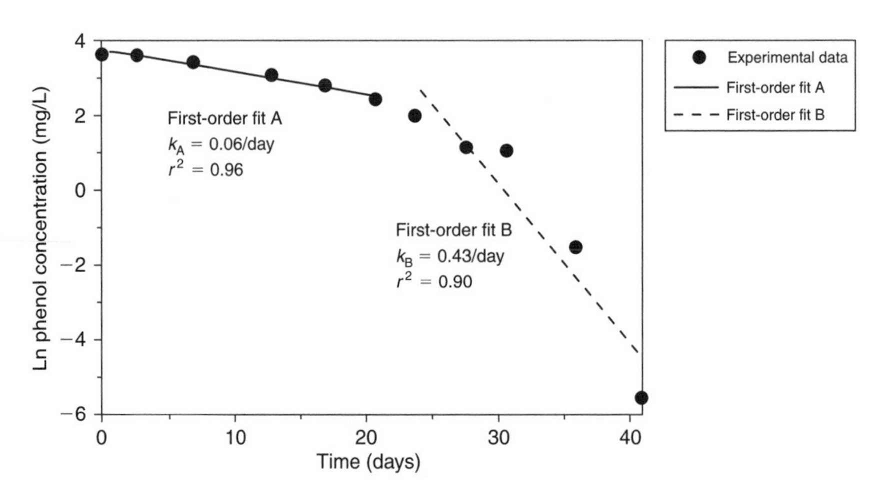

Figure 10.22. A logarithmic plot of phenol concentration vs time, showing two apparent first order rates (Bekins et al., 1998). *Note*: the logarithmic plot appears in Example 10.7 when line 20 in USER_GRAPH is changed to: 20 graph_y log(tot("Phenol") * 1 × 10^3).

Bekins et al. also plotted their data assuming first order degradation of phenol (Figure 10.22). For a first order rate, the concentration of phenol should decrease logarithmically with time, but the plot showed a non-linear trend of ln(phenol) *vs* time. The data were divided in two linear portions, each with a different rate constant. Interestingly, the plot shows that the apparent first order rate constant is greater for larger time than the initial rate constant (initially 0.06/day, finally 0.43/day). This behavior in logarithmic plots is typical for Monod type kinetics and will be noted when the degradation rate is monitored for both large and small concentrations of S compared to $k_{½}$.

If the bacterial mass does vary, the change of B can be coupled to substrate decrease and to biomass decay according to:

$$\frac{dB}{dt} = -Y\frac{dS}{dt} - k_{Bd}B \tag{10.51}$$

where k_{Bd} is the first order death rate of the bacteria (s^{-1}). Neglecting the death rate ($k_{Bd} = 0$), the growth can be included in Equation (10.44):

$$\frac{dS}{dt} = R_{\text{Monod}} = -\mu_{\max}\frac{(B_0 + Y(S_0 - S))}{Y}\frac{S}{k_{½} + S} \tag{10.52}$$

where B_0 is the initial bacterial mass (mol $CH_{1.4}O_{0.4}N_{0.2}$/L), and S_0 is the initial substrate concentration (mol C/L). $Y(S_0 - S)$ gives the cells which grew out of degradation of $S_0 - S$. This equation can be integrated analytically (Alexander and Scow, 1989), but it may be easier to look at a solution with PHREEQC as in Example 10.8.

EXAMPLE 10.8. *Xylene degradation with biomass growth*

Schirmer et al. (1999) measured breakdown of xylene (C_8H_{10}) in a batch experiment with aquifer sediment. The vial with pristine sediment contained 1.47 mg biomass/L, of which 0.2% was active in degrading aromatic hydrocarbons. When 16 mg xylene/L were added to the reaction vessel, the concentration dropped immediately to 8.6 mg/L which indicated that 46% of the chemical was sorbed to the sediment and evaporated into the headspace of the vial. In other words, the distribution coefficient $K_d = (16 - 8.6)/8.6 = 0.86$, and breakdown of the solute concentrations was retarded by $R = 1.86$. The data can be modeled with the Monod rate equation, with $B_0 = 7.5\times10^{-8}$mol $CH_{1.4}O_{0.4}N_{0.2}$/L, $\mu_{max} = 2.49\times10^{-5}$/s, $k_{½} = 7.45$ μmol xylene/L and $Y = 0.305$ (mol $CH_{1.4}O_{0.4}N_{0.2}$/mol xylene-C) (experimental conditions at 10°C, oxygen at 0.3 mM in excess). We extend the input file from Example 10.7 with a rate for biomass increase, and the experimental details from Schirmer et al. Note that Schirmer's parameters were recalculated from mg/L to mol/L (cf. Questions).

```
SOLUTION_MASTER_SPECIES; Xylene Xylene 0 106 106                # C8H10
SOLUTION_SPECIES; Xylene = Xylene; -log_k 0


RATES
 S_degradation
# dS/dt = -mu_max * (B/(Y * 8)) * (S / (k_half + S)) / R      (mol xylene/L/s)
# mu_max is maximal growth rate, 1/s.                          B is biomass, mol C/L.
# Y * 8 is yield factor, mol biomass-C/mol xylene.             S is xylene conc, mol/L.
# k_half is half saturation concentration, mol xylene/L. R is retardation, 1 + K_d
# K_d = q_xylene / c_xylene.
 -start
  1 mu_max = parm(1); 2 k_half = parm(2); 3 Y = parm(3); 4 R = 1 + parm(4)
 10 S = tot("Xylene")
 20 if S < 1e-9 then goto 60
 30 B = kin("Biomass")                       # kin(".i.") gives moles of "Biomass"
 40 rate = -mu_max * (B / (Y * 8)) * (S /(k_half + S)) / R
# 40 rate 5 -mu_max * (B / (Y * 8)) * (S /(k_half + S + S^2/8.65e-4)) / R
 50 dS = rate * time
 60 save dS                                  # d mol(C8H10)
 70 put(rate, 1)                             # Store dS/dt for use in Biomass rate
 -end


 Biomass
# dB/dt = dBg/dt - dBd/dt
# dBg/dt = - Y dS/dt is biomass growth rate. dBd / dt = k_Bd * B is biomass death rate.
# k_Bd is death rate coefficient, 1/s.
 -start
  1 Y = parm(1); 2 R = 1 + parm(2); 3 k_Bd = parm(3)
 10 rate_S = get(1) * R                 # Get degradation rate, multiply by Retardation
 20 B = m
 30 rate = -(Y * 8) * rate_S - k_Bd * B
 40 dB = rate * time
 50 save -dB                            # dB is positive, counts negative to solution
 -end


SOLUTION 1; -units mg/L; Xylene 8.6
KINETICS 1
                         # xylene, Schirmer et al., 1999. JCH 37, 69-86, expt. Figure 6b...
 S_degradation;  -formula Xylene 1;      -m0 0
  -parms 2.49e-5 7.45e-6 0.305 0.86     # mu_max, k_half, Y, K_d
 Biomass;  -formula C 0;-m0 0.75e-7     # 1.33e-7 with Haldane inhibition
  -parms 0.305 0.86 0                   # Y, K_d, k_Bd
  -steps 0.6e6 in 50
INCREMENTAL_REACTIONS
```

```
USER_GRAPH
 -headings time c_xylene Biomass
 -axis_scale y_axis 0 10 2 1; -axis_scale x-axis 0 6 1 0.5
 -axis_titles "Time / days" "mg / L"
 -start
 10 graph_x total_time / 86400
 20 graph_y tot("Xylene") * 106e3, kin("Biomass") * 22.6e3
 -end
END
```

Results of the calculation are shown in Figure 10.23, together with the experimental data. Quite conspicuous is the delay of xylene breakdown because the microbes which degrade the substance first must grow and increase in number. When the biomass is sufficient, xylene is quickly gone. The rate for S_degradation takes this into account by multiplying with $B/(Y \times 8)$ (mol xylene/L). Initially, B is small (B_0 = "m0" = 75 nmol/L in "Biomass", keyword KINETICS), but as xylene is used for energy and growth, B increases. Note how the degradation rate (dS/dt) is passed on to the biomass growth rate with the special BASIC functions "put" and "get". Because the degradation rate for solute xylene is retarded by sorbed and headspace xylene, it is multiplied with the retardation factor $R = (1 + K_d)$ to obtain the rate for total xylene. Line 50 of the Biomass rate definition "saves" negative moles, and because dB is positive, the kinetic reactant (mass m of Biomass) increases. Furthermore, note that the death rate of the microbes is set to zero in this example (parm(3) = 0 in the call for Biomass in KINETICS).

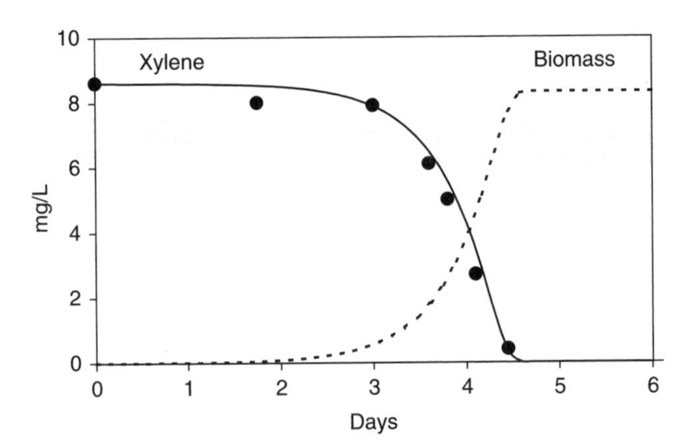

Figure 10.23. Microbial degradation of *m*-xylene in batch experiments (Schirmer et al., 1999). Datapoints and model line for solute xylene, dotted line for biomass.

The kinetic reactant "Biomass" is not a solute species and its reaction does not change the solution composition, which is indicated by the zero coefficient in "-formula C 0". In fact, these kinetic calculations could be done without any change in solution composition, only referring to the moles of kinetic reactants.

When Schirmer et al. added initially more xylene to the reaction vessel, the lag time increased. This was modeled with an inhibition factor of the form S^2/k_{HD} in the denominator of the Monod rate (with k_{HD} = 0.865 mmol S/L, the denominator becomes $(k_{½} + S + S^2/8.65 \times 10^{-4})$). Also, they used the initial concentration of B_0 = 133 nmol/L. If the initial biomass is made 100 μmol/L, and inhibition is absent, degradation starts immediately and the xylene concentration decreases linearly with time, similar to phenol in Figure 10.20.

When doing the following questions, it is helpful to compare the results of the simulations directly, fixing the chart of PHREEQC as can be modified in Edit, Preferences, Miscellaneous.

QUESTIONS:

Explain the increase of biomass quantitatively?

ANSWER: (8.6 mg xylene/L) / (106 g/mol xylene) × (8 mol C/mol xylene) × (R = 1.86) × (Y = 0.305) × (22.6 g/mol $CH_{1.4}O_{0.4}N_{0.2}$) = 8.32 mg biomass.

Change the yield factor from 0.305 to 0.5, explain the results?

ANSWER: degradation starts later, final biomass becomes 8.32 × 0.5 / 0.305 = 13.6 mg/L.

Change the yield factor back to 0.305 and make the initial biomass 100 μmol/L in the input file, compare the results.

ANSWER: degradation starts immediately with zero order rate.

Include the inhibition factor S^2 / k_{HD} and change m_0, compare the results.

ANSWER: Result is the same as in Figure 10.23.

What will be the effect of zero sorption of xylene ($K_d = 0$)?

ANSWER: solute xylene is exhausted quicker; final biomass = 8.32 / 1.86 = 4.47 mg/L.

And if you have bacterial decay ($k_{Bd} = 1 \times 10^{-6}$)?

ANSWER: xylene degradation starts later, biomass decreases when xylene is gone.

Model the retardation as sorption reaction, cf. Example 10.3.

Estimate the initial number of active bacterial cells in Schirmer's experiment?

ANSWER: (1.7×10^{-6} g/L) / (22.6 g/mol) / (10^{-14} mol C/cell) = 7.5×10^6 bacteria/L.

Schirmer expressed the yield factor as mg biomass/mg xylene. Find his Y?

ANSWER: 0.305 × (8 mol C/mol xylene) × (22.6 g/mol $CH_{1.4}O_{0.4}N_{0.2}$) / (106 g/molxylene) = 0.52

He also expressed $k_{½}$ and k_{HD} in mg xylene/L. Derive his values?

ANSWER: 7.45×10^{-3} × 106 = 0.79 mg/L; k_{HD} 0.865 × 106 = 91.7 mg/L

And lastly, obtain Schirmer's μ_{max}?

ANSWER: 2.49×10^{-5}/s, does not change.

Multiplicative Monod equations

In the experiments of Schirmer et al. (Example 10.8) oxygen was available in excess for the xylene degradation reaction, but this may not be the case in plumes of BTEX-chemicals in an aquifer. Here, oxygen must diffuse and disperse into the plume and its concentration will limit the breakdown rate. This dependency is modeled with the multiplicative Monod equation, in which the concentration of the electron acceptor appears as a multiplier in the same form as the electron donor (MacQuarrie et al., 1990; Schirmer et al., 2000; Prommer et al., 2000; Rittmann and McCarty, 2001). For example for oxygen:

$$\frac{dS}{dt} = -\left(\frac{k_{max} S}{k_{½} + S}\right)\left(\frac{k_{max,O_2} m_{O_2}}{k_{½,O_2} + m_{O_2}}\right) \tag{10.53}$$

where $k_{½,O_2}$ is the half-saturation concentration for oxygen (mol/L) and k_{max,O_2} is the maximal rate-multiplier for oxygen (−). In the experiment of Schirmer et al. the whole term was included in k_{max}. If another electron acceptor is present such as NO_3^-, another rate equation is written for the same substrate using NO_3^- instead of oxygen in the multiplicative part. If moreover, the oxidants are known to be utilized stepwise, yet another multiplier can be added which limits the reaction as long as the other oxidant is present. The equations then tend to become rather formidable (Essaid et al., 1995, 2003; Van Cappellen and Wang, 1996; Boudreau, 1996; MacQuarrie and Sudicky, 2001) but the logic is clear if simple steps are followed (Rittmann and McCarty, 2001). On the other hand, the equations seem to have lost the touch of the microbiological system in which different groups of bacteria utilize different substrates, each of them operating at a different pace that is probably dictated by the number of bacteria, and their growth and death rates.

Empirically, the usual oxidants in groundwater have large differences in their $k_{max,i}$, undoubtedly because the various bacterial groups are present in optimal form for the specific conditions of the aquifer. The large difference shows up in the persistence of nitrate when oxygen has been depleted, and of sulfate when nitrate has been utilized already completely. The multiplier that limits the reaction by the oxidant concentration (Equation 10.53) then is no longer necessary. Based on the degradation rates of organic carbon in aerobic soils and of nitrate reduction by organic carbon in aquifers, the following combination was obtained for the oxidant multiplier (Appelo and Parkhurst, 1998):

$$r_{Oxidants} = \left(\frac{m_{O_2}}{2.94 \times 10^{-4} + m_{O_2}} + \frac{0.01\, m_{NO_3^-}}{1.55\times10^{-4} + m_{NO_3^-}} + \frac{6.4 \times 10^{-5}\ m_{SO_4^{2-}}}{1\times10^{-4} + m_{SO_4^{2-}}} \right) \quad (10.54)$$

where m_i is molality. The degradation of natural organic carbon can now be calculated with:

$$\frac{dS_{OC}}{dt} = -k_1 \cdot S_{OC} \cdot r_{Oxidants} \quad (10.55)$$

where S_{OC} is organic carbon content (mol/kg soil), and k_1 is the first order decay constant/s. With $k_1 = 1.57\times10^{-9}$/s and $m_{O_2} = 0.3$ mM, Equation (10.55) gives a first order degradation rate of 0.025/yr, or 2.5% of organic carbon degrades each year, in accordance with estimates for aerobic soils in a temperate climate. The value for k_1 is for organic carbon in soils, and may be changed for older organic carbon (Equation 9.61). Note that Equation (10.54) is purely empirical, it does not account for the number of bacteria that process the 3 oxidants, which will differ from place to place, but it may give a first guess.

Monod rate coefficients

Values for rate coefficients have been noted in the literature, but these require careful consideration before comparisons can be made. Clearly, we would like to know the number of bacteria or the microbial mass and the yield factor Y since the rate is directly depending on it. However, these parameters are often not available, were assumed constant, or included in the value of k_{max} which is the easiest factor for describing the system. Moreover, the yield factor is usually expressed simply as mg dry biomass/mg substrate, with neglect of the different formula weights of the two (cf. Questions with Example 10.8). Finally, the oxidant species and its concentration will influence the overall rate, but usually these are only accounted for in models. A few values for μ_{max} (or k_{max}) and $k_{1/2}$ (for the organic compound) are noted in Table 10.4 together with the experimental conditions.

Table 10.4. Monod coefficients for degradation of organic chemicals in aquifers.

Chemical	μ_{max} (s^{-1})	$k_{1/2}$ (mol/L)	Reference
BTX	5×10^{-5}	2.2×10^{-4}	Goldsmith and Balderson, 1988, batch with contaminated soil, oxygen
xylene	1.06×10^{-4}	1.4×10^{-4}	Kelly et al., 1996, batch with contaminated soil, oxygen
BTX	5.6×10^{-5}	2×10^{-4}	Kelly et al., 1996, batch with contaminated soil, oxygen
	4.35×10^{-6}	2×10^{-4}	Eckert and Appelo, 2002, contaminated aquifer, 2.5 mM sulfate. k_{max} (in mol/L/s) decreases by 100 in less contaminated parts of the aquifer
toluene	5.69×10^{-6}	7.1×10^{-6}	MacQuarrie et al., 1990, Column with pristine sediment, oxygen
m-xylene	2.49×10^{-5}	7.45×10^{-6}	Schirmer et al., 1999. Batch with pristine sediment, oxygen (cf. Example 10.8)
phenol	1.16×10^{-7}	1.81×10^{-5}	Bekins et al., 1998. Inoculated sediment, methanogenic (cf. Example 10.7)

It has been noted that the various BTX compounds (benzene, toluene and xylene) show similar degradation rates under identical conditions (Kelly et al., 1996). A conclusion that can be drawn from Table 10.4 is that $k_{½}$ is small (as expected from Michaelis-Menten kinetics), and tends to be smaller in uncontaminated sediments, which indicates that the biomass is more prone to devour even small concentrations of the compound when nutrients are restricted as is usually the case in an uncontaminated environment. Where μ_{max} is given, the Monod rate must be multiplied with B/Y (the active biomass (mol C/L) divided by the yield factor (mol C in biomass/mol C in *S*)), to obtain the rate in mol carbon in *S*/L/s. To do so in a general sense the number of active bacterial cells (number/L) is multiplied with the average carbon content per cell (10^{-14} mol C/cell), and divided by the yield factor in mol C per mol C in the nutrient. In soils, the living biomass is about 5% of organic carbon content, or about 10^{10} cells per g soil with 3% *OC*. In lake sediments and shallow aquifers the concentration of cells was estimated to range from 10^5/g to 10^7/g, and in deep aquifers as many as 10^4–10^6 culturable organisms per gram sediment were obtained (Barns and Nierzwicki-Bauer, 1997). Of all the cells, only a part will be active in a specific process. Thus, for an aquifer,

$$B = (10^8 \text{ cells/kg}) \times (10^{-14} \text{ mol C/cell}) \times (6\,\text{kg/L for } \rho_b/\varepsilon)$$
$$\times (0.1 \text{ for an estimated active part}) = 6\times10^{-7} \text{ mol } CH_{1.4}O_{0.4}N_{0.2}/\text{L}.$$

Then, for methanogenic degradation of phenol,

$$k_{max} = (1.16\times10^{-7}/\text{s, Table 10.4}) \times (6\times10^{-7} \text{ mol } CH_{1.4}O_{0.4}N_{0.2}/\text{L})/$$
$$(Y = 0.05 \text{ (mol } CH_{1.4}O_{0.4}N_{0.2}/\text{mol C in phenol)} \times (6\,\text{mol C/mol phenol}))$$
$$\times (94 \times 10^3\,\text{mg/mol phenol}) \times (86400\,\text{s/day}) = 1.9\,\mu\text{g phenol/L/day}.$$

The degradation of phenol in anaerobic aquifers is indeed slow (Thornton et al., 2001).

QUESTION:
Compare k_{max} calculated above with the value of Bekins et al. (1998). *Hint*: recalculate B from 10 mg/L into mol C/L.

ANSWER: $B = 0.01 \;/\; 22.6 = 4.4\times10^{-4}$ mol C/L, gives in the inoculated sediment $k_{max} = 1.4$ mg phenol/L/day.

The yield factor

The yield factor Y in Equation (10.50) relates the growth of the bacteria with the reaction, and we noted that Y is small for methane producing reactions and increases in an aerobic milieu. Can we explain that, at least in a qualitative sense? For growth, the microbes need energy and chemicals for the synthesis of the organic molecules of the cell. Most experimental microbiological studies add nutrients and assure that the supply of chemicals is sufficient. With plenty of nutrients, it is the reaction energy released with substrate degradation that determines the energy available to the cells and thus the growth that can be sustained. The synthesis of a bacterial cell from pyruvate ($C_3H_4O_3$) as substrate, requires about 75 kJ/mol $CH_{1.4}O_{0.4}N_{0.2}$ (Rittmann and McCarty, 2001).

Since more energy is released when oxygen is used as electron acceptor than in the case of fermentation (Table 9.2; Figure 9.15), the yield factor for aerobic systems should be higher than for anaerobic conditions (Lensing et al., 1994; Rittmann and McCarty, 2001; Watson et al., 2003). If we compare the breakdown of pyruvate by fermenting and aerobic bacteria, the aerobic ones gain 1138 kJ/mol pyruvate and their fermenting friends only 115.6 kJ/mol pyruvate, or 9.8 times less (reaction energies from Rittmann and Mcarthy, 2001). Thus, also 9.8 times more aerobic bacteria can form. If the two microbial groups are present together and the reaction rates are comparable for the breakdown of pyruvate, then clearly, the oxygen reducers will quickly outcompete the fermenters in number and use up all the available pyruvate and oxygen in the proportion given by the reaction stoichiometry. First when oxygen is depleted, the fermenters stand a chance to consume the pyruvate substrate. Apparently, we have *growth of the fittest* who can snatch the energy (cf. Question).

The weak point in the theory is that the reaction rates of substrate breakdown by the different bacteria are variable and the parameters in the rates not exactly known. If the reaction rate is 9.8 times higher for the fermentation reaction than for aerobic oxidation, the energy advantage would be egalized and both groups grow with equal handicap. Various bacteria show only about two-fold growth increase when nitrate is added to an anaerobic culture with glucose as substrate, and up to four-fold increase when oxygen is added (Stouthamer, 1976). Thus, the observed growth-rate increase is smaller than expected from pure energy consideration, which can be related to low reaction efficiency or entropy loss (Heijnen and Van Dijken, 1992; VanBriesen, 2002). Moreover, the increase is twice larger for oxygen than for nitrate, which can not be explained simply by energetics, since the two oxidants provide nearly equal energy in the oxidation reaction. On the other hand, when chlorophenol, nitrate or sulfite were provided as electron acceptors to a bacterium that was able to degrade chlorinated aromatic hydrocarbons, the yield factor doubled almost uniformly compared with an anaerobic culture in pyruvate without electron acceptors (Mackiewicz and Wiegel, 1998). These oxidants give about equal energy. Also, VanBriesen (2002) compared yield factors from experiment with estimates based on reaction energies of various substrates with oxygen and found a good agreement (Figure 10.24). The observed yield (mol $CH_{1.4}O_{0.4}N_{0.2}$/mol C in the chemical) varied from 0.57 for glycerol ($C_3H_8O_3$, the most reduced substance in the comparison) to 0.22 for glyoxylate ($C_2H_2O_3$, the most oxidized substance). Rittmann and McCarty (2001, Table 3.1) list yield factors that vary from 0.7 for aerobic heterotrophic bacteria to 0.05 for methanogens and sufate reducers which use H_2 as electron donor, which shows the same trend of decreasing yield with smaller reaction energy.

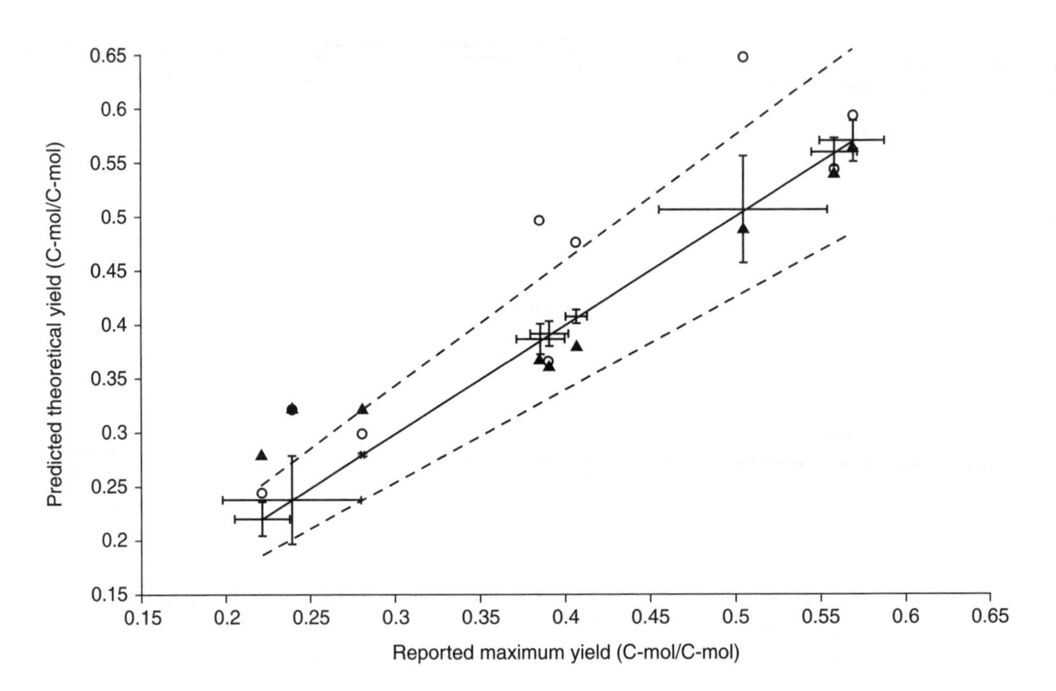

Figure 10.24. Comparison of predicted theoretical and reported maximum yield for various substrates in aerobic cultures. Triangles and circles are based on the methods proposed by Rittmann and McCarty (2001) and Heijnen and Van Dijken (1992), respectively. Experimental points are given on the solid line with 1:1 correlation, with error bars for reported yields. The dotted lines indicate 15% above and below reported yields. (VanBriesen, 2002).

The small yield factor for anaerobic processes translates in the slow start of fermentation compared to aerobic digestion in nature. Again, it is important to consider the timeframe when interpreting water quality patterns in an aquifer. In Chapter 9, we have related the water quality transitions in the plume below a waste site to oxidation reactions in the aquifer (Figure 9.18). Models typically attribute the water quality patterns to sequential biodegradation reactions in the aquifer and assume a constant composition of the leachate over time, that is given by the reduced, methanogenic water sampled directly below the waste site (Brun et al., 2002; Barry et al., 2002). However, it is likely that some time must pass before the leachate attains the methanogenic stage. The waste must be wet and sufficiently compacted so that oxygen diffusion is inadequate for aerobic degradation. It takes time before the bacterial population reaches the methanogenic stage, first a butyric acid stage must be ended that is characterized by high concentrations of organic acids and accompanied by high alkalinity concentrations. (These alkalinities are due to the organic acids, they are not from HCO_3^-). The spatial sequence in the aquifer has been observed as a time sequence in waste percolates that leached from waste sites and never did contact the aquifer below (Kjeldsen et al., 2002). The time sequence is related to the gas composition in the waste; it may take a decade before the methanogenic stage is reached (Farquhar and Rovers, 1973). Thus, instead of showing the reactions in the aquifer, the lateral concentration pattern in the plume probably represents the time variations in the composition of the waste site leachate, the initial, aerobic percolate having traveled farthest away from the site and the final, methanogenic stage being closest to the waste.

QUESTION:
Give the number of bacteria (n) after 24 hr for duplication times (d) of 1 hr (for aerobic bacteria) and 9.8 hr (for fermenting bacteria), starting with n_0?

ANSWER: $n / n_0 = 2^{t/d}$ yields 1.7×10^7 and 5, respectively.

10.5 KINETIC COMPLEXATION OF HEAVY METALS ON ORGANICS

We have noted that natural organic matter and organic chemicals can complex heavy metals and solubilize them (Section 7.5). Examples of concern are americium from nuclear waste bound to humic acid (Artinger et al., 1999) and gadolinium-pentatetic acid which has a medical application for magnetic resonance imaging (MRI) and can be traced in natural waters that receive effluents from hospitals (Knappe et al., 1999; Kummerer and Helmers, 2000). Natural organic matter, even in the dissolved state, forms large molecules with a charge-dependent potential which necessitates a special correction term when calculating complexation (discussed in Section 7.5). Complexation of heavy metals on smaller, artificial compounds such as NTA and EDTA which are used in industrial detergents and domestic washing-powders can be calculated in a more straightforward manner, but the kinetics of the complexation reactions are often slow (Hering and Morel, 1989). The kinetics of complex formation can be important if only a specific complex is biodegradable (VanBriesen et al., 2000). We will discuss an example of EDTA (ethylene diamine tetra acetate) in the Glatt river in Switzerland, which receives EDTA as the FeEDTA complex from sewage treatment plants. The FeEDTA forms because $FeCl_3$ is added to precipitate and coagulate phosphate and remove it from the waste water. In the river, iron is exchanged for Ca^{2+} and Zn^{2+}, in a slow, kinetic exchange (Xue et al., 1995).

The equilibrium speciation of EDTA for the river conditions shows that ZnEDTA and CaEDTA are the major complexes over the pH range 7.5–8.5 (Figure 10.25A). The total EDTA concentration in this figure is 0.06 μM ($10^{-7.22}$ M). Zn is extremely strongly complexed by EDTA and is able to displace Ca^{2+} from the complex already at a Zn/Ca (total concentration) ratio of 5×10^{-5} ($m_{Zn} = 10^{-7}$ M in Figure 10.25B). Lastly, the effect of increasing the EDTA concentration is illustrated in Figure 10.25C. The various complexes increase 1:1 with EDTA, except for ZnEDTA, which levels off when it is limited by the total Zn concentration.

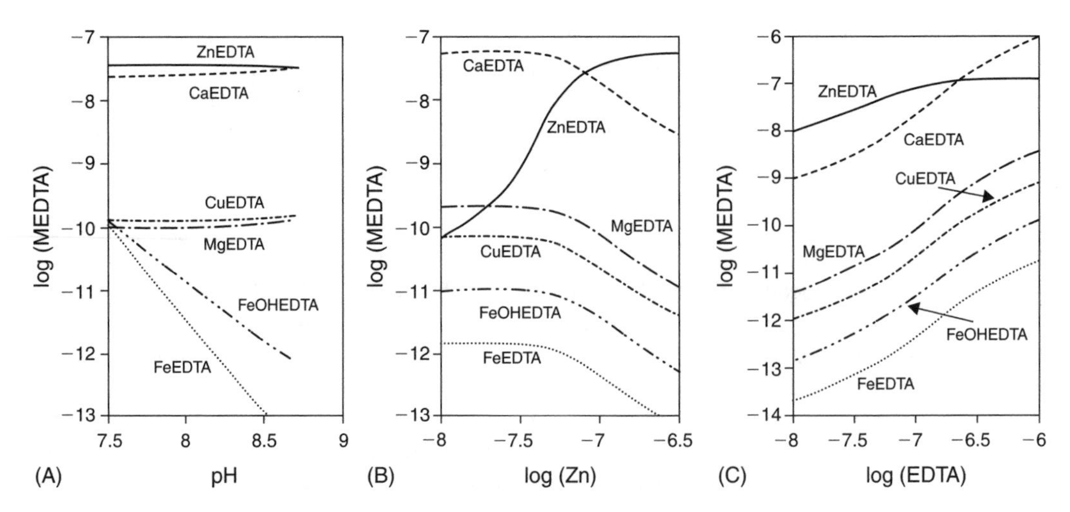

Figure 10.25. Equilibrium speciation of EDTA computed for the conditions of Glatt river water. The concentrations of the complexes are plotted as function of pH (A), of total Zn concentration (B), and of EDTA concentration (C). The initial water composition is given in Example 10.9. (Xue et al., 1995).

The data in the figure were calculated with Microql, and can be calculated with PHREEQC with the Minteq database. However, the constants must be adapted to fit the plotted concentrations and the alkalinity must be crossed out, as shown in Example 10.9.

EXAMPLE 10.9. *Speciation of EDTA in Glatt river water*

The speciation calculation for Glatt river water detailed in the PHREEQC input file turns out to be different from what is shown in Figure 10.25A, at pH = 8.25. However, a reasonable agreement can be obtained when a few complexation constants are adapted and the alkalinity is crossed out (apparently, the complexes with HCO_3^- are absent in Microql, used for Figure 10.25). Also, a complex of Cu^{2+} with DOC is included in the input file as was noted by Xue et al., but with a modified constant.

```
DATABASE minteq.dat
SOLUTION_MASTER_SPECIES; Lcu Lcu-2 0 1 1                          # DOC for Cu
SOLUTION_SPECIES
                                                                   # Complex Cu with DOC...
 Lcu-2 = Lcu-2; -log_k 0; Lcu-2 + Cu+2 = LcuCu; log_k 11.5        # Xue et al. used 13.6
                                                                # Modify Minteq database...
 Zn+2 + EDTA-4 = ZnEDTA-2;                 log_k 17.04 # Minteq 16.44
 Fe+3 + EDTA-4 = FeEDTA-;                  log_k 26.9  # Minteq 27.7
 FeOH+2 + EDTA-4 = FeOHEDTA-2;             log_k 21.49 # Minteq 21.99

SOLUTION 1 Glatt river water
 pH 8.25; pe 12 O2(g) -0.68;
 temp 25; # Alkalinity 3.97
 Na 1.2;         K 0.18;         Mg 0.71;       Ca 1.83
 Zn 170e-6;      Cu 31.5e-6;     Fe 0.2e-3      Ferrihydrite -1.69
 EDTA 60e-6;     NTA 20e-6;      Lcu 40e-6
END
```

which gives the speciation for EDTA (partial results):

```
-----------------------------Distribution of species------------------------------
                                                Log          Log          Log
       Species       Molality      Activity     Molality     Activity     Gamma
       EDTA          6.000e-08
       ZnEDTA-2      4.061e-08     2.940e-08     -7.391       -7.532      -0.140
       CaEDTA-2      1.909e-08     1.382e-08     -7.719       -7.860      -0.140
       CuEDTA-2      1.816e-10     1.314e-10     -9.741       -9.881      -0.140
       MgEDTA-2      1.178e-10     8.526e-11     -9.929      -10.069      -0.140
       FeOHEDTA-2    5.662e-12     4.099e-12    -11.247      -11.387      -0.140
       FeEDTA-       9.949e-13     9.177e-13    -12.002      -12.037      -0.035
```

The differences in the databases are not unusual and should be considered when doing these calculations. For example, Xue et al., use ferrihydrite with a solubility constant of $10^{-38.8}$, while the constant is $10^{-37.11}$ in MINTEQ.DAT. Thus, ferrihydrite is more soluble in the minteq database, which is mimicked in the input file by imposing subsaturation with $SI = -1.69$.

QUESTIONS:
Check the numbers in the column Log Molality with Figure 10.25A?
Find the concentration of $ZnEDTA^{-2}$ and $CaEDTA^{-2}$ with the minteq constants?
ANSWER: cross out the complex definitions, $\log m_{ZnEDTA^{-2}} = -7.648$; $\log m_{CaEDTA^{-2}} = -7.433$

The driving force for displacing iron from the FeEDTA complex is the low solubility of iron-hydroxide which gives a very small solute iron activity. The analyzed iron concentration in water can still be substantial because small, colloidal hydroxides may pass even the 0.05 μm filter that was used by Xue et al. However, the breakup of the FeEDTA complex is slow, and at least part of the analyzed iron will be present as the FeEDTA complex. The slow decline of the complex was followed over time after 1.25 μM FeEDTA had been added to a river water sample, as shown in Figure 10.26.

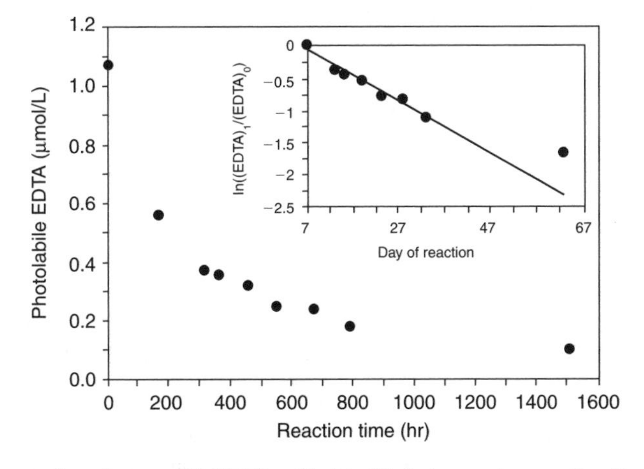

Figure 10.26. Concentration change of FeEDTA added to Glatt river water as a function of time. Inset shows logarithmic (first order) plot of the data.

The half-life of the first order reaction can be found to be 17.5 days from the inset at $\ln(0.5) = -0.69$. In that time, the river may have flown a distance of 2000 km, flushing the complex out from Switzerland down into the North Sea via the River Rhine (if it is not degraded in the meantime).

With Ca- and ZnEDTA as the major species in the Glatt river water, the exchange follows the reactions:

$$Ca^{2+} + FeOHEDTA^{2-} \rightarrow CaEDTA^{2-} + FeOH^{2+} \quad (10.56a)$$

$$Zn^{2+} + FeOHEDTA^{2-} \rightarrow ZnEDTA^{2-} + FeOH^{2+} \quad (10.56b)$$

$$Ca^{2+} + ZnEDTA^{2-} \rightarrow CaEDTA^{2-} + Zn^{2+} \quad (10.56c)$$

$$Zn^{2+} + CaEDTA^{2-} \rightarrow ZnEDTA^{2-} + Ca^{2+} \quad (10.56d)$$

and similar for the other complexes (their concentrations are less important). The exchanged iron precipitates:

$$FeOH^{2+} + 2H_2O \rightarrow Fe(OH)_3 + 2H^+ \quad (10.56e)$$

We can write kinetic equations for the formation of the EDTA complexes, for example for $ZnEDTA^{2-}$, Reaction (10.56b):

$$d[ZnEDTA^{2-}]/dt = k_{fw}[Zn^{2+}][FeOHEDTA^{2-}] - k_{bw}[ZnEDTA^{2-}][FeOH^{2+}] \quad (10.57)$$

where k_{fw} is the forward rate coeffcient ($k_{fw} = 10$/s, Xue et al., 1995) and k_{bw} the backward rate coefficient (s^{-1}). The backward rate coeffcient can be obtained from detailed balancing ($d[ZnEDTA^{2-}]/dt = 0$):

$$k_{bw} = k_{fw} \cdot K_{FeOHEDTA^{2-}} / K_{ZnEDTA^{2-}} = 10 \times 10^{21.49} / 10^{17.04} = 28184/s \quad (10.58)$$

where $K_{ZnEDTA^{2-}}$ is he association constant for the reaction:

$$Zn^{2+} + EDTA^{4-} \leftrightarrow ZnEDTA^{2-}; \quad K = 10^{17.04},$$

and similar for $K_{FeOHEDTA^{2-}}$ (cf. Example 10.9). Already, we can note that the backward rate coefficient is much higher than the forward counterpart.

The reaction of Zn^{2+} with $CaEDTA^{2-}$ (Reaction 10.56d) is added to (10.57) to obtain the total change of $[ZnEDTA^{2-}]$, and a rather formidable set of equations results. Some assumptions do permit to simplify the system and to derive analytical solutions for parts of the reactants (Xue et al. (1995), see also Atkins and de Paula (2002) for the basic explanations for doing these manipulations). Here, we will calculate the system numerically with PHREEQC (Example 10.10).

EXAMPLE 10.10. *Kinetic exchange of Fe(3)EDTA*
Calculate the kinetic exchange of Ca^{2+} and Zn^{2+} with 6×10^{-8} M $FeOHEDTA^{2-}$ added to Glatt river water. Use $k_{fw} = 10$/s for Zn/Fe and Ca/Fe exchange, and 1200/s for Ca/Zn exchange (Xue et al., 1995).

ANSWER:
We make an input file in which rates are defined for the EDTA complexes according to Equations (10.57) and (10.58). The rates for $CaEDTA^{2-}$ and $ZnEDTA^{2-}$ are explicitly written out, and the rate for $FeOHEDTA^{2-}$ follows by mass balance. For the complex activity, the moles of kinetic reactant are used. The mutual interaction of the complex reactions makes this a "stiff" set to solve with very different rates for the

two complexes. The Runge-Kutta procedure that is programmed in PHREEQC tends to take small timesteps to achieve the required tolerance, but the calculations can be quickened with the Ordinary Differential Solver that is implemented in PHREEQC-2.9. We note that the concentrations of EDTA are very low, so that the tolerance for the kinetic calculation must be adapted (default tolerance is 10^{-8} moles, which is decreased to 10^{-12} moles). We use a simplified Glatt river composition, and precipitate the exchanged iron as ferrihydrite.

```
PRINT; -status false
PHASES
 Ferrihydrite; Fe(OH)3 + 3H+ = Fe+3 + 3H2O; log_k 4.891
RATES
 CaEDTA
# E is short for EDTA...
# Reactions: FeOH-E + Ca = Ca-E + FeOH & Zn-E + Ca = Ca-E + Zn
#
# d(Ca-E)/dt = k_f1 * [FeOH-E] * [Ca+2] - k_b1 * [Ca-E} * [FeOH+2]
#   + k_f2 * [Zn-E] * [Ca+2] - k_b2 * [Ca-E} * [Zn+2]
#
# k_f1 = 10 /s. k_b1 = K_FeOH-E/K_Ca-E * k_f1
# k_f2 = 1200 /s. k_b2 = K_Zn-E/K_Ca-E * k_f2
#
# parms: 1= k_f1. 2= K_FeOH-E. 3= K_Ca-E. 4= k_f2. 5= K_Zn-E.
 -start
 10 k_b1 = parm(1) * parm(2) / parm(3)
 12 FeE = kin("FeEDTA")
 20 rate = parm(1)*FeE*act("Ca+2") - k_b1 * m * act("FeOH+2")
 30 k_b2 = parm(4) * parm(5) / parm(3)
 32 ZnE = kin("ZnEDTA")
 40 rate = rate + parm(4)*ZnE*act("Ca+2") - k_b2 * m * act("Zn+2")
 50 moles = rate * time
 60 save -moles
 70 put(moles, 1)
 -end

 ZnEDTA
# Reactions: FeOH-E + Zn = Zn-E + FeOH & Ca-E + Zn = Zn-E + Ca
# d(Zn-E)/dt = ... as for CaEDTA
#
# parms: 1= k_f1. 2= K_FeOH-E. 3= K_Zn-E. 4= k_f2. 5= K_Ca-E.
 -start
 10 k_b1 = parm(1) * parm(2) / parm(3)
 12 FeE = kin("FeEDTA")
 20 rate = parm(1)*FeE*act("Zn+2") - k_b1 * m * act("FeOH+2")
 30 k_b2 = parm(4) * parm(5) / parm(3)
 32 CaE = kin("CaEDTA")
 40 rate = rate + parm(4)*CaE*act("Zn+2") - k_b2 * m * act("Ca+2")
 50 moles = rate * time
 60 save -moles
 70 put (get(1) + moles, 1)
 -end

 FeEDTA
# Initially from waste stream, otherwise by mass balance...
 -start
 10 save get(1)
 -end
```

```
SOLUTION 1 Simplified Glatt river water
 pH 7.9; pe 12 O2(g) -0.68;
 Fe 0.2e-3 Ferrihydrite -1.69
# Run file with Ca or Zn crossed out, adapt step...
 Ca 1.83
# Zn 91.8e-6

KINETICS 1
 -step 432 in 50                              # for Ca, 0.12 hr
 # -step 21.6e6 in 50                         # for Zn, 6000 hr
 # -step 3.6e6 in 50                          # for Zn and Ca, 1000 hr

 CaEDTA; -formula Ca 1; m0 0; -parms 10 3.09e21 2.51e12 1200 1.1e17 6e-8
 ZnEDTA; -formula Zn 1; m0 0; -parms 10 3.09e21 1.1e17 1200 2.51e12 6e-8
 FeEDTA; -formula Fe 1; m0 6e-8
 -tol 1e-12
 -cvode true                                  # available in PHREEQC-2.9
INCREMENTAL_REACTIONS true

EQUILIBRIUM_PHASES 1; Ferrihydrite -1.69
USER_GRAPH
 -head hours FeEDTA CaEDTA ZnEDTA
 -axis_titles Hours mol/L
 -init false
 -start
 10 graph_x total_time / 3600
 20 graph_y kin("FeEDTA"), kin("CaEDTA"), kin("ZnEDTA")
 -end
END
```

The file should be run separately with either Ca or Zn in solution, and then with both ions present, with different timescales for the three runs ("-steps" under keyword KINETICS). The results are displayed in Figure 10.27. The exchange of Ca is complete in 0.1 hour, and follows for the most part a first order rate with a half-life of 0.014 hr (0.8 minutes). The exchange with Zn is much slower and takes 5000 hrs to complete (note the different time-axes in Figure 10.27). The difference is due to the 5×10^5 times smaller concentration (and activity) of Zn^{2+} compared to Ca^{2+}. Also, the shape for the ZnEDTA curve is different from the one for CaEDTA. It is more a second order reaction that is initially steeper as it depends both on the $FeOHEDTA^{2-}$- and the Zn^{2+}-activity; the driving activity of Zn^{2+} becomes less as it is taken up in the EDTA complex, while the activity of Ca^{2+} is barely changed by the small amount that is lost by the exchange reaction. Lastly, calculating the exchange with Ca^{2+} and Zn^{2+} together in solution shows intermediate kinetics and needs about 1000 hours for its completion, which is in the same range as shown in Figure 10.26.

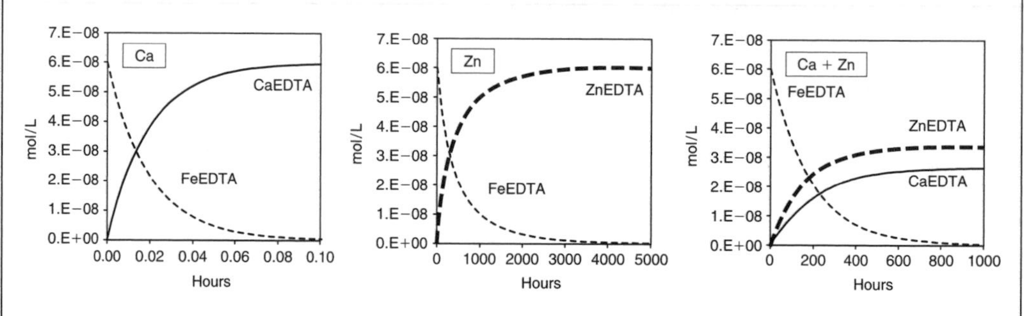

Figure 10.27. Kinetic exchange of Ca^{2+}, Zn^{2+} and both ions together, with $FeOHEDTA^{2-}$, calculated with PHREEQC. Note the different timescales in the three figures.

The mutual influence of various ions in kinetic complexation reactions is a common feature that will be noted when the driving force for the individual reactions is different because of concentration differences (as in Example 10.10), or because the rate coefficients are much dissimilar. The kinetics of complexation reactions are correlated with the water exchange rates of the water molecules in the primary hydration shell around the ion. Water exchange rates for many ions are illustrated in Figure 10.28 and show differences that span orders of magnitude. Correlation with the data in Figure 10.28 can be used to estimate the timelength that is necessary for attaining equilibrium in laboratory experiments in which complexation or sorption constants are to be determined (cf. Question).

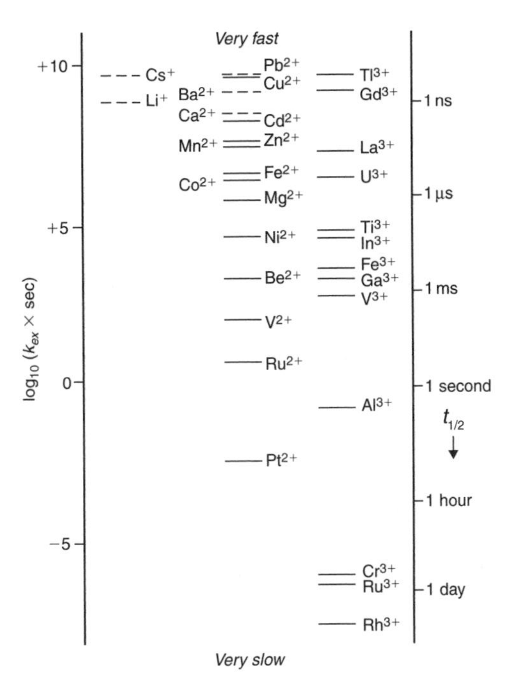

Figure 10.28. Half-lives for water exchange from the bulk solution with the primary hydration shell of various ions (Burgess, 1988).

QUESTION:
Estimate the half-life for sorption equilibrium of Fe^{2+} on the strong sites of goethite, cf. Section 7.6.1)?

ANSWER: The water exchange rate of Fe^{2+} is 1000 times smaller than of Pb^{2+} (Figure 10.28), hence we expect $t_{½} = 1000 \times 45\,\text{min} = 31$ days for Fe^{2+} (Figure 7.37).

PROBLEMS

10.1. Find K'_H of H_2O at 20 and 10°C from the PHREEQC database, with keywords REACTION_TEMPERATURE and USER_PRINT.

10.2. Find the different diffusion fluxes of $H_2{}^{18}O$ and $H_2{}^{16}O$ during evaporation, express the difference in the form of δ notation.

10.3. Model the naphthalene column experiment of Bayard et al., 2000, shown in Figure 10.13. Column length 7 cm, longitudinal dispersivity $\alpha_L = 0.7$ cm, flow rate 72 mL/hr, pore volume 21.6 mL. The soil has $f_{oc} = 0.014$ with $K_{oc} = 302$ L/kg, $\rho_b = 1.05$ g/cm^3, $\varepsilon = 0.58$. One third of the organic carbon reacts instantaneously, the rest with a first order exchange rate $k = 0.26$/hr. Injected were 85 *PV*'s with 9 mg naphthalene/L followed by elution with napthalene-free water.

In the second experiment, 4 g coaltar was mixed with the soil, $f_{oc} = 0.38$, $K_{oc} = 2026$ L/kg, hereof sorbed only 10% instantaneously naphthalene, and 90% with a first order exchange rate $k = 0.24$/hr. Injected were 110 *PV*'s with 9 mg naphthalene/L, again followed by elution with naphthalene-free water.

10.4. Use the PHREEQC input file made for Problem 5.19 to estimate the half-life of dehydration of HCO_3^- from 37 mM $NaHCO_3$ solution at pH = 8.3, when CO_2 is reduced to equilibrium with air.

10.5. Calculate the kinetic precipitation of calcite in your tea kettle, considering kinetic loss of $CO_{2(g)}$ from the heating water by the two-film diffusion model.
Data: The drinking water has temperature 10°C, is in equilibrium with calcite and $P_{CO_2} =$ 0.01 atm. It degasses kinetically to the atmospheric pressure of $10^{-3.5}$ atm during heating of 1 L in 2 minutes to 100°C. Consider only the water-film resistance, correct the diffusion coefficient for the temperature dependency of the viscosity (cf. Equation. 3.42) using:

$\eta_T = 10$^(−(1.37023 * (*tc* − 20) + 8.36×10^{-4} * (*tc* − 20)^2) / (109 + *tc*))

where *tc* is temperature in °C.

Already 10 g calcite exists in the kettle, with surface area $A = 2.4$ dm^2/dm^3. The specific weight of calcite is 2.5 g/cm^3.

10.6. Polycyclic aromatic hydrocarbons (PAH) were found in the soil below gasworks. Estimate the relative mobility of naphthalene and tetracene with respect to water flow in sand (0.1% OC; $\varepsilon = 0.3$) and peat (70% OC; $\varepsilon = 0.6$).

10.7. A waste-dump is situated as illustrated in the figure; Cd and TCE are leached from the waste with precipitation-excess in concentrations of resp. 1.12 and 7.1 mg/L, and enter the aquifer. Fronts and depths of leachates must be calculated, as well as a remedial pumping scheme, given the following hydrological parameters and aquifer-properties: $P = 0.3$ m/yr; porosity = 0.36; bulk density = 1.8 g/cm^3; dispersivity = 10% of flowlength. The aquifer is homogeneous.

a) Calculate distance and depth of the front for a conservative solute (i.e. without decay or adsorption) after 25 and 100 years; calculate the thickness of the plume at these fronts. Compare the distances, covered after 25 and 100 years.
b) TCE has an adsorption-isotherm $q = 7.4\ c^{0.4}$, where q is adsorbed TCE, mg/L groundwater, and c is solute TCE, mg/L.

Estimate a linear distribution-coefficient for Cd^{2+} from exchange equilibrium with Ca^{2+}, when $CEC = 10$ meq/kg, $Ca^{2+} = 10$ mmol/L, and the exchange coefficient $K_{Cd\backslash Ca} = 1$.

c) Calculate the fronts for Cd^{2+} and TCE for 25 and 100 years (distance and depth).
d) Calculate the distance over which 68% of the Cd-front has spread after 100 years as a result of dispersion, and give the concentration-profile in 3 points. Will the TCE-front be sharper or more spread-out?

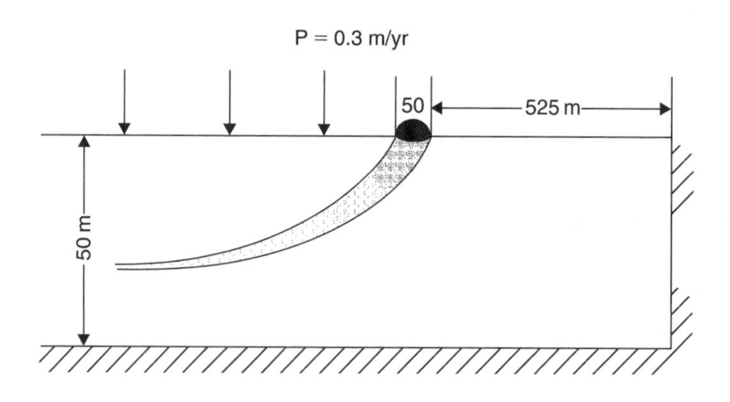

10.8. Model denitrification in an aquifer, using the rate Organic_C in phreeqc.dat. Input NO_3^- is 6 mM, Organic C is 0.1%, $\rho_b = 1.8\,g/cm^3$, $\varepsilon_w = 0.3$. Downward percolation velocity is 1 m/yr. At what depth is $NO_3^- < 0.4$ mM (=25 mg/L, the drinking water limit)? Consider the effect of dispersion, $\alpha_L = 1$ m?

REFERENCES

Ahmad, R., Kookana, R.S., Alston, A.M. and Skjemstad, J.O., 2001. The nature of soil organic matter affects sorption of pesticides. 1. Relationships with carbon chemistry as determined by ^{13}C CPMAS NMR Spectroscopy. *Env. Sci. Technol.* 35, 878–884.

Alexander, M. and Scow, K.M., 1989. Kinetics of biodegradation in soil. In B.L. Sawhney and K. Brown (eds), *Reactions and movement of organic chemicals in soils*. Soil Sci. Soc. Am. Spec. Pub. 22, 243–269.

Appelo, C.A.J. and Parkhurst, D.L., 1998. Enhancements to the geochemical model PHREEQC-1D transport and reaction kinetics. In G.B. Arehart and J.R. Hulston (eds), *Proc. 9th Water Rock Interaction Symp.*, 873–876, Balkema, Rotterdam.

Arnold, W.A. and Roberts, A.L., 2000. Pathways and kinetics of chlorinated ethylene and chlorinated actetylene reaction with Fe(0) particles. *Env. Sci. Technol.* 34, 1794–1805.

Artinger, R., Buckau, G., Geyer, S., Fritz, P., Wolf, M. and Kim, J.I., 1999. Characterization of groundwater humic substances: influences of sedimentary organic carbon. *Appl. Geochem.* 15, 97–116.

Atkins, P.W. and de Paula, J., 2002. *Atkins' Physical chemistry*, 7th ed. Oxford Univ. Press, 1149 pp.

Balkwill, D.L., Leach, F.R., Wilson, J.T., McNabb, J.F. and White, D.C., 1988. Equivalence of microbial biomass measures based on membrane lipid and cell wall components, adenosine triphosphate, and direct counts in subsurface aquifer sediments. *Microb. Ecol.* 16, 73–84.

Barns, S.M. and Nierzwicki-Bauer, S., 1997. Microbial diversity in modern subsurface, ocean, surface environments. *Rev. Mineral.* 35, 35–79.

Baedecker, M.J., Cozzarelli, I.M., Siegel, D.I., Bennett, P.C. and Eganhouse, R.P., 1993. Crude oil in a shallow sand and gravel aquifer. *Appl. Geochem.* 8, 569–586.

Barry, D.A., Prommer, H., Miller, C.T., Engesgaard, P., Brun, A. and Zheng, C., 2002. Modelling the fate of oxidisable organic contaminants in groundwater. *Adv. Water Resour.* 25, 945–983.

Bayard, R., Barna, L., Mahjoub, B. and Gourdon, R., 2000. Influence of the presence of PAHs and coaltar on naphthalene sorption in soils. *J. Contam. Hydrol.* 46, 61–80.

Bedient, P.B., Rifai, H.S. and Newell, C.J., 1994. *Groundwater contamination.* Prentice-Hall, Englewood Cliffs, 541 pp.

Bekins, B.A., Warren, E. and Godsy, E.M., 1998. A comparison of zero-order, first-order, and Monod biotransformation models. *Ground Water* 36, 261268.

Berg, J.M., Tymoczko, J.L. and Stryer, L., 2002. *Biochemistry*, 5th ed. Freeman, New York, 974 pp.

Boudreau, B.P., 1996. *Diagenetic models and their implementation*. Springer, Berlin, 414 pp.

Broholm, M.M., Tuxen, N., Rügge, K. and Bjerg, P.L., 2001. Sorption and degradation of the herbicide 2-methyl-4,6-dinitrophenol under anaerobic conditions in a sandy aquifer in Vejen, Denmark. *Env. Sci. Technol.* 35, 4789–4797.

Brun, A., Engesgaard, P., Christensen, T.H. and Rosbjerg, D., 2002. Modelling of transport and biogeochemical processes in pollution plumes: Vejen landfill, Denmark. *J. Hydrol.* 256, 228–247.

Burgess, J., 1988. *Ions in solution*. Ellis Horwood, Chichester, 191 pp.

Chapelle, F.H., 2001. *Groundwater microbiology and geochemistry*, 2nd ed. Wiley and Sons, New York, 477 pp.

Chiou, C.T., Peters, L.J. and Freed, V.H., 1979. A physical concept of soil-water equilibria for non-ionic organic compounds. *Science* 206, 831–832.

Chiou, C.T., Kile, D.E., Brinton, T.I., Malcolm, R.L., Leenheer, J.A. and MacCarthy, P., 1987. A comparison of water solubility enhancements of organic solutes by aquatic humic materials and commercial humic acids. *Env. Sci. Technol.* 21, 1231–1234.

Clausen, L. and Fabricius, I., 2001. Atrazine, isoproturon, mecoprop, 2,4-D and bentazone adsorption onto iron oxides. *J. Env. Qual.* 30, 858–869.

Conant, B.H., Gillham, R.W. and Mendoza, C.A., 1996. Vapor transport of trichloroethylene in the unsaturated zone: field and numerical modeling investigations. *Water Resour. Res.* 32, 9–22.

Curtis, G.P., Reinhard, M. and Roberts, P.V., 1986. Sorption of hydrophobic compounds by sediments. In J.A. Davis and K.F. Hayes (eds), *Geochemical process at mineral surfaces*. Adv. Chem. Series 323, 191–216.

Danckwerts, P.V., 1970. *Gas-liquid reactions*. McGraw-Hill, New York, 276 pp.

Eberhardt, C. and Grathwohl, P., 2002. Time scales of organic contaminant dissolution from complex source zones: coal tar pools vs. blobs. *J. Contam. Hydrol.* 59, 45–66.

Eckert, P. and Appelo, C.A.J., 2002. Hydrogeochemical modeling of enhanced benzene, toluene, ethylbenzene, xylene (BTEX) remediation with nitrate. *Water Resour. Res.* 38, DOI 10.1029/2001WR000692.

Essaid, H.I., Bekins, B.A., Godsy, E.M., Warren, E., Baedecker, M.J. and Cozzarelli, I.M., 1995. Simulation of aerobic and anaerobic biodegradation processes at a crude oil spill site. *Water Resour. Res.* 31, 3309–3327.

Essaid, H.I., Cozzarelli, I.M., Eganhouse, R.P., Herkelrath, W.N., Bekins, B.A. and Delinc, G.N., 2003. Inverse modeling of BTEX dissolution and biodegradation at the Bemidji, MN crude-oil spill site. *J. Contam. Hydrol.*, DOI 10.1016/S0169-7722(03)00034-2.

Farquhar, G.J. and Rovers, F.A., 1973. Gas production during refuse decomposition. *Water Air Soil Poll.* 2, 483–495.

Feenstra, S. and Guiguer, N., 1996. Dissolution of dense non-aqueous phase liquids (DNAPLs) in the subsurface. In J.F. Pankow and J.A. Cherry, *op. cit.*, 203–232.

Garbarini, D.R. and Lion, L.W., 1986. Influence of the nature of soil organics on the sorption of toluene and trichloroethylene. *Env. Sci. Technol.* 20, 1263–1269.

Gan, J., Yates, S.R., Becker, J.O. and Wang, D. 1998. Surface amendment of fertilizer ammonium thiosulfate to reduce methyl bromide emission from soil. *Env. Sci. Technol.* 32, 2438–2441.

Geller, J.T. and Hunt, J.R., 1993. Mass transfer from nonaqueous phase organic liquids in water-saturated porous media. *Water Resour. Res.* 29, 833–845.

Gillham, R.W. and O'Hannesin, S.F., 1994. Enhanced degradation of halogenated aliphatics by zerovalent iron. *Ground Water* 32, 958–967.

Goldsmith Jr., C.D. and Balderson, R.K., 1988. Biodegradation and growth kinetics of enrichment isolates on benzene, toluene, and xylene. *Water Sci. Technol.* 20, 505–507.

Grathwohl, P., 1990. Influence of organic matter from soils and sediments from various origins on the sorption of some chlorinated aliphatic hydrocarbons: implications on K_{oc} correlations. *Env. Sci. Technol.* 24, 1687–1693.

Haderlein, S.B. and Schwarzenbach, R.P., 1993. Adsorption of substituted nitrobenzenes and nitrophenols to mineral surfaces. *Env. Sci. Technol.* 27, 316–326.

Heijnen, J.J. and Van Dijken, J.P., 1992. In search of a thermodynamic description of biomass yields for the chemotrophic growth of microorganisms. *Biotechnol. Bioeng.* 39, 833–858.

Hering, J.G. and Morel, F.M.M., 1989. Slow coordination reactions in seawater. *Geochim. Cosmochim. Acta* 53, 611–618.

Hines, M.E., Crill, P.M., Varner, R.K., Talbot, R.W., Shorter, J.H., Kolb, C.E. and Harriss, R.C., 1998. Rapid consumption of low concentrations of methyl bromide by soil bacteria. *Appl. Env. Microbiol.* 64, 1864–1870.

Hussein, M. Jin, M. and Weaver, J.W., 2002. Development and verification of a screening model for surface spreading of petroleum. *J. Contam. Hydrol.* 57, 281–302.

Jellali, S., Benremita, H., Muntzer, P., Razakarisoa, O. and Schäfer, G., 2003. A large-scale experiment on mass transfer of trichloroethylene from the unsaturated zone of a sandy aquifer to its interfaces. *J. Contam. Hydrol.* 60, 31–53.

Jin, Y. and Jury, W.A., 1996. Characterizing the dependence of gas diffusion coefficient on soil properties. *Soil Sci. Soc. Am. J.* 60, 66–71.

Kan, A.T. and Tomson, M.B., 1990. Effect of pH concentration on the transport of naphthalene in saturated aquifer media. *J. Contam. Hydrol.* 5, 235–251.

Karickhoff, S.W., 1981. Semi-empirical estimation of sorption of hydrophobic pollutants on natural sediments and soils. *Chemosphere* 10, 833–846.

Karickhoff, S.W., 1984. Organic pollutant sorption in aquatic systems. *J. Hydraul. Eng.* 110, 707–735.

Kelly, W.R., Hornberger, G.M., Herman, J.S. and Mills, A.L., 1996. Kinetics of BTX biodegradation and mineralization in batch and column systems. *J. Contam. Hydrol.* 23, 113–132.

Kile, D., Wershawand, R. and Chiou, C.T., 1999. Correlation of soil and sediment organic matter polarity to aqueous sorption of nonionic compounds. *Env. Sci. Technol.* 33, 2053–2056.

Kjeldsen, P., Barlaz, M.A., Rooker, A.P., Baun, A., Ledin, A. and Christensen, T.H., 2002. Present and long-term composition of MSW landfill leachate: A review. *Crit. Rev. Env. Sci. Technol.* 32, 29–336.

Knappe, A., Sommer-von Jarmersted, C., Pekdeger, A., Bau, M. and Dulski, P. 1999. Gadolinium in aquatic systems as indicator for sewage water contribution. In H. Ármannsson (ed.), *Geochemistry of the earth's surface*, 187–190, Balkema, Rotterdam.

Kummerer, K. and Helmers, E., 2000. Hospital effluents as a source of gadolinium in the aquatic environment. *Env. Sci. Technol.* 34, 573–577.

Lensing, H.J., Vogt, M. and Herrling, B., 1994. Modeling of biologically mediated redox processes in the subsurface. *J. Hydrol.* 159, 125–153.

Liss, P.S., 1973. Processes of gas exchange across an air-water interface. *Deep Sea Res.* 20, 221–238.

Liss, P.S. and Slater, P.G., 1974. Flux of gases across the air-sea interface. *Nature* 247, 181–184.

Lyman, W.J., Reehl, W.F. and Rosenblatt, D.H., 1990. *Handbook of chemical property estimation methods. Am. Chem. Soc.*, Washington.

Mackiewicz, M. and Wiegel, J., 1998. Comparison of energy and growth yields for *Desulfitobacterium dehalogenans* during utilization of chlorophenol and various traditional electron acceptors. *Appl. Env. Microbiol.* 64, 352–355.

MacQuarrie, K.T.B., Sudicky, E.A. and Frind, E.O., 1990. Simulation of biodegradable contaminants in groundwater. 1. Numerical formulation in principal directions, *Water Resour. Res.* 26, 207–222.

MacQuarrie, K.T.B. and Sudicky, E.A., 2001. Multicomponent simulation of wastewater-derived nitrogen and carbon n shallow unconfined aquifers 1. Model formulation and performance. *J. Contam. Hydrol.* 47, 53–84.

Madigan, M.T., Martinko, J.M. and Parker, J., 2003. *Brock biology of microorganisms*, 10th ed. Prentice Hall, Englewood Cliffs, 1019 pp.

Madsen, L., Lindhardt, B., Rosenberg, P., Clausen, L. and Fabricius, I., 2000. Pesticide sorption by low organic carbon sediments. *J. Env. Qual.* 29, 1488–1500.

Mayer, K.U., Blowes, D.W. and Frind, E.O., 2001. Reactive transport modeling of an in situ reactive barrier for the treatment of hexavalent chromium and trichloroethylene in groundwater. *Water Resour. Res.* 37, 3091–3103.

McWhorter, D.B. and Kueper, B.H., 1996. Mechanics and mathematics of the movement of dense non-aqueous phase liquids in porous media. In J.F. Pankow and J.A. Cherry, *op. cit.*, 89–128.

Miljøprojekt, 1998. Status for lukkede boringer ved almene vandværker. Danish EPA.

Miller, C.T., Poirier-McNeill, M.M. and Mayer, A.S., 1990. Dissolution of trapped nonaqueous phase liquids: Mass transfer characteristics, *Water Resour. Res.* 26, 2783–2796.

Pankow, J.F. and Cherry, J.A. (eds), 1996. *Dense chlorinated solvents and other DNAPLs in groundwater.* Waterloo Press, Guelph, Ont.

Powers, S.E., Abriola, L.M., Dunkin, J.S. and Weber Jr, W.J. 1994. Phenomenological models for transient NAPL-water mass-transfer processes, *J. Contam. Hydrol.* 16, 1–33.

Prommer, H., Barry, D.A. and Davis, G.B., 2000. Numerical modelling for design and evaluation of groundwater remediation schemes. *Ecol. Model.* 128, 181–195.

Rittmann, B.E. and McCarty, P.L., 2001. *Environmental biotechnology*. McGraw-Hill, Boston, 754 pp.

Schirmer, M., Butler, B.J., Roy, J.W., Frind, E.O. and Barker, J.F. 1999. A relative-least-squares technique to determine unique Monod kinetic parameters of BTEX compounds using batch experiment. *J. Contam. Hydrol.* 37, 69–86.

Schirmer, M., Molson, J.W., Frind, E.O. and Barker, J.F., 2000. Biodegradation modelling of a dissolved gasoline plume applying independent laboratory and field parameters. *J. Contam. Hydrol.* 46, 339–374.

Schwarzenbach, R.P. and Westall, J.C., 1985. Sorption of hydrophobic trace organic compounds in groundwater systems. *Water Sci. Technol.* 17, 39–55.

Schwarzenbach, R.P., Gschwend, P.M. and Imboden, D.M., 1993. *Environmental organic chemistry*. Wiley and Sons, New York, 681 pp.

Stouthamer, A.H., 1976. *Yield studies in microorganisms*. Meadowfield Press, Durham, England, 88 pp.

Tiktak, A., Van den Berg, F., Boesten, J.J.T.I., Van Kraalingen, D., Leistra, M. and Van der Linden, A.M.A., 2000. *Manual of FOCUS PEARL version 1.1.1.* RIVM Rapport 711401008.

Thornton, S.F., Lerner, D.N. and Banwart, S.A., 2001. Assessing the natural attenuation of organic contaminants in aquifers using plume-scale electron and carbon balances: Model development with analysis of uncertainty and parameter sensitivity. *J. Contam. Hydrol.* 53, 199–232.

Troeh, F.R., Jabro, J.D. and Kirkam, D., 1982. Gaseous diffusion equations for porous materials. *Geoderma* 27, 239–253.

VanBriesen, J., 2002. Evaluation of methods to predict bacterial yield using thermodynamics. *Biodegradation* 13, 171–190.

VanBriesen, J., Rittmann, B.E., Xun, L., Girvin, D.C. and Bolton, H., 2000. The rate-controlling substrate of nitrilotriacetate for biodegradation by *Chelatobacter heintzii*. *Env. Sci. Technol.* 34, 3346–3353.

Van Cappellen, P. and Wang, Y., 1996. Cycling of iron and manganese in surface sediments. *Am. J. Sci.* 296, 197–243.

Verschueren, K., 2001. *Handbook of environmental data on organic chemicals,* 4th ed. Wiley and Sons, New York, 2416 pp.

Wanninkhof, R. and McGillis, W.R., 1999. A cubic relationship between air-sea CO_2 exchange and wind speed. *Geophys. Res. Lett.* 26, 1889–1892.

Watson, I.A., Oswald, S.E., Mayer, K.U., Wu, Y. and Banwart, S.A., 2003. Modeling kinetic processes controlling hydrogen and acetate concentrations in an aquifer-derived microcosm. *Env. Sci. Technol.* 37, 3910–3919.

Weaver, J.W., Charbeneau, R.J. and Lien, B.K., 1994. A screening model for nonaqueous phase liquid transport in the vadose zone using Green – Ampt and kinetic wave theory. *Water Resour. Res.* 30, 93–105.

Weber, W.J., LeBoeuf, E.J., Young, T.M. and Huang, W., 2001. Contaminant interactions with geosorbent organic matter: Insights drawn from polymer sciences. *Water Res.* 35, 853–868.

Weber, J.B. and Miller, C.T., 1989. Organic chemical movement over and through soil. In B.L. Sawhney and K. Brown (eds), *Reactions and movement of organic chemicals in soils*. Soil Sci. Soc. Am. Spec. Pub. 22, 305–334.

Xue, H., Sigg, L. and Kari, F.G., 1995. Speciation of EDTA in natural waters: Exchange kinetics of Fe-EDTA in river water. *Env. Sci. Technol.* 29, 59–68.

Yabusaki, S., Cantrell, D., Sass, B. and Steefel, C., 2001. Multicomponent reactive transport in an in situ zero-valent iron cell. *Env. Sci. Technol.* 35, 1493–1503.

Yeats, P.A. and Measures, C.I., 1998. The hydrographic setting of the second IOC contaminants baseline cruise. *Mar. Chem.* 61, 3–14.

Zachara, J.M., Ainsworth, C.C., Felice, L.J. and Resch, C.T., 1986. Quinoline sorption to subsurface materials: Role of pH and retention of the organic cation. *Env. Sci. Technol.* 20, 620–627.

11

Numerical Modeling

We like to address now how we can predict the fate of chemicals that leach from waste sites and appear to vanish in the invisible subsurface environment. We use numerical models since the combination of transport and chemistry often yields complicated concentration patterns which are difficult or impossible to calculate otherwise. Numerical solutions of the advection–reaction–dispersion (*ARD*) equation are approximative and may be incorrect if certain stability criteria are violated. These criteria are explained with 1D examples that illustrate how to calculate diffusion of seawater and advective flow in laboratory columns. They also apply in multidimensional models.

For modeling aquifers, the pattern of groundwater flowlines and travel times must be studied first. Inhomogeneities in an aquifer can be unraveled with environmental tracers and we consider an example of groundwater age dating with tritium/helium concentrations. Subsequently, cases are presented where hydrogeochemical modeling is used to learn what process is determining the concentration of organic and inorganic pollutants in groundwater. Often, chemical parameters are unknown and need to be estimated and optimized using literature data. Essentially, this chapter shows how the combination of modeling and predicting can help in improving our understanding.

In Chapter 3 we have combined the processes in one equation to model changes in concentration due to transport, reaction and dispersion:

$$\left(\frac{\partial c}{\partial t}\right)_x = -v\left(\frac{\partial c}{\partial x}\right)_t - \left(\frac{\partial q}{\partial t}\right)_x + D_L\left(\frac{\partial^2 c}{\partial x^2}\right)_t \tag{3.62}$$

where c is the solute concentration (mol/L), v is pore water flow velocity (m/s), D_L is hydrodynamic dispersion coefficient (m^2/s) and q is the concentration in the solid (mol/L of pore water).

Analytical solutions for Equation (3.62) are simple when q is a linear function of c, which permits replacement of $(\partial q / \partial t)$ by $K_d(\partial c / \partial t)$ so that we obtain

$$(1 + K_d)\left(\frac{\partial c}{\partial t}\right)_x = -v\left(\frac{\partial c}{\partial x}\right)_t + D_L\left(\frac{\partial^2 c}{\partial x^2}\right)_t \tag{11.1}$$

or:

$$\left(\frac{\partial c}{\partial (t/R)}\right)_x = -v\left(\frac{\partial c}{\partial x}\right)_t + D_L\left(\frac{\partial^2 c}{\partial x^2}\right)_t \tag{11.2}$$

where $(1 + K_d)$ is replaced by the retardation R, which defines the retarded time t/R. If the retardation is constant (i.e. the distribution coefficient K_d is independent of the concentration), then Equation (11.2) is identical with the simpler advection–dispersion equation (3.64). All the solutions for that equation are valid when the retarded time t/R is used: retarded chemicals show exactly the same amount of spreading at all points as a conservative chemical is doing, but arrive later depending on the retardation. For example, the formula which describes front spreading in an infinite column (Equation 3.65) becomes, with retardation:

$$c(x,t) = c_i + \tfrac{1}{2}(c_0 + c_i)\,\mathrm{erfc}\left(\frac{x - vt\,/\,R}{\sqrt{4D_L t\,/\,R}}\right) \tag{11.3}$$

We apply the usual simplification $D_L = \alpha_L v$, and $vt\,/\,R = x_{0.5}$ to get:

$$c(x,t) = c_i + \tfrac{1}{2}(c_0 + c_i)\,\mathrm{erfc}\left(\frac{x - x_{0.5}}{\sqrt{4\alpha_L x_{0.5}}}\right) \tag{11.4}$$

Thus, transport in an aquifer of a chemical that shows linear retardation can be calculated very simply, and completely analogous to a conservative compound. Let us compare the analytical solution with results from a numerical model in Example 11.1.

EXAMPLE 11.1. *Calculation of aquifer pollution by waste site leachate*
Figure 11.1 shows results of numerical calculation of pollutants leached from a waste site in an aquifer profile (Pickens and Lennox, 1976). Effects of dispersivity and different retardations are illustrated by iso-concentration lines (contours) of 0.1, 0.3, 0.5, 0.7, and 0.9 (fraction of final concentration). Compare the spread of the front with an analytical guess. Data: $vt = 415$ m; $\alpha_L = 10$ m; $\rho_b = 1.8$ g/cm^3; $\varepsilon = 0.3$; $K'_d = 0$, 0.1 and 1 mL/g. Initial concentration $c_i = 0$ mg/L, the waste site is a continuous source with $c_0 = 1$ mg/L in the leachate.

ANSWER:
Assume that a flowline, originating at the downstream end of the waste site is horizontal, and that transverse dispersion is absent. Travel distance of the $c\,/\,c_0 = 0.5$ concentration is $x_{0.5} = vt\,/\,R$. The concentration $c\,/\,c_0 = 0.1$ is at $x_{0.1}$, a distance further ahead that can be obtained from Equation (11.4): $c(x_{0.1}, t) = 0.1 = \tfrac{1}{2}\,\mathrm{erfc}[(x_{0.1} - x_{0.5})\,/\,\sqrt{(4\alpha_L\, x_{0.5})} = \tfrac{1}{2}\,\mathrm{erfc}[f(x)]$.

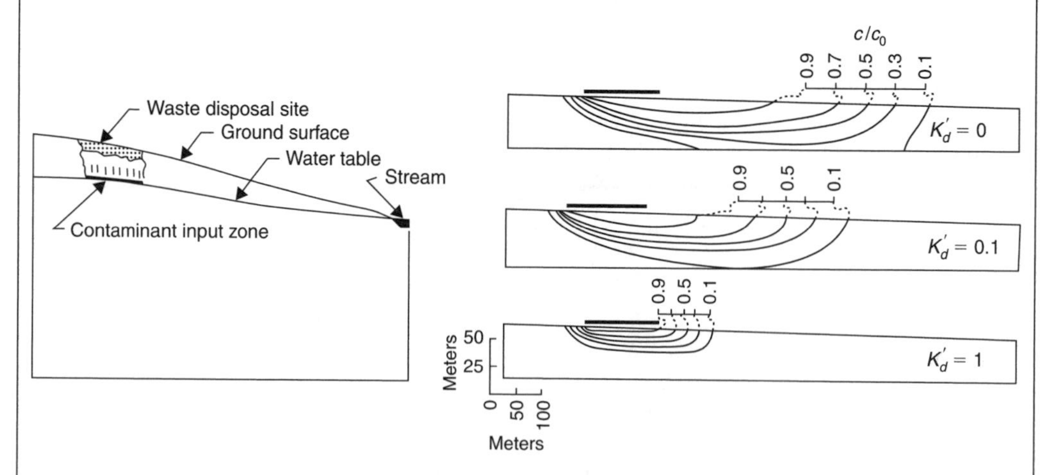

Figure 11.1. Effects of dispersivity and varying retardations on waste site leachate transport in an aquifer. Numerical results from Pickens and Lennox (1976). Approximate longitudinal spreading of 0.1 / 0.9, and 0.3 / 0.7 contour lines as calculated here is indicated by the bar.

If we know the inverse complementary error function of 0.2 (from Table 3.3), we obtain $f(x)$, and can calculate $x_{0.1}$. Thus, $f(x)$ = inverfc[0.2] = 0.906 = $(x_{0.1} - x_{0.5}) / \sqrt{(4\alpha_L\, x_{0.5})}$, which gives $x_{0.1}$ as function of α_L, vt and R. The concentration $c / c_0 = 0.9$, at $x_{0.9}$, is similarly found from $0.906 = -(x_{0.9} - x_{0.5}) / \sqrt{(4\alpha_L\, x_{0.5})}$. For $c / c_0 = 0.3$ and 0.7 the procedure is identical, with the value of inverfc[0.6] = 0.371. The results are summarized as:

	R	vt / R (m)	$\lvert x_{c/c_0} - (vt/R) \rvert$ (m)	
			0.1 / 0.9	0.3 / 0.7
$K'_d = 0$	1	415	117	47.8
$K'_d = 0.1$	1.6	259	92	37.8
$K'_d = 1.0$	7	59	44	18.1

11.1 NUMERICAL MODELING OF TRANSPORT

Simple functions may be unrealistic when geochemical problems must be tackled. For example, if precipitation of calcite is to be modeled along a flowline, then q_{Ca} is clearly a function of pH, $[HCO_3^-]$, and also of aqueous complexes and activity-coefficients. It is, in principle, possible to write out the transport equation (3.62) for each component, include equations for reactions among components, and solve the whole set with a differencing scheme. This approach has been developed mainly by Lichtner and co-workers (Lichtner, 1985; Miller and Benson, 1983; Lichtner and Seth, 1996; Steefel and Yabusaki, 1996 and Mayer et al., 2002). In another approach, transport (the A and D part of Equation 3.62) and chemistry (the R part) are solved separately in each timestep. The first method is the most general and may become the routine when equation solvers have advanced and still faster pc's become available. The second approach is less stringent in memory requirements (Yeh and Tripathi, 1989; Steefel and MacQuarrie, 1996) and is implemented in 1D in PHREEQC and recently also in 3D versions (PHAST, Parkhurst et al., 2004, or based on MODFLOW and MST3D in Alliances, and in PHT3D, Prommer, 2002).

11.1.1 *Only diffusion*

Let us split up the problem in three parts, corresponding to the three terms of the right-hand side of the advection-reaction-dispersion Equation (3.62). If we separate the dispersion term from the advection part, it is more appropriately termed diffusion. The change in concentration as result of diffusion is:

$$\left(\frac{\partial c}{\partial t}\right)_x = D_f \left(\frac{\partial^2 c}{\partial x^2}\right)_t \tag{11.5}$$

where we want to evaluate the change of c with time at some location x. It requires calculation of the second derivative of c with x. An approximation is obtained by combining two Taylor expansions of the function $f(x)$. First in forward sense:

$$f(x_n + \Delta x) = f(x_n) + f'(x_n)\Delta x + \frac{f''(x_n)}{2}(\Delta x)^2 + \frac{f'''(x_n)}{6}(\Delta x)^3 + \cdots \tag{11.6}$$

where f' is the first derivative, f'' is the second derivative, etc., and Δx is the step-size. A similar Taylor expansion in backward sense gives:

$$f(x_n - \Delta x) = f(x_n) - f'(x_n)\Delta x + \frac{f''(x_n)}{2}(\Delta x)^2 - \frac{f'''(x_n)}{6}(\Delta x)^3 + \cdots \tag{11.7}$$

Adding Equations (11.6) and (11.7) gives the finite difference approximation of the second derivative:

$$\frac{f(x_n + \Delta x) - 2f(x_n) + f(x_n - \Delta x)}{(\Delta x)^2} = f''(x_n) + O(\Delta x)^2 \tag{11.8}$$

The approximation of the second derivative of $f(x)$ at x_n is obtained by combining a backward and a forward step, and it is therefore termed a *central difference* equation. The error term $O(\Delta x)^2$ can be calculated with formulas derived in standard books on numerical techniques; here it is of the order of the square of the step-size Δx (Gerald and Wheatley, 1989).

The $f(x)$ used in the Taylor formula is nothing but the concentration $(c)^t_x$, i.e. the concentration at x, at time t. Thus, the second derivative in Equation (11.5) is approximated as:

$$\left(\frac{\partial^2 c}{\partial x^2}\right)_{t1} = f''(x)_{t1} = \frac{c^{t1}_{x-1} - 2c^{t1}_x + c^{t1}_{x+1}}{(\Delta x)^2} \tag{11.9}$$

We can also difference the first derivative with respect to time:

$$\left(\frac{\partial c}{\partial t}\right)_x = \left(\frac{\Delta c}{\Delta t}\right)_x = \frac{c^{t2}_x - c^{t1}_x}{\Delta t} \tag{11.10}$$

which amounts to taking steps forward in time. The solution for the diffusion Equation (11.5) is then obtained with a scheme that is central in space, and forward in time by combining Equations (11.9), (11.10) and (11.5):

$$\begin{aligned} c^{t2}_x &= c^{t1}_x + \frac{D_f \Delta t}{(\Delta x)^2}(c^{t1}_{x-1} - 2c^{t1}_x + c^{t1}_{x+1}) \\ &= mixf \cdot c^{t1}_{x-1} + mixf \cdot c^{t1}_{x+1} + (1 - 2mixf) \cdot c^{t1}_x \end{aligned} \tag{11.11}$$

where $mixf$ is $D_f \Delta t / (\Delta x)^2$.

The parameter $mixf$ is called a *mixing factor* since, according to Equation (11.11), diffusion consists of mixing of neighboring cells. When $mixf = ½$, Equation (11.11) simplifies to:

$$c^{t2}_x = \frac{c^{t1}_{x-1} + c^{t1}_{x+1}}{2} \tag{11.12}$$

On the other hand, when $mixf > ½$, a negative concentration will be obtained in the central cell, $c^{t2}_x < 0$, if the concentrations in the neighboring cells are zero, $c^{t1}_{x\pm1} = 0$. This is impossible of course, and *numerical instability* sets in, leading to *oscillations* in the results. Sometimes the instabilities occur already for $⅓ < mixf < ½$, especially when the concentrations change sharply from cell to cell.

Figure 11.2 shows such a case, with the central cell having the concentration $c^{t1}_x = 1$, surrounded by cells with $c^{t1}_{x\pm1} = 0$. Diffusion in time, of course, smears out the sharp transition and gives a smooth, Gaussian distribution. However, concentrations oscillate when calculated at $t = t2$ and $t3$ with $mixf = ½$. When the mixing factor is smaller than ⅓, such oscillations are prevented since never more is mixed out of a cell, than remains behind. This stability criterion thus requires that:

$$mixf = \frac{D_f \Delta t}{(\Delta x)^2} \leq ⅓, \quad \text{or} \quad \Delta t \leq \frac{(\Delta x)^2}{3D_f} \tag{11.13}$$

It is called the Von Neumann condition for the maximal size of the timestep (actually, Von Neumann specified $mixf \leqslant ½$). In case the simulation time t_n is larger than Δt, mixing is repeated a number of times ($Nmix$) until $\Sigma\Delta t = t_n$.

mixf = 1/2

t1	0	0	0	1	0	0	0
t2	0	0	1/2	0	1/2	0	0
t3	0	1/4	0	1/2	0	1/4	0
t4	1/8	0	3/8	0	3/8	0	1/8
x	**−3**	**−2**	**−1**	**0**	**1**	**2**	**3**

mixf = 1 / 3

t1	0	0	0	1	0	0	0
t2	0	0	1/3	1/3	1/3	0	0
t3	0	1/9	2/9	3/9	2/9	1/9	0
t4	1/27	3/27	6/27	7/27	6/27	3/27	1/27
x	**−3**	**−2**	**−1**	**0**	**1**	**2**	**3**

Figure 11.2. Stability in a central difference scheme depends on the mixing factor *mixf*; the scheme is stable when *mixf* < 1 / 3.

QUESTION:
Calculate the concentrations in Figure 11.2 at *t*3 for *mixf* = 1?

ANSWER: 0 1 −2 3 −2 1 0, note that the oscillations augment with time, but "mass balance" is conserved.

Boundary conditions may demand slightly different formulas for the mixing factor. When the concentration is a fixed c_0 at one boundary (for example when seawater diffuses into a fresh water sediment), a better approximation is obtained if we fix the concentration at that point. This can be arranged as illustrated in Figure 11.3. We halve the first cell, and calculate the concentration c_1 from central differences with $c_{1-½\Delta x}$ and $c_{1+½\Delta x}$. Clearly, $c_{1-½\Delta x} = c_0$ (Figure 11.3). The concentration $c_{1+½\Delta x}$ can be approximated as an average of two extremes. The first extreme is the midpoint-concentration between the first two cells (which is $½(c_1 + c_2)$), and the second one is the expected concentration when the line $c_0 - c_1$ is extrapolated (which gives $c_1 - (c_0 - c_1)$). The average of the two is:

$$c_{1+½\Delta x} = c_2/4 - c_0/2 + 5c_1/4 \tag{11.14}$$

Once more we use Equation (11.11) to calculate the concentration c_1 at the next timestep from c_1, $c_{1-½\Delta x}$ and $c_{1+½\Delta x}$, and obtain:

$$c_1^{t2} = ½ mixf^* c_0^{t1} + ¼ mixf^* c_2^{t1} + (1 - ¾\, mixf^*) c_1^{t1} \tag{11.15}$$

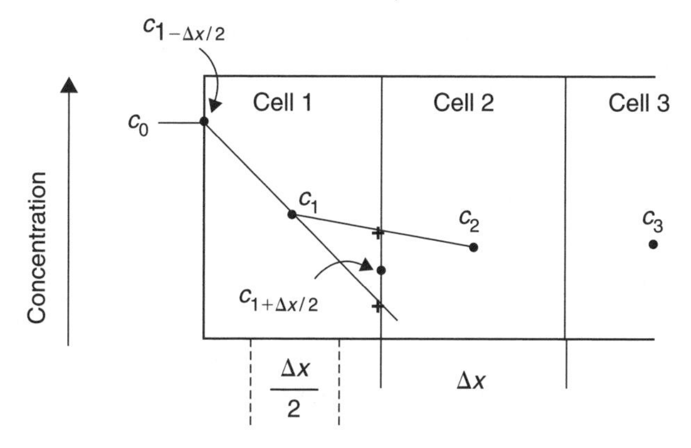

Figure 11.3. When concentrations are fixed at one boundary, the difference formulas at that point must be adapted.

The mixing factor for the halved cell is $mixf^* = D_f\Delta t \,/\, (\frac{1}{2}\,\Delta x)^2 = 4mixf$, and rewriting Equation (11.15) in terms of the mixing factor for the full cell yields:

$$c_1^{t2} = 2mixf\, c_0^{t1} + mixf\, c_2^{t1} + (1 - 3mixf)c_1^{t1} \tag{11.16}$$

We note that this boundary condition requires for generally stable calculations that $(1 - 3mixf) > \frac{1}{3}$, or $mixf < \frac{2}{9}$.

EXAMPLE 11.2. *A Pascal code to model Cl-diffusion from seawater into fresh-water sediment*
Pascal is a popular programming language somewhat similar to the C language. The Turbo Pascal compiler, which operates under DOS, can be downloaded from www.xs4all.nl/~appt, together with the programs in the examples in this Chapter. The following program calculates 100 years diffusion of seawater (Cl = 0.56 mol/L) into a 10 m thick fresh water sediment (Cl = 0.0 mol/L). The lower end is a closed boundary (a mirror for Cl-ions).

```
Program diffuse;                                    { brackets enclose comments }
                                                               { define variables...}
const  Ncel = 10;
var    i, j, Nmix                         : integer;
       Totx, Tott, Delx, De, mixf         : real;
       Ct1, Ct2                           : array[0..Ncel] of real;
begin
                                       { set variables and initial concentrations...}
  De := 1e-5 {cm2/s}; Totx := 1000.0 {cm}; Tott := 100 * 365 * 24 * 60 * 60.0 {s};
  Delx := Totx/Ncel;
  for i := 1 to Ncel do Ct1[i] := 0.0; Ct1[0] := 0.56;
                                                {calculate the mixing factor ...}
  mixf := De * Tott/(Delx*Delx);
  Nmix := 1 + trunc(mixf * 2/9);               {trunc(..) gives integer part }
  mixf := mixf/Nmix;
                                                              { now diffuse...}
for i:= 1 to Nmix do
  begin                                       { mix into Ct2-cells. first cell...}
    Ct2[1] := 2 * mixf * Ct1[0] + mixf * Ct1[2] + (1 - 3 * mixf) * Ct1[1];
    for j := 2 to Ncel - 1 do                                   { inner cells... }
      Ct2[j] := mixf * (Ct1[j-1] + Ct1[j+1]) + (1 - 2 * mixf) * Ct1[j];
    Ct2[Ncel] := mixf * Ct1[Ncel-1] + (1 - mixf) * Ct1[Ncel];          { .. end cell }
    for j := 1 to Ncel do                        { copy back into Ct1-cells... }
    Ct1[j] := Ct2[j];
end;

  writeln(' Depth below seabottom (m); Cl (mol/L) after', Tott/(31536e3):8:2,' years');
  writeln('                0 ;',Ct1[0]:8:4, Nmix:5, ' timesteps');
  for i:= 1 to Ncel do
    begin
      writeln('             ', (Delx * (i - 0.5)/100):9:3, Ct1[i]:8:4);
    end;
end.
```

Figure 11.4 shows a comparison of the results of this program with the analytical solution:

$$c_{x,t} = 0.56\,\mathrm{erfc}\left(\frac{x}{\sqrt{4D_e t}}\right) \tag{11.17}$$

The agreement is excellent, also for the first cell near the seawater boundary (usually this one gives some deviations).

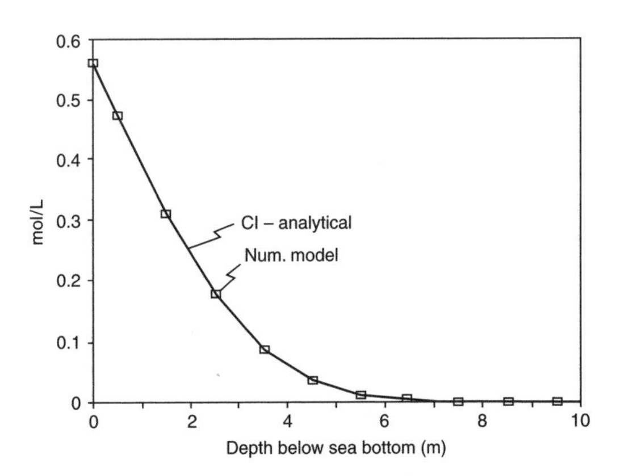

Figure 11.4. 100 years of seawater diffusion in fresh water sediment; comparison of central difference and analytical solution.

QUESTIONS:
What are the values of *Nmix* and the timestep Δt?
ANSWER: $Nmix = 15$, $\Delta t = 6.67$ years
What are *Nmix* and Δt in a 3 times finer grid (Ncel = 30, $\Delta x = 0.333$ m)?
ANSWER: 130, 0.77
Program the problem in a spreadsheet.
Refine the grid twice and compare the results (note how much easier this is done in Pascal or another programming language).
Calculate the problem with PHREEQC, refine the grid.
After seawater diffusion, continue the PHREEQC model with 50 years diffusion of fresh water.
Calculate the fractionation of $H_2{}^{16}O$ and $H_2{}^{18}O$ during 100 years diffusion. The initial $\delta^{18}O = -10‰$, $\delta^{18}O = 0‰$ in the boundary solution, cf. Chapter 3.
Draw/model the *c*/*x* curve for retarded chemicals, 1) $q = c$; 2) $q = c^2$; 3) $q = c^{0.5}$

We used the (known) concentrations at time $t1$ to calculate next timestep concentrations, but why not use these immediately in the central difference formula? Equation (11.11) would become:

$$\begin{aligned} c_x^{t2} &= c_x^{t1} + \frac{D_f \Delta t}{(\Delta x)^2}(c_{x-1}^{t2} - 2c_x^{t2} + c_{x+1}^{t2}) \\ (1 + 2mixf)c_x^{t2} &= c_x^{t1} + mixf \cdot c_{x-1}^{t2} + mixf \cdot c_{x+1}^{t2} \end{aligned} \tag{11.18}$$

Both the right hand side of this equation and $(1 + 2mixf)$ are positive, which means that we will always obtain a positive concentration c_x^{t2}, irrespective of the size of the timestep. Apparently, Equation (11.18) is *unconditionally stable*, and we can increase the timestep as much as we like.

The c_x^{t2}'s can be found by solving a set of linear equations. If the equations for the case of Example 11.2 are written out, we obtain:

$$\begin{array}{llllll} 1c_0^{t2} & +0c_1^{t2} & +0c_2^{t2} & +0c_3^{t2} & +0 & \cdots = c_0^{t1} \\ -2mixf\ c_0^{t2} & +(1+3mixf)c_1^{t2} & -mixf\ c_2^{t2} & +0c_3^{t2} & +0 & \cdots = c_1^{t1} \\ 0c_0^{t2} & -mixf\ c_1^{t2} & +(1+2\,mixf)\ c_2^{t2} & -mixf\ c_3^{t2} & +0 & \cdots = c_2^{t1} \\ \cdots & \cdots & \cdots & \cdots & \cdots & \cdots = \cdots \end{array} \tag{11.19}$$

in which we recognize a matrix form:

$$\begin{bmatrix} 1 & 0 & 0 & 0 & 0 & \cdots \\ -2mixf & (1+3mixf) & -mixf & 0 & 0 & \cdots \\ 0 & -mixf & (1+2mixf) & -mixf & 0 & \cdots \\ \cdots & \cdots & \cdots & \cdots & \cdots & \cdots \end{bmatrix} \begin{bmatrix} c_0^{t2} \\ c_1^{t2} \\ c_2^{t2} \\ \cdots \end{bmatrix} = \begin{bmatrix} c_0^{t1} \\ c_1^{t1} \\ c_2^{t1} \\ \cdots \end{bmatrix} \qquad (11.20)$$

which can be written as:

$$\boldsymbol{A} \cdot \boldsymbol{C}^{t2} = \boldsymbol{C}^{t1} \qquad (11.21)$$

There are as many equations as unknown c_x^{t2}'s and the vector $\boldsymbol{C}^{t2}$ can be obtained as follows. The matrix of the coefficients $\boldsymbol{A}$ is split up in a lower triangular $\boldsymbol{L}$ and an upper triangular $\boldsymbol{U}$ matrix, with the property that $\boldsymbol{L} \cdot \boldsymbol{U} = \boldsymbol{A}$. For example, with 4 cells in total and $mixf = 2$,

$$\overset{\boldsymbol{L}}{\begin{bmatrix} 1 & 0 & 0 & 0 \\ -4 & 7 & 0 & 0 \\ 0 & -2 & 31/7 & 0 \\ 0 & 0 & -2 & 65/31 \end{bmatrix}} \cdot \overset{\boldsymbol{U}}{\begin{bmatrix} 1 & 0 & 0 & 0 \\ 0 & 1 & -2/7 & 0 \\ 0 & 0 & 1 & -14/31 \\ 0 & 0 & 0 & 1 \end{bmatrix}} = \overset{\boldsymbol{A}}{\begin{bmatrix} 1 & 0 & 0 & 0 \\ 4 & 7 & -2 & 0 \\ 0 & -2 & 5 & -2 \\ 0 & 0 & -2 & 3 \end{bmatrix}} \qquad (11.22)$$

We now have to solve

$$\boldsymbol{L} \cdot \boldsymbol{U} \cdot \boldsymbol{C}^{t2} = \boldsymbol{C}^{t1}$$

which is easy because the matrices are triangular. First, we define $\boldsymbol{U} \cdot \boldsymbol{C}^{t2} = \mathbf{y}$, and solve y in $\boldsymbol{L} \cdot \boldsymbol{y} = \boldsymbol{C}^{t1}$, starting in the first row and substituting consecutively in the following rows. Then, $\boldsymbol{C}^{t2}$ is calculated from $\boldsymbol{U} \cdot \boldsymbol{C}^{t2} = \boldsymbol{y}$, now starting in the last row and working upwards to the first row. Gerald and Wheatley (1989) and Press et al. (1992) present calculation schemes for what is called *LU decomposition*.

EXAMPLE 11.3. *Implicit calculation of diffusion*

We redo the calculation of the Cl^- diffusion profile of Example 11.2, using only next-timestep concentrations and compare results for different timesteps. Only the 3 non-zero elements of each row of the coefficient matrix are stored.

```
Program diffuse2;
const   Ncel = 10;
var     i, j, Timesteps                          : integer;
        Totx, Tott, Delx, Delt, De, mixf, x      : real;
        Ct1, Ct2, y                              : array[0..Ncel] of real;
        A                                        : array[0..Ncel, 1..3] of real;
begin
                                    { set timesteps and initial concentrations...}
  Timesteps := 1; { ... also for Timesteps := 5, 15 }
  De := 1e-5 {cm2/s}; Totx := 1000.0 {cm}; Tott := 100 * 365 * 24 * 60 * 60.0 {s};
  Delx := Totx/Ncel;
  for i := 1 to Ncel do Ct2[i] := 0.0; Ct2[0] := 0.56;
                                                { calculate the mixing factor ...}
```

```
  Delt := Tott/Timesteps;
  mixf := De * Delt/(Delx * Delx);
                { fill coefficient matrix A, put the diagonal elements in A[.., 2] }
                                                                { boundary cells ...}
  A[0, 2] := 1; A[0, 3] := 0;
  A[1, 1] := -2 * mixf; A[1, 2] := 1 + 3 * mixf; A[1, 3] := -mixf;
  A[Ncel, 1] := -mixf; A[Ncel, 2] := 1 + mixf;
                                                                  { inner cells ... }
  for i := 2 to Ncel - 1 do
    begin
      A[i, 1] := -mixf; A[i, 2] := 1 + 2 * mixf; A[i, 3] := -mixf;
    end;
                  { decompose A in LU: store L in A[.., 1-2] and U in A[.., 3] ...}
  for i := 1 to Ncel do
    begin
      A[i-1, 3] := A[i-1, 3]/A[i-1, 2];
      A[i, 2] := A[i, 2] - A[i, 1] * A[i-1, 3];
    end;
                                                                   { now diffuse...}
  for i:= 1 to Timesteps do
    begin
      for j := 0 to Ncel do Ct1[j] := Ct2[j];
               { solve Ct2 in A.Ct2 = L.U.Ct2 = Ct1 ... First, find y in L.y = Ct1 }
      y[0] := Ct1[0];
      for j := 1 to Ncel do y[j] := (Ct1[j] - A[j, 1] * y[j-1])/A[j, 2];
                                                        { Now obtain Ct2 in U.Ct2 = y ...}
      Ct2[Ncel] := y[Ncel];
      for j := Ncel - 1 downto 1 do Ct2[j] := y[j] - A[j, 3] * Ct2[j+1];
    end;
  { writeln(' Depth .... etc. as in Example 11.2 }
end.
```

Figure 11.5 compares the results for various time discretizations. With 15 timesteps the accuracy of Example 11.2 is nearly reached. For 1 and 5 timesteps, the results are not as good, but they are smooth and without oscillations, and of course, obtained in less time.

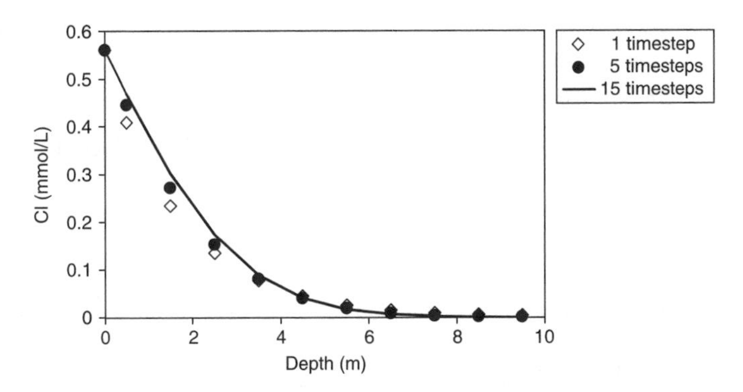

Figure 11.5. Diffusion of seawater Cl^- in a fresh sediment. Implicit calculation using different timesteps.

QUESTIONS:
Write the program with separate arrays for L and U, use L[0..Ncel, 1..2], U[0..Ncel].
Write the program, storing $\boldsymbol{y}$ in $\boldsymbol{C}^{t2}$ (this may be useful when computer memory is an issue).

Rather conspicuous in Figure 11.5 is that the concentrations are underestimated when few timesteps are used. The reason is that the gradient, at the end of a timestep, has decreased as result of diffusion. Accordingly, the flux is less than when the initial gradient is employed as in Example 11.2. For a better approximation, the two gradients may be combined, and a weighting factor can be applied:

$$c_x^{t2} = c_x^{t1} + \left(\frac{D_f \Delta t}{(\Delta x)^2}\right)\left((1-\omega)\quad(c_{x-1}^{t1} - 2c_x^{t1} + c_{x+1}^{t1}) \;+\; \omega(c_{x-1}^{t2} - 2c_x^{t2} + c_{x+1}^{t2})\right) \quad (11.23)$$

where ω is the weighting factor. When $\omega = 0$, the known concentrations at $t1$ are used, and the calculations are *explicit* (Example 11.2). When $\omega = 1$, the concentrations to be calculated are part of the algorithm, and the calculation scheme is *implicit* (Example 11.3). When $\omega = 0.5$, the two gradients are given equal weight, which makes the scheme central in time, commonly denoted as *Crank-Nicholson* and very often applied. Following the earlier discussion, the maximal timestep in the Crank-Nicholson scheme is:

$$\Delta t \le \frac{2(\Delta x)^2}{3\Delta_f} \quad (11.24)$$

or twice larger than in the explicit scheme (note that it is common to use $\Delta t \leqslant (\Delta x)^2/D$, but this may give oscillations for initially abrupt concentration fronts, cf. Figure 11.2). Actually, for a given number of cells and a fixed timestep, the most accurate results are obtained with the explicit method. However, for very precise calculations in which the timestep integration is varied depending on the accuracy of the computed result, the implicit method is often the most rapid. Variable timestepping can be done with Runge-Kutta routines that use intermediate steps to evaluate the accuracy of the integration and automatically decrease or increase the timestep to achieve a predefined exactness.

QUESTION:
Calculate the maximal timestep in a central-in-time scheme for $\omega = 0.45$.
ANSWER: $\Delta t < (\Delta x)^2/(1.65D)$
Program the seawater diffusion problem with the Crank-Nicholson scheme, using the outline from Example 11.3.

11.1.2 *Advection and diffusion/dispersion*

The second step in our modeling effort is to include transport due to advection:

$$\left(\frac{\partial c}{\partial t}\right)_x = -v\left(\frac{\partial c}{\partial x}\right)_t \quad (11.25)$$

The spatial derivative $(\partial c/\partial x)$ can be approximated by weighting concentrations from neighboring cells:

$$\frac{\partial c}{\partial x} = \frac{[(1-\alpha)c_x + \alpha c_{x+1}] - [(1-\alpha)c_{x-1} + \alpha c_x]}{\Delta x} \quad (11.26)$$

where α is the spatial weighting factor.

If we take $\alpha = 0$, then $(\partial c/\partial x) = (c_x - c_{x-1})/\Delta x$, which gives in Equation (11.25):

$$c_x^{t2} = c_x^{t1} - \frac{v\Delta t}{\Delta x}\left(c_x^{t1} - c_{x-1}^{t1}\right) \quad (11.27)$$

The equation indicates that c_x^{t2} may become negative when $v\Delta t > \Delta x$. Thus, in any case the timestep should be limited by

$$\Delta t \leq \frac{\Delta x}{v} \tag{11.28}$$

which is known as the *Courant condition*.

Let's take Δt such that $v\Delta t = \Delta x$. In that case is:

$$c_x^{t2} = c_x^{t1} - (c_x^{t1} - c_{x-1}^{t1}) = c_{x-1}^{t1} \tag{11.29}$$

Thus, we simply move along, pouring concentrations from one cell into the next one. Fronts with different concentrations on both sides, move neatly with the grid boundaries, and remain sharp. Such sharpness is blurred when the front movement does not correspond with the grid boundaries, i.e. when $v\Delta t < \Delta x$. In this case the mixing of old and new concentrations in a cell leads to a gradual smoothening of transitions, termed *numerical dispersion*. Figure 11.6 illustrates the process.

$v\,\Delta t = \Delta x$

t1	1	0	0	0	0
t2	1	1	0	0	0
t3	1	1	1	0	0
t4	1	1	1	1	0
x	**1**	**2**	**3**	**4**	**5**

$v\,\Delta t = ½\Delta x$

t1	1	0	0	0	0
t2	1	1/2	0	0	0
t3	1	3/4	1/4	0	0
t4	1	0.875	0.5	0.125	0
x	**1**	**2**	**3**	**4**	**5**

Figure 11.6. Numerical dispersion as result of time discretization.

Numerical dispersion can be counteracted by decreasing the gridsize (i.e. decrease Δx). However, this is a costly affair since smaller gridsize automatically decreases the timestep according to Equation (11.28), and thus *increases* the number of timesteps as well. Numerical dispersion can also be used to mimick physical (field) dispersion by choosing the discretization steps Δt in relation to Δx in such a way, that numerical dispersion is just equal to the physical dispersion which must be modeled (Van Ommen, 1985). However, front retardation as a result of reactions must also be taken into account. It can be shown (Herzer and Kinzelbach, 1989, Notodarmojo et al., 1991), that the numerical dispersivity amounts to:

$$\alpha_{num} = \frac{\Delta x}{2} - \frac{v\Delta t}{2R} \tag{11.30}$$

where α_{num} is numerical dispersivity (m), and R is the retardation.

The retardation can be calculated as the relative change of concentrations through reactions (sorption, precipitation, dissolution, etc.) within a cell:

$$R = 1 + \frac{\Delta q}{\Delta c} = 1 + \frac{c_{x,new} - c_{x,new+react}}{c_{x,new+react} - c_{x,old}} \tag{11.31}$$

where $c_{x,old}$ is the old concentration in a cell, $c_{x,new}$ is the concentration after the transport step, and $c_{x,new+react}$ is the concentration after reactions.

We recall here that (field) dispersivity forms part of the hydrodynamic dispersion coefficient (Chapter 3):

$$D_L = D_e + \alpha_L v \tag{3.73}$$

and must be included in the calculation of the mixing factor when advective flow is modeled in addition to diffusion. The actual value of α_L, to be comprised in D_L, can be corrected by subtracting the numerical dispersivity as result of transport and reaction:

$$\alpha_{mixf} = \alpha_L - \left(\frac{\Delta x}{2} - \frac{v\Delta t}{2R} \right) \tag{11.32}$$

and the mixing factor becomes:

$$mixf = \frac{D_e \Delta t}{(\Delta x)^2} + \frac{\alpha_{mixf} v\Delta t}{(\Delta x)^2} \tag{11.33}$$

Noting that *mixf* should be larger than zero, Equations (11.32) and (11.33) provide an estimate of maximum Δx, and hence of the minimum number of cells that is necessary for a numerical dispersion-free calculation. Take $D_e = 0$, and $v\Delta t = \Delta x$ (the latter is part of the sensible strategy to maintain $v\Delta t$ as large as possible, for efficient calculations and also to keep α_{mixf} positive in Equation (11.32), which means that in Equation (11.33), $\alpha_{mixf} / \Delta x > 0$. We combine with Equation (11.32), and obtain:

$$mixf > 0, \quad \text{if} \quad \Delta x < \frac{\alpha_L 2R}{R - 1} \tag{11.34a}$$

or with $Ncell = Totx / \Delta x$:

$$mixf > 0, \quad \text{if} \quad Ncell > \frac{Totx}{\alpha_L} \left(\frac{1 - 1/R}{2} \right) \tag{11.34b}$$

where *Totx* is the length of the (laboratory) column or flowline.

We have gained a measure for the maximal timestep (from $(\Delta t)_A \leqslant \Delta x / v$) and the minimal grid spacing (Equation (11.34) for numerical stability and arithmetical accuracy. But, what happens with the timestep criterion for diffusion ($(\Delta t)_D \leqslant (\Delta x)^2 / (3D_e)$, Equation 11.13)? That one relates the maximal timestep to the square of the cellsize, implying that when both timestep criteria are met for a given Δx, a grid refinement of 2 dictates that the advective timestep becomes *twice* smaller, while the diffusive timestep should be *four* times smaller. This conflict can be solved by operator splitting, taking multiple dispersion timesteps such that

$$\Sigma(\Delta t)_D = (\Delta t)_A \tag{11.35}$$

Thus, the mixing factor is defined as

$$mixf = \frac{D_L (\Delta t)_A}{n(\Delta x)^2} \tag{11.36}$$

where n is the smallest positive integer that keeps $mixf \leqslant 1/3$ (or $2/9$ for a fixed boundary condition). The dispersive timestep is then $(\Delta t)_D = (\Delta t)_A / n$, and n mixes are performed during an advective timestep. With the discussion so far, we are in the position to write a program for advective/dispersive transport with linear retardation.

EXAMPLE 11.4. *Model the linear retardation of γ-HCH in a 5 cm long, and 5 cm diameter laboratory column with a sandy* aquifer sample. The sand is low in organic carbon (0.05%), has porosity of 30%, and dispersivity is 7 mm. Injected in the column are 450 mL water containing 20 μg γ-HCH/L at a rate of 10 mL/hr. Assume that diffusion is negligible.

Preliminary calculations provide the following data. With $K_{oc} = 2 \times 10^3$ (Table 10.1) for γ-HCH, the distribution coefficient $K'_d = 5 \times 10^{-4} \cdot 2 \times 10^3 = 1$ L/kg, and the normalized distribution coefficient $K_d = (\rho_b/\varepsilon = 1.8 / 0.3$ kg/L)$\times$(1 L/kg) = 6. Hence the retardation $R = 7$. Pore volume (Porv) in the column is $\pi r^2 L \times \varepsilon = 29.5$ mL; pore water flow velocity (veloc) is 1.7 cm/hr.

From Equation (11.34) follows that *Ncell* (= Ncel) > (5 / 0.7 × (7 − 1) / 14) = 3.1, hence Ncel gets the integer value = 4. Then Δx (= Delx) = Totx / Ncel = 1.25 cm.

The maximal value of Δt (= Delt) is when $v\Delta t = \Delta x$, hence $\Delta t <$ (1.25 cm/1.7 cm/hr) = 0.735 hr. Total injection time of 45 hr is subdivided in 45/0.735 = (integer value) = 62 timesteps (= Nshift). Thus $\Delta t = 45/62 = 0.726$ hr. Finally, we calculate *mixf* = (1.7 cm/hr × 0.7 cm) × (0.726 hr) / (1.25 cm)^2 = 0.553. Hence, we take 2 dispersive timesteps (Nmix = 2) for each advective timestep, with Mixf = 0.553 / 2 = 0.276.

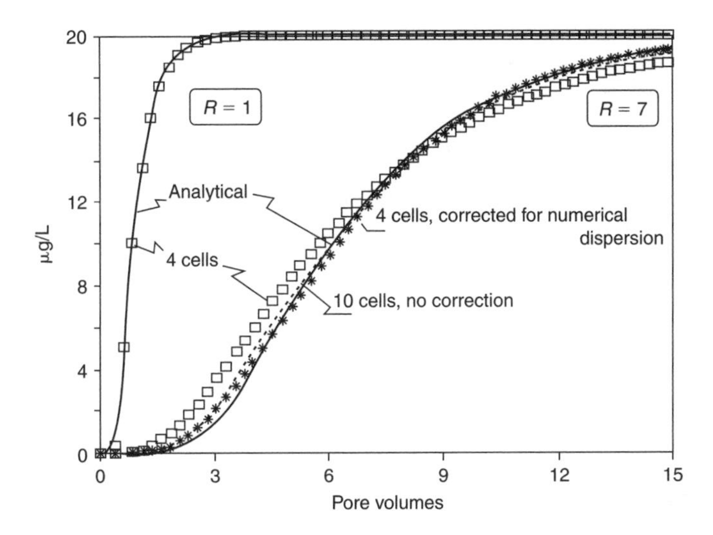

Figure 11.7. Breakthrough curves for conservative element, and retarded γ-HCH; comparison of finite difference and analytical solutions.

We subdivide our program into a number of subroutines:

dspvty	: calculates initial mixing factors mixf and dispersive timesteps.
dspcor	: corrects the mixing factors for mixing and retardation.
dffdsp	: mixes to mimick diffusion and dispersion.
flush	: performs advective transport.
distri	: distributes the chemical over solid (q) and water (c).
analyt	: calculates the analytical solution.
{MAIN}	: sets the variables, and calls the subroutines in the right order.

```
Program Columnperc;
Const Nel = 1;    Ncelmax = 80;
Var Numdisp                                     : boolean;
    i, j                                        : integer;
    Ncel, Nshift, Ish, Nmix, Mi                 : integer;
    Porv, Veloc, Tott, Delt, Totx, Delx         : real;
    Mixf, D, Disp, R, Rmax, Cort                : real;
```

```
    Cin                                    : array[0..Nel] of real;
    Ct1, Ct2, Ctold, Q, Mix                : array[0..Nel, 1..Ncelmax] of real;
    Pv                                     : array[1..200] of real;
    Cout                                   : array[0..Nel+2, 1..200] of real;
    file1                                  : text;

procedure dspvty;          { calculate mixing factor and dispersive timesteps... }
  begin
    Cort := 1 + 2/Ncel;        { Cort corrects for lack of end-cell mixing... }
    Mixf := (D + Disp * Veloc) * Delt/(Delx * Delx) * Cort;
    Nmix := 1 + trunc(3.0 * Mixf);
    Mixf := Mixf/Nmix;
                           { correct mixing factors for conservative element...}
    for j := 1 to Ncel do
      Mix[0, j] := Mixf - (1 - Veloc * Delt/Delx)/2/Nmix * Cort;
  end;

Procedure dspcor;                                { correct mixing factors ... }
Var Cdif : real;
  begin { Retardation is calculated as R ... }
    for j := 1 to Ncel do for i := 1 to Nel do
      begin
        { Ct2 ( = C_x, new) from flush. Ct1 ( = C_x,new+react ) from distri. }
        Cdif := Ct1[i,j] - Ctold[i,j];
        if abs(Cdif) > 1e-9 then R := (Ct2[i,j] - Ctold[i,j])/Cdif;
        if R < 1 then R := 1.0;
        if Rmax < R then Rmax := R;
        Mix[i,j] := Mixf - (1 - Veloc * Delt/R/Delx)/2/Nmix * Cort;
      end;
    end;

procedure dffdsp;                                    { diffuse and disperse... }
var Mixup, Mixdn : real;
  begin
                                                        { mix inner cells ... }
    for j:= 2 to Ncel-1 do for i:= 0 to Nel do
                             { average mixing factors of neighboring cells .. }
      begin
        Mixup := (Mix[i, j-1] + Mix[i, j])/2;
        Mixdn := (Mix[i, j] + Mix[i, j+1])/2;
        if Mixup < 0 then begin Mixup := 0; Numdisp := true; end;
        if Mixdn < 0 then Mixdn := 0;
        Ct2[i, j] :=
          (1.0 - Mixup - Mixdn) * Ct1[i, j] + Mixup * Ct1[i, j-1] + Mixdn * Ct1[i, j+1];
      end;
                                                     { mix boundary cells ... }
    for i:= 0 to Nel do
      begin
        Mixdn := (Mix[i, 1] + Mix[i, 2])/2;
        if Mixdn < 0 then Mixdn := 0;
        Ct2[i, 1] := (1 - Mixdn) * Ct1[i, 1] + Mixdn * Ct1[i, 2];
        Mixup := (Mix[i, Ncel-1] + Mix[i, Ncel])/2;
        if Mixup < 0 then Mixup := 0;
        Ct2[i, Ncel] := (1 - Mixup) * Ct1[i, Ncel] + Mixup * Ct1[i, Ncel-1];
      end;
    for j:= 1 to Ncel do for i:= 0 to Nel do Ct1[i, j]:= Ct2[i, j];
end;
```

```
procedure flush;                                              { advect... }
var x_in_cell : real;
  begin
    x_in_cell := Veloc * Delt/Delx;
    for i:= 0 to Nel do
      begin
        for j:= Ncel downto 2 do
          begin
            Ctold[i, j] := Ct1[i, j];              { ... keep old conc's for dspcor }
            Ct1[i, j] := x_in_cell * Ct1[i, j-1] + (1 - x_in_cell) * Ct1[i, j];
            Ct2[i, j] := Ct1[i, j];
          end;
                                                         { also the first cell... }
      Ctold[i, 1] := Ct1[i, 1];
      Ct1[i, 1] := x_in_cell * Cin[i] + (1 - x_in_cell) * Ct1[i, 1];
      Ct2[i, 1] := Ct1[i, 1];
    end;
  end;


procedure distri;
var Tot : real;
  begin
    for j := 1 to Ncel do
      begin
        Tot := Q[1, j] + Ct1[1, j];
        Ct1[1, j] := Tot/R;
        Q[1, j] := Tot - Ct1[1, j];
      end;
  end;


procedure analyt;
var P, a1, a2, a3, a4, a5, s, e, er1, er2 : real;


function erfc : real;
  begin
    e := 1/(1 + P * abs(s));
    erfc:= (e *(a1 + e *(a2 +e *(a3 + e *(a4 + e *a5))))) * exp(-s * s)
  end;


begin
     P := 0.3275911; a1 := 0.254829592; a2 := -0.284496736;
    a3 := 1.421413741; a4 := -1.453152027; a5 := 1.061405429;
    for i := 1 to Nshift do for j := 1 to 2 do
      begin
        if j = 1 then R := 1 else R := 7;
        s := (1 - Pv[i]/R)/2/sqrt(Disp/Totx * Pv[i]/R);
        if s > 0 then
          er1 := erfc else er1 := 2 - erfc;
        s := (1 + Pv[i]/R)/2/sqrt(Disp/Totx * Pv[i]/R);
        er2 := erfc * exp(Totx/Disp);
        Cout[Nel+j, i] := (er1 + er2) * Cin[j-1]/2;
    end;
  end;
```

```
{Main}
begin
                                                        { define column data... }
  D := 0.0 {cm2/s}; Disp := 0.7 {cm}; R := 7; Totx := 5.0 { cm };
  Ncel := 1 + trunc(Totx/Disp * (1 - 1/R)/2);
  if Ncel < 10 then Ncel := 10;
  Delx := Totx/Ncel { cm };
  Porv := pi * (2.5 * 2.5) * Totx * 0.3 { mL };
  Veloc := 10/Porv/3600 * Totx { cm/s };
  Nshift := 1 + trunc(450/Porv * Ncel);
  Delt := 45 * 3600/Nshift { s };
                                                 { set initial concentrations ... }
  for j := 1 to Ncel do for i := 0 to Nel do
    begin
      Ct1[i, j] := 0; Q[i, j] := 0;
    end;
    Cin[0] := 20; Cin[1] := 20 {ug/L};
                                                         { find mixing factors... }
  dspvty;
                                                         { now the experiment ...}
  for Ish := 1 to Nshift do
    begin
                                  { advect, distribute and correct mixing factors... }
      flush; distri; dspcor;
                                                      { disperse and distribute... }
      for Mi := 1 to Nmix do
        begin    dffdsp;    distri;    end;
                                      { find pore volume, assign concentrations... }
      Pv[Ish] := (Ish * Delt * Veloc/Delx + 0.5)/Ncel;
      for i := 0 to Nel do Cout[i, Ish] := Ct1[i, Ncel];
    end;
                                                   { Calculate analytical Cout... }
  analyt;
                                                     { write results to file... }
  assign(file1, 'ex11_4'); rewrite(file1);
  writeln(file1, ' PV R = 1(num) R = 7(num) R = 1(ana) R = 7(ana)');
  for j := 1 to Nshift do
    begin
      write(file1, Pv[j]:6:3);
      for i := 0 to Nel + 2 do write(file1, Cout[i,j]:9:3); writeln(file1);
    end;
  close(file1);
end.
```

You may note that the subroutine "dspcor" could have been called only once since in this example the mixing factor correction is constant (for a constant retardation). However, the same procedures are used in Example 11.5 where the retardation is variable, and we like to keep the algorithms general. The cellsize Δx also determines resolution of the outflow-profile, and it may be useful to include a statement that Ncel must have a lower value of e.g. 10 as in the program. Output of the example is shown in Figure 11.6 for Ncel of 4 and 10, and the figure includes the analytical solution (Equation 3.66):

$$c_{L,t} = \frac{c_i}{2}\left[\operatorname{erfc}\left(\frac{L - vt/R}{\sqrt{4D_L t/R}}\right) + \exp\left(\frac{vL}{D_L}\right)\operatorname{erfc}\left(\frac{L + vt/R}{\sqrt{4D_L t/R}}\right)\right] \tag{11.37}$$

In the procedure analyt, the arguments are all divided by $L(= \text{Totx})$ so that distance is expressed in pore volumes vt/L which are injected (or eluted).

Element 0 is a conservative substance that is transported by advection and dispersion only (e.g. Cl^-). The numerical calculation with only 4 cells shows already excellent agreement with the analytical solution. When HCH-elution is not corrected for numerical dispersion, an outflow-profile is calculated that is too diffuse compared with the analytical solution. Decreasing the grid-size by increasing Ncel to 10, gives a steeper curve and a better approximation, but even better results are obtained when the calculations with Ncel = 4 are corrected for numerical dispersion. The correction is, of course, highly profitable in terms of execution time compared to the other somewhat blunt (but easy) possibility of decreasing the cellsize.

There is a point with respect to boundary conditions that should be noted. Concentrations are calculated for the midpoint of each cell, including the two end-cells (a "cell-centered grid"). A "shift" moves concentrations in, and out of the column, without considering dispersion over the distance ½Δx to the column boundary. This neglect gives some deviation when few cells are used, but can be compensated for by multiplying the model mixing factor with (1 + 2/Ncel). Furthermore, the concentrations in the last cell need to be transported from the cell midpoint to the column end which requires a half shift, from the shifts.

QUESTIONS:
Include diffusion in the model, estimate the effective diffusion coefficient (cf. Equation 3.47).
Write a program that uses the implicit method for this column experiment ($\omega = 1$). Is the Courant condition essential when $\alpha = 0$?

For the overview, we write down the general finite difference solution for the *ARD* equation:

$$\begin{aligned} c_x^{t2} + q_x^{t2} = {} & c_x^{t1} + q_x^{t1} + \\ & (1-\omega)\left\{\frac{D_e \Delta t}{(\Delta x)^2}\left[c_{x+1}^{t1} - 2c_x^{t1} + c_{x-1}^{t1}\right] - \frac{v\Delta t}{\Delta x}\left[\alpha c_{x+1}^{t1} + (1-2\alpha)c_x^{t1} - (1-\alpha)c_{x-1}^{t1}\right]\right\} + \\ & \omega\left\{\frac{D_e \Delta t}{(\Delta x)^2}\left[c_{x+1}^{t2} - 2c_x^{t2} + c_{x-1}^{t2}\right] - \frac{v\Delta t}{\Delta x}\left[\alpha c_{x+1}^{t2} + (1-2\alpha)c_x^{t2} - (1-\alpha)c_{x-1}^{t2}\right]\right\} \end{aligned} \tag{11.38}$$

where the weighting factors α and ω signify different calculation schemes, with respect to

time $\omega = 0$, explicit
$\omega = ½$, central in time, *Crank-Nicholson*
$\omega = 1$, implicit

and with respect to

distance $\alpha = 0$, upwind when $v > 0$
$\alpha = ½$, central in space
$\alpha = 1$, upwind when $v < 0$.

A similar equation can be formulated for more dimensions (Zheng and Bennett, 2002). We have noted the pitfalls of the discretization, resulting in numerical oscillations and dispersion and inaccurate results for coarse grids. Also in more dimensions, the program to be used should be checked for consistency by calculating a straightforward 1D example and comparing the results with the analytical solution of the problem and the effects of grid refinement.

11.1.3 *Non-linear reactions*

The third step is to extend the model structure of Example 11.4 to more complicated chemical systems. Whenever a model is available for calculating the distribution of the chemicals over solution and solid in a batch system, it can be introduced in the procedure "distri" and then applied to model transport along a flowline. Let us consider an example of non-linear sorption, described by the Freundlich equation:

$$q = K_f c^n \tag{11.39}$$

where the exponent n determines the non-linearity of the sorption equilibrium. Newton's iterative method can be used to distribute the chemical over water (c) and solid (q) when the total quantity of a chemical (T) is known. Newton's method uses derivatives of the distribution function to obtain successive better approximations of the variables that determine the function. In our case we seek:

$$f(c) = c + q - T = c + K_f c^n - T = 0 \tag{11.40}$$

At a certain initial estimate c_1, the derivative of $f(c)$ can be approximated by:

$$\frac{f(c_2) - f(c_1)}{c_2 - c_1} = f'(c_1) \tag{11.41}$$

Now, the value of c_2 should be such that $f(c_2) = 0$; hence from Equation (11.41):

$$c_2 = c_1 - f(c_1) / f'(c_1) \tag{11.42}$$

This simple formula shows generally rapid convergence for smooth, continuous functions. The derivative $f'(C)$ can be obtained analytically, or numerically as in Example 11.5.

EXAMPLE 11.5. *Effect of the Freundlich exponent on breakthrough curves from a column*
Determine the effect for a range of n, varying from $n = 1$ to 0.2, all with identical $K_f = 1$.
The program of Example 11.4 needs only small extension for this problem. The procedure "distri" now calculates the distribution of the chemical over solid and solution iteratively, and stops when the mass-balance of Equation (11.40) is solved to be better than 10^{-6}. A function "frndl" calculates the amount of q for each concentration c. Since the value of the retardation used to correct numerical dispersion may vary, a preliminary run with 3 cells is made to determine its actual *maximal* value. This maximal value is then used to estimate the number of cells for a dispersion-free calculation. The parameter "numdisp" is set to "true" whenever numerical dispersion occurs, i.e. when averaging of two mixing factors would give a negative result. It proved necessary to increase this first estimate of Ncel by 1 (probably due to small concentrations which have a very high retardation when $n = 0.2$, and run farther ahead than is calculated in the 3-cell model).

```
Program Columnperc2; { ... copy from Example 11.4, add Var ... }
Var  Kf,  nf                                                : array[0..Nel] of real;
procedure dspvty; { ... }
Procedure dspcor; { ... }
procedure dffdsp; { ... }
procedure flush;  { ... }
function frndl (Kf, nf, C : real) : real;
  begin
    if C > 1e-10 then frndl := Kf * exp(nf * ln(C)) else frndl := 0;
  end;
```

```
procedure distri;
                        { distribution of Q and C according to Freundlich-isotherm .. }
Var C1, C2, T, Fc1, Fc2 : real;
    ic, ie, it : integer;
begin
  for ic := 1 to Ncel do for ie := 1 to Nel do
    begin
      T := Ct1[ie, ic] + Q[ie, ic];
                { .. T(otal) is redistributed over Ct1 and Q with Newton's method ... }
      C2 := T/(Kf[ie] + 1);
      it := 0;
      repeat
        C1 := C2;
        Fc1 := frndl(Kf[ie], nf[ie], C1) + C1 - T;
        C2 := C1 + 1e-9;
        Fc2 := frndl(Kf[ie], nf[ie], C2) + C2 - T;
        if abs(Fc1 - Fc2) > 0 then C2 := C1 - Fc1/((Fc2 - Fc1)/1e-9);
        if C2 < 1e-9 then C2 := 0.0;
        if C2 > T then C2 := T;
        it := it + 1;
        if it > 20 then begin
          writeln('FREUNDLICH Eqn, iteration > 20'); halt; end;
        until (abs(Fc1) < = 1e-6) or ((C1 + C2) = 0);
        Ct1[ie, ic] := C2; Q[ie, ic] := (T - C2);
      end;
    end;


procedure fill;
begin
  Delx := Totx/Ncel;
  for i := 0 to Nel do for j := 1 to Ncel do
    begin
      Ct1[i,j]:= 0; Q[i,j]:= 0; { .. or, if Ct1>0, then Q = frndl(.,.,.)}
    end;
  Cin[0]:= 1; Cin[1]:= 1 {ug/L};
end;


{Main}
begin
  D := 0.0 { cm2/s }; Disp := 0.5 { cm }; Totx := 5.0 { cm };
  Porv := pi * (2.5 * 2.5) * Totx * 0.3 { mL };
  Veloc := 10/Porv/3600 * Totx { cm/s };
  Tott := Totx/Veloc * 5; { Time for injection of 5 PV, s};
  Kf[1] := 1; nf[1] := 0.5;
                                    { initial run for estimating minimum no. of cells ...}
  Ncel := 3; Delt := Totx/Veloc/3 { s }; fill;
  Rmax := 1; dspvty; Numdisp := false;
  for Ish := 1 to 3 do begin flush; distri; dspcor; dffdsp; distri; end;
  if Numdisp = true then Ncel := 2 + trunc(Totx/Disp * (1 - 1/Rmax)/2);
                                                     { Now start realistic simulation ...}
  fill; Rmax := 1; Numdisp := false;
  Nshift := 1 + trunc(Tott * Veloc/Delx);
  Delt := Tott/Nshift {s};
  dspvty;
  for Ish:= 1 to Nshift do
```

```
    begin
      { ...cf Example 11.4... }
    end;
  { write to file1... }
end.
```

Output of the program is shown in Figure 11.8. The results have been calculated for 5 pore volumes entering the column with the chemical at a concentration Cin = 1, and 10 pore volumes flushing the column. Note that the fronts of increasing concentration are sharper when the Freundlich exponent n is smaller; a smaller n gives initially a more rapid elution, but leads to longer tails (cf. Section 3.4).

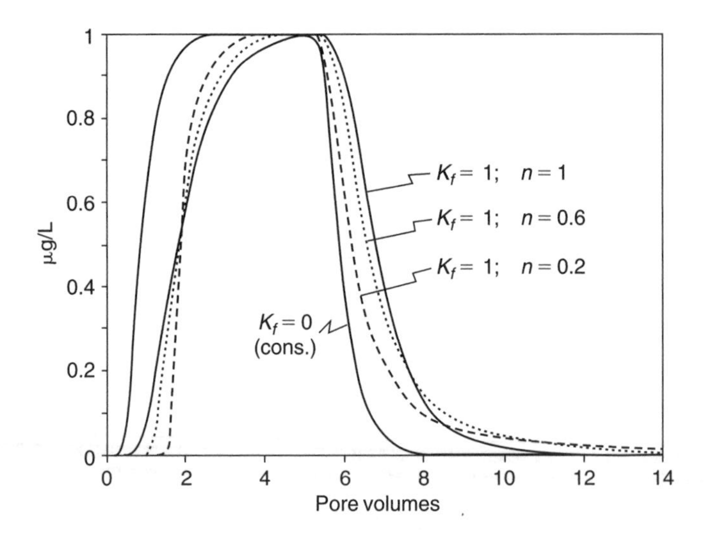

Figure 11.8. Effect of the exponent n in the Freundlich equation on column breakthrough curves.

PROBLEM:

Write a program for cation exchange of Na^+ and K^+ in a laboratory column (cf. Figure 6.26). The column length is 8 cm, dispersivity is 0.2 mm, the flow velocity is 100 m/yr, CEC = 1.1 mM, $c_{i,\text{Na}}$ = 1 mM, $c_{i,\text{K}}$ = 0 mM, $c_{0,\text{K}}$ = 1.1 mM, $c_{0,\text{Na}}$ = 0 mM.

11.2 EXAMPLES OF HYDROGEOCHEMICAL TRANSPORT MODELING

This section illustrates how transport models can help to bring order in the intricate geochemical reactions in aquifers and discusses practical examples from the literature in a step-by-step manner. First, an example of groundwater age dating with tritium/helium, showing the effects of differences in diffusion behavior and dispersion that must be understood to interpret vertical concentration profiles in aquifers. Second, the breakdown reactions of a NAPL (toluene) in an aquifer, and the reactions that occur when a site is remediated by adding a surplus oxidant to the system. Third, a pollution case of acid mine drainage from a uranium mine, where the retardation of the heavy metals determines the impact on the environment. Fourth, a system of in-situ iron removal where we learn to optimize a parameter from experimental data. Fifth, a closer look at the origin of the high As levels in Bangladesh and Bengal groundwaters in which we develop a concept that connects the concentration differences with geographical location. Sixth, a demonstration of isotopic modeling with PHREEQC.

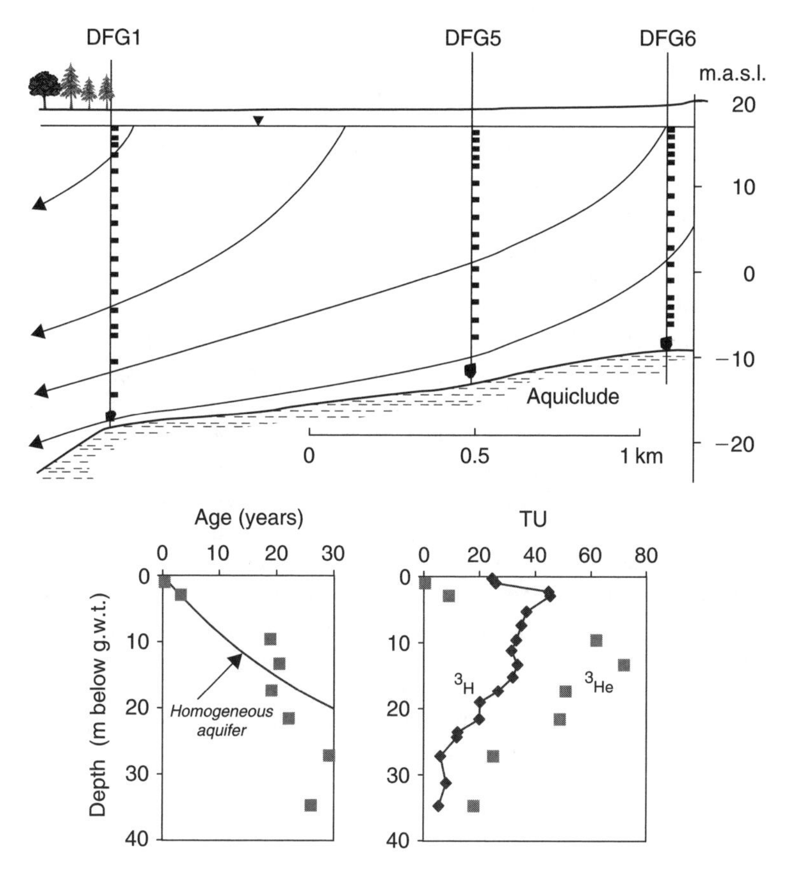

Figure 11.9. Regular flowlines in a phreatic aquifer (top) are contradicted by the age profile calculated from tritium/helium measurements in borehole DFG1 (modified from Leuchs, 1988).

11.2.1 *Tritium-Helium age dating*

The flowlines in a phreatic aquifer may follow the pattern shown in Figure 11.9 (modified from Leuchs, 1988, cf. Figure 9.14 for a chemical profile upstream in the same aquifer). Tritium and helium analyses of water from depth-specific sampling points were used for age-dating the groundwater and to obtain flow properties of the aquifer (Schlosser et al., 1988, 1989). To interpret the analyses, first consider the homogeneous case in which the age of water varies with depth according to (cf. Chapter 3):

$$t = \frac{D\varepsilon_w}{P} \ln\left(\frac{D}{D-d}\right) \tag{3.10}$$

where t is time (yr), D is thickness of the aquifer (here ≈35 m), ε_w is water-filled porosity (≈0.3), P is recharge (≈ 0.3 m/yr) and d is depth (m). Travel time through the unsaturated zone is about 1 year. In 1986, the year of tritium/helium sampling, the 1963 tritium peak is expected to be at 16.3 m depth below the groundwater table. In Figure 11.9, the tritium peak is located at 3 m, and the maximum of tritium + helium lies at 13.3 m depth. Schlosser et al. (1989) assumed that $P/\varepsilon_w = 0.375$ m/yr which would place the maximum at an intermediate depth of 7.3 m. However, a precipitation surplus (P) of (0.375 m/yr × 0.3) = 0.1 m/yr is unlikely small for the local conditions. Thus, there are discrepancies that must be explained.

We can calculate the age of the water according to:

$$t = -\ln(^3H / (^3H + ^3He)) / 0.0558 \text{ (yr)} \tag{3.20a}$$

The water in the borehole is about 20 years old over the 10–22 m depth range (Figure 11.9). The age is reasonable for $P = 0.3$ m/yr and the average depth, but the large spread suggests that waters of various ages are mixed over depth by dispersion. To derive the dispersion data, we follow Schlosser (1989) and calculate 1D concentration profiles for the variable tritium input in rain from 1950–1985, with ^3H to ^3He decay (Equation 3.20), a flux boundary condition for ^3H and either a flux or a constant concentration boundary for ^3He, and finally, a 4 times higher diffusion coefficient for ^3He than for ^3H. All these conditions can be included with little effort in the program given in Example 11.4 and we will discuss the results step-by-step (download tr_he.pas from the website).

First, to check the program, diffusion and dispersion are set to zero and the water age is calculated with depth following Equation (3.20a). The results are exactly identical to the analytical depth/time solution. Second, tritium and helium can be given the same diffusion coefficient of 10^{-9} m^2/s and flux boundary conditions. Now the calculated ages deviate slightly from the ideal (hydraulic) age (Figure 11.10). Above the tritium peak the ages are higher than the hydraulic age and below the peak they are lower. The deviations are due to the interplay of mixing (a linear process) and radioactive decay (an exponential process). As the result, the apparent age of a mixture of waters will come closer to the age of the tritium peak. Third, the diffusion coefficient of ^3He can be made 4 times larger than of ^3H (cf. Jähne et al., 1987 and data in Table 3.5). The generated ^3He now diffuses faster away from the peak than ^3H, and consequently, the 1963 water will appear to be younger while the surrounding water will be calculated to be older. In this case, the 1963 water is assigned a 5 years younger age, while the water at 23 m depth is aged by 13 years.

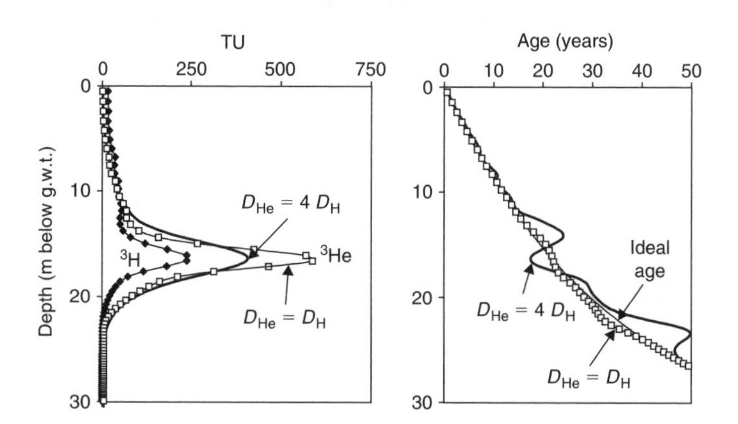

Figure 11.10. Tritium/Helium profiles and ages in a homogeneous phreatic aquifer with diffusion. ^3He has 4 times higher diffusion coefficient than ^3H which reduces the apparent age of water from the 1963 peak.

The analyzed ^3H and ^3He profiles are more disperse than in Figure 11.10 and as the fourth step dispersion is introduced in the model. A dispersivity $\alpha = 6$ m reduces the maximal tritium concentration to the analyzed 46 TU (Figure 11.11). However, the peak depths are different in Figures 11.9 and 11.11. Also, the model calculates a ^3He concentration of 30 TU at the groundwater table where a concentration of zero was analyzed. It suggests that ^3He may escape into the unsaturated zone, and as the fifth step, a constant boundary condition is defined for ^3He, $c_{^3He,d=0} = 0$. The profile now loses appreciable amounts of ^3He and the calculated age is considerably reduced.

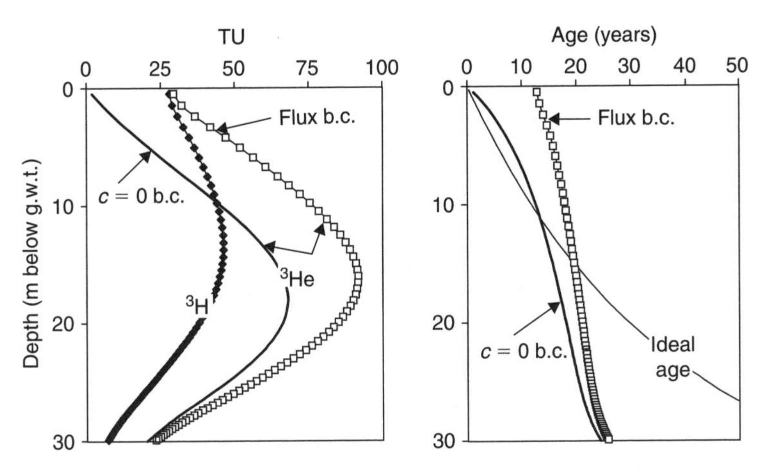

Figure 11.11. Tritium/Helium profiles and ages in a homogeneous phreatic aquifer with diffusion and dispersion. A flux boundary condition at zero depth for ^{3}H and ^{3}He is compared with a constant boundary condition for ^{3}He.

With the large dispersion, the difference in diffusion coefficients of ^{3}H and ^{3}He has no effect on the profiles. (This neatly shows the distinction between diffusion, that acts to separate the isotopes of a species, and dispersion, which mixes them).

How well do the model parameters represent the field conditions? A transverse vertical dispersivity of 6 m would imply a longitudinal dispersivity of 600 m (Chapter 3). However, borehole DFG1 lies about 2.3 km from the divide, which suggests that the longitudinal dispersivity should be 230 m or smaller.

We can also compare mass balances. The mass flow of tritium in a borehole is obtained by integrating over depth:

$$F_{\mathrm{TU}} = \int_{d=0}^{d=D} c_{^3\mathrm{H}+^3\mathrm{He}} \cdot v_{\mathrm{H_2O}} \, \mathrm{d}(d) \tag{11.43}$$

In the homogeneous aquifer, the flow velocity is the same at all depths:

$$v_{\mathrm{H_2O},x} = \frac{P \cdot x}{D\varepsilon_w} \tag{3.7}$$

where x is the distance between the borehole and the divide. The flow velocity increases linearly with x. Thus, in a homogeneous aquifer we can integrate the ^{3}H and ^{3}He concentrations over depth, multiply with $v_{\mathrm{H_2O},x}$ and divide by x, and the result is identical for all boreholes.

In the real aquifer the flow velocity probably varies with depth. Leuchs (1988) indicated that a 5 m thick gravel layer is present at 11 m depth in the otherwise sandy aquifer. If the gravel layer has a 6 times higher conductivity than the sand, it transmits 50% of the water in the profile. Consequently, it has a large share in the mass balance of Equation (11.43) and, actually, the mass balance is difficult to calculate if the variations in flow velocity are unknown. The gravel layer also explains the disperse distribution of tritium and helium in DFG1 since it mixes waters of various ages. The mixed water is spread out over depth again when the gravel layer ends, as sketched in Figure 11.12.

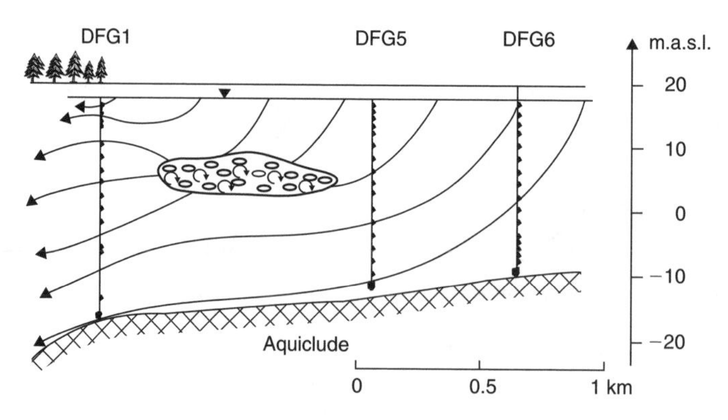

Figure 11.12. Water in the aquifer (Figure 11.9) mixes in a gravel layer and spreads out over depth when it ends.

With a few simple formulas we can estimate the flow properties of the gravel layer that may explain the observed age profile in DFG1. For example, if half of the profile (17.5 m) is homogenized, then the gravel layer has a conductivity that is $(D - z) / z$ larger than the rest, z being the thickness of the gravel layer. Mixing in the layer requires a flowpath L that can be estimated with Equation (3.56), neglecting the cross terms which increase α_{xy}:

$$\sigma^2 = 2D_L t = 2(D_e t + \alpha_T L) \tag{3.56a}$$

in which t and L are connected through:

$$t = \frac{D\varepsilon_w}{P} \ln\left(\frac{x}{x - L}\right) \tag{3.10c}$$

For example, with $D_e = 3\times10^{-10}\,\text{m}^2/\text{s}$, $x = 2200\,\text{m}$, $\alpha_T = 0.005\,\text{m}$, and $\sigma = z / 2 = 2.5\,\text{m}$, $L = 605\,\text{m}$. These relations are useful for initiating a hydrological model in which irregularities are tried out and distributed over the section to simulate observed age patterns. However, it will be difficult to find a 2D or 3D hydrological model in which helium and tritium can be given different diffusion coefficients and/or different boundary conditions at the water table.

Neglecting the difference in the transport properties of tritium and helium, we can calculate the profile of Figure 11.12 with PHAST (Parkhurst et al., 2004, files tr_he.chem.dat and tr_he.trans.dat from www.xs4all.nl/~appt). Figure 11.13 shows calculated tritium and tritium/helium age profiles at 300, 1900 and 2300 m distance from the divide for 1986. The 300 m profile is for the homogeneous aquifer and would be seen everywhere in case the gravel layer were absent. The 1900 m profile crosses the gravel layer at 11–16 m depth. Note that the peak concentration is much less than at 300 m and that the integrated concentration over depth is only half of what is found at 300 m. However, the fluxes calculated with Equation (11.43), divided by x, are identical. Lastly, the 2300 m profile is similar to the observed one in borehole DFG1 (Figure 11.9).

Inspection of the PHAST output files shows that the hydraulic heads in the cross section are down to the mm scale invariant over depth, implying that the flow velocity at $x = 1900$ m changes according to the hydraulic conductivity variation. Thus, when calculating the tritium flux according to Equation (11.43), the tritium peak concentrations between 11 and 16 m depth are multiplied by the higher velocity that is in direct proportion with the higher hydraulic conductivity of the gravel layer. We can also conclude that the velocity variations in the aquifer are difficult to discern from hydraulic head measurements which have only a mm-scale resolution. They can only be found by doing tracer tests.

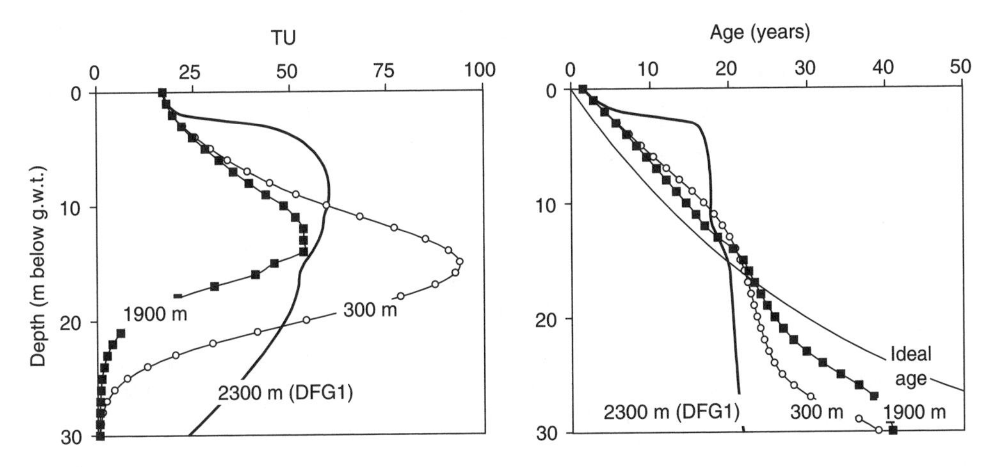

Figure 11.13. Tritium and tritium/helium age profiles in the aquifer of Figure 11.12 at varying distance from the divide, calculated in 2D with PHAST. The profile at 1900 m crosses a gravel layer and has a much smaller tritium content than at 300 and 2300 m (but the flux F_{TU} remains the same). The profile at 2300 m is for borehole DFG1 (Figure 11.9).

Finally, the modeled gravel layer is not the unique factor for explaining the analyzed profile, it may well be possible to find better matching results for the analyses of DFG1 with other hydrogeological configurations. Indeed, many profiles with environmental tracers such as tritium, tritium/helium and CFC's or SF6 have been reported and were compared with flow model results, but it appears difficult to generate a satisfactory model from many possible alternatives (Schlosser et al., 1988, 1989; Engesgaard et al., 1996; Szabo et al., 1996; Sheets et al., 1998; Shapiro et al., 1998; Shapiro, 2001; Zoellmann et al., 2001; Weissmann et al., 2002). Discrepancies are invariably attributed to improper conceptualization or non-uniqueness of the system. However, statistical optimization of the hydraulic conductivity variations using these environmental tracers has not been performed yet. On the other hand, the results presented in this section illustrate that water quality is rapidly homogenized over depth after a few km flow, which marks the problem of finding exact flowpaths as ill-posed from the start. We'll never find the exact answer.

How then, can we explain the presence of chemical variations with depth in many (saturated) profiles presented in previous chapters (Figures 1.7, 2.18, 3.2, 8.22, 9.14, 9.18, 9.20, 9.26, 9.44, 9.45, etc.)? Reinspection shows that the largest variety is found close to the point of infiltration where dispersion has not yet gained a dominant influence (Figures 3.2 and 9.18, DFG6 in Figure 9.14 is close to the divide, cf. Figure 11.9). In case a conservative parameter like Cl^- is already mixed over depth, the variations result from a chemical reaction with the aquifer (Figures 2.18, 8.22, 9.26 and 9.44). The changes can also indicate a diffusive source from another water, or from sediments with a different origin (Figures 3.28 and 9.45). Thus, it is the position in the flowfield of the aquifer, the reactivity of the aquifer with respect to the water quality, and the different properties of adjoining formations that determine contrasts in groundwater chemistry. And, that's why hydraulics, geology and geochemistry must be integrated when explaining water quality variations in aquifers.

11.2.2 *Toluene degradation in an aquifer*

Toluene, C_7H_8, and other BTEX compounds (benzene, toluene, ethyl-benzenes and xylene) have been spilled at thousands of sites and have contaminated groundwater. Yet, the actual spread of these contaminants is usually limited because the chemicals are degraded in the aquifer.

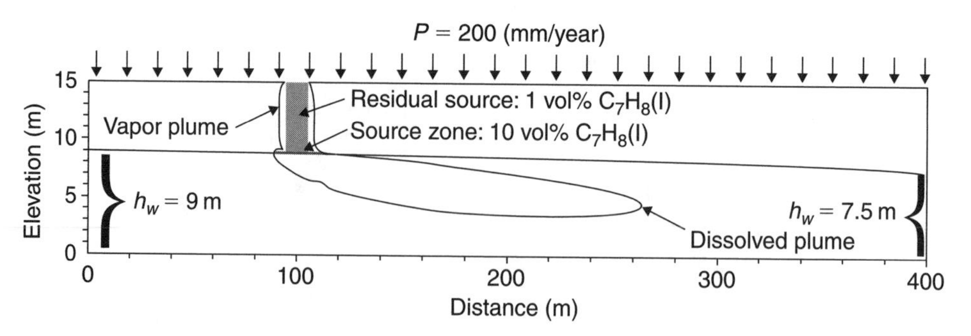

Figure 11.14. Cross section through a model aquifer polluted by toluene. The depicted plume extension is for conservative behavior of toluene. Note for the hand calculations in the text, that the left boundary ($x = 0$) is at 50 m from the divide (Mayer et al., 2002).

Breakdown occurs via microbially mediated oxidation with *dissolved* oxygen, nitrate, sulfate, and with *sedimentary* iron-oxyhydroxides as electron acceptors (MacQuarrie et al., 1990; Hutchins et al., 1991; Lovley et al., 1994; Chapelle et al., 1996; Schmitt et al., 1996; Barbaro and Barker, 2000; Cozarelli et al., 2001, Example 9.11). An interesting point is that the dissolved oxidants need to be mixed with the contaminant plume by dispersion and diffusion, while the solid and sorbed oxidants (and the associated bacteria which are fixed on the sediment particles) may react when the reducing pollutant flows by. In addition, BTEX can also be degraded by fermentation. The different degradation paths lead to various reaction products located distinctly within or around the plume.

The outline of an example case modeled by Mayer et al. (2002) is shown in Figure 11.14. The soil above a uniform, phreatic aquifer was polluted with toluene and a plume developed in the groundwater body. The concentrations of various chemicals, 10 years after onset of the pollution, are presented in Figure 11.15; we follow the numerical simulation of Mayer et al. with simple hand calculations. First, we calculate concentration and distance traveled by toluene, assuming conservative behavior. The effective solubility of toluene used by Mayer was 1.1 mM (this is less than the actual solubility to mimick the time-dependent dissolution from blobs, cf. Chapter 10). In 10 years time, $c_{0.5} = 0.55$ mM is expected to have traveled from 100 to 229 m for the conditions given in Figure 11.14, an average $h_w = 8.5$ m and a water-filled porosity $\varepsilon_w = 0.38$ ($h_w = D$ in Equation (3.9); note that $x = 100$ m in Figure 11.14 is 150 m from the divide). The position of $c_{0.16} = 0.18$ mM would be at $100 + 129 + \sqrt{(2 \times 129 \times 0.1)} = 234$ m for $\alpha_L = 0.1$ m. This plume extension fits the numerical result in Figure 11.14. However, Figure 11.15 shows that toluene has not advanced as far because of degradation.

The oxidation reaction of toluene,

$$C_7H_8 + 14H_2O \rightarrow 7CO_2 + 36H^+ + 36e^- \qquad (11.44a)$$

requires that 36 electrons per mol toluene are taken up by oxygen:

$$36e^- + 9O_2 + 36H^+ \rightarrow 18H_2O, \qquad (11.44b)$$

or by iron-oxyhydroxide:

$$36e^- + 36FeOOH + 108H^+ \rightarrow 36Fe^{2+} + 72H_2O, \qquad (11.44c)$$

or by methanogenic reduction of toluene:

$$36e^- + 1.8C_7H_8 + 36H^+ \rightarrow 12.6CH_4 \qquad (11.44d)$$

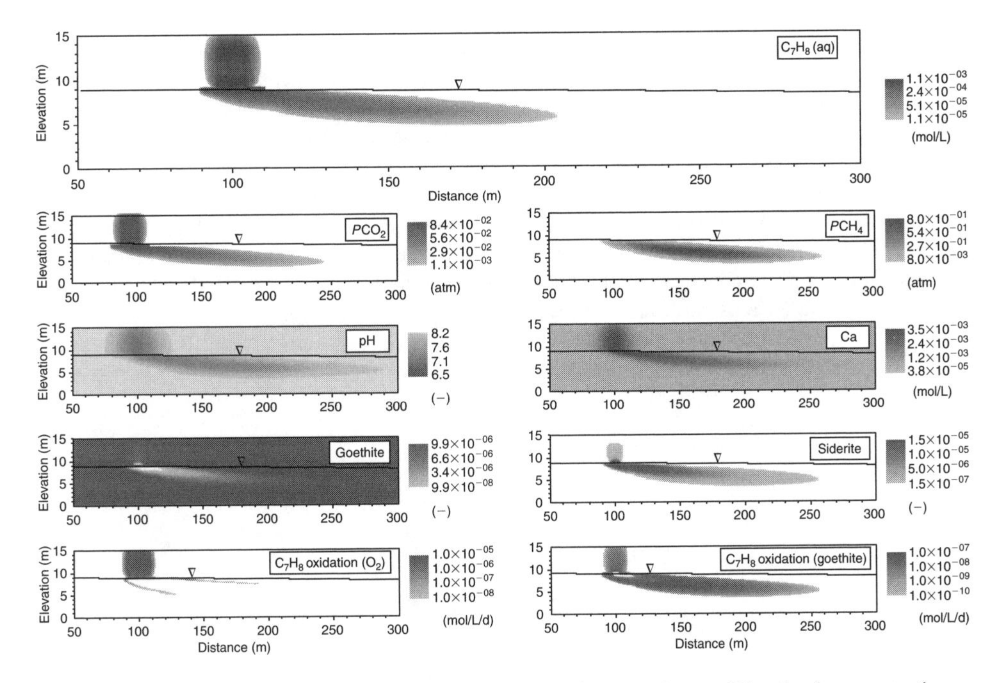

Figure 11.15. Simulation results of biodegradation of toluene in groundwater. Dissolved concentrations are in mol/L, gas pressures in atm, solids as volume fraction [−], and reaction rates in mol/L/day. (Mayer et al., 2002).

Protons are balanced when Reaction (11.44*a*) is combined with (11.44*b*) or (11.44*d*), so that the produced CO_2 acidifies the water. As a result, the pH decreases, calcite (when available) dissolves and the Ca^{2+} concentration increases (Figure 11.15). On the other hand, Reaction (11.44*c*) appears to consume considerable amounts of protons, but most of the released Fe^{2+} will precipitate in the form of siderite:

$$36e^- + 36FeOOH + 72H^+ + 36HCO_3^- \rightarrow 36FeCO_3 + 72H_2O \qquad (11.44e)$$

which still consumes acidity. Thus, the development of pH can indicate which oxidation reaction takes the lead.

The model results in Figure 11.15 show that pH is lower in the plume than in the surrounding groundwater, which indicates that oxidation by oxygen or methanogenic fermentation is the most important degradation reaction. It can be verified indeed in Figure 11.15, that the rate of oxidation by O_2 has been assumed to be 100 times larger than by goethite. Most of the oxidation takes place in the unsaturated zone where gaseous diffusion of oxygen is much more effective than in the aquifer, where only a small rim of the plume is reached by aqueous oxygen. Mayer et al. (2002) have considered various other reactions in their model, such as CH_4 oxidation, and they included also the less crystalline ferrihydrite, which reacts quicker than goethite.

QUESTIONS:
Recalculate the volume fractions of 10^{-5} goethite and 1.5×10^{-5} siderite in Figure 11.15 to mmol/L pore water? $\rho_{sid} = 3.96$ kg/L, $\rho_{go} = 4.3$ kg/L, $\varepsilon_w = 0.38$.

ANSWER: 1.27 mmol go/L, 1.35 mmol sid/L.

Calculate the residence of water in the unsaturated zone of the toluene case, Figure 11.14, when $\varepsilon_w = 0.1$?

ANSWER: ≈3.5 yr.

Use PHREEQC to calculate the pH, the CO_2 pressure and the Ca^{2+} concentration when 10^{-5} mol C_7H_8/L/d oxidizes with O_2 in the unsaturated zone? Calcite dissolves to equilibrium, temperature is 25°C.

ANSWER: pH = 5.78, $P_{CO_2} = 10^{0.36} = 2.29$ atm, $m_{Ca} = 13.0$ mM.

CO_2 escapes from water in the unsaturated zone, therefore repeat the calculations limiting $[P_{CO_2}]$ to 8.4×10^{-2}.

ANSWER: pH = 6.69, $m_{Ca} = 3.6$ mM.

Use the result of the previous calculation as starting point for toluene polluted groundwater, and calculate the composition when goethite oxidizes 10^{-7} mol C_7H_8/L/d over 10 yrs in 1 yr steps. Allow for precipitation of siderite. Plot the resulting pH, P_{CO_2}, Ca^{2+} and Fe concentrations with user_graph as a function of distance (use Equation (3.9), note that $x = 0$ is 50 m from the divide) and compare with Figure 11.15.

ANSWER:
```
SOLUTION 1; -temp 25
EQUILIBRIUM_PHASES 1
 Calcite; O2(g) -0.68; CO2(g) -1.075 0
REACTION
 C7H8 1; 0.0128
SAVE solution 2
END
USE solution 2
EQUILIBRIUM_PHASES 1
 Calcite; Siderite 0 0; Goethite
REACTION
 C7H8 1; 3650e-7 in 10
USER_GRAPH
 -head x Pco2/10 Ca Fe*100 pH
 -start
 10 graph_x 150 * exp(0.2 * step_no /(8.5 * 0.38)) - 50
 20 graph_y 10^SI("CO2(g)")/10, tot("Ca"), tot("Fe") * 100
 30 graph_sy -la("H+")
 -end
END
```

Write the redox equation for toluene oxidation with SO_4^{2-} as electron acceptor?

ANSWER: $36e^- + 4.5SO_4^{2-} + 45H^+ \rightarrow 4.5H_2S + 18H_2O$

In the PHREEQC input file, add 1 mM SO_4^{2-} to the initial, polluted groundwater and allow for siderite *and* pyrite precipitation. Plot the sulfate concentration and the amounts of siderite and pyrite that precipitate. Which phase is more insoluble, pyrite or siderite?

11.2.3 *Remediation of a BTEX polluted site*

The previous modeling example showed that a 2- or 3D model is necessary for evaluating the reactions at the plume's rim where water with oxygen or other oxidants admixes. Internally in the plume a 1D model may be applied along a flowline. The advantage of a 1D model is that simulated concentrations can be easily compared with observations in a single picture. A site below a former gasworks in Düsseldorf (Germany) was remediated by injecting water with nitrate, and the concentrations in the aquifer were measured in multilevel sampling points (Figure 11.16). The comparison of model and observed concentrations allowed to define a comprehensive reaction scheme in which the reactions of both mineral and organic compounds were combined (Eckert and Appelo, 2002).

Figure 11.17 shows a selection of observed concentrations and model results for conservative transport of the compound (grey lines) and with reactions (black lines) in the plume at 2 m distance from the injection well. The decrease of injected NO_3^- is initially accompanied by an increase of the SO_4^{2-} concentration due to reaction of FeS. This FeS formed earlier when the BTEX compounds reacted with SO_4^{2-} from groundwater (Wisotzky and Eckert, 1997).

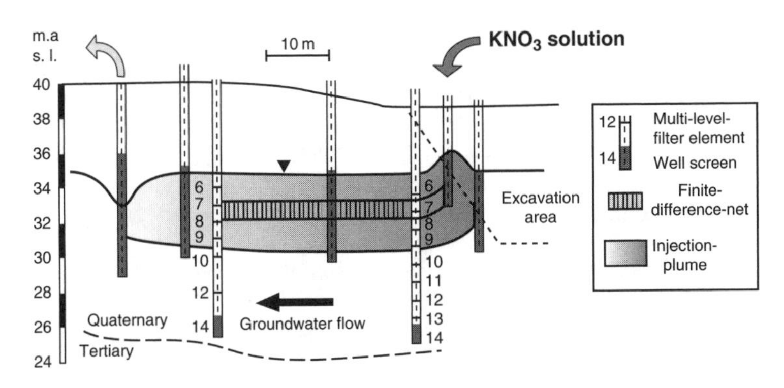

Figure 11.16. Cross section showing injection, abstraction and observation wells below a former gasworks site (Modified from Eckert and Appelo, 2002).

The oxidation of FeS was proved with $\delta^{34}S$, which is low in reduced sulfur compounds and indicates the source of dissolved SO_4^{2-}. When FeS is exhausted after about 50 days, the still present BTEX compounds give a small reduction of the NO_3^- concentration.

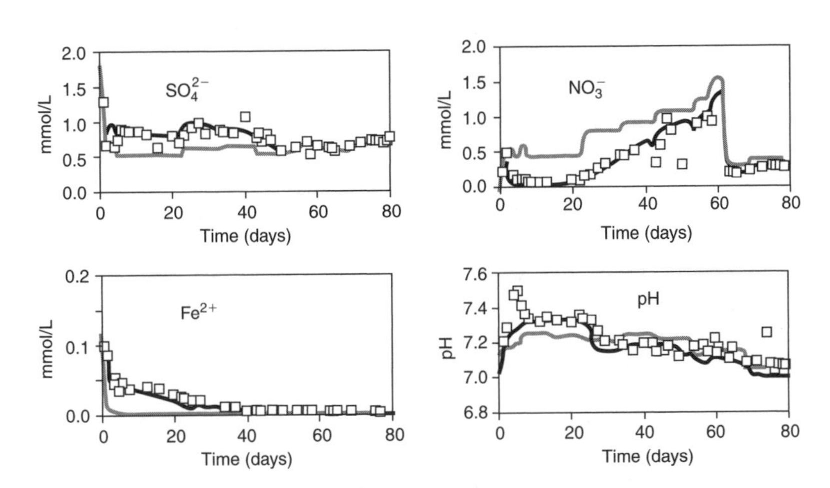

Figure 11.17. Selected concentrations in groundwater during remediation of a BTEX site with nitrate amended water. Squares: observed data; grey lines: conservative transport of injected water; black lines: model with reactions (Eckert and Appelo, 2002).

The data show the combined presence of NO_3^- and Fe^{2+} in the groundwater after 20 days, which is thermodynamically not feasible, and points to a kinetic oxidation of Fe^{2+} assisted by microbes or perhaps by solid surfaces. The kinetic reaction was modeled assuming that the rate was additive to the a-biological oxidation of Fe^{2+} with O_2 (cf. Example 4.12):

$$-\mathrm{d}m_{\mathrm{Fe}^{2+}} / \mathrm{d}t = m_{\mathrm{Fe}^{2+}} (2.91\times 10^{-9} + 1.33\times 10^{12}\ [\mathrm{OH}^-]^2\ (773\, m_{\mathrm{O}_2} + f_1\, m_{\mathrm{NO}_3^-})) \qquad (11.45)$$

where t is time (s) and m_i is the molality of i. The dependency of the oxidation rate on nitrate was added as ($f_1\ m_{\mathrm{NO}_3^-}$) and calibrated to $f_1 = 200$, or 4 times lower than for O_2. Actually, we don't know anything about the oxidation mechanism except that it happens in soils and with microbes.

Another interesting feature is the pH change that results from the redox reactions and which can be used for deciphering the reactions. When the oxidation is incomplete (Fe^{2+} is not oxidized), the reaction of NO_3^- with FeS increases the pH:

$$NO_3^- + 5/8 FeS. + H^+ \rightarrow 0.5N_2 + 5/8 SO_4^{2-} + 5/8 Fe^{2+} + 0.5H_2O \quad (11.46)$$

but when the oxidation is complete, the pH decreases:

$$NO_3^- + 5/9 FeS + 1/3 H_2O \rightarrow 1/2 N_2 + 5/9 SO_4^{2-} + 5/9 FeOOH + 1/9 H^+ \quad (11.47)$$

Whether or not iron-oxyhydroxide forms in the reaction does not really matter, the hydroxy-complexes of Fe^{3+} are already acidifying.

The fraction of iron that should oxidize for pH to remain neutral (as observed, Figure 11.7), and the amount of FeS that reacts can be calculated (cf. Question):

$$NO_3^- + 9/16 FeS + 1/4 H_2O \rightarrow 1/2 N_2 + 9/16 SO_4^{2-} + 1/2 FeOOH + 1/16 Fe^{2+} \quad (11.48)$$

or 1/9 of Fe from FeS should remain Fe^{2+}, which is close to observed in the initial 20 days (compare the reaction amounts of SO_4^{2-} and Fe^{2+} in Figure 11.7). Furthermore, initially the pH increases rather sharply due to the reaction of $(Fe,Ca)CO_3$:

$$x/5NO_3^- + Fe_xCa_{1-x}CO_3 + 7x/5H_2O + (1 - 9x/5)H^+ \rightarrow x/10N_2 + xFeOOH + (1 - x)Ca^{2+} + HCO_3^- \quad (11.49)$$

which increases pH when $x < 5/9$. The siderite formed earlier together with FeS by reaction of BTEX with iron-oxyhdroxide, cf. Figure 11.15.

QUESTIONS:

In Example 4.12, the rate of Fe^{2+} oxidation is expressed with P_{O_2} as variable. Find the conversion factor to use m_{O_2}, check with Equation (11.45)?

Include the empirical kinetic oxidation reaction of Fe^{2+} with NO_3^- in PHREEQC (use Example 9 from the PHREEQC manual as template). Is the reaction with NO_3^- more or less acid than with O_2?

ANSWER: less acid

Find the concentration of NO_3^- for which the Fe^{2+} oxidation rates with O_2 and NO_3^- are equal?

ANSWER: $m_{NO_3^-} = 773\, m_{O_2}/200$

Compare the changes in oxidation rates with time for O_2 and NO_3^- when the initial rates are the same?

ANSWER: NO_3^- maintains a larger rate because the reaction is less acid and the rate depends on OH^-.

Derive a pH-neutral oxidation reaction of pyrite with nitrate, similar to Equation (11.48). *Hint*: find the coefficients a and x in the reaction:

$$NO_3^- + aFeS_2 + (8a + 2x - 3)H_2O \rightarrow 1/2 N_2 + 2aSO_4^{2-} + (a-x)Fe^{2+} + xFeOOH$$

by balancing on H and charge.

ANSWER: $a = 6/14, x = 1/14$

11.2.4 *Acid drainage from a Uranium mine*

The former uranium mine "Königstein" is located South of Dresden (Germany), at a depth of about 250 m in Cretaceous deposits shown in cross section in Figure 11.18. The deposits consist of alternating (consolidated) sand and clay layers in which four aquifers are discerned.

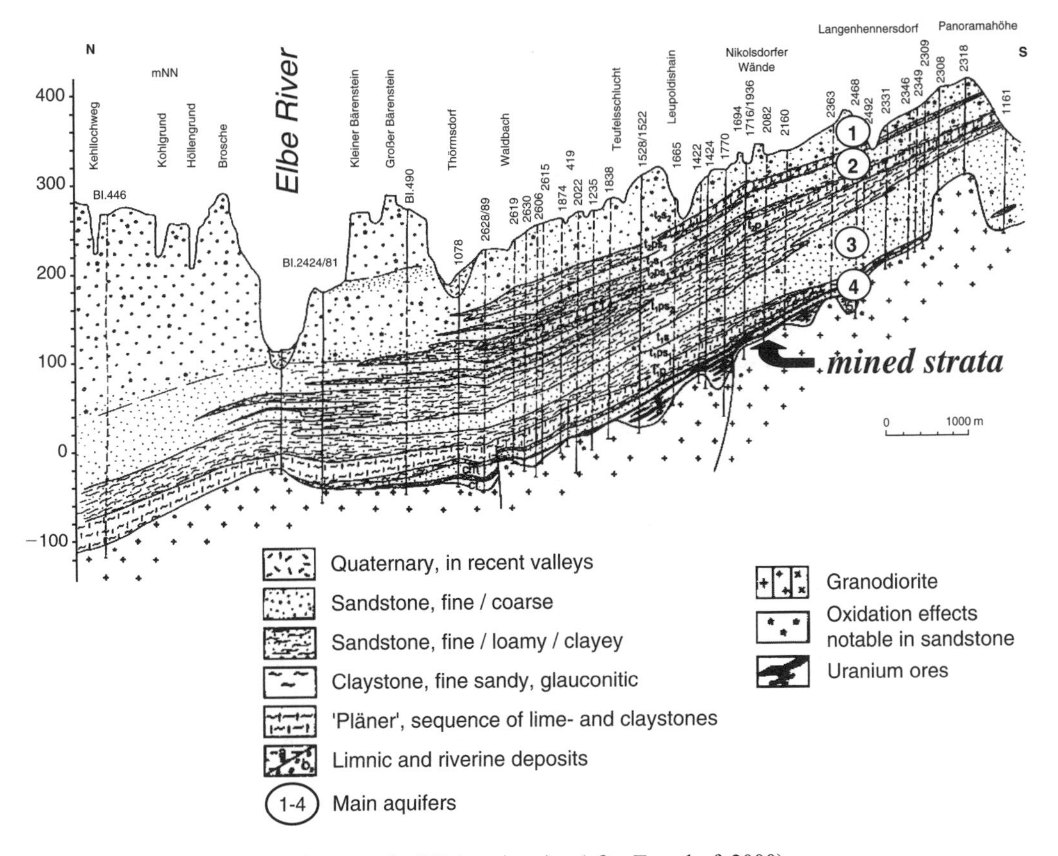

Figure 11.18. N–S cross section over the Königstein mine (after Tonndorf, 2000).

The fourth and deepest aquifer contains locally minable uranium, probably emplaced in the strata when U(6), advected by groundwaterflow, was reduced by organic matter and sulfides, to precipitate as uraninite (UO_2) and coffinite ($USiO_4$) in the form of rollover fronts (Tonndorf, 2000). The fourth aquifer is overlaid by an aquiclude and by the third aquifer, which under natural flow conditions is waterfilled for 50 m at the site of the mine.

However, the mine was pumped during operation, which created a pumpage cone in the third aquifer and oxidizing conditions in the mine, in which sulfides oxidized. In addition, sulfuric acid leaching of U formed part of the mining operations and the mine water still has, 10 years after closure, an extremely low pH and high concentrations of U and heavy metals. The mine water has been pumped and treated above ground so far, but the costs are excessive. When pumping ceases, the mine water will enter the third aquifer via a fault zone and flow towards the Elbe River where it will seep up in and near the river bed. We want to predict the concentrations of heavy metals in the plume and the seepage zones near the river, and to model the attenuation processes (Walter et al., 1994; Bain et al., 2001). The flowtime for conservative transport from the mine to the Elbe River is estimated to be about 15 years (data in Bain et al., 2001).

Ideally, we like to have column experiments to verify the fast sorption reactions of heavy metals and protons. The slow dissolution and precipitation reactions of various minerals that may occur over 15 years flowtime can only be extrapolated by using published kinetic rate data. The slow reactions of the resident alumino-silicates, sulfides, and other minerals are important, since they change the pH and pe of the mine water which in turn affect sorption and precipitation of the heavy metals.

Exchange and sorption of U, Cd^{2+} and Zn^{2+} as a function of pH

So far, the literature on the Königstein mine has neglected the sorption reactions that attenuate transport of the heavy metals, but we may apply the exchange and surface complexation models from Chapters 6 and 7 to this case. For lack of data, we use the default constants from the WATEQ4F database and assume the exchange capacity of the quartz rich sediments, $CEC = 50$ mmol/L, and the surface complexation on ferrihydrite, Hfo = 10 mmol/L. The dominant clay mineral in the sediments is kaolinite (or dickite), which has a *PZC* at pH $\approx$ 4.5. Accordingly, in the low-pH mine water proton exchange on X^- is important and included in the model. Using the mine water composition of Bain et al., we can estimate distribution coefficients ($K_d = q / c$) as a function of pH. We make an input file with the solution, add exchange and surface, and use NaOH as reactant to increase the pH from 2.3 in the mine water to about 10 (Table 11.1).

Table 11.1. PHREEQC input file for calculating distribution coefficients of U, Zn^{2+} and Cd^{2+} in mine water (Figure 11.19).

```
PRINT; -reset false
SOLUTION 1                                          # Mine water from Bain et al., 2001
 -temp 10;    pH 2.3
 Na 23.8;     K 0.1;      Mg 2.0;       Ca 11.6
 C 1.7e-4;    Cl 13.0;    P 0.08;       S(6) 52.8
 Al 6.5;      Cd 0.01;    Fe(3) 10.7;   Fe(2) 0.27;
 U(6) 0.18;   Zn 1.5;

SURFACE 1
 Hfo_w 2e-3 600 0.89; Hfo_s 5e-5; -equil 1          # (0.012 % FeOOH_am = 0.01 mol HFO/L)

EXCHANGE_SPECIES
 H+ + X- = HX; log_k 1.0; -gamma 9.0 0.0  # kaol has PZC = 4.6, sorbs protons
EXCHANGE 1
 X 50e-3; -equil 1                                  # (CEC = 50 meq/L)

REACTION 1; NaOH 1; 105e-3 in 100
INCREMENTAL_REACTIONS

USER_GRAPH  # plot q/c...

 -head pH U Zn Cd
 -start
 10 graph_x -la("H+")                               # find sorbed amounts...
 20 q_U = mol("Hfo_sOHUO2+2") + mol("Hfo_wOUO2+")
 30 q_Zn = mol("ZnX2") + mol("Hfo_sOZn+") + mol("Hfo_wOZn+")
 40 q_Cd = mol("CdX2") + mol("Hfo_sOCd+") + mol("Hfo_wOCd+")
                                                    # plot the distribution coefficients ...
 50 graph_y q_U / tot("U"), q_Zn / tot("Zn"), q_Cd / tot("Cd")
END
```

The results are shown in Figure 11.19. The distribution coefficient for Cd^{2+} is about 0.2 over most of the pH range, but increases for pH > 9.5. The sorbed concentration is entirely due to the exchange reaction, except at very high pH where surface complexation on Hfo becomes important. The distribution coefficient for Zn^{2+} is 1 in the low pH range. It increases to 2.5 at pH 8 by sorption on Hfo, and diminishes again at higher pH where aqueous complexes redistribute Zn^{2+} back into the solution.

At low pH, Zn^{2+} is only retarded by cation exchange, like Cd^{2+}, but why is the distribution coefficient 5 times higher for Zn^{2+} than for Cd^{2+}? Inspection of the database shows that both metals have the same exchange constant. However, for ZnX_2 an activity correction is defined, but not for CdX_2.

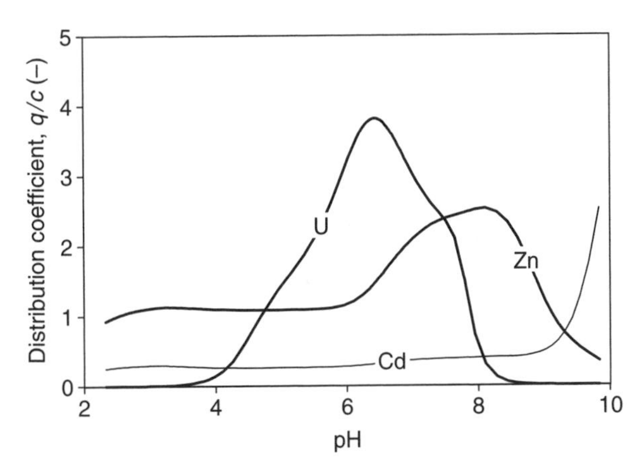

Figure 11.19. Distribution coefficients of trace metals as function of pH. The mine water composition is given in Table 1.1. The pH is increased by adding NaOH.

This explains most of the difference, while another part is due to aqueous complexes which are stronger for Cd^{2+} than for Zn^{2+} in this water. The distribution coefficient for U, which is only bound to Hfo (at least according to the WATEQ4F database), displays the usual competition among complexes in solution and on the surface that is characteristic for heavy metal sorption. At low pH, all U remains in solution and the distribution coefficient is 0. At intermediate pH, K_d increases to 4, but it dwindles again to 0 at higher pH. Kohler et al. (1996) noted that sorption of U(6) occurs on quartz, but it is small compared to Hfo.

Retardation in a multicomponent solution

In the previous section, the retardation of heavy metals was estimated from the distribution coefficient. Actually, the *slope* of the isotherm should be considered (Chapter 3), which is influenced by concentration changes of *all* the elements. For example for uranium:

$$\left(\frac{\partial c_{\mathrm{U}}}{\partial t}\right)_x = -v\left(\frac{\partial c_{\mathrm{U}}}{\partial x}\right)_t + D\left(\frac{\partial^2 c_{\mathrm{U}}}{\partial x^2}\right)_t - \left[\left(\frac{\partial q_{\mathrm{U}}}{\partial c_{\mathrm{U}}}\right)_{x,\mathrm{Cd},\mathrm{Zn},\ldots}\left(\frac{\partial c_{\mathrm{U}}}{\partial t}\right)_x\right] - \left[\left(\frac{\partial q_{\mathrm{U}}}{\partial c_{\mathrm{Zn}}}\right)_{x,\mathrm{U},\mathrm{Cd},\ldots}\left(\frac{\partial c_{\mathrm{Zn}}}{\partial t}\right)_x\right] - \ldots \tag{11.50}$$

The slopes $(\partial q_{\mathrm{U}}/\partial c_j)_{x,i \neq j}$ are termed *flushing factors* for U. A higher flushing factor means higher amounts sorbed, and thus more pore volumes for flushing the chemical. A flushing factor of 0 indicates zero effect on sorption and no effect on transport. For $c_j = c_{\mathrm{U}}$, the largest effect is expected on q_{U}, and indeed, $1 + \partial q_{\mathrm{U}}/\partial c_{\mathrm{U}}$ is equal to the retardation for U, as was defined in Chapter 3.

We can calculate the flushing factors for the Königstein example numerically, changing the concentration of an element in the initial solution a tiny bit and comparing the exchangeable and surface complexed concentrations before and after the addition. The result for U and Zn^{2+}, when changing the U concentration (adding 10^{-8} mol/L), is shown in Figure 11.20. The flushing factor is markedly different from the distribution coefficient for U, and the flushing factor for Zn^{2+} by U can even become negative. What is happening?

The distribution coefficient is the ratio of sorbed and solute concentrations, and it is assumed constant as if we have a linear isotherm. However, with the flushing factor we have calculated the *slope* of the isotherm. Apparently, the slope $\partial q/\partial c$ is larger than q/c for U, indicating that we have a concave isotherm at a fixed pH (the slope of the isotherm lies below the curve).

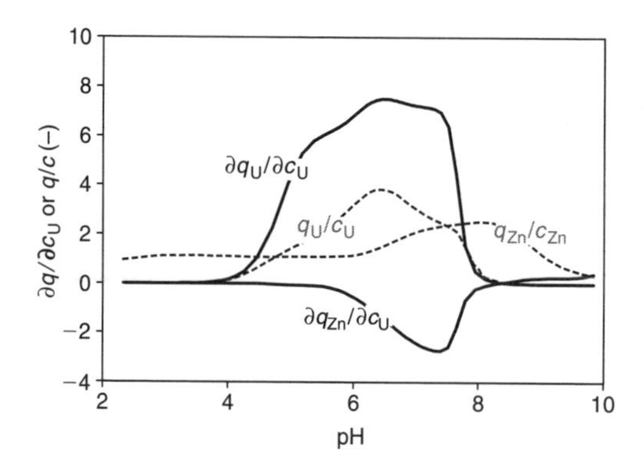

Figure 11.20. Flushing factors (slopes $(\partial q_i \,/\, \partial c_{\mathrm{U}})$) for U(6) and Zn^{2+} as function of pH (thick lines). The distribution coefficients $q \,/\, c$ from Figure 11.19 are drawn for comparison (dotted lines).

Consequently, the U(6) front will be sharpening when concentrations decrease. On the other hand, for Zn^{2+} the slope $(\partial q_{\mathrm{Zn}} \,/\, \partial c_{\mathrm{U}})_{i \neq \mathrm{U}}$ is negative in the pH range where U competes with Zn^{2+} for sorption sites. The negative flushing factor for Zn^{2+} by U indicates that Zn^{2+} is desorbed by U. It can be seen in Figure 11.20 that the displacing effect of U can be even larger than of Zn^{2+} itself, which suggests that the flushing factor for Zn^{2+} might become negative. However this does not happen. When the sorbed concentration of Zn^{2+} diminishes, also the flushing factor tends to zero and the retardation becomes 1 (retardation cannot be smaller than 1).

We can calculate a concentration profile with PHREEQC and see how these retardations apply along a flowline. Going from the mine to the Elbe River at 2000 m, Figure 11.21 shows the concentrations of SO_4^{2-}, Zn^{2+}, U(6) and pH when the mine water has traveled 1500 m. Sulfate shows almost conservative behavior (it is only very slightly sorbed by Hfo) and indicates the position of the acid front. Zinc is retarded by 2.1, in agreement with the calculated distribution coefficient of 1.1 when the pH is below 5 (Figure 11.19). In the low pH range, surface complexation of Zn^{2+} on Hfo is minor, and the retardation is not affected by U. The concentration of U shows an increase as if a heap of snow is pushed up in front of a snowplough. The typical shape is due to the pH variation in the front; uranium is sorbed at neutral pH, but as the pH decreases to below 4 in the acid mine water, U is desorbed again, and the decreasing concentration gives a sharp front.

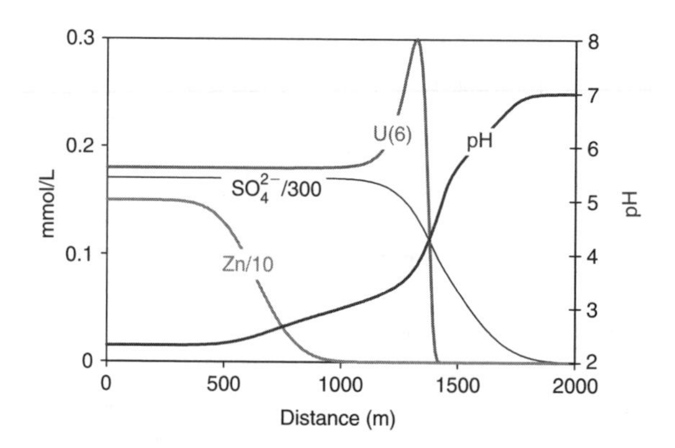

Figure 11.21. Calculated concentrations of SO_4^{2-}, Zn^{2+}, U(6) and pH along a flowline from the Königstein mine to the Elbe River. The acid front of the mine water is at 1500 m.

The flushing factors are needed in transport models when correcting for numerical dispersion (Example 11.4), or when correcting for reactions in implicit schemes. Kirkner and Reeves (1988) suggest to calculate the factors by numerical differentiation as we did with U. However, in general this is not easy since the difference between activity (which determines sorption) and concentration (which is transported) may influence the results. Furthermore, the flushing factors will show numerical roundoff errors which affect the charge balance of the transported solution after the correction. This may lead to non-convergence of the combined transport and geochemical model.

In most cases, the retardation ($1 + \partial q_i / \partial c_i$) will have the largest influence on transport of i, and can be calculated easily from changes in the solute concentration over time in a cell using Equation (11.31). Zysset et al. (1994) calculated its value after a half-time transport step, and corrected transport with it in the next half-time step in a procedure called Strang-splitting. After the correction, the model may iterate until the newly calculated retardation is equal to the previous one within a given tolerance (Yeh and Tripathi, 1989; Engesgaard, 1991). By neglecting the cross-terms ($\partial q_i \, / \, \partial c_j$), possible chromatographic peaks and troughs (cf. Chapter 6) are disregarded, but these are smeared out anyhow over the cell domain for which the geochemical model calculates the reactions. The code HydroGeoChem by Yeh et al. (2004) does consider the cross-terms (Yeh, pers. comm.).

An interesting point to consider is the retardation of Si, when the solution is in equilibrium with quartz, and pH is below 10. In this case, $dc_{Si} = 0$, and the retardation is infinite by Equation (11.31). The concentration of $H_4SiO_4^0$ is invariable when quartz is present, and we don't even need to calculate transport and reactions. This holds for other minerals as well, each mineral fixing the concentration of one component by the Phase Rule (Saalting et al., 1999).

QUESTIONS:

Calculate the distribution coefficient for mine water with activity correction for CdX_2 at low pH?

ANSWER: 0.8

What will be the flushing factor of U by Cd^{2+}?

ANSWER: minimal, Cd^{2+} is not much sorbed by Hfo

Calculate the flushing factors for Cd^{2+}, Zn^{2+} and U by U, Zn^{2+} and Cd^{2+} for the Königstein case? *Hint*: in the input file in Table 11.1, put q_i and c_i for each reaction step in a memory location, "60 put(q_U, 1, step_no); 70 put(tot("U"), 5, step_no)", etc. Then, in a subsequent simulation, add 10^{-8} mol U/L to the solution, calculate again the q_i's, and find $\Delta q_i / \Delta c_U$. Repeat, adding 10^{-8} mol/L of Zn^{2+} or Cd^{2+}.

Other effects on transport of heavy metals

Overall, Figures 11.19 and 11.20 show that retardation of the heavy metals will be small when acid mine drainage enters a silicate aquifer, but that it may become significant when the pH increases to 6. There are at least four mechanisms imaginable that can increase the pH, *viz.* sorption of protons, silicate dissolution (the concentration of Si is quite high in the mine water, cf. Bain et al., 2001), dilution by mixing with local recharge water, and reaction with calcite in confining layers. The assumed *CEC* and concentration of Hfo in the Königstein aquifer can take up 10 mmol H^+/L mine water at pH = 2.3, and hence can retard the acidity by a factor of $1 + (0.01 / 10^{-2.3}) = 3$ (cf. Figure 11.21 which shows that pH = 2.3 has only traveled 500 m). Equilibrium with goethite, pyrite and kaolinite will bring the pH to 4.6 and perhaps even higher when the iron-oxyhydroxide in the aquifer is more soluble than pure goethite (cf. Questions), but the reactions are kinetic. To assess the effect of dilution by mixing with local recharge water requires detailed hydrogeological information, and finally, the reaction with calcite may be feasible since large amounts of calcite are present in the confining layers (Tonndorf, 2000).

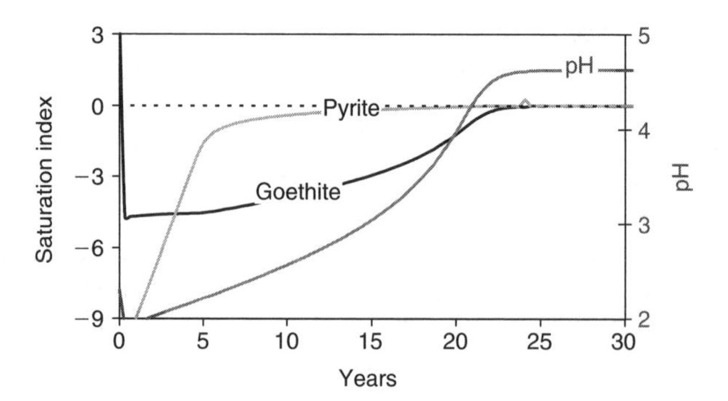

Figure 11.22. Kinetic reaction of pyrite and goethite with mine water increases the pH. The saturation indexes of pyrite and goethite indicate the time course of the reactions.

Kinetic mineral reactions

The reactivity of the kinetically reacting minerals in the aquifer with mine water can be explored with PHREEQC in a step-by-step approach starting with pyrite/goethite and then adding the silicate minerals (cf. Glynn and Brown, 1996). The major variables are the pH and the saturation indices of the minerals, which change during the reaction progress and end up at zero when final equilibrium is reached (Figures 11.22 and 11.23).

The mine water is supersaturated with respect to goethite, but undersaturated for pyrite. Dissolution of a tiny amount of pyrite lowers the pH and leads to subsaturation for goethite, which starts to dissolve:

$$FeS_2 + 14FeOOH + 26H^+ \rightarrow 15Fe^{2+} + 2SO_4^{2-} + 20H_2O \tag{11.51}$$

The kinetic dissolution rates for pyrite were discussed in Section 9.4.2, and for goethite given in Equation (9.82). Figure 11.22 shows how the pH increases concomitantly with the dissolution of goethite. Pyrite reaches saturation after approximately 12 years in the model, goethite needs 22 years and the pH is then 4.6.

The mine water is also subsaturated for kaolinite, a mineral that reacts quicker than goethite:

$$Al_2Si_2O_5(OH)_4 + 6H^+ \rightarrow 2Al^{3+} + 5H_2O + 2SiO_2 \tag{11.52}$$

where the released silica precipitates as cristobalite.

For kaolinite, we look up the rates defined by Sverdrup and Warfvinge (1995) and find:

```
Kaolinite
# Rate from Sverdrup and Warfvinge, 1995, Am Mineral. 31, 485.
# r in 1/6 kmol kaol/m2/s (is 1 kmol H+/m2/s)
 -start
 10 A = 3150 * m0                                   # m2 for 10 m2/g
 20 f_Al = 1 + tot("Al")/4e-6
 30 r = 10^-15.1 * act("H+")^0.7 * f_Al^-0.4 + 10^-17.6 * f_Al^-0.2
 40 moles = A * r/6 * 1e3 * (m/m0)^0.67 * (1 - SR("Kaolinite")) * time
 50 if moles < 0 then moles = moles / 10           # 10 times smaller precipitation rate
 60 save moles
 -end
```

Sverdrup and Warfvinge (1995) defined the rate in terms of proton buffering of 6 mol H^+ per mol kaolinite, and the rate was reformulated to moles of kaolinite. The rate contains inhibition factors that account for the decrease in rate by dissolved Al^{3+} as discussed in Chapter 8. Furthermore, the rate for precipitation was (arbitrarily) assumed to be 10 times smaller than for dissolution.

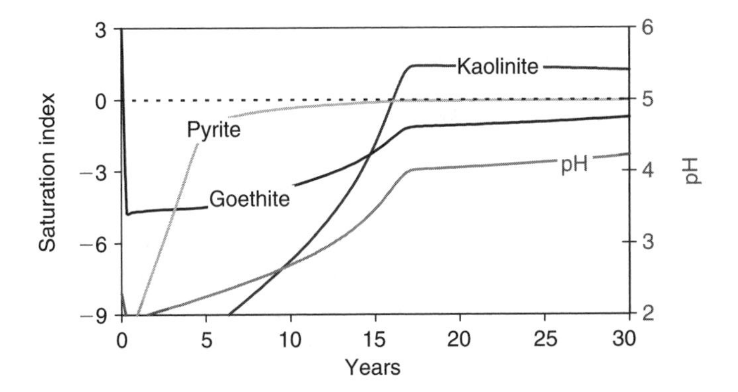

Figure 11.23. Kinetic reaction of pyrite, goethite and kaolinite. Initially, the pH increases about twice faster than without kaolinite, but the pH is lower after 30 years.

When kaolinite is added to the assemblage, the pH increases initially more rapidly to 4 after 16 years. However, the increase slows down when kaolinite reaches saturation. Kaolinite even becomes supersaturated because the pH continues to rise as goethite dissolves and it starts to precipitate (Figure 11.23). The precipitation of kaolinite buffers much of the pH increase from the goethite reaction and the approach to equilibrium of goethite is therefore retarded.

The calculations suggest that the pH will remain below 4.5 during the 15 years flowtime from the mine to the Elbe River. Thus, the pH buffering by mineral dissolution is inadequate to enhance sorption of the heavy metals.

QUESTIONS:

Calculate the pH of mine water in equilibrium with goethite, pyrite and kaolinite (WATEQ4F database)?

ANSWER: 4.6

And if the iron-oxyhydroxide in the aquifer has higher solubility, $SI_{goethite} = 2.0$?

ANSWER: 5.5

Find the distribution coefficient for U after the reaction with pyrite? Why is it low?

ANSWER: zero, surface complexation of U(4) is undefined in the WATEQ4F database.

Model the kinetic reaction of pyrite and goethite in mine water with PHREEQC over 30 yr. Use 100 mM pyrite ($A/V = 0.3$ m^2/mol/L) and 20 mM goethite ($A/V = 2000$ m^2/mol/L).

Add the kinetic reaction of kaolinite, as discussed before. The aquifer contains 5% kaolinite, surface area 10 m^2/g.

Add the kinetic reaction of K-feldspar (Table 8.7). The aquifer contains 5% K-feldspar and surface area 0.4 m^2/g.

Simulate an experiment with mine water used as influent in a 0.1 m column with pristine sediment in which pyrite and goethite react kinetically. Flowtime in the column is 1 day. The pristine water has $[P_{CO_2}] = 10^{-2}$ and is in equilibrium with calcite, pyrite and goethite ($SI_{goethite} = 2$).

A 2D transport model for Uranium

So far, we have assumed that uranium is present as U(6) (mainly UO_2SO_4 and UO_2^{2+}) in the mine water. However, it can be reduced to UOH^{3+} by H_2S in the aquifer:

$$UO_2^{2+} + 0.25H_2S + 0.5H^+ \rightarrow UOH^{3+} + 0.25SO_4^{2-} \tag{11.53}$$

The reduced form of U precipitates readily in various minerals, of which coffinite ($USiO_4$) is one of the most insoluble, and commonly observed in the ore at Königstein:

$$UOH^{3+} + H_4SiO_4 \rightarrow USiO_4 + H_2O + 3H^+ \tag{11.54}$$

Actually, it is probably this reaction scheme that emplaced the ore at Königstein (Tonndorf, 2000). When operating in the (oxic) mine water, it decreases the concentration of uranium by orders of magnitude from 0.1 mM to less than 0.1 μM. However, if the mine water enters directly the oxidized zone that was created during pumping of the mine, uranium is retarded only by surface complexation. In this case, only mixing by dispersion with water from the reduced confining layers may start the reduction reaction and precipitate coffinite. Let's look at the combination of transverse dispersion and chemical reaction with Alliances, an MT3D/PHREEQC combination developed at Andra, France.

A rectangular cross-section of the aquifer, 50 × 2000 m above the mine was discretized in 40 × 100 cells (Figure 11.24). Acid mine water enters the section from the lower left corner and is accompanied by an equal flow of natural water from the upper left corner. The water leaves the section in the middle right. The lower right quarter, from the mine until the outflow cell contains pyrite and reduced water, the rest of the cross-section is oxidized and contains water with 0.3 mM O_2.

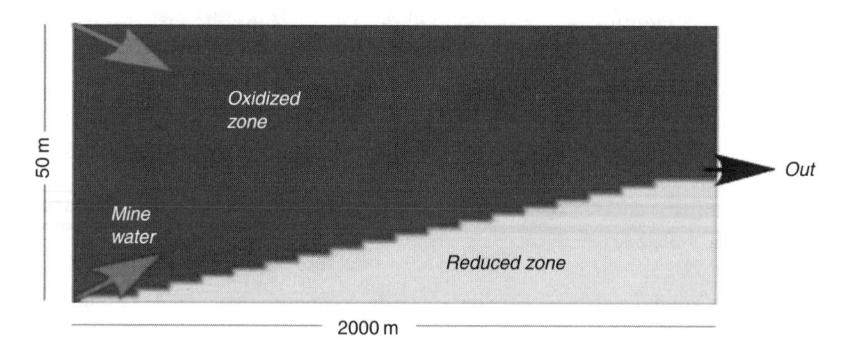

Figure 11.24. Outline of the model aquifer. The reduced zone contains pyrite.

The solute concentrations of U after 20 years indicate the loss towards the reduced zone where coffinite precipitates, while at the front the snowplough effect appears (Figure 11.25). The contact with pyrite is determined by the transverse dispersivity, here taken to be 1 m. As we note before, the translation towards the reduced zone enriched the strata with uranium in the geological past, and they will be fixing uranium again according to the model.

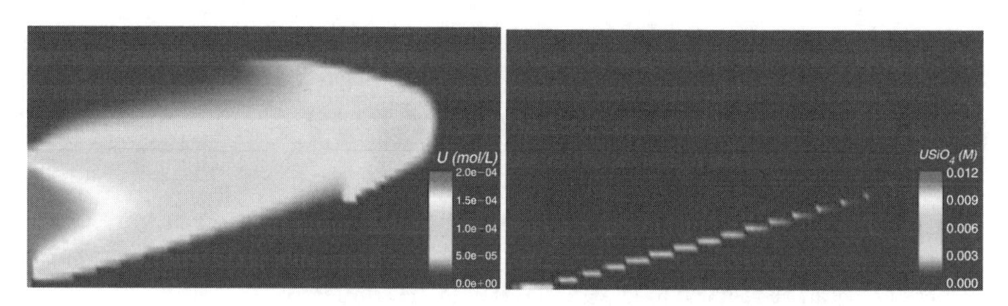

Figure 11.25. Modeled concentrations of U(6) in acid mine drainage entering an aquifer (left) and precipitation of $USiO_4$ (coffinite, right) at the border of the reduced zone.

11.2.5 *In-situ iron removal from groundwater*

The Fe^{2+} concentration of pumped groundwater can be lowered when first a volume of oxygenated water is injected in the aquifer. This process of in-situ iron removal is beneficial since above-ground treatment can be reduced or even omitted and the concentrations of other trace elements such as manganese and arsenic are diminished as well. The oxygen oxidizes Fe^{2+} to Fe^{3+}, which precipitates as iron-oxyhydroxide in the aquifer, but the reaction scheme and the contact mode of O_2 and Fe^{2+} have been disputed. At present, an intermediate reaction involving the desorption of Fe^{2+} from exchange and surface sorption sites seems the most plausible (Appelo et al., 1999). The desorbed Fe^{2+} reacts with O_2 according to:

$$O_2 + 4Fe^{2+} + 10H_2O \rightarrow 4Fe(OH)_3 + 8H^+ \tag{11.55}$$

When groundwater is pumped, Fe^{2+} from groundwater sorbs again to the exchange sites that lost their iron. The iron concentration increases as the exchange sites approach saturation with Fe^{2+} and when a limiting concentration is reached, pumping may be interrupted to start a new cycle by injecting another volume of oxygenated water. Figure 11.26 shows the iron concentration during the first and the seventh cycle in Schuwacht, the Netherlands. The arrival of Fe^{2+} occurs later in the 7th than in the 1st cycle. The delay is related to the transient nature of the process, and to the increase of the sorption capacity on iron-oxyhydroxide.

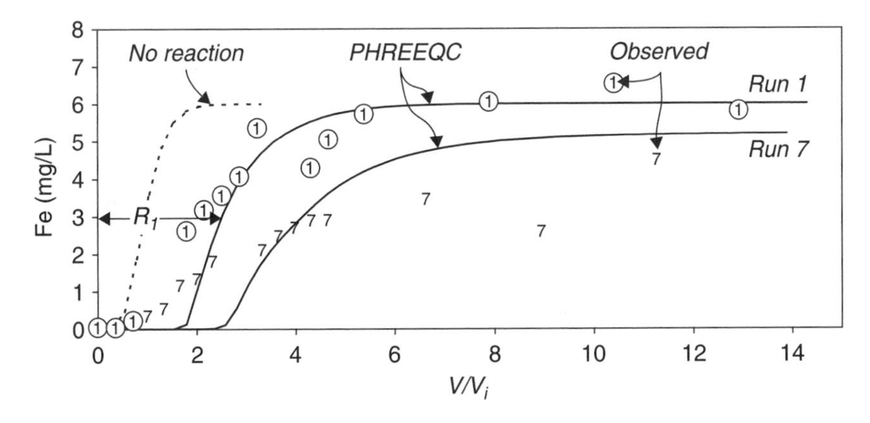

Figure 11.26. Concentrations of Fe during the 1st and 7th cycle of *in-situ* iron removal at Schuwacht. The x-axis gives the ratio of pumped to injected volume (Appelo and De Vet, 2003).

To understand what happens, consider the sequence of oxygen and iron profiles in Figure 11.27. During injection of $1000\,m^3$ aerated water, groundwater with dissolved iron is displaced into the aquifer. If sorption sites were absent, Fe^{2+} would simply move along with groundwater, and the reaction between oxygen and Fe^{2+} would be limited to small amounts at the front where mixing takes place. However, cations from the injected water exchange for sorbed Fe^{2+} and the oxidation of this iron consumes oxygen. Thus, the oxygen front lags behind the injected water front. When the operation is switched to pumping, first the injected volume is withdrawn, with some front spreading as a result of dispersion. Then, groundwater can be pumped with a reduced iron concentration, because Fe^{2+} is lost to the exchange sites when groundwater flows through the oxidized zone. After a fixed time, or when the iron concentration exceeds a limit, another volume of aerated water is injected and the next cycle of *in-situ* iron removal begins.

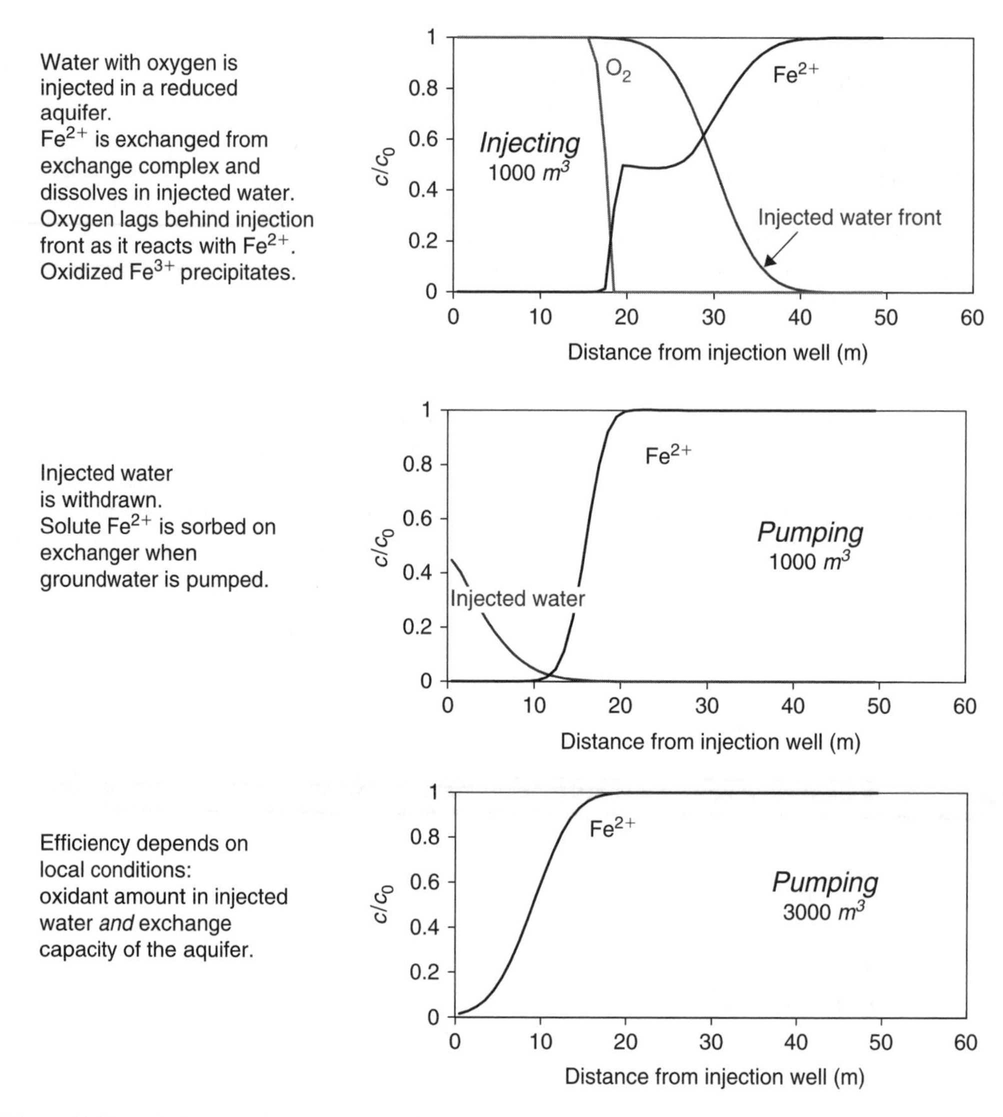

Figure 11.27. Oxygen- and iron-concentration profiles at three stages of an *in-situ* iron removal cycle.

The efficiency of the process can be calculated if we neglect dispersion and limit the sorption reactions to the zone where oxygen has penetrated, and then consider how much iron can be sorbed in that part of the aquifer. First, the position of the oxygen front at the end of the injection stage must be located. This position can be found from the Reaction (11.55) of oxygen with ferrous iron, in which all Fe^{2+} comes from sorbed iron, each mole consuming 1/4 mole of dissolved oxygen. Thus, the retardation of oxygen is:

$$R_{O_2} = 1 + \Delta q_{Fe} / 4\,\Delta m_{O_2} \tag{11.56}$$

where Δq_{Fe} is the difference in sorbed iron over the front (mol/L pore water) and Δm_{O_2} the difference in dissolved oxygen (mol/L). Sorbed iron in Equation (11.56) is in exchange equilibrium with dissolved iron in groundwater and is zero when oxygen is present. Likewise, oxygen is zero where Fe^{2+} is present. These reactions result in a sharp front that migrates through the aquifer.

For the case of linear flow, the distance between the oxygen front and the injection well is:

$$x_{O_2} = x_{inj} / R_{O_2} \tag{11.57}$$

where x_{inj} is the distance traveled by the injected water. The fraction of injected water from which oxygen is consumed is:

$$f_{inj} = (x_{inj} - x_{O_2}) / x_{inj} = 1 - 1 / R_{O_2} \tag{11.58}$$

Note that Equation (11.58) implies that oxygen may not be reacting at all when the retardation equals 1, i.e. when sorbed iron is zero ($\Delta q_{Fe} = 0$ in Equation 11.56). On the other hand, all oxygen is used up when sorbed iron is infinite. Thus, the efficiency of *in-situ* iron removal depends on the sorption capacity of the aquifer for iron, and it will be low in a coarse, gravelly aquifer.

Now during pumping, if native groundwater with dissolved iron returns and flows along the emptied sorption sites, Fe^{2+} sorbs again, and iron is retarded with respect to groundwater flow. The retardation equals (we assume a sharp front, but this depends on the counter-ions):

$$R_{Fe} = 1 + \Delta q_{Fe} / \Delta m_{Fe} \tag{11.59}$$

Equation (11.59) allows to calculate the volume of groundwater that can be pumped until the sorption sites are filled and the iron concentration of the native groundwater arrives at the well. The volume of water between the injection well and x_{O_2} is:

$$V_{inj}(1 - f_{inj}) = V_{inj} / R_{O_2} \tag{11.60}$$

and the volume of groundwater that can be pumped is:

$$V_{gw} = (V_{inj} / R_{O_2}) \times R_{Fe} \tag{11.61}$$

Note again, that Equation (11.61) implies that iron arrives immediately at the well when the ratio of sorbed and dissolved iron is very small, and, of course, that it will never arrive when the concentration of dissolved iron is zero. Thus, the efficiency of the process also depends on the iron concentration in the groundwater.

We can define the efficiency of *in-situ* iron removal as:

$$E = V_{gw} / V_{inj} \tag{11.62}$$

which, according to Equation (11.61), equals:

$$E = R_{Fe} / R_{O_2} \tag{11.63}$$

For a hand calculation, we can assume that Δq_{Fe} in Equation (11.59) is the same as was used for calculating the retardation of oxygen (Equation 11.56). If sorption on the precipitated iron-oxyhydroxide is known, it can be used to increase Δq_{Fe} in Equation (11.59).

Sorption of Fe^{2+} on ferrihydrite

For modeling the process, we need the surface complexation constant of Fe^{2+} to iron-oxyhydroxide. The constant can be estimated with linear free energy relations (*LFER*) given by Dzombak and Morel (cf. Table 7.5). We can also derive a value from experiments reported by Liger et al. (1999), who published the sorption edge of Fe^{2+} on amorphous iron-hydroxide *vs* pH (Figure 11.28). For fitting the sorption edge, we use a general non-linear parameter program (PEST) together with PHREEQC. PEST is popular among hydrologists for optimizing groundwater model parameters.

It fits model values with observations by varying one or more parameters in the input file of the simulation model. The output file generated by the simulation program is read by PEST and compared with the observations, and the parameters are optimized using the Marquardt algorithm. A shareware DOS version of PEST can be downloaded from www.xs4all.nl/~appt, together with guidelines for modeling the sorption edge shown in Figure 11.28.

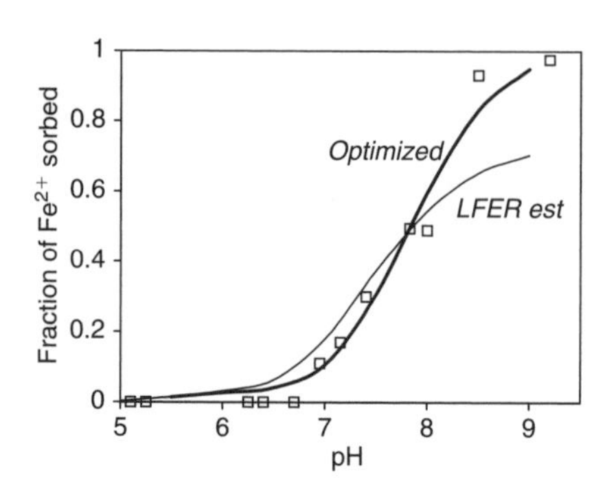

Figure 11.28. Sorption edge of Fe^{2+} on ferrihydrite in 0.1 N $NaNO_3$. Experimental points from Liger et al., 1999; fitted curves from model with surface complexation constants estimated with linear free energy relations (*LFER*) and optimized with PEST and PHREEQC.

Modeling an in-situ experiment

Let us try to model the iron concentrations shown in Figure 11.26. We consider a flowline of 15 m, inject 10 m^3 during 1 day, then pump at the same rate during 3 days and repeat the cycle three times. The water compositions are given in Table 11.2.

Table 11.2. Water compositions in a model for *in-situ* iron removal.

	Temp, °C	pH	pe	Ca^{2+}	Fe^{2+}	Alkalinity	Cl^-
Groundwater	10	7.0	0(?!)	3.0	0.1 mmol/L	6.2 mmol/L	0.001
Injected water	10	7.0	$P_{O_2} = 0.2$	3.0		6.0 mmol/L	0

The aquifer contains iron-oxyhydroxide (100 ppm Fe, or 10.7 mM Goethite, *SI* = 3.0), the *CEC* is 60 meq/L. How shall we set up the input file? Here is an outline:

```
SOLUTION_SPECIES
 H2O + 0.01e- = H2O-0.01; log_k -9.           # For program convergence
EQUILIBRIUM_PHASES 1-15                       # Define sediment
 Goethite 3.0 10.7e-3                         # Goethite equilibrium, SI, moles
SOLUTION 1-15                                 # Define groundwater, 15 cells
 -temp 10; pH 7.0;    pe 0.0 Goethite 3.0
 Ca 3; C(4) 6 charge; Fe 0.1; Cl 1e-3         # Cl added as tracer
END

EXCHANGE 1-15; X 0.06; -equil 1
SURFACE 1-15
 Hfo_w Goethite 0.2 5.3e4                     # Coupled to Goethite, proportion, m2/mol
```

```
 Hfo_s Goethite 1e-3                          # Coupled to Goethite, proportion
 -equil 1
END

PRINT; -reset false
SOLUTION 0                                    # Inject oxygenated water
 -temp 10; pH 7.0; pe 14.0 O2(g) -0.68
 Ca 3; C(4) 6 charge
TRANSPORT
 -cells 15; -length 1                         # Flowtube is 15 x 1 = 15 m.
 -disp 0.1                                    # Dispersivity, m
 -shifts 10 1                                 # Inject 10 m3
 -time 8640                                   # 1 shift = 2.4 hours residence time
 -punch_cells 1
END

 TRANSPORT                                    # Pump 30 m3
 -shifts 30 -1                                # No of displacements, direction backward
 USER_GRAPH
 -heading days Tracer Fe
 -axis_scale x_axis 0 3; -axis_scale y_axis 0 1.0; -axis_titles Days mmol/l
 -plot_concentration_vs time
 -start
 10 graph_x (sim_time + 4320)/(3600*24)
 20 graph_y tot("Cl")*1e6, tot("Fe(2)")*5e4
 -end
END

 TRANSPORT; -shifts 10 1; END                 # 2nd cycle
 TRANSPORT; -shifts 30 -1
END
 TRANSPORT; -shifts 10 1; END                 # 3rd cycle
 TRANSPORT; -shifts 30 -1
END
```

This simple example will produce the basic features of Figure 11.26. The increase of Fe^{2+} is retarded with respect to Cl^- (added as a tracer for conservative transport). The retardation increases in the second cycle because Fe^{2+} has not refilled all the exchange sites after 3 days of pumping groundwater. In subsequent cycles, the retardation increases because more sorption sites become available as more and more iron-oxyhydroxide precipitates.

A closer look at the actual data in Figure 11.26 shows that the Fe^{2+} concentrations increase quicker and more smoothly with V/V_0 than the model simulates. This discrepancy may be due to the heterogeneity of the aquifer (coarse, gravelly parts with a low exchange capacity have smaller retardation than loamy-sandy parts), to the kinetics of the sorption/exchange reaction, or perhaps to the oxidation-kinetics of Fe^{2+} to Fe^{3+}.

QUESTIONS:

What will be the effect of doubling the Ca^{2+} concentration?

ANSWER: Sorbed $q_{Fe^{2+}}$ is approximately halved, the retardation of iron is reduced, cf. Equation (11.59).

What is the effect, having Na^+ instead of Ca^{2+} as major cation?

ANSWER: the Fe^{2+} front during pumping is sharpening, it will steepen.

In-situ iron removal does not function when pyrite is present in the aquifer. Include pyrite in EQUILIBRIUM_PHASES 1–15 and explain the concentration patterns?

Behavior of Arsenic

It has been observed during *in-situ* iron removal (Appelo and De Vet, 2003) that the concentration of As increases temporarily above the pristine groundwater concentration in the initial pumping stage (Figure 11.29). Can we unravel the cause with PHREEQC?

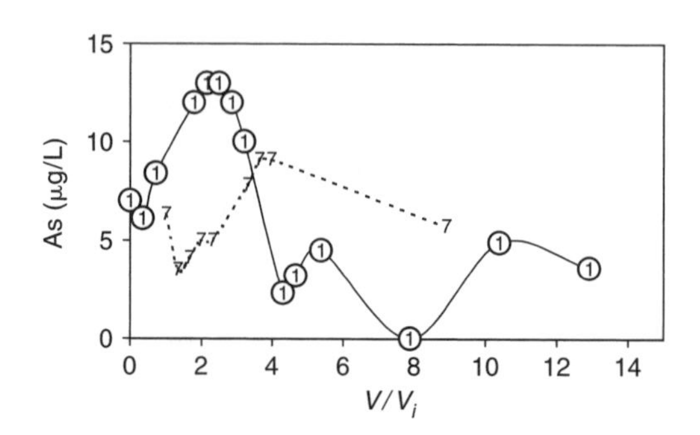

Figure 11.29. Observed concentrations of Arsenic in cycles 1 and 7 during *in-situ* iron removal (Appelo and De Vet, 2003).

Add 1 μM As to the groundwater (SOLUTION 1–15) and run again while displaying pH, As(3) and As(5). PHREEQC calculates that the As concentrations in pumped groundwater have decreased, but it is all As(5), while in groundwater the dominant species is As(3). It seems that As(3) is oxidized to As(5) which sorbs on Hfo. Let's uncouple the redox species to have only As(3) in the simulation, assuming that the oxidation of As(3) is slow and does not take place on the timescale of the experiment.
Make As(5) negligible with …

```
SOLUTION_SPECIES
H3AsO4 + 2H+ + 2e- = H3AsO3 + H2O; log_k 180.          # log_k = 18.897 in database
```

The As(5) concentration indeed becomes zero, but the As(3) concentration remains steady at 0.81 μM in the first cycle and is 0.79 μM in the second cycle. Apparently, the solute concentration is buffered by the surface complex on ferrihydrite. Appelo and De Vet did not analyze the speciation of As, but Rott et al. (1996) noted that As(3) in groundwater was almost completely oxidized during *in-situ* iron removal. Maybe we should model that As(3) is all oxidized and no longer present.
Make As(3) negligible with …

```
SOLUTION_SPECIES
H3AsO4 + 2H+ + 2e- = H3AsO3 + H2O; log_k -180.
```

Now the concentration of As(5) decreases initially, and then increases to 1.17 μM. The decrease coincides with the pH decrease that is associated with the precipitation of $Fe(OH)_3$ at the oxygen front. Inspection of the surface complexes of As(5) shows the dominant presence of $Hfo\_wOHAsO_4^{3-}$. This complex is enhanced when the surface gains positive charge with the pH decrease (hence, the As concentration in water decreases), and is released when the pH increases again. Thus, it seems that reactions involving redox and surface complexation yield the concentration peaks, while the competitive reactions with various other ions may also play a role (cf. Appelo and De Vet, 2003).

Usually, several knobs are tried and turned to improve the fit of the model with observed concentrations. The number of surface sites for ferrihydrite may be increased, or the K values may be varied, assuming that the natural oxyhydroxides are different from the laboratory synthesized material (Stollenwerk, 2003). The variety of bonding sites that can be discerned with spectroscopic techniques is indeed huge (Foster, 2003). On the other hand, the database contains K values that were selected based on experience in various laboratory and field situations and they provide a good starting point.

QUESTIONS:
Explain the distribution of As(3) and As(5) in the pumping stage of *in-situ* iron removal, with As(3) and As(5) present according to the database speciation?
ANSWER: Initially As(5) is dominant, but As(3) turns up when Fe^{2+} appears (cf. the redox in Figures 9.5 and 9.8).
Why is the Fe^{2+} front retarded when only As(5) is present?
ANSWER: Sorption of the negative As(5) complex decreases the positive charge on ferrihydrite, which enhances sorption of Fe^{2+}.
What is the effect of adding 10 μM P to the groundwater?
ANSWER: Sorption of As is reduced, the As concentrations diminish more quickly during pumping.

11.2.6 *Arsenic in Bangladesh groundwater*

A relatively large part of the human intake of arsenic stems from drinking water (Table 1.1) and concentrations not much above the WHO recommended limit of 10 μg/L may already be detrimental for health. High concentrations of As have been found in many aquifers around the world (Welch et al., 2000; Smedley and Kinniburgh, 2002), the saddest incidence being Bangladesh where millions of people are using groundwater with an excessive As content (BGS and DPHE, 2001; Kinniburgh et al., 2003). In the past, the Bangladesh people relied for drinking water on the great Himalayan rivers (Ganges, Brahmaputra and Meghna) which were reasonably clean until the population escalated in number and the natural clean-up proved insufficient for removing the increased viral and bacterial contamination to a harmless level. Groundwater was promoted as an alternative to surface water by various international organizations and the local government, and millions of shallow tubewells were installed. The change in the source of drinking water is believed to have been partly responsible for the reduction in infant mortality in Bangladesh since 1980 (Kinniburgh et al., 2003). However, it was not known that the groundwater in the area commonly contains As in concentrations that may far exceed the safe drinking water limit. It has taken some time before illnesses like skin and bladder cancer developed and the problem was diagnosed correctly, first in West Bengal (Chakraborty and Saha, 1987, cited by Smedley, 2003).

Based on the hydrological situation and confirmed by groundwater age determinations with tritium, the high arsenic concentrations stem from natural sediment-water interactions. This is rather enigmatic and a cause for intense debate since the solid concentrations lie in the common range of 1–10 mg/kg for sediments. One theory relates the release to oxidation of pyrite which may contain arsenic, in response to groundwater drawdown that facilitates the access of oxygen (Figure 11.30b). However, pyrite is not present everywhere and the sulfate concentrations in groundwater are low and indicate reduction and possibly pyrite formation, rather than oxidation (Kinniburgh et al., 2003).

Another theory relates the liberation of As to reductive dissolution of iron-oxyhydroxide that contains As in sorbed or occluded form (Bhattacharya et al., 1997; Nickson et al., 2000, Figure 11.30a). Oxalate extractions of the sediments, by which ferrihydrite is dissolved, show a clear correlation of As and Fe which makes iron-oxyhydroxide indeed a plausible source (Kinniburgh et al., 2003). In addition, As and Mg are correlated in the sediments, which suggests that biotite may also be important. Apparently, the iron-oxyhydroxide forms during weathering of primary minerals, and it takes up As in the soils in the upland or during erosion and transport with surface runoff. The As concentrations in river water can at times be relatively high (5–20 μg As/L).

One problem of the reductive dissolution theory is how the reductant contacts the oxyhydroxide. Organic matter is the likely reductant, but iron-oxyhydroxide is expected mostly in

A) Reductive dissolution of $Fe(OH)_3$

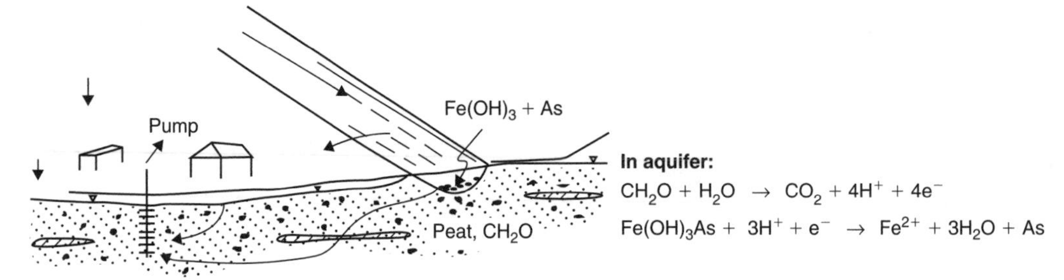

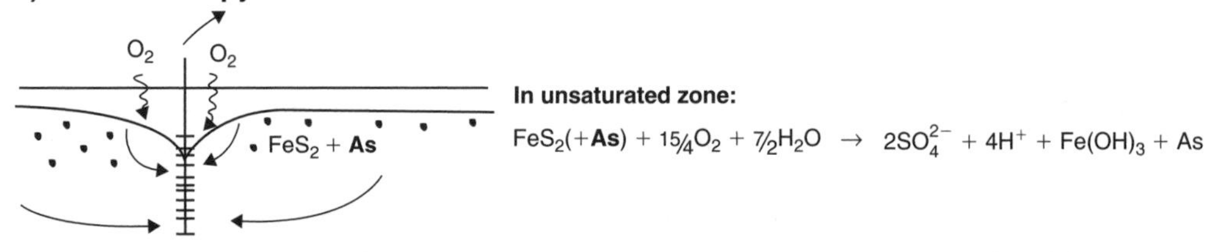

C) Displacement of As by HCO_3^-

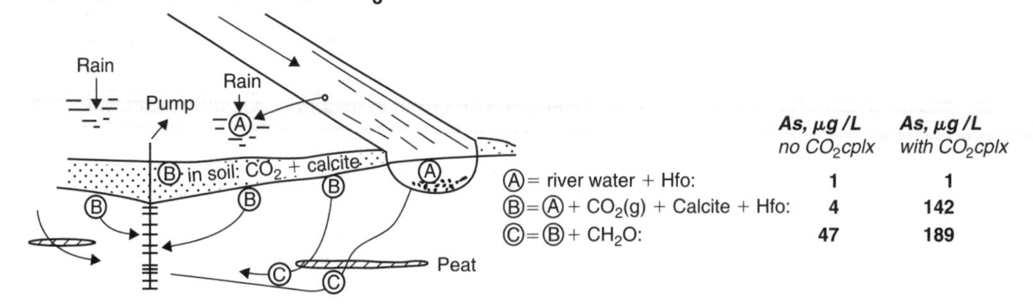

Figure 11.30. Mechanisms for explaining the release of As from sediment into groundwater in Bangladesh. (A) iron-oxyhydroxide is reduced by organic matter and releases sorbed As; (B) groundwater drawdown facilitates access of oxygen to pyrite that contains As which dissolves; (C) change in HCO_3^- concentration displaces As from iron-oxyhydroxide.

oxidized sediments where organic matter is sparse or refractory, meaning that organic matter must be transported with groundwater or by diffusion to the iron-oxyhydroxide. On the other hand, the As problem seems mostly confined to wells in Holocene aquifers, which may contain iron-oxyhydroxide and still reactive organic matter that were jointly sedimented in the river bed.

Let us calculate concentrations according to the reductive dissolution theory and compare these with correlations of As with HCO_3^- and Fe^{2+} found by Nickson et al., 2000 (Figure 11.31).

Make an input file with average river water, equilibrate with atmospheric oxygen and ferrihydrite, assume that only 1% of the analyzed goethite has the surface complexation properties of ferrihydrite and that it is 1000 times more soluble than goethite. Add CH_2O in steps, first reducing dissolved oxygen, then 1 mM ferrihydrite in 5 steps (i.e., add 0.05 mM CH_2O in these steps).

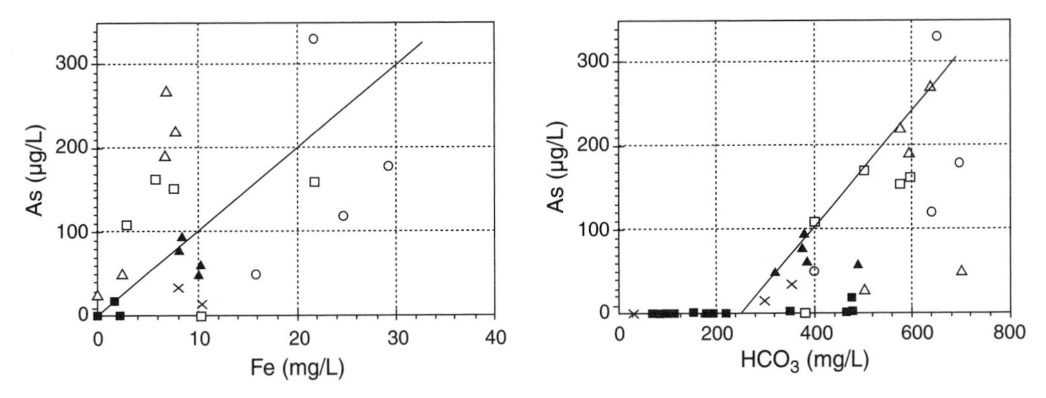

Figure 11.31. Correlations of As with Fe and HCO_3^- in Bangladesh groundwater (Nickson et al., 2000).

```
SOLUTION_SPECIES; H2O + 0.01e- = H2O-0.01; -log_k -9        # ... speed up convergence
PHASES; Ferrihydrite; FeOO2H3 + 3H+ = Fe+3 + 3H2O; log_k 2.0
SOLUTION 1 Ganges river water
 pH 8.39; pe 14 O2(g) -0.68
 -units mg/l
 Na 19.3; K 4; Mg 14.1; Ca 55.7; Si 16 as Si; P 0.1 as PO4
 Cl 9.6; Alkalinity 260 as HCO3; S(6) 4.4 as SO4
 As 1e-3                              # ...As is usually 10 times higher
EQUILIBRIUM_PHASES 1
                                      # 0.1% Fe in solid = 18 mmol/kg = 0.1 mol
                                      # FeOOH/L pore water
 Ferrihydrite 0 1e-3                  # assume only 1% has the properties of ferrihydrite
SAVE solution 1
END

SURFACE 1                             # iron hydroxide in riverbed
 Hfo_w Ferrihydrite 0.2 5.34e4; Hfo_s Ferrihydrite 0.5e-2; -equil 1
END

EXCHANGE 1                            # clay minerals in riverbed
 X 0.1; -equil 1
USE solution 1                        # for plotting initial conc's..

USER_GRAPH
 -axis_titles "mg HCO3/L", "mg / L", "pH"; -connect; -head Alk As*10 Fe S(6) pH
 -start; 10 graph_x Alk * 61e3
 20 graph_y tot("As") * 74.92e4, tot("Fe(2)") * 55.85e3, tot("S(6)") * 96e3
 30 graph_sy -la("H+"); -end
END

                                      # reduction of O2 and ferrihydrite by organic
                                      # matter in the aquifer...
USE surface 1; USE solution 1; USE exchange 1
EQUILIBRIUM_PHASES 2
 Ferrihydrite 0 1e-3;                 # Siderite 0 0
REACTION 2                            # reduce O2, ferrihydrite in 5 steps, then some SO4
 CH2O 1e-3; 0.2674 0.05 0.05 0.05 0.05 0.05 0.05 0.05
INCREMENTAL_REACTIONS
END
```

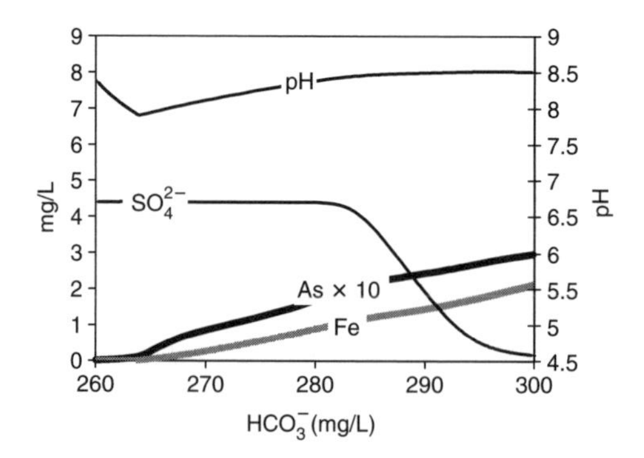

Figure 11.32. Release of As by reductive dissolution of iron-oxyhydroxide, with pH and Fe concentration as function of the alkalinity.

The result is shown in Figure 11.32. Overall, the As concentration increases as expected but there are less obvious aspects in the plot. In the first step, O_2 is reduced, and the pH decreases. Thereafter ferrihydrite is reduced and the pH increases again. Remarkably, the Fe concentration does not increase to 55.85 mg/L, expected when 1 mM ferrihydrite is reduced. Of course, part of Fe^{2+} will be taken up by the exchanger, but inspection of the PHREEQC output file reveals that ferrihydrite is not completely reduced. The remaining ferrihydrite will have As on its surface, not all available As is released yet into solution. Furthermore, sulfate is being reduced and also some methane is formed. Apparently, ferrihydrite is too stable to be reduced (cf. Section 9.6).

Decrease the stability of ferrihydrite in the input file by increasing *SI* to 3.0:

```
EQUILIBRIUM_PHASES 2
Ferrihydrite 3.0 1e-3 # Siderite 0 0
```

Indeed (inspect the PHREEQC graph and output), now the As and Fe^{2+} concentration reach a plateau value, all ferrihydrite is dissolved in the end. The As concentration becomes as high as 1190 μg/L. Only in the last but one step sulfate starts to be reduced, which is later than before and indicates that it has become more stable relative to ferrihydrite. Note also that the concentration of As increases more slowly because the proportion of As(5) is larger than in the previous case and because As(5) is sorbed more strongly than As(3) under the model conditions.

The concentration of Fe^{2+} is rather high (22 mg/L) and the solution is supersaturated with respect to siderite, so let's precipitate that mineral.

Uncomment the mineral Siderite in EQUILIBRIUM_PHASES 2, and run again.

The calculations show (cf. the PHREEQC graphs) that the Fe^{2+} concentration is low, that the alkalinity is smaller and that the pH increases to more than 9.0. All sorbed As is released into solution because ferrihydrite is converted completely to siderite. The lower Fe^{2+} concentration diminishes the stability of ferrihydrite, and practically the same results are obtained with $SI_{\text{ferrihydrite}} = 0$ (check!).

The release of As is in agreement with the reductive dissolution theory, but the simulated water compositions not altogether resemble the observed groundwaters qualities which have a pH slightly below 7, dissolved Fe^{2+} of 1–10 mg/L (in approximate equilibrium with siderite), and a much higher alkalinity together with a much higher CO_2 pressure and a much lower CH_4 pressure. It can be checked with PHREEQC that the reaction of organic matter with ferrihydrite cannot reproduce these concentrations. Apparently, organic matter reacts with O_2 to produce large amounts CO_2, i.e. the reaction takes place in the soil which, anyhow, is the source of most groundwater. So, let's have production of CO_2 and dissolution of calcite in the soil before the reduction of ferrihydrite starts.

Replace the last simulation of the input file with:

```
                                    # CO2 production and calcite dissolution in the soil...
USE surface 1; USE solution 1; USE exchange 1
EQUILIBRIUM_PHASES 2
 Ferrihydrite 0 1e-3; Calcite; O2(g) -0.68
```

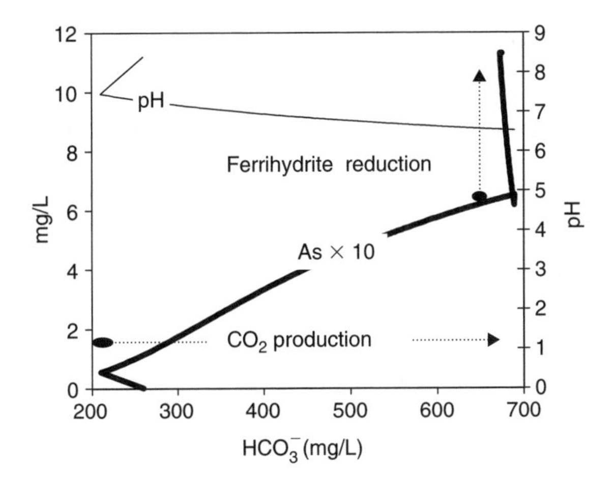

Figure 11.33. Release of As from ferrihydrite by displacement with HCO_3^- followed by reductive dissolution of ferrihydrite.

```
REACTION 2
 CH2O 1; 10e-3 in 10
INCREMENTAL_REACTIONS
SAVE solution 2; SAVE surface 2; SAVE exchange 2;
END                            # water B in Fig 11.30C
                               # reduction of O2 and FeOOH by organic matter...
USE surface 2; USE solution 2; USE exchange 2
EQUILIBRIUM_PHASES 3
 Ferrihydrite 0 1e-3; Calcite; Siderite 0 0
REACTION 3                     # reduce O2, ferrihydrite in 5 steps, then SO4...
 CH2O 1e-3; 0.2674 0.05 0.05 0.05 0.05 0.05 0.05 0.05
END                            # water C in Fig 11.30C
```

Remarkably, the concentration of As now increases to more than 600 μg/L when the alkalinity increases by the reaction of CO_2 and calcite (Figure 11.33). The increase is due to displacement by HCO_3^- which is sorbed by ferrihydrite (Appelo et al., 2002). Thus, iron-oxyhydroxide, transported by the river and in equilibrium with the low As and HCO_3^- concentration in surface water, is sedimented in the delta and loses part of its As when contacted by groundwater with a high HCO_3^- concentration (Figure 11.30c). Phosphate, which also has a higher concentration in groundwater than in surface water, has a similar effect. A further increase in As takes place when ferrihydrite is reduced (Figure 11.33). The Fe^{2+} concentration in groundwater indicates that reduction occurs, but the As lost by displacement alone is sufficient to explain the observed concentrations. Mixing of waters from different source areas with different CO_2 pressures, in combination with the different retardations of As, Fe^{2+}, and HCO_3^-, leads to the observed variations in groundwater compositions.

The HCO_3^- displacement theory predicts that As concentrations will be low in river water that infiltrates via the river bed into the aquifer and keeps its low alkalinity. Thus, bank filtration may be an option in Bangladesh to obtain safer groundwater with a lower As concentration.

QUESTIONS:
Show with reaction equations that the pH decreases and increases when CH_2O reduces O_2 and FeOOH, respectively?

ANSWER: $O_2 + CH_2O \rightarrow H^+ + HCO_3^-$;
$$4\,FeOOH + CH_2O + H_2O \rightarrow 4\,Fe^{2+} + HCO_3^- + 7\,OH^-$$

Show with a reaction equation that the pH increases more when siderite precipitates on reduction of FeOOH with CH_2O (for the same Fe^{2+} concentration).

ANSWER: $\frac{4}{3} CH_2O + \frac{16}{3} FeOOH \rightarrow \frac{4}{3} FeCO_3 + 4\,Fe^{2+} + 8\,OH^-$

What is the maximal As concentration in the model system, if 2 mM ferrihydrite were present initially?

ANSWER: $1190 \times 2 = 2380\,\mu g/L$

Calculate the distributions $(1 + K_d)$ and retardations $(1 + dq_i/dc_i)$ for As, Fe^{2+} and HCO_3^- for the model of Figure 11.33, at the last CO_2 production step (681.5 mg HCO_3^-/L) and the fourth reduction step (680.57 mg HCO_3^-/L).

ANSWER:

	$(1 + K_d)$			R		
(HCO_3^-)	As	Fe^{2+}	C	As	Fe^{2+}	C
(681.5)	1.75	1	1.01	1.38	1	1.0019
(680.6)	1.35	3.81	1.004	1.45	3.87	1.0009

(Note that the retardation of As is smaller than of Fe, and that R_{As} and R_C are interdependent)

Why is it impossible to have both goethite and ferrihydrite in equilibrium with a solution?

ANSWER: Try it out with PHREEQC

11.2.7 *Fractionation of isotopes*

Most simply, isotopic fractionation considers the ratio of two isotopes in two species or two phases at equilibrium. The ratio may grow and change 1) in separation processes where the product is removed from the reactant (Rayleigh fractionation), 2) with the progress of chemical reactions and 3) during diffusion. Rapidly, the calculations become complicated when these processes are conjuncted, but fortunately, PHREEQC can bring relief. It will be shown how to calculate oxygen fractionation during condensation of water vapor, and carbon fractionation in precipitating calcite. More examples are given by Appelo (2003).

When dN molecules of light $H_2{}^{16}O$ condense, they are accompanied by dN_i molecules of heavy $H_2{}^{18}O$, with the ratio given by

$$\frac{dN_i}{dN} = \alpha \frac{N_i}{N} \tag{2.7}$$

We have calculated the isotopic ratio in rain with the Rayleigh formula:

$$\frac{R}{R_0} = \left(\frac{N}{N_0}\right)^{(\alpha - 1)} \tag{2.10}$$

for a constant fractionation factor α (Example 2.1). To calculate the actual process, where vapor cools and condenses and α changes with temperature, requires numerical integration of Equation (2.7). The kinetic integrator of PHREEQC can do this in several ways, and one is given in the following input file. We calculate the condensation of 1 g $H_2{}^{16}O$ vapor over the temperature range from 25 to -20°C as a kinetic process (rate "condense"), which provides stepwise values of d(^{16}O). For each step, the amount d(^{18}O) is calculated in another rate ("Frn_18O") according to Equation (2.7). We take care to make the timestep large ("-step 6") to gain equilibrium among liquid and vapor $H_2{}^{16}O$ in the rate "condense". The result is shown in Figure 11.34.

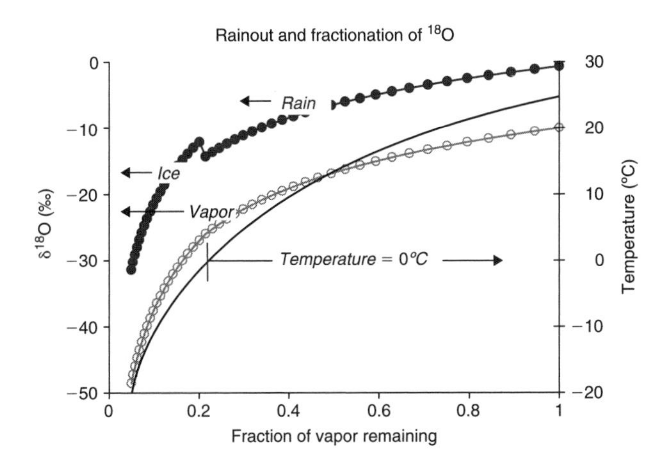

Figure 11.34. Fractionation of ^{18}O during cooling of water vapor from an initial temperature of 25°C to −20°C with condensation of rain and subsequently of ice. Initial $\delta^{18}O_{vapor} = -10$‰. The line without circles indicates temperature. Note the jump of ^{18}O in the transition rain/ice at 0°C, where the fractionation factor changes.

```
# Calculate 18O fractionation during condensation
RATES
 Condense
                                      # Condense 1 g water during cooling from 25 to -20 C
-start
 10 P_eq = SR("H2O(g)")                                # the vapor pressure
 20 n_eq = P_eq * 43.54165 / 0.082057 / TK             # n = PV/RT
 30 d16O = (m - n_eq) * time
 40 save d16O                                          # ... integrate
 50 put(d16O, 1)                                       # store d16O in temporary memory
-end

 Frn_18O
                                                       # d 18O/d 16O = aa_l_v * 18O/16O
 -start
  10 if TK > 273.15 then c = -2.0667e-3 else c = 1.0333e-3
  20 aa_l_v = exp(1.137e3/TK^2 - 0.4156/TK + c)        # Majoube, 1971
  30 d18O = aa_l_v * m / kin("Condense") * get(1)
  40 put(aa_l_v, 2)
  50 save d18O                                         # ...integrate
 -end

KINETICS 1
 Condense; -formula H2O 0; m0 55.506e-3
 Frn_18O; -formula H2O 0; m0 0.110177e-3               # d18O = -10 permil.
 -steps 6
INCREMENTAL_REACTIONS
REACTION_TEMPERATURE
 25 -20 in 46                                          # change temp in 1 degree steps
SOLUTION 1

PRINT
-reset false
USER_GRAPH
 -headings frac liq/ice vap temp
 -axis_titles "Fraction of vapor remaining" "delta 18O" "temp, C"
 -axis_scale y_axis -50 0
```

```
 -start
   2 if get(1)  = 0 then goto 60
  10 vap =  (kin("Frn_18O")/kin("Condense")/2.005e-3 - 1)*1e3
                                                            # in rain...
  20 liq = vap + 1e3*log(get(2))
  30 graph_x kin("Condense")/55.506e-3                      # ...Fraction of vapor remaining
  40 graph_y liq, vap
  50 graph_sy TC                                            # ... temp, C
  60 end
 -end
END
```

One particular aspect of the condensation problem is that actual reactions are never calculated, the rates add H_2O with a stoichiometric coefficient of zero as indicated with "-formula H2O 0" under keyword KINETICS. The integration could have been done as quickly with MATLAB or any other integrator, although it is helpful that PHREEQC has a database from which the water content of the vapor as a function of temperature can be extracted (lines 10 and 20 in "Condense").

QUESTIONS:
How many moles are in 1 g H_2O?
ANSWER: 55.506e-3 (cf. KINETICS 1)
What is the volume of 1 g H_2O as saturated vapor at 25°C?
ANSWER: 43.54165 L (cf. line 20 in the rate "Condense")
Add fractionation of deuterium to the input file, use $\ln \alpha^2\mathrm{H}_{l/g} = 52.612 \times 10^{-3} - 76.248/K + 24844/K^2$, K is temperature in Kelvin, initial $\delta^2\mathrm{H} = -80‰$.

Fractionation of Carbon during precipitation

The previous example could be done with any integrator, but PHREEQC is particularly useful for including the effects of a changing hydrogeochemistry in isotope calculations. For example, fractionation of ^{13}C in calcite that precipitates due to degassing of CO_2 is a function of pH because ^{13}C fractionates among the solute carbonate species and into CO_2 gas. During degassing, the pH increases. Again, this is a Rayleigh problem that needs numerical integration. The following input file calculates the isotopic composition for a system where CO_2 and calcite are both removed from solution (do not react back with the solution).

```
# Degas CaHCO3 solution, precipitate calcite, find d13C...
SOLUTION 1
 Ca 5 charge; C(4) 10
RATES
CO2(g)                                          # d(CO2) / dt = -k * (P_aq - P_gas)
# parms 1 = target log(P_CO2)_gas. 2 = reaction factor relative to calcite rate
 -start
  10 P_aq = sr("CO2(g)")
  20 dCO2 = parm(2) * (10^parm(1) - P_aq) * time
  30 save dCO2
  40 put(dCO2, 5)
 -end

Calcite
# lines 10 to 200 from PHREEQC.DAT database, add line
 210 put(moles, 4)                              # store amount of calcite precipitate
 -end
```

```
 C13
# Distribute 13C among solute species 1 = CO2aq, 2 = HCO3-, 3 = CO3-2
# and in calcite and CO2 gas.
 -start                                     # Define alpha's...
  10 aa_1_2 = exp-(9.866/TK - 24.12e-3)     # Mook et al., 1974
  20 aa_3_2 = exp(-0.867/TK + 2.52e-3)      # Mook et al., 1974
  30 aa_cc_2 = exp(-4.232/TK + 15.1e-3)     # Mook, 1986
  40 aa_g_2 = exp-(9.552/TK - 24.1e-3)      # Mook et al., 1974
                                            # Find H12CO3- and H13CO3-...
  50 aH = act("H+")
  60 K1 = aH * act("HCO3-") / act("CO2") / act("H2O")
  70 K2 = aH * act("CO3-2") / act("HCO3-")
  80 m13C = m / (aa_1_2 * aH/K1 + 1 + aa_3_2 * K2/aH)
  90 m12C = (tot("C(4)")) / (aH/K1 + 1 + K2/aH)
                                            # and the ratio 13C/12C in HCO3-...
  100 R2 = m13C / m12C
                                            # Fractionate 13C into calcite...
  110 d13C = aa_cc_2 * -get(4) * R2
                                            # and in CO2(g)...
  120 d13C_g = aa_g_2 * -get(5) * R2
                                            # Integrate...
  130 save d13C + d13C_g
                                            # for printout and graphics...
  140 put(R2, 2); 180 put(d13C, 10); 190 put(d13C_g, 20)
 -end

C13_cc; -start; 10 save -get(10); -end      # ... store 13C in calcite
C13_g; -start; 10 save -get(20); -end       # ... store 13C in gas

KINETICS 1
 CO2(g); m0 0; -parm -3.5 0.001             # parm(2) = 0.1 for equilib
 Calcite; m0 0; -parm 60 0.67
 C13; -formula C 0; m0 0.1112463e-3         # d13C = -10
 C13_cc; -formula C 0; m0 0
 C13_g; -formula C 0; m0 0
 -step 1e-5 3 5 10 15 20 30 45 60 100 150 300 600 1e3 5e3
INCREMENTAL_REACTIONS
USER_GRAPH
 -head cc d13C_cc _HCO3- _sol_tot _gas_tot pH
 -axis_titles "mmol Calcite" "d13C" "pH"
 -axis_scale x_axis 0 4 1 0.5
 -axis_scale y_axis -20 0 5 2.5
 -start
   2 if step_no = 0 then goto 50
  10 Rst = 0.011237
  20 graph_x kin("Calcite")*1e3
  30 graph_y (kin("C13_cc")/kin("Calcite")/Rst - 1)*1e3, (get(2)/Rst - 1)*1e3,\
        (kin("C13")/tot("C(4)")/Rst - 1)*1e3,\
        (kin("C13_g")/kin("CO2(g)")/Rst - 1)*1e3
  40 graph_sy -la("H+")
  50 end
 -end
END
```

Degassing of CO_2 is defined to be a function of the difference in CO_2 pressure of the solution and of the gas phase in the rate "CO2(g)". It provides stepwise values for the amounts of $^{12}CO_2$ lost to the gas phase. Furthermore, the rate for calcite dissolution and precipitation from the PHREEQC database is copied in the input file to render the amounts of ^{12}C that go into calcite. Like in the condensation problem, the values for d(^{12}C) are used to fractionate d(^{13}C) in the gas and the solid. First, the ratio $^{13}C / ^{12}C$ in HCO_3^- is calculated by fractionation over the solute species (Equation 5.13). Next, d(^{13}C) is obtained from the fractionation factors for gas-HCO_3^- and calcite-HCO_3^-. Figure 11.35 shows some results for varying relative rates for CO_2 lost to the gas-phase and the calcite.

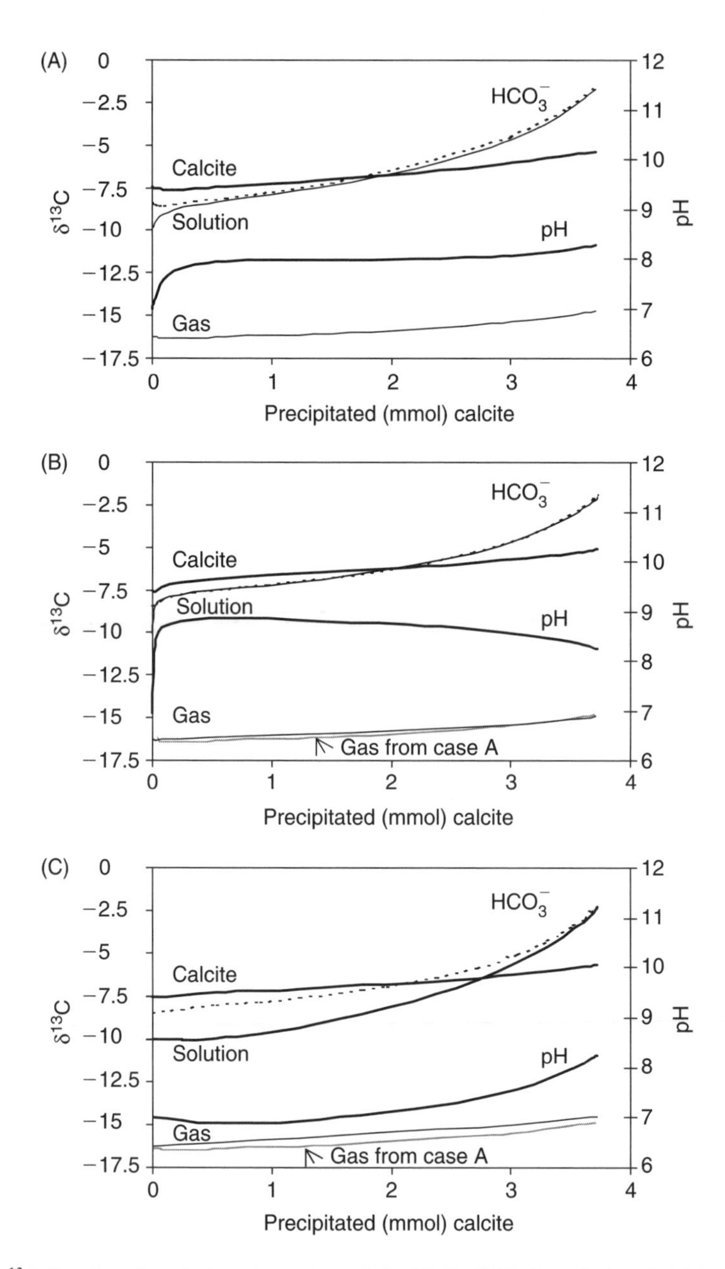

Figure 11.35. ^{13}C fractionation during degassing of 5 mM $Ca(HCO_3)_2$ solution. Initial pH = 7, solution $\delta^{13}C = -10$‰. In (A) the rates for degassing and for calcite precipitation approach equilibrium at the same pace. In (B), the rate for CO_2 degassing is 100 fold greater, in (C), the rate for calcite precipitation is 100 fold greater than in (a) which leads to subtle changes in the isotope compositions. In (B) and (C) the trace of $\delta^{13}C$ in gas from (A) is shown with a light-grey line for comparison.

For Figure 11.35A the two rates were set such that overall equilibrium with the final CO_2 pressure ($10^{-3.5}$ atm) and with calcite was reached at the same time. In this case, the pH increases rapidly to about 8 due to loss of CO_2, and subsequently more gradually because the precipitation of calcite keeps pace with the CO_2 loss. The $\delta^{13}C$ of the solution increases because the gas phase has low $\delta^{13}C \approx -15‰$. The average $\delta^{13}C$ of the solution, the gas phase and of calcite all increase during the approach to final equilibrium when the lighter isotope is fractionated into the gas phase which increases in mass and is removed from the system. Initially, $\delta^{13}C$ of the bicarbonate ion lies above the solution average because the lighter $CO_{2(aq)}$ forms a substantial part of total inorganic carbon (*TIC*), but as pH increases to 8.3, HCO_3^- becomes the major component of *TIC* and $\delta^{13}C$ of the solution and of HCO_3^- coincide. The line for calcite shows the change of $\delta^{13}C$ in the *average* solid, which increases from −7.5‰ to −5.4‰. However, the *actual* precipitate follows the line for bicarbonate and in the successive layers of the solid, $\delta^{13}C$ ranges from −7.5‰ to −0.7‰.

Subtle changes take place when the rate for CO_2 loss is much greater than for calcite precipitation, which is probably the most common condition for field situations (Figure 11.35B). With a 100-fold increase of the rate "CO2(g)", equilibrium among solution and the gas-phase is virtually attained at all stages (the factor entered as parm(2) is increased from 0.001 to 0.1). Now, the pH increases initially to 8.9, but then decreases as protons are released by the precipitation of calcite. The greater initial loss of carbon to the gas phase increases $\delta^{13}C$ in solution and in calcite. Therefore, a slightly lighter solution and gas are found in the end.

Conversely, if the rate of calcite precipitation is increased 100-fold, the pH decreases initially (Figure 11.35C). Again, $\delta^{13}C$ in HCO_3^- increases, this time because the fraction of $CO_{2(aq)}$ increases. As a result, the gas phase becomes heavier, and the final solution and calcite are lighter than in the two previous cases. Thus, although the final, *chemical* composition of the solution is identical for all the paths, the *isotope* composition differs and may be a tracer for the actual pathway.

PROBLEM:
Include degassing of CO_2 as a time dependent process with the two-film model (cf. Chapter 10). Calculate the fractionation of ^2H and ^{18}O during evaporation with the two-film model.

REFERENCES

Alliances, couplage chimie et transport avec MT3D/PHREEQC. Under development at Andra, France. Contact Alain.Dimier@Andra.fr

Appelo, C.A.J., 2003. Calculating the fractionation of isotopes in hydrochemical (transport) processes with PHREEQC-2. In H.D. Schulz and A. Hadeler (eds), *GEOPROC 2002*, 383–398, Wiley-VCH, Weinheim.

Appelo, C.A.J. and De Vet, W.W.J.M., 2003. Modeling in situ iron removal from groundwater with trace elements such as As. In A.H. Welch and K.G. Stollenwerk, *op. cit.*, 381–402.

Appelo, C.A.J., Drijver, B., Hekkenberg, R. and De Jonge, M., 1999. Modeling in situ iron removal from ground water. *Ground Water* 37, 811–817.

Appelo, C.A.J., Van der Weiden, M.J.J., Tournassat, C. and Charlet, L., 2002. Surface complexation of ferrous iron and carbonate on ferrihydrite and the mobilization of arsenic. *Env. Sci. Technol.* 36, 3096–3103.

Bain, J.G., Mayer, K.U., Blowes, D.W., Frind, E.O., Molson, J.W.H., Kahnt, R. and Jenk, U., 2001. Modelling the closure-related geochemical evolution of groundwater at a former uranium mine, *J. Contam. Hydrol.* 52, 109–135.

Barbaro, J.R. and Barker, J.F., 2000. Controlled field study on the use of nitrate and oxygen for bioremediation of a gasoline source zone. *Bioremediation J.* 4, 259–270.

BGS, DPHE, 2001. Arsenic contamination of groundwater in Bangladesh. In D.G. Kinniburgh and P.L. Smedley (eds), British Geological Survey (Technical Report, WC/00/19, 4 Volumes). British Geological Survey, Keyworth.

Bhattacharya, P., Chatterjee, D. and Jacks, G., 1997. Occurrence of arsenic-contaminated groundwater in alluvial aquifers from Delta Plains, eastern India. *Int. J. Water Resour. Devel.* 13, 79–92.

Chapelle, F.H., Bradley, P.M., Lovley, D.R. and Vroblesky, D.A., 1996. Measuring rates of biodegradation in a contaminated aquifer using field and laboratory methods. *Ground Water* 34, 691–698.

Cozarelli, I.M., Bekins, B.A., Baedecker, M.J., Aiken, G.R., Eganhouse, R.P. and Tucillo, M.E., 2001. Progression of natural attenuation processes at a crude-oil spill site: 1. Geochemical evolution of the plume. *J. Contam. Hydrol.* 53, 369–385.

Eckert, P. and Appelo, C.A.J., 2002. Hydrogeochemical modeling of enhanced benzene, toluene, ethylbenzene, xylene (BTEX) remediation with nitrate. *Water Resour. Res.* 38, DOI 10.1029/2001WR000692

Engesgaard, P., 1991. Geochemical modelling of contaminant transport in groundwater. Ph.D. thesis, Lyngby, 144 pp.

Engesgaard, P., Jensen, K.H., Molson, J., Frind, E.O. and Olsen, H., 1996. Large-scale dispersion in a sandy aquifer: Simulation of subsurface transport of environmental tritium. *Water Resour. Res.* 32, 3253–3266.

Foster, A.L., 2003. Spectroscopic investigations of arsenic species in solid phases. In A.H. Welch and K.G. Stollenwerk, *op. cit.*, 27–66.

Gerald, C.F. and Wheatly, P.O., 1989. *Applied numerical analysis*, 4th ed. Addison-Wesley, Reading, Mass., 679 pp.

Glynn, P. and Brown, J., 1996. Reactive transport modeling of acidic metal-contaminated groundwater at a site with sparse spatial information. In P.C. Lichtner, C.I. Steefel and E.H. Oelkers (eds), *Rev. Mineral.* 34, 377–438.

Herzer, J. and Kinzelbach, W., 1989. Coupling of transport and chemical processes in numerical models. *Geoderma* 44, 115–127.

Hutchins, S.R., Downs, W.C., Wilson, J.T., Smith, G.B., Kovacs, D.A., Fine, D.D., Douglass R.H. and Hendrix, D.J., 1991. Effect of nitrate addition on biorestoration of fuel-contaminated aquifer - field demonstration. *Ground Water* 25, 571–580.

Jähne, B., Heinz, G. and Deitrich, W., 1987. Measurement of the diffusion coefficients of sparingly soluble gases in water. *J. Geophys. Res.* 92, 10767–10776.

Kinniburgh, D.G., Smedley, P.L., Davies, J., Milne, C.J., Gaus, I., Trafford, J.M., Burden, S., Ihtishamul Huq, S.M., Ahmad, N. and Ahmed, K.M., 2003. The scale and causes of the groundwater arsenic problem in Bangladesh. In A.H. Welch and K.G. Stollenwerk, *op. cit.*, 211–257.

Kirkner, D.J. and Reeves, H., 1988. Multicomponent mass transport with homogeneous and heterogeneous chemical reactions: effect of the chemistry on the choice of numerical algorithm. 1. Theory. *Water Resour. Res.* 24, 1719–1729.

Kohler, M., Curtis, G.P., Kent, D.B. and Davis, J.A., 1996. Experimental investigation and modeling of uranium(VI) transport under variable chemical conditions. *Water Resour. Res.* 32, 3539–3551.

Leuchs, W., 1988. *Vorkommen, Abfolge und Auswirkungen anoxischer Redoxreaktionen in einem pleistozänen Porengrundwasserleiter*. Bes. Mitt. dt. Gewässerk. Jb. 52, 106 pp.

Lichtner, P., 1985, Continuum model for simultaneous chemical reactions and mass transport in hydrothermal systems. *Geochim. Cosmochim. Acta* 49, 779–800.

Lichtner, P.C. and Seth, M.S., 1996. *User's Manual for MULTIFLO: Part II—MULTIFLO 1.0 and GEM 1.0 Multicomponent-Multiphase Reactive Transport Model*. CNWRA 96–010, San Antonio, Texas.

Liger, E., Charlet, L. and Van Cappellen, P., 1999. Surface catalysis of Uranium(VI) reduction by iron(II). *Geochim. Cosmochim. Acta* 63, 2939–2955.

Lovley, D.R., Woodward, J.C. and Chapelle, F.H., 1994. Stimulated anoxic biodegradation of aromatic hydrocarbons using Fe(III) ligands. *Nature* 370, 128–131.

MacQuarrie, K.T.B., Sudicky, E.A. and Frind, E.O., 1990. Simulation of biodegradable contaminants in groundwater. 1. Numerical formulation in principal directions. *Water Resour. Res.* 26, 207–222.

Mayer, K.U., Frind, E.O. and Blowes, D.W., 2002. Multicomponent reactive transport modeling in variably saturated porous media using a generalized formulation for kinetically controlled reactions. *Water Resour. Res.* 38, 1174, DOI 10.1029/2001WR000862.

Miller, C.W. and Benson, L.V., 1983. Simulation of solute transport in a chemically reactive heterogeneous system: Model development and application. *Water Resour. Res.* 19, 381–391.

Nickson, R.T., McArthur, J.M., Ravenscroft, P., Burgess, W.G. and Ahmed, K.M., 2000. Mechanism of arsenic release to groundwater, Bangladesh and West Bengal. *Appl. Geochem.* 15, 403–413.

Notodarmojo, S., Ho, G.E., Scott, W.D. and Davis, G.B., 1991. Modeling phosphorous transport in soils and groundwater with two-consecutive reactions. *Water Res.* 25, 1205–1216.

Parkhurst, D.L., Kipp, K.L., Engesgaard, P. and Charlton, S.R., 2004. *PHAST—A program for simulating ground-water flow and multicomponent geochemical reactions.* http://wwwbrr.cr.usgs.gov/projects/GWC_coupled/phreeqc/index.html

Pickens, J.F. and Lennox, W.C., 1976. Numerical simulation of waste movement in steady groundwater flow systems. *Water Resour. Res.* 12, 171–180.

Press, W.H., Flannery, B.P., Teukolsky, S.A. and Vettering, W.T., 1992. *Numerical recipes in Pascal.* Cambridge Univ. Press, 759 pp.

Prommer, H., Barry, D.A. and Davis, G.B., 2000. Numerical modelling for design and evaluation of groundwater remediation schemes. *Ecol. Model.* 128, 181–195.

Prommer, H., 2002. A reactive multicomponent transport model for saturated porous media. www.pht3d.org

Rott, U., Meyerhoff, R. and Bauer, T., 1996. In situ treatment of groundwater with increased concentrations of iron, manganese and arsenic (in German). *Wasser-Abwasser* 137, 358–363.

Saalting, M.W., Ayiora, C. and Carrera, J., 1998. A mathematical formulation of reactive transport that eliminates mineral concentrations. *Water Resour. Res.* 34, 1649–1656.

Schlosser, P., Stute, M., Dörr, H., Sonntag, C. and Münnich, K.O., 1988. Tritium/^{3}He dating in shallow groundwater. *Earth Planet. Sci. Lett.* 89, 353–362.

Schlosser, P., Stute, M., Sonntag, C. and Münnich, K.O., 1989. Tritiogenic ^{3}He in shallow groundwater. *Earth Planet. Sci. Lett.* 94, 245–256.

Schmitt, R., Langguth, H.R., Püttmann, W., Rohns, H.P., Eckert P. and Schubert, J., 1996. Biodegradation of aromatic hydrocarbons under anoxic conditions in a shallow sand and gravel aquifer of the Lower Rhine Valley, Germany. *Org. Geochem.* 25, 41–50.

Shapiro, A.M., 2001. Effective matrix diffusion in kilometer-scale transport in fractured crystalline rock. *Water Resour. Res.* 37, 507–522.

Shapiro, S.D., Rowe, G., Schlosser, P., Ludin, A. and Stute, M., 1998. Tritium-helium 3 dating under complex conditions in hydraulically stressed areas of a buried-valley aquifer. *Water Resour. Res.* 34, 1165–1180.

Sheets, R.A., Bair, E.S. and Rowe, G.L., 1998. Use of H-3/He-3 ages to evaluate and improve groundwater flow models in a complex buried-valley aquifer. *Water Resour. Res.* 34, 1077–1089.

Smedley, P.L., 2003. Arsenic in groundwater - South and East Asia. In A.H. Welch and K.G. Stollenwerk, *op. cit.*, 179–209.

Smedley, P.L. and Kinniburgh, D.G., 2002. A review of the source, behaviour and distribution of arsenic in natural waters. *Appl. Geochem.* 17, 517–568

Steefel, C.I. and MacQuarrie, K.T.B., 1996. Approaches to modeling reactive transport in porous media. In P.C. Lichtner, C.I. Steefel and E.H. Oelkers (eds), *Rev. Mineral.* 34, 83–125.

Steefel, C.I. and Yabusaki S.B., 1996. *OS3D/GIMRT, Software for multicomponent-multidimensional reactive transport: User's Manual and Programmer's Guide*. PNL-11166, Richland.

Stollenwerk, K.G., 2003. Geochemical processes controlling transport of arsenic in groundwater: a review of adsorption. In A.H. Welch and K.G. Stollenwerk, *op. cit.*, 67–100.

Sverdrup, H.U. and Warfvinge, P., 1995. Estimating field weathering rates using laboratory kinetics. In A.F. White and S.L. Brantley (eds), *Rev. Mineral.* 31, 485–541.

Szabo, Z., Rice, D.E., Plummer, L.N., Busenberg, E. and Drenkard, S., 1996. Age dating of shallow groundwater with chlorofluorocarbons, tritium helium 3, and flow path analysis, southern New Jersey coastal plain. *Water Resour. Res.* 32, 1023–1038.

Tonndorf, H., 2000. *Die Uranlagerstätte Königstein.* Sächsisches Landesamt für Umwelt und Geologie, Dresden, 208 pp.

Van Ommen, H.C., 1985. The mixing cell concept applied to transport of non-reactive and reactive components in soils and groundwater. *J. Hydrol.* 78, 201–213.

Walter, A.L., Frind, E.O., Blowes, D.W., Ptacek, C.J. and Molson, J.W., 1994. Modelling of multicomponent reactive transport in groundwater. 2. Metal mobility in aquifers impacted by acidic mine tailings discharge. *Water Resour. Res.* 30, 3149–3158.

Weissmann, G.S., Zhang, Y., LaBolle, E.M. and Fogg, G.E., 2002. Dispersion of groundwater age in an alluvial aquifer system. *Water Resour. Res.* 38 (10), Art 1198, 14 pp.

Welch, A.H. and Stollenwerk, K.G. (eds) 2003. *Arsenic in groundwater, geochemistry and occurrence*. Kluwer, Boston, 475 pp.

Welch, A.H., Westjohn, D.B., Helsel, D.R. and Wanty, R.B., 2000. Arsenic in ground water of the United States: occurrence and geochemistry. *Ground Water* 38, 589–604.

Wisotzky, F. and Eckert, P., 1997. Sulfat-dominierter BTEX-Abbau im Grundwasser eines ehemaliges Gaswerksstandortes. *Grundwasser* 2, 11–20.

Yeh, G.T. and Tripathi, V.S., 1989. A critical evaluation of recent developments in hydrogeochemical transport models of reactive multichemical components. *Water Resour. Res.* 25, 93–108.

Yeh, G.T., Sun, J., Jardine, P.M., Burgos, W.D., Fang, Y., Li, M.-H. and Siegel, M.D., 2004. *HYDROGEOCHEM 5.0: A Three-Dimensional Model of Coupled Fluid Flow, Thermal Transport, and hydrogeochemical Transport through Variably Saturated Conditions*. ORNL/TM-2004/107.

Zheng, C. and Bennett, G.D., 2002. *Applied contaminant transport modeling*, 2nd ed. Wiley and Sons, New York, 621 pp.

Zoellmann, K., Kinzelbach, W. and Fulda, C., 2001. Environmental tracer transport (H-3 and SF6) in the saturated and unsaturated zones and its use in nitrate pollution management. *J. Hydrol.* 240, 187–205.

Zysset, A., Stauffer, F. and Dracos, T., 1994. Modeling of chemically reactive groundwater transport. *Water Resour. Res.* 30, 2217–2228.

APPENDIX A

Hydrogeochemical Modeling with PHREEQC

PHREEQC is used for simulating a variety of reactions and processes in natural waters or laboratory experiments. PHREEQC needs an input file in which the problem is specified via KEYWORDS and associated datablocks. The keywords are summarized here in *get-going* sheets. A full description of many alternatives for input and the mathematical backgrounds can be found in the manual of the program by Parkhurst and Appelo (1999).

PHREEQC was developed for calculating "*real world*" hydrogeochemistry and it does help to keep the connection with the physical situation palpable and concrete. Thus, we may start imagining a laboratory table with flasks with solutions, bottles with chemicals, minerals and exchangers, gascontainers, pipettes, a balance and a centrifuge with accessories, where we want to do some experiments. We use the following KEYWORDS:

SOLUTION *m–n* for the composition and quantity of solutions in flasks *m* to *n*. We can USE a specific solution, or MIX fractions of solutions in one flask.

Various reactants can be added to the flask:
EQUILIBRIUM_PHASES for a combination of minerals and/or gases which react reversibly to a prescribed equilibrium;
EXCHANGE for the capacity and composition of an exchanger;
SURFACE for the capacity and composition of surface complexers;
REACTION for stepwise adding or removing chemicals, minerals or water;
KINETICS for chemicals which react depending on time and composition of the solution;
GAS_PHASE for a combination of gases in a specified volume or at a given pressure;
SOLID_SOLUTIONS for adding solid solutions of minerals;
REACTION_TEMPERATURE for changing the temperature of the flask.

END is the signal for PHREEQC to calculate the composition of the solution and the reactants in what is termed a *simulation*. The compositions can be stored in computer memory with SAVE solution *no*, SAVE exchange *no*, etc., to be used later on in the same computer run (the same input file) with MIX or USE solution *no*, USE exchange *no*, etc., where *no* is a number or a range of numbers. The results can also be printed, punched in a spreadsheet file, or plotted in a graph.

For defining the output we use:
PRINT to limit printout to specific items and to suspend/resume print options;
SELECTED_OUTPUT to obtain results in spreadsheet type format;
USER_PRINT, USER_PUNCH and USER_GRAPH for defining tailor-made, specific output.

Flow and transport in the field and in laboratory columns can be modeled with keyword TRANSPORT for 1D dispersive/diffusive transport including mobile/immobile zones. For 3D transport, the code PHAST can be used from the US Geological Survey website.

Lastly, the chemical reactions which led to a given water quality can be recovered using keyword INVERSE_MODELING.

	PHREEQC Get-going sheet #
Getting Started	1
SOLUTION	2
Mineral Equilibration	3
PHREEQC Output	4
REACTION	5
MIX	6
EXCHANGE	7
Surface Complexation	8
Convergence; Titrating	9
TRANSPORT	10
KINETICS	11
SOLID_SOLUTION	12
GAS_PHASE	13
Databases	14
INVERSE_MODELING	15

Getting Started

The program PHREEQC, the manual and various example files can be downloaded via links in: www.xs4all.nl/~appt. Download Vincent Post's windows version and run psetup.exe to install the program and databases in c:\Program Files\phreeqc. Click on the PHREEQC icon to start. You'll be located in the input window.

Various options can be set, click Edit, Preferences (or ALT+e, e) and make your choices. The directories and the database file can be selected, click Calculations, Files (ALT+c, f).

To the right of the input window is a frame with PHREEQC keywords and BASIC statements. Click on + to expand the keywords, click again on + to expand keyword SOLUTION to see the identifiers (sub-keywords) temp, pH, pe, and more. Double clicking on a keyword or identifier will copy it to the input window.

Example files are in c:\Program Files\PHREEQC\Examples. Open U_GEX1, click Files, Open (ALT+f, o), U_GEX1. The file appears in the input window:

```
#            Graphs Ca vs F concentration in equilibrium with fluorite
#
SOLUTION 1                        # distilled water
END
EQUILIBRIUM_PHASES 1              # equilibrium with Fluorite, CaF2
 Fluorite
REACTION                          # Change Ca concentration for plot
 Ca(OH)2 1
 0.05 0.1 0.2 0.5 1 2 3 5 7 9 12 14 mmol
USE solution 1
USER_GRAPH                        # various chart definitions are omitted here
 10 graph_x tot("Ca")*1e3
 20 graph_y tot("F")*1e3
END
```

- Run the file, click the calculator icon, or Calculations, Start (ALT + c, s).
- Quickly, the progress window reports "Done". Press Enter, and you'll be transferred to the output tab. It contains a complete listing and many details of the calculated solution compositions.
- Click the Database tab. The basic data for calculating a speciation model appear, listed under keywords. The first one is keyword SOLUTION_MASTER_SPECIES which defines the elements and, among other things, their formula weights.
- Click the grid tab. It contains rows with the Ca and the F concentrations in equilibrium with fluorite as written by USER_GRAPH.
- Click the Chart tab. It shows a plot of the F *vs* Ca concentration in equilibrium with fluorite.

SOLUTION

The aqueous concentrations in a solution are defined with SOLUTION. The symbols for the elements are listed in the first column of SOLUTION_MASTER_SPECIES in the database.

```
                              # Information following a # is not read

SOLUTION 1  Speciate an analysis, calculate saturation indices
  temp 25.                    # temperature in degrees Celcius, default = 25°C
  pH 7.0                      # default pH = 7.
  pe 4.0                      # pe = -log(electron activity), default = 4.
  -units mmol/kgw             # default units mmol/kg water
  Ca 3.0                      # total dissolved Ca
  Na 1.0
  Alkalinity 3.8              # mmol charge / kgw
  S 1.0                       # total sulfur, mainly sulfate, S(6), at pe = 4
  N(5) 0.2                    # nitrogen in the form of nitrate, N(5)
  Cl 1.0
  water 1                     # kg water, default = 1 kg
  density 1                   # density, default = 1 kg/L
END

# Example 2.  define range of solution no's, and use mg/L units
#
SOLUTION 2-4                  # solutions 2, 3 and 4
  -units mg/L
  Ca 40                       # mol. weights listed under SOLUTION_MASTER_SPECIES
  Alkalinity 122 as HCO3      # HCO3 is used for recalculating to mol charge
END

# Example 3.  Adjust concentration or pH to charge balance
#
SOLUTION 4                    # default pH = 7, temp = 25, -units mmol/kgw
  Na 2.0
  Cl 1.3 charge               # adapt Cl to obtain charge balance
END

# Example 4.  Adjust pe to equilibrium with atmospheric oxygen
#
SOLUTION 6                    # see EQUILIBRIUM_PHASES (sheet # 3) for more info
  pe 10.0 O2(g)-0.68          # add PHASES name, SI = -0.68. P_gas = 10^-0.68
END

# Example 5.  Enter a suite of samples in spreadsheet format
# The data are tab delimited
#
SOLUTION_SPREAD
  -units mg/L
  -temp 11
Ca        Cl        S(6)      Temp      Alkalinity          pH        Li
                                        mg/L as HCO3                  ug/L
87.7      22.7      94.4      13        233.4               7.3
78.6      35.3      87.4      13.1      221.7               7.1       4
75.0      24.9      83.2                221.3               6.9
```

Note

The default temperature of 11°C will be given to the third sample.

Mineral Equilibration

A solution can be equilibrated with minerals assembled together in EQUILIBRIUM_PHASES, or the initial concentration of a component can be adapted when defining the SOLUTION.

```
# Example 1.  Equilibrate water with calcite and CO2
#
SOLUTION 1                      # default pH = 7, pe = 4, temp = 25, units = mmol/kgw
 Ca 3.0
 Alkalinity 4.0
 S(6) 1.0
EQUILIBRIUM_PHASES 1
 Calcite 0.0 10.0               # Mineral from database, SI = 0.0, amount = 10 mol
 CO2(g) -2.0                    # SI = - 2.0 or [P_CO2] = 10^-2.0, amount = 10 mol
END

# Example 2.  Equilibrate with Fluorite, adding gypsum.
#
SOLUTION 2
 F 1.5 mg/kgw                   # drinking water limit for F = 1.5 mg/L
EQUILIBRIUM_PHASES 2
 Fluorite 0.0 Gypsum            # obtain SI = 0. by adding Ca from gypsum
END                             # note the high SO4 concentration in the output...

# Example 3.  Equilibrate by changing an initial concentration
#
SOLUTION 3
 Ca 3.0 Calcite                 # adapt Ca to equilibrium with calcite
 Alkalinity 4.0
 S (6) 1.0
 F 0.1 Fluorite -1.0            # adapt F to SI = -1.0 for fluorite
END
```

Notes

The mineral names must be spelled exactly as entered in the database under PHASES (you can copy the name after selection with CTRL+c, then paste with CTRL+v).

Gas equilibria. The gas pressure can be varied, setting *SI*. The *SI* is equal to log(gas pressure/1 atm).

Example 4 illustrates the different effects of equilibrating by adjusting the initial concentration, and by the reversible dissolution reaction that is simulated with keyword EQUILIBRIUM_PHASES.

```
# Example 4.  Adjust the CO2 gas pressure to 10^-22.5 = 0.00316 atm
#
SOLUTION 1
 pH 7.0
 C(4) 1.0 CO2(g) -2.5           # Only TIC is adapted, pH remains 7.0
END

SOLUTION 2
 pH 7.0
EQUILIBRIUM_PHASES 2            # pH changes when carbonic acid dissolves
 CO2(g) -2.5 10.0
END
```

Solution 1 obtains a large electrical imbalance, while solution 2 remains electrically neutral.

PHREEQC Output

Using PRINT, you can modify the print to the output tab by setting output block identifiers true or false. These identifiers can be clicked and copied to the input tab after expanding keyword PRINT in the right window. The option -reset false is rather essential when doing transport simulations.

```
PRINT
 -reset false          # suspend all print output
 -totals true          # will print Solution composition, Description of solution
 -status true          # shows calculation progress
```

To write specific output in spreadsheet format ("punch" to a file), use SELECTED_OUTPUT. The selected information is written to the file after each simulation.

```
SELECTED_OUTPUT
 -file stine1.csv             # name of output file. OBLIGATORY
 -totals Fe Fe(2) Hfo_w       # total conc. in mol/kgw, MASTER_SPECIES element name
 -molalities Fe+2 CaX2        # molalities of species
 -activities Fe+2 H+          # log activities of species
 -eq Calcite                  # equilibrium_phases, note abbreviation -eq
```

Notes

Element names given under -totals should comply with the MASTER_SPECIES element names (first column in the databases).

Species names given under -mol and -act and mineral names must match the database names.

In transport calculations, -punch_cells and -punch_frequency can be used to further limit the punchout to the selected_output file.

USER_PUNCH, USER_PRINT and USER_GRAPH

The USER_ keywords give access to the BASIC compiler which is built in PHREEQC. You can punch out your own variables to the selected_output file with USER_PUNCH, print to the output file with USER_PRINT, and plot in PHREEQC for Windows with USER_GRAPH.

```
USER_PUNCH
 -headings mg_Na/l P_CO2               # column heading in file
 -start                                # start of BASIC lines
 10 punch tot("Na")*22.999e3
 20 punch 10^(SI("CO2(g)"))
 -end                                  # end of BASIC lines
END

USER_GRAPH                             # plot F and pH vs Ca concentration
 -start
 10 graph_x tot("Ca")*1e3              # mmol/L
 20 graph_y tot("F")*1e3
 30 graph_sy -la("H+")                 # pH on secondary y-axis
 -end
END
```

The BASIC lines can be seen as spreadsheet formulas, with special functions which give access to the aqueous model. These functions are explained under keyword RATES in the PHREEQC manual, and listed in the PHREEQC BASIC statements at the right of the input window.

REACTION

The keyword REACTION is used to stepwise add reactants or amounts of minerals to the solution.

```
# Example 1. Add 5 and 10 mmol NaCl to 1 liter distilled water.
#
SOLUTION 1         # Define start solution, just water
REACTION 1
 NaCl 1.1          # Reactant formula, stoichiometric coefficient in the reaction
 5e-3 10e-3        # Add 1.1*5e-3 mol NaCl and 1.1*10e-3 mol NaCl
# 10e-3 moles in 2 steps       # alternate way of writing the steps
END

# Example 2. Add Calcite and CO2 to water
#
SOLUTION 2
REACTION 2
 Calcite 1.0 CO2(g) 1.3          # mineral from database, stoichiometric coefficient
 2e-3          # Add 1*2e-3 = 2e-3 mol Calcite and 1.3*2e-3 = 2.6e-3 mol CO2
END

# Example 3.  Evaporate 99% of a HCl solution in 5 steps
#
SOLUTION 3
 pH 3.0 charge
 Cl 1.0
 water 1.0                      # 1 kg water = 55.51 mol
REACTION 3
 H2O -1.0
 54.951 moles in 5 steps
END
```

Notes

In example 1, the program recourses to the initial solution when adding the reaction steps. With keyword INCREMENTAL_REACTIONS true, the reaction steps are added cumulatively, thus giving 5×10^{-3} and 15×10^{-3} mol NaCl.

MIX

The effects of mixing of two or more groundwaters on the speciation and saturation state with respect to minerals can be calculated with keyword MIX. This keyword is also useful for simulating laboratory batch experiments.

```
# Example 1.  Calculate mixing corrosion in limestone water
#
SOLUTION 1 Calcite saturated, high CO2
 pH 7.1 charge
 Ca 2.0 Calcite
 C(4) 4.0 CO2(g) -1.5
SOLUTION 2 Calcite saturated, low CO2
 pH 7.1 charge
 Ca 2.0 Calcite
 C(4) 4.0 CO2(g) -2.5
END
MIX 1
 1 0.4                          # use fraction 0.4 of solution 1
 2 0.6                          # use fraction 0.6 of solution 2
EQUILIBRIUM_PHASES 1
 Calcite                        # find how much calcite dissolves in the mixture
END
```

```
# Example 2.  Evaporate 90% of 1 liter NaCl water, restore to 1 liter
#
SOLUTION 2
 Na 10
 Cl 10
REACTION 2
 H2O -1.0
 49.955                         # Evaporate 49.955 moles (= 0.9 * 55.506)
SAVE solution 3                 # leaves 5.5506 mol H2O = 0.1 liter
END
MIX 3
 3 10                           # Take fraction 10 of solution 3,
                                # is 10*5.5506 = 55.506 mol = 1 liter
END
```

```
# Example 3.  Displace exchangeable cations with 1 M NaCl
#
SOLUTION 0 1M NaCl
 Na 1e3
 Cl 1e3
END
EXCHANGE 1 Fill centrifuge tube with 1 g clay = 1 mmol CEC
 KX 0.02e-3; MgX2 0.09e-3; CaX2 0.4e-3
MIX 1
 0 0.02                         # Add 20 ml of 1M NaCl solution
END
```

EXCHANGE

The keyword EXCHANGE is used to define exchange properties and to calculate the composition of the exchanger.

```
# Example 1.  Calculate composition of an exchanger
#
SOLUTION 1
 Na 1.0
 K  0.2
 Cl 1.2
EXCHANGE 1
 X 30e-3                    # 30 mmol X-
 -equil 1                   # equilibrate with solution 1, --> 15 mmol NaX/L,
                            # 15 mmol KX/L
END
```

```
# Example 2.  Input measured exchanger
#
EXCHANGE 1
 CaX2 20e-3                 # 20 mmol exchangeable Ca
 NaX 10e-3                  # 10 mmol exchangeable Na
```

```
# Example 3.  Exchanger composition is calculated after reactions
#
SOLUTION 1-5
 Na 2.0
 C(4) 2.0 charge
EQUILIBRIUM_PHASES 1-5      # mineral equilibria for cells 1-5
 Calcite 0.0 0.01           # 0.01 mol Calcite, SI = 0.0
SAVE solution 1-5           # save equilibrated solutions
END

EXCHANGE 1-5                # exchangers for cells 1-5
 X 0.1                      # 0.1 mol X
 -equil 1
END
```

Notes

EXCHANGE defines moles of X^-, but the *concentration* depends on the amount of solution (default is 1 kg water). In example 1, the concentration is 30 mmol X^-/kg water.

In example 3, the exchanger is calculated with a mineral equilibrated solution, *after* SAVE and END. The calculation sequence is important here, the results are different when EXCHANGE 1–5 is included before the first END. Any PHREEQC simulation starts with equilibration of the exchanger (if defined). Next, the reactions are calculated together with the exchanger. Here, the initial solution contains only Na^+, and the exchanger would be pure NaX. If the solution is then equilibrated with calcite, Ca^{2+} displaces Na^+ from the exchanger, and the Na^+ concentration in solution 1–5 will become higher than 2 mM.

The moles of an exchanger may be coupled to minerals (eq) or kinetic reactants (kine).

```
EXCHANGE 1-5
 X Calcite eq 0.05          # X is coupled to calcite (an equilibrium phase)
                            # in proportion 0.05 mol X / mol calcite
```

Surface Complexation

PHREEQC contains the Dzombak and Morel database for surface complexation of heavy metal ions on hydrous ferric oxide (ferrihydrite, hfo). Ferrihydrite binds the metal ions and protons on strong and weak sites and develops a charge depending on the ions sorbed. The keyword SURFACE is used to define the surface complexation properties of sediments in much the same way as EXCHANGE, but since the surface may become charged, some precautions are in order.

```
# Example 1.  Calculate the composition of a surface complex
#
SOLUTION 1
 Na 1.; Cl 1.; Zn 1e-3                 # semicolon separates lines
SURFACE 1
 Hfo_w 2.4e-3 600 1.06                 # 2.4e-3 mol weak sites, sp. surface = 600 m2/g
                                       # gram ferrihydrite = 1.06
 Hfo_s 6e-5                            # strong site, 6e-5 mol sites
 -equil 1                              # equilibrate with solution 1
END
```

```
# Example 2.  Couple a surface with a mineral (eq) or kinetic reactant (kine)
#
SOLUTION 2                             # adjust pe to prevent (too much) Hfo dissolution
 pe 14 O2(g) -0.68                     # pe is for P_O2 = 10^-0.68 = 0.21 atm
 Na 1.; Cl 1.; Zn 1e-3
EQUILIBRIUM_PHASES 2
 Fe(OH)3(a) 0.0 1e-3
SAVE solution 2                        # Save solution 2 after equilibration
END
SURFACE 2
 Hfo_w Fe(OH)3(a) eq 0.2 5.34e4        # weak sites, coupled to ferrihydrite,
                                       # proportionality constant = 0.2,
                                       # sp. surface = 5.34e4 m2/mol
 Hfo_s Fe(OH)3(a) eq 0.005             # strong site, coupled to ferrihydrite,
                                       # proportionality constant = 0.005
 -equil 3
END
```

Notes

The default properties of Hfo are: 89 g/mol, 600 m^2/g, 0.2 mol weak sites/mol, 0.005 mol strong sites/mol.

Make sure to have reactions calculated for a solution, to save it and to END the simulation, before equilibrating a surface as is done in example 2 (see also discussion in get-going sheet *EXCHANGE*).

When Hfo is coupled to a mineral, the specific surface area is given in m^2/mol.

Two Tips

- In rare cases, mainly with redox problems with very low concentrations of the metal ions, PHREEQC has difficulties finding a solution. The first trick is to include a redox buffer in the input file:

```
SOLUTION_SPECIES
  H2O + 0.01e- = H2O-0.01; log_k -9.0
```

- Titration to a fixed pH is illustrated in the following example.

```
# Determine Cd sorption edge on Hfo
PHASES
  Fix_ph                          # Define pH fixing mineral
  H+ = H+; log_k 0                # pH = 10^(-SI)

SOLUTION 1
  pH 4.0; Na 100; Cl 100 charge; Cd 1e-3

SURFACE 1
  Hfo_w 0.2e-3 600 89e-3
  Hfo_s 0.05e-3
  -equil 1
END

USER_GRAPH
  -head pH %_cd_sor titrans
  -start
10 sor = mol("Hfo_sOCd+") + mol ("Hfo_wOCd+")           # 2 Cd surface complexes
20 graph_x -la("H+")                                    # pH on x-axis
30 graph_y 100 * sor/(sor + tot("Cd"))                  # % sorbed Cd on y-axis
# 40 graph_sy 1e6 * tot("Br")         # (to plot mol titrans on secondary y axis)
-end
END

USE solution 1; USE surface 1
EQUILIBRIUM_PHASES 1
  Fix_ph -4.01 Na(OH)0.999999Br0.000001         # NaOH for increasing pH, with tracer Br
END
  USE solution 1; USE surface 1
  EQUILIBRIUM_PHASES 1; Fix_ph -4.5 Na(OH)0.999999Br0.000001
END
# Etc. for pH up to 8...
```

TRANSPORT

Keyword TRANSPORT is used for modeling 1D flowlines and laboratory columns. The flowline is divided in a number of cells 1–*n* for which SOLUTION 1–*n* must be defined. With advective flow, a *shift* moves water from each cell into the next, higher- or lower-numbered cell, depending on the flow direction.

```
# u_gex3 -- Transport and ion exchange: CaCl2 flushes column with (Na,K)NO3
#
SOLUTION 1-16 Initial solution in column
 Na 1.0; K 0.2; N(5) 1.2                 # default pH, etc, note line separator ;
EXCHANGE 1-16
 X 1.1e-3; -equilibrate 1
END
SOLUTION 0 Displacing solution, CaCl2
 Ca 0.6; Cl 1.2
END
PRINT; -reset false                      # suspend output
TRANSPORT
 -cells 16
 -length 0.005               # cell length 0.005 m, column length 16*0.005 = 0.08 m
 -dispersivity 0.002                     # m
 -shifts 40                              # 40/16 = 2.5 pore volumes
 -flow_direction forward
 -time_step 720                          # 1 shift = 720 s. total time = 40*720 s
 -boundary_conditions flux flux          # at column ends
 -diffusion_coefficient 0.3e-9           # m2/s
 -punch 16                               # only punch/graph cell 16
USER_GRAPH
 -heading PV NO3 Cl Na K Ca;      -axis_titles Pore_Volumes mol/L
 -axis_scale x_axis 0 2.5;        -axis_scale y_axis 0 1.5e-3
 -plot_concentration_vs time
 -start
 10 graph_x (step_no + 0.5)/16    # transfer cell centered conc to column end
 20 graph_y tot("N"), tot("Cl"), tot("Na"), tot("K"), tot("Ca")
 -end
END
```

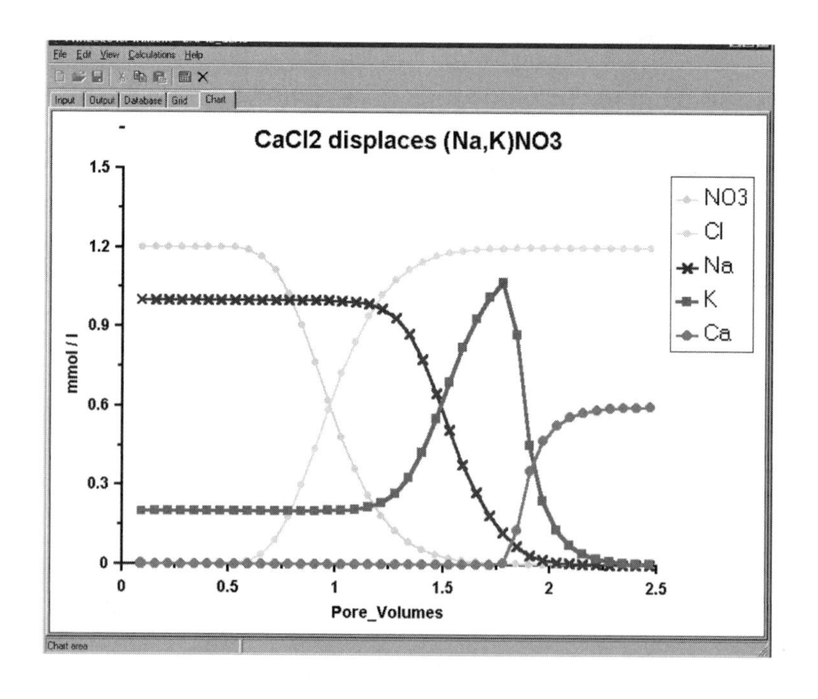

KINETICS

For kinetic calculations a rate equation must be programmed in BASIC using keyword RATES. The rate is called with keyword KINETICS which can also pass parameters to the rate. The BASIC functions are explained under keyword RATES in the PHREEQC manual.

```
#      Kinetic quartz dissolution
#
RATES
  Quartz                       # rate name
  -start
  #1 rem dQu/dt = -k * (1 - ΩQuartz). k = 10^-13.7 mol/m2/s (25 C)
  #2 rem parm(1) = A (m2), parm(2) = V (dm3) recalculate to mol/dm3/s
  10 moles = parm(1) / parm(2) * (m/m0)^0.67 * 10^-13.7 * (1 - SR("Quartz"))
  20 save moles * time         # integrate. save and time must be in rate definition
                               # moles count positive when added to solution
  -end
KINETICS             # Sediment: 100% qu, grain size 0.1 mm, por 0.3, rho_qu 2.65 kg/dm3
  Quartz                       # rate name
  -formula SiO2
  -m0 102.7                    # initial moles of quartz
  -parms 22.7 0.162            # parameters for rate eqn. Here:
# Quartz surface area (m2/kg sediment), water filled porosity (dm3/kg sediment)
  -step 1.58e8 in 10 steps     # 1.58e8 seconds = 5 years
  -tol 1e-8                    # integration tolerance, default 1e-8 mol
INCREMENTAL_REACTIONS true     # start integration from previous step
SOLUTION 1
USER_GRAPH
  -heading time Si; -axis_titles years mmol/L
  -axis_scale y_axis 0 0.12 0.02; -axis_scale x_axis 0 5
  -start
  10 graph_x total_time/3.1536e7
  20 graph_y tot("Si")*1e3
  -end
END
```

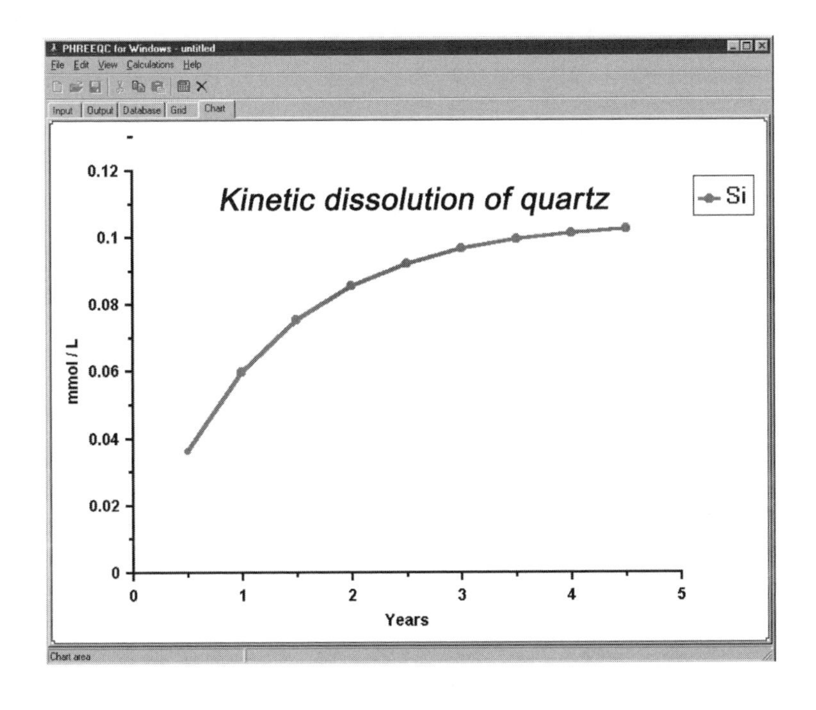

SOLID_SOLUTION

The solid solution option of PHREEQC can be used for modeling trace element behavior. The solubility of the trace component in a solid solution is generally much smaller than when the trace element forms a pure solid. Consequently the concentrations of the minor component are much reduced in solution, and the mobility of the trace element is diminished.

```
SOLUTION 1
   pH 7.5 Calcite               # K = 10^-8.48
   C(4) 4.0
   Ca 1.0
   Cd 1 ug/kgw Otavite          # Otavite = CdCO3, K = 10^-12.1
   SOLID_SOLUTION 1             # Note decrease of Cd concentration
   CaCdCO3
   -comp Calcite 0              # Ideal solid solution, initial amount = 0
   -comp Otavite 0
END

USE solution 1
SOLID_SOLUTION 1                # Regular solid solution with ideal properties
   CaCdCO3
   -comp1 Calcite 0
   -comp2 Otavite 0
# -distri D1 D2 x1 x2           # D1=D2=D_ideal = K_cc/K_ot for all x
   -distri 4168.69 4168.69 0.1 0.6
END
```

Ideal solid solutions can have as many components as desired in PHREEQC, regular solid solutions have only two components. The properties of regular solid solutions can be defined by various means (in the example via distribution coefficients).

This keyword is most useful for modeling liquid solutions (BTEX etc., see Chapter 10).

GAS_PHASE

This keyword is used to calculate equilibrium with a mixture of gases in a bubble. The bubble can have fixed volume and varying pressure, or it may have fixed total pressure and varying volume.

```
# Calculate CH4 gas formation under peat
#
SOLUTION 1
  pH 7.0 charge
  Ca 2.0 Calcite
  C(4) 2.0 CO2(g) -2.0             # P_CO2 = 0.01 atm
  S(6) 1.0
  N(0) 2.0 N2(g) 0.299             # P_N2 = 1.99 atm

REACTION 1
CH2O 1; 0.01                       # decompose 10 mmol organic carbon

GAS_PHASE
  -fixed_pressure
  -pressure 2.0                    # 2 atm, at 10 m depth
  -volume 1e-8                     # negligible initial volume, 1e-8 liter
  -temperature 25                  # default 25°C
  CH4(g) 0.0                       # gas name from PHASES, partial pressure
  H2S(g) 0.0
  CO2(g) 0.005
  N2(g) 0.995
END
```

Databases

Currently 4 databases are provided with PHREEQC, PHREEQC.DAT is the smallest. WATEQ4F.DAT has additional data for heavy metals, MINTEQ.DAT has a few more organic chemicals, and LLNL.DAT is a huge database with many minerals and large-range, temperature dependent equilibrium constants. The databases contain lists under keywords.

```
SOLUTION_MASTER_SPECIES                                  # always needed
#element      species      alk      gfw_formula         element_gfw
H             H+           -1.0     H                   1.008
H(0)          H2           0.0      H
H(1)          H+           -1.0     0.0
E             e-           0.0      0.0                 0.0
O             H2O          0.0      O                   16.0
O(0)          O2           0.0      O
O(-2)         H2O          0.0      0.0
# ...the above elements must always be present in the database
Ca            Ca+2         0.0      Ca                  40.08
Fe            Fe+2         0.0      Fe                55.847   # Fe+2 is primary redox species
Fe(+2)        Fe+2         0.0      Fe
Fe(+3)        Fe+3         -2.0     Fe
# The primary species reaction is repeated in SOLUTION_SPECIES, with log_k = 0
# etc...
SOLUTION_SPECIES
H+ = H+                                        # H+ primary species
              log_k        0.0
              -gamma       9.0      0.0        # params for activity coefficient
Fe+2 = Fe+2                                    # Fe+2 is the primary redox species for Fe
              log_k        0.0
              -gamma       6.0      0.0
Fe+2 = Fe+3 + e-                               # new species defined on the right of = sign
              log_k        -13.02
              delta_h      9.680    kcal       # temp. dependency from Van 't Hoff eqn
              -gamma       9.0      0.0
# etc...
PHASES                                         # Minerals
Calcite
   CaCO3 = CO3-2 + Ca+2
            log_k          -8.48
            delta_h        -2.297 kcal
            -analytic      -171.9065  -0.077993  2839.319  71.595 # log_k vs temp
# etc...
EXCHANGE_MASTER_SPECIES
   X          X-                               # element, species
EXCHANGE_SPECIES
   X- = X-                                     # reaction for master species with log_k = 0
            log_k          0.0
   Na+ + X-= NaX
            log_k          0.0
            -gamma         4.0      0.075
# etc...
SURFACE_MASTER_SPECIES                         # element, species
   Hfo_s    Hfo_sOH
   Hfo_w    Hfo_wOH
SURFACE_SPECIES
   Hfo_sOH = Hfo_sOH                           # reaction for master species with log_k = 0
            log_k          0.0
   Hfo_sOH + H+ = Hfo_sOH2+
            log_k          7.29                # = pKa1,int
# etc...
RATES
# etc...
END
```

INVERSE_MODELING

Inverse modeling aims to decipher the reactions and mixtures which have led to a given water quality.

```
# Find the reactions which produced solution 2 from solution 1.
#
PRINT; -reset false; inverse true
SOLUTION 1
Cl 1; Na 1
#  evaporate 90% of solution 1, then add 2 mM gypsum
SOLUTION 2
  Cl 10; Na 10; Ca 2; S(6) 2
END

INVERSE_MODELING
  -solutions 1 2            # Balance solution 2 from 1
   -phases                  # Using reactants from PHASES
  Gypsum
  Water                     # Water not in PHASES in database, is defined below
  -balances                 # Any elements not in phases
  Cl                        # Add 0 0.05 uncertainty for absolute balance of Cl
  Na
  Alkalinity 1              # Permitted uncertainty 100%


  -uncertainty 0.05         # Default uncertainty, 5%


  -range true

PHASES
  Water                     # Define water
  H2O = H2O
  log_k 0                   # log K not used
END
```

With the identifier -phases the possibly reacting phases are indicated. All reactants must have been defined either as EXCHANGE_SPECIES or under keyword PHASES. These "phases" are used only for mole balancing, their log K is unimportant when they are used in INVERSE_MODELING.

Elements which are not a part of the indicated phases, but are present in the solutions, must be listed with -balances, as done for Na, Cl and alkalinity in this example. Numbers after the elements indicate the permitted uncertainty for the solutions (in the order entered with identifier-solutions). A positive number indicates fractional uncertainty, a negative number means that the uncertainty, in mol/kgw, may increase up to the absolute value of the number. In the example, the permitted uncertainty for the alkalinity is set to 100% of the input concentration. The alkalinity is very small for these two solutions, and a balance can not be obtained without allowing for a large uncertainty.

Isotopes can be included in inverse models, the δ values are entered under SOLUTION.

APPENDIX B

ANSWERS TO PROBLEMS

CHAPTER 1

1.1a.

Na^+	K^+	Ca^{2+}	Mg^{2+}	$\Sigma+$	Cl^-	HCO_3^-	SO_4^{2-}	$\Sigma-$
2.70	0.002	4.05	0.74	7.49	−4.74	−3.00	−0.0	−7.74 meq/L

The difference: 7.49−7.74 = −0.25 meq/L = −1.7%
This is a fairly good analysis.

1.1b. Total Hardness = 4.05 Ca^{2+} + 0.74 Mg^{2+} = 4.79 meq/L
expressed as mg/L $CaCO_3$: 240 mg/L
expressed as °d (German Hardness): 13.5°d
Alkalinity (3.00 meq/L) equals 150 mg/L $CaCO_3$.
Hence, non-carbonate hardness is 90 mg/L. It derives from other sources than calcite-dissolution and will remain in the water when pH increases and calcite precipitates.

1.2.1. $\Sigma+ = 7.4$ meq/L, $\Sigma- = -6.0$ meq/L.
Comparison of these values with the measured *EC* of 750 μS/cm suggests that there is an anion-shortage. This shortage is probably due to a too low HCO_3^- value, since the calculated $[P_{CO_2}] = 10^{-3.7}$, which is below the atmospheric level.

1.2.2. $\Sigma+ = 3.6$ meq/L, $\Sigma- = -1.5$ meq/L.
In this case there is a cation-surplus and if we look at the *EC* (150 μS/cm), it seems that Ca^{2+} determination must be wrong since the contribution of measured Ca^{2+} concentration to the *EC* would be ≈ 280 μmho/cm. At the measured pH, the Al content is also rather high.

1.2.3. $\Sigma+ = 4.5$ meq/L, (Fe included as Fe^{2+}), $\Sigma- = -4.5$ meq/L.
Seems allright (*EC* = 450 μS/cm), but NO_3^- cannot occur together with Fe^{2+}. If a brown $Fe(OH)_3$ precipitate is present in the non-acidified sample, the anions must be wrong. Remember that the concentration of HCO_3^- decreases when Fe^{2+} becomes oxidized in the non-acidified sample:

$$Fe^{2+} + \tfrac{1}{4}O_2 + 2\,HCO_3^- + \tfrac{1}{2}H_2O \rightarrow Fe(OH)_3 + 2\,CO_2$$

Thus, alkalinity is best determined in the field!

1.3b. A = St. Heddinge waterworks
B = Maarum waterworks
C = Haralds mineralwater
D = Rainwater Borris

Concentrations in mmol/L

Sample		A	B	C	D
Na^+		0.783	19.139	0.391	0.313
K^+		0.161	0.358	0.026	0.016
Mg^{2+}		0.579	1.193	–	0.032
Ca^{2+}		3.019	1.048	–	0.035
NH_4^+		0.001	0.194	–	0.068
$\Sigma+$	(meq/L)	8.134	24.173	0.417	0.531
Cl^-		0.818	5.669	0.480	0.338
HCO_3^-		4.867	17.748	–	0.000
SO_4^{2-}		0.635	0.000	–	0.047
NO_3^-		0.565	0.032	0.081	0.073
$\Sigma-$	(meq/L)	−7.52	−23.450	−0.561	−0.505
EC	(μS/cm)	750	1900	88	

ad A: *E.B.* $= (8.134 - 7.520)/(8.134 + 7.520) \times 100 = 3.9\%$, which is reasonable but not perfect. Since $EC/100 = 750/100 = 7.50$ is in best agreement with $\Sigma-$, one of the cation concentrations must be a bit high, possibly Ca^{2+}.

ad B: *E.B.* $= (24.173 - 23.450)/(24.173 + 23.450) \times 100 = 1.5\%$. Based on the *E.B.*, this is a rather good analysis. However, $EC/100 = 19.00$ is not in good agreement with the sum of cations or anions, but we are at the upper limit where the simple relation $\Sigma+ = -\Sigma-$ (in meq/L) $= EC/100$ is justified.

ad C: The analysis is incomplete and the *E.B.* cannot be calculated. The hardness in german degrees is $1°dH = 17.8\,mg\ CaCO_3/L = 0.18\,mmol/L\ (Ca^{2+} + Mg^{2+}) = 0.36\,meq/L$. The $\Sigma+$ becomes then $0.78\,meq/L \sim 78\,\mu S/cm$ *EC*, which is in reasonable agreement with the measured *EC*.

ad D: *E.B.* $= (0.532 - 0.505)/(0.532 + 0.505) \times 100 = 2.6\%$.

1.3c.

Concentrations in mg/L.

Rain contribution	A	B	C
Concentration factor	2.42	16.75	1.42
Na^+	17.40	120.60	10.20
K^+	1.52	10.55	0.89
Mg^{2+}	1.89	13.07	1.11
Ca^{2+}	3.38	23.45	1.98
NH_4^+	2.97	20.60	1.74
Cl^-	29.00	201.00	17.00
SO_4^{2-}	10.83	75.04	6.35
NO_3^-	10.92	75.71	6.40
PO_4^{3-}	0.00	0.00	0.00

ad A: The contribution of rainwater is for most components small, except for Na which apparently is derived mainly from rainwater.

ad B: The concentration factor of 16.75 is much too high for Danish conditions and the high Cl^- concentration indicates a marine influence.

ad C: Most of the dissolved contents in this sample originates from rainwater.

1.3d. ad A: The high Ca^{2+} and HCO_3^- contents indicate the presence of $CaCO_3$ in the aquifer and the high NO_3^- content suggests agricultural activity in the recharge area.

ad B: The high Na^+ content, which is balanced by HCO_3^-, indicates that ion exchange has influenced this water composition.

ad C: This sample must originate from a reservoir with little reactive material, since most of the dissolved content is derived from rainwater.

CHAPTER 2

2.1. $x = I_0/(1.5E) = (0.025\,\text{m } H_2O \times 4\,\text{m/s} \times 3.15{\times}10^7\,\text{s/yr})/(1.5 \times 0.5\,\text{m } H_2O/\text{yr}) = 4200\,\text{km}$.

2.2. Integrate Equation (2.2) from 0 to 2500 km:

$$RP_{tot} = \frac{\int_0^{2.5\cdot 10^6} E\,x\,\mathrm{d}x}{\int_0^{2.5\cdot 10^6} (I_0 - 0.5\,E\,x)\,\mathrm{d}x} = \frac{E\,(2.5{\times}10^6)^2/2}{I_0 2.5{\times}10^6 - 0.5E\,(2.5{\times}10^6)^2/2} = 0.22$$

for $I_0 = 3.15{\times}10^6\text{m } H_2O \times \text{m/yr}$ and $E = 0.5\,\text{m } H_2O/\text{yr}$.

2.3a. In the following table, the concentrations of cations and anions are expressed in µeq/L. The estimated parameters are underlined.

	H^+	NH_4^+	Na^+	K^+	Ca^{2+}	Mg^{2+}	$\Sigma+$	Cl^-	NO_3^-	SO_4^{2+}	$\Sigma-$
Vlissingen	100	49	265	8	78	68	568	−297	−92	−170	−559
Deelen	112	119	30	3	32	10	306	−37	−90	−179	−306
Beek	26	143	23	11	100	10	313	−34	−87	−192	−313

Vlissingen: $E.B. = 0.8\%$

2.3b. The seawater-contribution to the total concentration of ion i may be calculated from $c_{i,\text{rain}} = (c_{i,\text{sea}}/c_{\text{Cl,sea}}) \times c_{\text{Cl,rain}}$.

	Vlissingen			Deelen			Beek		
	Total	Sea	%Sea	Total	Sea	%Sea	Total	Sea	%Sea
H^+	100	0	0	112	0	0	26	0	0
Na^+	265	254.5	96	30	31.7	>100	23	29.1	>100
K^+	8	5.6	70	3	0.7	23	11	0.6	6
Mg^{2+}	68	57.7	85	10	7.2	72	10	6.6	66
Ca^{2+}	78	10.5	13	32	1.3	4	100	1.2	1
NH_4^+	49	0	0	119	0	0	143	0	0
Cl^-	297	297	100	37	37	100	34	34	100
SO_4^{2-}	170	31	18	179	3.9	2	192	3.5	2
NO_3^-	92	0	0	90	0	0	87	0	0

2.3c. – The very low H^+, NO_3^- and NH_4^+ contents in seawater cannot contribute significantly to the rainwater composition. Instead, their sources in rainwater are industrial processes and the evaporation of ammonia.

- The seawater-contribution for the Na^+ concentration is negative for both Deelen and Beek. This might be an effect of industrial Cl^- production. In such cases Na^+ is preferred as conservative ion.
- Dust and fertilizer are likely sources for K^+ in rain.
- The seawater influence on the Ca^{2+} concentration is low and cement industry seems to be the source of high Ca^{2+} concentrations in rain at Beek.

2.4. Substitute δ for R: $\delta = (\delta_0 + 1000) f^{(\alpha-1)} - 1000$
differentiate:

$$d\delta/df = (\delta_0 + 1000)(\alpha - 1) f^{(\alpha-2)}$$

Hence:

$$\frac{d\,\delta^2H}{d\,\delta^{18}O} = \frac{(-95 + 1000)(1.0876 - 1)}{(-12.2 + 1000)(1.0099 - 1)}\, 0.9^{(1.09-1.01)} = 8.0$$

The deuterium excess is $(-95 + (1.0876 - 1) \times 1000) - 8 \times (-12.2 + (1.0099 - 1)) \times 1000) = 10.4$.

2.5. $^{18}O = 0.11117$: $\delta^{18}O = -0.065$‰; $^{18}O = 0.11118$: $\delta^{18}O = 0.025$‰; from $(0.05/10^3) \cdot R_{st} \cdot {}^{16}O = 5.56 \times 10^{-6}$ mol/L (i.e. 6 digits accurate).

2.6. From continuity for water flow, the stagnant, water filled porosity is $\varepsilon_{st} = \varepsilon_{rz} - \varepsilon_w = 0.25 - 0.089 = 0.161$. The Cl^- mass balance is $\varepsilon_{rz}c_{rz} = \varepsilon_w c_w + \varepsilon_{st}c_{st}$. Hence, $c_{st} = (0.25 \times 0.7 - 0.089 \times 0.19)/0.161 = 0.98$ mmol/L.

2.11. The answer follows the 3rd item in Example 2.1: $\delta_{H_2O} = -80 + 91.3f$. $f \approx 0.2$ at 0°C (Figure 2.10), $\delta_{H_2O} \approx -61.7$‰.

CHAPTER 3

3.1. 3 m/yr.

3.2. The infiltration reach of 200 m is distributed proportionally over depth, and obtains a thickness of $200/2000 \times 50 = 5$ m. Mean depth is $1000/2000 \times 50 = 25$ m; travel time is for water infiltrated at 1000 m: $t = 34.7$ yr, and for water infiltrated at 900 m: $t = 40$ yr.

3.3. At 1 km is $v = 20$, $v_D = 6$ m/yr. At 2 km is $v = 40$, $v_D = 12$ m/yr. Travel time 34.7 yr. Age of water at 10, 20, 30, 40 m depth is 11.2, 25.5, 45.8, 80.5 yr. The formula $t = d\varepsilon/P$ gives 10, 20, 30 and 40 years.

3.4. $x = x_0 \exp[Pt/(2D\varepsilon)]$ and $d = D(1 - \exp[-Pt/(D\varepsilon)])$

3.5. $v = -dx/dt = P(r^2 - x^2)/(2xD\varepsilon)$ gives $\ln[(x^2 - r^2)/(x_0^2 - r^2)] = Pt/(D\varepsilon)$.

3.6. 4 yrs.

3.7a. The gradient increases parabolically with distance (x^2) in the upper, phreatic aquifer, and is invariant in the lower aquifer.

3.7b. Flow is downward.

3.7c. A diverging bundle of lines in the D/x plane, with decreasing slope deeper down.

3.8. $x_0 = 333$ m. Travel time above the clay layer is from 666 to 1666 m in the upper phreatic aquifer, $t = (2D\varepsilon_w/(3P)) \cdot (\ln(1666) - \ln(666)) = 0.6D\varepsilon_w/P$. The travel time below the clay layer is, from mass balance, $t = V/Q = (1000D\varepsilon/3)/(333P) = D\varepsilon/P$, or 1.67 times longer. However, the hydraulic gradient increases at the end of the clay layer, and more water attempts to flow through the lower aquifer. A numerical model gives a 1.16 times longer travel time through the lower clay layer.

3.9. $\rho_b = 2.12$ g/cm^3; 2 ppm $= 2 \times 2.12/0.2 = 21.2$ mg/L.

3.10. $\varepsilon = 0.26$.

3.11. c_2 displaces c_1: at $PV = 1$, c shocks from 0.025 to 0.075 where $dq/dc = 0$. c then shocks to the end concentration 0.175 after $\Delta q/\Delta c = (0.9 - 0.2)/(0.17 - 0.075) = 7.3\ PV$.
c_1 displaces c_2: c shocks from 0.17 to 0.05 after $\Delta q/\Delta c = (0.9 - 0.4)/(0.17 - 0.05) = 4.17\ PV$, then a wave to $c = 0.025$, which ends at $dq/dc = 8$.

3.12. $\sigma = 25$ and 79 cm.

3.13. Integrate from 0 to 100 cm:

$$\int_0^{100} c_{x,t}\ dx = \frac{10}{\sqrt{4\pi Dt}} \int_0^{100} \exp\left(-\frac{x^2}{4\ Dt}\right) dx$$

Substitute $x^2 = 4Dt \cdot s^2$, and simplify:

$$\int_0^{100} c_{x,t}\ dx = \frac{10}{\sqrt{\pi}} \int_0^{\frac{100}{\sqrt{4Dt}}} \exp(-s^2) ds = 5\ \mathrm{erf}\left(\frac{100}{\sqrt{4Dt}}\right)$$

We find $\mathrm{erf}(100/\sqrt{4Dt}) = \mathrm{erf}(0.89) = 0.792$. Hence $5 \times 0.792 = 0.396$ g NaCl resides in between 0 and 100 cm.

3.15. $D_L = D_f \cdot \varepsilon + \alpha v$ gives 5×10^{-3}, 5×10^{-4} and 5.3×10^{-4} cm^2/s.

3.16. $\alpha_L = 0.02\ x$ and $0.06\ x$.

3.17. The temperature factor is $283 \times 0.891/(298 \times 1.3) = 0.65$. Thus $D_{NaCl} = 0.65 \times (1.33 \times 10^{-9} + 2.03 \times 10^{-9})/2 = 1.09 \times 10^{-9}$ m^2/s. $D_{Ca(HCO_3)_2} = 0.65 \times (0.79 \times 10^{-9} + 1.18 \times 10^{-9})/2 = 0.64 \times 10^{-9}$ m^2/s.

CHAPTER 4

4.1a. A: $SI = \log IAP - \log K = -11.08 + 10.57 = -0.51$
B: $SI = \log IAP - \log K = -10.07 + 10.57 = 0.50$
C: $SI = \log IAP - \log K = -8.78 + 10.57 = 1.79$

4.1b.

	I	γ_F	γ_{Ca}	*IAP*	*SI*
A:	0.026	0.86	0.54	−11.48	−0.91
B:	0.057	0.81	0.44	−10.61	−0.04
C:	0.199	0.75	0.31	−9.54	1.03

4.2a. 25°C $\log K_{villiaumite} = -0.49$

4.2b. 10°C $\log K_{villiaumite} = -0.50$

4.2c. $\log SI_{villiaumite}$
A: −5.41
B: −4.45
C: −2.67

4.3a. $SI_{gypsum} = \log IAP - \log K = -6.17 + 4.60 = -1.57$. Log *IAP* decreases when $CaSO_4^0$ is included in the calculation.

4.3b. 2.56 mmol gypsum dissolves, $[Ca^{2+}] = 3.24$ mmol/L and $[F^-] = 0.091$ mmol/L.

4.4. Analysis A, but it requires that fluorite is present in the rock.

4.5. $O_{2(g)} \leftrightarrow O_{2(aq)}$ $\Delta G_r^0 = 16.5 - 0 = 16.5$ kJ/mol
$\Delta G_r^0 = -RT \ln K \Rightarrow K = [O_{2(aq)}]/[P_{O_2}] = \exp(-16.5/2.48) = 10^{-2.89}$

At 25°C:
$[P_{O_2}] = 10^{-0.68} \rightarrow O_{2(aq)} = 0.27$ mmol/L $= 8.6$ mg/L.

Van't Hoff-equation:
$\log K_T = \log K_{25} + (-10{,}000/19.1) \times (1/298 - 1/T)$.

At 15°C:
(288 K): $\log K = -2.83 \rightarrow O_{2(aq)} = 0.31$ mmol/L $= 9.9$ mg/L.

At 5°C:
(278 K): $\log K = -2.76 \rightarrow O_{2(aq)} = 0.36$ mmol/L $= 11.5$ mg/L.

4.6. For smithsonite ($\log K = -10.0$) and calcite:

--User print--
aCa = 7.3848×10^{-4} aZn = 2.2293×10^{-5} aCa/aZn = 3.3126×10^{1}
aCa = 7.5910×10^{-4} aZn = 2.3148×10^{-7} aCa/aZn = 3.2793×10^{3}
$m_{Ca}/m_{Cd} = 16$ and 1600, respectively. The $ZnCl^+$ complex is not very important.

4.8b. $d(FeS_2)/dt = k\,(FeS_{surf})^2\,(S_{surf})\,P_{H_2S}$
$k = 2\times10^{-20}\,cm^6$ atm/mol/L/sec

4.8c. The second, since the rate depends on the dissolved sulfide concentration.

4.9. First calculate moles of *OC* in $0.1 \times 1 \times 1\,m^3$ soil: $0.1\,m^3 \times 1400\,kg/m^3 \times 0.4$ kg *OC*/kg soil/(12 g/mol C) = 4.67 kmol.
At steady state, *OC* is constant: $R = 4.67 \times 0.025 = 0.117\,kmol/yr/m^2$. For the whole Netherlands: $0.117 \times 3 \times 10^{10} = 3\times10^9$ kmol/yr.
Human emission is 10×10^3 kg/yr/person $\times 16\times10^6$ people/(12 kg/kmol) $= 1.3\times10^{10}$ kmol/yr, or about 5 times higher. Note that at steady state the soil absorbs as much C as it emits, thus increasing the contribution of human emission to atmospheric CO_2 even more.

4.10a.
```
RATES; Organic_C
    -start
    10 mSO4 = tot("S(6)")
    20 rate = 1.e-13 * mSO4 / (1.e-4 + mSO4)
    30 moles = rate * parm(1) * m * time
    40 save moles
    -end
KINETICS 1; Organic_C
  -formula CH2O; -m0 2.68; -parms 8e4; -step 2.16e6 in 25 #25 days
SOLUTION 1; S(6) 20; Na 40 charge
```

4.10b. For SO_4^{2-} reduction, $dOC/dt = -k\,OC$, $k = 8\times10^{-9}$/s = 0.25/yr.

CHAPTER 5

Laboratory exercise:
```
DATABASE phreeqc.dat
SOLUTION 1; EQUILIBRIUM_PHASES; CO2(g) -2.0
```

```
KINETICS 1
  Calcite; -parms 110.3 0.67; -step 8640 in 10 # only 1/10 day
INCREMENTAL_REACTIONS
USER_GRAPH
  -axis_titles "time / s" "Saturation Ratio"
  -start; 10 graph_x total_time
  20 graph_y sr("Calcite"); end
END
```

5.1. $\gamma_1 = 0.90$; $\gamma_2 = 0.67$. At pH = 7 is $[HCO_3^-] = 1/0.9$ times lower, $10^{-2.75}$. Similarly, $[CO_3^{2-}] = 10^{-5.88}$ and $TIC = 10^{-2.68}$. At pH = 10, the ionic strength is above 19, for beyond the range of the Debye-Hückel theory.

5.2. pH = 6.81; $[CO_3^{2-}] = 10^{-6.38}$. pH = 9.67; $[CO_3^{2-}] = (1.7 - 1.3)/2 = 0.2$ mmol/L.

5.3. $m_{CO_3^{2-}} = \alpha\ TIC$; $\alpha^{-1} = 10^{16.6}\ \gamma_2[H^+]^2 + [H^+]\ \gamma_2/(\gamma_1\ 10^{-10.3}) + 1$.
At pH = 10.5, $\alpha^{-1} = 2.7\times10^{-5} + 0.47 + 1 = 1.47$; $m_{CO_3^{2-}} = 1.70\times10^{-3.0}$ mol/L, $[CO_3^{2-}] = 1.14\times10^{-3}$.
At pH = 6.3, $\alpha^{-1} = 14144$; $m_{CO_3^{2-}} = 1.77\times10^{-7}$; $[CO_3^{2-}] = 1.18\times10^{-7}$.

5.4. $[H_2CO_3] = 10^{-3.5}$; $[H^+] = 10^{-4.9} = [HCO_3^-]$; $[CO_3^{2-}] = 10^{-10.3}$.

5.5. $Ca^{2+} = 1.58$ mmol/L.

5.6. $P_{CO_2} = 0.02$ atm.

5.7. $P_{CO_2} = 10^{-2.55}$ atm.

5.8. $k_1 \cdot [H^+] \approx 0$; $k_2 \cdot [H_2CO_3^*] = 10^{-9.8}$; $k_3 \cdot [H_2O] = 10^{-7.0}$; $R_{fw} = 10^{-7.0}$.
$[Ca^{2+}]_{eq.} = 0.5\times10^{-3}$; $R_b = k_4 \times 2 \times [Ca^{2+}]^2 = R_{fw}$; $k_4 = 0.2$.
95% saturation, or $m_{Ca^{2+}} = 4.75\times10^{-4}$; $t = 9159$ s = 2.5 hr.

5.9a. $[P_{CO_2}]$ can be calculated from $\log [P_{CO_2}] = 7.8 + \log HCO_3^- - pH$, which gives:
(1) −0.9 (2) −2.2 (3) −2.6 (4) −1.6

5.9b. Saturation with respect to calcite is calculated from:
$H^+ + CaCO_3 \leftrightarrow Ca^{2+} + HCO_3^-$; $\log K = 1.8$

$\log (IAP/K_{cc})$	In contact with calcite?	Open/closed?
(1) −5.7	no	–
(2) 0.5	yes	Open
(3) 0.9	yes	Partly closed
(4) −1.5	partly	

5.9c. (1) acid rain; (2) –; (3) SO_4^{2-} high; (4) K^+, NO_3^- high.

5.10a. CO_2 in water is analyzed as H_2CO_3.
$H_2CO_3 \leftrightarrow H^+ + HCO_3^-$; $K = 10^{-6.3}$
$[H^+] = 10^{-6.3} \times 66 \times 10^{-3}/6.38\times10^{-3} = 10^{-5.3}$; pH = 5.3

5.10b. Alkalinity will not change if CO_2 decreases. (Alkalinity = $HCO_3^- + 2\,CO_3^{2-} + OH^- - H^+$). CO_2-concentration will reach equilibrium with the atmosphere: $H_2CO_3 = 10^{-5}$ mol/L. Hence:

$$[H^+] = 10^{-6.3} \times 10^{-5}/6.38\times10^{-3} = 10^{-9.1}; \quad pH = 9.1$$

5.10c. $\Sigma+ = 4.45$ meq/L $\Sigma- = -7.78$ meq/L; Probably Fe precipitated in the non-acidified sample.

5.10d. To maintain charge balance an equivalent amount (in meq/L) of HCO_3^- must be removed from the solution:

$$Fe^{2+} + 2HCO_3^- + ¼O_2 + ½H_2O \leftrightarrow Fe(OH)_3 + 2CO_2$$

5.10e. Juvenile = first emergence at the Earth's surface.

5.11.1. Constant; if H_2CO_3 dissociates ($H_2CO_3 \leftrightarrow H^+ + HCO_3^-$), H^+ and HCO_3^- compensate each other's contribution to alkalinity.

5.11.2. Decrease; HCO_3^- reacts with added H^+: $HCO_3^- + H^+ \leftrightarrow H_2CO_3$

5.11.3. Decrease; Fe^{3+} will take up OH^-.

5.11.4. Constant.

5.11.5. Decrease; Mn^{2+} is oxidised by O_2, and releases H^+: $Mn^{2+} + H_2O + ½O_2 \leftrightarrow MnO_2 + 2H^+$.

5.11.6. Increase.

5.11.7. Increase.

5.11.8. Increase.

5.11.9. Constant.

5.12. 60 mg/L SiO_2 = 1 mmol/L ΣSi. At pH = 10.0 :
$[H_3SiO_4^-]/[H_4SiO_4^0] = (10^{-9.1})/(10^{-10.0}) = 10^{0.9} = 7.94$
$[H_3SiO_4^-] = 7.94/8.94 = 0.89$ mmol/L

When Alkalinity is titrated, $H_3SiO_4^-$ associates with H^+, to become $H_4SiO_4^0$. The contribution to the alkalinity is equivalent to the $H_3SiO_4^-$-concentration.

5.13. With Equation 5.24, we obtain Ca = 1.36 mol/m^3, or 0.41 mol/m^2/yr leaves the area. This is 41 g $CaCO_3$, or 20 cm^3/m^2/yr. Denudation rate is therefore 20 mm/1000 yr.

5.14. PHREEQC calculates Ca = 1.4 mM, which gives a flux of 0.14 mol/m^2/yr or 14 g $CaCO_3$/m^2/yr. 1 m^3 sand $\times \rho_b/\varepsilon$ = 1800 kg = 360 kg calcite. Decalcifation rate is (1 m^3/(360 $\times$ 10^3 g/14 g/m^2/yr) $\times$ 1000 mm/m) = 0.04 mm/yr.

5.15, 16. Download input files from www.xs4all.nl/~appt

5.17. See PHREEQC manual, p. 43.

5.18.
```
PRINT; -reset false; -user_pr true
SOLUTION 1; -temp 10
EQUILIBRIUM_PHASES 1; CO2(g) -1.5
RATES; Dolomite
  -start; 10 save -1.2e-10 * log( SR("Dolomite") + 1e-5) * time
  20 print SR("Dolomite"), total_time; -end
KINETICS 1; Dolomite; -step 1e8 in 100; INCREMENTAL_REACTIONS
END
```

For $[P_{CO_2}] = 10^{-3.5}$, $t = 2.6\times10^6$ s. For $[P_{CO_2}] = 10^{-1.5}$, $t = 1.4\times10^7$ s.

5.19. Download input file from www.xs4all.nl/~appt

5.19a. 85 s at 10°C, 15 s at 25°C.

5.19b. $10^{-1.54}$

5.20. 53, 46 and 15pmc, respectively.

5.21. Download input files from www.xs4all.nl/~appt

5.22.
```
DATABASE phreeqc.dat
 SOLUTION 1; KINETICS 1; Calcite; -parms 110.3 0.67; -step 864 in 10;
 INCREMENTAL_REACTIONS; PRINT; -reset false; -user_print true
 USER_PRINT; -start; 10 print total_time, "SR_cc", sr("Calcite"); end
 END # (about 5 min are sufficient).
```

CHAPTER 6

6.1. gfw = 389.91 g/mol. One packet of 20 units (with 24 oxygens) has 2×0.7 charge over 2 basal planes, i.e. $CEC = 2 \times 0.7 \times 1000 / 389.91 / 20 \times 1000 = 180$ meq/kg. Charge is $0.015\ q_e/\text{Å}^2 = 0.24\ \text{C/m}^2$.

6.2. gfw = 368.22 g/mol. $CEC = 217$ meq/kg. Charge = $0.28\ \text{C/m}^2$.

6.3. 10 times concentrated water: $\beta_{Na} = 0.0259$; $\beta_{Mg} = 0.134$; $\beta_{Ca} = 0.840$.
seawater: $\beta_{Na} = 0.584$; $\beta_{Mg} = 0.319$; $\beta_{Ca} = 0.097$.

6.4. $\gamma_1 = 0.85$; $\gamma_2 = 0.53$; $K_{Na\backslash Cd} = 0.286$ and 0.339.

6.5a. Sample 1 is rainwater (with Na/Cl-ratio of seawater) in which some calcite has dissolved. Sulfate is partly reduced. Sample 2 is slightly mixed with salt water (higher Cl^-). Cation-exchange is evident, Ca^{2+} is replaced by Na^+, causing undersaturation with respect to calcite. Renewed dissolution of calcite gives high alkalinity. Sample 3 has a low Na/Cl-ratio, sulfate partly reduced. Sample 4 is Seawater.

6.5b. A: $NaHCO_3$-water, interface moves downward. B: $CaCl_2$-water, interface moves upward.

6.5c.

	β_{Na}	β_K	β_{Mg}	β_{Ca}
Sample 1	0.0062	0.0014	0.0187	0.974
Sample 4	0.558	0.061	0.292	0.089

6.5d.

	Na-X	K-X	$Mg\text{-}X_2$	$Ca\text{-}X_2$	Σ
Sample 1	0.4	0.09	1.1	58.4	60.0 meq/L
Sample 4	33.5	3.7	17.5	5.3	60.0 meq/L

6.6a. Evaporation concentrates Nile-water in the first place. Using a "conservative" species, Cl^-, we can calculate the composition:

Concentration-factor $Cl_{groundwater} / Cl_{Nile} = 4.5 / .5 = 9.0$

	Na^+	K^+	Mg^{2+}	Ca^{2+}	HCO_3^-	Cl^-	SO_4^{2-}	pH
Nile-water	.5	.1	.4	.7	2.2	.5	.1	
9× concentrated	4.5	.9	3.6	6.3	19.8	4.5	.9	
groundwater	8	.1	.2	.7	2.7	4.5	1.0	8.1
Difference	3.5	−.8	−3.4	−5.9	−17.1	–	0.1	

The decrease of Ca^{2+} and HCO_3^- (and of some Mg) is an effect of calcite-precipitation. If we compare K and IAP of the reaction:

$$CaCO_3 + H^+ \leftrightarrow Ca^2 + HCO_3^-; \quad K = 63; \quad IAP = 136,$$

the water is clearly oversaturated. The increase of Na^+ is produced by cation exchange of Ca^{2+} and Mg^{2+} for Na^+ from the soil. The decrease of K^+ is an effect of uptake by vegetation or illite.

6.6b. The very high Cl-concentration in water from borehole 121 can only be explained by salt water intrusion. If we mix groundwater and seawater at the correct ratio, (contribution of seawater 0.473), we obtain:

	Na^+	K^+	Mg^{2+}	Ca^{2+}	HCO_3^-	Cl^-	SO_4^{2-}
Calculated	231	5	26	5.4	2.5	270	14
Analysed	180	2.5	35	15	7.5	270	1.5
Difference	−51	−2.5	9	10	5	–	−12

The decrease of Na^+ is an effect of cation exchange. Salt water intrudes in a fresh aquifer, Ca^{2+} will be removed from the exchange complex and replaced by Na^+. The decrease of SO_4^{2-} is an effect of reduction.

6.6c. Calculation of the composition in mg/L:

Na^+	K^+	Mg^{2+}	Ca^{2+}	HCO_3^-	Cl^-	SO_4^{2-}
4140	98	850	601	458	9571	144

The contribution of HCO_3^- to $TDS = 0.5 \times HCO_3^-$ (mg/L), since half of the HCO_3^- originates from $CO_{2(g)}$ (and would escape when the sample is evaporated to dryness). Hence: $TDS = 15{,}633$ mg/L.

6.7a. Na-X = 160.6; Mg-X_2 = 283.18; Ca-X_2 = 306.22 meq/L pore water.

6.7b. Total cations = 14.66 meq/L. Na^+ = 13.28; Mg^{2+} = 0.43; Ca^{2+} = 0.26 mmol/L.

6.7c. Exchanger composition in equilibrium with injected water is: Na-X = 56.35; Mg-X_2 = 81.1; Ca-X_2 = 612.54 meq/L pore water.

6.7d. Assume that Na-X is flushed from 160.6 to 56.35 meq/L, which needs (160.6 − 56.35)/(13.28 − 9.4) = 26.9 pore volumes = 6985 m^3 water.

6.8a. Na-X = 2.2, Mg-X_2 = 16.1, Ca-X_2 = 41.7 meq/L.

6.8b. Na^+ = 28.8, Mg^{2+} = 50.2, Ca^{2+} = 52.8 meq/L.

6.8c. The final composition of the exchanger is Na-X = 25.8, Mg-X_2 = 22.7, Ca-X_2 = 11.5 meq/L. Ca-X_2 requires (41.7 − 11.5)/(105.6 − 8.3) = 0.31 pore volumes. Note that the concentration of Ca^{2+} only reaches up to 17 mM, because dispersion smoothens the peak.

6.9.

$$K^V_{Na/I^{2+}} = \frac{2\,K_{Na/I^{2+}}}{\sqrt{1+\beta_{Na}}}$$

$$K^G_{Na/I^{2+}} = \frac{K_{Na/I^{2+}}}{\sqrt{1-\beta_{Na}}}$$

6.11. Smectite with 0.09 C m^{-2}.

x:	0	10	50	100 Å
ψ_0:	−0.393	−0.284	−0.078	−0.015 V
n_+/n_∞	4.3×10^6	6.3×10^4	20.6	1.8
m_+	43,000	63	0.206	0.018 mol/L
(Note that the concentrations at 0 and 10 Å are physically impossible)				
n_-/n_∞	2.3×10^{-7}	1.6×10^{-5}	0.05	0.55

6.12. *SAR* of irrigation water is 1.7; when 10 times concentrated, *SAR* changes to 5.5.

6.13. For illite (using activity of Ca^{2+} and K^+ in solution obtained with $[i] = \gamma_i m_i$):

	0.1	0.01	0.001	$10^{-4}N$	
$K^V_{Ca\backslash K}$	0.90	0.76	0.53	0.25	(Gapon)
$K^G_{Ca\backslash K}$	0.62	0.43	0.28	0.13	(Vanselow)
$K_{C\backslash K}$	1.09	0.81	0.55	0.25	(Gaines and Thomas)

and similarly for the resin:

	0.1	0.01	0.001	$10^{-4}N$	
$K^{V}_{\mathrm{Ca\backslash K}}$	2.68	2.41	2.37	1.80	(Gapon)
$K^{G}_{\mathrm{Ca\backslash K}}$	1.53	1.26	1.20	0.90	(Vanselow)
$K_{\mathrm{C\backslash K}}$	2.87	2.50	2.38	1.81	(Gaines and Thomas)

6.14. $CEC = (\beta_I^M \cdot i + \beta_J^M \cdot j + \beta_K^M \cdot k + \ldots) \cdot TEC$
$TEC = (\beta_I / i + \beta_J / j + \beta_K / k + \ldots) \cdot CEC$ and
$\beta_I \cdot CEC = i \cdot \beta_I^M \cdot TEC$ yield:

$$\beta_I = i \cdot \beta_I^M \cdot \frac{TEC}{CEC} = \frac{\beta_I^M \cdot i}{\beta_I^M \cdot i + \beta_J^M \cdot j + \beta_K^M \cdot k + \cdots}$$

$$\beta_I^M = \beta_I^M / i \cdot \frac{CEC}{TEC} = \frac{\beta_I / i}{\beta_I / i + \beta_j / j + \beta_K / k + \cdots}$$

6.15.
```
EXCHANGE_SPECIES; Ca+2 + 2X- = CaX2; log_k -10
                  Mg+2 + 2X- = MgX2; log_k -10 # remove database species
  0.5 Mg+2 + X- = Mg0.5X; log_k 0.3
  0.5 Ca+2 + X- = Ca0.5X; log_k 0.4
#or
  Ca+2 + Mg0.5X = Ca0.5X + Mg+2; log_k 0.1; -no_check; -mole_balance Ca0.5X
SOLUTION 1; Mg 1; EXCHANGE 1; X 2e-3; -equil 1
REACTION 1; Mg -1 Ca 1; 1.999e-3 in 20
USER_GRAPH; -start; 10 graph_x tot("Ca"); 20 graph_y mol("CaX2") + mol("Ca0.5X")/2
  -end
END
```
(compare the isotherms by crossing out the species)

6.16. Use:
```
EXCHANGE_SPECIES; K+ + X- = KX; log_k 0.43; -gamma 3.5 0.015
```

6.17. $m_{\mathrm{Na}} = 1\ \mu\mathrm{M}$ gives $q_{\mathrm{Na}} = 0.55\ \mu\mathrm{M}$, $R = 1.6$. $m_{\mathrm{Na}} = 10\,\mathrm{mM}$ gives $q_{\mathrm{Na}} = 5.2\,\mathrm{mM}$, $R = 1.5$.

CHAPTER 7

7.1. $\alpha / \ln 10 = 95$. $\kappa \varepsilon A_s = F^2 / (95RT \ln 10) = 1.7 \times 10^4\,\mathrm{F/kg}$. Apparently (cf. Example 7.5), the capacitance is not really constant at low ionic strength.

7.2. The ratio is 0.04, almost all sites are protonated, in agreement with Figure 7.21.

7.3. $\mathrm{p}K_{\mathrm{int}} = 2.8$; $\alpha / \ln 10 = 3 \cdot 1$; $K_{\mathrm{h\text{-}Al}} = 3 \cdot 10^{-3}$ give:

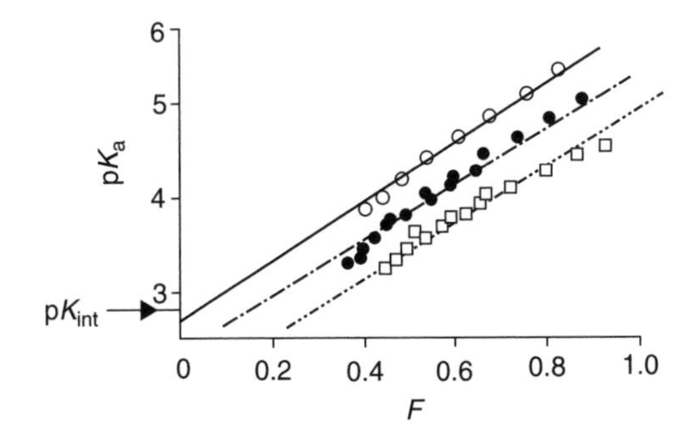

7.4.
```
SOLUTION 1; K 1 charge; Br 1; pH 5 # change Br concentration as necessary
SURFACE 1
  Hfo_w 8.5e-3 600 3.8    # fitted sites...
  -equil 1; -diff true
PRINT; -reset false; -user_print true
USER_PRINT
  -start
  10 print "mM Br = ", 1e3*tot("Br"), "R 5", 1 + edl("Br")/tot("Br")
  -end
END
```
For a fixed pH, protons sorb more at high ionic strength because ψ decreases. Hence higher charge on surface, and more Br^- needed in double layer for charge compensation. Charge increases approximately with $\sqrt{I}$. Hence, the distribution coefficient ($K_d = R - 1$) decreases by 3 for each tenfold increase of KBr concentration.

7.5.
```
SOLUTION 0; K 1 charge; Br 1; pH 5
SOLUTION 1-5; K 4 charge; Cl 4; pH 5 # fitted initial KCl concentration...
END
SURFACE 1-5; Hfo_w 8.5e-3 600 3.8; -equil 1; -diff true
PRINT; -reset false; -warnings false
TRANSPORT; -cells 5; -length 0.02; -shifts 25 1; -disp 0.002; -punch 5
USER_GRAPH; -head PV Br; -axis_scale x_axis 0 5 1; -plot time
  -start; 10 graph_x (step_no + 0.5)/cell_no; 20 graph_y tot("Br")/1e-3; -end
END
```

7.6.
```
SURFACE 1; Hfo_w 5.6e-4 60 2.5; -equil 1 # 10% active sites
SOLUTION 1; pe 14 O2(g) -0.68; Na 100; Cl 100
REACTION 1; H4AsO4 1; 0.267e-3
PHASES; Fix_ph; H+ = H+; log_k 0; EQUILIBRIUM_PHASES 1; Fix_ph -7.0 NaOH
USER_PRINT; -start;
  10 qAs5 = mol("Hfo_wH2AsO4") + mol("Hfo_wHAsO4-") + mol("Hfo_wOHAsO4-3")
  20 qAs3 = mol("Hfo_wH2AsO3"); 30 print '% As5 sorbed:', qAs5/ (tot("As(5)") + qAs5)
  40 print '% As3 sorbed:', qAs3 / (tot("As(3)") + qAs3)
  50 print 'mmol As5/kg sorbed', qAs5 / 2.5e-6, 'mmol As3/kg sorbed', qAs3 / 2.5e-6
  -end
END
SURFACE 1; Hfo_w 5.6e-4 60 2.5; -equil 1 # 10% active sites
SOLUTION 1; Na 100; Cl 100
REACTION 1; H3AsO3 1; 0.267e-3
EQUILIBRIUM_PHASES 1; Fix_ph -7.0 HCl
END
```
At high concentration of H_4AsO_4, the surface is rendered negative by Hfo_wHAsO4-, which diminishes sorption of the anion. At low concentration this effect is negligible; the higher chemical affinity of As(5) for ferrihydrite then gives higher sorption of As(5) than of As(3).

7.7. The half-life increases by 100 (becomes 3 days).

7.9. Approximately 4.5 hr.

7.10. Hfo_sOPb+ $= 4.473\times10^4$; Hfo_wOPb $= 3.132\times10^{-3}$.

7.11. cf. Figure 7.35.

7.14. $q_{Cd} = 5.7\times10^4\, c_{Cd}/(30.9 + c_{Cd})$, both q and c in μg/L pore water. For very small c_{Cd}, $K_d = q_{tot}/c_{Cd} = 5.7\times10^4/30.9 = 1844$. The distribution is linear if $c_{Cd} < 3.09$ μg/L. 75% of q_{tot} is reached when $c_l/(K_L + c) = 0.75$, or $c_{Cd} = 93$ μg/L.

7.15.
```
SOLUTION 1; Ca 2; Zn 2e-4
EXCHANGE 1; X 73.5e-3; -equil 1
USER_PRINT; -start; 10 print 'Kd_Zn = ', mol("ZnX2") / tot("Zn"); -end
END
```
gives K_d's: **a)** 18.1; **b)** 18.1; **c)** 9; **d)** 36.2; **e)** 18.3

7.16.
```
SOLUTION 1; Mg 2; Zn 2e-4; pH 7
SURFACE 1; Hfo_w 2e-4 600 0.088; Hfo_s 5e-6; -equil 1
USER_PRINT; -start
  10 print 'Kd_Hfo', (mol("Hfo_wOZn+") + mol("Hfo_sOZn+"))/tot("Zn"); -end
END
```
K_d's: **a)** 12.6; **b)** 8.8 (strong sites are full already); **c)** 11.8; **d)** 25.2 (note to double the grams of Hfo and weak and strong sites); **e)** 0.005

7.17. Mg^{2+} is sorbed on the weak sites, which increases the potential, rejects Zn^{2+}. Increase of the ionic strength increases the capacitance, decreases the surface potential. The chemical affinity becomes less important, H^+ displaces Zn^{2+} from the strong sites. Ca^{2+} is sorbed also on the strong sites and doubling the concentration has a much stronger effect than Mg^{2+}.

7.18.
```
SOLUTION 1; Pb 2; pH 3; Na 100; N(5) 100 charge
SURFACE 1; Hfo_w 20e-3 600 8.8; Hfo_s 5e-4; -equil 1
REACTION 1; NaOH 1; 28e-3 in 20
USER_GRAPH; -start; 10 graph_x -la("H+"); 20 graph_y (2e-3 - tot("Pb"))/2e-3;-end
END # and similar for Zn, Ca
```

7.19.
```
SOLUTION 1; Ca 0.33; Ni 1e-5; pH 3; N(5) 0.33 charge
EXCHANGE 1; X 8.7e-3; -equil 1
EXCHANGE_SPECIES; Ni+2 + 2X- = NiX2; log_k 0.6; H+ + X- = HX; log_k 4.5
REACTION 1; NaOH 1; 10e-3 in 20
SURFACE 1; Hfo_w 2e-5 600 8.8; Hfo_s 5e-7; -equil 1
SURFACE_SPECIES; Hfo_sOH + Ni+2 = Hfo_sONi+ + H+; log_k 0.37
USER_GRAPH; -start; 10 graph_x -la("H+1");
20 q_Ni = mol("NiX2") + mol("Hfo_sONi+") + mol("Hfo_wONi+")
30 graph_y q_Ni / (q_Ni + tot("Ni")); -end
END
```
The model sorption edge is steeper than in Figure 7.18, 10% sorption at low pH cannot be modeled, but see Bradbury and Baeyens (1999) who cancel the electrostatic contribution.

7.20.
```
SOLUTION 1; Ca 1; Cd 2e-4; Cl 2
EXCHANGE 1; X 55.7e-3; -equil 1
EXCHANGE_SPECIES; Cd+2 + 2X- = CdX2; -log_k 0.8; -gamma 0 0
USER_PRINT; -start; 10 print 'X... Kd_Cd = ', mol("CdX2") / tot("Cd"); -end
END
```
gives $K_d' = 24.1$ L/kg

7.21.
```
SOLUTION 1; -temp 10; pH 7 charge; Ca 1 Calcite; C(4) 1 CO2(g) -2.5; Mg 1 Dolomite
SURFACE 1; Hfo_w 8.4e-3 600 3.84; -equil 1
END
```
yields for carbonate $R = 1 + 3.6/3.9$ (mM/mM) $= 1.9$.

7.22. Download the files from www.xs4all.nl/~appt

7.23
```
SOLUTION 1; pH 7.67; Na 2.93; Ca 0.59; Alkalinity 3.85; F 0.26; As 1.3e-3
EQUILIBRIUM_PHASES 1; Goethite 3.0 0; Fluorite 0 0; Calcite
SURFACE 1; Hfo_w Goethite 0.2 5400; Hfo_s Goethite 2.5e-3
REACTION; FeSO4 1e-3 NaClO4 0.167e-3; 10 in 10
USER_GRAPH; -head step As F
-axis_titles Step "ug As/L or mg F/L"
-axis_scale y-axis 0 5
  -start; 10 graph_x step_no; 20 graph_y tot("As")*74.92e6, tot("F")*19e3; -end;
END
```
Already step 1 yields As $<$ 10 mg/L, and in step 3, $F^- = 2.5$ mg/L. Further reaction doesn't improve much. To arrive at step 3, the pill should contain: 456 mg $FeSO_4$, 61 mg $NaClO_4$ nd 39 mg Calcite.

CHAPTER 8

8.1. K-feldspar, quartz and gibbsite.

8.2. The base production by gibbsite dissolution ($Al(OH)_3$) is:

$$[OH^-] = 3[Al^{3+}] + 2[AlOH^2] + [Al(OH)_2^+]$$
$$[OH^-] = 3(6\times10^{-5}) + 2(7.7\times10^{-6}) + 6\times10^{-7} = 1.97\times10^{-4}$$
$$[H^+]_{initial} = 10^{-4.53} + 1.97\times10^{-4} = 2.26\times10^{-4} \text{ and } pH_{initial} = 3.65$$

8.3. Combining $K_{gibbsite}$ and $K_{jurbanite}$ gives:

$$2\log[OH^-] - \log[SO_4^{2-}] = -33.88 + 17.8 = -16.08$$

Combining with the dissociation of water yields:

$$2pH - \log[SO_4^{2-}] = 11.92$$

8.4a. The water chemistry can be explained by:
dissolving 0.016 mmol/L NaCl; 0.015 mmol/L gypsum; 0.014 mmol/L biotite; 0.175 mmol/L plagioclase; 0.115 mmol/L calcite and precipitating 0.033 mmol/L kaolinite; 0.081 mmol/L Ca-smectite

8.4b. A greater weathering of plagioclase than of biotite or calcite is inconsistent with the usual weathering sequence. However, Feth et al. (1964) stated rather explicitly that calcite is absent in the Sierra Nevada source rocks, and calcite (or gypsum) should in fact not be included in weathering balances for the area. Perhaps dry deposition is important.

8.5. $Na_{sea} = 0.206$ mM, $Na_{albite} = 0.137$ mM.
PHREEQC finds 2 models:
Albite 1.443×10^{-4}; Biotite 4.212×10^{-6}; K-feldspar 6.293×10^{-6}; Anorthite 5.546×10^{-5}; Gibbsite 1.657×10^{-5}; Kaolinite -1.411×10^{-4}; $CO_2(g)$ 1.381×10^{-3}
and
Albite 1.360×10^{-4}; Biotite 4.601×10^{-6}; K-feldspar 5.904×10^{-6}; Anorthite 5.569×10^{-5}; Kaolinite -1.289×10^{-4}; CO2(g) 1.381×10^{-3}
The expected composition of the plagioclase is $(Na_{0.71}Ca_{0.29})(Al_{1.29}Si_{3.7})O_8$.

CHAPTER 9

9.1. Find $H_2AsO_4^- = 0.58 \times As(5) = 0.68$ μM with the speciation factor. Using Equation (9.32), pe = 0.83, *Eh* = 0.05V.

9.2a. For example Fe_2O_3/MnO_2, $Fe_2O_3/MnCO_3$, $FeCO_3/MnCO_3$, $FeS_2/MnCO_3$.
For dissolved species Fe^{2+}/Mn^{2+}, Fe^{3+}/Mn^{2+}.

9.2b. For hematite, $Fe_2O_3 + 2e^- + 6H^+ \leftrightarrow 2Fe^{2+} + 3H_2O$, and $pe = 11.3 - \log[Fe^{2+}] - 3pH$

9.2c. A gradual decrease in pe or pH or both simultaneously.

9.2d. pH = 7.16; $[NO_3^-] = 110$ mg/L $= 10^{-2.75}$ M; $[NO_2^-] = 2.1$ mg/L $= 10^{-4.35}$ M.
From $NO_3^- + 2H^+ + 2e^- \leftrightarrow NO_2^- + H_2O$, $\log K = 28.3$
$\log[NO_2^-] - \log[NO_3^-] + 2pH + 2pe = 28.3 \rightarrow$ $pe = 7.79$
From $Mn^{2+} + 2H_2O \leftrightarrow MnO_2 + 4H^+ + 2e^-$, $\log K = -41.52$
$-4pH - 2pe - \log[Mn^{2+}] = -43.67$, $\log[Mn^{2+}] = -2.7$

9.3. *The PbO_2/PbO boundary*: Combining reactions 1) and 3) yields:

$PbO_2 + 2H^+ + 2e^- \leftrightarrow H_2O + PbO$ and
$\log K = 2\,pH + 2\,pe = \log K_1 - \log K_3 = 49.2 - 12.7 = 36.5$

The PbO/Pb boundary: Combining reactions 2) and 3) yields:

$PbO + 2H^+ + 2e^- \leftrightarrow Pb + H_2O$ and
$\log K = 2\,pH + 2\,pe = \log K_3 + \log K_2 = 12.7 - 4.26 = 8.44$

The PbO_2/Pb^{2+} boundary: For reaction 1) and $\log[Pb^{2+}] = -6$:

$\log K_1 = \log[Pb^{2+}] + 4\,pH + 2\,pe = 49.2$ and $4\,pH + 2\,pe = 49.2 + 6 = 55.2$

The PbO/Pb^{2+} boundary: For reaction 3) and $\log[Pb^{2+}] = -6$:

$\log K_3 = \log[Pb^{2+}] + 2\,pH = 12.7$ and $pH = ½(12.7 + 6) = 9.35$

The $Pb(OH)_3^-/Pb^{2+}$ boundary: For reaction 4):

$\log K_4 = -3\,pH + \log[Pb(OH)_3^-] - \log[Pb^{2+}] = -28.1$

If $\log[Pb(OH)_3^-] = \log[Pb^{2+}]$ then pH = 9.37. Accordingly, $Pb(OH)_3^-$ never becomes a dominant dissolved species and is therefore not included in the diagram.
The Pb/Pb^{2+} boundary: For reaction 2) and $\log[Pb^{2+}] = -6$:

$\log K = -\log[Pb^{2+}] + 2\,pe = -4.26$, and $pe = ½(-4.26 - 6) = -5.13$

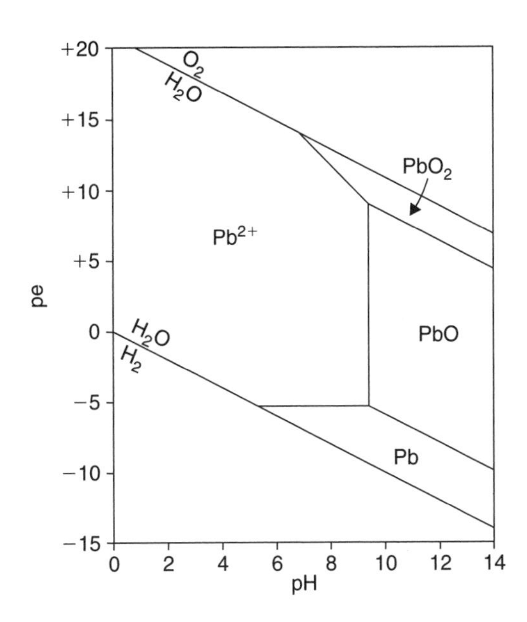

9.4a. Combination of several equations gives: $MnCO_3 + CO_2 + H_2O \leftrightarrow Mn^{2+} + 2HCO_3^-$; $K = 10^{-6.9}$
In water in contact only with $MnCO_3$: $[HCO_3^-] = 2[Mn^{2+}] \Rightarrow 4[Mn^{2+}]^3/[P_{CO_2}] = 10^{-6.9}$.
$[P_{CO_2}] = 10^{-1.5} \rightarrow Mn^{2+} = 1$ mmol/L (i.e. 1 mmol/L $MnCO_3$ will dissolve)

9.4b. Congruent dissolution: the mineral dissolves completely and no secondary phases are produced. Incongruent dissolution: the stoichiometry of dissolved substances does not correspond to their ratio in the dissolving phase, *i.e.* a secondary precipitate is formed.

9.4c. Rhodochrosite will dissolve according to 9.4a. Mn^{2+} oxidizes in the presence of O_2:

$$Mn^2 + H_2O + ½O_2 \leftrightarrow 2H^+ + MnO_2 \quad \text{(pyrolusite)}$$

meaning that rhodochrosite dissolves incongruently. The produced H^+ reacts with HCO_3^- and increases $[P_{CO_2}]$

9.5.

$$\begin{array}{lcl} KMnO_4 + 8H^+ + 5e^- & \leftrightarrow & Mn^{2+} + K^+ + 4H_2O \\ (CH_2O + H_2O & \leftrightarrow & CO_2 + 4H^+ + 4e^-) \cdot 5/4 \\ \hline KMnO_4 + 3H^+ + 5/4\,CH_2O & \leftrightarrow & Mn^{2+} + K^+ + 5/4\,CO_2 + 11/4\,H_2O \end{array}$$

9.6. Possible reactions are:

1. $H_2S_{(aq)} \leftrightarrow HS^- + H^+$; $K = 10^{-7}$
2. $H_2S_{(aq)} + 4H_2O \leftrightarrow SO_4^{2-} + 10H^+ + 8e^-$; $K = 10^{-40.6}$
3. $S_{(s)} + 4H_2O \leftrightarrow SO_4^{2-} + 8H^+ + 6e^-$; $K = 10^{-36.2}$
4. $(= 3 - 2)\ H_2S_{(aq)} \leftrightarrow S_{(s)} + 2H^+ + 2e^-$; $K = 10^{-4.4}$
5. $(= 4 - 1)\ HS^- \leftrightarrow S_{(s)} + H^+ + 2e^-$; $K = 10^{2.6}$
6. $(= 2 - 1)\ HS^- + 4H_2O \leftrightarrow SO_4^{2-} + 9H^+ + 8e^-$; $K = 10^{-33.6}$

Precipitation of S is possible when the pH < 6 a 7, and H_2S-concentration $> \pm 5$ mmol/L (at lower levels the stability-field of S will disappear).

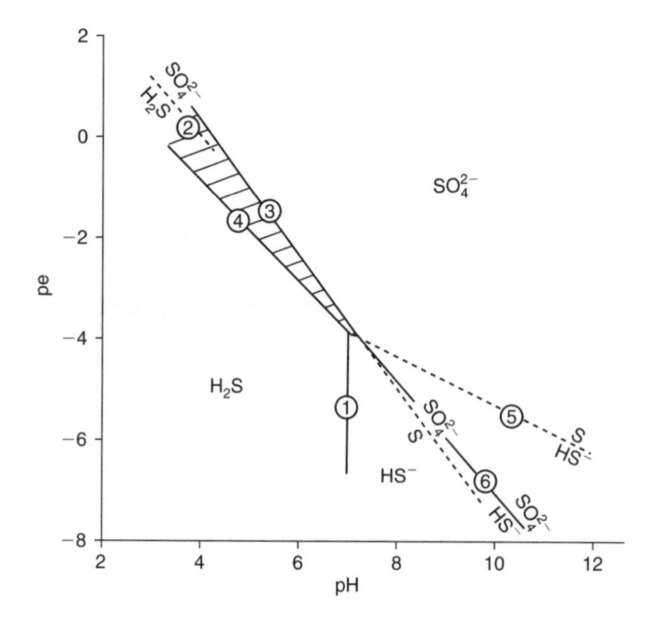

9.7b. Determine the fraction (fHfo_wOH$_2^+$) with PHREEQC for the experimental pH's (assuming that the surface properties of goethite and ferrihydrite are the same):

```
SOLUTION 1; pH 1.5; pe 10 O2(g) -0.68; Fe 0.1 Goethite; Cl 1 charge
SURFACE 1; Hfo_w 1e-3 600 444e-3; -equil 1; END
```

Then regress, and find that the *rate* $= 10^{-11.7}\,(f\mathrm{Hfo\_wOH_2^+})^{1.9}$ mol/m^2/s. (Note that the squared sum of errors is higher than when regressing on H^+, Equation (9.82)).

9.8. For 2 electrons in the reaction $H_2 \rightarrow 2H^+ + 2e^-$, K increases by $10^{-(2 \times -3/5.708)} = 11$, or H_2 must be 11 times higher than at equilibrium.

9.9. Use Equation (4.29) and find $K_{10}/K_{25} = 162$. Hence H_2 decreases by $162^{1/4} = 3.6$

9.10. The reaction: $4H_{2(g)} + 2HCO_3^- + H^+ \leftrightarrow CH_3COO^- + 4H_2O$; $\log K = 25.28$

add $8H^+ + 8e^- \leftrightarrow 4H_{2(g)}$; $\log K = 0$

Write in the input file: SOLUTION_SPECIES; 8e− + 2HCO3− + 9H+ = CH3COO− + 4H2O; log_k 25.28. (Note to write redox equations in PHREEQC with electrons). Plot on the y-axis mol("CH3COO−")*1e8. (Note that acetate is thermodynamically quite unstable).

CHAPTER 10

10.1. With USER_PRINT, print 10^SI("$H_2O(g)$")/(0.082 * TK)/(1000/18), which gives 1.73×10^{-5} and 9.44×10^{-6} L water/L air for 20 and 10°C

10.2. From Equation (3.61) find $DH_2{}^{16}O/DH_2{}^{18}O = 1.032$. Hence $\delta^{18}O_{vapor} = -32‰$ for kinetic fractionation by diffusion. (Compare with equilibrium fractionation, $\delta^{18}O_{vapor} \approx -10‰$).

10.3. Download input file from www.xs4all.nl/~appt

10.4. Download input file from www.xs4all.nl/~appt; the half-life is about 105 s.

10.5. Download input file from www.xs4all.nl/~appt

10.6. Naphthalene in peat: $\log K_d' = 3.0 + \log 0.7 = 2.85$; $K_d' = 700$ L/kg; bulk density of peat is $\rho_b = 1.0 \times 0.4 = 0.4$ kg/L. Hence $K_d = 700 \times 0.4/0.6 = 467$; velocity is 468 times smaller. Tetracene in sand: $\rho_b = 1.855$ kg/L; $R = 1 + \rho_b/\varepsilon \times 631 = 3904$.

10.7a. $x = 550 \exp(0.3\,t/50/0.36)$:

t (years)	x (m)	Flowlength ($= x - x_0$)	Depth (m)	Thickness (m)
25	834	284	17.0	3.0
100	2912	2362	40.6	0.9

Note that flowpath increases 9-fold, when time increases 4-fold.

10.7b. Assume $CEC = [\text{Ca-X}_2] = 10$ meq/kg. 1 kg = 556 mL = 200 mL H_2O, hence $CEC = 50$ meq/L = 25 mmol Ca^{2+}/L
$K_d(\text{Cd}) = [\text{Cd-X}_2]/[Cd^{2+}] = [\text{Ca-X}_2]/[Ca^{2+}] = 2.5$; $R(\text{Cd}) = 3.5$.
For TCE, the slope dq/dc decreases with c, hence we have a sharp front. $R_{TCE} = 1 + \Delta q/\Delta c = 1 + (16.2 - 0)/(7.1 - 0) = 3.3$

10.7c. $v_{H_2O} = R\,v_i = R\,dx_i/dt = P\,x/(D\varepsilon)$; $x = 550 \exp(0.3\,t/(R \times 50 \times 0.36))$;

t (Years)	x (M)	Flowlength ($= x - x_0$)	Depth (m)	Thickness (m)
25	620	70	5.6	4.0
100	885	335	18.9	2.8

10.7d. Flowlength after 100 yr = 335 m; $\alpha = 10\%$ of 335 m = 33 m. $\sigma^2 = 2Dt = 2\ \alpha x = 2 \times 0.1 \times (335)^2 = 22{,}110$; $\sigma = 148$ m. Conc. at 335 + 148 = 483 m is $(16/100)c_0 = 0.18$ mg Cd/L. At 335 m: $(1/2)c_0 = 0.55$ mg/L. At 335 − 148 = 187 m is $(84/100)c_0 = 0.94$ mg/L. The TCE-front will be sharper since a convex isotherm gives small concentrations a higher slope dq/dc, and hence more retardation.

10.8.

```
SOLUTION 1-40; -temp 10; pH 7.0 charge; Ca 1.0 Calcite; C(4) 2.0 CO2(g) -2.0
SOLUTION 0; -temp 10; pH 7.0 charge; pe 14.0 O2(g) -0.68; Ca 1.0 Calcite;
C(4) 2.0 CO2(g) -2.0
 N(5) 6
END
```

```
KINETICS 1-40; Organic_C; -formula C; -m0 0.5
EQUILIBRIUM_PHASES 1-40; Calcite
END
PRINT; -reset false; -status false
USER_GRAPH; -headings depth Ca NO3 Alk
  -start; 10 graph_x dist; 20 graph_y tot("Ca")*1e3, tot("N(5)")*1e3, Alk*1e3; -end
TRANSPORT; -cells 40; -shifts 30 1; -dispersivity 0.0; -diffc 0.0; -time_st 3.15e7;
-punch_frequency 30
END
```

For denitrifying 5.75 mM NO_3^-, 7.19 mM C are needed. 0.1% C = 500 mmol/L pore water. Reaction time of C is $t = -1/(5 \times 10^{-4}) \ln((5 - 7.19)/0.5) = 28.96$ yr, equals depth (m).

INDEX